Handbook of Muscle Disease

NEUROLOGICAL DISEASE AND THERAPY

Series Editor

William C. Koller

Department of Neurology
University of Kansas Medical Center
Kansas City, Kansas

1. Handbook of Parkinson's Disease, *edited by William C. Koller*
2. Medical Therapy of Acute Stroke, *edited by Mark Fisher*
3. Familial Alzheimer's Disease: Molecular Genetics and Clinical Perspectives, *edited by Gary D. Miner, Ralph W. Richter, John P. Blass, Jimmie L. Valentine, and Linda A. Winters-Miner*
4. Alzheimer's Disease: Treatment and Long-Term Management, *edited by Jeffrey L. Cummings and Bruce L. Miller*
5. Therapy of Parkinson's Disease, *edited by William C. Koller and George Paulson*
6. Handbook of Sleep Disorders, *edited by Michael J. Thorpy*
7. Epilepsy and Sudden Death, *edited by Claire M. Lathers and Paul L. Schraeder*
8. Handbook of Multiple Sclerosis, *edited by Stuart D. Cook*
9. Memory Disorders: Research and Clinical Practice, *edited by Takehiko Yanagihara and Ronald C. Petersen*
10. The Medical Treatment of Epilepsy, *edited by Stanley R. Resor, Jr., and Henn Kutt*
11. Cognitive Disorders: Pathophysiology and Treatment, *edited by Leon J. Thal, Walter H. Moos, and Elkan R. Gamzu*
12. Handbook of Amyotrophic Lateral Sclerosis, *edited by Richard Alan Smith*
13. Handbook of Parkinson's Disease: Second Edition, Revised and Expanded, *edited by William C. Koller*
14. Handbook of Pediatric Epilepsy, *edited by Jerome V. Murphy and Fereydoun Dehkharghani*
15. Handbook of Tourette's Syndrome and Related Tic and Behavioral Disorders, *edited by Roger Kurlan*
16. Handbook of Cerebellar Diseases, *edited by Richard Lechtenberg*
17. Handbook of Cerebrovascular Diseases, *edited by Harold P. Adams, Jr.*
18. Parkinsonian Syndromes, *edited by Matthew B. Stern and William C. Koller*
19. Handbook of Head and Spine Trauma, *edited by Jonathan Greenberg*
20. Brain Tumors: A Comprehensive Text, *edited by Robert A. Morantz and John W. Walsh*
21. Monoamine Oxidase Inhibitors in Neurological Diseases, *edited by Abraham Lieberman, C. Warren Olanow, Moussa B. H. Youdim, and Keith Tipton*
22. Handbook of Dementing Illnesses, *edited by John C. Morris*
23. Handbook of Myasthenia Gravis and Myasthenic Syndromes, *edited by Robert P. Lisak*

24. Handbook of Neurorehabilitation, *edited by David C. Good and James R. Couch, Jr.*
25. Therapy with Botulinum Toxin, *edited by Joseph Jankovic and Mark Hallett*
26. Principles of Neurotoxicology, *edited by Louis W. Chang*
27. Handbook of Neurovirology, *edited by Robert R. McKendall and William G. Stroop*
28. Handbook of Neuro-Urology, *edited by David N. Rushton*
29. Handbook of Neuroepidemiology, *edited by Philip B. Gorelick and Milton Alter*
30. Handbook of Tremor Disorders, *edited by Leslie J. Findley and William C. Koller*
31. Neuro-Ophthalmological Disorders: Diagnostic Work-Up and Management, *edited by Ronald J. Tusa and Steven A. Newman*
32. Handbook of Olfaction and Gustation, *edited by Richard L. Doty*
33. Handbook of Neurological Speech and Language Disorders, *edited by Howard S. Kirshner*
34. Therapy of Parkinson's Disease: Second Edition, Revised and Expanded, *edited by William C. Koller and George Paulson*
35. Evaluation and Management of Gait Disorders, *edited by Barney S. Spivack*
36. Handbook of Neurotoxicology, *edited by Louis W. Chang and Robert S. Dyer*
37. Neurological Complications of Cancer, *edited by Ronald G. Wiley*
38. Handbook of Autonomic Nervous System Dysfunction, *edited by Amos D. Korczyn*
39. Handbook of Dystonia, *edited by Joseph King Ching Tsui and Donald B. Calne*
40. Etiology of Parkinson's Disease, *edited by Jonas H. Ellenberg, William C. Koller, and J. William Langston*
41. Practical Neurology of the Elderly, *edited by Jacob I. Sage and Margery H. Mark*
42. Handbook of Muscle Disease, *edited by Russell J. M. Lane*
43. Handbook of Multiple Sclerosis: Second Edition, Revised and Expanded, *edited by Stuart D. Cook*

Additional Volumes in Preparation

Infections of the Central Nervous System, *edited by Karen L. Roos*

Handbook of Muscle Disease

edited by
Russell J. M. Lane
Charing Cross Hospital
Charing Cross and Westminster
Medical School
London, England

Marcel Dekker, Inc. New York • Basel • Hong Kong

Library of Congress Cataloging-in-Publication Data

Handbook of muscle disease / edited by Russell J. M. Lane.
p. cm. — (Neurological disease and therapy ; 42)
Includes index.
ISBN 0-8247-9494-X (hardcover : alk. paper)
1. Neuromuscular diseases. 2. Muscles—Diseases. I. Lane, Russell J. M.
II. Series [DNLM: 1. Muscular Diseases. W1 NE33LD v.42 1996 / WE
550 H2365 1996]
RC925.H36 1996
616.7'4—dc20
DNLM/DLC
for Library of Congress 96-21848
CIP

The publisher offers discounts on this book when ordered in bulk quantities. For more information, write to Special Sales/Professional Marketing at the address below.

This book is printed on acid-free paper.

Marcel Dekker, Inc.
270 Madison Avenue, New York, New York 10016

Current printing (last digit):
10 9 8 7 6 5 4 3 2 1

PRINTED IN THE UNITED STATES OF AMERICA

To Dianne,
Rebecca, and Anna

Series Introduction

Disease of muscle constitutes an important part of clinical neurology. A variety of pathological processes target muscle. It is the muscle that does the work of moving the limbs even though it is under precise control of neuronal mechanisms and subject to many diseases that attack the muscle directly. It is the depth and breadth of muscle disease that makes the *Handbook of Muscle Diseases* an essential volume. The diagnosis of muscle disease can be challenging even to the most seasoned clinician. Therefore, it is appropriate that in the *Handbook of Muscle Disease* Dr. Russell Lane has brought together a group of contributors who are experts in muscle disease and who can relate the many discoveries that have recently increased our understanding of muscle disease, in particular, advances in molecular genetics that have played a prominent role.

The first section of the book is dedicated to the technical aspects of the investigation of muscle disease. This is indeed important in the era of cost-saving medicine and will be helpful to the clinician in deciding which tests to order and which tests will have the highest likelihood of giving a diagnosis under a certain clinical circumstance. Many of these tests are expensive and need to be appropriately applied. The comprehensive discussion of congenital myopathies and muscular dystrophies follows. These are important diseases both for the pediatrician and for the physician seeing these patients later in life. Our understanding of these disorders and their molecular basis has increased dramatically. Currently, many of these disorders cannot be diagnosed through genetic testing.

Another group of diseases related to the muscle are related to altered membranes. These are discussed in Chapter 5. These conditions may be particularly difficult to diagnose. Many other disorders that can attack muscle, such as systemic disease, metabolic, mitochondrial, and inflammatory myopathies, are the topics of subsequent chapters. These diseases are discussed in depth regarding clinical features, diagnosis, pathophysiology, and treatment. The last section of the book concentrates on the clinical management of muscle disorders.

In keeping with the tradition of this handbook series, the *Handbook of Muscle Disease* provides a practical summary of muscle disease for the practicing physician. The book will serve as an excellent review text for clinicians seeking information about the diagnosis and management of patients with muscle diseases.

William C. Koller

Foreword

In the 1970s, when I was Professor of Neurology at the University of Newcastle upon Tyne and Dean of the Medical School, I was still able to continue, on a part-time basis, to direct the Muscular Dystrophy Research Laboratories there. As Dean, I soon became aware that among an exceptionally able group of medical students a young man called Russell Lane was an individual of outstanding ability, and this view was confirmed when, in the final examination in medicine, he achieved first-class honors, a distinction granted to the leading student of each class not more than once every three or four years. I was therefore happy when, after he had trained in internal medicine, he decided to move into neurology, and I was particularly pleased when it became clear that he was taking an interest in the field of neuromuscular disease. After a distinguished and varied career in Newcastle and in Durham, North Carolina, he obtained the consultant post he now holds in the Charing Cross and Westminister Medical School in London, where his interest in the neuromuscular field has continued to flower.

This remarkably comprehensive volume is a testimony to his interest and dedication to his chosen field of study, and I believe that it will be widely welcomed. The book is designed in an interesting format, beginning with a section on the investigation of a myopathy, followed by sections on the congenital myopathies, the muscular dystrophies, disorders of muscle membranes, myopathies in systemic and iatrogenic diseases, inherited metabolic myopathies, mitochondrial myopathies, inflammatory myopathies, miscellaneous disorders, and aspects of management. The book concludes with a diagnostic index of neuromuscular disorders.

Dr. Lane has recruited an impressive team of international authors of acknowledged ability and distinction and has, in my view, produced an exemplary reference work that will surely be welcomed and consulted regularly by all of those with an interest in this group of diseases. The last 20 years has seen a vast, even explosive, increase in our understanding of the pathogenesis of many of the conditions dealt with in these pages, and in some of the most tragic and disabling of them, the advances in molecular biology now occurring almost daily have brought within reach the prospect of effective amelioration through gene therapy. But other developments in biochemistry, electrophysiology, imaging, immunology, and microscopy in all forms, as well as new techniques of management and rehabilitation, have transformed the lives of many patients with these conditions; all of these aspects are covered in this outstanding work.

As editor myself of several books on neuromuscular disease, it is, I suppose, inevitable that one should look upon a new entrant into the field with mixed feelings; however, science and medical care continually advance, and my scrutiny of this book has convinced me that it adds immeasurably to our knowledge and understanding of this challenging and burgeoning topic. I am confident that it will stand the test of time.

John Walton, Kt., T.D., M.A., D.Sc., F.R.C.P.
(Lord Walton of Detchant)
President, World Federation of Neurology
Oxford, England

Preface

The last decade has witnessed remarkable developments in our understanding of muscle diseases, chiefly through advances in molecular genetics, biochemistry, and immunohistochemistry. These technologies have provided new insights into the pathogenesis of the dystrophies and other myopathies and will, it is hoped, allow increasing opportunities for rational treatment of these conditions in the future.

A number of excellent works are available on the clinical and pathological aspects of muscle disease, ranging from comprehensive multiauthor reference texts to clinical monographs. The aim of this book is, in part, to complement these standard texts and, in particular, to review recent developments in the molecular biology of muscle diseases.

The first section of the book includes discussions on the technical aspects of quantitative electromyography, spectroscopy, and myoimaging, together with an account of basic strategies in molecular genetics. Subsequent sections cover the whole range of muscle diseases, reviewing current opinion on diagnostic criteria, pathogenesis, pathology, and management. It was inevitable, in a discipline in which such rapid advances are being made, that significant developments in our understanding would occur during the production of this book; the classification of limb-girdle dystrophies and the genetic basis of the channelopathies are but two examples. Where appropriate, notes have been added to incorporate these latest observations.

Neuromuscular diseases are arguably the most practical of disciplines within neurology, and chapters are devoted to the elucidation and differential diagnosis of the symptoms and signs of muscle disease, measurement of muscle strength, and muscle biopsy techniques and analysis. The final section deals with various aspects of the general management of patients with neuromuscular diseases, including implications of different disorders on anesthetic practice, the role of physiotherapy, orthotics, and surgical intervention.

The Appendix comprises a practical Neuromuscular Diagnostic Index, based on ICD-10 and Read nomenclature and coding, for use in the clinic.

I should like to thank everyone who has contributed to this book—the authors from around the world for their patience, the publishers for their unflagging support, and my colleague, Dr. Richard Orrell, who coauthored several of my contributions. On a personal note, I should also like to take this opportunity to acknowledge and thank senior colleagues who fostered my interest in neurology and neuromuscular diseases, and who supported me in my early career, in particular, Lord Walton of Detchant and Dr. Peter Hudgson in Newcastle, Professor Allen D. Roses at Duke University, and Dr. Frank Clifford Rose and Dr. G. David Perkin at Charing Cross Hospital.

It is hoped that this book will provide a useful and stimulating resource for established physicians, nurses, therapists, and scientists working in the field, and also for those training in neurology, paediatrics, orthopedics, and physical medicine. I am, however, mindful of Sir Andrew Huxley's caution in his Sherrington Lectures*:

---

*Huxley AF. Reflections on Muscle. The Sherrington Lectures XIV. Liverpool: Liverpool University Press, 1980.

Many people no doubt would say that it is impossible that any substantial part of what now seems to be a well-established body of knowledge about muscle should come to be disregarded in a few decades' time, but to me it is no more impossible than it would have seemed to our predecessors of ninety or a hundred years ago, that the body of information about muscle gained with the light microscope in the 1870s and 1880s should disappear from human knowledge by 1930 or 1940.

One can only trust that those who read this work will find in it something of value. It is respectfully dedicated to patients with muscle disease and all those who care for them.

Russell J. M. Lane

Contents

Series Introduction *William C. Koller* v
Foreword *John Walton (Lord Walton of Detchant)* vii
Preface ix
Contributors xv

I. INVESTIGATION OF A MYOPATHY

1. Clinical Presentation: Symptoms and Sign of Muscle Disease and Their Interpretation **1**
Russell J. M. Lane and Geraint N. Fuller

2. Muscle Strength Measurement **19**
Mohammed Sanjak, Daryn Belden, Thomas Cook, and Benjamin R. Brooks

3. Exercise Tests **35**
Russell J. M. Lane

4. Electromyography **41**
Robin P. Kennett

5. Muscle Biopsy Techniques **53**
Richard H. T. Edwards, Malcolm J. Jackson, Timothy R. Helliwell, and J. E. Clague

6. Muscle Biopsy Analysis **61**
Margaret A. Johnson and Martin J. Barron

7. Electron Microscopy **81**
Caroline A. Sewry

8. Myoimaging **95**
Carl D. Reimers and Thomas J. Vogl

9. Magnetic Resonance Spectroscopy in the Investigation and Understanding of Muscle Disease **115**
Phillip R. J. Barnes, Doris J. Taylor, and David Hilton-Jones

10. Methods in Molecular Genetics Diagnosis of Muscle Disease **127**
Tracey Van-Tongeren, Nessa Carey, and Keith J. Johnson

II. CONGENITAL DYSTROPHIES AND RELATED DISORDERS

11. Congenital Muscular Dystrophies **139**
Bernard Echenne

12. Rigid Spine Syndrome 151
Luciano Merlini and Claudia Granata

13. Bethlem Myopathy 159
Gerrit Joost Jöbsis and Marianne de Visser

III. MORPHOLOGICALLY DEFINED CONGENITAL MYOPATHIES

14. Classification of Congenital Myopathies 165
Hans Hilmar Goebel and Anna Fidzianska

15. Core Diseases, Centronuclear Myopathies, and Nemaline Myopathy 177
Hans Hilmar Goebel and Anna Fidzianska

16. Congenital Fiber Type Disproportion 195
J. Michael Schröder

IV. THE MUSCULAR DYSTROPHIES

17. X- Linked Dystrophies: Clinical and Pathological Features 201
Frank L. Mastaglia and Byron A. Kakulas

18. Dystrophin: Structure, Localization, and Function 223
John B. Harris, Michael J. Cullen, and Ruth Vater

19. The Spectrum of Dystrophinopathies 235
Byron A. Kakulas

20. Autosomal Recessive Limb-Girdle Muscular Dystrophies: Clinical and Laboratory Findings 245
Mayana Zatz, Mariz Vainzof, Maria Rita Passos-Bueno, Jeanne Akiyama, and Suely K. N. Marie

21. Genetic and Linkage Analysis in Autosomal Recessive Limb-Girdle Muscular Dystrophy 257
Maria Rita Passos-Bueno, Mariz Vainzof, and Mayana Zatz

22. Dystrophin and DNA Findings in Duchenne and Becker Carriers 265
Mariz Vainzof, Maria Rita Passos-Bueno, and Mayana Zatz

23. Facioscapulohumeral Syndrome 275
W. J. Ken Cumming

24. Autosomal Dominant Limb-Girdle Muscular Dystrophy 287
Ashok Verma and Walter G. Bradley

25. Other Dystrophies with Autosomal Dominant Inheritance 297
Henning Schmalbruch

V. DISORDERS OF MUSCLE MEMBRANES—MYOTONIC DISORDERS AND PERIODIC PARALYSES

26. Myotonic Dystrophy 311
Russell J. M. Lane, Peggy Shelbourne, and Keith J. Johnson

27. Hypokalemic Periodic Paralysis 329
Rabi Tawil and Robert C. Griggs

28. Muscle Sodium Channel and Chloride Channel Diseases 339
Reinhardt Rüdel and Frank Lehmann-Horn

29. Neuromyotonia (Isaacs' Syndrome) 355
Ian K. Hart and John Newsom-Davis

VI. MYOPATHIES IN SYSTEMIC AND IATROGENIC DISEASES

30. Muscle Dysfunction in Endocrine Disease 365
Richard W. Orrell, Russell J. M. Lane, and Peter Hudgson

31. Toxic and Drug-Induced Myopathies 379
Russell J. M. Lane

32. Myopathies in Infections and Systemic Diseases 391
Richard W. Orrell and Russell J. M. Lane

VII. INHERITED METABOLIC MYOPATHIES

33. Glycogen Storage Diseases of Muscle 407
Corrado Angelini, Andrea Martinuzzi, and Lodovica Vergani

34. Lipid Storage Myopathies and Inborn Errors of Fatty Acid Oxidation 431
Jochen Schaefer, Sandra Jackson, and D. M. Turnbull

35. Myoadenylate Deaminase Deficiency 443
Richard L. Sabina

36. Malignant Hyperthermia 451
Beverley A. Britt

37. Sarcoplasmic Reticulum Ca^{2+}–Adenosine Triphosphatase Deficiency (Brody's Disease) 473
J. A. P. Hiel, P. J. H. Jongen, P. J. E. Poels, A. A. G. M. Benders, J. H. Veerkamp, N. Knoers, F. J. M. Gabreëls, and R. A. Wevers

VIII. MITOCHONDRIAL MYOPATHIES

38. Clinical Features of Mitochondrial Myopathies and Encephalomyopathies 479
Michio Hirano and Salvatore DiMauro

39. Genetics of Mitochondrial Myopathies 505
S. R. Hammans and A. E. Harding

40. Biochemical Investigation of Mitochondrial Respiratory Chain Abnormalities 519
J. M. Cooper and Anthony H. V. Schapira

41. Treatment of Mitochondrial Myopathies 533
R. C. A. Sengers, J. M. F. Trijbels, and W. Ruitenbeek

IX. INFLAMMATORY MYOPATHIES

42. Idiopathic Inflammatory Myopathies 539
Russell J. M. Lane and Peter Hudgson

43. Inclusion Body Myositis and Other Rimmed Vacuolar Myopathies 575
Michael Rose and Russell J. M. Lane

X. MISCELLANEOUS DISORDERS

44. Chronic Fatigue Syndrome 591
Russell J. M. Lane

45. Polymyalgia Rheumatica 595
Richard W. Orrell

46. Myositis Ossificans 599
Richard W. Orrell

47. Anthrogryposis Multiplex Congenita 603
Richard W. Orrell and Geraint N. Fuller

48. Compartment Syndromes 605
Richard W. Orrell

49. Myoglobinuria 607
Richard W. Orrell and Russell J. M. Lane

50. Nutritional Myopathies 613
Richard W. Orrell

51. Amyloid Myopathy and Intermediate Filament Myopathy 617
David J. Dick

52. HIV-Related Myopathies 623
Russell J. M. Lane

XI. ASPECTS OF MANAGEMENT

53. Anesthesia and Primary Muscle Disease 629
Nicholas P. Hirsch and Scott H. Russell

54. Management of Respiratory Problems in Neuromuscular Disease 639
Marlene Normand, John R. Bach, Ann Ford Fricke, Henry D. Owens, and Richard T. Moxley III

55. Cardiac Involvement in Neuromuscular Disease 655
Michael Swash and Martin S. Schwartz

56. Physical Therapy of Muscle Diseases 663
Sylvia Hyde

57. Surgical Care and Orthotics in Muscle Disease 671
John D. Hsu and Yuriko Lynn Yasuda

58. Genetic Counseling and Prenatal Diagnosis 683
Elizabeth M. Thompson

Appendix: Diagnostic Index of Neuromuscular Disorders 707
Russell J. M. Lane, Richard W. Orrell, and Hiba Tappouni

Index *729*

Contributors

Jeanne Akiyama, B.Sc. Department of Biology, Center for Myopathy, Institute of Biosciences, University of São Paulo, São Paulo, Brazil

Corrado Angelini, M.D. Department of Neurology, Regional Neuromuscular Center, University of Padua, Padua, Italy

John R. Bach, M.D. Kessler Institute, West Orange, New Jersey

Phillip R. J. Barnes, M.D. Department of Neurology, Radcliffe Infirmary, and John Radcliffe Hospital, Oxford, England

Martin J. Barron, B.Sc. (Hons.), M.Sc. Division of Neurobiology, University of Newcastle upon Tyne, Newcastle upon Tyne, England

Daryn Belden, M.D. Department of Neurology, ALS Clinical Research Center, University of Wisconsin Medical School, Madison, Wisconsin

A. A. G. M. Benders, M.D. Department of Biochemistry, University Hospital Nijmegen, Nijmegen, The Netherlands

Walter G. Bradley, D.M., F.R.C.P. Department of Neurology, University of Miami School of Medicine, Miami, Florida

Beverley A. Britt, M.D., F.R.C.P.(C) Department of Anesthesia and Pharmacology, Malignant Hyperthermia Investigation Unit, University of Toronto, and Toronto General Hospital, Toronto, Ontario, Canada

Benjamin R. Brooks, M.D. Department of Neurology, ALS Clinical Research Center, University of Wisconsin Medical School, Madison, Wisconsin

Nessa Carey, B.Sc., Ph.D. Department of Anatomy, Charing Cross and Westminster Medical School, University of London, London, England

J. E. Clague, M.D. University of Manchester, Manchester, England

Thomas Cook, Ph.D. Department of Biostatistics, University of Wisconsin Medical School, Madison, Wisconsin

J. M. Cooper, B.Sc., Ph.D. Department of Clinical Neurosciences, Royal Free Hospital School of Medicine, London, England

Michael J. Cullen, D. Phil. Muscular Dystrophy Group Research Laboratories, Newcastle General Hospital, Newcastle upon Tyne, England

W. J. Ken Cumming, B.Sc., M.D., F.R.C.P.I., F.R.C.P. Department of Neurology, Withington Hospital, Manchester, England

Marianne de Visser, M.D. Department of Neurology, Academic Medical Center, Amsterdam, The Netherlands

David J. Dick, M.D., F.R.C.P. Department of Neurology, Norfolk and Norwich Hospital, Norwich, and Department of Neurology, Addenbrooke's Hospital, Cambridge, England

Salvatore DiMauro, M.D. Department of Neurology, Columbia University College of Physicians and Surgeons, New York, New York

Bernard Echenne, M.D. Department of Neuropediatics, Gui de Chauliac Center, Montpellier, France

Richard H. T. Edwards, Ph.D., F.R.C.P. Department of Medicine, University of Liverpool, Liverpool, England

Anna Fidzianska, M.D. Department of Neurology, Warsaw University Medical Center, Warsaw, Poland

Ann Ford Fricke, R.N. Department of Neurology, Strong Memorial Hospital, University of Rochester, Medical Center, Rochester, New York

Geraint N. Fuller, M.A., M.D., M.R.C.P. Department of Neurology, Gloucester Royal Hospital, Gloucester, England

F. J. M. Gabreëls, M.D. Department of Child Neurology, Institute of Neurology, University Hospital Nijmegen, Nijmegen, The Netherlands

Hans Hilmar Goebel, M.D. Division of Neuropathology, Johannes Gutenberg University, Mainz, Germany

Claudia Granata, M.D. Rizzoli Orthopedic Institute and Spine Unit, Toniolo Hospital, Bologna, Italy

Robert C. Griggs, M.D. Department of Medicine, University of Rochester School of Medicine and Dentistry, Rochester, New York

S. R. Hammans, M.A., M.D., M.R.C.P. Wessex Neurological Centre, Southampton General Hospital, Southampton, England

A. E. Harding, M.D., F.R.C.P.† Department of Clinical Neurology, Institute of Neurology, London, England

John B. Harris, Ph.D., M.R.Pharm.S., F.I.Biol. Muscular Dystrophy Group Laboratories, Newcastle General Hospital, Newcastle upon Tyne, England

Ian K. Hart, Ph.D., M.R.C.P. Department of Neurology, University of Oxford, Oxford, England

Timothy R. Helliwell, M.D. M.R.C.Path. Department of Pathology, University of Liverpool, Liverpool, England

J. A. P. Hiel, M.D. Institute of Neurology, University Hospital Nijmegen, Nijmegen, The Netherlands

David Hilton-Jones, M.A., M.D., F.R.C.P.E. Department of Clinical Neurology, Radcliffe Infirmary, Oxford, England

†Deceased.

Michio Hirano, M.D. Department of Neurology, Columbia University College of Physicians and Surgeons, New York, New York

Nicholas P. Hirsch, M.B., B.S., F.R.C.A. Department of Anaesthesia, The National Hospital for Neurology and Neurosurgery, London, England

John D. Hsu, M.D., C.M., F.A.C.S. Department of Surgery, Rancho Los Amigos Medical Center, Downey, and Department of Orthopedics and Rehabilitation, University of Southern California, School of Medicine, Los Angeles, California

Peter Hudgson, F.R.C.P., F.R.A.C.P. Regional Neurological Center, Newcastle General Hospital, and University of Newcastle upon Tyne, Newcastle upon Tyne, England

Sylvia Hyde, F.C.S.P., M.B.A. Research and Development Center, Muskelsvindfonden, Arhus, Denmark

Malcolm J. Jackson, Ph.D., D.Sc., M.R.C.Path. Department of Medicine, University of Liverpool, Liverpool, England

Sandra Jackson, B.Sc. (Hons.), Ph.D. Department of Neurology, University of Newcastle upon Tyne, Newcastle upon Tyne, England

Gerrit Joost Jöbsis, M.D. Department of Neurology, Academic Medical Center, Amsterdam, The Netherlands

Keith J. Johnson, B.Sc., Ph.D. IBLS Laboratory of Genetics, University of Glasgow, Glasgow, Scotland

Margaret A. Johnson, Ph.D. Division of Neurobiology, University of Newcastle upon Tyne, Newcastle upon Tyne, England

P. J. H. Jongen, M.D. Institute for Neurology, University Hospital Nijmegen, Nijmegen, The Netherlands

Byron A. Kakulas, M.D. (Hon.Athens), M.D., (W.A.) F.R.A.C.P., F.R.C.Path., F.R.C.P.A. Royal Perth Hospital and Australian Neuromuscular Research Institute, Queen Elizabeth II Medical Centre, Perth, Western Australia, Australia

Robin P. Kennett, B.Sc., M.D., M.R.C.P. Department of Neurophysiology, Radcliffe Infirmary NHS Trust, Oxford, England

N. Knoers, M.D., Ph.D. Department of Human Genetics, University Hospital Nijmegen, Nijmegen, The Netherlands

Russell J. M. Lane, B.Sc., M.D., F.R.C.P. Regional Neurosciences Center, Charing Cross Hospital, and Charing Cross and Westminster Medical School, London, England

Frank Lehmann-Horn, M.D. Department of Applied Physiology, University of Ulm, Ulm, Germany

Suely K. N. Marie, M.D. Neuromuscular Group, University of São Paulo, São Paulo, Brazil

Andrea Martinuzzi, M.D. Ph.D. Department of Neurology, Regional Neuromuscular Center, University of Padua, Padua, Italy

Frank L. Mastaglia, M.D., F.R.A.C.P., F.R.C.P. Departments of Medicine and Neurology and Australian Neuromuscular Research Institute, Queen Elizabeth II Medical Centre, Perth, Western Australia, Australia

Luciano Merlini, M.D. Neuromuscular Laboratory, Rizzoli Orthopedic Institute, Bologna, Italy

Richard T. Moxley III, M.D. Neuromuscular Disease Center, University of Rochester Medical Center, Rochester, New York

John Newsom-Davis, M.D., F.R.C.P., F.R.S. Department of Neurology, University of Oxford, Oxford, England

Marlene Normand, M.D. Section of Neurology, University of Nebraska Medical Center, Omaha, Nebraska

Richard W. Orrell, B.Sc., M.B., Ch.B., D.C.H., M.R.C.P. Regional Neurosciences Center, Charing Cross Hospital, and Charing Cross and Westminster Medical School, London, England

Henry D. Owens, B.S., R.R.T. Department of Respiratory Care, University of Rochester Medical Center, Rochester, New York

Maria Rita Passos-Bueno, Ph.D. Center for Myopathy, Institute of Bioscience, University of São Paulo, São Paulo, Brazil

P. J. E. Poels, M.D. Institute of Neurology, University Hospital Nijmegen, Nijmegen, The Netherlands

Carl D. Reimers, M.D. Department of Clinical Neurophysiology, Georg-August-University of Göttingen, Göttingen, Germany

Michael Rose, M.D. Department of Neurology, King's College Hospital, London, England

Reinhardt Rüdel, Ph.D., M.Sc., M.B. Department of General Physiology, University of Ulm, Ulm, Germany

W. Ruitenbeek Department of Pediatrics, University Hospital Nijmegen, Nijmegen, The Netherlands

Scott H. Russell, M.B., B. S., F.R.C.A. Department of Anaesthesia, Queen Elizabeth Hospital, Birmingham, England

Richard L. Sabina, Ph.D. Department of Biochemistry, Medical College of Wisconsin, Milwaukee, Wisconsin

Mohammed Sanjak, Ph.D. Department of Neurology and ALS Clinical Research Center, University of Wisconsin Medical School, Madison, Wisconsin

Jochen Schaefer, M.D. Department of Neurology, University of Newcastle upon Tyne, Newcastle upon Tyne, England

Anthony H. V. Schapira, D.Sc., M.D., F.R.C.P. Department of Clinical Neurosciences, Royal Free Hospital School of Medicine, London, England

Henning Schmalbruch, M.D. Department of Medical Physiology, University of Copenhagen, Copenhagen, Denmark

J. Michael Schröder, M.D. Department of Neuropathology, Rheinisch-Westfälische Technische Hochschule Aachen, Aachen, Germany

Martin S. Schwartz, M.D., F.R.C.P. Department of Neurology, Atkinson Morley's Hospital, London, England

R. C. A. Sengers, M.D. Department of Pediatrics, University Hospital Nijmegen, Nijmegen, The Netherlands

Caroline A. Sewry, B.Sc., Ph.D. Neuromuscular Unit, Royal Postgraduate Medical School, and Muscle Cell Biology Group, MRC Clinical Sciences Centre, Hammersmith Hospital, London, England

Peggy Shelbourne, Ph.D. IBLS Laboratory of Genetics, University of Glasgow, Glasgow, Scotland

Michael Swash, M.D., F.R.C.P., F.R.C.Path. Department of Neurology, St. Bartholomew's and Royal London School of Medicine and Dentistry, Royal London Hospital, London, England

Hiba Tappouni Research Assistant, University of North Carolina, Chapel Hill, North Carolina

Rabi Tawil, M.D. Department of Neurology, University of Rochester School of Medicine and Dentistry, Rochester, New York

Doris J. Taylor, D.Phil. Biochemical and Clinical Magnetic Resonance Unit, University of Oxford and John Radcliffe Hospital, Oxford, England

Elizabeth M. Thompson, M.D., F.R.A.C.P. Centre for Medical Genetics, Women's and Children's Hospital, North Adelaide, South Australia, Australia

J. M. F. Trijbels, M.D. Institute of Pediatrics, University Hospital Nijmegen, Nijmegen, The Netherlands

D. M. Turnbull, M.B.B.S., Ph.D., M.D., F.R.C.P. Department of Neurology, University of Newcastle upon Tyne, Newcastle upon Tyne, England

Mariz Vainzof, Ph.D. Center for Myopathy, Institute of Biosciences, University of São Paulo, São Paulo, Brazil

Tracey Van-Tongeren, B.Sc.(Hons.), M.Phil. Department of Anatomy, Charing Cross and Westminster Medical School, University of London, London, England

Ruth Vater, B.Sc., M.Phil., Ph.D. Muscular Dystrophy Group Research Laboratories, Newcastle General Hospital, Newcastle upon Tyne, England

J. H. Veerkamp, Ph.D. Department of Biochemistry, University Hospital Nijmegen, Nijmegen, The Netherlands

Lodovica Vergani, M. D. Department of Neurology, Regional Neuromuscular Center, University of Padua, Padua, Italy

Ashok Verma, M.D., D.M. Department of Neurology, University of Miami School of Medicine, Miami, Florida

Thomas J. Vogl, M.D. Department of Radiology, Humboldt-University of Berlin, Berlin, Germany.

R. A. Wevers, Ph.D. Department of Neurology, University Hospital Nijmegen, Nijmegen, The Netherlands

Yuriko Lynn Yasuda, M.S.Ed., O.T.R., F.A.O.T.A. Department of Occupational Therapy, Rancho Los Amigos Medical Center, Downey, and University of Southern California, School of Medicine, Los Angeles, California

Mayana Zatz, Ph.D. Center for Myopathy, Institute of Biosciences, University of São Paulo, São Paulo, Brazil

1

Clinical Presentation

Symptoms and Signs of Muscle Disease and Their Interpretation

Russell J. M. Lane
Charing Cross Hospital and Charing Cross and Westminster Medical School, London, England

Geraint N. Fuller
Gloucester Royal Hospital, Gloucester, England

INTRODUCTION

This chapter provides a practical approach to the clinical assessment and interpretation of the symptoms and signs of muscle disease. While this is only the first step in making a specific diagnosis, which may depend on neurophysiological studies, myoimaging, muscle biopsy findings, and molecular genetic techniques discussed in later chapters, it is the initial clinical evaluation that determines the direction of these further investigations. The origins of the cardinal symptoms of muscle disease—weakness, fatigue, and myalgia—will be discussed in some detail, since an understanding of the mechanisms involved can aid interpretation of patients' symptoms. Since this book is devoted to disorders of muscle, symptoms relating to diseases of nerves and the neuromuscular junction, which must obviously be considered in differential diagnosis, will be mentioned only briefly. The chapter concludes with an account of some aspects of symptomatic treatment.

HISTORY

Presenting Features

Weakness

The most prominent symptom of muscle disease is usually weakness, which is understood by a physician to imply a reduction of static muscle strength. However, some patients use it to mean tiredness, fatigue, loss of energy, exercise intolerance, or even an alteration in sensation. Once it is clear that the patient does indeed experience true weakness, it is necessary to establish the rapidity of onset, its duration, and whether the weakness has been progressive or fluctuating. The distribution of the weakness and the resulting functional disability should also be noted. For example, when was the patient last able to jump and run, climb stairs unaided, or walk unaided, and what aids does the patient require to achieve these goals at present? Finally, the extent of resulting handicap should be documented. Has the disease resulted in inability to work or to perform activities of daily living?

Two points are particularly worth bearing in mind when considering a history of weakness: chronology of weakness and distribution of weakness.

Chronology of Weakness. Patients with chronic, slowly progressive weakness, or weakness affecting only certain muscle groups, may compensate or adapt very effectively and may present only when these compensatory mechanisms begin to fail. In defining "onset," therefore, it is always worth reviewing perinatal and developmental history, even if the weakness appears to be recent. If the patient is young, or a parent is available, it may be possible to determine whether the patient was a "floppy baby" or had difficulty in breathing, contractures, or other signs of weakness at birth. In infants and young children, details of the mother's pregnancy, such as polyhydramnios and the onset and vigor or fetal movements, may be relevant. At what age did the infant achieve the motor milestones of pulling the head up in the prone position, sitting and standing unsupported, crawling, and particularly walking? Could the patient run, jump, and hop as a child, and could he run as fast as his peers? Patients who had mild weakness in childhood may say that they "did not like sports" or always took a relatively less active role in games, such as goalkeeper or umpire. Such an approach may indicate that weakness perceived as being of relatively recent onset is, in fact, long-standing.

Distribution of Weakness. In terms of their distribution, muscle diseases can broadly be divided into the following categories.

1. Proximal/generalized myopathies. Muscle diseases commonly present with generalized weakness, particularly affecting proximal and axial weight-bearing muscles in a nonselective and symmetrical fashion. Initial problems may include ascending the stairs, getting out of low chairs, and working with the arms above the head. Quadriceps weakness tends to cause problems descending stairs, and patients may complain that their knees suddenly "give way." Common examples include polymyositis and limb-girdle syndrome.

2. Myopathies with distal weakness. Less commonly, muscle disorders may affect mainly the distal muscles, leading to problems such as a tendency to scuff the toes or trip over undulations in pavements due to footdrop, and difficulty in taking lids off bottles or in gripping pens, reflecting wrist and intrinsic hand muscle weakness. Myotonic dystrophy and inclusion body myositis are two conditions that commonly show this pattern of weakness, and neurogenic disorders are also important in the differential diagnosis.

3. Myopathies with selective muscle weakness. In some disorders, certain muscles in a body region may be strikingly weaker than others. For example, in facioscapulohumeral and scapuloperoneal dystrophies, weakness particularly affects muscle groups referred to in the names of the conditions, while in certain mitochondrial disorders, the extraocular muscles may appear to be exclusively involved. The basis of this selective involvement remains poorly understood.

Pathogenesis of Weakness

Weakness may result from lesions at any point in the "chain of command" from the motor cortex to the myofibrils. Evidence that the weakness is peripheral rather than central is usually clear from the clinical examination, although sometimes both central and peripheral abnormalities may be operative, as for example in amyotrophic lateral sclerosis and some mitochondrial disorders.

At the level of the muscle fiber, weakness may result from:

Alterations in the electrical and metabolic functions of the muscle membranes (e.g., nondystrophic myotonias and periodic paralyses)
Abnormalities of intermediary metabolism (e.g., the metabolic myopathies)
Disordered energy metabolism (e.g., mitochondrial myopathies)

More commonly, however, weakness results from widespread disruption of the myofibrillary and cytoskeletal architecture by inflammatory or degenerative processes, resulting from immunological attack (inflammatory myopathies) or defects in genes encoding structural proteins (muscular dystrophies), often resulting in replacement of muscle with fat and fibrous tissue. There are also a number of rare congenital abnormalities of muscle structure (congenital myopathies). Weakness may therefore be persistent and progressive, as in these latter disorders, or perhaps apparent only after provocation by exercise, as in some metabolic myopathies.

Fatigue, Tiredness, and Exercise Intolerance

"Fatigue" is a difficult complaint to interpret and, like weakness, is a term used by patients to describe a number of symptoms. In the strict physiological sense, fatigue is an inability to sustain force. Fatigue described as "tiredness" or "tired all the time," however, is an extremely common complaint in the general population (1) and is probably mainly of psychological origin, related to stress and relative physical inactivity. In such cases, patients will often find that the fatigue affects mental activities such as reading, as well as physical function, and may be present at rest, although exacerbated by activity. In such cases, there may be a perception of low or reduced output for the apparent effort expended in a physical task. Such disorders are sometimes referred to as effort syndromes, of which chronic fatigue syndrome (see Chapter 44) may be considered a particular form. The term "asthenia" or "neurasthenia," a weariness resulting from physical or mental effort, is also sometimes used to describe this condition. The influence of psychological or central neurological factors in the genesis of symptoms is important in such cases, although the mechanisms involved are currently unclear.

Attempts have been made—notably by Borg (2,3)—to quantify perceived effort in relation to force generated, work performed, and even mental tasks. "Tiredness" is a prominent symptom of depressive illness and also of anemia, metabolic disorders such as renal insufficiency, inflammatory disorders such as rheumatoid disease, and central nervous system diseases such as multiple sclerosis, although in these latter disorders there is also a demonstrable peripheral component to the fatigue state (4). This tiredness type of fatigue, however, is not usually a prominent feature in patients with neuromuscular diseases (5).

Pathogenesis of Fatigue

True muscular fatigue (inability to sustain force) is a universal experience following prolonged, repeated muscle contractions and occurs much more rapidly than normal in patients with neuromuscular disorders. Thus patients with myasthenia and other disorders of the neuromuscular junction may have relatively preserved static muscle strength but may develop weakness rapidly on repetitive muscle contraction. Indeed, patients with myopathies of any sort may first become aware of weakness only on exertion, with resulting loss of endurance, while static strength remains preserved initially. It should be noted that exercise intolerance may also result from exertional myalgia, which invokes other diagnostic possibilities (see below).

Peripheral mechanisms, rather than central nervous system events, are primarily responsible for "physiological" fatigue in normal subjects (6). Since the high "safety factor" for transmission at the normal neuromuscular junction ensures that events at this site are not usually important determinants of fatigue, the phenomenon ultimately reflects failure of the muscle fiber contraction mechanism itself. In this regard, fatigue can arise as a result of failure of propagation of the action potential from the neuromuscular junction to the sarcoplasmic reticulum (SR), evident at high rates of stimulation—high-frequency fatigue; or as a result of mechanical defects in excitation-contraction (e-c) coupling and the sliding filament mechanism itself—low-frequency fatigue (6). Examples of the former would include muscle membrane disorders such as the channelopathies (myotonias and periodic paralyses), while low-frequency fatigue is seen in normal subjects and patients with myopathies. Certain conditions, such as McArdle's disease, in which inborn metabolic errors result in impaired energy metabolism, may affect ionic processes involved in maintenance of membrane excitation as well as e-c coupling and myofibrillary contraction, resulting in both high- and low-frequency fatigue characteristics (7). Under normal conditions, however, the relatively brief time course of electrical events, compared to the much slower mechanical factors, determines that low-frequency fatigue mechanisms are paramount (8).

The Origin of Low-Frequency Fatigue

There appear to be two phases to the biochemical changes in muscle during low-frequency fatigue. During the first 20–30 sec. of a fatiguing contraction, phosphocreatine (PCr) levels fall while lactate and pH remain constant (the alactic phase). This is uninfluenced by occlusion of the circulation. When PCr levels have fallen to some 80% of resting levels, lactate, H^+, and phosphate ions accumulate, but

ATP levels do not normally fall (9). Accumulation of one or more of these metabolites must be involved in the genesis of fatigue since recovery does not occur if the circulation is occluded. Increasing [H^+] has been shown to inhibit cross-bridge formation between actin and myosin (reducing tension generation), to reduce action potential propagation (inhibiting recruitment of sarcomeres), and to inhibit Ca^{2+} reuptake by the SR, prolonging relaxation time (10–12). However, reductions in force of 50% or more can be induced by mild fatigue with minimal changes in pH, and low-frequency fatigue occurs in glycolytic disorders such as McArdle's disease, in which lactate is not produced and pH tends to *increase* during contraction (7). Thus, H^+ accumulation alone cannot explain physiological fatigue. Other possibilities include increased phosphate ions, particularly diprotonated phosphate ($H_2PO_4^{--}$), which has been correlated with force reductions in myofibrils (10–12). the nature of the fatiguing contractions is a further consideration. The metabolic cost of *eccentric* contractions (contractions opposing lengthening of muscle, as for example in the action of quadriceps when walking down stairs) has been shown to be 4–6 times *less* than that of concentric contractions, yet the fatigue generated is considerably greater (13). This appears to be due to mechanical disruption of sarcomeres produced by eccentric contractions and is associated with a marked rise in creatine kinase (CK) levels in the blood (14,15).

Central Fatigue

Central, neuronal factors may play a part in the genesis of fatigue under certain circumstances. The amplitude of the integrated electromyogram increases progressively during submaximal contraction of a muscle at constant force, due to recruitment of motor units of progressively increasing threshold. This tends to counteract the loss of force experienced by individual motor units as they fatigue. As fatigue progresses, the rate of development of mechanical twitch tension and relaxation reduces, but this is normally matched by a reduction in the motor unit firing rate to preserve the optimal muscle tension for the prevailing electromechanical and metabolic state of the muscle fibers (16,17). Abnormalities of this "matching," such as lack of motivation, or more complex disturbances, such as neurotransmitter abnormalities, may underlie central fatigue.

Recent studies have shown that amplitudes of motor evoked potentials elicited by transcranial magnetic stimulation are transiently decreased by exercise (18). This has been interpreted as reflecting decreased efficiency of central synaptic function, as a result of increased exercise-induced "central drive," causing continuous neurotransmitter mobilization and relative depletion, by analogy to events at the neuromuscular junction. Such decrements were not observed with electrical transcortical stimulation, which is believed to activate axons directly rather than via neurotransmitter release (18).

Muscle Pain

Patients experiencing myalgia may refer to muscle pain in a number of ways, but in general, pain or discomfort that is fairly persistent is usually described as an aching, soreness, tenderness, or stiffness (the latter particularly following exercise) while intermittent or paroxysmal pain is usually referred to as spasm or cramp. In either case, the pain may be diffuse or focal and usually has a dull, aching quality, which is often poorly localized, although it may be severe.

It is important to attempt to distinguish true myalgia from pain arising from joints, bones, and other connective tissues. For example, joint pain tends to be localized and worsened by moving the joint, but may be referred to a site remote from the pathology, such as the thigh muscles or knee in hip disease. Joint disease may also be associated with painful reflex muscle spasm, for example in cervical whiplash injuries. Bone pain tends to have a deep, boring, continuous quality, is often worse at night, and is usually unaffected by movement. The myofascial pain syndrome (fibromyalgia) often results in painful, focal areas of tenderness (trigger points) while polymyalgia rheumatica results in stiff painful muscles, particularly on rising, although the pathology resides in the blood vessels and connective tissues, rather than muscle (see Chapter 45). In addition, muscle pain may be a secondary feature of central nervous system disorders, causing changes in tone such as spasticity, dystonia, parkinsonian rigidity, tetanus, and stiff-man syndrome, as well as peripheral neurogenic disorders such as motor neuron disease and neuromyotonia.

Pathogenesis of Muscle Pain

Pain signals from muscles are transmitted centrally by thin myelinated A- and unmyelinated C-fibers, whose unencapsulated, multiple branched endings are distributed through the endomysium and other connective tissues, particularly in the regions of tendon insertions, fasciae, and aponeuroses. These nerve endings respond to mechanical and chemical stimuli; some appear to be unimodal, others polymodal. By contrast, fibers subserving Golgi tendon organs and muscle spindles do not transmit pain. The most significant pain fiber responses are induced by bradykinin, serotonin, histamine, K^+, and H^+ (19). Mechanical stimuli appear to produce brief, rapid-onset pain, while chemical nocioceptive signals are slow-onset and chronic. Examples of the former include muscle swelling from hematoma, hypertonia, reflex muscle spasm, cramp, and contracture, while myalgia due to necrotic, inflammatory, and metabolic disturbances is primarily chemically mediated (19).

Such pain mechanisms are common to many tissues and viscera. However, skeletal and cardiac muscle pain may also result from ischemia. Familiar examples include angina and intermittent claudication of the legs due to occlusive vascular disease and jaw claudication in cranial arteritis. Ischemic muscle pain is also a common experience in normal subjects (see below) and has thus been the object of considerable experimental investigation. Despite this, the origin of such pain remains unknown. It does not appear to be the result of ischemic hypoxia per se, since postexercise pain does not increase if ischemia is maintained by occlusion of blood supply to a muscle and is not augmented by ischemia prior to exercise. Lactic acid production also cannot be the sole determinant, since pain is a feature of contractures that occur in McArdle's disease and phosphofructokinase deficiency, in which lactate is not produced (19).

Since muscle pain induced by vascular occlusion resolves quickly with restoration of blood supply, a variety of small molecules that could be rapidly cleared from the muscle have been suggested as pain mediators (19). For example, adenosine, produced in abundance by ischemic exercise, has been shown to induce ischemia-like pain after intra-arterial injection (20); but levels are not increased in postexercise venous blood in McArdle's disease (21).

Finally, muscle pain may result from sarcomeric disruption produced by eccentric muscle contractions (14,15) (see above). This differs from mechanical pain due to muscle swelling or stretching in that the pain and tenderness develop progressively over the 24–48 hr following cessation of exercise, concomitant with a rise in CK levels, which may reach tens of thousands of units per liter (22).

Muscle Pain Syndromes

Myalgia can be classified into muscle pain occurring in normal individuals under predictable conditions and myalgia occurring in muscle diseases.

In normal subjects, the origin of myalgia is defined by its relationship to exercise. Muscle pain occurring during exercise has been shown to be due to ischemia and remits rapidly after cessation of exercise (19). Examples include forearm pain while carrying a heavy suitcase or repetitive isometric handgrips. The ischemia is thought to result from increasing intramuscular pressure through local tissue edema, causing occlusion of intramuscular blood vessels. Pain occurring after heavy or unaccustomed exercise, which may persist for days, has been shown to result from damage induced by eccentric muscle contractions, as described above (19).

Muscle diseases may present or be associated with a number of forms of muscle pain. The myalgic syndromes to be considered include diffuse and focal muscle pain, cramps, contractures, and the exertional muscle pain syndrome (EMPS, Table 1). Muscle stiffness may be due to myotonia, which can also be painful occasionally and will be considered briefly in this context.

Diffuse and Focal Muscle Pains. In general, muscle diseases causing myalgia are those in which there is either an active inflammatory or necrotic process or where there is impairment of muscle energy metabolism, in which case myalgia is usually related particularly or exclusively to exertion. These

Table 1 Causes of Myalgia

1. Diffuse and Focal Muscle Pain	
Diffuse myalgia	
With weakness	*Without weakness*
Inflammatory myopathy	Polymyalgia rheumatica
Muscle infections (e.g., viruses, *Toxoplasma*)	Fibromyalgia
	Denny-Brown syndrome
Necrotizing myopathy (drugs, alcohol)	Hypothyroidism
	Steroid withdrawal
Metabolic bone disease	
Focal muscle pain	
With swelling or induration	*Without swelling*
Hematoma	Claudication
Infection (e.g., pyomyositis)	Neurogenic (e.g., brachial plexus neuritis)
Nodular myositis	
Sarcoidosis	
Compartment syndrome	Restless legs
Acute alcoholic myopathy	Painful legs, moving toes
2. Cramps[a]	
a. In normal subjects	
Effort cramps	e.g., "stitch," calf or hamstrings in football players, swimmers, etc.
Rest cramps	Nocturnal cramps in the elderly
	Postexercise
	Pregnancy
Secondary to fluid balance and electrolyte disturbances	Sweating ("miners," etc.)
	Diarrhea
	Hemodialysis
	Diuretics
b. In neuromuscular diseases	
Neurogenic disorders	Motor neuron diseases
	Polyneuropathies
	Radiculopathies
Secondary to systemic metabolic diseases	Uremia
	Dysthyroidism
	Cirrhosis
	Hypoadrenalism
Fasiculation syndromes	Denny-Brown syndrome
	Other syndromes (see text)

[a]After Joekes (27).

disorders include infections of muscle, idiopathic inflammatory myopathies, metabolic and endocrine myopathies, and some drug-induced muscle diseases. The individual conditions are described in detail in other chapters.

In focal muscle pain, the presence of muscle swelling indicates infection, local inflammation, hematoma or infarction, or rarely a tumor. Focal muscle pain without swelling is less likely to be due to definable muscle disease.

Diffuse myalgia with associated weakness suggests inflammatory, metabolic, or toxic/drug-induced myopathy, while lack of weakness will suggest primary nonmyopathic disorders.

A history suggestive of myoglobinuria, the presence of weakness, and a raised serum CK level are important clues to the presence of primary muscle disorders in patients complaining of muscle pain. In the absence of such signs, no clear diagnosis may be forthcoming after full investigation (23). When exercise is the key provocative factor, such cases may reasonably be included under the heading of exertional muscle pain syndrome, which is discussed further below, but there are a number of other poorly understood clinical syndromes in which myalgia is a feature, including "growing pains" in children, restless legs (Ekbom's) syndrome, and Spillane's syndrome of painful legs and moving toes. However, pain mechanisms in such cases probably do not originate in the muscle.

Cramp. A cramp may be defined as a spasmodic, painful involuntary contraction of muscle. It is associated with a visible or palpable focal contraction of muscle fibers, often accompanied by secondary abnormal posturing of the affected joint and relieved by stretching the muscle. On electromyography, the contracted area exhibits high-frequency discharges of motor units, usually of normal morphology.

There is considerable evidence that cramps are neurogenic in origin. Thus, cramps are a feature of anterior horn cell diseases, radiculopathies, and polyneuropathies, and not of pure myopathies, and are frequently preceded by and associated with fasiculations, or less commonly myokymia, which are also thought to be neurogenic. Denny-Brown proposed that cramp derived from ectopic axonal discharges, chiefly from presynaptic end-plate terminals (24); thus, cramp can be induced by repetitive nerve stimulation distal to a peripheral nerve block, but then relieved by stretching the muscle, obviating the need to invoke any centrally mediated reflex mechanisms in the genesis of cramp (25). However, recent studies on patients with refractory cramps and myokymia suggested that the somatodendritic membrane of anterior horn cells in such cases might exhibit bistable polarization characteristics, in which inputs (via low-threshold afferents such as from muscle spindles) would shift membrane potential to a second stable state that allows repetitive discharge, causing cramp or myokymia depending on the extent of the motor neuron pool involved in the process (26). Such cramps could be relieved by strong afferent inputs to the motor pool, sufficient to excite high-threshold pain and tactile fibers (see below). Cramps may thus arise from both central (somatic) and peripheral (axonal) mechanisms.

Joekes (27) produced a simple classification of cramps, based on precipitating factors (Table 1). Briefly, cramp may occur in normal individuals under altered physiological circumstances, such as exercise and electrolyte losses, and in pathological situations, in neuromuscular diseases, and with certain drugs.

Ordinary, simple cramps are a common experience and usually of minor consequence. Therefore, we know little of their pathogenesis or associations. In younger individuals, cramps are most frequently related to severe exertion (effort cramps). Examples include "stitch" (cramps of the intercostal or abdominal wall muscles), swimmer's cramp (particularly in cold water), and cramp in track athletes and football players precipitated by sudden muscle contraction (of the calf muscles, for example) without a proper warmup period.

Rest cramps appear to be particularly common in the elderly, and night cramps in such patients may be severe and distressing (28). The propensity to cramping is increased in situations where there is alteration in the ionic milieu of nerve endings, for example hyponatremia due to replacement of sweat losses by drinking water ("stoker's cramp"), in metabolic disorders associated with low Na^+, Ca^{2+}, or Mg^{2+} such as dysthyroidism, uremia, cirrhosis, and also in pregnancy. Cramps can also be iatrogenic, following the use of certain drugs or as a result of hemodialysis for example, where the effect appears to be related to hypo-osmolality, since such cramps are reversed by hypertonic saline or glucose (29). Familial forms of cramp and myalgia have also been described (30,31) having similarities with isolated cases of idiopathic muscular pain fasciculation syndrome (32) (Denny-Brown and Foley syndrome), suggesting a common origin in a relatively benign peripheral nerve disorder. Fasciculations in isolation are also extremely common in normal subjects and were self-reported in 70% of 539 healthy medical personnel in one survey (33).

Cramps have to be distinguished clinically from entities such as dystonia (e.g., "writer's cramp"), from reflex muscle spasms, as in acute low back and neck pain, and from other central nervous system disorders characterized by motor unit hyperactivity, including tetanus, stiff-man syndrome and neuromyotonia (Chapter 29). Tetany induced by hypocalcemia and more commonly by stress-related hyperventilation may be associated with fatigue, myalgia, and cramps, which in extreme cases results in the familiar carpopedal spasm. The phenomenon of normocalcemic tetany or spasmophilia (latent tetany) probably has a similar origin.

Contracture. A contracture (physiological rigor) is clinically similar to cramp, but always results from repeated muscle contraction, and does not occur at rest. The fundamental distinction is that contracture, unlike a cramp, is electromyographically silent. The causes of contracture are given in Table 2. Its most familiar association is with enzyme defects in the energy-yielding glycolytic pathway, such as McArdle's disease and other rarer disorders of this type (see Chapter 33). Indeed, the phenomenon seems to be peculiar to glycolytic disorders and is not a feature of other metabolic myopathies such as lipid storage myopathies or mitochondrial disorders. It is therefore possible that lack of intracellular acidosis in exercise, characteristic of glycolytic defects, is an important predisposing factor. For example, contracture in experimental models in which CK activity or glycolysis is inhibited is not clearly related to fall in ATP or PCr concentrations, but may be related to high free ADP levels enhancing actin and myosin interaction (34).

Contracture also occurs with intrinsic defects in the muscle relaxation mechanism. This can be seen in paramyotonia congenita, in which the duration of muscle contraction may outlast the period of myotonic discharge (35) (Chapter 28), and in Brody's disease (Ca^{2+} ATPase deficiency of the SR, Chapter 37) in which there is impaired reuptake of Ca^{2+} after contraction, resulting in a prolonged active state of myofibrillary interaction (36). Finally, myoedema, the "mounding phenomenon" induced by percussion of the muscle belly, is also electrically silent. This may be seen in a number of conditions, notably cachexia, hypothyroidism (associated with prolonged tendon reflex relaxation times), and the so-called "rippling muscle syndrome" in which waves of contraction spread across the muscle after percussion (37). The mechanisms involved in such cases are not known, but again may relate to defects in Ca^{2+} handling by the SR.

Exertional Muscle Pain Syndrome. Some patients experience myalgia that is mainly or exclusively exercise-related, in which case a disorder of muscle energy metabolism must be considered. The circumstances in which exertional myalgia arises may provide a clue to the type of metabolic disturbance. Muscle contraction at high force and frequency depends on anaerobic glycolysis, and such activity rapidly causes fatigue, pain, and ultimately contracture in McArdle's disease and the other rare glycolytic diseases. In contrast, exertional myalgia in lipid storage myopathies, such as carnitine palmityl transferase (CPT) deficiency, tends to occur after prolonged steady-state exercise when fatty acids are vital for muscle energy metabolism. However, as mentioned, detailed investigation often fails to identify a cause for exertional myalgia. While psychogenic causes may be invoked in some cases, other patients sometimes show nonspecific abnormalities on investigation, such as mildly raised CK levels, myopathic EMG changes, and biopsy features suggestive of a muscle disease (38), and it seems likely that other specific metabolic and structural disorders of muscle remain to be discovered. Indeed, recent research suggests that some cases of EMPS may be due to entities such as Brody's disease and dystrophinopathy (39). Certain nonspecific histological and ultrastructural abnormalities are associated with EMPS more commonly than expected, including internalized capillaries (40) and tubular aggregates (41), but their pathophysiological significance is unclear. A deficiency of oxidative type 1 fibers, with a reduced type 1/2 fiber ratio, has also been associated with exertional myalgia (42).

Muscle Stiffness. Muscle stiffness is also a difficult symptom to interpret. It may indicate an increase in tone, due to spacticity or rigidity, myotonia, or one of the rare "stiff muscle syndromes"—

Table 2 Causes of Contracture

Abnormal glycolytic metabolism	e.g., McArdle's disease
Failure of relaxation	Paramyotonia
	Brody's disease
	Rippling muscle syndrome
	Myoedema

neuromyotonia and stiff-man syndrome. Alternatively, patients may use the term to refer to a form of muscle pain, particularly after exertion.

Myotonia is a condition of prolonged muscle contraction, with subsequent slowed relaxation, after activation, voluntarily or through mechanical or electrical stimulation. It results from inherited or acquired abnormalities of the electrical properties of the sarcolemma, which alter the resting membrane potential and time constants of depolarization and repolarization, resulting in repetitive discharges that give a characteristic EMG signal (myotonic runs, or "dive-bomber" potentials, see Chapter 4). Patients may not complain of typical myotonic symptoms, such as difficulty releasing objects, but may ascribe complaints of muscle stiffness to "arthritis," for example. Patients with weakness associated with myotonia, as in myotonic dystrophy, often find the myotonia to be only a minor problem and rarely seek treatment for it. In nondystrophic myotonias, and in periodic paralysis with myotonia, the myotonia may be severe and sometimes frankly painful. Typically, myotonia improves with exercise (the warmup phenomenon) although in paramyotonia congenita, symptoms are more common in the cold and worsen with exercise. The dystrophic and nondystrophic myotonias are reviewed in Chapters 26–28.

Systemic Review

Other aspects of the history may be helpful in some instances. Patients with exertional myalgia should be questioned regarding possible episodes of myoglobinuria, in which episodes of weakness, usually with marked myalgia, are associated with passing dark red-brown (cola-colored) urine. This is a sign of rhabdomyolysis and can be a feature of a number of metabolic, toxic, and inflammatory myopathies. It has also been described following status epilepticus, prolonged myoclonus, dystonia, and various forms of strenuous exercise including military training, marathon running, weight lifting, and body building, in individuals without evidence of muscle disease or metabolic defect in muscle (43).

Dysphagia and dysphonia may be features not only of a number of neurogenic disorders, but also of polymyositis and certain forms of muscular dystrophy. Breathlessness and respiratory difficulties, particularly at night, may be due to muscle disease, notably in acid maltase deficiency and polymyositis.

Skin rash is sometimes a feature of inflammatory muscle disease associated with autoimmune disorders and may be transient, as may arthralgia.

Family History

The family history may give vital clues in the diagnosis of muscle disease, by establishing that a disorder is inherited and showing the nature of the genetic trait. This may obviate the need for invasive investigations in other affected family members. It is essential to construct a detailed family tree in such cases, to include details of ages, causes of death, and birth order, and specific inquiries concerning consaguinity should be made. Such a device serves as an excellent *aide memoire* for patients and relatives and often prompts patients to seek supplementary information from other relatives when they leave the clinic.

Drug History

A careful drug history, and history of possible exposure to toxins, is essential in a proper assessment of muscle disease. It is now realized that muscle can be affected by a wide range of therapeutic and recreational agents, sometimes with severe or chronic consequences, and that drugs can mimic a number of naturally occurring neuromuscular diseases. This subject is discussed in detail in Chapter 31.

EXAMINATION

General Examination

Some forms of muscle disease are associated with physical signs in other systems. For example, a "heliotropic" erythematous, macular skin rash is characteristic of dematomyositis (Fig. 1), but may be subtle, particularly in children. Dilated nailfold capillaries may be revealed by transillumination in this and other connective tissue disorders, such as lupus, rheumatoid disease, and scleroderma, which have

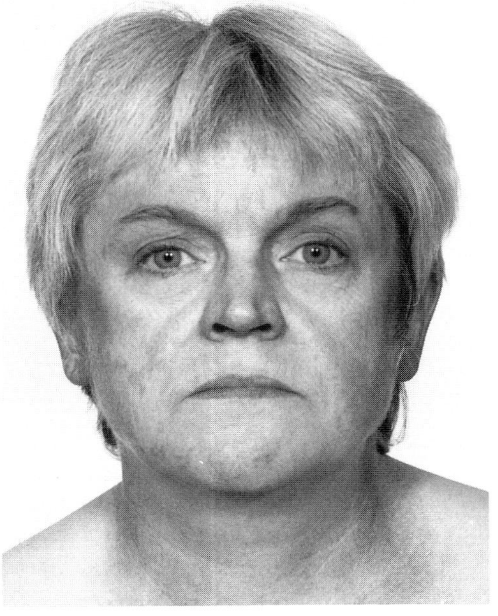

Figure 1 "Heliotrope cyanosis" skin rash in dermatomyositis.

their own additional clinical characteristics (Fig. 2). Patients with myotonic dystrophy may have a typical facies, at least in advanced cases. This includes premature frontal balding, bilateral ptosis, bilateral facial weakness, and wasting of the temporalis muscles and sternomastoid muscles (Fig. 3).

A wide range of systemic abnormalities may be found in patients with mitochondrial cytopathies, notably retinal pigmentary changes, deafness, and ataxia, while cardiac disease, in the form of dysrhythmias and subclinical electrocardiographic abnormalities, is found in these disorders and in myotonic and X-linked dystrophies, and occasionally in polymyositis.

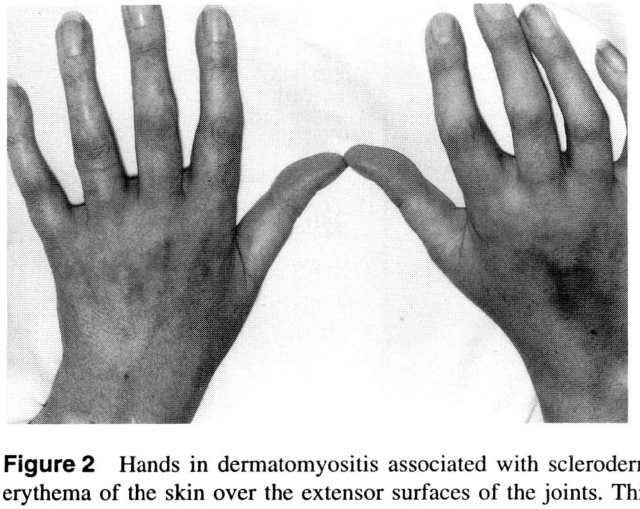

Figure 2 Hands in dermatomyositis associated with scleroderma. There is acrosclerosis with marked erythema of the skin over the extensor surfaces of the joints. This patient also had myasthenia and both conditions responded to thymectomy (see Ref. 45).

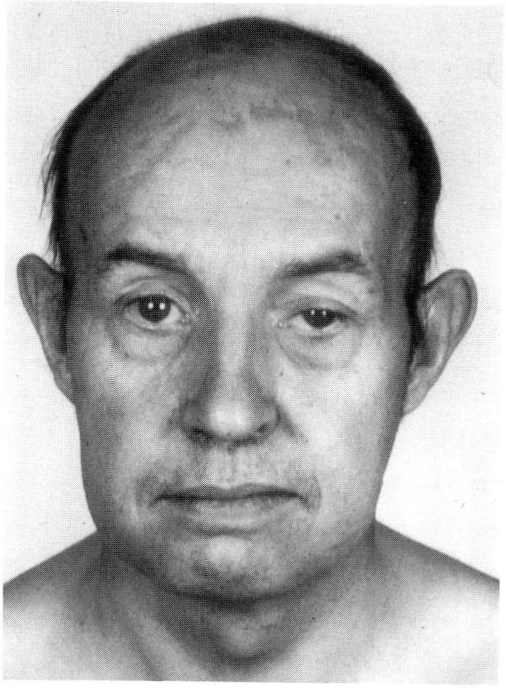

Figure 3 Facies in myotonic dystrophy. In this moderately affected patient, there was marked frontal balding, bilateral ptosis (more marked on the left than on the right); and facial weakness, and the sternomastoids were wasted and weak. (From Blackwood W, Perkin GD, Rose FC, Shawdon HH, Eds. *Slide Atlas of Neurology*. London: Gower Medical Publishing, 1983, by permission.)

Cranial Nerves

Retinitis pigmentosa, with pigmentary changes best seen in the periphery of the retina, may be seen in mitochondrial disorders. Ptosis with external ophthalmoplegia may be a feature of several diseases, including mitochondrial myopathies, oculopharyngeal muscular dystrophy, and certain congenital myopathies, as well as being a cardinal clinical sign of myasthenia. When marked, it may be associated with overactivity of the frontalis muscle with furrowing of the forehead, and in severe cases with neck extension, in an attempt to keep the eyes open. The temporalis muscles should be examined for evidence of wasting, which results in hollowing of the temples, while weakness of these muscles and of the masseters and pterygoids results in weakness of jaw opening and closure, which if asymmetrical results in trismus. Facial weakness is a feature of facioscapulohumeral muscular dystrophy (FSH) as well as myotonic dystrophy, but in FSH the lips adopt a characteristic patulous appearance (*bouche de tapir*), with loss of the normal cupid's bow, ptosis is unusual, and the sternomastoids are preserved (Fig. 4). Dysphonia may be evident, or develop, during history taking. In weakness of the bulbar musculature, the voice may take on an adenoidal quality, due to nasal escape, the weakness of the palate resulting in difficulties pronouncing hard "C" and "G" sounds. Tongue weakness, with lingual dysarthria, may be quite subtle and is often associated with dysphagia. Macroglossia is rare, but may occur in acid maltase deficiency (Fig. 5) and amyloid myopathy (Chapter 51). Tongue atrophy associated with denervation is much more common. When examining for tongue fasciculation, it is important to have the tongue at rest in the floor of the mouth, with the tip at the edge of the lips, and not protruded, since this may result in fine involuntary movements even in normal subjects. Bilateral fasciculation is almost invariably due to motor neuron diseases.

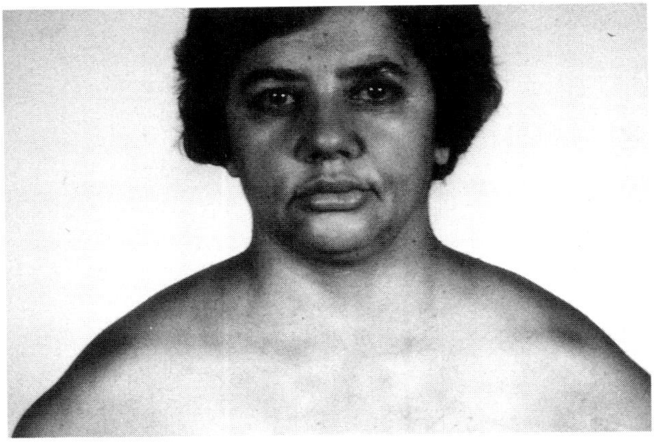

Figure 4 Facies in facioscapulohumeral muscular dystrophy. Striking facial weakness, with patulous lips, transverse smile and inability to whistle.

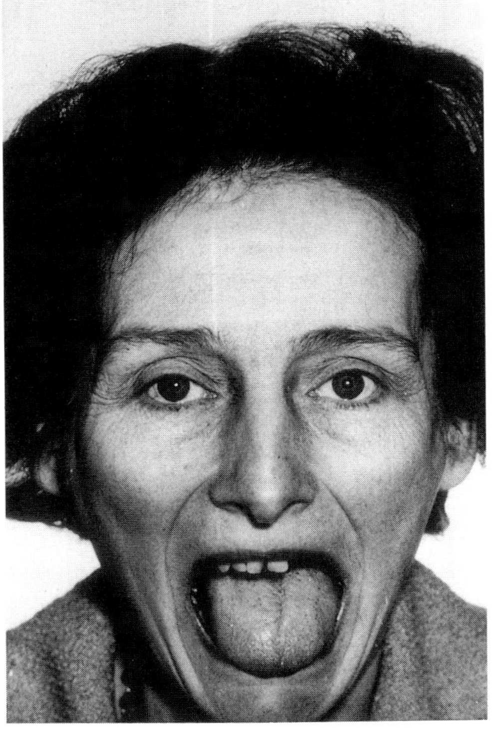

Figure 5 Macroglossia in acid maltase deficiency. The tongue is pale and enlarged, resulting from excess glycogen deposition.

The Limbs

The state of the motor system should be described under the following headings, relating to the classical pattern of examination of inspection, palpation, and percussion: bulk, tone, power, reflexes, and coordination. Finally, gait and posture should be noted.

Inspection: Bulk and Involuntary Movements

Muscle wasting (the term "atrophy" is often used to describe the effects of denervation) is common to most myopathies. Muscular hypertrophy is less common, but is a notable feature of myotonia congenita, X-linked dystrophies, hypothyroid myopathy, lipodystrophy, and occasionally in chronic denervation (Fig. 6).

Involuntary movements of muscle can be classified clinically in terms of the size of the elements involved. Fasiculations represent random, spontaneous contractions of small groups of muscle fibers. When associated with weakness and atrophy, they indicate denervation, as for example in motor neuron disease. Benign fasiculations are seen in normal subjects and patients with cramp-fasiculation syndrome (see above). Myokymia is a recurrent or persistent quivering or wave-like undulation of muscle fibers ("bag of worms") and again is essentially neurogenic. Localized myokymia, most often in the eyelid or periorbital muscles, is a common benign phenomenon related to fatigue and stimulants such as caffeine. When it affects a larger area of the face, it may indicate brainstem disease, for example in multiple sclerosis, while in limb muscles it can be seen in radiculopathy and peripheral nerve disease. Generalized myokymia, in association with muscle stiffness, is characteristic of neuromyotonia (see Chapter 29).

Palpation: Muscle Tone and Mobility

Muscle tone is typically normal or reduced in myopathies. Hypotonia is a particularly important sign in pediatric neuromuscular disease, and the "floppy infant" typically adopts a "frog leg" position and

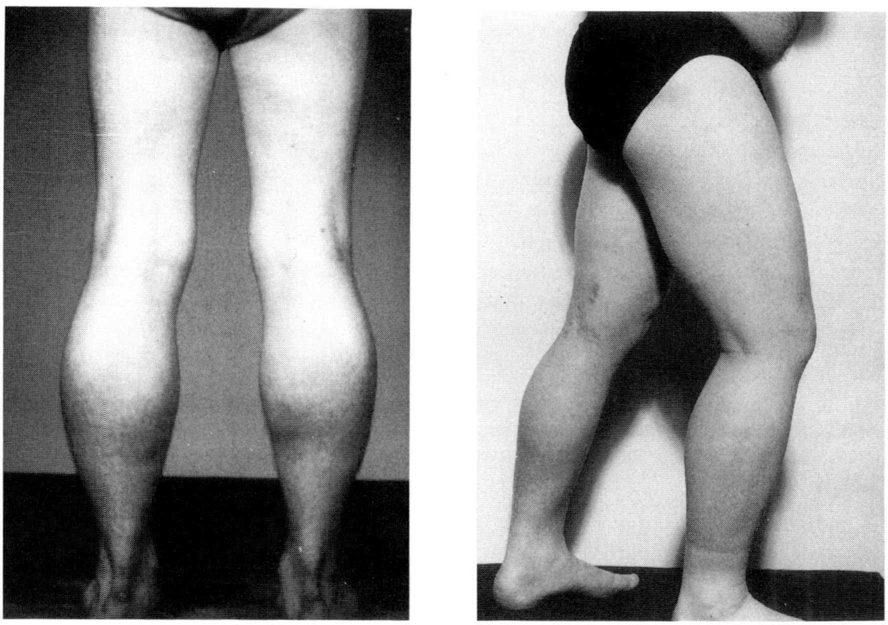

(a) (b)

Figure 6 Muscular hypertrophy in (a) Becker muscular dystrophy and (b) Hoffman syndrome (hypothyroid myopathy, see text).

demonstrates an abnormal head lag on being pulled from a horizontal position. Hypertonicity is much less common. It can be seen in neuromyotonia due to continuous muscle fiber activity and has a different quality from that found in pyramidal and extrapyramidal disease. Maneuvers to examine tone also allow for detection and measurement of contractures (which may be mobile or fixed deformities), and grip myotonia should also be sought (see Chapter 26).

Percussion

The tendon reflexes tend to become depressed, and may eventually be lost, in progressive myopathies in which muscle is replaced with fat and fibrous tissue, although in dystrophies the ankle jerks are often retained until late in the disease (a point of differentiation from neurogenic diseases). However, reflexes may be brisk or well preserved in inflammatory, hypocalcemic, and hyperthyroid myopathies.

Striking a muscle belly with a tendon hammer may also elicit useful physical signs. Fasciculations may become evident, while in cachectic states and hypothyroidism, myoedema (the "mounding phenomenon," produced by muscle fiber swelling in the percussed area) may be seen. Percussion myotonia can also be demonstrated. This is most commonly elicited by striking the thenar eminence, which results in a slow, sustained drawing of the thumb across the palm for a second or two, before it slowly returns to the rest position. Sometimes striking the thenar eminence or other muscles, such as the belly of a forearm muscle or tongue, may result in a persistent groove or dimple.

Strength Testing

Muscle strength is traditionally recorded using semiquantitative grading systems, such as the Medical Research Council (MRC) scale. Although this has been criticized in that it is nonlinear and, in its basic form, is insensitive, it requires no equipment, is simple to perform, and with minor modifications can be shown to give remarkably low intra- and interobserver variations, with reasonably linear characteristics (Fig. 7). Nevertheless, quantitative techniques using dynamometers are being used increasingly, notably in clinical trials, and this is discussed in detail in Chapter 2.

Posture and Gait

Proximal weakness, particularly of gluteus maximus, the paraspinals, and often the abdominal wall muscles, may result in hyperlordosis of the lumbar spine, with abdominal protrusion. Associated weakness of the quadriceps may be compensated by locking the knees in hyperextension, leading to a tendancy to toe-walk, while additional hip flexion weakness may result in the patient swinging the legs forward in a characteristic rolling, myopathic gait. Progressive weakness of axial muscles may result in a kyphoscoliosis. Such signs may be obvious when weakness is marked, but milder degress of weakness can sometimes be demonstrated. The Trendelenburg test, in which postural stability is examined by having the patient stand on each leg in turn, may reveal pelvic girdle (particularly gluteal) weakness, while toe- and heel-walking tests can demonstrate weakness of foot dorsi- and plantar flexion at an early stage. Proximal weakness can also be confirmed by looking for elements of Gower's sign, when a patient rises from the floor, or by noting difficulty on rising from a chair with the arms folded across the chest.

Creatine Kinase Measurement

Measurement of the serum CK level is such an important part of the investigation of a patient with muscle disease that it would seem appropriate to include it as part of the clinical assessment. However, CK is not always increased in muscle disease and may be raised in a variety of other situations, notably after exercise and in denervating conditions. If exercise is excluded, very high CK levels are seen in only a few conditions, notably rhabdomyolysis from any cause, and in inflammatory myopathies and X-linked dystrophies. In general, the elevation of the CK reflects the severity of the muscle damage, except where loss of muscle fibers is so severe that CK levels begin to fall, as in the late stages of Duchenne dystrophy.

Raised CK levels are sometimes a chance finding, in patients asymptomatic for neuromuscular disease. Detailed investigation of such cases, however, may lead to a number of diagnoses, including

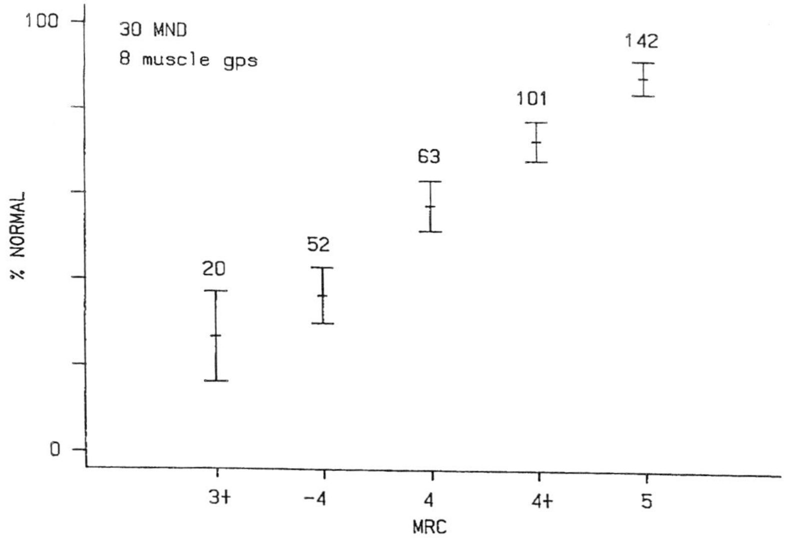

Figure 7 Charing Cross Strength Scale. Relationship between modified MRC grades and hand-held dynamometry readings. Hand-held dynamometry readings expressed as percent of mean for each muscle in a normal population. The mean and 95% confidence limits are shown, together with the number of muscles tested in each grade. Eight different muscle groups were tested in 30 motor neuron disease patients by the same observer. The grade definitions were: 0 = no contraction; 1 = flicker of contraction; 2 = contraction sufficient to move joint; 3- = just able to sustain contraction against gravity; 3 = can sustain contraction against gravity; 3+ = minimal active movement against resistance; 4 = moderate weakness; 4+ = minor weakness; 5 = normal. (From Ref. 55.)

polymyositis, inclusion body myositis, mitochondrial cytopathies, McArdle's disease, malignant hyperthermia state, and congenital myopathies (44). Recent studies have also shown that some patients with "idiopathic hyperCKemia" are female carriers of gene defects for Duchenne and Becker dystrophy, or males with clinically very mild dystrophinopathies (39).

Symptomatic Treatment of Weakness, Fatigue, and Myalgia

Some common neuromuscular diseases, particularly those that have an immunological basis, such as inflammatory demyelinating neuropathies, myasthenia, and inflammatory myopathies, can be greatly improved or even cured by the use of steroids, immunoglobulins, plasma exchange, cytotoxic drugs, radiotherapy, and, in some cases, thymectomy (45). Unfortunately, at present it is not possible to offer more than palliative and supportive treatment for weakness in genetically determined disorders, such as the dystrophies and metabolic myopathies, although gene therapy strategies are being developed and biochemical adjuvants may be of help in, for example, the mitochondrial disorders (see Chapter 41).

Improvement of strength and limb mobility through physiotherapy is the mainstay of general treatment for progressive muscular weakness, coupled with orthotic and orthopedic intervention in more advanced cases (see Chapters 56 and 57). Drugs such as anabolic steroids and clenbuterol can certainly be shown to increase muscle mass and strength in the short term (46) but have not so far gained general acceptance in the treatment of myopathies.

Fatigue characteristics of muscle can be altered through training, with sustained, low-impact routines promoting aerobic muscle metabolism and increasing the proportion of oxidative type 1 muscle fibers, while high-impact anaerobic regimes increase type 2 fiber complement (47).

The management of myalgia is still more problematic. When myalgia is due to defined inborn metabolic errors, avoidance of precipitating activities and "living within the limits" of the disease may be all that can be suggested. Dietary manipulations may be of some benefit, to circumvent lack of substrates resulting from the metabolic block, or to promote alternative energy-producing metabolic pathways.

Exertional myalgia and cramp will sometimes respond to calcium antagonists such as verapamil and nifedipine. Nifedipine was shown to alleviate cramp induced by exercise (48) and by hemodialysis (49), and verapamil improved symptoms in some patients with EMPS in a placebo-controlled trial (38). This drug also proved to be useful in patients with Brody's disease (36,50). Quinine, however, remains the mainstay of treatment for cramp and its effectiveness has been shown in a number of controlled trials (51,52). A variety of other drugs, including diazepam, dantrolene, phenytoin, and procaineamide, are also said to be beneficial (53). As mentioned above, certain cramp syndromes may result from central mechanisms involving anterior horn cell somatic membranes and may be relieved by suitable inhibitory inputs. Thus, transcutaneous nerve stimulation proved beneficial in one patient (54).

Management of individual muscle diseases is discussed in more detail in other chapters.

REFERENCES

1. Ridsdale L. Tired all the time. Br Med J 1991; 303:1490–1491.
2. Borg G. Perceived exertion as an indicator of somatic stress. Scand J Rehabil Med 1970; 2:92–98.
3. Borg GAV. Psychophysical basis of perceived exertion. Med Sci Sports Exerc 1982; 14:377–381.
4. Lenman AJR, Tulley FM, Vrobova G, Dimitrijevic MR, Towle JA. Muscle fatigue in some neurological disorders. Muscle Nerve 1989; 12:938–942.
5. Wesseley S, Powell R. Fatigue syndromes: a comparison of chronic "post-viral" fatigue with neuromuscular and affective disorders. J Neuro Neurosurg Psychiatry 1989; 52:940–948.
6. Edwards RHT. Human muscle function and fatigue. In: Porter J, Whelan R, eds. Human Muscle Fatigue: Physiological Mechanisms, Ciba Foundation Symposium No 82, London; Pitman Medical, 1981: 1–18.
7. Cooper RG, Stokes MJ, Edwards RHT. Myofibrillar activation failure in McArdle's disease. J Neuro Sci 1989; 93:1–10.
8. Hainault K, Duchateau J. Muscle fatigue, effects of training and disuse. Muscle Nerve 1989; 12:660–669.
9. Editorial. Fatigue. Lancet 1988; 2:546–547.
10. Wilson JR, McCully KK, Mancini DM, Boden B, Chance B. Relationship of muscular fatigue to pH and diprotonated Pi in humans: a ^{31}P-NMR study. J Appl Physio 1988; 64:2333–2339.
11. Wilkie DR. Muscular fatigue: effects of hydrogen ions and inorganic phosphate. Fed Proc 1986: 45:2921–2913.
12. Nosek TM, Fender KY, Godt RE. It is diprotonated inorganic phosphate that depresses muscle force in skinned skeletal muscle fibres. Science 1987; 236;191–193.
13. Newham DJ. The consequences of eccentric contractions and their relationship to delayed onset muscle pain. Eur J Appl Physiol 1988; 57:353–359.
14. Newham DJ, McPhail G. Mills KR, Edwards RHT. Ultrastructural changes after concentric and eccentric contractions of human muscle. J Neurol Sci 1983; 61:109–122.
15. Friden J. Muscle soreness after exercise: implications of morphological changes. Int J Sports Med 1984; 5:57–66.
16. Bigland-Ritchie B, Jones DA, Woods JJ. Excitation frequency and muscle fatigue: electrical responses during human voluntary and stimulated contractions. Exp Neurol 1979; 64:414–427.
17. Bigland-Ritchie B, Kukulka CG, Lippold OCJ, Woods JJ. The absence of neuromuscular transmission failure in sustained maximal voluntary contractions. J Physiol 1982; 330:265–278.
18. Brasil-Neto JP, Cohen LG, Hallett M. Central fatigue as revealed by postexercise decrement of motor evoked potentials. Muscle Nerve 1994; 17:713–719.
19. Mills KR, Newham DJ, Edwards RHT. Muscle pain. In: Wall PD, Melzek R, eds. Textbook of Pain. New York: Churchill Livingston, 1985: 319–330.
20. Sylven C, Jonzon B, Fedholm BB, Kaijser L. Adenosine injection into the brachial artery produces ischaemia-like pain or discomfort in the forearm. Cardiovasc Res 1988; 22:674–678.
21. Sinkeler SPT, Joosten EMG, Wevers RA, Binkhorst RA, Oerlemans FT, van Bennekom CA, Coerwinkel MM, Oei TL. Ischaemic exercise test in myoadenylate deaminase deficiency and McArdle's disease: measurement of plasma adenosine, inosine and hypoxanthine. Clin Sci 1986; 70:399–410.
22. Newham DJ, Jones DA, Edwards RHT. Large delayed plasma creatine kinase changes after stepping exercise. Muscle Nerve 1983; 6:380–385.
23. Mills KR, Edwards RHT. Investigative strategies for muscle pain. J Neurol Sci 1983; 58:73–88.

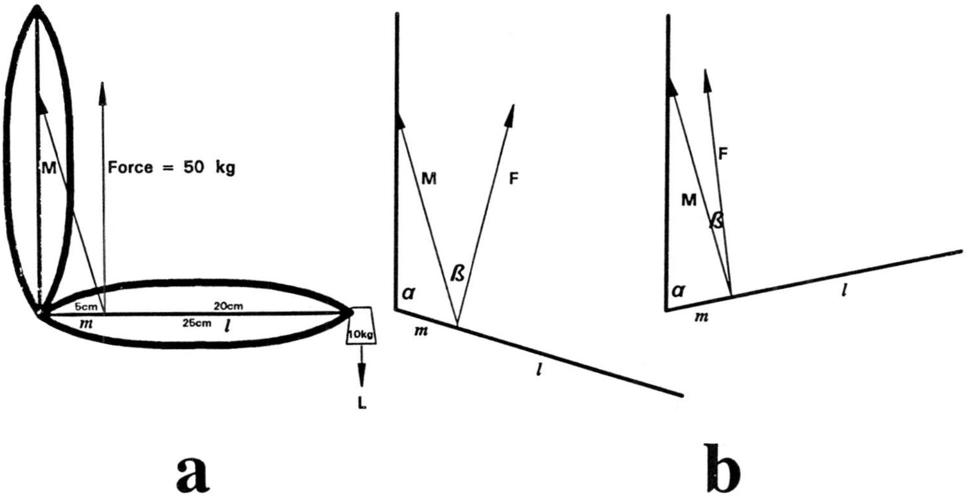

Figure 1 (a) Muscle strength measured by maximum weight that can be held isometrically for a specified length of time. Note that the force exerted by the muscle is much higher than the load lifted due to the unfavorable mechanical advantage of the skeletal system. (b) Effect of change in joint angle α during muscle strength measurement on the measured muscle force. The balance torque around the joint is equal to $M \bullet m = L \bullet l$ provided that muscle force (M) and the load (L) are acting at a right angle to the lever arm. The effective vector F is equal to M cos β.

or decelerates the load. The load that can be supported eccentrically by a muscle is greater than the load that can be lifted concentrically or held isometrically. It is easier to hold a specific weight than to lift it, and it is easier to lower it. In this regard, muscle tension may be expressed as strength, power, work, endurance (fatigue), rate of energy utilization (i.e., oxygen consumption during standardized work), velocity of contraction and tension development, relaxation rate, coordination of movement, or any other indices defined under controlled conditions, depending on the task to be performed (9–18).

Basic Principles

The force (tension) that a muscle can generate is proportional to the physiological cross-sectional area (CSA), which is determined by the number of sarcomeres arranged in parallel (19). The range through which a muscle shortens varies with its fiber length, determined by the number of sarcomeres arranged in series. Longer muscles are faster compared to shorter fibers contracting at the same rate (20) (Table 2).

The angle at which a muscle pulls (called the pennation angle, determined by the origin and insertion of the muscle) will influence the mechanical arrangements of the muscle lever system and subsequently muscle force and velocity. The larger the angle, the less is the tension and the velocity of shortening potential (4,20).

Table 2 Number of Sarcomeres in Selected Lower Limb Extensors and Flexors Obtained from Cadavers

Muscle group	Number of sarcomeres (10^4)
Quadriceps	3.12
Hamstrings	4.30
Dorsiflexors	2.93
Plantar flexors	1.53

Within a given muscle, the number and length of muscle fibers are determined by the design of muscle fiber arrangement relative to the muscle tendon (21). Pennate muscles can pack larger numbers of fibers at a greater density in the same volume than can be packed in fusiform muscle of comparable size and thickness. The total tension-length relationship of paralleled-fiber muscle is linear while that of pennated muscle is curvilinear (Fig. 2) (22). The paralleled-fiber muscle can exert a small constant increment in tension over a longer distance while the pennated muscle can increase the rate of tension development over a shorter distance. Muscle mass, length, and pennation angle measured in human muscle from cadavers demonstrate that the cross-sectional area of a muscle is directly proportional to its weight and inversely related to individual fiber length (20). The work a muscle can produce depends on the volume of the muscle, while its power depends on muscle volume and fiber type (4).

Measurement of Muscle Strength Under Static Conditions

Muscle strength is the maximum force that a group of muscles can exert isometrically in a single voluntary contraction. It is expressed as the maximum voluntary isometric contraction (MVIC) (23).

Muscle strength, under static conditions, is commonly assessed by fixed or hand-held mechanical (hydraulic, cable tensiometer, springs, etc.) or electronic (load cell strain gauge, piezoelectric transducer, etc.) dynamometers (Table 3).

Muscle strength under static condition may also be measured as the amount of weight (kgwt) a muscle can hold isometrically at certain angle (Fig. 1a).

Measurement of isometric muscle strength by dynamometer assumes a linear relationship between the load (kgwt) and the deformation of the sensing element device of the dynamometer. Therefore, it is essential to establish empirically this linearity between the actual weight (load) used and the force registered by the measuring device, within the desired range of measurement. Periodic calibration of the measuring device with variable loads is mandatory and cannot be overemphasized. Dynamometers that cannot be readily calibrated (like many of the hand-held dynamometers) should not be used.

Muscle force acts on a joint in a perpendicular fashion. Therefore, the isometric muscle strength test is a torque measurement. Subsequently, muscle force may need to be expressed in a moment force (torque) value such as Nm. In clinical trials of large patient populations, measuring the lever arm may be cumbersome and unnecessary if the point of application of force and the joint angle remain constant.

Since muscle strength depends on the existing mechanical advantage (Fig. 1b), the point of force application, angle, and direction should be the same on repeated serial measurements of muscle strength

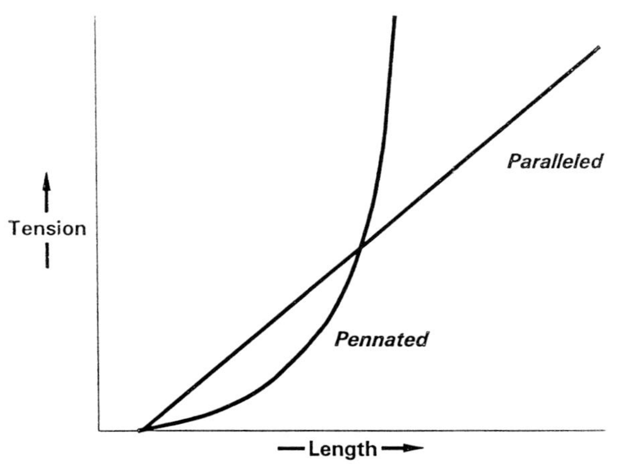

Figure 2 Total length-tension line of a parallel- versus pennated-fiber muscle.

Table 3 Measured Force Under Static Conditions

$Force_{kgf}$	$= K_{kgf/cm} \times \Delta_{cm}$	mechanical dynamometer
$Force_{kgf}$	$= K_{kgf/mV} \times \Delta_{mV}$	electronic dynamometer
$Force_{kgf}$	$= stress_{kgf/cm^2} \times$ cross-sectional $area_{cm^2}$	force platform
$Torque_{Nm}$	= force × moment arm	torque dynamometer
$Endurance_{kg\text{-}sec}$	= force × time	electronic dynamometer

K = constant (i.e., stiffness coefficient of a spring or strain gauge); Δ = deformation in the sensing device (i.e., changes in length of the spring in a mechanical dynamometer or change in voltage due to change in resistance of the sensing device (i.e., strain guage or quartz in piezoelectric) calibrated against standard weight; mV = millivolt; Nm = Newton-meter.

(i.e., in natural history studies or drug therapy clinical trials). The attachment of the dynamometer should be perpendicular (at 90 degrees) to the limb. Force obtained during strength testing is dependent on the distance of the dynamometer attachment from the joint axis of rotation. The shorter the distance between the point of force application and the center of joint rotation, the higher is the measured force. The calculated torque, however, will be the same.

Many factors will influence the mechanical advantage and joint angle during muscle strength measurements, such as lack of stabilization, change in joint and body position, deformation, contracture, spasticity due to disease progression, and gravitational as well as other forces acting on the joint (24–26). In designing MVIC testing for neuromuscular disease patients, one should select the position, angle, and direction that secure maximum stabilization, comfort, safety, and reproducibility, not necessarily the maximum mechanical advantage. The joint angle that produces optimal mechanical advantage for selected joints is given in Table 4 (24).

If the point of force application, angle, or direction of pull cannot be controlled, or if perpendicularity cannot be achieved, one should express muscle strength as torque by measuring the lever arm, or use a torque dynamometer that senses torsional twisting. One example in which the above condition might not be controlled is during repeated measurements of muscle force in children. With growth, the lever arm will change significantly as the anatomical definition of the point of force application is altered, especially with disease (e.g., children with Duchenne muscular dystrophy) (11,27–32). In older populations these assumptions may be easily met if the joint stays intact.

Measurement of Muscle Strength Under Dynamic Conditions

Muscle force or tension may be assessed under dynamic conditions as the maximal force a group of muscles can generate at a specified velocity (12). Muscle strength under dynamic conditions is expressed as net muscle moment (joint moment, joint muscle moment, tendon moment, joint torque), muscle power, muscle work, and muscle endurance (17,18) (Table 5). Muscle work, power, and

Table 4 Optimal Joint Angles at Which Different Muscle Groups Exert Maximum Force During Isometric Contraction

Muscle group	Angle	Muscle group	Angle
Shoulder adduction	110°	Hip adduction	110°
Shoulder abduction	180°	Hip abduction	180°
Shoulder flexion	180°	Hip flexion	180°
Shoulder extension	90°	Hip extension	90°
Elbow flexion	40°	Knee flexion	160°
Elbow extension	115°	Knee extension	115°
		Ankle dorsiflexion	125°
		Ankle plantar flexion	90°

Table 5 Measurement of Muscle Strength Under Dynamic Condition

Muscle moment expressed in Newton–meters

$M_{Nm} = I\alpha$ — M = net muscle moment; I = mass moment of inertia of segment (kgm^2); α = angular acceleration (rad/sec^2)

Muscle power expressed in watts

$P_w = M\omega$ — P = muscle power (watts, W); M = net muscle moment (Nm); ω = joint angular velocity (rad/sec)

Muscle work expressed in joules

$W_j = P \times time$ — (1 J = 1 Nm = 1 Ws)

Net muscle moment (M) is the sum of moments on the limb due to internal and external forces (muscle contraction, agonist and antagonist, friction between segments, gravity, ground reaction, arm pulling or pushing against dynamometer, leg pushing against ground, etc.).

Ergometry

Bench step

W (kgm/min) = body weight (kg) × bench height (m) × number of steps per minute

Treadmill

W (kgm/min) = body weight (kg) × velocity (m/min) × sin α

α is angle of the treadmill inclination

Cycle

W (kgm/min) = wheel resistance (kg) × distance (m) × rpm × time (min)

endurance can also be measured using Isokinetic dynamometer and other ergometers such as bench step, cycle, treadmill (Table 5), or as a rate of caloric output (kcal/min) (i.e., as the rate of oxygen consumption during standardized work on a treadmill, or cycle ergometers) (33). These techniques are widely applied in cardiopulmonary and sports medicine (33).

Accurate muscle strength evaluation under dynamic conditions requires detailed analysis of joint and muscle segments and application of mathematical inverse solution techniques (17,18). This procedure requires combining kinematics (displacement, velocity, and acceleration computed from high-speed photography and TV image analysis) and kinetics (force and moment of force computed with the use of ground reaction analysis, force platform, and in some cases EMG) techniques. These procedures are widely applied in gait, motion, and sport activity analyses (17,18,21).

APPLICATION OF MUSCLE STRENGTH MEASUREMENTS IN MUSCLE DISEASE

Measurements Under Static Conditions

The two most common methods of isometric muscle strength testing in clinical neurology are the manual muscle test (MMT) and the quantitative muscle strength test (QMT).

Manual Muscle Test

The MMT is an ordinal presentation of muscle strength. MMT has several standard forms all of which were shown to be reliable (high intrarater, moderate interrater) measures of muscle strength, and to be useful tools for clinical trial and natural history of muscle weakness in Duchenne muscular dystrophy, poliomyelitis, facioscapulohumeral (FSH) dystrophy, and amyotrophic lateral sclerosis (10,27,30,32, 34–42).

The major limitations with MMT, however, are lack of sensitivity to change and lack of correlation with absolute muscle strength. That is, no clinically useful prediction of QMT score from MMT score is possible, especially in the strong muscles (5,42–48). In clinical drug trial studies, the inability to detect a small physiological effect of a drug on muscle strength may result in potentially important, biologically significant effects of the drug being overlooked (49,50). This lack of sensitivity of MMT may affect the reliability and statistical power of the trial. For example, twice the number of patients is required for MMT to achieve the same power compared with QMT (43).

Quantitative Muscle Strength Test

The QMT measures the MVIC in an absolute unit of force, using a fixed or hand-held dynamometer, as described earlier. Fixed isometric muscle testing using an electronic base dynamometer (i.e., a load cell) is an ideal way to measure muscle strength of a variety of small and large, strong and weak muscles (11,43–45,49). In muscle disease, QMT was shown to be a reproducible, sensitive, and accurate measure of muscle strength in weak and strong muscles and even in muscles that were difficult to standardize in MMT (11,16,29,40,43,44,51–58). A standardized QMT has not been established. A proposed standardization of QMT procedures is given in Table 6.

Hand-held dynamometers have the advantage of portability, but may lead to an excursion with increasing force and thereby lead to underestimation of strong muscle and overestimation of weak muscle (57). Nevertheless, hand-held dynamometers correlate well with fixed dynamometers up to 30 kg force and have high intrareter and moderate interrated reliability (54).

An extensive review of muscle strength testing with various types of fixed and hand-held dynamometers, their validity, range and sensitivity, reliability, and testing procedures confirms their pivotal role in quantitating strength for scientific and clinical purposes (53).

The sensitivity of fixed dynamometer QMT in detecting disease progression has been assessed in a large-scale clinical trial in 728 ALS patient over 36 clinical centers [unpublished data, ACTS (ALS CNTF Treatment Study group)]. The time between test-retest was 1–3 weeks (mean = 15.1 days). The average

Table 6 Proposed Standardization Procedures for QMT of MVIC

Instrumentation
Fixed electronic load cell strain guage.
Force analog digitizing hardware and software—automated data acquisition system.
Comfortable testing table that helps stabilization and constant body position.
Position
Consistent joint angle and lever arm.
Consistent point of application of force.
Consistent direction of pull or push.
Consistent stabilization procedures.
Only muscles around the joint that can assume the predefined position of testing should be tested. Loss of this ability should be considered end-point.
Comfortable, painless, secure, and safe position.
Defintion of measurement
Define the number of contractions (i.e., 2 MVIC) and the rest period (3–4 sec contraction with 3–4 sec rest).
Define readout as average–not peak–of high sampling rate to avoid effect of jerking and instability within each trial. Report highest MVIC of two trials to eliminate the effect of fatigue or other errors.
Instructions
Inform patients of the intent of the test (i.e., MVIC).
Demonstrate action to patient.
Provide objective and consistent instruction that achieves consistent level of motivation.
Environmental factors
Private and quiet room–eliminates any factors that might affect instrument performance and patient cooperation.

change in strength across clinics between visits was –0.029 megascore units (SEM = 0.0059). The fact that we are able to detect a change in patient strength over a 2-week period suggests that the QMT is quite sensitive. In fact, we would still have 90% power to detect a change in patient status with only 313 patients over a 2-week period. Changes over a longer time interval would require even fewer patients to detect. Variability in test-retest data over different clinic sites might affect reliability in large clinical trials. However, while the elapsed time between measurement varies by clinic, we found little relationship between elapsed time and variability in combined megascore. Moreover, a plot of within-clinic variance against the quintile of the chi-square distribution shows that the variability between clinics is not practically significant, given the interpatient biological diversity in strength. In addition, there were no significant differences between clinics in the average change in this study between the two (baseline and initial visit) visits.

Measurements Under Dynamic Condition

Muscle torque, work, and power under dynamic conditions are widely measured using isokinetic dynamometers. Isokinetic muscle strength testing was shown to be reliable (high intrarater and moderate interater reliability) (8). An isokinetic dynamometer has been shown to be a sensitive method to quantitate deficit in the early stage of muscle weakness due to ALS (59,60), Parkinson's disease (61), multiple sclerosis (62), spastic hemiparesis (63), and various neuromuscular diseases (64). In ALS the isokinetic muscle strength measured at higher angular velocities significantly decreased despite normal isometric muscle strength (Fig. 3) (59). Isokinetic measurement may also be useful to study the loss of dynamic muscle strength over prolonged periods (Fig. 4, unpublished data). In many motor diseases, such as ALS (65) and poliomyelitis (66), symptomatic muscle weakness measured under static conditions only occurs after substantial neuronal loss. Our observation in ALS, and other neurological disorders, is that clinically significant loss of function in many muscle groups occurs only when isometric muscle strength declines to a level that is below the second percentile for gender-, age-, weight-, and height-matched normal control subjects (unpublished data). Therefore, dynamic muscle strength measurement is critical to quantitate presymptomatic muscle strength loss that may be secondary to upper motor neuron disease in such conditions. As the disease progresses, patients lose isokinetic muscle strength faster than isometric muscle strength. At this stage strength as measured by MMT would still be in the normal range.

Isometric Versus Dynamic Muscle Strength Measurements in Muscle Disease

The preceding discussion indicates that there are many ways to express muscle force or tension development. Muscle strength, as defined by MVIC, refers to one index that is easily, safely, and less expensively measured compared to dynamic strength. Under static conditions, muscle strength is evaluated under a well-defined point in the force-velocity-length relation (isotension, velocity, length, single joint) (Figs. 5 and 6) (17,22). This condition eliminates the extra force due to the elastic passive tension of muscle contractile elements and connective tissue that occurs during dynamic motion (22,67), thereby ensuring reproducibility. Since no significant change occurs in length during isometric contraction, the role of the noncontractile elastic element is not significant. Therefore, isometric strength measurement can better distinguish between contractile and noncontractile tissue pathology. In addition, measurement of muscle strength under dynamic conditions requires accurate control of velocity, which is very difficult, especially in neuromuscular disease. Since dynamic movements consist of muscular impulses that accelerate and decelerate muscle segments, there are always bursts, delays, absorption, and transmission of muscular energy between muscle groups (agonist and antagonist), body segments, and the joints involved in the motion (17,18,68). In many neuromuscular disorders, especially those of spastic nature, cocontractions of agonists and antagonists occur, leading to inefficiency (69,70). Mechanical efficiency, which is the work performed by the muscle (metabolic energy) in relation to the work done by the load, cannot be accurately estimated without measurement of oxygen uptake (70). Thus, one cannot easily localize weakness. In addition, neuromuscular disease slows movement, which leads to intermittent "near-isometric" holding of limb segments that cannot be quantified without EMG (68,71).

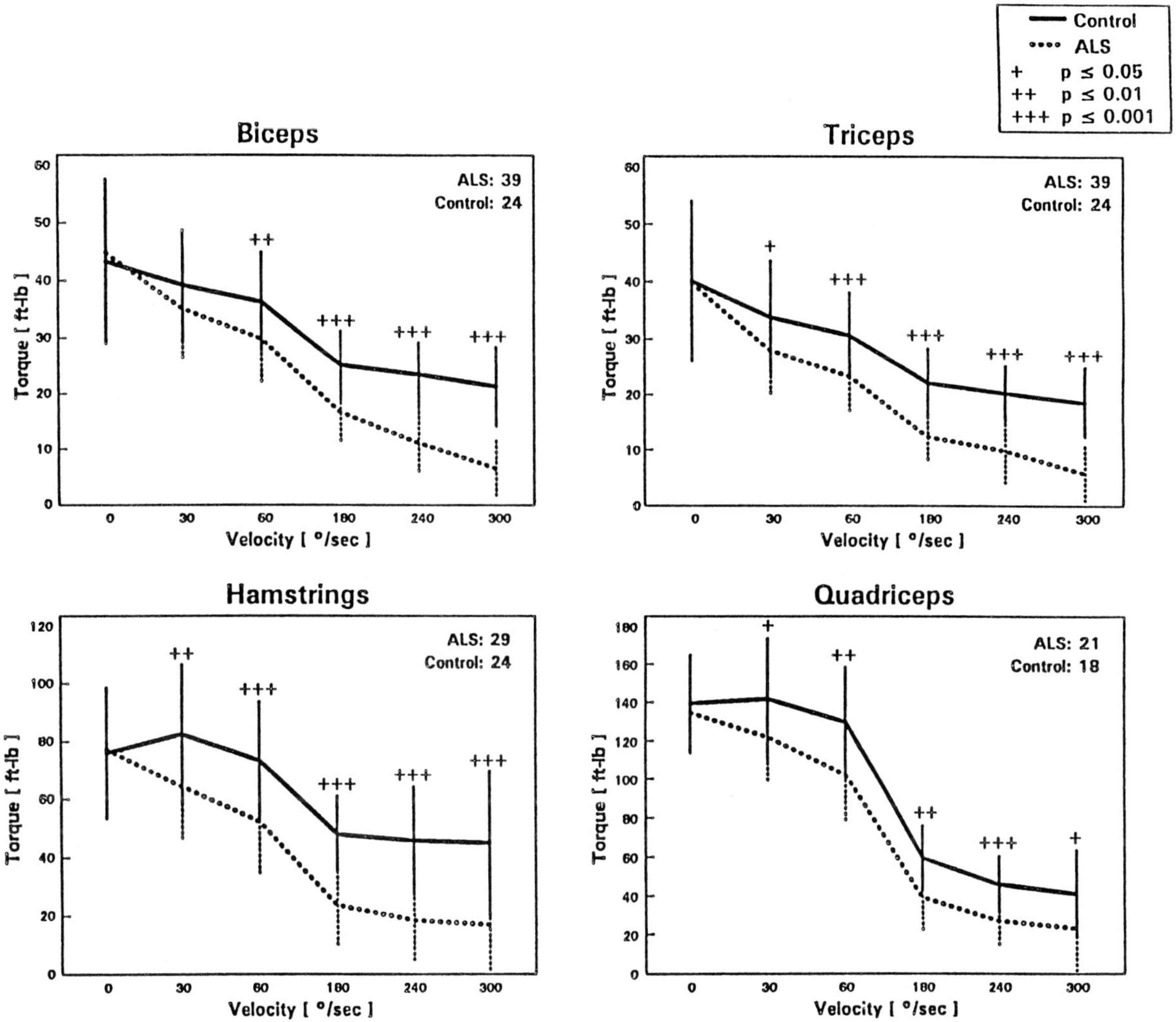

Figure 3 Peak dynamic force at various velocities in ALS vs. male age-matched control subjects within normal peak torque range at isometric measure).

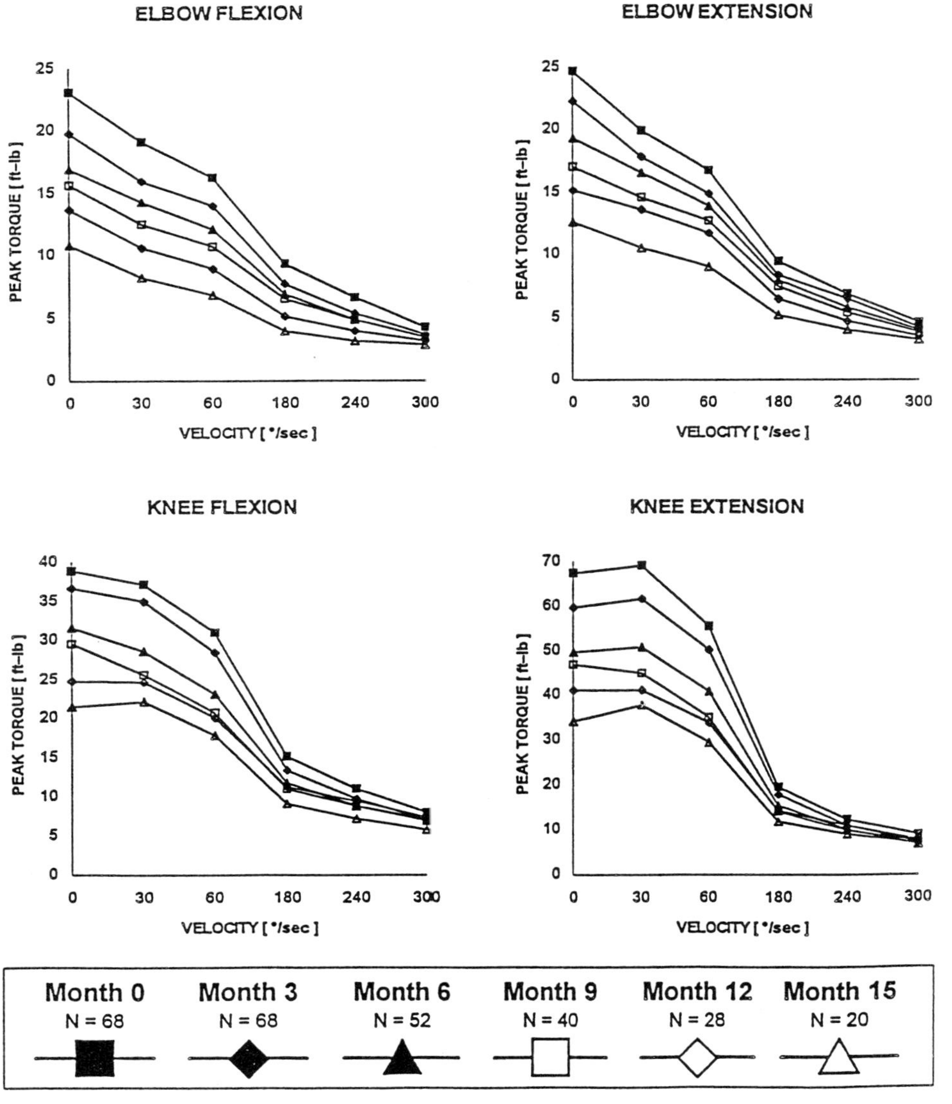

Figure 4 Peak torque during maximal isokinetic elbow and knee flexion and extension at different angular velocities obtained by serial evaluation of ALS patients.

Isometric muscle strength is superior to dynamic muscle strength measurement in neuromuscular disease in terms of fatigue and energy metabolism control. The energy supply for dynamic muscle contraction is complex and requires an efficient energy metabolism system (Fig. 7) (72). In contrast, MVIC is a simple task, commonly evaluated for 3–5 sec. This is an adequate time to reach MVIC without muscle fatigue, which may affect its measurement (72,73). In short-lasting MVIC, the ATP and phosphocreatine stores in the muscle can yield enough energy for the contraction, with additional oxygen present in the muscle (bound to myoglobin) for energy to be delivered aerobically (73).

In muscle disease, fatigue from central and peripheral origin is a common problem and may affect different muscles differently during MVIC (11,14,73). For example, in five consecutive measurements, MVIC in ALS declined consistently in right shoulder abductors, but did not change in right elbow

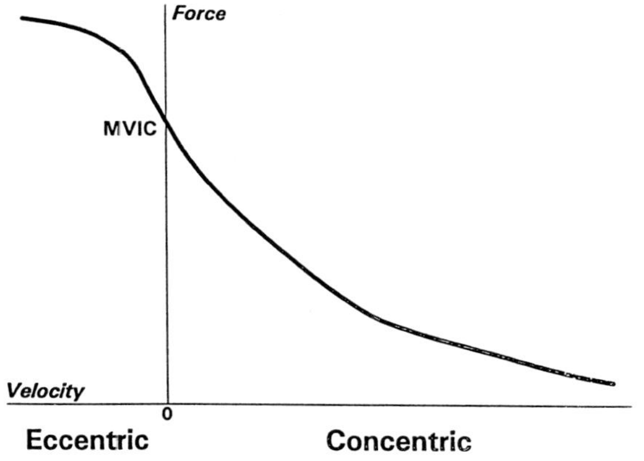

Figure 5 Force-velocity relationship of skeletal muscle. Maximum voluntary isometric contraction (MVIC) is measured isometrically at zero velocity during concentric or eccentric contraction, or dynamically during concentric or eccentric contraction at a given velocity.

extensors (74). This observation is important for selecting muscles that are less susceptible to fatigue in clinical trials, and it is important to take the highest, not the average of MVIC, of consecutive MVIC measurements.

FUTURE CONSIDERATIONS

Technological advances in force transducer design in terms of sensitivity and range, signal conditioning and digital filtering, analog to digital converter design, and low-cost, high-performance microprocessors provide fast and simple tools for muscle strength measurements. With the use of a microcomputer, one is able to instantaneously record, display, and quantitate the strength of a variety of small and large muscles in various primary and secondary muscle diseases. Establishing a data base on the natural history for individual muscle groups in different regions will provide a valuable tool to understand, manage, and treat many muscle diseases.

As mentioned earlier, MVIC measured in an absolute force unit has limited clinical interpretation

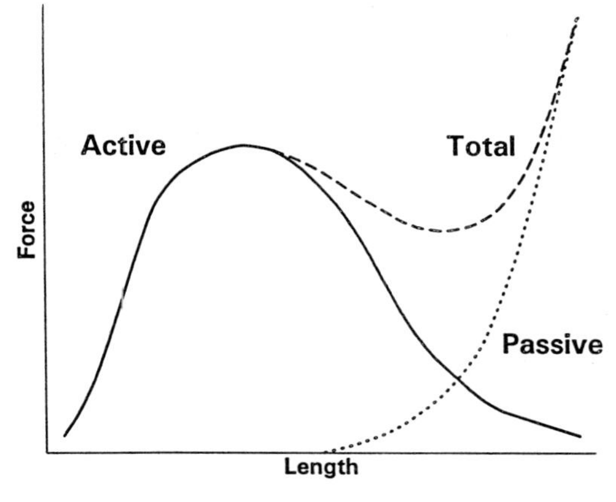

Figure 6 Effect of connective tissue components on the muscle force-length relationship.

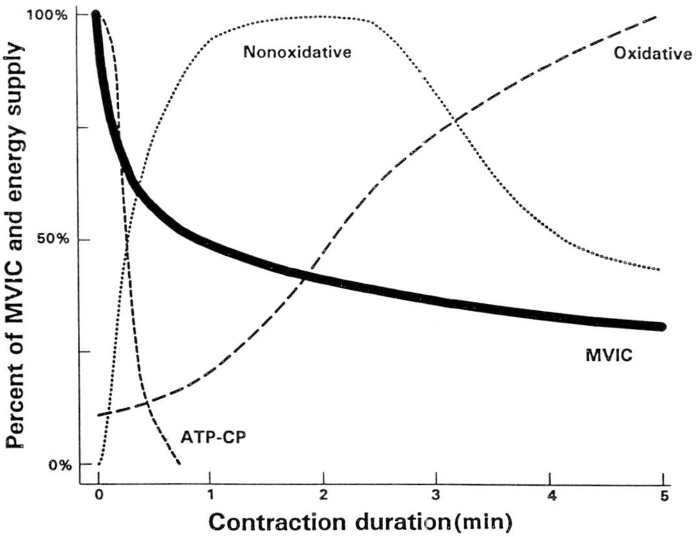

Figure 7 Energy supply of MVIC over time.

without standardized predicted values of MVIC for each muscle group measured, corrected for gender, age, weight, height, and level of physical training in a specific population. Construction of a multifactor prediction regression equation for each muscle group, measured at a standardized position, requires a large sample size (at least 30 in each decade to normalize the data) and is time-consuming. It is essential, however, if clinical inferences about the MVIC are to be made. Obtaining normative data from the literature is extremely difficult and limited (51). Most studies of MVIC are confined to one or a few muscle groups, using different angles, different procedures and instruments, and different populations. MVIC of 10 upper and 10 lower extremity muscle strength, collected from healthy normals, are shown in Figure 8.

Clinically, MVIC of each muscle group and the rate of its loss in muscle disease can be indexed as a decline in MVIC compared to predicted MVIC (such as those done for other physiological capacities, i.e., forced vital capacity, maximum oxygen consumption, etc.). Such a score (percent of predicted) can be useful for comparing loss of muscle strength across muscle groups or patient populations. It can also be combined into a single composite score (mean percent of predicted) over a number of muscle groups to yield an overall measure of muscle strength.

Z-score transformation (raw score– mean/standard deviation) of data with reference to a mean and standard deviation in a specific population, or by reference to a predicted value based on gender, age, and height, has been used in natural history and clinical trials in ALS and FSH dystrophy patients (42,51,75,76). The database, however, was not large enough to account for changes in strength in relation to gender, age, weight, and height. In clinical trials, especially of incurable muscle diseases, the main aim is to detect change, no matter how minor or in which muscle group. Combining scores of muscles that may respond to a treatment with nonresponding muscles may lead to inaccurate conclusions. Composite Z-scores, while achieving the goal of giving approximately equal weight to each muscle group, are difficult to interpret clinically. Furthermore, the mean of a Z-score is no longer a Z-score, so its natural statistical interpretation is lost. Since we have found that the coefficient of variation across a muscle group is nearly constant, the mean percent of predicted has similar statistical properties to the combined Z-score, while at same time having the benefit of a natural clinical interpretation. An additional benefit of percent predicted is that changes in strength over time are adjusted for baseline characteristics. If we consider the change in Z-scores that are based on baseline characteristics (42), the predicted scores cancel and do not influence the change in score.

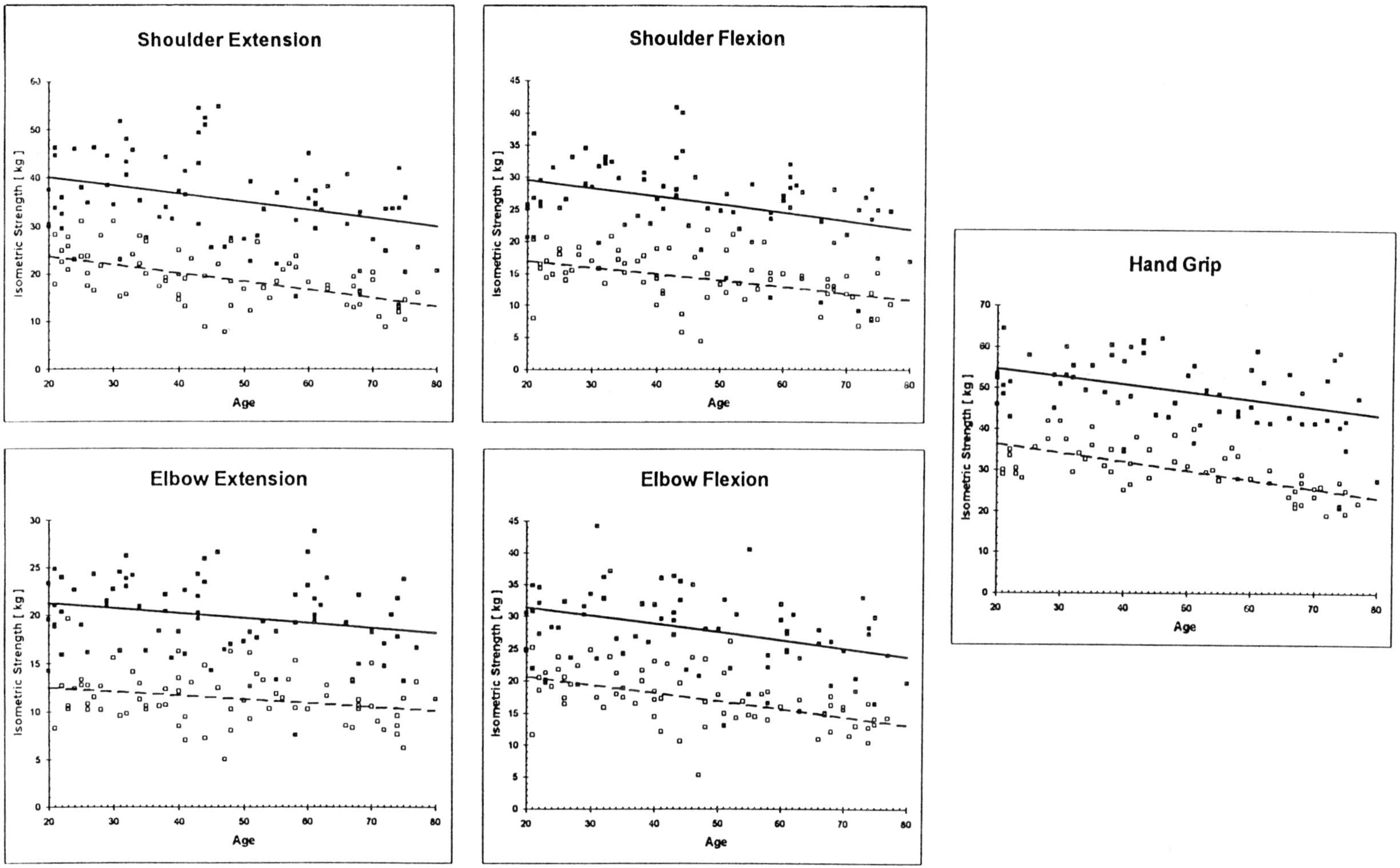

Figure 8 Normal values of maximal voluntary isometric contraction in cross-section of normal untrained male (■) and female (□) subjects across different ages (University of Wisconsin ALS Clinical Research Center, unpublished data).

REFERENCES

1. Dejong RN. The Neurologic Examination. Philadelphia: Harper & Row, 1979:279–328.
2. Burke RE, Edgerton VR. Motor unit properties and selective involvement in movement. Exerc Sport Sci Rev 1975; 3:31–81.
3. Henneman E, Olson CB. Relationship between structure and function in the design of skeletal muscle. J Neurophysiol 1965; 28:581–587.
4. Otten E. Concept and models of functional architecture in skeletal muscle. Exerc Sport Sci Rev 1988; 16:89–137.
5. Beasley WC. Quantitative muscle testing: principles and applications to research. Arch Phys Med Rehabil 1961; 42:398–425.
6. Lovett RW. Fatigue and exercise in treatment of infantile paralysis: study of one thousand eight hundred and thirty six. JAMA 1917; 69:168–176.
7. Kroemer KHE, Kroemer HJ, Kroemer-Elbert KE. Physiological basis of human factors/ergonomic. Adv Hum Factors/Ergonom 1986; 4:63–87.
8. Ostrennig LR. Isokinetic dynamometry: implication for muscle testing and rehabilitation. Exerc Sport Sci Rev 1986; 14:45–80.
9. Komi PV, Salonen M, Jarvinen M, Kokko O. In vivo registration of Achilles tendon forces in man. I. Methodological development. Int J Sports Med 1987; 8:3–8.
10. Edwards RHT. Studies of muscular performance in normal and dystrophic subjects. Br Med Bull 1980; 36:159–164.
11. Edwards RHT. Quantitative methods of assessing skeletal muscle function for therapeutic trials. Acta Cardiomiol III 1991; 2:185–198.
12. Knuttogen HC, Kramer WJ. Terminology and measurement in exercise performance. J Appl Sport Sci Res 1987; 1:1–10.
13. Kroemer KH. Human strength: terminology, measurement and interpretation of data. Hum Factors 1970; 12:297–313.
14. Milner-Brown HS, Mellenthin M, Miller RG. Quantfying human muscle strength, endurance and fatigue. Arch Phys Med Rehabil 1986; 67:530–535.
15. Wiles CM, Karni Y, Nicklin J. Laboratory testing of muscle function in the management of neuromuscular disease. J Neurol Neurosurg Psychiatry 1990; 53:384–387.
16. Wiles CM, Karni Y. The measurement of muscle strength in patients with peripheral neuromuscular disorders. J Neurol Neurosurg Psychiatry 1983; 46:1006–1013.
17. Winter DA. Biomechanics and Motor Control of Human Movement. New York: Wiley, 1991.
18. Zajac FE, Gordon ME. Determining muscle force and action in multi-articular movements. Exerc Sport Sci Rev 1989; 17:187–230.
19. Maughan RJ, Watson JJ, Weir J. Strength and cross-sectional area of human skeletal muscle. J Physiol 1983; 338:37–49.
20. Wickiewicz TL, Roy RR, Powell PL, Edgerton VR. Muscle architecture of the human lower limb. Clin Orthop Related Res 1983; 179:275–283.
21. Rasch PJ. Kinesiology and Applied Anatomy, 7th ed. Philadelphia: Lea & Febiger, 1989:41–65.
22. Arkin AM. Absolute muscle power: the internal kinesiology of muscle. Arch Surg 1941; 395–410.
23. Enoka RM. Muscle strength and its development: new perspectives. Sports Med 1988; 6:146–168.
24. Clarke HH, Bailey TL. Strength curves for fourteen joint movements. J Phys Ment Rehabil 1950; 4:12–16.
25. Chaffin DB. Ergonomic guide for the assessment of human static strength. Indust Hygi Assoc J 1975; 505–511.
26. Hills M, Bohannon RW. Shoulder extension strength is influenced by elbow position. Isokinet Exerc Sci 1992; 2:129–132.
27. Brooke MH, Griggs RC, Mendell JR, Fenichel GM, Shumate JB, Pellegrino RJ. Clinical trials in Duchenne dystrophy. I. The design of the protocol. Muscle Nerve 1981; 4:186–197.
28. Griggs RC, Moxley RT III, Mendell JR, Fenichel GM, et al. Randomized, double-blind trial of Mazindol in Duchenne dystrophy. Muscle Nerve 1990; 13:1169–1173.
29. Hosking GP, Bhat US, Dubowitz V, Edwards RHT. Measurements of muscle strength and performance in children with normal and diseased muscle. Arch Dis Child 1976; 51:957–962.
30. Kilmer DD, Abresch RT, Fowler WM. Serial manual muscle testing in Duchenne muscular dystrophy. Arch Phys Med Rehabil 1993; 74:1168–1171.
31. Mendell JR, Province MA, Moxley RT III, Griggs RC, Brooke MH, et al. Clinical investigation of Duchenne's muscular dystrophy: methodology for therapeutic trials based on natural history controls. Arch Neurol 1987; 44:808–811.
32. Scott OM, Hyde SA, Goddard C, Dubowitz R. Quantification of muscle function in children: a prospective study of Duchenne's muscular dystrophy. Muscle Nerve 1982; 5:291–301.

33. Astrand PO, Rodhal K. Textbook of Work Physiology, 2nd ed. New York: McGraw-Hill, 1977.
34. Daniels I, Worthington C. Muscle Testing: Techniques of Manual Examination. Philadelphia: Harper & Row, 1972.
35. Personius KE, Pandya S, King WM, Tawil T, McDeromott MP. Facioscapulohumeral dystrophy natural history study: standardization of testing procedures and reliability of measurements. Phys Ther 1994; 74:253–263.
36. Blair L. The role of the physical therapist in the evaluation studies of the poliomyelitis vaccine field trials. Phys Ther Rev 1955; 33:437–447.
37. Brooke MH, Florence JM, Heller SL, Kaiser KK, Phillips D, Gruber A, Babcock D. Controlled trial of thyrotropin-releasing hormone in amyotrophic lateral sclerosis. Neurology 1986; 36:146–151.
38. Dalakas MC, Elder G, Hallett M, Ravits J, Baker M, Paddopoulos N, Albrecht P Sever P. A long-term follow-up study of patients with post-poliomylitis neuromuscular symptoms. N Engl J Med 1986; 314:959–963.
39. Florence JM, Pandya S, King W, Robison J, Baty J, Miller P, Schierbecker J Signore L. Intrarater reliability of manual muscle test (MRC) grade in Duchenne's muscular dystrophy. Phys Ther 1992; 72:115–123.
40. Fowler MW, Gardner GW. Quantitative strength measurement in muscular dystrophy. Arch Phys Med Rehabil 1967; 48:629–644.
41. Mitsumoto H, Salgado E, Negroski D, Hanson MR, Salanga UD, Wilber JF, Wilbourn AJ, Brewer AC, Leatherman J. Amyotrophic lateral sclerosis: effects of acute intravenous and chronic subcutaneous administration of thyrotropin-releasing hormone in controlled trials. Neurology 1986; 36:152–159.
42. Tawil R, McDermott MP, Mendell JR, Kissel J, Griggs RL. Facioscapulohumeral muscular dystrophy (FSHD): design of natural history study and results of baseline testing. Neurology 1994; 44:442–446.
43. Aitkens S, Lord J, Bernauer E, Fowler WM, Lieberman JS, Peter B. Relationship of manual muscle testing to objective strength measurements. Muscle Nerve 1989; 12:173–177.
44. Andres PL, Thibodeau LM, Finison LJ, Munsat TL. Quantitative assessment of neuromuscular deficit in ALS. Neurol Clin 1987; 5:125–141.
45. Beasley WC. Instrumentation and equipment for quantitative clinical muscle testing. Arch Phys Med Rehabil 1956; 37:604–621.
46. Bohannon RW. Manual muscle test scores and dynamometer test scores of knee extension strength. Arch Phys Med Rehabil 1986; 67:390–392.
47. Van der Ploeg RJO, Oosterhuis HJGH, Reuvekamp J. Measuring muscle strength. J Neurol 1984; 231:200–203.
48. Wadsworth CT, Krishan R, Sear M, Harrold J, Neilsen DH. Interrater reliability of manual muscle testing and hand-held dynamometer muscle testing. Phys Ther 1987; 67:1342–1347.
49. Munsat TL, Anders PL, Skerry LM. The use of quantitative techniques to define amyotrophic lateral sclerosis. In: Munsat TL, ed. Quantification of Neurologic Deficit. Boston: Butterworths, 1989:129–142.
50. Lane RJM. Measurements of muscle strength. In: Rose FC, ed. Amyotrophic Lateral Sclerosis. New York: Demos, 1990:53–63.
51. Brooks BR, Sufit RL, Depaul R, Tan YD, Sanjak M, Robbins J. Design of clinical therapeutic trials in ALS. In: Munsat TL, ed. Advances in Neurology: Amyotrophic Lateral Sclerosis. New York: Butterworths, 1991:521–546.
52. Brussock CM, Haley SM, Munsat TL, Berndards DB. Measurement of isometric force in children with and without Duchenne's muscular dystrophy. Phys Ther 1992; 72:105–114.
53. Amundsen LR. Muscle Strength Testing: Instrumented and Noninstrumented Systems. New York: Churchill Livingstone, 1990.
54. Bohannon RW. Muscle strength testing with hand-held dynamometers. In: Amundsen LR, ed. Muscle Strength Testing: Instrumented and Non-Instrumented Systems. New York: Churchill Livingstone, 1990:69–88.
55. Laubach LL. Comparative muscular strength of men and women: a review of the literature. Aviat Space Environ 1976; 47:534–542.
56. Resnick JS, Mammel M, Mundale OM. Muscular strength as an index of response to therapy in childhood dermatomyositis. Arch Phys Med Rehabil 1981; 62:12–19.
57. Cook DJ, Soutter DS. Strength evaluation in neuromuscular disease. Neurol Clin 1987; 5:101–123.
58. Ioldings DM, Smith LK, Spender WA. Muscle testing. Part 2. Reliability in clinical use. Phys Ther Rev 1961; 41:249–256.
59. Sanjak M, Carey P, Belden DS, Sufit RL, Brooks BR. Isokinetic versus isometric muscle strength loss early in the course of ALS. Neurology 1992; 42:454.
60. Sufit RL, Clough JA, Schramm M, Conard J, Erickson L, Brooks BR. Isokinetic assessment in ALS. Neurol Clin 1987; 5:197–212.
61. Koller W, Kase S. Muscle strength testing in Parkinson's disease. Eur Neurol 1986; 25:130–133.
62. Armstrong LE, Winant DM, Swasey PR, Seidle ME, Carter AL, Gehlsen GM. Using isokinetic dynamometry to test ambulatory patients with multiple sclerosis. Phys Ther 1983; 63:1274–1279.

63. Nakamura R, Hosokawa T, Tsuji I. Relationship of muscle strength for knee extension to walking capacity in patients with spastic hemiparesis. Tohoku 1985; 145:335–340.
64. Griffin JW, McClure MH, Bertorini TE. Sequential isokinetic and manual muscle testing in patients with neuromuscular disease. Phys Ther 1986; 66:32–35.
65. Sobue G, Sahashi K, Takahashi A, Matsuoka Y, Muroga T, Sobue I. Degenerating compartment of motor neurons in ALS: possible process of motor neuron loss. Neurology 1983; 33:654–657.
66. Sharrard WJW. Poliomyelitis: The distribution of paralysis. In: Garling ER, Ross JP, eds. British Surgical Practice: Surgical Progress. London: 1956:83–103.
67. Cavagna GA, Dusman B, Magaria R. Positive work done by a previously stretched muscle. J Appl Physiol 1968; 24:21–32.
68. Quanbury AO, Winter DA, Reimer GD. Instantaneous power and power flow in body segments during walking. J Hum Movement Stud 1975; 1:59–67.
69. Lewis SF, Haller RG. Skeletal muscle disorders and associated factors that limit exercise performance. Exerc and Sport Sci Rev 1989; 17:67–113.
70. Sanjak M, Paulson D, Sufit RL, Reddan W, Beaulieu D, Erickson L, Shug A Brooks BR. Physiologic and metabolic response to progressive and prolonged exercise in amyotrophic lateral sclerosis. Neurology 1978; 37:1217–1220.
71. Falconer K, Winter DA. Quantitative assessment of co-contraction at the ankle joint in walking. Electromyogr Clin Neurophysiol 1985; 25:135–148.
72. Bergstrom J, Harris RC, Hultman E, Nordesjo LA. Energy rich phosphagens in dynamic and static work. In: Pernow B, Saltin B, eds. Muscle Metabolism During Exercise. New York: 1971:341–355.
73. Asmussen E. Muscle fatigue. Med Sci Sport 1979; 11:313–321.
74. Guiloff RJ, Eckland DJA. Observation on the clinical assessment of patients with motor neuron disease. Neurol Clin 1987; 5:171–192.
75. Munsat TL, Hollanders D, Andres PL, Finison L. Clinical trials in ALS: Measurement and natural history. In: Munsat TL, ed. Advances in Neurology: Amyotrophic Lateral Sclerosis. Boston, Butterworths, 1991:515–519.
76. Andres PL, Finison LJ, Conlon T, Thibodeau LM, Munsat TL. Use of composite scores (megascores) to measure deficit in amyotrophic lateral sclerosis. Neurology 1988; 38:405–408.

3
Exercise Tests

Russell J. M. Lane
Charing Cross Hospital and Charing Cross and Westminster Medical School, London, England

INTRODUCTION

Myalgia, fatigue, and cramps are common symptoms of patients with neuromuscular disorders and, in particular, patients with metabolic myopathies. Several forms of investigation may be appropriate to the analysis of such complaints, but the diagnostic protocol will often include some type of exercise test. In general terms, exercise tests are of two types; ischemic exercise tests (IET), used in the diagnosis of disorders of intermediary muscle metabolism, and incremental and steady-state exercise tests using a bicycle ergometer or treadmill to assess lactate production during or after exercise, for example in patients with suspected mitochondrial myopathy. Both forms of test are essentially screening tools, in that they may indicate that a metabolic abnormality is present but will not usually specify its nature.

ISCHEMIC EXERCISE TESTS

In 1951, McArdle reported a patient who experienced muscle pain and contracture on exertion and showed that his muscle failed to produce lactate when doing work under ischemic conditions (1). McArdle postulated from this observation that the patient's muscle disease resulted from an inborn error of glycolytic metabolism. McArdle's disease was the first metabolic myopathy to be defined, and the ischemic exercise test was subsequently used widely to identify patients with impaired lactate production due to this and other glycolytic disorders. In addition, the later identification of a metabolic myopathy due to myoadenylate deaminase deficiency (MADD) by Fishbein and colleagues (2), characterized by decreased ammonia production and thus an increased lactate-to-ammonia production rate during exercise, added a further dimension to the test, and measurements of plasma ammonia are now routinely performed as part of an IET.

In practice, performing a needle muscle biopsy for histochemical examination for MADD, myophosphorylase deficiency, and phophosfructokinase deficiency is relatively straightforward and may be diagnostic, obviating the need for an ischemic exercise test, but an abnormal IET result may indicate one of the less common metabolic myopathies, or perhaps the MADD carrier state (3).

Essentially, the IET is performed by taking a baseline blood sample from a forearm vein at rest, inflating a blood pressure cuff applied to the upper arm to well above systolic blood pressure, and asking the patient to exercise the forearm muscles (for example, by opening and closing the hand or squeezing a ball) for either a fixed period of time or until exhaustion or unbearable discomfort occurs. The cuff is then deflated and further venous samples are collected, immediately and at various time points after exercise, for plasma lactate and ammonia measurements. Many variations of this fundamental procedure have been described. A number of studies have attempted to standardize the test conditions in order to optimize lactate and ammonia production, and to increase the sensitivity of the procedure, and most laboratories have defined protocols and internal standards. Techniques relating lactate production to the amount of work performed (4,5), varying the duration of exercise (6,7), and using electrically stimulated muscle contractions rather than voluntary effort (8) have been reported.

In one recent study (9) a "standardized" IET, based on previous work on optimization of responses (10), compared isometric handgrip contractions at 80% of maximal grip strength to exhaustion, against a "nonstandardized" technique in which subjects opened and closed the fist for 2 min. Normal controls showed significantly greater increases in lactate and ammonia production with the standardized method, and it was noted that peak ammonia concentrations were reached later than peak lactate levels; sampling up to at least 7 min after release of the blood pressure cuff was recommended. It was also confirmed that a minimum increase in lactate concentration (of at least 2.5 mM) was required to demonstrate impaired ammonia production in cases of MADD. Failure to achieve such an increase may explain false positive IET results (i.e., low ammonia output through inadequate ischemic work) in normal subjects screened for MADD (9).

In another study (11), normal controls, patients with defined glycogen storage diseases and MADD, and patients with fatigue, weakness, or cramps of various causes were studied using a technique that measured the amount of work done by subjects during ischemic exercise to exhaustion (onset of fatigue, severe pain or cramp). In controls, lactate levels peaked within 3–5 min, but peak ammonia levels were not achieved until 7 min postexercise, confirming the observations in the study discussed previously. Serial baseline lactate and ammonia levels varied minimally, and there were no significant sex or age differences in terms of lactate or ammonia production. No relationship was found between peak lactate and ammonia levels and total work done, and the authors concluded that elaborate means to control and measure the work performed were unnecessary. They confirmed that the most important variables in correctly identifying patients with glycogen storage diseases and MADD were the maximum absolute increases in lactate and ammonia levels above baseline, rather than the proportional changes or variations in the molar ratios of lactate and ammonia. Using an ellipse defining the 95% confidence limits for maximal lactate and ammonia concentrations in normal subjects (Fig. 1), all three patients with myophosphorylase deficiency were identified by low lactate and increased ammonia production,

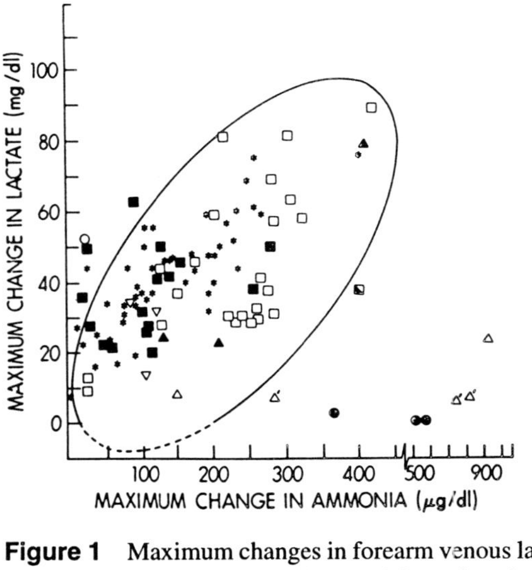

Figure 1 Maximum changes in forearm venous lactate versus ammonia concentrations in 23 normal men and women and in patients with complaints of weakness, fatigue, or cramps. ☐ - controls; ● - McArdle's disease; Δ - debrancher deficiency with enzyme deficiency in liver and muscle; ▲ - debrancher deficiency without enzyme defect in muscle; ∇ - glycogen storage disease I, II, or IX; ○ - MADD; ◩ - mitochondrial myopathy; ■ - other diagnosed muscle disease; * - undiagnosed or nonmyopathic diseases (including dystrophies, hypokalemic periodic paralysis, CPT deficiency, psychiatric disorders). (From Ref. 11, with permission.)

and three patients with debrancher enzyme deficiency confined to the liver were distinguished from three others with muscle involvement. However, results from a patient with phosphorylase kinase deficiency (glycogen storage disease type IX) were normal. The single patient with MADD showed the expected decreased ammonia production relative to lactate. Of 70 patients with fatigue, weakness, or cramps of unknown origin investigated by IET, 22 proved to have neurological or neuromuscular disorders, which included three with McArdle's disease, one with mitochondrial myopathy, and one with MADD, as suggested by their IET results.

In summary, under controlled conditions IET remains a useful screening test for metabolic myopathies, although false positive and negative results may occur.

INCREMENTAL AND STEADY-STATE EXERCISE TESTS

This type of exercise test has been employed principally in the investigation of patients with fatigue states and in the diagnosis and assessment of patients with possible mitochondrial cytopathies. Patients with a variety of defects of aerobic utilization of pyruvate via acetyl CoA, the citric acid cycle, and the mitochondrial respiratory chain have been described. A common consequence of such defects is the increased conversion of pyruvate to lactate by lactate dehydrogenase. Resting, baseline lactate levels in such patients may be raised, but often an exercise test will be required to demonstrate this abnormality. In general terms, such exercise tests examine lactate production in relationship to work done, an excessive rise indicating a "left shift" in the anaerobic threshold due to abnormalities in the control of muscle energy metabolism.

Many exercise test protocols have been reported, but in general these have not been standardized. In particular, rather indefinite work rates, such as "light work" or "exercise to exhaustion," have been employed, exercise periods have varied from minutes (12) to several hours (13), and no account has been taken of the age, weight, or sex of subjects. While such non-standardized tests may identify patients with gross defects, they seem unsatisfactory as screening tools.

Plasma lactate concentration depends on the balance between lactate production, principally from muscle, and its removal by tissues, principally the liver. During incremental aerobic exercise a work rate is reached beyond which plasma lactate increases sharply (Fig. 2). This point corresponds closely to the anaerobic threshold (AT) in terms of oxygen consumption (14), although it can be demonstrated that oxygen utilization and lactate production responses are in fact independent. AT has been measured in normal, untrained volunteers on the basis of ventilatory variables and lactate production, and tables are available that allow work rate and heart rate at the AT to be predicted for subjects with regard to age, weight, and sex (14,15).

In a recent study, normal subjects and patients with documented mitochondrial cytopathies were investigated using a braked bicycle ergometer technique, exercising at a constant work rate of 90% of the predicted anaerobic threshold (the subanaerobic threshold exercise test, SATET) for 15 min (15). Heart rates during exercise were recorded continuously, and the mean rate over the exercise period was compared with the expected rate at AT, based on earlier experimental data linking heart rate to oxygen consumption (14,15). Confidence limits for lactate concentrations at baseline, after the 15-min exercise period, and 30 min after exercise were established. All patients with mitochondrial disease were shown to have peak lactates exceeding these limits, with a false positive rate of only 7% (Fig. 3). In particular, two patients with mitochondrial disorder were correctly identified by the test, although both had normal magnetic resonance spectroscopy results and one a normal muscle biopsy. In a subsequent review, further evidence of the high sensitivity of the test was demonstrated by the identification of a patient with mitochondrial cytopathy of unusual phenotype, again with normal biopsy but with marked abnormalities of complex I and IV on biochemical studies of isolated mitochondria (16).

In these studies, it was also shown that (mean–expected) heart rates at AT were linearly related to peak lactate concentration (15), allowing an assessment of cardiovascular "fitness" in relationship to work rate, which can be of value in assessment of patients with "fatigue" states. In this regard, the technique has been used to look for evidence of abnormal muscle energy metabolism in patients with

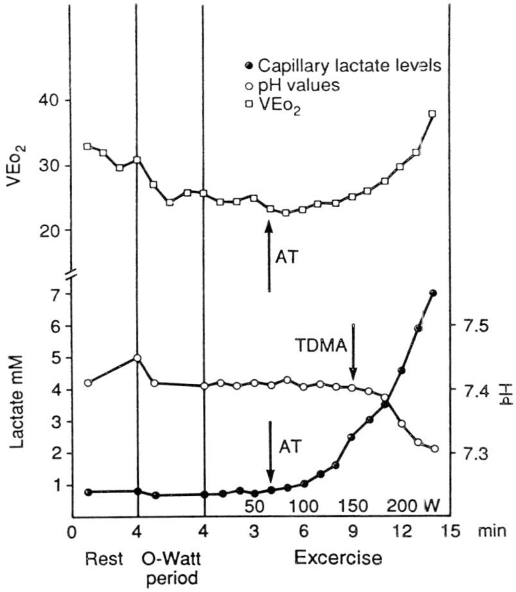

Figure 2 Relationship between venous lacatate level, ventilatory equivalent for oxygen (VE_{O_2}) and arterial pH during incremental exercise on a bicycle ergometer in a normal subject. During the initial stages, oxygen consumption declines as reduced shunting improves efficiency of gas transfer. At the anaerobic threshold (AT), oxygen consumption begins to increase; this corresponds closely with a rise in plasma lactate. pH remains stable until the rate of acid production exceeds the buffering capacity of the blood, at the threshold for decompensation for metabolic acidosis (TDMA). (Based on Ref. 14, with permission.)

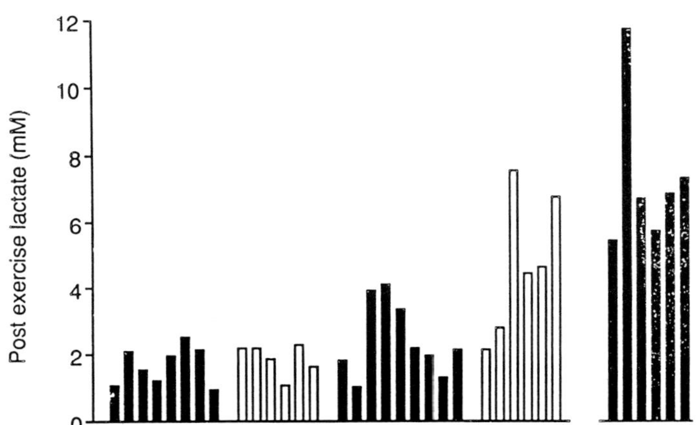

Figure 3 Postexercise venous plasma lactate levels in the SATET, in 29 control subjects and six patients with mitochondrial cytopathies. Two controls in the 35-50-year age group had levels exceeding the upper 99% confidence limit, but if an abnormal SATET was defined as an abnormal postexercise level *plus* either an abnormal baseline or 30-min postexercise level, all controls were normal and all patients with mitochondrial disease were abnormal. (From Ref. 15, with permission.)

chronic fatigue syndrome, which is discussed in detail in Chapter 44. Incremental treadmill exercise tests have also been used by several groups in attempting to elucidate the basis of fatigue in this disorder (17,18).

Exercise tests will continue to play a useful screening role in identifying patients with possible metabolic myopathies. In general, protocols should be devised to err on the side of false positive rather than false negative diagnosis, since patients who are in fact normal can then be reassured by further, specific investigations.

REFERENCES

1. McArdle B. Myopathy due to a defect in muscle glycogen breakdown. Clin Sci 1951; 10:13–33.
2. Fishbein WN, Armbrustmacher VW, Griffen JL. Myoadenylate deaminase deficiency: a new disease of muscle. Science 1978; 200:545–548.
3. Fishbein WN, Armbrustmacher VW, Griffen JL, Davis JI, Foster WD. Levels of adenylate deaminase, adenylate kinase and creatine kinase in frozen human muscle biopsy specimens relative to type I/II fibre distribution: evidence for a carrier state of myoadenylate deaminase deficiency. Clin Neurol 1984; 15:271–277.
4. McArdle B, Verel D. Responses to ischaemic work in the human forearm. Clin Sci 1956; 15:302–318.
5. Ionescu V, Luca N. Studies of carbohydrate metabolism in myasthenia gravis in conditions of ischemic exercise. Acta Neurol Scand 1966; 42:244–254.
6. Mercelis R, Martin JJ, Dehaene I, de Barsy T, Van den Bergh G. Myoadenylate deaminase deficiency in a patient with facial and limb-girdle myopathy. J Neurol 1981; 225:157–166.
7. Rumpf KW, Wagner H, Kaiser H, Meinck HM, Goebel HH, Scheler F. Increased ammonia production during forearm ischaemic work test in McArdle's disease. Klin Wochenschr 1981; 59:1319–1320.
8. Landen S, Wahren J. Blood flow, oxygen uptake and lactate production in the forearm during exercise induced by median nerve stimulation. Acta Physiol Scand 1969; 75:82–91.
9. Sinkelerr SP, Wevers R, Joosten EE, Binkhorst RA, Oei LT, Van't Hof MA, de Haan AF. Improvement of screening in exertional myalgia with a standardised ischaemic forearm test. Muscle Nerve 1986; 9:731–737.
10. Sinkeler SPT, Binkhorst RA, Daanen H, Wevers RA, Oei TL, Joosten EMG. The relationship between blood lactate and ammonia in ischaemic handgrip exercise. Muscle Nerve 1985; 8:523–527.
11. Coleman RA, Stajich JM, Pact VW, Percak-Vance MA. The ischaemic exercise test in normal adults and in patients with weakness and cramps. Muscle Nerve 1986; 9:216–221.
12. Morgan-Hughes JA, Darveniza P, Kahn SN, Landon DN, Sherratt RM, Land JM, Clark JB. A mitochondrial myopathy characterised by a deficiency in reducible cytochrome b. Brain 1977; 100:617–640.
13. Brooke MH, Carroll JE, Davis JE, Hagberg JM. The prolonged exercise test. Neurology 1979; 29:636–643.
14. Reinhard U, Muller PH, Schmulling RM. Determination of anaerobic threshold by the ventilation equivalent in normal individuals. Respiration 1979; 38:36–42.
15. Nashef L, Lane RJM. Screening for mitochondrial cytopathies: the sub-anaerobic threshold exercise test (SATET). J Neurol Neurosurg Psychiatry 1989; 52:1090–1094.
16. Lane RJM. Screening for mitochondrial cytopathies: the sub-anaerobic threshold exercise test (SATET). In: Methods of Toxicology, Vol 2. 1993:171–178.
17. Riley M, O'Brien CJ, Bell N, Nicholls DP, McCloskey DR. Aerobic work capacity in patients with chronic fatigue syndrome. Br Med J 1990; 301:953–956.
18. Gibson H, Carroll N, Clague JE, Edwards RHT. Exercise performance and fatiguability in patients with chronic fatigue syndrome. J Neurol Neurosurg Psychiatry 1993; 56:993–998.

4
Electromyography

Robin P. Kennett
Radcliffe Infirmary NHS Trust, Oxford, England

INTRODUCTION

The traditional role of clinical neurophysiology in the management of primary disorders of muscle has been in diagnosis, but the introduction of quantitative techniques and advanced electromyography (EMG) machines, which allow rapid acquisition and analysis of data, open the way for serial studies to assess disease progression and the response to treatment (1,2).

For diagnosis, the neurophysiological examination can assess the likely pathophysiological process affecting muscle, and also determine its distribution by the exploration of many sites. This usefully augments a muscle biopsy, which discloses anatomical and biochemical abnormalities in a restricted sample. In most cases, the most complete information will be obtained by correlation of neurophysiological and muscle biopsy data (3,4).

To formulate the possible pathophysiological process affecting muscle and the distribution of changes, a combination of neurophysiological techniques may be required, the exact combination being decided by the clinician during the examination. Nerve conduction studies and repetitive nerve stimulation tests play a relatively minor role in the assessment of patients with muscle disorders, and the main tool is electromyography, predominantly using standard electrodes such as the concentric needle. In some cases this may need to be enhanced by quantitative techniques including those employing special electrodes such as single fiber and macro EMG needles.

NERVE CONDUCTION STUDIES

These are used mainly to exclude alternative diagnoses, especially peripheral nerve disease, and in most primary myopathies motor and sensory nerve conduction are normal. However, abnormalities of nerve conduction have been reported in 45% of patients with mitochondrial myopathy (5) and in about 30% of patients with inclusion body myopathy, but a proportion of these were incidental nerve entrapment syndromes (6). In myotonic dystrophy, reduction is motor and sensory nerve conduction velocity probably reflects a generalized disorder of cellular function (7,8).

Measurement of compound muscle action potential (CMAP) amplitude may give a useful guide to the severity and distribution of muscle diseases. The CMAP of intrinsic hand muscle is reduced in distal myopathy and myotonic dystrophy and in a proportion of patients with acquired inflammatory myopathy, indicating severe and distal involvement (9). Measurement of CMAP amplitude is also valuable in diagnosis of the periodic paralyses where it falls progressively during the onset of weakness (10).

REPETITIVE NERVE STIMULATION (RNS)

These tests may be useful in differential diagnosis of neuromuscular transmission disorders that present with proximal muscle weakness. The Lambert-Eaton myasthenic syndrome is distinguished by a low resting CMAP amplitude, which shows further decrement with stimulation at low rates of 1–3Hz, but increment of more than 25% at higher rates (11). In myasthenia gravis the characteristic pattern on RNS is of progressive reduction in amplitude of the CMAP with the first four to five stimuli in a train at 3Hz followed by a mild increment (12).

A decrementing response to RNS is not confined to neuromuscular transmission disorders, but may be seen in muscle diseases, particularly the myotonic disorders (13,14). However, in this case decrement is more marked with high stimulation rates (above 10 Hz), and it is progressive during the early part of the train without showing mild increment after the 5th response.

ELETROMYOGRAPHY

Technical Considerations

Types of Electrodes (Fig. 1)

Routine electromyography most often makes use of the concentric needle electrode (CNE), although some centers prefer the monopolar electrode. The CNE consists of an active recording pickup threaded down the barrel of a hypodermic needle, from which it is electrically insulated, the shaft acting as the reference. The recording surface is beveled and therefore has directional properties (15). The recording area is typically 0.007 mm^2, although facial electrodes have smaller leading-off areas, which consequently affect motor unit action potential parameters, an important consideration when using these electrodes for quantitative purposes.

The single-fiber (SF) electrode also consists of a barrel insulated from a central active core, but the recording area is exposed at a side hole on the cannula. The smaller recording area of about 5×10^{-4} mm^2 enhances the selectivity of the recording electrode.

In contrast, the active recording site of the macroelectrode is the uninsulated tip of the cannula, usually over a distance of 15 mm, which is referenced to a distant subcutaneous electrode. The macroelectrode has a SF side port set 7.5 mm from the tip, which allows a muscle fiber action potential from the center of the unit to be used to trigger an averager. The averaged time-locked activity recorded from the uninsulated cannula referenced to the subcutaneous electrode is the macromotor unit action potential.

Recording Apparatus

Commercially available electromyographs are now built to the required technical standards (16). Electronic low-frequency filtering can enhance the spike components of a signal but reduces both duration and amplitude (Fig. 2). For this reason, wide band pass filter settings are required for quantitative analysis, and the same settings should be used for normal reference and patient data. One of the most important electronic devices is the trigger and delay line, which allows the oscilloscope to be triggered on a spike component of the recorded potential, and the sweep delayed so that preceding and subsequent linked signals can be identified. Ideally, the triggering point and delay time should be continuously variable. A loud speaker allows the examiner to hear signals generated at a distance from the recording electrode, which may not be visible on the screen. The electrode can then be repositioned to maximize the potential.

Concentric Needle Electromyography

The majority of routine electromyography is performed using this electrode. The most consistently studied features are insertional activity, spontaneous activity, motor unit action potential (MUAP)

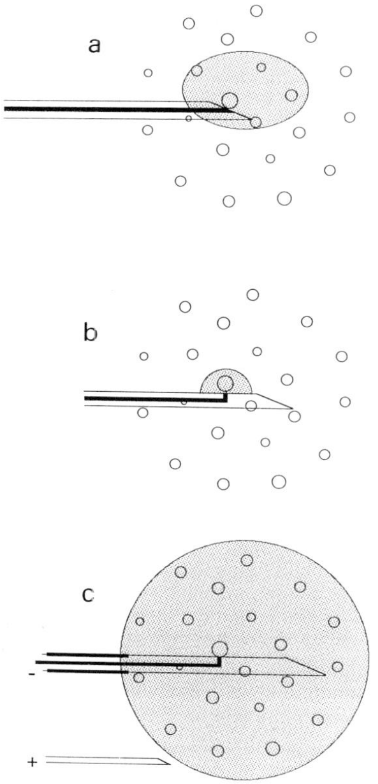

Figure 1 Recording properties of different EMG electrodes. The diagrams show muscle fibers of a motor unit shown in cross-section, and the pickup zone of the electrode is represented by the shaded area. The concentric needle electrode in (a) records from a group of fibers over about 1-mm radius from the leading-off area. The recording area of the single-fiber electrode in (b) is smaller, allowing selective recordings from one or a few fibers. A single muscle fiber from the center of the motor unit territory is recorded at the side port of the macro EMG electrode in (c). This is used to trigger an averager, and the time-locked electrical field generated by all the fibers in the unit recorded by the uninsulated tip of the cannula relative to a distant subcutaneous indifferent electrode is the macrounit potential.

characteristics, and the recruitment pattern. All of these may be assessed semiquantitatively, and some form the basis of more advanced quantitative techniques.

Insertional Activity

In normal muscles CNE insertion provokes a short burst of electrical activity, which is probably due to mechanically induced spontaneous depolarization of muscle fibers (17–19). The amount of this activity normally varies from one muscle to another, and at different sites in the same muscle. More prolonged bursts of insertional activity may be found in normal gastrocnemii, especially in athletic individuals (20). For this reason the assessment of the degree of insertional activity requires judgment and therefore plays only a limited role in diagnosis. Reduced insertional activity is seen in muscles largely replaced by connective tissue and at the height of attacks of periodic paralysis (10), and increased insertional activity occurs when muscle fibers are unusually excitable, as in acute myonecrosis (21).

The prolonged activity seen when an electrode is positioned in the end-plate zone of muscle should not be confused with abnormal activity. This "end-plate noise" is recognized as high-frequency

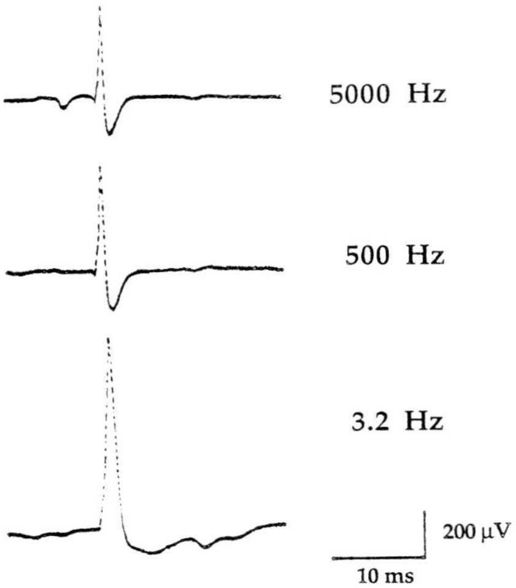

Figure 2 Effect of low-frequency filtering on the waveform and amplitude of a motor unit action potential recorded with a concentric needle electrode. The high frequency filter was set at 16 kHz in all cases.

discharges of low-amplitude spikes probably originating from the neuro-muscular junction (22). Recording in this site is often painful, and the activity disappears when the electrode is repositioned.

Abnormal insertional activity also includes myotonic and complex repetitive discharges. Myotonic discharges consist of simple monophasic components that recur initially at a frequency of about 70 Hz (23), but the discharge frequency and amplitude of individual components decline as the train proceeds, leading to the characteristic reduction in pitch heard through a loudspeaker. Typically myotonic discharges are activated by needle insertion or movement and may be enhanced by cooling. In myotonic disorders this activity is usually more pronounced distally and is detected in all patients, provided enough sites are sampled (24). Although the pathophysiological mechanism of myotonic discharges is thought to vary by condition, the basic abnormality is failure of muscle membrane repolarization beyond the threshold level, allowing recurrent spontaneous depolarizations until the transmembrane ionic fluxes render the fiber inexcitable (25–27).

Complex repetitive discharges (CRDs) are initiated by the same maneuvers as myotonic discharges, but the individual components consist of more numerous spikes and the discharge frequency is more variable between 5-and100/sec. Single-fiber EMG studies have shown low jitter values between the spike components of CRDs with increasing jitter in some before the discharge abruptly terminates (28). These findings have been interpreted as indicating ephaptic, or electrical, transmission between adjacent muscle fibers, and consequently CRDs are most often seen in conditions where large numbers of neighboring muscle fibers are hyperexcitable, the external electrical field generated by a single spontaneous depolarization in one fiber being sufficient to activate an adjacent fiber. This condition is seen in severe necrotizing myopathies, and CRDs are therefore a feature of muscular dystrophy and inflammatory myopathies (21).

Spontaneous Activity

Fibrillation potentials and positive sharp waves are spontaneous depolarizations of single muscle fibers (Fig. 3). In primary myopathies these most frequently occur because of functionally denervated segments of muscle fiber caused by necrosis or severe focal atrophy either at the end-plate zone or along the fiber's length (29,30). Physiological changes to muscle membranes after denervation, including the development of phasic changes in transmembrane potential, are thought to be responsible for spontaneous depolarization (31). The quantity of fibrillation potentials can be graded on a four-point scale depending on the number of areas at which activity is found and the profusion at a given site (20).

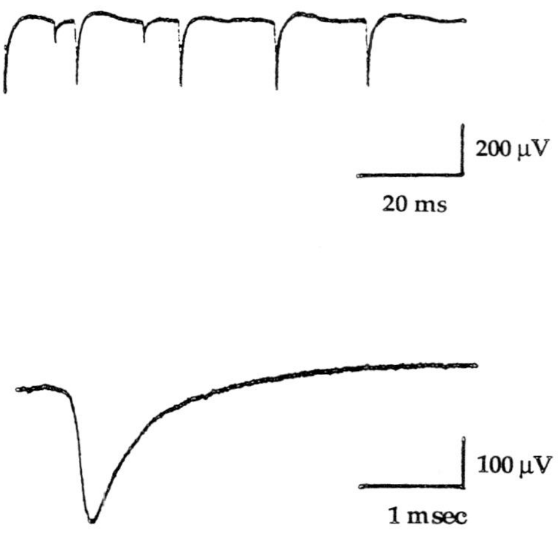

Figure 3 Characteristic waveform of positive sharp waves recorded with a concentric needle electrode.

As fibrillation potentials and positive sharp waves are a feature of functionally denervated muscle fibers, they appear when necrosis is a prominent pathological finding, as in Duchenne muscular dystrophy and in the inflammatory myopathies (21). They also occur when there is a disorder of muscle membrane function, as in paramyotonia on cold exposure (32,33), and at the start of attacks of periodic paralysis (10), but in this case they could represent volitional activation of motor units containing very few excitable muscle fibers.

Motor Unit Action Potentials (MUAP)

MUAP characteristics are usually assessed during light voluntary contraction when fatigue-resistant type 1 units are examined. Inspection of a real-time oscilloscope trace may give an impression about unit waveforms and amplitudes, but a trigger and delay line is required to give a more detailed semiquantitative analysis (Fig. 4). The MUAP parameters most often defined are duration, amplitude,

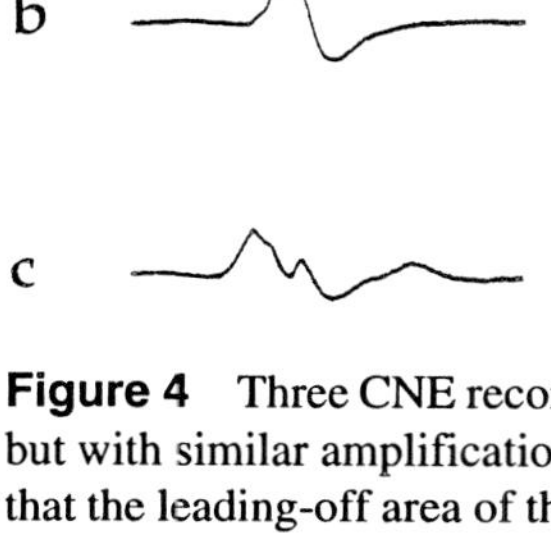

Figure 4 Three CNE recordings of the same motor unit action potential with different electrode positions but with similar amplification and filter settings. The maximum amplitude and fast rise time in (a) indicate that the leading-off area of the electrode is in the center of the motor unit territory. Moving to the periphery in (b) reduces amplitude and increases rise time, but further manipulation in (c) reveals a notched component. Only the potential in (a) should be used for quantitative analysis, but the irregularities in (c) may be used to assess the stability of the motor unit action potential.

waveform, and stability. Amplitude and duration are to some extent dependent variables, giving information about the summated electrical activity generated by the muscle fibers in the fraction of the motor unit in the pickup territory of the recording electrode. In normal MUAPs amplitude depends on summation of high-frequency spike components derived from two to 12 muscle fibers originating within a 1-mm radius of the recording electrode, whereas duration is determined by lower-frequency signals that arise at a greater distance from the electrode (34). Duration is therefore the most consistent measure of the overall size of a motor unit and correlates best with macrounit parameters. A reduction in the electrical field generated by individual muscle fibers, or fall in the number of fibers in a unit, common pathological processes in primary myopathies, will reduce both amplitude and duration of the MUAP. However, as the external electrical field is related to the diameter of a muscle fiber (35), a single hypertrophied fiber close to the recording electrode could give a MUAP of normal amplitude. In the extreme situation of muscle fiber atrophy and loss from a motor unit, the MUAP approximates the characteristics of a fibrillation potential. This is seen in end-stage myopathies.

The smooth waveform of a normal MUAP is due to the synchronicity of component single muscle fiber action potentials. This is achieved in normal muscle by a compact innervation zone (36) and uniform conduction velocities along nerve terminals and muscle fibers. The conduction velocity of muscle fiber action potentials is dependent on the external fiber diameter (37), and processes causing variation of muscle fiber diameters lead to desynchronization and increased irregularity or polyphasicity of the summated MUAP (38). The desynchronization is increased in necrotic conditions where the innervation zone is wide, and where long and newly myelinated nerve terminals cause a dispersion of conduction velocities along terminal axons. Therefore, in primary myopathies where there is a loss of muscle fibers by necrosis, or reduction in innervation ratio, or where there are normal numbers but increased variability in diameters, the MUAPs become polyphasic and of short duration and low amplitude. When there is chronic necrosis with reinnervation and hypertrophy, there may be polyphasic MUAPs with prolonged durations and high-amplitude components (Fig. 5). In some myopathies where both processes occur simultaneously, there is a bimodal distribution of MUAP durations.

Assessment of stability requires the use of a trigger and delay line and low-frequency filtering. Normal MUAP waveforms usually have smooth waveforms, but manipulation of the recording electrode in a motor unit territory will usually expose irregularities or polyphasic components (Fig. 4). When normal motor units fire recurrently, these irregular components are closely time-locked to the main spike, and little variability of waveform occurs; the unit is stable. Instability, i.e., variability of MUAP waveform with recurrent discharges, most often occurs when there is insecure neuromuscular transmission either at immature neuromuscular junctions during reinnervation or because of a reduced safety factor for transmission in myasthenic disorders. Complete failure of neuromuscular transmission results in "blocking" of some components of the MUAP. This failure of transmission may be at a poorly myelinated nerve terminal, or at the neuromuscular junction itself. Instability of MUAPs is therefore most likely to occur in myopathies where there is muscle fiber necrosis and is often seen in conjunction with fibrillation potentials or CRDs. However, when there is only minor necrosis, abnormal spontaneous activity may not be detected and instability of the MUAP may be the only indication of functional denervation/collateral reinnervation.

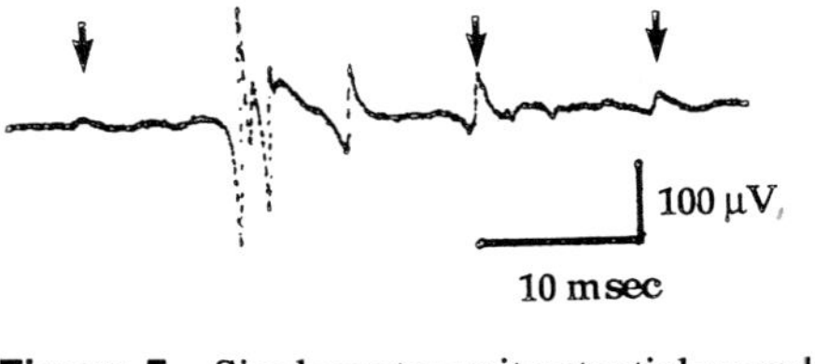

Figure 5 Single motor unit potential recorded from the biceps muscle of a patient with chronic polymyositis. Use of a trigger and delay line showed satellite potentials (arrows) linked to the main potential, giving the complex an overall duration of about 80msec.

Recruitment Pattern

During the earliest stages of voluntary contraction, the fatigue-resistant type 1 motor units become active, initially firing sporadically but then steadily at rates of 5–10 Hz as the force of contraction increases. Further tension is generated by increasing firing rates and the discharge of more units. Finally, near maximum force, there are bursts of larger type 2 units (Fig. 6). This orderly sequence is the recruitment pattern, which follows the "size principle" of smaller motor units being active before larger ones (39,40). The display of MUAPs on an oscilloscope screen is the interference pattern, and in normal muscles maximal contraction is associated with overlapping MUAPs, which obscure the baseline.

In myopathies, diseased muscle fibers may be unable to generate normal tension even at the end-stage of recruitment. However, the interference pattern may also appear to be complete at low forces if the low-threshold MUAPs are of unusually long duration. For instance, a single MUAP of 50-msec duration firing at a rate of 20 Hz will completely obscure an oscilloscope baseline. In end-stage myopathies there may be very few muscle fibers able to contribute to a MUAP, in which case the interference pattern is thin at maximal force of contraction.

Quantitative Electromyography

Quantitative techniques have evolved along two lines. The first uses signals obtained by the CNE, from which automatic MUAP or interference pattern analysis is performed. The second approach is to use special electrodes, giving rise to single-fiber EMG (SFEMG), macro EMG, scanning EMG, and multielectrode studies, of which only the first two are of established value in clinical practice.

Automatic MUAP Analysis

Individual MUAPs may be automatically identified and analyzed by a computer. Most often a "template" is used to recognize potentials of different waveform (41,42), but MUAPs selected by a trigger can be averaged before analysis (43). For statistical purposes a sample should consist of at least 20 MUAPs. After a sample has been collected, MUAP parameters are analyzed and mean values calculated or frequency histograms plotted. These techniques are therefore an extension of the semi-quantitative analysis of MUAP characteristics performed during routine concentric needle electromyography. As with routine studies, technical factors can significantly affect the results, the most important of which is ensuring that the CNE is positioned in the center of the motor unit territory, not at the periphery when artifactually low measurements of amplitude will be obtained.

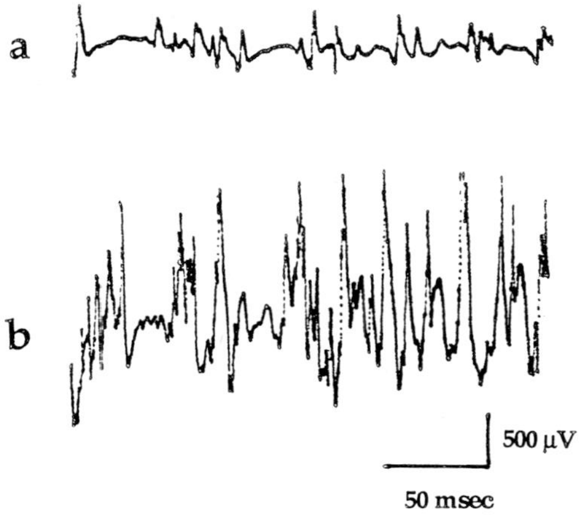

Figure 6 Pattern of motor unit recruitment, (a) with slight voluntary contraction, and (b) near maximal force.

A different approach is automatic decomposition of the interference pattern, which has the advantage of studying later recruited motor units (44,45).

Automatic Analysis of the Interference Pattern

The earliest quantitative analysis of the interference pattern assessed the number of phase changes (turns) with an amplitude greater than 100 μV, in 1 sec, and the mean amplitude of the potential between turns with the muscle contracting against a known force (46). One of the difficulties with this analysis is that amplitude (47) and possibly turns (48) are dependent on the force of contraction. Moreover, overlap of potentials as the interference pattern becomes complete can cause increased amplitude and decreased turns by electrical summation. This problem may be overcome by using a standard force relative to the maximum (47), or by disregarding force measurement. In one approach mean amplitude is plotted against turns/second at a variety of forces to give a "cloud" of data points (Stålberg's cloud) (49). Ninety-five percent confidence intervals for the edge of the cloud in normal muscle can be defined. Other parameters, excluding force measurement, include the peak ratio (the maximum ratio between turns and mean amplitude) (50) and the time interval between turns (51).

The problems encountered with interference pattern analysis are that it lacks diagnostic sensitivity, especially when there is little weakness and when higher forces are generated, that it is time consuming, that normal reference values are available for only a limited number of muscles, and that as yet it has not been shown to be able to distinguish between myopathies of different aetiology. It has been claimed that recent modifications using the peak ratio and time interval parameters increase diagnostic yield and make the technique more sensitive than MUAP analysis (52), but this is yet to be substantiated.

Single-Fiber EMG

The SFEMG parameters of greatest value are the fiber density (FD) and calculation of jitter, but analysis of complex potentials may also provide insight into pathophysiological processes.

The FD is the average number of spikes in the uptake area of the SF electrode. In normal muscles the majority of SF insertions record only a single action potential with amplitude greater than 200 μV and rise time faster than 300msec (53). In about a third of electrode positions, a second potential with these characteristics will be recorded and rarely are there three or more spikes, giving a normal FD of about 1.5 (54). To obtain the FD, the electrode is positioned to give maximal amplitude on a triggering spike and the number of linked potentials with the given characteristics is counted. The average number of spikes recorded after 20 repositionings of the electrode is the FD. Of great technical importance when calculating the FD is ensuring that the side port of the electrode is clean, as a proteinaceous deposit will severely affect recording conditions and give falsely low FD estimations.

The FD reflects the distribution of muscle fibers within a motor unit and is reduced when the number of muscle fibers capable of generating the electrical field required to fulfill the recording criteria falls (i.e., when the muscle fibers are atrophic). Increased FD may occur in myopathies by muscle splitting where a single muscle fiber generates two action potentials (55), by collateral reinnervation (56), and possibly by increased packing of muscle fibers due to atrophy (although this is concomitantly associated with reduced external electrical field), and by alteration of volume conduction of distant potentials by different tissue filtering characteristics produced by fibrosis and fatty infiltration (57). The FD is therefore increased in Duchenne muscular dystrophy, even in muscles that are not clinically affected, although it may fall later in the course of the disease (57). In polymyositis the FD is initially low or normal, but increases with disease chronicity (9).

Individual recordings can become complex with multiple spike components. The overall duration of these potentials and the time interval between spikes (expressed as mean interspike interval, MISI) may be of value in diagnosis, MISI increasing in myopathy as spikes tend to be widely dispersed throughout complexes.

The second important SFEMG parameter is the measurement of jitter. This is the small difference in the interpotential time interval between spike components on successive discharges (Fig. 7). Jitter

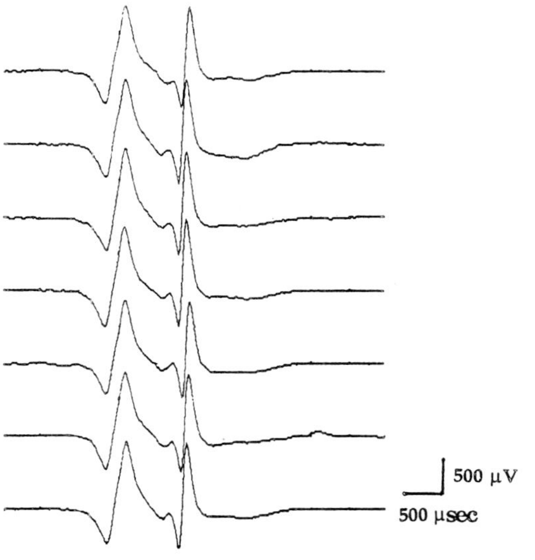

Figure 7 Raster display of two SFEMG spike potentials showing normal variability in interpeak latency (MCD 24μsec).

is usually calculated electronically and expressed as the mean time difference of interpotential interval between consecutive discharges (mean consecutive difference, MCD).

In myopathies producing collateral reinnervation, there can be an increased proportion of spikes with raised jitter values, as in axonal degeneration and neuromuscular transmission disorders. Raised jitter values in myopathy may also be due to increased variability of propagation velocity along abnormal muscle fibers (Fig. 8). Decreased jitter values (MCD below 5μsec) occur when two spikes are generated by a single neuromuscular junction, as in split muscle fibers or in ephaptic (electrical) transmission.

Thus, in Duchenne muscular dystrophy low jitter values are seen in 2–16% of recordings, and 20–40% of potentials show increased values (56).

Macro EMG

Whereas the SFEMG records highly selectively individual fibers of the motor unit, the macroelectrode aims to record all the electrical activity generated by the unit. To do this the triggering SF potential must be generated by a muscle fiber in the center of the motor unit territory (58). As with other quantitative techniques, at least 20 different potentials should be recorded from different sites within a muscle to give a valid sample (59). Macrounit amplitudes and durations are strongly positively correlated, and in myopathies the amplitudes are either normal or slightly reduced. Changes in macrounit waveform may be helpful with increasing complexity in conditions causing fractionation of motor units. The insight given by this technique into pathophysiological processes may be enhanced by correlation of macrounit parameters with the FD, which may be obtained at the same recording session.

CONCLUSION

The electrodiagnosis of primary muscle diseases is mainly dependent on needle electromyography. A semiquantitative assessment using a concentric needle electrode allows description of spontaneous activity, MUAP characteristics, and abnormalities of the recruitment pattern, often sufficient to reach

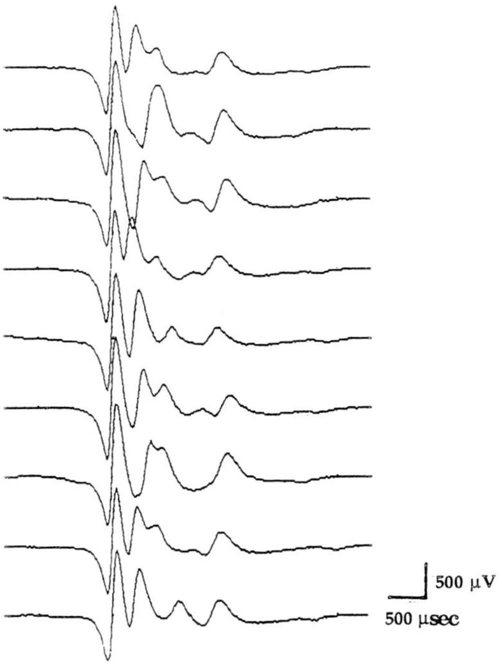

Figure 8 SFEMG recording of a complex potential in a chronic myopathy showing four linked potentials satisfying the criteria for inclusion into a fiber density calculation, with increased jitter between components (second potential MCD 92µsec).

diagnosis. This form of evaluation has the advantage of being quick to perform and permits sampling of many sites to give information about the distribution of changes.

Quantitative techniques can provide further information useful both diagnostically and for sequential review, but they are less popular as they are inclined to be time-consuming, even with modern EMG machines.

REFERENCES

1. Dorfman LJ, McGill KC. AAEE minimonograph #29: automatic quantitative electromyography. Muscle Nerve 1988; 11:804–818.
2. Stålberg E. Invited review: electrodiagnostic assessment and monitoring motor unit changes in disease. Muscle Nerve 1991; 14:293–303.
3. Hausmanowa-Petrusewicz I, Jedrzejowska H. Correlation between electromyographic findings and muscle biopsy in cases of neuromuscular disease. J Neurol Sci 1971; 13:85–106.
4. Buchthal F, Kamieniecka Z. The diagnostic yield of quantified electromyography and quantified muscle biopsy in neuromuscular disorders. Muscle Nerve 1982; 5:265–280.
5. Yiannikas C, McLeod JG. Peripheral neuropathy associated with mitochondrial myopathy. Electroencephalogr Clin Neurophysiol 1985; 61:S20.
6. Lotz BP, Engel AG, Nishino H, Stevens JC, Litchy WJ. Inclusion body myositis: observations in 40 patients. Brain 1989; 112:727–747.
7. Mongia SK, Lundervold A. Electrophysiological abnormalities in cases of dystrophia myotonica. Eur Neurol 1975; 13:360–376.
8. Olson ND, Jou M-F, Quast JE, Nuttall LFQ. Peripheral neuropathy in myotonic dystrophy. Arch Neurol 1978; 35:741–745.
9. Henriksson KG, Stålberg E. The terminal innervation pattern in polymyositis: a histological and SFEMG study. Muscle Nerve 1978; 1:2–13.

10. Gordon AM, Green JR, Lagunoff D. Studies on a patient with hypokalaemic familial periodic paralysis. Am J Med 1970; 48:185–195.
11. Eaton LM, Lambert EH. Electromyography and electric stimulation of nerves in diseases of motor unit: observations on myasthenic syndrome associated with malignant tumors. JAMA 1957; 163:1117–1124.
12. Sanders DB. Electrophysiologic study of disorders of neuromuscular transmission. In: Aminoff MJ, ed. Electrodiagnosis in Clinical Neurology, 2nd ed. London: Churchill Livingstone, 1986:307–332.
13. Brown JC. Muscle weakness after rest in myotonic disorders: an electrophysiological study. J Neurol Neurosurg Psychiatry 1974; 37:1336–1342.
14. Aminoff MJ, Layzer RB, Sutya S-M, Faden AI. The declining electrical response of muscle to repetitive nerve stimulation in myotonia. Neurology 1977; 27:812–816.
15. Nakao K, Nakanishi T, Tsubaki T. Action potentials recorded by coaxial needle electrodes in Ringer's solution. Electroencephalogr Clin Neurophysiol 1965; 18:412–414.
16. Guld C, Rosenfalck A, Willison RG. Technical factors in recording electrical activity of muscle and nerve in man. Electroencephalogr Clin Neurophysiol 1970; 28:399–413.
17. Weddell G, Feinstein B, Pattle RE. The electrical activity of voluntary muscle in man under normal and pathological conditions. Brain 1944; 67:178–257.
18. Kugelberg E, Petersen I. "Insertion activity" in electromyography. J Neurol Neurosurg Psychiatry 1949; 12:268–273.
19. Wiechers DO. Mechanically provoked insertional activity before and after nerve section in rats. Arch Phys Med Rehabil 1977; 58:402–405.
20. Daube JR. AAEM minimonograph #11: needle examination in clinical electromyography. Muscle Nerve 1991; 14:685–700.
21. Bohan A, Peter JB, Bowman RL, Pearson CM. Compute-assisted analysis of 153 patients with polymyositis and dermatomyositis. Medicine 1977; 56:255–286.
22. Wiederholt WC. "End-plate noise" in electromyography. Neurology 1970; 20:214–224.
23. McComas AJ, Mrozek K. The electrical properties of muscle fibre membranes in dystrophia myotonica and myotonia congenita. J Neurol Neurosurg Psychiatry 1968; 31:441–447.
24. Streib EW, Sun SF. Distribution of electrical myotonia in myotonic muscular dystrophy. Ann Neurol 1983; 14:80–82.
25. Lehmann-Horn F, Rudel R, Dengler R, Lorkovic H, Haass A, Ricker K. Membrane defects in paramyotonia with and without myotonia in a warm environment. Muscle Nerve 1981; 4:396–406.
26. Lehmann-Horn F, Rudel R, Ricker K, Lorkovic H, Dengler R, Hopf HC. Two cases of adynamia episodica hereditaria: in vitro investigation of muscle cell membrane and contraction parameters. Muscle Nerve 1983; 6:113–121.
27. Rudel R, Lehmann-Horn F. Membrane changes in cells from myotonia patients. Phys Rev 1985; 65:310–365.
28. Trontelj J, Stålberg E. Bizarre repetitive discharges recorded with single fibre EMG. J Neurol Neurosurg Psychiatry 1983; 46:310–316.
29. Desmedt JE, Borenstein S. Relationship of spontaneous fibrillation potentials to muscle fibre segmentation in human muscular dystrophy. Nature 1975; 258:531–534.
30. Streib EW, Wilbourn AJ, Mitsumoto H. Spontaneous electrical activity in polymyositis and dermatomyositis. Muscle Nerve 1979; 2:14–18.
31. Thesleff S, Ward MR. Studies on the mechanism of fibrillation potentials in denervated muscle. J Physiol 1975; 244:313–324.
32. Haass A, Ricker K, Rudel. Clinical study of paramyotonia congenita with and without myotonia in a warm environment. Muscle Nerve 1981; 4:388–395.
33. Subramony SH, Malhotra CP, Mishra SK. Distinguishing paramyotonia congenita and myotonia congenita by electromyography. Muscle Nerve 1983; 6:374–379.
34. Gath I, Stålberg E. The calculated radial decline of the extracellular action potential compared with in situ measurements in the human brachial biceps. Electroencephalogr Clin Neurophysiol 1978; 44:547–552.
35. Hakansson CH. Action potential recorded intra- and extracellularly from the isolated frog muscle fibre in Ringer's solution and in air. Act Physiol Scand 1957; 39:291–312.
36. Buchthal F, Guld C, Rosenfalck P. Multielectrode study of the territory of a motor unit. Acta Physiol Scand 1957; 39:83–104.
37. Hakansson CH. Conduction velocity and amplitude of the action potential as related to circumference in the isolated fibre of frog muscle. Acta Physiol Scand 1956; 37:14–34.
38. Nandedkar S, Sanders DB, Stålberg E. EMG of reinnervated motor units: a simulation study. Electroencephalogr Clin Neurophysiol 1988; 70:177-184.
39. Henneman E. Relation between size of neurons and their susceptibility to discharge. Science 1957; 126: 1345–1347.
40. Milner-Brown HS, Stein RB, Yemm R. The orderly recruitment of human motor units during voluntary isometric contraction. J Physiol 1973; 230:359–370.

41. Bergmans J. Computer assisted on line measurement of motor unit potential parameters in human electromyography. Electromyography 1971; 11:161–181.
42. Tanzi F, Taglietti V, Zucca G, Arrigo A, Moglia A, Sandrini G, et al. Computerized EMG analysis. Electromyogr Clin Neurophysiol 1979; 19:495–503.
43. Lang AH, Nurkkanen P, Vaahtoranta KM. Automatic sampling and averaging of electromyographic unit potentials. Electroencephalogr Clin Neurophysiol 1971; 31:404–406.
44. Mambrito B, De Luca CJ. A technique for the detection, decomposition and analysis of the EMG signal. Electroencephalogr Clin Neurophysiol 1984; 58:175–188.
45. McGill KC, Dorfman LJ. Automatic decomposition electromyography (ADEMG): validation and normative data in brachial biceps. Electroencephalogr Clin Neurophysiol 1985; 61:453–461.
46. Willison RG. Analysis of electrical activity in healthy and dystrophic muscle in man. J Neurol Neurosurg Psychiatry 1964; 27:386–94.
47. Fuglsang-Frederiksen A, Mansson A. Analysis of electrical activity of normal muscle in man at different degrees of voluntary effort. J. Neurol Neurosurg Psychiatry 1975; 38:683–694.
48. Liguori R, Dahl K, Fuglsang-Frederiksen A. Turns-amplitude analysis of the electromyographic recruitment pattern disregarding force measurement. 1. Method and reference values in healthy subjects. Muscle Nerve 1992; 15:1314–1318.
49. Stålberg E, Chu J, Bril V, Nandedkar S, Stålberg S, Ericsson M. Automatic analysis of the EMG interference pattern. Electroencephalogr Clin Neurophysiol 1983; 56:672–681.
50. Fuglsang-Frederiksen A, Lo Monaco M, Dahl K. Turns analysis (peak ratio) in EMG using the mean amplitude as a substitute of force measurement. Electroencephalogr Clin Neurophysiol 1985; 60:225–227.
51. Fuglsang-Frederiksen A, Scheel U, Buchthal F. Diagnostic yield of analysis of the pattern of electrical activity and individual motor units potentials in myopathy. J Neurol Neurosurg Psychiatry 1976; 39:742–750.
52. Liguori R, Dahl K, Fuglsang-Frederiksen A, Trojaborg W. Turns-amplitude analysis of the electromyographic recruitment pattern disregarding force measurement. II. Findings in patients with neuromuscular disorders. Muscle Nerve 1992; 15:1319–1324.
53. Stålberg E, Thiele B. Motor unit fibre density in the extensor digitorum communis muscle. J Neurol Neurosurg Psychiatry 1975; 38:874–880.
54. Gath I, Stålberg E. On the measurement of fibre density in human muscles. Electroencephalogr Clin Neurophysiol 1982; 54:699–706.
55. Hilton-Brown P, Stålberg E, Trontelj J, Milhelin M. Causes of the increased fiber density in muscular dystrophies studied with single fiber EMG during electrical stimulation. Muscle Nerve 1985; 8:383–388.
56. Hilton-Brown P, Stålberg E. The motor unit in muscular dystrophy, a single fibre EMG and scanning EMG study. J Neurol Neurosurg Psychiatry 1983; 46:981–995.
57. Stålberg E, Trontelj J. Single Fibre Electromyography. Old Woking, Surrey: Mirvalle Press, 1979.
58. Stålberg E. Macro EMG, a new recording technique. J Neurol Neurosurg Psychiatry 1980; 43:475–482.
59. Stålberg E, Fawcett PRW. Macro EMG in healthy subjects of different ages. J Neurol Neurosurg Psychiatry 1982; 45:870–878.

5
Muscle Biopsy Techniques

Richard H. T. Edwards, Malcolm J. Jackson, and Timothy R. Helliwell
University of Liverpool, Liverpool, England

J. E. Clague
University of Manchester, Manchester, England

INTRODUCTION

Muscle biopsy is an integral part of the investigation of any patient suspected of having a neuromuscular disorder and is useful in assessing muscle involvement in systemic illnesses such as connective tissue disease. A muscle biopsy can provide definitive diagnostic evidence in patients presenting with symptoms such as muscle weakness, cramps, or postexertional muscle pain. The use of closed biopsy techniques (1–3) has widened the application of this investigation and encouraged physicians to seek a tissue diagnosis for genuine, though vague, symptoms that might not be considered sufficient indication for open biopsy. The repeatability of closed techniques permits disease progress and treatment responses to be assessed (4), and the combination of closed muscle biopsy with the development of analytical methods has led to a better understanding of muscle cell structure and metabolism in health and disease (5,6).

SELECTION OF THE MUSCLE FOR BIOPSY

The muscle chosen for biopsy depends on the distribution of muscle weakness. A moderately affected muscle is most suitable for biopsy in dystrophic diseases. Minimally involved muscles may show no abnormality whereas severely involved muscles may not only be technically more difficult to biopsy, but may also be completely replaced by fat and connective tissue, and it may not be possible to distinguish end-stage neuropathic from myopathic processes. Tender areas in suspected vasculitis or myositis may be more suitable for biopsy. Imaging techniques including ultrasonography (7), radionuclide scanning (8), computerized tomography (9), and magnetic resonance imaging (10) have all been used to identify suitable biopsy sites.

Deltoid, supraspinatus, trapezius, biceps, and triceps brachii are suitable for biopsy in the upper limb; quadriceps femoris, gastrocnemius, and tibialis anterior in the lower limb. Needle biopsy has also been used to sample paraspinal muscles in the study of scoliosis. In our practice, where the vast majority of biopsies are now performed using the conchotome technique (11,12), we biopsy most commonly tibialis anterior, quadriceps femoris, gastrocnemius, and deltoid. It is important to remember to avoid biopsy in muscles that have been examined electromyographically as any needling can produce changes indistinguishable from inflammatory myopathy (13).

METHODS OF PERCUTANEOUS BIOPSY

We perform all our biopsies under local anesthesia and only rarely has any sedation been necessary for anxious patients. There is little justification for submitting patients with respiratory insufficiency to general anesthesia. Moreover, the particular hazards of anesthesia in muscle disease such as malignant hyperthermia and the susceptibility to respiratory complications in myotonic dystrophy patients should provide further argument against the use of general anesthesia.

We perform our biopsies at the bedside or in the day ward with one assistant in attendance. Routine clotting screen or platelet count is unnecessary if there is no history of bleeding or easy bruising.

CONCHOTOME BIOPSY TECHNIQUE

Routine aseptic precautions are taken. The skin is shaved and cleaned with chlorhexidine in alcohol or another suitable disinfectant. The biopsy site is infiltrated with 5 ml of 2% lignocaine, but it is important to avoid injecting the muscle itself. A 1-cm skin incision is made with a scalpel blade and the muscle fascia is also penetrated. The closed jaws of the conchotome are inserted through the incision into the muscle with the long axis of the jaws parallel to the muscle fibers. The jaws are opened and advanced 2–3 mm. In one movement the jaws are closed and the conchotome is twisted through 180° and withdrawn (Fig. 1). A sterile scalpel blade or needle can be used to deliver the sample to a piece of gauze for inspection. The procedure can be repeated several times through the same incision. Wound-closing procedures are similar to those for the needle technique. Adhesive strips are applied to appose the wound edges and an adhesive dressing is applied over the biopsy site. Pressure is then applied for 15 min to prevent bleeding and hematoma formation. We advise patients to rest for 30 min after the procedure and to avoid strenuous exercise for 24 hr. We ask patients to keep the dressing dry for 7 days after which the whole dressing can be removed. Conchotome sizes vary and we use the 4.5-mm size for younger subjects, the 6.5-mm size for larger subjects. The conchotome has not been found to require sharpening, maintaining its cutting edge for up to 4 years (9).

NEEDLE BIOPSY TECHNIQUE

For needle biopsy the same preparation procedures are followed. Several types of needles are available. A modified University College needle is shown in Figure 2. The needle is passed through a 5-mm skin incision and through the fascia with the aperture closed. When the muscle is entered, the aperture is opened by withdrawing the inner tube or "guillotine" and then closed, trapping a sample of muscle in the barrel. Up to three samples can be taken with each penetration. Low-grade suction can be applied through the needle, which sucks a larger sample into the aperture. The guillotine requires regular inspection and sharpening to maintain its cutting edge.

ADVANTAGES, COMPLICATIONS, AND LIMITATIONS OF CLOSED TECHNIQUES

Hematoma and wound infection are complications common to all techniques of biopsy. Closed techniques are clearly more advantageous with regard to patient comfort and scar formation. Both the needle and conchotome biopsy techniques are blind and sampling errors can occur, but this is also true of open biopsy. The conchotome technique extends the range of muscles that can be biopsied. Smaller muscles can be sampled as the conchotome technique does not require deep muscle penetration, as does needle biopsy. The conchotome technique also provides a larger sample, 50–150 mg, compared to needle biopsy samples with suction, which average 100 mg (9).

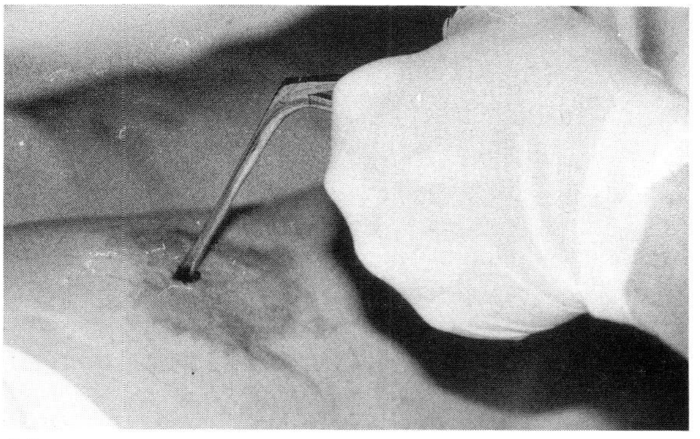

(a)

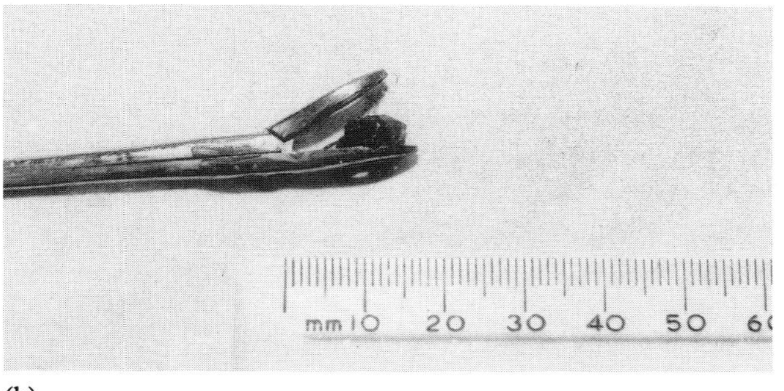

(b)

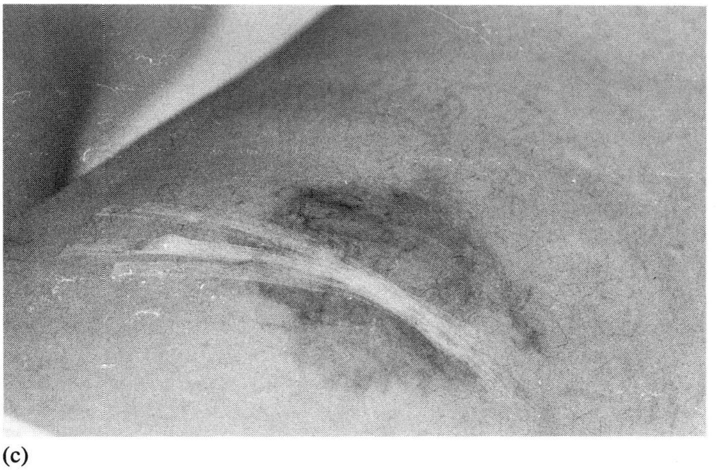

(c)

Figure 1 Conchotome biopsy procedure.

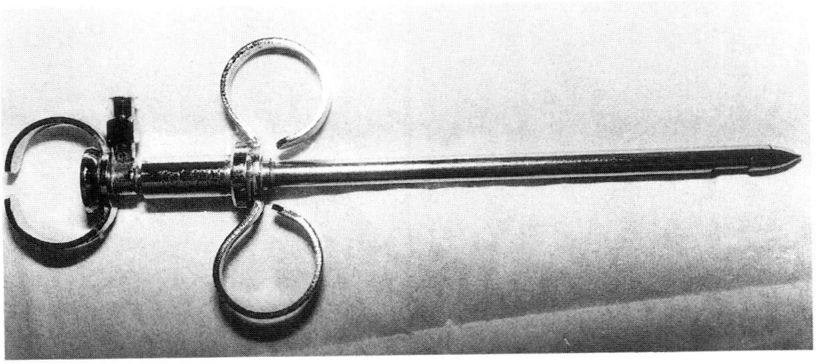

Figure 2 Modified University College biopsy needle.

OPEN BIOPSY

Open biopsy is a more extensive procedure that produces a larger and more unsightly scar. It does not have the repeatability of the closed procedures. It is now rarely indicated and might only be considered necessary in the following circumstances:

1. To obtain motor end-plates for functional and morphological studies in myasthenia gravis.
2. To obtain blood vessels of sufficient size to show vasculitis such as the changes seen in polyarteritis nodosa.
3. To obtain samples for functional and pharmacological examinations in the investigation of malignant hyperthermia. However, these techniques cannot exclude the risk of malignant hyperthermia and this indication for open biopsy is therefore not so strong.
4. To obtain some measurements of mitochondrial metabolism, e.g., ADP:O ratio.

TISSUE HANDLING AFTER BIOPSY

Transport to the Laboratory

Tissue should be transferred to the laboratory as rapidly as possible in a small plastic tube (5 ml) though a delay of up to 30 min will not compromise the morphology. Specimens may be kept moist with damp filter paper and the specimen tube kept in a container of ice water (0°C) if longer delays up to 1 hr are envisaged.

Tissue Allocation

Most histological studies are performed on frozen muscle sections, which are also suitable for molecular biological studies (Western blotting and DNA analyses can be carried out on a few tissue sections). A small piece of tissue is usually fixed for electron microscopy. Unless a particularly large specimen is available, routine formaldehyde fixation is not required as this limits the histochemical methods that can be applied.

For most biochemical analyses, 50–100 mg of muscle is placed in a plastic specimen tube and rapidly frozen in liquid nitrogen, although studies of intact mitochondria require fresh, unfrozen samples.

Tissue Dissection and Freezing

It is important for correct diagnosis to achieve high-quality, artifact-free sections. Artifactual tissue distortion can occur during the biopsy procedure, during freezing (due to ice crystal formation), and during cutting and staining, ice crystal formation occurs if the tissue block is too large to allow rapid cooling of the specimen center from the freezing fluid, if tissue is placed directly into liquid nitrogen, as bubbles form on the tissue surface and slow heat conduction, and if tissue thawing is allowed to occur during sectioning.

Diagnostic assessment of fiber sizes requires undistorted transverse sections of muscle fibers; longitudinal fibers are rarely useful.

For optimal freezing, tissue blocks should not exceed 5 × 4 × 4 mm. Larger specimens from open biopsy may need trimming with a sharp, mounted razor blade. Smaller conchotome or needle biopsy specimens (50 mg) should be carefully oriented on a 2-cm diameter cork disk using a dissecting microscope and forceps to obtain parallel fascicles, which can then be cut transversely (Fig. 3). Orientation should be performed rapidly to prevent drying. The oriented tissue is then surrounded by OCT mounting medium to support the block during freezing and cutting. The cork disk and tissue are immersed in isopentane cooled to -160°C in liquid nitrogen until boiling ceases (15–20 sec), rapidly transferred to liquid nitrogen for an additional 15–20 sec, wrapped in aluminum foil, and stored until required. Both the cork and the foil should be labeled with the patient details and biopsy date.

To prevent the formation of ice crystal artifact, it is important that the isopentane is sufficiently chilled. This can be recognized if the beaker of isopentane starts to freeze. If isopentane is not available, the biopsy can be rolled in starch (from surgical gloves), to improve heat transfer, before the specimen is placed directly into liquid nitrogen. Freezing in isopentane chilled with dry ice is not satisfactory and usually leads to some crystal formation; however, it will not preclude an accurate diagnosis of major types of muscle disease if optimal methods are unavailable.

Tissue Storage and Handling

The frozen tissue blocks are usually stored in a freezer at -70°C. This preserves tissue adequately for several years. There is a danger of desiccation with long-term freezer storage if specimens are not wrapped with foil or plastic film. To avoid this it may be preferable to store the tissue as sections on slides or coverslips in the freezer (wrapped in foil or plastic film), or to keep tissue in a vat of liquid

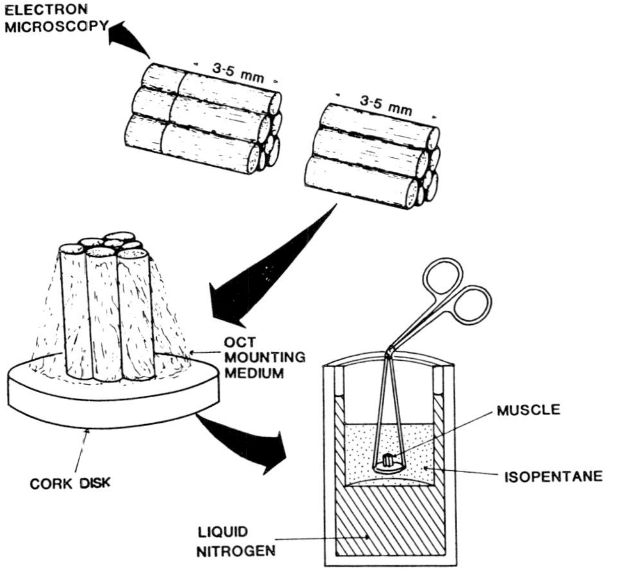

Figure 3 Preparation of muscle biopsy tissue for light microscopy (see text for details).

nitrogen. Sections are usually cut onto coverslips or slides in a cryostat at -18 to -20°C and 5–10 microns thickness; thinner sections are more difficult to cut but provide better morphological appearances for diagnostic work.

Electron Microscopy

For needle and conchotome specimens 1–2-mm cubes cut with a sharp razor are placed directly into fixative and hypercontraction artifact is rarely induced. If a large specimen from an open biopsy is available, a strip of muscle 0.2 cm in diameter and up to 1.5 cm long can be tied to a wooden splint to reduce hypercontraction during fixation. The fixative used needs to be optimized for each laboratory, but usually 2–6% glutaraldehyde in 0.1 M phosphate or 0.1 M cacodylate buffer at pH 7.2 is used. Tissue is fixed overnight and then rinsed in buffer below processing to resin.

BIOCHEMICAL ANALYSES

A wide range of biochemical analyses are possible on muscle biopsy samples. Those most frequently undertaken include analyses of key enzymes of the glycolytic pathways, mitochondrial citric acid cycle, and electron transport chains (14) or measurements of important intermediates in energy metabolism such as ATP and phosphocreatine (5). A substantial amount of data is also available on the cation content of normal and diseased muscle (15,16). In addition to the "traditional" biochemical analyses, the development of molecular techniques and the wide availability of cDNA probes to a large number of muscle genes have stimulated analyses of specific mRNA species in biopsy samples (17). More sensitive procedures, such as the polymerase chain reaction, are also being applied to the specific detection of very low levels of minor RNA species such as enteroviral RNA in muscle (18).

REFERENCES

1. Bergstrom, J. Muscle electrolytes in man: determined by neutron activation analysis on needle biopsy specimens: a study in normal subjects kidney patients and patients with chronic diarrhoea. Scand J Clin Lab Invest 1962; 14 (Suppl 68): 1–110.
2. Edwards RHT. Percutaneous needle biopsy of skeletal muscle in diagnosis and research. Lancet 1971; 2:593–595.
3. Edwards RHT, Young A, Wiles M. Needle biopsy of skeletal muscles in the diagnosis of myopathy and the clinical study of muscle function and repair. N Engl J Med 1980; 302:261–271.
4. Dubowitz V. Muscle Biopsy: A Practical Approach, 2nd ed. London: Bailliere Tindall, 1985:3–18.
5. Edwards RHT, Jones DA, Maunder C, Batra GJ. Needle biopsy for muscle chemistry. Lancet 1975; 1:736–740.
6. Edwards RHT, Round JM, Jones DA. Needle biopsy of skeletal muscle: a review of 10 years experience. Muscle Nerve 1983; 6:676–683.
7. Heckmatt JZ, Pier N, Dubowitz V. Real-time ultrasound imaging of muscles. Muscle Nerve 1988; 11:56–65.
8. Yonker RA, Webster EM, Edwards NL, et al. Technetium pyrophosphate muscle scans in inflammatory muscle disease. Br J Rheumatol 1987; 26:267–269.
9. Bulcke JA, Crolla D, Termote J-L, Baert A, Palmers Y, Van Den Bergh R. Computed tomography of muscle. Muscle Nerve 1981; 4:67–72.
10. Fujino H, Kobayashi T, Goto I, Onitsuka H. Magnetic resonance imaging of the muscles in patients with polymyositis and dermatomyositis. Muscle Nerve 1991; 14:716–720.
11. Henriksson KG. "Semi-open" muscle biopsy technique. A simple outpatient procedure. Acta Neurol Scand 1979; 59:317–323.
12. Dietrichson P, Coakley J, Smith PEM, Griffiths RD, Helliwell TR, Edwards RHT. Conchotome and needle percutaneous biopsy of skeletal muscle. J Neurol Neurosurg Psychiatry 1987; 50:1461–1467.
13. Engel WK. Focal myopathic changes produced by electromyographic and hypodermic needles. Arch Neurol 1967; 16:509–513.
14. Wagenmakers AJM, Kaur N, Coakley JH, Griffiths RD, Edwards RHT. Mitochondrial metabolism in myopathy and myalgia. Adv Myochem 1987; 1:219–230.
15. Bergstrom J. Percutaneous needle biopsy of skeletal muscle in physiological and clinical research. Scand J Clin Lab Invest 1975; 35:609–616.

16. Jackson MJ, Jones DA, Edwards RHT. Measurement of calcium and other elements in muscle biopsy samples from patients with Duchenne muscular dystrophy. Clin Chim Acta 1985; 147:215–221.
17. McConchie SM, Coakley J, Edwards RHT, Beynon RJ. Molecular heterogeneity in McArdle's disease. Biochim Biophys Acta 1991; 1096:26–32.
18. Gow J, Behan WMH, Clements GB, Woodall C, Riding M, Behan PO. Enteroviral RNA sequences detected by polymerase chain reaction in muscles of patients with postviral fatigue syndrome. Br Med J 1991; 302:692–696.

6
Muscle Biopsy Analysis

Margaret A. Johnson and Martin J. Barron
University of Newcastle upon Tyne, Newcastle upon Tyne, England

INTRODUCTION

The histopathological analysis of muscle biopsy material plays a pivotal role in the diagnosis of patients with neuromuscular disorders, but in order to be fully effective it should be integrated within an overall program of clinical and laboratory investigations. The precise role of histopathological assessment in the diagnostic process will vary according to the nature of the disorder under investigation. For example, the muscle biopsy findings would be definitive in the diagnosis of central core disease, whereas in the investigation of a lipid storage myopathy they would indicate the area in which biochemical studies should be directed but would be inconclusive in themselves. In this chapter our aim is to describe protocols designed to give the maximal amount of specific information on individual muscle disorders, but we shall also indicate those areas where other investigative techniques are essential to provide a definite diagnosis.

Today's muscle pathologists have at their disposal an increasing repertoire of histochemical and immunocytochemical techniques that are relevant to the investigation of neuromuscular disorders. Most laboratories use a basic histochemical screening protocol for the initial assessment of all biopsies and subsequent investigations will depend on the outcome of this initial screening procedure. Its main objective is to be able to arrive at a working diagnosis. This will involve being able to decide whether the observed histopathological features are predominantly (1) neurogenic (denervating), (2) myopathic, or (3) inflammatory. Recognition of features indicating a muscular dystrophy or a metabolic myopathy within the "myopathic"group should also be possible at this stage.

Further investigation will involve the use of "subprotocols" or more specific techniques; in the case of inflammatory myopathies and muscular dystrophies, these will be mainly immunocytochemical, whereas those relevant to the metabolic myopathies will be predominantly enzyme histochemical or organic histochemical methods. The application of this wide range of techniques is possible only if appropriately prepared tissue is available. In the case of muscle this means correctly frozen material. Provision for long-term storage is also important since new diagnostic techniques need to be tested using "archive" material. Stored biopsies may also enable retrospective diagnoses to be made as new techniques become available without recourse to rebiopsy of patients.

TISSUE PREPARATION AND STORAGE

Fresh (unfixed) frozen sections are essential for many enzyme histochemical techniques required in muscle pathology and for a wide range of immunocytochemical methods, especially those using monoclonal antibodies. Cryosection technology is also flexible and adaptable in that numerous fixation procedures or other posttreatments can be used on frozen sections, for example, to prevent loss of soluble tissue components, to optimize antigen reactivity, or to permeabilize cellular organelles.

Orientation of muscle for diagnostic work should normally be transverse since many histological features (muscle fiber diameter, fascicular architecture) need to be assessed on cross-sections. The height of a muscle block should not exceed its width, otherwise it will tend to be unstable, and in any case the height should not exceed 7–8 mm, otherwise good freezing will not be possible. This should be done using isopentane prefrozen in liquid nitrogen and used as it is melting, when the temperature is about -150°C. Direct immersion of muscle in liquid nitrogen causes ice-crystal damage and should be avoided. Small muscle samples (<2 mm^3) can be supported between thin slices of gel medium (15% gelatin plus equal part cryoembedding medium; OCT, Tissue Tek, Miles Inc.) prior to freezing. Orientation of needle or conchotome samples is best done using a dissection microscope.

Long-term storage of muscle biopsies should be at temperatures of -70°C or lower. Even at -70°C dehydration of tissue can be severe, so heat-sealed polythene packets should be used rather than vials or screw-cap containers. The gaseous phase of liquid nitrogen containers is often used for long-term storage, and here the temperature may vary from about -120°C to about -160°C depending on the type of container used.

ROUTINE HISTOCHEMICAL SCREENING PROTOCOL

The component techniques of a suggested routine screening protocol are listed in Table 1. Basic information on fiber types is needed in all instances, and the myofibrillar ATPase technique, with appropriate preincubation procedures, is used for this purpose. The general metabolic characteristics of the three major histochemical fiber types—type 1 (slow-twitch oxidative), type 2A (fast-twitch oxidative/glycolytic), and type 2B (fast-twitch glycolytic)—are summarized in Table 2. Whereas the reaction intensities of myofibrillar ATPase (mATPase) after alkaline preincubation approximate most closely those seen in vivo, acid preincubation procedures, which result in the "reversal"of mATPase characteristics, are used more extensively in diagnostic work since they give a particularly clear-cut differentiation of fiber types.

One of the major uses of mATPase methods is in the detection of abnormalities of fiber type distribution. The spatial distribution of fiber types in normal human muscle is random (Fig. 1A), but neurogenic disorders may show considerable uniform fiber type grouping (Fig. 1B) due to remodeling of motor units as a consequence of denervation and reinnervation (6). Selective atrophy of a particular fiber type may be an important diagnostic feature, as may abnormal predominance of one fiber type.

Succinate dehydrogenase (SDH) is used routinely as a "marker enzyme" for mitochondrial distribution and is the preferred method for this purpose. Demonstration of NADH dehydrogenase (NADH-tetrazolium reductase, NADH-TR) activity is often used instead. However, it is important to recognize that the enzyme activity demonstrated is not equivalent to the NADH-ubiquinone reductase (complex I) component of the respiratory chain but also includes NADH-cytochrome b_5 reductase

Table 1 Basic Histochemical Screening Protocol

Technique	Ref.	Detection of:
Myofibrillar ATPase	1	Uniform fiber type grouping; selective atrophy/hypertophy; fiber type predominance; "unstructured" cores
Succinate dehydrogenase	2	Mitochondrial abnormalities, "ragged-red" fibers; central cores, multicores; floccular fibers; "moth-eaten" fibers
Acid phosphatase	3	Lysosome distribution; excess in ceroid lipofuscinosis and type 2 glycogenosis; macrophage distribution
Periodic acid Schiff	4	Glycogen distribution/storage; "ring" fibers
Sudan Black B or Oil Red O	5	Lipid accumulation

Table 2 Histochemical Characteristics of Muscle Fiber Types

	Type 1	Type 2A	Type 2B	Type 2C
Myofibrillar ATPase, pH 10.2 preincubation	+	++	+++	++(+)
pH 4.6 preincubation	+++	+	++	+(+)
pH 4.3 preincubation	+++	–	–	+(+)
Succinate dehydrogenase	+++	++	+	
Sudan Black B	+++	++	+	
Myophosphorylase	+	++	+++	
Periodic acid Schiff	+	++	+++	

activity present in extramitochondrial sites. The diagnostic usefulness of SDH as a mitochondrial marker is not confined to the detection of "mitochondrial myopathies" (Fig. 2A). In several congenital myopathies, including central core disease and multicore disease, characteristic abnormalities of mitochondrial distribution are the major diagnostic features. In many chronic muscular dystrophies a "floccular" distribution of mitochondria is a common finding (Fig. 2B).

Acid phosphatase (AcPase) activity provides a histochemical marker of lysosomal distribution. In normal human muscle, lysosomes are sparsely located, usually adjacent to nuclei or elsewhere at the fiber periphery. Gross increases in muscle lysosomes are seen in Batten's disease (ceroid lipofuscinosis) and in the lysosomal glycogenosis Pompe's disease (α-1,4-glucosidase deficiency). Macrophages, with their high lysosomal content, show intense acid phosphatase activity, and this property is useful in identification of this cell type in cellular infiltrates. In necrotizing myopathies without an inflammatory basis, infiltrates will be almost purely macrophagic, whereas a significant lymphocytic component will be present in true inflammatory myopathies.

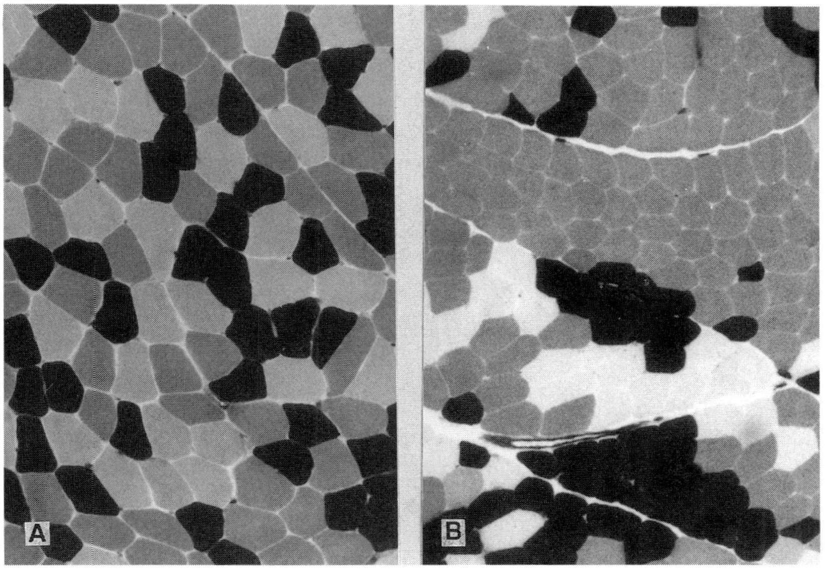

Figure 1 (A) Normal skeletal muscle showing random distribution of fiber types: type 1 (dark), type 2A (pale), type 2B (intermediate). Myofibrillar ATPase, after pH 4.6 preincubation; ×150. (B) Skeletal muscle from patient with a denervating disorder showing uniform fiber type grouping due to reinnervation. Myofibrillar ATPase, after pH 4.6 preincubation; ×150.

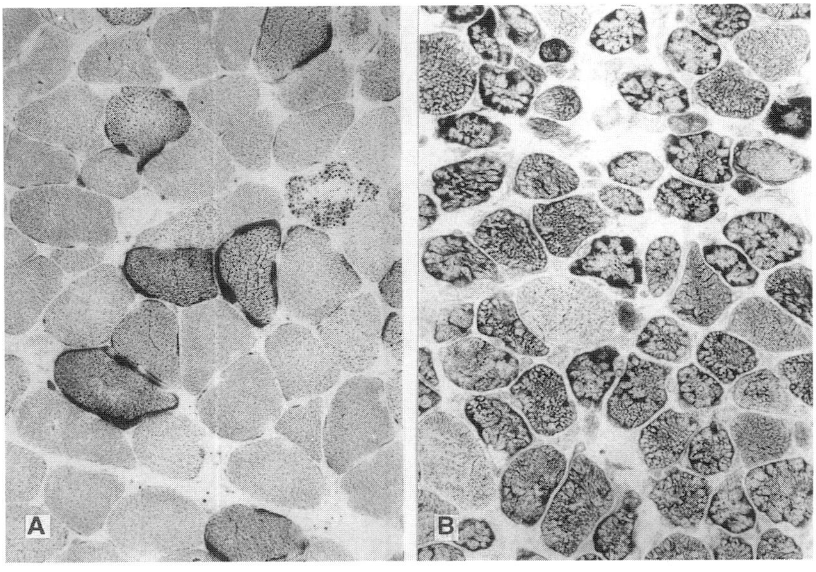

Figure 2 (A) Patient with a mitochondrial myopathy; biopsy findings include prominent subsarcolemmal mitochondrial aggregates. Succinate dehydrogenase; ×250. (B)Floccular distribution of mitochondria in the biopsy from a patient with a chronic limb-girdle syndrome. Succinate dehydrogenase; ×250.

The periodic acid-Schiff (PAS) reaction is used to monitor the amount of glycogen in skeletal muscle. Glycogen is highly diffusible, but prior fixation of sections in ethanolic formalin will retain normal amounts of cytosolic glycogen in situ. Some glycogenoses are characterized by gross vacuolation and glycogen storage, and in this context celloidinization of frozen sections may be necessary to retain all the stored glycogen.

Sudan Black B or Oil Red O methods are employed to demonstrate neutral lipid in muscle sections. In normal skeletal muscle, type 1 fibers contain larger and more numerous lipid droplets than are found in type 2 fibers, reflecting the relative importance of β-oxidation in the two fiber types, and if β-oxidation is impaired, more severe vacuolation due to lipid storage is encountered in type 1 fibers.

In addition to the histochemical techniques listed above, any routine screening protocol will include hematoxylin-and-eosin-stained sections, preferably serial to the histochemical preparations so that direct correlation with morphological abnormalities is possible. The hematoxylin–van Gieson technique is also useful for the selective staining of fibrous connective tissue components and provides a good indication of increases in endomysial and perimysial connective tissue. The Gomori trichrome method, modified for use on frozen sections (7), demonstrates the nemaline rods that are characteristic of rod-body myopathy.

The screening process outlined here will give sufficient information for the pathologist to decide which additional histochemical and/or immunocytochemical techniques are appropriate. Details of these will be given as the various categories of neuromuscular disease are considered in turn.

DENERVATING DISORDERS

Only a brief account will be given of the histopathological features of the spinal muscular atrophies and other denervating disorders since this volume is mainly concerned with primary myopathic disorders. However, in terms of differential diagnosis, infantile spinal muscular atrophies must be considered along with other causes of severe hypotonia in young children, and later-onset spinal muscular atrophies must be differentiated from various muscular dystrophies.

Infantile Spinal Muscular Atrophy (SMA)

The most severe form of infantile SMA (Werdnig-Hoffmann disease) is characterized by widespread denervation with no evidence of reinnervation. By the time the patient is 2–3 months old, a majority of muscle fibers show significant atrophy when compared with an expected size range of 8–16 μm at this age. A minority of motor units survive and their fibers show compensatory hypertrophy. The component fibers of these surviving units are invariably type 1 in histochemical profile, and this correlates with the lack of phasic activity in the severely weakened muscles. The territory of surviving motor units is dispersed, with fibers of atrophied units intervening. This feature allows them to be differentiated from reinnervated motor units, which are compact, in addition to being histochemically uniform.

In biopsies from patients with Werdnig-Hoffmann disease taken after 8 weeks of age, muscle fiber size distribution will be distinctly bimodal and the diagnosis straightforward. However, clinical and other pressures may lead to biopsy being undertaken prior to 8 weeks. In a fair proportion of such biopsies there will be no clear bimodal size distribution. In these circumstances it is difficult to differentiate infantile SMA from the more benign syndrome of congenital fiber type disproportion (CFTD), which is also characterized by hypotonia and where significant size differences are found between the major fiber types (8).

The "intermediate" form of SMA, in contrast to Werdnig-Hoffmann disease, can show considerable evidence of reinnervation with compact, histochemically uniform motor units. Most reinnervated units are type 1 in metabolic profile, and the capacity for phasic activity is severely restricted in these patients, although survival into the second decade is not unknown. A useful indication of ongoing reinnervation may be given if target fibers are present. These fibers show concentric zones of high and low mitochondrial oxidative activity, due, most probably, to the reorganization of organelles which accompanies fiber type interconversion.

Juvenile SMA (Kugelberg-Welander Disease)

Histochemical analysis is particularly important in more chronic forms of SMA where few atrophied fibers may be apparent on conventional histological examination, due to effective reinnervation. However, uniform fiber type grouping is often extensive and may involve groups of type 2A fibers or type 2B fibers, though type 1 fiber groups still tend to predominate. Whereas myopathic changes (acute necrosis, phagocytosis, fiber splitting) are rare in more severe forms of SMA, the prolonged survival and greater motor ability shown by Kugelberg-Welander patients is associated with greater stress being placed on surviving motor units. Under these circumstances, acute muscle fiber breakdown can become frequent and fiber splitting is a common reaction to hypertrophy or overload. As muscle fibers atrophy and are eventually lost altogether, increase in endomysial and perimysial fibrous connective tissue may be considerable. Destruction of muscle fascicles due to replacement of muscle fibers by fat and fibrous connective tissue may obliterate any clear evidence of uniform fiber type grouping. At this stage it may be very difficult to differentiate between a chronic SMA, complicated by secondary myopathic abnormalities, and a late-onset muscular dystrophy or limb-girdle syndrome.

Ancillary Techniques

The routine diagnostic protocol should be sufficient in most instances where denervating disorders need to be differentiated from other neuromuscular diseases. It is important, however, that estimates of atrophy and hypertrophy are made with reference to normal ranges of fiber size in adults and children (9,10). Detailed histographic analysis, even using computer-aided systems, is labor-intensive and rarely necessary for routine diagnostic purposes. Use of a stage micrometer, on the other hand, should be an invariable adjunct to reading muscle biopsies and provides a rapid method of estimating ranges of fiber size in individual biopsies. Quantitative techniques for the detection of nonrandom distribution of fiber types (uniform fiber type grouping) may be helpful in cases where grouping is not extensive (11,12).

MUSCULAR DYSTROPHIES

Histopathological analysis of this group of disorders is currently in an interesting phase, with specific molecular probes already available for the Xp21-linked muscle disorders (13) and the prospect of similar advances in other muscular dystrophies being not too far distant.

Duchenne Muscular Dystrophy (DMD)

The classic histopathological features of DMD include acute muscle necrosis, phagocytosis, regeneration, rounded over contracted (hyaline) fibers, endomysial and perimysial fibrous connective tissue increase, and eventual disorganization of fascicular architecture due to replacement by fat. None of these features is specific for DMD, but in practice pattern recognition of the characteristic combination of histopathological abnormalities is usually sufficient for accurate diagnosis. This can now be verified by dystrophin analysis, which has also proved invaluable in the differential diagnosis of Becker and other forms of muscular dystrophy.

Dystrophin Analysis

Dystrophin is the protein product associated with the Xp21 genetic locus. The muscle-specific form is located at the plasma membrane of skeletal muscle fibers (14) where it can be demonstrated by a range of immunocytochemical techniques. Dystrophin is a large molecule (c427 kDa), and monoclonal antibodies have been produced that recognize epitopes within various domains (N-terminus, rod-like central domain, C-terminus). Binding of antidystrophin primary antibody in muscle sections may be followed by a fluorochrome- or peroxidase-conjugated second antibody. For routine diagnostic use a peroxidase label is preferred (Fig. 3A) because the end-product is stable and slides can be stored for reference purposes. Hematoxylin counterstaining allows the general histological features of the tissue to be visualized in the same section.

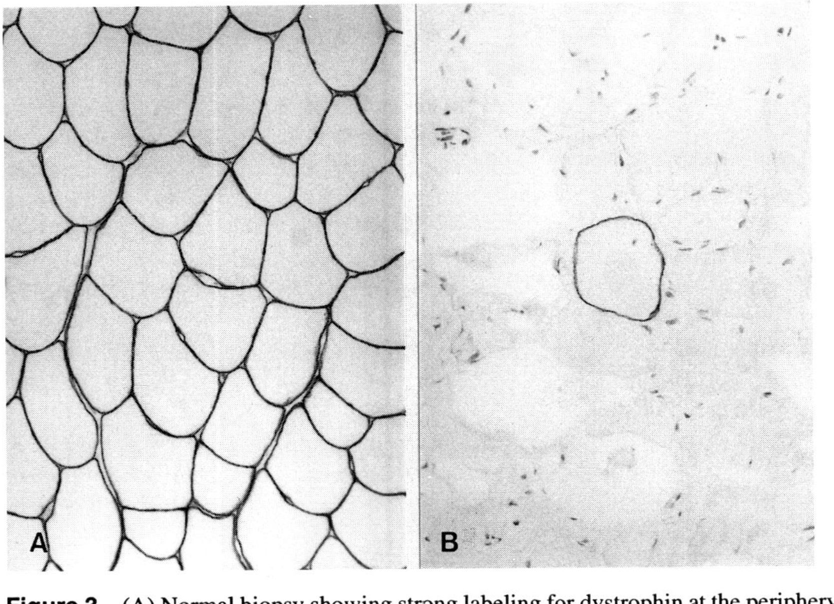

Figure 3 (A) Normal biopsy showing strong labeling for dystrophin at the periphery of all muscle fibers. Immunoperoxidase; ×250. (B) DMD patient showing absence of dystrophin labeling on all but one isolated muscle fiber. Immunoperoxidase; ×250. (C) BMD patient showing marked interfiber variation in intensity of labeling for dystrophin. Immunoperoxidase; ×250. (D) Manifesting carrier of DMD showing absence of labeling for dystrophin in some muscle fibers. Immunoperoxidase; ×250.

In muscle biopsies from DMD patients, three different labeling patterns are seen (15). Almost 50% of patients show no dystrophin labeling at all. However, it is also common to find occasional, strongly dystrophin-positive fibers, disposed singly or in small clusters (Fig. 3B). This type of labeling pattern is at least as frequent as the total absence of labeling. In a small minority of DMD biopsies, a third labeling pattern is seen, consisting of a faint to weak reaction on a high proportion of muscle fibers.

Becker Muscular Dystrophy (BMD)

Whereas the histopathological features of DMD are so distinctive that a process of pattern recognition usually suffices for diagnostic purposes, the same cannot be said of the more benign allelic condition, Becker muscular dystrophy. The histological features of BMD, both in early stages of the disorder and later, can prove a source of diagnostic confusion. Biopsies from early BMD (patients <10 years) are characterized by less severe fibrous connective tissue increase than in DMD, fewer hyaline fibers, and preservation of more normal fiber type proportions. The abnormally high proportion of type 1 fibers seen in many DMD biopsies does not occur in BMD.

However, a picture similar to that seen in early BMD occurs in biopsies from patients with severe autosomal recessive muscular dystrophy of childhood (SAR). Moreover, both early BMD and SAR generally show a particularly focal distribution of acutely necrotic fibers. These foci can be extensive, sometimes involving over 50 contiguous muscle fibers. Foci of grouped regenerating fibers are also seen, presumably representing a later stage in the process. At this stage the foci are easy to identify because of the 2C histochemical characteristics of the component fibers, seen in mATPase preparations after pH 4.3 preincubation.

In the absence of informative family histories, conventional histological investigation cannot clarify the diagnosis further, at least not with the certainty required for genetic counseling. The differentiation of early BMD from SAR by means of dystrophin analysis is therefore one of the most important contributions of this technique to diagnostic practice.

Emery-Dreifuss muscular dystrophy is X-linked with its gene locus situated on the long arm of

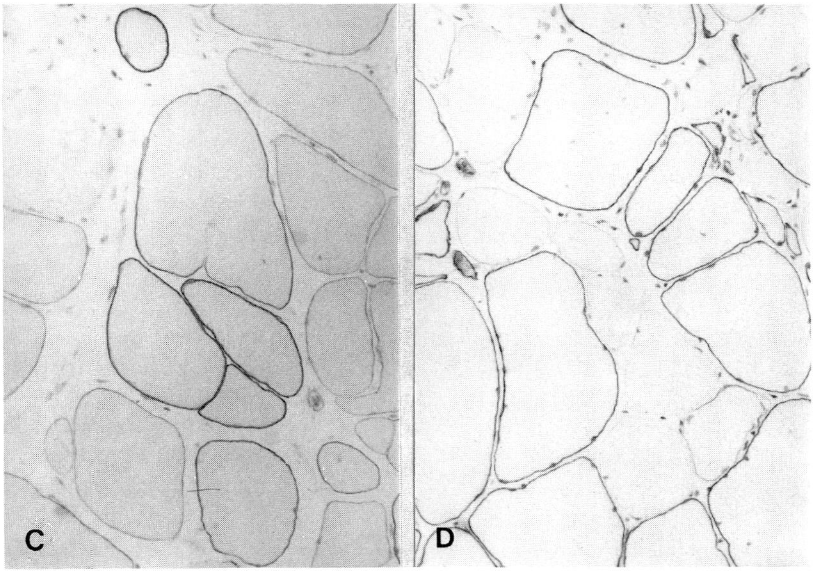

Figure 3 (Continued)

the X chromosome (16). Histopathological features are nonspecific, so dystrophin analysis is necessary to differentiate the condition reliably from BMD.

BMD patients who are biopsied in the second or third decades, or even later, show a pattern of histopathological abnormalities that is typical of many chronic muscular dystrophies and hence relatively nonspecific. The range of fiber diameters may be great; in severely affected biopsies atrophied fibers <10 μm are common and muscle hypertrophy can result in some fibers exceeding 200 μm. Features characteristic of chronically stressed muscle are seen, including fiber splitting and "ring fibers." These fibers show a corkscrew displacement of peripheral myofibrils, so a longitudinal myofibrillar banding pattern is seen in transverse sections. Abnormalities of mitochondrial distribution ("moth-eaten" fibers) are also common. However, all these features can also occur in various autosomally inherited limb-girdle syndromes, highlighting another important contribution of dystrophin analysis to the diagnostic process.

BMD patients, irrespective of the age at which biopsies are taken, show three major patterns of dystrophin labeling in muscle sections. (1) Labeling may show a severe uniform decrease with all muscle fibers affected. (2) Labeling may show variation between muscle fibers (interfiber variation, Fig. 3C) or within individual muscle fibers (intrafiber variation) in addition to significant overall decrease in labeling intensity. (3) Labeling may be near-normal in intensity, and immunocytochemical assessments (ICC) alone are insufficient grounds on which to base a diagnosis. In this latter category immunoblotting is essential to provide precise quantitation of any dystrophin deficiency (15).

Dystrophin Immunoblotting

An ideal investigative protocol for the muscular dystrophies will include dystrophin ICC and immunoblotting run in parallel since the techniques give information on different aspects of the protein and are complementary. Protein bands visualized by immunoblotting can be assessed quantitatively by means of densitometry, which is essential to detect BMD patients with near-normal dystrophin levels. The corollary to this is that, to show conclusively that non-Xp21 biopsies contain a normal amount of dystrophin, immunoblotting should also be done. In addition to information on dystrophin quantity, immunoblotting allows estimation of dystrophin size, which cannot be obtained by ICC.

Manifesting Carriers of DMD and BMD

A proportion of carriers of DMD or BMD will themselves show overt muscle disease. Such carriers may be misdiagnosed clinically as suffering from non-Xp21-linked limb-girdle muscular dystrophies or similar syndromes, and the histological abnormalities seen in biopsies include no specific features to differentiate them from non-Xp21-linked conditions. However, dystrophin ICC reveals a varying proportion of dystrophin-negative or dystrophin-deficient fibers (Fig. 3D) in most manifesting carriers (17). It is important to ascertain that absence of dystrophin labeling in isolated fibers is not due simply to necrosis, and it is recommended that labeling for β-spectrin (a muscle membrane protein independent of dystrophin) should be done on serial sections to monitor membrane integrity.

Limb-Girdle Syndromes (LGS)

The ability to discriminate between BMD and autosomally inherited limb-girdle syndromes and to identify female carriers of DMD and BMD has removed two major sources of confusion from the investigation of muscular dystrophies in adults. Prior to dystrophin analysis and in the absence of informative family histories, there was little in the way of specific histopathological features to help in differentiating autosomal recessive and autosomal dominant limb-girdle syndromes from BMD or from each other. As a group BMD patients show more severe ring fiber abnormalities than autosomal LGS patients. Also, as a group autosomal dominant LGS patients show more severe floccular abnormalities of mitochondrial distribution and more severe type 1 fiber atrophy than either BMD patients or patients with autosomal recessive LGS. However, discernible differences between groups of patients are of limited value in the diagnosis of individuals.

Facioscapulohumeral Muscular Dystrophy (FSH) and Scapuloperoneal Syndromes

The histopathology of FSH muscular dystrophy can vary from minimal abnormalities such as increased variation in fiber size to a much more severe picture, including acute necrosis, type 1 fiber atrophy, and abundant floccular muscle fibers. In addition, about one-third of FSH biopsies examined show significant inflammatory cell infiltration. On further immunocytochemical analysis this is generally found to be $T8^+$-cell-associated. It is possible that the inflammatory component results from sensitization of lymphocytes to muscle proteins externalized during primary myopathic degeneration. Biopsies from patients with scapuloperoneal syndromes without facial involvement show broadly similar histopathological features, including some inflammatory component in a proportion of cases.

Myotonic Muscular Dystrophy

The clinical diagnosis of this disorder is usually so straightforward that muscle biopsy is only rarely undertaken. Histopathological abnormalities include type 1 atrophy, which may be severe, and ring fibers, which are probably more common in this disorder than in any other. Biopsies from children with the congenital form of myotonic dystrophy generally show an abnormal predominance of type 1 muscle fibers.

Immunocytochemical Protocol for Biopsies from Patients with Suspected Muscular Dystrophies

A suggested ICC protocol is summarized in Table 3. Note that use of more than one antidystrophin antibody is recommended. A small proportion of DMD/BMD biopsies may give false negative results due to deletions that remove the epitope recognized by a particular antibody. Antibodies to different sectors of the molecule may enable the presence of limited amounts of dystrophin to be detected. Routine inclusion of β-spectrin labeling in the protocol ensures that scattered dystrophin-negative fibers that occur as a result of muscle necrosis are recognized as such, since they will also be β-spectrin-negative. Labeling for the chromosome 6-encoded dystrophin-related protein (DRP, utrophin) may be potentially useful in cases where dystrophin is minimally decreased and immunoblotting is unlikely to clarify the situation (e.g., suspected manifesting carriers). DRP labeling is generally seen in fibers where dystrophin is significantly decreased. However, some anti-DRP antibodies appear also to recognize a component in regenerating fibers, in non-Xp21-linked conditions as well as DMD/BMD (21).

Table 3 Immunocytochemical protocol for detection of Xp21-linked muscular dystrophies

Labelling	Ref.	Rationale for use
Dystrophin	18	Detection of dystrophin loss/decrease; involvement of specific dystrophin domains
Rod domain	19	
C-terminus	19	
N-terminus		
β-spectrin	20	Monitoring membrane integrity; detection of membrane loss due to necrosis.
Utrophin (DRP, dystrophin-related protein)	21	Confined to neuromuscular junctions in normal muscle; increased in fibers showing severe dystrophin decrease
NPA (no primary antibody) control		Monitoring specificity of labelling; detection of "background" reaction due to nonspecific binding of second antibody

CONGENITAL MYOPATHIES

Many of this group of disorders owe their initial recognition to histochemical and ultrastructural investigation; differential diagnosis still rests mainly with these two disciplines.

Central Core Disease and Multicore Disease

The cores of central core disease are extensive cylindrical areas within muscle fibers, which are devoid of mitochondria and hence readily detectable using techniques for mitochondrial marker enzymes such as SDH. In transverse sections, the cores may be single and centrally placed, or two or more may be seen within one fiber profile (Fig. 4A). In longitudinal sections, the cores may extend for long distances along muscle fibers (several hundred micrometers). If disruption of myofibrils is also seen within the cores, they are described as unstructured and will also be visible in mATPase-reacted sections as areas of low activity. However, structured cores, in which myofibrillar integrity and mATPase activity are retained, are more common. Central core formation appears to be associated with conversion to a type 1 histochemical profile and it is not uncommon to see total conversion to this fiber type.

Multicore disease is much less spectacular from a histopathological viewpoint. Areas of low mitochondrial density are often ill-defined in transverse sections and ultrastructural studies show that they rarely extend for more than a few sarcomeres longitudinally. The cores are almost always unstructured, i.e., myofibrillar integrity is also disrupted, and thus they are also detectable in mATPase preparations. Very little fiber atrophy is seen in either central core or multicore disease.

Centronuclear Myopathies

As the name suggests, this disorder is characterized by a high incidence of internal muscle nuclei. One form of the disorder also shows significant type 2 muscle hypotrophy; i.e., fibers have never attained a normal diameter. This fact can be deduced from ultrastructural studies that show no excessive amounts

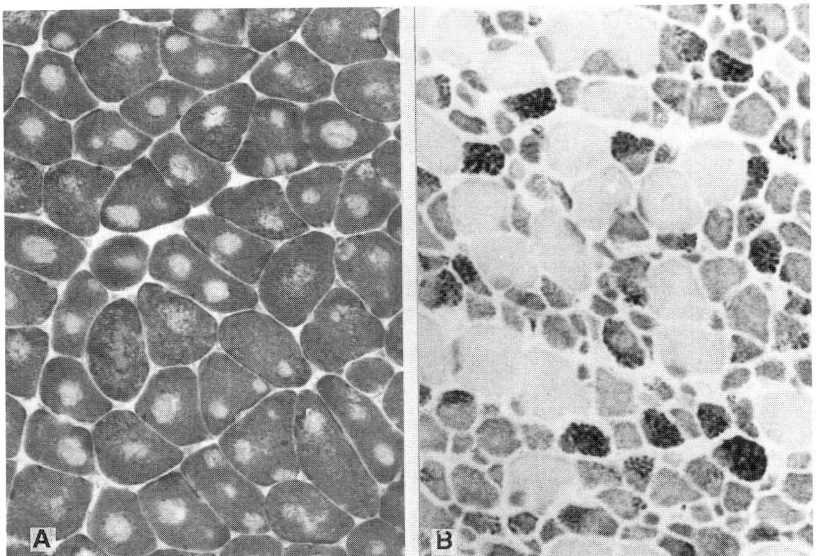

Figure 4 (A) Muscle biopsy from patient with central core disease showing absence of mitochondria from core-like areas within the fibers. Succinate dehydrogenase; ×200. (B) Nemaline myopathy in an infant aged 2 months showing the nemaline rods demonstrated by labeling for α-actinin. (Monoclonal antibody to α-actinin by courtesy of Dr. L. V. B. Nicholson, University of Newcastle upon Tyne.) Immunoperoxidase; ×600.

of basal lamina round the small-diameter fibers, whereas atrophied fibers would show this characteristic. The fact that many of the fibers with central nuclei are also of small diameter has, in the past, prompted the use of the term "myotubular myopathy" in connection with this disorder. However, affected fibers are histochemically differentiated into type 1 or, more rarely, type 2 whereas a true myotube is incompletely differentiated and would show 2C histochemical characteristics.

Nemaline Myopathy

The rods seen in nemaline myopathy are usually 1–5 μm in length and appear to be derived from proteins associated with the Z-line, principally α-actinin (22). "Streaming" or extension of Z-lines is often visible at the ultrastructural level, in addition to the presence of free nemaline rods. At light microscope levels, rods are reliably demonstrated in Gomori trichrome-stained sections and can also be demonstrated immunocytochemically using antibodies to α-actinin (Fig. 4B). Rod formation may be extensive, especially in severe neonatal forms of the disorder, and virtually all muscle fibers may be affected. In milder variants, nemaline rods are found primarily in type 1 fibers, which are also often atrophied.

INFLAMMATORY MYOPATHIES

Histological examination alone may be sufficient to establish the diagnosis of an inflammatory muscle disorder (Fig. 5A), but if the nature of any infiltrate is to be adequately characterized, immunocytochemical analysis is also necessary (23) (see Table 4). Useful cell marker techniques include those for the identification of B lymphocytes, $T4^+$ and $T8^+$ lymphocytes, and macrophages. In normal skeletal muscle, expression of MHC class I and class II antigens is confined to vascular endothelial cells with no expression on the surface of muscle fibers themselves. However, increased expression, especially of MHC-I, is an almost invariable finding in muscle fiber populations affected by inflammatory processes. Detection of neoantigens formed during the series of reactions involving complement (C5b-9, membrane attack complex, MAC) is also possible (24). The distribution of muscle capillaries can be localized by lectin binding (Ulex europaeus agglutinin-I, UEA-I), and this is useful for detection of areas of capillary depletion (25).

Juvenile Dermatomyositis (JDM)

This condition is a multisystem disorder in which much of the histopathology of muscle, skin, and intestinal tract may be accounted for on the basis of an underlying angiopathy (26). In muscle, preferential involvement of perifascicular muscle fibers in acute necrosis and/or atrophy (Fig. 5B) is an almost invariable finding, though the precise mechanisms are unclear. Inflammatory cell infiltration is predominantly perivascular and perimysial; foci of infiltration are often compact, with B lymphocytes as the prevalent cell type (Fig. 5C,D). This in itself is suggestive of humoral mechanisms of cell damage and muscle microvasculature would seem to be the primary target. Capillary loss is common in perifascicular areas, and muscle fibers of normal histological appearance may show central loss of mATPase activity, which is thought to be a prenecrotic change associated with partial or transient ischemia.

Even in biopsies from JDM patients with an extremely short clinical history (<1 month) where histological evidence of inflammatory cell infiltration or muscle necrosis may be minimal, expression of MHC-I on the surface of muscle fibers may be widespread. In inflammatory muscle disease in general, some indication of chronicity should be obtained from inspection of fiber diameters, which will be near-normal in acute cases, whereas wide variation will indicate a much longer time course.

Adult Dermatomyositis (AM)

Adult patients with dermatomyositis show a more variable histopathological picture. In some it is obvious that an underlying angiopathy is the main cause of muscle damage, and it is not unusual to see whole muscle fascicles affected by synchronous necrosis, their component fibers having lost all demonstrable

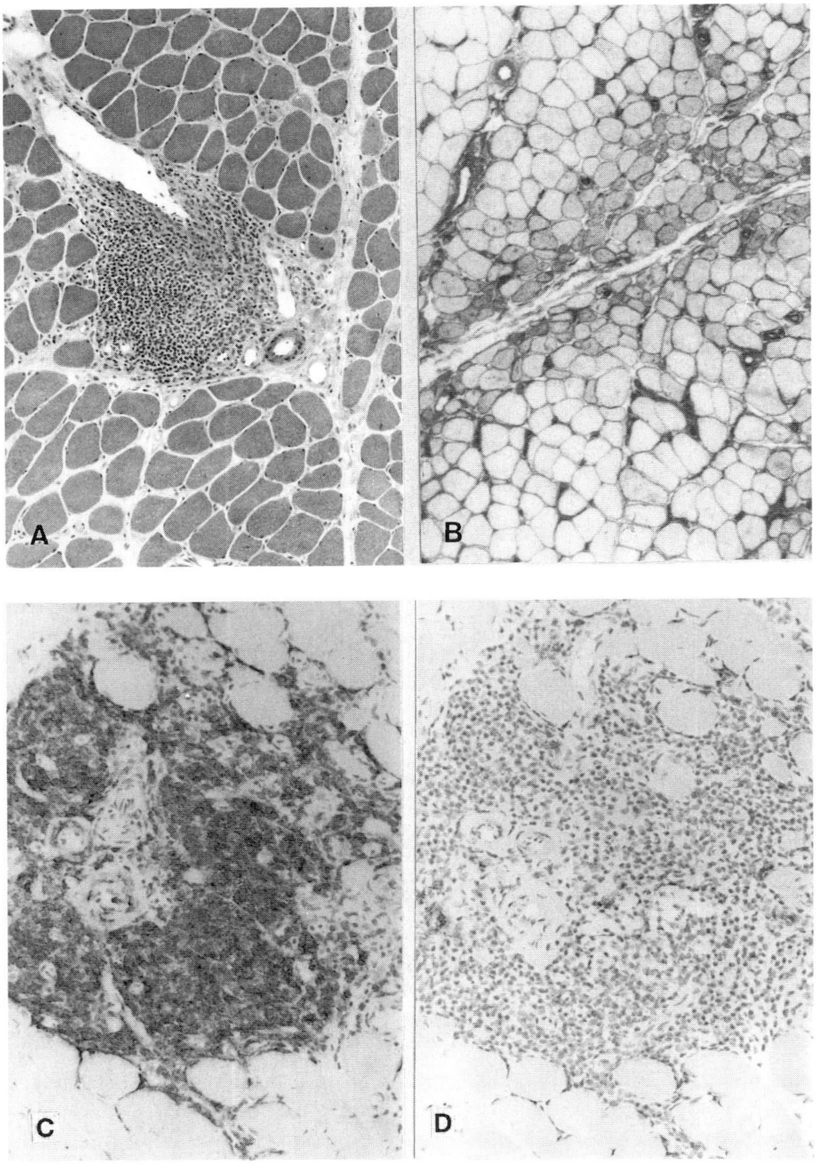

Figure 5 (A) Discrete perivascular inflammatory infiltrate in a muscle biopsy from a patient with mixed connective tissue disease. Hematoxylin-eosin; ×120. (B) Juvenile dermatomyositis in a patient aged 4 years; biopsy showed perifascicular atrophy and increased MHC-I expression in perifascicular areas. Immunoperoxidase; ×150. (C) Infiltrate from a patient with juvenile dermatomyositis showing marked predominance of B lymphocytes (CD22 expression). Immunoperoxidase, hematoxylin counterstain; ×220. (D) Serial section to (C) showing only sparse $T8^+$ lymphocytes (CD8 expression) in the infiltrate. Immunoperoxidase, hematoxylin counterstain; ×220.

Table 4 Immunocytochemical Protocol for Investigation of Inflammatory Myopathies

Labeling	Source	Dectection of:
Pan B cells (CD22)	D	Mature B cells, B-cell precursors
$T4^+$ cells (CD4)	O	T helper cells, MHC-II restricted
$T8^+$ cells (CD8)	D	T cytotoxic/repressor cells, MHC-I restricted
Macrophages (CD68)	D	Phagocytic activity
MHC-I	D	Molecule involved in antigen presentation to $T8^+$ cells
MHC-II	D	Molecule involved in antigen presentation to $T4^+$ cells
MAC (C5b-9)	B	Complement-mediated membrane lysis (membrane attack complex)
UEA-1 lectin	D	Distribution of muscle capillaries

B, Behring Diagnostics; D, Dakopatts A/S; O, Oxoid Ltd. (Unipath).

enzyme activity. Capillary loss in these areas may be severe and presumably restitution of vascular supply must occur as a prerequisite of muscle regeneration. ADM patients with this type of histopathology often show compact inflammatory cell foci with B lymphocytes predominating. Perifascicular fibers usually show strong MHC-I expression, which may be related to proximity to inflammatory cell foci.

Not all ADM biopsies show this type of picture. In some, inflammatory cell infiltration is more diffuse, occurring in endomysial rather than perimysial sites. Acutely necrotic fibers may be scattered in distribution, rather than clustered and/or perifascicular in location. In such cases histological and immunocytochemical features may be very similar to those seen in polymyositis, indicating that the presence or absence of skin involvement does not in itself imply a radically different histopathological picture.

Polymyositis (PM)

In PM, individual muscle fibers themselves appear to be the primary target of the disordered immune process and $T8^+$ (cytotoxic) lymphocytes are generally the prevalent cell type. The distribution of acutely necrotic muscle fibers is mainly random, which is consistent with a process of T-cell-mediated cytotoxicity, in contrast to the grouped, often perifascicular distribution of necrosis in JDM and some cases of ADM, which is dictated by vascular involvement.

In PM, uncomplicated either by malignancy or by connective tissue disorders, the majority of infiltrating cells are diffusely located, rather than perivascular. Early stages of necrosis are often accompanied by vacuolation of muscle fibers. MHC-I expression, known to be a prerequisite for muscle fibers to be targeted by $T8^+$ lymphocytes, is strongest adjacent to areas of infiltration and may be much weaker elsewhere.

If PM is associated with malignancy, disseminated fiber atrophy, with type 2B fibers selectively affected, may be seen. In PM complicated by rheumatoid disease, the constitution of the inflammatory infiltrate may be characterized by a predominance of B lymphocytes often in compact foci in perimysial sites.

Inclusion Body Myositis (IBM)

This type of inflammatory myopathy is often refractory to steroid treatment. The inclusion bodies that characterize the disorder are pleomorphic and may be 3–10 μm in size. They are frequently associated with basophilic-rimmed vacuoles. Ultrastructural studies reveal complex membranous whorled structures (myelin figures) and/or filamentous bodies within a single membrane. Typically the inflammatory infiltrate is sparse and $T8^+$ cells are dominant (27).

Other Inflammatory Conditions

The differential diagnosis of an inflammatory myopathy involves discrimination of several other conditions, many of them comparatively rare. Polymyalgia rheumatica patients present with muscle aches and a high ESR but often yield muscle biopsies with little specific pathology, though "moth-eaten" fibers with patchy mitochondrial distribution may be common. However, other patients with high ESR may show evidence of giant cell arteritis. If arteritic changes are sparsely distributed, the chance of a positive diagnosis is to some extent governed by thoroughness of sampling. At a minimum, sections should be taken from all tissue blocks available. This is equally applicable to the investigation of other arteritides, including polyarteritis nodosa, and in fact to the investigation of any condition with a patchy distribution.

METABOLIC MYOPATHIES

The routine histochemical screening protocol outlined earlier should give a reliable indication of the general nature of any metabolic disturbance. Additional histochemical techniques for the further investigation of metabolic myopathies are listed in Table 5.

Glycogenoses

The most severe glycogen storage is seen in the infantile form of α-1,4-glucosidase deficiency (type II glycogenosis, Pompe's disease), and vacuolation is frequently so severe that the stored glycogen can only be retained in tissue sections by means of prior celloidinization. In later-onset forms of this disorder, a large increase in lysosomal acid phosphatase is seen; in the infantile form, increased acid phosphatase is more diffusely localized. Although there is no specific histochemical method for the demonstration of α-1,4-glycosidase, the concomitant findings of glycogen storage (Fig. 6A) and increased acid phosphatase activity are virtually diagnostic of this disorder since it is the only known lysosomal glycogenosis.

Myophosphorylase deficiency (type V glycogenosis, McArdle's disease) is usually characterized by more moderate glycogen storage, often limited to subsarcolemmal vacuoles and clefts between myofibrils. The histochemical method for the demonstration of the enzyme allows a definite diagnosis to be made using tissue sections; a positive reaction is seen in smooth muscle cells of arterioles, since the disorder involves only the skeletal muscle isoenzyme, and this provides a useful internal control. If regenerating fibers are present, they will usually show some myophosphorylase activity since the fetal isoenzyme is also unaffected (Fig. 6B).

A specific histochemical method is also available for the diagnosis of phosphofructokinase deficiency. This is a multistep method, however, and parallel control incubation using fructose-1,6-diphosphate as substrate must be made to demonstrate the integrity of the metabolic pathway from aldolase onward. Rarer forms of glycogenosis, such as phosphoglucomutase deficiency and defects involving branching enzyme, need to be investigated using biochemical techniques.

Lipidoses

In disorders of lipid metabolism due to carnitine palmitoyl transferase (CPT) deficiencies, the degree of lipid accumulation may be minimal and thus negative histochemical evidence in no way rules out

Table 5 Histochemical Techniques for Investigation of Metabolic Myopathies

Technique	Ref.	Detection of:
Myophosphorylase	28	Type V glycogenosis
Phosphofructokinase	29	Type VII glycogenosis
Succinate dehydrogenase	2	Complex II deficiency
Cytochrome c oxidase	30	Complex IV deficiencies
AMP deaminase	31	Myoadenylate deaminase deficiency
NADH-tetrazolium reductase	2	Tubular aggregates in periodic paralysis, etc.

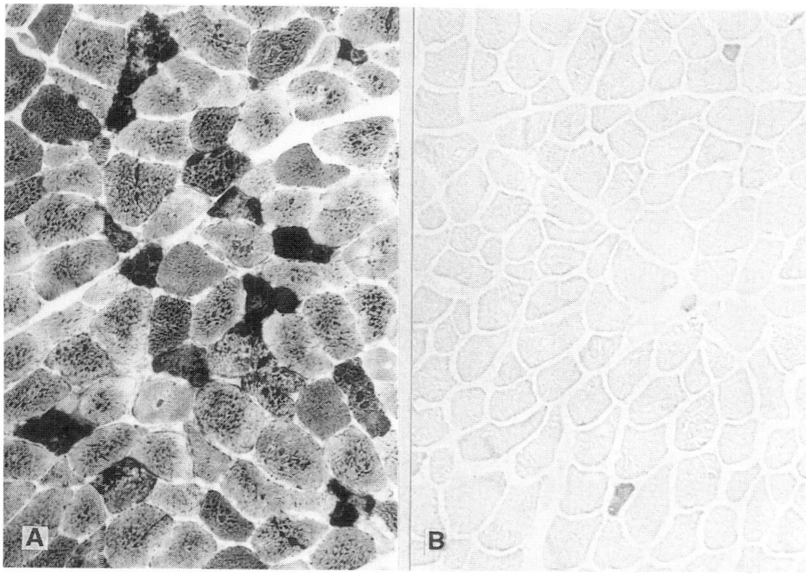

Figure 6 (A) Glycogenosis type 2 (α-1,4-glucosidase deficiency) in an adult patient showing variable glycogen storage in individual muscle fibers. Periodic acid-Schiff; ×250. (B) Glycogenosis type 5 (myophosphorylase deficiency) showing absence of phosphorylase activity from all muscle fibers except two small regenerating fibers (arrows). Phosphorylase; ×200.

these disorders. Defects of any of the acyl-CoA dehydrogenases (short-, medium-, or long-chain) give rise to more severe vacuolation and lipid storage, detectable by Oil Red O or Sudan Black B staining (Fig. 7A,B). No histochemical methods exist to differentiate the various types of acyl-CoA dehydrogenase deficiency from one another or from more distal defects of β-oxidation, i.e., those due to deficiency of electron transfer flavoprotein (ETF) or of its corresponding dehydrogenase.

Immunocytochemical methods, based on the use of antibodies specific for the different acyl-CoA dehydrogenases, have been used to study patients with deficiencies of these enzymes (32). However, this methodology suffers from the limitation that applies equally to all such ICC studies of enzymes. Absence of immunolabeling in tissue sections is usually a reliable indication that no enzyme protein is present. However, the presence of labeling, while showing that cross-reactive protein is there, can give no information on the functional integrity of the enzyme.

Respiratory Chain Abnormalities

Indication of an underlying respiratory chain abnormality is often given by finding fibers with abnormal subsarcolemmal accumulations of mitochondria ("ragged-red" fibers), demonstrable in Gomori trichrome-stained sections (33), or, more appropriately, using mitochondrial marker techniques. However, the absence of ragged-red fibers cannot be taken as evidence of normal respiratory chain function. For example, a proportion of patients with Leigh's disease show a severe deficiency of cytochrome c oxidase (CCO, complex IV) in skeletal muscle, yet the distribution of mitochondria is usually entirely normal. Fatal infantile CCO deficiency (34) associated with de Toni-Fanconi-Debre syndrome involves total absence of catalytic activity from all muscle fibers (Fig. 8A), whereas only a proportion of fibers show evidence of mitochondrial proliferation or clumping at light microscopic level.

Since CCO deficiency is associated with so many mitochondrial disorders (see Chapter 10), demonstration of CCO activity in tissue sections, using diaminobenzidine as electron donor, is an essential part of any protocol for the investigation of respiratory chain abnormalities. Histochemical

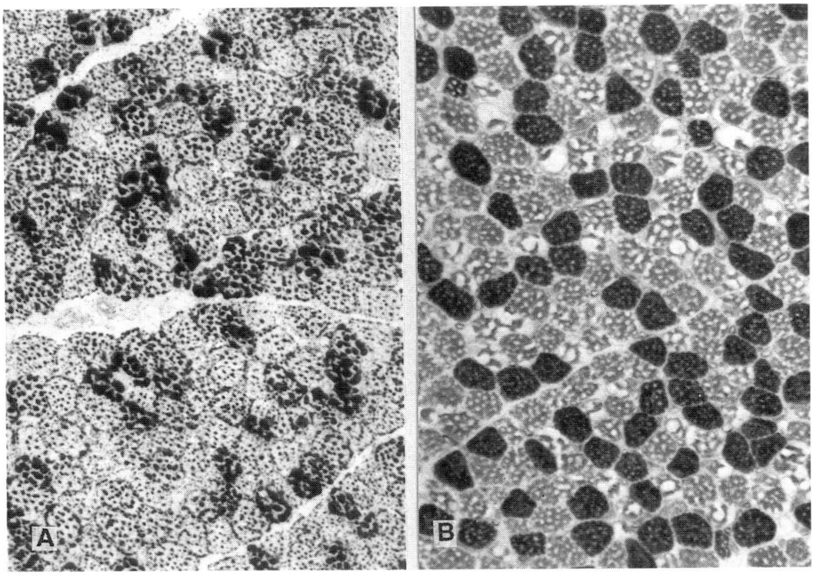

Figure 7 (A) Lipid storage in a muscle biopsy from an infant aged 6 months with ETF dehydrogenase deficiency. Sudan Black B; ×380. (B) Same patient as in (A) showing that vacuolation due to lipid storage is more severe in type 1 (pale) fibers. Myofibrillar ATPase, after pH 10.2 preincubation; ×380.

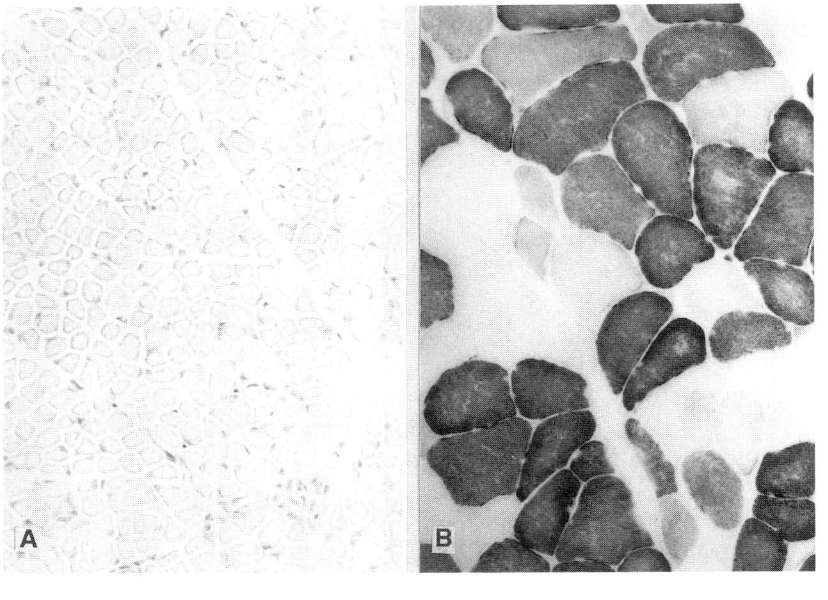

Figure 8 (A) Total cytochrome oxidase deficiency in an infant aged 3 months with de Toni-Fanconi-Debre syndrome. Cytochrome oxidase; ×380. (B) Mosaic of cytochrome oxidase–deficient fibers (asterisks) in a patient with chronic progressive external ophthalmoplegia and myopathy. Cytochrome oxidase; ×250.

monitoring is particularly important in differentiating the fatal infantile form of CCO deficiency from more benign infantile CCO defects, characterized by progressive increase in the proportion of muscle fibers showing functional CCO. This process can be adequately assessed using small-needle biopsy samples, a procedure that is invaluable when repeated biopsy is necessary.

A mosaic of CCO-deficient fibers (Fig. 8B) is seen in virtually all cases of Kearns-Sayre syndrome (KSS) or chronic progressive external ophthalmoplegia (CPEO), and this finding is a useful indicator of a probable abnormality of mitochondria DNA (mtDNA). Large-scale deletions of mtDNA are generally found in these patients (35) but are heteroplasmic; i.e., they affect only a proportion of total mtDNA. Segregation of mutated and wild-type mtDNA into individual muscle fibers or segments of fibers is thought to account for the mosaic of respiration-deficient fibers demonstrable histochemically.

Defects involving complex II (SDH) are rare but are demonstrable using standard tetrazolium salt reduction techniques, preferably in conjunction with quantitative estimation by means of microphotometry. No valid histochemical techniques exist for the demonstration of the catalytic activity of complexes I, III, or V. As mentioned previously, NADH-TR activity demonstrable histochemically is not identical to NADH-ubiquinone reductase (complex I) and includes NADH-cytochrome b_5 reductase activity and possibly other components as well.

Immunocytochemical techniques can be used to demonstrate possible absence or decrease of components of complex III (ubiquinol-cytochrome c reductase) since this complex contains relatively few subunits; this also applies to the investigation of complex IV (36). The use of subunit-specific antibodies to complex I components is impractical since there are >30 subunits. The required information on involvement of individual subunits can be acquired more appropriately by means of electrophoresis followed by immunoblotting using polyclonal antisera raised to the holocomplex.

Disorders of Purine Metabolism

Many patients with metabolic disturbances present with muscle cramps and pain on exercise. Two disorders of purine metabolism need to be considered in differential diagnosis. Myoadenylate deaminase deficiency may occur in patients who show minimal histopathological abnormalities. However, a specific histochemical technique is available for the demonstration of AMP deaminase activity, based on the reduction of a tetrazolium salt by ammonia in the presence of dithiothreitol. Xanthine oxidase (XO) deficiency may also give rise to muscle pain on exercise, but since XO levels are low even in normal skeletal muscle, the defect is usually demonstrated by means of jejunal biopsy.

Electrolyte Abnormalities

The hypokalemic, hyperkalemic, and normokalemic forms of periodic paralysis are dealt with in detail elsewhere (Chapter 7). Muscle biopsy may be uninformative if undertaken between attacks, although the hyperkalemic and normokalemic forms are more likely to show permanent myopathic features, including severe atrophy and type 1 fiber predominance. However, vacuolation is the major histopathological feature of the periodic paralyses; vacuoles contain little except water and electrolytes and are unlikely to be confused with lipid- or glycogen-containing vacuoles. "Tubular aggregates" are found in some patients with the hypo- or hyperkalemic forms. These consist of double-walled tubules, packed in hexagonal lattices, and show very high NADH–tetrazolium reductase activity but no SDH activity. The aggregates are often confined to type 2B fibers but are not specific to the periodic paralyses since they can also occur in cases of myotonia congenita and myasthenic syndromes.

REFERENCES

1. Brooke MH, Kaiser KK. Muscle fiber types: how many and what kind? Arch Neurol 1970; 23:369–379.
2. Pearse AGE. Histochemistry, Theoretical and Applied, 3rd ed. London: Churchill-Livingstone, 1972.
3. Lojda Z, Gossrau R, Schiebler TH. Enzyme Histochemistry. A Laboratory Manual. Berlin: Springer-Verlag, 1979.
4. McManus JFA. Histological and histochemical uses of periodic acid. Stain Technol 1948; 23:99–108.

5. Chiffelle TL, Putt FA. Propylene and ethylene glycol as solvents for Sudan IV and Sudan Black B. Stain Technol 1951; 26:51–56.
6. Edstrom L, Kugelberg E. Histochemical mapping of motor units in experimentally reinnervated skeletal muscle. Experimentia 1969; 25:1044–1052.
7. Engel WK, Cunningham GG. Rapid examination of muscle tissue. An improved trichrome method for fresh-frozen biopsy sections. Neurology 1963; 13:919.
8. Brooke MH. Congenital fiber type disproportion. In: Kabulas BA, ed. Clinical Studies in Myology. Amsterdam: Excerpta Medica, 1973:147–159.
9. Brooke MH, Engel WK. The histographic analysis of human muscle biopsies with regard to fiber types. I. Adult male and female. Neurology 1969; 19:221–233.
10. Brooke MH, Engel WK. The histographic analysis of human muscle biopsies with regard to fibre types. IV. Children's biopsies. Neurology 1969; 19:591–605.
11. Johnson MA, Polgar J, Weightman D, Appleton D. Data on the distribution of fibre types in thirty-six human muscles. An autopsy study. J Neurol Sci 1973; 18:111–129.
12. Lester JM, Silber DI, Cohen MH, Hirsch RP, Bradley WG, Brenner JF. The codispersion index for the measurement of fibre type distribution patterns. Muscle Nerve 1983; 6:581–587.
13. Hoffman EP, Brown BH, Kunkel LM. Dystrophin: the protein product of the Duchenne muscular dystrophy gene. Cell 1987; 51:919–928.
14. Zubrzycka-Gaarn EE, Bulman DE, Karpati G, Burghes AHM, Belfall B, Klamut HJ, Talbot J, Hodges RS, Ray PN, Warton RG. The Duchenne muscular dystrophy gene product is localized in sarcolemma of human skeletal muscle. Nature 1988; 333:466–469.
15. Nicholson LVB, Johnson MA, Gardner-Medwin D, Bhattacharya S, Harris JB. Heterogeneity of dystrophin expression in patients with Duchenne and Becker muscular dystrophy. Acta Neuropathol 1990; 80:239–250.
16. Emery AEH, Dreifuss FE. Unusual type of benign X-linked muscular dystrophy. J Neurol Neurosurg Psychiatry 1966; 29:338–342.
17. Arahata K, Ishihara T, Kamakura K, Tsukahara T, Ishiura S, Baba C, Matsumoto T, Nonaka I, Sugita H. Mosaic expression of dystrophin in symptomatic carriers of Duchenne's muscular dystrophy. N Engl J Med 1989; 320:138–142.
18. Nicholson LVB, Davison K, Falkous G, Harwood C, O'Donnell E, Slater CR, Harris JB. Dystrophin in skeletal muscle. I. Western blot analysis using a monoclonal antibody. J Neurol Sci 1989; 94:125–136.
19. Nicholson LVB, Johnson MA, Davison K, O'Donnell E, Falkous G, Barron M. Harris JB. Dystrophin or a "related protein" in Duchenne muscular dystrophy. Acta Neurol Scand 1992; 86:8–14.
20. Appleyard ST, Dunn MJ, Dubowitz V, Scott ML, Pittman SJ, Shotton DM. Monoclonal antibodies detect a spectrum-like protein in normal and dystrophic human muscle. Proc Natl Acad Sci USA 1984; 81:776–780.
21. Helliwell TR, Nguyen thi Man, Morris GE, Davies KE. The dystrophin-related protein, utrophin, is expressed on the sarcolemma of regenerating human skeletal muscle fibres in dystrophies and inflammatory myopathies. Neuromusc Disord 1992; 2:177–184.
22. Jockusch BM, Veldman H, Griffiths GW, Van Oost BA, Jennekens FGI. Immunofluorescence microscopy of a myopathy. α-Actinin is a major constituent of nemaline rods. Exp Cell Res 1980; 127:409–420.
23. Arahata K, Engel AG. Monoclonal antibody analysis of mononuclear cells in myopathies. I. Quantitation of subsets according to diagnosis and sites of accumulation and demonstration and counts of muscle fibers invaded by T cells. Ann Neurol 1984; 16:193–208.
24. Engel AG, Biesecker G. Complement activation in muscle fiber necrosis—demonstration of the membrane attack complex of complement in necrotic fibres. Ann Neurol 1982; 12:289–296.
25. Holthofer H, Virtanen I, Karniemi AL, et al. Ulex europaeus I lectin as a marker for vascular endothelium in human tissues. Lab Invest 1982; 47:60–66.
26. Banker BQ, Victor M. Dermatomyositis (systemic angiopathy) of childhood. Medicine 1966; 45:261–289.
27. Engel AG, Arahata K. Monoclonal antibody analysis of mononuclear cells in myopathies. II. Phenotypes of autoinvasive cells in polymyositis and inclusion body myositis. Ann Neurol 1984; 16:209–215.
28. Godlewsky HG. Are active and inactive phosphorylases histochemically distinguishable? J Histochem Cytochem 1963; 11:108–112.
29. Bonilla E, Schotland DL. Histochemical diagnosis of muscle phosphofructokinase deficiency. Arch Neurol 1970; 22:8–12.
30. Seligman AM, Karnovsky MJ, Wasserkrug HL, Hanker JS. Non-droplet ultrastructural demonstration of cytochrome oxidase activity with a polymerising osmiophilic reagent, diaminobenzidine (DAB). J Cell Biol 1968; 38:1–14.
31. Fishbein WN, Armbrustmacher VW, Griffin JL. Myoadenylate deaminase deficiency: a new disease of muscle. Science 1978; 200:545–548.
32. Farnsworth L, Shepherd IM, Johnson MA, Bindoff LA, Turnbull DM. Absence of immunoreactive enzyme protein in short chain acyl-CoA dehydrogenase deficiency. Ann Neurol 1990; 28:717–720.

33. Olson W, Engel WK, Walsh GO, Einaugler R. Oculocraniosomatic neuromuscular disease with "ragged-red" fibers. Arch Neurol 1972; 26:193–211.
34. DiMauro S, Mendell JR, Sahenk Z, Bachman D, Scarpa A, Schofield R, Reiner C. Fatal infantile mitochondrial myopathy and renal dysfunction due to cytochrome-c-oxidase deficiency. Neurology 1980; 30:795–804.
35. Holt IJ, Harding AE, Morgan-Hughes JA. Deletions of muscle mitochondrial DNA in patients with mitochondrial myopathies. Nature 1988; 331:717–719.
36. Johnson MA. Cytochemical and immunocytochemical investigation of respiratory complexes in individual muscle fibres of human skeletal muscle. In: Gorrod JW, Albaro O, Ferrari E, Papa S, eds. Molecular Basis of Neurological Disorders and their treatment. London: Chapman & Hall, 1991:40–56.

7
Electron Microscopy

Caroline A. Sewry
Royal Postgraduate Medical School and Hammersmith Hospital, London, England

INTRODUCTION

Electron microscopy plays an important role in accurately defining the ultrastructural changes associated with diseased muscle. Techniques such as immunocytochemistry, freeze fracture, and X-ray microanalysis are now more widely applied to diseased muscle and are adding an extra dimension to our understanding.

Skeletal muscle undergoes a variety of changes in response to disease or trauma, and the interpretation of these must take into account factors such as the small sample size, artifacts induced by preparation, and the nonspecificity of many of the features. It is therefore more appropriate to describe the wide variety of changes that occur in human neuromuscular disorders in relation to individual cellular components, rather than specific disorders. It is now recognized that a definitive diagnosis cannot be made by electron microscopy alone, and that clinical, histochemical, and electrophysiological data must also be considered. Nevertheless, consistent patterns are often seen and can be used as an aid to diagnosis. Citations from the literature have been kept to a minimum in this review and further details may be found in works by Sewry (1), Cullen and Landon (2), Cullen et al. (3,5), and Landon (4).

NORMAL MUSCLE

Muscle Cell Surface

Muscle fibers are multinucleated syncytia. In normal muscle nuclei are heterochromatic, do not divide, and are located at the periphery of the fibers. Each muscle fiber is enveloped in a surface membrane, the sarcolemma, composed of an inner plasma membrane and an outer basement membrane (Fig. 1). These are closely applied and move in parallel during contraction. The outer basement membrane consists of the electron-dense basal lamina (lamina densa) and a reticular outer layer of collagen fibrils, in particular type III collagen, proteoglycans, and fibronectin. The basal lamina is secreted by the fiber itself and its main constituents are laminin, type IV collagen, and glycoproteins. The plasma membrane is a proteolipid bilayer that contains a variety of structural, receptor, and metabolically active proteins. It is separated from the basement membrane by an electron-lucent region of about 10–15 nm, the lamina lucida (or lamina rara). Fine strands traverse the lamina lucida, connecting the plasma membrane with the basement membrane, and these probably enable the plasmalemma and basal lamina to move together during contraction.

The surface area of the plasmalemma is increased by an extensive system of caveolae that are in continuity with it via pores. Many caveolae also connect with the transverse tubular system (T system),

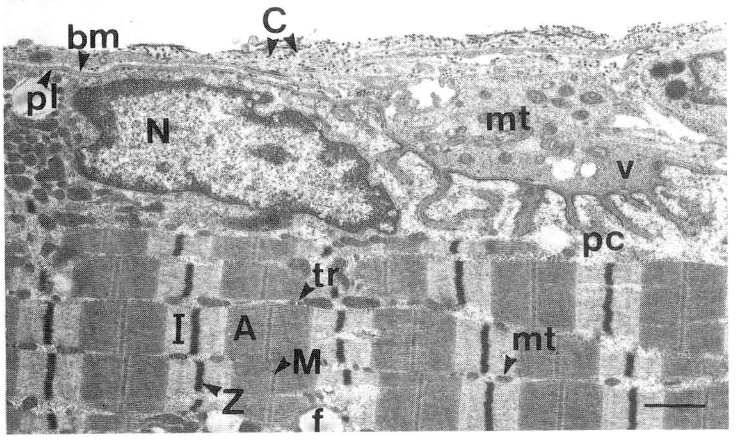

Figure 1 Longitudinal section of a muscle fiber through a neuromuscular junction showing the major components of a fiber. The nucleus (N) is peripheral, beneath the sarcolemma composed of the plasma membrane (pl) and basement membrane (bm). Collagen (C) is present in the reticular layer. Basement membrane extends into the postsynaptic clefts (pc), and the presynaptic axon contains mitochondria (mt) and numerous vesicles (v). The myofibrils show a striation pattern of dark A-bands (A) and light I-bands (I). The I-band is transected by the Z-line (Z) and the A-band by the M-line (M). A sarcomere is defined as the region between two Z-lines. Mitochondria lie at the level of the I-band and fat is associated with some (f). Triads can also be seen near the A/I junction (tr). Bar = 1 μm.

which is formed from extensions of the plasmalemma into the muscle fibers. The plasmalemma is associated with the myofibrils via components of the cytoskeleton, several of which have a costameric distribution and are located adjacent to the Z-line.

The neuromuscular junction is a specialized region of the sarcolemma where the nerve axons of the motor unit terminate (Fig. 1). Synaptic vesicles and mitochondria are the most conspicuous components at the axon terminal, but other structures such as coated vesicles, glycogen, neurofilaments, and microtubules can also be seen. The postsynaptic membrane is folded and has deep clefts (Fig. 1). It is separated from the presynaptic membrane by a synaptic space of about 60–80 nm, which contains basal lamina continuous with that of the muscle fiber and the Schwann cell at the axon terminal. An important component of the postsynaptic membrane is the acetylcholine receptor, which binds the acetylcholine released from the synaptic vesicles. Acetylcholine receptors are clustered at the crests of the folds, which often appear thicker and more electron-dense.

Immunocytochemical studies have shown that a variety of proteins are concentrated at the neuromuscular junction, including extracellular matrix proteins such as laminin, collagen IV, and fibronectin, and also cytoskeletal components such as desmin, dystrophin, utrophin (dystrophin-related protein), and β-spectrin (6–9), although the ultrastructural localization of some of these proteins has not yet been studied.

The myotendinous junction is also a specialized region of the sarcolemma. Here it becomes ridged and folded, and the plasmalemma has a thick electron-dense layer that is continuous with the Z-line of the myofibrils.

At the site of satellite cells the plasmalemma is indented (Fig. 2). Satellite cells are mononucleated cells that lie between the basal lamina and the plasmalemma and are thought to be a population of myoblastic stem cells. They are normally dormant but can be activated to divide and to differentiate and are the source of cells responsible for repair and regeneration of fibers. Inactive satellite cells have a small volume of cytoplasm and contain a few mitochondria, free ribosomes, rough endoplasmic reticulum, glycogen, microtubules, and intermediate filaments. Organized myofilaments are not a

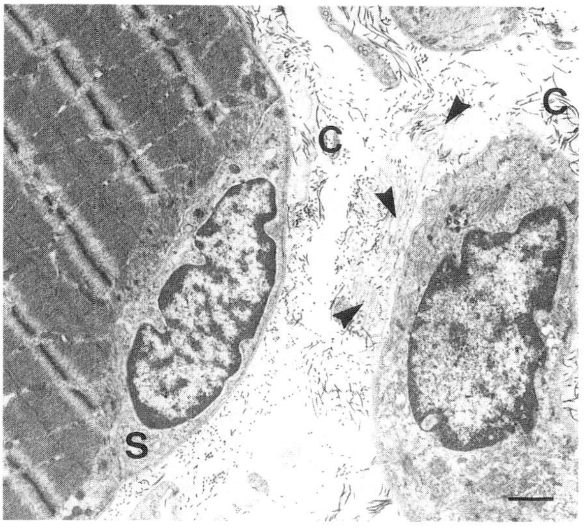

Figure 2 A satellite cell (S) beneath the basement membrane of a fiber, but separated from it by its own plasma membrane. The adjacent area shows extracellular collagen (C) and part of an atrophic fiber with redundant basement membrane (arrows). Bar = 1 μm.

feature of satellite cells. The number of satellite cells declines with age, and in adult human muscle they comprise about 3% of myonuclei. Their number is increased in diseased muscle, including situations associated with denervation and regeneration.

Myofibrils

About 85–90% of the volume of a muscle fiber is occupied by the myofilaments. The myofilaments are organized into bundles, the myofibrils, which are laterally aligned and have a regular arrangement of repeating components known as sarcomeres (Fig. 1). The regular alternation of myofilament proteins gives skeletal muscle its characteristic striated pattern. Each sarcomere consists of a dark anisotropic A-band, with a light isotropic I-band on either side, bisected by the dense Z-line. The center of the A-band is transected by a narrow band known as the M-line, which is flanked on either side by the pale H-zone. The number of recognizable striations on the M-line is fiber-type-dependent and varies from 3 to 5. A sarcomere is defined as the area between two Z-lines (Fig. 1). The length of the sarcomere varies with the degree of contraction, but the normal resting length is 2.5–3.0 μm. During contraction the length of the A-band remains constant at 1.5–1.6 μm but that of the I-band and H-zone is reduced. The myofibrils shorten by the I-filaments sliding towards the center of the A-band.

The major protein of the A-band is myosin and that of the I-band actin. In transverse section of the A-band, six actin molecules can be seen surrounding each myosin filament. Interacting with the actin filaments are tropomyosin and the troponins. The Z-line is organized into a lattice structure and is chiefly composed of α-actinin. Two other proteins, titin and nebulin, are also major constituents of the myofibrils. Both are very large proteins and are believed to provide elasticity to the myofibrils. Titin spans a considerable part of the sarcomere and is thought to make elastic connections with the Z-line at one end, and to be bound to the myosin filaments near the M-line at the other, where it has a role in regulating filament length and assembly.

Cytoskeleton

The cytoskeleton is composed of F-actin, intermediate filaments, and microtubules. Intermediate filaments, in particular desmin, connect the myofibrils to each other and to the plasmamembrane at the level of the Z-line. Desmin also connects the Z-line to the nucleus. Vimentin is also tightly associated with the nucleus and the plasmalemma, but it is not highly expressed in mature muscle fibers and is most prominent in developing and regenerating fibers. Desmin is also more highly expressed in immature fibers. Other components of the cytoskeleton, such as vinculin, ankyrin, and spectrin, are believed to attach the intermediate filaments to the plasma membrane.

Sarcoplasm

The space between the myofibrils and beneath the sarcoplasm is occupied by the cytoplasm of the fiber, the sarcoplasm. This contains organelles, such as mitochondria, the sarcoplasmic and T-tubule membrane systems, Golgi apparatus, and microtubules, as well as glycogen, ribosomes, lipid droplets, and lipofuscin. Glycogen granules occur throughout the sarcoplasm but are more numerous in the I-band than the A-band. Microtubules, Golgi apparatus, and free ribosomes are more commonly seen in perinuclear regions.

Mitochondria

Mitochondria are membranous structures concerned with energy production and calcium regulation. Each mitochondrion has an outer membrane and an inner membrane, folded into cristae, which contains several proteins and enzymes involved in electron transport and oxidative phopsphoryation. The space enclosed by the inner membrane contains a slightly electron-dense matrix that contains dense granules and calcium deposits. The size and shape of mitochondria are variable, but in normal muscle they are usually small and ovoid. Mitochondria are located in the sarcoplasm between the myofibrils and occur as pairs either side of the Z-line, adjacent to the I-band (Fig. 1). They also form subsarcolemmal aggregates of variable size. Mitochondria are more numerous in type 1, slow-contracting fibers, but in human muscle differences in mitochondrial volume between type 1 and type 2 fibers are less distinct than in other species.

Sarcoplasmic Reticulum and T-Tubule System

The sarcoplasmic reticulum (SR) and the T system form an interrelated network of membranes around the myofibrils and are concerned with the excitation, contraction, and relaxation of the fiber. The T system runs transversely across the fiber. It is developed from, and continuous with, the plasma membrane of the sarcolemma and is easily penetrated by fixatives and electron-dense substances such as lanthanum, horseradish peroxidase, and ferritin. The SR forms a complex network of membranes around each myofibril but there is no continuity with the T system, or with the extracellular space. At the junction of the A- and I-bands the SR interacts with a T tubule to form a triad (Fig. 1). Triads are composed of two lateral sacs, or terminal cisternae, of SR that lie on either side of a T tubule. The lateral sacs, in contrast to other parts of the SR and the T system, contain amorphous electron dense material.

Other Structures

Ultrathin sections also contain other structures in addition to the muscle fibers. *Capillaries* are frequently encountered and are covered by a prominent basement membrane. Pericytes are closely applied to the endothelial cells and have numerous pinocytotic vesicles. *Intramuscular nerves* are occasionally seen with both myelinated and unmyelinated axons. They are surrounded by perineural cells and basement membrane. *Fibroblastic-like cells* are also seen and are responsible for the secretion of the variable amount of collagen and fibronectin present between the muscle fibers.

ABNORMALITIES IN SKELETAL MUSCLE

Sarcolemma

Folding and irregularities in the surface membrane of the muscle fibers are frequently seen in a variety of neuromuscular disorders (Fig. 3). Some undulation, however, may be caused by fixation of the fiber in a

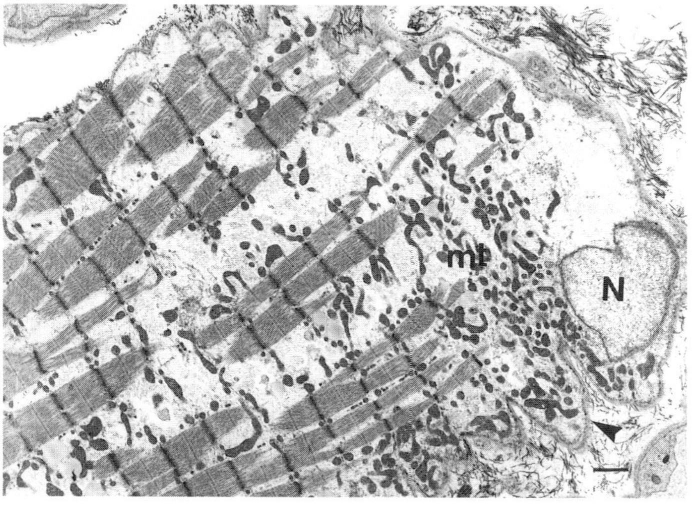

Figure 3 Abnormal loss and splitting of myofibrils. The space vacated is filled with glycogen, sarcoplasm, and mitochondria (mt) of varying size and shape. The sarcolemma is irregular and folded (arrow) and the nucleus is euchromatic. Bar = 1 μm.

contracted state. The basement membrane may separate from the underlying plasmalemma and be thrown into redundant folds, and occasionally glycogen may accumulate between the two membranes. In glycogen storage diseases this glycogen deposition may be excessive. Redundant basement membrane (Fig. 2) is considered a feature of atrophic fibers, in contrast to hypotrophic fibers, in which development and maturation are believed to be arrested, and where the basement membrane remains closely adherent. Duplication of the basement membrane is common in muscular dystrophies and occurs round regenerating fibers that secrete their own membrane while developing beneath the basement membrane of a necrotic fiber. Thickening of the basement membrane is also frequently observed, and both duplication and thickening occur around capillaries. The basement membrane is highly resistant to degradation and persists around necrotic fibers, even though the plasma membrane is lost. Focal lesions in the plasma membrane occur in a variety of disorders and allow the entry of extracellular markers to form "delta" lesions.

Nuclei

Changes in myonuclei include relocation internally in the fiber, changes in the chromatin distribution, irregularities in shape, and the presence of inclusions. The sarcoplasm around internal nuclei may show accumulations of mitochondria and glycogen, for example in centronuclear myopathy. Chromatin distribution is affected by the activity of the fiber and the degree of fiber damage. It can vary from very dispersed to highly condensed, as in pyknotic nuclei. Nucleoli in diseased muscle may be enlarged and in regenerating fibers they are particularly prominent.

Nuclear inclusions are rare but several reported cases show that they are commonly associated with inflammatory disorders. A variety of different types have been observed. Inclusions resembling myofilament components, such as actin-like filaments, or Z-line material and rods, or filaments thicker than myosin have been reported. In oculopharyngeal dystrophy 8.5-nm tubular filaments in at least 3% of nuclei have been considered a characteristic feature. In this condition the inclusions are largely confined to nuclei while in inclusion body myositis larger, 16–18-nm filaments occur in nuclei and the sarcoplasm. Some nuclear inclusions are thought to resemble viruses, but there is no direct evidence of a viral origin. In dermatomyositis, nuclear inclusions are frequently observed in sarcolemmal, interstitial cell, and endothelial cell nuclei.

Myofibrils

Selective and Nonselective Loss of Sarcomere Components. Loss and alterations in the myofibrils are the most common ultrastructural defect observed in all classes of genetic and acquired neuromuscular disorders. The degree of loss, splitting, and disruption of myofibrils is variable and its significance may be difficult to assess. Even in normal muscle, departures from the classical appearance frequently occur. Loss or disruption of myofibrils may be focal or widespread and affect whole fibers (Fig. 3). In necrotic fibers the myofilament structure is completely lost and the fiber is filled with amorphous granular material and invaded by macrophages. During earlier stages of fiber damage hypercontraction of myofibrils can occur. This may be focal or extensive, and areas of hypercontracted fibrils are interspersed with over-stretched myofibrils. Hypercontracted fibers are thought to correspond to the round, intensely stained fibers seen with the light microscope that are common in Duchenne muscular dystrophy.

Selective loss of the I-band, often accompanied by loss of the Z-line, is common, although care in interpretation is required if the plane of section does not pass through the whole myofibril. Selective loss of A-bands is more rare but has been reported as the predominant feature in some cases, including dermatomyositis, thrombotic thrombocytopenic purpura, congenital hypotonia, rhabdomyolysis, and steroid myopathy. It is, however, a nonspecific abnormality and isolated fibers with A-band loss may occur in a variety of conditions.

Cores and Targets

Disorganization and disruption of myofibrils is variable in degree and type. The normal striation pattern may show focal disruption and affect only a few sarcomeres, such as in minicores (Fig. 4), or it may extend the length of the fiber, as in central cores or target fibers. Cores of all types are characterized by an absence of mitochondria and a variable degree of Z-line disruption. In central core disease the cores are usually defined as structured or unstructured, depending on the extent of the disorganization. In structured cores the striation pattern is preserved but the core area is slightly contracted (Fig. 5), while in unstructured cores the striation pattern is lost and there is extensive Z-line streaming (Fig. 4). These pathologically distinct types of cores can also be distinguished histochemically with structured cores retaining their ATPase activity but unstructured cores loosing it. In most cases central cores are of one or the other type, but both types can occur in the same biopsy.

Target fibers occur in denervated muscle and some myopathic conditions. They resemble core fibers

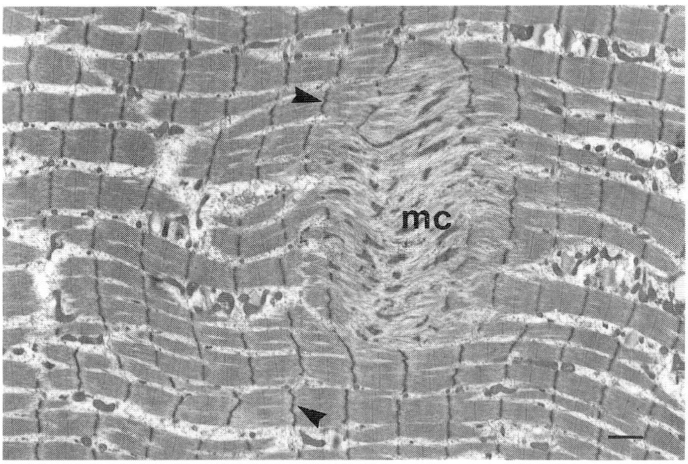

Figure 4 Z-line streaming, myofibrillar disruption, and absence of mitochondria in a minicore (mc). Z-line irregularities (arrows) are also seen in adjacent regions. Bar = 1 μm.

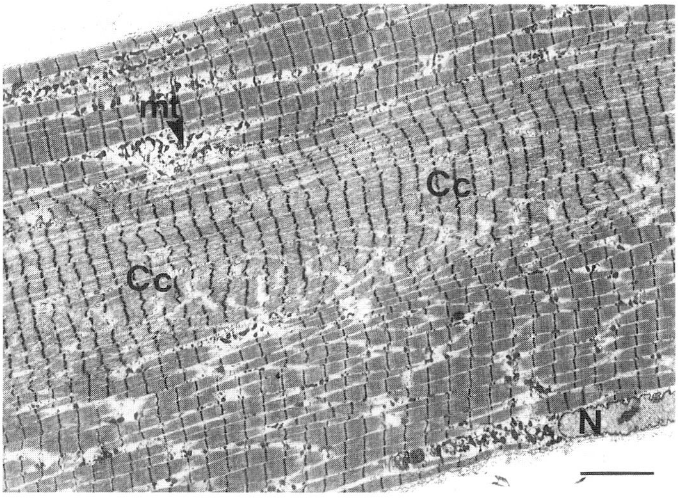

Figure 5 Longitudinal section through a fiber with a structured central core (Cc) showing contraction and misalignment of myofilaments in the central region and absence of mitochondria (mt). The subsarcolemmal nucleus (N) has a prominent nucleolus. Bar = 5 μm.

but have three concentric zones, at both the light and electron microscopic level. The outer zone has a relatively normal striation pattern, the intermediate zone shows mild Z-line irregularities and swelling of the sarcoplasmic reticulum, and the inner zone is markedly disrupted with extensive Z-line streaming, similar to that in unstructured cores. If the intermediate zone is difficult to distinguish, the fibers are known as targetoid. Zones of myofibrillar disruption have also been recognized in trilaminar fibers.

Ring Fibers

Myofibrillar disorientation may occur throughout a fiber or it may be restricted to one zone, such as in ring fibers (Fig. 6). These unusual fibers have one or more peripheral myofibrils at right angles to the

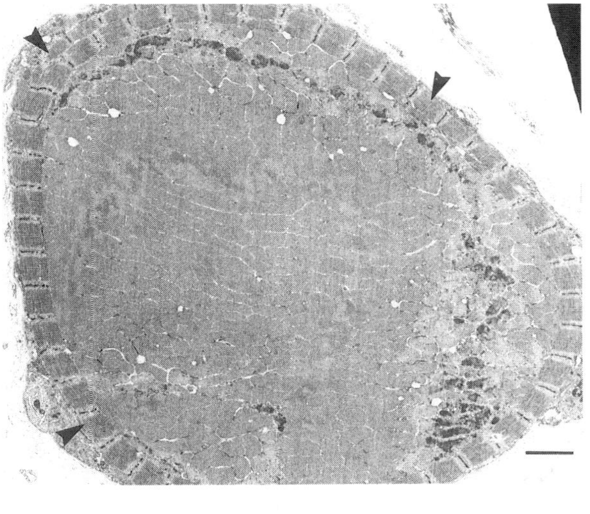

Figure 6 Ring fiber with a peripheral band of myofibrils (arrows) at 90° to the main axis of the fiber. Bar = 1 μm.

main axis of the fiber. Ring fibers occur in a variety of disorders but are particularly common in limb-girdle dystrophy and in myotonic dystrophy. In myotonic dystrophy ring fibers have a characteristic peripheral sarcoplasmic mass between the sarcolemma and the ring of myofibrils. The sarcoplasmic mass is devoid of myofibrils.

Z-line Abnormalities and Nemaline Rods. Z-line abnormalities are common and range from irregularities, to focal streaming, thickening, or duplication. The number of sarcomeres involved is variable and some Z-line streaming may occur in normal muscle. Rods, or nemaline bodies, are believed to be derived from Z-line material (Fig. 7). Biochemical and immunocytochemical evidence shows that they contain actin, tropomyosin, and α-actinin, a major constituent of the Z-line, and have peripheral desmin filaments. The filaments are arranged in a lattice-like pattern and have a periodicity similar to that of Z-lines. Rods are the predominant feature of nemaline myopathy, where they occur in a large population of fibers and are associated with a specific clinical phenotype. They may occur, however, in occasional fibers in a variety of disorders and are present in normal extraocular muscle and myotendinous regions. Nuclear rods have also been reported in a few unusual cases (10).

Cytoplasmic Bodies

Cytoplasmic bodies are also considered to be derived from the myofibrils. These have a dense circular or ovoid center with a peripheral zone of filaments radiating from it, and sometimes a pale halo between the two. The dark center resembles Z-line material and the radiating filaments are similar to actin, although desmin has been reported in these filaments. Cytoplasmic bodies may occur singly or in groups and continuity with myofibrils is sometimes seen. They are a nonspecific feature but are numerous in a progressive disorder known as "cytoplasmic body myopathy."

Other structures believed to be of myofibrillar origin are *filamentous bodies* (Fig. 8) and *concentric laminated bodies*, both of which are rare structures. Filamentous bodies are composed of actin-like filaments but other similar structures consisting of intermediate filaments have been reported. Concentric laminated bodies have whorled filamentous laminae and a center that often contains glycogen. They may show continuity with myofibrils, but their origin is uncertain and the size of the laminae is not the same as that of either thin or thick filaments.

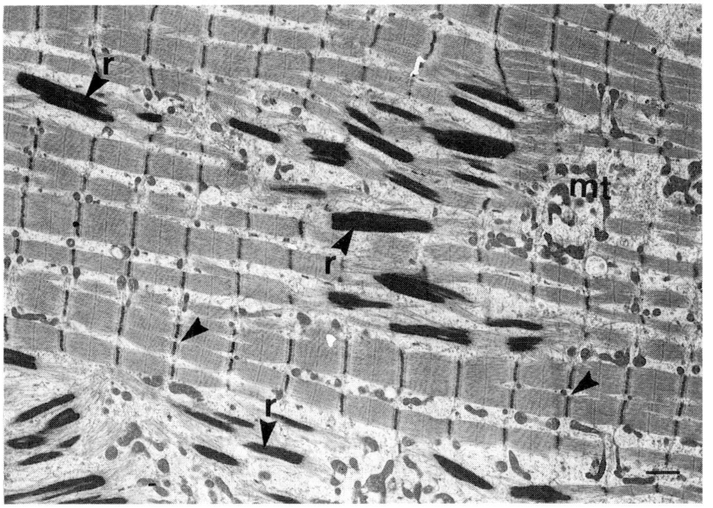

Figure 7 Rods (r) in a case of nemaline myopathy. Several rods show filaments attached to them and some show continuity with the myofilaments. The myofilaments are split (arrow) and some mitochondria (mt) are clustered in areas of myofibrillar loss. Bar = 1 μm.

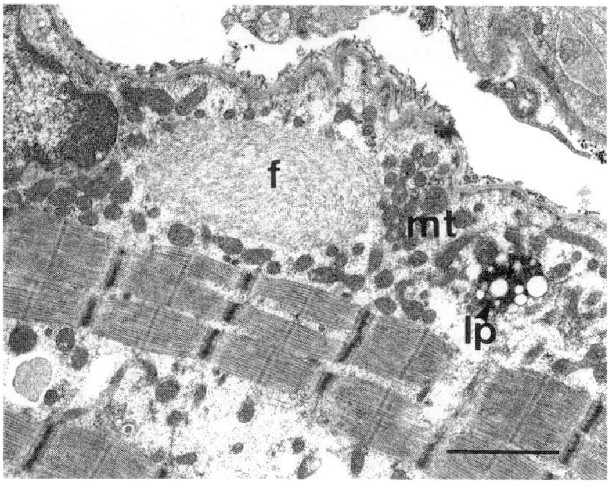

Figure 8 Subsarcolemmal filamentous body (f). Mitochondria (mt) and lipofuscin (lp) have also accumulated. Bar = 1 μm.

Mitochondria

The number, size, shape, and distribution of mitochondria vary considerably in diseased muscle (Fig. 3). Swelling of mitochondria occurs in many injured cells but may be artifactual in incorrectly fixed tissue. Areas of focal myofibrillar disruption are frequently accompanied by clustering of mitochondria and their shape is variable (Fig. 7). Focal loss of mitochondria may also occur.

Subsarcolemmal aggregates of mitochondria occur in normal muscle, but in lobulated fibers they are particularly prominent and form triangular areas corresponding to the abnormal accumulation of oxidative enzyme staining seen at the light microscope. Lobulated fibers are a nonspecific feature but they are rare in pediatric cases.

A variety of structural abnormalities have been observed in mitochondria. These may be the predominant feature in some disorders characterized by mitochondrial defects (Fig. 9). Structural

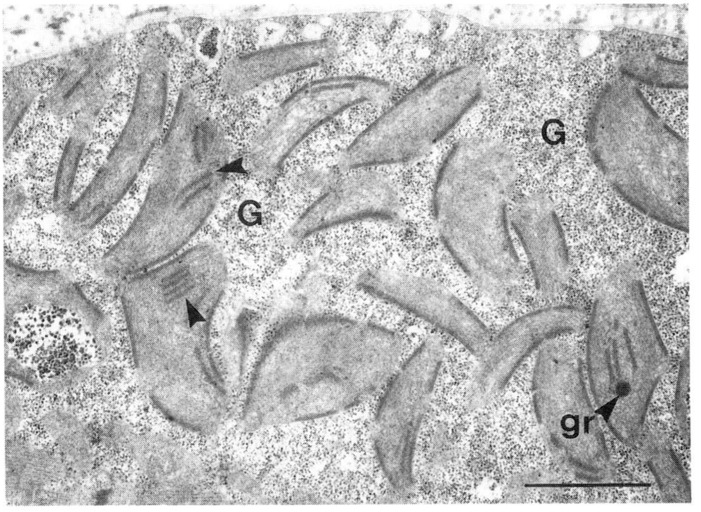

Figure 9 Cluster of abnormal subsarcolemmal mitochondria with rectangular inclusions (arrow) and an osmiophilic granule (gr). The mitochondria are surrounded by glycogen (G). Bar = 1 μm.

abnormalities are not, however, a universal feature of the mitochondrial myopathies and no specific biochemical defect is associated with a particular morphological abnormality. More than one structural abnormality may be seen in a single biopsy and these include crystalline inclusions, osmiophilic bodies, branching or concentric cristae, or giant mitochondria.

Membrane Systems

Dilation and proliferation of the sarcoplasmic reticulum (SR) and T system are common and in Duchenne dystrophy swelling of the SR is an early morphological abnormality. If fixation is inadequate, swelling of the membrane systems can be artifactual and it is then usually accompanied by disruption of the mitochondria. Marked dilation of the SR is believed to give rise to the vacuoles seen in periodic paralysis, and it is possible that the limiting membrane of autophagic vacuoles is also derived from the SR. Autophagic vacuoles contain sarcoplasm, mitochondria, and organelles undergoing degradation.

Displacement of triads from their position at the A/I-band junction may occur, and proliferation of the components of triads can give rise to tiers of alternating dense SR and translucent T-system tubules.

Proliferation of the T system is thought to give rise to honeycomb structures, which are common in areas of active degeneration and regeneration (Fig. 10). Honeycomb structures have similarities to triads with large, pale tubules, alternating with smaller, dense ones. Tubular aggregates are also believed to be derived from proliferation of the membrane system, and in some types connections with the SR can be seen (Fig. 11). Morphologically distinct types of tubules have been identified in tubular aggregates. One has a double-walled tubule, another a granular central core, and in a third type, filaments are regularly arranged around each tubule. The tubules often occur in parallel, in a hexagonal formation, but some haphazardly arranged tubules have been observed. These also differ from the above types in having an outer membrane enclosing more than one smaller tubule. Tubular aggregates occur in several neuromuscular disorders but are a common feature in periodic paralysis, where they are often confined to type 2 fibers and can be detected histochemically.

Storage Products and Particles

Glycogen and lipid both occur in normal muscle but excessive accumulations can occur in certain disease situations. Glycogen is frequently seen in areas of myofibrillar loss but it is particularly marked

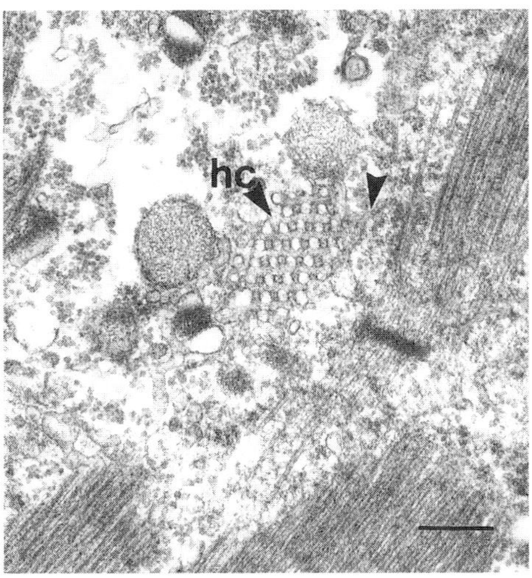

Figure 10 Honeycomb structure (hc) with alternating dense and light tubules. Connections to the reticulum system can be seen (arrow). Bar = 1 µm.

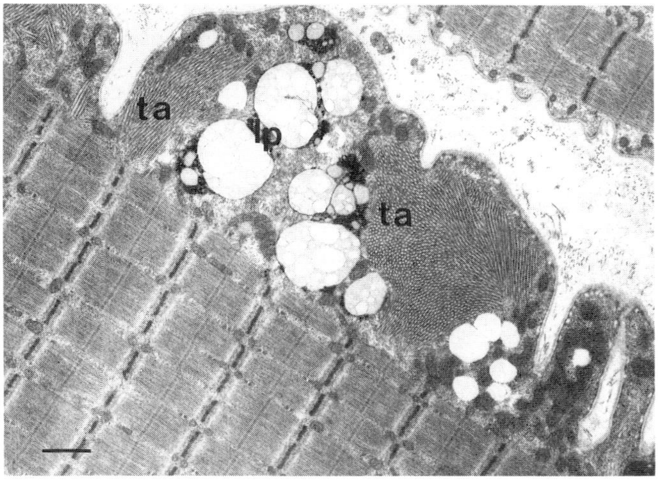

Figure 11 Subsarcolemmal tubular aggregate (ta) sectioned transversely and longitudinally. Lipofuscin (lp) is also seen but the myofibrillar pattern is well preserved. Bar = 1 μm.

in some glycogenoses. In type II (Pompe's) glycogenosis, the glycogen accumulates in membrane-bound sacs as well as in the cytoplasm, and there is often very little myofibrillar material.

Lipid is not retained by dehydration and embedding procedures, and clear, or lightly stained, areas represent its location. Lipid droplets occur in association with mitochondria and they are thus more numerous in type 1 fibers. Excessive accumulations occur in carnitine deficiency and sometimes in association with abnormal mitochondria.

Lipofuscin, the dense osmiophilic residue from lysosomal action, occurs in normal muscle but is more prevalent in diseased muscle. It is frequently found beneath the sarcolemmal and sometimes in the perinuclear region (Fig. 11).

Other Structural Abnormalities

A variety of unusual structures have been observed in diseased muscle. They are usually rare but occasionally they are the predominant feature of a clinical condition. For example, zebra bodies, fingerprint bodies, and reducing bodies have all given their name to specific disorders. Curvilinear bodies are accumulations of short, curved, membranous structures that occur in the brain, skeletal muscle, and other tissues in neuronal ceroid lipofuscinosis and in muscle of patients treated with chloroquine. Membranous whorls or myelin-like figures have dense material arranged in a whorl-like pattern and are often associated with autographic vacuoles. A variety of dense tubular structures of uncertain origin have also been reported.

ULTRASTRUCTURAL IMMUNOCYTOCHEMISTRY

The localization of antibodies at the ultrastructural level using preembedding, postembedding, or frozen sections is a rapidly expanding field. Although antibodies to a wide variety of proteins have been studied at the light level in human muscle, ultrastructural studies are still limited. The proteins localized include dystrophin, the defective gene product in Duchenne and Becker muscular dystrophy, β-spectrin, myosin, desmin, titin, α-actinin, extracellular matrix components, complement components, and the localization of lectins.

Dystrophin, using antibodies to the rod domain, and β-spectrin have both been localized to the cytoplasmic face of the sarcolemma (11,12). An antibody to the C-terminus of dystrophin, however, suggests that the C-terminus of dystrophin is inserted into the plasma membrane (13). This view is now supported by the association of dystrophin with a glycoprotein complex and the proposed link with laminin.

Fiber typing in human muscle has been demonstrated with monoclonal antibodies to slow myosin (14) but this has not been extended to pathological muscle.

Desmin has been shown to accumulate in some myopathic and cardiomyopathic conditions (15–17) and to be associated with a variety of structures in diseased muscle such as rods, cytoplasmic bodies, sarcoplasmic bodies, Mallory-like bodies, and cytoplasmic bodies. In Duchenne dystrophy the distribution of desmin is abnormal, in association with the myofibrillar disruption (18). Focal loss of labeling at the plasma membrane also occurs.

Immunogold labeling of titin in human dystrophic muscle has shown that extensive degradation occurs (18). Titin is degraded before desmin during necrosis and epitopes closer to the A/I-band junction appear to be lost before those in the A-band region. It is also suggested that lysis of titin may result in misalignment of the myofibrils. Loss of titin seems to be more common in Duchenne dystrophy than in other neuromuscular disorders.

The extracellular matrix components, collagen, laminin, and fibronectin are associated with the basement membrane and their localization at the ultrastructural level reflects that observed at the light level (19).

In myasthenia gravis circulating antibodies have been shown to localize to the neuromuscular junction (20). The terminal component of the complement pathway, C9, has also been identified at the neuromuscular junction in this disorder (21).

Lectins bind to specific carbohydrate groups and have been used in a similar manner to antibodies to detect abnormalities in dystrophic muscle. Concanavalin A was used several years ago to show that the plasma membrane is damaged in Duchenne dystrophy (22). The lectin *Ricinus communis* I (RCA I) recognizes α- and β-D-galactose residues and is also associated with the plasma membrane (23,24). In Duchenne dystrophy the plasma membrane fails to bind this lectin (23), but it is not yet known if this relates to the reduction of dystrophin-associated glycoproteins in dystrophic muscle. Biochemically, RCA I has recently been shown to recognize the α_1 (A), $\beta 1$ (B), and α_2 (M) subunits of laminin (25). During development of normal and dystrophic human muscle, RCA I binding to the plasma membrane increases from 12 to 20 weeks' gestation (26). The binding of RCA I to the plasma membrane in dystrophic fetuses suggests that its absence in muscle from patients with Duchenne dystrophy is a secondary phenomenon.

CONCLUSIONS

Electron microscopy has helped define the wide variety of structural changes associated with neuromuscular disorders. The nonspecificity of the abnormalities restricts the diagnostic value of the technique, but in selected cases electron microscopy has an important role in deciding whether a biopsy is abnormal, and in identifying specific structures such as rods and abnormal mitochondria. The advent of immunocytochemistry has already contributed to a greater understanding of the nature of ultrastructural changes, and its wider use will enhance this further.

REFERENCES

1. Sewry CA. Ultrastructural changes in diseased muscle. In: Dubowitz V. Muscle Biopsy: A Practical Approach. London: Baillière Tindall, 1985:129–183.
2. Cullen MJ, Landon DN. The ultrastructure of the motor unit. In: Walton JN, ed. Disorders of Voluntary Muscle, 5th ed. Edinburgh: Churchill Livingstone, 1988:27–73.
3. Cullen MJ, Hudgson P, Mastaglia FL. Ultrastructural studies of diseased muscle. In: Walton JN, ed. Disorders of Voluntary Muscle, 5th ed. Edinburgh: Churchill Livingstone, 1988:284–344.
4. Landon DN. Skeletal muscle—normal morphology, development and innervation. In: Mastaglia FL, Walton JN, eds. Skeletal muscle pathology, 2nd ed. Edinburgh: Churchill Livingstone, 1992:1–94.

5. Cullen MJ, Johnson MA, Mastaglia FL. Pathological reactions of skeletal muscle. In: Mastaglia FL, Walton JN, eds. Skeletal muscle pathology, 2nd ed. Edinburgh: Churchill Livingstone, 1992:123–184.
6. Sanes JR, Hall ZW. Antibodies that bind specifically to synaptic sites on muscle fiber basal lamina. J Cell Biol 1979; 97:357–370.
7. Askanas V, Bornemann A, Engel WK. Immunocytochemical localization of desmin at human neuromuscular junctions. Neurology 1990; 40:949–953.
8. Bewick GS, Nicholson LVB, Young C, O'Donnell E, Slater CR. Different distribution of dystrophin and related proteins at nerve-muscle junctions. NeuroReport 1992; 3:857–860.
9. Huard J, Fortier L-P, Dansereau G, Labrecque C, Tremblay JP. A light and electron microscopic study of dystrophin localisation at the mouse neuromuscular junction. Synapse 1992; 10:83–93.
10. Engel WK, Oberc MA. Abundant nuclear rods in adult onset rod disease. J Neuropathol Exp Neurol 1975; 34:119–132.
11. Cullen MJ, Walsh J, Nicholson LVB, Harris JB. Ultrastructural localisation of dystrophin in human muscle by gold immunolabelling. Proc R Soc Lond (Biol) 1990; 240:197–210.
12. Appleyard ST, Dunn MJ, Dubowitz V, Scott ML, Pittman SJ, Shotton DM. Monoclonal antibodies detect a spectrin-like protein in normal and dystrophic human muscle. Proc Natl Acad Sci USA 1984; 81:776–780.
13. Cullen MJ, Walsh J, Nicholson LVB, Harris JB, Zubrzycka-Gaarn EE, Ray PN, Worton RG. Immunogold labelling of dystrophin in human muscle, using an antibody to the last 17 amino acids of the C-terminus. Neuromusc Disord 1991; 1:113–120.
14. Semper AE, Fitzsimons RB, Shotton DM. Ultrastructural identification of type 1 fibers in human skeletal muscle; immunogold labelling of thin cryosections with a monoclonal antibody against slow myosin. J Neurol Sci 1988; 83:93–108.
15. Bertini E, Bosman C, Ricci E, Servidei S, Boldrini R, Sabatelli M, Salviati G. Neuromyopathy and restrictive cardiomyopathy with accumulation of intermediate filaments: a clinical, morphological and biochemical study. Acta Neuropathol 1991; 81:632–640.
16. Pellisier JF, Pouget J, Charpin C, Figarelli D. Myopathy associated with desmin type intermediate filaments. An immunoelectron microscopic study. J Neurol Sci 1989; 89:49–61.
17. Sabatelli M, Bertini E, Ricci E, Salviati G, Magi S, Papacci M, Tonali P. Peripheral neuropathy with giant axons and cardiomyopathy associated with desmin type intermediate filaments in skeletal muscle. J Neurol Sci 1992; 109:1–10.
18. Cullen MJ, Fulthorpe JJ, Harris JB. The distribution of desmin and titin in normal and dystrophic human muscle. Acta Neuropathol 1992; 83:158–169.
19. Dunn MJ, Sewry CA, Statham H, Dubowitz V. Studies on the extracellular matrix in diseased human muscle. In: Kemp RB, Hinchcliffe JR, eds. Matrices and Cell Differentiation. New York: Alan R Liss, 1984:213–231.
20. Engel AG, Lambert EH, Howard FM. Immune complexes (IgG and C3) at the motor end-plate in myasthenia gravis. Ultrastructural and light microscopic localization and electrophysiological correlations. Mayo Clin Proc 1977; 52:267–296.
21. Sahashi K, Engel AG, Lambert EH, Howard FM. Ultrastructural localisation of the terminal component (C9) at the motor endplate in myasthenia gravis. J Neuropathol Exp Neurol 1980; 39:160–172.
22. Bonilla E, Schotland DL, Wakayama Y. Duchenne dystrophy: focal alterations in the distribution of concanavalin A binding sites at the muscle cell surface. Ann Neurol 1978; 4:117–123.
23. Capaldi MJ, Dunn MJ, Sewry CA, Dubowitz V. Altered binding of *Ricinus communis* I lectin by muscle membranes in Duchenne muscular dystrophy. J Neurol Sci 1984; 63:129–142.
24. Capaldi MJ, Dunn MJ, Sewry CA, Dubowitz V. Binding of *Ricinus communis* I lectin to the muscle cell plasma membrane in diseased muscle. J Neurol Sci 1984; 64:315–324.
25. Wadsworth JDF, Okuno A, Strong PN. Assignment of laminin heavy chains using the lectin *Ricinus communis* agglutinin I. Biochem J 1993; 295:537–541.
26. Voit T, Sewry CA, Dunn MJ, Dubowitz V. Binding of *Ricinus communis* I lectin to developing dystrophic muscle in human fetus. J Neurol Sci 1988; 84:301–314.

8
Myoimaging

Carl D. Reimers
Georg-August-University of Göttingen, Göttingen, Germany

Thomas J. Vogl
Humboldt-University of Berlin, Berlin, Germany

INTRODUCTION

Limitations of standard diagnostic procedures such as electromyography and muscle biopsy, particularly with regard to serial observations, have provided an important impetus to the development of new noninvasive techniques for studying skeletal muscle pathology, notably ultrasound imaging (US, computerized X-ray (CT) scanning, and magnetic resonance imaging (MRI).

Ultrasound

Ultrasound has been used to depict skeletal muscles since 1968 (1). This technique is based on the interaction of high-frequency sound waves (frequency of more than 20 kHz) with the imaged matter; 3.5- to 10-MHz real-time transducers are routinely used for myosonography. These probes, in contrast to compound scanners, which were initially employed, facilitate the depiction not only of muscle structure, as do CT and MRI, but also of movements. Linear array probes are preferred to curved array or sector scanners because of their better tissue contact. The image quality is highly dependent on the frequency, intensity, and focus of the emitted waves and the image processing of the receiver, scan converter, and display device (Fig. 1a,b). Additional factors influencing the image are the angle between the transducer and the imaged tissue, the pressure on the tissue, and the acoustic coupling between the transducer and the tissue. The probe should be put on the tissue perpendicularly and without pressure because oblique angles result in false low echo intensities, and with undue pressure, false high echo intensities.

Computed Tomography

CT is an imaging modality for obtaining axial sections of the body using X-rays, which was first used in myoimaging in 1977 by O'Doherty et al. (20). During examination, an X-ray tube rotates perpendicularly to the body axis. The absorption of the radiation is measured by detectors. Linear attenuation coefficients are computed and reproduced on the screen as a gray scale display, each pixel representing a defined small volume of the imaged part of the body. The attenuation coefficient is measured in Hounsfield units (HU) in honor of Godfrey Hounsfield, the inventor of medical CT (3,4). The attenuation coefficient or radiodensity of water has been defined as 0 HU, that of air at -1000 HU, and that of bone at +1000 HU. Pure muscle tissue has an attenuation coefficient of approximately 54 HU (5). In myoimaging, structures of higher density in comparison to muscle are hyperdense, those of lower density hypodense. The standard examination protocol developed by Bulcke and Baert (6) and later modified by Rodiek and Küther (7) considers eight examination sites: prominentia laryngis, manubrium

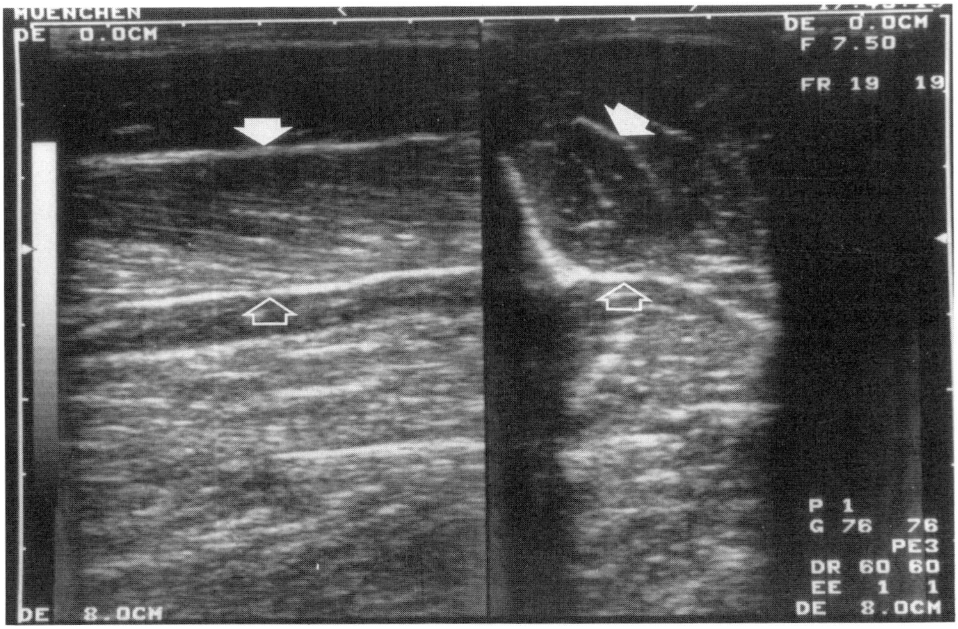

(a)

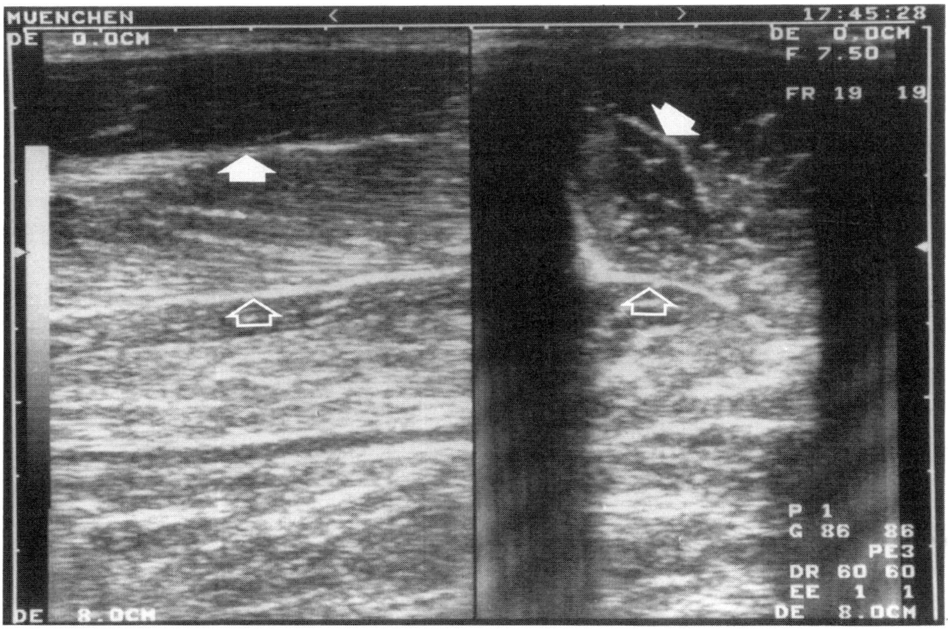

(b)

Figure 1 Longitudinal (left) and transverse (right) ultrasonograms (7.5 MHz linear array real-time electronic transducer) in a normal right tibialis anterior muscle (38-years-old man). (a) Gain 76 dB; (b) gain 86 dB. The deeper part of the muscle is more echogenic on the image created with a gain of 86 dB than with 76 dB. Arrows = aponeurosis; open arrows = membrana interossea cruris.

sterni, 15 cm proximal and 10 cm distal to the olecranon, navel, symphysis, and 15 cm proximal and 20 cm distal to the proximal edge of the patella.

Magnetic Resonance Imaging

MRI, the newest imaging technique, was first applied to myoimaging by Rodiek in 1985 (8). In a strong magnetic field (0.02–2 Tesla), paramagnetic atoms with unpaired electrons are stimulated by radiofrequency impulses. A portion of their nuclei assume a higher energy level. After the magnetic field is switched off, the stimulated nuclei return to their original state, and electromagnetic waves of the same energy and frequency as the stimulating impulses are produced. These so-called resonance signals are registered by coils and deliver information about the chemical and physical properties of the tissue. Their intensity depends on several parameters: the proton density, the T1-(spin-lattice)- and T2-(spin-spin)-relaxation time, the proton flow, and two freely adjustable examination parameters, the repetition time (TR) and the echo time (TE).

IMAGING OF HEALTHY MUSCULATURE

Ultrasound

Healthy muscle is echo-poor. Fasciae, the epimysium, aponeuroses, and bone echoes can easily be depicted as echogenic lines (Fig. 1a,b). The epimysium sometimes consists of two layers, which surround a thin layer of fat. The subcutaneous fat and the bone shadow also appear echo-poor, often separated by echogenic septa. On longitudinal scans perimysial fibroadipose septa are visible as thin echogenic lines separating the muscle bundles. On transverse scans these septa give the muscle a speckled appearance. During a short isometric contraction the muscle mass increases more or less depending on the tightness of its fascia. The overall echointensity decreases slightly. Fat planes between the muscles aid in the differentiation of neighboring muscles (9).

Compound Tomography

The fat layers between muscles are relatively hypodense. The zona compacta of bone is extremely hyperdense; bone marrow and subcutaneous fat are hypodense (Fig. 2).

Magnetic Resonance Imaging

Healthy muscle has moderate signal intensities on T1- and low signal intensities on T2-weighted images and appears gray. Septa between muscles with an increased connective tissue content present with even lower signal intensities; those with an increased adipose content are signal-intense (bright). The zona compacta has almost no signals (black); bone marrow and subcutaneous fat have high signal intensities on both T1- and T2-weighted images (Fig. 3a,b). After exercise, the T1 and T2 values increase, probably because of solute-induced water shifts caused by glycogenolysis. Perfusion is not required for this effect (10–12). The effect is significantly higher after concentric as compared to eccentric exercise (11).

THE DISEASED MUSCLE

Correlation of Histopathological and Imaging Findings

Ultrasound

Several studies have established a close relationship between the severity of mesenchymal abnormalities in the muscle biopsy and echogenicity (13,14). However, in the case of extensive fat infiltration echogenicity may decrease again (15). Semiquantitative correlation of lipid replacement and fibrosis with the corresponding echointensities showed that high echogenicity is caused by muscle lipomatosis, whereas muscle fibrosis does not considerably influence echogenicity (15,16). Muscle edema can cause

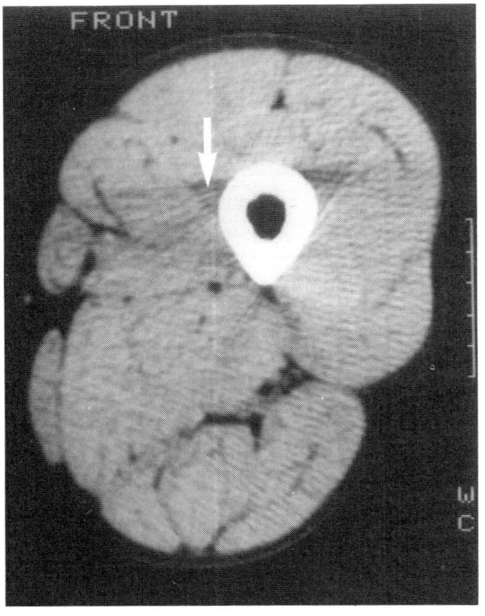

Figure 2 Normal computed tomogram of the thigh in a patient with glycogenosis type V (McArdle). Arrow = beam-hardening artifact.

low echointensities and muscle swelling. However, in the majority of cases the decrease of echointensity due to muscle edema is too small for the definite diagnosis of edema (16). Normal ultrasound findings cannot exclude neuromuscular disorders (16,17).

Computed Tomography

The Hounsfield value for adipose tissue is -106 HU and for fibrous tissue +74 HU (5). Decreased attenuation value on CT is related to the degree of muscular degeneration and replacement by fat as confirmed by histopathology (18). Unfortunately, not every histological lesion can be detected by muscle CT (19), and again, a normal muscle CT does not necessarily exclude muscle disease.

Magnetic Resonance Imaging

Adipose tissue is characterized by high signal intensities on both T1- and T2-weighted images (20) because of its short T1 and long T2 relaxation times. There is a curvilinear inverse relationship between the collagen content of tissue and T1 and T2 relaxation times (21). Thus, a high collagen content results in low signal intensity on T1- and high signal intensity on T2-weighted images, but an increase in muscle collagen content of greater than 60 mg/g of tissue is needed before noticeable changes in T1 or T2 images occur (21). In addition, muscle fibrosis is usually combined with fat replacement, so their effects overlap one another.

The tissue water content is linearly correlated to the spin density and the T1 and T2 relaxation times. The signal intensity on heavily T1-weighted spin-echo images, i.e., very short relaxation and echo times, is decreased. Sequences with longer relaxation times result in only slight changes in T1, which are often not sufficiently increased to counteract the increase in signal intensity from increased T2 time and spin density. Therefore, the net signal intensity on the T1-weighted image may be unchanged (20). Mild or moderate muscle edema can be overlooked when it occurs concomitantly with muscle lipomatosis, which also has a high signal intensity on T2-weighted images. Recently, so-called fat-suppressive sequences, for example short tau inversion recovery (STIR) sequences, have been introduced to enhance the detection of muscle edema. These sequences provide very sensitive depiction

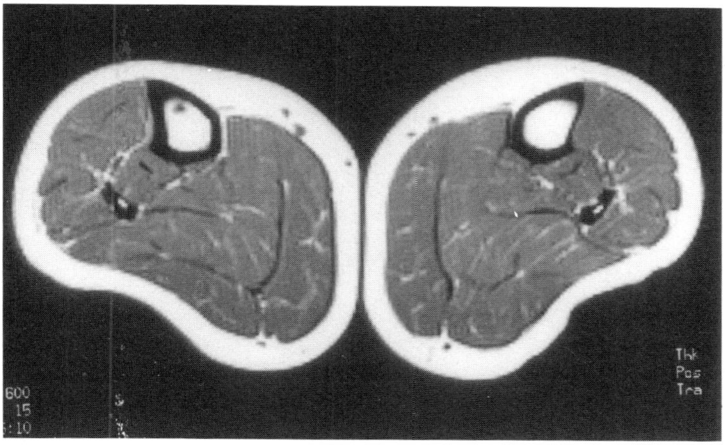

(a)

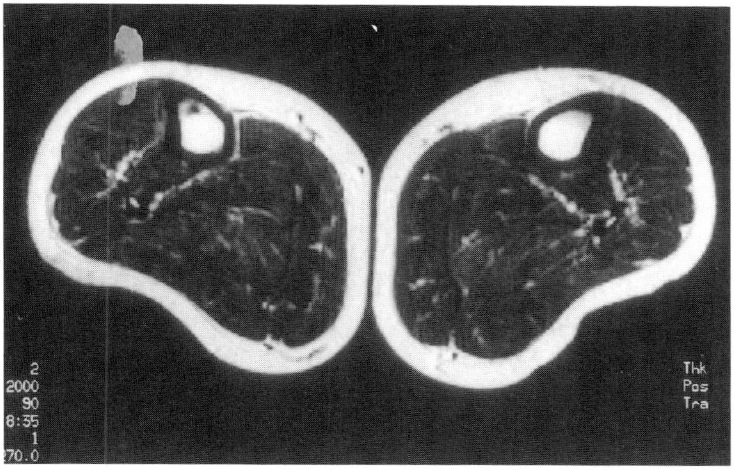

(b)

Figure 3 Normal spin echo sequence of the legs in a healthy female volunteer. (a) T1-weighted image (TR 600 msec/TE 15 msec); (b) T2-weighted image (TR 2100 msec/TE 90 msec).

of increased tissue water content, which, for example, is a typical feature of acute inflammatory myopathies (22–25). A further pitfall in the evaluation of MR images is that a relative increase of extracellular as opposed to intracellular fluid also increases the T1 and T2 relaxation times (26). This results in images that can be misinterpreted as muscle edema.

Inflammatory infiltrates result in increased STIR signal intensities (27). In rats, muscles with a larger proportion of fast-twitch fibers (type II fibers) are associated with longer T1 and T2 relaxation times than are those that have a greater proportion of slow-twitch fibers (type 1 fibers). This may be due to the larger extracellular fluid space of the slow-twitch fibers (26). In humans, the proportion of type I and type II fibers has an effect on T1 and T2 values, too. However, the relationship is reversed—higher proportions of slow-twitch fibers result in longer T1 and T2 values (28).

Table 1 summarizes the influence of intramuscular fat, connective tissue, increased water content, and calcifications on US, CT, and MR images.

Table 1 Correlation of Histopathological and Imaging Findings

Histopathological abnormality	Ultrasound	Computed tomography	Magnetic resonance imaging
Lipomatosis	Hyperechoic	Hypodense	High SI on T1 and T2
Connective tissue	No significant influence	Hyperdense	Low SI on T1 and high SI on T2
Increased total water content ("edema") or relative increase of extracellular space	Hyper- or hypoechoic	Hypodense	Low SI on T1, high SI on T2 and STIR
Calcification	Highly echogenic with acoustic shadow	Highly hyperdense	Very low SI on T1 and T2

SI = signal intensity; T1 = T1-weighted image; T2 = T2-weighted image; STIR = short tau inversion recovery.

Assessment of Muscle US, CT, and MRI Findings

Muscle disorders usually result in muscle fiber atrophy, rarely in hypertrophy, and in mesenchymal abnormalities such as lipomatosis, fibrosis, edema, and infiltration. In assessing the imaging results of muscles, the type, degree, and localization of the abnormalities must be determined.

Hypertrophy without mesenchymal abnormality is referred to as true hypertrophy (Figs. 4 and 5). Hypertrophy combined with abnormal mesenchymal structure is called pseudohypertrophy. Some neuromuscular diseases produce relatively focal muscle atrophy, hypertrophy, or mesenchymal abnormalities. Three categories of muscle degeneration can be distinguished (29): *intergroup selectivity*, for example selectivity of proximal or distal muscles or the distinction between the extensor and flexor groups; *intermuscular selectivity*, which results in the abnormality of individual muscles while others are normal (Figs. 4 and 5); and *intramuscular selectivity*, in which only parts of compound muscles, for example the quadriceps femoris or triceps surae muscles, are affected (29) (Fig. 5). Furthermore, in single muscles the distribution of mesenchymal abnormalities may be focal, multifocal, or homogenous. In myosonography, mesenchymal abnormalities cause obscured fascia, septa, and bone echoes. In the case of muscle atrophy, the thickness of the subcutaneous fat may be increased, especially in neuropathic diseases (13,30). To objectively quantify the degree of muscle atrophy and mesenchymal abnormalities an electronic caliper and computer-aided methods for measurements of echointensity (16,31,32), radiodensity (33,34), and relaxation times (35–37) should be employed. An additional method for quantifying ultrasonic abnormalities is to count the number of perimysial septa within 1 cm of muscle depicted by high-frequency sonography, as their number is increased (normal mean: 10.4 septa/cm; range: 8–13) in lower extremity muscles of many patients with neuromuscular diseases (38). Fasciculations can be depicted in B-mode real time ultrasound and documented using the M-mode-technique (39–42) (Fig. 6).

Advantages and Disadvantages of Different Imaging Techniques

MR imaging provides the best soft tissue contrast (8,43) and excellent differentiation between muscle parenchyma, fat, fibrosis, and edema. MRI permits multiplanar scanning. CT and MRI create cross-sections of the trunk or limbs, whereas most ultrasound images are limited to single muscles or small

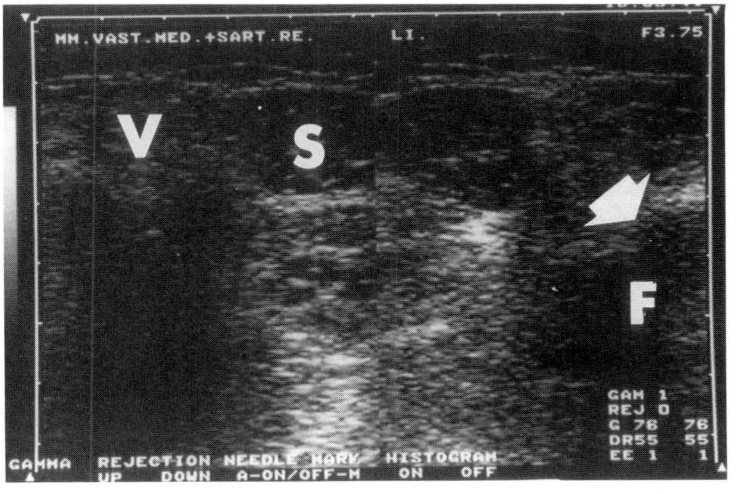

Figure 4 Transverse ultrasound of the vastus medialis and sartorius muscles in a patient with a benign X-linked muscular dystrophy of Becker type (right-sided muscles on left, and vice versa). True hypertrophy of the sartorius muscle (S), while the vastus medialis muscles (V) show marked atrophy and slightly increased echogenicity. Arrow = attenuated femur (F) echo.

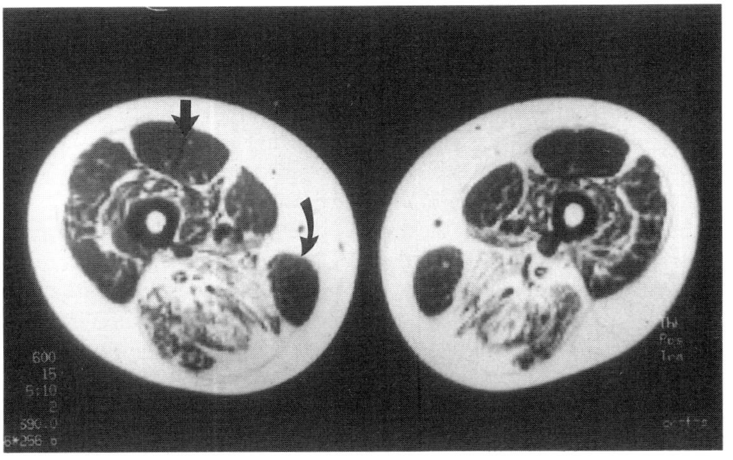

Figure 5 T1-weighted spin echo sequence (TR 600 msec/TE 15 msec) in a patient with autosomal recessive limb girdle muscular dystrophy. Compensatory hypertrophy of the rectus femoris (arrow), sartorius, and gracilis (curved arrow) muscles. Marked increase of signal intensities (fat) of the hamstrings and adductors, and less severe of the vasti muscles.

groups. Experimental ultrasonic CT can scan whole extremities (44). CT and MRI can image high-contrast objects as small as 0.5 mm, whereas spatial resolution of diagnostic ultrasound is about 1 mm (45). The sensitivity of MRI is superior to that of ultrasound (17,46) and CT (17,46). There is a controversy as to whether CT (46) or ultrasound (44,47) is more sensitive. CT uses ionizing X-rays and therefore should be employed cautiously, particularly in young patients. At present, real-time sonography is the only dynamic imaging technique that permits the depiction of muscle movements, such as fasciculations (Fig. 6).

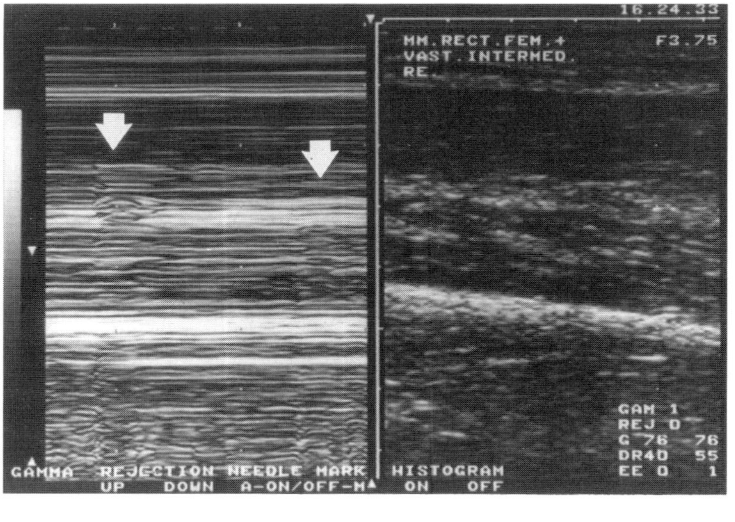

Figure 6 B- and M-mode (left) ultrasound of the right rectus femoris and vastus intermedius muscles in a patient with beginning amyotrophic lateral sclerosis. Normal echogenicity of the muscles; two fasciculations documented by the M-mode-technique (arrows).

Findings in Primary Muscle Disorders

General Remarks

Fasciculations, muscle edema, and calcifications are typical, but not pathognomonic, features of certain muscle diseases, irrespective of their degree and localization. *Fasciculations* can regularly be visualized by using real-time ultrasonography in spinal muscular atrophies (39–41) (Fig. 6), in motor neuron disease (39,41,42), in hereditary motor and sensory neuropathies (41), and in postpoliomyelitis muscular atrophy (42). *Muscle edema* is a typical feature of acute myositis, but can occasionally occur in metabolic myopathies (48), familial periodic paralysis (43), degenerative myopathies (facioscapulohumeral muscular dystrophy, myotonic dystrophy) (17,49), and neuropathies (49–51). *Muscle calcification* can be seen in dermatomyositis of children and rarely of adults (52,53), localized myositis ossificans (54) (Fig. 7), and myodysplasia ossificans progressiva (55). Clinical findings such as muscle *pseudohypertrophy* (16,56–58), asymmetry of muscle involvement, for example in facioscapulohumeral muscular dystrophy (59,60) (Figs. 8 and 9), postpoliomyelitis syndrome (43,56,61) (Fig. 10), spinal muscular atrophies (62), *muscle aplasia* (63), and *supernumery muscles* (64–66) can be documented by myoimaging.

The following sections deal with the typical myoimaging findings in particular neuromuscular diseases. Only observations derived from studies based on analysis of a sufficient number of patients were considered in this review. Up to now, nearly all studies dealt with imaging of paraspinal and lower extremity muscles. Reports on abnormalities of arm muscles are rare.

Several authors reported that primary myopathic and neuropathic disorders can be distinguished by means of myoimaging (Table 2). However, other authors have contested this diagnostic potential of myoimaging (72–74).

Findings in Special Diseases

Morphologically Defined Congenital Myopathies.

Central Core Disease. Ultrasound reveals markedly increased echogenicity of the thigh musculature (75,76). On CT scans of the thighs, hypodensities are depicted, particularly in the sartorius muscles (60,76).

Nemaline Myopathy. Cross-sectional areas of the muscles are below normal range and the amount of subcutaneous fat is increased. US shows mild to severe increase of echogenicity, with partial loss of fascial and bone echos, in the paraspinals, thighs, and lower legs (47,77), notably in the dorsiflexors of the feet (47). CT and MRI scans depict low density and increased signal intensity indicating increased adipose tissue content in the dorsiflexors of the feet (47).

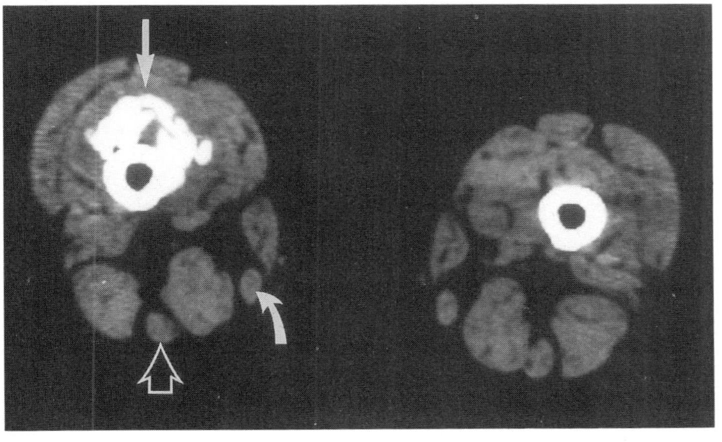

Figure 7 Computed tomograms of the thighs in a patient with localized myositis ossificans. Extremely hyperdense calcifications (arrow) of the right vastus intermedius muscle; atrophy of the gracilis (curved arrow) and semitendinosus (open arrow) muscles.

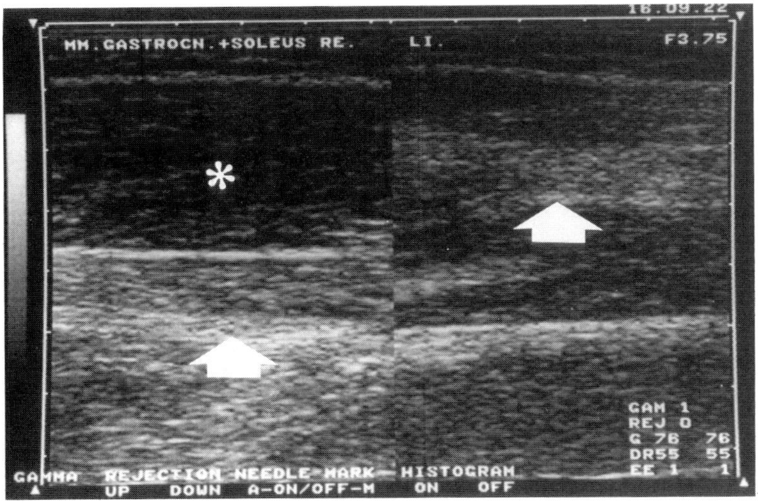

Figure 8 Longitudinal ultrasound scans of the gastrocnemius and soleus muscles in a patient with facioscapulohumeral muscular dystrophy. Hypertrophy of the right gastrocnemius muscle (shown on left; asterisk) whereas the left gastrocnemius and both soleus muscles (arrows) are atrophic. The soleus muscles additionally are abnormally echogenic.

Centronuclear Myopathy. Marked hypodensities of upper extremity muscles are shown on CT scans (60,73) (Fig. 11).

Muscular Dystrophies.

Congenital Muscular Dystrophy. Myosonography of affected children usually reveals considerably increased echointensities and obscured bone echoes in the lower extremities. However, in one study images of two newborns were normal (65).

Duchenne Type. There is controversy as to whether US reveals normal (67,78) or abnormal (thickened

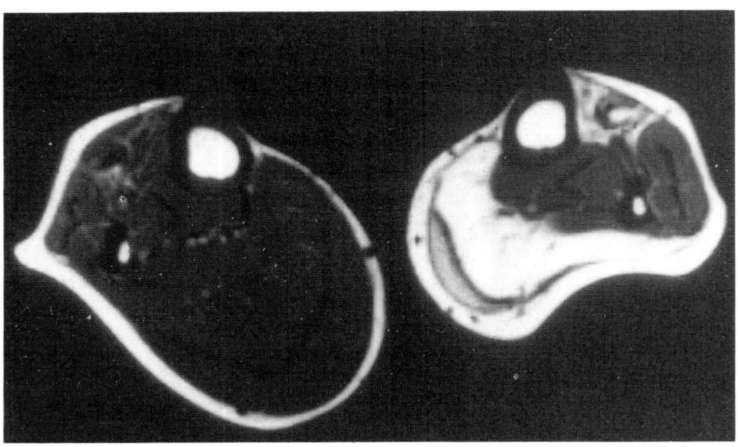

Figure 9 T1-weighted spin echo sequence (TR 600 msec/TE 15 msec) of the legs in a patient with facioscapulohumeral muscular dystrophy. Hypertrophy of the right calf (shown on left) whereas the left leg is atrophic, additionally presenting with markedly increased signal intensities (bright) of the tibialis anterior, extensor digitorum longus, and triceps surae muscles indicating fat replacement.

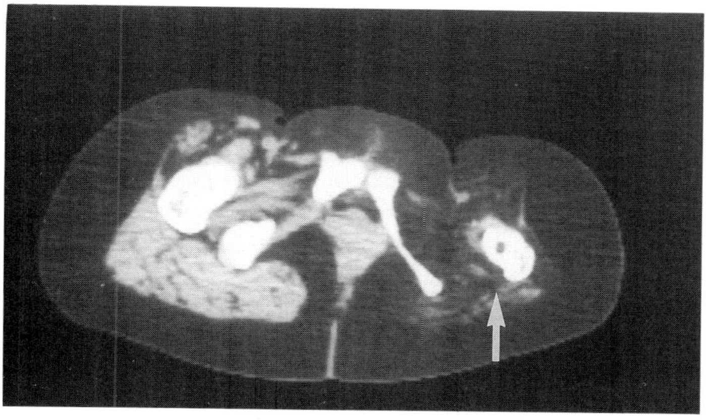

Figure 10 Axial computerized tomogram of the hip region in a patient with late postpoliomyelitis muscular atrophy. Considerable increase of subcutaneous fat tissue, marked atrophy and hypodensity (= fat replacement) of the right (shown on left) gluteal (arrow), obturatorius, sartorius, and rectus femoris muscles.

and hypoechoic) (79) muscles in preclinical stages. In later stages, muscle echogenicity is generally increased (67,69,78,80) and grows parallel to loss of muscle strength (81). The gastrocnemius muscle is more echogenic than the soleus (81). Bone and fascial echoes diminish (69,81). In early stages, CT reveals patchy hypodensities in some muscles. The vasti lateralis and medialis (68), adductors of the thighs, biceps femoris (34,68), glutei (34,82), the gastrocnemius (34,68,82,83), and the lateral part of the erector spinae muscles (34,84) are the first to be affected. The calves become hypertrophic or pseudohypertrophic (33,73,83). Gradually the low-attenuation areas coalesce, giving the muscles a fuzzy or washed-out appearance (68). The sartorius and gracilis (68,80,85), semitendinosus, and semimembranosus (83) muscles usually preserve their normal structure and enlarge until the final stages, when they also become hypodense.

The muscle size is normal, only slightly decreased or even increased in the early stages (67,80,86). MRI shows abnormally high T1 values in the thighs (36) and slightly decreased T1 values in the triceps surae muscles (37). T1 values rapidly decrease with the progression of the disease (36,37). In late stages, high signal intensities on both T1- and T2-weighted images, indicating muscle lipomatosis, can be seen with a similar distribution as in CT (87,88).

Becker Type. The distribution of abnormalities in the progressive muscular dystrophy of Becker type is similar to that in Duchenne type, but less pronounced. The glutei and thigh muscles with the exception of the sartorius and gracilis (88–90) (Fig. 4), adductor longus (88,90), and, in some patients, the semitendinosus (89,90), and semimembranosus (89) become atrophic. In contrast to the Duchenne type, fat infiltration in the calves is milder than in the paravertebral muscles (83).

Facioscapulohumeral type. The clearest abnormalities can be found in the biceps, triceps, and shoulder

Table 2 Myoimaging Features in Primary Myopathic and Neuropathic Disorders

Feature	Primary myopathy	Primary neuropathy	Ref.
Thickness of subcutaneous fat	Slightly increased	Markedly increased	13,30,67
Muscle atrophy	Late and less severe	Early and severe	13,30,6768
Increase of muscle echogenicity	Homogeneous	Inhomogeneous	69
Muscle involvement	Symmetrical	Asymmetrical	70
Selective muscle involvement	Present	Absent	70,71
Perimysium	Intact	Broken	60,70

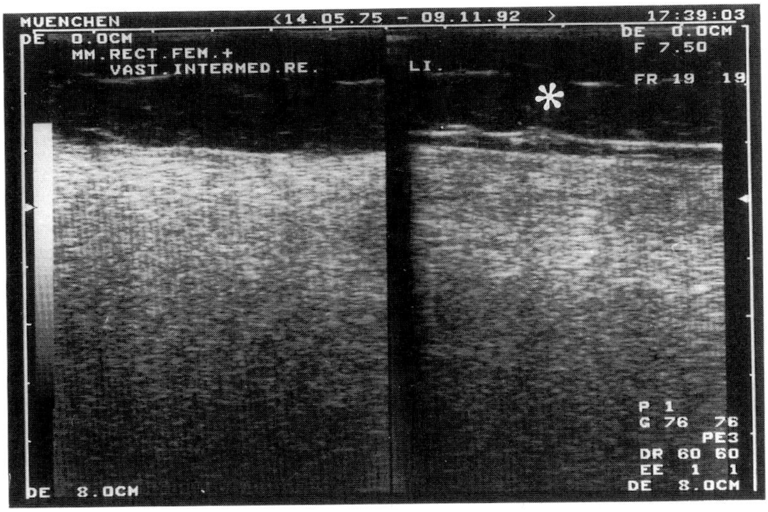

Figure 11 Longitudinal ultrasound scans of the rectus femoris and vastus intermedius muscles in a patient with centronuclear myopathy. Marked increase of muscles' echogenicity; perimysia, septa, and femur echoes no longer depictable. Asterisk = subcutaneous fat.

muscles (59). In the early stages of the disease, involvement of the buttocks, quadriceps femoris, hamstrings, and anterior and posterior leg muscles is already visible (59,86). The psoas muscle is hypertrophic in half the patients (86). The abnormalities are often asymmetrical (59,60) (Figs. 8 and 9).

Limb-Girdle Muscular Dystrophy. Muscle atrophies are rare (86). The majority of patients have lipomatosis in the paraspinal, gluteal, quadriceps femoris, semitendinosus, and semimembranosus muscles as well as in the medial head of the gastrocnemius (59,86,91). As in X-linked muscular dystrophies, the sartorius and gracilis muscles are usually hypertrophic or pseudohypertrophic (68,86, 91) (Fig. 5). The subcutaneous adipose is often thickened, particularly in women (59).

Oculopharyngeal Type. Hypodensities are visible in the paraspinal, hamstring, and triceps surae muscles (92).

Distal Types. In autosomal recessive distal muscular dystrophy, CT demonstrates preferential involvement of the lower leg flexors, hamstrings, adductors, and extensors of the thigh. The popliteus muscle usually remains unaffected. There is a stage showing hypertrophy of the sartorius, gracilis, adductor, rectus femoris, and hamstring muscles. CT abnormalities are clearly visible before muscle weakness and atrophy become clinically apparent (93). In rimmed vacuolar distal myopathy, low-density areas are usually noted in the tibialis anterior, extensor digitorum and hallucis, hamstring, adductor, and gracilis muscles. The gastrocnemius, soleus, tibialis posterior, and quadriceps femoris, and especially the vastus lateralis muscles are less often and less severely affected (94). In sporadic distal myopathy with early adult onset, hypodense areas are particularly common in the gastrocnemius, soleus, and gluteus minimus muscles (95).

Myotonic Disorders. US reveals atrophy of the tibialis anterior, triceps brachii, rectus femoris, and vastus intermedius muscles. Echointensities are highest in the tibialis anterior, brachioradialis, biceps brachii, peroneus longus, and rectus femoris muscles (17). Atrophy of the sternocleidomastoids is an early sign (96). With progressive disease, lipomatosis followed by atrophy of the extensor muscles of the spine (59,96), vastus intermedius (96–98), tibialis anterior (17,73,99), and the medial head of the gastrocnemius muscle (17,73,98) develops (Fig. 12). In later stages, the vastus lateralis and the lateral head of the gastrocnemius muscle, and finally the whole quadriceps femoris and triceps surae muscles,

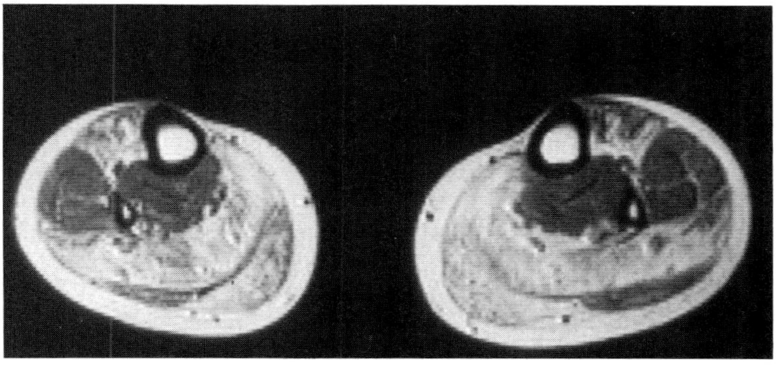

Figure 12 Axial T1-weighted spin echo sequence (TR 600 msec/TE 15 msec) of the legs in a patient with myotonic dystrophy. Markedly increased signal intensities (bright) of the tibialis anterior and soleus muscles and the medial head of the gastrocnemius muscles indicating fat replacement. The lateral head of the right gastrocnemius muscle (shown on left) shows slightly increased signal intensity.

are affected (96). The rectus femoris, adductor magnus and longus (17,96,98), and tibialis posterior (98) are usually spared or even hypertrophic, except in the final stages.

Familial Periodic Paralysis. Patients may show slightly increased muscle echointensities that do not correlate with the severity of the clinical signs, but do correlate with the patient's age and duration of the disease. In almost half the patients, thigh and calf muscles are partly replaced by adipose tissue, giving the muscles a moth-eaten appearance on CT. Those patients with weekly attacks present with decreased T1 values. The lesser thigh muscles are spared (80).

Inherited Metabolic Myopathies.

Type V (Phosphorylase Deficiency) and Type VII (Phosphofructokinase Deficiency). An increase of the cross-sectional area and the absence of MRI enhancement of skeletal muscles after exercise typify glycogenosis type V (McArdle's disease) (35,48). In both diseases, focal areas of low signal intensities on T1- and high signal intensities on T2-weighted and STIR images appear 24 hr after contractures, with complete recovery after several days or longer.

Inflammatory Myopathies.

Bacterial Myositis. Pyomyositis results in hypoechoic areas on US (100–102) and hypodense areas on CT (103–105) or diffuse muscle infiltration (102). They may be combined with ring enhancement (103,104), muscle swelling (101, 105,106), effacement of the fascial planes (104), involvement of the underlying bone (104), and subcutaneous edema (101,105). MRI reveals areas of high signal intensity on T2-weighted images, whereas T1-weighted images are normal or show only slightly increased signal intensity (106,107).

Parasitic Myositis. Wadia et al. (108) found multiple 5–7-mm-wide hypodense foci in muscular cysticercosis. The scolices were depicted as hyperdense, contrast-enhancing spots.

Poly-/dermatomyositis. In childhood dermatomyositis, echointensities are slightly increased, and bone and fascial echoes vanish (67,69,109,110). In adults, polymyositis increases echointensity. Muscle atrophy is predominantly found in thigh muscles, whereas in dermatomyositis the triceps brachii and forearm muscles are most clearly affected. Striking muscle atrophy is rare in dermatomyositis. Echointensities are lower in dermatomyositis as compared to polymyositis (16). In acute myositis, MRI shows focally or inhomogeneously increased signal intensities on T2-weighted spin echo and STIR sequences (22–25,111), predominantly in the vasti (24,111), glutei (22), adductors (22,112), biceps femoris, and gastrocnemius muscles (112). Focal and diffuse patterns of high signal intensities are of similar frequency (25). STIR sequences are more sensitive than T2-weighted images (111). The signal intensities on T1-weighted images are normal (22–24) or less increased than on T2-weighted images

(113) (Fig. 13a,b). The T1 values are increased in the rectus femoris and biceps femoris muscles (24). The signal intensity on STIR and T2-weighted images correlates with disease activity (114,115). They normalize after successful treatment (22,23).

Inclusion Body Myositis. The most obvious abnormality is severe generalized muscle atrophy in most patients with inclusion body myositis (16). In MRI focal increase of signal is usually found. The majority of patients show predominant involvement of the anterior muscle compartment of the thigh (25).

Granulomatous Myositis. In the nodular type of sarcoid myositis, US shows hypoechoic areas (57). CT demonstrates normal findings (57,116) or shows areas of low attenuation with a rim of contrast enhancement (117). MRI shows a center of low signal intensity on T1-weighted, proton-dense images (117). T2-weighted images have a border of high signal intensity (117,118), which is less predominant in T1-weighted images (117). In the myopathic type, hypertrophy of the triceps brachii and calf muscles can be observed. Echointensities are especially high in patients with insidious granulomatous polymyositis (16). CT findings may be normal (116,117) or show marked hypodensities (119). MRI findings may also be normal (117) or show high signal intensities on T2-weighted images (116,119).

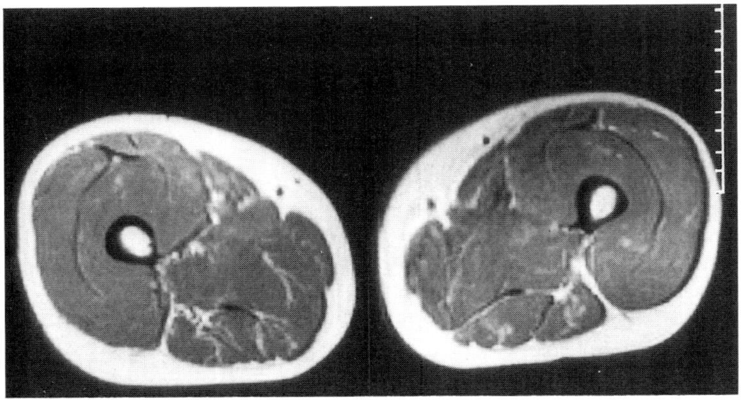

(a)

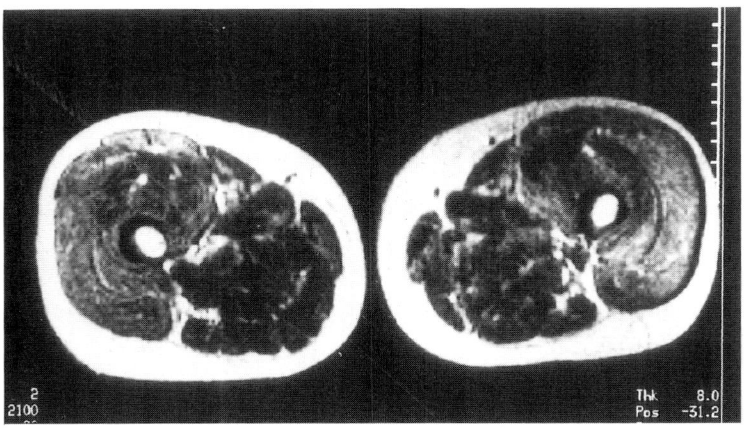

(b)

Figure 13 Spin echo sequences in a 63-year-old woman with subacute polymyositis. (a) T1-weighted image (TR 600 msec/TE 15 msec) without definite abnormality. (b) T2-weighted image (TR 2100 msec/TE 90 msec): considerably increased signal intensities in the vasti laterales and the right rectus femoris muscles (arrows) indicating muscle edema.

Other Myopathies.

Rhabdomyolysis. Rhabdomyolysis results in muscle swelling (46,120,121) and subcutaneous edema (121). About half the patients (46) have areas of low (122,123) or both de- and increased echogenicity (46). Focal low-attenuation areas are visible in CT (46,120–122). As a general rule, rhabdomyolysis generally leads to increased signal intensities on MRI with partial saturation, T2-weighted, and STIR sequences (46,124).

Myonecrosis. Acute myonecrosis results in hypodense areas (125,126), sometimes combined with a hyperemic contrast-enhancing border (125).

Impact of Myoimaging on the Diagnosis and Treatment of Neuromuscular Diseases

Myoimaging can document the extent of muscle atrophy and hypertrophy more precisely than any other technique. True hypertrophy can be distinguished from pseudohypertrophy. Myoimaging is the only noninvasive method that demonstrates muscle lipomatosis, fibrosis, and calcification. Myoimaging is thus a valuable tool for guiding electromyographic examinations (127–129), searching for the optimal biopsy site (130–132), and planning conservative or operative rehabilitation measures (133,134). It facilitates the documentation of the course of the disease (22,23,37,135,136) in most systemic myopathies. Several authors report that myoimaging is able to differentiate primary myopathies from neuropathies. In focal disease, for example in infectious myositis or rhabdomyolysis, imaging facilitates localization of the process. MRI, especially with fat-suppressive sequences, is a sensitive method for assessing muscle edema and the activity of inflammatory myopathies and thus may influence therapy. In summary, although there are no specific abnormalities, myoimaging can provide much information that may aid the differential diagnosis of neuromuscular disorders.

REFERENCES

1. Ikai M, Fukunaga T. Calculation of muscle strength per unit cross-sectional area of human muscle by means of ultrasonic measurement. Int Z angew Physiol einschl Arbeitsphysiol 1968; 26:26–32.
2. O'Doherty DS, Schellinger D, Raptopoulos V. Computed tomographic patterns of pseudohypertrophic muscular dystrophy: preliminary results. J Comput Assist Tomogr 1977; 1:482–486.
3. Hounsfield GN. A method of an apparatus for examination of a body by radiation such as X or gamma radiation. Patent specification 1283915. The Patent Office, London, 1968.
4. Hounsfield GN. Computed medical imaging. Nobel lecture, December 8, 1979. J Comput Assist Tomogr 1980; 4:665–674.
5. Grindrod S, Tofts P, Edwards R. Investigation of human skeletal muscle structure and composition by X-ray computerised tomography. Eur J Clin Invest 1983; 13:465–468.
6. Bulcke JAL, Baert AL. Clinical and Radiological Aspects of Myopathies. CT Scanning, EMG, Radioisotopes. Berlin: Springer, 1982:45–81.
7. Rodiek SO, Küther G. Technik computertomographischer Skelettmuskeluntersuchungen bei neuromuskulären Erkrankungen. Fortschr Röntgenstr 1985; 143:24–28.
8. Rodiek SO. Kernspintomographie der Skelettmuskulatur bei neuromuskulären Erkrankungen. Fortschr Röntgenstr 1985; 143:418–425.
9. Fornage B. Ultrasonography of Muscles and Tendons. Examination Technique and Atlas of Normal Anatomy of the Extremities. New York: Springer, 1989.
10. Archer BT, Fleckenstein JL, Bertocci LA, et al. Effect of perfusion on exercised muscle: MR imaging evaluation. JMRI 1992; 2:407–413.
11. Shellock FG, Fukunaga T, Mink JH, Edgerton VR. Acute effects of exercise on MR imaging of skeletal muscle: concentric vs eccentric actions. Am J Roentgenol 1991; 156:765–768.
12. Fleckenstein JL, Haller RG, Bertocci LA, Parkey RW, Peshock RM. Glykogenolysis, not perfusion, is the critical mediator of exercise-induced muscle modifications on MR images. Radiology 1992; 183:25–27.
13. Kamala D, Suresh S, Githa K. Real-time ultrasonography in neuromuscular problems of children. J Clin Ultrasound 1985; 13:465–468.
14. Lamminen A, Jääskeläinen J, Rapola J, Suramo I. High-frequency ultrasonography of skeletal muscle in children with neuromuscular disease. J Ultrasound Med 1988; 7:505–509.
15. Reimers K, Reimers CD, Wagner St, Paetzke I, Pongratz DE. Skeletal muscle sonography: a correlative study between echointensities and morphology. J Ultrasound Med 1993; 12:73–77.

16. Reimers CD, Fleckenstein JL, Witt ThN, Müller-Felber W, Pongratz DE. Muscular ultrasound in idiopathic inflammatory myopathies of adults. J Neurol Sci 1993; 116:82–92.
17. Schedel H, Reimers CD, Naegele M, Witt ThN, Pongratz DE, Vogl Th. Imaging techniques in myotonic dystrophy. A comparative study of ultrasound, computed tomography and magnetic resonance imaging of skeletal muscles. Eur J Radiol 1992; 15:230–238.
18. Ohiwa N, Kato T, Ando T, Mohri A, Yokoi K, Matsumoto H. CT findings of skeletal muscles in children with progressive muscular dystrophy. Brain Dev 1981; 13:156–159.
19. Vliet AM, Thijssen HOM, Joosten E, Merx JL. CT in neuromuscular disorders: a comparison of CT and histology. Neuroradiology 1988; 30:421–425.
20. Fleckenstein JL. Weatherall PT, Bertocci LA, et al. Locomotor system assessment by muscle magnetic resonance imaging. Magn Reson Q 1991; 7:79–103.
21. Scholz TD, Fleagle StR, Burns TL, Skorton DJ. Tissue determinants of nuclear magnetic resonance relaxation times. Effect of water and collagen content in muscle and tendon. Invest Radiol 1989; 24:893–898.
22. Hernandez RJ, Kein DR, Sullivan DB, Chenevert ThL, Martel W. Magnetic resonance imaging appearance of the muscles in childhood dermatomyositis. J Pediatr 1990; 117:546–550.
23. Fujino H, Kobayashi T, Goto I, Onitsuka H. Magnetic resonance imaging of the muscles in patients with polymyositis and dermatomyositis. Muscle Nerve 1991; 14:716–720.
24. Park JH, Vansant JP, Kumar NG, et al. Dermatomyositis: correlative MR imaging and P-31 MR spectroscopy for quantitative characterization of inflammatory disease. Radiology 1990; 177:473–479.
25. Fraser DD, Frank JA, Dalakas M, Miller FW, Hicks JE, Plotz P. Magnetic resonance imaging in the idiopathic inflammatory myopathies. J Rheumatol 1991; 18:1693–1700.
26. Polak JF, Jolesz FA, Adams DF. NMR of skeletal muscle differences in relaxation parameters related to extracellular/intracellular fluid spaces. Invest Radiol 1988; 23:107–112.
27. Fraser DD, Frank JA, Dalakas MC. Inflammatory myopathies: MR imaging and spectroscopy. Radiology 1991; 179:341–342.
28. Kuno S-y, Katsuta S, Inouye T, Anno I, Matsumoto K, Akisada M. Relationship between MR relaxation time and muscle fiber composition. Radiology 1988; 169:567–568.
29. Bulcke JAL, Baert AL. Clinical and Radiological Aspects of Myopathies. CT Scanning, EMG, Radioisotopes. Berlin: Springer, 1982:100–105.
30. Heckmatt JZ, Pier N, Dubowitz V. Assessment of quadriceps muscle atrophy and hypertrophy in neuromuscular disease in children. J Clin Ultrasound 1988; 16:177–181.
31. Forst R. Skelettmuskel-Sonographie bei neuromuskulären Erkrankungen. Stuttgart: Enke, 1986.
32. Heckmatt J, Rodillo E, Doherty M, Willson K, Leeman S. Quantitative sonography of muscle. J Child Neurol 1989; 4(Suppl):S101–S106.
33. Jones DA, Round JM, Edwards RHT, Grindrod SR, Tofts PS. Size and composition of the calf and quadriceps muscles in Duchenne muscular dystrophy. J Neurol Sci 1983; 60:307–322.
34. Gellerich I, Koch RD. Computertomographische Befunde bei maligner progressiver Muskeldystrophie Duchenne (DMD). Psychiatr Neurol Med Psychol (Leipzig) 1990; 42:282–290.
35. Fleckenstein JL, Haller RG, Lewis SF, et al. Absence of exercise-induced MRI enhancement of skeletal muscle in McArdle's disease. J Appl Physiol 1991; 71:961–969.
36. Matsumura K, Nakano I, Fukuda N, Ikehira H, Tateno Y, Aoki Y. Proton spin-lattice relaxation time of Duchenne dystrophy skeletal muscle by magnetic resonance imaging. Muscle Nerve 1988; 11:97–102.
37. Nagao H, Morimoto T, Sano N, et al. Magnetic resonance imaging of skeletal muscle in patients with Duchenne muscular dystrophy—serial axial and sagittal section studies. No To Hattatsu 1991; 23:39–43.
38. Dock W, Happak W, Grabenwöger F, Toifl K, Bittner R, Gruber H. Neuromuscular diseases: evaluation with high-frequency sonography. Radiology 1990; 177:825–828.
39. Reimers CD, Müller W, Schmidt-Achert M, Heldwein W, Pongratz DE. Sonographische Erfassung von Faszikulationen. Ultraschall 1988; 9:237–239.
40. Rott H-D, Lang C, Engelhardt A, Grahmann F. Sonographische Befunde bei pseudomyopathischer spinlar Muskelatrophie vom Typ Kugelberg-Welander. Ultraschall Klin Prax 1988; 3:25–28.
41. Reimers CD, Naegele M, Fenzi G, et al. Comparative Study of Ultrasound, MRI, CT, and Myopathological Findings in Generalized Neuromuscular Diseases. In: Nadjmi M, ed., Imaging of Brain Metabolism, Spine and Cord, Interventional Neuroradiology, Free Communications, XVth Congress of the European Society of Neuroradiology, Würzburg, September 13th–17th, 1988. Berlin: Springer, 1989:627–630.
42. Walker FO, Donofrio PD, Harpold GJ, Ferrell WG. Sonographic imaging of muscle contraction and fasciculations: a correlation with electromyography. Muscle Nerve 1990; 13:33–39.
43. Kaiser WA, Schalke BCG. Kernspintomographie bei generalisierten Skelettmuskelerkrankungen. Untersuchungsmodalitäten und Indikationsempfehlungen. Röntgenpraxis 1989; 42:338–345.
44. Rott H-D. Muskeldystrophie Duchenne: Konduktorinnendiagnostik mit bildge benden Verfahren. In: Stuhler Th, Feige A, eds. Ultraschalldiagnostik des Bewegungsapparates. Berlin: Springer, 1987:81–88.

45. Bushong SC, Archer BR. Diagnostic Ultrasound. Physics, Biology, and Instrumentation. St. Louis: Mosby Year Book, 1991:69.
46. Lamminen AE, Hekali PE, Tiula E, Suramo I, Korhola OA. Acute rhabdomyolysis: evaluation with magnetic resonance imaging compared with computed tomography and ultrasound. Br J Radiol 1989; 62:326–330.
47. Wallgren-Pettersson C, Kivisaari L, Jääskeläinen J, Lamminen A, Holmberg C. Ultrasonography, CT, and MRI of muscles in congenital nemaline myopathy. Pediatr Neurol 1990; 6:20–28.
48. Fleckenstein JL, Peshock RM, Lewis SF, Haller RG. Magnetic resonance imaging of muscle injury and atrophy in glycolytic myopathies. Muscle Nerve 1989; 12:849–855.
49. Schedel H, Reimers CD, Vogl T, Lissner J. Kernspintomographische Darstellung des Muskeloedems bei neuromuskulären Erkrankungen. Radiol Diagn 1992; 33:29–32.
50. Shabas D, Gerard G, Rossi D. Magnetic resonance imaging examination of denervated muscle. Comput Radiol 1987; 11:9–13.
51. Fleckenstein JL, Watumull D, Conner K, et al. Denervated human skeletal muscle: MRI evaluation. Radiology 1993; 187:213–218.
52. Randle HW, Sander HM, Howard K. Early diagnosis of calcinosis cutis in childhood dermatomyositis using computed tomography (letter). JAMA 1986; 256:1137–1138.
53. Coakley JH, Smith PEM, Jackson MJ, Edwards RHT, Carty AT. Myositis ossificans nonprogressiva—reversible muscle calcification in polymyositis. Br J Rheumatol 1989; 28:443–445.
54. Kramer FL, Kurtz AB, Rubin C, Goldberg BB. Ultrasound appearance of myositis ossificans. Skel Radiol 1979; 4:19–20.
55. Reimers CD, Vogl T, Pongratz DE. Bildgebende Verfahren bei neuromuskulären Erkrankungen Med Klin 1992; 87:469–478.
56. Bertorini TE, Igarashi M. Postpoliomyelitis muscle pseudohypertrophy. Muscle Nerve 1985; 8:644–649.
57. Kayanuma K, Uono M. A case of muscular sarcoidosis of palpable nodules type with pseudohypertrophy—comparison of computerized tomography and ultrasound imaging of skeletal muscles. Clin Neurol 1987; 27:760–766.
58. de Visser M, Hoogendijk JE, Ongerboer de Visser BW, Verbeeten Jr BJ. Calf enlargement in hereditary motor and sensory neuropathy. Muscle Nerve 1990; 13:40–46.
59. Jiddane M, Gastaut JL, Pellissier JF, Serratrice G, Salamon G. CT of primary muscle diseases. Am J Neuroradiol 1983; 4:773–776.
60. Herson D, Larde D, Ferry M, Brunet P, Fardeau M. Apport diagnostique du scanner X en pathologie musculaire. Rev Neurol (Paris) 1985; 141:482–489.
61. Rodiek SO. CT-MR-Tomographie und MR-Spektroskopie bei neuromuskulären Erkrankungen. Stuttgart: Enke, 1987:22–26.
62. Sambrook M, Rickards D, Cumming WJK. CT muscle scanning in the evaluation of patients with spinal muscular atrophy (SMA). Neuroradiology 1988; 30:487–496.
63. Wright AR, Milner RH, Bainbridge LC, Wilsdon JB. MR and CT in the assessment of Poland syndrome. J Comput Assist Tomogr 1992; 16:442–447.
64. Romanus B, Lindahl S, Stener B. Accessory soleus muscle. A clinical and radiographic presentation of eleven cases. J Bone Joint Surg 1986; 68:731–732.
65. Buschmann WR, Cheung Y, Jahss MH. Magnetic resonance imaging of anomalous leg muscles: accessory soleus, peroneus quartus and flexor digitorum longus accessorius. Foot Ankle 1991; 12:109–116.
66. Sanger JR, Krasniak CL, Matloub HS, Yousif NJ, Kneeland JB. Diagnosis of an anomalous superficialis muscle in the palm by magnetic resonance imaging. J Hand Surg 1991; 16:98–101.
67. Heckmatt JZ, Pier N, Dubowitz V. Real-time ultrasound imaging of muscles. Muscle Nerve 1988; 11:56–65.
68. Hawley BJ, Schellinger D, O'Doherty DS. Computed tomographic patterns of muscles in neuromuscular diseases. Arch Neurol 1984; 41:383–387.
69. Fischer AQ, Carpenter DW, Hartlage PL, Carroll JE, Stephens S. Muscle imaging in neuromuscular disease using computerized real-time sonography. Muscle Nerve 1988; 11:270–275.
70. Calò M, Crisi G, Martinelli C, Colombo A, Schoenhuber R, Gibertoni M. CT and the diagnosis of myopathies. Neuroradiology 1986; 28:53–57.
71. Bulcke JAL, Baert AL. Clinical and Radiological Aspects of Myopathies. CT Scanning, EMG, Radioisotopes. Berlin: Springer, 1982:114–116.
72. Forst R. Skelettmuskel-Sonographie bei Neuromuskulären Erkrankungen. Stuttgart: Enke, 1986:138.
73. Mielke U, Wussow W. Computed tomography of calf muscles in progressive muscular dystrophy. In: Gerstenbrand F, Grcevic N, Aichner F, eds. Stuttgart: Fischer, 1985:175–182.
74. Nordal HJ, Dietrichson P, Eldevik P, Gronseth K. Fat infiltration, atrophy and hypertrophy of skeletal muscles demonstrated by X-ray tomography in neurological patients. Acta Neurol Scand 1988; 77:115–122.
75. Forst R. Skelettmuskel-Sonographie bei neuromuskulären Erkrankungen. Stuttgart: Enke, 1986:114–115.
76. Arai Y, Sumida S, Osawa M, Hirasawa K, Kawai M, Shishikura K. Skeletal muscle CT scan and ultrasound imaging in two siblings with central core disease. No To Hattatsu 1990; 22:55–60.

77. Forst R. Skelettmuskel-Sonographie bei neuromuskulären Erkrankungen. Stuttgart: Enke, 1986:113–114.
78. Forst R, Casser H-R. 7 MHz-real-time-Sonographie der Skelettmuskulatur bei Duchenne-Muskeldystrophie. Ultraschall 1985; 6:336–340.
79. von Rohden L. Neuromuskuläre Erkrankungen. In: Hofmann V, ed. Ultraschalldiagnostik in Pädiatrie und Kinderchirurgie. Leipzig: VEB Georg Thieme, 1989:269–271.
80. Alanen A, Kantola I, Komu M. New imaging methods in hereditary neuro-muscular diseases. Acta-Cardiomiol 1989; 1:33–45.
81. Aizawa H, Kozimo S, Takagi A. Ultrasound imaging of muscles in Duchenne muscular dystrophy. Clin Neurol 1989; 29:49–53.
82. Kawai M, Kunimoto M, Motoyoshi Y, Kuwata T, Nakano I. Computed tomography in Duchenne type muscular dystrophy—morphological stages based on the computed tomographical findings. Clin Neurol 1985; 25:578–590.
83. Kawai M. A computed tomographic study of calf pseudohypertrophy in the three types of progressive muscular dystrophy (Duchenne, Becker and Fukuyama). Brain Dev 1985; 17:241.
84. Imhof H, Toifl K. CT staging of neuromuscular diseases. In: Gerstenbrand F, Grcevic N, Aichner F, eds. Neuroimaging. Stuttgart: Fischer, 1985:171–174.
85. Horikawa H, Konagaya M, Takayanagi T, Otsuji H. Quantitative analysis of muscular wastings of lower limbs in Duchenne muscular dystrophy by computed tomography. Muscle Nerve 1986; 9(Suppl):242.
86. Serratrice G, Salomon G, Jiddane M, Gastatut JL, Pellissier JF, Pouget J. Resultats du scanner X musculaire dans 145 cas de maladies neuro-musculaires. Rev Neurol 1985; 141:404–412.
87. Schreiber A, Smith WL, Ionasescu V, Zellweger H, Franken EA, Dunn V, Ehrhardt J. Magnetic resonance imaging of children with Duchenne muscular dystrophy. Pediatr Radiol 1987; 17:495–497.
88. Kuriyama M, Hayakawa K, Konishi Y, et al. MR imaging of myopathy. Comput Med Imag Graph 1989; 13:329–333.
89. Bulcke JAL, Baert AL. Clinical and Radiological Aspects of Myopathies. CT Scanning, EMG, Radioisotopes. Berlin: Springer, 1982:148–149.
90. de Visser M, Verbeeten B. Computed tomography of the skeletal musculature in Becker-type muscular dystrophy and benign infantile spinal muscular atrophy. Muscle Nerve 1985; 8:435–444.
91. Murphy WA, Totty WG, Carrol JE. MRI of normal and pathologic skeletal muscle. Am J Roentgenol 1986; 146:565–574.
92. Medici M, de Tenyi A, Vincent O, Tome F. Muscle computed axial tomography features in oculopharyngeal dystrophy. Muscle Nerve 1986; 9(Suppl):244.
93. Mizusawa H, Kobayashi F, Nakanishi T. Distribution of skeletal muscle involvement in autosomal recessive distal muscular dystrophy: a clinical and computed tomographic study. Clin Neurol 1987; 27:177–184.
94. Mizusawa H, Kurisaki H, Takatsu M, Inoue K, Mannen T, Toyokura Y, Nakanishi T. Rimmed vacuolar distal myopathy: a clinical, electrophysiological, histopathological and computed tomographic study of seven cases. J Neurol 1987; 234:129–136.
95. de Visser M. Computed tomographic findings of the skeletal musculature in sporadic distal myopathy with early adult onset. J Neurol Sci 1983; 59:331–339.
96. Bulcke JAL, Baert AL. Clinical and Radiological Aspects of Myopathies. CT Scanning, EMG, Radioisotopes. Berlin: Springer, 1982:127–135.
97. Rickards D, Isherwood I, Hutchinson R, Gibbs A, Cumming WJK. Computed tomography in dystrophia myotonica. Neuroradiology 1982; 24:27–31.
98. Damian MS, Bachmann G, Herrmann D, Dorndorf W. Magnetic resonance imaging of muscle and brain in myotonic dystrophy. J Neurol 1993; 240:8–12.
99. Gellerich I, Müller D, Koch RD. Computertomographische Untersuchungen bei myotonen Dystrophien. Psychiat Neurol med Psychol (Leipzig) 1986; 38:378–383.
100. Yagupsky P, Shahak E, Barki Y. Non-invasive diagnosis of pyomyositis. Clin Pediatr Phila 1988; 27:299–301.
101. Quillin SP, McAlister WH. Rapidly progressive pyomyositis. Diagnosis by repeat sonography. J Ultrasound Med 1991; 10:181–184.
102. Belli L, Reggiori A, Cocozza E, Riboldi L. Ultrasound in tropical pyomyositis. Skel Radiol 1992; 21:107–109.
103. Schlech WF III, Moulton P, Kaiser AB. Pyomyositis: tropical disease in a temperate climate. Am J Med 1981; 71:900–902.
104. Tumeh SS, Butler GJ, Maguire JH, Nagel JS. Pyogenic myositis: CT evaluation. J Comput Assist Tomogr 1988; 12:1002–1005.
105. Piper C, Henrich R, Wendt B, von Egidy H. Nichttropische Pyomyositis in Deutschland. Med Klin 1990; 85:707–714.
106. Yuh WTC, Schreiber AE, Montgomery WJ, Ehara S. Magnetic resonance imaging of pyomyositis. Skel Radiol 1988; 17:190–193.
107. Schedel H, Reimers CD, Vogl T, Lissner J. Ultrasound and MR imaging of acute myositis. Eur Radiol 1992; 2:70–72.

9

Magnetic Resonance Spectroscopy in the Investigation and Understanding of Muscle Disease

Phillip R. J. Barnes
Radcliffe Infirmary and John Radcliffe Hospital, Oxford, England
Doris J. Taylor
University of Oxford and John Radcliffe Hospital, Oxford, England
David Hilton-Jones
Radcliffe Infirmary, Oxford, England

INTRODUCTION

It has long been an aim of physicians, biomedical researchers, and patients alike to develop noninvasive, safe, and painless techniques for investigating biochemical and physiological processes and their disturbance in disease. The successful application of nuclear magnetic resonance (NMR) to living tissues in the last 20 years or so provides three examples where this aim has been achieved. Magnetic resonance imaging (MRI) can provide structural information of quite startling quality without the use of ionizing radiation and has revolutionized radiological practice. In addition, the development of functional imaging techniques based on magnetic resonance (1) promises to provide information superior in both temporal and spatial resolution to that from positron emission tomography (PET) and without the use of isotopes. Finally, magnetic resonance spectroscopy (MRS) can provide information about the relative chemical composition of tissues without recourse to invasive sampling as well as enabling the determination of intracellular pH. This allows repeated studies to follow progression of diseases and sequential studies to follow changes due to acute stresses such as exercise or ischemia.

In this chapter we introduce the reader to some of the basic principles of MRS and demonstrate its power as a noninvasive method of assessing biochemical processes in vivo.

WHAT IS MAGNETIC RESONANCE SPECTROSCOPY?

Physical Basis

Magnetic resonance signals are generated by the interaction of extremely strong magnetic fields and radiofrequency energy pulses. The high-field magnet aligns those atomic nuclei that possess the quantum mechanical property of spin and thus behave like minute magnets. The strength of the magnetic field will influence this relationship according to the equation $\omega = \gamma B_0$ where ω is the angular velocity, γ is a constant for each nucleus called the magnetogyric ratio, and B_0 is the strength of the magnetic field. If radiofrequency energy is applied at the resonant frequency, then the nuclei will absorb the energy, and then, when the pulse is terminated, the energy will be released again and can be detected by a receiver coil. The precise magnetic field experienced by an individual nucleus, and therefore the precise frequency at which that nucleus will absorb and release energy, is dependent on its local chemical environment. For example, the frequencies at which the phosphorus nuclei in ATP and phosphocreatine absorb energy are slightly different. This allows a frequency spectrum to be produced for a given sample containing a particular NMR visible nucleus. The relative area under each peak is directly proportional to the amount of that compound or group of compounds in a sample.

MRS Visible Nuclei

Nuclei that may be of use in the study of biological phenomena include ^{1}H, ^{13}C, ^{15}N, ^{17}O, ^{19}F, ^{23}Na, ^{31}P, and ^{87}Rb. In practice, however, only ^{1}H, ^{13}C, ^{19}F, and ^{31}P have been used widely in human studies in vivo, and nearly all work on human muscle has used ^{31}P.

Relationship of MRS and MRI

MRS and MRI both use high-field magnets and sophisticated computational methods, but the precise techniques used are different (2). Imaging is often possible on systems used primarily for spectroscopy, and conversely, some imaging units will have purchased the hardware and additional software for spectroscopy. In general, however, MRS is not available in most imaging units and the expertise in data analysis is clearly quite different from that required for the interpretation of MRI.

SPECTROSCOPIC INFORMATION FROM NORMAL SKELETAL MUSCLE

Phosphorus MRS

Resting Muscle

^{31}P NMR has been the most widely used technique in the study of muscle. A spectrum from the resting gastrocnemius of a healthy volunteer is shown in Figure 1. Seven peaks are visible and are identified

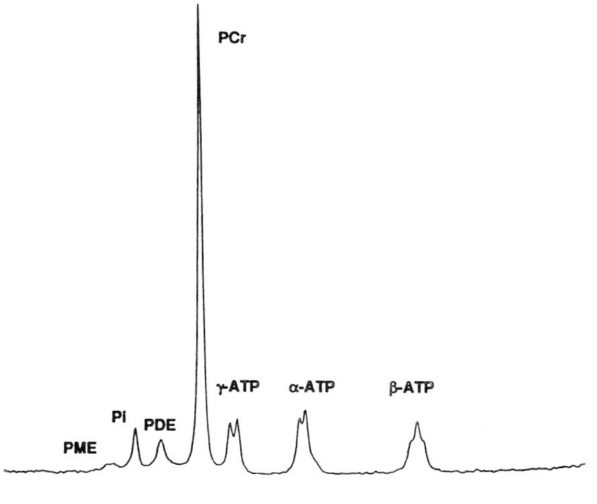

Figure 1 ^{31}P NMR spectrum from the gastrocnemius of a healthy volunteer. The major peaks are (from left to right) pooled phosphomonoesters (PME), inorganic phosphate (Pi), pooled phosphodiesters (PDE), phosphocreatine (PCr), and three peaks corresponding to the γ, α, and β phosphate groups of the ATP molecule. Spectrum acquired using a 5-cm surface coil in a 2.0-T superconducting magnet (average of 256 scans).

in the figure legend. The dominant peak is that due to phosphocreatine (PCr) while the three peaks to the right of this correspond to the γ, α, and β phosphate groups, respectively, of the ATP molecule. Note that although the areas under the peaks of the γ and β resonances are approximately equal, that of the α peak is greater. This is due to additional signal at the same resonant frequency arising from the phosphate in $NAD(P)^+$ and NAD(P)H. To the left of PCr the main peak is due to inorganic phosphate (P_i), and there are two smaller peaks corresponding to poorly defined pools of phosphomonoesters [including glycolytic intermediates such as fructose-6-phosphate and nucleotide monophosphates such as inosine monophosphate (IMP)] and phosphodiesters (including glycerol-3-phosphocholine and glycerol-3-phosphoethanolamine).

It is possible to measure intracellular pH directly from the ^{31}P NMR spectrum. In the physiological pH range the resonant frequency of PCr remains constant but that of P_i depends on the relative proportions of HPO_4^{2-} and $H_2PO_4^-$. That, in turn, depends on pH (the pK_a for the equilibrium reaction between the two is 6.72). Although the two phosphate species have a significant chemical shift (frequency) difference, they are in very rapid exchange, so a single P_i peak is observed corresponding to the weighted mean of the two. The pH may then be calculated from the chemical shift of the Pi peak relative to that of PCr.

One important phosphorus-containing compound that is notably absent from the spectrum in Figure 1 is adenosine diphosphate (ADP). Although ADP is present at a total concentration of 1–2 mM, about 90% of the total ADP is bound to myofibrils. Binding restricts molecular motion, which means that this fraction is not NMR-visible. It is possible, however, to estimate the level of free ADP (which is the important fraction of ADP in regulating metabolic processes) from the creatine kinase equilibrium expression

$$[ADP] = \{[ATP][\text{total creatine}]\}/\{[PCr][H^+][K_{eq}]\} \quad (1)$$

where K_{eq} is the equilibrium constant for the reaction. This also allows the calculation of the phosphorylation potential (PP), which is a measure of the free energy available from ATP hydrolysis and thus of the energetic state of the cell (3). Phosphorylation potential is given by the expression PP = $[ATP]/([ADP][P_i])$.

Exercise and Recovery

Although interesting information may be derived from resting muscle, far more is available if the muscle is studied during exercise and in the recovery period following exercise. Several approaches have been used but two have provided most of the information on disease states. The first is to exercise at a number of steady states (where workload is constant and levels of PCr and P_i remain steady) and to relate the levels of high-energy phosphate to work rate in a "work-energy cost transfer function" (4). The resulting data theoretically give a measure of the effectiveness of oxidative phosphorylation during exercise. One disadvantage of this approach is that it is often difficult to obtain the required steady state in patients with muscle disease. The other method is to perform some form of incremental exercise, where the workload is gradually increased and changes in high-energy phosphates are followed throughout. This approach has been shown to produce equivalent changes in PCr and P_i, for a given workload, to steady-state exercise (5) and has the advantage of being quicker and more straightforward to perform. No matter which protocol is used, PCr falls during exercise and P_i rises. In the incremental techniques pH usually falls due to proton production from glycogenolysis (Fig. 2a–c), and in normal muscle there is a characteristic relationship between changes in pH and PCr utilization, as shown in Figure 2d. In the steady-state technique, protocols are designed to prevent any fall in pH since only in the absence of a change in $[H^+]$ do the changes in Pi and PCr represent an index of mitochondrial function (because [ADP] is affected by the change in pH via the creatine kinase equilibrium—see Eq. 1).

During recovery from exercise, glycogenolysis stops so the rate of PCr resynthesis provides a measure of the rate of mitochondrial oxidative phosphorylation, because PCr resynthesis directly follows net ATP production as a result of the creatine kinase equilibrium (Eq. 1). In skeletal muscle the rate of PCr resynthesis in vivo appears to be tightly controlled by the concentration of free ADP (6)

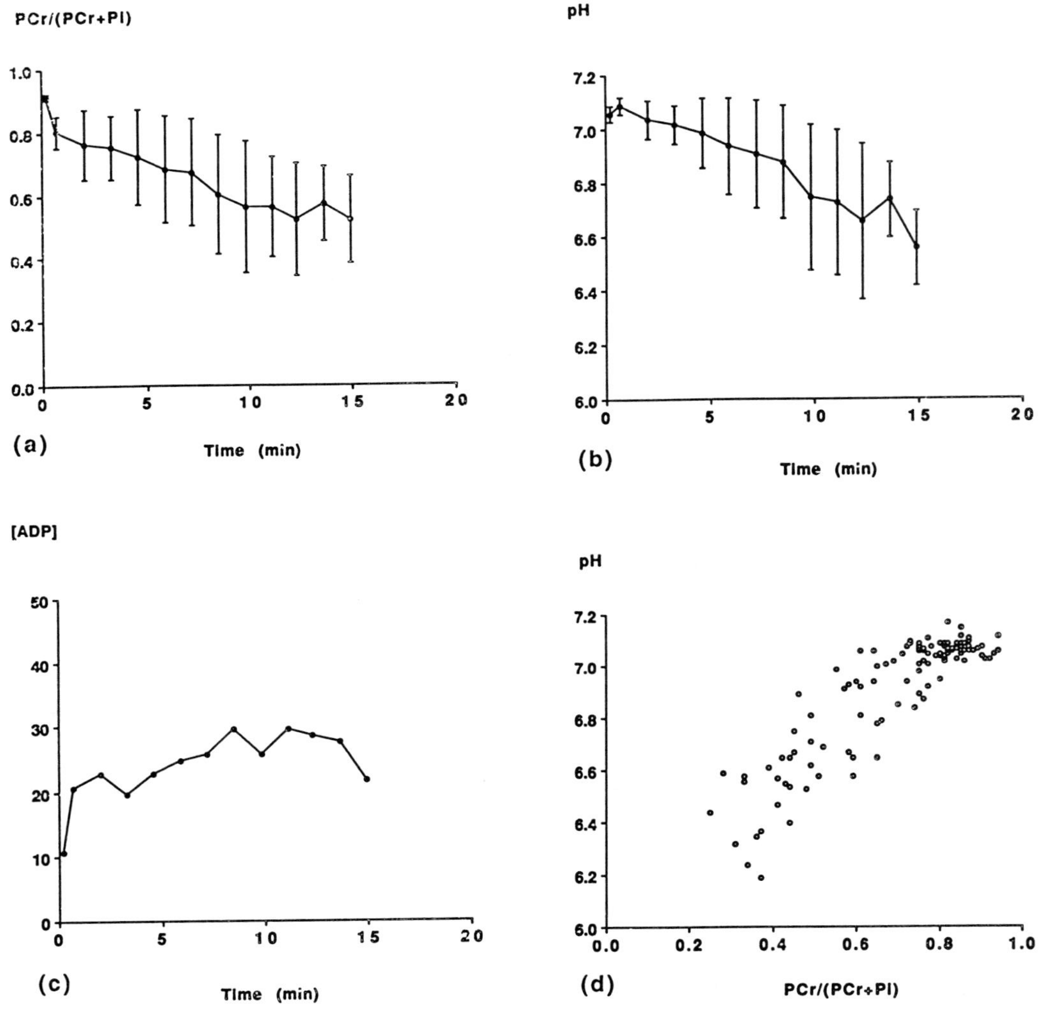

Figure 2 Changes in (a) PCr [expressed as PCr/(PCr+Pi)], (b) pH_i, and (c) calculated ADP concentration (in μM) in the *flexor digitorum superficialis* muscle of healthy volunteers during aerobic exercise, and the relationship of PCr/(PCr+Pi) and pH_i (d) during this protocol. Exercise consists of pulling a handle attached to a weight, which is increased in small increments throughout exercise ("incremental exercise"; see text). Error bars represent the standard deviations of the means and have been omitted from calculated [ADP] for clarity.

in the same way as ATP synthesis in vitro (7). It is possible to derive a figure for the apparent V_{max} for oxidative phosphorylation and this is a sensitive measure of primary or secondary mitochondrial dysfunction (8). Recovery kinetics of P_i and pH also may be calculated. Finally, more complex calculations have recently been developed to estimate initial rates of glycolysis during exercise and rates of proton extrusion from the cell, all using basic data from exercise studies (9), and these are beginning to be useful in the study of disease.

MRS Using Other Nuclei

Relatively little use has been made of ^1H in the evaluation of muscle disease largely because quantitative studies are technically difficult. The only other nucleus of clear use in human muscle work is the carbon isotope ^{13}C. This is only a minor isotope occurring with a natural abundance of 1.1%. Added to its natural NMR properties, this means that its detection is only 0.24% as sensitive as that of ^{31}P. Under some circumstances exogenous ^{13}C-labeled compounds (such as ^{13}C-glucose) (10) have been administered and their metabolism followed, but this is less acceptable for routine patient studies and is expensive. It is also possible, however, to acquire "natural abundance" ^{13}C spectra by using longer collection periods and ^{13}C-glycogen is detectable in this way (11). The time required for adequate signal-to-noise resolution means that dynamic studies (such as during exercise) are not feasible at present.

All of the studies reviewed in this chapter relate to ^{31}P MRS unless specifically stated to the contrary.

MRS STUDIES IN HUMAN MUSCLE DISEASE

Primary Muscle Diseases

Mitochondrial Diseases

The major strength of ^{31}P MRS is the understanding of bioenergetics, so the study of mitochondrial diseases is clearly of interest.

^{31}P spectra from resting muscle in patients with mitochondrial myopathies are often abnormal (12–18). There are decreases in the ratios of PCr/Pi and PCr/ATP (12–18), which probably reflect loss of PCr, since [ATP] has been normal in those cases in which it has been assayed (15,17). Similarly high Pi/ATP ratios are often seen and suggest raised Pi (12,15,17,18). In addition, resting intracellular pH is sometimes higher than normal, despite the high serum lactate seen in many patients (15,17). The overall energetic status of the resting cell may be derived from the phosphorylation potential, which is lower in most patients because both [Pi] and [ADP] are raised (15).

During exercise the poor energetic state of the cell is again clear. There is rapid depletion of PCr and accumulation of Pi[15], while in studies where steady-state work rates are used, the slope of the plot of work rate against Pi/PCr (the "work-energy cost transfer function"; see above) is low, which may itself be interpreted as evidence of mitochondrial dysfunction (14,16). The fall in pH_i during exercise is much less than would be expected for the PCr loss, however, despite the systemic lactic acidosis that occurs (15,16), and it seems that the muscle has increased its ability to handle a proton load (9,15). This may be a sensible adaptation for two reasons. First, it allows the clearance of the excess protons produced by high glycolytic flux in the face of reduced mitochondrial pyruvate or fatty acid utilization. Second, consideration of the creatine kinase equilibrium shows that high pH_i and low [PCr] together will cause higher levels of [ADP]. Since the main regulator of mitochondrial ATP synthesis in skeletal muscle appears to be the free [ADP] (6,7), mechanisms that produce higher-than-normal levels of [ADP] would be an appropriate adaptation.

Studies of recovery after exercise confirm that pH_i recovery is rapid (9,15), suggesting increased proton-extruding capacity, while recovery of levels of [PCr], [Pi], and [ADP] to the resting state is slow. It is possible to analyze recovery data to produce an apparent mitochondrial V_{max} for ATP synthesis (8) and this is also, not surprisingly, low in mitochondrial diseases (G. J. Kemp and D. J. Taylor, unpublished observations).

A few groups have looked at the effect of potential treatments for mitochondrial myopathies upon energetics in individual cases. Although these are mainly anecdotal, improvements in bioenergetics have been reported following treatment of complex III deficiency with vitamins K_3 and C (14), cytochrome c oxidase deficiency with ubiquinone (coenzyme Q) (19,20), and MELAS syndrome with nicotinamide and riboflavin (21) or with ubiquinone and idebenone (22). A more recent study of coenzyme Q_{10} therapy in 16 patients has shown no evidence of any bioenergetic or clinical improvement, however (23).

Several other conditions have been postulated or proven to be due to mitochondrial dysfunction and some of these have been studied by ^{31}P MRS. *Leber's hereditary optic neuropathy* is usually associated with one of two mitochondrial DNA (mtDNA) point mutations. A study of three siblings, with a mutation at base pair 11778, showed high pH_i at rest in all patients but otherwise resting muscle was normal, but the behavior of PCr/Pi during exercise and recovery was consistent with mitochondrial dysfunction in muscle (24). Unfortunately, the pH response during exercise and recovery was not given and PCr and Pi kinetics were not shown individually. In *Parkinson's disease* (PD) there is now good evidence that there is a specific deficiency of NADH CoQ_1 reductase (complex I) in the substantia nigra of affected patients (25), but conflicting evidence regarding respiratory chain function in skeletal muscle (26–28). A study of patients with PD using ^{31}P MRS has shown no evidence of any significant reduction in mitochondrial function in skeletal muscle in vivo (D. J. Taylor, P. R. J. Barnes, and G. J. Kemp, unpublished observations). In the acquired myopathy associated with *zidovudine* (AZT) therapy for human immunodeficiency virus (HIV) infection, ^{31}P MRS data support the hypothesis that it is due to some form of mitochondrial dysfunction, presumably because of interference with mtDNA synthesis (29).

Glycogen Storage Diseases of Muscle

Short-term exercise relies heavily on utilization of muscle glycogen stores, and genetic defects affecting the breakdown of glycogen cause significant muscle symptoms and produce marked changes in skeletal muscle bioenergetics as studied by ^{31}P MRS.

Myophosphorylase Deficiency (McArdle's Disease) and Debrancher Enzyme Deficiency. In these two disorders glycogen breakdown is impaired but the glycolytic breakdown of glucose to pyruvate is normal.

In resting muscle pH_i and the ratios of high-energy phosphates are usually normal (30–32). The response to exercise is characteristic, however. There is rapid breakdown of PCr and a rise in Pi, suggesting defective energy production by the cell (30,31). The pH_i does not fall, but in fact will usually rise during exercise (30,31). This is presumably because although lactic acid production is low due to failure of glycogenolysis, PCr hydrolysis continues and is a proton-*consuming* process. Recovery following exercise is normal in these patients, indicating that mitochondrial function is unimpaired (30–32).

Besides providing bioenergetic correlates of some of the clinical features of these diseases, such as rapid fatigue and subnormal lactate responses to exercise, they also shed some light on other features. One of these is the "second-wind" phenomenon, in which if the level of exercise is reduced by the patient at the point where fatigue begins to become noticeable, then he or she often will be able to continue exercising and even increase the workload. Ross and Radda (33) noted an increase in the level of PCr and fall in pH_i in one patient who experienced this phenomenon and suggested that mobilization of alternative fuels was the explanation. Non-MRS studies have shown that the second-wind may be prevented by inhibitors of fatty acid mobilization (34). Others have failed to find bioenergetic correlates of this phenomenon in patients who have experienced it clinically, however (31). Also the electrically silent contractures seen in McArdle's disease had been thought to be due to ATP depletion causing a rigor-like state in the muscle, but several ^{31}P MRS studies have shown that ATP depletion does not occur even during contractures (29,32,35), thus confirming biochemical studies (36).

The glycogenoses have been the one group of muscle diseases where ^{13}C MRS has proved useful to date. ^{13}C-glycogen storage has been quantitated in normal muscle (11) and excess storage in McArdle's patients has been clearly shown (37). This finding could open the way to noninvasive diagnosis of abnormal glycogen storage (but not of specific types as yet).

Glycolytic Disorders. These disorders differ slightly from the glycogenoses described above. Although both groups are associated with subnormal acid production and rapid PCr depletion, the glycolytic disorders additionally affect the metabolism of exogenous glucose. They result in the accumulation of sugar phosphate intermediates proximal to the site of the block and these appear as phosphomonoesters (PMEs) in the ^{31}P spectrum. They have been described in phosphofructokinase (PFK) deficiency (31,38,39), phosphoglycerate kinase (PGK) deficiency (31), and a case of partial phosphoglycerate mutase (PGAM) deficiency (40).

Again in contrast to the glycogenoses, recovery of PCr levels following exercise is abnormal in each of the disorders described (31,38–40). It is thought that this is not simply due to low supplies of pyruvate arising from low flux though the glycolytic pathway becoming the rate-limiting factor for flux through the tricarboxylic acid cycle, but rather that some of the phosphate required for ATP synthesis (and thus PCr resynthesis) is "trapped" within the PME pool, which is only slowly metabolized. Support for this point of view comes from observations that in these disorders the sum of PCr, Pi, and PME remains constant throughout exercise (suggesting there is net transfer of some of the phosphate groups from PCr to sugar phosphates via ATP) (39).

Treatments. Various therapists have been suggested for McArdle's disease, and the effects of some of these assessed by ^{31}P MRS. Intravenous glucose infusion resulted in reduced PCr depletion and normal acidification in a few studies (32,35,41), in contrast to a lack of effect in PFK deficiency (39). A similar improvement was, surprisingly, seen in the case of partial PGAM deficiency described above (40), but this is hard to interpret since it was only a partial defect. A more practical therapy that has been suggested for McArdle's disease is a high-protein diet, and a single report has shown both clinical and bioenergetic improvement after 6 weeks of such a diet, but not with acute infusions of amino acids (41).

The Muscular Dystrophies

Xp21 Dystrophies. The possibility that abnormal muscle bioenergetics might contribute to the weakness and wasting seen in several muscular dystrophies has led to their examination by ^{31}P MRS.

In early studies of *Duchenne muscular dystrophy* (DMD) a reduction in the ratio PCr/ATP and an increase in Pi/ATP were noted (38,42), with an abnormally large peak in the phosphodiester region. The pH_i was also found to be abnormally alkaline. In a study on gastrocnemius, it was found that the changes in high-energy phosphates became more severe with age, in parallel with worsening of the dystrophy (43). This is consistent with the findings that patients with *Becker muscular dystrophy* (BMD) tend to have changes similar to but less marked than those seen in DMD (8,44,45).

With the exception of a single patient described by Griffiths and colleagues (46), there have been no published studies of exercise and recovery in DMD, largely because the patients are too weak to exercise by the time they are old enough to be sufficiently cooperative. Such studies have been performed upon carriers of Xp21 dystrophies and patients with BMD (8,45). As mentioned above, BMD patients had similar resting abnormalities to those in DMD; carriers, however, had normal levels of high-energy phosphates and a normal phosphorylation potential, but pH_i was slightly but significantly raised (8). During exercise BMD patients depleted PCr more rapidly than controls but pH_i fell less (45). Similar results were found in carriers, whether clinically affected or not (8,45). The available evidence suggests that glycolytic proton production is normal (8) and that improved proton extrusion is most likely to account for the subnormal pH_i drop (9). Although it has been suggested, based on the relationship between Pi/PCr and work rate in exercise, that mitochondrial function may be impaired in BMD and Xp21 carriers (45), analysis of recovery data does not support this and apparent mitochondrial V_{max} is found to be normal in carriers at least (8).

A single study of the effect of attempted treatment of DMD with allopurinol or ribose showed neither function improvement nor any change in the ^{31}P MRS findings (46).

Other Dystrophies. In a study of a family with an autosomal dominant *oculopharyngeal dystrophy*, resting forearm flexor muscle showed low values of PCr/(PCr+Pi) and high pH_i (47). During exercise there was rapid PCr depletion and a marked fall in pH_i. Following exercise PCr recovered normally but pH_i was slow to recover. These studies are useful in showing that there is widespread involvement in what is clinically a very focal dystrophy, and also in showing that not all dystrophies show a pattern of reduced acidosis during exercise.

In *myotonic muscular dystrophy* (MyD) resting ratios of Pi/ATP, PME/ATP, and PDE/ATP were all elevated in forearm flexor muscle but pH_i was normal (48). During exercise there was rapid depletion of energy stores but the change in pH_i was small. Findings during exercise are similar to those in mitochondrial myopathies, but recovery of PCr after exercise showed no evidence of mitochondrial dysfunction. Interestingly, preliminary studies of the less clinically affected gastrocnemius muscle in MyD showed raised Pi/ATP, PDE/ATP, and pH_i, while in exercise energy stores were rapidly depleted but with a normal pH_i change (49). The reasons for these differences between the two muscle groups are not yet clear.

Finally ^{31}P MRS has been used to demonstrate that the cold-induced paralysis of *paramyotonia congenita* is not associated with any bioenergetic change in muscle from its resting state, which was itself no different from that in controls (50).

Miscellaneous Muscle Problems

Chronic Fatigue Syndrome. An early study of a single patient with a chronic fatigue syndrome following severe varicella infection showed extremely marked acidification during exercise, occurring early in exercise and being disproportionate to the PCr depletion (51). A later study showed the phenomenon of early acidification in five of six cases, but this was disproportionate to PCr depletion in only two (including the previously reported single case) (52). Nevertheless, abnormal acidification was not seen in a group of seven patients with chronic fatigue studied by another group (53), and in a detailed analysis of 46 patients seen in the Oxford laboratory we have found no consistent abnormality of bioenergetics or pH_i control (54). Thus, although the clinical syndrome of chronic fatigue may include a minority of patients with abnormal bioenergetics (51,52,54), there is no diagnostic abnormality (54).

Malignant Hyperpyrexia. This syndrome comprises a genetic predisposition to muscle hypermetabolism and rhabdomyolysis triggered by certain anesthetic agents. Recent genetic linkage work suggests that the genetic defect lies in the gene for the ryanodine receptor, and excessive release of Ca^{2+} from the sarcoplasmic reticulum is thought to underlie the clinical features (55,56). Magnetic resonance studies of patients with malignant hyperpyrexia show excessive acidification during exercise and some continuing acidification followed by a slow recovery of pH_i postexercise (57,58). Changes in PCr were not much different from controls and postexercise recovery was normal. In a more recent study, the resting ratio of Pi/PCr was slightly reduced, and there was slow recovery of PCr/Pi following exercise (59). The slow PCr/Pi recovery does not necessarily imply impaired mitochondrial function, however, since PCr recovery is expected to be slow when the pH is lower than normal, partly because the concentration of ADP also will be low (see above). These results are consistent with excessive activation of glycogenolysis and glycolysis by raised Ca_i^{2+} and are rather similar to those found in a case of subtotal Ca^{2+} ATPase deficiency (60).

Systemic Diseases and Muscle

Many diseases have secondary clinical effects on skeletal muscle and several of these have been investigated using ^{31}P MRS. The principal conclusions of these studies are summarized in Table 1. One such disease of particular clinical importance to the myologist is hypothyroid myopathy, since the myopathy may be the only clinical manifestation of the disease and we will briefly enlarge on the bioenergetics of this.

Hypothyroidism

Resting skeletal muscle from patients with hypothyroidism showed raised Pi/ATP and PCr/ATP in the ^{31}P spectrum and reduced PCr/Pi (61–63). This is in keeping with the in vitro finding of a 25% reduction in the level of ATP in these patients (64). Resting pH_i was significantly reduced (62). During exercise PCr depleted more rapidly than in controls but following exercise it was resynthesized normally although Pi disappearance was slow (62). The pH_i response to exercise is unique: during early exercise pH_i increased more than in controls (suggesting reduced acid production), but later in exercise the pH_i fell more than in the controls (62). This is consistent with a delay in the activation of glycogenolysis (62). A suggestion of reversible mitochondrial dysfunction due to hypothyroidism was made by Argov and his colleagues on the basis of slow PCr/Pi recovery following exercise (61). This observation probably reflects slow Pi recovery with normal PCr recovery, however, making this possibility unlikely (62). All of the changes observed were reversed by thyroxine treatment (61,62).

Table 1 Bioenergetic Effects of Some Common Systemic Diseases on Skeletal Muscle as Studied Using ^{31}P Magnetic Resonance Spectroscopy

Condition	Resting muscle	Exercise	Recovery	Conclusion	Ref.
Hypothyroidism	PCr/ATP↑, Pi/ATP↑, PCr/Pi↓ pH_i↓	pH_i↑ then ↓, PCr↓	pH_i slow Pi slow PCr normal (PCr/Pi slow)	Delayed activation of glycogenolysis ?Reduced proton-extruding capacity	61–63
Hyperthyroidism	Normal	PCr↓, pH_i↓[65] Normal[61]	Normal[65] PCr fast[61]	↑ATP use in exercise ? Excessive reliance upon glycolysis	61,65
Uremia	Pi/ATP↑	Reduced exercise duration; pH_i and PCr both ↓ Improved by erythropoetin	PCr slow (not improved by erythropoetin)	Impaired oxidative metabolism ? due to uremia per se	66,67
Respiratory failure	Normal	PCr/Pi↓, pH_i↓	PCr slow	Impaired oxidative metabolism due to hypoxia	68
Heart failure	Pi/ATP↑	PCr↓, pH_i↓ Blood flow normal	PCr slow pH_i slow	?Peripheral factor causing abnormal metabolic control and impaired oxidative metabolism	69–71
Peripheral vascular disease	pH_i↑ in severe disease	PCr↓, pH_i↓ ADPi	PCr slow pH_i slow	Imparied oxidative metabolism due to low O_2 supply Poor washout of H^+	72

CONCLUSIONS

MRS is a valuable tool for studying primary and secondary muscle diseases, and results such as those discussed in this chapter have added significantly to the understanding of bioenergetics and the control of pH in various disease states. It thus remains largely a powerful research tool, but it may also be a useful adjunct to standard clinical investigations such as open biopsy and exercise testing in certain circumstances since characteristic patterns of responses to exercise are seen in some conditions such as McArdle's disease. MRS could also prove useful in the monitoring of the response to new therapies for muscle diseases including those for diseases not thought of as primarily metabolic myopathies, but where there are reproducible MRS abnormalities.

REFERENCES

1. Kwong KK, Belliveau JW, Chesler DA, et al. Dynamic magnetic resonance imaging of human brain activity during primary sensory stimulation. Proc Natl Acad Sci USA 1992; 89:5675–5679.
2. Andrew ER, Bydder G, Griffiths J, Iles R, Styles P. Clinical Magnetic Resonance Imaging and Spectroscopy. Chichester, England: Wiley, 1990.
3. Veech RL, Lawson JWR, Cornell NW, Krebs HA. Cytosolic phosphorylation potential. J Biol Chem 1979; 254:6538–6547.
4. Chance B, Leigh JS Jr, Clark BJ, et al. Control of oxidative metabolism and oxygen delivery in human skeletal muscle: a steady state analysis of the work/energy cost transfer function. Proc Natl Acad Sci USA 1985; 82:8384–8388.
5. Moonen CTW, Radda GK, Yonge RP. Gradually increasing dynamic work mimics prolonged steady state exercise in man—a ^{31}P-magnetic resonance spectroscopy (m.r.s.) study. J Physiol (Lond) 1987; 382:168P.
6. Kemp GJ, Taylor DJ, Radda GK. Control of phosphocreatine resynthesis during recovery from exercise in human skeletal muscle. NMR Biomed 1993; 6:66–72.
7. Chance B, Williams GR. Respiratory enzymes in oxidative phosphorylation. III. The steady state. J Biol Chem 1955; 217:409–427.
8. Kemp GJ, Taylor DJ, Dunn JF, Frostick SP, Radda GK. Cellular energetics of dystrophic muscle. J Neurol Sci 1993 (in press).
9. Kemp GJ, Taylor DJ, Styles P, Radda GK. The production, buffering and efflux of protons in human skeletal muscle during exercise and recovery. NMR Biomed 1993; 6:73–83.
10. Shulman GI, Rothman DL, Jue T, Stein P, DeFronzo RA, Shulman RG. Quantitation of muscle glycogen synthesis in normal subjects and subjects with non-insulin-dependent diabetes by ^{13}C nuclear magnetic resonance spectroscopy. N Engl J Med 1990; 322:223–228.
11. Avison MJ, Rothman DL, Nadel E, Shulman RG. Detection of human muscle glycogen by natural abundance ^{13}C NMR. Proc Natl Acad Sci USA 1988; 85:1634–1636.
12. Gadian DG, Radda GK, Ross BD, et al. Examination of a myopathy by phosphorus nuclear magnetic resonance. Lancet 1981; 2:774–775.
13. Radda GK, Bore PJ, Gadian DG, et al. ^{31}P NMR examination of two patients with NADH-CoQ reductase deficiency. Nature 1982; 295:608–609.
14. Eleff S, Kennaway NG, Buist NRM, et al. ^{31}P-NMR study of improvement in oxidative phosphorylation by vitamins K_3 and C in a patient with a defect in electron transport at complex III in skeletal muscle. Proc Natl Acad Sci USA 1984; 81:3529–3533.
15. Arnold DL, Taylor DJ, Radda GK. Investigation of human mitochondrial myopathies by phosphorus magnetic resonance spectroscopy. Ann Neurol 1985; 18:189–196.
16. Argov Z, Bank WJ, Maris J, Peterson P, Chance B. Bioenergetic heterogeneity of human mitochondrial myopathies; phosphorus magnetic resonance spectroscopy study. Neurology 1987; 37:257–262.
17. Matthews PM, Allaire C, Shoubridge EA, Karpati G, Carpenter S, Arnold DL. In vivo muscle magnetic resonance spectroscopy in the clinical investigation of mitochondrial disease. Neurology 1991; 41:114–120.
18. Matthews PM, Berkovic SF, Shoubridge EA, et al. In vivo magnetic resonance spectroscopy of brain and muscle in a type of mitochondrial encephalomyopathy (MERRF). Ann Neurol 1991; 29:435–438.
19. Nishikawa Y, Takahashi M, Yorifuji S, et al. Long-term coenzyme Q_{10} therapy for a mitochondrial encephalomyopathy with cytochrome c oxidase deficiency: a ^{31}P NMR study. Neurology 1989; 39:399–403.
20. Bendahan D, Desnuelle C, Vanuxem D, et al. ^{31}P NMR spectroscopy and ergometer exercise test as evidence for muscle oxidative performance improvement with coenzyme Q in mitochondrial myopathies. Neurology 1992; 42:1203–1208.
21. Penn AMW, Lee JWK, Thullier P, et al. MELAS syndrome with mitochondrial $tRNA^{Leu(UUR)}$ mutation:

correlation of clinical state, nerve conduction, and muscle ^{31}P magnetic resonance spectroscopy during treatment with nicotinamide and riboflavin. Neurology 1992; 42:2147–2152.
22. Ihara Y, Namba R, Kuroda S, Sato T, Shirabe T. Mitochondrial encephalomyopathy (MELAS): pathological study and successful therapy with coenzyme Q_{10} and idebenone. J Neurol Sci 1989; 90:263–271.
23. Matthews PM, Ford B, Dandurand RJ, et al. Coenzyme Q_{10} with multiple vitamins is generally ineffective in treatment of mitochondrial disease. Neurology 1993; 43:884–890.
24. Cortelli P, Montagna P, Avoni P, et al. Leber's hereditary optic neuropathy: genetic, biochemical and phosphorus magnetic resonance spectroscopy study in an Italian family. Neurology 1991; 41:1211–1215.
25. Schapira AHV, Mann VM, Cooper JM, et al. Anatomic and disease specificity of NADH CoQ_1 reductase (complex I) deficiency in Parkinson's disease. J Neurochem 1990; 55:2142–2145.
26. Bindoff LA, Birch-Machin M, Cartlidge NE, Parker WD Jr, Turnbull DM. Mitochondrial function in Parkinson's disease. Lancet 1989; 2:49.
27. Shoffner JM, Watts RL, Juncos JL, Torroni A, Wallace DC. Mitochondrial oxidative phosphorylation defects in Parkinson's disease. Ann Neurol 1991; 30:332–339.
28. Mann VM, Cooper JM, Krige D, Daniel SE, Schapira AHV, Marsden CD. Brain, skeletal muscle and platelet homogenate mitochondrial function in Parkinson's disease. Brain 1992; 115:333–342.
29. Weissman JD, Constantinitis I, Hudgins P, Wallace DC. ^{31}P magnetic resonance spectroscopy suggests impaired mitochondrial function in AZT-treated HIV-infected patients. Neurology 1992; 42:619–623.
30. Ross BD, Radda GK, Gadian DG, Rocker G, Esiri M, Falconer-Smith J. Examination of a case of suspected McArdle's syndrome by ^{31}P nuclear magnetic resonance. N Engl J Med 1981; 304:1338–1342.
31. Duboc D, Jehenson P, Tran Dinh S, Marsac C, Syrota A, Fardeau M. Phosphorus NMR spectroscopy study of muscular enzyme deficiencies involving glycogenolysis and glycolysis. Neurology 1987; 37:663–671.
32. Argov Z, Bank WJ, Maris J, Chance B. Muscle energy metabolism in McArdle's syndrome by in vivo magnetic resonance spectroscopy. Neurology 1987; 37:1720–1724.
33. Ross BD, Radda GK. Application of ^{31}P n.m.r. to inborn errors of muscle metabolism. Biochem Soc Trans 1983; 11:627–630.
34. Pernow BB, Havel RJ, Jennings DB. The second wind phenomenon in McArdle's syndrome. Acta Med Scand 1967; 472(Suppl):294–307.
35. Lewis SF, Haller RG, Cook JD, Nunnally RL. Muscle fatigue in McArdle's disease studied by ^{31}P-NMR: effect of glucose infusion. J Appl Physiol 1985; 59:1991–1994.
36. DiMauro S, Bresolin N. Phosphorylase deficiency. In: Engel AG, Banker BQ, eds. Myology. New York: McGraw-Hill, 1986:1585–1602.
37. Jehenson P, Duboc D, Bloch G, Fardeau M, Syrota A. Diagnosis of muscular glycogenosis by in vivo natural abundance ^{13}C NMR spectroscopy. Neuromusc Disord 1991; 1:99–101.
38. Edwards RHT, Wilkie DR, Dawson MJ, et al. Clinical use of magnetic resonance in the investigation of myopathy. Lancet 1982; 1:722–731.
39. Argov Z, Bank WJ, Maris J, Leigh JS Jr, Chance B. Muscle energy metabolism in human phosphofructokinase deficiency as recorded by ^{31}P nuclear magnetic resonance spectroscopy. Ann Neurol 1987; 22:46–51.
40. Argov Z, Bank WJ, Boden B, Ro Y-I, Chance B. Phosphorus magnetic resonance spectroscopy of partially blocked muscle glycolysis. An in vivo study of phosphoglycerate mutase deficiency. Arch Neurol 1987; 44:614–617.
41. Jensen KE, Jakobsen J, Thomsen C, Henriksen O. Improved energy kinetics following high protein diet in McArdle's syndrome. A ^{31}P magnetic resonance study. Acta Neurol Scand 1990; 81:499–503.
42. Newman RJ, Bore PJ, Chan L, et al. Nuclear magnetic resonance studies of forearm muscle in Duchenne dystrophy. Br Med J 1982; 284:1072–1074.
43. Younkin DP, Berman P, Sladky J, Chee C, Bank WJ, Chance B. ^{31}P NMR studies in Duchenne muscular dystrophy: age-related metabolic changes. Neurology 1987; 37:165–169.
44. Newman RJ. An in vivo study of muscle phosphate metabolism in Becker's dystrophy by ^{31}P NMR spectroscopy. Metabolism 1985; 34:737–740.
45. Barbiroli B, Funicello R, Iotti S, Montagna P, Ferlini A, Zaniol P. ^{31}P-NMR spectroscopy of skeletal muscle in Becker dystrophy and DMD/BMD carriers. Altered rate of phosphate transport. J Neurol Sci 1991; 109:188–195.
46. Griffiths RD, Cady EB, Edwards RHT, Wilkie DR. Muscle energy metabolism in Duchenne dystrophy studied by ^{31}P-NMR: controlled trials show no effect of allopurinol or ribose. Muscle Nerve 1985; 8:760–767.
47. Zochodne DW, Koopman WJ, Witt NJ, et al. Forearm P-31 nuclear magnetic resonance spectroscopy studies in oculopharyngeal muscular dystrophy. Can J Neurol Sci 1992; 19:174–179.
48. Taylor DJ, Kemp GJ, Woods CG, Edwards JH, Radda GK. Skeletal muscle bioenergetics in myotonic dystrophy. J Neurol Sci 1993 (in press).
49. Barnes PRJ, Kemp GJ, Taylor DJ, Radda GK. Bioenergetics of the gastrocnemius muscle in myotonic dystrophy. J Neurol Neurosurg Psychiatry 1993; 56:728.

50. Lehmann-Horn F, Hopfel D, Rudel R, Ricker K, Küther G. In vivo P-NMR spectroscopy: muscle energy exchange in paramyotonia patients. Muscle Nerve 1985; 8:606–610.
51. Arnold DL, Bore PJ, Radda GK, Styles P, Taylor DJ. Excessive intracellular acidosis in skeletal muscle on exercise in a patient with post-viral exhaustion/fatigue syndrome, a ^{31}P nuclear magnetic resonance study. Lancet 1984; 1:1367–1369.
52. Arnold DL, Bore PJ, Radda GK, Styles P, Taylor DJ. Enhanced intramuscular acidosis during exercise by patients with postviral exhaustion/fatigue syndrome. J Neurol Sci 1985; 35(Suppl 1):165.
53. Kent-Braun J, Sharma KR, Weiner MW, Massie B, Miller RG. Central basis of muscle fatigue in chronic fatigue syndrome. Neurology 1993; 43:125–131.
54. Barnes PRJ, Taylor DJ, Kemp GJ, Radda GK. Skeletal muscle bioenergetics in the chronic fatigue syndrome. J Neurol Neurosurg Psychiatry 1993; 56:679–683.
55. MacLennan DH, Duff C, Zorzato F, et al. Ryanodine receptor gene is a candidate for predisposition to malignant hyperthermia. Nature 1990; 343:559–561.
56. Fujii J, Otsu K, Zorzato F, et al. Identification of a mutation in porcine ryanodine receptor associated with malignant hyperthermia. Science 1991; 253:448–451.
57. Kozak-Reiss G, Gascard JP, Redouane-Bénichou K. Dépistage de l'hyperthermie maligne anesthétique par les tests de contracture musculaire et par la spectroscopie RMN. Ann Fr Anest Reanim 1986; 5:584–589.
58. Webster DW, Thompson RT, Gravelle DR, Laschuk MJ, Driedger AA. Metabolic response to exercise in malignant hyperthermia-sensitive patients measured by ^{31}P magnetic resonance spectroscopy. Magn Reson Med 1990; 15:81–89.
59. Olgin J, Rosenberg H, Allen G, Seestedt R, Chance B. A blinded comparison of noninvasive, in vivo phosphorus nuclear magnetic resonance spectroscopy and the in vitro halothane/caffeine contracture test in the evaluation of malignant hyperthermia susceptibility. Anesth Analg 1991; 72:36–47.
60. Taylor DJ, Brosnan MJ, Arnold DL, et al. Ca^{2+}-ATPase deficiency in a patient with an exertional muscle pain syndrome. J Neurol Neurosurg Psychiatry 1988; 51:1425–1433.
61. Argov Z, Renshaw PF, Boden B, Winokur A, Bank WJ. Effects of thyroid hormones on skeletal muscle bioenergetics. In vivo phosphorus-31 magnetic resonance spectroscopy study of humans and rats. J Clin Invest 1988; 81:1695–1701.
62. Taylor DJ, Rajagopalan B, Radda GK. Cellular energetics in hypothyroid muscle. Eur J Clin Invest 1992; 22:358–365.
63. Kaminsky P, Robin-Lherbier B, Brunotte F, et al. Energetic metabolism in hypothyroid skeletal muscle, as studied by phosphorus magnetic resonance spectroscopy. J Clin Endocrinol Metab 1992; 74:124–129.
64. Wiles CM, Young A, Jones DA, Edwards RHT. Muscle relaxation rate, fibre-type composition and energy turnover in hyper- and hypo-thyroid patients. Clin Sci 1979; 57:375–384.
65. Kaminsky P, Robin-Lherbier B, Walker P, et al. Muscle bioenergetic impairment in hyperthyroid man: a study by ^{31}P NMR spectroscopy. Acta Endocrinol (Copenh) 1991; 124:271–277.
66. Thompson CH, Kemp GJ, Taylor DJ, Ledingham JGG, Radda GK, Rajagopalan B. Effect of chronic uraemia on skeletal muscle metabolism in man. Nephrol Dial Transplant 1993; 8:218–222.
67. Nishida A, Kubo K, Nihel H. Impaired muscle energy metabolism in uremia as monitored by ^{31}P-NMR. Nippon Jinzo Gakkai Shi 1991; 33:65–73.
68. Payen JF, Wuyam B, Reutenauer H, et al. Impairment of muscular metabolism in chronic respiratory failure. A human ^{31}P-MRS study. NMR Biomed 1991; 4:41–45.
69. Wilson JR, Fink L, Maris J, et al. Evaluation of energy metabolism in skeletal muscle of patients with heart failure with gated phosphorus-31 nuclear magnetic resonance. Circulation 1985; 71:57–62.
70. Massie BM, Conway M, Yonge R, et al. ^{31}P Nuclear magnetic resonance evidence of abnormal skeletal muscle metabolism in patients with congestive heart failure. Am J Cardiol 1987; 60:309–315.
71. Massie BM, Conway M, Rajagopalan B, et al. Skeletal muscle metabolism in heart failure: abnormalities during ischaemic exercise. Clin Res 1987; 35:303A.
72. Hands LJ, Bore PJ, Galloway G, Morris PJ, Radda GK. Muscle metabolism in patients with peripheral vascular disease investigated by ^{31}P nuclear magnetic resonance spectroscopy. Clin Sci 1986; 71:283–290.

10

Methods in Molecular Genetics Diagnosis of Muscle Disease

Tracey Van-Tongeren and Nessa Carey
Charing Cross and Westminster Medical School, University of London, London, England

Keith J. Johnson
University of Glasgow, Glasgow, Scotland

INTRODUCTION

Advances in the techniques of molecular biology within the past two decades have brought about a revolution in the field of human genetics. Rapid progress in the development and application of new techniques have led to elucidation of the defects underlying a range of human genetic disorders. For an increasing number of diseases this information is being used to generate a clearer understanding of normal cellular and physiological processes. This is perhaps particularly true of the neuromuscular disorders as relatively little was previously known of the functions and interactions of expressed genes in this system. By improving our knowledge of the diseases at the molecular level it will ultimately be possible to offer the prospect, not just of improved diagnosis and, if appropriate, prenatal screening, but of the rational design of therapeutic regimes that may be drug-based or, in the near future, genetic engineering approaches to therapy.

Neurological and neuromuscular disorders with a major genetic component are listed in Table 1. In certain of these, for example Duchenne muscular dystrophy (1–3), myotonic dystrophy (4–8), and Huntington's disease (9), the genetic defect underlying at least some cases of the disease has been identified. In other disorders, a possible chromosomal localization for the gene has been obtained but the gene itself and the mutation that leads to disease have yet to be identified. Taken together these diseases constitute a major human health problem, with enormous social and economic consequences. They are also a very heterogeneous group of disorders, and accurate discrimination between some of these diseases on clinical criteria may be problematic; one of the major advantages of a successful genetic approach to these conditions will be improvements in diagnostic accuracy.

The majority of human genetic disorders that have been studied at the molecular level, including the neuromuscular diseases, are termed "single-gene disorders," i.e., the observed phenotype is caused by a mutation in one gene. It is important to realize, however, that very similar phenotypes may arise through different single-gene defects in different families. This is particularly true for diseases in which a sequential metabolic pathway is interrupted. For example, mutations of various enzyme genes in the urea cycle (10) may lead to very similar clinical features. This is also true of certain neuromuscular disorders, such as the Charcot-Marie-Tooth heritable peripheral neuropathies (hereditary sensorimotor neuropathies) and myotonic disorders, as described later. However, for a disease manifesting within a particular pedigree and segregating with a Mendelian mode of inheritance, the phenotype is usually a consequence of a mutation in only one gene. This distinguishes these disorders from polygenic conditions in which the phenotype depends on the interaction of a number of different genes, for

Table 1 Types of Mutations Underlying Human Neuromuscular Diseases

Mutation type	Example of Disease
Large deletions	Duchenne muscular dystrophy (42)
	Becker muscular dystrophy (42)
Single-codon deletions	Charcot-Marie-Tooth type 1B (46,47)
	Duchenne muscular dystrophy (42)
Point mutations	Hyperkalemic periodic paralysis (61)
	Paramyotonia congenita (61)
Expansion of repetitive elements	Huntington's disease (9)
	Myotonic dystrophy (4–8)
	Kennedy's disease (62)
	Spinocerebellar ataxia (63)
Contraction of repetitive elements	Facioscapulohumeral dystrophy (55)
Duplications	Charcot-Marie-Tooth type 1A (44,45)

example high blood pressure or cleft lip and palate. Such complex traits are much more difficult to analyze successfully at the genome level, and this review will concentrate primarily on the single-gene neuromuscular disorders.

ISOLATION OF GENES INVOLVED IN HUMAN NEUROMUSCULAR DISORDERS

Candidate Gene Strategy

An approach that has been successfully applied to the study of human genetic disorders in the past is the candidate gene strategy. This is applicable when the observed phenotype is a clear and obvious consequence of a relatively simple biochemical dysfunction, for example some of the hemoglobinopathies (11), hemophilias (12), and certain of the metabolic disorders such as phenylketonuria (13). In these cases it is possible to predict from the aberrant protein the gene most likely to be involved in the disease. The protein sequence can be used to predict the sequence of the gene, which can be isolated and analyzed comparatively between patients and unaffected individuals to determine the precise nature of the mutation.

Positional Cloning ("Reverse Genetics")

Although the candidate gene approach has been successful for certain genetic diseases, its usefulness is limited by the paucity of information on the way in which the eventual phenotype is manifested. In complex tissues such as the neuromuscular system, a vast number of gene products interact. The expression of certain genes may occur only at particular points in embryonic, fetal, or adult life, and a small alteration in the abundance or activity in what may be a relatively minor protein in terms of total cellular synthesis may lead to gross alterations in final cellular biochemistry and physiology. Under these circumstances the positional cloning strategy (previously known as "reverse genetics") is the most appropriate technique. The basis of positional cloning is the identification of a disease gene in the complete absence of any information about the gene product or its role in the disease process. It must be appreciated that although conceptually quite straightforward, such gene-mapping exercises represent an enormous commitment of both resources and time. For example, in both myotonic dystrophy and Huntington's disease, 10 years elapsed between the initial assignment of the disease gene to a chromosome (14,15) and the identification of the mutated gene itself. Two of the major genetic diseases in which the genes involved have been successfully isolated by a positional cloning strategy are

Duchenne muscular dystrophy and cystic fibrosis, and the major steps in the searches of these genes have been thoroughly reviewed by a number of authors (16,17).

Chromosomal Assignment

An essential prerequisite for any positional cloning strategy is the availability of well-ascertained pedigrees in which cotracking of the disease with genetic markers can be achieved. Ideally such pedigrees will contain a large number of informative meioses [such as the extensive Venezuelan cohort studied in the Huntington's disease collaboration (9)] and careful clinical assessment is essential. This may be less than straightforward due to the problems inherent in distinguishing between different diseases with relatively similar presentation. For example, the nondystrophic myotonias are a complex and heterogeneous group of disorders that require extremely rigorous clinical criteria for accurate diagnosis. Inaccurate diagnosis may hamper gene-mapping efforts if two different disease entities are mistakenly included in the same study (heterogeneity). The rarity of certain neuromuscular diseases may also make gene mapping difficult as there may be few pedigrees in which there are sufficient family members presenting with the disease for linkage to be established.

The first stage in a positional cloning strategy is localization of the gene to a particular chromosome. In certain cases this may be relatively straightforward; for example, the almost total restriction of Duchenne muscular dystrophy to males indicated the likelihood of an X-linked recessive disorder. In some past instances it proved possible to assign a disease locus to a chromosome based on coinheritance in a family of the disease with a known protein marker, most commonly a red blood cell antigen. For example, myotonic dystrophy was known to cosegregate in families with the Lutheran and other blood group antigens, and when this linkage group was shown to map to chromosome 19 the disease gene could be coassigned with it (14). However, such association of a disease with an easily detectable protein marker is fortuitous and rare, and methods have now been developed for studying inheritance of the disease with markers detectable directly on the chromosome.

Although single-gene disorders rarely lead to observable cytogenetic changes [the fragile X syndrome of mental retardation is a rare exception (18)], in some circumstances cytogenetic alterations have led to a focusing of the mapping effort on a specific region of the human genome. For example, the efforts to map the Duchenne muscular dystrophy gene were boosted by the identification of rare female patients with partial translocations of the short arm of the X chromosome (19). This focused the gene-mapping program on the region around the translocation breakpoint, thereby decreasing markedly the amount of the chromosome that had to be analyzed in the search for the mutated gene.

DNA Markers: RFLPs, Tandem Repeats, and Genetic Mapping

Unfortunately, in the majority of genetic diseases no cytogenetic information is available to indicate where a mutation is likely to lie and a genome-wide search must be initiated. Broadly speaking, this involves analyzing the coinheritance of the disease and DNA markers in an affected pedigree. Markers that cosegregate with the disease are deemed to be linked to the disease locus, and the closer together the two lie, the less likely they are to be separated by homologous recombination during meiosis (thereby deviating from Mendel's law of independent assortment). The degree of linkage can be expressed mathemetically (resulting in the lod score, which is the logarithm of the odds of two loci being within measurable distance of one another) to gain a statistical measure of the tightness of linkage. This is usually carried out using sophisticated computer programs. (An excellent discussion of lod scores can be found in Ref. 19a). The extent of recombination between two loci can be used to estimate the genetic distance between them, a recombination fraction of 10% representing a genetic distance of 10 centiMorgans, which can be approximated very roughly as 10 million base pairs. This estimate is very approximate as certain regions of the genome are more prone to recombination than others (20). Ordering the relative positions of such markers leads to the generation of a *genetic map*, based on recombination intervals.

The original genetic markers employed in gene-mapping projects were *restriction fragment length polymorphisms* (RFLPs) (21,22). These are polymorphic alterations in DNA sequence that lead to an alteration in a restriction enzyme recognition site, which can be detected by Southern blotting, as shown in Figure 1. Although these have proved useful, their versatility is limited as many of these polymorphisms may be present at a low frequency in the population, and this frequency will tend to remain constant if mating is random. This is predicted by the Hardy-Weinberg equilibrium, which demonstrates mathematically that, in the absence of external influences such as selection or migration, the frequency of a genetic allele in the population will remain constant from generation to generation. This decreases the likelihood that the polymorphism will be informative in a pedigree as it is probable that family members will be homozygous for the common allele. Only heterozygotes are informative in gene linkage studies. In many cases the polymorphism may have only two possible isoforms, which again will limit the extent to which it is useful in gene tracking studies.

Huge strides are now being made in positional cloning problems through the availability of highly polymorphic markers, based on repetitive DNA sequences. These are *tandem repeats* of short lengths

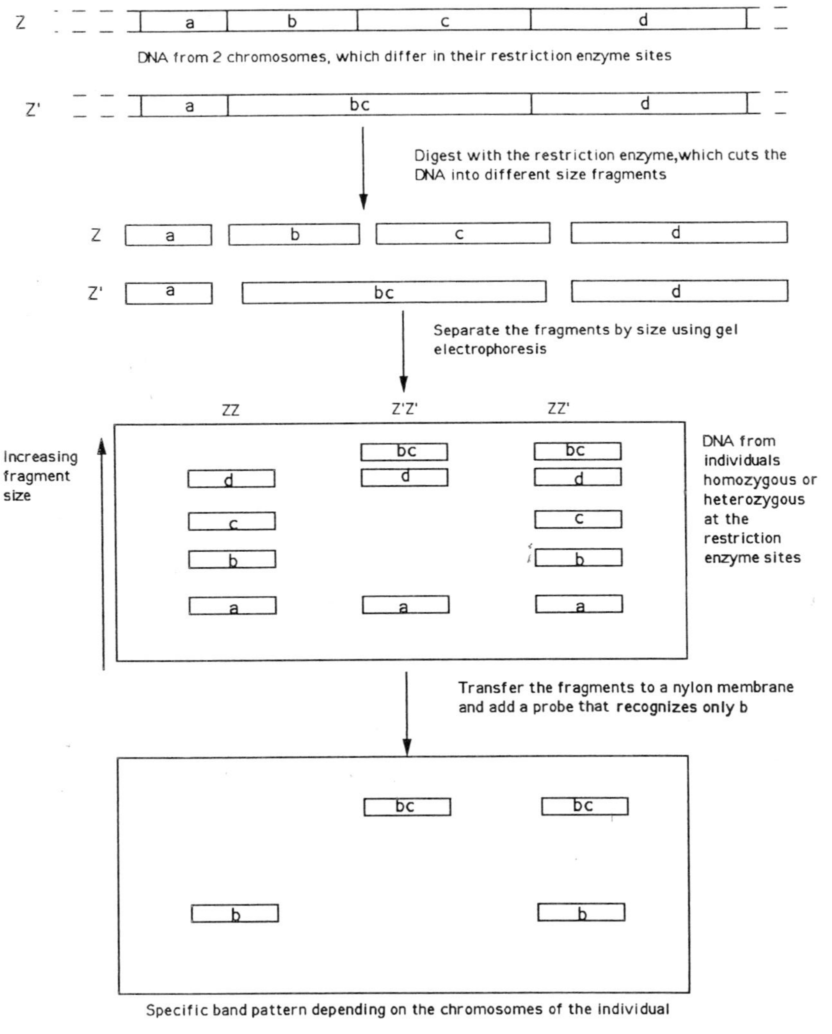

Figure 1 Schematic of the major steps in Southern blotting.

of DNA, which are usually stably inherited but exist in a wide variety of isoforms in the population through variation in the repeat number. This vastly increases the likelihood that meioses within a pedigree will be informative, as there is an increased chance of all the chromosomes in a cross being distinguishable, thereby permitting more informative tracking of the cosegregation of disease and marker. The original highly variable repeat was based on a 33-bp repeat in the human myoglobin gene (23), and a number of other such *minisatellite repeats* were subsequently identified, leading to the technique known as DNA fingerprinting.

Although in some cases analysis of a combination of such probes can lead to absolute discrimination between individuals, they have disadvantages in gene-mapping strategies. These polymorphisms are analyzed by Southern blotting, which is relatively time-consuming and expensive, and they require the availability of probes for the repetitive sequences. For positional cloning they are being supplanted by the microsatellite repeats, based on smaller repetitive elements, especially CA repeats (24). These are *dinucleotide repeats*, which occur throughout the genome. They are highly polymorphic in the population; i.e., at a particular CA repeat locus there is a high likelihood that in two unrelated individuals there will be a difference in the number of repeats, but they are inherited stably. This makes them ideal for linkage analysis. They also have the advantage that they can be analyzed by the polymerase chain reaction (PCR) using primers that flank the repeat locus. This removes the need for a Southern-blotting step as radiolabeled PCR products can be directly sized accurately on denaturing acrylamide gels, and there are none of the problems that can arise if a probe for a particular locus has not been made widely available. The genome-mapping projects have resulted in the generation of a huge number of genetic markers spanning the entire genome, and this has decreased enormously the length of time required for the initial localization of disease loci, since in previous positional cloning efforts a great deal of the research had to focus on the generation of de novo flanking markers.

Sometimes a gene-mapping program may initially succeed only in excluding areas of the genome as candidate regions. In Friedreich's ataxia 80% of the genome had been excluded before linkage to chromosome 9 was finally established (25). To focus a gene-mapping strategy, flanking markers (i.e., a set of markers that lie telomeric and centromeric to the disease locus) are required (Fig. 2). This allows the gene to be approached from two directions, with the generation of more closely linked markers, which can be ordered on the chromosome, thereby defining more accurately the region that must be searched for the disease locus.

Gene Searching Within a Defined Interval

When markers very close to the gene have been isolated, very few meioses will demonstrate recombination between the markers and the disease due to the rarity of crossing-over in a small genetic interval, although in terms of gene searching this region may still be as large as one million base pairs [1 megabase (Mb)]. Under these circumstances it may be appropriate to apply a directed positional cloning approach by screening the region for genes of a class that may be involved in the etiology of the disease. In paramyotonia congenita and hyperkalemic periodic paralysis this approach was successful, with the demonstration of linkage of these disorders to a sodium channel gene (SCN4A) on human chromosome 17 (26,27). If, however, absolutely no convincing physiological evidence for a candidate gene is available, the entire region must be analyzed. This analysis will focus on exonic sequences, delineating the genes in the region and identifying mutations that are present only in the affected patients.

For a positional cloning strategy to succeed, it requires, in addition to informative family resources, effective means of cloning and manipulating large regions of human DNA to generate a comprehensive *physical map* of the region. A physical map specifies the physical location and order of physical features (usually short segments of known sequence called sequence-tagged sites, or STSs) on a chromosome, and the information can be combined with that of the genetic map, thereby allowing far more rapid "placement" of disease loci on a chromosome. Large technical advances have been made on both these fronts, which are starting to alter the rate-limiting steps in this approach to gene mapping. Hybrid cell lines have been established that contain one or a part of one human chromosome in a rodent cell line (28). Great progress has also been made in cloning large stretches of DNA, an essential prerequisite for

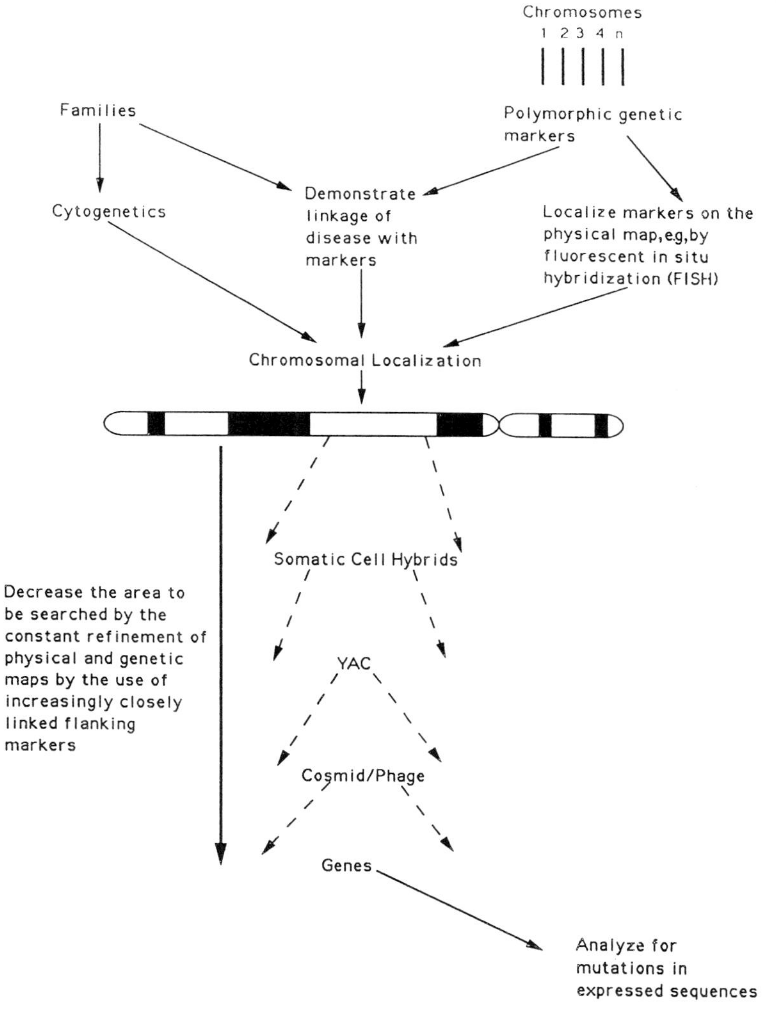

Figure 2 Schematic of the major steps in positional cloning of a disease gene. A simplified view of the positional cloning of a disease gene is shown. By constant refinement of the linkage data for affected pedigrees, and increasingly precise localization of the target region of the genome by positioning of "anonymous" polymorphic markers from the genetic map onto the physical map, successively smaller regions of the genome are pinpointed for in-depth analysis. By proceeding in this way it is possible to isolate a disease gene in the complete absence of any information about its identity or normal cellular function.

this kind of approach as otherwise the sheer number of clones that have to be handled in the generation of linkage data becomes too cumbersome. In the yeast artificial chromosome (YAC) system (29) linear DNA stretches of up to 1 Mb have been cloned, and sets of overlapping clones (contigs) are now becoming available that cover a whole chromosome, for example chromosome 19. Using this approach, a physical map of the human genome has recently been reported (30).

No one cloning system is suitable for all requirements. For example, YACs have a tendency toward instability, which may create difficulties in the precise ordering of genetic markers. The P1 system (31) in *Escherichia coli* is now capable of cloning fragments up to 100 kb in size, providing a useful link between the YACs and more conventional cloning vectors, cosmids (up to 45 kb of insert) and bacteriophage λ (up to 25 kb). Improved methodology for cloning large stretches of chromosomes as efficiently as possible have also been developed [chromosome jumping (32,33)] allowing large stretches of the genome to be covered in a positional cloning strategy as rapidly as possible.

When a possible disease region has been identified, a number of strategies can be employed to isolate the disease gene itself. cDNA libraries from tissues likely to express the gene can be screened with genomic probes from the isolated region to obtain a limited number of candidate genes. cDNAs are artificially generated using an mRNA template and the enzyme reserve transcriptase to generate DNA clones that contain only expressed (i.e., exonic) sequences of genes. An increasing number of cDNA libraries from a wide variety of tissues are now available, again improving the resources required for a successful positional cloning strategy. Any cDNA clones isolated from the candidate region would be sequenced to attempt to assign putative function to the gene product that may correlate with the observed disease phenotype, although this is not sufficient to confirm that this is actually the disease gene. Genomic DNA digests from different species (zoo blots) can also be used to confirm that the probe detects single-copy sequences that are evolutionary conserved. Detection of such sequences increases the likelihood that a functionally significant gene has been isolated, as only such genes are likely to be significantly conserved during evolution.

A summary of the major steps in a positional cloning strategy is shown in Figure 2.

MUTATIONS IN HUMAN NEUROMUSCULAR DISORDERS

To demonstrate that an isolated cDNA is the gene underlying a disease a number of criteria must be met. Mutations must be present that cosegregate with the disease in affected family members but are absent in unaffected members (this may be complicated by nonpenetrance of a trait, but this should be minimal in true single-gene disorders), and introduction of the normal gene or its product into cells displaying the abnormal phenotype should result in restoration of normal function. It should be realized that these criteria may not be as straightforward as they sound. For example, if the number of informative meioses and the number of pedigrees are limited, it may be difficult to exclude genetically the possibility that a suspected mutation in a candidate gene is anything other than a rare polymorphism. Problems may also arise if the same disease phenotype is due to different mutations, albeit in the same gene, in different families. It is also not always feasible to demonstrate functional restoration by a gene for technical reasons, particularly if the defect at the cellular level is not well characterized, or the phenotype is not due to a loss of function mutation. Although great progress has been made in the production of transgenic mice into which mutated versions of genes can be introduced, or in which the normal murine genes are inactivated to mimic loss of gene function, these may not always present with the same phenotype as is found in humans. This is true even for well-characterized disease genes, as has been discovered for the mdx mouse model of Duchenne muscular dystrophy (34) and the mouse model of Lesch-Nyhan syndrome in which inactivation of the hypoxanthine-guanine phosphoribosyl transferase (HGPRT) gene results in no discernible phenotype (35,36), despite clear evidence of the involvement of loss of function of this gene in the human disease (37).

A wide range of types of mutations have been shown to underlie human neuromuscular disorders. Examples of some of the different types of mutation are shown in Table 1. It is important to recognize, however, the sheer complexity in genetic terms of the human neuromuscular disorders. Very similar phenotypes may arise from mutations in different genes, or a variety of mutations in the same gene may result in either similar or disparate symptomatology. At one end of the scale are diseases with pronounced genetic homogeneity, for example myotonic dystrophy in which greater than 95% of cases are due to expansion of a trinucleotide repeat at the 3′ end of the gene (38). At the other end of the range are diseases such as paramyotonia congenita, in which four independent mutations have been identified in the SCN4A sodium channel gene (39,40) in the small number of families so far analyzed (allelic heterogeneity), and malignant hyperthermia susceptibility (MHS), in which extensive genetic and allelic heterogeneity are present (41). The mutation that leads to Duchenne muscular dystrophy can also be extremely variable (42). In approximately two-thirds of cases this is a partial deletion of the dystrophin gene, but it can be difficult to correlate directly the extent of the deletion with the severity of clinical symptoms. This appears also to be influenced by factors including whether the reading frame is maintained after the deletion, and the position of the deleted segment in the gene. Other cases of Duchenne muscular dystrophy are caused by point mutations (42). Diseases with very similar phenotypes are now being shown to be genetically very heterogeneous. Some cases of hyperkalemic periodic paralysis fail to show linkage to the SCN4A gene (43), and the disease complex Charcot-Marie-Tooth heritable peripheral neuropathies have been shown to be due to duplications in a gene on chromosome 17p11.2 (type 1A) (44,45), codon deletion, or a single amino acid change in the peripheral myelin protein gene P_0 on chromosome 1q21-23 (type 1B) (46,47). Other cases of Charcot-Marie-Tooth syndrome are X-linked (48) and also show allelic heterogeneity, with a variety of point mutations, including one that results in the creation of a premature stop codon in the connexin32 gene (49). In Charcot-Marie-Tooth type 2, 50% of families show linkage to a locus on chromosome 1p35-36 (50), and Charcot-Marie-Tooth 4A has been shown to map to chromosome 8q13-21.1 (51). Interestingly, different mutations in one gene may lead to either dominant or recessive modes of inheritance, as shown for the skeletal muscle chloride channel gene on chromosome 7 in myotonia congenita (52,53) (the dominant Thomsen myotonia congenita or Becker's recessive generalized myotonia). Such differences in the mutations underlying human neuromuscular disorders have important consequences for genetic diagnosis and counseling.

DISEASE DIAGNOSIS

Vast improvements in genetic counseling for inherited neuromuscular disorders have resulted from the advances made in molecular genetics in the past two decades, particularly in the fields of accurate diagnosis and prenatal screening, although the identification of a disease gene may not in itself always lead to a straightforward clinical assessment.

Screening for both carrier status and prenatal testing has been improved, even in cases where the mutation itself has not been identified, by the increased availability of linked RFLPs or other markers for disease loci. This allows prediction of whether an individual has inherited the chromosome that is carrying the defective gene. However, this strategy is limited by its inability to determine conclusively that there has been no recombination at the disease allele, and because certain families may not be informative for this study either because the number of meioses is too small for such predictive testing or because the appropriate markers are not sufficiently informative because key individuals are homozygous for the marker. This will be a particular problem in consanguineous pairings.

Where possible, direct mutation detection is the preferred diagnostic option. In a few diseases, notably sickle cell anemia (54), the point mutation that leads to the phenotype also creates an RFLP, but such cases are rare. In Duchenne muscular dystrophy Southern blotting is used to detect deletions in the dystrophin gene, and a similar approach can be used in diagnosis of facioscapulohumoral muscular dystrophy (55). The trinucleotide repeat disorders can be diagnosed readily by analysis of the expansion in repeat length, either by the polymerase chain reaction or by Southern blotting for the larger expansions. For diseases in which there is a plethora of possible mutations, the approach adopted for diagnosis is of necessity more complex. Family studies may be undertaken to determine which mutation is segregating in the pedigree, but this may not be feasible if the number of accessible family members is limited. In Duchenne muscular dystrophy a method has been developed in which reverse transcription of dystrophin mRNA is followed by polymerase chain reaction with primers developed to allow analysis of a variety of possible mutations (56). Methods are being devised that allow simultaneous screening of DNA from an individual for a number of the most common known mutations, for example using the multiplex polymerase chain reaction (57). Other strategies, such as single-strand conformation polymorphism analysis (SSCP) (58) or denaturing gradient gel electrophoresis (DGGE) (59), allow detection of single base changes in the absence of data about such a mutation, and this may yet prove to be the analytical method of choice in diseases in which the incidence of new mutations is high. This is a particular problem in diseases caused by defects in very large genes, for example the neurofibromatosis type 1 gene (60). Care must therefore be exercised when deciding on the approach for genetic screening of patients presenting with genetic neuromuscular disorders, especially those known to show marked genetic heterogeneity, in order to optimize resource use and enhance the likelihood of reaching a meaningful diagnosis.

CONCLUSION

The application of molecular genetic techniques is revolutionizing the field of human neuromuscular diseases, altering its status from a largely phenomenological and descriptive sphere of study to one in which precise, well-defined genetic aberrations can be shown to underlie the presenting phenotype. The study of how such a variety of mutations result in disease symptoms provides scientists with a tool for understanding both normal and abnormal tissue function and also offers the prospect of better patient care through improved diagnosis and prenatal screening. As a large number of neuromuscular diseases arise through the appearance of new mutations (that is, mutations that have not previously been observed in a particular family), the ability to dissect the underlying defect through molecular genetics offers the greatest hope yet for the rational design of therapeutic regimes that may begin to improve the prospects of patients.

REFERENCES

1. Den Dunnen JT, Bakker E, Klein-Breteler EG, et al. Direct detection of more than 50% of Duchenne muscular dystrophy mutations by field inversion gel electrophoresis. Nature 1987; 329:640–642.
2. Forrest SM, Cross GS, Speer A, et al. Preferential deletion of exons in Duchenne and Becker muscular dystrophies. Nature 1987; 329:634–640.
3. Koenig M, Hoffman EP, Bertelson CJ, et al. Complete cloning of the Duchenne muscular dystrophy (DMD) cDNA and preliminary genomic organisation of the DMD gene in normal and affected individuals. Cell 1987; 50:509–517.
4. Brook JD, McCurrach ME, Harley HG, et al. Molecular basis of myotonic dystrophy: expansion of a trinucleotide (CTG) repeat at the 3′ end of a protein kinase family member. Cell 1992; 68:799–808.
5. Buxton J, Shelbourne P, Davies J, et al. Detection of an unstable fragment of DNA specific to individuals with myotonic dystrophy. Nature 1992; 355:547–548.
6. Fu Y-H, Pizzuti A, Fenwick RG, et al. An unstable triplet repeat in a gene related to myotonic muscular dystrophy, myotonin protein kinase. Science 1992; 255:1256–1258.
7. Harley HG, Brook JD, Rundle SA, et al. Expansion of an unstable DNA region and phenotypic variation in myotonic dystrophy. Nature 1992; 355:545–547.
8. Mahadevan M, Tsilfidis C, Sabourin L, et al. Myotonic dystrophy mutation. An unstable CTG repeat in the 3′ untranslated region of a candidate gene. Science 1992; 255:1253–1255.
9. The Huntington's Disease Collaborative Research Group. A novel gene containing a trinucleotide repeat that is expanded and unstable in Huntington's disease chromosomes. Cell 1993; 72:971–983.
10. Brusilow SW. Disorders of the urea cycle. Hosp Pract 1985; 20:65–72.
11. Collins FS, Weissman SM. The molecular biology of human haemoglobin. Prog Nucl Acid Res Mol Biol 1984; 31:315–458.
12. Giannelli F, Green PM, High KA, et al. Haemophilia B: database of point mutations and short additions and deletions—second ed. Nucleic Acids Res 1991; 19(Suppl):2193–2219.
13. Robson KJH, Chandra T, MacGillivray RTA, Woo SLC. Polysome immunoprecipitation of phenylalanine hydroxylase mRNA from rat liver and cloning of its cDNA. Proc Natl Acad Sci USA 1982; 79:4701–4705.
14. Whitehead A, Solomon E, Chambers S, et al. Assignment of the structural gene for the third component of complement to chromosome 19. Proc Natl Acad Sci USA 1982; 79:50221–5025.
15. Gusella JF, Wexler NS, Conneally PM, et al. A polymorphic DNA marker genetically linked to Huntington's disease. Nature 1983; 306:234–238.
16. Monaco AP. Molecular human genetics and the Duchenne/Becker muscular dystrophy gene. In: Partridge T, ed., Molecular and Cell Biology of Muscular Dystrophy, 1st ed. London: Chapman & Hall, 1993:1–11.
17. Collins FS. Cystic fibrosis: molecular biology and therapeutic implications. Science 1992; 256:774–779.
18. Verkerk AJMH, Pieretti M, Sutcliffe JS, et al. Identification of a gene (FMR-1) containing a CGG repeat coincident with a breakpoint cluster region exhibiting length variation in fragile X syndrome. Cell 1991; 65:905–914.

19. Boyd Y, Buckle V, Holt S, et al. Muscular dystrophy in girls with X;autosome translocations. J Med Genet 1986; 23:484–490.
19a. Davies KE, Read AP. Molecular Basis of Inherited Disease, 2nd ed. New York: Oxford University Press, 1992:27.
20. Hartley D, Davies KE, Drayna D, White RL, Williamson R. A cytological map of the human X chromosome—evidence for non-random recombination. Nucleic Acids Res 1984; 12:5277–5285.
21. Kan YW, Dozy AM. Polymorphism of DNA sequence in the β-globin gene region. Proc Natl Acad Sci USA 1978; 75:5631–5635.
22. Solomon E, Bodmer WF. Evolution of sickle variant gene. Lancet 1979; 1:923.
23. Jeffreys AJ, Wilson V, Thein SL. Individual specific "fingerprints" of human DNA. Nature 1985; 316:76–79.
24. Hearne CM, Ghosh S, Todd JA. Microsatellites for linkage analysis of genetic traits. Trends Genet 1992; 8:288–294.
25. Chamberlain S, Shaw J, Rowland A, et al. Mapping of mutation causing Friederich's ataxia to human chromosome 9. Nature 1988; 3334:248–250.
26. Fontaine B, Khurana TS, Hoffman EP, et al. Hyperkalemic periodic paralysis and the adult muscle sodium alpha subunit gene. Science 1990; 250:1001002.
27. Ebers GC, George AL, Barchi RL, et al. Paramyotonia congenita and hyperkalemic periodic paralysis are linked to the adult muscle sodium channel gene-locus. Am J Hum Genet 1991; 49:851–854.
28. Weiss MC, Green H. Human-mouse hybrid cell lines containing partial complements of human chromosomes and functioning human genes. Proc Natl Acad Sci USA 1976; 58:1104–1111.
29. Burke DT, Carle GF, Olsen MV. Cloning of large segments of exogenous DNA into yeast by means of artificial chromosome vectors. Science 1987; 236:806–812.
30. Cohen D, Chumakov I, Weissenbach J. A first-generation physical map of the human genome. Nature 1993; 366:698–701.
31. Sternberg, N. Bacteriophage P1 cloning system for the isolation, amplification and recovery of DNA fragments as large as 100 kilobase pairs. Proc Natl Acad Sci USA 1990; 87:103–107.
32. Collins FS, Weissman SM. Directional cloning of DNA fragments at a large distance from an initial probe. Proc Natl Acad Sci USA 1984; 6812–6816.
33. Poustka A, Pohl TM, Barlow DP, Frischauf A-M, Lehrach H. Construction and use of human chromosome jumping libraries from Not1-digested DNA. Nature 1987; 325:353–355.
34. Bulfield G, Siller WG, Wright PAL, Moore KJ. X chromosome-linked muscular dystrophy (mdx) in the mouse. Proc Natl Acad Sci USA 1984; 81:1189–1192.
35. Hooper M, Hardy K, Handyside A, Hunter S, Monk M. HPRT-deficient (Lesch-Nyhan) mouse embryos derived from germline colonization by cultured cells. Nature 1987; 326:292–295.
36. Kuehn MR, Brafley A, Robertson EJ, Evans MJ. A potential animal model for Lesch-Nyhan syndrome through introduction of HPRT mutations into mice. Nature 1987; 326:295–298.
37. Seegmiller JE, Rosenbloom FM, Kelley WN. Enzyme defect associated with a sex-linked human neurological disorder and excessive purine synthesis. Science 1967; 155:1682–1684.
38. Redman JB, Fenwick RG, Fu Y-H, et al. Relationship between parental nucleotide GCT repeat length and severity of myotonic dystrophy in offspring. JAMA 1993; 269:1960–1965.
39. McClatchey AI, Van den Bergh P, Pericak-Vance MA, et al. Temperature-sensitive mutations in the III-IV cytoplasmic loop region of the skeletal muscle sodium channel gene in paramyotonia congenita. Cell 1992; 68:769–774.
40. Ptacek LJ, George AL Jr, Barchi RL, et al. Mutations in the S4 segment of the adult skeletal muscle sodium channel cause paramyotonia congenita. Neuron 1992; 8:891–897.
41. Johnson K. Malignant hyperthermia hots up! Hum Mol Genet 1993; 2:849.
42. Hoffman EP. Genotype/phenotype correlations in Duchenne/Becker dystrophy. In: Partridge T, ed. Molecular and Cell Biology of Muscular Dystrophy, 1st ed. London: Chapman & Hall, 1993:12-36.
43. Wang J, Zhon J, Todorovic SM, et al. Molecular genetic and genetic correlations in sodium channelopathies: lack of founder effect and evidence for a second gene. Am J Hum Genet 1993; 53:1074–1084.
44. Lupski JR, Montes de Oca-Luna R, Slaugenhaupt S, et al. DNA duplication associated with Charcot-Marie-Tooth disease type 1a. Cell 1991; 66:219–232.
45. Raeymaekers P, Timmerman V, Nelis E, et al. Duplication in chromosome 17p11.2 in Charcot-Marie-Tooth neuropathy type 1a (CMT1a). Neuromusc Disord 1991; 1:93–97.
46. Hayasaka K, Hinoro M, Sato W, et al. Charcot-Marie-Tooth neuropathy type 1b is associated with mutations of the myelin P_0 gene. Nature Genet 1993; 5:31–34.
47. Kulkens T, Bolhuis PA, Wolterman RA, et al. Deletion of the serine 34 codon from the major peripheral myelin protein P_0 gene in Charcot-Marie-Tooth disease type 1B. Nature Genet 1993; 5:35–39.
48. Gal A, Mucke J, Theile H, Wieacker PF, Ropers HH, Wienker TF. X-linked dominant Charcot-Marie-Tooth disease: suggestion of linkage with a cloned DNA sequence from proximal Xq. Hum Genet 1985; 70:38–42.

49. Ben Othmane K, Middleton LT, Loprest LJ, et al. Localization of a gene (CMT2A) for autosomal-dominent-Charcot-Marie-Tooth disease type 2 to chromosome –1p and evidence of genetic heterogeneity. Genomics 1993; 17(2):370–375.
50. Ben Othmane K, Hentati F, Lennon F, et al. Linkage of a locus (CMT4A) for autosomal recessive Charcot-Marie-Tooth disease to chromosome 8q. Hum Mol Genet 1993; 2:1625–1628.
51. Bergoffen J, Scherrer SS, Wang S, et al. Connexin mutations in X-linked Charcot-Marie-Tooth disease. Science 1993; 262:2039–2042.
52. Koch MC, Steinmeyer K, Lorenz C, et al. The skeletal muscle chloride channel in dominant and recessive human myotonia. Science 1992; 257:797–800.
53. George AL Jr, Crackower MA, Abdalla JA, et al. Molecular basis of Thomsen's disease (autosomal dominant myotonia congenita). Nature Genet 1993; 3:305–310.
54. Geever RF, Wilson LB, Nallaseth FS, et al. Direct identification of sickle cell anaemia by blot hybridisation. Proc Natl Acad Sci USA 1981; 78:5081–5085.
55. Wijmenga C, Hewitt JE, Sandkuijl LA, et al. Chromosome 4q DNA rearrangements associated with facioscapulohumeral muscular dystrophy. Nature Genet 1992; 2:26–30.
56. Roberts RG, Bobrow M, Bentley DR. Point mutations in the dystrophin gene. Proc Natl Acad Sci USA 1992; 89:2331–2335.
57. Dorkin HL, Parad RB, Witt D, Klinger KW. Multiplex PCR amplification from the CFTR gene using DNA prepared from buccal brushes/swabs. Hum Mol Genet 1993; 2:159–163.
58. Sekiya T. Detection of mutant sequences by single-strand conformation polymorphism analysis. Mutat Res 1993; 288:79–83.
59. Costes B, Fanen P, Goossens M, Ghanem M. A rapid, efficient and sensitive assay for simultaneous detection of multiple cystic fibrosis mutations. Hum Mutat 1993; 2:185–191.
60. Rodenhiser DI, Ainsworth PJ, Coulter-Mackie MB, Singh SM, Jung JH. A genetic study of neurofibromatosis type 1 (NF1) in south-western Ontario.II. A PCR based approach to molecular and prenatal diagnosis using linkage. J Med Genet 1993; 30:363–368.
61. Placek LJ, Johnson KJ, Griggs RC. Genetics and physiology of the myotonic muscle disorders. N Engl J Med 1993; 328:482–489.
62. La Spada AR, Wilson EM, Lubahn DB, Harding AE, Fishbeck KH. Androgen receptor gene mutations in X-linked spinal and bulbar muscular atrophy. Nature 1991; 352:77–79.
63. Orr HT, Chung M-Y, Banfi S, et al. Expansion of an unstable trinucleotide CAG repeat in spinocerebellar ataxia type 1. Nature Genet 1993; 4:221–226.

11
Congenital Muscular Dystrophies

Bernard Echenne
Gui de Chauliac Center, Montpellier, France

INTRODUCTION

The first description of a congenital muscular dystrophy (CMD) was made by Batten in 1903, concerning three patients with proximal muscular weakness since birth, associated with joint contractures. The disease was not progressive. In a fourth patient, Batten subsequently showed that the muscle was replaced by fat and fibrous tissue. In 1908, Howard described multiple joint contractures in a newborn, related to dystrophic muscular lesions, which he called "dystrophia muscularis congenita." The first cases of CMD associated with central nervous system (CNS) involvement were described by Fukuyama et al. (1). Later, other cases were reported outside Japan, frequently associated with cerebral and ophthalmological abnormalities (2–4).

This group of diseases or syndromes seems heterogeneous, but recent discoveries concerning biochemical and genetic markers will probably simplify the classification in the near future (Table 1).

The classification of CMD is currently based on clinical and neuroimaging criteria, together with common "dystrophic"—and nonspecific—changes observed on muscle biopsy.

Clinically, CMD are characterized by signs that are present from birth or that appear during the first weeks or months of life, or at least during the first year: marked hypotonia; generalized or proximal muscular weakness and delayed motor milestones; frequent joint contractures affecting knees, hips, feet, elbows, or shoulders; perinatal cyanosis or respiratory difficulties. Drinking or sucking problems are frequently noted.

The severity of the disease varies markedly from one patient to another. In most cases, a slow improvement of motor abilities is observed during the first years, but later, generally after 5–6 years, a slowly progressive worsening occurs, with increasing muscle weakness and joint stiffness or deformation (5,6). This progression of motor disability leads to severe motor handicaps—most patients are never able to walk and some die [15% in the study of Serratrice et al. (7)]. However in some cases, the disease remains benign, with only slight motor disability.

The common feature in all cases is the "dystrophic" aspect of muscles. The morphological criteria include:

random variation of muscle fiber size, predominantly fiber atrophy
marked fibrous tissue proliferation, the fat infiltration being initially less severe
nuclear multiplication and centralization
absence of necrosis, or only moderate mononuclear cell infiltration
absence of regeneration
absence of specific structural abnormalities (Figs. 1 and 2)

Table 1 Classificiation of Congential Muscular Dystrophies

A.	Congenital muscular dystrophy without central nervous system involvement	
B.	Congenital muscular dystrophy with central nervous system involvement	
	1.	Fukuyama-type CMD
	2.	The cerebro-ocular dysplasia-muscular dystrophy syndromes Walker-Warburg syndrome Muscle-eye and brain syndrome
	3.	Overlap syndromes between FCMD, WWS, and MEB
	4.	CMD with cerebral hypomyelinization
C.	Miscellaneous forms	
D.	CMD with absence of dystrophin	
E.	The congenital inflammatory myopathies	

Ophthalmological examination, determination of the intellectual level, and neuroimaging investigations [cerebral computerized-tomography (CT) scan and especially magnetic resonance imaging (MRI)] are necessary for differentiating the different types of CMD.

CLASSIFICATION

CMD Without CNS Involvement

Both boys and girls are affected. Most cases are sporadic, but frequently, a recessive autosomal mode of inheritance can be found. Exceptionally, an autosomal dominant heredity has been demonstrated in some benign forms (8).

Hypotonia, muscular weakness, and joint contractures are most often present from birth or appear during the first months of life. The muscular weakness predominates at the proximal level; cervical and facial muscles are usually involved, but ocular movements are normal. Clubfeet and hip subluxation are frequent. The development of a kyphoscoliosis can be troublesome as the child grows older. Some clinical variants have been described: the atonic-sclerotic type of Ullrich, in which infants are floppy,

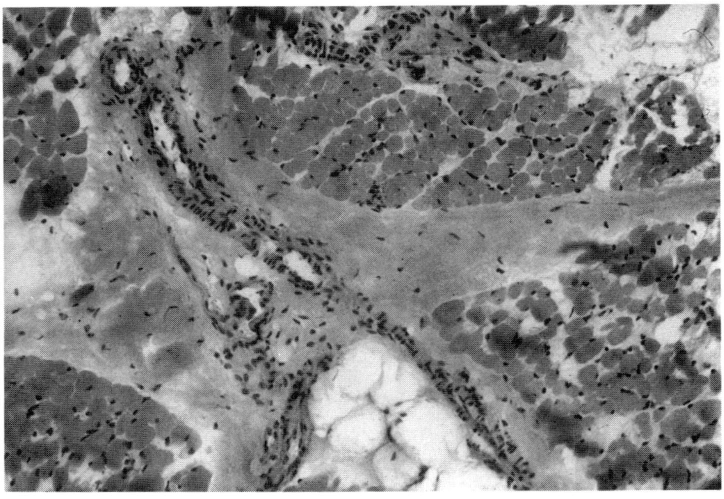

Figure 1 Dystrophic changes: variation of fiber size, intense fibrous tissue proliferation, fat infiltration, absence of necrosis. Hematoxylin and eosin ×40.

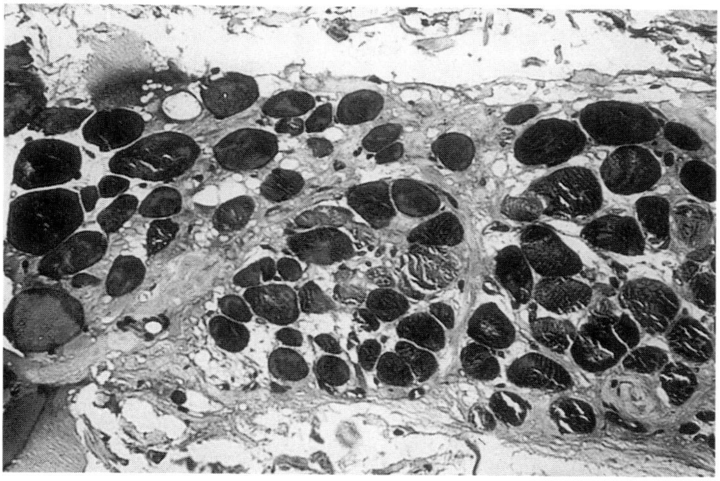

Figure 2 Marked variation of muscle fiber size, with marked fibrosis. Gomori trichrome stain ×60.

with contractures of the proximal joints, but with laxity and hyperextensibility of hands and fingers and the "stick man" dystrophy, described by Goebel et al. (9) and by Brooke (10), in which the patient's limbs are extremely thin, almost skeletal, resembling children's drawings of "stick men." In spite of a considerable muscle atrophy, there is almost no deterioration of gross motor function, as opposed to most of the patients with CMD. Typically, the children are weak and floppy during the first year of life; the motor milestones are often, but not always, delayed. Then toe walking appears, with contractures limited to the heel cord or extended to other joints, such as the knees, hips, elbows, and posterior cervical region, with a picture of rigid spine syndrome (11). The muscles are extremely thin and deep tendon reflexes are diminished or absent. The intellectual level is normal.

Irrespective of the clinical manifestations and rate of progression, certain findings are frequently observed during evolution, including progressive joint contractures, tendinous shortenings, spinal deviation, and chewing and swallowing difficulties secondary to the involvement of facial muscles. There is no cardiomyopathy. However, the severity of the disease is variable. Some patients have only a mild motor disability, with nonprogressive muscular weakness. But in many others, the disease is severe. Some patients, although hypotonic from birth, can have a Duchenne-like progression and die in their teens. Most patients, after a slow improvement during the first years, will worsen during the second decade of life, because of the worsening of joint contractures, the occurrence of spine and chest deformities, and the slow progression of muscle disease (Fig. 3).

The serum creatine kinase is increased during early years, but not to the extent seen in cases of Duchenne dystrophy. There is no relationship between the clinical evolution and the enzyme levels. The electromyogram (EMG) is normal or shows myopathic changes. The cerebral CT scan and MRI are normal.

On muscle biopsy, significant fibrous tissue proliferation with random variability in the size of fibers is the most significant findings. There is often a type 1 fiber predominance, as there is in many congenital muscle diseases. In the "stick man" form, in addition to dystrophic changes, a nonrandom distribution in the variously sized fibers, mimicking denervation, can be found (10). When investigated, dystrophin has always been found to be normal (12).

CMD with CNS Involvement: The Cerebromuscular Syndromes

CMD and brain abnormalities, associated with mental retardation and/or signs of CNS dysfunction, are found in at least three conditions: the Fukuyama-type CMD, the muscle-eye and brain syndrome of

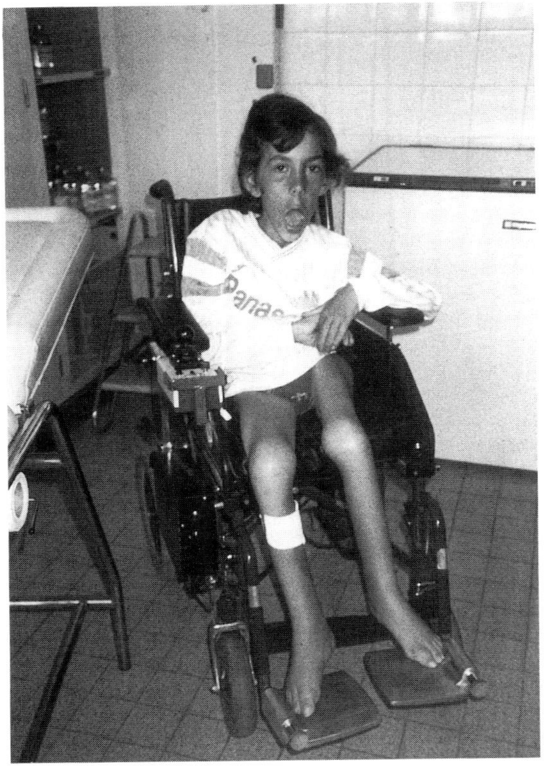

Figure 3 CMD without CNS involvement in 11-year-old boy: joint contractures, spinal and chest deformities, marked facial muscles involvement with chewing difficulties.

Santavuori, and the Walker-Warburg syndrome, with many overlaps between these different forms. Several subgroups, based on clinical or neuroimaging characteristics, can also be identified. It must be emphasized that two-thirds of CMD patients examined neuroradiologically have abnormal cerebral CT and/or MRI findings (6,13).

The Fukuyama-Type CMD (FCMD)

FCMD is an autosomal recessive disorder endemic to Japan, but rarely observed in other countries. The prevalence in Japan varies from 1×10^{-5} to 5.6×10^{-5}. The main clinical features are (14):

- Early onset, usually at birth, at least before the 9th month of life
- Generalized hypotonia and muscular weakness
- Involvement of facial muscles
- Joint contractures present at birth or appearing during the first months, often with an arthrogryposis-like character
- Mental retardation, often severe (IQ between 30 and 50), apparently nonprogressive
- Microcephaly
- Febrile convulsions and epilepsy in approximately 50% of cases
- Severe motor disability, most patients never being able to stand up or walk
- Increase of serum enzymes, but at a lower level than in Duchenne dystrophy; myopathic pattern on EMG and dystrophic changes on muscle biopsy
- Slow worsening of motor functions after 5–6 years

Variations in the severity of the disease may occur, but rare patients who become able to walk show a degradation later, with loss of motor acquisitions. Histologically, the muscle changes are nonspecific and correspond to severe dystrophic lesions; they become worse with time when biopsies are performed twice in the same patient. Dystrophin is almost always present. The cerebral CT scan and MRI are nearly always abnormal, with hypodensity of cerebral white matter on CT scan in 91% of cases (15) and with hyperintense signal on MRI, slight ventricular dilatation, enlargement of pericerebral spaces, and variable cortical atrophy. On MRI, localized or more diffuse areas of pachygyria and polymicrogyria are often seen, but it is impossible to ascertain whether this can always be demonstrated. In fact, severe neuronal migration defects are a constant feature of anatomical investigations, with more or less extensive polymicrogyria (77% of cases) or pachygyria (38%), meningeal adherences (92%), neuronal heterotopias (46%), myelin pallor or white matter spongiosis being the most common anomalies. The myelin pallor seems to correspond to a dysmaturation state since the white matter hypodensity disappears on repeated CT scanning after the age of 3 years (15).

Another important finding concerns the ophthalmological abnormalities, which were considered infrequent for a long time. However, recent studies have shown that they are more frequent than expected. Primary vitreous remnants, disc coloboma, and adhesions of the anterior chamber are not usually seen, but other anomalies, such as myopia, optic atrophy, and cortical blindness, are frequent (16).

The Cerebro-ocular-dysplasia Muscular Dystrophy Syndromes

These syndromes are characterized by the association of CMD with severe neuronal migration defects and developmental ocular anomalies. In our opinion, these syndromes correspond to phenotypic variations of FCMD.

The most severe form corresponds to Walker-Warburg syndrome transmitted as an autosomal recessive trait. The clinical onset occurs at birth, with severe encephalopathy, hypotonia, blindness, hydrocephalus, lack of motor or mental progression, and death (17,18). The ophthalmological anomalies consist of microphthalmy and ocular anterior chamber dysgenesis with Peters syndrome (cataract, retinal dysplasia with persistance of primitive vitreous, optic atrophy, and coloboma). From a neuropathological point of view, important lesions are encountered, which correspond to pathological factors acting at different ages of pregnancy: meningeal thickening with cerebral adherences, pachygyria and lissencephaly, neuronal heterotopia, lack of cortical stratification, hydrocephalus, agenesis of the vermis and cerebellar microgyria.

When performed, the muscular biopsy shows nonspecific dystrophic changes.

It must be emphasized that in type I lissencephaly (Miller Dicker syndrome), muscle dystrophic changes are not seen (study performed in eight personal cases, unpublished data).

Less severe is the muscle-eye and brain (MEB) syndrome described by Santavuori (2–19). In these patients, the pedigree data suggest a genetic disorder, presumably of autosomal recessive inheritance.

The clinical findings consist of muscular hypotonia and weakness appearing at birth or during the first months, delayed milestones, dysmorphic facial appearance (high and prominent forehead with wide fontanelle, narrow calcarium, flatness of midfacies with short nose and philtrum, macrocephaly), mental delay, early joint contractures, spasticity of limbs observed after age 5 years, and in some cases, hydrocephalus, epilepsy, initially brisk reflexes, with later absent reflexes with the development of spasticity when brisk reflexes could again be found.

The motor progression is very limited and most patients do not progress, with early death [4/19 between 6 and 16 years (19)].

In addition, severe visual failure is typical, with severe myopia (usually > -10 diopters), congenital or infantile glaucoma, retinal dystrophy, nystagmus, uncontrolled eye movements, and choroidal hypoplasia. The electroretinogram is often flat, with abnormal visual evoked potentials of very high amplitude (>50 μV). The results of cerebral CT scan are close to similar of FCMD: white matter hypodensity, ventricular dilatation, even hydrocephalus, corpus callosum agenesis, and absence of the septum pellucidum. The muscular dystrophic lesions are nonspecific. As with FCMD, the disease is slowly progressive.

Overlap Syndromes

In our opinion, different phenotypes of cerebro-ocular muscle disorders do exist, and are probably due to the same gene defect or to related mechanisms. Many overlaps concern Walker-Warburg syndrome, MEB syndrome, and FCMD: the muscular dystrophic lesions are the same; the ophthalmological anomalies and the cortical neuronal organization disorders of the brain are identical, with more or less severe lesions. Similar architectonic findings in brain necropsy studies have been demonstrated in FCMD and Walker-Warburg syndrome. Some reported cases are difficult to classify (6,20,21), and the two syndromes (FCMD and Walker-Warburg syndrome) have been reported by Yoshioka et al. (22) in three members of the same family.

Most of the published cases correspond to a genetically determined disease, transmitted by recessive autosomal inheritance. The different syndromes may be allelic with variable phenotypes. Recently, Matsumura et al. (23) reported that dystrophin-associated proteins have abnormally low expression in FCMD, despite near-normal or normal expression of dystrophin (this defect especially concerns the 43-kDa protein). This expression is normal in cases of CMD without CNS involvement. FCMD has recently been linked to chromosome 9 (9q31–9q33) (TODA et al., Nature Genet. 5:283, 1993) and Japanese authors have shown that FCMD and Walker-Warburg syndrome could be different phenotypes of the same gene mutation (Yoshioka et al, unpublished data).

In some rare cases, the role of a viral agent inducing a prolonged and potentially reproducing fetopathy has been the subject of discussion. Such cases of CMD could correspond to a fetal myositis occurring during pregnancy, to neonatal hypotonia and joint contractures, with or without eye and brain lesions (24). During the past 15 years, we observed two such patients in whom, during the first weeks of life, very high levels of CK were found, with muscle lesions characterized by areas of necrosis, unusual inflammatory cell proliferation, absence of fat, and only slight fibrosis. A second biopsy performed 1 year later showed typical dystrophic changes, with marked fibrosis, disappearance of necrosis, and decrease of mononuclear cell infiltration. The tissue cultures and the serological investigations remained negative.

CMD with Cerebral Hypomyelination

It came as a surprise some 14 years ago when CNS changes in the form of white matter translucency on cerebral CT scans were reported in some patients with CMD (3,4,25). The MRI investigations show that such patients have a diffuse alteration of myelin, as evidenced by marked differentiation between gray and white matter, and by increased signal intensity of white matter on the T2-weighted images. The changes are diffuse in the centrum semiovale and patchy in the external capsule and cerebellum. The internal capsule, corpus callosum, basal ganglia, thalami, and brainstem are completely spared (26). The occurrence of this finding in CMD without mental retardation or signs of CNS dysfunction seems to define a particular form of CMD, or "occidental type" of CMD (27,28). Its incidence is unknown. We found it in eight of 14 patients during the last 10 years. At least 90 cases have been reported from the West or reported from Japan as "non-Fukuyama CMD" (3,4,6,10,13,26–33, personal unpublished data) (Figs. 4, 5, 6).

The family pedigrees and frequent consanguinity suggest an autosomal recessive inheritance. Weakness noted at birth, early onset of contractures, delayed motor milestones, normal intelligence, normal ophthalmological examination, absence of epilepsy, and often macrocephaly are usually seen. The motor disability is severe, and after a moderate improvement during the first years, a slow worsening occurs, with progressive muscle atrophy and joint stiffness, chest deformation, and kyphoscoliosis.

The unique finding is an abnormal myelination on neuroimaging studies, without evidence of brain dysgenesis. Although abnormal myelination is also the most characteristic change for FCMD, the situation seems different, since the neuroimaging anomalies remain stable over time. The nature and

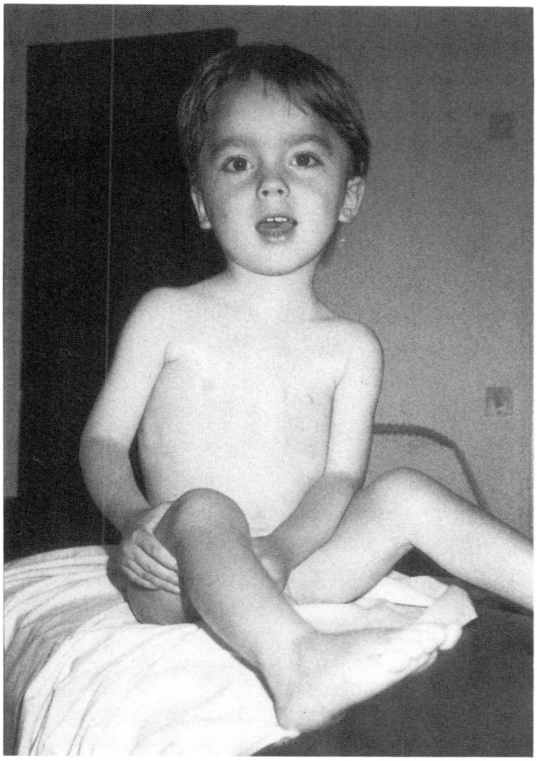

Figure 4 "Occidental-type" CMD. Normal intelligence and clinical appearance.

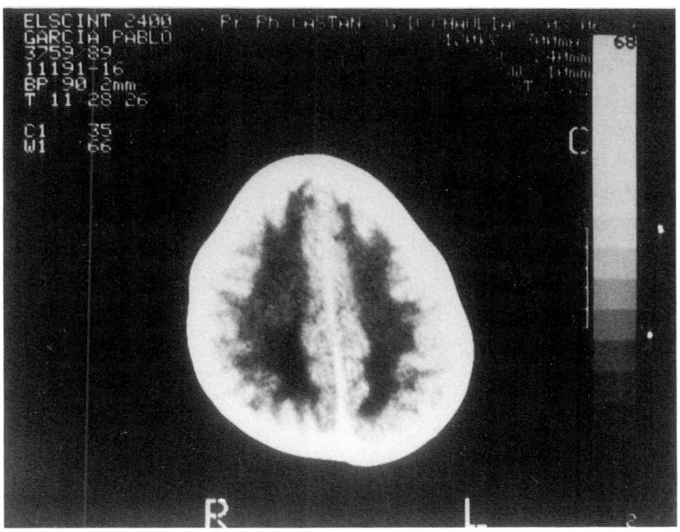

Figure 5 CT scan of the same patient as in Figure 5. Isolated white matter hypodensity.

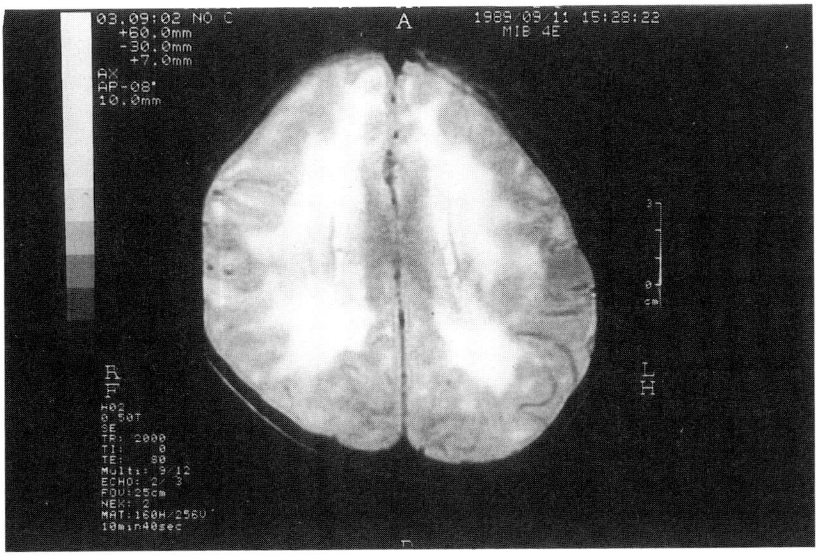

Figure 6 Cerebral MRI, with increased signal intensity of white matter with the T2-weighted images.

significance of white matter changes remain obscure. There are no signs of impaired cognitive functions or disturbed personality in these children, regardless of the white matter changes. Some affected patients are now in their thirties, and no change toward normal myelination has been observed in the patients examined twice. So the possibility of myelin dysmaturation, as in FCMD, seems excluded. To our knowledge, none of the patients with isolated neuroimaging white matter abnormalities have died, and neuropathological correlation is lacking up to now.

The majority of patients constitute a relatively homogeneous group that is commoner in the West (26). However, a normal intellectual level, with or without epilepsy, may coexist with cerebral cortical dysplasia. In addition, MRI, and of course CT scan investigations, may show only white matter changes and misdiagnose restricted areas of cortical dysgenesis. We believe that this occidental type of CMD is heterogeneous, which is clearly demonstrated by some reported cases in the literature:

A normal intelligence level with absence of CNS dysfunction occurred in some of the patients described by Egger et al. (29), Stern et al. (30), and Trevisan et al. (31). However, the apparently isolated white matter hypodensity on CT scan was associated with necropsy-proven focal areas of micropolygyria or pachygyria with cortical disorganization. One of Nogen's patients, in spite of a normal IQ, had hydrocephalus (4). CMD with white matter anomaly, macrocrania, and epilepsy, but with normal intelligence, has also been reported (3,32).

In the same family, the severity of the disease may vary from one patient to another: encephalopathy, epilepsy, and CMD with abnormal cerebral white matter are observed in one sibling, whereas in the other, the CMD with neuroimaging anomaly remains isolated (30,31). In addition, FCMD may coexist with the occidental type in the same family, and even with CMD without cerebral involvement (31).

Thus, it seems inadequate to classify CMD only on neuroimaging and clinical criteria, such as intellectual level, head circumference, or epilepsy. It would be better to regroup these syndromes under the term "cerebromuscular syndromes," until biological investigations perhaps demonstrate a common or different genetic markers.

Two different forms of occidental-type CMD may now be characterized: the merosin-positive and the merosin-negative forms. As a matter of fact, a specific absence of merosin, the laminin M-chain, has been demonstrated in the classical non-Japanese forms of CMD (Tome et al., C. R. Acad. Sci. Paris 317–351, 1994) and these merosin-negative CMD have been linked to a group of markers within the 6q2 region (Hilaire et al., Human Mol. Genet. 3:1657, 1994), confirming that FCMD and merosin-negative CMD are not allelic. The merosin-negative phenotype appears to correspond to the most severely affected children presenting with neonatal hypotonia, joint stiffness, markedly delayed motor development, raised serum CK, and general changes of the white matter on MRI.

Miscellaneous Forms

Some cases remain difficult to classify. Some patients have a CMD, a borderline intellectual level, and a cerebellar atrophy or hypoplasia on neuroimaging, without white matter changes (31,33,34). We described a CMD in two siblings with progressive evolution, macrocephaly, facial dysmorphism, epilepsy, borderline intellectual level, and cerebral white matter hypodensity on CT scan, but with only white matter spongiosis seen on necropsy, without any cortical dysplasia or neuronal heterotopia, very different from the FCMD anatomical studies (35). One of Brooke's patients (10) and those reported by Vles et al. (36) may correspond to the same syndrome.

CMD with Absence of Dystrophin

In CMD, as in Duchenne dystrophy (DMD), cytochemical and immunocytochemical studies have shown increased calcium-overloaded fibers, consistent with plasma membrane abnormality. Immunocytochemistry of dystrophin, quantitative analysis of the immunocytochemical staining patterns, and immunoblot analysis of dystrophin have shown that in FCMD, a high percentage (17–43%) of muscle fibers immunoreacted abnormally for dystrophin, with absent or abnormal staining (partially deficient, fluffy, or intense). This was related to the increased number of regenerating type 2C fibers, and it was hypothesized that the immunocytochemical abnormalities of FCMD muscles might reflect a primary abnormality of the muscle fiber membrane or might be a secondary consequence of generalized fiber damage (37). However, Arikawa et al. (37) reported two patients with apparent FCMD with complete absence of dystrophin, in one of whom an intragenic deletion was detected, suggesting the possibility of an early onset of DMD.

We believe that this diagnosis has to be considered in every case of CMD. We observed a boy with hypotonia since birth, neonatal motor limb hypomobility, and swallowing difficulties, without joint contractures. The milestones were delayed, the child sitting without support at age 12 months and walking alone at 30 months. At age 7 months, he had a marked generalized hypotonia with areflexia and macrocrania (>98th percentile), and the diagnosis of myopathy was suspected because of a high level of CK (1757 IU, N < 250). At age 3 years 9 months, he walked alone with rolling gait and had generalized muscular weakness and hypotonia, without joint contractures. Mental delay was severe (IQ, 40 MacCarthy scale), and he had a dysmorphic appearance (macrocrania, high arched palate, involvement of facial muscles). EEG, cerebral CT scan, and MRI were normal. Marked dystrophic changes were seen on the muscle biopsy. In a follow-up, we observed stabilization of the mental deficiency, the occurrence of epileptic seizures after age 9 years, and moderate and slow improvement of motor functions up to 10 years, followed by a slow worsening (Fig. 7). The CK remained markedly increased (6000 to 11,200 IU). Dystrophin was completely absent on a second biopsy, which showed a worsening of the muscular dystrophic process. A deletion between exons 44 and 51 was found on cDNA analysis in the patient and also in his mother (38).

We believe that in CMD patients, some rare cases of early-onset DMD or of a dystrophinopathy different from DMD do exist. Such cases are characterized by the severity of the mental deficiency, which remains stable, by the discrepancy between the early onset and the stability of the disease, at least during the first 10 years, and by some unusual clinical findings such as macrocrania, epilepsy, facial dysmorphic features, slimness of the limbs, generalized amyotrophy, and lack of early orthopedic complications; some rare identical cases have been observed by others [J. C. Netter and G. Ponsot in France, unpublished data, and Topaloglu et al. in Turkey (39)].

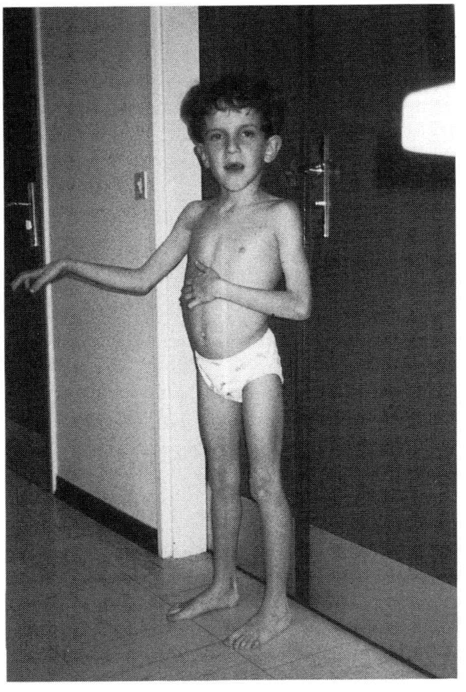

Figure 7 Neonatal dystrophinopathy: clinical aspect of a 10-year-old boy.

The Congenital Inflammatory Myopathies

As mentioned above, identification of this condition as a separate nosological entity remains under discussion. Clinically, the presentation is identical to CMD: muscular hypotonia and weakness with or without arthrogryposis, every elevated serum CK activity, myopathic EMG, and muscle fiber damage on biopsy, with inflammatory cell infiltration. CNS involvement may exist, with variable features similar to Walker-Warburg syndrome.

The differential diagnosis with CMD is anatomical. Necrosis and interstitial inflammatory cells must be clearly abundant and the loss of fibers intense. Staining for immunoreactive class 1 major histocompatibility complex gene products should be positive in the sarcolemma of most muscle fibers (24). In these very rare cases, treatment with steroids may be helpful, but results are variable (40).

CONCLUSION

Genetic markers are beginning to be available for the congenital muscular dystrophies, and we will rapidly progress in knowledge of biochemical abnormalities that could explain the clinical disorders. Treatments are only symptomatic. They include physical therapy and orthopedic procedures, and in some cases, antiepileptic drugs are needed, or neurosurgery for hydrocephalus. Range-of-motion exercises to prevent contractures and preventive treatments of spinal deformities are indicated at an early stage.

REFERENCES

1. Fukuyama Y, Kawazura M, Haruna H. A peculiar form of congenital progressive muscular dystrophy: report of fifteen cases. Pediatr Univ Tokyo 1960; 4:5–8.
2. Santavuori P, Leisti J, Kruus S. Muscle-eye and brain disease: a new syndrome. Neuropädiatrie 1977; 8:550–553.

3. Bernier JP, Brooke MH, Naidich TP, Carroll JE. Myoencephalopathy: cerebral hypomyelination revealed by CT scan of the head in a muscle disease. Trans Am Neurol Assoc 1979; 104:224–246.
4. Nogen AG. Congenital muscle disease and abnormal findings on computerized tomography. Dev Med Child Neurol 1980; 22:658–663.
5. MacMenamin JB, Becker LE, Murphy EG. Congenital muscular dystrophy: a clinicopathologic report of 24 cases. J Pediatr 1982; 5:692–697.
6. Leyten QH, Gabreels FJM, Renier WO, Ter Laak HJ, Sengers RCA, Mullaart RA. Congenital muscular dystrophy. J Pediatr 1989; 115:214–221.
7. Serratrice G, Cros D, Pellissier JF, Gastaut JL, Pouget J. Dystrophie musculaire congénitale. Rev Neurol 1980; 136(6–7):445–472.
8. Leyten QH, Gabreels FJM, Joosten EMG, et al. An autosomal dominant type of congenital muscular dystrophy. Brain Dev 1986; 8:533–537.
9. Goebel HH, Lenard HG, Langenbeck U, Mehl B. A form of congenital muscular dystrophy. Brain Dev 1980; 2:387–400.
10. Brooke MH. A Clinician View of Neuromuscular Diseases. Baltimore: Williams & Wilkins 1986.
11. Echenne B, Astruc J, Brunel D, Pages M, Baldet P, Martinazzo G. Congenital muscular dystrophy and rigid spine syndrome. Neuropediatrics 1983; 14:97–101.
12. Uchino M, Araki S, Miike T, Teramoto H, Nakamura T, Yasutake T. Localization and characterization of dystrophin in muscle biopsy from Duchenne muscular dystrophy and various neuromuscular disorders. Muscle Nerve 1989; 12:1009–1016.
13. Pihko H, Louhimo T, Valanne L, Donner M. CNS in congenital muscular dystrophy without mental retardation. Neuropediatrics 1992; 23:116–122.
14. Fukuyama Y, Ozawa M, Susuki H. Congenital progressive muscular dystrophy of the Fukuyama type: clinical, genetic, and pathological considerations. Brain Dev 1981; 3:1–29.
15. Yoshioka M, Saiwai S. Congenital muscular dystrophy (Fukuyama type). Changes in the white matter low density on CT. Brain Dev 1988; 10:41–44.
16. Tsutsumi A, Uchida Y, Osawa M, Fukuyama Y. Ocular findings in Fukuyama type congenital muscular dystrophy. Brain Dev 1989; 11:413–419.
17. Pagon RA, Clarren SK, Milam DF, Hendrickson AE. Autosomal recessive eye and brain anomalies: Warburg syndrome. J Pediatr 1983; 103:542–546.
18. Bordarier C, Aicardi J, Gouttieres F. Congenital hydrocephalus and eye abnormalities with severe development brain defects: Warburg's syndrome. Ann Neurol 1984; 16:60–65.
19. Santavuori P, Somer H, Sainio K, et al. Muscle-eye and brain disease (MEB). Brain Dev 1989; 11:143–153.
20. Korinthenberg R, Palm D, Schalke W, Klein J. Congenital muscular dystrophy, brain malformation and ocular problems (muscle, eye and brain disease) in two German families. Eur J Pediatr 1984; 142:64–68.
21. Heyer R, Ehrich J, Goebel HH, Christen HJ, Hanefeld F. Congenital muscular dystrophy with cerebral and ocular malformations (cerebro-oculo muscular syndrome). Brain Dev 1986; 8:614–619.
22. Yoshioka M, Kuroki S, Nigami H, Kawai T, Nakamura H. Clinical variations within sibships in Fukuyama-type congenital muscular dystrophy. Brain Dev 1992; 14:334–337.
23. Matsumura K, Nonaka I, Campbell KP. Abnormal expression of dystrophin associated proteins in Fukuyama-type congenital muscular dystrophy. Lancet 1993; 341:521–522.
24. Shevell M, Rosenblatt B, Silver K, Carpenter S, Karpati G. Congenital inflammatory myopathy. Neurology 1990; 40:1111–1114.
25. Echenne B, Arthuis M, Billard C, et al. Congenital muscular dystrophy and cerebral CT scan anomalies. Results of a collaborative study of the Société de Neurologie Infantile. J Neurol Sci 1986; 75:7–22.
26. Cook JD, Gascon GG, Haider A, et al. Congenital muscular dystrophy with abnormal radiographic pattern. J Child Neurol 1992; 7(Suppl):S51–S63.
27. Castro-Gago M, Pena-Guitian J. Congenital muscular dystrophy of a non-Fukuyama-type with characteristic CT images. Brain Dev 1988; 10:60 (letter).
28. Topaloglu H, Yalaz K, Renda Y, et al. Occidental type cerebro-muscular dystrophy: a report of eleven cases. J Neurol Neurosurg Psychiatry 1991; 54:226–229.
29. Egger J, Kendall BE, Erdohazi M, Lake BD, Wilson J, Brett EM. Involvement of the central nervous system in congenital muscular dystrophies. Dev Med Child Neurol 1983; 26:32–42.
30. Stern LM, Albertyn L, Manson JI. Fukuyama congenital muscular dystrophy in two Australian female siblings. Dev Med Child Neurol 1990; 32:808–819.
31. Trevisan CP, Carollo C, Segalla P, Angelini C, Drigo P, Giordano R. Congenital muscular dystrophy: brain alterations in an unselected series of Western patients. J Neurol Neurosurg Psychiatry 1991; 54:330–334.
32. Streib EW, Lucking CH. Congenital muscular dystrophy with leucoencephalopathy. Eur Neurol 1989; 29:211–215.
33. Kohrman MH, Picchietti DL, Wollmann R, Chelmicka-Schorr EE. A variant of Fukuyama congenital muscular dystrophy in a non-Japanese child. Pediatr Neurol 1986; 2:290–293.

34. Knubley WA, Bertorini T. Congenital muscular dystrophy with cerebellar atrophy. Dev Med Child Neurol 1988; 30:378–380.
35. Echenne B, Pages M, Marty-Double C. Congenital muscular dystrophy with cerebral white matter spongiosis. Brain Dev 1984; 6:491–495.
36. Vles JSH, Dekrom MCTFM, Visser R, Höweler CJ. Two Dutch siblings with congenital muscular dystrophy (Fukuyama type). Clin Neurol Neurosurg 1983; 85(3):175–180.
37. Arikawa E, Ishihara T, Nonaka I, Sugita H, Arahata K. Immunocytochemical analysis of dystrophin in congenital muscular dystrophy. J Neurol Sci 1991; 105:79–87.
38. Echenne B, Chevron MP, Pons F, et al. Congenital muscular dystrophy with absence of dystrophin. Pediatr Neurol 1992; 8:372 (abstract).
39. Topaloglu H, Dinger P, Gögüs S, Ayter S, Topcu M. Unusual case of Duchenne muscular dystrophy. Pediatr Neurol 1992; 8:410 (abstract).
40. Kinoshita M, Nishina M, Koya N. Ten years follow-up study of steroid therapy for congenital encephalomyopathy. Brain Dev 1986; 8:280–284.

12
Rigid Spine Syndrome

Luciano Merlini
Rizzoli Orthopedic Institute, Bologna, Italy
Claudia Granata
Rizzoli Orthopedic Institute and Toniolo Hospital, Bologna, Italy

INTRODUCTION

Rigid spine syndrome (RSS) was first described in 1973 by Dubowitz (1) as a benign myopathic condition in which the predominant feature is a marked limitation in flexion of the whole dorsolumbar and cervical spine due to contracture of the spinal extensors. The myopathy tends to be nonprogressive except for the development of scoliosis and other joint contractures. The condition has an early onset but the neonatal period is normal.

More recently (2–20) it became clear that the most important (16) and constant feature of RSS was the development of a restrictive ventilatory syndrome often so severe to be life-threatening or to require mechanical ventilation (12,14–16,19,20). The delay in recognizing the restrictive ventilatory impairment as the hallmark of RSS and the fact that contracture of spinal exstensors is present in other neuromuscular disorders (15) led both to revisions of the diagnosis of RSS (21,22) and to denial that the syndrome exists (7,9,23,24). It is thus important to reestablish precise criteria of diagnosis distinguishing this condition from the many others that have only the "rigid spine" sign (15) in common with it.

CLINICAL FEATURES

The features that establish the diagnosis of RSS are the presence of all the following:

1. Early marked limitation of flexion of cervical and dorsolumbar spine and, frequently, a progressive scoliosis in the teens
2. Mild, nonprogressive proximal muscle wasting and weakness but severe involvement of the axial, especially paravertebral, muscles
3. Severe restrictive respiratory syndrome, which may develop in childhood, in adolescence, or later
4. Muscle biopsy showing nonspecific myopathic features sometimes with marked proliferation of connective and adipose tissue

Additional features are:

1. Moderately elevated serum creatine kinase (CK)
2. Contractures of other joints, especially at the elbows and ankles, with resulting flexion of the elbow and equinus of the feet.

The syndrome is usually recognized during infancy or childhood. The child is not floppy and does not have multiple contractures at birth. In some there is delay in reaching the motor milestones, such as standing or walking (25). Signs of muscle weakness, such as waddling gait and difficulty in climbing stairs and running, appear during infancy or early childhood. The contractures develop early and sometimes precede or mask the motor signs. Not infrequently, the presenting symptom is a nonpainful limitation in flexion of the neck (Fig. 1) and trunk or equinus of the feet. Some patients develop normally until about 7 years of age, when limb weakness or spine rigidity is recognized. The limitation of movement, particularly flexion of the spine, is the main feature, and weakness is of secondary importance and relatively nonprogressive (1). The rigid spine is explained by a shortening of the extensors of the spine, which results in the characteristic posture. The neck is held in hyperextension with the trunk tilted stiffly forward so that the arms are suspended anterior to the pelvis (26). Contractures are usually present also at the elbows and ankles, resulting in limited extension of the elbow and equinus of the foot. Scoliosis is frequent in patients who develop rigid spine during childhood. Once scoliosis appears, it tends to worsen (Fig. 2) especially during adolescence, leading in extreme cases to loss of balance of the trunk so that standing becomes impossible. Muscular strength is selectively affected. The loss of strength is marked in the axial muscles and affects the flexors (especially sternocleidomastoid and abdominals) more than the extensors of the spine. In the limbs, shoulder, and pelvic girdles muscular power is good (MRC grade 4 or 5) even though the muscles are very thin.

Respiratory function is invariably impaired (15) in RSS and its severity governs the prognosis. Some patients have a 40–60% reduction of vital capacity (VC), which tends to remain stationary: these have a fair prognosis and are often symptomless, although hypoventilation can be demonstrated during sleep (20). Others have a more severe restrictive syndrome or one that tends to worsen. These patients develop symptoms of chronic nocturnal hypoventilation (dyspnea, nightmares, morning headache, diurnal sleepiness, palpitation, increased fatigue, loss of weight), which may precede ventilatory decompensation with hypercapnia and hypoxia even in the daytime to the point of cardiocirculatory collapse and death (9,11,15,27–29).

Cardiomyopathy is not a common feature of RSS, although a few reports have shown this association (5,9,14,27). Complete heart block resulted in the death of a young man (5) and of a 7½-year-old girl with hypertrophic cardiomyopathy (27). In RSS the cardiac changes such as right ventricular hypertrophy and right-sided heart failure are most likely secondary to pulmonary hypertension following the severe restrictive respiratory failure (9,14).

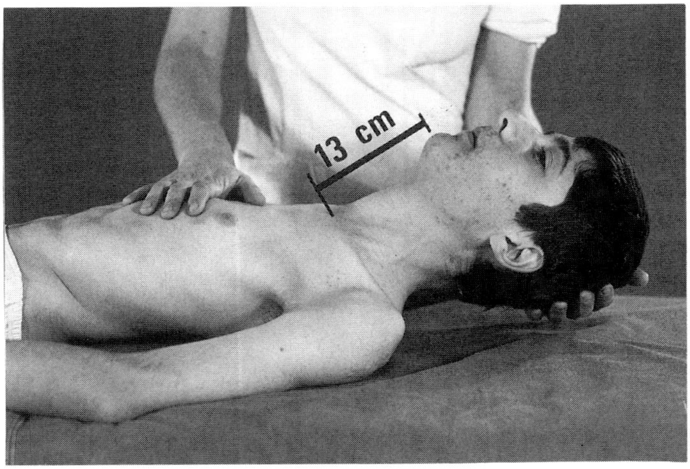

Figure 1 Measurement of the chin-sternum distance (in centimeters) to define the severity of rigid spine at the cervical level. The patient is lying on his back and his neck and head are passively flexed as much as possible.

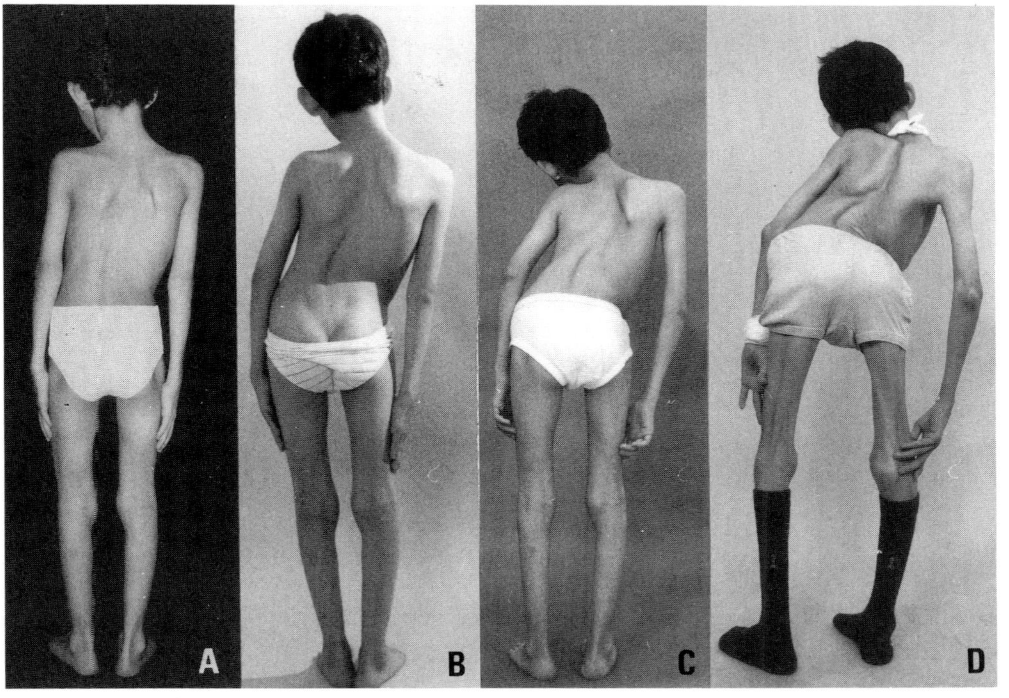

Figure 2 Boy with rigid spine syndrome who had a progressive worsening of scoliosis from the ages of 11 years (A), 13 years (B), 15 years (C), and 17 years (D). During this period there was no progression of weakness, but the restrictive respiratory syndrome deteriorated steadily, and he had to sleep in an iron lung from the age of 13 years and subsequently had a tracheotomy (D).

LABORATORY FINDINGS

The CK may be normal but more often it is elevated, from moderately up to 10 times the upper limit of normal (1,3,9,15,25,30). The electromyographic recordings show a myopathic pattern (1,25,30), and nerve conduction velocity is normal (1).

Muscle computerized tomography (CT) (16,19,31,32) shows atrophy and lower density of the paravertebral muscles (Fig. 3), especially at the cervical and upper thoracic levels, while the intercostal muscles are spared (16). X-ray (Fig. 4) shows marked limitation of the cervical spine, both in flexion and in extension (15,33), and the peculiar combination ("alligator sign") (33) of marked C2–C7 extension and C1–C2 flexion in the neutral position of the neck.

PATHOLOGICAL FEATURES

Muscle biopsy of the spinal (1,25) and the limb muscles (1,3,9,25) shows a "myopathic" pattern with variation in fiber size, some degenerating fibers, and proliferation of connective tissue and adipose tissue. Inflammatory changes are usually absent. Fiber type predominance is sometimes present (6,10,26,30,34–38). Electron microscopic studies have not shown any distinguishing features except for the presence of Z-band streaming (6). Autopsy in an 8-year-old boy who died of pneumonia showed no abnormality in the spinal cord or peripheral nerves and an almost complete replacement by connective tissue (29) of the neck extensor muscles.

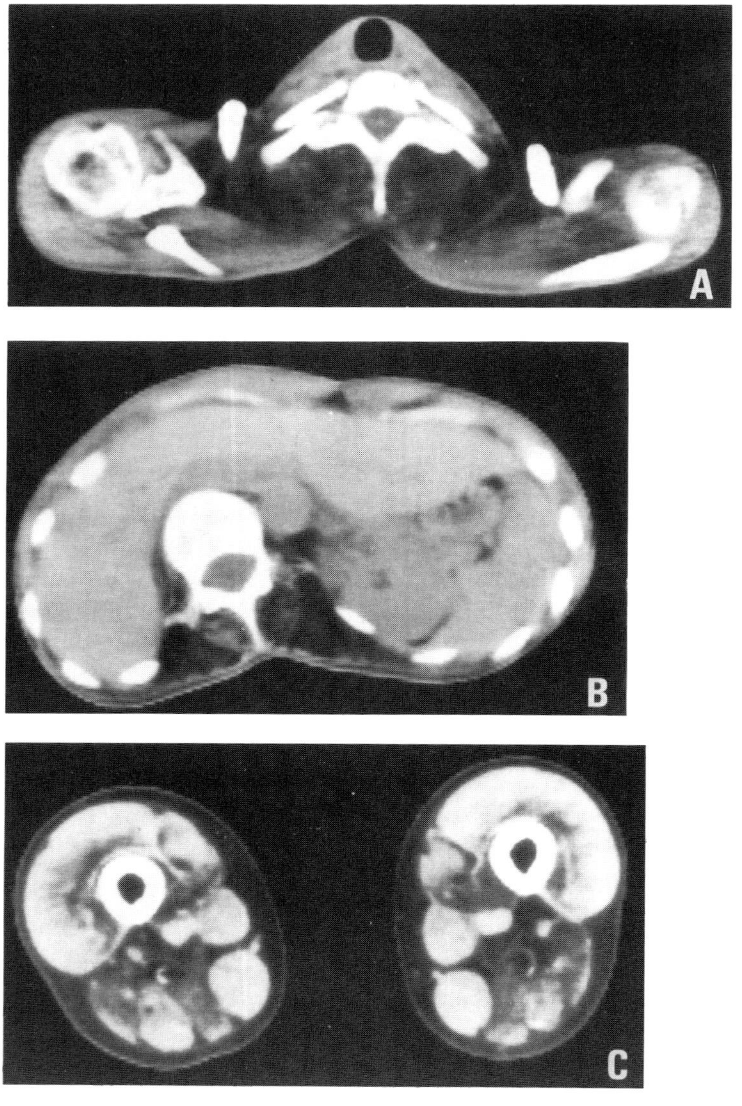

Figure 3 Muscle CT scans of a 17-year-old boy with rigid spine syndrome. The paravertebral muscles at the cervical (A), thoracic (not shown), and lumbar (B) levels are atrophic and show end-stage replacement by lower-density connective tissue and fat. The thigh muscles (C) are comparatively better preserved with sparing of vastus lateralis, sartorius, and gracilis muscles.

PATHOGENESIS

The limitation of spine flexion is due to a fibrous shortening of the paravertebral muscles (1,16,19,26, 31). As it cannot be modified under anesthesia, this rules out muscle spasm (25). The typical RSS posture is partly functional: to look forward, a patient with hyperextended rigid cervical spine has to keep his trunk bent forward.

The severe restrictive respiratory dysfunction is mechanical in origin. It is due to extreme flattening of the chest (11,16,39) and to the contracture of costovertebral joints, which markedly limit thoracic

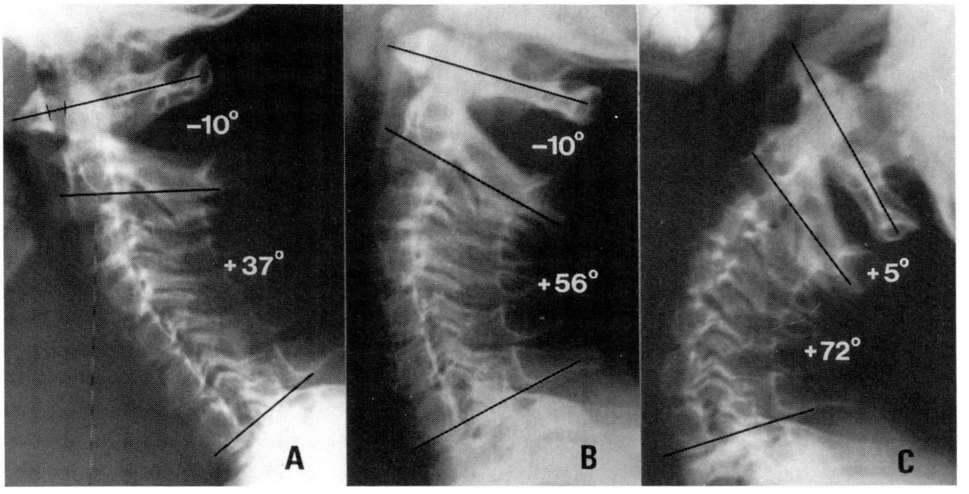

Figure 4 Radiographic study of the cervical spine in a 10-year-old boy with rigid spine syndrome. Lateral X-rays were taken in three postures: (A) maximal flexion, (B) neutral position, and (C) maximal extension. The angular displacement is measured between three lines: the first is a line of the atlas that runs through the center of the anterior arch and the posterior tubercle, the second is the line of the axis from the lower margin outline of the transverse process to the lowest point of the spinolaminar profile, and the third is the line of the inferior border of the C7 body and the lowest point of the spinolaminar profile. The angular displacement of C1–C7 from maximal extension (C) to maximal flexion (A) is 50° ([72° + 5°]–[-10° + 37°]) (normal, up to 120°). The range of extension is 31° ([72° + 5°]–[-10° + 56°]) (normal, up to 65°), and the range of flexion is 19° ([-10° + 56°]–[-10° + 37°]) (normal, up to 55°). In the neutral position (B), the vertebrae from C2 to C7 are in extension (spinous processes are closed together) partially compensated by an almost complete flexion of C1–C2. Because of this flexion, the laminae of C1 and C2 are wide open posteriorly, looking like the open jaws of an alligator ("alligator sign"). (From Ref. 15 and Ref. 33.)

movements during respiration (6,11,16). No atrophy of intercostal muscles is evident on CT scan (16), so atrophy and weakness of the respiratory muscles are not important factors in the respiratory failure, but the thin and fixed thorax is important (16).

Subjects with RSS are as a rule short in stature and underweight (2,6,10,11). We (2) reported a boy and a girl both aged 14 with a height and weight below the 5th percentile and a severe RSS (36% and 22% of VC, respectively). The hyposomia and thinness are possibly due to chronic hypoventilation (3,5,6,10,11,16), as correction of the latter allows some catchup growth and weight gain in growing youngsters and weight gain in the others (16).

GENETIC FEATURES

Most cases of RSS are sporadic, and at first there was a distinct predominance of boys. Later several descriptions of the syndrome were reported in females (2,3,5,6,9–11,15,16,18,20,27,28,31,34,35,39–48) and within families (9,11,15,31,37,39–41,47,49,50). Aver'ianov et al. (40) reported three familial cases with recessive inheritance.

TREATMENT

Orthopedic treatment is usually given to correct the equinus of the feet. Spine surgery for scoliosis has been done only occasionally (2,9,33). In one of our patients (2) this surgery afforded no permanent benefit, as it was followed by recurrence (15). Giannini et al. (33) reported a 10-year-old boy in whom they corrected the hyperextension of the neck by opening the C2–C7 interspinous spaces through a capsulotomy and stabilized it with bone grafts. The boy with the straightened cervical spine was then able to look forward without any compensatory posture.

Since it has been understood that the major effect of the disease is development of a life-threatening restrictive respiratory failure (2–20), the recommended measures are monitoring of the respiratory function with spirometry (20), BGA (19), and sleep polysomnography (13–15,19,20). Mechanical ventilation at night (12,14–16,19,20) affords protection from respiratory failure and improves the quality of life.

DIFFERENTIAL DIAGNOSIS

The diagnosis of RSS is chiefly one of exclusion since many other muscular disorders may present limitation of flexion of the neck or trunk due to shortening of the trunk extensors (15). Emery-Dreifuss muscular dystrophy (EDMD), congenital muscular dystrophy, and congenital myopathies must be considered in the differential diagnosis. EDMD is an X-linked condition characterized by early contractures, humeroperoneal muscle wasting and weakness, and cardiomyopathy with conduction defect (23,51). Cases that were initially described as RSS (30) have subsequently been revised to EDMD (21,22). Limited flexion of neck and trunk is reported in half of EDMD cases (52) but is not so marked as in RSS, whereas flexion of the elbows and equinus of the feet is much more severe. Scoliosis is recorded in only one member of EDMD family number 1 (53) but not in other families (15). A severe restrictive respiratory syndrome is a constant feature of RSS but not of EDMD (51,52), whereas the reverse is true for cardiomyopathy with conduction defect. The recent discovery of the EDMD gene (Toniolo, in press) will improve the diagnosis of this condition. Congenital muscular dystrophy (54) is a condition in which the child shows muscle weakness, hypotonia, and multiple contractures at birth, delayed motor milestones, and striking muscle histopathological changes similar to those of muscular dystrophy (29). Both scoliosis and rigid spine may be present in this condition (10,41,55,56), but the presentation at birth and the severity of the muscle biopsy findings help in the differentiation from RSS. Rigid spine is also present in several congenital myopathies such as central core disease (57), minicore (58,59), and centronuclear disease (15), in which muscle biopsy is diagnostic.

In summary, the rigid spine sign must be distinguished from RSS. Rigid spine is a clinical sign (15) that is present in several neuromuscular disorders in which the replacement of the spinal extensor muscles by fibrous and adipose tissue results in their shortening and inhibits the flexion of the spine. RSS is a particular clinical entity characterized by mild, nonprogressive proximal weakness and severe involvement of the trunk muscles that leads to fixation and loss of movement of the spine and the thoracic cage, ending in rigid spine, scoliosis, and severely restrictive respiratory failure.

ACKNOWLEDGMENT

This work was supported by Italian Telethon (project no. 340).

REFERENCES

1. Dubowitz V. Rigid spine syndrome: a muscle syndrome in search of a name. Proc Roy Soc Med 1973; 66:219–220.
2. Savini R, Cervellati S, Merlini L. La sindrome del rachide rigido. Prog Patol Vertebr 1979; 2:183–189.
3. Prost A, Mussini JM, Audrian M, Bainvel JV, Mathè JF, Cottin S. Syndrome de la colonne raide. Une observation. Rev Rhum 1980; 47:49–52.
4. Abousamra S, de Vernejoul MC, Huchon G, et al. [Pleuropulmonary disease and rigid spine syndrome (author's transl)]. Ann Med Interne Paris 1981; 132:202–204.
5. Mussini JM, Gray F, Hauw JJ, Piette AM, Prost A. Rigid spine syndrome: histological examinations of male and female cases. Acta Neuropathol 1981; 7:(Suppl VII):331–333.

6. Mussini JM, Mathè JF, Prost A, Gray F, Labat JJ, Fève JR. Le syndrome de la colonne raide. Un cas féminin. Rev Neurol 1982; 138:25–37.
7. Serratrice G, Pellissier JF, Pouget J, Gastaut JL. Le syndrome de la colonne vertébrale rigide et ses frontières nosologiques. Deux observations. Presse Méd 1984; 13:1129–1132.
8. Jusic A. Rigid spine snydrome associated with tent-like configuration of the thoracic cage, cerebral lesion and delayed sexual maturation in three patients. J Neurol 1985; 232:32–37.
9. Poewe W, Willeit H, Sluga E, Mayr U. The rigid spine syndrome—a myopathy of uncertain nosological position. J Neurol Neurosurg Psychiatry 1985; 48:887–893.
10. Bertini E, Marini R, Sabetta G, et al. The spectrum of the so-called rigid spine syndrome: nosological considerations and report of three female cases. J Neurol 1986; 233:248–253.
11. Van Munster ETL, Joogsten EM, Van Munster-Uijtdehaage MAM, Kruls HJA, Ter Laak HJ. The rigid spine syndrome. J Neurol Neurosurg Psychiatry 1986; 49:1292–1297.
12. Winter JH, Neilly JB, Henderson AF, et al. Life-threatening respiratory failure due to a previously undescribed myopathy. Q J Med 1986; 61:1171–1178.
13. Cirignotta F, Mondini S, Zucconi M. Breathing impairments during sleep in rigid spine syndrome. Sleep Res 1987; 16:417 (abstract).
14. Efthimiou J, McLelland J, Round J, Gribbin HR, Loh L, Spiro SG. Diaphragm paralysis causing ventilatory failure in an adult with the rigid spine syndrome. Am Rev Respir Dis 1987; 136:1483–1485.
15. Merlini L, Granata C, Ballestrazzi A, Marini ML. Rigid spine syndrome and rigid spine sign in myopathies. J Child Neurol 1989; 4:274–282.
16. Morita H, Kondo K, Hoshino K, Maruyama K, Yanagisawa N. Rigid spine syndrome with respiratory failure. J Neurol Neurosurg Psychiatry 1990; 53:782–784.
17. Onodera O, Yamazaki M, Atsumi T, Miyatake T, Izumi T. [A case of rigid spine syndrome with rimmed vacuole]. Rinsho Shinkeigaku 1990; 30:516–519.
18. Bein T, Lenhart FP, Berger H, et al. [Rupture of the trachea during difficult intubation]. Anaesthesist 1991; 40:456–457.
19. Akiyama Y, Aimoto Y, Nishimura M, Takai S, Kawakami Y. Rigid spine syndrome with selective respiratory muscle weakness. Respiration 1992; 59:48–51.
20. Kawata A, Suga M, Miyamoto K, Hirose K, Tanabe H. Rigid spine syndrome and nocturnal alveolar hypoventilation. Intern Med 1993; 32:638–640.
21. Voit T, Krogman O, Lenard HG, et al. Emery-Dreifuss muscular dystrophy: disease spectrum and differential diagnosis. Neuropediatrics 1988; 19:62–71.
22. Dubowitz V. A Colour Atlas of Muscle Disorders in Children. London: Wolfe Medical, 1989:38.
23. Rowland LP, Ferell M, Olarte M, Hays A, Singh N, Wanar FE. Emery-Dreifuss muscular dystrophy. Ann Neurol 1979; 5:111–117.
24. Vita G, Girlanda P, Santoro M, Messina C. Is rigid spine syndrome a distinct clinical entity? Muscle Nerve 1983; 6:458–460.
25. Dubowitz V. Muscle Disorders in Childhood. Philadelphia: Saunders, 1978:239.
26. Seay AR, Ziter FA, Patajan JH. Rigid spine syndrome. Arch Neurol 1977; 34:119–122.
27. Colver AF, Steer CR, Godman MJ, Uttley WS. Rigid spine syndrome and fatal cardiomyopathy. Arch Dis Child 1981; 56:143–148.
28. Pavone L, Gullotta F, La Rosa M, Grasso A. Rigid spine syndrome. Some evidence of varying pathological patterns. Helv Paediatr Acta 1983; 38:367–372.
29. Dubowitz V. Muscle Biopsy: A Practical Approach. London: WB Saunders, 1985:616.
30. Goebel HH, Lenard HG, Gorke W, Kunze K. Fibre type disproportion in the rigid spine syndrome. Neuropaediatrie 1977; 8:467–477.
31. Vanneste JA, Augustijn PB, Stam FC. The rigid spine syndrome in two sisters. J Neurol Neurosurg Psychiatry 1988; 51:131–135.
32. Matsumoto W, Umehara F, Fukunaga H, Okadome T, Osame M. [A case of scapuloperoneal atrophy with rigid spine having lobulated fibers in muscle biopsy]. Rinsho Shinkeigaku 1991; 31:417–421.
33. Giannini S, Ceccarelli F, Granata C, Capelli T, Merlini L. Surgical correction of cervical hyperextension in rigid spine syndrome. Neuropediatrics 1988; 19:105–108.
34. Lossner VJ, Oertel G, Ziegan J. Rigid spine-Syndrom. Seltene Kombination mit einer neuromuskularen Affektion. Beitr Orthop Traumatol 1981; 28:429–433.
35. Dougados M, Tourbet A, Amor B. Le syndrome de la colonne vertébrale rigide. Rev Rhum 1986; 53:265–267.
36. Tardieu M, Schovart D, Metral S, Landrieu P. [The rigid spinal column syndrome. Aspects of hypotrophy and predominance of type I muscular fibers]. Arch Fr Pediatr 1986; 43:115–118.
37. Takase Y, Yamamoto K, Nogaki H, Fukusako T, Sasabe F, Morimatsu M. Rigid spine syndrome associated with cardiomyopathy: clinical and nosological considerations. Jpn J Med 1990; 29:555–559.
38. Tardieu M, Schovart D, Metral S, Landrieu P. Syndrome de la colonne raide. Aspect d'hypotrophie et prédominance des fibres musculaires de type I. Arch Fr Pediatr 1986; 43:115–118.

39. Zheng HM, Zhou GZ, Ding SJ. Rigid spine syndrome. Report of 4 cases. Chin Med J Engl 1989; 102:702–708.
40. Aver'ianov IuN; , Il'ina NA, Vinogradova NV, Potomskaia LZ, Sokolina NA. [Rigid spine syndrome]. Zh Nevropatol Psikhiatr 1981; 81:1457–1461.
41. Echenne B, Astruc J, Brunel D, Pages M, Baldet P, Martinazzo G. Congenital muscular dystrophy and rigid spine syndrome. Neuropediatrics 1983; 14:97–101.
42. Ben Hamida M, Hentati F, Ben Hamida C, et al. [Case of the rigid spine syndrome in a female patient]. Tunis Med 1986; 64:645–648.
43. Drndarski I, Todorovic S. [The rigid spine syndrome]. Srp Arh Celok Lek 1989; 117:519–530.
44. Emery AE. Emery-Dreifuss muscular dystrophy and other related disorders. Br Med Bull 1989; 45:772–787.
45. Tanaka K, Yoshimura T, Muratani H, Kira J, Itoyama Y, Goto I. Familial myopathy with scapulohumeral distribution, rigid spine, cardiopathy and mitochondrial abnormality. J Neurol 1989; 236:52–54.
46. Todorovic S. Rigid spine syndrome and progressive external opththalmoplegia in a 15-year-old girl. Dev Med Child Neurol 1989; 31:811–815.
47. Palmucci L, Mongini T, Doriguzzi C, Maniscalco M, Schiffer D. Familial autosomal recessive rigid spine syndrome with neurogenic facio-scapulo-peroneal muscle atrophy. J Neurol Neurosurg Psychiatry 1991; 54:42–45.
48. Wehnert M, Machill G, Grimm T, et al. Evidence supporting tight linkage of X-linked Emery-Dreifuss muscular dystrophy to the factor VIII:C gene. Hum Hered 1991; 41:364–369.
49. Di Iorio G, Lus G, Cutillo C, Cecio A, Cotrufo R. Histopathological heterogeneity and cytopathological similarity of findings in different muscles of two brothers affected by rigid spine syndrome. J Neurol Sci 1989; 94:107–114.
50. Ricci E, Bertini E, Boldrini R, et al. Late onset scleroatonic familial myopathy (Ullrich disease): a study of two sibs. Am J Med Genet 1988; 31:933–942.
51. Yates JRW. European Workshop on Emery-Dreifuss muscular dystrophy 1991. Neuromusc Disord 1991; 1:393–396.
52. Merlini L, Granata C, Dominici P, Bonfiglioli S. Emery-Dreifuss muscular dystrophy: report of five cases in a family and review of the literature. Muscle Nerve 1986; 9:481–485.
53. Thomas PK, Calne DB, Elliot CF. X-linked scapuloperoneal syndrome. J Neurol Neurosurg Psychiatry 1972; 35:208–215.
54. Lazaro RP, Fenichel GM, Kilroy AW. Congenital muscular dystrophy: case reports and reappraisal. Muscle Nerve 1979; 2:349–355.
55. Lenard HG, Goebel HH. Congenital muscular dystrophies and unstructured congenital myopathies. Brain Dev 1980; 2:119–125.
56. Serratrice G, Cros D, Pellissier JF, Gastaut JL, Pouget J. Dystrophie musculaire congénitale. Rev Neurol 1980; 136:445–472.
57. Merlini L, Mattutini P, Bonfiglioli S, Granata C. Nonprogressive central core dissease with severe congenital scoliosis: a case report. Dev Med Child Neurol 1987; 29:106–109.
58. Ben Hamida M, Hentati F, Ben Hamida C. Maladie a multiminicores au cours d'un syndrome de la colonne raide. Rev Neurol 1987; 143:284–289.
59. Merlini L, Granata C, Govoni E. Miopatia congenita tipo minicore con rachide rigido. Arch Ortop Reumatol 1983; 96:455–459.

13
Bethlem Myopathy

Gerrit Joost Jöbsis and Marianne de Visser
Academic Medical Center, Amsterdam, The Netherlands

INTRODUCTION

Myopathies with limb-girdle distribution of weakness and autosomal dominant inheritance constitute a disparate group of disorders (1). Due to the development of sophisticated diagnostic methods, a classification has emerged that distinguishes well-defined disorders such as congenital and metabolic myopathies. However, there still exists a wide variety of autosomal dominant myopathies that cannot be easily classified. Within this broad category, the autosomal dominant myopathy described by Bethlem and van Wijngaarden in three unrelated Dutch families in 1976 is considered a distinct nosological entity (2). This benign myopathy is characterized by the following salient clinical features: slow progression of limb-girdle weakness from childhood onward with periods of arrest for several decades; contractures of fingers, elbows, and ankles; and absence of cardiac involvement. Subsequently, several other families with a similar condition have been identified, both by Bethlem and associates (3,4) and by others (5–10). In 1988 the name "Bethlem myopathy" was proposed by Mohire et al. (5).

CLINICAL FINDINGS

There exists some variation in the literature regarding the age of onset. In Bethlem's original article onset was around the fifth year of life, weakness being the first symptom. From subsequently published kindreds it has become apparent that onset can be either in infancy with slightly delayed developmental milestones or in early childhood (3,5,7). Somer et al. reported a Finnish family that displayed all the characteristics of Bethlem myopathy except for a wide variation in the age of onset ranging from early infancy to the fifth decade. Most patients of this kindred had an onset in adolescence (8). In addition to a new kindred, Boers reevaluated the previously published Dutch families in detail with regard to presenting symptoms (4). Reliable data could be acquired from 13 of 27 patients. There was a convincing history of reduced fetal movements during pregnancy in two. In seven there had been floppiness in the neonatal period that had gradually decreased during the first year of life. In two kindreds an extension contracture of the ankles was present at birth in all affected individuals but one. Atypical crawling due to proximal paresis was seen in eight and inability to walk or to stand without support before the age of 1½ years was noted in seven children. A positive Gowers sign was present after 2 years of age in nearly all.

During childhood there is difficulty in arising from a squat, clumsy or waddling gait, easy tripping, and an inability to run. Typically insidious progression is observed in the first decade, later followed by a relative increase in muscle power. During adolescence the positive Gowers sign, which is invariably present in childhood, disappears. In some female patients muscle strength deteriorates during pregnancy. After the third decade muscle weakness tends to worsen gradually, hallmarked by the reappearance of the Gowers sign. However, in most patients ability to work is preserved until old age. In their sixth or seventh decade some patients become wheelchair-dependent for out-of-doors transport (unpublished observations), whereas in others a walking cane suffices. Life expectancy is unaffected.

There is generalized slight atrophy of the musculature and diffuse mild weakness, proximal more severe than distal. Extensors are weaker than flexors. The muscles innervated by the cranial nerves are preserved. In virtually all kindreds hypertrophic calves are not present, except for one individual from a Finnish pedigree (8) and multiple sibs from two French families (6). Hypertrophy of the extensor digitorum brevis muscle has been reported in one pedigree (10). Flexion contractures of the interphalangeal joints of the four last fingers (Fig. 1), elbows (Fig. 2), and ankles (pes equinovarus) are present in nearly all patients. The former, caused by a shortening of the flexor digitorum profundus muscle, is most striking at full wrist extension. Contractures of the metacarpophalangeal joints, wrists, knees, hips, and shoulders are encountered in many patients. In a few families contractures of the neck are present in some patients whereas spine rigidity seems even rarer (5,10). There is no relationship between the contractures and the severity of the muscular weakness. One author reports a slow progression with age (10). Congenital torticollis due to contracture of the sternocleidomastoideus muscle and kyphoscoliosis are observed in a few patients of some kindreds. Deep tendon reflexes are either preserved, decreased, or absent, irrespective of the severity of weakness.

ADDITIONAL INVESTIGATIONS

Serum creatine kinase (CK) activity is usually either normal or slightly elevated (2–5 times the upper limit). In one study a moderate increase in CK activity (up to 15 times the upper limit) was noted in three young individuals (10).

Generally, the electromyogram (EMG) shows on voluntary activity short-duration, small-amplitude, rapidly recruiting potentials with increased polyphasia. In most patients there is no spontaneous muscle activity at rest (2,5); however, in some patients some fibrillation potentials, positive sharp waves, and high-frequency bizarre discharges are present (4,8). Satellite potentials and long-duration motor unit action potentials have been noted by some authors (4,8). In a minority of cases the EMG is entirely normal or shows a neurogenic picture with mildly prolonged insertional activity and reduced number of increased-amplitude polyphasic motor units (4–8). Motor conduction velocities are normal.

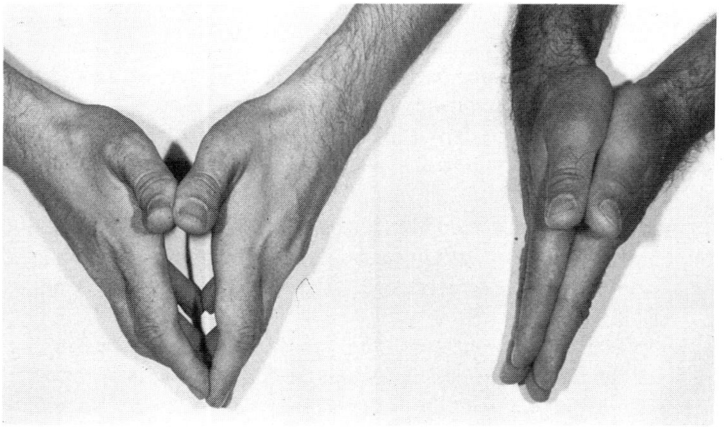

Figure 1 Flexion contractures of the fingers in a patient with Bethlem myopathy (left) compared to a normal subject (right).

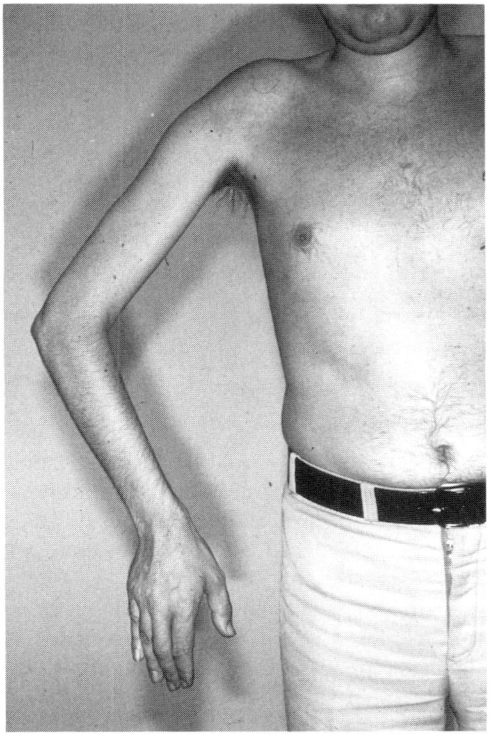

Figure 2 Flexion contracture of the elbow in addition to mainly proximal muscle wasting.

Computerized tomography scans of the musculature show diffuse replacement of muscle tissue by fat, proximal more than distal and increasing with age (4). Unexpected early involvement of the lumbar paravertebral muscles was reported by Merlini et al. (10). On magnetic resonance imaging, which has been done in one published family, selective involvement of the quadriceps femoris muscle was noted, whereas the gracilis and sartorius muscles and the hamstrings were relatively spared (8).

Cardiac examinations, including echocardiography, are unrevealing (11).

PATHOLOGICAL FEATURES

Muscle biopsy specimens show nonspecific myopathic changes, including variation in muscle fiber diameter with a moderate increase in fibers with internal nuclei, a marked increase in fatty tissue, and a small increase in connective tissue. Lobulated or moth-eaten type 1 fibers have been noted in a few patients of some pedigrees (2,7,8). Occasional necrosis and phagocytosis were reported in one study (10); otherwise structural abnormalities are absent. Electron microscopy reveals no specific changes in the muscle fibers.

Three autopsies have been reported showing no convincing abnormality of brain, spinal cord, or peripheral nerves (2,5). The distribution of pathological features in skeletal muscles is similar to that found clinically. Histopathological findings are comparable to those of muscle biopsies. Except for changes consistent with atherosclerotic coronary artery disease, microscopic examination of the heart is normal (5).

THERAPY

No specific therapy is known, but patients may benefit from reassurance about the natural course of this benign myopathy and from general supportive measures commonly used in rehabilitation medicine. Achilles tendon shortening and torticollis might be relieved by surgery.

GENETIC FEATURES

The underlying biochemical defect of Bethlem myopathy remains unknown. From the published pedigrees it can be inferred that the mode of inheritance is autosomal dominant with complete penetrance and virtually constant expression. Recently, genetic linkage has been established in six Dutch pedigrees with markers on the distal end of the long arm of chromosome 21 (12).

DIFFERENTIAL DIAGNOSIS

Bethlem myopathy shares many features with Emery-Dreifuss muscular dystrophy (EDD) and the rigid spine syndrome (RSS), especially the slowly progressive weakness of limb-girdle musculature, an onset in infancy or early childhood, and widespread contractures (13,14). However, due to cardiac involvement both EDD and RSS run a less benign course. EDD has an X-linked recessive inheritance but autosomal dominant forms do exist (15). RSS usually occurs sporadically and mostly in males.

Several congenital myopathies have a limb-girdle distribution of weakness with slow or absent progression (16). Flexion contractures sometimes occur. Muscle biopsy will lead to the correct diagnosis.

Limb-girdle muscular dystrophy with autosomal dominant inheritance usually has its onset after adolescence (1), although exceptions to this rule occur (17,18). Progression is more rapid than in Bethlem myopathy and contractures are absent (1,19).

Although Bethlem myopathy is apparently a distinct nosological entity, some reported kindreds share many, but not all, features with Bethlem myopathy, thus making a definite classification rather difficult. The Finnish family reported by Somer et al. is somewhat atypical because of its wide variation in onset and hypertrophic calves in one sib (8). One of the two reported French kindreds displays calf hypertrophy in all affected sibs and has an adult onset (6). In the other kindred age of onset is variable and calf hypertrophy is present in some individuals. Since all affected individuals described belong to one generation, there is no evidence for clear autosomal dominant inheritance.

A family described by Bailey et al. showed onset of limb-girdle weakness in early infancy with delayed developmental milestones and very slow progression, contractures of elbows, heel cords, and knees, pes equinovarus, spine rigidity and thoracic scoliosis, absence of cardiac involvement, and a normal or moderately elevated CK (20). The findings of EMG and muscle biopsy were consistent with a myopathy. Inheritance was suggestive of an autosomal dominant trait with incomplete penetrance. A peculiar facies with frontal bossing and low-set ears in multiple cases and brisk tendon reflexes with ankle subclonus were features atypical of Bethlem myopathy.

Schmalbruch et al. described a family with benign limb-girdle weakness with onset in early childhood and autosomal dominant inheritance, Achilles tendon shortening, and torticollis in some individuals (21). Although not mentioned in the clinical description, contractures of the elbows can be observed from the photograph of one patient. However, the muscle biopsies showed features atypical of Bethlem myopathy, i.e., a necrotizing myopathy with pronounced regeneration and formation of aberrant myofibrils (*ringbinden*) and prominent fibrosis.

REFERENCES

1. Shields RW. Limb girdle syndromes. In: Engel AG, Franzini-Armstrong C, eds. Myology, vol 2. New York: McGraw-Hill, 1994:1258–1274.
2. Bethlem J, van Wijngaarden GK. Benign myopathy, with autosomal dominant inheritance: a report on three pedigrees. Brain 1976; 99:91–100.

3. Arts WF, Bethlem J, Volkers WS. Further investigations on benign myopathy with autosomal dominant inheritance. J Neurol 1978; 217:201–206.
4. Boers JM. Een congenitale myopathie met contracturen. Thesis, University of Amsterdam, 1984.
5. Mohire MD, Tandan R, Fries TJ, Little BW, Pendlebury WW, Bradley WG. Early-onset benign autosomal dominant limb-girdle myopathy with contractures (Bethlem myopathy). Neurology 1988; 38:573–580.
6. Serratrice G, Pellissier JF. Benign myopathy affecting mainly the girdles with dominant autosomal inheritance in two families. Rev Neurol 1988; 144:43–46.
7. Tachi N, Tachi M, Sasaki K, Imamura S. Early-onset benign autosomal-dominant limb-girdle myopathy with contractures (Bethlem myopathy). Pediatr Neurol 1989; 5:232–236.
8. Somer H, Laulumaa V, Paljärvi L, Partanen J, Lamminen A, Pihko H, Sariola H, Haltia M. Benign muscular dystrophy with autosomal dominant inheritance. Neuromusc Disord 1991; 1:267–273.
9. Nielsen JF, Jakobsen J. A Danish family with limb-girdle muscular dystrophy with autosomal dominant inheritance. Neuromusc Disord 1994; 4:139–142.
10. Merlini L, Morandi L, Granata C, Ballestrazzi A. Bethlem myopathy: early-onset benign autosomal dominant myopathy with contractures. Description of two new families. Neuromusc Disord 1994; 4:503–511.
11. de Visser M, de Voogt WG, la Rivière GV. The heart in Becker muscular dystrophy, facioscapulohumeral dystrophy, and Bethlem myopathy. Muscle Nerve 1992; 15:591–596.
12. Jöbsis GJ, Barth PG, Wolterman RA, Baas F, Bolhuis PA, de Visser M. Bethlem myopathy: clinical and genetic aspects. Clin Neurol Neurosurg 1994; 96;199.
13. Merlini L, Granata C, Dominici P, Bonfiglioli S. Emery-Dreifuss muscular dystrophy: report of five cases in a family and a review of the literature. Muscle Nerve 1986; 9:481–485.
14. Poewe W, Willeit H, Sluga E, Mayr U. The rigid spine syndrome: a myopathy of uncertain nosological position. J Neurol Neurosurg Psychiatry 1985; 48:887–893.
15. Gilchrist JM, Leshner RT. Autosomal dominant humeroperoneal myopathy. Arch Neurol 1986; 43:734–735.
16. Fardeau M, Tomé FMS. Congenital myopathies. In: Engel AG, Franzini-Armstrong C, eds. Myology, vol 2. New York: McGraw-Hill, 1994:1487–1532.
17. Panegyres PK, Mastaglia FL, Kakulas BA. Limb girdle syndromes; clinical, morphological and electrophysiological studies. J Neurol Sci 1990; 95:201–218.
18. Miller G, Beggs AH, Towfighi J. Early onset autosomal dominant progressive muscular dystrophy presenting in childhood as a Becker phenotype, the importance of dystrophin and molecular genetic analysis. Neuromusc Disord 1992; 2:121–124.
19. Gilchrist JM, Pericak-Vance M, Silverman L, Roses AD. Clinical and genetic investigation in autosomal dominant limb-girdle muscular dystrophy. Neurology 1988; 38:5–9.
20. Bailey RO, Dentinger MP, Toms ME, Hans MB. Benign muscular dystrophy with contractures: a new syndrome? Acta Neurol Scand 1986; 73:439–443.
21. Schmalbruch H, Kamieniecka Z, Fuglsang-Frederiksen A, Trojaborg W. Benign congenital muscular dystrophy with autosomal dominant heredity: problems of classification. J Neurol 1987; 234:146–151.

14
Classification of Congenital Myopathies

Hans Hilmar Goebel
Johannes Gutenberg University, Mainz, Germany

Anna Fidzianska
Warsaw University Medical Center, Warsaw, Poland

INTRODUCTION

How did the concept of congenital myopathies develop?

The term "congenital myopathies" (CM) was introduced into the scientific literature some 35 years ago (57) when "central core disease" (CCD) was first described as "a new congenital non-progressive myopathy." However, this report did not immediately result in the clear delineation of this group of muscular disorders; even 13 years later they were simply referred to as "new" myopathies (17). Reviewing the evolution of the concept of CM as a nosological group of neuromuscular diseases reveals apparent difficulties in finding appropriate designations for individual CM, a *conditio sine qua non* for the definition of either a nosological entity or a group of similar diseases. For example, the "new congenital non-progressive myopathy" (57) was finally termed "central core disease" only a few years after the original description (2,19), while the subgroup of CM associated with centrally placed nuclei, centronuclear myopathy (CNM), was initially designated "myotubular myopathy" (60), but 5 years later renamed "familial centronuclear myopathy" (55). Again, nemaline myopathy (NM), also sometimes called rod myopathy, was described 7 years after the first report of CCD (58), but a simultaneous report on the same disorder failed to gain equal acceptance, largely because the terms nemaline and rod bodies were not used, but rather a new term, "myogranules," was coined (11). This latter name never gained nosological acceptance. Similarly, the terms "multicore" (22) and "minicore" (14) have been used more widely than "focal loss of cross striations," introduced by Engel (23) to describe identical lesions. Curiously, multicores had earlier been described by the same authors (21) as "multifocal degeneration of muscle fibres."

Use of the term "congenital" (57,58) paved the way for recognition of CM, and they are now firmly established and accepted as a separate group of childhood neuromuscular conditions. Although the term "congenital" refers to the presence of a neuromuscular condition present at the time of birth, there are exceptions to this rule: First, there are other well-defined neuromuscular conditions present at birth but not included as forms of CM. Second, there are neuromuscular conditions that are listed with the CM because of structural abnormalities in the skeletal muscle that are identical or very similar to those seen in muscle specimens of childhood CM, which, however, have a late onset and are often sporadic. Even the term "new" myopathies, possibly introduced in an attempt to circumvent the difficulties in accepting the "congenital" aspects of CM (17), was misleading, and it also encompassed early glycogenoses, e.g., infantile type II or Pompe disease. The term acknowledged the recent recognition of these congenital myopathies, but not necessarily their recent emergence as new entities, since the CM are defined in terms of morphological abnormalities in muscle fibers, and these have been recognized only since the advent and application of recent morphological techniques, principally enzyme histochemistry and electron microscopy. We may assume that CM existed before the era of these modern myopathological techniques, but were perhaps not identified because of the rather primitive and now obsolete manner of handling muscle tissue. It is remarkable, however, that central cores can be readily recognized in routine trichrome preparations, central nuclei can be identified in hematoxylin-eosin sections, and nemaline bodies or rods had been detected many years earlier in muscle specimens stained with phosphotungstic acid (PTAH) by the late Dr. Reye at the Royal Alexandra Hospital for Children in Sydney in 1961 (28). The enormous task of systematically screening muscle tissues from children with neuromuscular disorders who would have presented with symptoms of CM, to provide accurate retrospective diagnoses, has yet to be performed.

DEFINITION

What is a congenital myopathy?

Defining an individual CM or the group of CMs generates the same difficulties as accurately defining other neuromuscular diseases. For example, biochemical abnormalities in either muscle or generalized metabolism define metabolic myopathies; abnormalities in nerve supply constitute neurogenic processes; and functional impairment of the neuromuscular junction is responsible for myasthenic syndromes. In other neuromuscular conditions, problems of definition and classification are more complex. For instance, myoglobinuria can be classified among the metabolic diseases even if the cause for the myoglobinuria cannot be explained by any known metabolic effect. Defining a heterogeneous group of disorders such as muscular dystrophies becomes still more difficult. For instance, limb-girdle muscular dystrophy is part of the limb-girdle syndrome, nosologically heterogeneous itself, while distal muscular dystrophy is a form of distal myopathy. Even in classical muscular dystrophies such as the Duchenne type, problems of classification in individual patients may be encountered when Duchenne muscular dystrophy is redefined as a defect of dystrophin, since a few patients afflicted with congenital muscular dystrophy of the Fukuyama type, based on clinical course and morphological findings, have also been found to lack dystrophin completely (1) or partially (1,35).

A CM is generally defined as a primary myopathy, commencing in early childhood, more often of a hereditary nature, slowly progressive, characterized by the nonspecific features of type I fiber predominance and/or type I fiber atrophy (Fig. 1) together with specific structural lesions.

Disease-specific or disease-typical morphological abnormalities in the skeletal muscle are a crucial feature for definition of CM, and their presence in reasonable number and consistency both in an individual patient and in other patients with the same type of CM represents an important aspect of definition. Such CM, marked by disease-specific features, have to be distinguished from those neuromuscular conditions where the same morphological hallmarks may occur, but either less frequently or in association with an obviously different neuromuscular disease. For example, rods may appear in a variety of neuromuscular conditions but as an infrequent finding, and they may be seen rather consistently in a myopathy linked to infection with the human immunodeficiency virus (HIV) (8,15,59). On the other hand, definition of the morphological lesion that is considered specific or typical of a CM may not be without difficulty.

For example, various forms of "cores" have been described in the past (28), but the precise separation of "cores" from target or targetoid phenomena remains controversial, the former being a feature of CCD, a CM, the latter of neurogenic processes.

Clinical details, genetic data, and morphological features on muscle biopsy are of principal importance in diagnosing CM, while electrophysiological findings and laboratory investigations, such as creatine kinase levels, are of less significance. Neuroimaging techniques or ultrasonography have also not added substantial new data to the evaluation of patients with CM. Clinical and genetic as well as morphological findings not only allow classification of a patient's CM, but also identify the individual CM because, apart from CM-specific morphological features, certain CM not infrequently show certain characteristic clinical symptoms and signs (28). For instance, extraocular muscle weakness, especially

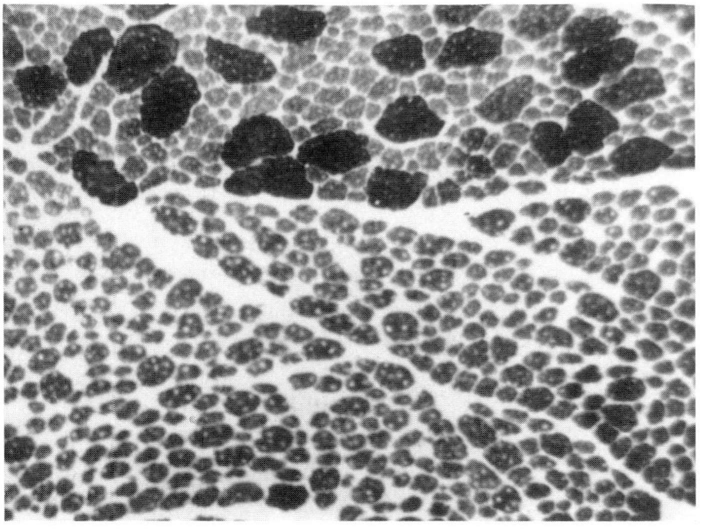

Figure 1 Type I fiber predominance and type I fiber atrophy, a typical nonspecific feature in many congenital myopathies, congenital fiber type disproportion. ATPase pH 9.3. × 220.

ptosis, is a prominent clinical sign in centronuclear myopathy, but it has also been observed in multicore disease, nemaline myopathy, and congenital fiber type disproportion. Facial involvement has been seen in patients with almost any CM. Moreover, weakness of respiratory muscles resulting in respiratory insufficiency is often associated with CM, particularly in the infantile forms of nemaline myopathy and centronuclear/myotubular myopathy, but it may also be seen in adults, who, for instance, suffer from cytoplasmic-body myopathy. In addition to skeletal muscles, cardiac muscle may be affected in a number of CM, such as minicore disease, nemaline myopathy, and granulofilamentous myopathy, and rod-like features have also been encountered in postmortem cardiac muscle in nemaline myopathy. That patients with central core disease may be prone to develop malignant hyperthermia has repeatedly been shown.

Progress in clarifying the etiopathogenesis of CM has been extremely slow. The concept of immaturity of muscle fibers, applied to explain certain morphological features in, e.g., myotubular myopathy and in a few other CM, e.g., NM, has been based on new immunomorphological techniques, in that proteins of fetal muscle fibers (desmin, vimentin) (50) and neural cell adhesion molecule (N-CAM) can now be demonstrated.

Even in adult-onset myotubular/centronuclear myopathy, desmin and vimentin show abnormal distribution (44). In nemaline myopathy, a fast type of myosin heavy chain was found defective (56), while α-actinin variants in muscle fibers of nemaline myopathy corresponded to those in controls (33). Fetal myosin was also seen in nemaline myopathy and centronuclear myopathy (54). However, these data are still too scant to allow more generalized statements, but they emphasize the growing need for and application of immunohistochemical techniques to biopsied muscle specimens in CM. Recently, even dystrophin, a cytoskeleton protein of the muscle fiber the pathology of which is usually related to Duchenne and Becker muscular dystrophies, has been found in abnormal amounts in an apparently hitherto undescribed CM (47).

Certain familial myopathies, which have not been classified with any commonly accepted category of neuromuscular diseases and often become apparent early in childhood, may, therefore, appropriately be assigned to the CM. Lysosomal glycogen storage disease without acid maltase deficiency (49) has now repeatedly been documented as a nosological entity of male patients afflicted by a myopathy and cardiomyopathy, autophagic vacuoles, and intralysosomal accretion of glycogen, but with normal biochemical activity of acid maltase. The X-linked (congenital) myopathy with excessive autophagy

(39) may be closely related to this lysosomal myopathy, but, in a single child, the X-linked character of this disorder may not be obvious (31). Bethlem myopathy (3,43), also considered a benign muscular dystrophy (53), has an early onset as a benign autosomal-dominant limb-girdle myopathy without specific myopathological features. Congenital myopathy as a sign of associated neuromuscular lesions (63) may often be part of syndromes (34,51,62). However, investigations into neuromuscular conditions, i.e., clinical symptoms and myopathology, as components of syndromes and other genetic abnormalities have not been rigorously pursued.

If such a myopathy is not named specifically, but rather called "childhood myopathy" (42) and only later characterized by delayed fiber type differentiation and cessation of fiber growth (13), it is of similar dubious nosological significance as "an unusual familial muscle disorder" (12), which was characterized by clinical heterogeneity, but conspicuous autosomal-dominant inheritance and highly atrophic muscle fibers that suggested limited denervation. Classification of such nondescript (congenital) myopathies may be quite difficult, and subsequent awareness of it may, therefore, become lost, and thus its nosological connotation may largely remain unclarified.

PROBLEMS IN CLASSIFICATION

How can we delineate individual congenital myopathies?

Starting from a mere enumeration of different forms of CM as the basis for their classification, we first separated CM with structured lesions from those without structured lesions, calling them "structured types" and "nonstructured types" (Fig. 2) (25,26,41). We have since attempted to group different forms of CM according to the underlying morphological lesion, as shown in Table 1 (28). This classification is based on structural lesions that are either derived from common normal constituents of the muscle fiber, or from those that cannot be considered derivatives of structural components of the normal muscle fiber. This, however, does not mean that CM listed within the same category, e.g., lesions that are supposed to be related to or derived from the Z-disc, can be considered linked to each other pathogenetically or etiologically. But the simultaneous occurrence, within lesioned sarcomeres, of Z-disc abnormalities and core-like features (4,52,61) suggests certain morphogenetic links.

While the frequency of structural lesions within the muscle tissue of an individual patient with the

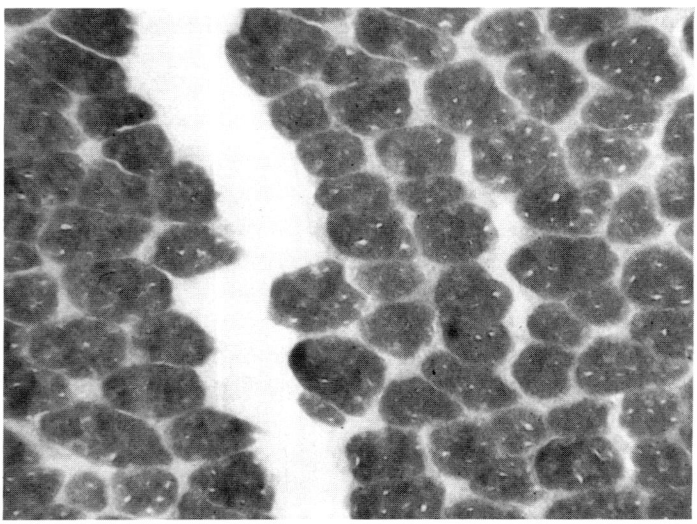

Figure 2 Type I fiber uniformity, typical of "nonstructured types" of congenital myopathies. ATPase pH 10.4. ×200.

Table 1 Neuromuscular Disorders Present at Birth or in Early Infancy

Infantile spinal muscular atrophy
Infantile polyneuropathy
Hereditary motor and sensory neuropathy type III
Infantile dystrophic myotonia
Congenital myopathies
Congenital muscular dystrophies
Congential myasthenic syndromes
Infantile type II glycogenosis
Mitochondrial myopathies (e.g., cytochrome-c-oxidase deficiency)

respective CM is important for classification, the frequency of an individual CM within the total spectrum of CM may also be considered significant. As no precise epidemiological data are known for CM, frequency data are probably most certain for the rarest forms of CM, assuming that ascertained cases do not remain unreported. Thus, to our knowledge, sarcotubular myopathy (37) has been described only once in two brothers, and no further cases have since been reported or have come to the authors' attention (F. Jerusalem, personal communication, 1991). It seems that either familial or multiple sporadic cases of a CM are necessary to recognize it as a nosological entity among the CM. Spheroid body myopathy (SBM) (29) is such a case in point. Although it was originally reported as an autosomal-dominant myopathy, affecting several generations within the same kinship with identical morphological lesions in the muscle fibers, no independent cases were subsequently reported for a long time. But recently, another member of this family was found to be affected and to harbor the spheroid bodies in his diseased muscle (Fig. 3), and a firm link to another family briefly reported earlier with myofibrillar inclusion bodies (10) could be established (A. N. D'Agostino, personal communication, 1992), identifying within the kinship with SBM (29) an Indiana branch and an Oregon branch (10). Recently, however, unrelated patients with SBM have been described (16). In the meantime, the concept of spheroid body myopathy—or cytoplasmic body myopathy, referring to the cytoplasmic body (Fig. 4) as an inclusion body similar to the spheroid bodies—has been expanded, largely by experimental induction of spheroid-cytoplasmic complexes (SCC) (9) and the observation of such SCC as a component of a CM in humans (32).

The common denominator of such pathological inclusion bodies within muscle fibers is the abnormal aggregation of desmin (Fig. 5), the physiological intermediate filament of skeletal muscle fibres, in all these lesions (20,24,32,45,48). With regard to these desmin-related myopathies among the CM, which may, one day, be called "desminopathies" by analogy to the "dystrophinopathies," a new principle of classification beyond purely descriptive morphological aspects, such as Z-band abnormalities, core lesions, and so forth, has now been introduced into the classification of CM. This recent development reflects not only a gradual abandonment of the purely descriptive concept of CM, but also one that may prove to be fruitful for further elucidation of the pathogenesis of CM.

Considering the nosological importance of structural abnormalities in CM, which are often pathognomonic, one could justifiably reclassify oculopharyngeal muscular dystrophy (OPMD) as a late-onset autosomal-dominant CM because of the disease-specific intranuclear filaments of 10 nm, which, to our knowledge, have not been encountered in any other neuromuscular condition. OPMD certainly meets the criterion of heredity, because it is autosomal-dominantly transmitted, and of slow rather than rapid progression, and of a congenital nature as the, so far unknown, gene defect is congenitally present—by definition of a hereditary disease. The late onset of OPMD would not militate against it being classified as a CM as, for instance, spheroid body myopathy has also been observed as a late-onset form among autosomal-dominant neuromuscular conditions (29).

The late onset of clinical symptoms and signs associated with structural lesions typical of a CM has been used as evidence against the concept of the congenital nature of CM. Weight to this criticism is added by the observation that such cases often occur sporadically. However, late onset of clinical symptoms

Table 2 Congenital Myopathies

Structured types

- I. Abnormalities of sarcomeres
  - A. Central core disease
    - Multicore disease
    - Minicore disease
  - B. Myofibrillar lysis myopathy
    - Selective myosin degeneration myopathy
    - Cap disease
  - C. Trilaminar fiber myopathy
- II. Z-disc abnormalities
  - A. Nemaline/rod myopathy
  - B. Mitochondria-jagged Z-line myopathy
  - C. Z-band plaque myopathy
  - D. Desmin-related forms
    - Cytoplasmic body myopathy
    - Spheroid body myopathy
    - Mallory body–like myopathy
    - Sarcoplasmic body myopathy
    - Granulofilamentous body myopathy
    - Abnormal accumulation of desmin and dystrophin
- III. Nuclear abnormalities
  - A. Myotubular myopathy
    - Centronuclear myopathy
  - B. Nucleodegenerative myopathy
  - C. Nemaline myopathy with intranuclear rods

Mixed myopathies

- IV. Abnormal inclusions
  - A. Fingerprint body myopathy
  - B. Zebra body myopathy
  - C. Reducing body myopathy
  - D. Tubulomembranous inclusion myopathy
  - E. Cylindrical spirals myopathy
- V. Abnormalities of organelles
  - A. Sarcotubular myopathy
    - Tubular aggregate myopathy
    - Honeycomb myopathy
  - B. Rimmed vacuole myopathy
    - Granulovacuolar lobular myopathy
    - Lysosomal glycogen storage disease with normal acid maltase activity
- VI. Miscellaneous
  - A. Abnormal myomuscular junction myopathy
  - B. Minimal change myopathy
  - C. Nonspecific congenital myopathies in syndromes

Unstructured types

- Congenital fiber type disproportion
- Type I fiber predominance
- Uniform type I fiber myopathy
- Type IIA fiber uniformity and smallness
- Microfiber myopathy
- Type II muscle fiber hypoplasia
- Type I myofiber hypotrophy
- Bethlem myopathy
- Childhood myopathy (42)

Congenital myasthenic syndromes
Familial myasthenia
Congenital endplate acetylcholinesterase deficiency
Slow channel syndrome
Congenital end-plate AChR deficiency
Partially characterized congenital myasthenic syndromes
AChR deficiency and increased affinity for *d*-tubucurarine
Decreased MEPP amplitude without AChR deficiency
Possible defect in ACh synthesis, mobilization storage
Familial limb-girdle myasthenia

---

may not necessarily mean late formation of the respective structural lesions. Too little is known about the evolutionary dynamics of myopathological features in CM; most such cases have been designated as examples of CM exclusively from muscle biopsy findings, not clinical evidence. Moreover, repetitive muscle biopsies performed in the same patient are extremely rare, and perhaps only when identical lesions can be documented by subsequent muscle biopsies should the designation of a CM be confirmed. For example, is zebra body myopathy (40) now eliminated from the list of CM because a subsequent muscle biopsy specimen from the same patient did not contain zebra bodies (B. D. Lake, personal communication, 1990)? Lesions may be focal or infrequent, resulting in apparently different findings at different times in biopsy tissues from various muscles in the same patient. Structural lesions may also lag behind enzyme histochemical abnormalities, as seen by us in a father and several of his children with CCD, the father and one daughter showing cores and type I fiber uniformity, an affected son only showing type I fiber uniformity, but no cores (25). As long as we know so little about the pathogenesis and molecular aspects of CM, late presentation may still be compatible with congenital onset. While sporadic occurrence of late-onset CM is seemingly incompatible with the hereditary nature of many CMs, sporadic CMs in childhood have also been described, e.g., reducing body myopathy (7) and congenital fiber type

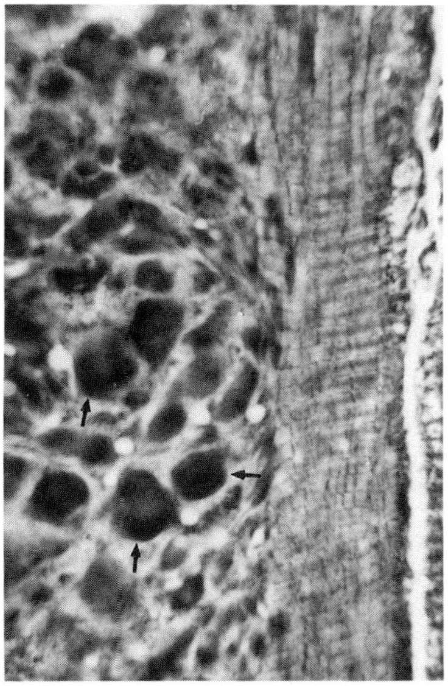

Figure 3 Numerous spheroid bodies (arrows) in spheroid body myopathy, modified trichrome strain, phase contrast. ×1120. (Courtesy of Jans Muller, M.D., Indianapolis.)

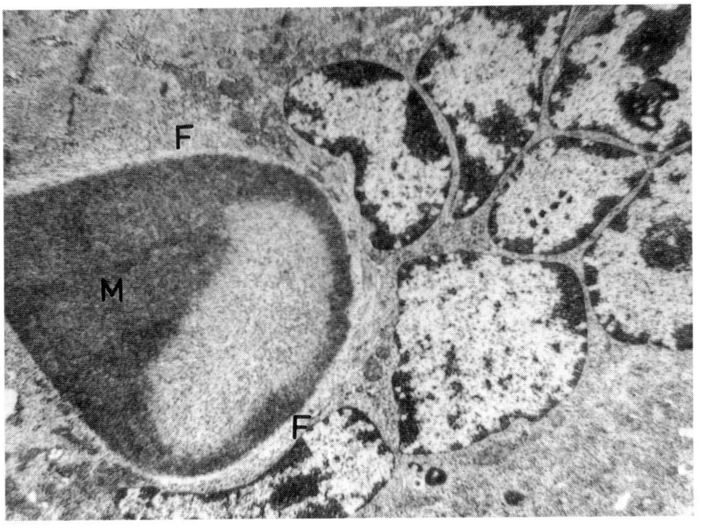

Figure 4 Cytoplasmic body consisting of an electron-dense matrix (M) and filaments (F) in the periphery and center of this inclusion, cyoplasmic body myopathy. ×17,600.

disproportion (6). It is likely that the familial nature of late-onset disorders may be more difficult to confirm than that of early-onset diseases, particularly when families are small, prenatal diagnosis is not undertaken or is not possible, and possibly affected family members are unavailable for adequate study.

Precise classification of a CM becomes even more difficult if myopathological features usually attributed to one CM are found together with those of another well-defined CM, such as reducing bodies and cytoplasmic bodies (Fig. 6) (38), central nuclei and minicores (30), or other mixed myopathies (18).

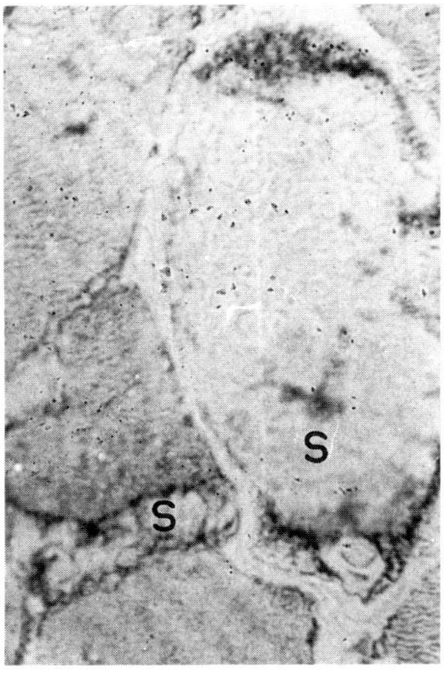

Figure 5 Spheroid bodies (S) are surrounded by increased amounts of immunoreactive desmin, spheroid body myopathy. PAP technique. ×600.

Table 3 Protein Abnormalities and Congenital Myopathies

α-Actinin	Nemaline myopathy
Vimentin	Myotubular myopathy
Desmin	Myotubular myopathy
(desminopathies?)	Cytoplasmic body myopathy
	Sarcoplasmic body myopathy
	Spheroid body myopathy
	Mallory body–like myopathy
	Granulofilamentous myopathy
	Central core disease
α-B Crystallin	Spheroid body myopathy
	Cytoplasmic body myopathy
	Granulofilamentous myopyathy
Dystrophin	"Congenital myopathy"
	Central core disease?

Based on myopathology, the cardinal factor in defining CM, mild aberrations from classical features may suggest variants, e.g., intranuclear rods together with intrasarcoplasmic rods as a possible variant of nemaline myopathy (36,46), since intranuclear rods are not common in nemaline myopathy. Nosological connotation becomes even more obscure when intranuclear rods without intrasarcoplasmic nemaline bodies occur (Fig. 7) in a young child's neuromuscular disease.

CONCLUSION

This chapter deals with the problems of classifying CM, and other chapters discuss the most frequent forms, CCD, CNM, NM, and congenital fiber-type disproportion. Details of other congenital myopathies have been reviewed recently by this author and others (5,25–28).

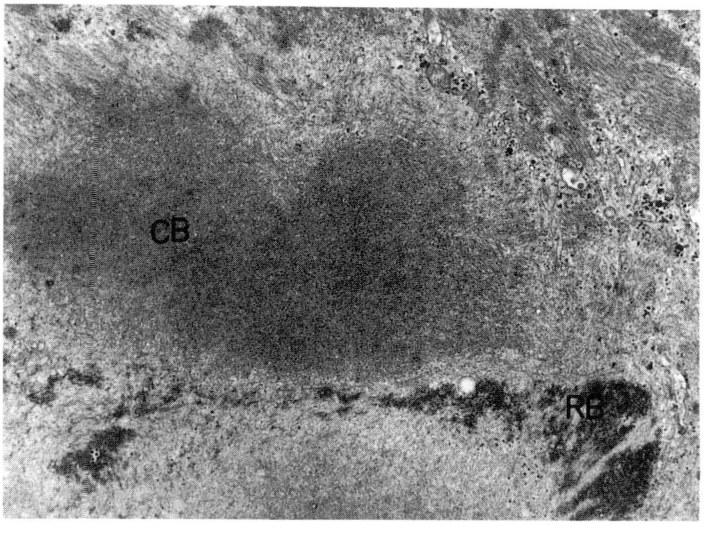

Figure 6 Both a cytoplasmic body (CB) and a darker reducing body (RB) are present as an example of a "mixed" congenital myopathy. ×16,380.

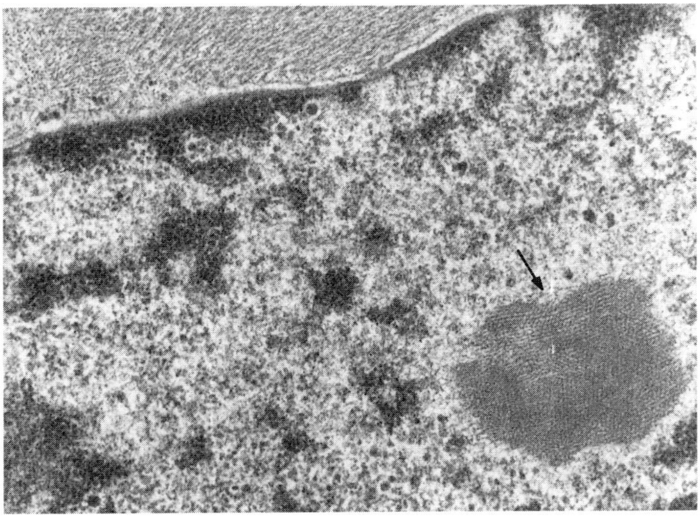

Figure 7 Intranuclear nemaline/rod body (arrow) in a muscle fiber. ×38,000.

ACKNOWLEDGMENTS

Preparation of this chapter was financially supported by the Deutsche Gesellschaft zur Bekämpfung der Muskelkrankheiten e.V., Freiburg, Germany, and by a special grant from the Federal Ministry of Health, Bonn, Germany, to Dr. A. Fidzianska (122-4072-12 P 6/10). The manuscript was edited and typed by Mrs. A. Wöber; Mr. W. Meffert kindly provided the figures.

REFERENCES

1. Arikawa E, Arahata K, Nonaka I, Sugita H. Dystrophin analysis in congenital muscular dystrophy. Neurology 1990; 40(Suppl 1):206 (abstract).
2. Bethlem J, Posthumus-Meyjes FE. Congenital nonprogressive central core disease of Shy and Magee. Psychiatr Neurol Neurochir (Amst) 1960; 63:246.
3. Bethlem J, van Wijngaarden GK. Benign myopathy with autosomal dominant inheritance: a report on three pedigrees. Brain 1976; 99:91–100.
4. Bethlem J, Arts WF, Dingemans KP. Common origin of rods, cores, miniature cores, and focal loss of cross-striations. Arch Neurol 1978; 35:555–566.
5. Bodensteiner J. Congenital myopathies. Neurol Clin 1988; 6:499–518.
6. Brooke MH. Congenital fiber type dysproportion. In: Kakulas BA, ed. Clinical Studies in Myology. Part 2. Amsterdam: Excerpta Medica, 1973:147–159.
7. Brooke MH, Neville HE. Reducing body myopathy. Neurology 1972; 22:829–840.
8. Cabello A, Martínez-Martín P, Gutiérrez-Rivas E, Madero S. Myopathy with nemaline structures associated with HIV infection. J Neurol 1990; 237:64–66.
9. Chou SM, Mizuno Y. Induction of spheroid cytoplasmic bodies in a rat muscle. Neurology 1984; 34 (Suppl 1):82.
10. Clark JR, D'Agostino AN, Wilson J, Brooks RR, Cole GC. Autosomal dominant myofibrillar inclusion body: clinical, histologic, histochemical, and ultrastructural characteristics. Neurology 1978; 28:399.
11. Conen PE, Murphy EG, Donohue WL. Light and electron microscopic studies of "myogranules" in a child with hypotonia and muscle weakness. Can Med Assoc J 1963; 89:983–986.
12. Cumming WJK, Kristmundsdottir F, Mahon M. An unusual familial muscle disorder. J Neurol Neurosurg Psychiatry 1989; 52:266–269.
13. Cumming WJK, Mahon M, Kristmundsdottir F. Childhood myopathy with delayed fibre type differentiation and cessation of fibre growth. J Neurol Sci 1990; 98:101.
14. Currie S, Noronha M, Harriman D. "Minicore" disease. IIIrd International Congress on Muscle Diseases. Amsterdam: Excerpta Medica, 1974:12.

15. Dalakas MC, Pezeshkpour GH. Neuromuscular diseases associated with human immunodeficiency virus infection. Ann Neurol 1988; 23(Suppl):S38–S48.
16. Dickoff DJ. Adult onset of inherited myopathies. Prog Clin Neurosci 1988; 1:65–80.
17. Dubowitz V. The "new" myopathies. Neuropädiatrie 1969; 1:137–148.
18. Dubowitz V. The Floppy Infant. Philadelphia: JB Lippincott, 1980.
19. Dubowitz V, Pearse AGE. Oxidative enzymes and phosphorylase in central core disease of muscle. Lancet 1960; 2:23.
20. Edström L, Thornell L-E, Eriksson A. A new type of hereditary distal myopathy with characteristic sarcoplasmic bodies and intermediate (skeletin) filaments. J Neurol Sci 1980; 47:171–190.
21. Engel AG, Gomez MR. Congenital myopathy associated with multifocal degeneration of muscle fibers. Am Neurol Assoc 1966; 9:222–223.
22. Engel AG, Gomez MR, Groover RV. Multicore disease: a recently recognized congenital myopathy associated with multifocal degeneration of muscle fibers. Mayo Clin Proc 1971; 46:666–681.
23. Engel WK. Central core disease and focal loss of cross-striations. In: Goldensohn ES, Appel SH, eds. Scientific Approaches to Clinical Neurology. Philadelphia: Lea & Febiger, 1977; 2:1555–1571.
24. Fidzianska A, Goebel HH, Osborn M, Lenard HG, Osse G, Langenbeck U. Mallory body-like inclusions in a hereditary congenital neuromuscular disease. Muscle Nerve 1983; 6:195–200.
25. Goebel HH. Neuropathological aspects of congenital myopathies. Prog Neuropathol 1986; 6:231–262.
26. Goebel HH. Congenital myopathies. In: Adachi M, Sher HJ, eds. Neuromuscular Disease. New York, Tokyo: Igaku-Shoin, 1990:197–244.
27. Goebel HH. Congenital myopathies. Acta Paediatr Jpn 1991; 33:247–255.
28. Goebel HH, Lenard HG. Congenital myopathies. In: Rowland LP, DiMauro S, eds. Handbook of Clinical Neurology. Amsterdam: Elsevier, 1992; 18:331–367.
29. Goebel HH, Muller J, Gillen HW, Merritt AD. Autosomal dominant "spheroid body myopathy." Muscle Nerve 1978; 1:14–26.
30. Goebel HH, Meinck HM, Reinecke M, Schimrigk K, Mielke U. Centronuclear myopathy with special consideration of the adult form. Eur Neurol 1984; 23:425–434.
31. Goebel HH, von Loh S, Gehler J. Childhood neuromuscular disease with rimmed vacuoles. Eur J Pediatr 1986; 144:557–562.
32. Halbig L, Goebel HH, Hopf HC, Moll R. Spheroid-cytoplasmic complexes in a congenital myopathy. Rev Neurol (Paris) 1991; 147:300–307.
33. Hashimoto K, Shimizu T, Nonaka I, Mannen T. Immunochemical analysis of α-actinin of nemaline myopathy after two-dimensional electrophoresis. J Neurol Sci 1989; 93:199–209.
34. Ionasescu V, Patil S, Hart M, Rhead W, Smith W. Multiple congenital anomalies syndrome with myopathy in chromosome 16 abnormality. Am J Med Genet 1987; 26:189–194.
35. Jay V, Becker L, Bulman D, Ray P. Dystrophin and DNA analysis in the diagnosis of muscular dystrophy in childhood: a review of 51 cases including 7 cases of congenital muscular dystrophy. J Neuropathol Exp Neurol 1992; 51:355 (abstract).
36. Jenis EH, Lindquist RR, Lister RC. New congenital myopathy with crystalline intranuclear inclusions. Arch Neurol 1969; 20:281–287.
37. Jerusalem F, Engel AG, Gomez MR. Sarcotubular myopathy: a newly recognized, benign, congenital, familial muscle disease. Neurology 1973; 23:897–906.
38. Hübner G, Pongratz D. Granularkörperchenmyopathie (sog. Reducing body myopathy). Virchows Arch (A) 1981; 392:97–104.
39. Kalimo H, Savontaus M-L, Lang H, et al. X-linked myopathy with excessive autophagy: a new hereditary muscle disease. Ann Neurol 1988; 23:258–265.
40. Lake BD, Wilson J. Zebra body myopathy: clinical, histochemical and ultrastructural studies. J Neurol Sci 1975; 24:437–446.
41. Lenard HG, Goebel HH. Congenital muscular dystrophies and unstructured congenital myopathies. Brain Dev 1980; 2:119–125.
42. Mahon M, Kristmundsdottir F, Cumming WJK, Noronha MJ. Sequential studies of a childhood myopathy: a clinical, histochemical and morphometric investigation. Neuropathol Appl Neurobiol 1989; 15:3–12.
43. Mohire MD, Tandan R, Fries TJ, Little BW, Pendlebury WW, Bradley WG. Early-onset benign autosomal dominant limb-girdle myopathy with contractures (Bethlem myopathy). Neurology 1988; 38:573–580.
44. Misra AK, Menon NK, Mishra SK. Abnormal distribution of desmin and vimentin in myofibers in adult onset myotubular myopathy. Muscle Nerve 1992; 15:1246–1252.
45. Osborn M, Goebel HH. The cytoplasmic bodies in a congenital myopathy can be stained with antibodies to desmin, the muscle-specific intermediate filament protein. Acta Neuropathol (Berl) 1983; 62:149–152.
46. Paulus W, Peiffer J, Becker I, Roggendorf W, Schumm F. Adult-onset rod disease with abundant intranuclear rods. J Neurol 1988; 235:343–347.

47. Prelle A, Moggio M, Comi GP, et al. Congenital myopathy associated with abnormal accumulation of desmin and dystrophin. Neuromusc Disord 1992; 2:169–175.
48. Rappaport L, Contard F, Samuel JL, et al. Storage of phosphorylated desmin in a familial myopathy. FEBS Lett 1988; 231:421–425.
49. Riggs JE, Schochet SS, Gutmann L, Shanske S, Neal WA, DiMauro S. Lysosomal glycogen storage disease without acid maltase deficiency. Neurology 1983; 33:873–877.
50. Sarnat HB. Myotubular myopathy: arrest of morphogenesis of myofibres associated with persistence of fetal vimentin and desmin. Can J Neurol Sci 1990; 17:109–123.
51. Schmitt HP. Involvement of the larynx in a congenital "myopathy," unilateral aplasia of the arytenoid, micrognathia, and malformation of the brain—a new syndrome? Virchows Arch (A) 1978; 381:85–96.
52. Seitz RJ, Toyka KV, Wechsler W. Adult-onset mixed myopathy with nemaline rods, minicores, and central cores; a muscle disorder mimicking polymyositis. J Neurol 1984; 231:103–108.
53. Serratrice G, Pellissier JF. Deux familles de myopathies bénignes prédominant sur les ceintures d' hérédité autosomique dominante. Rev Neurol (Paris) 1988; 144:43–46.
54. Sewry CA, Rodillo E, Heckmatt JZ, Dubowitz V. Immunocytochemical studies of congenital centronuclear and nemaline myopathies. J Neurol Sci 1990; 98:100.
55. Sher JH, Rimalovski AB, Athanassiades TJ, Aronson SM. Familial centronuclear myopathy: a clinical and morphological study. Neurology 1967; 17:727–742.
56. Shimizu T, Hashimoto K, Nonaka I, Mannen T. Structural proteins in nemaline myopathy. J Neurol Sci 1990; 98:101.
57. Shy GM, Magee KR. A new congenital non-progressive myopathy. Brain 1956; 79:610–621.
58. Shy GM, Engel WK, Somers JE, Wanko Th. Nemaline myopathy: a new congenital myopathy. Brain 1963; 86:793–810.
59. Simpson DM, Bender AN. Human immunodeficiency virus-associated myopathy: analysis of 11 patients. Ann Neurol 1988; 24:79–84.
60. Spiro AJ, Shy GM, Gonatas NK. Myotubular myopathy. Arch Neurol 1966; 14:1–14.
61. Vallat JM, de Lumley L, Loubet A, Leboutet MJ, Corvisier N, Umdenstock R. Coexistence of minicores, cores, and rods in the same muscle biopsy. Acta Neuropathol (Berl) 1982; 58:229–232.
62. Vanek J, Janda J, Amblerová V, Losan F. Freeman-Sheldon syndrome: a disorder of congenital myopathic origin? J Med Genet 1986; 23:231–236.
63. Watanabe K, Kumagai T, Wakayama Y, Hara K, Yamada H. Congenital myopathy and communicating hydrocephalus—a possible pathogenetic combination. Brain Dev 1982; 4:455–462.

15

Core Diseases, Centronuclear Myopathies, and Nemaline Myopathy

Hans Hilmar Goebel
Johannes Gutenberg University, Mainz, Germany

Anna Fidzianska
Warsaw University Medical Center, Warsaw, Poland

INTRODUCTION

The three classical congenital myopathies, discovered in the late 1950s and early 1960s following the introduction of new morphological techniques into myopathology (electron microscopy and enzyme histochemistry), are *central core disease* (CCD), *centronuclear/myotubular myopathy* (CNM), and *nemaline myopathy* (NM). These three neuromuscular conditions, or groups of conditions (if one considers clinical variants such as neonatal and adult forms as possibly distinct nosological entities), differ in that NM shows conspicuous ultrastructural features, the nemaline bodies or rods, whereas CCD and CNM are best recognized at the light microscopic level, by enzyme histochemistry and well-established histological techniques.

CORE DISEASES

Central Core Disease

CCD, the first of the congenital myopathies described (1), is a member of the "core" diseases (2), which also encompass multicore disease, minicore disease (MCD), and "focal loss of cross-striations," the latter three possibly defining a single entity.

Clinical Features

CCD occurs in an autosomal-dominant fashion and is the most common of the congenital myopathies. Clinically (Fig. 1a), patients with CCD present with congenital or early childhood muscle weakness and marked muscle hypotonia, as floppy babies, or with developmental delay in standing and walking. However, a patient may be clinically normal and may be diagnosed only by systematic biopsy studies of family members. Muscle weakness is proximal or generalized, with relatively little muscle atrophy, more of the lower than of the upper extremities, with impaired deep tendon reflexes without evidence of neuropathy. Skeletal abnormalities, possibly related to prenatal muscle weakness, may include kyphoscoliosis, foot deformities, congenital hip dislocation, contractures, and hypermobile joints (3). Severe scoliosis may be a prominent sign (4) although there is usually little progression. Respiratory insufficiency may cause considerable handicap (5). Several cardiac abnormalities have been noted in CCD (3). A few patients have a normal electromyographic (EMG) pattern, while in others, short-duration, low-amplitude

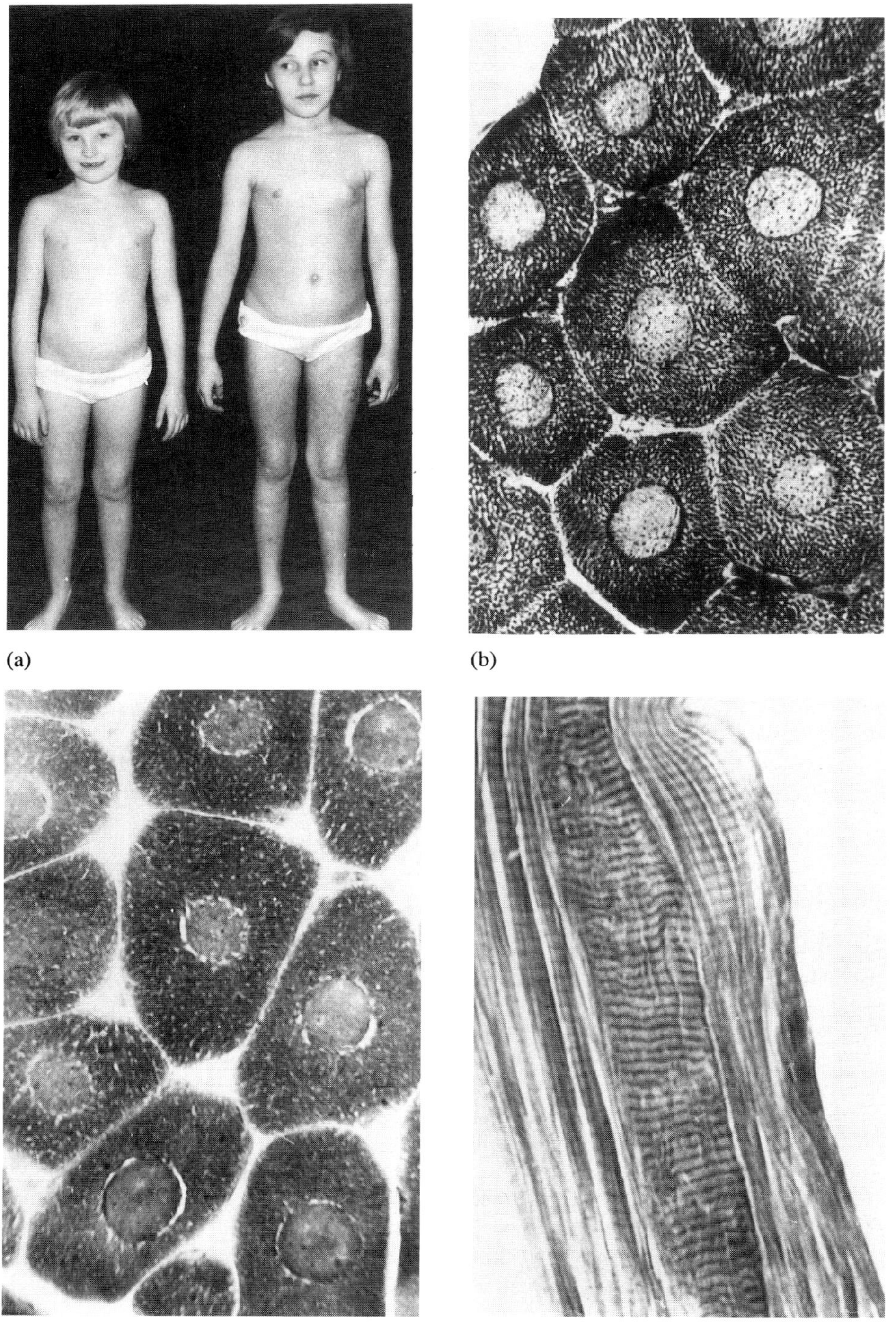

(a) (b) (c) (d)

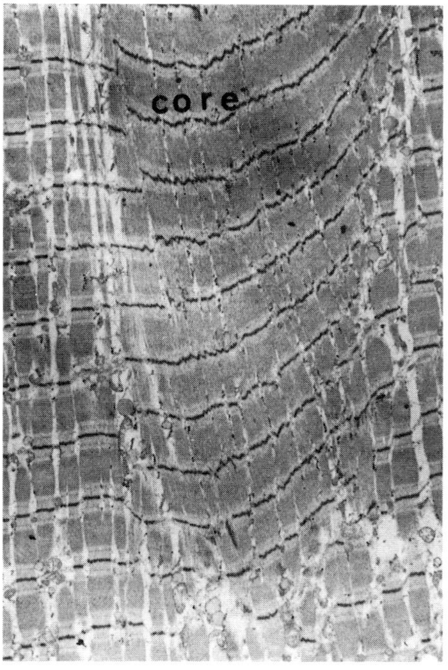

(e)

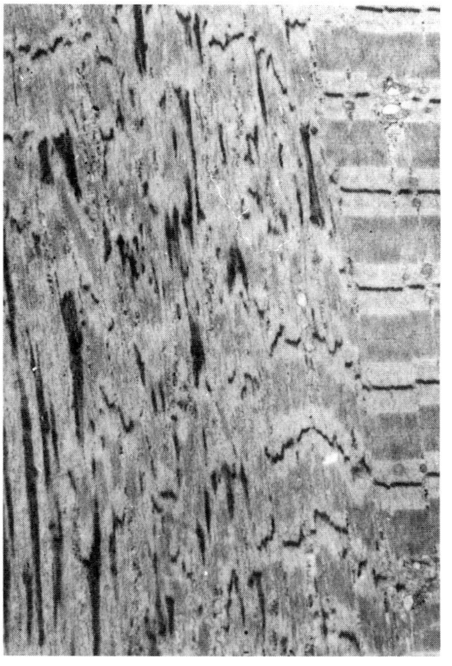

(f)

Figure 1 Central core disease. (a) Two sisters with CCD. (b) Single central cores in type I fibers (NADH, ×502). (c) Structured cores displaying enzyme activity (ATPase pH 4.3, ×495). (d) A structured core, in longitudinal section, reveals contracted sarcomeres, which are regular and in register, compared to adjacent myofibrils the sarcomeres of which are in different register (modified trichrome stain, ×1680). (e) Contraction of sarcomeres in a structured central core is evident as is lack of mitochondria compared to peripheral sarcomeres (×5710). (f) Unstructured core reveals lack of mitochondria and disruption of the sarcomeric pattern (×5000).

polyphasic motor unit potentials and reduced interference pattern on maximal effort may be present. Single-fiber EMG has shown a slightly increased fiber density as in other benign myopathies (6).

Genetics

Numerous reports emphasize the association of CCD with susceptibility to malignant hyperthermia (7–10). The King syndrome, which encompasses susceptibility to malignant hyperthermia and congenital facial and skeletal abnormalities, is also marked by a slowly progressive myopathy but is not a core disorder. A variant of the King syndrome, with cleft palate, has also been reported (11). Thus, every patient suspected or proven to have CCD should be investigated for a susceptibility to malignant hyperthermia. Although malignant hyperthermia may be associated with a variety of neuromuscular disorders, its occurrence in CCD is more frequent than expected. It is thus perhaps not unexpected that one of the genes for malignant hyperthermia has been localized to the same region on chromosome 19 as that for CCD (19q12–13.2), and it has been suggested that these two genes may be allelic (12,13). More recently two mutations within the ryanodine receptor *(RYR I)* gene, Ag 163 Cys and Ile 403 Met, were detected in two separate CCD kinships (14).

Pathology

Congenital myopathies are characterized by morphological abnormalities in most instances, the characteristic feature of CCD being the core (Fig. 1b,c,d), a centrally or eccentrically located lesion within the muscle fiber, only later named as such (15) in material from the original family reported by Shy and Magee (1). Cores are lesions of the sarcomeres and are divided into structured (Fig. 1e) and unstructured forms (Fig. 1f). The form first described was a structured core, i.e., aberrantly running central and eccentric myofibrils within the majority of myofibers (1) that consist of contracted sarcomeres surrounded laterally by sarcomeres of wider register. Each of the five biopsied, clinically affected members of the original family displayed these structured or normally striated cores (1), resulting in the concept that some patients or families with CCD have only structured cores whereas others have unstructured ones (16). Unstructured cores (Fig. 1f) are marked by disruption of the sarcomeric pattern with excessive streaming of the Z-band. Common to both structured and unstructured cores is the paucity or absence of mitochondria within the core region, as revealed by NADH-tetrazolium reductase (Fig. 1b) and cytochrome-c-oxidase preparations.

The pathogenesis of cores in human CCD has not been fully clarified. It has been proposed that during the 24th–28th week of gestation, myotubes of different maturation fuse with centrally or peripherally located, less mature myotubes, which lag in maturation even after fusion of the myotubes has occurred, thereby producing a central or peripheral core with sarcomeres that register differently from adjacent areas derived from other myotubes (17). It is suggested that unstructured cores simply represent destruction of the sarcomeres, a possibly nonspecific lesion similar to target and targetoid lesions and "streaming of the Z-band." According to this concept (17), structured and unstructured cores would follow fundamentally different pathogenetic principles, the former being that of a developmental error, the latter that of degeneration of sarcomeres. However, it is conceivable that even a fundamentally aberrant structured core might "deteriorate" into an unstructured one; the occurrence of both structured and unstructured cores within the same muscle biopsy specimen and even the same muscle fiber has previously been documented and illustrated (2,18). The reverse evolution of structured cores from previous unstructured ones, however, appears difficult to conceive.

Cores have been produced experimentally by tenotomy, but differ in fine structure from naturally occurring cores, and also by local tetanus toxin injection (19) and after immobilization of the limb in the shortened position (20). Increased amounts of both desmin and dystrophin may be observed within cores as a result of abnormal protein metabolism, which has previously been suggested to originate in early myogenesis (17).

Some developmental disturbance is further suggested by the frequently encountered fiber type uniformity, histochemically of 2c fibers (17), which may occasionally be the only sign of myopathology in a CCD family (2). In other instances, type I fiber predominance and type I fiber atrophy are characteristic light microscopic features (21).

Other types of cores have been described including "reversed," "atypical," "segmental," and multi-/minicores (2). It is not unusual to see, in addition to central cores, minicores that are marked by absence of mitochondria like central cores and by circumscribed disruption of sarcomeres by Z-band streaming. Structured mini- or multicores have not been clearly identified, but the occurrence of both central and minicores in the same biopsy has been reported (22). Indeed, in certain instances both cores and nemaline bodies or rods have been found in the same biopsy and individual muscle fibers (22,23). Furthermore, the morphological lesions within myofibers are not confined to abnormalities of the sarcomeres, (i.e., central cores, minicores, and rods); the sarcoplasmic reticulum and transverse tubules may also be involved, in both core and noncore regions, but more conspicuously in the former (21). It is interesting that abnormalities of the sarcotubular system are not confined to patients at risk for malignant hyperthermia (21). On that basis, it is not perhaps surprising that the father of a family with CCD, whose three clinically affected children all had numerous central cores, showed a wealth of tubular aggregates within muscle fibers, thought to be derived from the sarcoplasmic reticulum (24). It thus seems that in CCD, not only the sarcomeres but also other major organelles such as mitrochondria, transverse tubules and the sarcoplasmic reticulum may be severely affected, although the relationship between these changes is currently not clear.

Minicore/Multicore Disease

While minicores/multicores have been observed in CCD (and may also be associated with aggregates of nemaline bodies in rod myopathy), a separate congenital myopathy termed MCD seems to exist, in which very small, multiple cores abound within muscle fibers (Fig. 2b,c). Like unstructured cores, these cores also lack mitochondria. Another term used to describe these appearances is "focal loss of cross-striations," and it is now accepted that, based on clinical and morphological features, minicore myopathy and multicore myopathy are identical (25).

Clinical Features

Clinically (Fig. 2a), this disorder is a slowly progressive or nonprogressive myopathy of onset in early childhood, which is not marked by disease-specific clinical findings. Weakness of extraocular muscles may be a feature, as may respiratory insufficiency due to progressive muscle weakness (26). Rigid spine has also been described in MCD (27). An associated susceptibility to malignant hyperthermia as seen in CCD patients has not been clearly documented in patients with MCD, however. When cardiomyopathy is associated (28), patients may die early from cardiac failure. Interestingly, patients with congenital minicore myopathy and cardiomyopathy may show similar lesions of sarcomeres, mitochondria, and the transverse tubular system in skeletal and myocardial fibers (29).

Genetics

The gene for MCD has not yet been localized and is apparently not at the CCD locus at 19q13. Genetic transmission has been reported to be autosomal recessive (28), autosomal dominant (25), or apparently sporadic.

Pathology

In CCD, smaller cores of the unstructured type, often encountered without or together with central cores, suggest that unstructured CCD may develop from mini/multicore disease, and that instances of this latter congenital myopathy may actually represent forms of CCD biopsied at a very early stage in the course of the disease, or variants of CCD in which an arrest of evolution from multiple small cores to larger central or eccentric cores has occurred. Currently, however, to assume such an evolutionary process is speculative, partly based on the dearth of subsequent biopsy studies from a patient with mini/multicore disease. Indeed, circumstantial evidence militates against mini/multicore disease being a mere variant of CCD: (1) CCD is a benign, slowly progressive or nonprogressive disease while mini/multicore disease has been found to be rapidly fatal in a few patients; (2) the gene locus for CCD has been placed at 19q13 while the gene locus for mini/multicore disease is still unknown; (3) unequivocal patients with CCD and mini/multicore disease in the same family have not been reported; (4) as will be discussed later, minicores have been reported occasionally in conjunction with centronuclear myopathy, but central cores have not.

Thus, it appears that CCD and mini/multicore disease represent at least two different nosological entities. Mini/multicores might conceivably progress to form unstructured central cores by fusion, but this is at present unproven because of the essentially nonprogressive nature of the condition. Conversely, mini/multicores may be so large or central cores may be so small that differentiation of these two types may be difficult in individual instances. Other characteristic features of congenital myopathies, type I fiber predominance and/or type I fiber atrophy, are also seen in mini/multicore myopathy (25). Central nuclei may also be present (30), and precise nosological differentiation of centronuclear myopathy with minicores from mini/multicore myopathy with central nuclei may become difficult.

CENTRONUCLEAR/MYOTUBULAR MYOPATHY

This congenital myopathy may actually consist of three nosological entities: CNM of childhood, CNM of the adult, and myotubular myopathy (MM).

The first case of what was later designated as CNM was unfortunately originally described as MM

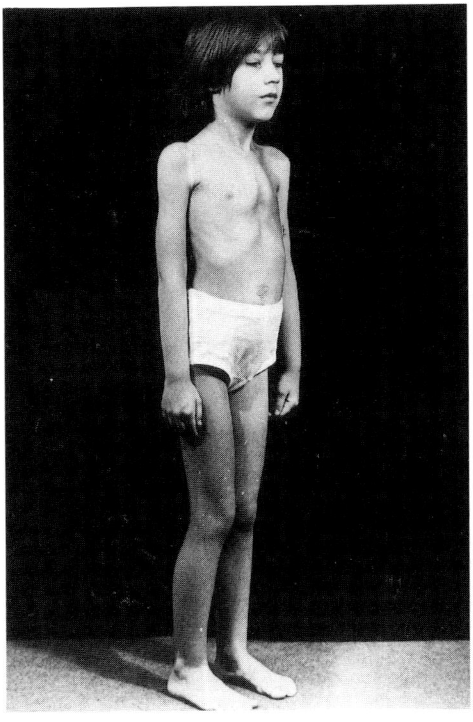

(a)

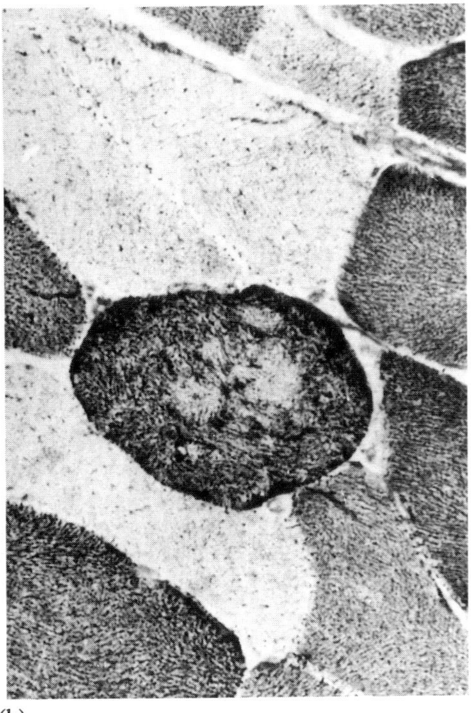

(b)

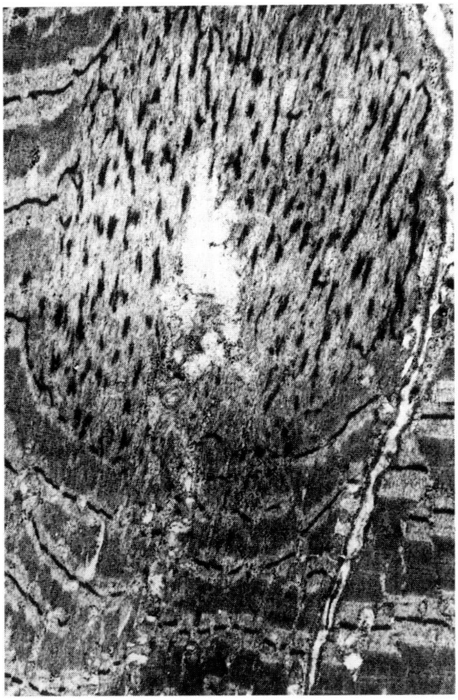

(c)

Figure 2 Minicore disease. (a) Nine-year-old girl with minicore disease. (b) Central type I fiber contains several small, irregular mini/multicores (NADH, ×526). (c) Circumscribed unstructured mini/multicore (×3532).

(31). Spiro et al. in 1966 reported a 12-year-old boy whose muscle fibers were marked by centrally located nuclei resembling myotubes, which led the authors to suggest developmental arrest as a cause in this boy's disease (31). Since the subsequent report on the same entity, 1 year later (32), which not only contested the pathogenetic concept, but also replaced its name by the descriptive term "centronuclear myopathy," CNM has consistently been described as a genetic disorder, with autosomal dominant, autosomal recessive, and sporadic forms. Loci for abnormal genes of these CNM have not yet been identified, however.

We propose to call the rapidly progressive, and often fatal, X-linked infantile form *myotubular myopathy*, whereas the non-X-linked, hereditary (autosomal recessive or autosomal dominant) and sporadic forms should be designed *centronuclear myopathy*, the common denominator of each condition being the morphological feature of centrally located single nuclei within muscle fibers. However, the terms myotubular and centronuclear have been used interchangeably even recently, e.g., "X-linked CNM" (33) and "adult-onset MM" (34).

X-Linked Neonatal Myotubular Myopathy

Clinical Features

This severe neonatal myopathy, which was first described (35) 3 years after the original report by Spiro et al. (31), affects newborn boys, who present with profound hypotonia (floppy infant) and respiratory insufficiency requiring intubation and artificial ventilation. Not infrequently, this has to be maintained until death, which often occurs within days or weeks and occasionally directly after birth. In other instances, survival for a few years is possible, as in one of our patients who died at age 4 after lifelong respirator support. An older male sibling who had died immediately after birth, of respiratory insufficiency, probably also had MM. This early mortality in (often unrecognized) MM is not a rare feature (36). Some children manage to breathe spontaneously for a while (37). The pregnancy often seems to be complicated by polyhydramnios (36,38–40) or preterm termination (36,39). Deep-tendon and Moro reflexes are weak or absent.

Additional clinical symptoms in individual patients include ptosis, joint contractures, a high-arched palate, and cryptorchidism (39). Limb movements are often much reduced. Certain skeletal deformities, such as contractures of arms and/or legs, clubfeet, micrognathia (41), and even arthrogryposis multiplex congenita (42), indicate intrauterine muscle weakness.

Genetics

Recently, using DNA markers St 14, cpX 67, DX 13, and pSt 35-691, X-linked MM has been localized at Xq 28 permitting prenatal diagnosis (33,38). Nuclear haploidy (in addition to diploidy, i.e., mosaicism) has also been documented (43).

Pathology

The morphological hallmark is rounded muscle fibers with a central, rather large nucleus (Fig. 3a) while peripheral nuclei are absent. This large nucleus is surrounded by a halo free of sarcomeres but filled with sarcoplasmic contents, which results in strong staining reactions for oxidative enzymes (Fig. 3b). Conversely, the periphery of these muscle fibers is often devoid of oxidative enzyme activities. Thus, typing of muscle fibers, based on oxidative enzyme preparations, is difficult and often inadequate. ATPase preparations, conversely, may reveal a checkerboard pattern or a type I fiber predominance (38). Intrafusal fibers are exempt from this central nucleation process. At the electron microscopic level (Fig. 3c), no additional abnormalities may be noted and the centrally located nuclei may be surrounded by sarcomeres or sarcomere-free sarcoplasm.

The original suggestion of immaturity of muscle fibers in MM (31) has recently been reiterated by the demonstration of abnormal amounts of desmin and vimentin as they appear in fetuses between the 8th and 15th weeks of gestation (36,44). Embryonic myosin was occasionally, and fetal myosin was more often, encountered in respective myofibers (45). The concept of persistence of fetal muscle fibers, i.e., maturational arrest of myofibers in MM, is also supported by neurohistochemical findings including the

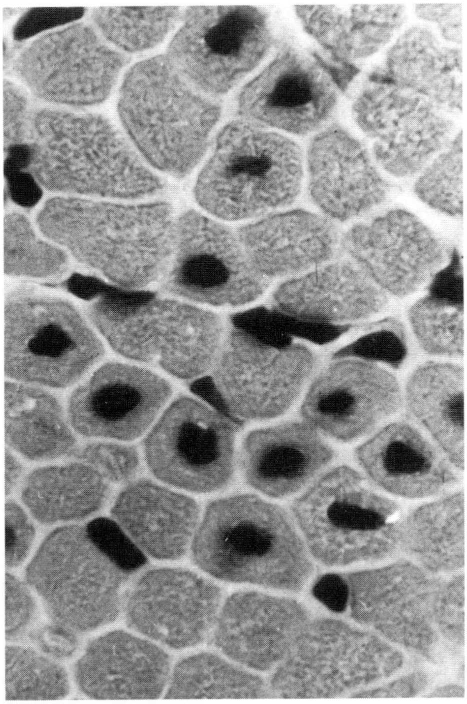

(a)

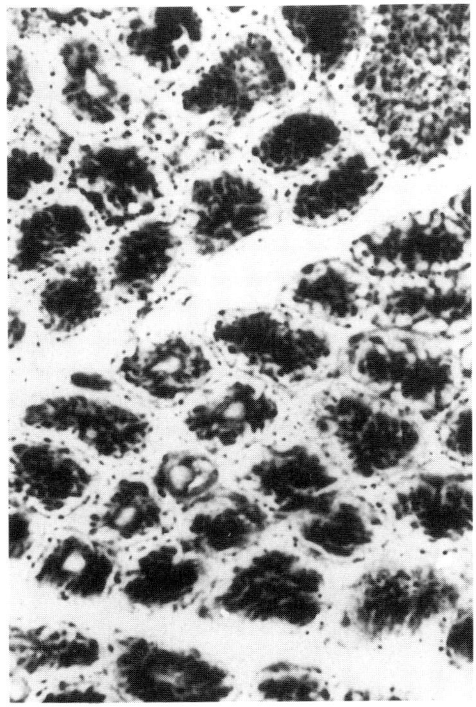

(b)

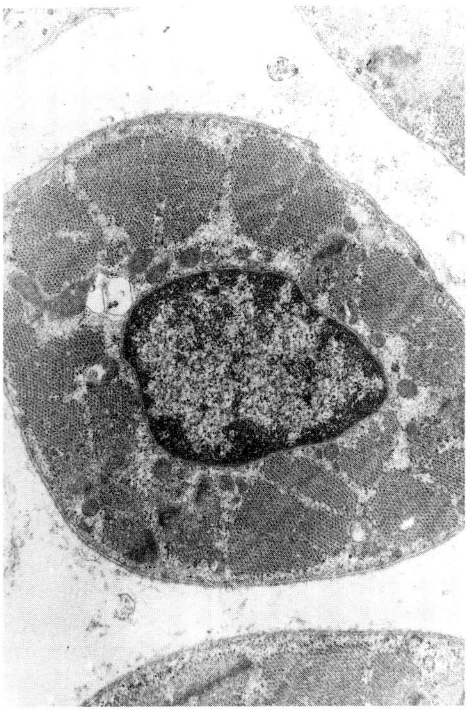

(c)

presence of the prenatal-myosin heavy-chain isoform (41). However, the strong reactivity to antibodies to desmin and vimentin in myofibers corresponded to a fetal stage of 8–15 weeks (44), whereas the presence of the prenatal myosin heavy-chain isoform and the pattern of MHC expression corresponded to that of 19–20 weeks' fetal stage (41). These findings indicate that, although the common denominator is maturational arrest or immaturity, individual fetal parameters correspond to different prenatal stages, the dynamics of which are still ill-understood. Small myofibers may also react positively for N-CAM. Immaturity of myofibers has also been shown using X-ray analysis, which demonstrated fetal patterns of sodium, chloride, and potassium distribution (46). Evidence of a neurogenic process, however, has not been found at peripheral-nerve or spinal cord levels (37,44). Muscle specimens from carrier mothers (women having delivered more than one affected son) may be normal (40) or may show a nonspecific abnormal variation in muscle-fiber diameters (2, 8) and central nuclei (44).

Centronuclear Myopathy

Clinical Features

This group of congenital myopathies encompasses the childhood form, which may follow autosomal recessive or autosomal dominant (Fig. 4a) modes of inheritance or may occur sporadically, as well as the late-onset or adult type. The childhood form, which affects girls and boys equally, is often characterized by ptosis or even total ophthalmoplegia (41) and facial weakness, in addition to generalized or proximal limb muscle weakness. Muscle hypotonia and reduced or absent deep tendon reflexes may be present. Occasionally, postnatal asphyxia may require intubation (47). Other patients show normal early motor development, but retarded motor development later (48). The childhood form of CNM may also be only mildly progressive or run a protracted course with diagnosis being delayed until adulthood (49). In a number of such patients, subtle myopathic changes, i.e., short-duration, polyphasic motor unit potentials, by concentric needle EMG have been observed, but a reduced interference pattern during maximal effort and the presence of fibrillation potentials at rest, suggesting a neurogenic component, has also been seen frequently (49).

Genetics

To our knowledge, genetic data equivalent to the Xq 28 form of MM have not been identified for childhood CNM. Thus, prenatal diagnosis is still not feasible. Similarly, prenatal myosin heavy-chain isoforms, found in X-linked MM, could not be confirmed in childhood CNM (41).

Pathology

The muscle is usually characterized by predominance of type I fibers (Fig. 4b), small type I fibers, and central nuclei of varying frequency. The number of central nuclei may increase with age (50) and both perinuclear and peripheral halos may be observed. Cytoplasmic bodies have also been seen in muscle fibers of CNM (49). Moreover, a considerable number of type 2c fibers, indicative of immaturity, may also be seen (50). Then, unequivocal diagnostic and nosological differentiation between X-linked MM and childhood CNM of non-X-linked type in a boy may be difficult. Nosological differentiation may be even further complicated if type I fiber predominance and type 2 hypotrophy but no increase in central nuclei are found in girls with clinical features identical to those of X-linked MM (51). Lo et al. (51) consider the congenital myopathy with type I hypotrophy without central nuclei to be a member of the CNM group (51). CNM may be associated with numerous minicores (52). This coexistence of CNM and minicores was documented in two cousins as a persistent feature in serial muscle biopsy specimens (52). Likewise, minicores had previously been reported in other patients with CNM (53).

Figure 3 Myotubular myopathy. (a) Several muscle fibers contain large single central nuclei, hematoxylin-eosin (×1330). (b) Muscle fibers are marked by a dense, centrally located substrate and a light peripheral rim with several muscle fibers showing another light central region, the site of the central nucleus (NADH, ×1493). (c) Cross-sectioned muscle fiber contains a single central unremarkable nucleus and no further ultrastructural abnormalities (×13,178).

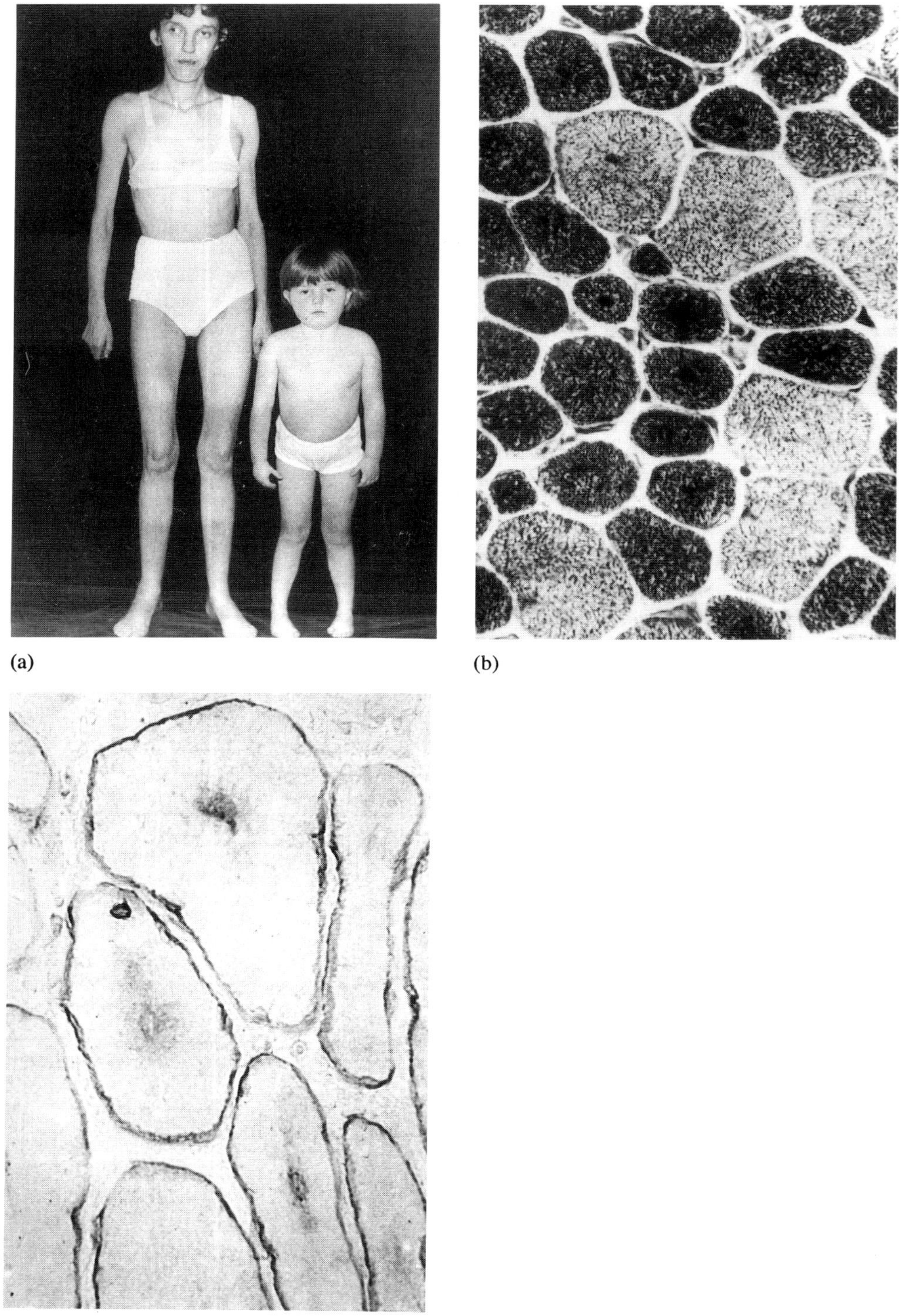

(a) (b) (c)

Figure 4 Autosomal dominant centronuclear myopathy. (a) Mother and daughter with AD centronuclear myopathy. (b) Predominance and selective atrophy of type I fibers with a few central dark dots. The perinuclear area is rich in enzyme activity. (Muscle biopsy tissue from the daughter, NADH, ×511). (c) Antibody against the C-terminus of dystrophin marks the periphery and the centre of muscle fibers (×540). (Courtesy of Dr. A. Engel, Mayo Clinic.)

However, since small patches of "streaming of the Z-band" (marked by circumscribed disruption of the sarcomeres and lack of mitochondria within these regions) and minicores may only differ in size, such minicores will have to be a constant myopathological feature in CNM to be considered a significant component of this disorder.

Late-Onset Centronuclear Myopathy

A number of patients who have myopathological changes of CNM present clinical symptoms, usually extraocular weakness without ptosis, during adolescence or even adulthood. CNM often appears sporadic, but an autosomal-dominant mode of inheritance has been reported (54). Occasionally, the calf muscles appear hypertrophic.

Myopathology is similar to that of the childhood form, marked by type I fiber hypotrophy and predominance. The immaturity of centrally nucleated myofibers may be evident from the presence of perinuclear vimentin, desmin, and dystrophin (34,54) (Fig. 4c), even in sporadic late-onset CNM. A neurogenic origin of the muscle lesions has been suggested by electromyography and myopathology (52) for this late-onset or adult type of CNM although unequivocal morphological abnormalities of peripheral nerves or spinal cords have not been reported.

NEMALINE MYOPATHY

NM is now firmly established as a separate congenital myopathy characterized by the abundant formation of rods derived from the Z-bands. NM was the second entity to be described within the spectrum of the newly delineated congenital myopathies (55), central core disease having been described earlier (1). However, in the same year, rods were also described in a child with hypotonia and muscle weakness, but called myogranules (56). These simultaneous publications, and an addendum in the former paper (55) concerning two additional unrelated children affected by NM, emphasized the relative frequency of NM within the spectrum of neuromuscular conditions in children. Therefore, it is not surprising that rods had actually been discovered some time before, in muscle fibers of a weak young child, by Dr. Reye, a distinguished pathologist at the Royal Alexandra Hospital for Children in Sydney, Australia, but not further reported (2).

NM (derived from the Greek word νημα, *thread*) may appear in several clinical forms:

1. A slowly progressive or nonprogressive childhood form, usually autosomal dominantly inherited
2. A neonatal or early-infantile, often rapidly fatal, form, usually of autosomal recessive inheritance
3. An adult form

Whether NM with intranuclear rods in isolation represents a separate clinical form remains open to debate.

Nemaline Myopathy in Children

Clinical Features

The most frequent and earliest-described clinical form of NM (Fig. 5a) is that occurring in infancy or early childhood. This often presents with signs of the floppy infant (55), with hypotonia and proximal or generalized muscle weakness, but patients with more distal muscle weakness and reduced muscle bulk have also been described, running a slowly progressive or nonprogressive course (57). Early onset of facial muscle weakness may result in a myopathic facies, with a high-arched palate—a physiognomic feature, however, not confined to NM. Severe respiratory insufficiency may develop (58). The feet may be flat, and tendon reflexes reduced. Creatine kinase in the serum is usually mildly elevated or normal. Serial electromyographic studies may reveal normal features until the age of 3 years, succeeded by progressive changes marked by increasing myopathic features and even neuropathic phenomena after the age of 9 years. These first appear in distal muscles and spread during the course of the disease, indicating continual degeneration and regeneration (59). No sensory deficits or abnormalities of nerve conduction velocities have been noted (59).

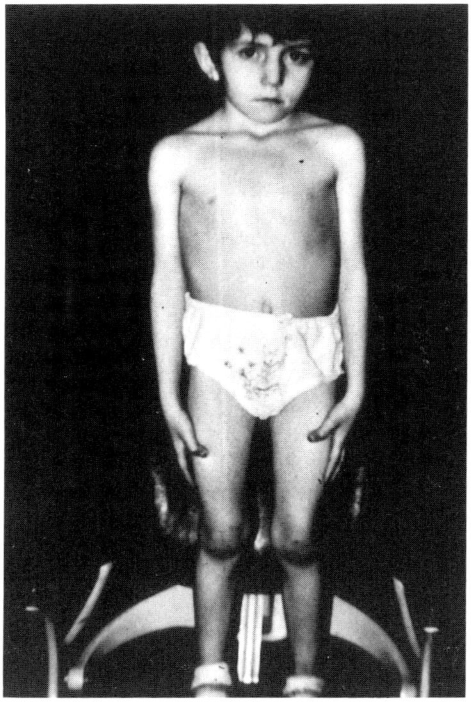

(a)

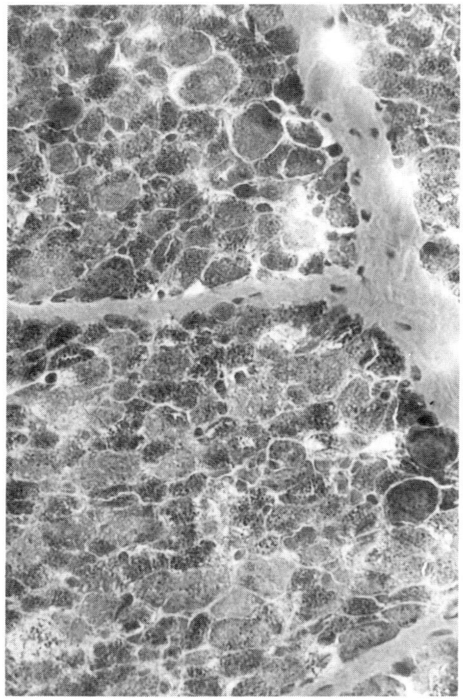

(b)

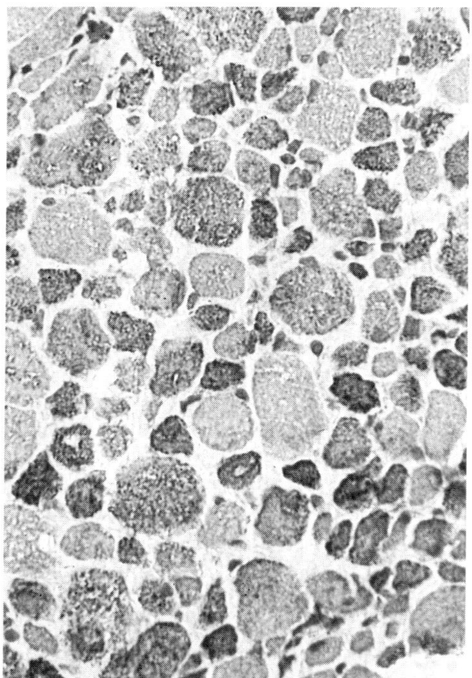

(c)

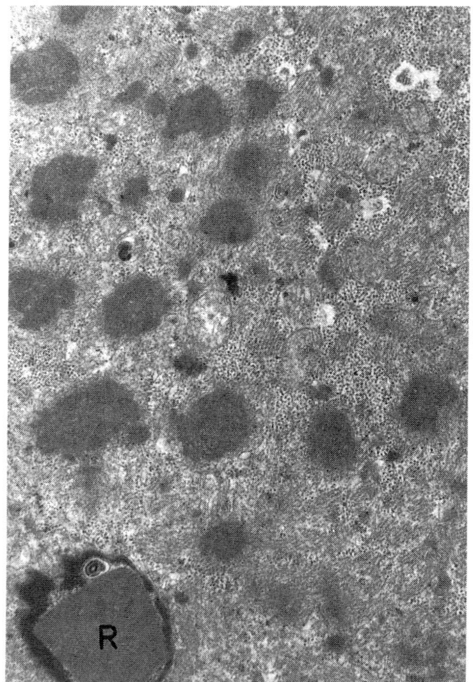

(d)

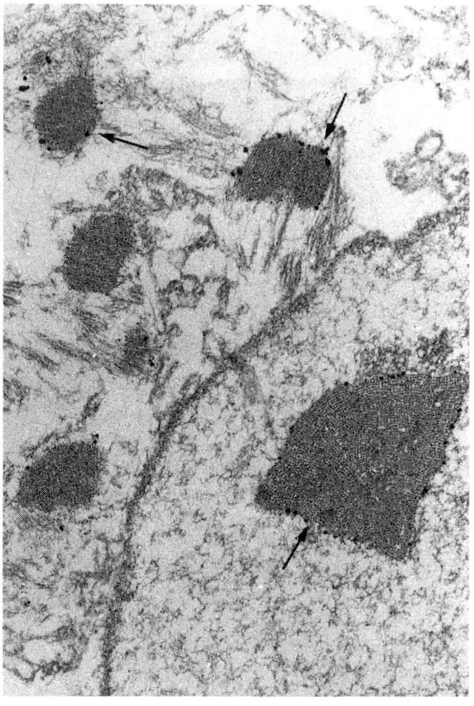

(e)

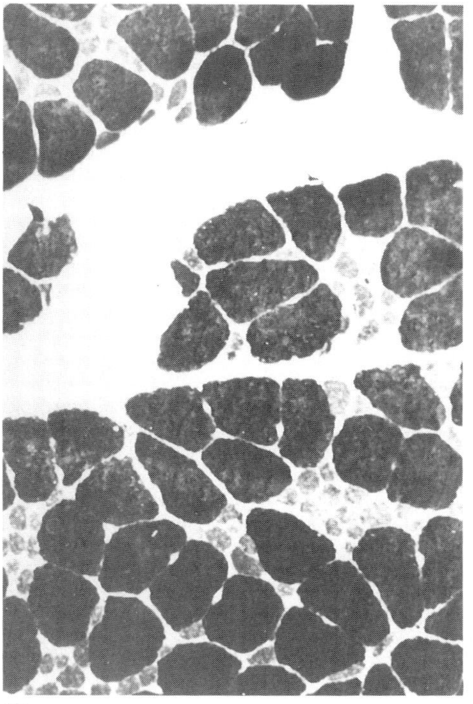

(f)

Figure 5 Nemaline myopathy. (a) Young Turkish girl with slowly progressive nemaline myopathy. (Courtesy of Dr. M. Caliskan, Istanbul.) (b) Innumerable myofibers contain aggregates of rods (modified trichrome stain, ×400). (c) Numerous small myofibers react intensely with an antibody against α-actinin indicating aggregates of rods (×400). (d) Aggregate of sarcoplasmic rods and nucleus with single large intranuclear rod (R) (×15,675). (e) Sarcoplasmic rods and intranuclear rod are decorated with immune gold grains (arrows) using antibody against α-actinin. Note crisscross pattern of intranuclear rod (×27,500). (f) Predominance and selective type I fiber atrophy (alkaline ATPase preparation, ×400).

Neonatal and Early-Infantile Form

A clinically more severe form, often rapidly fatal, may be distinguished. These children were not only floppy at birth, but also often require artificial ventilation because severe respiratory insufficiency with inability to breathe normally is a frequent cause of death, with or without bronchopneumonia. Moreover, we have recently observed a baby girl who died in infancy and had only *intranuclear* rods. Intranuclear rods in very young children may indicate an unfavorable prognosis, since two other unrelated infants, one of whom died at the age of 3½ months due to respiratory failure, had both intranuclear and intrasarcoplasmic rods (H. H. Goebel, unpublished observations).

Adult-Onset Nemaline Myopathy

Adult-onset NM may be considered a separate entity, being sporadic and occasionally simulating polymyositis. It too is characterized by weakness of trunk and proximal muscles and is associated with a myopathic EMG pattern. This form should be distinguished from familial NM with autosomal dominant inheritance, in which some affected family members may develop clinical signs during adulthood (60). Such patients may be similar to the large kinship reported to have an abnormal gene, *NEM 1*, on chromosome 1 (61) (see below).

Genetics of Nemaline Myopathy

NM is characterized not only by different clinical forms, but also by different modes of inheritance. Both autosomal dominant and autosomal recessive inheritance have been described (62) although curiously, in Finland, only an autosomal recessive trait could be defined, based on complete ascertainment of patients with NM (63). Linkage to APOA2, a marker localized to 1q21–q23, was established in a single family showing late-onset autosomal dominant NM (61). The candidate gene for the adult form of NM *(NEM I)* is now provisionally assigned to 1p13–1q25 (61). It should be noted that the presumptive gene *NEM I* differs from the two genes on chromosome 1 coding for α-actinin. Although NM and one familial form of Charcot-Marie-Tooth (CMT) disease have been seen within the same family, and both genes are located on chromosome 1, a consistent association of NM with this CMT gene has not been reported. Genes for the other clinical forms of NM have not yet been localized. Thus, with the rare exception of *NEM I*, prenatal diagnosis in NM cannot be based on regular genetic studies. However, the observation of intranuclear rods in several affected infants with NM suggests that fetal muscle biopsy might be considered for histological/electron microscopic evaluation, although, to our knowledge, this approach has not yet been successful.

Pathology

Irrespective of the clinical form, the myopathological features of NM are uniform. Nemaline bodies, or rods (Fig. 5b), are the morphological hallmark and are derived from enlargement and extension of the Z-bands. The ultrastructure of rods is identical to that of Z-bands, and α-actinin, a component of the Z-band, is also a component of rods (Fig. 5c) (64,65). Electrophoretic analysis revealed an acidic variant of α-actinin in type I and type IIA myofibers, but a basic variant in type IIB myofibers (65). The rods often assemble in clusters beneath the sarcolemma, but they may also be encountered inside the myofiber, and a trend had been noted to find central rods more often in older patients (66). In several adult and infant patients, rods have also been encountered within nuclei (67–69). We recently encountered three such infants with intranuclear rods (Fig. 5d,e), one having intranuclear rods in isolation, but two also having intrasarcoplasmic rods. These three children had the severe congenital form, which proved fatal in two patients. Rods have also been encountered together with abnormal mitochondria (70) and rimmed vacuoles (71), as well as with cores (22). The spontaneous association of rods and cores has an experimental equivalent in that both lesions can be produced in type I muscle fibers after tenotomy (72). Rod-like structures have also occasionally been seen in cardiac myocytes of patients with NM and cardiomyopathy (73–75). However, these inclusions often did not look like fully formed rod bodies.

Rods have also been observed as a nonspecific myopathological feature in other neuromuscular disorders such as dermatomyositis (76), in CMT (77) (see above), and in a few other conditions (78). Although typical in ultrastructure, they were few in number. Atrophic muscle fibers, severely depleted of myofilaments, may often retain Z-bands (with adjacent thin myofilaments) as a residual state, and even show some enlargement of the Z-bands.

NM is often characterized by type I fiber atrophy and type I fiber predominance, a morphological pattern of fiber type disproportion (Fig. 5f). There may be areas of such a fiber type disproportion without rods, alternating with rod-enriched areas within the same biopsy sample (18) (Fig. 5f). This histochemical pattern without rods may help in identifying affected family members in NM (79). Occasionally, deficiency of type IIB, or both type IIA and IIB myofibers may be seen (66), suggesting a deficient differentiation of type II myofibers in NM. Some of the myofibers in childhood NM may actually be fetal in nature (80), and muscle fiber type transformation (from type II to type I fibers) has been observed in NM (81). Thus, in addition to rod formation, there are a variety of associated or independent histochemical and ultrastructural abnormalities in NM muscle fibers.

Rods have been identified in biopsied muscle tissue of heterozygous parents among some clinically affected family members of patients with autosomal recessive NM (62). Other heterozygous parents lacked rods, but showed abnormal histochemical patterns (82).

Cardiac involvement in NM has been found in several adults (73,83) and children (74,84), but not in patients with congenital NM (85). However, few morphological studies of rod formation in cardiac myocytes have been undertaken (73,74,83), some having found rod-like material (73), others not (83,86).

ACKNOWLEDGMENTS

The preparation of this paper was supported by the Deutsche Gesellschaft für Muskelkranke e.V., Freiburg, Germany, and by a special grant from the Federal Ministry of Health, Bonn, Germany, to Dr. A. Fidzianska (122-407-12 P 6/10). The manuscript was edited and typed by Mrs. A. Wöber, and Mrs. I. Warlo kindly provided the figures.

REFERENCES

1. Shy GM, Magee KR. A new congenital non-progressive myopathy. Brain 1956; 79:610–621.
2. Goebel HH, Lenard HG. Congenital myopathies. In: Rowland LP, DiMauro S, eds. Handbook of Clinical Neurology. Amsterdam: Elsevier, 1992; 18(62):331–367.
3. Shuaib A, Paasuke RT, Brownell AKW. Central core disease. Clinical features in 13 patients. Medicine 1987; 66:389–396.
4. Merlini L, Mattutini P, Bonfiglioli S, Granata C. Non-progressive central core disease with severe congenital scoliosis: a case report. Dev Med Child Neurol 1987; 29:106–109.
5. Soza M, Hoppe A, Gonzalez S. Central core disease (CCD): severe and progressive form. J. Neurol Sci 1990; 98:337 (abstract).
6. Rowinska-Marcinska K, Strugalska MH, Hausmanowa-Petrusewicz I. Fiber density in congenital muscle fibre type disproportion. Electromyogr Clin Electrophysiol 1990; 30:475–481.
7. Denborough MA, Dennett X, Anderson RMcD. Central-core disease and malignant hyperpyrexia. Br Med J 1973; 1:272–273.
8. Frank JP, Harati Y, Butler IJ, Nelson TE, Scott CI. Central core disease and malignant hyperthermia syndrome. Ann Neurol 1980; 7:11–17.
9. Eng GD, Epstein BS, Engel WK, McKay DW, McKay R. Malignant hyperthermia and central core disease in a child with congenital dislocating hips. Arch Neurol 1978; 35:189–197.
10. Krivosic-Horber R, Krivosic I. Myopathie à axe central (central core disease) associée à une sensibilité à l'hyperthermie maligne. Presse Med 1989; 18:828–831.
11. Stewart CR, Kahler SG, Gilchrist JM. Congenital myopathy with cleft palate and increased susceptibility to malignant hyperthermia: King syndrome? Pediatr Neurol 1988; 4:371–374.
12. Kausch K, Lehmann-Horn F, Janka M, Wieringa B, Grimm T, Müller CR. Evidence for linkage of the central core disease locus to the proximal long arm of human chromosome 19. Genomics 1991; 10:765–769.
13. Haan EA, Freemantle CJ, McCure JA, Friend KL, Mulley JC. Assignment of the gene for central core disease to chromosome 19. Hum Genet 1990; 86:187–190.
14. Quane KA, Healy JMS, Keating KE, et al. Mutations in the ryanodine receptor gene in central core disease and malignant hyperthermia. Nature 1993; 3:51–55.
15. Greenfield JG, Cornman T, Shy GM. The prognostic value of the muscle biopsy in the "floppy infant." Brain 1958; 81:461–484.
16. Neville HE, Brooke MH. Central core fibers: structured and unstructured. In: Kakulas BA, ed. Basic Research in Myology. Proceedings of the Second International Congress on Muscle Diseases (part 1). Amsterdam: Excerpta Medica, 1973:497–511.
17. Fidzianska A, Niebrój-Dobosz I, Badurska B, Ryniewicz B. Is central core disease with structural core a fetal defect? J Neurol 1984; 231:212–219.
18. Goebel HH. Neuropathological aspects of congenital myopathies. Prog Neuropathol 1986; 6:231–262.
19. Chou SM, Chou TM, Mori M. The "core" myofibres induced by local tetanus and tenotomy in rats. J Neuropathol Exp Neurol 1981; 40:300.
20. Bruce-Gregorios J, Chou SM. Core myofibres and related alterations induced in rats' soleus muscle by immobilization in shortened position. J Neurol Sci 1984; 63:267–275.
21. Hayashi K, Miller RG, Brownell AKW. Central core disease: ultrastructure of the sarcoplasmic reticulum and T-tubules. Muscle Nerve 1989; 12:95–102.
22. Seitz RJ, Toyka KV, Wechsler W. Adult-onset mixed myopathy with nemaline rods, minicores, and central cores: a muscle disorder mimicking polymyositis. J Neurol 1984; 231:103–108.
23. Vallat JM, de Lumley L, Loubet A, Leboutet MJ, Corvisier N, Umdenstock R. Coexistence of minicores, cores, and rods in the same muscle biopsy. Acta Neuropathol (Berl) 1982; 58:229–232.
24. Castro L, Coelho T, Pinheiro AV, Guimarães A. Tubular aggregates in a case of central core disease. J Neurol Sci 1990; 98:337 (abstract).
25. Paljärvi L, Kalimo H, Lang H, Savontaus M-L, Sonninen V. Minicore myopathy with dominant inheritance. J Neurol Sci 1987; 77:11–22.

26. Fardeau M. Some orthodox or non-orthodox considerations on congenital myopathies. In: Ellinson RJ, Murray NMF, Halliday AM, eds. The London Symposia (EEG Suppl. 39). Amsterdam: Elsevier, 1987: 85–90.
27. Ben Hamida M, Hentati F, Ben Hamida C. Maladie à multiminicores au cours d'un syndrome de la colonne raide. Rev Neurol (Paris) 1987; 143:284–289.
28. Shuaib A, Martin JME, Mitchell LB, Brownell AKW. Multicore myopathy: not always a benign entity. Can J Neurol Sci 1988; 15:10–14.
29. Davis DG, Nelson KR, Markesbery WR. Congenital myopathy and cardiomyopathy with identical ultrastructural changes. Arch Neurol 1990; 47:1141–1144.
30. Edström L, Mair WGP, Wroblewski R, Hovmöller M, Malm G. Type distribution of muscle fibres and their ultrastructure related to intracellular elemental composition as revealed by energy dispersive X-ray microanalysis. A study of multicore myopathy. J Neurol Sci 1986; 76:31–48.
31. Spiro AJ, Shy GM, Gonatas NK. Myotubular myopathy. Persistence of fetal muscle in an adolescent boy. Arch Neurol 1966; 14:1–14.
32. Sher JH, Rimalovski AB, Athanassiades TJ, Aronson SM. Familial centronuclear myopathy: a clinical and pathological study. Neurology 1967; 17:727–742.
33. Liechti-Gallati S, Wolff G, Ketelsen U-P, Braga, S. Prenatal diagnosis of X-linked centronuclear myopathy by linkage analysis. Pediatr Res 1993; 33:201–204.
34. Misra AK, Menon NK, Mishra SK. Abnormal distribution of desmin and vimentin in myofibers in adult onset myotubular myopathy. Muscle Nerve 1992; 15:1246–1252.
35. Van Wijngaarden GK, Fleury P, Bethlem J, Meijer AEFH. Familial "myotubular" myopathy. Neurology 1969; 19:901–908.
36. Sarnat HB. Myotubular myopathy: arrest of morphogenesis of myofibres associated with persistence of fetal vimentin and desmin. Four cases compared with fetal and neonatal muscle. Can J Neurol Sci 1990; 17:109–123.
37. Oldfors A, Kyllerman M, Wahlström J, Darnfors C, Henriksson KG. X-linked myotubular myopathy: clinical and pathological findings in a family. Clin Genet 1989; 36:5–14.
38. Lehesjoki A-E, Sankila E-M, Miao J, Somer M, Salonen R, Rapola J, De la Chapelle A. X-linked neonatal myotubular myopathy: one recombination detected with four polymorphic DNA markers from Xq28. J Med Genet 1990; 27:288–291.
39. Bucher HU, Boltshauser E, Briner J, Gnehm HE, Janzer RC. Severe neonatal centronuclear (myotubular) myopathy: an X-linked recessive disorder. Helv Paediatr Acta 1986; 41:291–300.
40. Keppen LD, Husain MM, Woody RC. X-linked myotubular myopathy: intrafamilial variability and normal muscle biopsy in a heterozygous female. Clin Genet 1987; 32:95–99.
41. Sawchak JA, Sher JH, Norman MG, Kula RW, Shafiq SA. Centronuclear myopathy heterogeneity: distinction of clinical types by myosin isoform patterns. Neurology 1991; 41:135–140.
42. Miller A, Solimano A, Norman MG. Arthrogryposis multiplex congenita and hypotonia in a male neonate. Pediatr Neurosci 1987; 13:272–277.
43. Metze K, Schnabel R, Mellin W, Gullotta F, Hiddemann W. Heterogeneity of centronuclear myopathy shown by flow cytometry of nuclear DNA. Neuropathol Appl Neurobiol 1993; 19:182–184.
44. Sarnat HB. Vimentin and desmin in maturing skeletal muscle and developmental myopathies. Neurology 1992; 42:1616–1624.
45. Sewry CA, Rodillo E, Heckmatt JZ, Dubowitz V. Immunocytochemical studies of congenital centronuclear and nemaline myopathies. J Neurol Sci 1990; 98:100 (abstract).
46. Edström L, Wroblewski R, Mair WGP. Genuine myotubular myopathy. Muscle Nerve 1982; 5:604–613.
47. Pavone L, Mollica F, Grasso A, Pero G. Familial centronuclear myopathy. Acta Neurol Scand 1980; 62:41–54.
48. Müller B, Mostacciuolo ML, Danieli GA, Grimm T. Problems in genetic counseling in a family with an "atypical" centronuclear myopathy. Am J Med Genet 1989; 32:417–419.
49. Serratrice G, Pellissier JF, Bes A, Arne-Bes MC. Un cas de myopathie centro-nucléaire de l'adulte de topographie facio-scapulo-péronière. Rev Neurol (Paris) 1987; 143:693–697.
50. Sasaki T, Shikura K, Sugai K, Nonaka I, Kumagai K. Muscle histochemistry in myotubular (centronuclear) myopathy. Brain Dev 1989; 11:26–32.
51. Lo WD, Barohn RJ, Bobulski RJ, Kean J, Mendell JR. Centronuclear myopathy and type-1 hypotrophy without central nuclei. Distinct nosologic entities? Arch Neurol 1990; 47:273–276.
52. Goebel HH, Meinck HM, Reinecke M, Schimrigk K, Mielke U. Centronuclear myopathy with special consideration of the adult form. Eur Neurol 1984; 23:425–434.
53. Lee Y-S, Yip WCL. A fatal congenital myopathy with severe type I fibre atrophy, central nuclei and multicores. J Neurol Sci 1981; 50:277–290.
54. Ferrer X, Vital C, Coquet M, Deleplanque B, Ellie E, Lagueny A, Julien J. Myopathie centronucléaire autosomique dominante. Rev Neurol (Paris) 1992; 148:622–630.
55. Shy GM, Engel WK, Somers JE, Wanko T. Nemaline myopathy. A new congenital myopathy. Brain 1963; 86:793–810.
56. Conen PE, Murphy EG, Donohue WL. Light and electron microscopic studies of myogranules in a child with hypotonia and muscle weakness. Can Med Assoc J 1963; 89:983–986.

57. Wallgren-Pettersson C. Congenital nemaline myopathy. A clinical follow-up study of twelve patients. J Neurol Sci 1989; 89:1–14.
58. Sasaki M, Yoneyama H, Nonaka I. Respiratory muscle involvement in nemaline myopathy. Pediatr Neurol 1990; 6:425–427.
59. Wallgren-Pettersson C, Sainio K, Salmi T. Electromyography in congenital nemaline myopathy. Muscle Nerve 1989; 12:587–593.
60. Shimomura C, Nonaka I. Nemaline myopathy: comparative muscle histochemistry in the severe neonatal, moderate congenital, and adult-onset form. Pediatr Neurol 1989; 5:25–31.
61. Laing NG, Majda BT, Akkari PA, et al. Assigment of a gene (NEM I) for autosomal dominant nemaline myopathy to chromosome 1. Am J Hum Gent 1992; 50:576–583.
62. Arts WF, Bethlem J, Dingemans KP, Eriksson AW. Investigations on the inheritance of nemaline myopathy. Arch Neurol 1978; 35:72–77.
63. Wallgren-Pettersson C, Kääriäinen H, Rapola J, Salmi T, Jääskeläinen J, Donner M. Genetics of congenital nemaline myopathy: a study of 10 families. J Med Genet 1990; 27:480–487.
64. Jockusch B, Veldman H, Griffith GW, Van Oost BA, Jennekens FGI. Immunofluorescence microscopy of a myopathy. Exp Cell Res 1980; 127:409–420.
65. Hashimoto K, Shimizu T, Nonaka I, Mannen T. Immunochemical analysis of α-actinin of nemaline myopathy after two-dimensional electrophoresis. J Neurol Sci 1989; 93:199–209.
66. Wallgren-Pettersson C, Rapola J, Donner M. Pathology of congenital nemaline myopathy. J Neurol Sci 1988; 83:243–257.
67. Jenis EH, Lindquist RR, Lister RC. New congenital myopathy with crystalline intranuclear inclusions. Arch Neurol 1969; 20:281–287.
68. Engel WK, Oberc MA. Abundant nuclear rods in adult-onset rod disease. J Neuropathol Exp Neurol 1975; 34:119–132.
69. Barohn RJ, Jackson CE, Kagan-Hallet KS. Neonatal nemaline myopathy with abundant intranuclear rods. Neuromusc Disord 1994; 4:513–520.
70. Fukunaga H, Osame M, Igata A. A case of nemaline myopathy with ophthalmoplegia and mitochondrial abnormalities. J Neurol Sci 1980; 46:169–177.
71. Takeuki Y, Iwami H, Inoue F, et al. Rimmed vacuoles in biopsied muscle of nemaline myopathy. Acta Neuropathol (Berl) 1985; 68:253–255.
72. Karpati G, Carpenter S, Eisen AA. Experimental core-like lesions and nemaline rods. Arch Neurol 1972; 27:237–251.
73. Meier C, Voellmy W, Gertsch M, Zimmermann A, Geissbühler J. Nemaline myopathy appearing in adults as cardiomyopathy. Arch Neurol 1984; 41:443–445.
74. Buonocore G, Balestri P, Toti P, Bagnoli F. A new case of severe congenital nemaline myopathy. Acta Pediatr 1993; 82:1082–1084.
75. Ishibashi-Ueda H, Imakita M, Yutani C, et al. Congenital nemaline myopathy with dilated cardiomyopathy: an autopsy study. Hum Pathol 1990; 21:77–82.
76. Danon MJ, Giometti CS, Manaligod JR, Perurena OH, Skosey JL. Adult-onset nemaline rods in a patient treated for suspected dermatomyositis. Arch Neurol 1981; 38:761–766.
77. Danon MJ, Sarpel G, Manaligod JR. Nemaline rod myopathy and Charcot-Marie-Tooth disease. Arch Neurol 1980; 37:123–126.
78. Jerusalem F, Goetze H, Mumenthaler M. Zur diagnostischen Spezifität von nemaline- Strukturen. Z Neurol 1971; 200:148–157.
79. Bender AN, Willner JP. Nemaline (rod) myopathy: the need for histochemical evaluation of affected families. Ann Neurol 1978; 4:37–42.
80. Nonaka I, Tojo M, Sugita H. Fetal muscle characteristics in nemaline myopathy. Neuropediatrics 1983; 14:47–52.
81. Miike T, Ohtani Y, Tamari H, Ishitsu T, Une Y. Muscle fiber type transformation in nemaline myopathy and congenital fiber type disproportion. Brain Dev 1986; 8:526–532.
82. Wallgren-Pettersson C. Congenital nemaline myopathy. A longitudinal study. In: Spring E, ed. Commentationes Physico-mathematicae. Vol. 111, Dissertationes No. 30. Helsinki: Finnish Society of Sciences and Letters, 1990.
83. Stoessl J, Hahn AF, Malott D, Jones DT, Silver MD. Nemaline myopathy with associated cardiomyopathy. Arch Neurol 1985; 42:1084–1086.
84. Van Antwerpen L, Gospe SM, Dentinger MP. Nemaline myopathy associated with hypertrophic cardiomyopathy. Pediatr Neurol 1988; 4:306–308.
85. Airenne A-L, Wallgren-Pettersson C. Normal cardiac contractility in patients with congenital nemaline myopathy. Neuropediatrics 1988; 19:115–117.
86. Shafiq SA, Dubowitz V, Peterson HdeC, Milhorat AT. Nemaline myopathy: report of a fatal case, with histochemical and electron microscopic studies. Brain 1967; 90:817–828.

16
Congenital Fiber Type Disproportion

J. Michael Schröder
Rheinisch-Westfälische Technische Hochschule Aachen, Aachen, Germany

INTRODUCTION

Muscle fiber type disproportion (FTD), according to Brooke's (4) definition, is characterized by (1) hypertrophy of type 2 muscle fibers, (2) predominance of type 1 fibers with normal or slightly reduced diameters, and (3) a difference between the mean diameter of these main fiber types of more than 12%. Although FTD was first regarded as a congenital myopathy (4–9), our data and those of others (cf. 2, 20,21,31,32,39) indicate that it is a heterogeneous syndrome caused by a variety of disorders, including peripheral neuropathy. In this respect, it is important to distinguish between pure congenital fiber type disproportion (CFTD), a common cause of benign congenital hypotonia, and the histopathological feature of FTD which can be observed in a number of conditions.

GENETICS

Half the patients studied by Brooke (4) had relatives with similar symptoms. In others only distant family members were affected, so autosomal recessive inheritance was supposed; in one patient, however, the father and mother were affected as well, and autosomal dominant inheritance was suggested. This was also true for two patients studied by Fardeau et al. (21), the father of whom showed the same characteristic fiber type disproportion in his muscle biopsy, although he was not conscious of any muscle weakness. However, he had a small, narrow face, slight proximal muscle wasting, and depressed tendon reflexes in the arms. A similar disease was recorded in concordant twins (13). A balanced chromosomal translocation t(10;17) (p 11.2;q25) in a case with congenital fiber type disproportion and arthrogryposis multiplex congenita has been described and the translocation breakpoints considered as candidate regions for a myopathy gene (25).

CLINICAL ASPECTS

The children studied by Brooke (4) appeared to be hypotonic ("floppy babies"). The disease presented shortly after birth. In half the cases there were contractures of different muscles in the lower limbs. One patient had torticollis due to contracture of the sternocleidomastoid muscle. Half the patients showed congenital hip dislocation, which could be unilateral or bilateral. The extent of weakness varied from case to case. All muscles of the trunk and of the extremities were involved, although in some patients the lower extremities were more affected than the upper ones. Usually there was only a delay of the motor milestones without prominent paralysis. In some cases an initial increase of the weakness during the first year was observed; in none of the children, however, was there a progression of the disease after the age of 2 years, and in some improvement occurred. During the first year, there were repeated infections of the respiratory tract. Twelve of 14 patients had 30% lower weight than normal children at this age. In eight patients body size was reduced by more than 10%. Eight patients had a high-arched palate; in six patients kyphoscoliosis developed during growth. Deformities of the feet, such as flat or high-arched feet were frequent. The serum creatine kinase level was usually normal, and the electromyogram was either normal or "neuropathic." Other cases, however, showed complex high-amplitude potentials (32).

Clancy et al. (12) emphasized that the clinical picture in only three of their five cases followed the description of Brooke (4), whereas in two other cases there was only slight hypotonia in the legs; in another case there was severe retardation of the motor and mental development together with disturbances of growth, deformity of the head, and scoliosis. The diagnosis would not have been suggested, even after the results of the muscle biopsy were known, on the basis of the clinical findings alone. Some cases were unusually severe (32) or fatal (11,42). In a study of congenital fiber type disproportion in eight children, the presence of severe weakness at birth and profound respiratory muscle weakness were considered as predictors of a high rate of mortality in infancy and poor developmental outcome in the survivors (45).

MUSCLE BIOPSY

By Brooke's criteria, type 1 muscle fibers should be smaller than type 2 fibers. The difference in the mean fiber diameter should be at least 12%, but is usually much more (15). In general, the mean caliber of the type 1 fibers is not significantly below normal values for the corresponding age group, and thus there is no general type 1 fiber atrophy. Type 2 fibers, however, are considerably enlarged, especially type 2B fibers. Grouping of fiber types or groups of atrophic muscle fibers were not a feature. In eight of the 14 patients studied there were more type 1 fibers than type 2 fibers; in four patients there were exceptionally few type 2B fibers and in one patient remarkably few type 2A fibers. Occasionally central nuclei were observed. Some fibers appeared to be "moth-eaten" or "whorled"; occasional *ringbinden* and nemaline bodies were also noted. Large sectors of sarcomeric disarray in small fibers were noted in a more severely affected case (32) similar to the case with central core disease illustrated in Figure 1.

DIFFERENTIAL DIAGNOSIS AND NOSOLOGICAL HETEROGENEITY OF FTD

It is important to distinguish patients with congenital fiber type disproportion (CFTD) from those with Werdnig-Hoffmann disease because their prognosis is significantly better. The disease should also be distinguished from the various forms of congenital muscle dystrophy. Facioscapulohumeral muscular dystrophy, which may show large type 2 fibers, should also be considered. Fiber type disproportion is also seen in myotonic dystrophy, in spinocerebellar disorders (9), in fetal alcohol syndrome, globoid cell leukodystrophy, infantile acid maltase deficiency (33), in rigid spine syndrome (27,40), Marden-Walker syndrome (24), Lowe's or Mobius' syndrome, hypothyroidism, and hydrocephalus (41).

Small type 1 fibers may be encountered in many congenital myopathies, including nemaline myopathy (14,30,36), centronuclear myopathies (28,34), and fingerprint myopathy (22). Myotubular myopathy was observed in one member, and fiber type disproportion in another in the same family (31). Several patients initially reported as having CFTD showed rods in their muscle biopsies (4,10 case 2,24). The histological picture of FTD can be considered a "mixed bag" of cases (5,15,20) and includes a variety of myopathies (see also 3,18,19,26,29,35,37,38,41,43,44).

Computer retrieval of our cases classified as FTD from a data file of more than 5000 muscle biopsies revealed 102 cases (Table 1). Thus FTD represents a relatively frequent histopathological feature although much less frequent than selective type 2 muscle fiber atrophy. Table 1 illustrates the considerable variety of disorders, including peripheral neuropathy, which had FTD as a feature

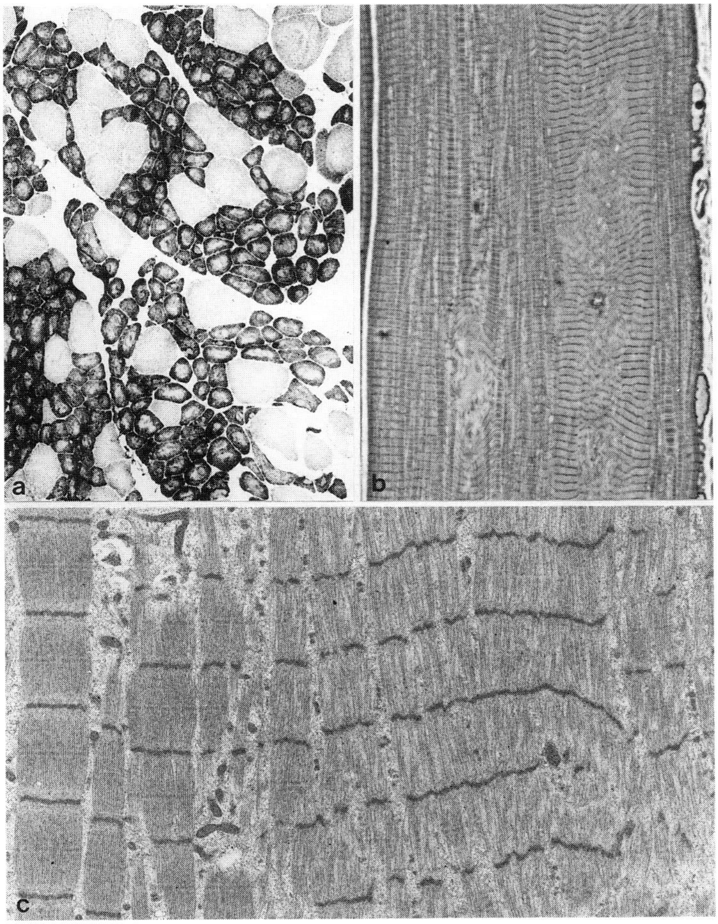

Figure 1 Quadriceps femoris muscle biopsy showing congenital muscle fiber type disproportion with central cores in a 27-year-old adipose woman. (a) NADH reaction showing large fibers with low oxidative activity and a mean diameter of about 100 μm, and small fibers showing high activity with a mean diameter of 40 μm. Most of the small fibers have a light center. ×120. (b) Semithin section stained with paraphenylenediamine. There are several foci of myofibrillar disarray and clearing where the A-bands are more severely affected than the Z-bands. ×1050. (c) At the electron microscopic level, myofibrils on the left appear to be normal, whereas on the right myofibrils are more or less severely depleted of myosin filaments including their M-lines, in an area where the number of mitochondria is reduced. Uranylacetate and lead citrate, ×1050.

in our series. In one of our patients, who died because of nocturnal apnea at the age of 19 years, several months after the muscle biopsy had been performed, we noted microscopic hydromyelia at the thoracic level (39). His brother, who later developed neurofibromatosis, also died of nocturnal apnea and had similar fiber type disproportion (J. Bohl, personal communication). Thus FTD should be regarded as a histopathological feature that is nonspecific and may be associated with a variety of congenital and noncongenital muscle disorders. Nevertheless, pure fiber size disproportion at an early age (in Table 1 below 30 years of age) is highly suggestive of a congenital syndrome, and there are numerous cases of benign congenital hypotonia in which CFTD is the only histopathological abnormality.

Table 1 Muscle Fiber Type Disproportion: Clinical and/or Histological Diagnoses in 102 Successive Cases

Pure fiber type disproportion	
Probably congenital fiber type disproportion (CFTD) ≤ 30 years of age	33
Fiber type disproportion (FTD) ≥ 30 years of age	26
Well-defined myopathic disorders	
Mitochondrial myopathies	3
Myotonic dystrophy	2
Malignant hyperthermia	2
Central core and minicore myopathy	2
Limb girdle dystrophy	1
Neurogenic disorders	
Neurogenic atrophy	7
Neuropathy, axonal type	7
Neuropathy, demyelinating type	3
Hydromyelia and syringomyelia	2
Neurofibromatosis	1
Inflammatory disorders	
Dermatomyositis	4
Vasculitis	2
Various disorders	
Glycogenosis	3
Crampus syndrome	2
Eales syndrome	1
Wallenberg syndrome	1

PATHOGENESIS

Although FTD as a histopathological syndrome is heterogeneous and seen in a variety of disorders, such as those listed in Table 1, the final pathway of its pathogenesis and histochemical manifestation could be similar. Muscle weakness in these disorders might be more, or less, well compensated by hyperactivity and overload of the remaining fibers. It is remarkable in this context that 74% of the patients in our series are male and only 26% female. If this type of overload causing hyperactivity is generally of short duration, it would mimic training conditions such as weight lifting in which selective hypertrophy of type 2 muscle fibers is known to occur (17). But other causes, such as undefined neurogenic disturbances of the peripheral reflex arc, should also be taken into consideration.

REFERENCES

1. Argov Z, Gardner-Medwin D, Johnson MA, Mastaglia FL. Patterns of muscle fiber-type disproportion in hypotonic infants. Arch Neurol 1984; 41:53–57.
2. Berger M, Schröder JM, Gibbels E, Thomas E. Fasertypendisproportion, eine Form der sogenannten kongenitalen, nichtprogressiven Myopathien. Fortschr Neurol Psychiatr 1979; 47:466–478.
3. Bethlem J, van Wijngaarden GK, Meijer AEFH, Hülsmann WC. Neuromuscular disease with type I fiber atrophy, central nuclei, and myotube-like structures. Neurology 1969; 19:705–710.
4. Brooke MH. A neuromuscular disease characterized by fiber types disproportion. In: Kakulas BA, ed. Proceedings of the Second International Congress on Muscle Diseases, Perth, Australia, November 1971. Amsterdam: Excerpta Medica, ICS No. 294, Part 2, 1973:147–159.
5. Brooke MH, Carroll JE, Ringel SP. Congenital hypotonia revisited. Muscle Nerve 1979; 2:84.
6. Brooke MH, Engel WK. The histographic analysis of human muscle biopsies with regard to fiber types. Neurology 1969; 19:221–223.
7. Brooke MH, Engel WK. Diseases of the upper and lower motor neuron. Neurology 1969; 19:378–393.

8. Brooke MH, Engel WK. Myotonias, mysthenia gravis, and hypokalemic periodic paralysis. Neurology 1969; 19:469–477.
9. Brooke MH, Engel WK. Children's biopsies. Neurology 1969; 19:591–605.
10. Caille B, Fardeau M, Harpey JP, Lafourcade J. Hypotonie congénitale avec atteinte sélective des fibres musculaires de type 1. Arch Franc Pédiatr 1971; 28:205.
11. Cavanagh NPC, Lake BD, McMenamin P. Congenital fibre type disproportion myopathy; a histological diagnosis with an uncertain clinical outlook. Arch Dis Child 1979; 54:735.
12. Clancy RR, Kelts KA, Oehlert JW. Clinical variability in congenital fiber type disproportion. J Neurol Sci 1980; 46:257–266.
13. Curless RG, Nelson MB. Congenital fiber type disproportion in identical twins. Ann Neurol 1977; 2:455–459.
14. Dahl DS, Klutzow FW. Congenital rod disease. Further evidence of innervational abnormalities as the basis for the clinicopathologic features. J Neuro Sci 1974; 23:371–385.
15. Dubowitz V. The floppy infant. In: Clinics in Developmental Medicine no. 76, 2nd ed. London: SIMP with Heinemann Medical, Lippincott, Philadelphia, 1980.
16. Dubowitz V, Brooke MH. Muscle Biopsy: A Modern Approach. London: WB Saunders, 1973:475 pp.
17. Edström L, Ekblom B. Differences in sizes of red and white muscle fibres in vastus lateralis of musculus quadriceps femoris of normal individuals and athletes. Relation to physical performance. Scand J Clin Lab Invest 1972; 30:175–181.
18. Eisler T, Wilson JH. Muscle fibre-type disproportion. Report of a family with symptomatic and asymptomatic members. Arch Neurol 1978; 35:823.
19. Engel WK, Wanko T, Fenichel GM. Nemaline myopathy: a second case. Arch Neurol 1964; 11:22–39.
20. Fardeau M. Congenital myopathies. In: Mastaglia FL, Lord Walton of Detchant, eds. Skeletal muscle pathology, 2nd ed. London: Butler & Tanner, 1992:258.
21. Fardeau M, Harpey JP, Caille B. Disproportion congénitale des differents types de fibre musculaire, avec petitesse relative des fibres de type I. Rev Neurol 1975; 131:745.
22. Fardeau M, Tomé FMS, Derambure S. Familial fingerprint body myopathy. Arch Neurol 1976; 33:724–725.
23. Farkas-Bargeton E, Aicardi J, Chevrie JJ, Thieffry S. Apport des techniques histo-enzymologiques à l'étude des hypotonies congénitales. Rev Neurol 1968; 119:513.
24. Garcia-Alix A, Blanco D, Cabanas F, Garcia-Sanchez P, Pellicer A, Quero J. Early neurological manifestations and brain anomalies in Marden-Walker syndrome. Am J Med Genet 1992; 44:41–45.
25. Gerdes AM, Petersen MB, Schroder HD, Wulff K, Brondum-Nielsen K. Congenital myopathy with fiber type disproportion: a family with a chromosomal translocation t(10;17) may indicate candidate gene regions. Clin Genet 1994; 45:11–16.
26. Glick B, Shapira Y, Stern A, Blank A, Rosenmann E. Congenital muscle fiber-type disproportion myopathy: a follow-up study of twenty cases. Ann Neurol 1984; 16:405.
27. Goebel HH, Lenard HG, Gorke W, Kunze K. Fibre type disproportion in the rigid spine syndrome. Neuropädiatrie 1977; 8:467–477.
28. Inokuchi T, Umezaki H, Santa T. A case of type 1 muscle fibre hypotrophy and internal nuclei. J Neurol Neurosurg Psychiatry 1975; 38:475–482.
29. Jaffé M, Shapira J, Borochowitz Z. Familial congenital fiber type disproportion (CFTD) with an autosomal recessive inheritance. Clin Genet 1988; 33:33.
30. Kinoshita M, Satoyoshi E. Type I fiber atrophy and nemaline bodies. Arch Neurol 1974; 31:423–425.
31. Kinoshita M, Satoyoshi E, Kumagai M. Familial type I fiber atrophy. J Neurol Sci 1975; 25:11–17.
32. Lenard HG, Goebel HH. Congenital fibre type disproportion. Neuropaediatrie 1975; 6:220–231.
33. Martin JJ, Clara R, Ceutrick C, Joris C. Is congenital fibre type disproportion a true myopathy? Acta Neurol Belg 1976; 76:335–344.
34. Meyers KR, Golomb HM, Hansen JL, McKusick VA. Familial neuromuscular disease with myotubes. Clin Genet 1974; 5:327.
35. Mizuno Y, Komiya K. A serial muscle biopsy study in a case of congenital fiber-type disproportion associated with progressive respiratory failure. Brain Dev 1990; 12:431–436.
36. Radu H, Ionescu V. Nemaline (neuro)myopathy. Rod-like bodies and type I fibre atrophy in a case of congenital hypotonia with denervation. J Neurol Sci 1972; 17:53.
37. Radu H, Ionescu V, Radu A, Paler V, Rosu AM, Marian A. Hypotrophic type I muscle fibres with central nuclei, and central myofibrillar lysis preferentially involving type II fibres. Eur Neurol 1974; 11:108–127.
38. Sarnat HB. Vimentin and desmin in maturing skeletal muscle and developmental myopathies. Neurology 1992; 42:1616–1642.
39. Schröder JM. Pathologie der Skelettmuskulatur. Spez. Pathol. Anat. 1982; 15:272–274.
40. Seay AR, Ziter FA, Petajan J. Rigid spine syndrome. A type I fiber myopathy. Arch Neurol 1977; 34:119–122.
41. Simon R, Mateos F, Seijo M, Balseiro J, Madero S, Cabello A. Congenital fiber-type disproportion: analysis of a series of 11 cases. Neurologia 1991; 6:281–286.

42. Spiro AJ, Horoupian DS, Snyder DR. Biopsy and autopsy studies of congenital muscle fibre type disproportion: a broadening concept. Neurology 1977; 27:405.
43. Sulaiman AR, Swick HM, Kinder DS. Congenital fibre type disproportion with unusual clinico-pathologic manifestations. J Neurol Neurosurg Psychiatry 1983; 46:175.
44. ter Laak HJ, Jaspar HHJ, Gabreels FJM, Breuer TJM, Sengers RCA, Joosten EMG, Stadhouders AM, Gabreels-Festen AAWM. Congenital fibre type disproportion: a morphometric study (4 cases). Clin Neurol Neurosurg 1981; 83:67.
45. Torres CF, Moxley RT. Early predictors of poor outcome in congenital fiber-type disproportion myopathy. Arch Neurol 1992; 49:855–856.

17
X-Linked Dystrophies: Clinical and Pathological Features

Frank L. Mastaglia
Australian Neuromuscular Research Institute, Queen Elizabeth II Medical Centre, Perth, Western Australia, Australia

Byron A. Kakulas
Royal Perth Hospital and Australian Neuromuscular Research Institute, Queen Elizabeth II Medical Centre, Perth, Western Australia, Australia

INTRODUCTION

The most important forms of X-linked dystrophy are the Duchenne and Becker varieties, both of which are now known to be associated with mutations in the dystrophin gene at the Xp21 locus and deficiency of dystrophin. Discovery of the dystrophin gene and the finding of mutations in the gene in patients with Duchenne and Becker dystrophy confirmed the allelic nature of these two conditions, which may sometimes coexist in the same family (1). In addition, the discovery of dystrophin and the subsequent development of immunohistochemical methods for its localization within the muscle fiber, as well as more sensitive immunoblotting techniques for the protein, have provided more specific methods for establishing the diagnosis of these conditions. It has also been recognized that mutations in the gene may be associated with a wider spectrum of clinical expression than was previously appreciated and these conditions are now collectively referred to as the *dystrophinopathies.*

The other X-linked dystrophies are much rarer and, less is known about their molecular genetics and pathology.

X21 DYSTROPHIES

Incidence

Duchenne muscular dystrophy (DMD), the most rapidly progressive and lethal form of dystrophy, is also the most common variety, having an incidence of roughly 1 in 3500 live male births and a reported prevalence of about 50–70 x 10^{-6} total male population in most surveys (2). With the development of more specific diagnostic methods such as dystrophin immunolabeling and cDNA probes, it has been recognized that the more benign Becker variety (BMD) is more common than was previously thought. In a study of the north of England the incidence was found to be roughly one in 18,000 live male births and the prevalence of 23.8 x 10^{-6} (3).

Inheritance

The inheritance pattern is that of an X-linked recessive trait, with males manifesting the typical clinical condition. Heterozygous females, who are usually unaffected, are carriers and may pass on the condition to 50% of their sons and the carrier state to 50% of their daughters. In other words, there is a 1 in 4 risk that a carrier will have an affected child.

Female DMD carriers are usually clinically unaffected, but some show minor nonprogressive clinical abnormalities, elevated serum CK levels, or mild electromyographic changes or histological muscle abnormalities. Rarely, carriers with incomplete lyonization of the maternal X chromosome may have more severe progressive muscle disease (*manifesting carriers*). On average, about one-third of new cases of DMD occur in families with previously affected males in the same or in previous generations, while in about two-thirds or more of cases there is no previous family history of the condition (2, 4). Of the two-thirds of sporadic cases, about one-third are born to previously unknown carriers or to mothers with a new mutation, while the remainder are boys with new mutations (4, 5). Some workers (6) have reported a higher carrier frequency in DMD families and have suggested that the rate of new mutations is lower. In rare instances, the Duchenne phenotype may occur in females with an XO,X/XX or X/XX/XXX karyotype or with an X-autosome translocation (at Xp21), or in one of a pair of monozygotic twins due to lyonization of the paternal X chromosome (4).

In BMD, unlike DMD, most affected males are able to have children and all their daughters will be carriers. A higher proportion of mothers of affected males are carriers and it has been estimated that only about 10% of cases are due to new mutations (4).

Molecular Genetics

The dystrophin gene locus is large, being about 2500 kb in length and consisting of over 70 exons. The large size of the gene is thought to account for the very high mutation rate for DMD, which has been estimated to be of the order of one in 10,000 (7). Up to 70% of DMD cases and over 80% of BMD cases have indentifiable deletions in the gene (2, 8). In contrast, a small proportion (5–10%) of DMD cases have a duplication and the remaining cases are presumed to have point mutations in the gene. Deletions may occur at any position in the gene but are most frequent in the central (exons 45–55) and proximal (first 20 exons) regions (9). The position and size of deletions (and duplications) correlates poorly with the clinical phenotype, which depends more on whether or not the deletion disrupts the nucleotide reading frame. In general, *out-of-frame* deletions are associated with the severe DMD phenotype and a virtually complete absence of muscle dystrophin, while *in-frame* deletions are associated with the milder and more variable BMD phenotype and with a dystrophin molecule that is truncated but that is still presumed to be partly functional (10). Although there are exceptions, deletions in certain regions of the gene have been associated with certain phenotypes. For example, deletions involving exons 45–75 have been associated with typical mild BMD with preservation of walking until over the age of 40 years. Deletions in the central region of the gene, even when large, may also be associated with a mild clinical phenotype or with a syndrome of myalgia and muscle cramping without weakness or even with asymptomatic elevation of the serum CK level (2).

Rarely, deletions in the Xp21 region that extend beyond the dystrophin gene may be associated with overlapping syndromes of muscular dystrophy and glycerol kinase deficiency, congenital adrenal hypoplasia, chronic granulomatous disease, and the McLeod syndrome, in which there is reduced antigenicity of the Kell blood group (2). The McLeod syndrome may also be associated with a mild subclinical myopathy (11) in which the muscle dystrophin content is normal (12).

Clinical Phenotypes

Duchenne Dystrophy

The clinical phenotype resulting from complete deficiency of dystrophin is remarkably uniform, being characterized by progressive wasting and weakness particularly of proximal limb-girdle muscles usually becoming apparent before the age of 5 years, with inability to walk by the age of 8–12 years and death from respiratory or cardiac failure in the late 20s or early 30s. The earliest signs are usually noted only after the child begins to walk, most cases presenting between the ages of 3 and 6 years because of an abnormal gait, inability to run, or falls. In most cases motor development is retarded, about 50% of patients still being unable to walk at the age of 18 months and some even at the age of 3 years (13). Early clues to the diagnosis in the young child include hypotonia, prominence of the calves, and abnormalities of gait, which may become apparent when the child tries to run, to jump, or to climb stairs (14). In some cases of DMD the onset is delayed until 8–9 years of age, but such a later onset may suggest that the patient has the more benign BMD.

The pattern of muscle involvement is very consistent, being bilaterally symmetrical and involving the girdle and proximal limb muscles most severely, particularly in the lower limbs. A characteristic feature is the selectivity of muscle involvement, the quadriceps, iliopsoas, and glutei being particularly involved in the lower limbs and the latissimus dorsi, serrati, sternocostal head of pectoralis major, biceps, triceps, and brachioradialis in the upper limbs. Weakness of the knee and hip extensors results in the classical Gowers sign when the child attempts to rise from the floor, and weakness of the hip extensors leads to the development of an accentuated lumbar lordosis. At an earlier stage, even before the Gowers sign can be elicited, affected boys may be unable to rise from a sitting position on the floor if asked to do so with the arms folded (2). The gait subsequently becomes waddling in character owing to the weakness of the hip abductors and there is often a tendency to walk on the toes. Pseudohypertrophy of the calves, which is due largely to the accumulation of adipose and connective tissue (15), is a distinctive feature in over 90% of cases, the calf muscles having a firm, doughy, or woody consistency on palpation. Other muscle groups, such as the deltoids, wrist extensors, spinati, masseters, and the tongue, may also be enlarged in some patients (2).

As the disease advances there is progressive atrophy and weakness of the proximal limb and girdle muscles and the differential pattern of muscle involvement becomes progressively less obvious. The deep tendon reflexes are depressed, with the exception of the ankle jerks, which are preserved even in the late stages of the disease. Kyphoscoliosis and joint deformities are prone to develop, particularly once the patient becomes confined to a wheelchair, unless spinal bracing and an active program of physiotherapy is instituted. Although the weakness is progressive, there may be periods of apparent arrest. Periods of inactivity associated with orthopedic procedures or with viral or other intercurrent illnesses may lead to rapid deterioration and loss of the ability to walk. Weakness of the intercostal muscles and the development of thoracic deformities lead to progressive deterioration in pulmonary function and a predisposition to respiratory infections in the later stages of the disease when respiratory failure often develops (16). Diaphragmatic function is relatively preserved, which may explain why respiratory failure develops only late in the course of the disease and chronic alveolar hypoventilation is uncommon (2).

Becker Dystrophy

The age of onset and rate of progression are much more variable in BMD than in DMD. Most patients present between the ages of 5 and 15 years but the onset may not be until the 3rd or 4th decade of life in some cases. As in DMD, early motor milestones are often delayed and many patients give a history of never being able to run properly or to participate in school sports. Muscle cramps are often an early symptom. As in DMD, the pattern of muscle weakness is selective, with particular involvement of the sternocostal portion of the pectoralis major, latissimus dorsi, serratus anterior, biceps, triceps, and brachioradialis in the upper limbs and of the hip flexors, medial vasti, and later of the adductors and abductors of the thighs and anterior tibial muscles in the lower limbs (17). Muscle hypertrophy is a prominent feature in most cases, involving particularly the calves and sometimes also the lateral vasti, anterior tibial, deltoid, triceps, and forearm extensor muscles. The knee jerks and upper limb reflexes become depressed and are often lost as the disease progresses, while the ankle jerks are preserved even in the later stages (18). There is often a lordotic posture and waddling gait as in DMD. Pes cavus is not uncommon but muscle contractures and spinal deformities do not usually occur.

The rate of progression of the disease is variable. Most patients are still able to walk in the 20s or even later and most survive beyond the age of 20 and some even into the 7th or 8th decades (18). In the most severe cases with a clinical phenotype intermediate between that of typical DMD and BMD, patients become disabled at an earlier age and may be wheelchair-bound in the teens. In general, such patients have the lowest amounts of dystrophin in their muscle (about 10%) (10). In general, patients with deletions in the central rod domain of the dystrophin gene are more likely to have the more benign form of disease whereas deletions in the amino-terminal or carboxyl-terminal domains appear to be associated with more severe clinical phenotypes although there is a considerable amount of clinical variability particularly in patients with deletions near the center of the rod domain (10).

Other Phenotypes

Following the discovery of dystrophin, a number of variant phenotypes from the classical DMD and BMD forms have been recognized in patients with lesser degrees of dystrophin deficiency. These include patients with weakness confined to the quadriceps muscles (18), patients with a clinical phenotype resembling limb-girdle dystrophy or spinal muscular atrophy (20), as well as patients with myoglobinuria or with myalgia, fatigue, and muscle cramps (21), or even some patients with asymptomatic elevation of the serum CK level (*hyperCKemia*) (10). Such symptoms may also occur in female carriers, particularly those with dystrophin levels less than 50–60% of normal, and may be associated with some degree of muscle weakness and more frequently with elevation of the serum CK level (10, 22).

Involvement of Other Tissues

Cardiac Muscle

Cardiac involvement is most frequent in DMD in which fibrosis, particularly of the left ventricular myocardium and conducting system, is invariably found at postmortem (23, 24). During life, asymptomatic sinus tachycardia, arrhythmias, mitral valve prolapse, and other nonspecific murmurs are common as are electrocardiographic abnormalities, which are present in most cases and include conduction defects, tall R waves in right precordial leads, and deep Q waves in the limb and left precordial leads (2). Echocardiography shows characteristic changes in the contractility of the free posterior left ventricular wall, which increase with the passage of time (25). Sudden death may occasionally occur as a result of an arrythmia, but progressive cardiac failure is unusual until the terminal stages of the disease. Cardiomyopathy may also occur in patients with BMD but is less frequent than in DMD, although it is a prominent feature in some families and may lead to cardiac failure even at a relatively early stage of the disease (26, 27). Electrocardiographic abnormalities may be present in up to 50% of cases (10). There are also reports of isolated cases or families with cardiomyopathy as the only or major manifestation of a dystrophinopathy. A number of these patients have had deletions involving exon 1 or the muscle-promotor region of the dystrophin gene (28, 29).

Central Nervous System

About one-third of patients with DMD have some degree of mental retardation and an IQ less than 75 (17). The IQ distribution of DMD patients is about one standard deviation lower than that of a normal population, and mean IQs are lower than in control groups of patients with spinal muscular atrophy or other conditions causing physical incapacity (2, 30). Characteristically, there is impairment of verbal abilities, which is nonprogressive, although in some patients both verbal and performance skills are impaired (31). Electroencephalographic abnormalities of a nonspecific type may be found in up to 50% of DMD cases, while cranial computerized tomography and magnetic resonance imaging studies may show evidence of mild cerebral atrophy particularly in patients with low IQ scores (32, 33). Neuropathological changes that have been found in some cases and that may underlie the intellectual impairment include neuronal heterotopias (34) and dendritic abnormalities (35). Attempts to correlate the occurrence of mental retardation with the size and extent of mutations in the dystrophin gene have not been conclusive, although in one study of 66 DMD patients with mental retardation twice as many were found to have deletions involving exon 52 than any other portion of the gene (36).

Other Tissues

There is some evidence that the smooth muscle of the gastrointestinal tract and bladder may also be affected in DMD, although this is rarely symptomatic. Apart from this, there is no convincing evidence that any other tissues are functionally affected by the genetic defect (2).

Pathology

Duchenne Muscular Dystrophy

The Fetus. Several studies of the skeletal musculature of male fetuses at risk for DMD or BMD are reported in the literature. Overall it seems that in the fetus lack of dystrophin is compatible with normal histogenesis of skeletal muscle since the abnormalities that have been described are relatively minor. Vassilopolous et al. (37, 38) measured the size of muscle fiber nuclei in eight male fetuses at risk for DMD and in eight boys with DMD aged between 2 and 6 years. They found an overall slight but significant increase in nuclear size. In another study Mahoney et al. (39) reported excessive variation in muscle fiber diameter, hyalinization of larger fibers with disorganization of myofibrillar architecture, and interruptions in the muscle cell membrane. In 13 male fetuses at risk for DMD, Emery (40) also reported excessive variation in muscle fiber diameter and hyaline change in 2% of fibers. None of these workers found necrosis or phagocytosis in fetal DMD muscle. However, an important observation in view of later events was the finding of high intracellular calcium levels in fetal muscle by Emery and Burt in 1980 (41). Despite the lack of necrosis, elevated serum CK levels (up to 9140 IU) are reported in DMD fetuses, presumably as a result of "leaky membranes" (41). In contrast to these subtle morphological changes, dystrophin is totally lacking in aborted male fetuses of DMD families.

Neonates. Skeletal muscle studies in DMD neonates are also few. Bradley et al. (42) could find little evidence of degeneration or regeneration in a 2½-week-old boy with elevated serum CK activity (1433 IU). However, in another report of a 2-month-old DMD boy the muscle lesions were well developed, with polyfocal necrosis, phagocytosis, and regeneration (43). In the 2½-week-old boy studied by Bradley et al. (42) only hyalinized fibers were found, but by 3 years of age florid histological changes had developed. Bradley et al. (42) found the most severe degree of necrosis to be present at 5–6 years of age. A feature of the early DMD lesions is the presence of "fields" of necrosis affecting many adjoining muscle fibers. Regeneration is also prominent in the early stages (Fig. 1) and regenerating fibers may be grouped in early lesions (Fig. 2). Lymphocytic infiltration is variable and may be prominent in individual patients. The finding of CD8+ (cytotoxic-suppressor) T cells in close proximity to and focally invading muscle fibers (44), and of cells with the ultrastructural features of activated lymphocytes in muscle from affected individuals and from at risk fetuses (45), suggests that sensitized T cells may play a part in causing the ongoing muscle damage in DMD, as does the observation of class 1 MHC antigen expression on muscle fibers (46).

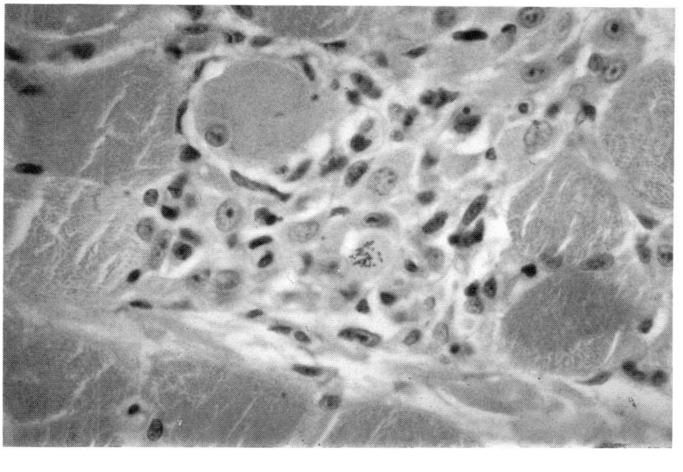

Figure 1 Muscle regeneration in DMD. In the center of the field there is a mitotic figure within a myoblast. Other myoblasts show vesicular nuclei and prominent nucleoli. Macrophages and round cells are also present. H&E, x600.

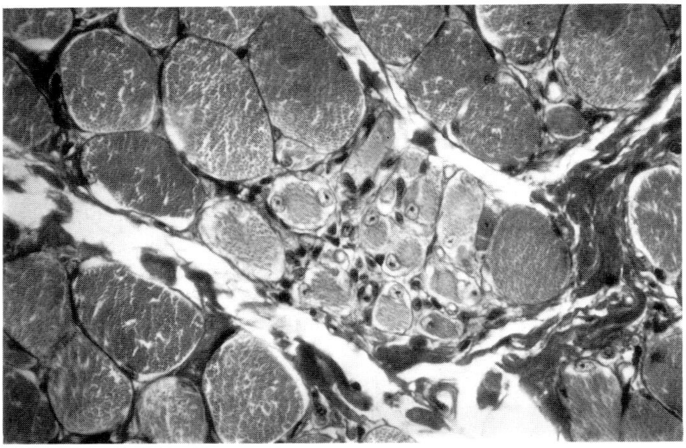

Figure 2 A group of regenerating fibres is present near the center of the field. These show vesicular nuclei, prominent nucleoli, and reduced myofibrillar density. Fibrous tissue is increased. Trichrome stain, X600.

Boyhood and Later Stages. By the time the Gowers sign develops at 5 or 6 years of age the muscle lesions are well advanced, especially in the larger muscles such as the gastrocnemius and vastus lateralis. At this stage there is overall disorganization of muscle architecture, which is the "hallmark" of DMD and a helpful sign for the diagnostic pathologist (Fig. 3). At this stage 50% or more of muscle fibers may be lost and are replaced by fat and fibrous tissue. The remaining muscle fibers show excessive variation in size and shape, with atrophy, hypertrophy, irregular fiber outlines, splitting, myofibrillar whorls, central nucleation, and hyalinization. These myopathic features form a general background added to which is focal necrosis, phagocytosis by macrophages, and regenerative changes. In regenerating myotubes pyknosis of nuclei and poorly defined or disorganized protomyofibrils indicate that regeneration may be abnormal in DMD (47, 48). Histochemical differentiation of muscle fibers is poor even in the early stages of DMD, with many type 2C fibers, which represent regenerating muscle fibers. Both type 1 and type 2 fibers are prone to necrosis.

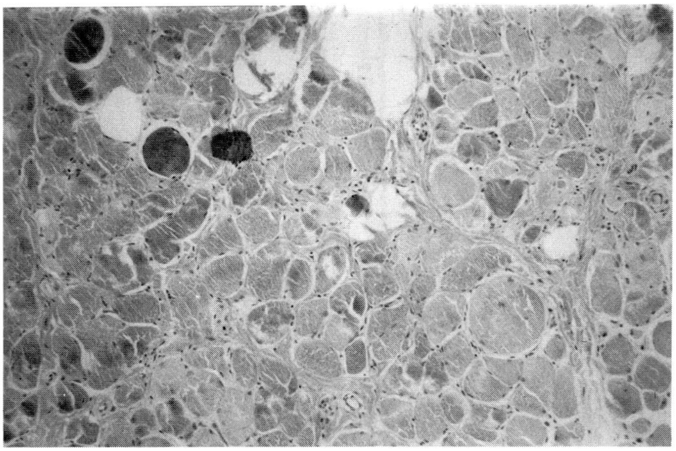

Figure 3 Overall disorganization of muscle architecture in DMD with excessive variation in size and shape of residual fibers. Large round fibers, some hyalinized (top left), and fibrous tissue condensation with some fat increase is shown. H&E, x320.

Skeletal muscle lesions that have progressed to the end-stage show almost total loss of muscle fibers, which are replaced by dense fibrous tissue and fat (Fig. 4). Necrosis is rare at this stage. A particular feature of the lesions of DMD is the sparing of muscle spindles, which remain essentially normal. It is presumed that the small size of the intrafusal fibers protects them from necrosis, a feature in common with the extraocular muscles.

An unexplained aspect of the myopathology of DMD is the different rate of progression of the lesions in different muscles. In some muscles the disorder is rapid and aggressive while in others it is protracted and delayed. In general, the larger limb-girdle muscles and the gastrocnemii are most affected, approaching the end-stage when the patient is 11 or 12 years of age. This coincides with the boy becoming wheelchair-bound. Other smaller muscles, such as those of the head and neck, are relatively spared. The tongue, intercostal muscles, and diaphragm may show focal necrosis in boys surviving beyond 20 years (49). In a necropsy survey of 19 DMD patients the minimal changes were found in the masseter, temporalis (Fig. 5), and tongue, consisting of polyfocal necrosis and regeneration. The external ocular muscles escaped altogether whereas the sternomastoids showed moderate change. The respiratory muscles, diaphragm, and intercostal muscles showed intermediate-stage lesions consisting of active necrosis and regeneration with slight muscle fiber outfall and the usual nonspecific myopathic changes, such as central nucleation and irregular size and shape of muscle fibers with fat and fibrous tissue increase. The distal muscles of the hands and feet were also found to be relatively spared compared with the quadriceps, gastrocnemii, triceps, deltoid, and biceps, all of which showed end-stage changes in this necropsy study (49).

Nicholson et al. (50) have suggested that there are two subtypes of DMD based on their analysis of 30 boys with DMD correlated with dystrophin content. The mean age of loss of mobility of DMD in boys with completely absent dystrophin labeling on tissue sections was 7.9 years, whereas the mean age for those with some labeling was 9.9 years.

Electron Microscopy. Numerous electron microscopic studies of DMD muscle biopsies have been reported (51–56). Of particular relevance was the observation by Mokri and Engel (52) of gaps in the muscle membrane in view of the later finding of absence of dystrophin in the subsarcolemmal region in DMD. Such gaps were believed to allow the ingress of calcium ions leading to hypercontraction of sarcomeres. The term "delta lesion" was used by Cullen and Fulthorpe (57) to describe the triangular sarcolemmal lesions in relation to hypercontracted sarcomeres in the vicinity (Fig. 6). The delta lesion is believed to signify the beginning of necrosis (Fig. 7). Necrosis is

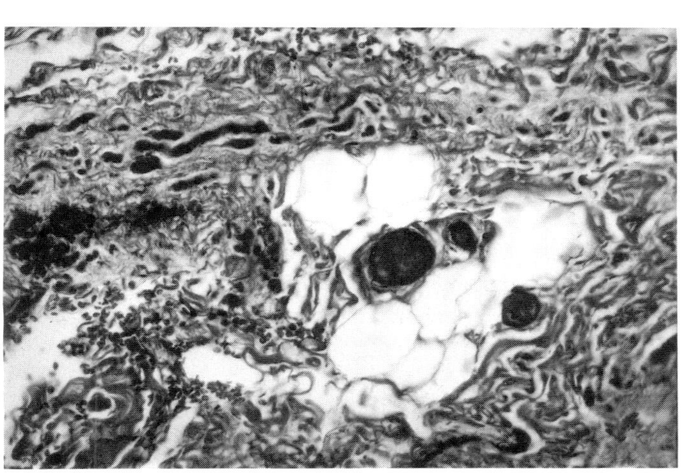

Figure 4 End-stage lesion in DMD with few remaining darkly stained muscle fibers among dense connective tissue and fat. Trichrome stain, x320.

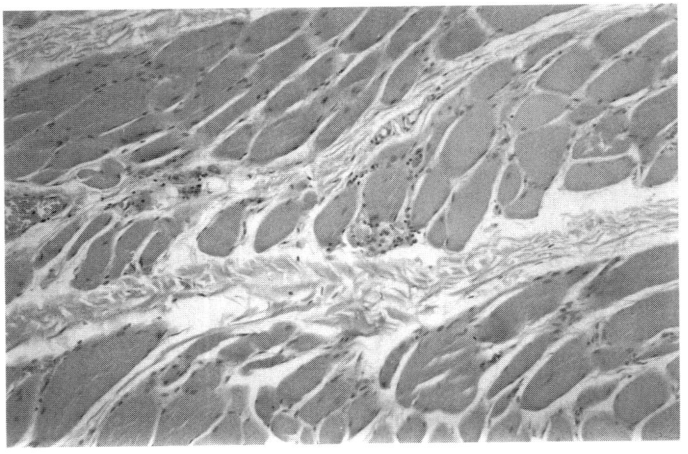

Figure 5 Temporalis muscle showing minimal changes at autopsy in a 26-year-old man with DMD. Focal necrosis is present in the center of the field. H&E, x120.

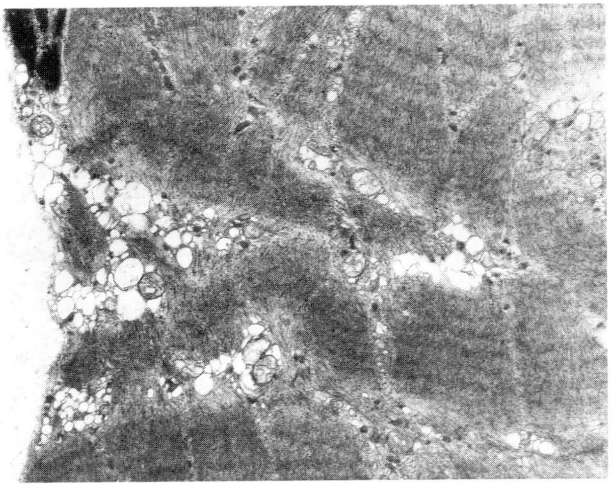

Figure 6 The "delta" lesion in DMD. Myofibrils are lost within a triangular area with the base at the sarcolemma (left). This zone is occupied by many vesicles and a small clump of homogenized myofibrils. Elsewhere myofibrils are hypercontracted, indicative of early necrosis. EM, x6000.

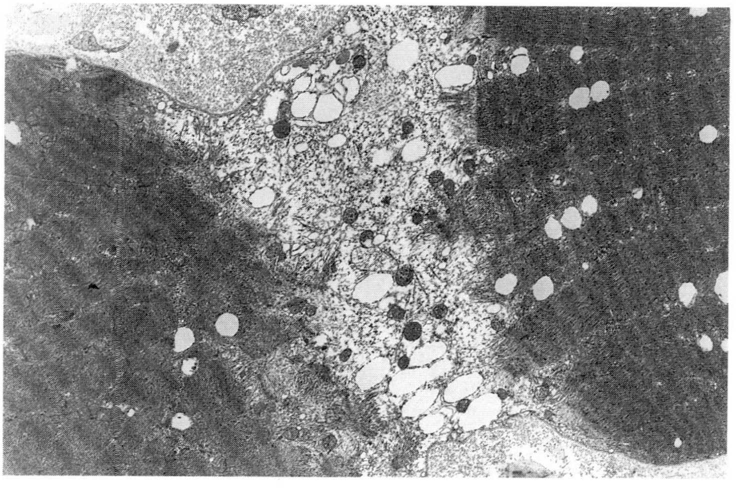

Figure 7 DMD showing a slightly more advanced lesion than in Figure 6. There is a central zone of rarefaction with hypercontracted and homogenized sarcomeres in the vicinity representing early necrosis. EM, x6000.

characterized by Z-band smearing, swelling of the sarcoplasmic reticulum and mitochondria, myelin figures, and rarefaction of myofibrils, which may become "homogenized." The I-band is destroyed prior to the A-band. Lipofuscin bodies and fat droplets are found in the perinuclear and subsarcolemmal zones. Other electron microscopic findings consist of reduced numbers of mitochondria and dilation of the sarcotubular systems (56, 57). Regenerative changes consist of enlarged nuclei with prominent nucleoli, polyribosomes in the sarcoplasm, proliferation of the Golgi apparatus, and newly formed sarcomeres. Redundant and duplicated basal laminae may signify past necrosis and regeneration (58). In DMD regenerating muscle fibers may show degenerative features such as Z-band streaming, disorganization of myofilaments, increased fat and lipofuscin content, and myelin figures (48, 56).

Becker Muscular Dystrophy

The lesions in BMD are similar to those of DMD except for their slower tempo of evolution. There is no histopathological feature that may be said to be distinctive of either type. In muscle biopsy diagnosis the only hint that the disorder may be the milder Becker form is the slightly better histochemical differentiation of type 1 and 2 muscle fibers in the vastus lateralis.

Dystrophin Analysis

Dystrophin is normally distributed uniformly in the subsarcolemmal zone of skeletal muscle fibers (Fig. 8). Characteristically immunocytochemistry or Western blot analysis shows a total absence of dystrophin in DMD. In BMD variable amounts of dystrophin are preserved, muscle fibers usually showing a discontinuous subsarcolemmal distribution of the protein (Figs. 9 and 10). However, in both DMD and BMD a few smaller muscle fibers completely encircled by dystrophin may be found and have been termed "revertant" fibers (see below) (Fig. 11).

In general, the amount of dystrophin present in BMD is inversely proportional to the clinical severity of the disorder. There are several subtypes of BMD each milder than the other and each showing a corresponding degree of preservation of dystrophin (59). According to Hoffman et al. (59), dystrophin of 3% or less is indicative of DMD, while dystrophin of 3–15% is associated with the more severe form of BMD. Milder BMD patients have 20% or greater amounts of dystrophin. The "Hoffman rule" is a useful guide in muscle biopsy diagnosis, although there are many exceptions to the rule. The basis for these exceptions has not been established in all cases.

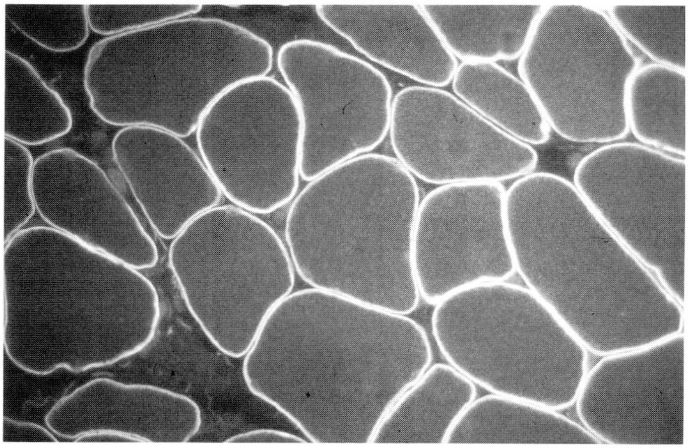

Figure 8 Distribution of dystrophin in the subsarcolemmal region of a normal control muscle. Immunofluorescence, x320.

Special Features

The heart is affected in both DMD and BMD. It is presumed that dystrophin is totally absent in the heart in DMD and partially deficient in BMD. However, opportunities for cardiac biopsy have not arisen and information is therefore sparse. In DMD, at autopsy, cardiac muscle fibers are lost and are replaced by dense connective tissue, which surrounds islands of surviving myofibers (Fig. 12). The changes are most severe in the left ventricular myocardium and conducting system.

The pathological basis for the moderately reduced IQ levels found in approximately two-thirds of DMD patients, but not in BMD, is not well understood. It is known that a form of dystrophin of slightly lower molecular weight is present in the developing brain, but its function is unknown. If dystrophin is required for normal neuronal migration, this may explain the reported finding of minor degrees of

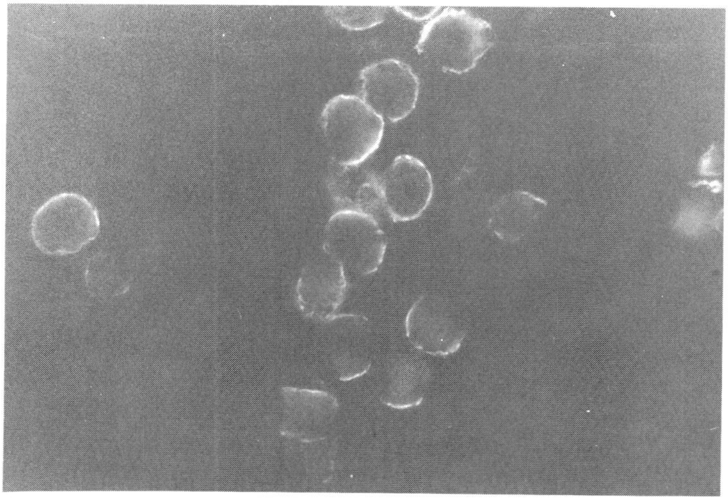

Figure 9 Dystrophin immunocytochemistry in a BMD fetus. There are a few muscle fibers partly surrounded by dystrophin. x480.

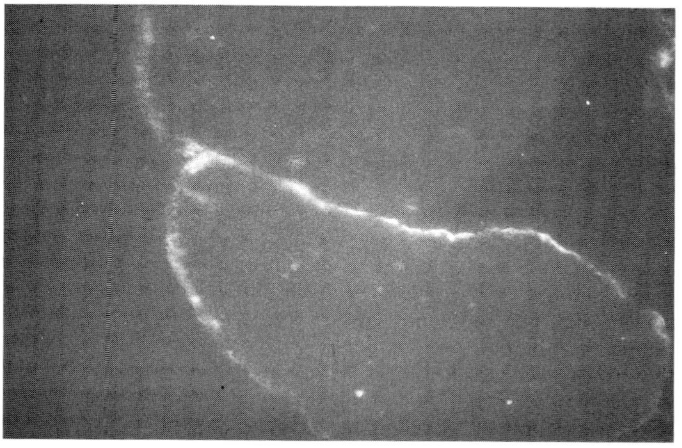

Figure 10 Dystrophin immunocytochemistry in a BMD patient showing incomplete circumferential staining of two muscle fibers. x600.

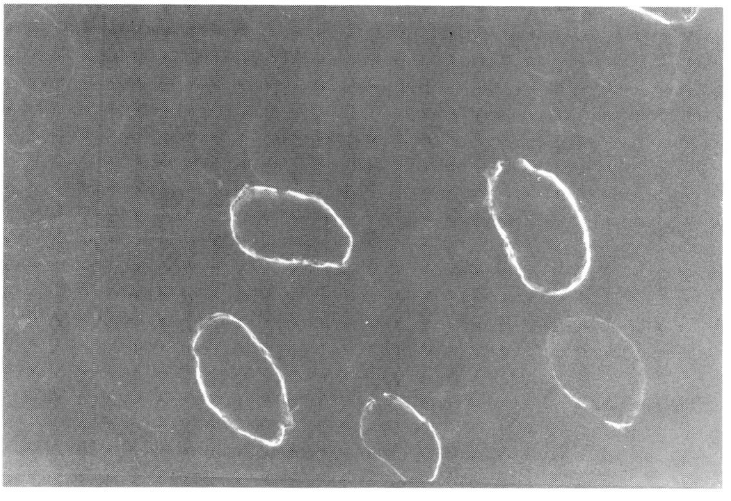

Figure 11 Immunocytochemistry in DMD showing a number of dystrophin-positive muscle fibers. These have been termed "revertant fibers." x300.

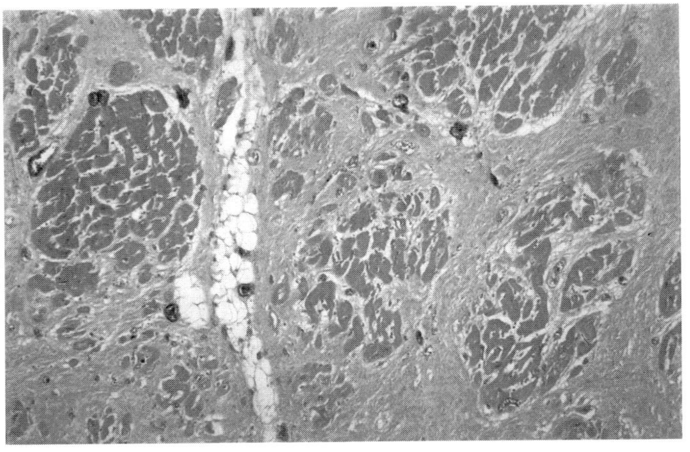

Figure 12 The heart in DMD showing islands of residual myocardial fibers surrounded by dense fibrous tissue. Adipose tissue is also present near the center of the field. H&E, x320.

pachygyria and of neuronal heterotopia in a smaller number of DMD patients (34). However, such findings are subtle and have not been confirmed by others (60).

Pathogenesis

The discovery of dystrophin provides a basis for the formulation of a system of pathogenesis for DMD and BMD. It is now presumed that the abnormalities of the muscle fiber membrane shown by freeze fracture (61) are due to the lack of dystrophin.

It has long been known that continuing polyfocal muscle fiber necrosis results eventually in outfall of muscle fibers, presumably from exhaustion of regeneration, and fibrous tissue increase, which is secondary and passive. The original work leading to this well-accepted principle of myopathology was first undertaken in a small marsupial, the Rottnest Island Quokka (*Setonix brachyurus*) (62–64). Further, abnormalities in the later stages of regeneration in DMD are believed to be responsible for acceleration of the "necrosis/regeneration" cycle causing earlier outfall of muscle fibers than would occur otherwise (47, 48, 56).

Given that the pathogenesis of the muscle lesions to the end-stages is well understood in DMD, the initial cause of the polyfocal necrosis remains to be explained. At first sight this seems to be straightforward, as the presumed function of dystrophin, as an element of the cytoskeleton, is to provide stability for the muscle membrane. The muscle fiber is subject to considerable mechanical deformation with high tension during contraction. A function of dystrophin is to prevent membrane disruption under such conditions. On first approximation this concept would explain the origin of the "delta" lesion as being due to membrane rupture. Following breakdown of the membrane, calcium, among other electrolytes, enters the sarcoplasm in large quantities and is thought to activate calcium-sensitive proteases precipitating necrosis (56). However, this rather simple view may not be the full explanation as excessive calcium levels have been reported even in cultured human dystrophic myotubes (65). As a result of the calcium trigger, there is hypercontraction and homogenization of sarcomeres followed by cyclical necrosis and regeneration with progression toward the end-stage. In general, these principles of pathogenesis are readily accepted. However, many details remain unexplained, both at the molecular level and grossly, such as the variable topographic distribution in the severity of the muscle lesions. The escape of smaller-diameter muscle fibers, as in the muscles of the hands, feet, head, and neck, is not easily explained by a simple mechanical stress. Interaction of the dystrophin molecule with other elements of the cytoskeleton may provide the answer, but little information is available in this respect.

Furthermore, the surprising variability of the Xp21 phenotypes now termed the *dystrophinopathies*, some with neurogenic features, also requires explanation.

Although dystrophin is believed to be purely concerned with mechanical stability of the membrane, other metabolic functions such as calcium regulation may also exist. Four glycoproteins and one nonglycosylated protein interact with dystrophin (66), and in the *mdx* mouse components of this complex are greatly reduced (67). It has also been found that a mutation of the 50-kDa glycoprotein causes an autosomal recessive form of muscular dystrophy in humans and in the hamster (68, 69).

In conclusion, a system of pathogenesis based on the knowledge of dystrophin is now being suggested for DMD and BMD. However, this formulation will require considerable further development before a full explanation for all the effects of dystrophin deficiency become known.

Diagnosis

Confirmation of the diagnosis in these conditions initially involves the demonstration of an elevated serum CK level and confirmation of a primary myopathic as opposed to a neurogenic process by electromyography and muscle biopsy and, ultimately, the demonstration of dystrophin deficiency or molecular alteration by immunohistochemistry or immunoblotting, or by the demonstration of a deletion or other mutation in the dystrophin gene. The serum CK is always markedly elevated (up to 300 times normal) at the time of presentation and even in the preclinical stages, already being high at birth, reaching a peak at 1–2 years, and gradually declining thereafter, but reaching normal levels only in the late stages of the disease (2). Markedly elevated CK levels, comparable to those found in DMD, are also found in the early stages of BMD and fall with increasing age (17).

The marked elevation of the serum CK level in affected boys at birth has been the basis for the development of screening programs for newborn males in a number of countries (70). By detecting isolated cases such programs make it possible to prevent the birth of secondary cases in families. However, in the case of DMD it has been estimated that this would prevent only 15–20% of new cases, and the value of such screening programs has therefore been questioned. It has been argued that almost equal detection and prevention rates could be achieved by more limited screening of boys with delayed motor milestones who are still not walking at the age of 18 months (13, 71). Neonatal screening programs will assume much greater importance when an effective treatment for these conditions becomes available (70).

The gold standard for establishing the diagnosis of an Xp21 dystrophy (dystrophinopathy) is the immunohistochemical demonstration of complete or partial dystrophin deficiency in a biopsy, which should ideally be taken from a muscle that is only moderately affected clinically. Complete or virtually complete absence of dystrophin is diagnostic of DMD while patchy preservation of dystrophin would suggest BMD (see above). Dystrophin immunohistochemistry may therefore be of prognostic as well as diagnostic value, particularly in young boys in whom it may not be possible to determine on clinical or other grounds whether they have DMD or BMD at the time of presentation. The importance of using a panel of antibodies to different regions of the dystrophin molecule from the N- to the C-terminus is now recognized (72, 73). Immunoblotting for dystrophin in muscle tissue is a more sensitive technique, which may demonstrate more subtle changes in the molecular size or quantity of dystrophin (74), but is necessary only when the immunohistochemical findings are normal or equivocal. Screening for a deletion (or duplication) in the dystrophin gene can be carried out on DNA obtained from leukocytes but is not necessarily diagnostic as not all cases of DMD or BMD have deletions and deletions may occasionally be found in the absence of muscle disease (4).

Prenatal diagnosis in at-risk males is possible by carrying out linkage studies or by demonstrating a deletion at Xp21 in DNA obtained from cultured amniotic fluid cells or from a chorionic villus biopsy. The latter may be performed as early as 10 weeks of gestation and a prenatal diagnosis can therefore be made at an earlier stage than with amniocentesis. It is also possible to perform fetal muscle biopsy in the second trimester of pregnancy and to perform dystrophin immunohistochemistry using a series of antibodies to different regions of the dystrophin molecule. However, the technique of fetal muscle biopsy is more difficult than that of chorionic biopsy and has a higher risk of fetal loss (2).

Carrier Detection

Confirmation of the carrier status in the mothers or female relatives of affected males followed by appropriate genetic counseling is the most important way of preventing new cases of muscular dystrophy. In a small proportion of female carriers (5–10%), including some without a prior family history of X-linked dystrophy, there is clinically detectable muscle involvement. While this is usually mild, in some carriers it may be comparable in severity to that in males with DMD (73). The serum CK level is helpful in identifying carriers and is elevated in up to 80% of definite carriers. The levels are highest just before the age of puberty, subsequently fall, and may be normal in adult life.

More recent advances in DMD/BMD carrier detection include linkage studies using DNA probes to detect polymorphic DNA sequences (RFLPs) within and flanking the gene locus as well as more direct methods for identifying the mutation, including DNA dosage analysis, detection of junction fragments, in situ hybridization, DNA sequencing, and amplification of lymphocyte mRNA (2, 4). If the family mutation is known, carrier status can be determined with 100% accuracy using DNA analysis (4). Immunohistochemical staining for dystrophin in biopsy specimens may show a mosaic pattern of dystrophin-positive and -negative muscle fibers in biopsies from manifesting carriers (73, 75) but is usually normal in symptomless carriers, as is the abundance and molecular size of dystrophin. Variability in dystrophin immunostaining and a reduction in the amount of dystrophin or the demonstration of dystrophin molecules of two different sizes may, however, be found in a proportion of younger DMD/BMD carriers (76, 77).

Management

General Management

The major objectives are to prevent the development of joint and spinal deformities, to prolong ambulation as long as possible, and to maintain the best possible level of general health. A program of active physical exercise may improve muscle strength and function and is also psychologically beneficial (2). The development of contractures at the weight-bearing joints in the lower limbs, which is one of the important factors leading to loss of the ability to ambulate, can be retarded by passive stretching exercises if initiated early in the course of the disease, while the judiciously timed fitting of leg braces and release of lower limb contractures by tenotomy will often prolong the period of ambulation and independence. In the more rapidly progressive DMD, the prevention of scoliosis is of particular importance, especially when the child is no longer able to walk. Particular attention should be directed to posture when the patient becomes confined to a wheelchair, and if necessary, moulded and fitted back supports can be used. If scoliosis does develop, a spinal brace should be provided. Spinal surgery may be necessary in some patients with rapidly evolving scoliosis (78). Even short periods of immobility and confinement to bed should be avoided insofar as possible as they may lead to marked functional deterioration, which may be difficult to reverse. Dietary control is important to prevent excessive weight gain.

Impairment of pulmonary function, due to the development of spinal and chest deformities and progressive weakness of the intercostal muscles, is a major source of morbidity and, by predisposing to respiratory infection and the development of respiratory failure, is a frequent cause of death in DMD. Regular breathing exercises, postural drainage, avoidance of smoking (active or passive), vaccination against influenza at the beginning of winter, prompt treatment of respiratory infections with appropriate antibiotics, and prevention of scoliosis and chest deformities are therefore important aspects of the management of these patients (2). When symptoms of respiratory insufficiency develop, consideration needs to be given to the use of some form of assisted ventilation, particularly at night when hypoventilation is most likely to occur (79).

Attention also needs to be directed to the early detection of cardiac arrhythmias and signs of cardiac failure, which are treated by conventional measures.

Particular care is required during general anesthesia, when serious arrhythmias and cardiac arrest may occur, and also postoperatively, when acute rhabdomyolysis and myoglobinuria may occur and may lead to acute renal failure. Succinylcholine has been implicated in most of the reported cases of postoperative myoglobinuria in DMD patients and should be avoided (2).

Drug Trials

Corticosteroids are the only drugs that have been shown to influence the clinical course of DMD in properly conducted, randomized clinical trials. In a number of trials, prednisone in doses of 0.75–1.5 mg/kg/day has been found to increase muscle strength, function, and muscle mass in boys with DMD for periods of up to 2 years and even longer (80, 81). Patients with BMD may also respond to prednisone (84). The use of the drug is limited by the potentially serious adverse effects that may develop with prolonged therapy. Although equally effective in the short term, alternate-day treatment with prednisone (2.5 mg/kg) was not found to have a sustained effect after the first 6 months of treatment (81). Deflazacort, which has fewer side effects than prednisone and may have anabolic properties, has also been found to be effective and warrants further evaluation (83, 84). Corticosteroid treatment does not influence the amount of dystrophin in the dystrophic muscle, and the beneficial effects may be due to an immunosuppressive action as biopsy studies showed a significant reduction in T-cell numbers and in numbers of muscle fibers invaded by mononuclear cells in prednisone-treated patients (85).

Myoblast Transfer

A number of experimental approaches aimed at overcoming the deficiency of dystrophin in dystrophic muscle are being investigated (64, 86). The injection of normal myoblasts into dystrophic muscles has been shown to lead to the expression of dystrophin in some muscle fibers both in the *mdx* mouse and in humans, presumably as a result of fusion of the injected myoblasts with dystrophic muscle fibers (86, 87). However, the results of preliminary experiments in which myoblasts were injected into various limb muscles in boys with DMD have not been encouraging so far (2, 88).

Gene Therapy

Recent work in tissue culture and in the *mdx* mouse suggests that direct gene therapy may be a feasible method of correcting the gene defect in DMD/BMD. It has been shown that dystrophin is expressed in cell cultures transfected with a human dystrophin cDNA (89, 90) and in vivo in the dystrophin-deficient *mdx* mouse after the intramuscular injection of a dystrophin minigene (91, 92). However, much more work is necessary particularly on a systemic method of delivery and on targetting the gene to the myocardium as well as the skeletal muscles.

EMERY-DREIFUSS MUSCULAR DYSTROPHY

This unusual and distinctive form of X-linked muscular dystrophy was first described by Dreifuss and Hogan in 1961 (93) and subsequently by Emery and Dreifuss in 1966 (94) in a large family from Virginia in the United States and is now recognized as a separate entity from the Xp21 dystrophies (95).

Clinical Features

The condition is characterized by slowly progressive wasting and weakness, which involves mainly the scapulohumeral muscles in the upper limbs and the anterior tibial and peroneal groups in the lower limbs. Another characteristic feature is the development at a relatively early stage of muscle contractures, which result in a distinctive pattern of limb deformities and a pathognomonic posture with flexion of the elbows, equinovarus deformities at the ankles, and limitation of neck flexion due to contracture of the posterior neck muscles. The onset is usually around the age of 4–5 years and a waddling gait and increased lumbar lordosis usually develop by the teens. The clinical severity is variable within families, ranging from individuals who are asymptomatic to those who are severely disabled and unable to walk in adult life, although the latter are in the minority (95, 96). Other notable features include the absence of pseudohypertrophy, with the exception of the extensor digitorum brevis muscles, and the absence of mental retardation.

Of particular importance is the frequent occurrence of myocaridal involvement, usually not manifesting until the second or third decade of life, and leading to disorders of cardiac conduction, arrhythmias, and sudden death in a high proportion of affected individuals (95, 96). Cardiac involvement may occur even in individuals with relatively mild skeletal myopathy and is characteristically associated with abnormalities of atrial conduction and the development of atrial paralysis. Manifestations range from low-amplitude P waves, sinus bradycardia, and first-degree heart block in asymptomatic individuals to atrial fibrillation, atrial flutter, accelerated junctional rhythms, and more severe degrees of atrioventricular conduction block, which may lead to syncopal episodes, thromboembolic events, and sudden death (95, 96). More generalized cardiomyopathy and congestive failure may develop in the later stages of the condition (95, 97).

Diagnosis

The clinical features are usually quite distinctive, particularly the pattern of muscle involvement, contractures, and cardiac manifestations. The serum CK level may be normal or elevated, sometimes markedly so (98), while the aldolase level is usually normal (99). The electromyographic findings are those of a primary myopathic process with low-amplitude, short-duration polyphasic motor unit potentials and some associated spontaneous activity (98).

Muscle biopsy shows typical dystrophic changes of variable severity, including isolated myofiber necrosis, regenerating fibers that may be grouped, atrophic and hypertrophic fibers with fiber splitting, increased perimysial and endomysial connective tissue, and normal intramuscular nerves (98). Fiber type changes are variable and include type 2 atrophy or hypertrophy, type 2 predominance, and type 1 atrophy or predominance (100). Dystrophin is present in normal amounts and has a normal distribution and molecular weight (2, 98).

Related Conditions

It now seems likely that the condition described in the literature as X-linked scapuloperoneal myopathy (101–103) is the same condition as Emery-Dreifuss muscular dystrophy (2). It has also been suggested that some cases described in the literature under the title of the rigid spine syndrome may represent formes frustes of Emery-Dreifuss muscular dystrophy (104, 105).

Genetics

The gene for Emery-Dreifuss muscular dystrophy has been localized to the distal Xq28 region close to the F8C and red/green color vision loci (106–108). The finding of close linkage with color blindness in both Emery-Dreifuss muscular dystrophy and the scapuloperoneal syndrome supports the identity or allelic nature of the two disorders (109).

Carrier Detection

Identification of carriers is important not only for prevention, but also because they may have cardiac involvement and are at risk of developing cardiac arrhythmias with advancing age. Sudden death has also been recorded in carriers (97). The degree of phenotypic expression in carriers is variable, some having only subtle clinical abnormalities while occasional females may demonstrate a typical male phenotype and others may have only cardiac involvement (98). The serum CK level may be elevated, particularly in younger women, but the majority of carriers have normal CK levels (98).

Management

Complete cardiological investigation is mandatory both in affected males and in female carriers in view of the risk of serious arrhythmias and sudden death, and the insertion of a cardiac pacemaker may be lifesaving. Other measures include active and passive muscle-stretching exercises to try to forestall the development of muscle contractures. Some patients will require tenotomies to overcome the shortening of the Achilles tendons. The spinal deformity is nonprogressive and does not usually require surgical intervention (98).

OTHER FORMS OF X-LINKED MYOPATHY

An unusual form of hereditary myopathy characterized by slowly progressive weakness mainly of the legs, with no associated cardiac or neural involvement, has been described in five members of a Finnish family in a pedigree pattern consistent with X-linked recessive inheritance (110, 111). The serum CK level was elevated and electron microscopic examination showed evidence of increased autophagic activity in muscle fibers. Linkage studies have excluded the DMD locus and have tentatively assigned the gene for this condition to Xq28 (111). However, further studies are required to confirm this location and to determine whether the condition is allelic with Emery-Dreifuss dystrophy, which has been assigned to the same region.

ACKNOWLEDGMENTS

The authors are grateful to Mrs. S. Moncrieff, who prepared the manuscript, to Mr. R. Johnson and Mr. P. Jacobsen for preparation of the figures, and to Dr. Nigel Laing for his helpful advice and constructive criticism.

REFERENCES

1. Wilton SD, Johnsen RD, Pedretti JR, Laing NG. Two distinct mutations in a single dystrophin gene: identification of an altered splice-site as the primary Becker muscular dystrophy mutation. J Med Genet 1993; 46:563–569.
2. Emery AEH. Duchenne Muscular Dystrophy, 2nd ed. Oxford: Oxford Medical Publications, 1993.
3. Bushby KMD, Thambyayah M, Gardner-Medwin D. Prevalence and incidence of Becker muscular dystrophy. Lancet 1991; 337:1022–1024.
4. Laing NG. Molecular genetics and genetic counselling for Duchenne/Becker muscular dystrophy. In: Partridge T, ed. Molecular and Cell Biology of Muscular Dystrophy. London: Chapman Hall, 1993:37–84.
5. Moser H. Duchenne muscular dystrophy: pathogenetic aspects and genetic prevention. Hum Genet 1984; 66:17–40.
6. Lane RJM, Robinow M, Roses AD. The genetic status of mothers of isolated cases of Duchenne muscular dystrophy. J Med Genet 1983; 20:1–11.
7. Cummings MR. Human Heredity, Principles and Issues. St. Paul, MN: West Publishing Co, 1991.
8. Bushby KMD, Gardner-Medwin D. The clinical, genetic and dystrophin characteristics of Becker muscular dystrophy. 1. Natural history. J Neurol 1992 (in press).
9. Baumbach LL, Chamberlain JS, Ward PA, Farwell NJ, Caskey CT. Molecular and clinical correlations of deletions leading to Duchenne and Becker muscular dystrophies. Neurology 1989; 465–474.
10. Hoffman EP. Genotype/phenotype correlations in Duchenne/Becker dystrophy. In: Partridge T, ed. Molecular and Cell Biology of Muscular Dystrophy. London: Chapman Hall, 1993:12–36.
11. Swash M, Schwartz MS, Carter ND, Heath R, Leak M, Rogers K. Benign X-linked myopathy with acanthocytes (McLeod syndrome). Its relationship to X-linked muscular dystrophy. Brain 1983; 106:717–733.
12. Danek A, Witt TN, Stockmann HBAC, Weiss BJ, Schotland DL, Fischbeck KH. Normal dystrophin in McLeod myopathy. Ann Neurol 1990; 28:720–722.
13. Gardner-Medwin D, Bundey S, Greer S. Early diagnosis of Duchenne muscular dystrophy. Lancet 1978; 1:1102.
14. Gardner-Medwin D. Recognising and preventing Duchenne muscular dystrophies. Br Med J 1983; 287:1083–1084.
15. Jones DA, Round JM, Edwards RHT, Grindwood SR, Tofts PS. Size and composition of the calf and quadriceps muscle in Duchenne muscular dystrophy: J Neurol Sci 1983; 60:307–322.
16. Newson-Davis J. The respiratory system in muscular dystrophy. Br Med Bull 1980; 36:135–138.
17. Walton JN, Gardner-Medwin D. Progressive muscular dystrophy and the myotonic disorders. In: Walton JN, ed. Disorders of Voluntary Muscle. Edinburgh: Churchill Livingstone, 1981:481.
18. Bradley WG, Jones MZ, Fawcett PRW. Becker-type muscular dystrophy. Muscle Nerve 1978; 1:111–132.

19. Sunohara N, Arahata K, Hoffman EP, et al. Quadriceps myopathy: forme fruste of Becker muscular dystrophy. Ann Neurol 1990; 28:634–639.
20. Laing NG, Mears ME, Thomas HE, et al. Differentiation of Becker muscular dystrophy from limb-girdle muscular dystrophy and Kugelberg-Welander disease using a cDNA probe. Med J Aust 1990; 152:270–271.
21. Gospe SM, Lazaro RP, Lava NS, Grootscholten PM, Scott MO, Fischbeck KH. Familial X-linked myalgia and cramps: a nonprogressive myopathy associated with a deletion in the dystrophin gene. Neurology 1989; 39:1277–1280.
22. Minetti C, Chang HW, Medori R, et al. Dystrophin deficiency in young girls with sporadic myopathy and normal karyotype. Neurology 1991; 1288–1292.
23. Perloff JK, De Leon AC, O'Doherty D. The cardiomyopathy of progressive muscular dystrophy. Circulation 1966; 33:625–648.
24. Heymsfield SB, McNish T, Perkins JV, Felner JM. Sequence of cardiac changes in Duchenne muscular dystrophy. Am Heart J 1978; 95:283–294.
25. Goldberg SJ, Stern LZ, Feldman L, Allen HD, Sahn DJ, Valdes-Cruz LM. Serial two-dimensional echocardiography in Duchenne muscular dystrophy. Neurology 1982; 32:1101–1105.
26. Katiyar BC, Somani PN, Misra S, Chaterji AM. Congestive cardiomyopathy in a family of Becker's X-linked muscular dystrophy. Postgrad Med J 1977; 53:12–15.
27. Siegel IM. Cardiomyopathy: presenting symptom of progressive muscular dystrophy. JAMA 1972; 222: 1060.
28. Muntoni F, Cau M, Ganau A et al. Brief report: deletion of the dystrophin muscle-promotor region associated with X-linked dilated cardiomyopathy. N Engl J Med 1993; 329:921–925.
29. Yoshida K, Ikeda SI, Nakamura A, et al. Molecular analysis of the Duchenne muscular dystrophy gene in patients with Becker muscular dystrophy presenting with dilated cardiomyopathy. Muscle Nerve 1993; 16:1161–1166.
30. Karagan NJ. Intellectual functioning in Duchenne muscular dystrophy: a review. Psychol Bull 1979; 86:250–259.
31. Karagan NJ, Zellweger HU. Early verbal disability in children with Duchenne muscular dystrophy. Dev Med Child Neurol 1978; 20:435–441.
32. Yoshioka M, Okuno T, Honda Y, Nakana Y. Central nervous system involvement in progressive muscular dystrophy. Arch Dis Child 1980; 55:589–594.
33. Al-Qudah AA, Kobayashi J, Chuang S, et al. Etiology of intellectual impairment in Duchenne muscular dystrophy. Pediatr Neurol 1990; 6:57–59.
34. Rosman NP, Kakulas BA. Mental deficiency associated with muscular dystrophy. A neuropathological study. Brain 1966; 89:769–788.
35. Jagadha V, Becker LE. Brain morphology in Duchenne muscular dystrophy: a Golgi study. Pediatr Neurol 1988; 4:87–92.
36. Rapaport D, Passos-Bueno MR, Brandão L, et al. Apparent association of mental retardation and specific patterns of deletions screened with probes Cf56a and Cf23a in Duchenne muscular dystrophy. Am J Med Genet 1991; 39:437–441.
37. Vassilopoulos D, Emery AEH. Muscle nuclear changes in fetuses at risk for Duchenne muscular dystrophy. J Med Genet 1977; 14:13–15.
38. Vassilopoulos D, Lumb EM, Emery AEH. Muscle nuclear size in neuromuscular disease. J Neurol Neurosurg Psychiatry 1976; 39:159–162.
39. Mahoney MJ, Haseltine FP, Hobbins JC, Banker BQ, Caskey CT, Golbus MS. Prenatal diagnosis of Duchenne's muscular dystrophy. N Engl J Med 1977; 18:968–973.
40. Emery AEH. Muscle histology and creatine kinase levels in the foetus in Duchenne muscular dystrophy. Nature 1977; 266:472–473.
41. Emery AEH, Burt D. Intracellular calcium and pathogenesis and antenatal diagnosis of Duchenne muscular dystrophy. Br Med J 1980; 355–357.
42. Bradley WG, Hudgson P, Larson PF, Papapetropoulos TA, Jenkison M. Structural changes in the early stages of Duchenne muscular dystrophy. J Neurol Neurosurg Psychiatry 1972; 35:451–455.
43. Hudgson P, Pearce GW, Walton JN. Preclinical muscular dystrophy: histopathological changes observed on muscle biopsy. Brain 1967; 90:565–576.
44. Arahata K, Engel AG. Monoclonal antibody analysis of mono-nuclear cells in myopathies. I. Quantitation of subsets according to diagnosis and sites of accumulation and demonstration and counts of muscle fibers invaded by T cells. Ann Neurol 1984; 16:193–208.
45. Fidzianska A, Hoebel HH, Kosswig R, Burck U. "Killer" cells in Duchenne disease: ultrastructural study. Neurology 1984; 34:295–303.
46. Appleyard ST, Dubowitz V, Dunn MJ, Rose ML. Increased expression of HLA ABC class I antigens by muscle fibres in Duchenne muscular dystrophy, inflammatory myopathy, and other neuromuscular disorders. Lancet 1985; 1:361–363.

47. Mastaglia FL, Kakulas BA. Regeneration in Duchenne muscular dystrophy: a histological and histochemical study. Brain 1969; 92:809–818.
48. Mastaglia FL, Papadimitriou JM, Kakulas BA. Regeneration of muscle in Duchenne muscular dystrophy. An electron microscopic study. J Neurol Sci 1970; 11:425–44.
49. Kakulas BA, Guo Yu Pu. Quantitative assessment of muscle lesions in 19 Duchenne necropsies. Muscle Nerve 1986; 9 (No 5S, Suppl):206 (abstract 39.23).
50. Nicholson LVB, Johnson MA, Bushby KMD, Gardner-Medwin D. Functional significance of dystrophin positive fibres in Duchenne muscular dystrophy. Arch Dis Child 1993; 68:632–646.
51. Kakulas BA, Adams RD. Diseases of the Muscle. Pathological Foundations of Clinical Myology, 4th ed. Philadelphia: Harper & Row, 1985.
52. Mokri B, Engel AG. Duchenne dystrophy: electron microscopic findings pointing to a basic or early abnormality in the plasma membrane of the muscle fiber. Neurology 1975; 25:1111–1120.
53. Mastaglia FL, Kakulas BA. The natural history of human muscle diseases studied by means of serial biopsy. Proc Aust Assoc Neurol 1968; 5:537–543.
54. Papadimitriou JM, Kakulas BA. Ultrastructure of skeletal muscle in muscular dystrophy, the carrier state and other human myopathies. Proc Aust Assoc Neurol 1968; 5:87–94.
55. Papadimitriou JM, Mastaglia FL, Kakulas BA. The natural history of Duchenne muscular dystrophy: an ultrastructural study. Proc Aust Assoc Neurol 1969; 6:87–92.
56. Cullen MJ, Mastaglia FL. Morphological changes in dystrophic muscles. Br Med Bull 1980; 36:145–152.
57. Cullen MJ, Fulthorpe JJ. Stages in fibre break-down in Duchenne dystrophy. An electron microscopic study. J Neurol Sci 1975; 24:179–200.
58. Schmalbruch HH. The muscular dystrophies. In: Mastaglia FL, Walton JN, eds. Skeletal Muscle Pathology. Edinburgh: Churchill Livingstone, 1982:235–265.
59. Hoffman EP, Fischbeck KH, Brown RH, et al. Characterization of dystrophin in muscle-biopsy specimens from patients with Duchenne's or Becker's muscular dystrophy. N Engl J Med 1988; 318(21):1363–1368.
60. Dubowitz V, Crome L. The central nervous system in Duchenne muscular dystrophy. Brain 1969; 92:805–808.
61. Wakayama Y, Okayasu H, Shibuya S, Kumagai T. Duchenne dystrophy: reduced density of orthogonal array subunit particles in muscle plasma membrane. Neurology 1984; 34:1313.
62. Kakulas BA, Adams RD. Principles of myopathology as illustrated in the nutritional myopathy of the Rottnest Quokka. Ann NY Acad Sci 1966; 138:90–101.
63. Kakulas BA. Man, Marsupial and Muscle. The Investigation of Muscular Paralysis Occurring in the Rottnest Island Quokka (*Setonix brachyurus*). Seattle: University of Washington Press, 1982.
64. Kakulas BA, Mastaglia FL. Pathogenesis and Therapy of Duchenne and Becker Muscular Dystrophy. New York: Raven Press, 1990.
65. Fong P, Turner PR, Denetclaw WF, Steinhardt RA. Increased activity of calcium leak channels in myotubes of Duchenne human and mdx mouse origin. Science 1990; 250:673.
66. Campbell KP, Kahl SD. Association of dystrophin and an integral membrane glycoprotein. Nature 1989; 338:259.
67. Ohlendieck K, Campbell KP. Dystrophin-associated protein are greatly reduced in skeletal muscle from mdx mice. J Cell Biol 1991; 115:1685.
68. Matsumura K, Tomé FMS, Collin H, et al. Deficienty of the 50K dystrophin-associated glycoprotein in severe childhood autosomal recessive muscular dystrophy. Nature 1992; 359:320.
69. Roberds SL, Ervasti JM, Anderson RD, et al. Disruption of the dystrophin-glycoprotein complex in the cardiomyopathic hamster. J Biol Chem 1993; 268:11496.
70. Van Ommen GJB, Scheuerbrandt G. Neonatal screening for muscular dystrophy. Consensus recommendation of the 14th Workshop sponsored by the European Neuromuscular Center (ENMC). Neuromusc Disord 1993; 3:231–239.
71. Crisp DE, Ziter FA, Bray PF. Diagnostic delay in Duchenne's muscular dystrophy. JAMA 1982; 247:478–480.
72. Muntoni F, Mateddu A, Cianchetti C, et al. Dystrophin analysis using a panel of antidystrophin antibodies in Duchenne and Becker muscular dystrophy. J Neurol Neurosurg Psychiatry 1993; 56:26–31.
73. Bushby KMD, Goodship JA, Nicholson LVB, Johnson MA, Haggerty ID, Gardner-Medwin D. Variability in clinical, genetic and protein abnormalities in manifesting carriers of Duchenne and Becker muscular dystrophy. Neuromusc Disord 1993; 3:57–64.
74. Morandi L, Mora M, Bernasconi P, et al. Very small dystrophin molecule in a family with a mild form of Becker dystrophy. Neuromusc Disord 1993; 3:65–70.
75. Arahata K, Ishihara T, Kamakura K, et al. Mosaic expression of dystrophin in symptomatic carriers of Duchenne's muscular dystrophy. N Engl J Med 1989; 320:138–142.
76. Clerk A, Rodillo E, Heckmatt JZ, Dubowitz V, Strong PN, Sewry CA. Characterisation of dystrophin in carriers of Duchenne muscular dystrophy. J Neurol Sci 1991; 102:197–205.

77. Vainzof M, Passos-Bueno MR, Pavanello RCM, Schreiber R, Satz M. A model to estimate the expression of the dystrophin gene in muscle from female Becker muscular dystrophy carriers. J Med Genet 1992; 29:476–479.
78. Rideau Y, Glorion B, DeLaubier A, Tarle O, Bach J. The treatment of scoliosis in Duchenne muscular dystrophy. Muscle Nerve 1984; 7:281–286.
79. Heckmatt JZ, Loh L, Dubowitz V. Night-time nasal ventilation in neuromuscular disease. Lancet 1990; 335:579–582.
80. Mendell JR, Moxley RT, Griggs RC, et al. Randomized, double-blind six-month trial of prednisone in Duchenne's muscular dystrophy. N Engl J Med 1989; 320:1592–1597.
81. Fenichel GM, Mendell JR, Moxley RT, et al. A comparison of daily and alternate-day prednisone therapy in the treatment of Duchenne muscular dystrophy. Arch Neurol 1991; 48:575–579.
82. Rifai Z, Welle S, Moxley RT, et al. Mechanism of action of prednisone in Duchenne dystrophy. Neurology 1992; 42:1428.
83. Mesa LE, Dubrovsky AL, Corderi J, Marco P, Flores D. Steroids in Duchenne muscular dystrophy—Deflazacort trial. Neuromusc Disord 1991; 1:261–266.
84. Griggs RC, Fischbeck K. X-linked muscular dystrophies. In: Rowland LP, DiMauro S, eds. Handbook of Clinical Neurology. Amsterdam: Elsevier, 1992:117–143.
85. Kissel JT, Burrow KL, Rammohan KW, et al. Mononuclear cell analysis of muscle biopsies in prednisone-treated and untreated Duchenne muscular dystrophy. Neurology 1991; 41:667–672.
86. Partridge TA, Morgan JE, Coulton GR, Hoffman EP, Kunkel LM. Conversion of *mdx* myofibres from dystrophin-negative to -positive by injection of normal myoblasts. Nature 1989; 337:176–179.
87. Partridge TA. Invited review: myoblast transfer: a possible therapy for inherited myopathies. Muscle Nerve 1991; 14:197—212.
88. Dubowitz V. Transferring myoblasts in Duchenne dystrophy. Br Med J 1992; 305:844–845.
89. Dickson G, Love DR, Davies KE, Wells KE, Piper TA, Walsh FS. Human dystrophin gene transfer: production and expression of a functional recombinant DNA-based gene. Hum Genet 1991; 88:53–58.
90. Lee CC, Pearlman JA, Chamberlain JS, Caskey CT. Expression of recombinant dystrophin and its localization to the cell membrane. Nature 1991; 349:334–336.
91. Acsadi G, Dickson G, Love DR, et al. Human dystrophin expression in *mdx* mice after intramuscular injection of DNA constructs. Nature 1991; 352:815–818.
92. Ragot T, Vincent N, Chafey P, et al. Efficient adenovirus-mediated transfer of a human minidystrophin gene to skeletal muscle of *mdx* mice. Nature 1993; 361:647–650.
93. Dreifuss FE, Hogan GR. Survival in X-chromosomal muscular dystrophy. Neurology 1961; 11:734–741.
94. Emery AEH, Dreifuss FE. Unusual type of benign X-linked muscular dystrophy. J Neurol Neurosurg Psychiatry 1966; 28:338–342.
95. Hopkins LC, Warren ST. Emery-Dreifuss muscular dystrophy. In: Rowland LP, DiMauro S, eds. Handbook of Clinical Neurology. Amsterdam: Elsevier, 1992:145–160.
96. Merlini L, Granata C, Dominici P, Bonfiglioli S. Emery-Dreifuss muscular dystrophy: report of five cases of a family and review of the literature. Muscle Nerve 1986; 9:481–485.
97. Fishbein MC, Siegel RJ, Thompson CE, Hopkins LC. Sudden death of a carrier of X-linked Emery-Dreifuss muscular dystrophy. Ann Intern Med 1993; 119:900–905.
98. Specht LA. Case record of the Massachusetts General Hospital. N Engl J Med 1992; 327:548–557.
99. Bialer MG, Bruns DE, Kelly TE. Muscle enzymes and isoenzymes in Emery-Dreifuss muscular dystrophy. Clin Chem 1990; 36:427–430.
100. McKee AC. Case records of the Massachusetts General Hospital. N Engl J Med 1992; 327:548–557.
101. Thomas PK, Calne DB, Elliott CJ. X-linked scapuloperoneal syndrome. J Neurol Neurosurg Psychiatry 1972; 35:208–215.
102. Rotthauwe HW, Kowalewski S. Gutartige recessiv X-chromosomal verebte Muskeldystrophie. I. Utersuchungen bei Merkmalsträgern. Hum-genet 1966; 3:17–29.
103. Waters DD, Nutter DO, Hopkins LC, Dorney ER. Cardiac features of an unusual X-linked humeroperoneal neuormuscular disease. N Engl J Med 1975; 293:1017–1022.
104. Colver AF, Steer CR, Godman MJ, Uttley WS. Rigid spine syndrome and fatal cardiomyopathy. Arch Dis Child 1981; 56:148–151.
105. Takase Y, Yamamoto K, Nogaki H, Fukusako T, Sasabe F, Morimatsu M. Rigid spine syndrome associated with cardiomyopathy—clinical and nosological considerations. Jpn J Med 1990; 29:555–559.
106. Consalez GG, Thomas NST, Stayton TCL, et al. Assignment of Emery-Dreifuss muscular dystrophy to the distal region of Xq28: the results of a collaborative study. Am J Hum Genet 1991; 48:468–480.
107. Kress W, Müller E, Kausch K, et al. Multipoint linkage mapping of the Emery-Dreifuss muscular dystrophy gene. Neuromusc Disord 1992; 2:111–115.
108. Yates JRW, Waner JP, Smith JA, et al. Emery-Dreifuss muscular dystrophy: linkage to markers in distal Xq28. J Med Genet 1993; 30:108–111.

109. Goldblatt J, Schram LJ, Wallis G, Oswald A, Beighton P. Emery-Dreifuss syndrome and X-linked muscular dystrophy with contractures: evidence for homogeneity. Clin Genet 1989; 35:1–4.
110. Kalimo H, Savontaus M-L, Lang H, et al. X-linked myopathy with excessive autophagy: a new hereditary muscle disease. Ann Neurol 1988; 23:258–265.
111. Saviranta P, Lindlof M, Lehosjoki A-E, et al. Linkage studies in a new X-linked myopathy, suggesting exclusion of DMD locus and tentative assignment to distal Xq. Am J Hum Genet 1988; 42:84–88.

18

Dystrophin: Structure, Localization, and Function

John B. Harris, Michael J. Cullen, and Ruth Vater
Newcastle General Hospital, Newcastle upon Tyne, England

INTRODUCTION

Dystrophin, a protein with a molecular mass of approximately 427 kDa, comprises approximately 3685 residues and is the product of a 14-kilobase transcript. The gene coding for dystrophin contains between 75 and 80 exons and is located on the short arm of the X chromosome. In the majority of cases of Duchenne-type muscular dystrophy, deletions in the gene introduce a stop codon into the reading frame resulting in the production of an abnormal protein that is, presumably, rapidly destroyed. The muscle fibers of these patients are, therefore, virtually devoid of dystrophin. Abnormalities in the appropriate promotor region or point mutations could be the cause of the major deficiency in dystrophin in those patients in whom no major deletion can be detected. In the less severe Becker-type muscular dystrophy, the abnormality in the gene does not disrupt the reading frame, thus allowing the production of a variant of dystrophin that may be smaller than normal. In many cases, the gene product is also of rather low abundance, possibly because the "abnormal" proteins are more susceptible to degradation. In general, the protein produced in patients with Becker-type muscular dystrophy has an intact N- and C-terminal. The protein is usually correctly localized to the plasma membrane and the consequent muscle pathology is less severe. A few patients (less than 5%) exhibit duplications of exons. These patients produce a protein of slightly larger than normal mass.

Although the abnormality in the gene and its protein product is of obvious relevance to the expression of Duchenne- and Becker-type muscular dystrophies, the precise relationship between the genetic abnormality, the relative or near-total absence of dystrophin, and the progressive degeneration of the skeletal muscle is far from clear. A great deal of attention has, therefore, been paid to the structure, localization, and function of dystrophin in normal tissue.

Although dystrophin is expressed in skeletal and cardiac muscle and, at lower abundance, in smooth muscle, it is becoming increasingly apparent that there are a number of variants of dystrophin. Five distinct promoters have been identified in the gene. Three promoters produce a full-length protein, but from distinct first exons. The two other promoters are located much closer to the 5′ region and produce much smaller homologs of dystrophin. The various transcripts of the "dystrophin gene" exhibit a considerable degree of tissue- and cell-type specificity, and most are not expressed in skeletal muscle (1).

An additional variant of dystrophin, known variously as dystrophin-related protein (DRP) or utrophin, is elaborated on chromosome 6. It is expressed in skeletal muscle, where it appears to be preferentially localized to the neuromuscular junction (1,2).

In this chapter, we concentrate on the structure, localization, and function of the 427-kDa native dystrophin, commenting briefly on the other variants of dystrophin as appropriate.

THE STRUCTURE OF DYSTROPHIN

The structure of dystrophin has been elucidated as a result of detailed studies based on its presumed amino acid sequence, the effects of controlled proteolysis, and the analysis of rotary-shadowed images of the isolated protein. There is general agreement that four domains may be identified in the protein. The first is the actin-binding N-terminal domain, the second the rod domain, the third the cysteine-rich domain, and the fourth the C-terminal domain (Fig. 1).

The N-terminal domain consists of approximately 240 residues. It is structurally, and possibly functionally, homologous with the F-actin-binding region of α-actinin and β-spectrin (3,4).

The rod domain is by far the largest of the four domains. Its most important characteristic is its organization into between 24 and 26 repeating units of 109 residues. These repeat units are similar in many respects to the repeat units of 106 residues of the cytoskeletal protein β-spectrin; the repeats in β-spectrin constitute a series of α-helical components of a central rod, and it is predicted that the repeat units of dystrophin likewise form a central rod of α-helical repeats.

The central rod domain is further subdivided by four segments, each containing a number of proline residues. One of these segments appears at the beginning of the rod domain, one at the end, one between repeat elements 3 and 4, and one between elements 19 and 20. The high content of proline is significant because the presence of proline residues is typically associated with a break in an α-helix and the introduction of an element of flexibility into the molecule. These proline-rich regions in dystrophin have been calculated to extend for between 45 and 75 residues and may act as "hinges" in an otherwise rigid molecule.

The cysteine-rich domain, comprising approximately 150 residues, shows some homology with a number of Ca^{2+}-binding regions in physiologically important proteins such as calmodulin and slime-mold α-actinin (5).

The C-terminal domain, comprising the terminal 420 residues of dystrophin, shows no homology with any other class of protein other than the autosomal DRP (6).

Dystrophin has been localized to the plasma membrane (see below), and it has been shown that the C-terminal region (comprising the cysteine-rich and C-terminal domains) is associated with a complex of at least six proteins, called the dystrophin-associated glycoproteins (DAGs). This complex consists of four or more integral membrane proteins [25, 35, 43, and 50 kDa (adhalin)], a cytoplasmic protein triplet of 50 kDa (syntrophin), and a 156-kDa protein that is located externally and links the plasma membrane to the basal lamina by binding to laminin (7–11). The 156- and 43-kDa proteins (α- and β-dystroglycan) are encoded by a single mRNA; the mature proteins are produced following posttranslational cleavage and glycosylation of the gene product, a 97-kDa polypeptide (11). Which of the two domains of the C-terminal region of dystrophin is specifically associated with the complex is not yet known, although recent evidence suggests that the cysteine-rich and the first part of the

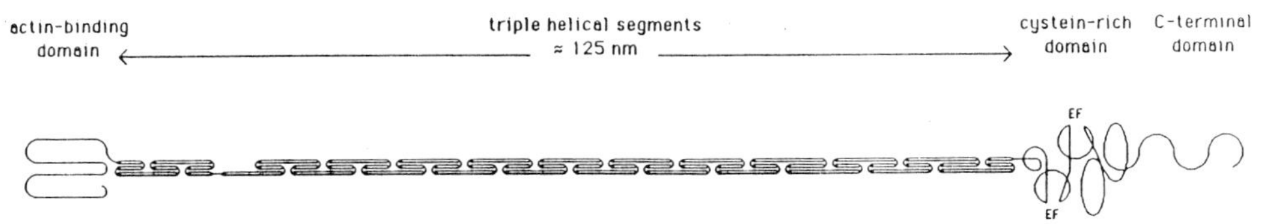

Figure 1 Schematic diagram showing the predicted structure of the dystrophin molecule. The cysteine-rich domain includes potential calcium binding sites (EF hands). (Adapted from Ref. 5.)

C-terminal domains represent the binding site (12). The latter half of the C-terminal domain does not appear to be important in the binding of the dystrophin-associated glycoproteins.

There is considerable debate about whether dystrophin is monomeric or polymeric. Isolated molecules of dystrophin tend to aggregate into polymers of various kinds (13–15), and although it is tempting to speculate that some form of polymerization reflects the in vivo organization of dystrophin, there is no definitive evidence for such an association. Nonetheless, a "model" that attempts to accommodate most of the known and some of the proposed features of dystrophin and its relationship to the DAGs is shown in Figure 2.

THE LOCALIZATION OF DYSTROPHIN

The first approach used to study the localization of dystrophin was subcellular fractionation. Skeletal muscles of rabbit and mouse were used to prepare subcellular fractions enriched with membrane of the transverse tubular system (T system). Subsequent analysis led to the observation that this fraction was relatively rich in dystrophin and thus to the concept that dystrophin was primarily localized to the T system at the triads (16,17). More recent work has shown that the fractions of the T system were contaminated with fractions of plasma membrane, and that the data were, therefore, unreliable.

Immunolabeling of tissue sections has a number of advantages over cell fractionation when the localization of an antigen is to be determined. The most important is that whereas tissue is broken up during homogenization and fractionation, the basic structure of the tissue is not destroyed during the preparation of tissue sections. This, in turn, introduces several other advantages: It is possible to prepare both transverse and longitudinal sections and thus build up a three-dimensional view of the distribution of the antigen. It is possible to apply the technique to single teased muscle fibers and to isolated "sheets" of plasma membrane. Finally, it is possible to use immunological techniques in combination with electron microscopy to determine accurately the cellular distribution of the antigen under study. Immunofluorescence, immunoperoxidase, and immunogold labeling have all been used in attempts to determine the localization of dystrophin.

The first published light-microscopic study of dystrophin labeling utilized immunofluorescence (18,19), and this is still the most widely used technique. There is general agreement that antibodies to

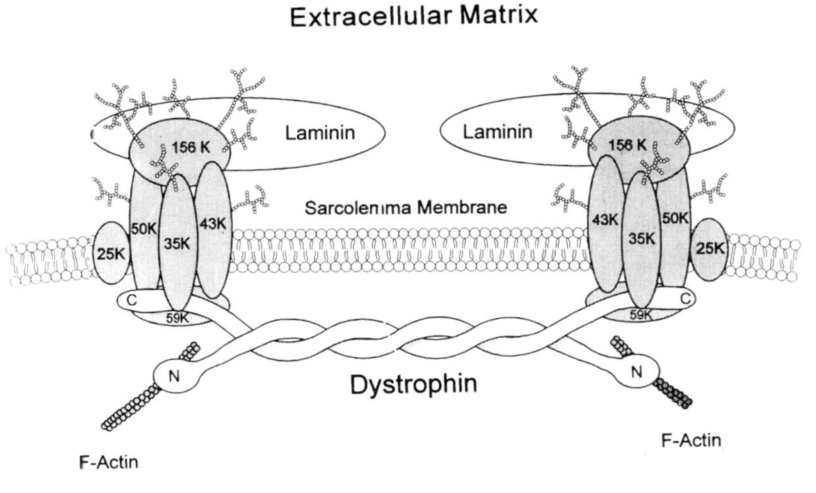

Figure 2 Schematic model of dystrophin and the dystrophin-associated glycoproteins at the plasma membrane. 156K = 156-kDa glycoprotein; 50K = 50-kDa glycoprotein; 43K = 43-kDa glycoprotein; 35K = 35-kDa glycoprotein; 25K = 25-kDa protein; 59K = 59-kDa protein. This model is frequently revised—see Note Added in Proof. (Reproduced with permission from Ref. 57.)

dystrophin label the plasma membrane very specifically (Fig. 3) and do not label the T system or triads. Immunoreactivity is more intense at the neuromuscular junction than at the extrajunctional regions of the muscle fiber, but it is not clear whether this reflects a genuine difference or is an artifact introduced by postjunctional folding. There also appears to be a concentration of dystrophin at the myotendinous junction of skeletal muscle fibers (20,21), a finding that may have functional significance (see below).

Although there is now general agreement that there is no localization of dystrophin to the T tubules in mature fibers, Bornemann and Schmalbruch (22) have reported that regenerating muscle exhibits dystrophin labeling at the triads. This raises a number of, as yet, unresolved questions on the expression of dystrophin (or its related proteins) in developing muscle.

The sarcolemma, first described in 1840 by Bowman (23), is essentially a histological term for a structure that can be divided into four distinct components: (1) the plasma membrane, peripheral to which lies (2) the basal lamina and (3) the reticular lamina. Internal to the plasma membrane and closely associated with it is (4) the cortical cytoskeleton (= membrane cytoskeleton). To distinguish with which of these components dystrophin is associated, immunolabeling at the electron microscope level is required.

This can be best achieved using either immunoperoxidase or immunogold labeling. Unfortunately, immunoperoxidase tends to smear in the section, so the exact labeling site is often poorly defined unless great care is taken to obtain relatively low-intensity labeling (24). Colloidal gold has several advantages as an ultrastructural marker: it is highly electron-dense, it is discrete, so its distribution can be quantified, and the colloidal particles can be prepared in different sizes, thus allowing the concomitant labeling of more than one antigen.

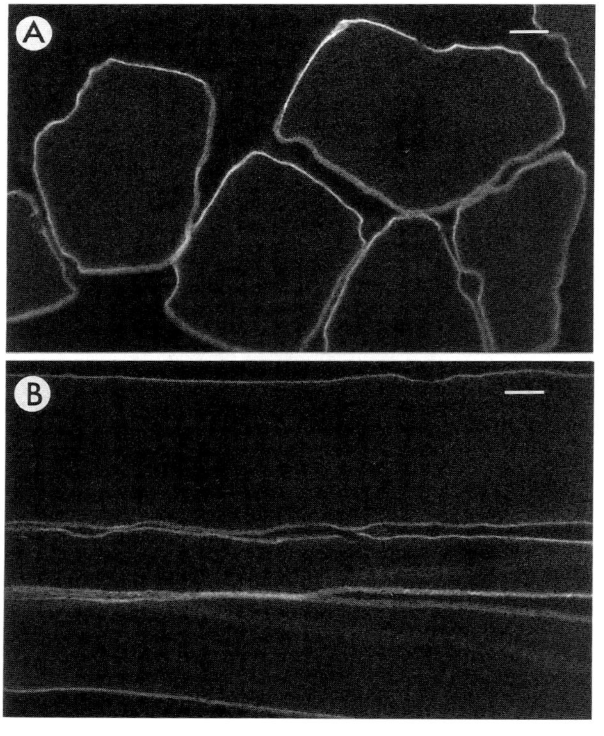

Figure 3 (A) Transverse section showing immunofluorescence (rhodamine labeling) of a group of myofibers after incubation with the dystrophin antibody Dy4/6D3. The labeling is restricted to the periphery of the fibers. (B) Longitudinal section labeled for dystrophin in the same way. Muscle = rat soleus. Bar = 10 μm.

When monoclonal antibodies to the rod portion of dystrophin are labeled with colloidal gold, the gold particles are found mostly in the cortical cytoskeleton, very close to the cytoplasmic face of the plasma membrane (25,26) (Fig. 4). When monoclonal antibodies to the last 17 amino acids of the C-terminus are used, the particles lie over the plasma membrane (27,28). These data are entirely consistent with the view that the cysteine-rich and the first half of C-terminal domains bind to the complex of dystrophin-associated glycoproteins (12). It is generally thought unlikely that the C-terminal domain penetrates or crosses the lipid bilayer of the plasma membrane because it is insufficiently lipophilic, although it is of interest that Suwa et al. (29) claim to have identified a fragment of dystrophin (amino acids 3107–3200 in the cysteine-rich domain) that may contain four helices that could act as transmembrane domains.

Where distances between labeling sites (nearest neighbor) in the plane of the plasma membrane have been measured, a range of values has been recorded: 100–125 nm in the mouse (26), 100–140 nm in humans (25,27), 100–200 nm (175 nm mean) in the rat (30), and 100–120 nm at the rat myotendinous junction (20). The rather wide range in values found in each set of measurements is not unexpected when it is considered that the thickness of the cryosection (70–100 nm) is probably less than the repeat distance of the molecules, that the efficiency of labeling is unlikely to exceed 60%, and that there is a lot of inherent "noise" in the technique (31). The size of the dystrophin molecule in vivo has not yet been established. The predicted length of dystrophin (100–125 nm) (5,32) is close to both the nearest-neighbor measurements and the calculated length of dystrophin obtained from rotary-shadowed images of dystrophin purified from rabbit skeletal muscle (13) and chicken gizzard smooth muscle (14). If, as we believe, the monoclonal antibodies used in this work react with a single epitope, it would

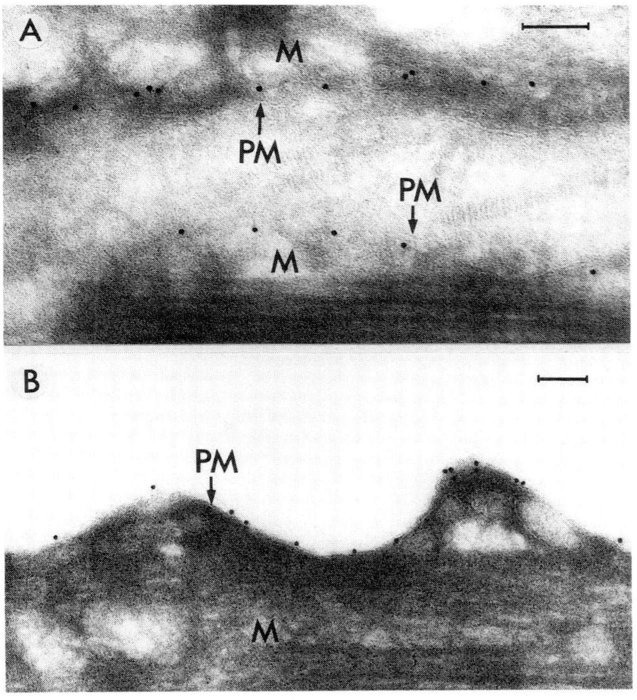

Figure 4 (A) Edges of two adjacent myofibers (M) immunogold-labeled after incubation with the dystrophin rod antibody Dy4/6D3. The 10-nm gold particles lie internal to the plasma membranes (PM). Muscle = human quadriceps. Bar = 100 nm. (B) Edge of a myofiber (M) immunogold-labeled after incubation with an antibody Dy8/6C5 to the C-terminus of dystrophin. The gold particles lie on or very close to the plasma membrane (PM). Muscle = human quadriceps. Bar = 100 nm.

suggest that dystrophin forms a network of monomers. When nearest-neighbor distances were compared in transverse and longitudinal sections of the same muscle, no significant differences were found (25). This implies that there is no strong axial component to the arrangement of dystrophin at the plasma membrane.

The relatively even distribution of labeling sites at the plasma membrane seen under the electron microscope suggests that dystrophin is arranged as a uniform lattice at the surface of the muscle fiber. Under the light microscope, however, dystrophin labeling, whether by fluorescence or by peroxidase, can sometimes appear discontinuous, with a periodicity corresponding to one sarcomere length (Fig. 5). A number of investigators have now described this "costameric" distribution of dystrophin in some detail (33–36). The term "costamere" was introduced by Pardo et al. (37) approximately 10 years ago to describe vinculin-rich bands that encircled the muscle fibers at the level of the I-band, and that were thought to mark sites of attachment between the myofibrils and the sarcolemma. Subsequently, spectrin and α-actinin have also been shown to be components of costameres. In the recent descriptions of dystrophin in costameres it has been described as colocalizing with vinculin (33,36), spectrin (33), and α-actinin (35), although one report noted inconsistencies in the colocalization with vinculin (34). Two of the papers describe an increased concentration of dystrophin opposite the I-band and a reduction in concentration exactly opposite the Z-line (33,35). An increased concentration opposite the M-line (33,36) and fine longitudinal connections (33,35) have also been described.

While an association with costameres cannot be refuted, it is intriguing that in none of the ultrastructural immunogold studies of dystrophin (25–27,30,38) has such a discontinuous distribution been observed. Labeling is as dense opposite the A-band as opposite the I-band, and there is no visible reduction opposite the Z-line or elevation opposite the M-line. It has been suggested (33) that immunogold labeling is too sparse for a costameric pattern to be detected, but we maintain that the density is indeed sufficient for any differences to be detectable. If the distribution of dystrophin in vivo is costameric, it implies that there is some redistribution of dystrophin during processing (perhaps at the fixation stage), before the sections are viewed under the electron microscope. At the moment we are unable to explain the discrepancy between the light microscopic and electron microscopic results.

Other Tissues

Dystrophin is also found in cardiac muscle, in similar concentrations as in skeletal muscle, and in smaller concentrations in smooth muscle, the central nervous system, and the outer plexiform layer of the retina. In cardiac muscle, dystrophin is situated at the plasma membrane as in skeletal muscle, but is also found

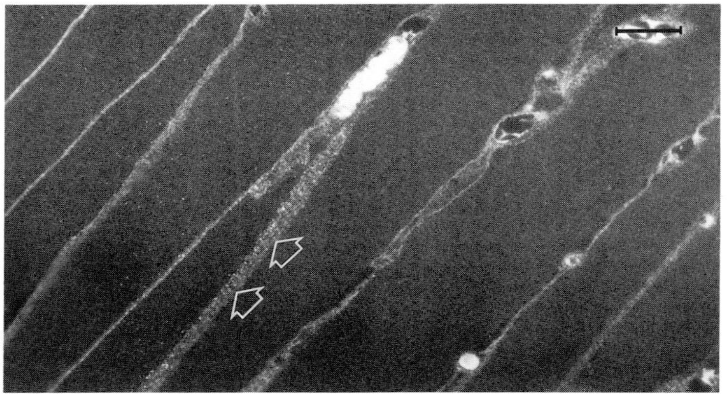

Figure 5 Immunofluorescence (rhodamine) labeling of a group of myofibers after incubation with the dystrophin antibody Dy8/6C5. Where the plasma membrane lies oblique in the section (arrows) there are suggestions of a sarcomeric repeat to the labeling. Muscle = dog peroneus. Bar = 20 μm. (Courtesy of Mr. J. Fulthorpe.)

on the T-tubule membranes. (In cardiac muscle the T tubules are broad invaginations of the plasma membrane that are lined by basal lamina.) Dystrophin is not found on that part of the plasma membrane incorporated into the intercalated disks (39). In smooth muscle, dystrophin is found predominantly in the vesicle-rich areas of plasma membrane but not in the areas of smooth membrane that are associated with the attachment plaques. Within the parenchyma proper, DRP is located at the inner plasma face of astrocyte foot processes at the abluminal aspect of the blood-brain barrier (40).

DAGs are normally found in the plasma membrane when dystrophin is present and with a similar apparent spacing to dystrophin (Fig. 6). When dystrophin is absent, as from the extrajunctional plasma membrane of skeletal muscle from patients with Duchenne dystrophy or from the *mdx* mouse, the DAGs are often greatly reduced (11). The DAGs are also present in dystrophic mouse and human muscle fibers at the neuromuscular junction where they associate with locally expressed DRP (41). This colocalization of DRP and DAGs suggests a similar association as that between dystrophin and the DAGs.

Those homologs of dystrophin not found in skeletal muscle (e.g., the homolog known as Dp71, one of the C-terminal transcripts) are specifically localized to a number of major organs including kidney, lung, and pancreas (for details see Ref. 1).

THE FUNCTION OF DYSTROPHIN

The widespread occurrence of dystrophin (and its alternative transcripts and homologs) among vertebrates and in numerous cell types suggests that, despite its general exiguity, it is a protein of considerable importance to the organization, development, maintenance, and function of the cell. Whether its role is the same in all cell types is unclear but improbable given the diversity of cell types, contractile and noncontractile, in which it is expressed (42).

In skeletal muscle, the principal isoform of dystrophin is the 427-kDa protein, localized to the plasma membrane. Its function is not known, but it is an important constituent of the cytoskeleton. Thus, although dystrophin constitutes only 0.001–0.002% of the total protein of the muscle fiber, it constitutes 5% of the proteins of the membrane cytoskeleton (9). The location and relatively high abundance of dystrophin at the plasma membrane have suggested to many that its principal function is to confer mechanical strength and support to the plasma membrane, particularly protecting the membrane against stress-induced damage resulting from cycles of mechanical activity. There are several superficial attractions to this theory, but little definitive evidence in its favor. For example, the plasma membrane is not a major load-bearing structure in skeletal muscle. More significantly, experiments designed specifically to test the hypothesis that dystrophin-deficient muscle fibers are prone to damage have provided highly equivocal data (7,43,44).

Menke and Jockusch (45) reported that myotubes grown from explants of dystrophin-deficient

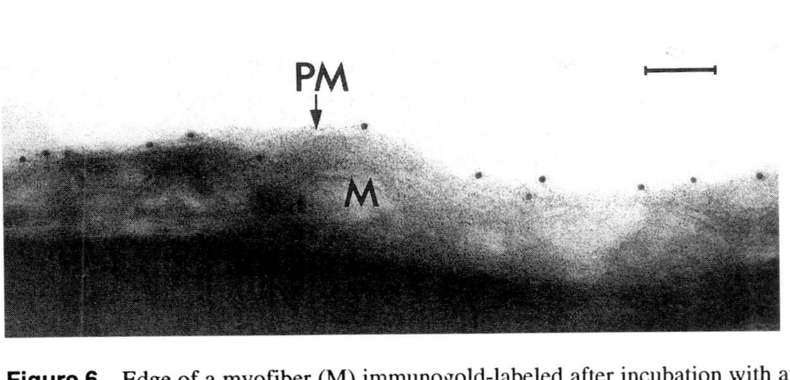

Figure 6 Edge of a myofiber (M) immunogold-labeled after incubation with an antibody to the 43-kDa dystrophin-associated glycoprotein. The 10-nm gold particles lie on or very close to the plasma membrane. Muscle = human quadriceps. Bar = 100 nm.

mouse and human muscles were more sensitive to hyposmotic shock than normal myotubes. Similarly, Franco and Lansman (46) used suction electrodes applied to the plasma membrane of myotubes and determined the negative pressure required to rupture the plasma membrane. They could find no differences between dystrophin-deficient and normal myotubes. Weller et al. (44) used a technique to ensure that *mdx* and normal mice were required to use the muscle of their lower limbs to perform eccentric contractions (a condition where a muscle is forced to lengthen at the same time as it is bearing increasing tension). Such contractions can be intensely damaging. In both normal and dystrophin-deficient muscle fibers, the contractions caused a significant increase in the number of physically damaged muscle fibers, but although there were, overall, more necrotic fibers in the dystrophin-deficient muscles both before and after exercise, the increase in the number of necrotic fibers was much larger in control than in dystrophin-deficient muscles (12-fold compared with fourfold).

The concentration of dystrophin at the myotendinous junction in normal muscle (20) has led to speculation that it is involved in protecting myotendinous junctions from tension-induced damage (21). The hypothesis implies that the absence of dystrophin would lead to "weak" myotendinous junctions, easily damaged when skeletal muscles were required to bear tension. There is, however, no evidence that this is a site of weakness in dystrophin-deficient fibers.

It is commonly believed that many of the pathological features of dystrophic muscle arise because of an increase in internal free Ca^{2+}. Such an increase, it is suggested, would activate a number of Ca^{2+}-dependent proteases, lead to Ca^{2+}-induced damage to the mitochondria, and stimulate hypercontraction (see, for example, Ref. 47). Fong et al. (48) have reported that in dystrophic, dystrophin-deficient muscle fibers, Ca^{2+}-leak channels (specific channels that are *not* activated by either stretch, mechanical activity, or membrane potential) are open more frequently than normal and could thus be implicated in the accumulation of free $[Ca^{2+}]_i$. Franco and Lansman (46) also identified a Ca^{2+}-selective channel that was open more frequently than normal in dystrophic muscle, although in this case, the channels appeared to open in response to stretch. It is not clear whether the Ca^{2+} channels identified by these two groups of workers were the same, but were exhibiting slightly different physical properties because of differences in experimental techniques and conditions, but it has been suggested by each group that dystrophin (and possibly other homologs) could bind to and stabilize transmembrane ion channels. It is important to note that two substantial studies (49,50) have failed to confirm the elevated internal Ca^{2+} levels reported by Fong et al. (48).

The relationship between dystrophin, the DAGs, and the actin filaments of the cytoskeleton is of great potential significance with regard to the function of dystrophin because it may be an indication that dystrophin has a major role in linking the extracellular matrix with the cortical cytoskeleton, a role that could be particularly important if dystrophin was found to be a part of the costamere; in that case, dystrophin would be involved in stabilizing the myofibrils within the muscle fiber and linking them via the cytoskeleton to the plasma membrane.

There is, of course, no reason why the loss or absence of dystrophin should not lead to any or all of the pathological features discussed above—loss of Ca^{2+} homeostasis, mechanical weakness of plasma membranes, fragility of myofibrils, and weakness of the myotendinous junction—but given the diversity of the pathological features associated with X-linked muscular dystrophy, the number of dystrophin homologs and isoforms, and the variety of cell types involved (contractile and noncontractile), the effects of the loss of "dystrophin" will differ from cell type to cell type.

It is also possible that the association between the absence of dystrophin and cellular pathology is indirect and not causal. For example, several recent studies on muscle from patients with Duchenne muscular dystrophy, in which an internally truncated form of dystrophin was expressed, have shown that despite the severe phenotype, dystrophin was still localized to the plasma membrane (51–53). Perhaps other key proteins, and their overall organization in the cytoskeleton, are particularly important. It is, for example, well established that in many disorders of skeletal muscle—including the X-linked and autosomal recessive forms of muscular dystrophy—there is a reduction in the expression of the DAGs (11,54,55). The converse has also been reported—the absence of key glycoproteins has been shown to be associated with an absence of dystrophin, even when the gene for dystrophin is fully transcribed (57). It might be argued from such data that the loss of one of a number of proteins could lead to an unpredictable instability of the cytoskeleton and the onset of a variable pathological response.

The role of dystrophin in central neurons is particularly enigmatic. It is well established now that many patients with X-linked muscular dystrophy have a deficit in cognitive function. The deficit is, however, highly variable, is totally unrelated to any other measure of disability or severity of the disease, and is nonprogressive (56). It has been speculated that the deficit is related to the low levels of dystrophin in certain central neurons, including the cerebellum, hippocampus, and neocortex (see Ref. 1 for a review), but the data are not compelling.

In conclusion, much is known of the molecular biology and structure of dystrophin, its isoforms and homologs. There is continuing discussion about its localization and organization in skeletal muscle and many other cell types. As yet, however, very little is known about the function of dystrophin or its related proteins.

NOTE ADDED IN PROOF

Since this article was written, several alternative theoretical models of the dystrophin-glycoprotein complex have been published. For a recent scheme the reader is referred to: Campbell KP, Three muscular dystrophies: loss of cytoskeleton-extracellular matrix linkage. Cell 1995; 80:675–679.

REFERENCES

1. Ahn AH, Kunkel LM. The structural and functional diversity of dystrophin. Nature Genet 1993; 3:283–291.
2. Bewick GS, Nicholson LVB, Young C, Slater CR. Relationship of a dystrophin-associated glycoprotein to junctional acetylcholine receptor clusters in rat skeletal muscle. Neuromusc Disord 1993; 3:503–506.
3. Byers TJ, Husain-Chishti A, Dubreuil RR, Branton D, Goldstien LSB. Sequence similarity of the amino-terminal domains of Drosophila beta spectrin to α-actinin and dystrophin. J Cell Biol 1989; 109:1633–1641.
4. Hemmings L, Kuhlman PA, Critchley DR. Analysis of the actin-binding domains of α-actinin by mutagenesis and demonstration that dystrophin contains a functionally homologous domain. J Cell Biol 1992; 116:1369–1380.
5. Koenig M, Monaco AP, Kunkel L. The complete sequence of dystrophin predicts a rod-shaped cytoskeletal protein. Cell 1988; 53:219–228.
6. Love DR, Hill DF, Dickson G, Spurr NK, Blyth BC, Marsden RF, Walsh FS, Edwards YF, Davies KE. An autosomal transcript in skeletal muscle with homology to dystrophin. Nature 1989; 339:55–58.
7. Campbell KP, Kahl SD. Association of dystrophin and an integral membrane glycoprotein. Nature 1989; 338:259–262.
8. Yoshida M, Ozawa E. Glycoprotein complex anchoring dystrophin to sarcolemma. J Biochem 1990; 108:748–752.
9. Ohlendieck K, Campbell KP. Dystrophin constitutes 5% of membrane cytoskeleton in skeletal muscle. FEBS Lett 1991; 283:230–234.
10. Ohlendieck K, Ervasti JM, Snook JB, Campbell KP. Dystrophin-glycoprotein complex is highly enriched in isolated skeletal muscle sarcolemma. J Cell Biol 1991; 112:135–148.
11. Ibraghimov-Beskrovnaya O, Ervasti JM, Leveille CJ, Slaughter CA, Sernett SW, Campbell KP. Primary structure of dystrophin-associated glycoproteins linking dystrophin to the extracellular matrix. Nature 1992; 355:696–702.
12. Suzuki A, Yoshida M, Yamamoto H, Ozawa E. Glycoprotein-binding site of dystrophin is confined to the cysteine-rich domain and the first half of the carboxyl-terminal domain. FEBS Lett 1992; 308:154–160.
13. Murayama T, Sato O, Kimura S, Shimizu T, Sawada H, Maruyama K. Molecular shape of dystrophin purified from rabbit skeletal muscle myofibrils. Proc Jpn Acad 1990; 66:96–99.
14. Pons F, Augier H, Heilig R, Leger J, Mornet D, Leger JJ. Isolated dystrophin molecules as seen by electron microscopy. Proc Natl Acad Sci USA 1990; 87:7851–7855.
15. Sato O, Nonomura Y, Kimura S, Maruyama K. Molecular shape of dystrophin. J Biochem 1992; 112:631–638.
16. Hoffman EP, Knudson CM, Campbell KP, Kunkel LM. Subcellular fractionation of dystrophin to the triads of skeletal muscle. Nature 1987; 330:754–756.
17. Knudson CM, Hoffman EP, Kahl SD, Kunkel LM, Campbell KP. Characterization of dystrophin in skeletal muscle triads. J Biol Chem 1988; 163:8480–8485.
18. Sugita H, Arahata K, Ishiguro T, Suhara Y, Tsukahara T, Ishiura S, Eguchi C, Nonaka I, Ozawa E. Negative

staining of Duchenne muscular dystrophy (DMD) and *mdx* muscle surface membrane with antibody against synthetic peptide fragment predicted from DMD cDNA. Proc Jpn Acad B 1988; 64:37–39.
19. Zubrzycka-Gaarn EE, Bulman DE, Karpati G, Burghes AHM, Belfall B, Klamut HJ, Talbot J, Hodges RS, Ray PN, Worton RG. The Duchenne muscular dystrophy gene product is localised in sarcolemma of human skeletal muscle. Nature 1988; 333:466–469.
20. Samitt CE, Bonilla E. Immunocytochemical study of dystrophin. Muscle Nerve 1990; 13:493–500.
21. Tidball JG, Law DJ. Dystrophin is required for normal thin filament-membrane associations at myotendinous junctions. Am J Pathol 1991; 138:17–21.
22. Bornemann A, Schmalbruch H. Desmin and vimentin in regenerating muscle. Muscle Nerve 1992; 15:14–20.
23. Bowman W. On the minute structure and movements of voluntary muscle. Phil Trans Roy Soc Lond 1840; 130:457–501.
24. Carpenter S, Karpati G, Zubrzycka-Gaarn EE, Bulman DE, Ray PN, Worton RG. Dystrophin is localized to the plasma membrane of human skeletal muscle fibers by electron-microscopic cytochemical study. Muscle Nerve 1990; 13:376–380.
25. Cullen MJ, Walsh J, Nicholson LVB, Harris JB. Ultrastructural localization of dystrophin and human muscle using gold immunolabelling. Proc Roy Soc Lond B 1990; 240:197–210.
26. Watkins SC, Hoffman EP, Slayter HS, Kunkel LM. Immunoelectron microscopic localization of dystrophin in myofibres. Nature 1988; 333:863–866.
27. Cullen MJ, Walsh J, Nicholson LVB, Harris JB, Zubrzycka-Gaarn EE, Ray PN, Worton RG. Immunogold labelling of dystrophin in human muscle using an antibody to the last 17 amino acids of the C-terminus. Neuromusc Disord 1991; 1:113–119.
28. Cullen MJ, Watkins SC. Ultrastructure of muscular dystrophy: new aspects. Micron 1993; 24:287–307.
29. Suwa M, Mitaku S, Kuroda Y. Theoretical analysis of amino acid sequence of human dystrophin. Biochem Biophys Res Commun 1993; 191:782–789.
30. Vater R, Cullen MJ, Nicholson LVB, Harris JB. The fate of dystrophin during the degeneration and regeneration of the soleus muscle of the rat. Acta Neuropathol 1992; 83:140–148.
31. Harris JB, Cullen MJ. Ultrastructural localization and the possible role of dystrophin. In: Kakulas BA, MacHowell J, Roses AD, eds. Duchenne Muscular Dystrophy: Animal Models and Genetic Manipulation. New York: Raven Press, 1992:19–40.
32. Cross RA, Stewart M, Kendrick-Jones J. Structural predictions for the central domain of dystrophin. FEBS Lett 1990; 262:87–92.
33. Porter GA, Dymtrenko GM, Winkelmann JC, Bloch RJ. Dystrophin colocalizes with β-spectrin in distinct subsarcelemmal domains in mammalian skeletal muscle. J Cell Biol 1992; 117:997–1005.
34. Masuda T, Fujimaki N, Ozawa E, Ishikawa H. Confocal laser microscopy of dystrophin localization in guinea pig skeletal muscle fibers. J Cell Biol 1992; 119:543–548.
35. Straub V, Bittner RE, Leger JJ, Voit T. Direct visualization of the dystrophin network on skeletal muscle fiber membrane. J Cell Biol 1992; 119:1183–1191.
36. Minetti C, Beltrame F, Marcenaro G, Bonilla E. Dystrophin at the plasma membrane of human muscle fibers shows costameric localization. Neuromusc Disord 1992; 2:99–109.
37. Pardo JV, Siliciano JD, Craig SW. A vinculin-containing cortical lattice in skeletal muscle: transverse lattice elements ("costameres") mark sites of attachment between myofibrils and sarcolemma. Proc Natl Acad Sci USA 1983; 80:1081–1088.
38. Squarzoni S, Sabatelli P, Maltarello MC, Cataldi A, Diprimio R, Maraldi NM. Localization of dystrophin COOH-terminal domain by the fracture-label technique. J Cell Biol 118:1401–1410.
39. Byers TJ, Kunkel LM, Watkins SC. The subcellular distribution of dystrophin in mouse skeletal, cardiac and smooth muscle. J Cell Biol 1991; 115:411–421.
40. Lidov HGW, Byers TJ, Watkins SC, Kunkel LM. Localization of dystrophin to postsynaptic regions of central nervous system cortical neurons. Nature 1990; 348:725–728.
41. Matsumura K, Ervasti JM, Ohlendieck K, Kahl SD, Campbell KP. Association of dystrophin-related protein with dystrophin-associated proteins in *mdx* mouse muscle. Nature 1992; 360:588–591.
42. Meng G, Kress W, Scherpf S, Bettecken T, Feichtinger W, Schempp W, Schmid M, Müller CR. A comparison of the dystrophin gene structure in primates and lower vertebrates. In: Angelini C, Danieli GA, Gontanari D, Eds. Muscular Dystrophy Research: From Molecular Diagnosis Towards Therapy. Amsterdam: Elsevier, 1991:23–30.
43. Baatsen PHWW, Moens P, Marechal G. Structural basis for force decline in EDL-muscle from mdx-mice after eccentric contraction. J Musc Cell Motil 1993; 14:255.
44. Weller B, Karpati G, Carpenter S. Dystrophin deficient *mdx* muscle fibres are preferentially vulnerable to necrosis induced by experimental lengthening contractions. J Neurol Sci 1990; 100:9–13.
45. Menke A, Jockusch H. Extent of shock-induced membrane leakage in human and mouse myotubes depends on dystrophin. J Cell Sci 1995; 108:727–733.

46. Franco A, Lansman JB. Calcium entry through stretch-inactivated ion channels in *mdx* myotubes. Nature 1990; 344:670–673.
47. Bodensteiner JB, Engel AG. Intracellular calcium accumulation in Duchenne dystrophy and other myopathies—a study of 567,000 muscle fibers in 114 biopsies. Neurology (Minneapolis) 1978; 28:670–677.
48. Fong P, Turner PR, Denetclaw WF, Steinhardt RA. Increased activity of calcium leak channels in myotubes of Duchenne human and *mdx* mouse origin. Science 1990; 250:673–676.
49. Rivet-Bastide M, Inbert N, Cognard G, Duport G, Rideau Y, Raymond G. Changes in resting ionized calcium level and in calcium transients during in vitro development of normal and Duchenne muscular dystrophy cultured skeletal muscle measured by laser cytofluorimetry using Indo-1. Cell Calcium 1993; 14:563–571.
50. Pressmar J, Brinkmeier H, Seewald MJ, Naumann T, Rüdel R. Intracellular Ca^{2+} concentrations are not elevated in resting cultured muscle from Duchenne (DMD) patients and in MDX mouse muscle fibres. Pflügers Arch 1994; 426:499–505.
51. Récan D, Chafey P, Leturcq F, Hugnot J-P, Vincent N, Tomé F, Collin H, Simon D, Czernichow P, Nicholson LVB, Fardeau M, Kaplan J-C, Chelly J. Are cysteine-rich and COOH-terminal domains of dystrophin critical for sarcolemmal localization? J Clin Invest 1992; 89:712–716.
52. Hoffman EP, Garcia CA, Chamberlain JS, Angelini C, Lupski JR, Fenwick R. Is the carboxyl-terminus of dystrophin required for membrane association? A novel, severe case of Duchenne muscular dystrophy. Ann Neurol 1991; 30:605–610.
53. Helliwell TR, Ellis JM, Mountford RC, Appleton RE, Morris GE. A truncated dystrophin lacking the C-terminal domains is localized at the muscle membrane. Am J Hum Genet 1992; 50:508–514.
54. Matsumura K, Tomé FMS, Collin H, Azibi K, Chaouch M, Kaplan J-C, Fardeau M, Campbell KP. Deficiency of the 50K dystrophin-associated glycoprotein in severe childhood autosomal recessive muscular dystrophy. Nature 1992; 369:320–322.
55. Ervasti JM, Campbell KP. Membrane organization of the dystrophin-glycoprotein complex. Cell 1991; 66:1121–1131.
56. Nicholson LVB, Johnson MA, Bushby KD, Gardner-Medwin D, Curtis A, Ginjaar IB, den Dunnen JT, Welch JL, Butler TJ, Bakker E, van Ommen G-J, Harris JB. Integrated study of 100 patients with Xp21 linked muscular dystrophy using clinical, genetic, immunochemical and histopathological data. Part 2. Correlations within individual patients. J Med Genet 1993; 30:737–744.
57. Matsumura K, Campbell KP. Deficiency of dystrophin-associated proteins: a common mechanism leading to muscle cell necrosis in severe childhood muscular dystrophies. Neuromusc Disord 1993; 3:109–118.

19

The Spectrum of Dystrophinopathies

Byron A. Kakulas
Royal Perth Hospital and Australian Neuromuscular Research Institute, Queen Elizabeth II Medical Centre, Perth, Western Australia, Australia

INTRODUCTION

A vast improvement in the understanding of Duchenne muscular dystrophy (DMD) and its allele Becker muscular dystrophy (BMD) followed the molecular discoveries of Kunkel and his group (1). In a series of reports from 1986 onward, Kunkel (2), his co-workers Hoffman (3,4), Monaco (5), Koenig (6), and others delineated the DMD gene product, named it dystrophin, and suggested its function as a structural protein of skeletal muscle. It was soon recognized to be a member of the greater spectrin family of cytoskeletal proteins. Suprisingly and quite consistently, the mutation responsible for the majority of DMD and BMD cases was shown to be a deletion of DNA (7). It was also found that the amount of dystrophin present was in general inversely proportional to the severity of the disease. In DMD dystrophin was usually totally absent, whereas in BMD it was present but in reduced quantities, roughly corresponding to the clinical severity of the disorder (8).

DYSTROPHIN

The structure, localization, and function of dystrophin are described in another chapter. Briefly, the DMD gene consists of 2.5 million base pairs with more than 80 exons represented by 14 kDa of mRNA. The product, dystrophin, is a very large protein with an *MW* of 427 kDa containing 3685 amino acids. Dystrophin is present in very small amounts in skeletal muscle accounting for 0.002% of total muscle protein (4,5). It exists as a large oligomeric complex with four glycoproteins, one of which is greatly reduced in DMD, the loss of which may be one of the first stages of the molecular pathogenesis of DMD (9). Dystrophin occurs in heart, brain, and smooth muscle where it has a molecular weight of 405 kDa. Multiple forms of dystrophin generated by different splicing of the transcript were described by Koenig et al. (10).

Dystrophin belongs to the greater spectrin family with extensive α-helical domains and a structural similarity to the cytoskeleton proteins spectrin and α-actinin. It has an actin-binding site and is believed to have a similar role in the cytoskeleton to α-actin and spectrin. In an analysis of the repeat domain of dystrophin, Koenig and Kunkel (2) identified four potential hinge segments that they believe confer flexibility to the membrane-associated network of dystrophin and provide resilience to the membrane. The molecular biology of dystrophin is fully reviewed by Gutmann and Fischbeck (11).

Sugita et al. (12) demonstrated the subsarcolemmal location of dystrophin and its loss in DMD and *mdx* mice using antibodies to a synthetic peptide fragment predicted from DMD cDNA. This location to the membrane was also shown by others (13,14).

The function of dystrophin is not yet fully known. It is believed mainly to have a physical or mechanical role as a membrane stabilizer. However, small quantities are also found at the junction of the terminal cisternae of the sarcoplasmic reticulum and T system, where it may be involved with calcium metabolism. High concentrations are found at the myoneural junction. Dystrophin is believed to confer flexibility and stability on the plasma membrane and may have a role in excitation contraction coupling and regulation of calcium movement, as discussed by Harris and Cullen (15).

MOLECULAR BASIS OF DMD AND BMD

There is no particular site of predilection for the DNA deletion in DMD and the severity of the disease (16). It was postulated and soon verified by Monaco et al. (17) that the absence of dystrophin in DMD was due to the deletions causing a "frameshift" disorder so that no dystrophin is produced. In BMD the majority of deletions are near the center of the gene involving exons 45–47, which are an "in-frame," so a truncated protein is produced (17). This explanation for difference in severity between DMD and BMD is now known as the "reading frame" hypothesis of Monaco (17).

Hoffman et al. (8) suggest that DMD and BMD can be divided into four clinical categories according to the amount of dystrophin present. DMD (with the patient becoming wheelchair-bound at the age of 11 years) occurs when the quantity of dystrophin is less than 3% of normal; in severe BMD (when the patient becomes wheelchair-bound from 13 to 20 years of age) dystrophin quantities are between 3 and 10%; and moderate or mild BMD (with ambulation beyond the age of 20 years) is related to a dystrophin quantity of 20% or more. These guidelines have proved useful in the clinical investigation of sporadic cases. Nevertheless, the correlation between dystrophin quantity and the severity of the disease is not strict (18) and there are many exceptions. Some patients with abundant dystrophin may still show the aggressive DMD phenotype, and in some milder cases, patients have no demonstrable dystrophin (19). Some of the DMD phenotypes with dystrophin present are believed to involve the glycoprotein-binding site (20) of the N-terminus. Others may be due to a disorder of the glycoprotein itself (9).

These cases are quite apart from autosomal recessive pseudohypertrophic muscular dystrophy, and other phenotypes with clinical similarities to DMD or BMD are coded on other genes. As an example, some patients whose disorder resembles BMD have autosomal dominant inheritance (21).

DYSTROPHIN IN THE PATHOGENESIS OF DMD

The discovery that dystrophin is an element of the membrane cytoskeleton conforms with well-established concepts of the pathogenesis of DMD where a weakened membrane has for many years been considered to predispose to focal necrosis, the first observable lesion. Membrane rupture during contraction results in muscle fiber necrosis due to the entry of large amounts of calcium into the sarcoplasm so that proteases are excessively activated. A cycle of repeated necrosis and abortive regeneration eventually leads to exhaustion of regeneration, so fat and fibrous tissue replacement is passive. These formulations were first derived from the investigation of the Rottnest Island Quokka from which principles of myopathology were derived (22). When applied to the human necrobiotic myopathies, the principles soon became the standard concept of mechanisms of pathogenesis of the necrobiotic myopathies including DMD (23,24). In the context of pathogenesis, the spectacular molecular genetic discoveries of Kunkel et al. (1,6) were quickly followed by theoretical discussions and practical experiments directed toward a curative treatment for DMD (25).

CLINICAL APPLICATION OF THE MOLECULAR ADVANCES

Application of the discoveries to the clinical situation has also been rewarding. Major improvements have resulted from the use of DNA technology in the diagnosis of DMD and BMD. Dystrophin immunocytochemistry and Western immunoblots are applied to muscle biopsies, and recombinant

DNA technology is used on blood samples for DNA deletion analysis. Thus, for the first time in the history of myopathology, specific tests are now available for the diagnosis of DMD and BMD.

Prior to the molecular genetic revelations underlying DMD, female carrier diagnosis depended on indirect methods such as slightly elevated serum creatine kinase levels, electromyography, and muscle biopsy. Genetic counseling was based on "approximate" information. Nevertheless, significant benefits, including prevention, were achieved in this way (26).

The recent use of cDNA methods for carrier detection and genetic counseling have vastly improved the situation, providing accurate and direct diagnosis in most cases. In those DMD or BMD families that have deletions the affected male fetus can be rapidly identified at 8 weeks of pregnancy using DNA from chorionic villus biopsies.

The availability of the multiplex polymerase chain reaction (PCR) has made even more rapid diagnosis possible. Similarly, the use of microsatellites has enhanced the practicality of linkage analysis for 35% or so of families who do not have DNA deletions. We have also identified a point mutation, an A-to-C substitution at the junction of the exon 19 and the intron 19 splice site in a BMD family, which has made antenatal diagnosis possible for the female carriers in the pedigree (27).

THE DYSTROPHINOPATHIES

In the first molecular reports the gene responsible for DMD was referred to as the "Duchenne gene." However, this may be considered an inappropriate name since the gene produces a normal cytoskeletal protein, dystrophin. Therefore, the term "dystrophin gene" is more accurate. It is believed that the term "dystrophinopathies" is superior to the previously applied title Xp21.2 myopathies as it is derived from the known gene product.

Soon after the molecular discoveries were made, the *mdx* mouse, which has been widely used as a model for BMD, was found to lack dystrophin altogether. This was shown to be due to a point mutation in the homologous gene (28). Another animal equivalent, a golden retriever dog with muscular dystrophy (GRMD), also completely lacks dystrophin (29) as the result of a point mutation (30). Other animal equivalents are the cat and the Rottweiller dog (31). Because the eponymous terms "Duchenne" and "Becker" refer to the human diseases, the comparative animal disorders cannot be so named. For this reason they were first called the "dystrophinopathies."

With the observation that some other human neuromuscular disorders previously considered to be unrelated to DMD were also due to abnormalities of the dystropin gene, the term "dystrophinopathy" was soon applied to the wider range of human and animal diseases. Examples were patients previously thought to have spinal muscular atrophy (32–34), limb-girdle muscular dystrophy (35), quadriceps myopathy (36), and even congenital muscular dystrophy (37).

The full range of the human dystrophinopathies has not yet been determined. Our experience is continuing to evolve and it is not yet known how many clinical variants may be associated with abnormalities of dystrophin. Nor is it known how many phenotypes exist that are similar to DMD and BMD that are due to abnormalities of other genes (8,21). A classification of the human dystrophinopathies as they stand is given in Table 1 and the group is described below.

Duchenne Muscular Dystrophy

It has long been observed that the clinical features and natural history of DMD are almost exactly alike in every patient regardless of ethnicity or location. The young "infant Hercules," who develops a waddling gait and muscular weakness by 5 years of age, together with enlarged calves, proximal limb-girdle weakness, lordosis, loss of ambulation at 11 or 12 years, and death before 30 years of age, is the DMD stereotype. It is now known that in DMD dystrophin is usually totally lacking or present in amounts less than 3% of normal. The few dystrophin-containing fibers that may be found in muscle biopsies are so-called "revertant" fibers, which occur in regenerating fibers as the myopathy evolves and are believed to be due to overriding of the stop codon. In DMD there is no particular site of

Table 1 The Spectrum of Dystrophinopathies

Animal	
The *mdx* mouse	
The *mdx* dog (golden retriever and Rottweiler)	
The *mdx* cat	
Human	
Duchenne muscular dystrophy (DMD)	
The manifesting DMD carriers	
Becker muscular dystrophy (BMD)	
Atypical BMD	
Miscellaneous	
Quadriceps myopathy	
Congenital muscular dystrophy (Fukuyama type)	
Cardiomyopathies	
Myalgia and cramps	
Creatine kinase aemias	
Asymptomatic individuals	

predilection for the mutation, which may be in any part of the gene. The most common type of mutation (responsible for 60–70% of cases) is a DNA deletion. In DMD the deletions (or other mutations) have in common their "out of frame" character and no dystrophin is produced.

Rare DMD phenotypes do exist in which normal or near-normal amounts of dystrophin are found. These DMD patients may have abnormalities of the C-terminus or the glycoprotein to which dystrophin is attached (9). One such patient reported was a particularly severe DMD phenotype, but biochemically his muscle contained substantial amounts of dystrophin more in keeping with BMD. The dystrophin gene in the patient was deleted for exons encoding the carboxyl-terminal domain of the dystrophin protein and some of the penultimate cysteine-rich domain. This observation supports the hypothesis that an intact carboxyl terminus is crucial for correct dystrophin function. Such patients are rare (38).

The Manifesting DMD Carrier

Laboratory tests for dystrophin have shown that symptomatic, i.e., "manifesting" female carriers, are not as uncommon as once believed. In the past, sporadic females, with mild clinical features, have been confused with limb-girdle muscular dystrophy (LGMD) or autosomal recessive pseudohypertrophic muscular dystrophy. The female carrier of DMD is heterozygous for the recessive dystrophin gene and is believed to develop dystrophic changes because of excessive "lyonization" (i.e., embryonic inactivation) of her normal X chromosome. A proportion of her muscle fibers are deficient in dystrophin and others are normal, forming a mosaic (39). These females usually have a mild BMD phenotype.

There are other situations in which a girl may have a DMD-like picture. These females usually have a chromosomal disorder in which on X chromosome is inactivated or lost. The best-known examples are X-autosomal translocations at Xp21.2, causing inactivation of the gene and Turner's syndrome. The subject of dystrophin findings in carriers is discussed more fully in another chapter.

Becker Muscular Dystrophy (BMD)

The clinical features and natural history of BMD are similar to those of DMD except for a more protracted course. In contrast to DMD, where the clinical picture is more or less stereotyped and the progression of the disease highly predictable, there is much greater variability in the natural history of BMD. Subtypes of BMD are recognized that closely correlate with the amount and quality of dystrophin expressed in skeletal muscle as described above (8). Severe, intermediate, and mild forms are known. Those patients relatively lacking in dystrophin have more severe disease than those approaching normal quantities (40). However, the correlation is not strict and many exceptions are reported. For instance, Patel et al. describe a BMD patient with no demonstratable dystrophin (19). In this patient no dystrophin was detected with antibodies to the 60-kDa and 90-kDa antibodies. Gel electrophoresis and Western blotting showed complete absence of dystrophin. There were no DNA deletions. The patient was in his 20s and still able to walk (19).

In BMD deletions of DNA are found in 60–70% of families, with almost half the cases involving exons 45–53. The "in frame" mutation hypothesis of Monaco explains the presence of a truncated dystrophin protein in BMD (17). Some families have been reported to have both DMD and BMD phenotypes. In one such instance two distinct mutations accounting for the phenotypic difference were identified (41).

Atypical Becker Muscular Dystrophy

Pseudo Spinal Muscular Atrophy

It is now recognized that some patients who were originally diagnosed as having spinal muscular atrophy (SMA) of the late childhood or adolescent Kugelberg-Welander type have a dystrophinopathy, as demonstrated by molecular laboratory methods (32). Thus some sporadic males with proximal muscle weakness, enlarged calves, and a protracted clinical course have now been shown to be BMD variants even though they may have neuropathic features in their muscle biopsies and neurogenic features in their electromyogram (EMG). These cases were often incorrectly labeled as the Kugelberg-Welander type of SMA, but they now can be shown to be true BMD variants on the basis of immunocytochemistry or Western immunoblots on muscle biopsies and/or with deletions of DNA from white blood cells or elsewhere.

Our own experience includes a young man aged 17 years with "arrested SMA" who had a neurogenic EMG, neuropathic muscle biopsy, and normal serum creatine kinase. He was also found to have a deletion of the DMD gene by cDNA probe 9–14 band c indicating an exon greater than 50. Calf hypertrophy is not always present in the atypical BMD cases. For instance, five brothers with features of muscular dystrophy without calf hypertrophy were shown to have DNA deletions (42).

In clinical practice it was long accepted that cases of sporadic male patients with chronic neurogenic proximal weakness were spinal in origin even though moderately elevated serum creatine kinase levels were usually present. The creatine kinase elevation was explained on the basis of denervated muscle being more prone to nonspecific injury. However, in an early study Bradley et al. drew attention to the prominence of neurogenic features in EMGs of BMD patients (43). This study was undertaken when the sick motor neuron theory of McComas (44) was in vogue. Why neurogenic features are prominent in some BMD patients given the subsarcolemmal mechanical role of dystrophin is not known. A role for dystrophin at the level of the motor unit cannot be discounted.

Pseudo Limb-Girdle Muscular Dystrophy

The same experience has occurred with limb-girdle muscular dystrophy (LGMD) patients. When LGMD patients have been investigated using the molecular genetic methods, a significant number, even females (manifesting carriers), have been shown to have abnormalities of the dystrophin gene and are therefore variants of BMD and are true dystrophinopathies (33,35).

Arikawa et al. (35), In their analysis of 41 patients with a clinical diagnosis of LGMD, found five males with an abnormal dystrophin pattern by immunofluorescence diagnostic for BMD and two female LGMD patients with dystrophin patterns consistent with manifesting carriers of DMD. They report, therefore, that 17% of their limb-girdle population showed a dystrophinopathy. Five of the same 41 patients, including two of 24 females, had DNA deletions for the dystrophin gene.

Similarly, Norman et al. (45) found that four of 15 LGMD patients tested showed dystrophin gene deletions. Their study "emphasises the clinical overlap between limb girdle muscular dystrophy and dystrophinopathies and reinforces the necessity of dystrophin protein gene studies for the accurate clinical diagnosis of isolated cases of muscular dystrophy" (45).

It should be emphasized that not all sporadic male patients with SMA or LGMD are dystrophinopathies. In their analysis Hoffman et al. (8) found that about 60% of patients with the clinical features of BMD but with no family history were true Becker patients. It is noteworthy that in the pseudo SMA and LGMD patients the exon deletions are in the same area of the gene as the majority of BMD cases involving exons 45–47 using cDNA 8 (46).

In our experience about 15% of SMA and LGMD patients have been shown to have abnormalities of the dystrophin gene. That is, there is a common phenotype between the atypical dystrophinopathies and true SMA or LGMD patients.

Miscellaneous Dystrophinopathies

Quadriceps Myopathy

Soon after the seminal molecular discoveries were made, Sugita and his group, using Southern blots and/or immunocytochemistry, found that some patients with quadriceps myopathy may also have deletions and/or abnormalities of dystrophin. Thus four male patients with quadriceps myopathy were described by Arikawa et al. (47) and all showed abnormalities of dystrophin by immunofluorescence and immunoblot tests.

Congenital Muscular Dystrophy

Similarly, some patients with the "Fukuyama" type of congenital muscular dystrophy (FCMD) have been shown to have abnormalities of the dystrophin gene (37). Arikawa et al. (37) investigated 36 patients with FCMD and found that in 34 there was unusual immunostaining pattern for dystrophin with occasional negative or partially deficient fibers. Dystrophin was totally absent in two of the 36 patients with FCMD. An intragenic deletion was found in one FCMD patient between exons 51 and 53, which may be an asymptomatic region of the gene. They noted that spectrin immunostaining in FCMD was identical to that for dystrophin when examined in consecutive cryostat sections. Those fibers that stained intensely for dystrophin also stained intensely for spectrin and those that stained negative for dystrophin also stained negative for spectrin. The relationship between a dystrophinopathy and FCMD remains enigmatic. The deletion in one FCMD patient may be incidental, and the true relationship between FCMD and a dystrophinopathy is yet to be established.

Cardiomyopathies

Predominant cardiomyopathy may be another dystrophinopathy variant. It is well known that patients with BMD may develop overt cardiac failure. However, in some BMD patients the cardiomyopathy is the clinically dominant feature and there may be little or no evidence of muscular weakness (48). Gold et al. (48) reported three families in whom the initial tentative diagnosis was either metabolic myopathy or a familial metabolic cardiomyopathy. The patients' muscle symptoms consisted of exertional cramps, and physical examination revealed well-developed muscles and only minimal weakness, e.g., MRC grade 5- for hip flexion. Cardiac involvement caused congestive cardiomyopathy. The molecular weight of dystrophin was reduced and there was faint diffuse straining by immunocytochemistry. One patient had muscle pain and cramps, especially in the thighs. Therefore, the findings were similar to BMD. The reason for variable clinical cardiac complications in patients with BMD is unknown. Dystrophin is expressed in the heart, and it is assumed that in some BMD patients reduced quantities are responsible for a disorder of myocardial function.

Myalgia and Cramps

It has also been reported that certain patients with simple myalgia and cramps, with little or no clinical weakness, may have deletions of the dystrophin gene (49). According to Hoffman et al. (3), deletions in the proximal portion of the gene tend to cause severe cramps and myalgia and loss of the middle portion, a mild phenotype, the only sign being elevation of creatine kinase (49). Gospe et al. (50) also report predominant familial myalgia and cramps associated with a deletion in the dystrophin game.

The "creatine kinase–emias"

Patients with the McLeod syndrome may have very high levels of creatine kinase without muscle weakness. Other patients without the Kel blood factor abnormalities or acanthocyatosis may also have unexplained elevations of creatine kinase (51). In the clinic the usual approach to patients with elevated serum creatine kinase and no other abnormalities has been to consider them as possible malignant hyperthermia susceptibles and to recommend tests for that disorder. However, it has become apparent that some of these patients ("CK-emias") have abnormalities of the dystrophin gene (51).

Glycolytic Disorders

Patients with glycerol kinase deficiency may also have deletions of the 3′ end of the dystrophin gene and are thus associated with myopathies. Darras and Francke (52) have shown that all five patients with myopathy and glycerol kinase complex deficiency showed DNA deletions extending through the 3′ end of the gene. Their patients without myopathy had no deletion in the dystrophin cDNA. In a recent report Rose et al. (53) described a man with a clinically benign myopathy who had muscle glycogen storage and a dystrophin gene deletion. None of the known glycogen storage metabolic defects were identified. It was suggested that the deletion played some as yet unknown role in the pathogenesis.

Asymptomatic Dystrophinopathies

Some patients with deletions may be totally asymptomatic (54). It is believed that in such cases the reduction in size of dystrophin affects the rod domain and the molecule is able to function normally. Koenig et al. (46) refer to asymptomatic individuals with "in frame" deletions in the region of exons 31–44. A normal male with an inherited deletion of one exon within the DMD gene has also been reported (55). England et al. (56) reported very mild muscular dystrophy associated with a deletion of 46% of dystrophin. The dystrophin loss in their 61-year-old ambulant patient covered 5106 base pairs, resulting in the production of a truncated polypeptide localized correctly in the muscle cell. The deletion in the family included exons 17–48, removing 5106 bp off the coding sequence. The report further demonstrates the individual heterogeneity that exists within the BMD phenotype.

CONCLUSION

The term "dystrophinopathy," which describes a wide range of neuromuscular disorders, has come into use as a result of the finding of dystrophin gene or protein abnormalities in a variety of patients, not all of whom conform to the DMD or BMD stereotypes. Some of these patients were previously considered to have spinal muscular atrophy or LGMD (57). Others were even more variable, being previously considered to have quadriceps myopathy, congenital muscular dystrophy, or only myalgia or cramps. With the increasing use of molecular genetic methods for diagnostic purposes it is probable that more atypical dystrophinopathies will become known. A suggestion is provided by the finding of abnormalities by Western immunoblots or by immunocytochemistry in two patients with asymmetrical facioscapulohumeral muscular dystrophy in our clinic.

With the application of the new diagnostic methods excess dystrophin has also been recognized (58). Excess dystrophin may also be nonspecifically associated with autophagosomes. Whether conditions showing dystrophin accumulation in the sarcoplasm should be called "dystrophinopathies" is debatable, and they are not included in the classification.

It is salutary to be at a point in history when a classification for DMD, BMD, and related neuromuscular diseases may be presented in molecular terms. As we enter the era of clinical molecular genetics, our horizons are expanded, genetic dogmas overturned, and our appreciation of the molecular basis for many neuromuscular disorders vastly improved.

ACKNOWLEDGMENTS

The assistance of Drs. G. Hammond-Tooke, S. S. Gubbay, Theo Kyriakides, Sungkom Jongpiputvanich, and Peter Silberstein in the clinic and Nigel Laing, David Chandler, and Russell Johnsen in the laboratory is gratefully acknowledged.

REFERENCES

1. Kunkel LM, Monaco AP, Middlesworth W, Ochs HD, Latt SA. Specific cloning of DNA fragments absent from the DNA of a male patient with an X chromsome deletion. Proc Natl Acad Sci USA 1985; 82:4778–4782.
2. Koenig M, Kunkel LM. Detailed analysis of the repeat domain of dystrophin reveals four potential hinge segments that may confer flexibility. J Biol Chem 1990; 265(8):4560–4566.
3. Hoffman EP, Knudson CM, Campbell KP, Kunkel LM. Subcellular fractionation of dystrophin to the triads of skeletal muscle. Nature 1987; 330:754–758.
4. Hoffman EP, Brown RH Jr, Kunkel LM. Dystrophin: the protein product of the Duchenne muscular dystrophy locus. Cell 1987; 51:919–928.
5. Monaco AP. Dystrophin, the protein product of Duchenne/Becker muscular dystrophy gene. Trends Biol Sci 1989; 14:412–415.
6. Koenig M, Hoffman EP, Bertelson CJ, Monaco AP, Feener C, Kunkel LM. Complete cloning of the Duchenne muscular dystrophy (DMD) cDNA and preliminary genomic organization of the DMD gene in normal and affected individuals. Cell 1987; 50(3):509–517.
7. Koenig M, Beggs AH, Moyer M, et al. The molecular basis for Duchenne versus Becker muscular dystrophy: correlation of severity with type of deletion. Am J Hum Genet 1989; 45:498–506.
8. Hoffman EP, Kunkel LM, Angelini C, Clarke A, Johnson M, Harris JB. Improved diagnosis of Becker muscular dystrophy by dystrophin testing. Neurology 1989; 39:1011–1017.
9. Ervasti JM, Ohlendieck K, Kahl SD, Gaver MG, Campbell KP. Deficiency of glycoprotein component of the dystrophin complex in dystrophic muscle. Nature 1990; 345:315–319.
10. Koenig M, Feener CA, Hoffman EP, Kunkel LM. Multiple forms of dystrophin are generated by differential splicing of the transcript. In: Kakulas BA, Mastaglia FL, eds. Pathogenesis and Therapy of Duchenne ands Becker Muscular Dystrophy. New York: Raven Press, 1990:25–32.
11. Gutmann DH, Fischbeck KH. Molecular biology of Duchenne and Becker's muscular dystrophy: clinical applications. Ann Neurol 1989; 26:189–194.
12. Sugita H, Arahata K, Ishiguro T, et al. Negative immunostaining of Duchenne muscular dystrophy (DMD) and mdx mouse muscle surface membrane with antibody against synthetic peptide fragment predicted from DMD cDNA. Proc Jpn Acad 1988; 64:210–212.
13. Bonilla E, Samitt CE, Miranda AF, et al. Duchenne muscular dystrophy: deficiency of dystrophin at the muscle cell surface. Cell 1988; 54:447–452.
14. Zubrzycka-Gaarn EE, Bulman DE, Karpati G, et al. The Duchenne muscular dystrophy gene product is localized in sarcolemma of human skeletal muscle. Nature 1988; 333:466–469.
15. Harris JB, Cullen MJ. Chpt 3 Ultrastructural localisation of the possible role of dystrophin. In: Kakulas BA, McC Howell J, Roses AD, eds. Duchenne Muscular Dystrophy: Animal Models and Genetic Manipulation. New York: Raven Press, 1992:19–40.
16. Baumbach LL, Chamberlain JS, Ward PA, Farwell NJ, Caskey CT. Molecular and clinical correlations of deletions leading to Duchenne and Becker muscular dystrophies. Neurology 1989; 39:465–474.
17. Monaco AP, Bertelson CJ, Liechti-Gallati S, Moser H, Kunkel LM. An explanation for the phenotypic differences between patients bearing partial deletions of the DMD locus. Genomics 1988; 2(1):90–95.
18. Gillard EF, Chamberlain JS, Murphy EG, et al. Molecular and phenotypic analysis of patients with deletions within the deletion rich region of the Duchenne muscular dystrophy (DMD) gene. Am J Hum Genet 1989; 45:507–520.
19. Patel K, Leevers S, Abbs S, et al. Absence of dystrophin in Becker muscular dystrophy. Lancet 1989; 1:47.
20. Campbell KP, Kahl SD. Association of dystrophin and an integral membrane glycoprotein. Nature 1989; 338:259–262.
21. Miller G, Beggs AH, Towfighi J. Early onset autosomal dominant progressive muscular dystrophy presenting in childhood as Becker phenotype—the importance of dystrophin and molecular genetic analysis. Neuromusc Disord 1992; 2(2):121–124.
22. Kakulas BA. Experimental muscle diseases. In: Jasmin G, Cantin M, eds. Methods and Achievements in Experimental Pathology 1975; 7:109–131.
23. Kakulas BA, Adams RD. Principles of myopathology as illustrated in the nutritional myopathy of the Rottnest Island Quokka. Ann NY Acad Sci 1966; 138:90–101.
24. Kakulas BA, Adams RD. Diseases of Muscle: Pathological Foundations of Clinical Myology, 4th ed. Philadelphia: Harper & Row, 1985.
25. Kakulas BA. A Consideration of therapeutic interventions in the light of the muscle pathology and likely pathogenesis of the XP21.2 muscular dystrophies. In: Kakulas BA, Mastaglia FL, eds. Pathogenesis and Therapy of Duchenne and Becker Muscular Dystrophy. New York: Raven Press, 1990:47–57.
26. Hurse P, Kakulas BA, Gubbay SS, Knight JO. Carrier detection and genetic counselling in muscular dystrophy. In: Kakulas BA, ed. Clinical Studies of Myology. Amsterdam: Excerpta Medica, 1973; Part 2:638–643.
27. Wilton SD, Johnsen RD, Pedretti JR, Laing NG. Two distinct mutations in a single dystrophin gene:

identification of an altered splice site as the primary Becker muscular dystrophy mutation. Am J Med Genet 1993; 46:563–569.
28. Sicinski P, Geng Y, Ryder-Cook AS, Barnard EA, Darlison MG, Barnard PJ. The molecular basis of molecular dystrophin in the mdx mouse: a point mutation. Science 1989; 244:1578–1580.
29. Kornegay, JN. Golden retriever muscular dystrophy: the model and relevance to developmental therapeutics. In: Kakulas BA, McC.Howell J, Roses AD, eds. Duchenne Muscular Dystrophy: Animal Models and Genetic Manipulation. New York: Raven Press, 1992:105–108.
30. Sharp NJH, Kornegay JN, Van Camp SD, et al. Exon skipping during dystrophin mRNA processing in the canine homologue of Duchenne muscular dystrophy. In: Kakulas BA, McC Howell J, Roses AD, eds. Duchenne Muscular Dystrophy: Animal Models and Genetic Manipulation. New York: Raven Press, 1992:19–40.
31. Cooper BJ. The *xmd* dog: molecular and pheotypic characteristics. In: Kakulas BA, McC.Howell J, Roses AD, eds. Duchenne Muscular Dystrophy: Animal Models and Genetic Manipulation. New York: Raven Press, 1992:105–108.
32. Lunt PW, Cumming WJK. DNA probes in differential diagnosis of Becker muscular dystrophy and spinal muscular atrophy. Lancet 1989; 1:46–47.
33. Laing NG, Mears ME, Thomas HE, et al. Differentiation of Becker muscular dystrophy from limb girdle muscular dystrophy and Kugelberg-Welander disease using cDNA probe. Med J Aust 1990; 152:270–271.
34. McDonald TD, Medori R, Younger DS, et al. Becker muscular dystrophy or spinal muscular atrophy? Dystrophin studies resolve conflicting results of electromyography and muscle biopsy. Neuromusc Disord 1991; 1(3):195–200.
35. Arikawa E, Hoffman EP, Kaido M, Nonaka I, Sugita H, Arahata K. The frequency of patients with dystrophin abnormalities in a limb-girdle patient population. Neurology 1991; 41:1491–1496.
36. Sunohara N, Arahata K, Hoffman EP, et al. Quadriceps myopathy: forme fruste of Becker muscular dystrophy. Ann Neurol 1990; 28:634–639.
37. Arikawa E, Ishihara T, Nonaka I, Sugita H, Arahata K. Immunocytochemical analysis of dystrophin in congenital muscular dystrophy. J Neurol Sci 1991; 105:79–87.
38. Hoffman EP, Garcia CA, Chamberlain JS, Angelini C, Lupski JR, Fenwick R. Is the carboxyl-terminus of dystrophin required for membrane association? A novel, severe case of Duchenne muscular dystrophy. Ann Neurol 1991; 30(4):605–610.
39. Arahata K, Ishihara T, Kamakura K, et al. Mosaic expression of dystrophin in symptomatic carriers of Duchenne muscular dystrophy. N Engl J Med 1989; 320:138–142.
40. Hoffman EP, Fischbeck KH, Brown RH, et al. Characterization of dystrophin in muscle-biopsy specimens from patients with Duchenne's or Becker's muscular dystrophy. N Engl J Med 1988; 318;21:1363–1368.
41. Laing NG, Layton MG, Johnsen RD, Chandler DC, Mears ME, Goldblatt J, Kakulas BA. Two distinct mutations in a dystrophin gene: chance occurrence or pre-mutation? Am J Med Genet 1992; 42:688–692.
42. de Visser M, Bakker E, Defesche JC, Bolhuis PA, van Ommen GJ. An unusual variant of Becker muscular dystrophy. Ann Neurol 1990; 27:578–581.
43. Bradley WG, Jones MZ, Mussini JM, Fawcett PRW. Becker-type muscular dystrophy. Muscle Nerve 1978; 1:111–132.
44. McComas AJ, Sica REP, Campbell MJ. "Sick" motoneurones. A unifying concept of muscle disease. Lancet 1971; 1:3221–325.
45. Norman A, Cokley J, Thomas N, Harper P. Distinction of Becker from limb-girdle muscular dystrophy by means of dystrophin cDNA probes. Lancet 1989; 1:466–468.
46. Koenig M, Beggs AH, Moyer M, et al: The molecular basis for Duchenne versus Becker muscular dystrophy: correlation of severity with type of deletion. Am J Hum Genet 1989; 45:498–506.
47. Arikawa E, Arahata K, Sunohara N, Ishiura S, Nonaka I, Sugita H. Immunocytochemical analysis of dystrophin in muscular dystrophy. In: Kakulas BA, Howell J, Roses AD, eds. Duchenne Muscular Dystrophy: Animal Models and Genetic Manipulation. New York: Raven Press, 1992:81–88.
48. Gold R, Kress W, Meurers B, Meng G, Reichmann H, Müller CR. Brief Communication: Becker muscular dystrophy of unusual disease courses combined approach to dystrophin analysis. Muscle Nerve 1992; 15:214–218.
49. Beggs AH, Hoffman EP, Snyder JR, et al. Exploring the molecular basis for variability among patients with Becker muscular dystrophy: dystrophin gene and protein studies. Am J Hum Genet 1991; 49:54–67.
50. Gospe SM, Lazaro RP, Lava NS, Grootscholten DM, Scott MO, Fischbeck KH. Familial X-linked myalgia and cramps: a nonprogressive myopathy associated with a deletion in the dystrophin gene. Ann Neurol 1989; 26(3):466.
51. Rowland LP. Clinical concepts of Duchenne muscular dystrophy: the impact of molecular genetics. Brain 1988; 111:479–495.
52. Darras BT, Francke U. Myopathy in complex glycerol kinase deficiency patients is due to 3′ deletions of the dystrophin gene. Am J Hum Genet 1988; 43:126–130.

53. Rose MR, Howard RS, Genet SA, McMahon J, Whitfield A, Morgan-Hughes JA. A case of myopathy associated with dystrophin gene deletion and abnormal glycogen storage. Muscle Nerve 1993; 16:57–62.
54. Koh J, Bartlett RJ, Pericak-Vance MA, Speer MC, Yamaoka LH, et al. Inherited deletion at Duchenne dystrophy locus in normal male. Lancet 1987; 2:1154–1155.
55. Nordenskjold M, Nicholson L, Edstrom L, et al. A normal male with an inherited deletion of one exon within the DMD gene. Hum Genet 1990; 84:207–209.
56. England SB, Nicholson LVB, Johnson MA, et al. Very mild muscular dystrophy associated with the deletion of 46% of dystrophy. Nature 1990; 343:180–182.
57. Laing NG, Mears ME, Chandler DC, et al. Two years experience in a comprehensive carrier screening and prenatal diagnostic laboratory of the diagnosis of Duchenne and Becker muscular dystrophies. Med J Aust 1991; 154:14–18.
58. Prelle A, Moggio M, Comi GP, et al. Congenital myopathy associated with abnormal accumulation of desmin and dystrophin. Neuromusc Disord 1992; 2(3):169–175.

20

Autosomal Recessive Limb-Girdle Muscular Dystrophies

Clinical and Laboratory Findings

Mayana Zatz, Mariz Vainzof, Maria Rita Passos-Bueno, Jeanne Akiyama, and Suely K. N. Marie
University of São Paulo, São Paulo, Brazil

INTRODUCTION

The term "limb-girdle muscular dystrophy" (LGMD) was used by Walton and Natrass (1) to describe a condition in which the onset of muscular weakness occurs in the second or third decade, without facial involvement, and with slow progression leading to a stage of severe disability about 20 years after onset. The inheritance is usually autosomal recessive (AR) affecting both sexes equally, with an estimated incidence of about 1/100,000 (2,3), which is 10–15 times less frequent than Xp21 Duchenne muscular dystrophy (DMD). During the last few years, it has been shown that many of the cases that had been diagnosed as LGMD were in fact other forms of neuromuscular diseases, such as X-linked Becker muscular dystrophy (BMD), Kugelberg-Welander–type spinal muscular atrophy (SMA), the scapuloperoneal syndrome, and even DMD-manifesting carriers. A severe form of AR childhood muscular dystrophy was also identified (4–7), which is clinically very similar to Xp21 DMD. This condition, also reported as "Duchenne-like muscular dystrophy" (DLMD) or severe childhood autosomal recessive muscular dystrophy (SCARMD), affects both sexes equally, with onset in childhood and rapid progression leading to severe disability in the second decade. It is currently known that the occurrence of severe AR forms among patients diagnosed as X-linked DMD is not rare (8,9), in particular among Arab populations with a high degree of inbreeding (10).

The localization of the DMD/BMD gene and identification of its product, the protein dystrophin, brought great advances in the clinical differentiation of the AR and the Xp21 muscular dystrophies, in particular in families with no affected female patients. In the early nineties, it was shown that there is genetic heterogeneity for the mild autosomal forms of LGMD, with at least one dominant gene (LGMD1, located at 5q; 11) and more than one gene responsible for the AR forms (12). More recent investigations in Brazilian patients indicate that genetic heterogeneity exists for the severe "Duchenne-like" autosomal muscular dystrophies as well (13,14).

The differential diagnosis between the different types of LGMD is extremely important for genetic counseling and for delineating the basic defect responsible for the pathology of each type. Therefore, in this chapter and the next we will describe clinical, laboratory, and genetic findings, including DNA and dystrophin analysis in patients mostly belonging to large families in which AR inheritance has been confirmed, in an attempt to better characterize the different subgroups of AR LGMD.

CLINICAL AND LABORATORY FINDINGS

Data from 91 patients (38 males and 53 females), belonging to 28 families who have recently been reassessed, were presented to an International Workshop on the LGMD (organized by Dr. Kate Bushby from the University of Newcastle upon Tyne) (15,16) and comared with those from several countries in Europe, the United States, and North Africa. Following this workshop, 14 additional Brazilian families were analyzed. Data from the AR forms, including 22 large families previously reported (17), are summarized below.

The Mild Forms

The characteristic features are the following:

1. The age of onset varies (even among patients from the same family). LGMD usually occurs in the second or third decade, ranging from 5 to 45 years of age, but has been reported to occur even in the late fifties in the population from Japan (18).

2. The rate of progression is usually slow but also variable, with some period of apparent stabilization. The majority of our patients related the onset of clinical signs to difficulties with running, climbing stairs, and frequent falls. Involvement of upper limbs (difficulty raising the arms) is usually noticed later, but some patients reported onset of weakness in lower and upper limbs simultaneously. In some rare patients the myopathy is limited or affects predominantly the quadriceps, and reaches the upper limbs only many years later (19). The loss of ambulation is also variable, from 20 to 50 years of age in most of our patients, on average 20–30 years after onset. Some patients show mild clinical features and are never confined to wheelchair, in contrast to others such as those from the Amish community in whom loss of ambulation may occur in the early teens (20). Longevity does not seem to be much reduced in the Brazilian patients with the mild form of LGMD. Only one of our patients with the mild form died suddenly at age 28, but his older sister, aged 45, is only mildly affected, illustrating the intrafamilial variability the may be found in the severity of the phenotype.

3. On examination, the muscle involvement is variable, as suggested by the clinical history; in some patients the pelvic musculature is involved first, while in others scapular and pelvic muscles are simultaneously involved. In patients from two of our families an early distal involvement was observed. The facial muscles and the small muscles of the hands are not affected or are only minimally affected. Early contractures are rare but may be apparent in later stages of the disease.

4. There is a high inter- and intrafamilial variability for the presence or absence of calf hypertrophy (Fig. 1). It is rare in the patients from Réunion Island, but frequent among the Brazilians (where it is observed in about two-thirds of the cases) and in the Amish population, although again not consistent within families. In some of our patients, we observed asymmetry for the presence or absence of calf hypertrophy (Fig. 2).

5. Tendon reflexes are usually depressed and may disappear with the progression of the disease with the exception of the brachioradialis and Achilles, which are usually retained longer or even until the late stages of the disease.

6. Serum creatine kinase (CK) activity may be only moderately elevated (two- to 10-fold), as reported in the patients from Réunion Island (21). In the majority of our patients, serum CK activities are increased 10- to 30-fold above the upper normal limit and decrease with the progression of the disease (in particular after loss of ambulation), but were found to be extremely high (up to 150-fold) in some young, still asymptomatic Brazilian patients.

7. Electromyography is characteristic of a myopathic process, with the exception of one of our patients in whom it was found to be myopathic at age 9 and typical of a neurogenic process when repeated at age 22.

8. Muscle histopathological analysis may show a typical myopathic pattern (Fig. 3a) or only discrete alterations such as focal degeneration in preclinical cases (Fig. 3b). The more frequent abnormalities are: variation of fiber size, proliferation of endomysial and perimysial connective tissue, degeneration and regeneration, and increase in internal nuclei. In some of our patients the mus-

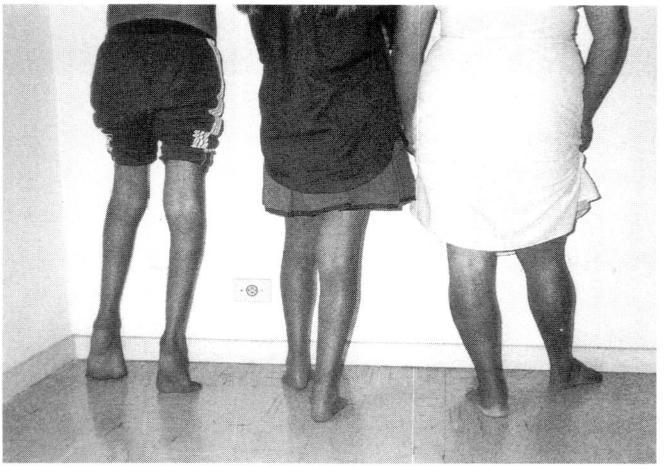

Figure 1 Intrafamilial variability for the presence or absence of calf hypertrophy in three affected sibs: (left to right), 18-year old male, 13-year old female, and 28-year old female.

cle biopsy showed a mixed pattern with regions suggestive of a neurogenic process (Fig. 3c). Interestingly, as reported in other studies (4) we did not observe a direct correlation between the degree of muscle degeneration and the severity of the phenotype. Some patients who were only mildly affected had pathological changes as severe as others who were in an advanced stage of the disease (Fig. 4).

9. The intellectual capacity is usually normal, although one of our female patients (aged 44) had a schizotypical personality (with hallucinations and paranoia) while her older LGMD affected sister had a normal personality.

Dystrophin Analysis

Dystrophin was assessed in 60 Brazilian patients with mild features (25 males and 35 females), through immunohistochemistry (IF) and Western blot (WB) analysis, using at least two antibodies, against the

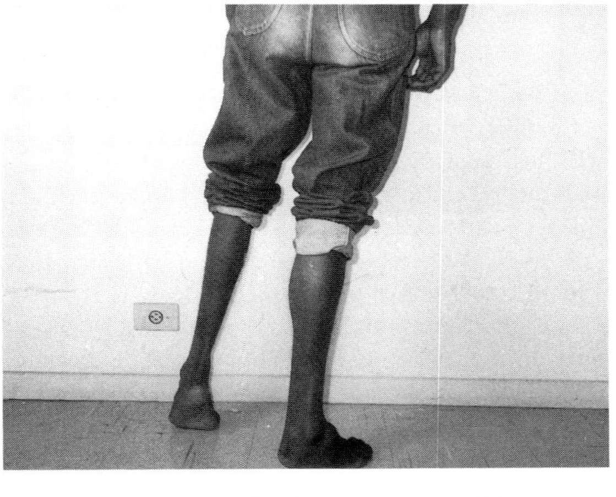

Figure 2 Assymetry for the presence of calf hypertrophy in a 23-year-old affected male. Both his affected older sister (Fig. 4, left) and younger brother (not shown) have pronounced calf hypertrophy bilaterally.

(a)

(b)

(c)

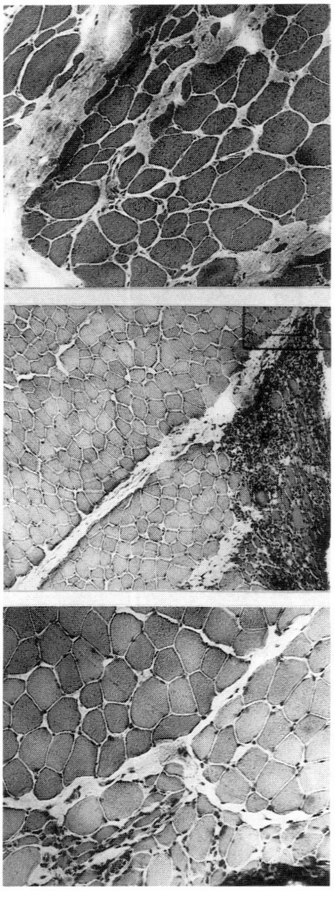

Figure 3 Muscle histopathology from patients in families with multiple affected cases. (a) Characteristic myopathic pattern showing variability in fiber diameter, perimysial and endomysial connective tissue proliferation, splitting, and internal nuclei. (b) Discrete alterations with focal areas of degeneration, necrosis, and inflammatory infiltration. (c) Mixed pattern showing small groups of atrophic fibers suggestive of a neurogenic process.

N- and the C-terminal regions. Most cases showed a normal IF pattern although occasionally some fibers were not totally positive. In the majority of patients, WB analysis showed a protein of normal molecular weight (MW) and quantity. Recently, Hoffman et al. (22) assessed dystrophin in 505 females with neuromuscular diseases and found that about 10% of the isolated cases have a dystrophinopathy as the cause of their disease. In another dystrophin study of 41 patients (24 females and 17 males) who had been diagnosed as LGMD in Japan (23), the authors identified seven patients who had been misclassified: five male patients had a dystrophin of reduced MW, diagnostic of BMD (which was confirmed through DNA analysis), and two females were manifesting carriers of DMD. All others had a dystrophin of normal MW and abundance. However, we have observed that in muscles with a high degree of degeneration, the quantitative analysis may show apparently reduced dystrophin even after correction with reference to the myosin abundance, causing difficulties in differential diagnosis between LGMD and Xp21 Becker dystrophy in isolated male patients with no detectable DNA deletion in the dystrophin gene. Identification of the defective primary product responsible for each form of LGMD will be fundamental to resolve this problem.

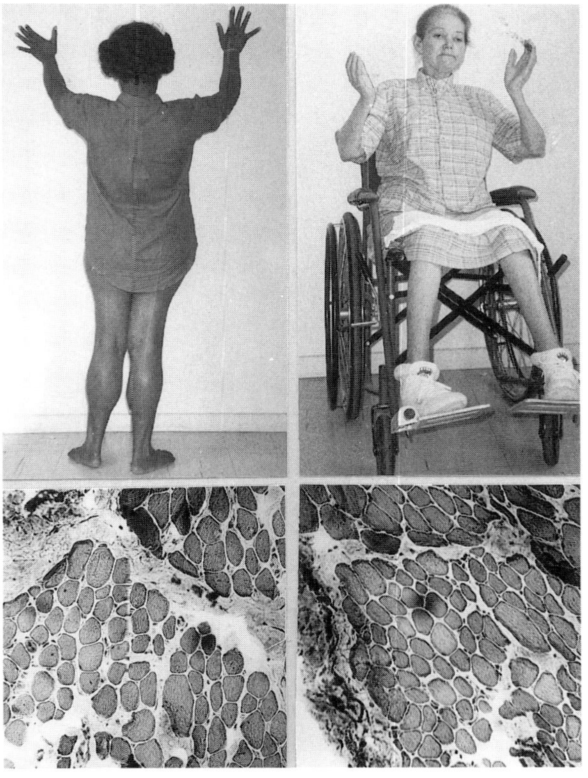

Figure 4 Muscle biopsies with similar pathological changes, although the patient on the right is much more severely affected.

The Severe Forms

The severe forms of LGMD, also classified as DLMD or SCARMD, are apparently not rare in some populations from Switzerland (24) and may be even more prevalent than X-linked DMD in highly inbred Arab populations from North Africa (such as Kuwait, Lybia, Saudi Arabia, Sudan, Tunisia, and Qatar) (25). In Great Britain and North America, the AR forms seem to be 20 times less common than the X-linked forms (24), which is not very different from our previously estimated proportion of 6.8% of families with DMD inherited as an AR trait in the Brazilian population (8,9).

The clinical picture is severe, and in male patients it may be indistinguishable from X-linked DMD. The age of onset in the North African form ranges from 3 to 12 years of age (in 75% of the cases until age 9), and loss of ambulation may occur from 10 to 31 years of age, but in the majority of cases it occurs before age 20 (10). The first symptoms are difficulties in running, climbing stairs, and rising from the floor. Calf hypertrophy is common, and serum CK may be very high in the early stages (up to 100 times normal) but declines rapidly with progression of the disease.

In Brazilian patients, the onset occurs between 3 and 8 years of age with confinement to wheelchair from age 9 to 16 years, which is apparently more severe than the North African form. Early toe walking is not a prominent feature as reported previously by Gardner-Medwin and Johnston in patients from Great Britain (5). Serum CK is extremely elevated in the early stages of the disease, in the same range observed for DMD (up to 150 times normal) (26), but also decreases sharply with progression of the disease.

The proximal pelvic girdle musculature is most severely affected, but in the majority of our patients

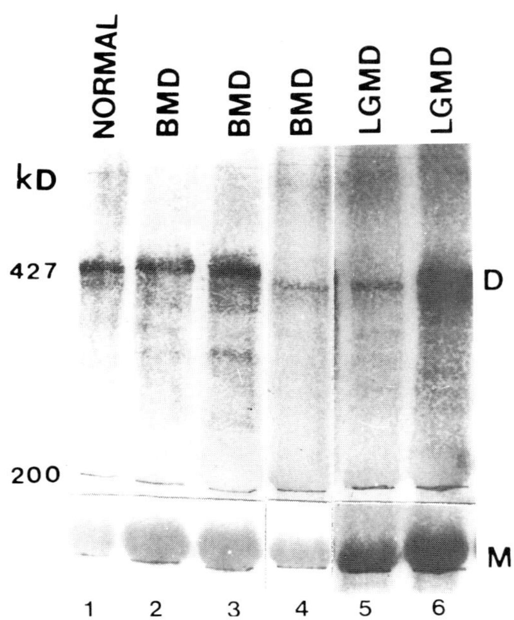

Figure 5 Dystrophin Western blot pattern in three patients with X-linked Becker muscular dystrophy (who have no detectable DNA deletions) and 2 LGMD patients, illustrating the difficulties for differential diagnosis through dystrophin quantitative analysis, in particular comparing patients from lanes 4 and 5.

we have observed that involvement of the scapular girdle musculature occurs earlier than in Xp21 DMD. In most cases, the muscles of the face and distal muscles of the upper limbs are not affected even in the late stages of the disease.

Tendon reflexes are usually depressed and disappear rapidly with progression of the disease. The electromyogram is myopathic and muscle biopsy shows a marked variation in fiber size, an increased proportion of internal nuclei, some necrotic fibers, and increased fat and connective tissue.

The intellectual capacity is usually normal, in contrast to DMD, where 30–50% of the patients are mentally impaired (4,27,28).

Dystrophin and Associated Glycoproteins

Dystrophin assessment in 10 Brazilian patients (including two sisters) with DLMD phenotype showed positive labeling with IF (Fig. 6), but frequently with a "fuzzy" pattern, and a band of normal MW with WB, but which can again be quantitatively diminished. Interestingly, two boys who had been previously diagnosed as X-linked DMD were reclassified as DLMD based on dystrophin analysis. An intriguing problem is how to explain the genetic heterogeneity for the DMD phenotype, that is, how different genes (X-linked and AR) result in clinical pictures that are so similar.

A model to explain such findings has been proposed by Campbell and his collaborators (29,30). According to this model, dystrophin, the protein product of the DMD gene, is a cytoskeletal protein tightly associated with a large oligomeric complex of sarcolemmal glycoproteins that provide a linkage to the extracellular matrix. These authors have recently demonstrated (31) the deficiency of one of the associated glycoproteins, the 50-kDA or dystrophin-associated glycoprotein (DAG), in two siblings from North Africa. Dysfunction of the dystrophin-glycoprotein complex due to deficiency of the 50DAG is presumed to lead to muscle cell necrosis in DLMD.

Recently, the 50DAG-associated protein was analyzed using IF by Campbell's group of investigators, in 10 patients from nine unrelated Brazilian families, all with a DLMD phenotype. Interestingly,

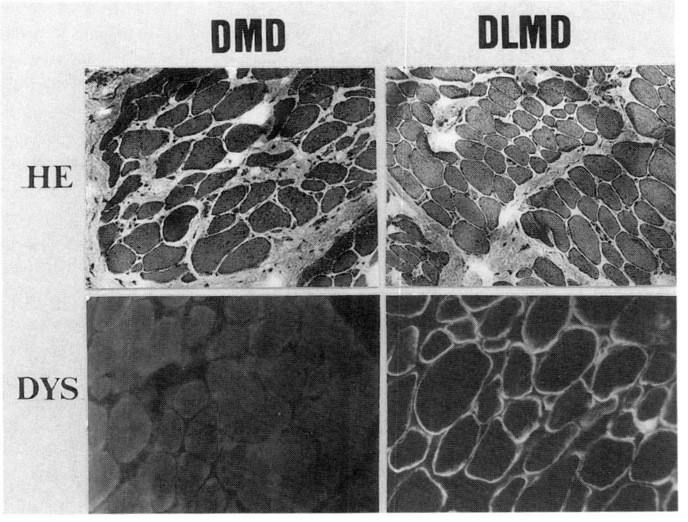

Figure 6 Muscle from an Xp21 DMD patient (left) and from a patient with AR DLMD (right) comparing histopathological findings (HE) and dystrophin (DYS).

six patients (including the two boys who had previously been diagnosed as Xp21) showed deficiency for 50DAG, but in four, from three unrelated families, the protein was apparently normal (Fig. 7).

It is now known that the deficiency of the 50DAG may represent the primary or a secondary effect of a defective gene, and these data support the hypothesis of genetic heterogeneity, as discussed further in the next chapter. In patients with a similar phenotype, the deficiency of the 50DAG glycoprotein in some cases and of an apparently normal 50DAG in others could be due to allelic heterogeneity (leading to the absence of or to the production of an abnormal protein as occurs in the Xp21 DMD and BMD). It could be due to nonallelic (or locus) heterogeneity; that is, the production of a normal 50DAG or its deficiency (which could be due to the primary or secondary effect of a defective gene) would be controlled by different genes. A third possibility would be the existence of both allelic and locus heterogeneity, as occurs for example in the mucopolysaccharidoses (32).

Unusual Families

In three additional unrelated Brazilian families, 16 patients (six males and 10 females) attracted our attention because of the presence of early contractures of elbows, knees, and ankles, clinical features similar to those of X-linked Emery-Dreifuss muscular dystrophy. The rate of progression in these patients was slow, serum CK was moderately increased (two- to 14-fold), and muscle biopsy showed a myopathic pattern.

In two other large families, two forms of muscular dystrophy were reported to segregate in the same pedigree. In a large inbred Palestinian family (33), there were 10 cases of congenital muscular dystrophy (CMD) and nine cases of adult LGMD, both displaying AR inheritance. The patients with CMD had generalized muscular weakness and hypotonia at birth without arthrogryposis or central nervous system involvement and then a relatively benign evolution with stabilization of the clinical picture and variable degree of severity. The other nine patients had a clinical picture of adult LGMD but with an unusual tendency to stabilization of symptoms. In another large consanguineous kindred from the Larsmo Islands, some patients had a severe limb-girdle–type muscular dystrophy while others had a mild late-onset distal myopathy (34). According to the author, pedigree analysis was compatible with the possibility that the mild-onset distal myopathy is caused by a dominant gene and that the more severe LGMD may be expressed in homozygotes.

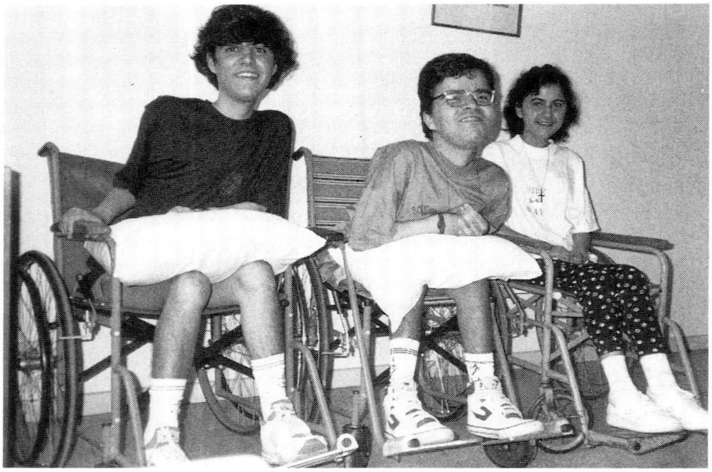

(a)

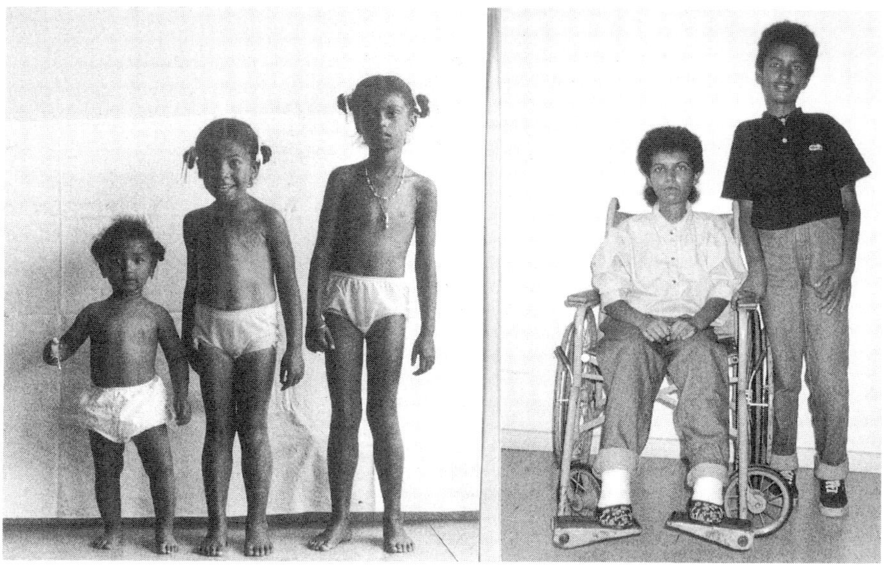

(b)

Figure 7 (a) Deficiency of the 50DAG glycoprotein in three affected sibs with DLMD, aged (from left to right) 21, 19, and 17 years. They were confined to wheelchair at age 10, 11, and 13, respectively. (b) (Left) Normal 50DAG glycoprotein in three affected sisters with DLMD, aged 15 months, 5 and 7 years, and (right) aged 18 and 12 years. The two oldest were confined to wheelchair at age 12 and the second died of heart failure at age 14.

COMPARISON BETWEEN THE DIFFERENT SUBTYPES

As discussed in the next chapter, results of linkage analysis have already shown that there is genetic heterogeneity for both the mild and severe forms of AR LGMD. One gene, located at 15q, has been identified as responsible for one of the mild forms, the LGMD1 type (21,35), but it is already known that the majority of the cases are caused by another gene (or other genes). Since 15q-linked and nonlinked families have been found in the Brazilian population, we compared patients with both forms in an attempt to find any clinical or laboratory test that would discriminate between them. Such comparison was not possible for the severe forms of DLMD because none of the Brazilian families tested so far showed linkage to the 13q probes, as recently reported for Tunisian patients (36).

Unfortunately, the comparison between 10 patients from the two families linked to 15q probes and 36 from 10 families that were excluded or unlikely to be controlled by the 15q gene showed no discriminating clinical or histopathological features. In all patients there was variability in the age of onset and rate of progression, presence or absence of calf hypertrophy, and involvement of muscle groups. Serum CK activities in both groups were highly variable. Although they were apparently more elevated, on average, among nonlinked ($x^2 = 300.0 \pm 396.1$) than among linked forms ($x^2 = 183.1 \pm 180.2$, $t = 1.47$; $p = 0.07$, these differences were not statistically significant. A notable finding was that in two female patients aged 5 and 7, respectively, from non-15q families, serum CK activities were increased 170-fold above the normal upper limit, in the same range found for young Xp21 DMD patients.

FUTURE CLASSIFICATION

Based on results from linkage analysis, it is possible to anticipate that a new classification will be required for the AR forms, but as illustrated below, it is not known yet how many genes are involved in the pathogenesis of LGMD.

1. The mild forms
    - The clinically similar forms
        - 15q LGMD2
        - Non-15q LGMD (one or more genes?)
    - The unusual forms (allelic heterogeneity? nonallelic heterogeneity? or both allelic and nonallelic heterogeneity?)
        - AR form with early contractures?
        - Myopathy limited to the quadriceps?
        - The Palestinian form?
        - The Finland form?
2. The severe forms:
    - The 13q form
    - The non-13q form (one or more genes?)

ACKNOWLEDGMENTS

The authors are extremely grateful for the collaboration of Dr. Rita C. M. Pavanello, Dr. Ivo Pavanello, Dr. Ana Maria Tsanaclis, Marta Canovas, Sabine Eggers, Constancia Urbani, and all the staff from ABDIM; to all the patients who collaborated for this study; and to FAPESP, PADCT, ABDIM, and CNPq for financial support.

NOTE ADDED IN PROOF

After the submission of the present chapter, two new LGMD genes have been identified as summarized in the most recent LGMD workshop report (37). One gene, located on chromosome 2p, is responsible for a mild progression (38,39). On the other hand, the gene that codes for the 50DAG-associated protein (adhalin), localized at 17q (40), may lead to a severe or a mild phenotype (41,42). In addition, there is evidence for at least two other ARLGMD genes, one causing a mild and the other a severe DLMD phenotype. Interestingly, three patients from a Brazilian family that showed positive lod scores at the 13q locus had a mild phenotype (43), in contrast to the Tunisian DLMD families who have a severe Duchenne-like course (44).

The most recent proposal for a new nomenclature and the respective phenotypes for ARLGMD (37) is the following:

LGMD2A (15q): mild phenotype
LGMD2B (2p): mild phenotype
LGMD2C (13q): (previously SCARMD)—usually severe phenotype
LGMD2D (17q): phenotype ranging from severe to mild

REFERENCES

1. Walton JN, Natrass FJ. On the classification, natural history and treatment of myopathies. Brain 1954; 2: 189.
2. Morton NE, Chung CS. Formal genetics of muscular dystrophy. Am J Hum Genet 1959; 11:360–379.
3. Yates JRW, Emery AEH. A population study of adult onset limb-girdle muscular dystrophy. J Med Genet 1985; 22:250–257.
4. Dubowitz V. Muscle Disorders in Childhood. London: WB Saunders, 1978:282.
5. Gardner-Medwin D, Johnston HM. Severe muscular dystrophy in girls. J Neurol Sci 1984; 64:79–87.
6. Somer H, Voutilainen A, Knuutila S, Kaitila I, Rapole J, Leinonen H. Duchenne-like muscular dystrophy in two sisters with normal karyotypes: evidence for autosomal recessive inheritance. Clin Genet 1985; 28:151–156.
7. Yoshioka M, Itagaki Y, Saida K, Nishitani Y. Clinical and genetic studies of muscular dystrophy in young girls. Clin Genet 1986; 29:137–142.
8. Zatz M, Passos-Bueno MR, Rapaport D. Estimate of the proportion of Duchenne muscular dystrophy with autosomal inheritance. Am J Med Genet 1989; 32:407–410.
9. Vainzof M, Pavanello RCM, Pavanello I, et al. Screening of male patients with autosomal recessive Duchenne dystrophy through dystrophin and DNA studies. Am J Med Genet 1991; 39:38–41.
10. Ben Hamida M, Fardeau M, Attia N. Severe childhood muscular dystrophy affecting both sexes and frequent in Tunisia. Muscle Nerve 1983; 6:469–480.
11. Speer MC, Yamaoka LH, Gilchrist JH, et al. Confirmation of genetic heterogeneity in limb-girdle muscular dystrophy: linkage of an autosomal dominant form to chromosome 5q. Am J Hum Genet 1992; 50:1211–1217.
12. Passos-Bueno MR, Richard I, Vainzof M, et al. Evidence of genetic heterogeneity for the autosomal recessive adult forms of limb-girdle muscular dystrophy following linkage analysis with 15q probes in Brazilian families. J Med Genet 1993; 30:385–387.
13. Passos-Bueno MR, Oliveira JRM, Bakker E, et al. Genetic heterogeneity for Duchenne-like muscular dystrophy. Hum Mol Genet 1993; 2(11):1945–1947.
14. Zatz M, Matsumura K, Vainzof M, Passos-Bueno MR, Pavanello RCM, Marie KS, Campbell KP. Deficiency of the 50Kda dystrophin-associated glycoprotein in Brazilian patients with severe childhood autosomal recessive muscular dystrophy (SCARMD). J Neurol Sci 1994; 123:122–128.
15. Bushby K. Report on the 12th ENMC sponsored International Workshop—the "limb-girdle" muscular dystrophies. Neuromusc Disord 1992; 2:3–5.
16. Clarke A. Report of ENMC workshop on the limb-girdle muscular dystrophies. J Med Genet 1992; 29:753–755.
17. Passos-Bueno MR, Vainzof M, Pavanello RCM, Pavanello-Filho I, Lima MABO, Zatz M. Limb-girdle syndrome: a genetic study of 22 large Brazilian families: comparison with X-linked Duchenne and Becker dystrophies. J Neurol Sci 1991; 103:65–75.
18. Arahata K, Beggs AB, Honda H, et al. Preservation of the C-terminal of dystrophin molecule in the skeletal muscle from Becker muscular dystrophy. J Neurol Sci 1991; 101:143–156.
19. Walton JN, Gardner-Medwin D: The muscular dystrophies. In: Walton JN, ed. Disorders of Voluntary Muscle, 5th ed. Edinburgh: Churchill Livingstone, 1988:519–568.
20. Jackson CE, Strehler DA. Limb-girdle muscular dystrophy: clinical manifestations and detection of preclinical disease. Pediatrics 1968; 41:494–502.

21. Beckmann J.S, Richard I, Hillaire D, et al. A gene for limb-girdle muscular dystrophy maps to chromosome 15 by linkage. CR Acad Sci Paris Ser III 1991; 312:141–148.
22. Hoffman EP, Arahata K, Minetti C, et al. Dystrophinopathy in isolated cases of myopathy in females. Neurology 1992; 42:967–975.
23. Arikawa E, Hoffman EP, Kaido M, Nonaka I, Sugita H, Arahata K. The frequency of patients with dystrophin abnormalities in a limb-girdle patient population. Neurology 1991; 41:1491–1496.
24. Emery AEH. Duchenne Muscular Dystrophy. New York: Oxford University Press, 1987:315.
25. Farag TI, Teebi AS. Duchenne-like muscular dystrophy in the Arabs. Am J Med Genet 1990; 37:289–290.
26. Zatz M, Rapaport D, Vainzof M, et al. Serum creatine-kinase (CK) and pyruvate-kinase (PK) activity as a function of clinical evolution in Duchenne (DMD) and Becker (BMD) muscular dystrophies. J Neurol Sci 1991; 102:190–196.
27. Bortolini ER, Zatz M. Investigation on genetic heterogeneity in Duchenne muscular dystrophy. Am J Med Genet 1986; 24:111–117.
28. Rapaport D, Passos-Bueno MR, Brandão L, Love D, Vainzof M, Zatz M. Apparent association of mental retardation and specific patterns of deletions screened with probes cf56a e cf23a in Duchenne muscular dystrophy. Am J Med Genet 1991; 39:437–441.
29. Ervasti JM, Campbell KP. Membrane organization of the dystrophin-glycoprotein complex. Cell 1991; 66:1–20.
30. Ibraghimov-Beskrovnaya O, Ervasti JM, Leveille CJ, Slaughter CA, Sernett SW, Campbell KP. Primary structure of dystrophin-associated glycoproteins linking dystrophin to the extracellular matrix. Nature 1992; 355:696–702.
31. Matsumura K, Tome FMS, Collin H, et al. Deficiency of the 50K dystrophin-associated glycoprotein in severe childhood autossomal recessive muscular dystrophy. Nature (Lond) 1992; 359:320–322.
32. Thompson MW, McInnes RR, Willard HF. Genetics in Medicine, 5th ed. Philadelphia: WB Saunders, 1991.
33. Mahjhen I, Vannelli G, Bushby K, Marconi G.P. A large inbred palestinian family with two forms of muscular dystrophy (to be published).
34. Udd B. Limb-girdle type muscular dystrophy in a large family with distal myopathy: homozygous manifestation of a dominant gene? J Med Genet 1992; 29:383–389.
35. Young K, Williams P, Foroud CE, et al. Confirmation of linkage of limb-girdle muscular dystrophy to chromosome 15. HGM11 1991, Abstract 26966.
36. Othmane KB, Ben Hamida M, Pericak-Vance MA, et al. Linkage of Tunisian autosomal recessive Duchenne-like muscular dystrophy to the pericentromeric region of chromosome 13q. Nature Genet 1992; 315–317.
37. Bushby KMD, Beckman JS. The limb-girdle muscular dystrophies—proposal for a new nomenclature. Neuromusc Disord 1995; 5(4):337–343.
38. Bashir R, Strachan T, Keers S, Stephenson A, Mahjneh I, Marconi G, Nashef L, Bushby KMD. A gene for autosomal recessive limb-girdle muscular dystrophy maps to chromosome 2p. Hum Mol Genet 1994; 3:455–457.
39. Passos-Bueno MR, Bashir R, Moreira ES, Vainzof M, Marie SK, Vasquez L, Iughetti P, Bushby K, Bakker E, Zatz M. Confirmation of the 2p locus for mild autosomal recessive limb-girdle muscular dystrophy in three families allows refinement of the candidate region. Genomics 1995; 27:192–195.
40. Roberds SL, Leturcq F, Allamand V, Piccolo F, Jeanpierre M, Anderson RD, Lim LE, Lee JC, Tomé FMS et al. Missense mutations in the adhalin gene linked to autosomal recessive muscular dystrophy. Cell 1994; 78:625–633.
41. Romero NB, Tomé FMS, Leturcq F, El Kerch F, Azibi K, Bachner L, Anderson RD, Roberds SL, Campbell KP, Fardeau M, Kaplan JC. Genetic heterogeneity of severe childhood autosomal recessive muscular dystrophy with adhalin (50-kDa dystrophin-associated glycoprotein) deficiency. CR Acad Sci Paris 1994; 317:70–76.
42. Passos-Bueno MR, Moreira ES, Vainzof M, Chamberlain J, Marie SK, Pereira L, Akiyama J, Roberds SL, Campbell KP, Zatz M. A common missense mutation in the adhalin gene in three unrelated Brazilian families with a relatively mild form of autosomal recessive limb-girdle muscular dystrophy. Hum Mol Genet 1995; 4(7):1163–1167.
43. Passos-Bueno MR, Moreira ES, Marie SK, Bashir R, Vasquez L, Love DR, Vainzof M, Iughetti P, Oliveira JR, Bakker E, Strachan T, Bushby K, Zatz M. Main clinical features for the three mapped autosomal recessive limb-girdle muscular dystrophy and estimated proportion of each form in 13 Brazilian families (submitted).
44. Ben Othmane KB, Speer MC, Stauffer J, Beel S, Middleton L, Ben Hamida C, et al. Evidence for linkage desiquilibrium in chromosome 13-linked Duchenne-like muscular dystrophy (LGMD2C). Am J Hum Genet 1995; 57:732–734.

21
Genetic and Linkage Analysis in Autosomal Recessive Limb-Girdle Muscular Dystrophy

Maria Rita Passos-Bueno, Mariz Vainzof, and Mayana Zatz
University of São Paulo, São Paulo, Brazil

INTRODUCTION

Limb-girdle muscular dystrophy (LGMD) is a progressive muscular dystrophy with weakness beginning in the limb or shoulder girdle, progressing later to the distal muscles, usually with no facial muscle involvement. The clinical course is variable from very severe forms with onset in childhood and very rapid progression (1–4) to mild types with onset in the second or third decade and a much slower course (1,5). In the majority of cases the inheritance is autosomal recessive (AR). Sporadic cases and families with autosomal dominant inheritance are also observed (6–8).

Different patterns in the mode of genetic transmission as well as variability in the age of onset and in the clinical course suggest genetic heterogeneity for the LGMD phenotype. The only way to assess the problem of genetic heterogeneity is through mapping and cloning of the genes that cause LGMD and identification of their products. This strategy will be extremely important for the understanding of the process responsible for muscle degeneration as well as for diagnostic purposes.

BRAZILIAN LGMD FAMILIES

Description of the Families

Since 1969, 123 unrelated families have been referred to our center with a diagnosis compatible with LGMD, while 826 were considered likely to have Xp21 muscular dystrophies: 715 with Duchenne muscular dystrophy (DMD) and 111 with Becker muscular dystrophy (BMD). Therefore, in the population seen in our center, LGMD seems to be approximately as frequent as BMD but 5–6 times less frequent than DMD. However, it is possible that among the earlier cases referred with a diagnosis of LGMD who were not reassessed, there were some BMD as well as spinal muscular atrophy (SMA) families.

The differential diagnosis has been based on family history, clinical examination, electromyography and/or muscle biopsy with a myopathic pattern, and elevated serum creatine kinase (CK) and/or pyruvate kinase (PK). For the last 3 years, dystrophin analysis (by imunocytochemistry and Western blot) as well as DNA studies have been performed in all families in which the diagnosis of Xp21 muscular dystrophy could not be excluded on clinical or genetic grounds alone.

A study including clinical and genetic data from 22 large Brazilian families with patients affected with LGMD that had been reevaluated has been published (9). Following this report, other families were recognized or reinvestigated, and we were able to confirm the diagnosis of LGMD in 126 patients from 42 unrelated families. Twenty patients (eight males and 12 females) from eight unrelated families showed severe progression, clinically indistinguishable from the severe Xp21 DMD, and were classified as having "Duchenne-like" muscular dystrophy (DLMD) or severe childhood autosomal recessive muscular dystrophy (SCARMD). Interestingly, among these, two isolated affected boys had been referred to our center with a diagnosis of Xp21 DMD and were shown to be affected by DLMD based on dystrophin analysis (4,10). The remaining 106 patients had a milder clinical course.

Segregational Analysis

The sex ratio among affected patients was 59 males and 67 females when all 42 families were taken together, with 49 males and 54 females among the 34 families with mild LGMD, 10 males and 13 females having the severe form. Although there is a bias of ascertainment toward families with affected females, these proportions did not differ from the 1:1 ratio expected for AR inheritance (DLMD: $\chi^2 = 0$; $p = 1.0$; mild LGMD: $\chi^2 = 0.02$; $p = 0.89$).

In our previous studies, we had observed a high consanguinity rate among the LGMD families; 68% (13/19) for the mild LGMD (9) and 53% (8/15) for the DLMD (3). This high consanguinity rate, which has been confirmed in the present sample (17 in 34 for the mild and four in eight for DLMD), is in accordance with a rare AR gene [q = 0.004 (6)]. However, among nonconsanguineous families there was a high degree of racial admixture, including patients from two unrelated families with the less severe mild LGMD, who were descended from a Mongoloid father and a Negroid mother in one and a Mongoloid father and a Caucasian mother in the other. These findings indicate that at least one common recessive gene responsible for mild LGMD is segregating in different ethnic groups.

Segregational analysis, considering multiple incomplete ascertainment (11), showed that the proportion of affected individuals among the 42 families was 28.4% (89/313). This proportion is not significantly higher than the theoretically expected 0.25 for AR inheritance ($Z = 1.22$; $p = .011$).

STRATEGIES FOR IDENTIFICATION OF GENES CAUSING THE PHENOTYPE OF LGMD

In the last 10 years, numerous Mendelian human diseases, such as Huntington's disease (12) and cystic fibrosis (13) among others, have been genetically mapped through the use of restriction fragment length polymorphisms (RFLPs), and more recently using microsatellite DNA markers (14). However, this strategy is time-consuming, labor-intensive, and expensive. Therefore, other approaches, such as the use of candidate genes, that could speed up the process of mapping disease genes and decrease the costs are of utmost interest. Hyperkalemic periodic paralysis (HYPP), an autosomal dominant disorder characterized by episodes of muscle weakness (due to depolarization of the muscle cell membranes associated with elevated serum potassium), is an example. Since electrophysiological studies have implicated the adult muscle sodium channel in this disorder, the locus for the alpha subunit of the adult muscle sodium channel was tested as a candidate gene for HYPP; tight linkage was observed in a large family with this disorder allowing localization of the HYPP gene at 17q (15).

Another approach, which has contributed to the localization of several disease genes, is based on identification of chromosomal rearragements, such as the DMD girls with X/autosomal translocations that allowed the mapping of the DMD gene to Xp21 (14). However, this strategy cannot be used routinely since it depends on luck to find such rare cases.

Therefore, as described below, the search for the LGMD genes has been done mainly through two approaches: candidate genes and screening with random markers through the whole genome. In addition, a Brazilian LGMD family with an atypical inheritance has been tested for unidissomy in an attempt to map this condition.

Selection of Families: The Problem of Genetic Heterogeneity

A serious obstacle to linkage studies in human diseases is genetic heterogeneity, which may be allelic, nonallelic (locus), or both. For example, the acute and chronic forms of SMA are caused by mutations at the same gene at 5q13 (16); that is, there is allelic heterogeneity. On the other hand, tight linkage has been found in some SMA pedigrees but it is entirely absent in others due to locus heterogeneity (17).

Nonallelic heterogeneity represents a great problem for linkage analysis because evidence of linkage to a locus in one family (leading to positive lod scores) will be offset by evidence against linkage in another family (leading to negative lod scores).

Therefore, to overcome the problem of genetic heterogeneity, several approaches should be used. The use of a single pedigree large enough to suffice for linkage analysis, whenever possible, represents an ideal situation. Gusella et al. (12), for example, analyzed a single extended Venezuelan pedigree segregating for Huntington's disease, which contained hundreds of individuals. Linkage analysis for chronic SMA was initiated based on a blind search through the genome in three large, highly informative families (16). With this approach, linkage was detected after testing for 115 DNA markers, which has been confirmed in other families. Simultaneously, this same gene was mapped after testing of 132 polymorphic DNA markers in 24 multiplex, but small (only two individuals affected per genealogy), SMA families of European, Middle Eastern, and African ancestry (18).

A related approach for avoiding heterogeneity is to restrict attention to a single population that has been relatively isolated from a genetic standpoint.

However, for many genetic diseases it will be difficult to collect enough families with multiple affected individuals or families in such circumstances.

Another potentially powerful approach, homozygosity mapping, has recently been described for mapping rare recessive diseases, by using DNA of inbred affected children. In such cases, one way to find a disease locus would be by searching for a region that is consistently homozygous by descent in inbred affected individuals (19).

The LGMD phenotype, a relatively rare disease, may be caused by locus and/or allelic heterogeneity.

Linkage analysis in Brazilian LGMD families was started in 1989 by testing candidate genes. For this approach, we selected families with a very similar clinical course that were classified in two groups according to the severity of the disease: DLMD and mild LGMD.

More recently, we have started a blind search through the genome using microsatellites. Four large consanguineous families have been selected for this study.

Linkage Analysis

Unusual Pedigrees

Morton and Chung (6) estimated that the frequency for the AR LGMD gene is about 0.004. In fact, the high consanguinity rate observed among the Brazilian families (see above) is in accordance with a rare autosomal recessive gene.

Among the 123 LGMD families studied, there is one (family 21 in Ref. 9) in which there are two affected female patients born from two independent nonconsanguineous marriages. Chromosomal analysis revealed a normal karyotype in both, excluding an X-autosome translocation or any other chromosomal abnormality. Considering the gene frequency as $q = 0.004$, we have estimated that the probability for such an occurrence is approximately 0.0013 (2pq or $8/1000 \times 2/3 \times 1/4$), which might be explained by chance taking into account the size of our population.

Alternatively, the clinical signs in the LGMD patients belonging to the above genealogy could be caused by an autosomal dominant AD gene not fully penetrant, since asymptomatic obligate heterozygotes have been reported for the AD forms (20). However, testing this hypothesis will be possible only after identification of the mutation or the product of the gene for AD LGMD.

A third possibility, which has been demonstrated for some diseases (14), is the occurrence of affected cases with AR inheritance as a result of uniparental unidissomy, in which only one of the parents was found to be heterozygous for a defective AR gene. If true in the present family, the demonstration of parental unidissomy would allow the assignment of the defective gene to one particular chromosome. To test this hypothesis, a total of 27 pair of primers (one informative marker per chromosome) were tested in this family, but no evidence of unidissomy was observed (E. Bakker, personal communication).

However, unidissomy can be more frequent than previously expected. Therefore, screening for uniparental dissomy among isolated cases belonging to large sibships could be an interesting approach in an attempt to contribute to the mapping of disease genes.

Testing Candidate Genes

Dystrophin Homologs. As pointed out in the last chapter, the AR LGMD shares remarkable clinical similarity with the allelic Xp21 muscular dystrophies, Duchenne and Becker. Therefore, it has been suggested that any protein that shares structural or functional homology with the Xp21 product, i.e., dystrophin, may cause a phenotype similar to the Xp21 myopathies.

Using a cDNA probe from the carboxy-terminal domain of dystrophin, Love et al. (21) detected a homologous sequence that exhibits nucleic acid and predicted homologies with dystrophin of approximately 65% and 80%, respectively, and that is mapped to chromosome 6q24 (22). The protein encoded by this sequence has been called utrophin or dystrophin-related protein (DRP). Recently, Tinsley et al. (23) have cloned and sequenced the whole DRP cDNA, and it was observed that the homology between DRP and dystrophin extends over their entire length. DRP is localized in normal adult muscle primarily at the neuromuscular junction (24,25) and it is much more abundant in fetal and regenerating muscle (24–26). In the absence of dystrophin and DMD patients, DRP is also present in the sarcolemma (24,25). Therefore, DRP may play a similar role to dystrophin in early development (24,25) and it might functionally replace dystrophin in adult tissue. In addition, DRP is associated with an identical or antigenically similar complex of sarcolemmal proteins (27). The similarity between these proteins led to the suggestion that DRP might be a strong candidate for one of the autosomal recessive muscular dystrophies (28,29).

To test this hypothesis, 19 large Brazilian families with affected patients with LGMD were selected for linkage analysis using three probes flanking the 6q24 region (MYB, ESR, TCP1). Sixteen of the 19 families were informative: 14 in which the affected patients had a milder phenotype and two in which the patients had DLMD. It was possible to exclude DRP as a candidate gene for the milder form of LGMD at $0 = 0.10$. The lod scores for the DLMD and these markers were also negative, but more informative families would be necessary for a definite exclusion (30). Further evidence suggesting that DRP is not the gene responsible for the DLMD phenotype has been provided by linkage analysis in DLMD families from Tunisia (31) as well as anti-DRP studies in muscle from DLMD patients (25).

These results suggest that DRP is not the gene for the mild or severe forms of LGMD, but in view of the difficulties of conclusive results from linkage analysis studies in small genealogies, it is not possible to rule out this gene as responsible for LGMD in some families. Therefore, testing the anti-DRP in muscle from any affected patient with LGMD (mild and severe forms) might provide important information.

Glycoproteins Associated with Dystrophin. Dystrophin is associated with a complex of muscle membrane glycoproteins (156DAG, 59DAP, 50DAG, 43DAG, and 35DAG) that provide a linkage to the extracellular matrix protein laminin (32,33). The absence of dystrophin leads to a dramatic reduction of the dystrophin-associated proteins in the sarcolemma of patients with DMD. Therefore, deficiency of one of these glycoproteins can lead to decrease of dystrophin and consequently results in a phenotype similar to DMD or even BMD. The recent reports showing deficiency of 50DAG in muscle from DLMD patients support this hypothesis (10,27).

The dystrophin glycoproteins 43K and 156K, called dystroglycan, are expressed in a variety of muscle and nonmuscle tissues (34). Therefore, the gene that codifies for this protein is a candidate gene for muscular dystrophies with a phenotype similar to the Xp21 myopathies. Since human dystroglycan cDNA has been isolated and assigned to chromosome 3p21 (35), we have studied the 3p21 marker D3S11 in four families with a mild form of LGMD to test this hypothesis. The lod scores were negative, suggesting that mutations on this glycoprotein are not the primary cause of this form of LGMD. However, other markers mapped at this region have to be tested for a definitive exclusion.

Screening Markers Along the Genome

A Mild Form of LGMD Is Caused by a Gene at 15q. Using markers from the whole genome in families with patients affected with the mild form of LGMD from a geographic isolate (Réunion Island) in the Indian Ocean, Beckmann et al. (36) found positive linkage with the 15q1.1 marker D15S25. The LGMD form linked to this 15q gene has been classified as LGMD2 and confirmed to be responsible for the clinically similar condition in families from the Amish population as well (37).

In a study including 12 Brazilian families with patients affected with mild LGMD, only two were shown to be linked to the 15q gene; seven were definitely excluded and the remaining three (which had negative lod scores) are probably unlinked, supporting genetic heterogeneity for the mild form of LGMD (38). Markers from the 15q region have also been tested in Brazilian and Tunisian families with the severe form of LGMD; in both groups the linkage analysis showed negative lod scores, excluding this gene as responsible for the DLMD phenotype (31,39).

In addition to confirming that the mild form of LGMD is genetically heterogenous, linkage analysis with 15q markers suggests that: (1) the mild and severe forms of LGMD are probably caused by different genes, and (2) the gene on chromosome 15q is apparently rare in the Brazilian population, but it will be important to assess its relative frequency in other ethnic groups.

A Gene at 13q that Causes DLMD. Othmane et al. (31) tested 135 different microsatellites in three highly inbred DLMD families from Tunisia and found a significant lod score ($z = 9.15$ at $0 = 0.03$) with the 13q12 locus D13S115 and with two additional 13q12 markers (D13S143 and D13S120). Therefore, it was concluded that the gene responsible for the primary DLMD defect in the Tunisian population lies in the pericentromeric region of chromosome 13q.

We have tested the marker D13S115 in eight Brazilian families: four with the non-15q mild form of LGMD and four with affected patients with DLMD. Negative lod scores were observed for both the mild and the severe forms. The exclusion of the 13q12 gene as responsible for the non-15q mild LGMD in Brazilian families suggests again that the mild and severe forms of LGMD are likely to be determined by different genes.

On the other hand, patients from the four DLMD families that were also excluded with 13q probes were also tested for 50DAG through imunocytochemistry. Interestingly, in three of the families, 50DAG were negative (although the other dystrophin-associated glycoproteins were normal), but in one family a normal 50DAG reaction was observed (35). Therefore, these linkage results suggested for the first time that this gene is not mapped at 13q.

PRELIMINARY CONCLUSIONS

1. Although the autosomal recessive LGMD shows a great spectrum of clinical variability, apparently no family has been reported with patients clinically classified as DLMD and mild LGMD belonging to the same genealogy, which supports genetic heterogeneity for the severe and mild phenotypes.
2. So far, two genes that cause LGMD have been mapped through the use of random probes along the genome: one at 15q related to the mild phenotype and another at 13q associated with the DLMD phenotype.
3. Markers from the 15q region showed that only a small proportion (approximately 16%, or two of 12) of the tested Brazilian families with the mild phenotype are linked to the 15q gene.
4. DRP and dystroglycan, two candidate genes for the LGMD phenotype, have been excluded as responsible for DLMD and mild LGMD in Brazilian families.
5. In four large Brazilian families, the DLMD phenotype is not caused by mutations of the gene located at 13q.
6. Results of linkage analysis have already demonstrated that there is locus heterogeneity for both the mild and the severe forms. However, only after mapping and cloning of all the responsible genes will it be known if there is also allelic heterogeneity and how many genes are involved in the AR forms of LGMD.

NOTE ADDED IN PROOF

As mentioned in Chapter 20, at least four genes are associated with the LGMD phenotype; two of these have already been cloned: the 15q gene, which codes for calpain (40), and the 17q, which codifies for adhalin (41). More recently, based on linkage analysis using the closest available informative markers, we have determined the relative proportion for each of the mapped four autosomal recessive LGMD genes among 13 Brazilian families with mild LGMD, as follows: 33% of them are caused by mutations at the 15q gene, 33% at the 2p gene, 17% by mutations in the adhalin gene, and less than 10% may be due to mutations at the 13q locus. In addition, based on the finding that a large family is unlinked to any of these loci, we suggested that there is at least one other gene responsible for the mild LGMD phenotype (42). On the other hand, the linkage analysis among DLMD families revealed that there is at least another gene (non-13q or non-17q) causing deficiency of adhalin as a secondary effect (43).

ACKNOWLEDGMENTS

The authors are very grateful to Eloisa S. Moreira, Marta Canovas, Simone Campiotto, Reinaldo I. Takata, Jõao Ricardo M. Oliveira, and Dr. Rita Pavanello; to Dr. E. Bakker and his collaborators; and to FAPESP, CNPq, PADCT, and ABDIM for financial support.

REFERENCES

1. Shields RW Jr. Limb girdle syndrome. In: Engel AG, Banker BQ, eds. Myology. Vol II. New York: McGraw-Hill, 1986.
2. Ben Hamida M, Fardeau M, Attia N. Severe childhood muscular dystrophy affecting both sexes is frequent in Tunisia. Muscle Nerve 1983; 6: 469–480.
3. Zatz M, Passos-Bueno MR, Rapaport D. Estimate of the proportion of Duchenne muscular dystrophy with autosomal recessive inheritance. Am J Med Genet 1989; 32:407–410.
4. Vainzof M, Pavanello RCM, Pavanlello-Filho I, et al. Screening of male patients with autossomal recessive Duchenne dystrophy through dystrophin and DNA studies. Am J Med Genet 1991; 39:38–41.
5. Walton JN, Gardner-Medwin D. The muscular dystrophies. In: Walton JN, ed. Disorders of Voluntary Muscle, 5th ed. Edinnburgh: Churchil Livingstone, 1988:519–568.
6. Morton NE, Chung CS. Formal genetics of muscular dystrophy. Am J Hum Genet 1959; 11:360–379.
7. Chutkow JG, Heffner RR, Kramer AA, Edwards JA. Adult-onset autosomal dominant limb-girdle muscular dystrophy. Ann Neurol 1986; 30:240–248.
8. Gilchrist JM, Perick-Vance MA, Silverman L, Roses AD. Clinical and genetic investigation in autosomal dominant limb-girdle muscular dystrophy. Neurology 1988; 38:5–9.
9. Passos-Bueno MR, Vainzof M, Pavanello RCM, Pavenllo-Filho I, Lima MABO, Zatz M. Limb-girdle syndrome: a genetic study of 22 large Brazilian families: comparison with X-linked Duchenne and Becker dystrophies. J Neurol Sci 1991; 103:65–75.
10. Zatz M, Matsumara K, Vainzof M, Passos-Bueno MR, Pavanello RCM, Marie SK, Campbell KP. Deficiency of the 50kda dystrophin associated glycoprotein in Brazilian patients with severe childhood autosomal recessive muscular dystrophy (SCARMD). J Neurol Sci 1994; 123:122–128.
11. Emery AEH. Methodology in Medical Genetics, 2nd ed. Edinburgh: Churchill Livingstone, 1976:157.
12. Gusella JF, Wexler NS, Conneally PM, et al. A polymorphic DNA marker genetically linked to Huntington's disease. Nature 1983; 306:234–238.
13. Tsui L-C, Buchwald M, Barker D, et al. Cystic fibrosis locus defined by a genetically linked polymorphic DNA marker. Science 1985; 230:1054–1057.
14. Thompson MW, McInnes RR, Willard HF. Genetics in Medicine, 5th ed. Philadelphia: WB Saunders, 1991.
15. Fontaine B, Khurana TS, Hoffman EP, et al. Hyperkalemic periodic paralysis and the adult muscle sodium channel α-subunit gene. Science 1990; 250:1000–1002.
16. Brzustowicz LM, Lehner T, Castilla LH, et al. Genetic mapping of chronic childhood-onset spinal muscular atrophy to chromosome 5q11.2–13.3. Nature 1990; 344:540–541.
17. Dubowitz V. Chaos in the classifications of the spinal muscular atrophies of childhood. Neuromusc Disord 1991; 1:7780.

18. Melki J, Abdelhak S, Sheth P, et al. Gene for chronic proximal spinal muscular atrophies maps to chromosome 5q. Nature 1990; 344:767–768.
19. Lander ES: Mapping complex genetic traits in humans. In: Davies KE, ed. Genome Analysis. Oxford: IRL Press, 1988:171–189.
20. DeCoster W, DeReuck JD, Thiery E. A late autosomal dominant form of limb-girdle muscular dystrophy. Eur Neurol 1974; 12:159–172.
21. Love D, Hill DF, Dickson G, Spurr NK, Byth BC, Marsden RF, Walsh FS, Edwards YH, Davies KE. An autosomal high molecular weight transcript in skeletal muscle with homology to dystrophin. Nature 1989; 339:55–58.
22. Buckle VJ, Guenet JL, Simon-Chazottes D, Love DP, Davies KE. Localization of a dystrophin-related autosomal gene to 6q24 in man and to mouse chromosome 10 in the region of the dystrophia muscularis (dy) locus. Hum Genet 1990; 85:324–326.
23. Tinsley JM, Blake DJ, Roche A, et al. Primary structure of dystrophin-related protein. Nature 1992; 360:591–593.
24. Love DR, Morris GE, Ellis JM, et al. Tissue distribution of the dystrophin-related gene product and expression in the *mdx* and *dy* mouse. Proc Natl Acad Sci USA 1991; 88:3243–3247.
25. Khurana TS, Watkins SC, Chafey P, et al. Immunolocalization and developmental expression of DRP in skeletal muscle. Neuromusc Dis 1991; 1:185–194.
26. Takemitsu M, Ishiura S, Koga R, Kamakura K, Arahata K, Nonaka I, Sugita H. Dystrophin-related protein in the fetal and denervated skeletal muscles of normal and mdx mice. Biochem Biophys Res Commun 1991; 180:1179–1186.
27. Matsumara K, Tome FMS, Collin H, et al. Deficiency of the 50K dystrophin-associated glycoprotein in severe childhood autossomal recessive muscular dystrophy. Nature 1992; 359:320–322.
28. Mandel JL. The gene and its product. Nature 1989; 339:584–586.
29. Bakker E. Duchenne muscular dystrophy. Carrier detection and prenatal diagnosis by DNA analysis—new mutation and mosaicism. Ph.D. thesis, Leiden, 1989.
30. Passos-Bueno MR, Terwilliger J, Ott J, Vainzof M, Love DR, Davies KE, Zatz M. Linkage analysis in families with autosomal recessive limb-girdle muscular dystrophy (LGMD) and 6q probes flanking the dystrophin-related sequence. Am J Med Genet 1991b; 38:140–146.
31. Othmane KB, Hamida MB, Pericak-Vance MA, et al. Linkage of Tunisian autosomal recessive Duchenne-like muscular dystrophy to the pericentromeric region of chromosome 13q. Nature Genet 1992; 2:315–317.
32. Ervasti JM, Ohlendieck, K, Kahl SD, Gaver MG, Campbell KP. Deficiency of a glycoprotein component of the dystrophin complex in dystrophic muscle. Nature 1990; 345:315–319.
33. Ervasti JM, Campbell KP. Membrane organization of the dystrophin-glycoprotein complex. Cell 1991; 66:1–20.
34. Ibraghimov-Beskrovnaya O, Ervasti JM, Leveille CJ, Slaughter CA, Sernett SW, Campbell KP. Primary structure of dystrophin-associated glycoproteins linking dystrophin to the extracellular matrix. Nature 1992; 355:696–702.
35. Ibraghimov-Beskrovnaya O, Milatovich A, Ozcelik T, Yang B, Francke U, Campbell KP. Dystroglycan: tissue distribution, human muscle cDNA, genomic structure and chromosome mapping. Am J Hum Genet 1992; 51:A130.
36. Beckmann JS, Richard I, Hillaire D, et al. A gene for limb-girdle muscular dystrophy maps to chromosome 15 by linkage. CR Acad Sci Paris Ser III 1991; 312:141–148.
37. Young K, Foroud T, Williams P, Jackson CE, Beckmann JS, Cohen D, Conneally PM, Tischfield J, Hodes ME. Confirmation of linkage of limb-girdle muscular dystrophy, type-2, to chromosome 15. Genomics 1992; 13:1370–1371.
38. Passos-Bueno MR, Richard I, Vainzof M, et al. Evidence of genetic heterogeneity for the autosomal recessive adult forms of limb-girdle muscular dystrophy following linkage analysis with 15q probes in Brazilian families. J Med Genet 1993; 30:385–387.
39. Passos-Bueno MR, Bakker E, Marie SKN, Pavanello RC, Vainzof M, Carvalho AS, Cohen D, Beckmann J, Zatz M. Exclusion of the 15q locus as a candidate gene for severe childhood autosomal recessive Duchenne-like muscular dystrophy in Brazilian families. Hum Mol Genet 1993; 2:201–202.
40. Richard I, Broux O, Allamand V, Fougerousse F, Chiannilkulchai N, Bourg N, Brenguier L, Devaud C, Pasturaud P, Roudaut C, Hillaire D, Passos-Bueno MR, Zatz M, Tishfield JA, Fardeua M, Jackson CE, Cohen D, Beckmann J. A novel mechanism leading to muscular dystrophy: mutations in calpain 3 cause limb-girdle muscular dystrophy type 2A. Cell 1995; 81:27–40.
41. Roberds SL, Leturcq F, Allamand V, Piccolo F, Jeanpierre M, Anderson RD, Lim LE, Lee JC, Tomé FMS, Romero NB, Fardeua M, Beckmann JS, Kaplan J-C, Campbell KP. Missense mutations in the adhalin gene linked to autosomal recessive muscular dystrophy. Cell 1994; 78:625–633.
42. Passos-Bueno MR, Moreira ES, Marie SK, Bashir R, Vasquez L, Love DR, Vainzof M, Iughetti P, Oliveira

JR, Bakker E, Strachan T, Bushby K, Zatz M. Main clinical features for the three mapped autosomal recessive limb-girdle muscular dystrophies and estimated proportion of each form in 13 Brazilian families. 1995 (submitted).

43. Passos-Bueno MR, Moreira ES, Vainzof M, Chamberlain J, Marie SK, Pereira L, Akiyama J, Roberds SL, Campbell KP, Zatz M. A common missense mutation in the adhalin gene in three unrelated Brazilian families with a relatively mild form of autosomal recessive limb-girdle muscular dystrophy. Hum Mol Genet 1995; 4:1163–1167.

22

Dystrophin and DNA Findings in Duchenne and Becker Carriers

Mariz Vainzof, Maria Rita Passos-Bueno, and Mayana Zatz
University of São Paulo, São Paulo, Brazil

INTRODUCTION

Duchenne muscular dystrophy (DMD) and its allelic less severe form, Becker muscular dystrophy (BMD), are caused by a defective gene located at Xp21. At least two-thirds of the DMD/BMD cases are inherited through carrier mothers who have a 50% chance of transmitting the abnormal allele to their offspring (1,2). Females who are heterozygotes for the DMD/BMD gene carry a normal dystrophin allele on one X chromosome and a mutated gene on the other. Since females have two X chromosomes, they have twice the chance of having a mutation as compared to males. As a result of lyonization (3),it is expected that in most female heterozygotes, the X chromosome bearing the mutant dystrophin gene remains active in approximately 50% of the myogenic precursor cells whereas in the other 50% the normal X chromosome remains active. Since one normal gene is enough to guarantee normal cell function, most DMB/BMD carriers are asymptomatic. However, approximately 8% of DMD heterozygotes have some clinical signs, ranging from calf hypertrophy to severe muscle weakness, as a result of uneven lyonization (1).

Since the early sixties many investigations have been reported in asymptomatic and clinically affected carriers of the DMD/BMD gene. The most important are summarized below.

SERUM ENZYMES AND DNA STUDIES

For many years, serum enzyme determinations were the best method for detecting the presence of the DMD/BMD gene in potential carriers. About 50–70% of the female heterozygotes for the DMD/BMD gene have elevated serum creatine kinase (CK) and pyruvate kinase (PK) levels (2,4). Moreover, some histopathological muscle abnormalities can also be found in about 70–75% of carriers (5).

Before the introduction of DNA technology, heterozygosity risks were estimated by combining data from the pedigree with results of serum enzyme determinations through Bayesian calculations, which was an inadequate method since 30–50% of the obligate carriers may have normal serum CK and PK. However, results of serum CK determinations may still be informative as an adjunct test, in particular when they are found to be elevated.

Localization and cloning of the dystrophin gene (6–9) allowed a significant improvement in carrier detection and prenatal diagnosis through the use of DNA-linked markers and by direct analysis of the abnormal gene (10–12). With the use of these techniques, it is currently possible to assess carrier status with 98% accuracy in over 95% of cases (13–15).

If the proband has a molecular deletion, as is found in about 60–70% of cases, it is possible to directly assess whether his mother carries the same deletion through analysis of the intensity of the band following quantitative polymerase chain reaction (PCR) (16) and Southern blot analysis (17,18), which requires a careful comparison of signal intensities of deleted and nondeleted exons. If the deletion is detected in the mother, it is possible to offer direct prenatal diagnosis if a male fetus is identified in a subsequent pregnancy.

When a mutation/deletion (or duplication) cannot be detected directly, genetic markers within or near the gene are used to track the defective gene (18–21). Although the method is highly reliable in 95% of the cases, it is subject to error because of the natural recombination events that occur between the genetic markers and the disease gene. The accuracy with which carrier status can be determined is greatly enhanced using a pair of flanking markers, the error determined in this case by the frequency of double recombination between the markers (13,15). Currently, multiplex PCR assays for polymorphisms are making linkage analysis much faster and simpler (20,21).

We have recently assessed the carrier status in 100 females from families with patients affected with DMD. Only one family was not informative. In 79 cases it was possible to define heterozygosity risks in potential carriers through the presence ($n = 24$) or absence ($n = 55$) of a DNA deletion along the dystrophin gene. Through the use of DNA polymorphic markers, in 12 families it was possible to classify seven females at risk as possible DMD carriers and 11 as unlikely to be carriers.

GERMLINE MOSAICISM

A mutation at a mitotic stage of germ lineage development may occur and be observed in more than one offspring of a male or female whose somatic cells lack the mutation (13,22,23). At least 19 cases have been reported so far for DMD, in which two or more affected sibs had a deletion (or duplication) that was not detected in their mother based in DNA analysis from peripheral blood (revision in Ref. 24).

Our observation of a significantly higher proportion of women with increased serum CK activities among obligate DMD carriers (who had at least one affected son and one affected brother) than among mothers of two or more affected sons in only one generation gave further support to the existence of germline mosaicism (25). Therefore, if a woman without a family history of DMD or BMD has a single affected son, and if she is unlikely herself to be a carrier, her risk of having another affected son is not negligible because of the possibility of germline mosaicism, which may be as high as 14% (14,18,24).

However, most interestingly, based on a collaborative study of 473 patients (by the University of São Paulo and University of Leiden), new data have shown that these risks differ according to the site of the deletion. Proximal deletions were much more frequent among germinal mosaic cases (six among seven cases) while distal deletions were more frequent among isolated sporadic cases (24 among 38 cases). These findings have important implications for estimates of recurrence risks for DMD isolated cases, in which the mutation is not detected in the probands' mothers through DNA analysis in lymphocytes. For example, if the proband has a deletion in the proximal region of the dystrophin gene, the recurrence risk for another affected son is 15%, while if the deletion is detected in the more distal region, this risk is only 2% (24).

DYSTROPHIN FINDINGS IN DMD/BMD CARRIERS

Asymptomatic DMD and BMD Carriers

Until recently, it was not possible to identify simplex cases of dystrophinopathy in females, but with the advent of molecular diagnosis several reports have been published in females at risk for DMD.

Immunohistochemical studies (IF) have shown a dystrophin mosaic pattern in DMD carriers with clinical manifestations. However, among those with no clinical signs, negative fibers or those with a patchy IF are usually absent or infrequent (26–28) although an altered dystrophin IF in some BMD asymptomatic carriers has been reported (29–31). Among 21 adult Xp21 asymptomatic carriers (16 DMD, 11 with elevated serum CK, and five BMD, two with elevated serum CK) studied at our center, we observed a positive IF pattern of dystrophin in all, which did not differ from normal controls, suggesting that abnormalities in dystrophin immunostaining are not easily observed in carriers with no clinical signs (32).

Based on dystrophin analysis in clonal myoblasts derived from one DMD carrier, Hurko et al. (33) suggested this approach as an alternative to established methods of determining carrier status. Although this method might be a sensitive test to identify DMD/BMD heterozygotes, it is time-consuming and expensive and certainly would not be appropriate for routine carrier detection.

In studies of *mdx* mice and the *xmd* dog, young female carriers have shown an IF dystrophin mosaic pattern, which becomes more uniformly positive with aging. This observation has been explained by the fusion of dystrophin-competent satellite cells into negative fiber segments during regeneration or by intracellular diffusion of dystrophin (34–36). To investigate if the same apparent decrease in dystrophin negative fibers with aging observed in *mdx* female heterozygotes also occurs in carriers of the DMD/BMD gene, we have studied the muscle from 10 young daughters of obligate DMD carriers, aged 6–17 years, all with grossly increased serum CK activities. Some dystrophin-negative fibers were seen in only one 9-year-old possible DMD carrier (32). On the other hand, a clear dystrophin mosaic pattern was reported in two young carriers aged 3 and 4 years, which was not seen in their carrier mothers (28). These results suggest that the mosaic pattern seen in young *mdx* and *xmd* female dogs is not frequent in young DMD female carriers, unless they are very young.

Quantitative dystrophin studies have also shown that small reductions in dystrophin levels cannot be detected in the majority of DMD carriers through dystrophin Western blot (WB) analysis (37,38). However, some BMD carriers (those related to patients with an abnormal dystrophin band) can be identified through a qualitative WB since their muscle contains both normal and abnormally sized dystrophin (Fig. 1) (39–42).

Such females provide a good model for studying dystrophin expression in carriers. Since muscle fibers are syncytial cells, each myofiber contains nuclei derived from different mononuclear myoblasts. Sarcolemmal distribution of an internally deleted dystophin was analyzed in BMD carriers who also had two bands (one normal and one altered) detected through WB. The results suggested that the two types of nuclei were randomly distributed (those in which the active X chromosome contained the normal allele and those in which the X chromosome contained the defective allele), resulting in normal and abnormal dystrophin molecules that were so intimately mixed that dystrophin-incompetent fibers could not be distinguished in the skeletal muscle from these Xp21 carriers (43).

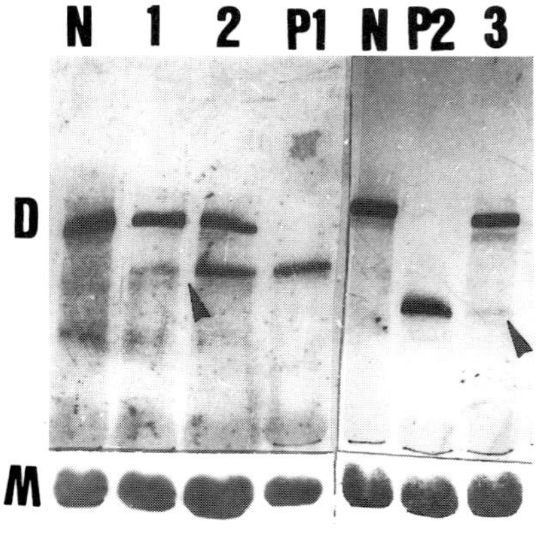

Figure 1 Dystrophin Western blotting in three BMD carriers (1, 2, and 3) as compared to their two related BMD patients (P1 and P2) from two different families. The three carriers show an additional dystrophin band whose MW is similar to the one found in the affected patients from the same family (arrows). (C-terminal Dy8/6C5 antibody courtesy of Dr. Louise V. B. Nicholson.)

Further support for the hypothesis of a mechanism of compensation (probably due to the diffusion of dystrophin within the myofiber from dystrophin-positive to dystrophin-negative nuclear regions) was recently reported in a t(X;12) DMD female carrier in whom there was a discordance between the amount of dystrophin (5% of normal) detected through WB and the percentage of labeled fibers (36%) detected through IF (44). Such a mechanism of compensation could explain the absence of a clear mosaic pattern in the majority of asymptomatic DMD carriers.

Manifesting DMD Carriers

Several examples have been reported of manifesting carriers of the DMD gene, who had a normal X-chromosome constitution, as an effect of uneven lyonization (45,46). The traditional clinical criteria for identifying a manifesting carrier are a family history of DMD, proximal limb weakness, high serum CK levels, and myopathic changes on both muscle biopsy and electromyography (46–48).

Recent publications on dystrophin IF analysis have shown that most manifesting DMD carriers have a mosaic pattern of dystrophin-positive and -negative fibers (Fig. 2A) (26–28,37,32). Depending on the proportion of negative fibers, a reduced abundance of dystrophin may or may not be observed

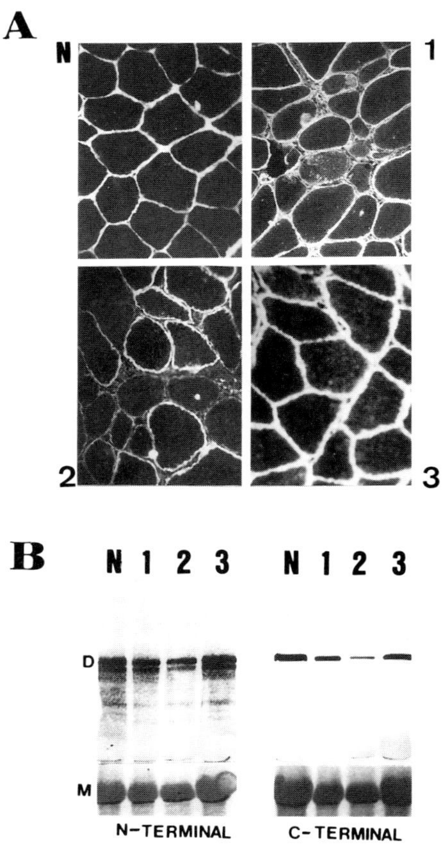

Figure 2 Dystrophin studies in three DMD carriers. (A) Dystrophin immunofluorescence in: (N) a normal control; (1) manifesting DMD carrier with 12% of totally or partially negative fibers; (2) manifesting carrier with about 50% mosaic pattern; (3) obligate asymptomatic DMD carrier. (B) Western blotting in the same carriers using N-terminal antidystrophin antibody, 60 kDa, and C-terminal 1461. D, 427-kDa dystrophin band; M, myosin content. (Antibodies courtesy of Dr. E. P. Hoffman, Dr. L. M. Kunkel, Dr. E. E. Zubrzycka-Gaarn, Dr. D. Bulman, P. N. Ray, and Dr. R. G. Worton.)

by WB in such carriers (28,49,37), and an accurate quantification depends on correlation with the myosin content of the muscle extract (Fig. 2B).

Before the discovery of dystrophin, the differential diagnosis between the X-linked forms and the autosomal recessive forms of limb-girdle muscular dystrophy (LGMD) was very difficult, particularly among isolated patients with no family history. Therefore, manifesting DMD carriers were often diagnosed as having AR LGMD. Two independent dystrophin studies have shown that two of 24 (50) and 25 of 210 (49) patients who had been diagnosed as LGMD (based on clinical course, high serum CK activity, and myopathic histopathology) were indeed manifesting carriers of the DMD gene. Therefore, about 10–15% of isolated affected female myopathic patients may be DMD manifesting carriers. The identification of such carriers is extremely important for genetic counseling, since they have a 50% chance of having an affected son while patients with AR LGMD have a negligible chance of having affected offspring (as long as there is no consanguineous marriage).

Manifesting carriers of DMD have occasionally been described in the same family (45,46,51–53). In two of six families with manifesting carriers studied at our center, more than one female had clinical signs and dystrophin alterations. Such familial aggregation suggests either environmental factors or that the process of lyonization is genetically influenced (29).

One finding that attracted our attention was that although mental impairment is rare among asymptomatic carriers, the majority of manifesting DMD carriers studied at our center are mentally retarded. Deletions in the dystrophin brain promoter have recently been excluded as responsible for mental retardation in DMD/BMD (54,55). Therefore, the mechanisms responsible for mental impairment in a proportion of DMD-affected patients and in symptomatic carriers are still under investigation.

In conclusion, both the percentage and the distribution of dystrophin-negative myofibers seem to be important markers of clinical severity in DMD-affected carriers (43,49). The manifestation of muscle weakness probably depends on two contrasting variables: the tendency for the muscle to become dystrophin-positive and the progressive histopathological changes of dystrophin-deficient muscle leading to fiber loss and clinical weakness (49).

Severely Affected Girls

Severe clinical signs in female heterozygotes for X-linked recessive genes may occur due to numerical or structural abnormalities in one of the X chromosomes, as reported in females with an XO sex chromosomal constitution in a large proportion of cells (56–59) or with a structurally abnormal X chromosome, in particular X/A translocations (60–64). We have recently studied four female patients with a clinical course and dystrophin findings comparable to those of Xp21 DMD.

Patient 1, a 6-year-old girl, had an XO chromosomal constitution and a frame-shifting DNA deletion of exon 44 in the dystrophin gene. Dystrophin analysis revealed a negative pattern through IF and a faint 250-kDa band through WB (Fig. 3), thus confirming the XO chromosomal constitution in her muscle cells.

Patient 2, a 13-year-old-girl with a mosaic XO/XX/XXX chromosomal constitution (36% 45,X cells in lymphocytes and 64% in fibroblasts), had been previously reported (59). She inherited the DMD gene from her mother, who is an obligate carrier. Her menarche at age 12, and an apparently normal sexual development, suggest the existence of a cell line with at least two X chromosomes in her ovaries. However, whe was confined to a wheelchair at age 9 with a clinical course as severe as that of an affected boy, which suggested that the majority of her cell lines in muscle carried the X chromosome with the defective gene. Indeed, dystrophin analysis, performed several years later, showed a negative dystrophin pattern through both IF and WB, confirming this hypothesis (Fig. 3). DNA analysis in this girl did not detect any deletion along the dystrophin gene. By contrast, a report on dystrophin analysis in another female with a XO/XX/XXX chromosomal constitution showed a mosaic dystrophin pattern, with only 9% of negative fibers and a mild phenotype (49). These observations indicate that the degree of muscle weakness in such unusual cases depends on the distribution and proportion of the different cell lines.

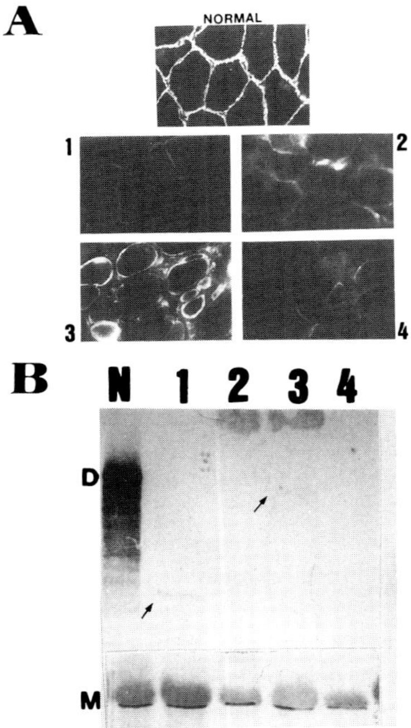

Figure 3 Immunofluorescence (A) and western blot (B) of four severely affected girls (1) 45,X; (2) XO/XX/XXX; (3) X/22; (4) affected girl with normal karyotype. IF showed a DMD pattern, with occasional partially labeled fibers. Western blot: arrows indicate the faint band in patients 1 and 3. N, Normal control; D, 427-kDa dystrophin band; M, myosin content.

Patient 3, a 13-year-old girl, was one of the three DMD-affected girls described from our center who had an X/autosome translocation (X/6, X/2, and X/22) (64,65). This girl, carrier of an X/22 translocation, became wheelchair-bound at age 10 and is mentally retarded (she was never able to attend normal school). DNA analysis did not detect any molecular deletion, and dystrophin analysis showed a negative pattern with occasional positive fibers through IF and a very faint 427-kDa band through WB (Fig. 3). Similar findings were recently reported in three other girls with translocations (49).

In patient 4, a 10-year-old girl also severely affected and mentally handicapped, DNA analysis revealed a deletion encompassing exons 12–13, but no chromosomal abnormality, suggesting skewed X-chromosome inactivation (with the majority of her muscle cells carrying the active X chromosome with the defective gene). Dystrophin analysis showed a pattern similar to the one found in DMD boys (Fig. 3).

An interesting explanation for the observation of severely affected girls with two normal X chromosomes is summarized by Walton and Gardner-Medwin (66), based on the report of a few remarkable pairs of identical twin girls, in which one twin has a severe phenotype (typical of a DMD-affected boy) while the other is entirely unaffected (67–71). It has been suggested that in such cases, early random X inactivation of the normal X chromosome in some cells and of the X containing the mutant DMD gene in other cells might contribute to the causation of the twinning process in the early carrier zygote. The resultant twins, both heterozygotes for the DMD gene, might have opposite phenotypes, one normal and the other typical of DMD. If this explanation is correct, one might expect early fetal loss of one twin to result occasionally in apparently isolated, severely affected DMD carriers, with normal karyotype.

CONCLUSIONS

Until an effective treatment is available for DMD/BMD, prevention of new cases through the identification of female carriers and prenatal diagnosis is of upmost importance. As a result of recent investigations at the molecular level, it is now possible to correctly classify a carrier in the majority of the cases. Although cases arising from new mutations cannot be prevented, the application of recent advances to genetic counseling and prenatal diagnosis has decreased the incidence of familial cases (49). In addition, investigations in asymptomatic and manifesting female carriers are extremely important for enhancing our comprehension of the mechanisms responsible for Xp21 muscular dystrophies.

NOTE ADDED IN PROOF

The presence of biased and nonbiased patterns of X inactivation has been reported in recent studies, correlating with dystrophin expression in DMD-manifesting carriers. However, Matthews et al. (72) observed no consistent relationship between X inactivation and the proportion of dystrophin-negative fibers or the phenotype in manifesting carriers, while Pegoraro et al. (73) reported that X-inactivation patterns correlate with muscle dystrophin content and clinical phenotype. In addition, a biochemical dystrophin normalization in muscle from female carriers was observed (73).

ACKNOWLEDGMENTS

The authors are extremely grateful for the collaboration of Dr. Rita C. M. Pavanello, Dr. Ivo Pavanello Filho, Dr. Ana Maria Tsanaclis, Dr. Sueli K. M. Marie, Marta Canovas, Reinaldo I. Takata, Simone Campioto, and Antonia M. Cerqueira and for financial suport from Fundacao de Amparo a Pesquisa do Estado de Sao Paulo (FAPESP), Conselho Nacional de Desenvolvimento Cientifico e Tecnologico (CNPq), and Associacao Brasileria de Distrofia Muscular (ABDIM).

REFERENCES

1. Emery AEH. Duchenne Muscular Dystrophy. Oxford: Oxford Medical Publications, 1987.
2. Zatz M. Otto PA. Evaluation of carrier detection rate for Duchenne and Becker muscular dystrophies using serum creatine-kinase (CK) and pyruvate-kinase (PK) through discriminant analysis. Am J Med Genet 1986; 25:219–231.
3. Lyon MF. Gene action in the X-chromosome of the mouse *(Mus musculus)*. Nature 1961; 190:372–373.
4. Zatz M, Shapiro LJ, Campion DS, Oda E, Kaback MM. Serum pyruvate-kinase (PK) and creatine-phosphokinase in progressive muscular dystrophies. J Neurol Sci 1978; 36:349–362.
5. Mauder-Sewry CA, Dubowitz V. Needle muscle biopsy for carrier dection in Duchenne muscular dystrophy. J Neurol Sci 1981; 49:305–324.
6. Koenig M, Hoffman EP, Bertelsom CJ, Monaco AP, Feener C, Kunkel LM. Complete cloning of the Duchenne muscular dystrophy (DMD) cDNA and preliminary genomic organization of the DMD gene in normal and affected individuals. Cell 1987; 50:509–517.
7. Monaco AP, Bertelson CJ, Liechti-Gallati S, Moser H, Kunkel LM. An explanation for the phenotypic differences between patients bearing partial deletions of the DMD locus. Genomics 1988; 2:90–95.
8. Forrest SM, Cross GS, Flint T, Speer A, Robson KJH, Davies KE. Further studies of gene deletions that cause Duchenne and Becker muscular dystrophies. Genomics 1988; 2:109–114.
9. Hoffman EP, Brown RH, Kunkel LM. Dystrophin: the protein product of the Duchenne muscular dystrophy locus. Cell 1987; 51:919–928.
10. Bakker E, Bonten EJ, de Lange LF, et al. DNA probe analysis for carrier detection and prenatal diagnosis of Duchenne muscular dystrophy: a standard diagnostic procedure. J Med Genet 1986; 23:573–580.
11. Lindolf M, Kaariainen H, Davies KE, de la Chapelle A. Carrier detection and prenatal diagnosis in X linked muscular dystrophy using restriction fragment length polymorphisms. J Med Genet 1986; 23:560–572.
12. Hodgson S, Walker A, Cole C, Hart K, Johnson L, Jubowitz V, Bobrow M. The application of linkage analysis to genetic counselling in families with Duchenne or Becker muculaar dystrophy. J Med Genet 1987; 24:152–159.

13. Bakker E, Hoffer MH, Goor N, et al. Prenatal diagnosis and carrier detection of Duchene muscular dystrophy with closely linked RFLP's. Lancet 1985; 1:655–658.
14. Bakker E, Van Broeckhoven C, Bonten EJ, Van de Vooren MJ, Veenema H, Van Hul W, Van Ommen GJB, et al. Germline mosaicism and Duchenne muscular dystrophy mutations. Nature 1987; 328:554–556.
15. Worton RG, Gillard EF. Muscular dystrophy, molecular genetics. Encycloped Hum Biol 1991; 5:191–204.
16. Abbs S, Yau SC, Clark S, Matthew CG, Bobrow M. A convenient multiplex PCR system for the detection of dystrophin gene deletions: Comparative analysis with cDNA hybridization shows mistypings by both methods. J Med Genet 1991; 28:304–311.
17. Monaco AP, Bertelson CJ, Colletti-Feener C, Kunkel LM. Localisation and cloning of Xp21 deletion breakpoints involved in muscular dystrophy. Hum Genet 1987; 75:221–227.
18. Bakker E, Veenema H, Den Dunnen JT, Van Broeckhoven C, Grootscholten PM, Bonten EJ, Van Ommen GJB, et al. Germinal mosaicism increases the recurrence risk for new Duchenne muscular dystrophy mutations. J Med Genet 1989; 26:553–559.
19. Roberts RG, Cole CG, Hart KA, Bobrow M, Bentley DR. Rapid carrier and prenatal diagnosis of Duchenne and Becker muscular dystrophy. Nucleic Acids Res 1989; 17:811.
20. Beggs AH, Kunkel LM. A polymorphic CACA repeat in the 3′ untraslated region of dystrophin. Nucleic Acids Res 1990; 18:1931.
21. Feener CA, Boyce FM, Kunkel LM. Rapid detection of CA polymorphisms in cloned DNA: application to the 5′ region of the dystrophin gene. Am J Hum Genet 1991; 48:621–627.
22. Darras BT, Francke U. A partial deletion of the muscular dystrophy gene transmitted twice by an unaffected male. Nature 1987; 329:556–558.
23. Wood S, McGillivray BC. Germinal mosaicism in Duchenne muscular dystrophy. Hum Genet 1988; 78:282–284.
24. Passos-Bueno MR, Bakker E, Kneppers ALJ, et al. Different mosaicism frequencies for proximal and distal Duchenne muscular dystrophy (DMD) mutations indicate difference in etiology and recurrence risk. Am J Hum Genet 1992; 51:1150–1155.
25. Passos-Bueno MR, Lima MABO, Zatz M. Estimate of germinal mosaicism in Duchenne muscular dystrophy. J Med Genet 1990; 27:727–728.
26. Bonilla E, Schmidt B, Samitt CE, et al. Normal and dystrophin-deficient muscle fibres in carrier of the gene for Duchenne muscular dystrophy. Am J Pathol 1988; 133:440–445.
27. Morandi L, Mora M, Gusoni E, Tedeschi S, Cornelio F. Dystrophin analysis in Duchenne and Becker muscular dystrophy carriers: correlation with intracellular calcium and albumin. Ann Neurol 1990; 28:674–679.
28. Clerk A, Rodillo E, Heckmatt JZ, Dubowitz V, Strong PN, Sewry CA. Characterisation of dystrophin in carriers of Duchenne muscular dystrophy. J Neurol Sci 1991; 102:197–205.
29. Glass IA, Nicholson LVB, Watkiss E, et al. Investigation of a female manifesting Becker muscular dystrophy. Med Genet 1992; 29:578–582.
30. Voit BT, Stuettgen P, Cremer M, Goebel HH. Dystrophin as a diagnostic marker in Duchenne and Becker muscular dystrophy. Correlation of immunofluorescence and Western blot. Neuropediatrics 1991; 22:152–162.
31. Minetti C, Bonilha E. Mosiac expression of dystrophin in carriers of Becker's muscular dystrophy and the X-linked syndrome of myalgia and cramps. N Engl J Med 1992; Oct 8:1100.
32. Vainzof M, Pavanello RCM, Pavanello I, et al. Dystrophin Immunoflurescence Pattern in Manifesting and Asymptomatic carriers of Duchenne's and Becker muscular dystrophies of different ages. Neuromusc Disord 1991; 3:177–183.
33. Hurko O, Hoffman EP, McKee L., et al. Dystrophin analysys in clonal myoblasts derived from a Duchenne muscular dystrophy carrier. Am J Hum Genet 1989; 44:820–826.
34. Watkins SC, Hoffman EP, Slayter HS, Kunkel LM. Dystrophin distribution in heterozygote *MDX* mice. Muscle Nerve 1989; 12:861–868.
35. Karpatti G, Gaarn EEZ, Carpenter S, Bulmam DE, Ray PN, Worton RG. Age-related conversion of dystrophin-negative to -positive fiber segments of skeletal but not cardiac muscle fibers in heterozygote *mdx* mice. J Neuropathol Exp Neurol 1990; 2:96–105.
36. Cooper BJ, Gallagher EA, Smith CA, Valentine BA, Winand NJ. Mosaic expression of dystrophin in carriers of canine X-linked muscular dystrophy. Lab Invest 1990; 2:171–178.
37. Arahata K, Ishihara T, Kamakura K, et al. Mosaic expression of dystrophin in symtomatic carrier of Duchenne'muscular dystrophy. N Engl J Med 1989; 320:138–142.
38. Beggs AH, Kunkel LM. Improved diagnosis of Duchenne/Becker muscular dystrophy. J Clin Invest 1990; 85:613–619.
39. Angeline C, Beggs AH, Hoffman EP, Fanin M, Kunkel LM. Enormous dystrophin in a patient with Becker muscular dystrophy. Neurology 1990; 40:808–812.
40. Beggs AH, Hoffman EP, Snyder JR, et al. Exploring the molecular basis for variability among patients with Becker muscular dystrophy: dystrophin gene and protein studies. Am J Hum Genet 1991; 49:54–67.

41. Vainzof M, Passos-Bueno MR, Pavanello RCM, Schreiber R, Zatz M. A model to estimate the expression of the dysrophin gene in muscle from Becker female muscular dystrophy carriers. J Med Genet 1992; 29:476–479.
42. Chevron MP, Tuffery S, Echenne B, et al. Becker muscular dystrophy: demonstration of the carrier status of a female by immunobloting and immunostaining. Neuromusc Disord 1992; 1:47–50.
43. Vainzof M, Nicholson LVB, Bulman DE, Tsanaclis AM, Passos-Bueno MR, Pavanello RCM, Zatz M. Sarcolemmal distribution of abnormal dystrophin in Xp21 carriers. Neuromusc Disord 1993 (in press).
44. Wenger SL, Steele MW, Hoffman, EP, Barmada MA, Wessel HB. X inactivation and dystrophin studies in a t(x;12) female: evidence for biochemical normalization in Duchenne muscular dystrophy carriers. Am J Med Genet 1992; 43:1012–1015.
45. Zatz M, Levisky RB, Levy JA, Valente BO, Gianotti M, Frota-Pessoa O. Clinical symptoms in a female carrier of Duchenne muscular dystrophy. J Hum Genet 1973; 21:297–305.
46. Moser H, Emery AEH. The manifesting carrier in Duchenne muscular dystrophy. Clin Genet 1974; 5:271–284.
47. Yoshioka M. Clinically manifesting carriers in Duchenne muscular dystrophy. Clin Genet 1981; 20:6–12.
48. Barkhaus PE, Gilchrist JM. Duchenne muscular dystrophy manifesting carriers. Arch Neurol 1989; 46:673–675.
49. Hoffman EP, Arahata K, Minetti C, et al. Dystrophinopathy in isolated cases of myopathy in females. Neurology 1992; 42:967–975.
50. Arikama E, Hoffman EP, Kaido M, Nonaka I, Sugita H, Arahata K. The frequency of patients with dystrophin abnormalities in a limb-girdle patient population. Neurology 1991; 41:1491–1496.
51. Frouhar SP, Spiro R, Lubs ML. Familial carrier manifestations in families with X-linked Duchenne muscular dystrophy. Am J Hum Genet 1975; 27:37A (abstract).
52. Falcao-Conceicao DN, Goncalves-Pimentel MM, Baptista ML, Ubatuba S. Detection of carrier of X-linked gene for Duchenne muscular dystrophy by levels of creatine kinase. J. Neurol Sci 1983; 62:171–180.
53. Reddy BK, Anandavalli TE, Reddi OS, X-linked Duchenne muscular dystrophy in an unusual family with manifesting carrier. Hum Genet 1984; 67:460–2.
54. Den Dunnen JT, Casula L, Makover A, et al. Deletion of dystrophin brain and muscle promoters is compatible with normal intellectual capability and may cause mild dystrophy. Neuromusc Disord 1991; 5:327–331.
55. Rapaport D, Passos-Bueno MR, Takata IR, et al. A deletion including the brain promoter of the Duchenne muscular dystrophy gene is not associated with mental retardation. Neuromusc Disord 1992; 2:117–120.
56. Walton JN. The inheritance of muscular dystrophy. Acta Genet Statist Med 1957; 7:318–320.
57. Ferrier P, Bamatter F, Klein D. Muscular dystrophy (Duchenne) in a girl with Turner's syndrome. J Med Genet 1965; 2:38–46.
58. Jalbert PC, Mouriquand C, Beaudoing A, Jaillard M. Myopathie progressive de type Duchenne et mosaique XO/XX/XXX. Considerations sur la genese de la fibre musculaire striee. Ann Genet (Paris) 1966; 9:104–107.
59. Bortolini ER, Silva DM, Chequer RS, Viana-Morgante AM, Zatz M. Brief Clinical Report: Duchenne muscular Dystrophy in a girl with a 45,X/46,XX/47,XXX chromosome constitution. Am J Med Genet 1986; 25:239–243.
60. Greenstein RM, Reardon MP, Chan TS. An X/autosome translocation in a girl with Duchenne muscular distrophy (DMD). Evidence for DMD gene localization. Pediatr Res 1977; 11:457.
61. Verellen C, Freund M, deMeyer R, Laterre C, Scholberg B, Frederic J. Progressive muscular dystrophy of the Duchenne type in a young girl associated with an aberration of chromosome X. Excerpta Med Int Cong Ser 1977; 426:42.
62. Canki N, Dutrillaux B, Tivadar I. Dystrophie musculaire de Duchenne chez une petite fille porteuse d'une translocation t(X;3) (p21;q13) de novo. Ann Genet 1979; 22:35–39.
63. Lindenbaum RH, Clarke G, Patel C, Moncrieff M, Hughes JT. Muscular dystrophy in a X;1 translocation female suggestes that Duchenne locus is on X chromosome short arm. J Med Genet 1979; 16:389–92.
64. Zatz M, Vianna-Morgante AM, Campos P, Diament AJ. Translocation (X;6) in a female with Duchenne muscular dystrophy: implications for the localisation of the DMD locus. J Med Genet 1981; 18:442–447.
65. Vianna-Morgante AM, Batista DAS, Levisky RB, Zatz M. X autosome translocations in females with X-linked recessive diseases. 7th International Congress of Human Genetics (Berlin) 1986; part I:97 (abstract).
66. Walton J, Gardner-Medwin D. The muscular dystrophies. In: Walton J, ed. Disorders of Voluntary Muscle, 5th ed. New York: Churchill Livingstone, 1988:519–568.
67. Burn J, Povey S, Boyd Y, Munro EA, West L, Harper K, Thomas D. Duchenne muscular dystrophy in one of monozygotic twin girls. J Med Genet 1986; 23:494–500.
68. Chutkow JG, Hyser CL, Edwards JA, Heffner RR Jr, Czyrny JJ. Monozygotic female twin carriers discordant for the clinical manifestations of Duchenne muscular dystrophy. Neurology 1987; 37:1147–1151.
69. Bonilla E, Younger DS, Chang HW, et al. Partial dystrophin deficiency in monozygous twin carriers of the Duchenne gene discordant for clinical myopathy. Neurology 1990; 40:1267–1270.
70. Richards CS, Watkins SC, Hoffman EP, Schneider NR, Milsark IW, Katz KS, Cook JD, Kunkel LM, Cortada

JM. Skewed X inactivation in a female MZ twin resulta in Duchenne muscular dystrophy. Am J Hum Genet 1990; 46:672–681.

71. Lupski JR, Garcia CA, Zoghbi HY, Hoffman EP, Fenwick RG. Discordance of muscular dystrophy in monozygotic female twins: Evidence supporting asymmetric splitting of the inner cell mass in a manifesting carrier of Duchenne dystrophy. Am J Med Genet 1991; 40:354–364.

72. Matthews PM, Benjamin, D, Van Bakel I, et al. Muscle X-inactivation patterns and dystrophin expression in Duchenne muscular dystrophy carriers. Neuromusc Disord 1995; 5:209–220.

73. Pegoraro E, Schimake RN, Garcia C, et al. Genetic and biochemical normalization in female carriers of Duchenne muscular dystrophy: evidence for failure of dystrophin production in dystrophin-competent myonuclei. Neurology 1995; 45:677–690.

23
Facioscapulohumeral Syndrome

W. J. Ken Cumming
Withington Hospital, Manchester, England

INTRODUCTION

The facioscapulohumeral (FSH) syndrome encompasses a group of disorders having the same phenotype, with weakness, of widely variable severity and age at onset, affecting particularly the facial muscles (eye and mouth closure), the serratus anterior, the humeral muscles, and the anterior tibial group in the majority of cases. The autosomal dominant (dystrophic) form is associated with a gene defect at 4q35, as are many apparently "sporadic" cases (1).

Although Landouzy and Déjerine (2) are credited with the original descriptions of FSH syndrome, it is likely that Duchenne, in his original papers in the *Archives of General Medicine* (3) concerning the differentiation between neurogenic and myopathic processes underlying clinical subgroups, also described patients with what would now be considered to be FSH syndrome. He went on to say, *"Par la face on elle atrophie quelques muscles, principalement l'orbiculaire des levres et les zygomatiques"* (4,5). Davidenkow (6) later pointed out that the condition could be quite benign and restricted in its clinical manifestations. The familial nature of the disease was underlined by the report of Tyler and Stephens (7), in which they described a family of 1249 individuals in six generations, of which 159 were known to have FSH disease. These authors were the first to recognize so-called "subclinical" involvement, where some family members had very limited manifestations of the disease. The autosomal-dominant nature of the disorder was later clarified by Chung and Morton (8).

Walton and Nattrass (9), in their classic paper on classification and natural history of the myopathies, denoted facioscapulohumeral dystrophy as one of the three major forms of dystrophic neuromuscular disease, the others being the Duchenne and limb-girdle forms. They characterized the condition as having an onset within the first 30 years of life, being very slowly progressive, and not normally affecting life span, and in which arrested progression was not unusual. They emphasized that the face was usually involved first, followed by the shoulder girdles, and finally the pelvic muscles. They noted that asymmetry of muscle involvement was not unusual.

Gardner-Medwin and Walton (1) more recently defined the characteristics of FSH syndrome as follows:

1. Expression in either sex
2. Transmission usually as an autosomal-dominant trait
3. Onset at any age from childhood until adult life
4. Frequent occurrence of aborted or mildly affected cases
5. Initial involvement of the face and shoulder girdle muscles with subsequent spread to the muscles in the lower limbs

6. Rarity of muscular hypertrophy
7. Infrequent occurrence of muscle contractures and skeletal deformity
8. Insidious progress of the disease with prolonged periods of apparent arrest

With the advent of molecular genetics, and therefore the need for well-defined criteria for each neuromuscular disease, the European Neuromuscular Centre held a workshop on the diagnostic criteria for facioscapulohumeral dystrophy (FSHD) (10), which has now been expanded (11,12). This group defined the criteria as follows:

1. Onset of disease in facial or shoulder muscles, with sparing of the extraocular, pharyngeal, and lingual muscles and the myocardium
2. Facial weakness in more than 50% of affected family members
3. Autosomal-dominant inheritance in familial cases
4. Evidence of *myopathic* disease on electromyography (EMG) and muscle biopsy in at least one affected member, without features specific to alternative diagnoses

However, while this most recent attempt at definition is clearly important with respect to the autosomal-dominant-inherited form of FSH dystrophy, it does not address the problem of alternative pathological diagnoses (FSH syndrome). FSH dystrophy is clinically and genetically distinct from scapuloperoneal syndrome (13), while scapuloperoneal dystrophy appears to be either allelic with (14) or a restricted form of FSHD (15).

MUSCLE PATHOLOGY

The interpretation of muscle pathology is difficult in a condition that shows slow clinical evolution and must be considered with regard to the muscle biopsied, the duration of disease at the time of biopsy, and the strength of the muscle biopsied. It is also important that muscle from which the biopsy is taken is affected, but not severely so (16).

In true FSHD, there will be evidence of a myopathic process with variation in the fiber diameter, central nucleation, occasional necrotic fibers and regenerating fibers, and variable increase in endomysial and perimysial connective tissue (17). Within so-called "dystrophic" muscle, there may be alterations of oxidative enzyme activity (moth-eaten fibers), and occasionally "lobulated" fibers are seen, caused by aberrant myofibers that appear to partition the muscle fiber (18). The presence of focal areas of mononuclear cell infiltration, in a pervivascular or perimysial distribution, is well recognized (19), occurring in both familial and sporadic cases. These occur in up to 70% of cases (20), being mild in the majority (>50 cells/1000 fibers) and marked (>600 cells/1000 fibers) in one-third.

However, there is undoubtedly pathological heterogeneity in the FSH syndrome (21–24). A neurogenic form of the condition has been described (25–27), and there have been several subsequent individual cases and families described in which the condition appears to have a neurogenic basis (28). To some extent this classification into neurogenic versus myopathic forms depends on the criteria for the diagnosis of a neurogenic disorder, and some of the previously reported cases must be considered suspect in that regard, in that in some instances only EMG studies were undertaken (29) while in others the histochemical analysis could not be considered adequate. However, in the author's series of 62 patients with FSH syndrome (22), the largest group (45 patients) showed evidence of neurogenic pathology, according to standard criteria (16,30) (Table 1).

Within this group of patients, at the time of biopsy 54 patients showed an FSH distribution of weakness only, and the other eight showed evidence of additional lower limb weakness. In all instances the quadriceps was biopsied whether it was clinically involved or not, even though it could be shown to be involved subclinically on CT scanning in all cases (Fig. 1) (31). The diagnosis of a neurogenic abnormality was based on the presence of scattered and grouped atrophy and fiber type grouping (Fig. 2) (16,30). Seven patients showed clear-cut dystrophic abnormalities. In three patients, the only

Table 1 Pathological Features in 62 Quadriceps Biopsies in FSH Syndrome

Neurogenic	Dystrophic	Dystrophic + inflammatory infiltrate	Lobulated	Myositis	IBM
45					
	9				
		3			
			3		
				1	
					1

abnormality demonstrated was the presence of lobulated fibers, and each of these patients had an age of onset of greater than 40 years (Fig. 3a). Three patients had evidence of a dystrophic process with, in addition, an inflammatory infiltrate (Fig. 3b). One patient showed the classical features of polymyositis (including repeat biopsy 5 years later), and one patient showed features of inclusion body myositis. These findings indicating pathological heterogeneity are in keeping with those of Mussini and Le Noan (32) and others who have described the FSH phenotype in association with polymyositis (33), mitochondrial myopathies (34–36), inflammatory infiltrates (19), congenital myopathies (37), familial inclusion body myositis (38), and accumulation of desmin filaments (39). This pathological heterogeneity is also in keeping with evidence of clinical heterogeneity (40,41). It has been suggested that retinal abnormalities are ubiquitous in FSH (42); however, most would accept that only *some* families [10% in Padberg's series (43,44)] and sporadic cases show evidence of retinal and/or hearing abnormalities (Coats syndrome) (45–52), which thus defines this subgroup of patients as part of the FSH syndrome.

It has been shown that while patients with a single pathological substrate (denervation) may have phenotypes of limb girdle, FSH, or scapuloperoneal type, there is in fact no real difference between these patients on the basis of the muscles involved as assessed by computerized-tomography scanning (31). It is therefore important to distinguish FSH dystrophy from FSH syndrome (35,53).

FACIOSCAPULOHUMERAL DYSTROPHY

It is unusual for FSHD to appear in infancy (54,55). The usual age of onset has been described as being between 7 and 20 years (56), between 7 and 27 years (9), and between 3 and 44 years (22).

Since the onset of weakness is very slow, many patients do not accept that they have a clinical problem until well in to their twenties or even thirties. Given that the weakness starts in the face and is recognized to be inherited, the slight drooping of the eyelids and the slightly transverse smile are accepted as being normal facial characteristics for that family, and the scapular winging, which often develops in the early teens, is more often considered an aberration, rather than being of clinical importance. It is usual to find that the patient has never been able to whistle or use a straw, because of the weakness of the facial muscles, or if they have done so, these functions have been lost early. It is often only when closure of the eyelids fails to cover the sclerae that patients seek clinical attention, and initial approaches are often made via the department of ophthalmology. However, in studying families with FSHD, it is clear that weakness of the orbicularis oculi on formal testing is a very early sign. The weakness of the shoulder girdle muscles, in particular the scapular fixators, also occurs early but rarely brings the patient to clinical attention until arm abduction becomes impaired because of loss of scapular fixation. The muscles involved are the latissimus dorsi, the lower portion of the trapezius, the rhomboid, and the serratus anterior. This gives the classic appearance at rest where the scapulae are lying more laterally than normal and rise upward and forward, producing a forward-sloped appearance to the shoulders (Fig. 4a). The clavicles are virtually horizontal; eventually the upper borders of the scapulae rise into the substance of trapezius, and the condition may initially be mistaken for trapezius hypertrophy (Fig. 4b).

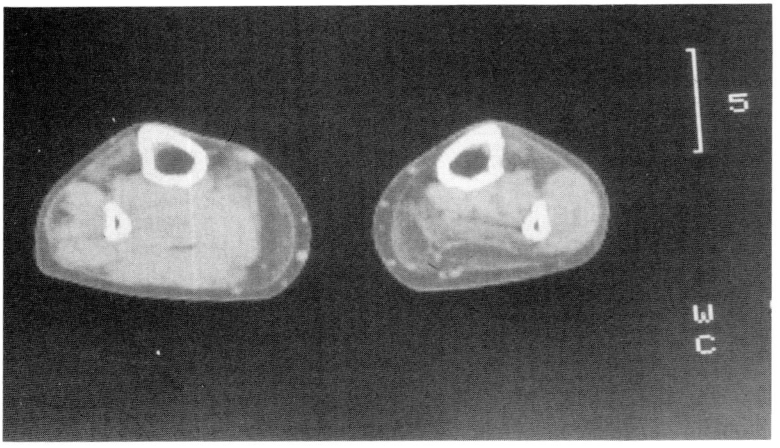

(a)

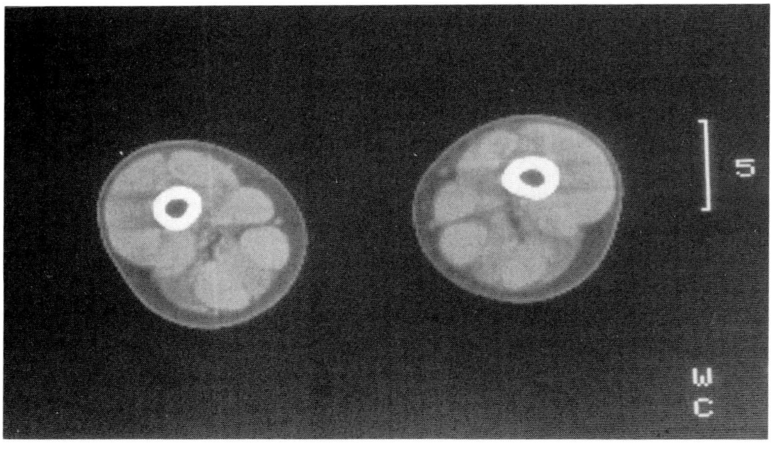

(b)

Figure 1 Computerized-tomography muscle scan of 28-year-old man without clinical evidence of lower limb weakness. (a) Showing evidence of muscle involvement. (b) Thigh muscles showing less marked, but evident, muscle involvement.

There is evidence for selectivity of muscle involvement in the early stages in that although the serratus anterior may be markedly weak, the supraspinatus and infraspinatus are often only mildly involved, but the sternocostal head of the pectoralis major is almost always involved. This may be associated with a degree of pectus excavatum.

When the patient is elevating the arm, the first 25° of arm elevation is carried out by the supraspinatus muscle. The remaining movement to bring the arm to the horizontal position is undertaken by the deltoid. To get beyond that position, for example, in reaching to a high shelf and using the arm for attention to the face or head, the scapula has to be rotated; this requires that it be fixed to the chest wall for the development of appropriate mechanical advantage. It is when that function is impaired or lost that the patient often seeks clinical attention. The timing of that will depend on how much the patient raises his or her arms above the head, either in work or at home. Most patients learn "trick" movements whereby they throw their arm forward into the air with a quick movement and then support it with the other arm. One patient coped perfectly well in his job as an electronics technician in the navy

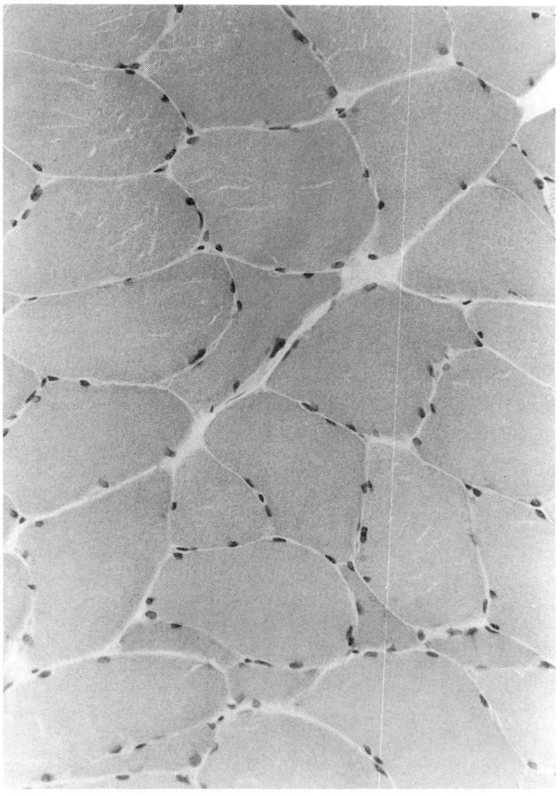

Figure 2 Quadriceps biopsy: scattered and groups of angular fibers with evidence of grouping on H&E. ×62.

and came to clinical attention when he was sent on an officer's course that required him to undertake physical activity under supervision, which revealed his scapular winging and inability to use the arm above the head. In Padberg's series (44,57), over 80% of the patients sought clinical attention because of weakness in the upper limbs.

The weakness that develops in the lower limbs (58) typically affects the anterior tibial compartment in the first instance, leading to footdrop. However, up to 50% of patients never develop clinically significant impairment of lower limb function although it is demonstrable on clinical examination. When the lower limb weakness spreads outside the anterior tibial group, it then classically involves the pelvic girdle, leading to a proximal myopathy, which may be asymmetrical or symmetrical. When severe, this proximal lower limb weakness may require use of or dependence on a wheelchair.

Kazakov et al. (59,60), in a study of 200 cases of FSH, divided their patients into two groups depending on the lower limb involvement. In the larger group, the late lower limb involvement primarily affected the anterior tibial and peroneal muscles, while in the smaller group, the proximal pelvic girdle muscles were more severely affected. They believed that these groups of patients were genetically distinct (61). However, pathologically it was not possible to rigorously exclude neurogenic disease in the late proximal lower limb group.

Although the differentiation between the patients with FSHD, on the basis of their lower limb pattern of involvement, is clinically well defined, long-term follow-up studies with adequate pathological investigation will be required to demonstrate whether this group has a separate pathological substrate.

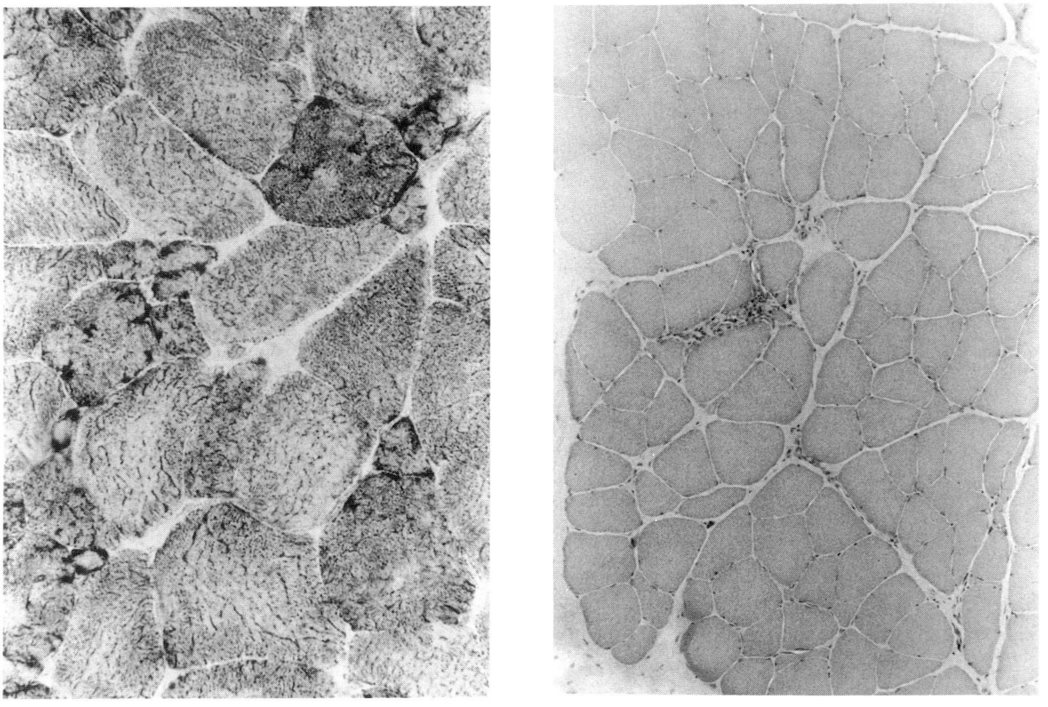

(a) (b)

Figure 3 (a) Quadriceps biopsy showing lobulated fibers. NADHTr. ×100. (b) Quadriceps biopsy showing inflammatory infiltrates. H&E. ×32.

There appears to be a subgroup of FSHD patients who progress rapidly from facial, through upper limb, anterior tibial, and then pelvic girdle involvement, which may result in the necessity for ambulatory aids at an early stage, whereas other patients do not seem to progress beyond minimal anterior tibial weakness. It has been suggested that these two lower limb variants of the condition run true in families. To be certain of this, however, will require extended longitudinal studies within families, with sequential graded neuromuscular testing; such studies are now being devised (62).

There was some evidence, albeit soft, to suggest truncal weakness, and hence involvement of the axial musculature, in one series (63).

It is clear that the life span of FSHD patients is not greatly affected. In Padberg's (44,57) series, the mean age at death was 64 years in men and 70 years in women, compared to expected figures of 65 years and 67 years, respectively.

In the more rapidly progressive cases, death may occur from respiratory failure in midlife, and rarely the infantile form (64,65) may be severe enough to be fatal in childhood (66). Eventual severity of the disease shows an inverse correlation with age at first recognized onset (67). Lunt and associates (15,58), in their extensive studies of FSHD, have shown that, in familial cases, penetrance for clinical signs is 95% by age 20, and that 19% of patients require a wheelchair by age 40.

GENETICS

Autosomal-dominant inheritance is the expected mode of inheritance in this condition. Sporadic cases must be considered to be part of familial FSHD until all family members have been examined and minor degrees of neuromuscular involvement have been excluded (68). This may involve measurement of

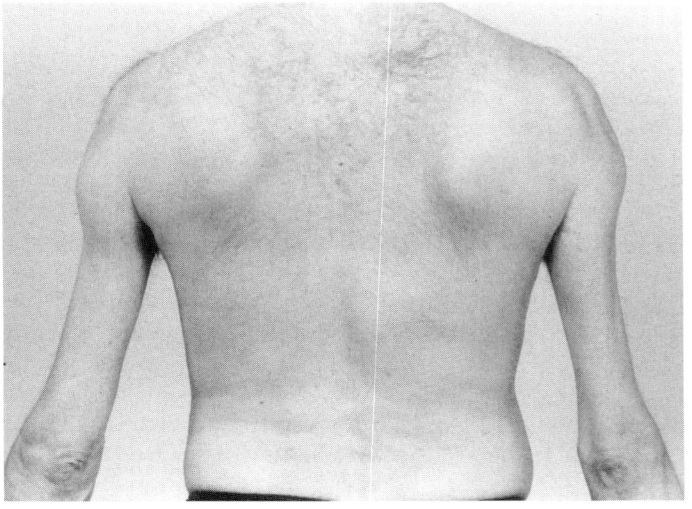

(a)

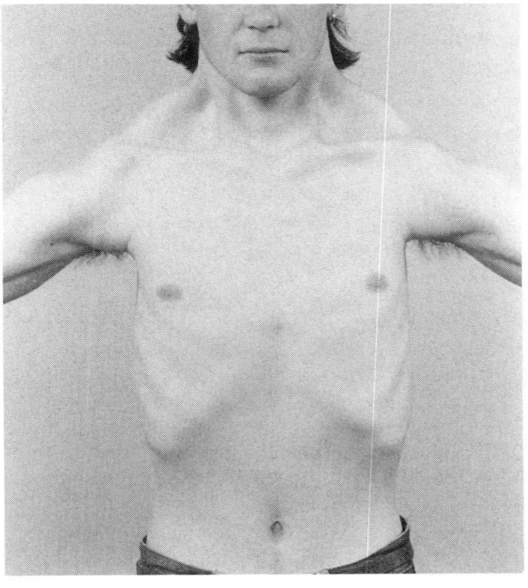

(b)

Figure 4 (a) Typical upper limb girdle appearances of FSH. (b) Overriding of the scapulae on arm elevation simulating trapezius hypertrophy.

serum creatine kinase, EMG studies, and muscle biopsy. Only in the absence of evidence of subclinical disease by these investigations may the case be considered to be truly sporadic. Autosomal-recessive inheritance and sporadic cases may occur (69,70); however, X-linked inheritance has not been described in FSHD. Germinal mosaicism has been described (71).

The laboratory investigations of patients with FSHD are those used to investigate any neuromuscular condition. Some elevation of the serum creatine kinase will be found in the majority of individuals together with an abnormal EMG showing polyphasic motor units of low amplitude and short duration in up to 50% of patients.

MOLECULAR GENETICS

The genetic locus for FSHD has now been identified and is on the long arm of chromosome 4, at 4q35 (72–94). The studies published to date are those of autosomal-dominant FSH disease and presumably therefore of FSH dystrophy, although pathological correlations are conspicuous by their absence. Such studies have been used for prenatal diagnosis and counseling (95–97). However, even among patients who conform to the standard autosomal dominant pattern of inheritance, there is now evidence of genetic heterogeneity (44,98–100).

At present the studies in molecular biology have advanced to the point where the locus for FSH disease has been narrowed, but as yet the responsible gene has not yet been cloned (101–105). At 4q35 there is a tandemly arranged, 3.2-kb repeat DNA sequence, and FSHD has been associated with deletions within this region leading to variable size of the resulting fragment (74) between 15 and 40 kb. This disruption of the normal pattern may perturb the "coding" sequence, although this view has been challenged (93).

In familial cases the fragment size ranged from 19 to 30 kb, was constant within the family, and was associated with a median age of onset of 18 years (range 8–23 years), whereas in de novo cases the fragment size was generally smaller (12–24 kb) and was associated with an earlier age of onset and more rapid progression, which was continued into successive generations in some families. Age at onset of wheelchair use was again correlated with fragment size; the smaller the fragment, the earlier the use of a wheelchair (15,106).

These results have major clinical implications and suggest that large fragments (>30 kb) cosegregating with 4q35 markers will lead to a milder phenotype with later age of onset and vice versa, thus permitting more refined clinical and genetic counseling (107). Pathological correlations with the presence or absence of a 4q35 abnormality and its size should resolve the still unanswered problems outlined above (20).

THERAPY

There is no specific therapy that will alter the natural history of the underlying disorder. Sporadic reports suggest that patients who have inflammatory infiltration or an apparent necrobiotic myopathy may benefit from steroid therapy. However, no long-term studies confirm this belief, and Munsat and Bradley (108) clearly showed that long-term treatment in one of their patients was of no benefit in slowing progression. Scapular fixation may be of benefit in selected patients (109), provided deltoid strength is preserved. However, it is the author's experience that patients will only tolerate stabilization of one scapula.

Patients may be reassured that the disease is usually only very slowly progressive and that life span is not affected.

It has generally been considered that the heart is spared in FSH disease and, in particular, FSH dystrophy (11,110,111). However, cardiac involvement may occur occasionally (112–114). It is difficult to determine whether this cardiac involvement indicates a further clinical subdivision of FSHD or whether, as in inflammatory myopathy, the more assiduously heart disease is sought, the more often it will be found.

REFERENCES

1. Gardner-Medwin D, Walton JN. The muscular dystrophies. In: Walton JN, Karpati G, Hilton-Jones D, eds. Disorders of Voluntary Muscle, 6th ed. Edinburgh: Churchill Livingstone, 1994:543–594.
2. Landouzy L, Déjerine J. De la myopathie atrophique progressive (myopathie héréditaire débutant dans l'enfance, par la face sans altération du système nerveux). C R Seances Acad Sci 1884; 98:53.

3. Duchenne GB. Récherches sur la paralysie musculaire pseudohypertrophique, ou paralysie myosclérosique. Arch Gen Med 1868; 11:5, 179,305,421,552.
4. Munsat TL. Facioscapulohumeral dystrophy and the scapuloperoneal syndrome. In: Engel AG, Banker BQ, eds. Myology. New York: McGraw-Hill, 1986:1251–1266.
5. Munsat TL, Serratrice G. Facioscapulohumeral and scapuloperoneal syndromes. In: Vinken PJ, Bruyn GW, Klawans HL, eds. Handbook of Clinical Neurology. Amsterdam: Elsevier, 1992:161–177.
6. Davidenkow S. Über die Veretbung der dystrophia musculorum progressiva und über Unterformen. Arch Rassenbiol 1930; 18:22.
7. Tyler FH, Stephens FE. Studies in disorders of muscles. II. Clinical manifestations and inheritance of facioscapulohumeral dystrophy in a large family. Ann Intern Med 1950; 32:640.
8. Chung CS, Morton NE. Discrimination of genetic entities in muscular dystrophy. Am J Hum Genet 1959; 11:339.
9. Walton JN, Nattrass FJ. On the classification, natural history and treatment of the myopathies. Brain 1954; 77:169.
10. Padberg GW, Lunt PW, Koch M, Fardeau M. Diagnostic criteria for fascioscapulohumeral muscular dystrophy. Neuromusc Disord 1991; 1:231–234.
11. Padberg GW, Lunt PW, Koch M, Fardeau M. Facioscapulohumeral muscular dystrophy. In: Emery AEH, ed. Diagnostic Criteria for Neuromuscular Disorders. Baarn, The Netherlands: European Neuromuscular Centre, 1994:19–24.
12. Lunt PW. Report of the Sixth International Workshop on Facioscapulohumeral Muscular Dystrophy: San Francisco, 11 November 1992; and current guidelines for clinical application of DNA rearrangements at locus D4S810. Neuromusc Disord 1994; 4:83–86.
13. Bachmann H, Ziegan J, Wagner A, Rhoden L, Krebs P. Clinical and genetic differentiation of facioscapulohumeral and scapuloperoneal neuromuscular syndromes. J Neurol Sci 1990; 98(Suppl):194.
14. Jardine PE, Upadhyaya M, Maynard J, Harper P, Lunt PW. A scapular onset muscular dystrophy without facial involvement: possible allelism with facioscapulohumeral muscular dystrophy. Muscle Nerve 1994; 4:477–482.
15. Lunt PW, Jardine PE, Koch MC, et al. Correlation between fragment size at D4F104S1 and age at onset or at wheelchair use, with a possible generational effect, accounts for much phenotypic variation in 4q35-facioscapulohumeral muscular dystrophy (FSHD). Hum Mol Genet 1995; 4:951–958.
16. Cumming WJK, Fulthorpe J, Hudgson P, Mahon M. Colour Atlas of Muscle Pathology. London: Mosby Year Book, 1994.
17. Schmalbruch H. The muscular dystrophies. In: Mastaglia FL, Lord Walton of Detchant, eds. Skeletal Muscle Pathology, 2nd ed. Edinburgh: Churchill Livingstone, 1992:283–318.
18. Bethlem J, Van Wijngaarden GK, De Jong J. The incidence of lobulated fibres in the facioscapulohumeral type of muscular dystrophy and the limb-girdle syndrome. J Neurol Sci 1973; 18:351.
19. Munsat TL, Piper D, Cancilla P, Mednick J. Inflammatory myopathy with facioscapulohumeral dystrophy. Neurology 1972; 22:335.
20. Arahata K, Ishihara T, Fukunaga H, et al. Inflammatory response in facioscapulohumeral muscular dystrophy (FSHD): Immunocytochemical and genetic analyses. Muscle Nerve. 1995; 18 (Suppl 2):S56–S66.
21. Apostolski S, Dozic S, Trikic R, Cvetkovic D, Gospavic J. The heterogeneity of FSH syndrome. J Neurol Sci 1990; 98(Suppl):194.
22. Kristmundsdottir F, Mahon M, Cumming WJK. Fascioscapulohumeral syndrome: pathological heterogenity. J Neurol Sci 1990; 98(Suppl):195.
23. Lunt P. Facioscapulohumeral disease—heterogeneity or homogeneity? J Neurol Sci 1990; 98(Suppl):196.
24. Arahata K, Lee JH, Goto K, et al. Muscle histology of facioscapulohumeral muscular dystrophy (FSHD): is genetic heterogeneity exists for FSHD patients associated with inflammatory changes? Muscle Nerve 1994; S29.
25. Fenichel GM, Emery ES, Hung P. Neurogenic atrophy simulating facioscapulohumeral dystrophy. A dominant form. Arch Neurol 1967; 17:257.
26. Furukawa T, Toyokura Y. Chronic spinal muscular atrophy of facioscapulohumeral type. J Med Genet 1976; 13:285–289.
27. Kazakov VM. Spinal Facioscapuloperoneal (or facioscapuloshin) type of muscular atrophy simulating facioscapuloperoneal type of muscular dystrophy: report of a case. Acta Cardiomiol 1992; 4:179–195.
28. Lin M-Y, Nonaka I. Facioscapulohumeral muscular dystrophy: muscle fiber type analysis with particular reference to small angular fibers. Brain Dev 1991; 13:331–338.
29. Fawcett PRW, Kennett RP, Schofield IS, Johnson MJ. A neurophysiological investigation of the facio-scapulo-peroneal (FSH) syndrome. J Neurol Sci 1990; 98(Suppl):195.
30. Jennekens FGI. Neurogenic disorders of muscle. In: Mastaglia FL, Lord Walton of Detchant, eds. Skeletal Muscle Pathology, 2nd ed. Edinburgh: Churchill Livingstone, 1992:563–598.

31. Sambrook P, Rickards D, Cumming WJK. CT muscle scanning in the evaluation of patients with spinal muscular atrophy (SMA). Neuroradiology 1988; 30:487–496.
32. Mussini J-M, Le Noan H. Inflammatory changes in facio scapulo humeral dystrophy. J Neurol Sci 1990: 98(Suppl):196.
33. Bates D, Stevens JC, Hudgson P. "Polymyositis" with involvement of the facial and distal musculature. One form of the facioscapulohumeral syndrome. J Neurol Sci 1973; 19:105–108.
34. Hudgson P, Bradley WG, Jenkinson M. Familial mitochondrial myopathy. Part I. Clinical, electrophysiological and pathological findings. J Neurol Sci 1972; 16:343–370.
35. Slipetz DM, Aprille JR, Goodyear PR, Rozen R. Deficiency of complex III of the mitochondrial respiratory chain in a patient with facioscapulohumeral disease. Am J Hum Genet 1991; 48·502–510.
36. Worsfold M, Park DC, Pennington RJ. Familial "mitochondrial" myopathy: a myopathy associated with disordered oxidative metabolism in muscle fibers. II. Biochemical findings. J Neurol Sci 1973; 19:261–274.
37. Serratrice G, Pellissier JF, Bes A, Arne-Bes MC. [A case of centronuclear myopathy in adults with facio-scapulo-peroneal topography]. Rev Neurol (Paris) 1987; 143:693–697.
38. McKee D, Karpati G, Carpenter S, Johnson W. Familial inclusion body myositis (IBM) mimics facioscapulohumeral dystrophy (FSHD). Neurology 1992; 42:302.
39. Askanas V, Alvarez RB, McFerrin J, Serdaroglu P, Lew MF, Engel WK. Facio-scapulo-humeral (FSH) syndrome with abnormal desmin accumulation. Neurology 1992; 42:229.
40. Ohkubo R, Nakagawa M, Yamada H, et al. Clinical heterogeneity and molecular analysis of facioscapulohumeral muscular dystrophy (FSHD) in southern part of Japan. Muscle Nerve 1994; S178.
41. Padberg G, Frants RR, Brouwer OF, Wijmenga C, van Deutekom J, van Geel M. Facioscapulohumeral dystrophy: clinical picture in relation to molecular genetics. Muscle Nerve 1994; S29.
42. Fitzsimons RB, Gurwin EB, Bird AC. Retinal vascular abnormalities in facioscapulohumeral muscular dystrophy: a general association with genetic and therapeutic implications. Brain 1987; 110:631–648.
43. Padberg GW, Brouwer OF, De Keizer RJW, et al. On the significance of retinal vascular disease and hearing loss in facioscapulohumeral muscular dystrophy. Muscle Nerve. 1995; 18 (Suppl 2):S73–S80.
44. Padberg GW, Frants RR, Brouwer OF, Wijmenga C, Bakker E, Sandkuijl LA. Facioscapulohumeral muscular dystrophy in the Dutch population. Muscle Nerve 1995; 18 (Suppl 2):S81–S84.
45. Brouwer OF, Padberg GW, Ruys CJM, Brand R, De Laat JAPM, Grote JJ. Hearing loss in facioscapulohumeral muscular dystrophy. Neurology 1991; 41:1878–1881.
46. Desai UR, Sabates FN. Long-term follow-up of facioscapulohumeral muscular dystrophy and Coats' disease. Am J Ophthalmol 1990; 110:568–569.
47. Frants RR, Brouwer OF, Ruys CJM, et al. Hearing loss in facioscapulohumeral muscular dystrophy. J Neurol Sci 1990; 98(Suppl):197.
48. Padberg GW, Brouwer OF, de Keizer RJW, Wijmenga C. Retinal vascular disease and perception deafness in facioscapulohumeral muscular dystrophy. J Neurol Sci 1990; 98(Suppl):196–197.
49. Gieron MA, Korthals JK, Kousseff BG. Facioscapulohumeral dystrophy with cochlea hearing loss and tortuosity of retinal vessels. Am J Med Genet 1988; 22:143–147.
50. Matsuzaka T, Sakuragawa N, Terasawn K, Kuwabara H. Facioscapulohumeral dystrophy associated with mental retardation, hearing loss, and torturosity of retinal arterioles. J Child Neurol 1986; 1:218–223.
51. Voit T, Lamprecht A, Lenard HG, Goebel HH. Hearing loss in facioscapulohumeral dystrophy. Eur J Pediatr 1986; 145:280–285.
52. Mitsuyoshi I, Saida K, Itagaki Y, Iwamura K, Nishitani H. Facioscapulohumeral muscular dystrophy associated with sensorineural hearing loss, tortuosity of retinal vessels and abnormal capillary proliferations in biopsied muscle. Muscle Nerve 1994; S179.
53. Serratrice G, Pellissier JF, Bes A, Arne-Bes MC. Centronuclear myopathy in adults. A facio-scapulo-peroneal case. Rev Neurol 1987; 143:693–697.
54. Hanson PA, Rowland LP. MÖbius syndrome and facioscapulohumeral muscular dystrophy. Arch Neurol 1971; 24:31.
55. Brouwer OF, Padberg GW, Bakker E, Wijmenga C, Frants RR. Early onset facioscapulohumeral muscular dystrophy. Muscle Nerve 1995; 18(Suppl 2):S67–S72.
56. Buchwald D, Garrity D. Comparison of patients with chronic fatigue syndrome, fibromyalgia, and multiple chemical sensitivities. Arch Intern Med 1994; 154:2049–2053.
57. Padberg GW. Facioscapuloperoneal disease. Doctoral Thesis, University of Leiden, 1982.
58. Lunt P, Compston DAS, Harper PS. Estimation of age dependent penetrance in facioscapulohumeral muscular dystrophy by minimising ascertainment bias. J Med Genet 1989; 26:755–760.
59. Kazakov VM, Bogorodinsky DK, Znoyko ZV, Skorometz AA. The facio-scapulo-limb (or the facioscapulohumeral) type of muscular dystrophy: clinical and genetic study of 200 cases. Eur Neurol 1974; 11: 236.
60. Kazakov VM, Rudenko DI. Clinical variability of facioscapulohumeral muscular dystrophy in Russia. Muscle Nerve. 1995; 18(Suppl 2):S85–S95.

61. Kazakov V, Rudemko D, Pavlov IP. Clinical variability of facioscapulohumeral muscular dystrophy in Russia. Muscle Nerve 1994; S179.
62. Tawil R, McDermott MP, Mendell JR, Kissel J, Griggs RC. Facioscapulohumeral muscular dystrophy (FSHD): design of natural history study and results of baseline testing. Neurology 1994; 44:442–446.
63. Awerbuch GI, Nigro MA, Wishnow R. Beevor's sign and facioscapulohumeral dystrophy. Arch Neurol 1990; 47:1208–1209.
64. Brouwer OF, Padberg GW, Wijmenga C, Frants RR. Facioscapulohumeral muscular dystrophy in early childhood. Arch Neurol 1994; 51:387–394.
65. Brouwer OF, Padberg G, Wijmenga C, Frants RR. Early onset Facioscapulohumeral muscular dystrophy. Muscle Nerve 1994; S177.
66. McGarry J, Garg B, Silbert S. Death in childhood due to facio-scapulo-humeral dystrophy. Acta Neurol Scand 1983; 68:61–63.
67. Lunt PW, Harper PS. Genetic counselling in facioscapulohumeral muscular dystrophy. J Med Genet 1991; 28:655–664.
68. Tawil R, Storvick D, Feasby TE, Weiffenbach B, Griggs RC. Extreme variability of expression in monozygotic twins with FSH muscular dystrophy. Neurology 1993; 43:345–348.
69. Moser von H, Weismann U, Richterich R, Rossi E. Progressive Muskeldystrophie. VIII Haufigkeit, Klinik und Genetik der Typen I und III. Schweiz Med Wochenschr 1966; 96:169.
70. Griggs RC, Tawil R, Storvick D, Mendell JR, Altherr MR. Genetics of facioscapulohumeral muscular dystrophy: new mutations in sporadic cases. Neurology 1993; 43:2369–2372.
71. Upadhyaya M, Maynard J, Osborn M, Jardine P, Harper PS, Lunt P. Germinal mosaicism in facioscapulohumeral muscular dystrophy (FSHD). Muscle Nerve. 1995; 18(Suppl 2):S45–S49.
72. Wijmenga C, Frants RR, Brouwer OF, Moerer P, Weber JL, Padberg GW. Location of facioscapulohumeral muscular dystrophy gene on chromosome 4. Lancet 1990; 336:651–653.
73. Wijmenga C, Padberg GW, Moerer P, et al. Mapping of facioscapulohumeral muscular dystrophy gene to chromosome 4q35-qter by multipoint linkage analysis and in situ hybridization. Genomics 1991; 9:570–575.
74. Sarfarazi M, Wijmenga C, Upadhyaya M, et al. Regional mapping of facioscapulohumeral muscular dystrophy gene on 4q35: combined analysis of an international consortium. Am J Hum Genet 1992; 51:396–403.
75. Upadhyaya M, Lunt P, Sarfarazi M, Broadhead W, Farnham J, Harper PS. The mapping of chromosome 4q markers in relation to facioscapulohumeral muscular dystrophy (FSHD). Am J Hum Genet 1992; 51:404–410.
76. Wijmenga C, Sandkuijl LA, Moerer P, et al. Genetic linkage map of facioscapulohumeral muscular dystrophy and five polymorphic loci on chromosome 4q35-qter. Am J Hum Genet 1992; 51:411–415.
77. Weiffenbach B, Bagley R, Falls K, et al. Linkage analyses of five chromosome 4 markers localizes the facioscapulohumeral muscular dystrophy (FSHD) gene to distal 4q35. Am J Hum Genet 1992; 51:416–423.
78. Gilbert JR, Stajich JM, Speer MC, et al. Linkage studies in facioscapulohumeral muscular dystrophy (FSHD). Am J Hum Genet 1992; 51:424–427.
79. Mathews KD, Mills KA, Bosch EP, et al. Linkage localization on facioscapulohumeral muscular dystrophy (FSHD) in 4q35. Am J Hum Genet 1992; 51:428–431.
80. Mills KA, Buetow KH, Xu Y, et al. Genetic and physical mapping on chromosome 4 narrows the localization of the gene for facioscapulohumeral muscular dystrophy (FSHD). Am J Hum Genet 1992; 51:432–439.
81. Wijmenga C, Hewitt JE, Sandkuijl LA, et al. Chromosome 4q DNA rearrangements associated with facioscapulohumeral muscular dystrophy. Nature Genet 1992; 2:26–30.
82. Upadhyaya M, Jardine P, Maynard J, et al. Molecular analysis of British facioscapulohumeral dystrophy families for 4q DNA rearrangements. Hum Mol Genet 1993; 2:981–987.
83. Altherr MR, Mathews KD, Winokur ST, Weiffenbach B, Schutte BC. Efforts toward understanding the molecular basis of facioscapulohumeral muscular dystrophy. Muscle Nerve 1994; S28.
84. Frants R, van Deutekom J, Rommens J, et al. Towards a transcription map of the FSHD gene region on 4q35. Muscle Nerve 1994; S28.
85. Hewitt JE, Lyle R, Clark LN, et al. Analysis of the tandem repeat locus D4Z4 associated with facioscapulohumeral muscular dystrophy. Hum Mol Genet 1994; 3:1287–1295.
86. Jardine PE, Upadhyaya M, Harper PS, Lunt PW. A clinical study of new mutation cases of facioscapulohumeral muscular dystrophy. Muscle Nerve 1994; S178.
87. Zatz M, Marie SK, Passos-Bueno MR, et al. High proportion of new mutations and possible anticipation following molecular genetic studies in Brazilian facioscapulohumeral muscular dystrophy (FSHD) families. Muscle Nerve 1994; S178.
88. Zatz M, Marie SK, Passos-Bueno MR, et al. High proportion of new mutations and possible anticipation in Brazilian facioscapulohumeral muscular dystrophy families. Am J Hum Genet 1995; 56:99–105.
89. Lee JH, Goto K, Matsuda C, Arahata K. Characterization of a tandemly repeated 3.3-kb KpnI unit in the facioscapulohumeral muscular dystrophy (FSHD) gene region on chromosome 4q35. Muscle Nerve 1995; 18(Suppl 2):S6–S13.

90. Wijmenga C, Dauwerse HG, Padberg GW, et al. FISH mapping of 250 cosmid and 26 YAC clones to chromosome 4 with special emphasis on the FSHD region at 4q35. Muscle Nerve 1995; 18(Suppl 2):S14–S18.
91. Van Deutekom JCT, Hofker MH, Romberg S, et al. Search for the FSHD gene using cDNA selection in a region spanning 100 kb on chromosome 4q35. Muscle Nerve 1995; 18(Suppl 2):S19–S26.
92. Lee JH, Goto K, Sahashi K, Nonaka I, Matsuda C, Arahata K. Cloning and mapping of a very short (10-kb) EcoRI fragment associated with facioscapulohumeral muscular dystrophy (FSHD). Muscle Nerve 1995; 18(Suppl 2):S27–S31.
93. Altherr MR, Bengtsson U, Markovich RP, Winokur ST. Efforts toward understanding the molecular basis of facioscapulohumeral muscular dystrophy. Muscle Nerve 1995; 18(Suppl 2):S32–S38.
94. Bakker E, Wijmenga C, Vossen RHAM, et al. The FSHD-linked locus D4F104S1 (p13E-11) on 4q35 has a homologue on 10qter. Muscle Nerve 1995; 18(Suppl 2):S39–S44.
95. Upadhyaya M, Lunt PW, Sarfarazi M, et al. DNA marker applicable to presymptomatic and prenatal diagnosis of facioscapulohumeral disease. Lancet 1990; 336:1320–1321.
96. Passos-Bueno MR, Wijmenga C, Takata RE, et al. No evidence of genetic heterogeneity in Brazilian facioscapulohumeral muscular dystrohy families (FSHD) with 4q markers. Hum Mol Genet 1993; 2:557–562.
97. Wijmenga C, Frants RR, Hewitt JE, et al. Molecular genetics of facioscapulohumeral muscular dystrophy. Neuromusc Disord 1993; 3:487–491.
98. Iqbal Z, Roper H, Pericak-Vance MA, et al. Genetic heterogeneity in facioscapulohumeral disease. Am J Hum Genet 1992; 51:A191.
99. Gilbert JR, Stajich JM, Wall S, et al. Evidence for heterogeneity in facioscapulohumeral muscular dystrophy (FSHD). Am J Hum Genet 1993; 53:401–408.
100. Lunt P, Jardine P, Upadhyaya M, Harper P. Genetic counselling and molecular definition of phenotype in facioscapulohumeral muscular dystrophy. Muscle Nerve 1994; S29.
101. Fischbeck KH, Garbern JY. Facioscapulohumeral muscular dystrophy defect identified. Nature Genet 1992; 2:3–4.
102. Van Deutekom JCT, Wijmenga C, Van Tienhoven EAE, et al. FSHD associated DNA rearrangements are due to deletions of integral copies of a 3.2 kb tandemly repeated unit. Hum Mol Genet 1993; 2:2037–2042.
103. Wijmenga C, Van DJCT, Hewitt JE, et al. Pulsed-field gel electrophoresis of the D4F104S1 locus reveals the size and the parental origin of the facioscapulohumeral muscular dystrophy (FSHD)-associated deletions. Genomics 1994; 19:21–26.
104. van Deutekom JCT, Wijmenga C, van Tienhoven EAE, et al. FSHD associated DNA rearrangements are due to deletions of integral copies of a 3.2kb tandemly repeated unit. Hum Mol Genet 1993; 2:2037–2042.
105. Wijmenga C, Hofker MH, Padberg GW, Frants RR. Genetic mapping of facioscapulohumeral muscular dystrophy. In: Partridge T, ed. Molecular and Cell Biology of Muscular Dystrophy. London: Chapman Hall, 1993:111–138.
106. Griggs RC, Tawil R, McDermott M, et al. Monozygotic twins with facioscapulohumeral dystrophy (FSHD): implications for genotype/phenotype correlation. Muscle Nerve 1995; 18(Suppl 2):S50–S55.
107. Lunt PW, Jardine PE, Koch M, et al. Phenotypic-genotypic correlation will assist genetic counseling in 4q35-facioscapulohumeral muscular dystrophy. Muscle Nerve 1995; 18(Suppl 2):S103–S109.
108. Munsat TL, Bradley WG. Serum creatine phosphokinase levels and prednisone treated muscle weakness. Neurology 1977; 27:96.
109. Bunch WH, Siegel IM. Scapulothoracic arthrodesis in facioscapulohumeral muscular dystrophy. Review of seventeen procedures with three to twenty-one-year follow-up. J Bone Joint Surg 1993; 75A:372–376.
110. De Visser M, De Voogt WG, La Riviere GV. The heart in Becker muscular dystrophy, facioscapulohumeral dystrophy, and Bethlem myopathy. Muscle Nerve 1992; 15:591–596.
111. Bensaid J. ECG differentiation of muscular dystrophy types. Am Heart J 1993; 125:1820–1821.
112. Topaloglu H, Celiker A, Oezme S, et al. Facioscapulohumeral syndrome with cardiomyopathy. Postgrad Med J 1990; 66:1088–1089.
113. Stevenson WG, Perloff JK, Weiss JN, Anderson TL. Facioscapulohumeral muscular dystrophy: evidence for selective, genetic electrophysiologic cardiac involvement. J Am Coll Cardiol 1990; 15:292–299.
114. Shen EN, Madsen T. Facioscapulohumeral muscular dystrophy and recurrent pacemaker lead dislodgment. Am Heart J 1991; 122:1167–1169.

24

Autosomal Dominant Limb-Girdle Muscular Dystrophy

Ashok Verma and Walter G. Bradley
University of Miami School of Medicine, Miami, Florida

INTRODUCTION

Although Duchenne is mainly known for his descriptions of the X–linked childhood muscular dystrophy that bears his name, it is certain that in his series of papers, he was also describing dystrophies with other modes of inheritance and with other patterns of weakness. In one family reviewed by Nevin (1), Duchenne described a relatively benign limb-girdle myopathy in a father-son pair, suggesting a Mendelian dominant trait; the weakness began in the fifth decade in the father. In 1884, Erb (2) described a juvenile form of muscular dystrophy that primarily affected the shoulder-girdle musculature, spared facial muscles, and had a benign clinical course. Early descriptions of Leyden (3) and Möbius (4) resembled the type described by Erb except that hip-girdle musculature was early and predominantly involved. These descriptions were mostly in sporadic cases, however, and at the time of these early reports the muscular dystrophies were certainly confused with other proximal myopathies, spinal muscular atrophies, and other neurogenic atrophies.

Limb-girdle muscular dystrophy (LGMD) as a nosological entity was first defined in 1954 by Walton and Nattrass (5). These authors reviewed their personal experience with 105 cases of muscular dystrophies of various types and proposed a new classification based on clinical and genetic features. Their concept of LGMD was broad, and it possibly included among LGMDs other neuromuscular disorders with predominant proximal limb weakness which could not be fitted under other categories. Nevertheless, analysis of the pedigrees of most families with LGMD showed that the disease was transmitted by an autosomal recessive (AR) gene (6,7). We have previously reviewed AR LGMD (8). From these early published series of autosomal LGMD, occasional families that did not show recessive inheritance can be culled to make a case of possible autosomal dominant (AD) LGMD. For example, in one family (L4) of Walton et al. (6), the affected cases occurred in more than one generation, suggesting an AD inheritance, although the authors suggested that the affected parent might be a homozygote for the AR trait. There was no consanguinity in this family. In another early series based on a population study of LGMD, Stevenson (9,10) came across two cases in a two-generation family (family D29). We recently reported a six-generation family of LGMD (11). Many more multigeneration kindreds of LGMD have been documented in recent decades (12–21).

The purpose of this chapter is to review the published reports of familial muscular dystrophy that have the following characteristics: (1) *AD inheritance*: Both males and females are affected by inheritance from an affected parent, and each child (or other first-degree relative) of an affected individual has a 50% risk of inheriting the disease. Father-to-son transmission distinguishes AD-pattern from X-linked dominant inheritance. LGMD confined only to females in multiple sibships (22–24) may be a form of X-linked dominant myopathy and is excluded in this review. (2) *Primary muscle degeneration*: Walton (25) defined muscular dystrophy as genetically determined primary degenerative myopathy. A number of diseases with apparently primary degeneration of skeletal muscles, such as nemaline (26,27) and other disfigurative fiber myopathies (28–34), have been recognized since this definition was advanced. Many of them show an AD pattern of inheritance (26–32) and may affect predominantly proximal muscles. However, we have not included these so-called "congenital myopathies" because their characteristic ultrastructural or histochemical features distinguish them from the muscular dystrophies. Also, frequently they have a different age of onset and are less progressive than the muscular dystrophies. (3) *Syndrome of limb-girdle weakness*: A detailed outline of limb-girdle pattern of weakness has been described by Bradley (8) elsewhere. Briefly, the weakness may begin either in pelvic girdle musculature or in muscles of the pectoral girdle and then gradually spread to involve both limb girdles. Eventually, other limb muscles may become affected, but it is the specific tendency for girdle muscles to become weak and wasted early in the disease, with relative sparing of other muscles, which produces the characteristic clinical picture of LGMD. Within the girdle and limb musculature, some muscles are selectively or predominantly involved earlier and preferentially compared to other adjacent muscles. The facial and distal limb muscles are spared until the late stages in LGMD.

GENETIC FEATURES

The AD LGMD are inherited in each family as single-gene conditions in a conventional Mendelian pattern. The underlying concept of single dominant gene inheritance is that a specific disease entity is the ultimate expression of a specific basic defect determined by a mutant gene.

An affected individual with AD LGMD may not always have a clinically affected parent, for a number of reasons. First, an apparently sporadic case may arise from a new mutation. Parenthetically, we note that if a dominant disorder causes death before reproductive age or severely impairs the ability to marry and reproduce, then the cases of such disease would arise only as a result of a new mutation. For example, it can be argued that sporadic congenital muscular dystrophy with early case fatality may be a single gene dominant disorder. The fact that this is rare is not incompatible with this suggestion since every single case would have to arise from a new mutational event.

Second, an apparently sporadic case can result from decreased disease penetrance. Penetrance is a phenomenon that refers to the clinical expression, or lack of it, of a mutant gene. Third, the disease expressivity, i.e., magnitude and severity of weakness, may be too low to become clinically manifest in another family member. Detailed clinical examination, extended follow-up, and laboratory investigations frequently reveal previously unknown cases in families with AD LGMD. Fourth, information about relatives may simply be unavailable, unappreciated, or false. Finally, in an occasional apparently sporadic case, the putative father may not be the actual biological father or may have died before the onset of symptoms.

A review of the reported pedigrees of AD LGMD shows considerable clinical variability among the affected members of different families. It is possible that a single disease locus undergoes more than one mutation (intra-allelic mutations), as has been recognized at the dystrophin-gene locus. Different intra-allelic mutations (allelic heterogeneity) lead to different alterations in gene product, which in turn may produce somewhat different clinical consequences. Conversely, mutations located at different genetic loci (locus heterogencity), involving defects in different gene products, can also lead to clinically indistinguishable phenotypes. Resolution of the basis of heterogeneity among inherited diseases of muscle has just begun to unfold.

The cellular defect(s) in AD LGMD is currently unknown. Some large multigeneration families of AD LGMD provide the opportunity to obtain significant DNA linkage data by using modern tools of molecular genetics. The linkage studies with expressed biological markers (6,12,16,20,35) have so far not been successful in defining a linked marker. Schneiderman et al. (12) reported a family in whom another AD trait involving leukocytes, the Pelger-Hüet anomaly, was found in some family members.

Linkage between LGMD and the Pelger-Hüet anomaly revealed a lod score of 0.35, which fails to indicate linkage between the two disorders.

By the positional cloning approach, one form of AD LGMD has been genetically linked to chromosome 5q22.3–31.3 and another to the telomeric region of chromosome 21q (35a,b). No flanking markers have been reported to date in these forms of AD LGMD and the molecular pathology underlying their dystrophic process is unknown. On the other hand, there have been spectacular advances in understanding the molecular genetic basis of the autosomal-recessive LGMD in recent years. At least three different genes coding separate components of the dystrophin-associated glycoprotein [α (adhalin)-, β-, and λ-sarcoglycan)] and abnormalities of merosin (chromosome 6q22–23) and a muscle-specific proteolytic enzyme, "calpain 3," are now known to cause AR LGMD (35c,d,e).

CLINICAL FEATURES

Dominantly inherited LGMD is distinctly rare. An early study of 214 cases of LGMD by Chung and Morton (36) indicated that 59% of the cases were AR while the rest were sporadic. One case each in the collections of 15 LGMD families by Walton et al. (6) in Northeast England and of 16 families by Stevenson (9,10) in Northern Ireland were probably of AD type. Yates and Emery (37), in a population study of adult-onset LGMD in the Lothian region of Scotland, collected 11 affected subjects in 10 families, nine sporadic and one set of two brothers. Bradley (8) studied 30 cases with as certain a diagnosis of LGMD as possible. In only six patients was there a family history of similar disease. Three patients had a Becker-type X-linked muscular dystrophy, and in two the condition was inherited as an AR trait. One patient had an affected paternal uncle, though his father was clinically normal. Panegyres et al. (21), in a review of 18 LGMD cases in 12 families, described two cases with an AD mode of inheritance. In one, facial muscles were involved, although the clinical profile was different from typical facioscapulohumeral muscular dystrophy (FSHMD). The reports of families with AD LGMD (1,6,9–21,38,39) are summarized in Table 1.

A full comprehension of the clinical picture of AD LGMD is hampered by the lack of longitudinal studies, in addition to the relatively small number of well-documented cases and the difficulty in determining the age of disease onset. Further, a considerable inter- and even intra-familial clinical heterogeneity exists in AD LGMD. Despite these limitations, an overall composite clinical picture of AD LGMD can be pieced together and several clinical subgroups can be identified. The childhood-onset AD LGMD (11,15,16,40) with prominent muscle contractures and benign clinical course is clearly a separate group. Juvenile-onset and early-adult-onset AD LGMD (13,19–21,38) may have a relatively progressive clinical course leading to wheelchair confinement 20–30 years after onset. Late-age-onset AD LGMD (1,14,38) in general causes only limited disability and presumably normal life span of the afflicted individuals. Some other clinical variants are discussed below.

Among the infantile- and childhood-onset cases, Bethlem and Van Wijngaarden (15) described affected members in three European families who started with pelvic girdle weakness and several decades later developed striking contractures of biceps, long flexors of the forearm, and calf muscles. The duration of illness in index cases at the time of evaluation was 34, 40, and 45 years, respectively, and all patients were ambulatory without assistance. Similar examples with what has been termed Bethlem myopathy were reported by Arts et al. (16) in a Polish family, by Mohire et al. (11) in a French-Canadian family, and by Sasabe et al. (40) in a Japanese family. We believe these constitute an entity distinguishable on clinical grounds from the other families with AD LGMD.

The initial weakness affects the pelvic-girdle muscles, especially the hip flexors, in the majority. However, the pattern of initial involvement may not be uniform even in the affected members of the same family. Bacon and Smith (13) described a familial patient who had shoulder muscle weakness at age 18 followed 10 years later by weakness in the thighs. Another member of this family had weakness of thigh muscles at age 21 followed by weakness in the upper arms in the 5th decade. Eventually both upper and lower limb-girdle muscles are involved in almost all cases of AD LGMD. The pattern of weakness is similar to that of AR LGMD (8). Hip flexors, glutei, quadriceps, and hamstrings in lower

Table 1 Families Reported with Autosomal Dominant Limb-Girdle Muscular Dystrophy

Source	Affected generations/ # affected members	Onset	Distribution of weakness	Clinical course	Serum CK	EMG	Biopsy	Remarks
Duchenne (quoted by Nevin, 1936) (1)	2/3	48 yr in father	PG,SG	—	—	—	—	Father and two sons affected
Sezary et al. (1909) (quoted by Nevin, 1936) (1)	2/2	6th decade	SG,A	—	—	—	—	Father and son affected
Barnes (1932) (38)	6/61	2nd–6th decade	PG,SG,A Late in Hands	Very slow	NA	Myogenic	Dystrophic	Four phases: hypertrophy, pseudohyper-trophy, atrophy, and terminal
Nevin (1936) (1)	2/2	5th, 6th decade	PG,SG,A	Very slow	NA	NA	Dystrophic	Father and daughter affected
Hamptman and Thannhauser (1941) (39)	3/9	1st decade	PG,A,T	Extremely slow	NA	NA	NA	Webbed neck, shortening of neck and back muscles, biceps, and hamstrings
Stevenson (1953) (Family D29) (9,10)	2/2	1st decade	PG,A	Very slow	NA	NA	NA	Father and daughter affected
Walton (1955) (Family L4) (6)	2/3	2nd, 3rd decade	PG,SG	Very slow	NA	NA	NA	Mother and two sons affected
Scheiderman et al. (1969) (12)	4/18	1st, 3rd decade	PG,SG	Extremely slow	NL	Myogenic	Myopathic	Pelger-Hüet anomaly of leukocytes in some members
Bacon and Smith (1971) (13)	3/6	2nd, 3rd decade	SG,PG,A,T	Extremely slow	NL	NA	Myopathic	—
De Coster et al. (1974) (14)	3/9	5th decade	PG	Very slow	NL-1-2+	Myogenic	Dystrophic	—

Bethlem and Van Wijngaarden (1976) (15)								
Family A	4/12	1st decade	PG,SG,A	Extremely slow	NL-1+	Myogenic	Dystrophic	Contractures of biceps, forearm and calf muscles
Family B	4/21	1st decade	PG,SG,A	Extremely slow	NL-1+	Myogenic	Dystrophic	–do–
Family C	5/12	1st decade	PG,SG,A	Extremely slow	NL-1+	Myogenic	Dystrophic	–do–
Arts et al. (1978) (16)	4/6	1st decade	PG,SG,A	Extremely slow	NL-1+	Myogenic	Dystrophic	Contractures of muscles, torticollis in one
Hastings et al. (1980) (17)								
Family A	3/3	1st decade	PG,SG,A	Extremely slow	1-2+	Myogenic	Dystrophic	Pseudohyper-trophy of calf
Family B	2/2	1st decade	PG,SG	Extremely slow	1-2+	Myogenic	Dystrophic	Pseudohyper-trophy of calf
Fenichel et al. (1982) (18)	2/3	1st decade	PG,A	Very slow	NL	Mixed	Myopathic	Cardiac involvement, sudden death in two cases
Chutkov et al. (1986) (19)	2/7	2nd–6th decade	PG,SG,A,T	Slow	NL-1+	Myogenic	Dystrophic, rimmed vacuoles; ragged red fibers	—
Mohire et al. (1988) (11)	6/33	1st decade	PG,SG,A,T	Extremely slow	NL-1-2+	Myogenic	Dystrophin	Contractures at elbow, hand, and ankles
Gilchrist et al. (1988) (20,35a)	7/34	2nd-4th decade	PG,SG	Slow	NL-1-2+	Myogenic	Myopathic	Linkage to chro-mosome 5q
Panegyres et al. (1990) (21)								
Family A	3/4	1st–3rd decade	PG,SG	Slow	NL-1-2+	Myogenic	Dystrophic	—
Family B	3/3	3rd decade	SG	Extremely slow	NL	Myogenic	Normal	—

Distribution of weakness: PG, pelvic girdle; SG, shoulder girdle; A, arms; T, thighs. Serum CK: 1+ (mild), 2+ (moderate) elevations; NL, normal; NA, not available.

limbs and pectorals and serratus anterior, biceps, brachioradialis, and triceps in upper limbs are the commonly affected muscles. It is the functional impairment of the pelvic-girdle muscles that usally brings the patient to seek medical advice.

The most striking clinical feature of all the reported cases of AD LGMD is the relatively benign clinical course. Examination of presymptomatic affected members shows that the disease often exists for many years before the patient seeks medical attention. One study (13) reported a postmortem examination of a 48-year-old man with LGMD who died of unrelated cause and who had not sought medical attention for a myopathy that was first detected during an army medical examination at age 18. Genealogical analysis led to the discovery of other affected members who were ambulatory well beyond 60 years of age and who had a normal life span. In two families reported by Hastings et al. (17), the fathers clinically had only a forme fruste myopathy and abnormal laboratory findings; muscle strength was normal.

A number of patients with AD LGMD show atypical features with regard to the nature and distribution of muscle weakness or muscle fiber morphology. Barnes (38) described an interesing six-generation family with 61 affected members, of whom 33 were alive and evaluated. At an early stage, there was true hypertrophy of muscles, often with increased strength, but in midadult life, muscle weakness appeared progressively through a pseudohypertrophic phase to an atrophic phase. During these phases, the weakness was particularly proximal, but in the latest stages, some atrophy in the hand muscles also developed. The oldest patient in the first generation died at age 87. Longevity in other affected members was not affected by the muscle disease. Pseudohypertrophy of calf muscles was also noted in affected persons of two families reported by Hastings et al. (17). The characteristic histological features in addition to the dystrophic fibers were the central cores and internalized capillaries in type I fibers. These capillaries were not ascribed to the phenomenon of fiber splitting. An AD LGMD with humeropelvic distribution of weakness and cardiac involvement (18) similar to the Emery-Dreifus form of X-linked dystrophy (41,42) has been reported.

In one family of Panegyres et al. (21), the weakness was confined to the pectoral-girdle musculature. De Costner et al. (14) reported another family with pure lower limb-girdle weakness, the symptoms beginning in the 5th decade. The clinical progression in these families was very slow and disability slight. Chutkow et al. (19) described LGMD in seven cases of a two-generation family. The weakness variably started in the 2nd to 6th decade in lower limbs and then ascended. The index case was wheelchair-bound at age 72 years, 30 years after onset of LGMD. Muscle biopsies showed marked dystrophic changes, rimmed vacuoles in all seven cases, and ragged red fibers in two cases. Rimmed vacuoles, one of the morphological findings in inclusion body myositis, have also been reported in other familial myopathies (43,44).

Although very slow progression is the most typical course of AD LGMD, periods of disease acceleration may be observed (12,19,38). This may be more apparent than real or may coincide with another, unrelated illness (12).

Determination of serum creatine kinase (CK) is the most widely used enzymatic test for the detection of muscular dystrophy, including preclinical cases (11,12,16). In AD LGMD, CK is normal or elevated to a mild to moderate extent (Table 1). This is in contrast to AR LGMD where CK is usually very high even in preclinical childhood cases (45). In general, CK in AD LGMD decreases with the age of the patient and with the duration of illness. It is often in the normal range after several decades of symptoms.

The electromyographic (EMG) changes in AD LGMD are typically myopathic (11,12–17,19–21). Nerve conduction studies are normal.

The pathological changes (11–21,40) in AD LGMD are similar to those observed in other muscular dystrophies (33). The abnormalities include fiber size variation, central nucleation, and individual muscle fiber necrosis and regeneration. Endomysial connective tissue is generally increased. Ultrastructural studies disclose many nonspecific changes typical of primary muscle disease, namely, focal disintegration of myofilaments, alteration of the Z-disk, sarcoplasmic masses, autophagic vacuoles, and foci of glycogen granules and lipid droplets.

Table 2 Differential Diagnosis of Autosomal Dominant Myopathies with Predominat Limb-Girdle Weakness

Disorder	Onset	Distribution of weakness	Clinical course	Serum CK	EMG	Biopsy	Remarks
LGMD	1st–6th decade	LG	Slow to extremely slow	NL-1-2+	Myogenic	Dystrophic	—
FSHMD	1st–3rd decade	Face, LG,P	Slow	NL-1+2+	Myogenic	Dystrophic, occ. inflm. reaction	Facial weakness early and prominent, gene located on chromosome 4 in one type
Myotonic dystrophy	Congenital, 1st–3rd decade	Face, jaw, neck, distal, LG	Rapid in congenital	NL-1+	Mogenic, myotonia+	Dystrophic, nuclear chain, ringed fiber, sarcoplasmic masses	Other system involvement, myotonin kinase gene on chromosome 17
Nemaline rod myopathy	Congenital	LG	Nonprogressive	NL	Myogenic	Rod bodies	May go undetected until later life
Central core disease	Congenital	LG	Nonprogressive	NL	Myogenic	Central cores, type 1 predominance	Pes cavus, scoliosis
Centronuclear myopathy	Congenital	Face, LG	Slow	NL	Myogenic	Central nuclei, type 1 predominance	Ptosis, ophthalmoplegia, seizures occasionally
Congenital fiber-type disproportion	Congenital	LG	Nonprogressive	NL	NL, myogenic	Small type 1 fibers	Muscle contractures
Hypokalemic periodic paralysis	Childhood to adult life	Generalized in attacks Late LG	Episodic paralysis, late slow progression	NL	NL (between attacks)	Normal, late vacuolar myopathy	Low serum K^+ in attacks, some with hyperthyroidism
Hyperkalemic periodic paralysis	Childhood to adult life	Generalized in attacks Late LG	Episodic paralysis, late slow progression	NL	NL (between attacks)	Normal, late vascular myopathy	Elevated serum K+ in attacks, some with myotonia

Disorder: LGMD, limb-girdle muscular dystrophy; FSHMD, facioscapulohumeral muscular dystrophy. Weakness: LG, limb girdle; P, personeal. Serum CK: creatine kinase; NL, normal; 1+ (mild), 2+ (moderate) elevations.

DIFFERENTIAL DIAGNOSIS

The diseases entering into the differential diagnosis of LGMD include many neuromuscular disorders [see list compiled by Bradley (8)]. A positive family history differentiates LGMD from polymyositis and a number of other acquired disorders of skeletal muscle, such as endocrine and metabolic myopathies. A clear AD pattern of inheritance further differentiates AD LGMD from X-linked dystrophies, such as Becker's, and from AR limb-girdle syndromes, such as AR LGMD, inherited metabolic myopathies, and spinal muscular atrophies (8,46). Familial quadriceps myopathy (47) has been reported, and there is evidence that the disease may eventually spread to other muscles, but its AR pattern differentiates it from AD LGMD.

Myotonic dystrophy (MyD) and facioscapulohumeral muscular dystrophy (FSHMD) are inherited in an AD pattern and are generally associated with significant proximal muscle weakness. MyD is usually easy to diagnose on the basis of characteristic facies, ptosis, and clinical and electromyographic myotonia (48). The distribution of muscle weakness in MyD is often diffuse, and frequently there is significant distal as well as facial, jaw, and neck muscle weakness. The MyD gene (myotonin kinase gene) has been localized to chromosome 19, and the molecular basis of the disease has recently been delineated (49,50). FSHMD may occasionally occur in abortive forms without significant facial involvement (8,51). The findings of other cases with classical facial involvement in the family help eliminate the diagnosis of LGMD. The topography of shoulder muscle weakness in FSHMD is quite selective, leading to a very characteristic clinical picture (51). In both FSHMD of the jump type (52) and the scapuloperoneal syndrome (51), peroneal muscle weakness at a relatively early stage helps separate these cases from AD LGMD.

An increasing number of myopathies with specific pathological features are now recognized under the heading of congenital myopathies. These are usually present from birth, though symptoms may not arise until later in life. Many of these congenital myopathies run in families in an AD pattern (26–32,34). The clinical course can be slowly progressive and similar to AD LGMD. However, in the congenital myopathies, the muscle involvement is diffuse and nonselective, and the morphological features on ultrastructural and histochemical studies differentiate them from AD LGMD (Table 2).

Familial periodic paralysis (53), both hypo- and hyperkalemic forms, are inherited with an AD pattern. Patients with familial hypokalemic periodic paralysis may in later life develop a progressive proximal vacuolar myopathy, Goldflam myopathy (53–55). A similar picture has occasionally been observed in cases of hyperkalemic periodic paralysis (53,56,57). The characteristic attacks of weakness in periodic paralysis clearly separate these families from progressive AD LGMD.

Although there is currently no specific treatment for AD LGMD, or for other forms of muscular dystrophies, certain therapeutic measures are helpful in maximizing residual muscle power and preserving function. Fortunately, many AD LGMDs have onset in late age, are only slowly progressive, and hence lead to limited disability. Principal goals of therapy in such cases are to prevent contractures and deformities and to maintain upright posture and independent ambulation. Both active exercise against graded resistance and passive stretching exercises are important.

Genetic counseling in families with AD LGMD should take into account the possibility of considerable variation in the severity of the disease from one individual to another. It may be difficult to convince family members that the risk of a relatively severely affected offspring may be as high for mildly as for severely affected persons. Once there is knowledge of the molecular genetic basis of LGMD, accurate genetic counseling, including prenatal diagnosis and disease prevention, will become available.

REFERENCES

1. Nevin S. Two cases of muscular degeneration occurring in late adult life, with a review of the recorded cases of late progressive muscular dystrophy (late progressive myopathy). Q J Med 1936; 5:51–68.
2. Erb W. Über die "juvenile form" der progressiven muskeltrophie und ihre beziehung zur sogenannten pseudohypertrophie der muskeln. Deutsch Arch Klin Med 1884; 34:467–519.

3. Leyden E. Klinik der Rückenmarks-krankheiten. Vol 2. Berlin: Hirschwald, 1876:531.
4. Möbius PJ. Über die hereditären nervenkrankheiten Volksmanns. Samml Klin Notiz 1879; 171:1505–1531.
5. Walton JN, Nattrass FJ. On the classification, natural history, and treatment of the myopathies. Brain 1954; 77:169–231.
6. Walton JN, Race RR, Philip U. On the inheritance of muscular dystrophy. Ann Hum Genet 1955; 20:1–38.
7. Walton JN. The inheritance of muscular dystrophy: further observations. Ann Hum Genet 1956; 21:40–58.
8. Bradley WG. Limb-girdle syndromes. In: Vinken PJ, Bruyn GW, Ringel SP, eds. Handbook of Clinical Neurology. Vol 40. Diseases of Muscle, Part I. Amsterdam: North Holland, 1979:433–467.
9. Stevenson AC. Muscular dystrophy in Northern Ireland. I. An account of the condition in fifty-one families. Ann Eugen 1953; 18:50–93.
10. Stevenson AC. Muscular dystrophy in Northern Ireland. II. An account of nine additional families. Ann Hum Genet 1955; 19:159–164.
11. Mohire MD, Tandan R, Fries TJ, Little BW, Pendelbury WW, Bradley WG. Early-onset benign autosomal dominant limb-girdle myopathy with contractures (Bethlem myopathy). Neurology 1988; 38:573–580.
12. Schneiderman LJ, Sampson WI, Schoene WC, Haydon GB. Genetic studies of a family with two unusual autosomal dominant conditions: muscular dystrophy and Pelger-Hüet anomaly. Am J Med 1969; 46:380–393.
13. Bacon PA, Smith B. Familial muscular dystrophy of late onset. J Neurol Neurosurg Psychiatry 1971; 34:93–97.
14. De Costner W, De Reuck JD, Thiery E. A late autosomal dominant form of limb-girdle muscular dystrophy. Eur Neurol 1974; 12:159–172.
15. Bethlem, J. Van Wijngaarden GK. Benign myopathy with autosomal dominant inheritance. A report on three pedigrees. Brain 1976; 99:91–100.
16. Arts WF, Bethlem J, Volkers WS. Further investigations on benign myopathy with autosomal dominant inheritance. J Neurol 1978; 217:201–206.
17. Hastings BA, Groothuis DR, Vick NA. Dominantly inherited pseudohypertrophic muscular dystrophy with internalized capillaries. Arch Neurol 1980; 37:709–714.
18. Fenichel GM, Sul YC, Kilroy AW, Blouin R. An autosomal-dominant dystrophy with humeropelvic distribution and cardiomyopathy. Neurology 1982; 32:1399–1401.
19. Chutkow JG, Heffner RR, Kramer AA, Edwards JA. Adult-onset autosomal dominant limb-girdle muscular dystrophy. Ann Neurol 1986; 20:240–248.
20. Gilchrist JM, Pericak-Vance M, Silverman L, Roses AD. Clinical and genetic investigation in autosomal dominant limb-girdle muscular dystrophy. Neurology 1988; 38:5–9.
21. Panegyres PK, Mastaglia FL, Kakulas BA. Limb girdle syndromes. Clinical, morphological and electrophysiological studies. J Neurol Sci 1990; 95:201–218.
22. Henson TE, Muller J, DeMyer WE. Hereditary myopathy limited to females. Arch Neurol 1967; 17:238–247.
23. Hertrich O. Mitteilung uber eine sippe mit dominant vererblicher, wahrscheinlich weiblich geschlechtsgebundener progressiver muskeldystrophie des Schultergurteltyps. Nervenarzt 1957; 28:325–327.
24. Heyck H, Laudahn G. Die progressiv-dystrophischen Myopathien. Berlin: Springer-Verlag, 1969:58–60.
25. Walton JN. Clinical aspects of human muscular dystrophy. In: Bourne GH, Golarz MN, eds. Muscular Dystrophy in Man and Animals. New York: Hafner, 1963:263–321.
26. Shy GM, Engel WK, Somers JE, Wanko, T. Nemaline myopathy: A new congenital myopathy. Brain 1963; 86:793–810.
27. Arts WF, Bethlem J, Dingemans KP, Eriksson AW. Investigations on the inheritance of nemaline myopathy. Arch Neurol 1978; 35:72–77.
28. Shy GM, Magee KR. A new congenital non-progressive myopathy. Brain 1956; 79:610–620.
29. Byrne E, Blumbergs PC, Hallpike JF. Central core disease. Study of a family with five affected generations. J Neurol Sci 1982; 53:77–83.
30. Kinoshita M, Satoyoshi E, Kumagai M. Familial type 1 fiber atrophy. J Neurol Sci 1975; 25:11–17.
31. Bradley WG, Price DL, Watanabe CK. Familial centronuclear myopathy. J Neurol Neurosurg Psychiatry 1970; 33:687–693.
32. McLeod JG, Baker DC, Lethlean AK, Shorey CD. Centronuclear myopathy with autosomal dominant inheritance. J Neurol Sci 1972; 15:375–387.
33. Dubowitz V. Muscle Biopsy: A Practical Approach, 2nd ed. London: Bailleire-Tindall, 1985.
34. Banker BQ. The congenital myopathies In: Engel AG, Banker BQ, eds. Myology. Vol 2. New York: McGraw-Hill, 1986:1527–1581.
35. Stevenson AC, Cheeseman EA, Huth MC. Muscular dystrophy in northern Ireland. III. Linkage data with particular reference to autosomal limb girdle muscular dystrophy. Ann Hum Genet 1955; 19:165–173.
35a. Speer MC, Yamaoka LH, Gilchrist JM, et al. Confirmation of genetic heterogeneity in limb-girdle muscular dystrophy: linkage of an autosomal dominant form to chromosome 5q. Am H Hum Genet 1992; 50:1211–1217.

35b. Bushby KM, Beckmann JS. The limb-girdle muscular dystrophies—proposal for a new nomenclature. Neuromusc Disord 1995; 5:337–343.
35c. Campbell KP. Loss of three muscular dystrophies: disruption of cytoskeleton-extracellular matrix linkage. Cell 1995; 80:675–679.
35d. Campbell KP. Adhalin gene mutations and autosomal recessive limb-girdle muscular dystrophy. Ann Neurol 1995; 38:353–355.
35e. Worton R. Muscular dystrophies: diseases of the dystrophin-glycoprotein complex. Science 1995; 270:755–756.
36. Chung CS, Morton NE. Discrimination of genetic entities in muscular dystrophy. Am J Hum Genet 1959; 11:339–359.
37. Yates JRW, Emery AEH. A population study of adult onset limb-girdle muscular dystrophy. J Med Genet 1985; 22:250–257.
38. Barnes S. A myopathic family; with hypertrophic, pseudohypertrophic, atrophic and terminal (distal in upper extremities) stages. Brain 1932; 55:1–46.
39. Hamptman A, Thannhauser SJ. Muscular shortening and dystrophy. A heredofamilial disease. Arch Neurol Psychiatry 1941; 46:654–664.
40. Sasabe F, Takase Y, Fukusako T, Yamamoto K, Morimatsu M. Early-onset benign autosomal dominant limb-girdle myopathy with contractures (Bethlem myopathy) in a Japanese family. Rinsho-Shinkeigaku 1992; 32:138–142.
41. Dreifuss FE, Hogan GR. Survival in X-chromosomal muscular dystrophy. Neurology 1961; 11:734–737.
42. Emery AEH, Dreifuss FE. Unusual type of benign X-linked muscular dystrophy. J Neurol Neurosurg Psychiatry 1966; 29:338–342.
43. Tomè FMS, Fardeau M. Nuclear inclusions in oculopharyngeal dystrophy. Acta Neuropathol (Berl) 1980; 49:85–87.
44. Sunohara N, Nonaka I, Kamei N, Satoyoshi E. Distal myopathy with rimmed vacuole formation: a follow-up study. Brain 1989; 112:65–83.
45. Jackson CE, Strehler DA. Limb-girdle muscular dystrophy: clinical manifestations and detection of preclinical disease. Pediatrics 1968; 41:495–502.
46. Bradley WG, Jones MZ, Mussini JM, Fawcett PRW. Becker-type muscular dystrophy. Muscle Nerve 1978; 1:111–132.
47. Espir MLE, Mathews WB. Hereditary quadriceps myopathy. J Neurol Neurosurg Psychiatry 1973; 36:1041–1045.
48. Harper PS. Myotonic disorders. In: Engel AG, Banker BQ, eds. Myology. Vol 2. New York: McGraw-Hill, 1986:1267–1296.
49. Mahadevan M, Tsilfidis C, Sabourin L, et al. Mytonic dystrophy mutation: an unstable CTG repeat in the 3′ untranslated region of the gene. Science 1992; 255:1253–1255.
50. Brook JD, McCurrach ME, Harley HG, et al. Molecular basis of myotonic dystrophy: expansion of a trinucleotide (CTG) repeat at the 3′ end of a transcript encoding a protein kinase family member. Cell 1992; 68:799–808.
51. Munsat TL. Facioscapulohumeral dystrophy and the scapuloperoneal syndrome. In: Engel AG, Banker BQ, eds. Myology. Vol 2. New York: McGraw-Hill, 1986:1251–1266.
52. Kazakov VM, Bogorodinsky DK, Skorometz AA. The myogenic scapulo-peroneal syndrome. Muscular dystrophy in the K. kindred: clinical study and genetics. Clin Genet 1976; 10:41–50.
53. Brooke MH. Periodic paralyses In: Bradley WG, Daroff RB, Fenichel GM, Marsden CD, eds. Neurology in Clinical Practice. Vol 2. Boston: Butterworth-Heinemann, 1991:1871–1873.
54. Odor DL, Patel AN, Pearce LA. Familial hypokalemic periodic paralysis with permanent myopathy. J Neuropathol Exp Neurol 1967; 26:98–114.
55. Pearson CM. The periodic paralysis. Differential features and pathological observations in permanent myopathic weakness. Brain 1964; 87:341–353.
56. Bradley WG, Taylor R, Rice DR, et al. Progressive myopathy in hyperkalemic periodic paralysis. Arch Neurol 1990; 47:1013–1017.
57. Bradley WG. Ultrastructural changes in adynamia episodica hereditaria and normokalaemic familial periodic paralysis. Brain 1969; 92:379–390.

25

Other Dystrophies with Autosomal Dominant Inheritance

Henning Schmalbruch
University of Copenhagen, Copenhagen, Denmark

OCULOPHARYNGEAL DYSTROPHY

This disorder was first recognized as a hereditary nosological entity in French Canada in 1915 (1), although sporadic patients with ptosis and dysphagia were described as early as 1878 (for references, see Ref. 2). Until 1969, 249 cases had been found in Canada alone, which could all be traced back to a French immigrant who came to Quebec in 1634 (3). The disease is, however, not limited to Canada, and families have been described in many different countries (for references, see Refs. 2,4).

The first muscle affected is usually the levator palpebrae. Some patients never develop other symptoms than ptosis of the eyelid; in others, the disease spreads and dysphagia and dysarthria occur. In spite of its name, oculopharyngeal dystrophy is not restricted to ocular and pharyngeal muscles; not only other extraocular muscles, but also facial, limb-girdle, and even distal limb muscles become involved to variable degrees. Skeletal muscles may be histologically affected without showing clinically significant atrophy and wasting. The course of the disease is slow, and weakness of the limb muscles is rarely disabling. Lid drop eventually obstructs vision (2,5). Dysphagia may give rise to serious complications such as malnutrition and aspiration pneumonia (6). Onset is usually beyond the 4th decade, but it may be much earlier (2). One of our patients was investigated at age 5 years because of distinct ptosis and a family history of oculopharyngeal dystrophy with four affected generations. Heredity is autosomal dominant with complete penetrance but variable expression. Cases may be mistaken as sporadic if relatives have only inconspicuous ptosis.

A form of oculopharyngeal dystrophy with pronounced affection of the distal muscles has been described (7–10). We have seen one sporadic case with this form. Goto et al. (11) observed two patients with oculopharyngodistal dystrophy who also had cardiomyopathy; cardiomyopathy is, however, not described as a consistent feature of oculopharyngeal dystrophy.

Polymyositis, the oculocraniosomatic syndrome, myotonic dystrophy, myasthenia gravis, bulbar paralysis, and even stenosis of the esophagus in patients with senile ptosis should be considered for differential diagnosis (2). In a Japanese family with familial spinal muscular atrophy, Matsunaga et al. (12) found several members with ophthalmoplegia and dysphagia as the only symptoms.

There is little doubt that oculopharyngeal dystrophy is a primary myopathy. Autopsy studies did not reveal any changes in brainstem nuclei or spinal anterior horn cells pointing toward a neurogenic origin of the disease (7,13). All motor nuclei of the extraocular muscles of a female from a Swedish kinship who died at the age of 73 years were normal; there were, however, severe hypoxic changes with complete loss of Purkinje cells (own unpublished observation). Nonspecific changes of the central or peripheral nervous system that have been attributed to the disease (14) probably reflect old age, severe malnutrition, or long-lasting hypoxia. No electrophysiological evidence for neurogenic involvement has been found in skeletal muscles; investigations of extraocular muscles are hampered by the fact that the diagnostic criteria established for limb muscles do not apply, and that only few laboratories have experience with these muscles.

Muscle biopsies from limb muscles reveal relatively modest myopathic changes, depending on the stage of the disease (Figs. 1, 2, 3). The variability of fiber diameter is increased, and very small fibers are scattered throughout the muscle (Fig. 1a). These fibers usually have peripheral nuclei and may on cross-section appear angulated or band-like (Figs. 1b, 2a). The number of central myonuclei is initially small, but later increases (Fig. 2). Then small and immature fibers with central nuclei are also found. Central nuclei indicate regeneration following necrosis, but necrotic fibers (Fig. 3b) are rare, corresponding to the small, usually only borderline increase of the serum creatine kinase (CK). Fibre type grouping or selective involvement of one of the fiber types has not been found. Loss of muscle fibers leads to fatty infiltration, which, however, is modest in limb muscles. Fibrosis does not occur. Atrophic angulated fibers become numerous in advanced stages, and groups of such fibers (Fig. 2a) may lead to the suspicion of a neurogenic lesion. At that stage, however, "rimmed" vacuoles are prominent (Fig. 2b).

These rimmed vacuoles are autophagic and contain myelin figures and cell debris. The basophilic content is dissolved during routine histology and becomes attached to the surrounding membrane. Semithin sections of plastic-embedded material and electron micrographs (Figs. 3a, 4a) show accumulations of lipid bodies instead of rimmed vacuoles; i.e. rimmed vacuoles are artifacts. Dubowitz (15, Fig. 9.198-9) describes as characteristic for the vacuoles in oculopharyngeal dystrophy that filaments

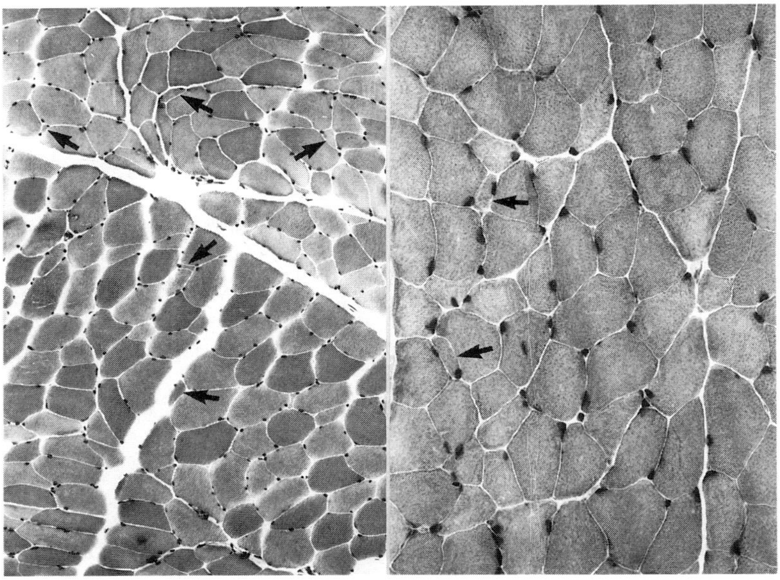

(a) (b)

Figure 1 Oculopharyngeal dystrophy. (a) Female aged 29 years with ptosis and dysphagia, but no ophthalmoplegia and no weakness of other muscles. The father had only ptosis and the grandmother ptosis and dysphagia. Myasthenia gravis was excluded. Brachial biceps muscle, H&E. Small fibers are scattered across the muscle (arrows), the number of central myonuclei is not increased. (89X.) (b) Five-year-old son of the patient shown in (a). The child had moderate ptosis, a weak voice, and sometimes difficulty in swallowing solid food. No other muscles were affected. Medial vastus muscle, H&E. Few central nuclei and scattered atrophic fibres (arrows) are the only abnormalities. Electron microscopy showed small autophagic vacuoles, which were not seen by light microscopy. (135X.)

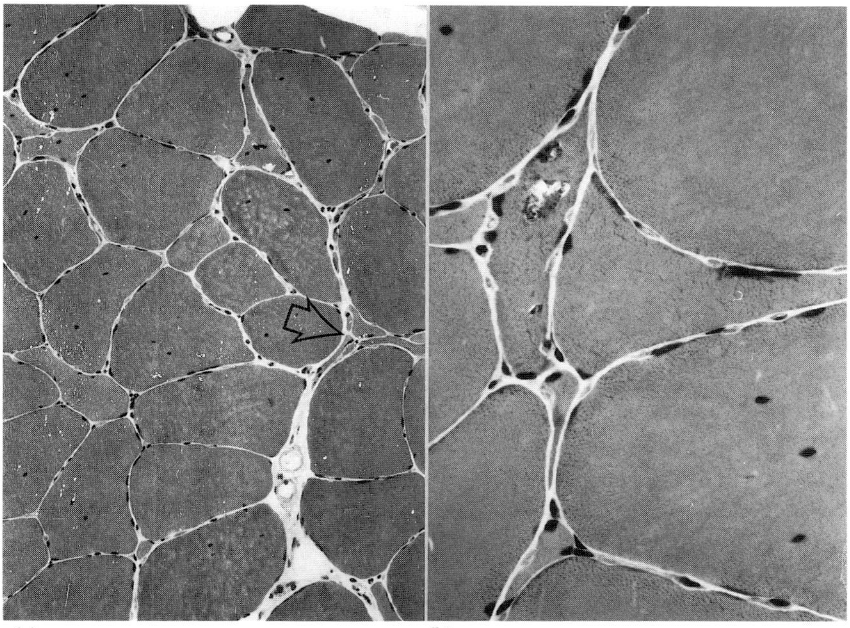

Figure 2 Oculopharyngeal dystrophy. Female aged 63 years. The disease started at the age of 48 years with ptosis and dysphagia; thereafter weakness of proximal and distal limb muscles and a hoarse voice developed. Numerous relatives had ptosis, ptosis and dysphagia, or ptosis, dysphagia, and walking difficulties. Medial vastus muscle, H&E. (a) The muscle fibers are hypertrophic or atrophic and contain multiple central nuclei. A group of angulated or band-like fibers (open arrow) may be mistaken as neurogenic change. (120X.) (b) Atrophic fibers, one with a large autophagic vacuole. Note central nuclei in large fibers. (360X.)

arise from their center of periphery and extend into the mounting medium above the frozen section; I have never seen this phenomenon, and electron microscopy does not show filaments that might be identifiable by light microscopy. Autophagocytosis occurs in both fiber types; it is most pronounced in atrophic fibers and may eventually lead to cell death. The biopsy from a child (Fig. 1b) showed autophagocytosis only in electron micrographs, not in light micrographs. Autophagocytosis, i.e., rimmed vacuoles, is not specific for oculopharyngeal dystrophy and occurs in a variety of other neuromuscular diseases (e.g., distal myopathy, see below).

Tomé and Fardeau (16,17) and Martin et al. (18) observed, in electron micrographs of all patients they studied, that a small proportion of myonuclei contained hollow filaments of 8.5 nm diameter (Fig. 4b). These intranuclear filaments differ from those in inclusion body myositis and also from filamentous inclusions seen in infantile polymyositis. The incidence of nuclei with inclusions may be small and their absence cannot be used as an exclusion criterion.

Fibers with abnormal mitochondria (called "ragged-red fibers" because they stain reddish with Gomori's trichome stain and their surface appears ruffled because clusters of subsarcolemmal mitochondria are most intensely stained) are absent. This distinguishes oculopharyngeal dystrophy from those ocular myopathies that are comprised by the term "oculocraniosomatic syndrome." The content of autophagic vacuoles sometimes stains reddish with the trichrome stain, and heavily affected fibers may be mistaken as ragged-red fibers. Reports of ragged-red fibers in extraocular muscles must be judged with caution, because these muscles normally contain fibers extremely rich in mitochondria, and the architecture of extraocular muscle fibers differs from that of other skeletal muscle fibers (19).

At autopsy, extraocular and pharyngeal muscles and the tongue show loss and fatty replacement

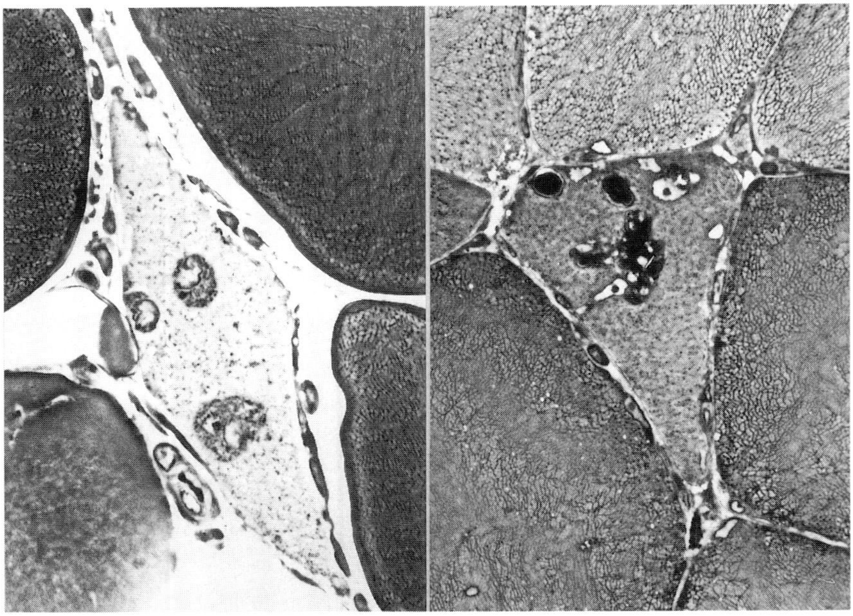

(a) (b)

Figure 3 Oculopharyngeal dystrophy. Same patient and muscle as in Figure 2. Epoxysections stained with *p*-phenylenediamine. (a) Necrotic muscle fiber containing three phagocytes. (610X.) (b) Autophagic vacuole containing lipid remnants. Note internal nuclei in the atrophic fiber. (610X.)

of muscle fibers (14). The pharyngeal muscles of our autopsied case (Fig. 5a) showed fatty replacement as well; fibers with large autophagic vacuoles were prominent both in pharyngeal muscles and in limb muscles. Fat infiltration of the brachial biceps muscle was distinct but not excessive. The histological picture of the tongue muscle was striking: small groups of isolated muscle fibers and large, apparently intact nerves run through a tissue consisting entirely of fat cells (Fig. 5b). The superior and lateral rectus muscles were normal; fibrosis or fat infiltration was lacking; all intramuscular nerve branches contained normal myelin sheaths. No autophagic vacuoles were found in the persisting fibers of the tongue and in the lateral rectus muscle. The histological changes seen in biopsies from the brachial biceps and medial vastus muscles taken 10 years before death (Figs. 2 and 3) were identical with those at autopsy but were much less pronounced.

DISTAL MYOPATHY

Myopathia Distalis Hereditaria Tarda Welander

Adult-onset neuromuscular disorders in which weakness and wasting of distal muscles are the primary symptoms are usually considered neurogenic if myotonic dystrophy has been excluded. Nevertheless, since Gowers' (20) description of a sporadic case, a man aged 18 years, with distal myopathy that affected the extensors of the hand and feet and the facial muscles, several authors have presented sporadic but also familial cases that had been classified as distal myopathy. The first thorough investigation was that by Welander (21), who found 249 patients (149 men and 100 women) with distal muscle atrophy in 72 Swedish kinships. The disease started after the age of 30 years (usually 40–60 years) with weakness of the extensor muscles of the hands. Progression, if any, was slow. Many patients were so little affected that they attributed the symptoms to beginning arthrosis or hard work and did not seek medical help. Later, the extensors of the feet became weak as well. Signs of

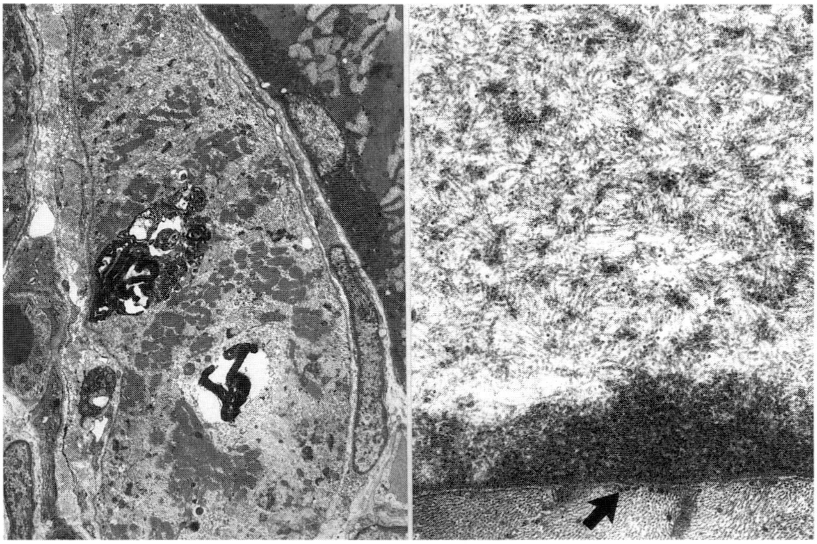

(a) (b)

Figure 4 Oculopharyngeal dystrophy. Same patient and muscle as in Figure 2. Electron micrographs. (a) Atrophic muscle fiber containing two autophagic vacuoles that are filled with myelin figures. (2500X.) (b) Myonucleus with intranuclear filaments. The nuclear membrane (arrow) and thin actin filaments are visible below. Normal chromatin is seen attached to the nuclear membrane. (25,600X.)

involvement of proximal limb muscles were found in about 15% of the patients; cranial muscles were usually not affected, but six patients had weakness of ocular muscles. None of the patients was severely disabled, and only about 10% had stopped working. Inheritance was autosomal dominant. Intermarriages between affected pedigrees were not uncommon due to the high frequency of intermarriage in remote regions of Sweden. Nine patients, all products of intermarriages, presented with "grossly atypical" disease; proximal limb muscles and flexor muscles of the hands and feet became involved at an early stage of the disease, and the patients were more severely incapacitated than "typical" cases; half of them were completely helpless. The grossly atypical cases were regarded as homozygous for the gene (21).

The disease was assumed to be myogenic, because the electromyogram disclosed a myopathic picture, and because muscle biopsies showed myopathic changes (21). Welander's monograph (21), published in 1951, contains eight light micrographs all of which this author, with present knowledge, would classify as compatible with neuropathy. In 1975, Edström (22) described biopsies from 13 Swedish patients that showed morphological changes similar to myotonic dystrophy (type 1 atrophy, chains of myonuclei, sarcoplasmic masses, *ringbinden*). Other authors found histological and electrophysiological signs suggesting a neuropathy with loss of small terminal nerve twigs (23–26). Recent studies from the same Swedish laboratories showed that patients with Welander myopathy had sensory deficits, that electromyograms of the anterior tibial and extensor digitorum communis muscles were myopathic, and that the muscle fibers contained "rimmed" vacuoles (27) and cytoplasmic and intranuclear filaments of 15–18 nm thickness as in inclusion body myositis (28,29). The absence of inflammatory infiltrations was noted as the distinguishing feature between distal myopathy and inclusion body myositis. Some members of typical families presented with clinical symptoms and electrophysiological abnormalities already in their twenties when muscle biopsies were still normal (27). Borg et al. (27) suggest that a peripheral nerve lesion precedes or coincides with the muscle affection. An interesting hypothesis has been put forward by Udd and associates (30–32), who studied

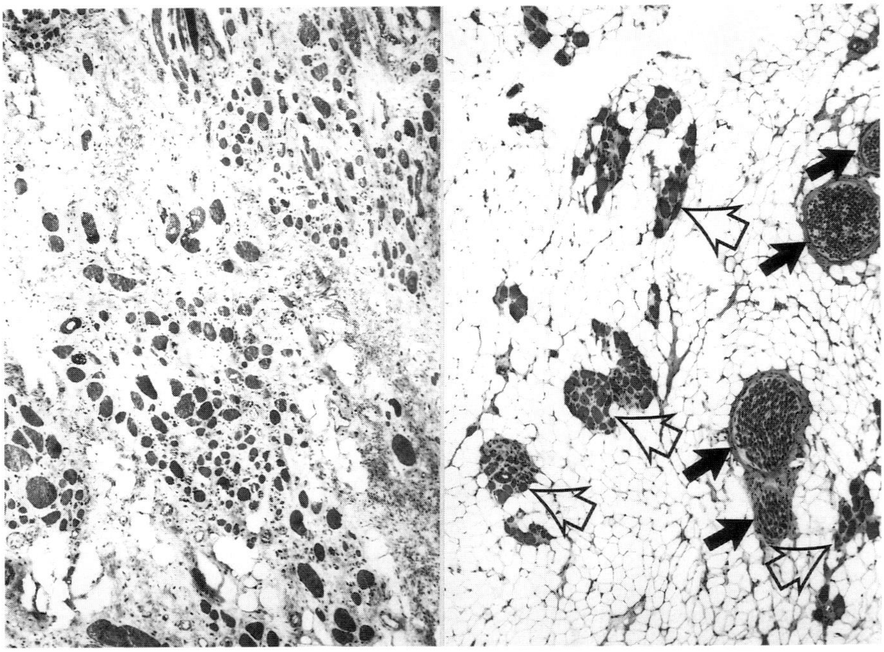

(a) (b)

Figure 5 Oculopharyngeal dystrophy. Same patient as in Figure 2, who died from pneumonia at the age of 73 years; autopsy. (a) Pharyngeal muscles. Few muscle fibers widely varying in size are scattered within fat and loose connective tissue. Many of these fibers contain rimmed vacuoles (not visible at this magnification). (37X.) (b) Tongue. Most muscle fibers have been replaced by fat cells; the remaining fibers are clustered in small groups (open arrows). Large nerve branches (arrows) are intact. (37X.)

a large, consanguineous Finish family. Some patients had limb-girdle dystrophy with little affection of distal muscles, others had distal myopathy. The patients with distal myopathy, however, differed from those studied by Welander (21) in that symptoms started in the lower limbs rather than in the hands. The authors suggest that patients with distal myopathy are heterozygotic for limb-girdle muscular dystrophy, which usually is considered to be recessively inherited. This would mean that Welander's "grossly atypical" patients might have had "typical" limb-girdle dystrophy, and that patients with distal myopathy were manifesting heterozygotes.

The nosological status of the Welander myopathy is still uncertain: the results of different studies are contradictory and confusing. Edström (22), in his first paper, describes histopathological changes as in myotonic dystrophy but no autophagic vacuoles; in this context, it is interesting that Schotland and Rowland (7) emphasized that the distinction between "distal myopathy" and "myotonic dystrophy without myotonia" may be difficult. Autophagic (rimmed) vacuoles have become the central issue in recent morphological studies of Welander myopathy. The same vacuoles occur in other forms of distal myopathy (see below) and in a large variety of chronic neuromuscular disorders. One may suspect that autophagocytosis is a common feature of all forms of destructive myopathy that do not show segmental necrosis initiated by membrane defects and calcium influx as in Duchenne dystrophy (33), and that all these disorders eventually develop rimmed vacuoles. An experimental model for this process is the toxic myopathy caused by chloroquine or chlorphentermine, substances that induce autophagocytosis and eventually cause breakdown (necrosis) of skeletal muscle fibers (34). The classification of distal myopathies by means of morphological analysis of the vacuoles may hence be futile. It is striking that Lindberg et al. (28), in vacuoles both in Welander myopathy and in inclusion body myositis, found immunoreactivity for mumps virus. This is an artifact (28) that had previously given rise to the hypothesis that inclusion body myositis is due to infection by a defective mumps virus (35).

I cannot contribute my own results to the discussion, because we have not seen biopsies from patients with typical Welander myopathy in Copenhagen over a period of 20 years, and none of the referring clinicians could remember having seen such a patient. This remains unexplained, because the disease apparently is quite common in Sweden. Patients with confirmed distal myopathy have become rare also in other laboratories as diagnostic methods have improved (36). The only case with unequivocal distal myopathy we have seen was a 54-year-old woman who over a period of 16 years had experienced progressive weakness of the hand and finger extensors; there was no family history. This patient also had hypothyroidism, and tubular aggregates were found in a biopsy from the extensor digitorum communis muscle.

Non-Scandinavian Forms of Distal Myopathy

Several non-Swedish patients with distal myopathy have been described. Markesbery et al. (37) and Bachmann and Wagner (38) observed patients with a late-onset familial form with dominant inheritance; the three patients described by Markesbery et al. (37) differed from the classical Welander form in that distal leg muscles were initially affected; eventually proximal limb muscles also became involved. Bulbar muscles were spared. The course of the disease in these familial cases was relatively benign. Muscle biopsies showed a wide spectrum of myopathic changes including vacuoles. One of six patients developed cardiomyopathy. In addition, a progressive form affecting young adults has been described (39–42). Also, in these patients weakness started in distal leg muscles and then spread to hand and to proximal muscles. One 26-year-old man had complete heart block (42). Serum creatine kinase (CK) was elevated, and the muscles showed pronounced autophagocytosis, excessive Z-streaming, and muscle fiber necrosis. Most early-onset cases were sporadic (43), but the two patients of Kuhn and Schröder (41) were brothers produced by a consanguineous marriage. Inheritance was therefore assumed to be autosomal recessive (41,43). Markesbery and Griggs (43) consider three different forms of distal myopathy as definite entities: the classical Welander form, a non-Scandinavian adult form with late onset, and an adult form with early onset.

Families with autosomal dominant distal myopathy with infantile or juvenile onset have also been described. The infantile form causes bilateral footdrop and weakness of the extensors of the fingers; onset is evident at the age of 2 years with no obvious progression after the age of 18 years (44,45). The distal myopathy with infantile onset observed by Bautista et al. (46) has alternatively been characterized as congenital myopathy with fiber type disproportion. The juvenile form (47) starts between the age of 5 and 15 years, affects the extensors and flexors of the hands and feet simultaneously, is slowly progressive, and becomes stationary at age 50 years.

Two other forms of distal myopathy with autosomal-recessive inheritance have been found in Japan. "Autosomal-recessive distal muscular dystrophy" (48) affects the gastrocnemius muscle more than the peroneal muscles, CK is elevated, and the muscles contain necrotic fibers. "Distal myopathy with rimmed vacuole formation" (49–54) affects the peroneal muscles more than the gastrocnemius muscle, CK is normal or borderline, and biopsies show pronounced autophagocytosis but no necrosis.

All reviews cite older reports of sporadic or familial cases of distal myopathy; attempts to relate these—seen from today—incomplete reports to modern descriptions are usually speculative. Some of these patients might in reality have had motoneuron disease, Charcot-Marie-Tooth disease, or myotonic dystrophy.

DESMIN-STORAGE MYOPATHY

In 1943 Milhorat and Wolff (55,56) described an American family with dominantly inherited distal myopathy starting in middle age in the lower limbs. The disease was progressive, and several patients died with signs of cardiovascular disease; autopsies disclosed pronounced muscular wasting and no involvement of the nervous system. We (57a) had the opportunity to investigate members of subsequent generations of this family. The pedigree now comprises five generations with 99 members of whom 26 have or had the disease. Morphology disclosed that it was identical with the desmin-storage myopathy described by Fardeau et al. (57) and Rappaport et al. (58) in France. Gait disturbance due to distal leg weakness started in early to middle adulthood; all extremities and bulbar, respiratory, and facial muscles became affected within 5–10 years and the patients became wheelchair-bound. Cardiac involvement was frequent, and arrhythmia, conduction block, or congestive heart failure was often the cause of premature death. CK was at most 3–5 times normal. Morphological or electrophysiological signs of neuropathy were absent. The course of the disease was apparently more benign in the French than in the American family.

The initial lesion is probably the deposition of a granulofilamentous material, which was found already in modestly affected muscles (Figs. 6 and 7). These deposits have been demonstrated to contain desmin (58,59) (Fig. 6d). The stored desmin was phosphorylated (58) and hence did not polymerize into intermediate filaments (60,61). The muscles initially showed modest myopathic changes (Fig. 6a), but eventually became fibrotic and their fibers were replaced by fat cells (Fig. 6b,c). Autophagocytosis was a relatively early change (Fig. 6a), and it became most prominent in advanced cases (Figs. 6c, 7d). The biopsy of one of our patients had, without knowledge of the family history, been classified as inclusion body myositis; in another patient the diagnosis was motoneuron disease. Sporadic cases of desmin-storage myopathy with the same morphological features with or without cardiomyopathy have been reported (59,62). Familial and sporadic cases in which the clinical picture was entirely dominated by the cardiomyopathy have also been described (63,64). The morphology of the deposits, however, varied in different patients, and it is disputable whether all had the same disease (see below). Liu and Gumbinas (65) observed already in 1974 a young man with untractable cardiomyopathy and generalized muscle atrophy including bulbar muscles; at autopsy they found in heart and skeletal muscles, by electron microscopy, deposits of the same type that have later been found in patients with desmin-storage myopathy (57) (Fig. 7c); in addition, axonal filamentous spheroids were present in various parts of the central nervous system.

The matter is complicated by the fact that Edström et al. (66) encountered a family with an obviously different form of desmin-storage disease. The disorder presented as dominantly inherited distal myopathy starting in the flexor muscles of the hands during the third or fourth decade; it later spread to proximal muscles but never to bulbar muscles. The patients became wheelchair-bound within 10 years, and most of them developed cardiomyopathy. The muscle fibers contained an excess of desmin (termed skeletin by the authors), which, however, formed typical intermediate filaments in the sarcoplasm. Most fibers also contained dense bodies filled with filaments; these bodies were membrane-bound, showed autofluorescence like lipofuscin granules (secondary lysosomes), and reacted with antidesmin. Myopathic changes were pronounced in muscle biopsies, the electromyogram was myopathic, and the CK was about twice the upper limit of normal. The disease differed from the Welander form of distal myopathy by being more progressive and by affecting flexor muscles more than extensor muscles; it differed from the myopathy described by Fardeau et al. (57) and from the family investigated by us (57a) in that symptoms started in the upper rather than in the lower limbs, that bulbar muscles were never affected, and that desmin was stored as intermediate filaments and in dense bodies (secondary lysosomes?) rather than as granulofilamentous deposits. Hence, based on morphology and on clinical data, we are faced with at least two forms of desmin-storage myopathy. A third form may be an autosomal recessive congenital myopathy with cardiomyopathy with antidesmin-reactive cytoplasmic bodies (67,68). The patient seen by Liu and Gumbinas (65) (see above) may belong to the third group. The picture is complicated by the indiscriminate use of the terms "dense bodies," "granulofilamentous material," and "intermediate filament accumulation" for bundles of filaments, membrane-bound structures containing filaments (lysosomes), and fine-granular deposits without enclosing membrane, and the fact that patients with accumulations of unidentified filaments surmised to contain desmin have been reviewed as reports of desmin-storage myopathy (59,62). "Desmin-storage myopathy" probably covers an unknown number of different disorders. Desmin is a normal constituent of the intermediate filaments of muscle cells; an excess of normal intermediate filaments may in some cases have been reactive and hence unspecific.

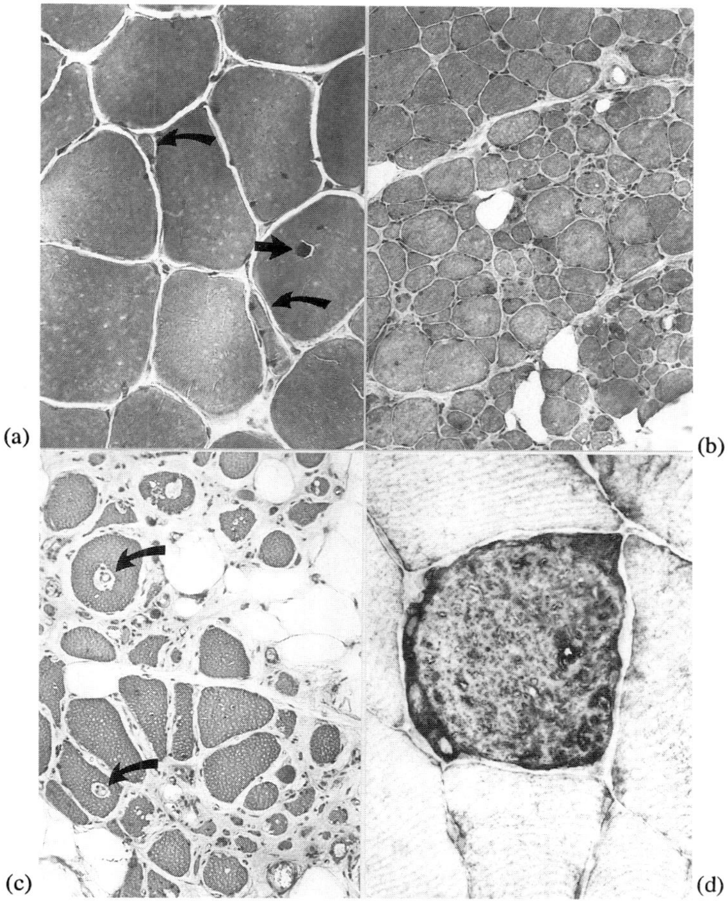

Figure 6 Familial desmin-storage myopathy. Light micrographs of frozen sections stained with Gomorri's trichome method (a,b), H&E (c), and antidesmin (d). (a) Thirty-five-year-old man, quadriceps muscle, modest changes. The muscle fibers are of normal size, but two fibers (curved arrows) are atrophic. One central nucleus and a stained inclusion (arrow), which corresponds to an autophagic vacuole, are seen. (250X.) (b) Forty-four-year-old female, deltoid muscle, advanced stage. The muscle fibers vary in size, the endomysium is fibrotic, and fat cells have replaced lost fibers. (100X.) (c) Forty-two-year-old man, vastus muscle. Fibrotic end-stage muscle, the fibers have partly been replaced by fat cells; few large and many small muscle fibers are seen. Several fibers are vacuolated; two fibers contain internal capillaries (curved arrows), which have been trapped by lateral fusion of myotubes during regeneration. (100X.) (d) Forty-nine-year old female, medial vastus muscle. Immunostaining for desmin (avidin-biotin-peroxidase). Intensely stained fiber with subsarcolemmal and intracellular reaction product. The surrounding fibers show normal Z-line-associated reactivity for desmin. (290X.)

ADDENDUM

Since this chapter was finished (May 1993), 12 additional patients with desmin-storage myopathy, often associated with cardiomyopathy, have been reported (69,70). Inheritance was usually dominant. The present author found typical desmin storage in muscle biopsies from a mother who had already died from the disease and from her daughter (without cardiomyopathy); both patients had for years been diagnosed and treated as polymyositis. I also know of two other cases in Denmark (both with

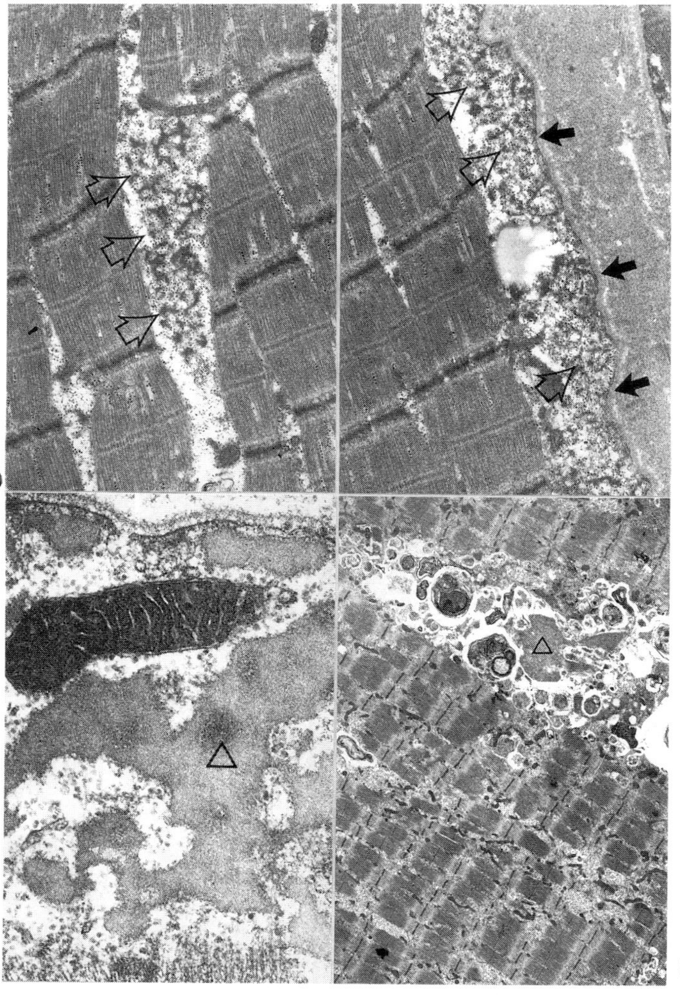

Figure 7 Familial desmin-storage myopathy, electron micrographs. (a,b) Same muscle as shown Figure 6a. An electron-dense material forms reticular meshworks (open arrows), which are deposited between the myofibrils (a) and beneath the sarcolema (b) (arrows). (10,000X.) (c) Same muscle as shown in Figure 6b. High magnification of the granulofilamentous material (triangle) shown in (a,b). It is intermingled with glycogen granules and is not enclosed by a membrane. (35,000X.) (d) Same patient as Figure 6c. Large autophagic vacuoles containing myelin figures and membrane whorls and also granulofilamentous material (triangle). (2200X.)

cardiomyopathy). This suggests that desmin-storage myopathy may not be as rare as originally believed. The morphology in some cases led to the assumption (70,71) that desmin-storage myopathy is related to spheroid body myopathy (72).

ACKNOWLEDGMENTS

The author's work has been supported by grants from the Danish MRC, Velux Fonden, Lundbeck Fond, Aage Haensch Fond, and Inger Bonnén Fond. The skillful technical assistance of Mrs. Marianne Bjærg and Mrs. Lis Hansen is gratefully acknowledged. Dr. Horowitz (Albany, NY) and the author thank Drs. H. Vinters (UCLA, Los Angeles) and S. Lipper (Long Island Jewish Medical Center, New York) for providing biopsy material of three patients with familial desmin-storage myopathy.

REFERENCES

1. Taylor EW. Progressive vagus-glossopharyngeal paralysis with ptosis. A contribution to the group of family diseases. J Nerv Ment Dis 1915; 42:129–139.
2. Bastiaensen LAK. Chronic Progressive External Ophthalmoplegia. Alphen a/d Ryn-Netherlands; Stafleu Publishers, 1978:386.
3. Barbeau A. Oculopharyngeal muscular dystrophy in French Canada. In: Brunette JR, Barbeau A, eds. Progress in Neuro-Ophthalmology. Vol. 2. Amsterdam: Excerpta Medica, 1969:3.
4. Tomé FMS, Fardeau M. Ocular myopathies. In: Engel AG, Banker BQ, eds. Myology. Vol 2. New York: McGraw-Hill, 1986:1327–1347.
5. Victor M, Hayes R, Adams RD. Oculopharyngeal muscular dystrophy. A familial disease of late life characterized by dysphagia and progressive ptosis of the eyelids. N Engl J Med 1962; 267:1267–1272.
6. Kiel DP. Oculopharyngeal muscular dystrophy as a cause of dysphagia in the elderly. Am Geriatr Soc 1986; 34:144–147.
7. Schotland DL, Rowland LP. Muscular dystrophy. Features of ocular myopathy, distal myopathy, and myotonic dystrophy. Arch Neurol 1964; 10:433–445.
8. Satoyoshi E, Kinoshita M. Oculopharyngodistal myopathy. Report of four families. Arch Neurol 1977; 34:89–92.
9. Vita G, Dattola R, Santoro M, Messina C. Familial oculopharyngeal muscular dystrophy with distal spread. J Neurol 1983; 230:57–64.
10. Scrimgeour EM, Mastaglia FL. Oculopharyngeal and distal myopathy: a case study from Papua New Guinea. Am J Med Genet 1984; 17:763–771.
11. Goto I, Kanazawa Y, Kobayashi T, Murai Y, Kuroiwa Y. Oculopharyngeal myopathy with distal and cardiomyopathy. J. Neurol Neurosurg Psychiatry 1977; 40:600–607.
12. Matsunaga M, Inokuchi T, Onishi A, Kuroiwa Y. Oculopharyngeal involvement in familial neurogenic muscular atrophy. J Neurol Neurosurg Psychiatry 1973; 36:104–111.
13. Little BW, Perl DP. Oculopharyngeal muscular dystrophy. An autopsied case for the French-Canadian kindred. J Neurol Sci 1982; 53:145–158.
14. Schmitt HP, Krause K-H. An autopsy study of a familial oculopharyngeal muscular dystrophy (OPMD) with distal spread and neurogenic involvement. Muscle Nerve 1981; 4:296–305.
15. Dubowitz V. Muscle Biopsy: A Practical Approach, 2nd ed. London: Bailliére Tindall, 1985:720.
16. Tomé FMS, Fardeau M. Nuclear inclusions in oculopharyngeal dystrophy. Acta Neuropathol (Berl) 1980; 49:85–87.
17. Tomé FMS, Fardeau M. Nuclear changes in muscle disorders. In: Jasmin G, Simard R, eds. Methods and Achievements in Experimental Pathology. Vol 12. Basel: Karger, 1986:261–296.
18. Martin J-JR, Ceuterick CM, Mercelis RJ. Nuclear inclusions in oculopharyngeal muscular dystrophy. Muscle Nerve 1982; 5:735–737.
19. Schmalbruch H. Skeletal Muscle. Berlin: Springer, 1985:440.
20. Gowers WR. A lecture on myopathy and a distal form. Br Med J 1902; 2:89–92.
21. Welander L. Myopathia distalis tarda hereditaria. Acta Med Scand 1951; 141 (Suppl 265):1–124.
22. Edström L. Histochemical and histopathological changes in skeletal muscle in late-onset hereditary distal myopathy (Welander). J Neurol Sci 1975; 26:147–157.
23. Stålberg E, Ekstedt J. Signs of neuropathy in distal hereditary myopathy (Welander). Electroencephalogr Clin Neurophysiol 1969; 26:343.
24. Thornell L-E, Edström L, Billeter R, Butler-Browne GS, Kjörell U, Whalen RG. Muscle fibre type composition in distal myopathy (Welander). An analysis with enzyme- and immuno-histochemical, gel-electrophoretic and ultrastructural techniques. J Neurol Sci 1984; 65:269–292.
25. Borg K, Borg J, Lindblom U. Sensory involvement in distal myopathy (Welander). J Neurol Sci; 1987; 80:323–332.
26. Borg K, Solders G, Borg J, Edström L, Kristensson K. Neurogenic involvement in distal myopathy (Welander). Histochemical and morphological observations on muscle and nerve biopsies. J Neurol Sci 1989; 91:53–70.
27. Borg K, Åhlberg G, Borg J, Edström L. Welander's distal myopathy: clinical, neurophysiological and muscle biopsy observations in young and middle aged adults with early symptoms. J Neurol Neurosurg Psychiatry 1991a; 54:494–498.
28. Lindberg C, Borg K, Edström L, Hedström A, Oldfors A. Inclusion body myositis and Welander distal myopathy: a clinical, neurophysiological and morphological comparison. J Neurol Sci 1991; 103:76–81.

29. Borg K, Tomé FMS, Edström L. Intranuclear and cytoplasmic filamentous inclusions in distal myopathy (Welander). Acta Neuropathol (Berl) 1991; 82:102–106.
30. Udd B, Kaarianen H, Somer H. Muscular dystrophy with separate clinical phenotypes in a large family. Muscle Nerve 1991; 14:1050–1058.
31. Udd B. Limb-girdle type muscular dystrophy in a large family with distal myopathy: homozygous manifestation of a dominant gene? J Med Genet 1992; 29:383–389.
32. Udd B, Rapola J, Nokelainen P, Arikawa E, Somer H. Nonvacuolar myopathy in a large family with both late adult onset distal myopathy and severe proximal muscular dystrophy. J Neurol Sci 1992; 113:214–221.
33. Schmalbruch H. The muscular dystrophies. In: Mastaglia FL, Lord Walton, eds. Skeletal Muscle Pathology, 2nd ed. Edinburgh: Churchill Livingstone, 1992:283–318.
34. Schmalbruch H. The early changes in experimental myopathy induced by chloroquine and chlorphentermine. J Neuropathol Exp Neurol 1980; 39:65–81.
35. Chou SM, Örvell C. Localization of mumps virus structural proteins with monoclonal antibodies in muscles of inclusion body myositis. Ann Neurol 1987; 22:129–130.
36. Walton JN, Gardner-Medwin D. Progressive muscular dystrophy and the myotonic disorders. In: Walton JN, ed. Disorders of Voluntary Muscle, 4th ed. Edinburgh: Churchill Livingstone, 1981:481–524.
37. Markesbery WR, Griggs RC, Leach RP, Lapham LW. Late onset hereditary distal myopathy. Neurology 1974; 24:127–134.
38. Bachmann H, Wagner A. Myopathia distalis tarda hereditaria. Ergebnisse einer genealogischen, klinischen, klinisch-chemischen und elektromyographischen Familienuntersuchung. Psychiatr Neurol Med Psychol (Leipzig) 1980; 32:20–28.
39. Markesbery WR, Griggs RC, Herr B. Distal myopathy: electron microscopic and histochemical studies. Neurology 1977; 27:727–735.
40. Miller RG, Blank NK, Layzer RB. Sporadic distal myopathy with early adult onset. Ann Neurol 1979; 5:220–227.
41. Kuhn E, Schröder JM. A new type of distal myopathy in two brothers. J Neurol 1981; 226:181–185.
42. Krendel DA, Gilchrist JM, Bossen EH. Distal vacuolar myopathy with complete heart block. Arch Neurol 1988; 45:698–699.
43. Markesbery WR, Griggs RC. Distal myopathies. In: Engel AG, Banker BQ, eds. Myology. Vol 2. New York: McGraw-Hill, 1986:1313–1325.
44. Magee KR, DeJong RN. Hereditary distal myopathy with onset in infancy. Arch Neurol 1965; 13:387–390.
45. Van der Does de Willebois AEM, Bethlem J, Meyer AEFH, Simons AJR. Distal myopathy with onset in early infancy. Neurology 1968; 18:383–390.
46. Bautista J, Rafel E, Castilla JM, Alberca R. Hereditary distal myopathy with onset in early infancy. Observation of a family. J Neurol Sci 1978; 37:149–158.
47. Biemond A. Myopathia distalis juvenilis hereditaria. Acta Psychiatr Neurol Scand 1955; 30:25–38.
48. Miyoshi K, Kawai H, Iwasa M, Kusaka F, Nishino H. Autosomal recessive distal muscular dystrophy as a new type of progressive muscular dystrophy. Seventeen cases in eight families including an autopsied case. Brain 1986; 109:31–54.
49. Nonaka I, Sunohara N, Ishiura S, Satoyoshi E. Familial distal myopathy with rimmed vacuole and lamellar (myeloid) body formation. J Neurol Sci 1981; 51:141–155.
50. Nonaka I, Sunohara N, Satoyoshi E, Terasawa K, Yonemoto K. Autosomal recessive distal muscular dystrophy: a comparative study with distal myopathy with rimmed vacuole formation. Ann Neurol 1985; 17:51–59.
51. Kumamoto T, Fukuhara N, Nagashima M, Kanda T, Wakabayashi M. Distal myopathy. Histochemical and ultrastructural studies. Arch Neurol 1982; 39:367–371.
52. Mizusawa H, Kurisaki H, Takatsu M, et al. Rimmed vacuolar distal myopathy: a clinical, electrophysiological, histopathological and computed tomographic study of seven cases. J Neurol 1987; 234:129–136.
53. Mizusawa H, Kurisaki H, Takatsu M, et al. Rimmed vacuolar distal myopathy. An ultrastructural study. J Neurol 1987; 234:137–145.
54. Sunohara N, Nonaka I, Kamei N, Satoyoshi E. Distal myopathy with rimmed vacuole formation. A follow-up study. Brain 1989; 112:65–83.
55. Milhorat AT, Wolff HG. Studies in diseases of muscle. XII. Heredity of progressive muscular dystrophy; relationship between age of onset of symptoms and clinical course. Arch Neurol Psychiatry 1943; 49:641–654.
56. Milhorat AT, Wolff HG. Studies in diseases of muscle. XIII. Progressive muscular dystrophy of atrophic distal type; report of a family; report of autopsy. Arch Neurol Psychiatry 1943; 49:655–664.
57. Fardeau M, Godet-Guillain J, Tomé FMS, et al. Une nouvelle affection musculaire familiale, définie par l'accumulation intra-sarco-plasmique d'un matérial granulo-filamentaire dense en microscopie électronique. Rev Neurol (Paris) 1978; 131:411–425.
57a. Horowitz SH, Schmalbruch H. Autosomal dominant distal myopathy with desmin storage: a clinicopathologic and electrophysiologic study of a large kinship. Muscle Nerve 1994; 17:151–160.

58. Rappaport L, Contard F, Samuel JL, et al. Storage of phosphorylated desmin in a familial myopathy. FEBS Lett 1988; 2341:421–425.
59. Pellissier JF, Puget J, Charpin C, Figarella D. Myopathy associated with desmin type intermediate filaments: an immunoelectron microscopic study. J Neurol Sci 1989; 89:49–61.
60. Geisler N, Weber K. Phosphorylation of desmin in vitro inhibits formation of intermediate filaments: identification of three kinase A sites in the aminoterminal head domain. EMBO J 1988; 7:15–20.
61. Inagaki M, Gonda Y, Matsuyama M, Nishizawa K, Nishi Y, Sato C. Intermediate filament reconstitution in vitro. The role of phosphorylation on the assembly-disassembly of desmin. J Biol Chem 1988; 263:5970–5978.
62. Telerman-Toppet N, Bauherz G, Noël S. Auriclo-ventricular block and distal myopathy with rimmed vacuoles and desmin storage. Clin Neuropathol 1991; 10:61–64.
63. Porte A, Stoeckel M-E, Sacrez A, Batzenschlager A. Unusual familial cardiomyopathy with storage of intermediate filaments in the cardiac muscle cells. Virchows Arch A Pathol Anat Histol 1980; 386:43–58.
64. Tanaka M, Kawahara Y, Motomiya T, et al. Cardiomyopathy characterized by abnormal accumulation of desmin-type intermediate filaments in cardiac mucle fibers. A case report and review of the literature. Acta Pathol Jpn 1989; 39:266–273.
65. Liu MH, Gumbinas M. Axonal filamentous spheroids associated with cardiomyopathy with "targetoid fibers." 1. Clinical, histologic, and electron microscopic studies. Neurology 1974; 24:547–554.
66. Edström L, Thornell L-E, Eriksson A. A new type of hereditary distal myopathy with characteristic sarcoplasmic bodies and intermediate (skeletin) filaments. J Neurol Sci 1980; 47:171–190.
67. Goebel HH, Lenard H-G, Langenbeck U, Mehl B. A form of congenital muscular dystrophy. Brain Dev 1980; 2:387–400.
68. Fidzianska A, Goebel HH, Osborn M, Lenard HG, Osse G, Langenbeck U. Mallory body-like inclusions in a hereditary congenital neuromuscular disease. Muscle Nerve 1983; 6:195–200.
69. Vajsar J, Becker LE, Freedom RM, Murphy EG. Familial desminopathy: myopathy with accumulation of desmin-type intermediate filaments. J Neurol Neurosurg Psychiatry 1993; 56:644–648.
70. Nakano S, Waclawik AJ, Engel AG. Myopathy with desmin excess. 1. Clinical and ultrastructural data. Muscle Nerve 1994; (Suppl 1):S214 (abstract).
71. Fardeau M, Tomé FMS. Congenital myopathies. In: Engel AG, Franzini-Armstrong C, eds. Myology, 2nd ed. Vol 2. New York: McGraw-Hill, 1994:1487–1532.
72. Goebel HH, Muller J, Gillen HW, Merritt AD. Autosomal dominant "speroid body myopathy." Muscle Nerve 1978; 1:14–26.

26
Myotonic Dystrophy

Russell J. M. Lane
Charing Cross Hospital and Charing Cross Hospital and Westminster Medical School, London, England

Peggy Shelbourne and Keith J. Johnson
University of Glasgow, Glasgow, Scotland

INTRODUCTION

Myotonic dystrophy, also referred to as myotonia atrophica, dystrophia myotonica (DM), myotonic muscular dystrophy and Steinert's disease, is "a genetically determined disorder in which a characteristic pattern of dystrophic muscle disease is accompanied by myotonia and by specific abnormalities of other systems." While this definition, from Harper's outstanding monograph on the subject (1), encompasses the elements of the condition as found typically in middle-aged patients, it belies the extaordinarily wide manifestations of the disorder, ranging from asymptomatic individuals, through cases with minimal features such as cataract and asymptomatic myotonia, to congenital cases with severe weakness but without clinical myotonia, and patients with prominent additional features such as mental retardation, dysphagia, and cardiopulmonary disease. DM is truly a multisystem disorder and may present to clinicians in a number of specialties.

Delineation of DM arose from the study of hereditary myotonic disorders, principally in Germany at the turn of the century. Descriptions appeared of atypical cases of *myotonia congenita*, the first of these conditions to be reported, with muscle wasting and additional features such as testicular atrophy and cataract. The mental impairment and other nonneuromuscular manifestations displayed by some patients, and the dominant mode of inheritance of the disease, were established by studies in the first half of this century (2). More recent research has focused on the unusual molecular genetics of the disease, an example of "dynamic mutation" by which the heterogeneity of phenotypic expression among individuals of a kinship appears to be determined.

Clinical Spectrum of the Myotonias

The differential diagnosis of myotonic disorders includes certain forms of nondystrophic myotonic disorders and periodic paralyses [the sodium channel disorders and the various forms of congenital myotonia (chloride channelopathies), see Chapter 27] and rare disorders such as chondrodystrophic myopathy (Schwartz-Jampel syndrome), as well as myotonic dystrophy. To this list can now be added the recently described proximal myotonic myopathy (PROMM) (3), a genetically distinct disorder which has a phenotype that includes certain features common to DM, such as autosomal dominant inheritance and cataracts, but in which the facial and ocular muscles are spared, the weakness is proximal rather than distal, and cardiac involvement appears to be less frequent.

EPIDEMIOLOGY OF MYOTONIC DYSTROPHY

DM is a worldwide disease. Since cases may go undiagnosed or present atypically with nonneurological features, prevalence figures are not entirely reliable, but a reasonable estimate is around 5/100,000, with an estimated heterozygote frequency of around 13/100,000. However, "hot spots" of high disease frequency have been noted, such as in Saguenay Province of northeast Quebec, where the prevalence is around 200/100,000 (4). Genealogical studies have subsequently shown that the disease in this region developed in 10–14 generations from a couple who had settled there in 1657 (5). In Guipuzcoa, in the Basque region of Spain, the prevalence was reported to be 26.5/100,000 (6) and in Mallorca, 11–12/100,000 (7)—again, significantly greater than expected. In contrast, the prevalence appears lower in ethnic African, Thai, Cantonese, and Oceanic populations (8).

CLINICAL FEATURES

The main features of DM are summarized in Table 1.

Myotonia

The cardinal sign of DM is myotonia, but unlike patients with congenital myotonia, DM patients rarely complain of or seek treatment for this problem, and it may be asymptomatic in mild cases. It is often described as "stiffness" and may be attributed to "arthritis." Occasionally, DM comes to light for the first time when myotonia causes a problem, such as inability to throw down a baseball bat after striking the ball or to release the trigger of a gun. *Grip myotonia* is demonstrated by asking the patient to relax the hand rapidly after a forceful grasp (Fig. 1). This phenomenon may be less marked after repeated contraction and relaxation ("warm-up phenomenon") and is usually worse in the cold. However, changes in action potential characteristics and force generation in the cold in DM cases do not seem to differ from normal (9). *Percussion myotonia* is usually demonstrable even if grip myotonia is absent. Striking the thenar eminence with a percussion hammer results in the thumb being drawn across the palm, followed by slow relaxation and return to the initial position. Percussion myotonia in other muscles, such as the forearm and tongue, may also be demonstrable and may be accompanied by the production of an indentation or groove in the muscle belly by the hammer blow. Even when such clinical signs are absent or difficult to elicit, as when muscle wasting and weakness are very severe, myotonia can almost invariably be demonstrated by electromyography (EMG), which elicits the characteristic "dive bomber" potentials (myotonic runs, see Chapter 4). In the earliest stages of the disease, however, electrical myotonia may be scarce or absent in certain muscles and extensive EMG screening may be required. In such cases, myotonia is most likely to be detected in small, distal muscles and facial muscles.

The basis of the myotonia in DM is at present less clear than in the nondystrophic myotonias. Earlier microelectrode studies of isolated fibers in DM showed little abnormality in membrane cable properties or in chloride or potassium conductance, and there was substantial variability between patients. However, a reduced resting membrane potential has commonly been observed and some defect in sodium conductance has been suggested to underly the myotonia of DM (1). Unfortunately, biochemical studies of muscle and nonmuscle cell membranes over a number of years, looking for abnormalities of lipid composition, protein phosphorylation, and ATPase activities, have given conflicting results (1). Recent studies on the flux of protons, ^{23}Na, and ^{13}C in erythrocyte membranes using multinuclear NMR have revealed increased water permeability and reduced Na^+ pumping, together with an increased rate of glycolysis (10), and studies of sarcolemmal Na/K-ATPase and sarcoplasmic reticulum Ca-ATPase in skeletal muscle and cultured muscle cells from DM patients showed reduced activities compared to controls, due to a reduction of approximately 40% in the molecular concentrations of these enzymes in the membranes (11). Behrens et al. (12) have also shown that the bee venom apamin, which binds to a slow Ca^{2+}-activated K^+ channel, reduces myotonia in thenar muscles when injected directly into the tissues. These apamin-binding sites appear after denervation and in fetal muscle and their appearance may be related to the development of myotonia in DM.

Table 1 Clinical and Laboratory Features of DM

	Clinical features	Laboratory tests
Neuromuscular	Myotonia	Myotonic runs on EMG
	Weakness	
	Reduced reflexes	Axonal neuropathy in some cases
Eye	Cataract	Slit lamp
	Ptosis	
	Ophthalmoparesis	
	Retinal pigmentation	ERG
		VEPs
Endocrine	Testicular atrophy	FSH, testosterone
	Diabetes	Insulin resistance
	Pituitary dysfunction	GH, prolactin
		Impaired TRH response
	Hyperparathyroidism	PTH
Skin	Frontal balding	
	Pilomatrixoma	
Cardiovascular	Hypotension	Low blood pressure
	Syncope, palpatations	ECG abnormalities
	Sudden death	
	Mitral valve prolapse	Echocardiograph
Respiratory	Aspiration pneumonia	
	Somnolence	
	Sleep apnea	Polycythemia
Gastrointestinal	Dysphagia	
	Pseudo-obstruction	Volvulus, megacolon
	Cholelithiasis	on ultrasound, etc.
CNS	Mental retardation	Neuropsychometry
		Atrophy on CT
		White matter lesions on MRI
		Hypometabolism on SPECT, PET
Immune system		Reduced immunoglobulins,
		T-cell functions
		Reduced C3, 4

ERG = electroretinogram; PTH = parathormone assay.

Muscular Weakness and Wasting

The second, and physically most disabling, feature of DM is progressive muscular weakness and wasting. This is most marked in the cranial musculature and distal limb muscles, although any muscle can be affected; muscles of the pelvic girdle, hamstrings, gastrocnemius, and soleus tend to be least involved. The tendon reflexes are normal to reduced or absent. Patients typically show bilateral ptosis, weakness of muscles of facial expression (rather "unsmiling," lugubrious expression), and jaw muscle weakness, which can lead to temperomandibular joint dysfunction (see Chapter 1, Fig. 3). Facial muscle wasting may produce a "hatchet-like" facies (Fig. 2a). Palatal, tongue, and laryngeal weakness may result in nasal dysarthria and eventually aspiration pneumonia. Sternomastoid weakness and wasting is often prominent at an early stage (Fig. 2b), and this, together with extraocular muscle involvement when present and relative preservation of neck extensors and shoulder girdle muscles, distinguishes DM from facioscapulohumeral muscular dystrophy. In the limbs, weakness of grip and wrist function

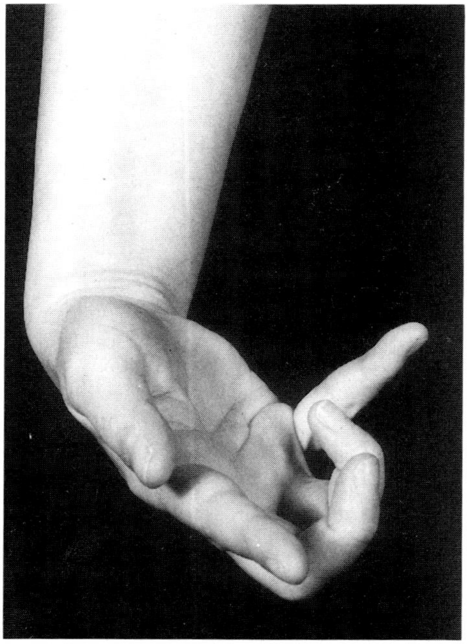

Figure 1 Grip myotonia. The patient is attempting to relax her hand and straighten her fingers after clenching her fist.

and footdrop are early signs. Wasting of the hand and foot muscles is common, but clawhand and pes cavus do not develop. Unlike limb-girdle dystrophies, the proximal weight-bearing muscles tend to be relatively well preserved, and most patients with DM, even with marked muscle involvement, retain mobility.

Classification of Disease Severity

A clinical subdivision of DM based on age of onset and phenotypic manifestations has been proposed (13). *Congenital DM* is characterized by hypotonia, respiratory and feeding difficulties at birth, mental retardation, and a mother or sibling with definite DM; *early childhood DM* encompasses cases with the cardinal features of DM and marked developmental delay, but no clear abnormalities at birth; *classical DM* cases present in adult life with symptoms related to the disease; and *minimal DM*, commonly affecting older patients, is characterized by at least two of the following: cortical cataracts, myotonic runs on EMG, and positive family history showing autosomal-dominant inheritance—to which might now be reasonably added the demonstration of abnormal expansion of the CTG repeat sequence at the DM locus (see below). The ability to identify obligate carriers, who may have no clinical or laboratory stigmata of the disease, suggests that a further category of *asymptomatic DM* is also justified. Mathieu et al. (14) have proposed a five-point muscular disability rating scale for DM, based on distal to proximal progression of muscular involvement in the condition, which proved useful in genotype/phenotype correlation studies (Table 2).

Muscle Pathology

Skeletal muscle in DM may be completely normal histologically, and a normal biopsy does not rule out the diagnosis. Indeed, few centers would now consider biopsy a mandatory investigation in a patient where the diagnosis was probable on clinical and genetic grounds, and the recent development of

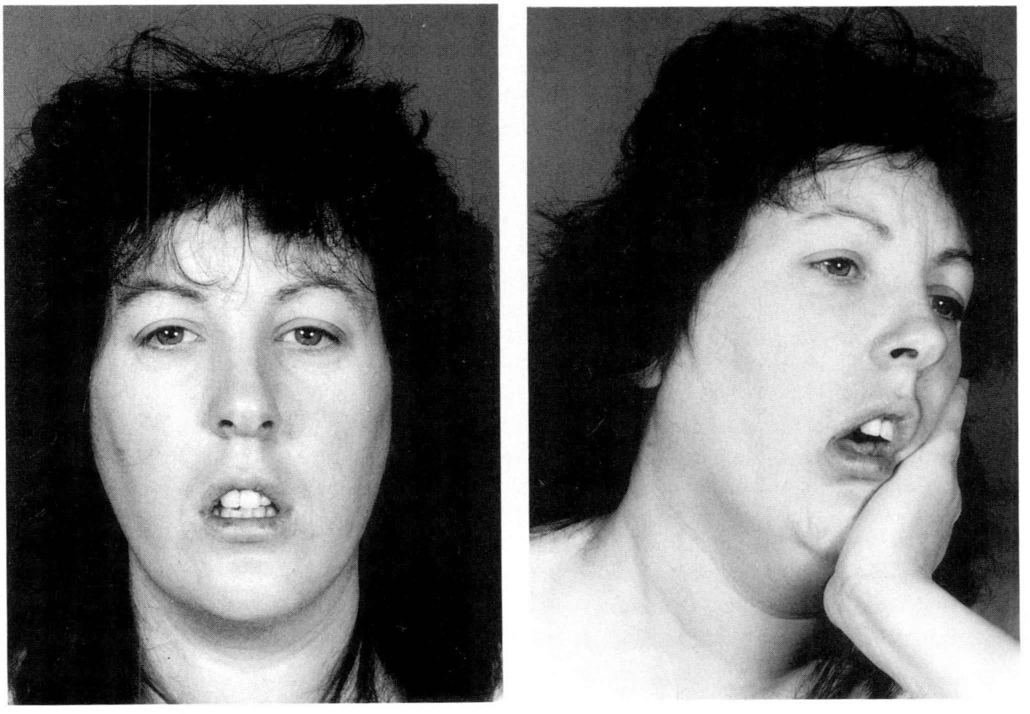

(a) (b)

Figure 2 (a) Facies in a young, female patient. Hatchet-shaped face, "carp mouth." (b) Weakness and wasting of sternomastoid.

specific genetic tests for the disease now makes other diagnostic tests largely superfluous. When the biopsy is abnormal, the changes are relatively nonspecific. The most common finding, best appreciated on longitudinal sections, is an increase in internal nuclei, frequently arranged in long chains down the center of fibers. The appearances are reminiscent of centronuclear myopathy (Chapter 15) and could represent a similar abnormality of muscle fiber maturation. Ringed fibers (*ringbinden*, striated annulets)

Table 2 Classification of Disease Severity in DM

Grade	Clinical description
1	No clinical neuromuscular impairment (diagnosed by laboratory tests)
2	Minimal signs (myotonia and cranial signs, such as jaw and temporal wasting, sternamostoid wasting/weakness, ptosis, nasal speech) No significant distal weakness
3	Distal weakness (no proximal weakness, except triceps)
4	Severe proximal weakness (uses wheelchair for some trips)

Source: After Mathieu et al. (14).

are also seen occasionally. These represent bundles of myofibrils running in spiral fashion around conventionally orientated fibrils beneath a common basement membrane, and it has been suggested that they are formed as a result of myofibrillary rupture caused by a variety of factors, which can include fiber splitting, abnormal relations between myofibrils and cytoskeleton during development and repeated myotonic contracture (15). It should be noted that, in contrast with other forms of dystrophy, fiber necrosis, degenerative and regenerative changes, cellular infiltration, and replacement of muscle by fat and fibrous tissue are not features of DM muscle.

Other features, such as sarcoplasmic masses, small angular fibers, and changes in fiber diameter and architecture, are inconstant and too nonspecific to be of diagnostic value (16). In the congenital form of DM, however, a number of rather distinctive changes have been noted, including marked muscle hypoplasia, particularly of type 1 fibers, and fibers with an unusual peripheral ring, devoid of mitochondria when seen in transverse section (17). The persistence of chains of central nuclei reminiscent of fetal myotubes, and the poor histochemical differentiation of fibers in congenital DM, suggest that the myopathy of DM may represent a fundamental defect in muscle maturation (18).

Mitochondrial abnormalities may be observed on electron microscopy, and magnetic resonance spectroscopy of DM patients with significant exercise intolerance has revealed changes similar to those seen in mitochondrial myopathy at rest and during exercise (see Chapter 9), except that intramuscular acidification was less marked and recovery characteristics for phosphate, phosphocreatine, and ATP were normal (19). However, consistent abnormalities of mitochondrial DNA have not been found despite earlier claims of a high incidence of such mutations (20).

Symptoms and Signs in Nonmuscular Tissues

The Eye

The early recognition of the association between lens opacities and DM was important, since it confirmed that disease expression was not confined to skeletal muscle and showed that otherwise asymptomatic relatives could be linked to apparently isolated cases of DM through autosomal-dominant inheritance (1). The earliest changes on slit lamp examination appear to be specific to DM, although recent studies indicate that the cataracts in PROMM are identical (3). These comprise punctate, iridescent, white and/or multicolored opacities distributed in the subcapsular regions, with sparing of the central lens region, progressing to rosette-type cataracts with radiating spokes and reticulation of the cortex (Fig. 3), and finally to mature cataract indistinguishable from that seen in other conditions,

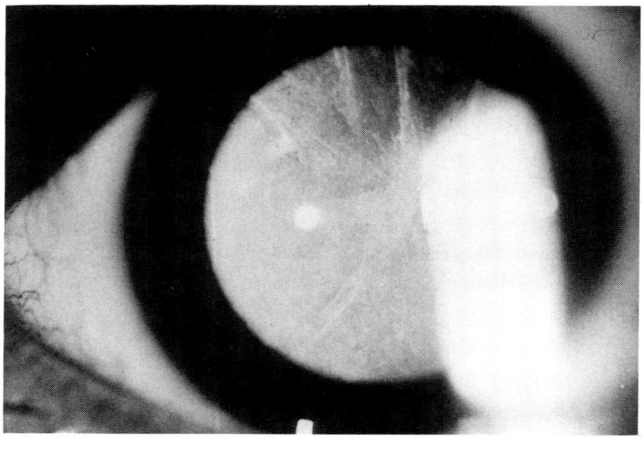

Figure 3 Cataract in myotonic dystrophy. Posterior lens opacities with characteristic "radiating spokes" pattern. (From *Atlas of Clinical Neurology*, 2nd ed. Wolfe Publishing; 1993, with permission.)

with resulting visual impairment. Specific lens opacities can be observed in virtually all adult cases of DM by slit lamp. They may be the only manifestation of the disease in elderly subjects, but are less common in younger patients, and rare in children under 10 and in congenital cases. Diabetes associated with DM may accelerate the maturation of cataracts.

The etiopathogenesis of cataract in DM is unknown, but cholesterol-lowering drugs such as triparanol and diazocholesterol, which have been used to induce myotonia experimentally, can also induce cataract (1).

Retinal changes, including peripheral pigmentary degeneration similar to retinitis pigmentosa, and butterfly-shaped pigmentary macular lesions, are less familiar features than cataract, but are progressive, although they do not seem to cause much problem with acuity. Abnormalities of the electroretinogram may occur, and abnormal visual evoked responses in a high proportion of cases imply damage to central visual pathways (21).

Ptosis in DM has already been mentioned and can be an early feature. Extraocular muscle weakness is also common but often asymptomatic. Intraocular pressure is often reduced, probably due to degeneration of the ciliary muscle (1).

Endocrine Disorders

The first indication that the manifestations of DM extended beyond skeletal muscle was the observation by Steinert of *testicular atrophy* in four of his original patients (1). The testicular atrophy results from atrophy, hyalinization, and fibrosis of the seminiferous tubules, with sparing of the interstitial cells. Thus, FSH levels may be increased while LH levels are normal in the early stages. Testosterone levels fall and LH levels may rise as the disease progresses (22). Male fertility is moderately reduced as a result of these changes. There is little evidence of gonadal dysfunction in females, however. Since testosterone deficiency can cause a myopathy (see Chaper 29), trials of testosterone therapy in DM have been conducted, unfortunately without demonstrating benefit (23).

The incidence of clinical, symptomatic *diabetes mellitus* in DM is not known with certainty, but is probably about four times that of the normal population, around 6–7%. The incidence of glucose intolerance is significantly higher, however, around 16% (1). The basis of this glucose intolerance appears to be "insulin resistance," and this phenomenon has been the subject of considerable investigation. Some studies have reported increased basal insulin and proinsulin levels and an exaggerated insulin response to glucose, but with little resulting effect on plasma glucose, and a normal glucose response to exogenous insulin (24). It has been suggested that compensatory pancreatic B-cell hyperactivity causes the "hyperinsulinism" and results in increased ratios of proinsulin to insulin (25). However, most nonobese DM patients have less severe whole-body and more severe forearm muscle insulin resistance, with normal insulin levels (26), and others have shown a normal ratio of proinsulin to insulin (27). The structures of insulin and insulin receptors in DM are normal (28), and it seems likely that insulin resistance probably ultimately results from abnormalities of cell membrane structure and composition, as reflected in abnormal membrane fluidity, which might hamper the function of insulin receptors (29).

Disorders of pituitary function have been reported occasionally, but there is no consistent abnormality. The pituitary fossa may be small, and changes in the skull X-ray reminiscent of acromegaly may be seen. Studies of bone density indicate a generalized tendancy to hyperostosis in the disease (30). Prolactin levels have been variably reported to be reduced, increased, or normal (1). Growth hormone levels are also increased in some patients, and other evidence of hypothalamic-pituitary axis dysfunction includes the finding of occasional hypernatremia secondary to defective arginine-vasopressin secretion in response to hyperosmolar challenge, and impaired TSH responses to TRH (1). In general, however, such abnormalities do not appear to lead to clinically significant effects.

Pituitary tumours have been found in some patients. While these may be coincidental, they might arise secondarily to chronic hypogonadism or to excess hypothalamic function (1).

Recent studies have also indicated increased parathyroid function in DM (31), and DM was associated with neurocristopathy (hyperparathyroidism, neurofbromatosis, and multiple endocrine adenomatosis) in females in three different families (32).

Premature *frontal balding*, which may affect females as well as males, is often cited as a cardinal feature of DM but is probably overemphasized, and its origin is unclear; there is little evidence that it is hormonally determined. Another rare, but apparently specific, dermatological association of DM is *pilomatrixoma* (calcifying epithelioma of Malherbe), but again, the nature of the association is unclear (1).

Cardiovascular Disorders

The most common cardiovascular complications associated with DM are conduction defects and systemic hypotension. Electrocardiographic (ECG) abnormalities are common in DM and were found in 202/236 (85%) of cases in one early study, although symptoms occurred in only 16% (33). The most frequent abnormalities are forms of sinoatrial and bundle branch block, but serious atrial and ventricular dysrhythmias and complete heart block can develop (26,33–38). These abnormalities are related to the rather selective degeneration and fibrosis of the conducting system that occurs (26,36). In contrast, cardiac failure due to cardiomyopathy and coronary artery disease appears to be uncommon.

There is some relationship between the prevalence of ECG abnormalities and the severity of neuromuscular symptoms (34). Sudden death related to dysrhythmias is well documented and has led to suggestions that serial studies using techniques such as ambulatory monitoring and His bundle electrocardiography should be considered, with a view to prophylactic pacemaker implantation in vulnerable cases (35,36). In one study, 17 DM patients asymptomatic for cardiac disease were followed over an 8-year period; 15/17 had ECG abnormalities at baseline and all had conduction defects at 8 years, while two patients suffered cardiac deaths (37). Thus, lack of cardiac symptoms does not preclude progressive heart disease in DM. A similar high frequency of ECG and echocardiographic abnormalities, again largely asymptomatic, was found in pediatric cases in one study (38). A permanent pacemaker may be required in both pediatric and adult cases, and it would seem reasonable to perform annual ECGs on DM patients and to consider more detailed investigations if there are presyncopal or syncopal symptoms, or if AV block or widened QRS complexes are evident (36).

It is worth noting that procainamide and quinidine, sometimes used for the treatment of myotonia, have been shown to prolong the P-R interval, while phenytoin, which is equally efficacious, does not.

Mitral value prolapse is also a common finding on echocardiography in DM patients, but pathological studies indicate that it is functional rather than structural, and does not appear to predispose to embolic cerebral infarction (1).

Hypotension was the first cardiovascular abnormality to be noted in DM and has been confirmed in more recent studies (39). The hypotension is largely asymptomatic and could indeed confer biological advantage. It probably relates to the widespread abnormality of smooth muscle function in DM (see below).

Respiratory Disorders

Respiratory disorders are largely secondary to bulbar and respiratory muscle weakness in adult cases, the most common manifestation being aspiration pneumonia, but anesthesia in DM poses hazards at any stage (40) (see also Chapter 52). Pulmonary immaturity and respiratory muscle weakness may contribute significantly to infant mortality in the congenital form, where diaphragmatic myotonia developing during inspiration, as well as weakness, may play a role (1). In addition, some patients appear to have alveolar hypoventilation, with resulting hypercapnia and secondary polycythemia, which some have suggested is out of proportion to the impairment expected on the basis of weakness and myotonia. This may imply abnormalities of central respiratory drive; chemoreceptor responses have been reported to be both normal and abnormal (1). Whether such respiratory disorder explains fully or in part the hypersomnia sometimes found in DM patients is a matter of debate. In one recent study, neuropsychological function and sleep characteristics studied by polysomnography were examined in eight patients. While significant apneic events were recorded in half, there was no correlation between severity of respiratory disorder and diurnal congenitive deficit (41).

Gastrointestinal and Other Smooth Muscle Disorders

Reduction in intraocular tension and systemic hypotension have been mentioned previously and have been attributed to abnormal smooth muscle function. Gastrointestinal symptoms may also arise on this basis.

The most common is dysphagia, which is due to abnormal striated muscle function in the pharynx and upper esophagus, and in some patients to smooth muscle dysfunction more distally (1,26). Cases of gastric volvulus, megacolon, and intestinal pseudo-obstruction requiring colonic resection have also been reported (42). Anal sphincter function may be defective in congenital and adult cases, and poor contraction of the gall bladder may predispose to cholelithiasis. Uterine dysfunction may cause problems in pregnancy and parturition (1), but curiously, urinary bladder function is generally unaffected.

Nervous System Disorders

A number of early studies of DM patients commented on changes in mental function. Patients were typically described as showing apathy and inertia, and even being hostile and unreliable, but more recent controlled investigations have failed to confirm any particular personality characteristics in DM (1). Several studies have shown reduction in mean IQ and educational level in populations and groups of DM patients, and it seems likely that a degree of retardation and cognitive impairment is a feature of the disease, although there is a wide variation in terms of severity (1). Psychiatric diseases such as depression and psychosis have not been considered to be prominent features of the condition, although this view has been challenged (43).

Abnormalities of VEPs in relation to eye disorder have already been mentioned, and several studies have also found evidence of central disorders using short-latency brainstem and long-latency auditory event-related potentials (44), and brainstem auditory evoked resonses and studies of vestibulo-ocular reflexes (45,46), indicating abnormalities in central sensory processing of visual and vestibuloauditory function. However, no abnormalities of central motor conduction velocities were found in one study, suggesting sparing of the corticospinal tracts (47), and abnormalities of somatosensory evoked responses seem uncommon (46).

Early radiological studies of the brain indicated that progressive cerebral atrophy was a feature of DM, compatible with autopsy studies showing reduced brain weight (1). CT scanning has shown occasional examples of basal ganglia calcification, which may in part be due to deposits in thalamostriate blood vessels (48). The recent advent of MRI imaging has shed much needed light on the nature of brain involvement in DM and its relationship with cognitive dysfunction. Several MRI studies have shown white matter lesions in addition to cerebral atrophy (49–52), and some have found a significant association between these lesions and cognitive impairment, constituting a subcortical dementia (52). Positron emmission tomographic (PET) studies in DM have shown reduced cerebral glucose utilization (53,54), and a recent study of cerebral blood flow using single photon emission computerized tomography (SPECT) found significant abnormalities in cerebral perfusion, notably in the temporal and frontal cortex, which correlated with reduction in IQ; blood flow was also significantly reduced on xenon 133 clearance studies (55). Many of the patients studied complained of somnolence, and it was commented that the pattern of hypoperfusion changes seen in the DM patients was also typical of sleep apnea (see above). Interestingly, cognitive and blood flow abnormalities were more marked in cases inherited through the maternal line. In relation to these observations, cognitive dysfunction followed the onset of heart block but improved with pacing in one case, suggesting defective cerebrovascular autoregulation and a critical perfusion state (48).

Pathological studies of brain have revealed abnormalities of gyral architecture and cellular organization, suggesting disturbance of neuronal migration during brain development, but such changes are nonspecific. Curious inclusions in thalamic neurons have been reported in several studies. While they are said to occur in normal subjects, they were remarkably common in DM cases (1).

Recently, the rather selective distribution of Alzheimer's neurofibrillary tangles in structures comprising the limbic system, in the relative absence of senile plaques, has been noted (56). Case control studies indicate that their frequency is age-related (57), and these selectively placed lesions appearing in the presenium may provide an insight into the rather curious neuropsychological and cognitive dysfunction in this disease.

Abnormalities of peripheral nerve and the neuromuscular junction have been reported in DM. Clinically, tendon reflexes may become depressed or absent and while sensation is preserved, abnormal thermal thresholds have been demonstrated. Reduction in motor and sensory nerve conduction velocities may be found, particularly affecting the most rapidly conducting fibers, compatible with an axonal peripheral neuropathy (58). In keeping with this, sural nerve biopsy may show reduced density of myelinated fibers. Nevertheless, pathological studies have not shown strong evidence for a significant neurogenic component to muscle fiber atrophy in the disease.

There is no good evidence of autonomic involvement in the condition. While abnormalities such as reduced blood pressure responses to postural change and handgrip and pupillary contraction rate can be demonstrated, these appear to be due to neuromuscular defects rather than to defective autonomic reflexes (59).

Hearing loss is not uncommon in DM and may be severe. In adults this usually takes the form of a high-frequency sensorineural deafness, which appears to have both peripheral and central components in its etiology, while in congenital and childhood cases conductive deafness related to recurrent otitis media and craniofacial anomalies appears to be the most important factor (1).

Immunological Abnormalities

A number of studies have demonstrated reduction in levels of immunoglobulins and defects in cell-mediated immunity in DM (1). Reduced levels of C3 and C4 complement components may also be found, and autoantibodies to smooth muscle were found in almost half the cases studied recently by one group (60).

CONGENITAL MYOTONIC DYSTROPHY

DM presenting at birth or in infancy and childhood deserves special consideration, not only because of the particular clinical features and management problems posed, but also in view of the genetics underlying such manifestations and their influence on our understanding of pathogenesis. The extensive studies of Harper and his colleagues now present a clear picture of the congenital form and its prognosis (1), and the genetic aspects are discussed later in this chapter.

Incidence and Diagnostic Features

While congenital DM was initially believed to be rare, it is now evident that in the past, many patients with congenital DM died at birth or were misdiagnosed variously as having "cerebral hypotonia," Moebius syndrome, or one of the other congenital neuromuscular diseases. Furthermore, the diagnosis may not be made at birth if the infant is relatively mildly affected, but present in infancy or childhood as a result of delay in motor milestones or mental retardation. The true incidence is thus unknown.

Some diagnostic clues may be available from the history. It is now established that, with very rare exceptions (see below), congenital DM cases are born to affected mothers, while patients presenting in adult life are more likely to have affected fathers. Mothers of congenital DM cases may have been aware of reduced fetal movements and may have encountered problems such as polyhydramnios (due to impaired fetal swallowing) and difficulties in labor due to pelvic muscle and uterine dysfunction (see above). Furthermore, while examination of the newborn infant may be inconclusive for DM, clinical and electrophysiological studies on the mother usually confirm that she is affected.

The phenotype in typical cases is characteristic. There is severe bilateral facial muscle weakness, which progresses with age with the development of a "tented" appearance of the upper lip (carp mouth). The jaw hangs open due to jaw muscle weakness, which together with severe pharyngeal and laryngeal weakness, discussed above, results in drooling and problems with sucking and swallowing. The limbs are profoundly hypotonic and weak, but the reflexes are usually preserved. Clinical myotonia is absent, and electrical myotonia may also be undetectable in children under 1 year of age and is found in only about 10% in those under 5 years. Half the patients have talipes, usually of equinovarus type, and some have more generalized joint contractures typical of arthrogryposis.

The most immediate consequence of the condition is usually respiratory distress. This is often multifactoral, being due to weakness of diaphragmatic and intercostal muscles, pharyngeal and laryngeal weakness predisposing to aspiration pneumonia, pulmonary immaturity due to lack of respiratory movements in utero, and probably to cerebral factors including brain damage due to intrauterine hypoxia, possibly cerebral haemorrhage, and cerebral disease related to expression of the DM gene. Chest X-ray may show consolidations due to uncleared secretions and aspiration, together with elevation of the right hemidiaphragm and very thin ribs, also found in other congenital myopathies.

Improvements in resuscitation and intensive care facilities have improved the immediate prognosis for such infants, and prospective studies have shown that patients who survive, while showing severe delay in motor development, do eventually walk and have an independent physical existence. Neuroimaging studies in congenital DM have shown that ventricular dilatation is commonly present at birth but is apparently nonprogressive, at least over the first few years of life, and that periventricular and subcortical white matter lesions discussed above are also present (61).

The outlook for patients with congenital DM is poor. A recent survey of 115 congenital cases found that there was a 25% risk of death before 18 months of age and only a 50% chance of survival into the midthirties (62). Furthermore, congenital DM patients are often very handicapped, although more by virtue of their mental retardation than physical disability. Approximately, 60–70% of patients have mental retardation, with IQs usually in the 40–80 range, with less than 10% having normal intelligence.

PATHOGENESIS OF MYOTONIC DYSTROPHY

Molecular Genetics of DM

While DM was first delineated at the turn of the century, its familial nature was not recognized until Fleischer's study in 1922, which, as mentioned above, showed that otherwise unaffected individuals in a pedigree could be linked to affected cases by the presence of cataract (1). Autosomal-dominant inheritance was not established until the detailed studies of Bell (1947) and Thomasen (1948) (1). Subsequent linkage studies assigned the gene for DM to the long arm of chromosome 19 (2), and further genetic and physical maps defined the locus to the band 19q13.3 between the flanking markers ERCC1 and D19S51 (63,64) (Fig. 4). DNA from the intervening region was screened for mutations in candidate genes, with the eventual identification of a DNA fragment of larger and more variable size in DM patients than in the normal population—an example of a "dynamic mutation" (65).

CTG Repeat Size and Phenotypic Variation

The mutation in DM was shown to be due to a marked increase in copy number of a tandemly arrayed trinucleotide (CTG) repeat sequence. Normal individuals usually have between five and 37 repeats, while DM patients typically have more than 50 (corresponding to an expansion of >0.15 kb). In general, there appears to be a relationship between disease severity and size of the repeat sequence. Patients with 50–80 repeats are minimally affected, while patients with severe disease, such as the congenital form, may have expansions of several kilobases (65,66) (Fig. 5). The mutation also shows somatic instability and may vary in size from tissue to tissue, being larger, for example, in muscle than leukocytes (67,68). Precisely how this relates to severity of disease expression in different tissues is not yet known.

This novel mechanism of mutation is now known to be shared by several other disorders, including

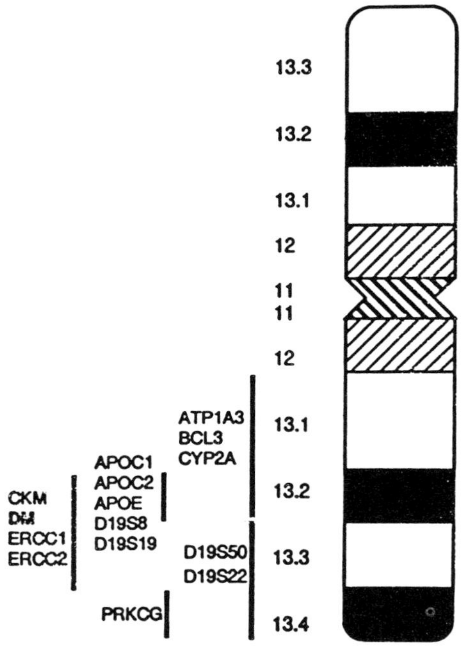

Figure 4 Gene map for MyD. Positions of genes, denoted by Gene Database symbols and polymorphic DNA markers, on 19q13.2–13.3. DM is now known to map distal to the muscle creatine kinase gene (*CKM*) and the excision repair genes (*ERCC1* and *ERCC2*) and proximal to the protein kinase C gamma subunit gene (*PRKCG*) and the anonymous DNA markers D19S50 and D19S22. (Detailed maps and their methods of derivation can be found in Refs. 63 and 64.)

fragile-X mental retardation, X-linked spinobulbar muscular atrophy (Kennedy's syndrome), Huntington's disease, dentatorubral-pallidoluysian atrophy (DPRLA), one form of dominantly inherited cerebellar ataxia (SCA1), and Machado-Joseph disease; undoubtedly, other examples will be discovered.

Intrafamilial Phenotypic Heterogeneity: The Phenomenon of "Anticipation"

When inheritance of the CTG repeat region is traced through individual families, it usually increases in size in successive generations. This provides an explanation for "anticipation," the tendency of the disease to be more severe in the offspring of successive generations (69,70) (Fig. 6). Rarely, however, the reverse phenomenon may be observed, with the inheritance of a *smaller* copy number or, exceptionally, a sequence size within the normal range (70). This intergenerational contraction of repeat size may result in individuals who are minimally affected or have no stigmata of the disease and probably accounts for instances of "incomplete penetrance" in DM families.

Maintenance of the DM Gene Pool

Autosomal-dominant diseases associated with a significant loss of reproductive fitness, as in DM, are usually associated with a high mutation rate, neccessary to maintain the gene pool. However, relatively few founder mutations have been detected in DM. Studies of CTG repeat sizes at the DM locus in the *normal* population found a major common allele of CTG_5 in about 40% of chromosomes and three other major alleles ($CTG_{11,12,13}$) and one minor allele (CTG_{14}) in an additional 50%. It was postulated that larger expansions ($CTG_{>19}$) present in the remaining 10% of the population constituted a pool of

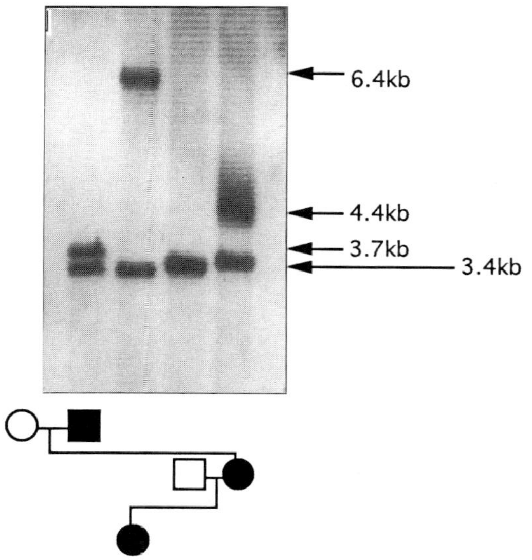

Figure 5 Southern blot of genomic DNA samples, digested with the restriction enzyme BglI and probed with p5B1.4 (70) from individuals in a family with DM. Filled symbols represent affected cases and show two fragments, one of 3.4 kb corresponding to the normal allele, containing between five and 35 copies of the CTG repeat, and additional higher-molecular-weight bands representing approximately 100 CTGs (3.7 kb), 230 CTGs (4.4 kb), and 1000 CTGs (6.4 kb), respectively. Note that the 4.4-kb band shows an elongated smear indicating marked heterogeneity of the repeat number.

potentially unstable alleles that could increase into the DM range on transmission (71). The dynamics of transmission of CTG expansions is also influenced by the sex of both parent and offspring. It has recently been shown that the intergenerational contraction of repeat size noted above nearly always occurs with *paternal* transmission. By corollary, it is uncommon for fathers to pass on expansions of larger than 1.5 kb, while maternal transmission shows no such restrictions (72). This may be reflected in the fact that younger, more severely affected patients are usually the offspring of affected mothers, while older, less severely affected patients have affected fathers. In addition, selective transmission of larger, potentially unstable CTG expansions, in preference to wild-type, from fathers to sons has been shown (73). This phenomenon, of "meiotic drive" at the DM locus, resulting in male-to-male transmission of unstable CTG repeats, may be the basis whereby the gene pool is maintained (74). Thus the prevalence of diseases like DM that are characterized by dynamic mutations may be determined by the frequency of such critical size repeat sequences in different populations. These potentially unstable, intermediate-size alleles, which are larger than found in most of the normal population but not large enough to cause disease, appear to be commoner in races with higher disease incidence (75). This meiotic instability also appears to underlie the almost exclusive association between the severe congenital form of DM and maternal transmission; affected females tend to transmit larger mutations (72,73). The recent reports of rare paternal transmission of the congenital form argue against an nongenetic "maternal factor" in congenital DM (76,77).

While affected individuals in a family can have quite variable sizes of repeats, families may show clustering of organ-specific complications such as symptomatic cardiac disease, pilomatrixomas, and smooth muscle involvement (66). In addition, it was noted in one study of affected sibling pairs that the later-born sibs appeared to be more severely affected than the older sibs, this effect increasing with the age difference between the sibs. It was speculated that this might be a reflection of increasing maternal age (78).

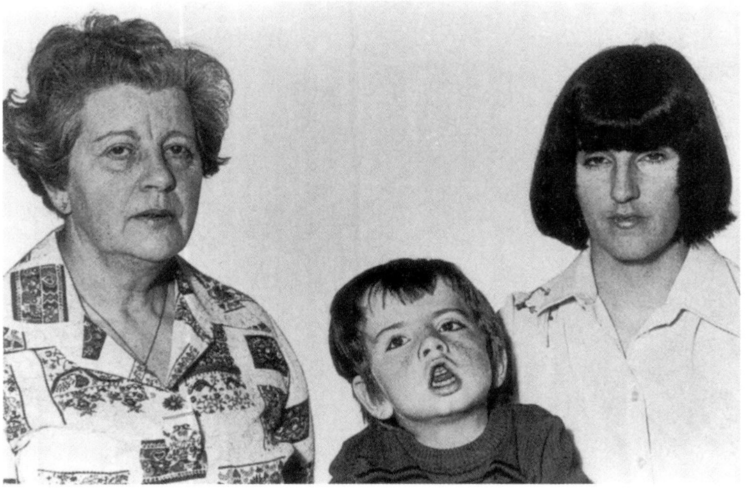

Figure 6 Anticipation in myotonic dystrophy. Three generations of one family. The grandmother was minimally affected, with electrical myotonia and cataracts; her daughter had classical DM, and there was a congenitally affected son.

Diagnosing DM by Molecular Genetics

The identification of the DM mutation and development of specific genetic tests for the disease have revolutionized the diagnosis of DM, particularly in very mild or apparently isolated cases. Thus, it has been demonstrated that such tests are more sensitive and specific than any combination of clinical assessments, such as slit-lamp examinations and electromyography (79) (Fig. 5). Such tests have proved to be of particular value in genetic counseling. Affected mothers have a 10% risk of having a severely affected child and a 40% risk if the mother already has a CMD child (80), the risk being greatest if the mother has clinical signs of the disease at the time of pregnancy. Direct mutation detection by chorionic villus sampling has proved sensitive and accurate in diagnosing DM prenatally (81,82), but there are potential pitfalls in terms of predicting disease severity, in that there is no simple, clear relationship between CTG repeat size and phenotype. The observed size of CTG repeat in lymphocytes or biopsy sample may not be representative of all tissues. The instability of the dynamic mutation also raises the possibility of mosaicism, in which a low copy number is present at the DM locus in somatic cells, but a larger mutation determining a more severe phenotype is present in gonadal cells (83).

Pathophysiology of DM

Several studies have shown that the CTG repeat region lies in a gene that has strong homology to the cAMP-dependent protein kinase family (the DMPK gene) (84), which has now been fully sequenced (85). Reports of DMPK mRNA expression in tissues have been inconsistent, some laboratories finding decreases in mRNA transcript (86) or complete absence in congenital cases (87), while others have reported increases (88). The CTG expansion has no effect on the coding sequences of the DMPK gene, and the function of this gene is unknown. Studies of the protein product using antipeptide antibodies have been hampered by lack of specificity of these reagents, and it is likely that the protein species localized to date are cross-reacting species unrelated to the DMPK gene (unpublished personal observations).

One possibility under investigation is that the CTG expansion has an incremental effect on the function of several genes in the region of the DM locus. In this model, CTG expansion into the DM range would first affect gene(s) determining the development of cataract, the manifestation of the disease in the most mildly affected individuals. Further expansion would cross a threshold determining disease expression in muscle, either through a differential tissue sensitivity of the gene determining cataract formation or by remote effects on other genes in the region (a field effect); there is a high gene density in the DM region of chromosome 19. Further expansions would occur, with increasing phenotypic expression at earlier ages in successive generations (anticipation) until a threshold was crossed involving a number of genes, resulted in the congenital form. These genetic effects would be modified by factors such as heterogeneity of CTG repeat length in different tissues and critical timing of expression of affected genes during development.

In summary, the discovery of the genetic mutation for DM has greatly increased diagnostic precision, but the mechanism by which the CTG expansion results in disease is unknown. Our understanding would be aided considerably through the study of transgenic animals expressing the mutation, but there are formidible technical problems involved and it is clear that there is still much to be done before we have satisfactory explanations for the mysteries that collectively comprise the DM phenotype.

REFERENCES

1. Harper PS. Myotonic Dystrophy, 2nd ed. Philadelphia: WB Saunders, 1989.
2. Shaw DJ, Harper PS. Myotonic dystrophy: developments in molecular genetics. Br Med Bull 1989; 45:745–759.
3. Ricker K, Koch MC, Lehmann-Horn F, Pongratz D, Speich N, Reimers K, Schneider C, Moxley RT III. Proximal myotonic myopathy (PROMM): clinical features of a multisystem disorder similar to myotonic dystrophy. Arch Neurol 1995; 52:25–31.
4. Bouchard G, Roy R, Declos M, Kouladjian K, Mathieu J. Spreading of the gene for myotonic dystrophy in Saguenay (Quebec). J Genet Hum 1988; 36:221–237.
5. Mathieu J, De Braekeleer M, Prevost C. Genealogical reconstruction of myotonic dystrophy in the Saguenay-Lac-Saint-Jean area. Neurology 1990; 40:839–842.
6. Lopez-de-Munain A, Blanco A, Emparanza JI, Poza JJ, Marti-Masso JF, Cobo A. Prevalence of myotonic dystrophy in Guipuzcoa (Basque country, Spain). Neurology 1993; 43:1573–1576.
7. Burcet J, Canellas F, Cavaller G, Vich M. Epidemiological study of myotonic dystrophy on the island of Mallorca. Neurologia 1992; 7:61–64.
8. Ashizawa T, Epstein HF. Ethnic distribution of the myotonic dystrophy gene. Lancet 1991; 338:642–643.
9. Ricker K, Hertel G, Langscheid K, Stodieck G. Myotonia is not aggravated by cooling: force and relaxation of the adductor pollcics in normal subjects and in myotonia as compared to paramyotonia. J Neurol 1977; 216:9–20.
10. Kuwabara T, Yuasa T, Ohno T, Yamamuro M, Miyatake T. Study on the erythrocytes from myotonic dystrophy with multi-nuclear NMR. Muscle Nerve 1991; 14:57–63.
11. Benders AA, Timmermans JA, Oosterhof A, Ter-Laak HJ, Van Kuppevelte TH, Wevers RA, Veerkamp JH. Deficiency of Na^+/K^+-ATPase and sarcoplasmic reticulum Ca^{++}-ATPase in skeletal muscle and cultured muscle cells of myotonic dystrophy patients. Biochem J 1993; 293:269–274.
12. Behrens MI, Jahl P, Serani A, Vergara F, Alvarez O. Possible role of apamin-sensitive K^+ channels in myotonic dystrophy. Muscle Nerve 1994; 17:1264–1270.
13. Jaspert A, Fahsold R, Grehl H, Claus D. Myotonic dystrophy: correlation of clinical symptoms with the size of the CTG trinucleotide repeat. J Neurol 1995; 242:99–104.
14. Mathieu J, de Braekleer M, PrCvost C, Boily C. Myotonic dystrophy: clinical assessment of muscular disability in an isolated population with presumed homogenous mutation. Neurology 1992; 42:203–208.
15. Sawicke E. Origin of the ring muscle fibres in neuromuscular diseases. Neuropatol-Polska; 1991; 29:29–40.
16. Casanova G, Jerusalem F. Myopathology of myotonic dystrophy: a morphometric study. Acta Neuropathol 1979; 4:231–240.
17. Farkas E, Tome FMS, Fardeau M, Arsenio-Nunes ML, Dreyfus P, Driebler MF. Histochemical and ultrastructural study of muscle biopsies in 3 cases of dystrophia myotonica in the newborn child. J Neurol Sci 1974; 21:273–288.
18. Argov Z, Gardner-Medwin D, Johnson MA, Mastaglia FL. Congenital myotonic dystrophy: fibre type abnormalities in two cases. Arch Neurol 1980; 18:693–696.
19. Taylor DJ, Kemp GJ, Woods CG, Edwards JH, Radda GK. Skeletal muscle bioenergetics in myotonic dystrophy. J Neurol Sci 1993; 116:193–200.
20. Thyagarajan D, Byrne E, Noer S, Lertrit P, Uttanophol P, Kapsa R, Marzuki S. Significance of mitochondrial DNA deletions in myotonic dystrophy. Acta Neurol Scand 1993; 87:32–36.

21. Kimizuka Y, Kiyosawa M, Tamai M, Takase S. Retinal changes in myotonic dystrophy: clinical and follow-up evaluation. Retina 1993; 13:129–135.
22. Harper P, Penney R, Foley TP Jr, Migeon CJ, Blizzard RM. Gonadal function in males with myotponic dystrophy. J Clin Endocrinol Metab 1972; 35:852–856.
23. Griggs RC, Pandya S, Florence JM, Brooke MH, Kingston W, Miller JP, Chutkow J, Herr BE, Moxley RT. Randomised controlled trial of testosterone in myotonic dystrophy. Neurology 1989; 39:219–222.
24. Piccardo MG, Pacini G, Rosa M, Vichi R. Insulin resistance in myotonic dystrophy. Enzyme 1991; 45:14–22.
25. Krentz AJ, Clark PM, Cox L, Williams AC, Nattrass M. Hyperproinsulinaemia in patients with myotonic dystrophy. Diabetologia 1992; 35:1170–1172.
26. Moxley III RT. Myotonic muscular dystrophy. In: Rowland LP, DiMauro S, eds. Handbook of Clinical Neurology. Vol 18: Myopathies. Amsterdam: Elsevier, 1992:209–259.
27. Poffenbarger PL, Pozefsky T, Soeldner JS. The direct relationship of proinsulin-insulin hypersecretion to basal serum levels of cholesterol and triglyceride in myotonic dystrophy. J Lab Clin Med 1976; 87:384–396.
28. Livingston JN, Moxley III RT. Phenotype, genotype and insulin resistance in myotonic dystrophy. Diabetes Rev 1994; 2:29–42.
29. Tong P, Thomas T, Berrish T, Humphriss D, Barriocanal L, Stewart M, Walker M, Wilkinson R, Alberti KGMM. Cell membrane dynamics and insulin resistance in non-insulin dependent diabetes mellitus. Lancet 1995; 345:357–358.
30. Rodriguez JR, Castillo J, Leira R, Pardo J, Lema M, Noya M. Bone abnormalities in myotonic dystrophy. Acta Neurol Scand 1991; 83:360–363.
31. Yoshida H, Oshima H, Saito E, Kinoshita M. Hyperparathyroidism in patients with myotonic dystrophy. J Clin Endocrin Metab 1988; 6:488–492.
32. Rosenberg NL, Diliberti JH, Andrews AM, Buist NR. Myotonic dystrophy and hyperparathyroidism: association with neurofibromatosis and multiple endocrine adenomatosis type 2a. J Neurol Neurosurg Psychiatry 1988; 51:1578–1580.
33. Church SC. The heart in myotonia atrophica. Arch Intern Med 1967; 119:176–181.
34. Forsberg H, Olofsson BO, Andersson S, Henriksson A, Bjerle P. 24-hour electrocardiography study in myotonic dystrophy. Cardiology 1988; 75:241–249.
35. Hawley RJ, Milner MR, Gottdiener JS, Cohen A. Myotonic heart disease: a clinical follow-up. Neurology 1991; 41:259–262.
36. Editorial. The heart in myotonic dystrophy. Lancet 1992; 339:528–529.
37. Melacini P, Buja G, Fasoli G, Angelini C, Armani M, Scognamilio R, Dalla-Volta S. The natural history of cardiac involvement in myotonic dystrophy: an eight year follow-up in 17 patients. Clin Cardiol 1988; 11:231–238.
38. Morgenlander JC, Nohiria V, Saba Z. EKG abnormalities in pediatric patients with myotonic dystrophy. Pediatr Neurol 1993; 9:124–126.
39. O'Brien T, Harper PS, Newcombe RG. Blood pressure and myotonic dystrophy. Clin Genet 1983; 23:422–426.
40. Russell SH, Hirsch NP. Anaesthesia and myotonia. British J Anaesth 1994; 72:210–216.
41. Broughton R, Stuss D, Kates M, Roberts J, Dunham W. Neuropsychological deficits and sleep in myotonic dystrophy. Can J Neurol Sci 1990; 17:410–415.
42. Brunner HG, Hamel BC, Rieu P, Howeler CJ, Peters FT. Intestinal pseudo-obstruction in myotonic dystrophy. J Med Genet 1992; 791–793.
43. Brumback RA. Magnetic resonance imaging and clinical correlates of intellectual impairment in myotonic dystrophy. Arch Neurol 1990; 47:253.
44. Raggazoni A, Pinto F, Taiuti R, Silveri MC. Myotonic dystrophy: an electrophysiological study of cognitive deficits. Can J Neurol Sci 1991; 18:300–306.
45. Verhagen WI, Ter-Bruggen JP, Huygen PL. Oculomotor, auditory and vestibular responses in myotonic dystrophy. Arch Neurol 1992; 49:954–960.
46. Cosi V, Bergamaschi R, Versino M, Callieco R, Sandrini G, Ruiz L. Multimodal evoked potentials in myotonic dystrophy (MyD). Neurophysiol Clin 1992; 22:41–50.
47. Cruz-Martinez A. Peripheral nerve conduction and central motor conduction after magnetic stimulation of the brain in myotonic dystrophy. Electromyogr Clin Neurophysiol 1992; 32:295–297.
48. Roche SW, Schwarz A, Lane RJ, Bridges LR. Reversal of cognitive deficit by cardiac pacing in a young man with dystrophia myotonica. J Roy Soc Med 1991; 84:625–626.
49. Glantz RH, Wright RB, Huckman MS, Garron DC, Siegel IM. Central nervous system magnetic resonance imaging findings in myotonic dystrophy. Arch Neurol 1988; 45:36–37.
50. Huber SJ, Kissel JT, Shuttleworth EC, Chakeres DW, Clapp LE, Brogan MA. Magnetic resonance imaging and clinical correlates of intellectual impairment in myotonic dystrophy. Arch Neurol 1989; 46:536–540.
51. Damian MS, Bachmann G, Herrmann D, Dorndorf W. Magnetic resonance imaging of muscle and brain in myotonic dystrophy. J Neurol 1993; 240:8–12.

52. Bachmann G, Damian MS, Schilling G. The clinical significance of 'white-matter-lesions' in myotonic dystrophy. Radiol J CEPUR 1993; 13:15–20.
53. Fiorelli M, Duboc D, Mazoyer BM, Blin J, Eymard B, Fardeau M, Samson Y. Decreased cerebral glucose utilisation in myotonic dystrophy. Neurology 1992; 42:91–94.
54. Mielke R, Herholz K, Fink G, Ritter D, Heiss WD. Positron emission tomography in myotonic dystrophy. Psychiatry Res 1993; 50:93–99.
55. Chang L, Anderson T, Migneco OA, Boone K, Mehringer CM, Villanueva-Meyer J, Berman N, Mena I. Cerebral abnormalities in myotonic dystrophy. Arch Neurol 1993; 50:917–923.
56. Yoshimura N. Alzheimer's neurofibrillary changes in the olfactory bulb in myotonic dystrophy. Clin Neuropathol 1990; 9:234–239.
57. Kiuchi A, Otsuka N, Namba Y, Nakano I, Tomonaga M. Presenile appearance of abundant Alzheimer's neurofibrillary tangles without senile plaques in the brain in myotonic dystrophy. Acta Neuropathol (Berlin) 1991; 82:1–5.
58. Logullo F, Censori B, Danni M, Del-Pesce M, Di-Bella P, Provinciali L. Peripheral neuropathy in myotonic dystrophy: electrophysiological and clinical features. Electromyogr Clin Neurophysiol 1992; 32:515–520.
59. Pierangeli G, Lugaresi A, Contin M, Martinelli P, Montagna P, Parchi P, Verlicchi A, Cortelli P. Autonomic nervous system function in myotonic dystrophy. Ital J Neurol Sci 1992; 13:589–592.
60. Fiorito S, Pocobelli D, Antonioni G, Picelli A, Fragola PV, Autore C, Magrini L, Cannata D. Autoantibodies in myotonic dystrophy. Allergy Immunol (Paris) 1989; 21:379–380.
61. Tanabe Y, Iai M, Tamai K, Fujimoto N, Sugita K. Neuroradiological findings in children with congenital myotonic dystrophy. Acta Paediatr 1992; 81:613–617.
62. Reardon W, Newcombe R, Fenton I, Sibert J, Harper PS. The natural history of congenital myotonic dystrophy: mortality and long term clinical aspects. Arch Dis Child 1993; 68:177–181.
63. Johnson K, Shelbourne P, Davies J, Buxton J, Nimmo E, Siciliano MJ, Bachinski LL, Anvret M, Harley H, Rundle S, Miki T, Brunner H, Williamson R. A new polymorphic probe which defines the region of chromosome 19 containing the myotonic dystrophy locus. Am J Hum Genet 1990; 46:1073–1081.
64. Tsilfidis C, MacKenzie AE, Shulter G, Leblond S, Bailly J, Johnson K, Williamson R, Siegel-Bartelt J, Korneluk RG. D19S51 is closely linked with and maps distal to the myotonic dystrophy locus on 19q. Am J Hum Genet 1991; 49:961–965.
65. Buxton J, Shelbourne P, Davies J, Jones C, Van Tongren T, Aslanidis C, de Jong P, Jansen G, Anvret M, Riley B, Williamson R, Johnson K. Detection of an unstable fragment of DNA specific to individuals with myotonic dystrophy. Nature 1992; 355:547–548.
66. Brunner HG, Nillesen W, Van Oost BA, Jansen G, Wieringa B, Ropers HH, Smeets HJ. Presymptomatic diagnosis of myotonic dystrophy. J Med Genet 1992; 29:780–784.
67. Anvret M, Ahlberg G, Grandell U, Hedberg B, Johnson K, Edstrom L. Larger expansions of the CTG repeat in muscle compared to lymphocytes from patients with myotonic dystrophy. Hum Mol Genet 1993; 2:1397–1400.
68. Thornton CA, Johnson K, Moxley RT III. Myotonic dystrophy patients have larger CTG expansions in skeletal muscle than in leucocytes. Ann Neurol 1994; 35:104–107.
69. Shelbourne P, Winqvist E, Kunert E, Davies J, Leisti J, Thiele H, Bachmann H, Buxton J, Williamson R, Johnson K. Unstable DNA may be responsible for the incomplete penetrance of the myotonic dystrophy phenotype. Hum Mol Genet 1992; 1:467–473.
70. Shelbourne P, Davies J, Buxton J, Anvret M, Blennow E, Bonduelle M, Schmedding E, Glass I, Lindenbaum R, Lane R, Williamson R, Johnson K. Direct diagnosis of myotonic dystrophy using a disease-specific marker. N Engl J Med 1993; 328:476–480.
71. Imbert G, Kretz C, Johnson K, Mandel J-L. Origin of the expansion mutation in myotonic dystrophy. Nature Genet 1993; 4:72–76.
72. Cobo AM, Baiget M, Lopez De Munain A, Poza JJ, Emparanza JI, Johnson K. Sex-related difference in intergenerational expansion of myotonic dystrophy gene. Lancet 1993; 341:1159–1160.
73. Gennarelli M, Dallapiccola B, Baiget M, Martorell L, Novelli G. Meiotic drive at the myotonic dystrophy locus. J Med Genet 1994; 31:980–982.
74. Carey N, Johnson K, Nokelainen P, Peltonen L, Savontaus M-L, Juvonen V, Anvret M, Grandell U, Chotai K, Robertson E, Middleton-Price H, Malcolm S. Meiotic drive at the myotonic dystrophy locus? Nature Genet 1994; 6:117–118.
75. Burke JR, Ikeuchi T, Koide R, Tsuji S, Yamada M, Pericak-Vance MA, Vance J. Dentatorubral-pallidoluysian atrophy and Haw River syndrome. Lancet 1994; 344:1711–1712.
76. Nakagawa M, Yamada H, Higuchi I, Kamanishi Y, Miki T, Johnson K, Osame M. A case of paternally inherited congenital myotonic dystrophy. J Med Genet 1994; 31:397–400.
77. Ohya K, Tachi N, Chiba S, Sato T, Kon S, Kikuchi K, Imamura S, Yamagata H, Miki T. Congenital myotonic dystrophy transmitted from an asymptomatic father with a DM-specific gene. Neurology 1994; 44:1958–1960.

78. Andrew PI, Wilson J. Relative disease severity in siblings with myotonic dystrophy. J Child Neurol 1992; 7:161–167.
79. Reardon W, Harley HG, Brook JD, Rundle SA, Crow S, Harper PS, Shaw DJ. Minimal expression of myotonic dystrophy: a clinical and molecular analysis. J Med Genet 1992; 29:770–773.
80. Koch MC, Grimm T, Harley HG, Harper PS. Genetic risks for children of women with myotonic dystrophy. Am J Hum Genet 1991; 48:1084–1091.
81. Myring J, Meredith AL, Harley HG, Kohn G, Norbury G, Harper PS. Specific molecular prenatal diagnosis for the CTG mutation in myotonic dystrophy. J Med Genet 1992; 29:785–788.
82. Smeets HJM, Nillesen WM, Los F, Busch HF, Korneluk RG, Wieringa B, Brunner HG. Prenatal diagnosis of myotonic dystrophy by direct mutation analysis. Lancet 1992; 340:237–238.
83. Shelbourne P, Winqvist R, Kunert E et al. Unstable DNA may be responsible for the incomplete penetrance of the myotonic dystrophy phenotype. Hum Mol Genet 1992; 1:467–475.
84. Brook JD, McCurrach ME, Harley HG et al. Molecular basis of myotonic dystrophy: expansion of a trinucleotide (CTG) repeat at the 3′ end of a transcript encoding a protein kinase family member. Cell 1992; 68:799–808.
85. Mahedevan M, Tsilfidis C, Sabourin L et al. Myotonic dystrophy mutation: an unstable CTG repeat in the 3′ untranslated region of a candidate gene. Science 1992; 255:1253–1256.
86. Fu Y-H, Friedman DL, Richards S et al. Decreased expression of myotonin-protein kinase messenger RNA and protein in adult form of myotonic dystrophy. Science 1993; 260:235–238.
87. Hoffman-Radvanyi H, Lavedan C, Rabes J-P, Savoy D, Duros C, Johnson K, Junien C. Myotonic dystrophy: absence of CTG enlarged transcript in congenital forms, and low expression of the normal allele. Hum Mol Genet 1993; 2:1263–1266.
88. Sabourin LA, Mahadevan MS, Narang M, Lee DS, Surh LC, Korneluk RG. Effect of the myotonic dystrophy (DM) mutation on mRNA levels of the DM gene. Nature Genet 1993; 4:233–238.

27

Hypokalemic Periodic Paralysis

Rabi Tawil and Robert C. Griggs
University of Rochester School of Medicine and Dentistry, Rochester, New York

INTRODUCTION

The primary periodic paralyses are a group of uncommon disorders characterized by episodic attacks of weakness. They are classically divided into hypokalemic and hyperkalemic forms based on serum potassium levels during attacks (1–3). The hyperkalemic form includes a clinically heterogeneous group of disorders now known to be caused by allelic mutations of the sodium channel α-subunit on chromosome 17 (4–7). Hypokalemic periodic paralysis (HypoKPP), the most common form of periodic paralysis, is the focus of the present chapter.

GENETICS

HypoKPP is an autosomal dominant disorder with variable penetrance in women (8–10). Sporadic cases are relatively common and are clinically indistinguishable from familial cases (11). Estimates of disease prevalence range from 0.4/100,000 in Finland (10) to 1.25/100,000 in Denmark (9) but are likely underestimates of the true prevalence, given the disease's variable penetrance. Recently, HypoKPP has been linked to the skeletal muscle dihydropyridine receptor where point mutations have been identified predicting histidine to arginine substitutions in the S4 segments of receptor domains II and IV (12,13).

CLINICAL FEATURES

The age of onset is usually adolescence with a range of 6–25 years (8,10,12). Rarely, ages of onset of 60–70 years have been described (9). Spontaneous attacks of weakness occur most frequently late at night and patients often awake with weakness. The most consistently reported precipitating factors include large carbohydrate or sodium-rich meals and rest following exertion (2,8,9). Less commonly, attacks occur with exposure to cold, alcohol ingestion, or emotional stress (8,9). Prodomal symptoms of stiffness, heaviness of the limbs, and diaphoresis are common (8,9). Weakness, which often starts proximally in the lower extremities, can remain mild and focal or progress to flaccid quadriplegia (8,9,14). Ocular, bulbar, and respiratory muscles are usually spared but can occasionally be involved, with life-threatening consequences (9,10,15). Oliguria and rarely anuria are noted during the attacks (9,16).

Examination during an attack shows flaccid weakness with hyporeflexia proportional to the degree of weakness (1,9). Lid-lag myotonia is a variable sign and is the only myotonic phenomenon seen with HypoKPP (17). Attacks vary in duration from 2 to 72 hr (2,9) and occur at a frequency ranging from one per day to one to two within a lifetime (1,9). With age, attack frequency usually diminishes; after years of attacks, persistent, often progressive, weakness supervenes (2,8,9). The percentage of affected individuals developing persistent weakness is not clear. All affected members in two separate families were found to have persistent weakness (18,19), whereas in a study of 21 affected individuals from seven different families in Denmark, none were said to have persistent weakness (9). Interattack weakness can also be seen as early in the course of the disease as 8 years of age (20,21).

Cardiac involvement in periodic paralysis is common in a form of hyperkalemic periodic paralysis (22–25). Cardiac dysrhythmias have been reported in HypoKPP (26,27), but the clinical and laboratory features make the diagnosis unclear (28). Bradycardia and sinus arrhythmia that occur during HypoKPP attacks are attributable to hypokalemia (9,10,29,30). In one autopsy case study, while light microscopy showed extensive vacuolization and fibrosis in the skeletal muscles, the cardiac muscle was normal (19).

Mortality during paralytic attacks has been reported to be as high as 10% by Talbot, based on a survey of the literature until 1941 (31). Mortality may reflect inaccurate diagnoses or less sophisticated medical care (9). Recent reports of death are rare and happen in the setting of severe, prolonged untreated attacks mostly as a result of aspiration rather than cardiac complications (9,15,19).

SECONDARY HYPOKALEMIC PARALYSIS

In many patients with secondary hypokalemic weakness the underlying condition is clinically evident and HypoKPP need not be considered in the differential diagnosis. Some, however, have no other evident medical problems and present with recurrent episodes of transient weakness indistinguishable from HypoKPP. Most secondary forms of hypokalemic paralysis result from severe chronic potassium depletion (32). The one exception is thyrotoxic periodic paralysis, which is thought to result from alterations in the muscle membrane permeability (33).

Thyrotoxic periodic paralysis occurs in 1.9% of thyrotoxic Japanese patients (34) and in a similar percentage of thyrotoxic Chinese patients (35). Although much rarer outside East Asia, it has been described in patients of various ethnic backgrounds in the United States and occurs in about 0.2% of thyrotoxic patients (33). The clinical presentation is often indistinguishable from HypoKPP except for subtle evidence of hyperthyroidism. Paralytic episodes often antedate the occurrence of clinical hyperthyroidism (33). Correction of the hyperthyroid state almost invariably eliminates the paralytic attacks (33). Unlike HypoKPP they do not respond to acetazolamide; treatment with β-blockers prevents attacks (36).

A number of potassium-wasting conditions cause secondary hypokalemic weakness. Drugs and toxins that have been implicated include diuretics (37), amphotericin B (38), laxatives (39), alcohol (40), and glycyrrhizin (41). The latter, an aldosterone agonist, is also present in licorice, snuff, and chewing tobacco all of which have also been reported to cause hypokalemic weakness (42–44). Renal conditions that can present with recurrent hypokalemic weakness include primary or secondary forms of renal tubular acidosis (45–48) and Bartter's syndrome (49). Gastrointestinal potassium loss with weakness has been reported in infectious enteritides (50,51) and rectal villous adenomas (52). Finally, adrenal cortical overactivity in the form of primary hyperaldosteronism can result in recurrent hypokalemic weakness (53).

The incidence of death during an attack of secondary hypokalemic paralysis may be higher than in HypoKPP (54,55). This discrepancy may be due to differences in total-body potassium content. Whereas in secondary forms of hypokalemic paralysis, severe potassium depletion exists, hypokalemia in HypoKPP is the result of a transient intracellular shift (54).

DIAGNOSIS

The diagnosis of HypoKPP depends on documenting hypokalemia during a spontaneous or induced attack and a consistent clinical profile. A positive family history is helpful but not essential. All sporadic cases, especially those with disease onset above 30 years, should be screened for secondary forms of hypokalemic paralysis. Such patients should be questioned specifically about their use of potassium-wasting drugs or toxins such as diuretics, laxatives, or the excessive intake of licorice. Thyroid function tests should be checked with T4, T3, and a sensitive TSH assay. If thyrotoxicosis is excluded, hypokalemia secondary to gastrointestinal loss, mineralocorticoid excess, or renal potassium wasting should be considered.

To document hypokalemia during a spontaneous attack, blood should be sampled early following the onset of an attack since serum potassium often returns to normal levels before the resolution of weakness. If spontaneous attacks are frequent, provocative testing may be necessary. Hypokalemia can be induced by the oral administration of 1.5 g/kg of glucose (maximum 100 g). If no weakness is induced and serum potassium remains >3 mg/dl, intravenous glucose 1.5–3.0 g/kg (maximum 200 g) is infused over 1 hr along with intravenous insulin 0.1 U/kg given at 30 and 60 min (2). Serial glucose, electrolytes, and electrocardiograms (ECG) as well as muscle strength testing are performed at 30-min intervals for 5 hr. Such tests must not be performed in patients with major cardiac, renal, or other complicating illness, and a physician must be in attendance during testing. An alternative, though more time-consuming, method described by Johnsen (56) consists of giving hourly oral doses of glucose (50 g/150 ml water) until an attack is precipitated, usually after 8–10 doses. Equivocal results with provocative testing should be interpreted carefully. Hyperkalemic periodic paralysis patients can become weak if made sufficiently hypokalemic (28,57) and, conversely, HypoKPP patients may weaken with significant hyperkalemia (58). It is extremely important to remember that patients with HypoKPP are often hypokalemic late in the course of a spontaneous attack.

Electrophysiological studies are useful adjunctive tests in the diagnosis of HypoKPP. Documenting myotonia by electromyography virtually rules out HypoKPP. The exercise nerve conduction protocol described by McManis et al. demonstrates significant and sustained drop in compound muscle action potential amplitudes following isometric exercise (59). This procedure, with a sensitivity of 71%, is useful in establishing the diagnosis of periodic paralysis in questionable cases (59). It does not, however, differentiate between the primary and secondary forms of periodic paralysis or between the various forms of primary periodic paralyses (59). Engel et al. combined nerve conductions with intra-arterial injection of epinephrine (2 μg/min × 5 min) (60). A decrement of >30% in the amplitude of the compound muscle action potential after epinephrine injection was specific for HypoKPP (60).

PATHOLOGY

While muscle biopsies are occasionally normal early in the course of HypoKPP, changes are consistently seen following repeated bouts of weakness and in patients with persistent weakness (8). The most characteristic change is the presence of well-demarcated central vacuoles often of considerable size (Fig. 1), some filled with PAS-staining material (8,61,62). In addition, nonspecific myopathic changes, such as variability in fiber size, increased central nuclei, fibrosis, and occasionally necrotic fibers, occur (8,19,61–63). Vacuoles evolve from dilated sarcoplasmic reticulum containing cellular debris to a remodeled vacuole containing sarcoplasmic invaginations (8).

PATHOPHYSIOLOGY

Skeletal muscle weakness in HypoKPP, as in other forms of periodic paralysis, is the result of transient inexcitability of muscle fibers (64). The mechanism leading to this loss of normal excitability remains uncertain. Both the contractile apparatus (65) and the neuromuscular junction (66) appear to be normal in HypoKPP.

Until muscle wasting occurs, total-body potassium in HypoKPP is normal (67). During spontaneous or glucose-insulin-induced attacks, there is a shift of potassium into muscle fibers (66,68), which

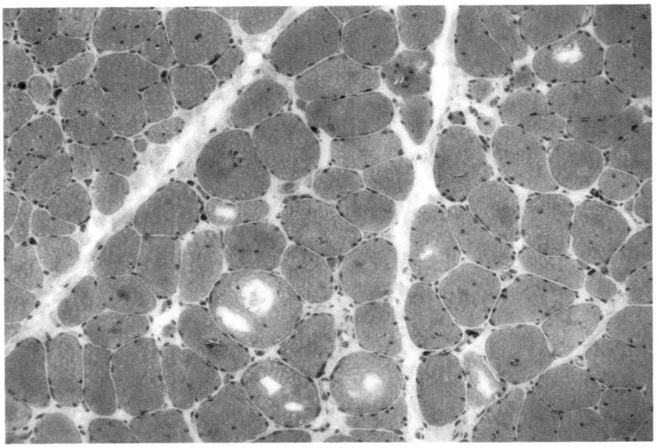

Figure 1 Vastus lateralis muscle from a patient with hypokalemic periodic paralysis and persistent weakness. Central vacuoles at different stages of development are seen in several fibers. (Hematoxylin and eosin.)

may be accompanied by an influx of water. The resultant intravascular volume contraction (69) may explain the oliguria observed during paralytic attacks (67).

Carbohydrate metabolism has been investigated extensively in HypoKPP. Engel et al. (70) found no disturbance of glycogen metabolism on biochemical studies of HypoKPP muscle biopsies. Johnsen et al. found a slightly higher-than-normal insulin response in HypoKPP patients challenged with glucose (71) and decreased insulin binding to monocytes (72). Hoffman et al., on the other hand, found increased insulin receptor binding and concluded that HypoKPP patients may have a supraphysiological response to normal fluctuations in blood insulin (73). More recently, an obese HypoKPP patient was studied using the euglycemic insulin clamp technique (74). The expected resistance to the action of insulin on glucose was observed, but there was a greater-than-expected hypokalemic response with resulting muscle weakness (74). HypoKPP patients may therefore have a disproportionate sensitivity to the action of insulin on potassium, independent of its action on glucose (74).

In vitro studies have consistently demonstrated the presence of lower-than-normal resting potentials in HypoKPP muscle (73,75). In addition, HypoKPP muscle fibers have a paradoxical response to both insulin (73) and decreased extracellular potassium concentrations (75). Whereas normal muscle became hyperpolarized, HypoKPP muscle was depolarized and became inexcitable; this response was not inhibited by sodium channel blockade with tetrodotoxin (75). The recent in vitro demonstration of improved contraction force in a HypoKPP muscle preparation in response to a potassium channel opener suggests the presence of a defective potassium channel (76).

Investigations into the effects of the recently described HypoKPP-associated dihydropyridine receptor mutations indicate a loss of receptor function (77). However, it is not yet clear how this loss of function relates to the hypokalemia-induced paralytic attacks (77).

TREATMENT

The treatment of HypoKPP is either acute, aiming to limit the severity and duration of acute paralytic attacks, or prophylactic, aiming to reduce attack frequency and prevent or reverse persistent weakness.

Treating Acute Attacks

The efficacy of oral potassium replacement in decreasing attack severity and duration is universally accepted although the extent of benefit has not been subjected to randomized controlled study. Oral potassium salts (0.25 mEq/kg body wt) are given every 30 min until weakness improves. Serum

Table 1 Treatment of Hypokalemic Periodic Paralysis

Treatment of acute paralytic attacks	
I.	Oral potassium salts: 25 mEq/30 min until weakness improves (not to exceed 200 mEq over 12 hr)
II.	Intravenous potassium (to be used only if oral administration is precluded): 0.05–0.1 mEq/kg body weight KCl in 5% mannitol given as a bolus; monitor with serial electrocardiograms and serum potassium levels
Prevention of paralytic attacks	
I.	Diet: avoidance of carbohydrate and sodium loads
II.	Pharmacological measures
	A. Slow-release potassium supplements 20–40 mEq given at bedtime
	B. Acetazolamide 125–1000 mg/day
	C. Dichlorphenamide 50–200 mg/day
	D. Triemterene 25–100 mg/day[a]
	E. Spironolactone 25–100 mg/day[a]

[a]Potassium supplements to be given cautiously.

potassium should be checked before doses in excess of 200 mEq in a 12-hr period are administered. If use of the oral route is precluded because of vomiting or bulbar weakness, the intravenous route can be used with caution and under serial ECG monitoring. Intravenous potassium, however, should not be given in glucose or normal saline, which will often cause a fall in plasma potassium (78,79). The preferred diluent is 5% mannitol, and 0.05–0.1 mEq/kg body weight of KCL is given in bolus form (2,80). The total dose of potassium needed to normalize serum potassium levels is difficult to calculate since HypoKPP patients are not potassium-depleted and their urine output may be reduced during an attack. Life-threatening hyperkalemia can occur with high-dose oral and intravenous potassium administration (58; Dalakas, personal communication). Therefore, serial serum levels are essential in guiding potassium administration and preventing overtreatment.

Prophylactic Treatment

Dietary manipulations such as low-carbohydrate and low-salt diets have had mixed results in prevention of paralytic attacks (16,60,81,82). Potassium, taken at bedtime, is used by many patients as prophylaxis (21). Whether potassium actually prevents early-morning attacks or just blunts their intensity is open to question. Potassium-sparing diuretics (16) have been used with some success, but neither β-blockers (83) nor diazoxide (84) effectively reduced attack rates.

Acetazolamide, a carbonic anhydrase inhibitor (CAI), is the preferred drug in the prophylactic treatment of HypoKPP. It has been shown to be effective both in reducing attack frequency and in improving persistent weakness (21,85). Dichlorphenamide, a more potent CAI, appears to be beneficial in patients whose response to acetazolamide is waning (86,87). A multicenter randomized, placebo-controlled trial is currently underway to test the efficacy of dichlorphenamide in the treatment of all forms of periodic paralysis. The CAIs, because of smaller average doses, are better tolerated in the periodic paralyses than in other disorders (88). However, the frequency of nephrolithiasis with acetazolamide is as high as in other patients treated with the drug (88). In patients responding inadequately to CAIs, the addition of a potassium-sparing diuretic may result in an increased therapeutic response.

Exacerbation of paralytic attacks by acetazolamide occurs in about 10% of HypoKPP patients (89); in such cases, potassium-sparing diuretics are the drugs of choice. HypoKPP patients worsened by acetazolamide may represent a distinct subgroup of HypoKPP.

The mechanism of action of the CAIs in the treatment of HypoKPP is uncertain. There is conflicting evidence as to the role of the metabolic acidosis induced by such treatment (86,90–92). A direct effect on muscle carbonic anhydrase is unlikely since acetazolamide has little or no effect on carbonic anhydrase III, the predominant muscle isoenzyme (93). There is, however, indirect evidence that CAIs may act on the muscle membrane (91).

A new class of medications, the potassium channel openers, developed to treat cardiac arrhythmias may prove beneficial in HypoKPP (76). In vitro studies have shown that cromakalin, a potassium channel opener, can improve the contraction force of fibers from patients with HypoKPP, potentially by enhancing membrane potassium conductance (76).

Whether prevention of paralytic attacks can retard or prevent the development of a progressive myopathy remains unanswered. Structural changes similar to those seen in HypoKPP can be seen in secondary forms of hypokalemic paralysis (94). It therefore seems reasonable to assume that successful prophylactic treatment may limit structural damage to muscle. Poor correlation between attack rate and the development of persistent weakness may appear to contradict such a contention (18). However, "fixed" weakness is often partially reversible (21,86), indicating that it may represent a state of persistent paralytic "attack." Prophylactic treatment should then be given both to patients with attacks and to those with persistent weakness and few or no attacks.

ANESTHESIA

Successful anesthetic management of HypoKPP patients depends on avoidance of known precipitating factors and careful intra- and postoperative monitoring. Thus, large carbohydrate meals should be avoided on the night before surgery and low preoperative potassium levels should be corrected (95,96). Intraoperative hypothermia, the use of glucose and sodium-containing solutions, and the use of cardioplegic solutions containing glucose and insulin should be avoided (95,97). Whether HypoKPP patients are more susceptible to the effects of muscle relaxants is not clear (95,98). However, neuromuscular blockade should be avoided as it could complicate recovery from a perioperative paralytic attack.

REFERENCES

1. Gamstrop I. Disorders characterized by spontaneous attacks of weakness connected with changes of serum potassium. Prog Clin Biol Res 1989; 306:175–195.
2. Griggs RC. Periodic paralysis. Semin Neurol 1983; 3:285–293.
3. Riggs JE. Periodic paralysis. Clin Neuropharmacol 1989; 12:249–257.
4. Fontaine B, Khurana TS, Hoffman EP, et al. Hyperkalemic periodic paralysis and the adult muscle sodium channel α-subunit gene. Science 1990; 250:1000–1002.
5. Ptácek LJ, Johnson KJ, Griggs RC. Genetics and physiology of the myotonic muscle disorders. N Engl J Med 1993; 328:482–489.
6. Ptácek LJ, George AL, Griggs RC, et al. Identification of a mutation in the gene causing hyperkalemic periodic paralysis. Cell 1991; 67:1021–1027.
7. Rojas CV, Wang JZ, Schwartz LS, Hoffman EP, Powell BR, Brown RH. A methionine to valine mutation in human hyperkalemic periodic paralysis. Nature 1991; 353:387–389.
8. Engel AG. Periodic paralysis. In: Engel AG, Banker BQ, eds. Myology. New York: McGraw-Hill, 1986: 1843–1870.
9. Johnsen T. Familial periodic paralysis with hypokalemia: experimental and clinical investigations. Dan Med Bull 1981; 28:1–27.
10. Kantola IM, Tarssanen LT. Familial hypokalemic periodic paralysis in Finland. J Neurol Neurosurg Psychiatry 1992; 55:322–324.
11. Ropers HH, Szliwowski HB. Periodic hypokalemic paralysis transmitted by an unaffected male with negative family history: a delayed mutation? Hum Genet 1979; 48:113–116.
12. Ptacek L, Tawil R, Griggs RC, et al. Dihydropyridine receptor mutations cause hypokalemic periodic paralysis. Cell 1994; 77:863–868.
13. Jukat-Rott K, Lehmann-Horn F, Elbaz A, et al. A calcium channel mutation causing hypokalemic periodic paralysis. Hum Mol Genet 1994; 3:1415–1419.
14. Schipperheyn JJ, Wintzen AR, Buruma OJS. Periodic paralysis. In: Rowland LP, DiMauro S, eds. Handbook of Clinical Neurology. Amsterdam: Elsevier, 1992:457–477.

15. Ionasescu V, Schochet SS, Powers JM, Koob K, Conway TW. Hypokalemic periodic paralysis. Low activity of sarcoplasmic reticulum and muscle ribosome during an induced attack. J Neuro Sci 1974; 21:419–429.
16. De Graeff J, Lameijer LDF. Periodic paralysis. Am J Med 1965; 39:70–80.
17. Resnick JS, Engel WK. Myotonic lid lag in hypokalemic periodic paralysis. J Neurol Neurosurg Psychiatry 1967; 30:47–51.
18. Buruma OJS, Bots GTAM, Lodewyk NW. Familial hypokalemic periodic paralysis: 50-year follow-up of a large family. Arch Neurol 1985; 42:28–31.
19. Links TP, Zwarts MJ, Wilmink JT, Molenaar WM, Oosterhuis HJ. Permanent muscle weakness in familial periodic paralysis. Brain 1990; 113:1873–1889.
20. Dyken M, Zeman W, Rusche T. Hypokalemic periodic paralysis. Children with permanent myopathic weakness. Neurology 1969; 19:691–699.
21. Griggs RC, Engel WK, Resnick JS. Acetazolamide treatment of hypokalemic periodic paralysis: prevention of attacks and improvement of persistent weakness. Ann Intern Med 1970; 73:39–48.
22. Andersen ED, Krasilnikopff PA, Overvad H. Intermittent muscular weakness, extrasystoles and multiple development anomalies: a new syndrome? Acta Pediatr Scand 1971; 60:559–564.
23. Gould RJ, Steeg CN, Eastwood AB, Penn AS, Rowland LP, De Vivo D. Potentially fatal cardiac dysrhythmia and hyperkalemic periodic paralysis. Neurology 1985; 35:1208–1212.
24. Lisak RP, Lebeau J, Tucker SH, Rowland LP. Hyperkalemic periodic paralysis and cardiac arrhythmia. Neurology 1972; 22:810–815.
25. Yoshimura T, Kaneuji M, Okuno T, et al. Periodic paralysis with cardiac arrhythmia. Eur J Pediatr 1983; 140:338–343.
26. Stubbs WA. Bidirectional ventricular tachycardia in familial hypokalemic periodic paralysis. Proc Roy Soc Med 1976; 69:223–224.
27. Levitt LP, Rose LI, Dawson D. Hypokalemic periodic paralysis with arrhythmia. N Engl J Med 1972; 286:253–254.
28. Rowland LP, Lisak RP. Periodic paralysis: K^+ levels during arrhythmia. N Engl J Med 1972; 287:50 (letter).
29. Fukuda K, Ogawa S, Yokozuka H, Handa S, Nakamura Y. Long-standing bidirectional tachycardia in a patient with hypokalemic periodic paralysis. J Electrocardiogr 1988; 21:71–76.
30. Kramer LD, Cole JP, Messenger JC, Ellestad MH. Cardiac dysfunction in a patient with familial hypokalemic periodic paralysis. Chest 1979; 75:189–192.
31. Talbot JH. Periodic paralysis. Medicine 1940; 20:85–143.
32. Layzer RB. Mineral and electrolyte disorders. In: Layzer RB, ed. Neuromuscular Manifestations of Systemic Disorders. Philadelphia: FA Davis, 1985:47–78.
33. Ober KP. Thyrotoxic periodic paralysis in the United Stated: report of 7 cases and review of the literature. Medicine 1992; 71:109–120.
34. Okinaka S, Shizume K, Iino S, et al. The association of periodic paralysis and hyperthyroidism in Japan. J Clin Endocrinol Metab 1957; 17:1454–1459.
35. McFazdean AJS, Yeung R. Periodic paralysis complicating thyrotoxicosis in Chinese. Br Med J 1967; 1:451–455.
36. Conway MJ, Seibel JA, Eaton RP. Thyotoxicosis and periodic paralysis: improvement with beta blockade. Ann Intern Med 1974; 81:332–336.
37. Cohen T. Hypokalemic muscle paralysis associated with administration of chlorthiazide. JAMA 1959; 170:2083–2085.
38. McChesney JA, Marquardt JF. Hypokalemic paralysis induced by amphotericin B. JAMA 1964; 189:1029.
39. Basser LS. Purgatives and periodic paralysis. Med J Aust 1979; 1:47–48.
40. Dominiczak AF, Semple PF, Fraser R, Lever AF, Wallace AM. Hypokalemia in alcoholics. Scot Med J 1989; 34:489–494.
41. Hayashi R, Maruyama T, Maruyama K, Yanagawa S. Tako K, Yanagisawa N. Myotonic and repetitive discharges in hypokalemic myopathy associated with glycyrrhizin-induced hypochloremia. J Neurol Sci 1992; 107:74–77.
42. Blachley JD, Knochel JP. Tobacco chewer's hypokalemia: licorice revisited. N Engl J Med 1980; 302:784 (letter).
43. Nielsen I, Pedersen RS. Life-threatening hypokalemia caused by licorice ingestion. Lancet 1984; 1:1305.
44. Valeriano J, Tucker P, Kattah J. An unusual cause of hypokalemic muscle weakness. Neurology 1983; 33:1242–1243.
45. Owen EE, Verner Jr. JV. Renal tubular disease with muscle paralysis and hypokalemia. Am J Med 1960; 28:8–21.
46. Peltz S, Hashmi S. Severe hypokalemia and ventilator-dependent generalized paralysis. NY State J Med 1991; 91:32–33.
47. Poux JM, Peyronnet P, Le Meur Y, Favereau JP, Charmes JP, Leroux-Robert. Hypokalemic quadriparesis and respiratory arrest revealing primary Sjögren's syndrome. Clin Nephrol 1992; 37:189–191.

48. Pun K, Wong C, Tsui EY, et al. Hypokalemic periodic paralysis due to Sjögren syndrome in Chinese patients. Ann Intern Med 1989; 110:405–406.
49. Cannon PJ, Leeming JM, Sommers SC, Winters RW, Laragh JH. Juxtaglomerular cell hyperplasia and secondary hyperaldosteronism (Bartter's syndrome): a reevaluation of pathophysiology. Medicine 1968; 47:107–131.
50. Orman RA, Lewis JB. Flaccid quadriparesis associated with *Yersinia* enterocolitis-induced hypokalemia. Arch Intern Med 1989; 149:1193–1194.
51. Reid A, Draper I. Flaccid quadriparesis due to salmonella enteritis. Scot Med J 1980; 25:129–130.
52. Keyloun VE, Grace WJ. Villous adenoma of the rectum associated with severe electrolyte imbalance. Am J Dig Dis 1967; 12:104.
53. Conn JW. Aldosteronism in man. JAMA 1963; 183:871–878.
54. Links ThP, Smit AJ, Oosterhuis HJGH. Potassium and sudden death. Lancet 1991; 338:1457.
55. Nimmannit S, Malasit P, Chaovakul V, Susaengrat W, Nilwarangkur S. Potassium and sudden unexplained nocturnal death. Lancet 1990; 336:116–117.
56. Johnsen T. A new standard and effective method of inducing paralysis without administration of exogenous hormone in patients with familial periodic paralysis. Acta Neurol Scand 1976; 54:167–172.
57. Layzer RB, Lovelace RE, Rowland LP. Hyperkalemic periodic paralysis. Arch Neurol 1967; 16:455–472.
58. Sestoft L. Direct transition from hypo- to hyperkalemic paralysis during potassium treatment of familial periodic paralysis. Dan Med Bull 1967; 14:157–160.
59. McManis PG, Lambert EH, Daube JR. The exercise test in periodic paralysis. Muscle Nerve 1986; 9:704–710.
60. Engel AG, Lambert EH, Resevear JW, Tauxe WN. Clinical and electrophysiologic studies in a patient with primary hypokalemic periodic paralysis. Am J Med 1965; 38:626–640.
61. Biczyskowa W, Fidzianska A, Jedrzejowska H. Light and electron microscopic study of the muscles in hypokalemic periodic paralysis. Acta Neuropathol 1969; 12:329–338.
62. Weller RO, McArdle B. Calcifications within muscle fibers in the periodic paralyses. Brain 1971; 94:263–272.
63. McDonald RD, Rewcastle NB, Humphrey JG. Myopathy and hypokalemic periodic paralysis: an electron microscopic study. Arch Neurol 1969; 20:565–585.
64. Rüdel R, Ricker K. The periodic paralyses. Trends Neurosci 1985; 8:467–470.
65. Engel AG, Lambert EH. Calcium activation of electrically inexcitable muscle fibers in primary periodic paralysis. Neurology 1969; 19:851–858.
66. Grob D, Johns RJ, Liljestrand A. Potassium movement in patients with familial periodic paralysis. Am J Med 1957; 23:356–375.
67. Shy GM, Wanko T, Bawley PT, Engel AG. Studies in familial periodic paralysis. Exp Neurol 1961; 3:53–121.
68. Zierler KL, Andres R. Movement of potassium into skeletal muscle during spontaneous attack in familial periodic paralysis. J Clin Invest 1957; 36;730–737.
69. Gordon AM, Green JR, Lagunoff D. Studies on a patient with hypokalemic familial periodic paralysis. Am J Med 1970; 48:185–195.
70. Engel AG, Potter CS, Rosevear JW. Studies on carbohydrate metabolism and mitochondrial respiratory activities in primary periodic paralysis. Neurology 1967; 17:329–336.
71. Johnsen T, Houston ME, Saltin B. Skeletal muscle characteristics and carbohydrate metabolism after glucose loading in hypokalemic periodic paralysis. Scand J Clin Lab Invest 1978; 38:549–556.
72. Johnsen T, Beck-Neilsen H. Insulin receptors, insulin secretion and glucose disappearance rate in patients with periodic hypokalemic paralysis. Acta Endocrinol 1979; 90:272–282.
73. Hoffman WW, Adornato BT, Reich H. The relationship of insulin receptors to hypokalemic periodic paralysis. Muscle Nerve 1983; 6:48–51.
74. Minaker KL, Meneilly GS, Flier JS, Rowe JW. Insulin-mediated hypokalemia and paralysis in familial hypokalemic periodic paralysis. Am J Med 1988; 84:1001–1006.
75. Rüdel R, Lehman-Horn F, Ricker K, Kuther G. Hypokalemic periodic paralysis: in vitro investigation of muscle fiber membrane parameters. Muscle Nerve 1984; 7:110–120.
76. Grafe P, Quasthoff S, Strupp M, Lehman-Horn F. Enhancement of K^+ conductance improves in vitro the contraction force of skeletal muscle in hypokalemic periodic paralysis. Muscle Nerve 1990; 13:451–457.
77. Hoffman EP, Lehmann-Horn F, Rudel R. Overexcited or inactive: ion channels in muscle disease. Cell 1995; 80:681–686.
78. Griggs RC, Resnick J, Engel WK. Intravenous treatment of hypokalemic periodic paralysis. Arch Neurol 1983; 40:539–540.
79. Kunin AS, Surawicz B, Sims EAH. Decrease in serum potassium concentrations and appearance of cardiac arrhythmias during infusion of potassium with glucose in potassium-depleted patients. N Engl J Med 1962; 266:228–233.
80. Moreno M, Murphy C, Goldsmith C. Increase in serum potassium resulting from the administration of hypertonic mannitol and other solutions. J Lab Clin Med 1969; 73:291–298.

81. McQuarrie I, Ziegler MR. Hereditary periodic paralysis. Effects of various types of diet on occurrence of paralytic attacks. Metabolism 1952; 1:129–144.
82. Rowley PT, Kilman B. The effect of sodium loading and depletion on muscular strength and aldosterone excretion in familial periodic paralysis. Am J Med 1960; 28:376–385.
83. Johnsen T. Treatment of familial periodic hypokalemia with propranolol (Inderal). Acta Neurol Scand 1977; 56:613–619.
84. Johnsen T. Trial of the prophylactic effect of diazoxide in the treatment of familial periodic hypokalemia. Acta Neurol Scand 1977; 56:525–532.
85. Resnick JS, Engel WK, Griggs RC, Stam AC. Acetazolamide prophylaxis in hypokalemic periodic paralysis. N Engl J Med 1968; 278:582–586.
86. Dalakas MC, Engel WK. Treatment of "permanent" muscle weakness in familial hypokalemic periodic paralysis. Muscle Nerve 1983; 6:182–186.
87. Sander C. Treatment of hypokalemic periodic paralysis with diclofenamid. Monatsschr Kinderheilkd 1988; 136:149–150.
88. Tawil R, Moxley III RT, Griggs RC. Acetazolamide induced nephrolithiasis: implications for treatment of neuromuscular disorders. Neurology 1993 (in press).
89. Torres CF, Griggs RC, Moxley RT et al. Hypokalemic periodic paralysis exacerbated by acetazolamide. Neurology 1981; 31:1423–1428.
90. Jarrell MA, Greer M, Maren TH. The effect of acidosis in hypokalemic periodic paralysis. Arch Neurol 1976; 33:791–793.
91. Riggs JE, Griggs RC, Moxley III RT. Dissociation of glucose and potassium arteriovenous differences across the forearm by acetazolamide. Arch Neurol 1984; 41:35–38.
92. Vroom FQ, Jarrell MA, Maren TH. Acetazolamide treatment of hypokalemic periodic paralysis: probable mechanism of action. Arch Neurol 1975; 32:385–392.
93. Peyronnard J, Charron LF, Messier J, Lavoie J, Faraco-Cantin F, Dubreuil M. Histochemical localization of carbonic anhydrase in normal and diseased human muscle. Muscle Nerve 1988; 11:108–113.
94. Comi G, Testa D, Cornelio F, Canal N. Potassium depletion myopathy: a clinical and morphological study. Muscle Nerve 1985; 8:17–21.
95. Ellis FR. Inherited muscle disease. Br J Anaesth 1980; 52:153–164.
96. Melnick B, Chang J, Larson CE, Bedger RC. Hypokalemic familial periodic paralysis. Anesthesiology 1983; 58:263–265.
97. Lema G, Urzua J, Moran S, Canessa R. Successful anesthetic management of a patient with hypokalemic familial periodic paralysis undergoing cardiac surgery. Anesthesiology 1991; 74:373–375.
98. Horton B. Anesthetic experiences in a family with hypokalemic periodic paralysis. Anesthesiology 1977; 47:308–310.

28

Muscle Sodium Channel and Chloride Channel Diseases

Reinhardt Rüdel and Frank Lehmann-Horn
University of Ulm, Ulm, Germany

INTRODUCTION

The diseases described in this chapter are characterized by a disturbance of the excitation mechanism of the membranes of the skeletal muscle fibers. A more or less pronounced hyperexcitability that is experienced by the patient as muscle stiffness (symptom of myotonia) may be followed by hypoexcitability causing muscle weakness. In the diseases called *myotonias* hyperexcitability dominates and muscle weakness is apparent only for seconds, if at all. In the diseases called *periodic paralyses* myotonia may become noticeable only at the beginning of an episode of weakness, or never develop; here the hypoexcitability is the predominant feature, causing muscle weakness or even complete paralysis of large muscle groups for many minutes, or even several hours.

The symptoms of muscle stiffness and weakness—although they seem direct opposites—are caused by the same process, i.e., temporary depolarization of the skeletal muscle fiber membrane. Slight depolarization may lead to hyperexcitability; stronger depolarization leads to inactivation of the sodium channels and thus to hypoexcitability.

Electrophysiology and molecular biology have shown that genetically determined defects of the sodium channel or the chloride channel proteins may underlie the depolarization.

The altered sodium channels do not always inactivate (close) properly and thus excitation may be repetitive, causing myotonia. The imperfect inactivation may lead to slowly progressive depolarization between subsequent action potentials or to very slowly progressing depolarization without action potentials. In either case an episode of paralysis may result.

The alteration of the chloride channels leads to a reduced membrane conductance causing reduced stability of the membrane potential. The instability is highest during voluntary muscle activation following rest. Repetitive activity then ensues, giving rise to myotonia. The excessive activity may also lead to slowly progressive depolarization between action potentials. Weakness may then be experienced. It usually lasts only for some seconds, but may nevertheless hamper movement more than the stiffness.

This chapter is confined to these specific channelopathies; other, more extensive reviews exist for both the muscle sodium channel (1) and chloride channel (2) diseases. Hypokalemic periodic paralysis and myotonic dystrophy, although belonging to the group of periodic paralyses and myotonias, are excluded by this definition and are dealt with separately in other chapters.

CLASSIFICATION

Recently, a consortium of experts agreed to the following classification of the muscle channelopathies (3).

Muscle sodium channel diseases
- Hyperkalemic periodic paralysis (HyperPP; MIM* 170500)
- Normokalemic periodic paralysis (NormoPP; MIM 170600)
- Paramyotonia congenita (PC; MIM 168300)
- Sodium channel myotonia (myotonia fluctuans)

Muscle chloride channel diseases
- Dominant myotonia congenita (Thomsen) (MIM 160800)
- Recessive myotonia congenita (Becker) (MIM 255700)

DEFINITION OF MYOTONIC SIGNS

The literal translation of the Greek word *myotonia* is muscle stiffness. The stiffness may decrease with continued muscle activity, a phenomenon termed *warm-up*, or it may increase with muscle activity, a phenomenon termed *paradoxical myotonia*. Myotonic muscles react to a blow with the percussion hammer by contracting for a brief period and may also become indented. This is called *percussion myotonia*. Myotonic muscles have a lowered electrical threshold and an increased tendency to react to direct current with a prolonged contraction. The combination of these mechanical and electrical abnormalities is called the *myotonic reaction*. This reaction was used to diagnose a myotonic disease in the early days of electrophysiology and has now been replaced by the electromyogram (EMG) where the muscles exhibit *myotonic runs*, i.e., repetitive activity. In very mild cases myotonic stiffness might not be clinically present, yet the EMG might reveal myotonic runs. This is termed *latent myotonia*.

Myotonia is by definition myogenic as opposed to neurogenic. The myotonias are thus delineated from *neuromyotonia*, spontaneous motor unit activity originating from hyperexcitability of the terminal motor nerve branches (4).

SPECIFICATION OF THE MUSCLE SODIUM CHANNEL DISEASES

Familial Hyperkalemic Periodic Paralysis

This disease resembles clinically the much more common primary hypokalemic periodic paralysis, but differs by the fact that the serum potassium rises rather than falls during an attack. The disease is transmitted with complete penetrance in both sexes, although incomplete penetrance was reported for two rare mutations (3,5). Sporadic cases have also been reported (6,7), and the existence of a de novo mutation was proven in a patient whose genetically confirmed father and mother did not carry the defective gene (8).

Attacks of generalized weakness usually begin to occur in the first decade of life. They commonly start in the morning, last 15 min to an hour, and then disappear spontaneously. Immobility or intake of potassium, e.g., during a provocative test, may provoke an attack of weakness, and prior strenuous work aggravates the attack. The generalized weakness is usually accompanied by a significant increase of serum potassium (up to 5–7 mM). Sometimes the level of serum potassium remains within the upper range of the norm, and only rarely does it rise so high that the function of the heart is impaired. The recovery can be improved by moderate exercise. Slight weakness may continue for several days. Between attacks the serum potassium is normal. Usually the frequency of attacks declines during the second half of life.

EMG reveals the existence of a variant with myotonia and a variant in which not only clinical, but also electrical, myotonia is absent (9–11). There is also a paramyotonic variant of hyperkalemic periodic paralysis, a condition also called paralysis periodica paramyotonica (12,13). In some patients, a chronic progressive myopathy (14) may develop, which seems to be genetically determined (15).

---

*MIM-Mendelian inheritance in man (V. A. McKusick, Johns Hopkins University Press, Baltimore).

Table 1 Clinical, Diagnostic, and Therapeutic Features of the Muscle Sodium Channel Diseases

Disorder	Hyperkalemic periodic paralysis (HyperPP)	Paramyotonia congenita (PC)	Sodium channel myotonia (Myotonia fluctuans)
Inheritance	Autosomal dominant; *de novo* mutations are known	Autosomal domiannt	Autosomal dominant; *de novo* mutations are known
Symptoms	Generalized attacks of weakness; occasionally the weakness becomes gradually permanent	At the beginning of exposure to cold muscular stiffness; on prolonged exposure to cold, muscular weakness	Generalized muscle stiffness without weakness and without cold sensitivity
First occurrence of symptoms	At the end of the 1st decade or beginning of 2nd decade	At birth	Early childhood
Frequency of attacks	A few times per year to daily	Upon every exposure to a cold environment	Depending on severity, occasionally or permanently
Severity of attacks	Occasionally generalized paralysis, frequently moderate weakness, often only in single muscle groups	Stiffness and weakness becoming more severe, the heavier the work in the cold; the lower the temperature, the more severe the weakness	Stiffness becomes more severe, the heavier the work; low temperature has no influence on muscle stiffness
Duration of attacks	Usually 1/2 to 1 hr, occasionally several hours	Stiffness only at the beginning of the work in a cold environment; weakness may last for hours after rewarming.	30 min to 2 hr
Time of day when symptoms appear	Early to late morning, occasionally at any time of day	At any time when exposed to cold environment	Any time during and after heavy exercise
Triggering factors	Muscular excercise with subsequent rest (attack 20–30 min later), cold, hunger	Cooling and heavy muscular work, particularly the conmbination of the two	Muscular exercise with subsequent rest; K^+ loading; succinyl-choline anesthesia
Diagnostic provocation	Musuclar exercise (ergometer or running upstairs) Administration of K^+	Cooling of lower arm for 30 min in water of 12–15°C; repeated forceful closure of the fist.	Muscle exercise and rest
Concomitant symptoms at the beginning of an attack	Myotonia, when present, is increased; cases without myotonia might present lid lag; paresthesias are common	Sensations of tension in the muscles	Sensations of tension in the muscles

Table 1 (*continued*)

Disorder	Hyperkalemic periodic paralysis (HyperPP)	Paramyotonia congenita (PC)	Sodium channel myotonia (Myotonia fluctuans)
K^+ in the serum during an attack	4.5–8 mM; after the attack occasionally <3 mM	3.5–4.5 mM	3.5–4.5 mM
Electromyogram	In cases with myotonia: bursts of fibrillation-like potentials in all muscles; during severe attacks no spontaneous activity	At the beginning of cooling severe fibrillation-like spontaneous activity; during the weakness no spontaneous activity	Bursts of fibrillation-like potentials in all muscles; in severe cases permanently present
Therapy	Preventive: frequent meals, acetazolamide, dihydrochlorothiacide; at the beginning of attack: movement, salbutamol spray, injection of calcium gluconate	Preventive: keeping the muscles warm, mexiletine; no therapeutic action is known for relief of weakness	Mexiletine depot or acetozolamide

Normokalemic Periodic Paralysis

This very rare condition resembles hyperkalemic periodic paralysis in many respects, but differs from it in that even during serious attacks of paralysis the serum potassium is not increased (16). Its existence as a nosological entity was questioned on the basis of molecular genetics (3). Large doses of sodium improve the weakness but glucose administration has no effect. There are no consistent changes in serum electrolytes, but increased sodium excretion and potassium retention occur during the attacks. The urinary potassium retention, the lack of a beneficial effect of glucose, and failure of the serum potassium to increase in attacks distinguish this disease from primary hyperkalemic periodic paralysis.

Resistance to potassium loading distinguishes another kinship (17) from the one first described (16). Thus heterogeneity may exist for this condition.

Paramyotonia Congenita

The hallmarks of the disease, as first described by Eulenburg (18) and independently by Rich (19), and later specified in a large number of families by Becker (20), are (1) paradoxical myotonia, i.e., a muscle stiffness that appears during muscle exercise and increases with continued exercise, (2) severe aggravation of this exercise-induced stiffness by cold, (3) a predilection of the myotonia for face, neck, and the long muscles of the hands, and (4) weakness after protracted exercise and exposure to cold. Some paramyotonia patients also have attacks of hyperkalemic muscle weakness [paralysis periodica paramyotonica (12,13)]. The condition is transmitted with complete penetrance.

Paramyotonic symptoms are present at birth and remain basically unchanged for the entire lifetime, though the attacks of weakness and hyperkalemia begin to appear in adolescence, if at all. In the cold, the face may appear mask-like, and the eyes cannot be opened for several seconds. Working in the cold makes the fingers so stiff that the patient becomes unable to move them within minutes. The stiffness quickly gives way to weakness. After warming, the hands may not regain strength for several hours. As a rule, the legs are less affected. Under warm conditions many patients have no complaints. Muscle pain, muscle atrophy, and hypertrophy are not typical for the disease.

Sodium Channel Myotonia

This relatively new classification is the consequence of recent genetic information (21). Becker (13) had investigated many families with nondystrophic dominant myotonia and proposed many subtypes for what he thought was myotonia congenita. Molecular biology revealed that some of these families in fact had paramyotonia, but not the classic type characterized by cold-induced weakness, but by cold- and exercise-induced stiffness. Other families whose condition was linked to the gene encoding the muscle sodium channel did not present the typical symptoms of paramyotonia as defined above. Rather, the clinical signs of afflicted persons resembled those of myotonia congenita with the peculiarity that the stiffness tended to vary from day to day. The condition was therefore called *myotonia fluctuans* (22,23). The patients never have muscle weakness and are not substantially sensitive to cold in regard to muscle stiffness. Their muscle stiffness is provoked by exercise, but often it occurs with some delay during rest after heavy exercise. The stiffness would then last half an hour to 2 hr. On many days or even for weeks afflicted persons experience no muscle stiffness at all. Ingestion of potassium causes severe myotonia, but no weakness. The potassium-induced myotonia can be prevented by administration of acetazolamide; hence in early descriptions, this variant was named "acetazolamide-responsive myotonia" (24). In some families the muscle stiffness is permanently very severe. The EMG reveals continuous myotonic activity. Because of the persisting myotonia this variant was described as "permanent myotonia" (21).

PATHOPHYSIOLOGY OF THE SODIUM CHANNEL DISEASES

Microelectrode recordings of the membrane potential in excised muscle fiber bundles reflected the different clinical features in vitro in that the depolarization could be most readily elicited by cooling when the muscle was from a paramyotonia patient and by increased extracellular potassium when the muscle was from a patient with hyperkalemic periodic paralysis (10,11). Patch clamp studies of sodium channels in whole muscle preparations, cultured cells, and expression systems revealed an increased number of openings and/or an increase in the time constant of fast inactivation, no matter whether the patient's diagnosis was hyperkalemic periodic paralysis, paramyotonia congenita, or sodium channel myotonia (21,25–28). The shift in the inactivation pattern toward a more frequent occurrence of late sodium channel openings explained the sustained membrane depolarization and the hyperexcitability of the muscle fiber membrane discovered in many earlier electrophysiological tests (10,29–31), although it should be mentioned that similar alterations in the inactivation of the sodium channels have also been detected in various myotonic diseases where the mutation is clearly not in the gene encoding the sodium channel (32).

A possible explanation for the long-lasting depolarization in hyperkalemic periodic paralysis could be the following. Since in the patients one of the alleles of the sodium channel gene is mutated and the other is normal, two kinds of sodium channels are expressed in their muscle fiber membranes, one that inactivates (= closes) normally, and one that fails to inactivate properly. The physiological effect of hyperkalemia induced by potassium intake or by work is a slight depolarization of the muscle fiber membrane. In the defective muscle this depolarization causes the abnormally inactivating sodium channels to open (27,28,30). Sodium ions continue to enter the muscle fibers through the noninactivating channels, thus reinforcing and sustaining the depolarization. This in turn causes inactivation of the normally functioning sodium channels (i.e., those expressed by the normal gene), leading to inexcitability of the muscle fibers, which is the immediate reason for the paralysis (30,33). In association with this influx of sodium, water is shifted into the muscle fibers. This causes hemoconcentration, and the resulting increase of serum potassium will affect those muscle fibers that had had high resting potential. This then may result in paralysis of the whole muscle. The attack is most likely terminated by increased excretion of potassium by the kidneys. Furthermore, the Na/K pump may be maximally activated by the increased concentration of intracellular sodium and extracellular potassium. When serum potassium returns to normal values, the defective sodium channels are likely to close, allowing the membrane to repolarize and become excitable again.

In paramyotonia and in sodium channel myotonia the mechanism must be rather similar. The opposite clinical signs of muscle stiffness and/or paralysis are determined by a relatively small difference in the membrane potential, as a small permanent sarcolemmal depolarization may cause hyperexcitability and a moderate depolarization may cause inexcitability. Obviously, the variability of signs between family members is determined not only by the altered channel function, but also by a number of additional physiological conditions.

SPECIFICATION OF THE MUSCLE CHLORIDE CHANNEL DISEASES

Dominant Myotonia Congenita (Thomsen)

The prime feature of the disease is generalized myotonia (2). The symptom is either congenital or appears during the first or second decade of life. The basis of the myotonia is hyperexcitability of the sarcolemma, which causes electrical and mechanical afteractivity. This inability of the skeletal muscles to relax at will is experienced by the patient as muscle stiffness. The stiffness is provoked by voluntary muscle contraction and is particularly pronounced after a period of rest. The highest probability for the stiffness to occur is when a rested patient performs a particularly strenuous movement. The relaxation of the first contraction may still be relatively unimpeded, for the myotonia first increases, reaches a maximum during the first three to four contractions, and then disappears. Sometimes the stiffness is associated with the symptom of transient weakness that depends on the patient's past activity in a similar way as the stiffness. The muscles show percussion myotonia, the myotonic reaction, and myotonic runs in the EMG. In contrast to an often-quoted opinion, cold is not a complicating major factor, nor is pain a typical feature. Episodes of long-lasting weakness do not occur. The muscles do not develop significant structural changes.

Table 2 Clinical, Diagnostic, and Therapeutic Features of the Muscle Chloride Channel Diseases

Disorder	Myotonia congenita (Thomsen)	Myotonia congenita (Becker)
Inheritance	Autosomal dominant	Autosomal recessive
Symptoms	Muscle stiffness, particularly after rest, muscle function improving with continuing exercise (warm-up)	Muscle stiffness as in dominant form; additionally marked transient weakness after rest, which also improves with exercise
First occurrence of symptoms	From birth to early childhood	Occasionally in early childhood, usually first decade of life
Pitfalls for correct diagnosis	There are cases of Na channel myotonia with very similar symptoms; diagnosis based on molecular biology is then necessary for correct diagnosis The EMG does not provide a safe differential diagnosis for dominant and recessive myotonia congenita	The phenomenon of anticipation in myotonic dystrophy may lead to a familial constellation suggesting recessive inheritance and thus lead to the spurious diagnosis of recessive myotonia congenita Misinterpretation of transient weakness may lead to spurious diagnosis of myotonic dystrophy
Therapy	Mexiletine depot, if necessary	Mexiletine depot, if necessary

The onset of myotonia is usually recognized in early childhood, in mild cases in late childhood. The myotonia is generalized, though the legs are often most affected, causing the children to fall frequently. Hands, arms, and cranial muscles may also be severely involved, so the patients are less skilled in grasping objects. Sometimes chewing becomes difficult.

Members of the same family can be affected to a different degree. In a few kinships the myotonia is regularly very mild and clinically almost nonexistent. This was considered a special form and termed *myotonia levior* (34). For the individual patient, the myotonic signs are rather constant over the entire lifetime. The symptom may increase somewhat in women during pregnancy, but this never creates a major problem. Hypothyroidism may cause aggravation of the myotonia.

Some patients show muscle hypertrophy and hence have an "athletic" appearance. Muscle strength is normal or even greater than normal, and these patients can be quite successful in sports where strength is more important than speed. Occasionally, a slight contracture of the calves may be found limiting dorsiflexion of the feet. Lid lag is usually present, and in some patients percussion blepharospasm myotonia can be observed after forceful closure of the eyes. The muscle stretch reflexes are normal.

Recessive Myotonia Congenita (Becker)

Basically, the clinical picture of recessive myotonia congenita resembles that of the dominant form, but some slight differences exist. In many patients myotonia does not become apparent until the age of 10–14 years, or even later, though in a few children myotonia was already found at the age of 2–3 years. The severity of the myotonia may slowly increase for some years, but this progression rarely goes beyond the age of 25–30 years.

On average the myotonic signs are more severe in recessive than in dominant myotonia congenita, so patients having the recessive form are more handicapped in daily life. Part of their disability consists in a severe myotonic stiffness, in particular in their leg muscles. Yet, even more disabling is a peculiar transient weakness affecting mostly the muscles of the hands and arms (35). This is best demonstrated when such a patient closes his fist with all strength after having the fingers kept at rest for 5–10 min. Finger force vanishes almost completely within a few seconds of contraction. For example, when the patient wants to lift a heavy suitcase, the handle will slide out of his hand. In contrast, a patient with the dominant form would be unable to release the handle for a short period of time. Only with repeated muscle contractions does force gradually return within 20–60 sec. This transient weakness is often generalized, causing the patient much trouble, e.g., when he wants to get from a supine to an upright position after rest or sleep.

The leg and buttock muscles often show a pronounced hypertrophy, and hyperlordosis is common. In contrast, the neck, shoulder, and arms may be rather poorly developed, especially in old age. This results in a characteristic disproportionate figure.

The serum creatine kinase (CK) is normal in most patients. Occasionally there is a borderline elevation in dominant and a two- or threefold elevation in recessive myotonia congenita. Needle EMG shows typical myotonic runs in every skeletal muscle. Muscle biopsy will not provide much relevant information for a diagnosis, although there might be a type II fiber deficiency and hypertrophied fibers and central nuclei. Isolated tubular aggregates may be found with the electron microscope. As a point of caution, it might be mentioned that in older patients with a severe degree of the recessive form, morphological changes may be such that the false consideration of myotonic dystrophy may arise.

PATHOPHYSIOLOGY OF CHLORIDE CHANNEL DISEASES

The electrophysiological mechanisms leading to the lowered electrical threshold and the increased tendency to react to a constant stimulus with a prolonged contraction were first explained by Bryant (36), who studied the membrane parameters of skeletal muscles excised from an animal model, the myotonic goat. The resting potential was normal and the action potential had a normal fast rising phase, but its falling phase had a prolonged afterdepolarization. The fibers lacked accommodation to sustained depolarizing current, so repetitive firing occurred along with a slow depolarization of the membrane. Upon slow depolarization to about –70 mV, the fibers became inexcitable. The most intriguing result was a very high membrane resistance of these fibers. Since in skeletal muscle 80% of the resting membrane conductance is chloride conductance, Bryant investigated how replacing chloride in the extracellular medium by less permeant anions affected normal fibers and found that they behaved remarkably like myotonic fibers in a chloride-containing medium. Not only was the membrane resistance equivalent to that of the myotonic fibers, but the fibers had the typical myotonic excitability, e.g., low threshold, lack of accommodation, repetitive firing, and myotonic afterdischarge. This led to the proposal that the basis for myotonia was the congenital absence or reduction of the chloride conductance (36). Lipicky and Bryant (37–40) went on to show that the sarcolemmal chloride conductance was indeed lower than normal in myotonia congenita patients, and this was later confirmed by other groups (41,42).

On the basis of further studies on myotonic goat muscle, the "low chloride conductance theory of myotonia" was put forward, which explains the hyperexcitability as a direct consequence of a reduced chloride conductance in the T system of the myotonic muscle fibers (43,44). During physiological activity potassium leaves the interior of the fibers and accumulates in the lumen of the T tubules. In a normal fiber with a high chloride conductance, tubular potassium accumulation causes depolarization of about 0.1 mV per action potential, not enough to be of physiological importance. In myotonic fibers, in which chloride conductance is lacking, the accumulation causes about 1 mV depolarization per action potential. This depolarization decays within seconds. In other words, while in normal fibers the chloride conductance tends to clamp the membrane potential to the chloride equilibrium potential assumed to be near –80 mV (45), in myotonic fibers this clamp is ineffective, leading to the afterdepolarization and subsequent repetitive activity.

Variation of the parameters for a mathematical membrane model that included the T system showed that the mere lowering of the chloride conductance did not cause repetitive firing (46). In fact, appropriate adjustment of the sodium channel activating kinetics was necessary for a correct simulation of the experimentally observed lack of accommodation. Interestingly, abnormalities in the properties of the sodium channels were noted in myotonic muscle, from both the goat (47) and humans (32,42,43).

Variation of the degree of artificial block of the chloride conductance of normal rat muscle showed that for myotonia to occur, the fraction of the chloride conductance may not be larger than 30% of the total membrane conductance (the rest being mainly potassium conductance) (48). Determination of this fraction in a larger number of patients having either dominant or recessive myotonia congenita showed that this fraction ranged from 0 to 80% (32,42). Thus the hyperexcitability in myotonia congenita must have an additional, as yet not quite understood mechanism.

GENETICS OF THE MUSCLE CHANNELOPATHIES

Mutations in SCN4A, the Gene Encoding the α Subunit of the Skeletal Muscle Sodium Channel Protein, Located on Chromosome 17q22

Fifteen different mutations (Fig. 1) have been discovered in SCN4A at the time of writing of this chapter (February 1994). Four of them lead to hyperkalemic periodic paralysis, five to paramyotonia congenita, and six to sodium channel myotonia. Six polymorphisms of SCN4A leading to harmless amino acid substitutions were found as a by-product of the search for mutations in affected families.

The various mutations causing the same particular sodium channel disease are not distributed over the gene in a predictable or easily understandable, regular manner. Some of them are in the part of the gene encoding the inactivation gate (substituting amino acids Gly-1306, Thr-1313) or its acceptor at the inner mouth of the pore (Thr-704, Ser-906, Val-1589, Met-1592). Others are located in the part encoding the IVth channel repeat (Met-1360, Arg-1448) believed to be involved in channel inactivation (1). Obviously not only the site of replacement, but also the properties of the replaced and the replacing amino acids, play a very important role. This is best seen in amino acid Gly-1306 where three different replacements lead to clinical pictures that differ substantially in severity. Another such example with three known replacements is amino acid Arg-1448.

Met-1592-Val, the first-detected SCN4A mutation (8) results in the form of *hyperkalemic periodic paralysis* that is associated with myotonia and does not lead to permanent weakness and is the second most commonly encountered mutation in all investigated families (49,50). In contrast, the most common Thr-704-Met mutation may cause both the myotonic and the nonmyotonic form (3,15). Independently

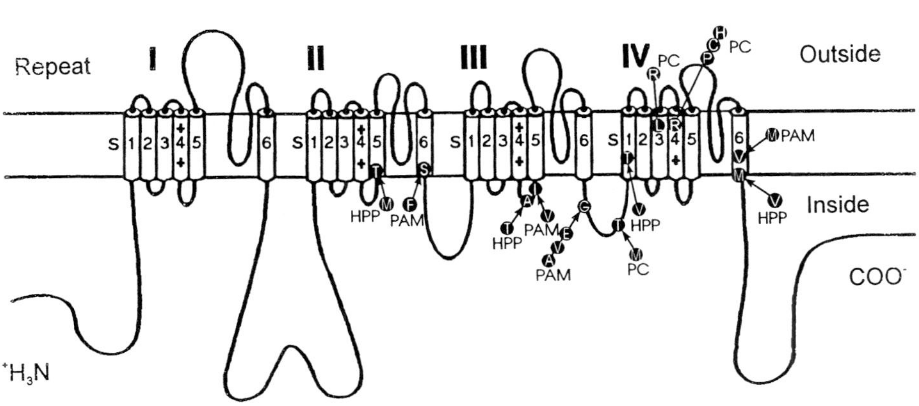

Figure 1 Scheme of the primary structure of the α subunit of the adult skeletal muscle Na^+ channel consisting of four repeats, which contain six transmembrane segments each. Amino acids are shown for which substitutes were deduced from point mutations found in hyperkalemic periodic paralysis (HPP), paramyotonia congenita (PC), and sodium channel myotonia (potassium-aggravated myotonia, PAM) (T = threonine 704, S = serine 804, A = alanine 1156, G = glycine 1306, T = threonine 1313, M = methionine 1360, L = leucine 1433, R = arginine 1448, V = valine 1589, M = methionine 1592). Two different mutations were found for R1448, three different mutations for G1306 (see text).

of the presence of myotonia, progressive myopathy is commonly found in older and sometimes even in younger patients. The rare third and fourth mutations causing hyperkalemic periodic paralysis (Ala-1156-Thr and Met-1360-Val) are characterized by incomplete clinical penetrance, although "nonaffected" family members show electrical myotonia in the EMG indicating the presence of defective sodium channels (3,5).

An interesting, as yet unexplained finding is that a family that was convincingly diagnosed as having hyperkalemic periodic paralysis did not show linkage to SCN4A. Genetic heterogeneity is a possible explanation (51).

In only one family with the clinical diagnosis of *normokalemic periodic paralysis* has a mutation been reported so far (3), namely the common Thr-704-Met substitution that leads to hyperkalemic periodic paralysis in other families. This finding suggests that the disease is not a nosological entity but a variant of the hyperkalemic form caused by additional factors not yet understood. A molecular biological study of the original family described by Poskanzer and Kerr (16) should clarify the situation.

In most families having *paramyotonia congenita*, substitutions were detected for amino acid Arg-1448, which is part of the transmembrane segment S4 of repeat IV. This segment is thought to act as voltage sensor because of its positive charges and may be important for the mechanism of coupling between channel activation and inactivation. These mutations reduce the number of positive charges. The Arg-1448-His mutation is found in families having cold-induced and potassium-induced stiffness combined with spontaneous episodic attacks of weakness (50,52). The Arg-1448-Cys mutation is typical for families in which affected members suffer from cold-induced stiffness (without weakness) and spontaneous episodic attacks of weakness. The Arg-1448-Pro mutation causes the most cold-sensitive paramyotonia form in which even slight cooling leads to paralysis (E. P. Hoffman, personal communication). The mutation Leu-1433-Ala was discovered in a typical paramyotonia congenita family characterized by muscle stiffness exacerbated in the cold, the stiffness giving way to flaccid paralysis when the cooled muscles were exercised (50). Another mutation, Thr-1313-Met, located in the inactivation gate, causes muscle stiffness aggravated by cold (53).

Six mutations at four different loci are responsible for *sodium channel myotonia*. One of these loci, closely located to the Thr-1313-Met mutation, is in the region coding for the inactivation gate. It is particularly remarkable because three different point mutations are known that lead to three different replacements for the same nucleotide. These mutations result in different amino acid substitutes for one (Gly-1306) of a pair of glycines 1306/07 supposed to act as the hinge of the inactivation gate. Length, ramification, and charge of the side chains in the substitutes correlate well with both the degree of membrane hyperexcitability and the clinical phenotype. Glutamic acid, an amino acid with a long side chain, causes permanent myotonia, the most severe form of sodium channel myotonia (21), formerly reported as the myogenic type of Schwartz-Jampel syndrome (54). Valine, an amino acid with a side chain of intermediate size, was the substitute in patients with moderate exercise-induced myotonia (21,53), and alanine, distinguished by a short side chain, resulted in myotonia fluctuans, the most moderate form of sodium channel myotonia (21). This form can also be produced by the Ser-804-Phe mutation (5,55). For the "acetazolamide-responsive myotonia" (24) an Ile-to-Val substitution was detected in the intracellular loop between segments S4 and S5 of repeat III (56).

Another mutation, Val-1589-Met, causes cold- and potassium-induced stiffness in absence of weakness (57). This substitution is located next to the helical position of the opposite substitution Met-1592-Val, the latter causing hyperkalemic periodic paralysis (8).

Mutations in HUMCLC, the Gene Encoding the Skeletal Muscle Chloride Channel Protein, Located on Chromosome 7q32

Five point mutations and one deletion (Fig. 2) have been described in HUMCLC, and the number is supposed to increase with the analysis of further families. Two of the point mutations lead to dominant myotonia congenita; the other three point mutations and the deletion lead to the recessive form. One of the mutations causing the dominant form is a Gly-230-Glu substitution in the extracellular linker between the 3rd and 4th of the putative 12 transmembrane domains of the channel protein (58); the

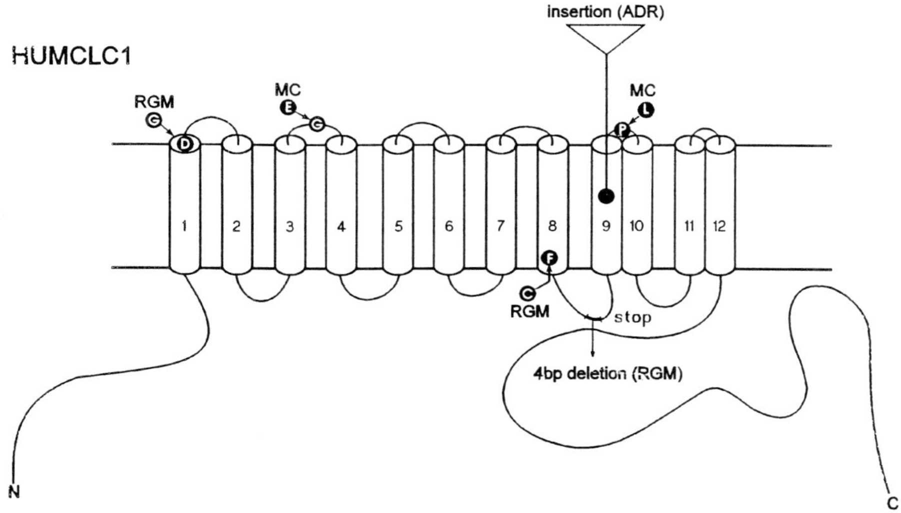

Figure 2 Scheme of the primary structure of the adult skeletal muscle chloride channel consisting of 12 transmembrane segments. Amino acids are shown for which substitutes were deduced from point mutations found in dominant (MC) and recessive (GM) myotonia congenita. A four-base deletion in the genetic region coding for transmembrane segment 9 causes an early stop codon very close to the site where the insertion of a transposon in the homologous mouse gene is responsible for generation of a mouse model of recessive myotonia, the so-called ADR mouse.

other is a Pro-480-Leu substitution in the extracellular linker between the 9th and 10th domains (59). The mutations found in both alleles of the patients with the recessive form are a Phe-413-Cys substitution in the 8th and an Asp-136-Gly substitution in the 1st transmembrane domain (60,61).

As with SCN4A, the various mutations are not distributed over the gene in a predictable or easily understandable, regular manner. Two of the disease-causing mutations are located at the extracellular side of the membrane-spanning domains, removing (Asp-136-Gly) or adding (Gly-230-Glu) a negative charge.

Several patients with recessive myotonia congenita were found to be heterozygous with respect to a known mutation (e.g., Phe-413-Cys), the second mutation being expected to exist in the parts of HUMCLC for which primers are not yet available. It seems that compound heterozygosity is common, as described for other recessive diseases (62). Consequently it is an open question as to whether some amino acid substitutions discovered in normal individuals are harmless polymorphisms or could lead to recessive myotonia congenita.

DIFFERENTIAL DIAGNOSIS OF THE MUSCLE CHANNELOPATHIES

Molecular genetics now allows specialized centers to diagnose patients with a muscle channelopathy. Abnormalities in the mobility of single-stranded DNA fragments of either SCN4A or HUMCLC can be found by screening the genomic DNA of such patients. Primer sequences have been published for all 24 exons of SCN4A (63) and for several exons of HUMCLC (58). The mutation underlying the disease of a proband can be detected by sequencing the DNA eluted from an aberrant band.

Differentiation between the various types of channelopathies is important for a patient's prognosis and therapy. For example, in the case of hyperkalemic periodic paralysis, progressive myopathy is expected only in the case of the most frequent mutation, Thr-704-Met (15). Furthermore, clinical differentiation between some forms may be impossible, e.g., between Thomsen myotonia and myotonia fluctuans or, more importantly, between Becker myotonia and "sporadic" myotonic dystrophy in patients who have no dystrophy at the time of examination. Moreover, the genotype of sporadic cases with a de novo mutation and of asymptomatic members of myotonia families can be determined for adequate genetic counseling.

THERAPY

Regular and frequent food ingestion, a low-potassium diet, and avoidance of stenuous work help to prevent attacks of hyperkalemic periodic paralysis. Many patients are able to prevent or abort attacks by continuing slight exercise and/or the ingestion of carbohydrates at the onset of weakness (e.g., 2 g glucose per kg body weight by mouth). However, severe attacks may fail to respond to these measures (9,62). Patients should be advised to rise early and have a good breakfast.

Some patients are able to diminish the extent and the duration of an attack by prompt oral intake of a thiazide diuretic or acetazolamide or by inhalation of a β-adrenergic agent. The beneficial effect of the former is probably based on their capacity to lower the serum potassium. The effect of β-adrenergic agents is probably medicated via stimulation of the Na/K pump (65–67).

For many patients prevention of attacks by the continuous administration of a thiazide diuretic (68–70) or acetazolamide (71,73) is recommended. These diuretics lower serum potassium and are very effective in mild cases. The diuretics are used in modest dosages at intervals from twice daily to twice weekly (68,71,73). Thiazide diuretics are preferable because of the possible complications of acetazolamide therapy (74). The dosage should be kept as low as possible, e.g., 25 mg of hydrochlorothiazide daily or every other day. The drug should not lower the serum potassium below 3.3 mM or the serum sodium below 135 mM (71). In severe cases, 50 or 75 mg of hydrochlorothiazide should be taken daily very early in the morning. In very severe cases this therapy may be ineffective. In some patients small doses of fludrocortisone (0.1 mg daily) combined with acetazolamide have been helpful (75). β-Adrenergic treatment may also prevent attacks (76).

Myotonic stiffness responds well to drugs that reduce the increased excitability of the cell membrane. These are the blockers of the sodium channels, i.e., local anesthetics, antiarrhythmic drugs, and related agents. They suppress myotonic runs by decreasing the number of available sodium channels and have no known effect on chloride channels Mexiletine (1–3 times 200 mg/day), quinine, procaine amide, and phenytoin are most commonly used. Tocainide (77) has beneficial effects similar to those of mexiletine, but bone marrow depression and interstitial pulmonary fibrosis have been reported with a few cardiac patients receiving the drug. Hence it is no longer recommended. Dantrolene, a blocker of calcium release from the sarcoplasmic reticulum, also reduces myotonic stiffness (78).

Paramyotonia or myotonia congenita patients usually need no treatment. They rarely ask for it and reject taking medication regularly. Young patients sometimes wish to use antimyotonic drugs for a few days when they want to participate in sports or dancing. Mexiletine in the depot form, given in the morning and in the afternoon, is the medication of first choice. As the resorption differs from patient to patient, the dosage should be adapted to the serum levels of the drug. Older patients should have an electrocardiogram, and the cardiologist should be consulted. Therapy with mexiletine will reduce the myotonic stiffness as well as the transient weakness, though the latter not to a remarkable degree. At the end of a longer treatment period, the dosage of mexiletine should be gradually reduced over several days. Otherwise a rather severe "rebound" of myotonic stiffness and muscle weakness may occur. Phenytoin and carbamazepine have considerably less antimyotonic effect.

In paramyotonic hyperkalemic periodic paralysis, mexiletine and hydrochlorothiazide can be given together to prevent both cold-induced stiffness and weakness and attacks of generalized weakness and hyperkalemia (70).

REFERENCES

1. Rüdel R, Ricker K, Lehmann-Horn F. Genotype-phenotype correlations in human skeletal muscle sodium channel diseases. Arch Neurol 1993; 50:1241–1248.
2. Rüdel R, Lehmann-Horn F. Membrane changes in cells from myotonia patients. Physiol Rev 1985; 65:310–356.

3. Lehmann-Horn F, Rüdel R, Ricker K. Workshop report: non-dystrophic myotonias and periodic paralyses. Neuromusc Disord 1993; 3:161–168.
4. Newsom-Davis J. New concepts in neuromuscular disorders. IX Int Congr Electromyography and Clin Neurophysiol, Jerusalem, 1992:60.
5. McClatchey AI, McKenna-Yasek D, Cros D, et al. Novel mutations in families with unusual and variable disorders of the skeletal muscle sodium channel. Nature Genet 1992; 2:148–152.
6. Dyken ML, Timmons GD. Hyperkalemic periodic paralysis with hypocalcemia episode. Arch Neurol 1963; 9:508–519.
7. Riggs JE, Moxley RT, Griggs RC, Horner FA. Hyperkalemic periodic paralysis: an apparent sporadic case. Neurology 1981; 31:1157–1159.
8. Rojas CV, Wang J, Schwartz L, Hoffman EP, Powell BR, Brown RH Jr. A Met-to-Val mutation in the skeletal muscle sodium channel alpha-subunit in hyperkalemic periodic paralysis. Nature 1991; 354:387–389.
9. Gamstorp I. Adynamia episodica hereditaria and myotonia. Acta Neurol Scand 1963; 39:41–58.
10. Lehmann-Horn F, Rüdel R, Ricker K, Lorkovic H, Dengler R, Hopf HC. Two cases of adynamia episodica hereditaria: in vitro investigation of muscle cell membrane and contraction parameters. Muscle Nerve 1983; 6:113–121.
11. Ricker K, Camacho L, Grafe P, Lehmann-Horn F, Rüdel R. Adynamia episodica hereditaria: what causes the weakness? Muscle Nerve 1989; 10:883–891.
12. Lundberg PO, Stalberg E, Thiele D. Paralysis periodica paramyotonica. A clinical and neurophysiological study. J Neurol Sci 1974; 21:309–321.
13. Becker PE. Myotonia Congenita and Syndromes Associated with Myotonia. Stuttgart: Thieme, 1977.
14. Bradley W, Taylor R, Rice D, et al. Progressive myopathy in hyperkalemic periodic paralysis. Arch Neurol 1989; 47:1013–1017.
15. Ptáček LJ, George AL Jr., Griggs RC, Tawil R, Kallen RG, Barchi RL, Robertson M, Leppert MF. Identification of a mutation in the gene causing hyperkalemic periodic paralysis. Cell 1991; 67:1021–1027.
16. Poskanzer DC, Kerr DNS. A third type of periodic paralysis with normokalemia and favorable response to sodium chloride. Am J Med 1961; 31:328–341.
17. Meyers KR, Gilden DH, Rinaldi CF, Hansen JL. Periodic muscle weakness, normokalemia, and tubular aggregates. Neurology 1972; 22:269–279.
18. Eulenburg A. Über einer familiäre, durch 6 Generationen verfolgbare Form congenitaler Paramyotonie. Neurol Zentralbl 1886; 5:265–272.
19. Rich EC. A unique form of motor paralysis due to cold. Med News 1894; 65:210–213.
20. Becker PE. Paramyotonia congenita (Eulenburg). In: Fortschritte der allgemeinen und klinischen Humangenetik. Stuttgart: Thieme, 1970:1–131.
21. Lerche H, Heine R, Pika U et al. Human sodium channel myotonia: slowed channel inactivation due to substitutions for a glycine within the III/IV linker. J Physiol 1993; 470:13–22.
22. Ricker K, Lehmann-Horn F, Moxley RT. Myotonia fluctuans. Arch Neurol 1990; 47:268–272.
23. Lennox G, Purves A, Marsden D. Myotonia fluctuans. Arch Neurol 1992; 49:1010–1011.
24. Trudell RG, Kaiser KK, Griggs RC. Acetazolamide responsive myotonia congenita. Neurology 1987; 37:488–491.
25. Cannon SC, Strittmatter SM. Functional expression of sodium channel mutations identified in families with periodic paralysis. Neuron 1993; 10:317–326.
26. Cummins TR, Zhou J, Sigworth FJ et al. Functional consequences of a Na^+ channel mutation causing hyperkalemic periodic paralysis. Neuron 1993; 10:667–678.
27. Lehmann-Horn F, Iaizzo PA, Hatt H, Franke Ch. Altered gating and reduced conductance of single sodium channels in hyperkalemic periodic paralysis. Pflügers Arch 1991; 418:297–299.
28. Cannon SC, Brown RH Jr, Corey DP. A sodium channel defect in hyperkalemic periodic paralysis: potassium-induced failure of inactivation. Neuron 1991; 6:619–626.
29. Lehmann-Horn F, Rüdel R, Dengler R, Lorkovic H, Haass A, Ricker K. Membrane defects in paramyotonia congenita with and without myotonia in a warm environment. Muscle Nerve 1981; 4:396–406.
30. Lehmann-Horn F, Küther G, Ricker K, Grafe P, Ballanyi K, Rüdel R. Adynamia episodica hereditaria with myotonia: a non-inactivating sodium current and the effect of extracellular pH. Muscle Nerve 1987; 10:363–374.
31. Lehmann-Horn F, Rüdel R, Ricker K. Membrane defects in paramyotonia congenita (Eulenburg). Muscle Nerve 1987; 10:633–641.
32. Iaizzo PA, Franke CH, Spittelmeister W, Ricker K, Rüdel R, Lehmann-Horn F. Altered sodium channel behaviour causes myotonia in dominant myotonia congenita. Neuromusc Disord 1991; 1:47–53.
33. Creutzfeldt OD, Abbot BC, Fowler WM, Pearson CM. Muscle membrane potentials in episodic adynamia. Electroencephalogr Clin Neurophysiol 1963; 15:508–519.
34. DeJong JGY. Myotonia levior. In: Kuhn E, ed. Progressive Muskeldystrophie - Myotonie - Myasthenie. Heidelberg: Springer, 1966:255–259.

35. Ricker K, Haass A, Hertel G, Mertens HG. Transient muscular weakness in severe recessive myotonia congenita. Improvement of isometric muscle force by drugs relieving myotonic stiffness. J Neurol 1978; 218:253–262.
36. Bryant SH. Cable properties of external intercostal muscle fibres from myotonic and non-myotonic goats. J Physiol 1969; 204:539–550.
37. Lipicky RJ, Bryant SH. Ion content, potassium flux and cable properties of myotonic, human, external intercostal muscle. Trans Am Neurol Assoc 1971; 96:34–38.
38. Lipicky RJ, Bryant SH, Salmon JH. Cable parameters, sodium, potassium, chloride, and water content and potassium efflux in isolated external intercostal muscle of normal volunteers and patients with myotonia congenita. J Clin Invest 1971; 50:2091–2102.
39. Lipicky RJ, Bryant SH. A biophysical study of the human myotonias. In: Desmedt JE, ed. New Developments in Electromyography and Clinical Neurophysiology. Vol 1. Basel: Karger, 1973:451–463.
40. Lipicky RJ. Studies in human myotonic dystrophy. In: Rowland LP, ed. Pathogenesis of Human Muscular Dystrophy. Amsterdam: Excerpta Medica, 1977:729–738.
41. Rüdel R, Ricker K, Lehmann-Horn F. Transient weakness and altered membrane characteristic in recessive generalized myotonia (Becker). Muscle Nerve 1988; 11:202–211.
42. Franke C, Iaizzo PA, Hatt H, Spittelmeister W, Ricker K, Lehmann-Horn F. Altered Na channel activity and reduced Cl conductance cause hyperexcitability in recessive generalized myotonia (Becker). Muscle Nerve 1991; 14:762–770.
43. Adrian RH, Bryant SH. On the repetitive discharge in myotonic muscle fibres. J Physiol 1974; 240:505–515.
44. Bryant SH. Altered membrane properties in myotonia. In: Bolis L, Hoffman LP, Leaf A, eds. Membranes and Diseases. New York: Raven Press, 1976: 197–206.
45. DeCoursey TE, Bryant SH, Owenburg KM. Dependence of membrane potential on extracellular ionic concentrations in myotonic goats and rats. Am J Physiol 1981; 240:C56–C63.
46. Adrian RH, Marshall MW. Action potentials reconstructed in normal and myotonic muscle fibres. J Physiol 1976; 258:125–143.
47. Bryant SH. Physical basis of myotonia. In: Schotland DL, ed. Disorders of the Motor Unit. New York: Wiley, 1982:381–389.
48. Barchi RL. An evaluation of the chloride hypothesis. Arch Neurol 1975; 32:175–180.
49. Feero WG, Wang J, Barany F, et al. Hyperkalemic periodic paralysis. Rapid molecular diagnosis and genotype/phenotype correlations for 12 families. Neurology 1993; 43:668–673.
50. Ptáček LJ, Gouw L, Kwiecicski H, et al. Sodium channel mutations in paramyotonia congenita and hyperkalemic periodic paralysis. Ann Neurol 1993; 33:300–307.
51. Wang J, Zhou J, Todorovic SM, et al. Molecular genetic and genetic correlations in sodium channelopathies: lack of founder effect and evidence for a second gene. Am J Hum Gen 1993; 52:1074–1084.
52. Ptáček LJ, George AL jr., Barchi RL, et al. Mutations in an S4 segment of the adult skeletal muscle sodium channel cause paramyotonia congenita. Neuron 1992; 8:891–897.
53. McClatchey AI, van den Bergh P, Pericak-Vance MA, et al. Temperature-sensitive mutations in the III-IV cytoplasmic loop region of the skeletal muscle sodium channel gene in paramyotonia congenita. Cell 1992; 68:769–774.
54. Spaans F, Theunissen P, Reekers A, Smit L, Veldman H. Schwartz-Jampel syndrome. Part I. Clinical, electromyographic, and histologic studies. Muscle Nerve 1990; 13:516–527.
55. Ricker K, Moxley RT, Heine R, Lehmann-Horn F. Myotonia fluctuans, a third type of muscle sodium channel disease. Arch Neurol 1994; 51:1095–1102.
56. Griggs RC, personal communication, 1994.
57. Heine R, Pika U, Lehmann-Horn F. A novel SCN4A mutation causing myotonia aggravated by cold and potassium. Hum Mol Genet 1993; 2:1349–1353.
58. George AL, Crackover MA, Abdalla JA, Hudson JA, Ebers GC. Molecular basis of Thomsen's disease (autosomal dominant myotonia congenita). Nature Genet 1993; 3:305–310.
59. Steinmeyer K, Lorenz C, Pusch M, Koch MC, Jentsch TJ. Multimeric structure of ClC-1 chloride channel revealed by mutations in dominant myotonia congenita (Thomsen). EMBO J 1994; 13:737–743.
60. Koch MC, Steinmeyer K, Ricker K, et al. Tight linkage of recessive and dominant forms of human myotonia to skeletal muscle chloride channel gene. Science 1992; 257:797.
61. Heine R, George AL, Pika U, Deymeer F, Rüdel R, Lehmann-Horn F. Proof of a non-functional muscle chloride channel in recessive myotonia congenita (Becker) by detection of a 4 base pair deletion. Hum Mol Genet 1994; 3:1123–1128.
62. Cutting GR, Kasch LM, Rosenstein BJ, Tsui L-C, Kazazian HH Jr, Antonarakis SE. Two patients with cystic fibrosis, nonsense mutations in each cystic fibrosis gene, and mild pulmonary disease. N Engl J Med 1990; 323:1685–1689.
63. George AL, Iyer GS, Kleinfield R, Kallen RG, Barchi RL. Genomic organization of the human skeletal muscle sodium channel gene. Genomics 1993; 15:598–606.

64. Sagild U. Hereditary Transient Paralysis. Copenhagen: Ejnar Munksgaards, 1959.
65. Clausen T, Wang P, Orskov H, Kristensen O. Hyperkalemic periodic paralysis: relationship between changes in plasma water, electrolytes, insulin and catecholamines during attacks. Scand J Clin Lab Invest 1980; 40:211–220.
66. Clausen T. Regulation of active Na^+–K^+ transport in skeletal muscle. Physiol Rev 1986; 66:542–580.
67. Ballanyi I, Grafe P. Changes in intracellular ion activities induced by adrenalin in human and rat skeletal muscle. Pflügers Arch 1988; 411:283–288.
68. Samaha FJ. Hyperkalemic periodic paralysis. Arch Neurol 1965; 12:145–154.
69. Layzer RB, Lovelace RE, Rowland LP. Hyperkalemic periodic paralysis. Arch Neurol 1967; 16:455–472.
70. Ricker K, Böhlen R, Rohkamm R. Different effectiveness of tocainide and hydrochlorothiazide in paramyotonia congenita with hyperkalemic episodic paralysis. Neurology 1983; 33:1615–1618.
71. McArdle B. Adynamia episodica hereditaria and its treatment. Brain 1962; 85:121–148.
72. Riggs JE, Moxley RT, Griggs RC, Horner FA. Hyperkalemic periodic paralysis: an apparent sporadic case. Neurology 1981; 31:1157–1159.
73. Carson MJ, Pearson CM. Familial hyperkalemic periodic paralysis with myotonic features. J Pediatr 1964; 64:853–865.
74. Riggs JE, Griggs RC. Diagnosis and treatment of the periodic paralyses. In: Klawans HL, ed. Clinical Neuropharmacology. Vol 4. New York: Raven Press, 1979:123–138.
75. Streeten DHP, Dalakos TG, Fellerman H. Studies on hyperkalemic periodic paralysis. Evidence of changes in plasma Na and Cl and induction of paralysis by adrenal glucocorticoids. J Clin Invest 1971; 50:142–155.
76. Bendheim PE, Reale EO, Berg BO. Beta-adrenergic treatment of hyperkalemic periodic paralysis. Neurology 1985; 35:746–749.
77. Rüdel R, Dengler R, Ricker K, Haass A, Emser W. Improved therapy of myotonia with the lidocaine derivative tocainide. J Neurol 1980; 222:275–278.
78. Eberstein A, Goodgold J. Relief of myotonia by dantrolene in experimental myotonia. Exp Neurol 1978; 62:804–809.

29

Neuromyotonia (Isaacs' Syndrome)

Ian K. Hart and John Newsom-Davis
University of Oxford, Oxford, England

INTRODUCTION

Neuromyotonia is a descriptive term for a disorder in which hyperexcitability of peripheral nerves leads to spontaneous and continuous muscle activity. This overactivity can be triggered by voluntary muscle contraction and is associated with characteristic electromyographic (EMG) abnormalities. Clinically, patients often present with a typical syndrome that includes muscle stiffness, cramps, myokymia, and delayed muscle relaxation.

It is useful to divide cases of neuromyotonia into two main groups. The first consists of cases that arise in certain inherited diseases, and the second contains cases that are acquired, including those that develop in association with various types of noninherited peripheral neuropathy. The acquired form is the most common and is the main subject of this chapter.

Our knowledge of neuromyotonia is based on a series of case reports that each describe a small number of patients. Although only about 120 cases in total have been reported, the syndrome has at various times been called undulating myokymia (1), the syndrome of continuous muscle activity (2), Isaacs' or Isaacs-Mertens syndrome, and quantal squander (3), as well as neuromyotonia (4), which is now the most generally accepted term. Both the rarity of the syndrome and the confusing terminology have led to some dispute over diagnostic criteria and made research difficult. As a result, the etiology of acquired neuromyotonia has remained obscure. Although a few cases have been reported to occur after exposure to environmental toxins, the majority have been labeled as idiopathic.

Recently, however, evidence has emerged that autoimmune mechanisms may play an important role in the etiology of acquired neuromyotonia. Here we review this evidence and discuss the diagnostic clinical and laboratory features of the disease.

DIAGNOSIS

Until more is known about the pathogenesis of neuromyotonia, diagnosis is necessarily based on a combination of clinical and EMG features. Most of these features are common to all forms of neuromyotonia.

Clinical Features

At presentation motor symptoms predominate. In the fully developed syndrome patients have muscle stiffness, cramps, myokymia, slow muscle relaxation after contraction (pseudomyotonia), increased sweating, and weakness (5,6). However, there is a wide range of both clinical severity and rate of progression, and not all of these symptoms may be present in all cases. Age of onset ranges from

Table 1 Clinical Associations of Neuromyotonia

Acquired diseases	
	Thymoma ± myasthenia gravis
	Bronchial carcinoma
	Acute inflammatory demyelinating peripheral neuropathy
	Chronic inflammatory demyelinating peripheral neuropathy
	Addison's disease
	Penicillamine therapy
Inherited diseases	
	Hereditary peripheral neuropathies
	Schwartz-Jampel syndrome
	Hereditary paroxysmal ataxias
	Distal spinal muscular atrophy

childhood to the eighth decade, and the condition tends to be chronic, although spontaneous remission has occurred in a few patients (7).

Stiffness, cramps, and myokymia are the most common presenting features, and although they may occur at rest, they are often triggered by muscle contraction. Characteristically, myokymia continues during both sleep and general anesthesia (2). The slow relaxation of muscle (for example, handgrip) is often worsened by cold, but despite its resemblance to true myotonia, the response to percussion is normal (2). Increased sweating is thought to be caused by the increase in the basal metabolic rate and body temperature that results from increased muscle activity (2,8). Evidence of primary autonomic dysfunction is confined to isolated cases (9).

These motor features tend to be most prominent in the limb and trunk muscles, but may also involve the face and tongue. Dyspnea, which is due to involvement of the respiratory muscles (2,10,11), and stridor, due to laryngeal muscle overactivity (11,12), may also develop. Neuromyotonia has occasionally been reported in association with ischemia-induced carpopedal spasm (11,13,18) and masseter spasm elicited by percussion (16). In these cases serum calcium, magnesium, and electrolyte levels and acid-base balance were all normal.

Examination may reveal weakness (usually mild) that can be proximal, distal, or generalized. Tendon reflexes may be reduced or absent even if there is no other evidence of peripheral neuropathy (8). Patients with severe long-standing neuromyotonia may develop muscle hypertrophy (19), pes cavus (2), or distal muscle wasting associated with contractures (2,8,10,20). Onset of acquired neuromyotonia in childhood may be associated with growth retardation (2,8), which may be corrected by treatment of the muscle overactivity (8).

Some patients have sensory symptoms, including paresthesia (10,11,15,16,21,22), painful dysesthesia (20,23), and numbness (18,22), although these are rarely prominent features of the clinical presentation. These findings suggest that in neuromyotonia the sensory as well as the motor nerves may be hyperexcitable (15). Electrophysiological or histological evidence of a peripheral neuropathy was found in only a minority of these cases (10,15,20,22).

Central disorders, including hallucinations and altered behavior (9,24), and personality change (25) have been seen in some patients with acquired neuromyotonia. This association is known as *chorée fibrillaire de Morvan.* There is one report of a patient with intellectual impairment in association with generalized cerebral atrophy on pneumoencephalography (21).

Laboratory Features

Before discussing the characteristic electrophysiological features of neuromyotonia, it is essential to consider the site of origin of the abnormal activity. Isaacs showed that the muscle discharges in neuromyotonia are generated in the peripheral motor nerves (2). He described two patients in whom muscle overactivity continued during both thiopentone-induced general anesthesia and complete nerve block, but was abolished by junctional neuromuscular blockade with either curare or succinylcholine. In one patient discharges were also unaffected by spinal anesthesia. These findings have since been confirmed in many other cases of neuromyotonia (both acquired and inherited) and suggest that the origin of the spontaneous action potentials in these cases is the distal part of the motor nerves and/or within its terminal nerve branches (8–10,12,18–20,22,26–29). This latter possibility would explain the fibrillation potentials that are found in some cases (see below). By contrast, in other patients the peripheral nerve activity has been reduced (8,11,14,16,28,30,31) or abolished (8,21,22,27,28,32,33) by peripheral nerve block, and in one case epidural anesthesia stopped the spontaneous discharges (22), indicating a more proximal site of origin.

Taken together, these findings suggest that the abnormal discharges may be generated at different sites throughout the length of the nerve. In some cases these sites are distal and in others proximal, perhaps even including the anterior horn cell.

EMG Findings

Diagnosis is confirmed by EMG. Concentric needle EMG shows distinctive doublet, triplet, multiplet, or continuous motor unit discharges that have a high (40–200 sec^{-1}) intraburst frequency (1,2). These are called neuromyotonic discharges (Fig. 1) and are present even in cases where there is no visible myokymia (25). The bursts occur at irregular intervals (1–30 sec), and voluntary or electrical activation of the nerve often elicits prolonged afterdischarges of a similar type. In addition, fibrillation potentials (discharges of single muscle fibers usually of less than 100 μV amplitude) and fasciculations (single motor unit discharges at low and irregular frequency) are sometimes present. The peripheral nerve origin of the abnormal impulses can be confirmed by performing blocks at the level of the nerve and neuromuscular junction.

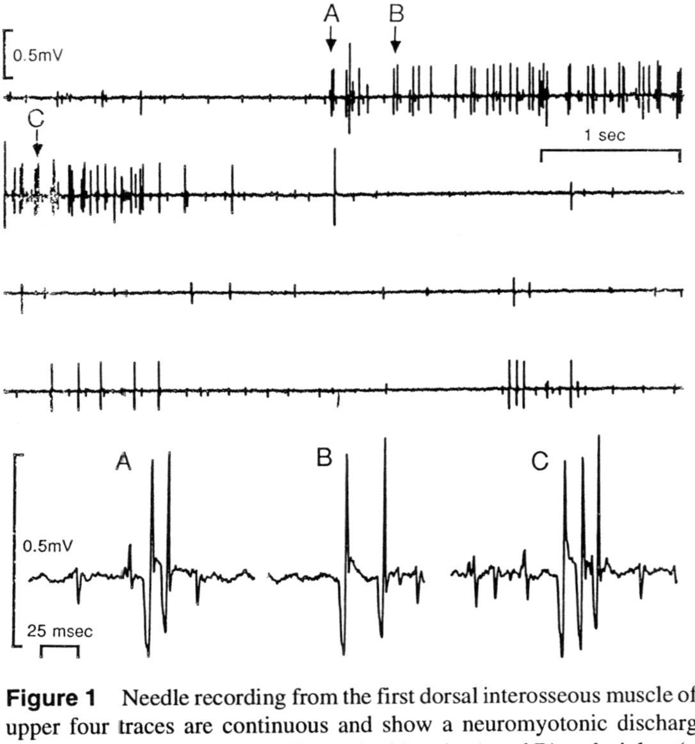

Figure 1 Needle recording from the first dorsal interosseous muscle of a patient with neuromyotonia. The upper four traces are continuous and show a neuromyotonic discharge lasting about 4 sec. Within this discharge, single motor units fire as doublets (at A and B) and triplets (at C). (From Ref. 25.)

Nerve Conduction Studies

Many cases of acquired neuromyotonia show no convincing evidence of peripheral neuropathy (2,4,8,25,30). In other cases, however, there may be evidence, often subtle, of damage to both motor and sensory fibers (10,11,25,35) or to motor fibers alone (34). The abnormalities are nonspecific and may suggest primary involvement of either the axon (11,20,25) or the myelin (35,37). Occasionally, neuromyotonia has been associated with either acute (32) or chronic (36) inflammatory demyelinating polyneuropathy.

Histology

Morphological data are limited, variable, and confusing. Moreover, the value of sural nerve biopsy, in what is predominantly a motor syndrome, is doubtful. Examination of the sural nerve in acquired neuromyotonia by both the light and electron microscope may be normal or show nonspecific signs of axonal damage (20,25,33) or myelin degeneration (10,13,35), or both (11,37). In two cases there was a selective loss of the larger myelinated axons, although the full range of both myelinated and nonmyelinated fibers was affected to some extent (33). Intramuscular nerve biopsies may show either preterminal (17,34,38) or intraterminal collateral sprouting (18), although these changes are not disease-specific. Preterminal sprouting is a sign of nerve degeneration from any cause, and intraterminal sprouting occurs in both myasthenia gravis and botulinum intoxication, as well as in neuromyotonia (18).

Muscle biopsy is not more helpful as the reported abnormalities are patchy and vary from case to case (2,13,28,31,33). Most, if not all, of these findings are probably caused by prolonged nerve firing and sustained muscle contraction.

Cellular infiltrates have not been found in either nerve or muscle except in rare cases of inflammatory demyelinating polyneuropathy with neuromyotonia (36). IgG deposits were absent in the one case where staining was carried out (11).

Other Findings

Cerebrospinal fluid (CSF) has not been analyzed in most cases of neuromyotonia. However, increased CSF total IgG has been found in a few patients (24,28). More recently, three cases of acquired neuromyotonia were reported to have abnormal CSF oligoclonal bands on electrophoresis, with increased IgG bands mainly in the fast γ regions (26). In the latter three cases serum immunoelectrophoresis was normal. When measured, both CSF total protein level and cell count have usually been normal (2,16,33). CSF γ-aminobutyric acid level was increased in one case (27).

Immune complexes have been found in two Japanese patients with acquired neuromyotonia (29,39). In both cases there was an abnormal creatine kinase (CK) isoenzyme band on serum electrophoresis that consisted of CK3 isoenzyme linked to IgA. In one patient the complex was found on the motor end-plates and muscle membranes (29).

Biochemical tests are unhelpful in diagnosis. In some cases serum levels of muscle proteins, including lactate dehydrogenase, aldolase, myoglobin, and CK (without an abnormal isoenzyme band), may be moderately raised (10,25,39), probably as a result of muscle damage due to overactivity. Serum calcium, magnesium, electrolytes, and acid-base balance are all normal.

Differential Diagnosis

As mentioned above, neuromyotonia may develop in association with various inherited diseases including several types of hereditary neuropathy (15,40–44), distal spinal muscular atrophy (45), Schwartz-Jampel syndrome (46), and various forms of hereditary paroxysmal ataxia (47,48). In these cases, although the symptoms and EMG findings may be identical to those of the acquired form, the family history and the characteristic features of the associated syndromes will help in diagnosis.

In the vast majority of patients with acquired neuromyotonia there is no obvious precipitating cause. However, in isolated cases direct damage to the peripheral nerves by herbicide (20) or insecticide

Table 2 Differential Diagnosis of Neuromyotonia

Focal myokymia
Benign myokymia
Nerve damage due to
ischemia
demyelination
tumor invation
entrapment
ionizing radiation
snake venom
Stiff-man syndrome
Cramp-fasciculation syndrome
Brody's disease (Ca^{2+}-ATPase deficiency)

poisoning (37) or chronic alcohol (49) or toluene abuse (50) has been suspected. Therefore, a history of chemical and drug exposure should be taken in every case.

Neuromyotonia is one of several conditions that are characterized by visible myokymia (51). In benign myokymia, which may also be associated with muscle cramps, the EMG shows a single polyphasic action potential that fires repetitively but varies in shape and form from impulse to impulse (1). Neuromyotonic discharges are not seen. Myokymia, usually localized, may also result from damage to the peripheral nerves caused by radiation, demyelination, entrapment, tumour invasion, ischemia, and timber rattle snake envenomation (52) or to the lower motorneurons within the central nervous system (CNS) by tumors, ischemia, and demyelination. In all these conditions the EMG may show doublet or multiplet motor unit discharges that differ from neuromyotonia in that they are localized and that the individual bursts of discharges occur rhythmically or semirhythmically.

Muscle stiffness and spasm are prominent features of the stiff-man syndrome (28,53). Unlike neuromytonia, however, the abnormal muscle activity in this CNS disease stops during sleep and does not display the high-frequency burst discharges.

The cramp-fasciculation syndrome presents with similar clinical and EMG findings to neuromyotonia, although fibrillation potentials and fasciculations are not found (54). Despite this, it is possible that at least some of these cases may be mild variants of neuromyotonia.

PATHOGENESIS

Lack of understanding of the etiology of neuromyotonia has resulted in most cases of the disease being classed as idiopathic. However, recent work has provided evidence that autoimmune mechanisms may be involved in the pathogenesis of at least some of these cases (25,55).

The stimulus for the investigation of the immune system in neuromyotonia came from the reported clinical associations of the disorder. Neuromyotonia has been described in six patients with thymoma (9,11,56,57). This is clearly not a chance association as these six cases represent about 15% of all the cases of neuromyotonia that have been reported in the past 20 years. Thymoma has long been recognized to be associated with myasthenia gravis (57) and with other immunologically mediated conditions (58). Three of the cases of neuromyotonia with thymoma had clinical evidence of myasthenia gravis (11,56,57) and two others had raised anti-acetylcholine receptor (AchR) antibody titers (9).

A patient with seropositive rheumatoid arthritis developed neuromyotonia after several years of treatment with penicillamine (59). The muscle overactivity responded rapidly to phenytoin and was abolished after penicillamine was withdrawn. Penicillamine is known to precipitate myasthenia gravis and the production of anti-AchR antibodies (60) and has also been implicated in a case of dermatomyositis (61).

Sporadically, neuromyotonia has been reported in association with other autoimmune diseases.

These include acute (32) and chronic (36) inflammatory demyelinating polyneuropathy and a case of Addison's disease with a mixed sensory-motor demyelinating neuropathy (35). In the latter case, muscle overactivity ceased after corticosteroid therapy. In these patients, one possible explanation for the neuromyotonia might be that it was produced by immune mechanisms similar to those that are thought to cause the underlying disease.

The finding that neuromyotonia has developed in three patients with bronchial carcinoma suggests that neuromyotonia can be a paraneoplastic disorder (16,21,62). Although histology was available in only one of these cases, it was of the small cell type (62). It is well known that this tumor associates with several neurological paraneoplastic syndromes that are considered to be autoimmune-mediated (63). In the Lambert-Eaton myasthenic syndrome (LEMS), IgG autoantibodies to tumor calcium channel antigens are thought to cross-react with nerve-terminal calcium channels (64,65). This leads to channel down-regulation and a reduction in transmitter release. By analogy, a different tumor antigen in paraneoplastic neuromyotonia might trigger the production of antibodies that increase peripheral nerve excitability.

There is now clinical evidence that a humoral factor may be important in the pathogenesis of neuromyotonia. In five patients it was found that plasma exchange, which depletes circulating immunoglobulins, resulted in clinical improvement and a reproducible reduction in the frequency of the abnormal muscle discharges (25,55,66,67).

Experimental evidence for a humoral factor is provided by studies where plasma or IgG from patients with neuromyotonia has been injected into mice (25,55). The injected animals, when compared with controls, showed (1) a significant enhancement of tubocurarine resistance at the neuromuscular junction, and (2) a significant increase in the quantal content of the end-plate potential (68). The simplest explanation for all these findings is that a circulating factor, probably an IgG, interferes with ion channels that regulate peripheral nerve excitability. Ion channels seem especially vulnerable to attack by autoantibodies: acetylcholine receptors are the target in myasthenia gravis and calcium channels in LEMS.

This hypothesis raises the question of what is the target ion channel in neuromyotonia? The best candidate would seem to be neuronal voltage-gated potassium channels. These channels are present in peripheral nerves (69,70), especially in the nodal and paranodal regions, and a decrease in their activity would lead to both an increase in quantal content and the generation of repetitive discharges, effects seen when these channels are blocked by drugs such as 4-aminopyridine (71,72). Other suggestions for target antigens in neuromyotonia, including muscle AchR (9) and neuronal sodium channels, seem less probable.

It appears that the spontaneous impulses in neuromyotonia can be generated at sites that lie within the blood-nerve barrier, as well as at the nerve terminal. Until recently it was thought this barrier, and its CNS counterpart, protected neurons from immunological attack. However, this seems not to be the case as there is now evidence that autoimmune mechanisms may be directly involved in the pathogenesis of CNS diseases such as the stiff-man syndrome (73) and various paraneoplastic neurological conditions (74). In neuromyotonia the report of abnormal CSF oligoclonal bands may be important (25).

The significance of the IgA-CK immune complexes (found in two patients, as described above) to the pathogenesis of neuromyotonia is not clear (29,39). It seems possible, however, that these complexes are a consequence of the release of enzymes from muscle damaged by overactivity, rather than a primary cause of nerve hyperexcitability.

TREATMENT

Phenytoin (2) and carbamazepine (4) are the first-line treatments for the symptoms of all types of neuromyotonia. Phenytoin (200–400 mg/day) controls most symptoms in most patients within a few days. If this drug is ineffective, carbamazepine (200–600 mg/day) should be tried. Some resistant cases may be helped by a combination of both these drugs (8). In isolated cases, sodium valproate (41) or dantrolene (27) has been reported to alleviate symptoms. Other drugs, including baclofen, benzodiazepines, carbonic anhydrase inhibitors, and quinine, are generally unhelpful.

Plasma exchange should be considered in the minority of patients with disabling symptoms of acquired neuromyotonia that fail to respond to drugs. Useful improvement (albeit temporary) in symptoms, confirmed by serial EMG studies, has been seen in five patients treated in this way (25,55,66,67).

There are few data on the use of immunosuppressive drugs. The results of a pilot study, in which three patients who responded to plasma exchange were given azathioprine (2.5 mg/kg/day) and prednisolone (up to 80 mg on alternate days) for up to 8 months, were inconclusive (25). Further trials are needed to determine the usefulness of long-term immunosuppressive therapy.

CONCLUSION

The term neuromyotonia describes a characteristic syndrome of clinical and EMG abnormalities that results from hyperexcitability of peripheral motor (and probably sensory) nerves. Although rare, it needs to be distinguished from other causes of muscle overactivity as the symptoms of neuromyotonia can usually be controlled by drugs. From the evidence available, it seems likely that neuromyotonia may be induced by several mechanisms that damage the peripheral nerve, including inherited abnormalities and perhaps chemical toxicity. Recently, preliminary studies have suggested that autoimmunity may be involved in the pathogenesis of at least some acquired cases. This work has already led to the exploration of new treatments and may eventually result in the identification of the molecular defect (or defects) that underlies the disease.

REFERENCES

1. Denny-Brown D, Foley JM. Myokymia and the benign fasciculation of muscular cramps. Trans Assoc Am Physicians 1948; 61:88–96.
2. Isaacs H. A syndrome of continuous muscle-fibre activity. J Neurol Neurosurg Psychiatry 1961;24:319–325.
3. Isaacs H. Quantal squander. S A J Lab Clin Med 1964; 10:93–95.
4. Mertens HG, Zschocke S. Neuromyotonie. Klin Wochenschr 1965; 43:917–925.
5. Kimura J. Neuromuscular diseases characterized by abnormal muscle activity. In: Electrodiagnosis in Diseases of Nerve and Muscle: Principles and Practice, 5th ed. Philadelphia: FA Davis, 1983:549–565.
6. Rowland LP. Cramps, spasms and muscle stiffness. Rev Neurol (Paris) 1985; 141:261–273.
7. Isaacs H, Heffron JJ. The syndrome of "continuous muscle-fibre activity" cured: further studies. J Neurol Neurosurg Psychiatry 1974; 37:1231–1235.
8. Irani PF, Purohit AV, Wadia NH. The syndrome of continuous muscle fiber activity. Acta Neurol Scand 1977; 55:273–288.
9. Halbach M, Homberg V, Freund H-J. Neuromuscular, autonomic and central cholinergic hyperactivity associated with thymoma and acetylcholine receptor-binding antibody. J Neurol (Berlin) 1987; 234:433–436.
10. Harik SI, Baraka AS, Tomeh GF, Mire-Salman J, Kronful A, Afifi AK. Autonomous peripheral nerve activity causing generalized muscle stiffness and fasciculations: report of a case with physiological, pharmacological, and morphological observations. Johns Hopkins Med J 1976; 139:49–60.
11. García-Merino A, Cabella A, Mora JS, Liaño H. Continuous muscle fiber activity, peripheral neuropathy, and thymoma. Ann Neurol 1991; 29:215–218.
12. Jackson DL, Sataya-Murti S, Davis L, Drachman DB. Isaacs' syndrome with laryngeal involvement: an unusual presentation of myokymia. Neurology 1979; 29:1612–1615.
13. Welch LK, Appenzeller O, Bicknell JM. Peripheral neuropathy with myokymia, sustained muscle contraction, and continuous motor unit activity. Neurology 1972; 22:161–168.
14. Negri S, Caraceni T, Boiardi A. Neuromyotonia. Report of a case. Eur Neurol 1977; 16:35–41.
15. Lance JW, Burke D, Pollard J. Hyperexcitability of motor and sensory neurons in neuromyotonia. Ann Neurol 1979; 5:523–532.
16. Partanen VSJ, Soininen H, Saksa M, Riekkinen P. Electromyographic and nerve conduction findings in a patient with neuromyotonia, normocalcemic tetany and small-cell lung cancer. Acta Neurol Scand 1980; 61:216–226.
17. Coërs C, Telermann-Toppet N, Durdu J. Neurogenic benign fasciculations, pseudomyotonia and pseudotetany. A disease in search of a name. Arch Neurol 1981; 38:282–287.

18. Oda K, Fukushima N, Shibasaki H, Ohnishi A. Hypoxia-sensitive hyperexcitability of the intramuscular nerve axons in Isaacs' syndrome. Ann Neurol 1989; 25:140–145.
19. Zisfein J, Sival M, Aron AM, Bender AN. Isaacs' syndrome with muscle hypertrophy reversed by phenytoin therapy. Arch Neurol 1983; 40:241–242.
20. Wallis WE, Van Poznak A, Plum F. Generalized muscular stiffness, fasciculations, and myokymia of peripheral nerve origin. Arch Neurol 1970; 22:430–439.
21. Walsh JC. Neuromyotonia: an unusual presentation of intrathoracic malignancy. J Neurol Neurosurg Psychiatry 1976; 39:1086–1091.
22. Hosokawa S, Shinoda H, Sakai T, Kato M, Kuroiwa Y. Electrophysiological study on limb myokymia in three women. J Neurol Neurosurg Psychiatry 1987; 50:877–881.
23. Wakayama Y, Ohbu S, Machida H. Myasthenia gravis, muscle twitch, hyperhidrosis, and limb pain associated with thymoma: proposal of a possible new myasthenic syndrome. Tohoku J Exp Med 1991; 164:285–291.
24. Fischer-Perroudon C, Trillet M, Mouret J, et al. Investigations polygraphiques et métaboliques d'une insomnie durable avec hallucinations. A propos de l'etude anatomo-clinique d'une observation de chorée fibrillaire de Morvan. Rev Neurol (Paris) 1974; 130:111–125.
25. Newsom-Davis J, Mills KR. Immunological associations of acquired neuromyotonia (Isaacs' syndrome). Report of five cases and literature review. Brain 1993; 116:453–469.
26. Barron SA, Heffner RR. Continuous muscle fiber activity. A case with unusual clinical features. Arch Neurol 1979; 36:520–521.
27. Sakai T, Hosokawa S, Shibasaki H et al. Syndrome of continuous muscle-fiber activity: increased CSF GABA and effect of dantrolene. Neurology 1983; 33:495–498.
28. Valli G, Barbieri S, Cappa S, Pellegrini G, Scarlato G. Syndromes of abnormal muscular activity: overlap between continuous muscle fibre activity and the stiff man syndrome. J Neurol Neurosurg Psychiatry 1983; 46:241–247.
29. Nagashima T, Kamegai M, Hirose K, et al. Creatine kinase (CK)-linked IgA in Isaacs' syndrome. An immune complex disease? J Neurol Sci 1985; 67:269–276.
30. Nakanishi T, Sugita H, Shimada Y, Toyokura Y. Neuromyotonia. A mild case. J Neurol Sci 1975; 26:599–604.
31. Lütschg J, Jerusalem F, Ludin HP, Vassella F, Mumenthaler M. The syndrome of "continuous muscle fiber activity." Arch Neurol 1978; 35:198–205.
32. Vasilescu C, Alexianu M, Dan A. Muscle hypertrophy and a syndrome of continuous motor unit activity in prednisone-responsive Guillain-Barré polyneuropathy. J Neurol (Berlin) 1984; 231:276–279.
33. Ono S, Munakata S, Nagao K, Shimizu N, Shimizu N. The syndrome of continuous muscle fibre activity: light and electron microscopic studies in muscle and nerve biopsies. J Neurol (Berlin) 1989; 236:377–381.
34. Lublin FD, Tsairis P, Streletz LJ, et al. Myokymia and impaired muscular relaxation with continuous motor unit activity. J Neurol Neurosurg Psychiatry 1979; 42:557–562.
35. Vilchez JJ, Cabello A, Benedito J, Villarroya T. Hyperkalaemic paralysis, neuropathy and persistent motor unit discharges at rest in Addison's disease. J Neurol Neurosurg Psychiatry 1980; 43:818–822.
36. Joy JL, Allen RF, Sunwoo IN, Oh SJ, Thomas TD. Isaacs' syndrome associated with chronic inflammatory demyelinating polyneuropathy. Muscle Nerve 1990; 13:868 (abstract).
37. Black JT, Garcia Mullin R, Good E, Brown S. Muscle rigidity in a newborn due to continuous peripheral nerve hyperactivity. Arch Neurol 1972; 27:413–425.
38. Isaacs H. Continuous muscle fibre activity in an Indian male with additional evidence of terminal motor fibre abnormality. J Neurol Neurosurg Psychiatry 1967; 30:126–133.
39. Kuwasaki N, Shoji H, Tominaga H, Kaji M, Nonaka K. A case of Isaacs' syndrome—continuous muscle fiber activity syndrome. No To Shinkei 1986; 38:531–536.
40. Auger RG, Daube JR, Gomez MR, Lambert EH. Hereditary form of sustained muscle activity of peripheral nerve origin causing generalized myokymia and muscle stiffness. Ann Neurol 1984; 15:13–21.
41. Vasilescu C, Alexianu M, Dan A. Neuronal type of Charcot Marie-Tooth disease with a syndrome of continuous motor unit activity. J Neurol Sci 1984; 63:11–25.
42. McGuire SA, Tomasovic JJ, Ackerman N Jr. Hereditary continuous muscle fibre activity. Arch Neurol 1984; 41:395–396.
43. Hahn AF, Parkes AW, Bolton CF, Stewart SA. Neuromyotonia in hereditary motor neuropathy. J Neurol Neurosurg Psychiatry 1991; 54:230–235.
44. Serratrice G, Pouget J, Pellissier JF, Cros D. Carbamazepine sensitive neuromyotonia and Charcot Marie-Tooth disease of the neuronal type. Rev Neurol (Paris) 1989; 145:867–868.
45. Castilla JM, Alberca R, Chichon I, Gil-Neciga E, Rafel E. Continuous muscle activity and distal spinal muscular atrophy. Eur Neurol 1991; 31:156–159.
46. Taylor RG, Layzer RB, Davis HS, Fowler WM. Continuous muscle fiber activity in the Schwartz-Jampel syndrome. Electroenceph Clin Neurophysiol 1972; 33:497–509.

47. Gancher ST, Nutt JG. Autosomal dominant episodic ataxia: a heterogeneous syndrome. Mov Disord 1986; 1:239–253.
48. Vaamonde J, Artieda J, Obeso JA. Hereditary paroxysmal ataxia with neuromyotonia. Mov Disord 1991; 6:180–182.
49. Vasilescu C, Florescu A. Peripheral neuropathy with a syndrome of continuous motor unit activity. J Neurol (Berlin) 1982; 226:275–282.
50. Devathasan G, Low D, Teoh PC, Wan SH, Wong PK. Complications of chronic glue (toluene) abuse in adolescents. Aust NZ J Med 1984; 14:39–43.
51. Gutmann L. AAEM minimonograph #37: facial and limb myokymia. Muscle Nerve 1991; 14:1043–1049.
52. Brick JF, Gutmann L, Brick J, Apelgren KN, Riggs JE. Timber rattlesnake venom-induced myokymia: evidence for peripheral nerve origin. Neurology 1987; 37:1545–1546.
53. Moersch FP, Woltman HW. Progressive fluctuating muscular activity and spasm. Mayo Clin Proc 1956; 31:421–426.
54. Tahmoush AJ, Alonso RJ, Tahmoush GP, Heiman-Patterson TD. Cramp-fasciculation syndrome: a treatable hyperexcitable peripheral nerve disorder. Neurology 1991; 41:1021–1024.
55. Sinha S, Newsom-Davis J, Mills K, Byrne N, Lang B, Vincent A. Autoimmune aetiology for acquired neuromyotonia (Isaacs' syndrome). Lancet 1991; 338:75–77.
56. McComas AJ. Myasthenia gravis preceded by quantal squander myokymia. In: Neuromuscular Function and Disorders, 2nd ed. London: Butterworths, 1977:315–317.
57. Namba T, Brunner NG, Grob D. Myasthenia gravis in patients with thymoma, with particular reference to onset after thymectomy. Medicine 1978; 57:411–433.
58. Souadjian JV, Enriquez P, Silverstein MN, Pepin J-M. The spectrum of diseases associated with thymoma. Arch Intern Med 1974; 134:374–379.
59. Reeback J, Benton S, Swash M, Schwartz MS. Penicillamine-induced neuromyotonia. Br Med J 1979; 279:1464–1465.
60. Vincent A, Newsom-Davis J. Acetylcholine receptor antibody characteristics in myasthenia gravis. II. Patients with penicillamine-induced myasthenia or idiopathic myasthenia of recent onset. Clin Exp Immunol 1982; 49:266–272.
61. Fernandes L, Swinson DR, Hamilton EBD. Dermatomyositis complicating penicillamine treatment. Ann Rheum Dis 1977; 36:94–95.
62. Waerness E. Neuromyotonia and bronchial carcinoma. Electromyogr Clin Neurophysiol 1974; 14:527–535.
63. Vincent A, Lang B, Newsom-Davis J. Autoimmunity to the voltage-gated calcium channel underlies the Lambert-Eaton myasthenic syndrome, a paraneoplastic disorder. Trends Neurosci 1989; 12:496–502.
64. Roberts A, Perera S, Lang B, Vincent A, Newsom-Davis J. Paraneoplastic myasthenic syndrome IgG inhibits $45Ca2^{+}$ flux in a human small cell carcinoma line. Nature 1985; 317:737–739.
65. Chalk CH, Murray NMF, Newsom-Davis J, O'Neill JH, Spiro SG. Response of the Lambert-Eaton myasthenic syndrome to treatment of associated small-cell lung carcinoma. Neurology 1990; 40:1552–1556.
66. Bady B, Chauplannaz G, Vial C, Savet J-F. Autoimmune aetiology for acquired neuromyotonia. Lancet 1991; 338:1330.
67. Heindenreich, F. Personal communication.
68. Shillito P, Lang B, Newsom-Davis J, Bady B, Chauplannaz G. Evidence for an autoantibody mediated mechanism in acquired neuromyotonia. J Neurol Neurosurg Psychiatry 1992; 55:1214 (abstract).
69. Bostock H, Baker M. Evidence for two types of potassium channel in human motor axons in vivo. Brain Res 1988; 462:354–358.
70. Röeper J, Schwarz JR. Heterogeneous distribution of fast and slow potassium channels in myelinated rat nerve fibres. J Physiol 1989; 416:93–110.
71. Kocsis JD, Bowe CM, Waxman SG. Different effects of 4-aminopyridine on sensory and motor fibers: pathogenesis of paresthesias. Neurology 1986; 36:117–120.
72. Fu WM, Lin RH, Lin-Shiau SY. (1987) Study on the neuromuscular action of 4-aminopyridine in the mouse diaphragm. Arch Int Pharmacodynam Ther 1987; 289:198–211.
73. Solimena M, Folli F, Denis-Donini S et al. Autoantibodies to glutamic acid decarboxylase in a patient with the stiff-man syndrome, epilepsy, and type I diabetes mellitus. N Engl J Med 1988; 318:1012–1020.
74. Posner J, Furneaux H. Paraneoplastic syndromes. In: Waksman BH, ed. Immunological Mechanisms in Neurology and Psychiatric Disease. New York: Raven Press, 1990:187–219.

30

Muscle Dysfunction in Endocrine Disease

Richard W. Orrell and Russell J. M. Lane
Charing Cross Hospital and Charing Cross and Westminster Medical School, London, England

Peter Hudgson
Newcastle General Hospital and University of Newcastle upon Tyne, Newcastle upon Tyne, England

INTRODUCTION

As one of the largest tissues in the body, muscle is particularly susceptible to the metabolic and trophic effects of the endocrine system. Abnormalities of muscle have been reported with virtually all forms of endocrine dysfunction, including iatrogenic causes. Some of these abnormalities are subclinical however, and sometimes only detectable at a biochemical or ultrastructural level; but endocrine myopathies can cause major disability. Fortunately, such muscular disability is usually eminently treatable by correcting the underlying endocrine abnormality, so the importance of recognizing an endocrine cause for a myopathy is clear. Most of the initial investigations required to diagnose an endocrine myopathy are readily performed, and failure to make the diagnosis and offer the appropriate treatment is most likely to result from a lack of appreciation of the possibility of an endocrine basis for the symptoms and signs.

The most common abnormalities of muscle function are found in disorders of thyroid (hypothyroidism and hyperthyroidism) and the corticosteroid hormones (Cushing's disease, Addison's disease, and iatrogenic). These are dealt with first, followed by muscle dysfunction associated with epinephrine (pheochromocytoma), growth hormone, insulin, parathormone, and testosterone.

THYROXINE

Hyperthyroid Myopathy

More than 75% of patients with hyperthyroidism have evidence of muscle weakness on clinical examination. although this is a major symptom in only around 5% (1,2). The features are those of a proximal myopathy, with wasting, and the weakness may be out of proportion to the degree of muscle wasting. Prominent shoulder girdle wasting with winging of the scapulae may be a specific feature of the condition (3). Cramps, myalgia, and fasciculation may be noted (4,5), and the reflexes are normal or brisk. This combination of signs may occasionally lead to a clinical picture resembling amyotrophic lateral sclerosis (6) (Fig. 1,a and b).

Serum creatine kinase (CK) is usually normal or low (7), probably related to increased clearance, although levels may be very high when associated with rhabdomyolysis in a thyroid storm (8). Electromyography (EMG) shows features of a myopathy (9,10), and nearly all thyrotoxic patients have EMG features suggestive of a myopathy in proximal muscles. The muscle biopsy appearances are

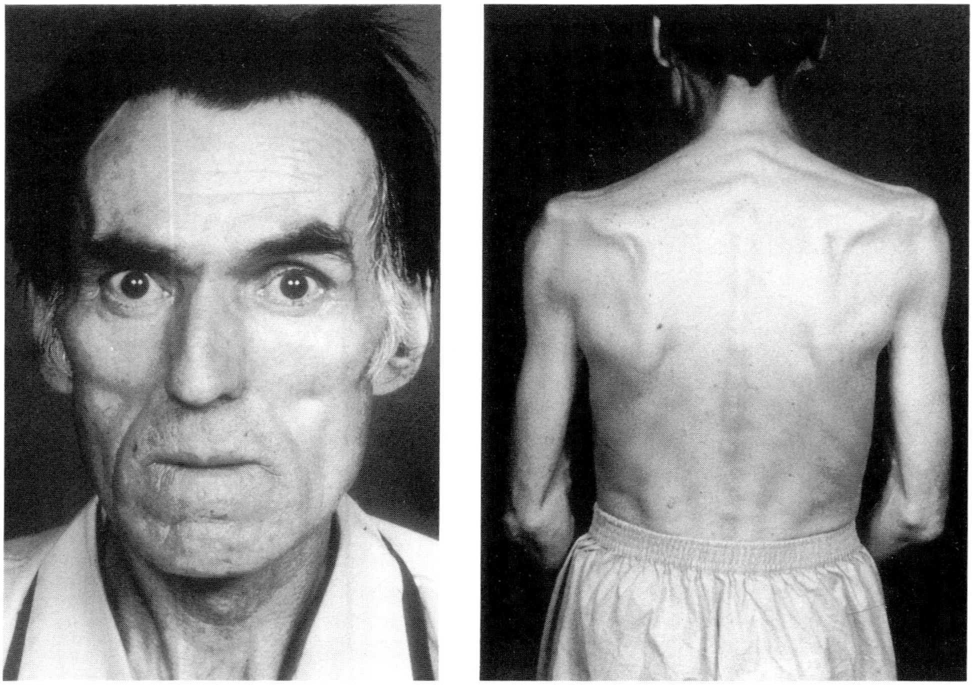

(a) (b)

Figure 1 (a) Thyrotoxic myopathy. Gaunt facial features, with slight proptosis and lid retraction. (b) Striking muscle wasting around the shoulder girdle. A degree of scapular winging was apparent on specific testing.

generally nonspecific and include atrophy of both fiber types (11,12). Experimental studies in hyperthyroid rats showed conversion of type 1 to type 2 fibers (13).

As with many endocrine myopathies, the pathogenesis of the myopathy is uncertain. Thyroxine influences many metabolic processes and in particular has a catabolic effect on muscle protein. There appears to be a reduced efficiency of contraction of the muscle (12,14), which may explain the disproportionate weakness to the degree of wasting. When the thyroxine excess is corrected, there is improvement in muscle power, usually before restoration of muscle bulk (2). The improvement may take several months, and propranolol may correct the muscle weakness more rapidly (15). In hyperthyroid patients with sudden generalized weakness, and bulbar palsy, associated myasthenia gravis should be considered (see below).

Hypothyroid Myopathy

Up to 40% of patients with hypothyroidism have clinical evidence of muscle weakness (2). Stiffness, pain, and cramps in the muscle may be present, especially on exercise or in cold weather (16). A delayed relaxation or "pseudomyotonia" is present in around a quarter of patients, and a triad comprising proximal weakness with raised CK, slurring dysarthria, and slow relaxing reflexes may be the presenting feature of the endocrine disorder (17).

There is a spectrum of presentation of muscle disease in hypothyroidism, the extreme presentation being Hoffmann's syndrome (18,19). Hoffmann's original patient was an adult with muscle hypertrophy, weakness, slow movements, painful spasms, and delayed muscle relaxation, which did not improve with repetition. (A patient with Hoffmann's syndrome, illustrating the muscular hypertrophy, is shown in Fig. 6b, Chapter 1.) The clinical sign of "myoedema" may be present—a sustained, electrically silent, focal muscle contraction producing mounding is present on tapping the muscle with a tendon hammer. A similar syndrome of muscle hypertrophy (but without cramps) associated with hypothyroidism, when seen in children with congenital hypothyroidism, is referred to as Kocher-Debré-Sémélaigne syndrome (20,21).

The pathogenesis of the muscle disease remains uncertain, with no consistent findings on muscle biopsy to explain the muscle hypertrophy. Serum CK is raised (7), often to around 10 times normal, and EMG may show myopathic features and sometimes fibrillations, positive waves, and other spontaneous activity (22,23). The muscle symptoms and signs generally recover completely when the hypothyroidism is corrected.

Thyrotoxic Periodic Paralysis

A periodic paralysis, clinically similar to hypokalemic periodic paralysis, may be associated with thyrotoxicosis (1). The condition is most common in thyrotoxic patients of Japanese or Chinese origin and is usually sporadic (24). It is more common in males. Proximal limb and trunk weakness may be precipitated by exercise, cold, or high carbohydrate intake. The paralysis may last from hours up to a week and may occur several times in a week (24). Serum potassium level may be low during an attack, and an attack may be precipitated by infusion of insulin and glucose (25). Normalization of thyroid status prevents the attacks, and individual attacks may respond to oral potassium administration. The cause of the paralysis is uncertain, but may relate to an increase in the number of muscle Na-K pumps and altered membrane excitability (26). As in hypokalemic periodic paralysis, vacuolar dilatation of the sarcoplasmic reticulum may be seen on electron microscopy (24,27). This condition is discussed more fully in Chapter 27.

Thyroid-Associated Ophthalmopathy (Exophthalmic Graves' Disease)

Exophthalmos and ocular myopathy may be associated with autoimmune thyroid disease. The majority of patients are biochemically hyperthyroid, but others are hypothyroid or euthyroid (28). Patients usually present with exophthalmos, which is often painful, and diplopia. The optic nerve may become compressed leading to loss of vision, and this complication appears to be related to the degree of enlargement of the extraocular muscles (29). The swelling is due to edema of the orbital contents and extraocular muscles as a result of glycoprotein accumulation and inflammatory changes (26,30). The autoimmune process appears to involve incompletely defined antibodies specific for extraocular muscles, with presumed shared epitopes on eye muscle and thyroid cells (31,32); alternatively, the extraocular muscles may be exposed to a high level of thyroid antibodies as a result of selective delivery through lymphatic channels (33).

At the earliest stages, lid retraction may respond to adrenergic blocking eye drops (guanethidine) and taping of the eyelids at night. More severe edema may respond to local injection of steroids. Any biochemical disturbance of thyroid hormone should be treated. Systemic steroids and cyclosporin may also be beneficial (34), but surgical decompression of the orbit may be necessary, especially to prevent compressive optic neuropathy and exposure keratitis (35).

Thyroid-associated ophthalmopathy must be distinguished from orbital myositis, a term that includes the entity of "orbital pseudotumor" and some examples of Tolosa-Hunt syndrome (36). Other causes of extraocular muscle enlargement include tumor infiltration (e.g., lymphoma) and local vascular malformations or venous congestion (37). Computerized tomography (CT) or magnetic resonance imaging (MRI) of the orbit usually demonstrates symmetrical and bilateral extraocular muscle enlargement in thyroid disease (Fig. 2), while in orbital myositis unilateral involvement is more typical and there may be involvement of only a single muscle, or more general asymmetrical involvement (Fig. 3); irregularity of the muscle outline with involvement of the tendon insertion on examination of the sclera is characteristic. Clinically, orbital myositis may also be distinguished by the presentation, with acute orbital pain, and responsiveness to corticosteroids (38). The erythrocyte sedimentation rate is also often raised, and there may be eosinophilia, but there is no abormality of thyroid function.

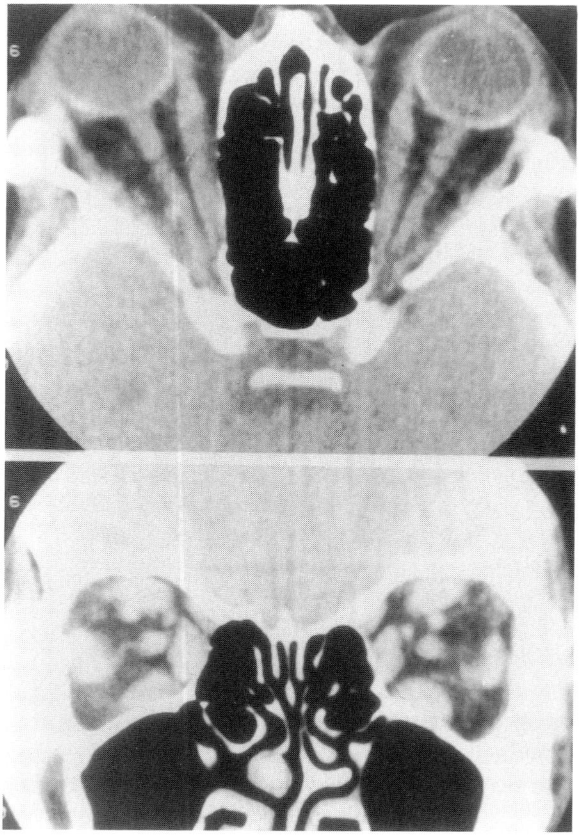

Figure 2 CT scan of the orbit showing hypertrophy of the extraocular muscle in a patient with Graves' disease.

Thyroid Disease and Myasthenia Gravis

There is an increased incidence of myasthenia gravis in patients with thyroid disease, both hypothyroidism and hyperthyroidism (39). The association is important to recognize, as the disturbance of thyroid function may exacerbate the clinical features of myasthenia gravis, and specific treatment is required for both conditions (40,41) (Fig. 4).

CORTICOSTEROIDS

Cushing's Syndrome and Steroid Myopathy

Similar myopathic features are found in patients with Cushing's disease, ectopic production of ACTH, and as a complication of corticosteroid administration. The clinical features are usually of a painless, symmetrical proximal myopathy, affecting the legs more than the arms, together with marked muscle wasting and the general systemic features of glucocorticoid excess (42) (Fig. 5, a and b).

Up to 80% of patients with Cushing's disease develop a myopathy (43). The muscle weakness and wasting resolve with correction of the glucocorticoid excess. Patients on chronic corticosteroid treatment may also develop a chronic, painless proximal myopathy (44). When monitored specifically for muscle strength, most patients develop some reduction in strength within a few weeks of starting treatment (45). It may be difficult to distinguish the myopathy attributed to corticosteroid administration

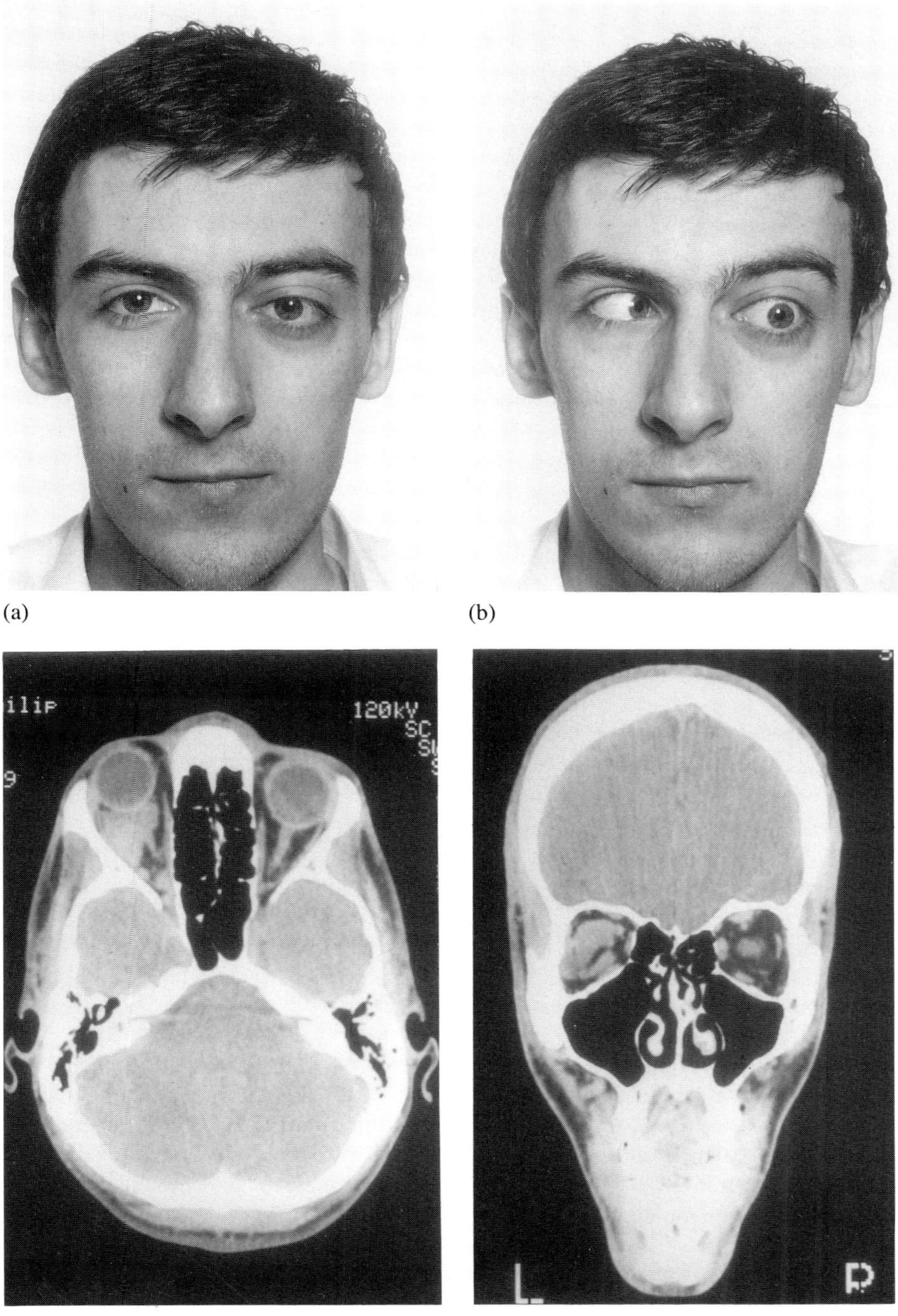

(a) (b)

(c) (d)

Figure 3 (a and b) Orbital myositis. This young man developed a painful left eye, with evident proptosis, and diplopia principally on attempted gaze to the left. (c and d) CT scan showing axial (c) and coronal (d) views of the orbits. There is proptosis of the globe on the left, with severe irregular enlargement of the extraocular muscles, particularly of the lateral rectus.

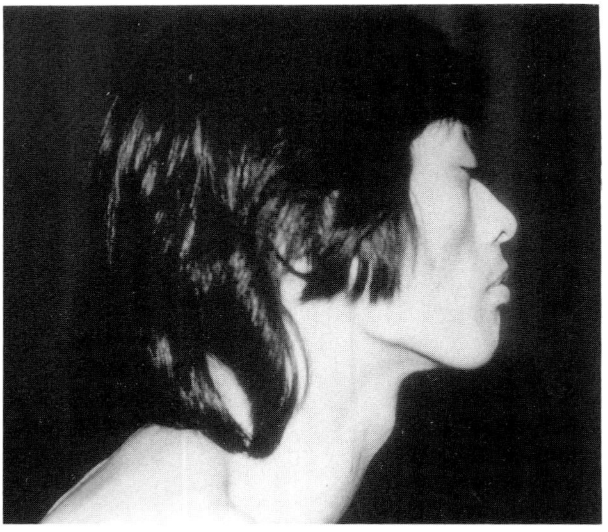

Figure 4 This young Chinese man was clinically and biochemically thyrotoxic, with a prominent, smooth goiter, but also showed striking bilateral ptosis and fatigable limb weakness, which responded to edrophonium. He was shown to have coincident myasthenia gravis. He also suffered episodes of periodic paralysis. After restoration of normal thyroid function, he responded well to thymectomy with eventual resolution of all neuromuscular symptoms.

in patients who are receiving the medication for another underlying myopathic disorder (See also Chapter 31.)

EMG may show myopathic features (42), serum CK is usually normal (46), and muscle biopsy typically shows selective type 2 fiber atrophy with increased muscle glycogen (47,48). Electron microscopy may show mitochondrial aggregates and vacuolation (49). As in many of the endocrine myopathies, these features do not seem sufficient to fully account for the degree of muscle weakness, and it may be relevant that glucocorticoids also affect muscle carbohydrate and protein metabolism. As in thyrotoxicosis, the muscle wasting may be disproportionately severe to the relative preservation of muscle strength, and it has been suggested that this may reflect potentiation of excitation-contraction coupling (26).

Laboratory investigations for the diagnosis of Cushing's disease should include the dexamethasone suppression test and measurement of urinary free cortisol (50). Dexamethasone, 1 mg, is taken at midnight and serum cortisol measured at 9 A.M., when the suppressed cortisol level should be less than 50 nmol/L in the normal subject. This test has high sensitivity but low specificity, with a false positive rate of 12.5% and false negative rate of 2% (51), and further investigation may be required beyond the initial screen. A 24-hr urine collection for urinary free cortisol should also be performed, but may be less sensitive.

Following treatment of Cushing's syndrome by hypophysectomy, recovery in muscle strength may occur, but is variable and may be slow and incomplete (52). Treatment of iatrogenic steroid myopathy is limited by the need to continue to treat the underlying disease. The corticosteroid should be given at the lowest dose possible, and alternate-day administration may help. A nonfluorinated corticosteroid should be given whenever possible, as the fluorinated corticosteroids (triamcinolone, betamethasone, and dexamethasone) are more likely to produce weakness. Starvation or protein deprivation may exacerbate a corticosteroid myopathy, and adequate nutrition should be given. Physical inactivity may also potentiate muscle wasting, and intensive physical therapy may partially prevent this (53).

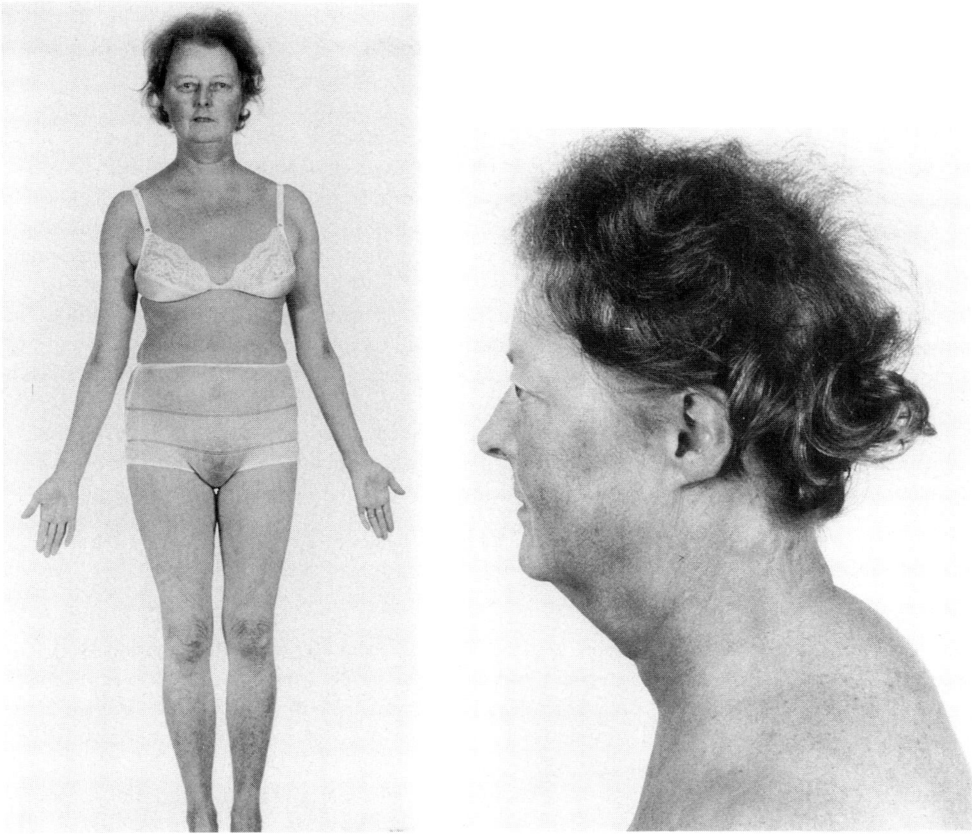

(a) (b)

Figure 5 (a) Cushing's disease. This patient presented with difficulty climbing stairs and lifting heavy objects above the head, and fatigue. She had a typical "moonface" with facial erythema, a centripetal fat distribution, but no striae. She had a marked symmetrical proximal weakness, with depressed reflexes and normal CK. (b) Marked "buffalo hump" neck appearance and facial erythema.

An acute, severe myopathy may develop within days of administration of high doses of corticosteroids, particularly intravenous steroids (54). This may be an acute necrotizing myopathy in patients with sepsis and a less severe myopathy in those with uncomplicated asthma (55). This "acute necrotizing myopathy of intensive care" may be triggered by other factors related to the underlying disease (commonly asthma) including immobility, neuromuscular blockade for mechanical ventilation, and sepsis (54). The features of critical illness polyneuropathy are usually not present. It has been suggested that the steroids have a priming effect on the muscle, and other factors such as nondepolarizing blocking agents act as a trigger for the myonecrosis. Withdrawal of the trigger may enhance recovery (54). (See also Chapter 31.)

It is probable that ACTH excess alone may produce a proximal myopathy, as demonstrated in patients with Nelson's syndrome following adrenalectomy (with corticosteroid replacement therapy), who subsequently develop a myopathy (56). The myopathy is particularly associated with those patients who develop excessive pigmentation (56).

Muscle weakness may also be a presenting symptom in primary hyperaldosteronism (Conn's syndrome) and is a common feature of the condition (57). The weakness, which is commonly episodic, may be attributable to hypokalemia, and tetany may result from alkalosis. The muscle symptoms resolve when the underlying disorder is treated.

Glucocorticoid Deficiency

Around 25–50% of patients with adrenal insufficiency have symptoms of generalized weakness, muscle cramps, and fatigue (58). Adrenal insufficiency may result from malignant or infectious destruction of the adrenal gland, autoimmune adrenal failure, adrenal hemorrhage, or ACTH deficiency (isolated or in more generalized hypopituitarism). The myopathic features are similar regardless of etiology. Addison's disease may cause additional weakness of the respiratory muscles (59) and precipitate myasthenia gravis (60).

EMG and CK are usually normal, and muscle biopsy shows no specific features (26,58). The myopathic features result from impaired muscle carbohydrate metabolism, disturbance of water and electrolyte balance, and impaired blood flow with exercise-induced hypotension (26). The muscle symptoms recover when the glucocorticoid deficiency is corrected.

A form of hyperkalemic periodic paralysis may be seen in adrenal insufficiency. Intake of potassium or exercise may precipitate a flaccid quadriplegia, which reverses when the serum potassium is lowered (61). Treatment is by glucocorticoid (and sometimes mineralocorticoid) replacement.

EPINEPHRINE

Pheochromocytoma

A focal myositis of skeletal muscle has been described in a patient with pheochromocytoma (62). The presentation was acute, with muscle tenderness and weakness in the arms and legs. It was proposed that the myositis was a result of a large discharge of catecholamine from the tumor. There was a marked rise in serum CK, and the patient died of renal failure. There was no evidence of the myocarditis that may be associated with pheochromocytoma (63).

Around 25% of patients (8–58% in different series) with pheochromocytoma complain of weakness (64,65), although this appears to be relatively nonspecific and not necessarily associated with muscle disease.

CALCITONIN

A myopathy has been associated with medullary carcinoma of the thyroid with hypercalcitoninemia (66). The clinical features of a 19-year-old girl with a medullary carcinoma of the thyroid secreting calcitonin were described. She had been "floppy" since birth, with features of proximal limb weakness. Additional features of Marfan's syndrome, mucosal neuromas on the tongue and eyelid, and skin pigmentation aggravated by sunlight were present. EMG showed myopathic features and serum CK was normal. Muscle biopsy showed histological features similar to a nemaline myopathy, with most fibers displaying a high level of oxidative enzyme activity, correlated with large numbers of mitochondria, and increased sarcoplasmic glycogen. The patient had a complex developmental disorder, with a myopathy classified as benign congenital hypotonia. It is not clear whether there was any causal relationship between the calcitonin excess and the myopathy.

GROWTH HORMONE

Acromegaly

Around 50% of patients with acromegaly (increased growth hormone) have proximal muscle weakness with pain and reduced exercise tolerance (67). The muscle weakness is slowly progressive, with minimal muscle wasting.

EMG may show myopathic features (67), and there may be additional features of nerve entrapment, including carpal tunnel syndrome. Features of muscle biopsy may include hypertrophy or atrophy of both fiber types (68), with excess deposition of lipofuscin and glycogen, loss of myofibrils, and increase in satellite cells (67).

The weakness and fatigability are out of proportion to the degree of visible muscle wasting. Possible mechanisms for the reduced power of muscle contraction include decreased sarcolemmal excitability and reduced myofibrillar ATPase activity, and the fatigability may be due to impaired carbohydrate metabolism or reduced blood flow in the muscle (26). The myopathy responds to correction of the growth hormone excess, usually by surgical removal of a pituitary adenoma, local irradiation, or bromocriptine.

Hypopituitarism

Growth hormone deficiency is usually associated with pituitary failure. In adults, causes include infarction, local tumor, head injury, and meningitis. These legions may lead to a more generalized disturbance of pituitary function, and many of the symptoms may be attributed to loss of thyroid and adrenocortical hormones. Growth hormone deficiency may contribute to the symptoms, which are of significant weakness and fatigue, out of proportion to any muscle wasting (26). The symptoms may respond to correction of thyroxine and adrenal corticosteroids, but in isolated growth hormone deficiency, specific replacement may be required (69).

In children, before puberty, the hypopituitarism is usually idiopathic or due to a craniopharyngioma. These children require growth hormone replacement to achieve normal muscle development and growth (69).

INSULIN

Insulin resistance may be associated with several endocrine disorders, including hyperthyroidism and hypothyroidism. The lack of the anabolic effect of insulin may contribute to muscle atrophy in some of these conditions. An unusual group of disorders, which may be associated with insulin resistance and sometimes myopathy, are the lipodystrophies.

Lipodystrophy

The primary feature of lipodystrophies is a reduction in, or loss of, subcutaneous fat. Acquired partial lipodystrophy affects mainly women, usually commencing in the first or second decade, characteristically with loss of facial fat, and sometimes additional loss of fat over the arms, chest, and abdomen, but with normal or increased fat deposition over the legs (70). Sometimes the lack or excess of subcutaneous fat may be misinterpreted as muscle wasting or hypertrophy. Some lipodystrophic patients have been described with clinical, electrophysiological, and histological features of a myopathy (71,72). The muscle symptoms may include aching and fatigue of the legs with marked proximal leg weakness and may be associated with insulin resistance (72). Clinical and metabolic studies suggest that partial lipodystrophy may be an incomplete variant of acquired generalized lipodystrophy.

Insulin resistance, with diabetes mellitus, may be associated with acquired generalized lipodystrophy. In one report, a 36-year-old woman with lipotrophic diabetes (or insulin-resistant diabetes and generalized lipodystrophy) had additional features of muscle weakness (73). Insulin resistance in lipodystrophy has been attributed to an insulin receptor defect (74) and also to a prereceptor abnormality (75). Abnormal muscle histology, including hypertrophy of type I and type 2 fibers, with accumulation of lipid droplets between the myofibrils has been described in acquired partial lipodystrophy (72). The pathogenesis of the myopathy in this condition is still poorly defined. Other aspects of insulin resistance are addressed in Chapter 26.

Diabetic Neuromyopathy

"Diabetic amyotrophy" is primarily a neuropathy (76), with two main clinical patterns: an abrupt onset of generalized weakness in one leg, with pathological evidence of infarction of proximal major nerve trunks; and an insidious, progressive weakness and wasting of proximal muscles, usually bilaterally, presumed to be due to a proximal intramuscular crural neuropathy attributed to metabolic effects (77).

Ischemic infarction of the thigh muscles may occur in poorly controlled diabetes mellitus (78). The acute onset of pain and tenderness and edema of the thigh is associated with a palpable mass, usually in the quadriceps or hamstring muscle groups. Serum CK is normal, and focal muscle abnormalities may be seen on CT or MRI. Muscle biopsy demonstrates widespread muscle infarction with necrosis, with evidence of arteriolar occlusion. The symptoms resolve spontaneously and treatment should be symptomatic, without surgical exploration if possible (79). The condition may recur in the same or opposite thigh. (See also Chapter 42.)

PARATHORMONE

Hyperparathyroidism

Patients with primary hyperparathyroidism, due to parathyroid adenoma, may complain of generalized weakness and muscle stiffness (80). There may be proximal muscle weakness and wasting, especially of the lower limbs, tongue fasciculation, and sometimes hyperreflexia. EMG may show myopathic features (80,81), the serum CK is usually normal (80,82), and muscle biopsy may be normal or show nonspecific features (80,83). Parathyroidectomy corrects the hormonal disturbance and relieves the symptoms (80,84). The marked muscle involvement is rarely seen now, owing to earlier biochemical diagnosis and treatment (82).

Secondary hyperparathyroidism may be found in patients with chronic renal failure. They may develop a myopathy that predominantly affects the lower limbs (85). Other metabolic abnormalities, including carnitine deficiency found in patients who develop renal failure, may contribute to the myopathy (86). The myopathy associated with metabolic bone disease and chronic renal failure is considered further in Chapter 32.

Hypoparathyroidism and Pseudohypoparathyroidism

Hypoparathyroidism is a deficiency of parathormone, usually resulting from surgical excision of the parathyroid glands or other local disease. In pseudohypoparathyroidism the parathormone level is normal or increased, but there is a defect in the cellular response to parathormone. The most common muscle symptoms and signs are those of tetany, resulting from hypocalcemia and hypomagnesemia, causing hyperexcitability of the nerve axons. This may result in diffuse muscle cramps and carpopedal spasm. Acute treatment is with intravenous calcium, and sometimes magnesium. Chronic treatment includes oral supplements of calcium, vitamin D, and magnesium if indicated.

A myopathy rarely occurs in parathyroid hormone deficiency, but has been described in both hypoparathyroidism (87) and pseudohypoparathyroidism (88). In hypoparathyroidism the muscle weakness may be mild, CK mildly elevated, and the muscle biopsy shows normal or atrophic fibers. The symptoms resolve with calcium and vitamin D supplements.

TESTOSTERONE

The effect of androgens on muscle size and function is well recognized in animals, where administration may cause increased muscle bulk, and castration or pituitary lesions lead to muscle wasting and weakness. Males generally have an increased muscle mass and strength compared to females, and the increase in circulating androgens at puberty presumably accounts in part for the increase in muscle mass.

Generalized muscle weakness may be found in men with primary and secondary hypogonadism (89), but where it is severe, a proximal, symmetrical weakness is typical. In such patients, the presenting complaints are most commonly impotence and loss of libido, although in one study 9/28 patients presented with generalized weakness, 19/28 having symptoms of muscle weakness on direct questioning (89).

Secondary testosterone deficiency may occur as a result of pituitary or testicular disease and as a result of aging (90). A myopathy may be associated with this deficiency, and in such instances testosterone replacement may produce significant functional improvement (91). A convenient method of administration is as a monthly intramuscular injection of a depot preparation allowing slow release of testosterone. Significant risks are associated with the administration of testosterone, however, including prostatic cancer, benign prostatic hypertrophy, and sleep apnea (92). Using conventional replacement levels of androgens, there appears to be little long-term gain in muscle mass in injured, undernourished, or elderly patients without testosterone deficiency (93). It is not clear whether androgen usage by athletes actually has a positive benefit on muscle function (94).

REFERENCES

1. Swanson JW, Kelly JJ, McConahey WM. Neurologic aspects of thyroid dysfunction. Mayo Clin Proc 1981; 56:504–512.
2. Kissel JT, Mendell JR. The endocrine myopathies. In: Rowland LP, DiMauro S, eds. Handbook of Clinical Neurology, vol 18 (62): Myopathies. Amsterdam: Elsevier, 1992:527–551.
3. Ramsay ID. Muscle dysfunction in hyperthyroidism. Lancet 1966; 2:931–934.
4. Feibel JH, Campa JF. Thyrotoxic neuropathy (Basedow's paraplegia). J Neurol Neurosurg Psychiatry 1976; 39:491–497.
5. McComas AJ, Sica REP, McNabb AR, Goldberg WM, Upton ARM. Evidence for reversible motoneurone dysfunction in thyrotoxicosis. J Neurol Neurosurg Psychiatry 1974; 37:548–558.
6. Serradell AP, Gonzalez JR, Torres JMC, Trull JL, Bielsa JO, Elola AU. Syndrome de sclerose laterale amyotrophique et hyperthyroidie: guerison sous antithyroidiens. Rev Neurol 1990; 146:219–220.
7. Docherty I, Harrop JS, Hine KR, Hopton MR, Matthews HL, Taylor CJ. Myoglobin concentration, creatine kinase activity, and creatine kinase B subunit concentrations in serum during thyroid disease. Clin Chem 1984; 30:42–45.
8. Bennett WR, Huston DP. Rhabdomyolysis in thyroid storm. Am J Med 1984; 77:733–735.
9. Ramsay ID. Electromyography in thyrotoxicosis. Q J Med 1965; 34:255–267.
10. Puvanendran K, Cheah JS, Naganthan N, Wong PK. Thyrotoxic myopathy: a clinical and quantitative analytic electromyographic study. J Neurol Sci 1979; 42:441–451.
11. Korenyi-Both A, Korenyi-Both I, Kayes BC. Thyrotoxic myopathy: pathomorphological observations of human material and experimentally induced thyrotoxicosis in rats. Acta Neuropathol (Berlin) 1981; 53:237–248.
12. Wiles CM, Young A, Jones DA, Edwards RHT. Muscle relaxation rate, fibre-type composition and energy turnover in hyper- and hypothyroid patients. Clin Sci 1979; 57:375–384.
13. Ianuzzo D, Patel P, Chen V, O'Brien P, Williams C. Thyroidal trophic influence on skeletal muscle myosin. Nature 1977; 270:74–76.
14. Zurcher RM, Harber FF, Grunig BE, Frey FJ. Effect of thyroid dysfunction on thigh muscle efficiency. J Clin Endocrinol Metab 1989; 69:1082–1086.
15. Pimstone N, Marine N, Pimstone B. Beta-adrenergic blockade in thyrotoxic myopathy. Lancet 1968:2:1219–1220.
16. Wilson J, Walton JN. Some muscular manifestations of hypothyroidism. J Neurol Neurosurg Psychiatry 1959; 22:320–324.
17. Wise MP, Blunt S, Lane RJM. Neurological presentations of hypothyroidism: the importance of slow relaxing reflexes. J Roy Soc Med 1995; 88:272–274.
18. Hoffman J. Weitear Beitrag zur Lehre von der Tetaine. Deutsche Z Nervenheilk 1897; 9:278.
19. Klein I, Parker M, Sherbert R, Ayyar DR, Levey GS. Hypothyroidism presenting as muscle stiffness and pseudohypertrophy: Hoffmann's syndrome. Am J Med 1981; 70:891–894.
20. Kocher T. Zur Verhutung der Creatinismus und cretinoider Zustande nach neuen Forschungen. Deutsche Z Chir 1892; 29:556.
21. Debré F, Sémélaigne G. Syndrome of diffuse muscular hypertrophy in infants causing athletic appearance: its connection with congenital myxoedema. Am J Dis Child 1935; 50:1351–1361.
22. Venables GS, Bates D, Shaw DA. Hypothyroidism with true myotonia. J Neurol Neurosurg Psychiatry 1978; 41:1013–1015.
23. Scarpalezos S, Lygidakis C, Papageorgiou C, Maliara S, Koukoulommati. Neural and muscular manifestations of hypothyroidism. Arch Neurol 1973; 29:140–144.

24. Satoyoshi E, Murakami K, Koine H, et al. Periodic paralysis in hyperthyroidism. Neurology 1963; 13:746–752.
25. McFadzean AJS, Yeung R. Periodic paralysis complicating thyrotoxicosis in Chinese. Br Med J 1967; 1:451–455.
26. Kaminski HJ, Ruff RL. Endocrine myopathies (hyper- and hypofunction of adrenal, thyroid, pituitary, and parathyroid glands and iatrogenic corticosteroid myopathy). In: Engel AG, Frazini-Armstrong C, eds. Myology, 2nd ed. New York: McGraw-Hill, 1994:1726–1753.
27. Engel AG. Electron microscopic observations in primary hypokalemic and thyrotoxic periodic paralysis. Mayo Clin Proc 1966; 41:797–808.
28. Salvi M, Zhang ZG, Halgert D, Woo M, Liberman A, Cadarso L, Wall JR. Patients with endocrine ophthalmopathy not associated with overt thyroid disease have multiple thyroid immunologic abnormalities. J Clin Endocrinol Metab 1990; 70:89–94.
29. Hallin E, Feldon SSE. Graves' ophthalmopathy. II. Correlation of clinical signs with measures derived from computed tomography. Br J Ophthalmol 1988; 72:678–682.
30. Riley FC. Orbital pathology in Graves's disease. Mayo Clin Proc 1972; 47:975.
31. Ahman A, Baker JR, Weetman AP, et al. Antibodies to porcine eye muscle in patients with Graves' ophthalmopathy: identification of serum immunoglobulins directed against unique determinants by immunoblotting and enzyme-linked immunosorbent assay. J Clin Endocrinol Metab 1987; 64:454–460.
32. Hiromatsu Y, Fukazawa H, Guinard F, et al. A thyroid cytotoxic antibody that cross-reacts with an eye muscle cell surface antigen may be the cause of thyroid associated ophthalmolopathy. J Clin Endocinol Metab 1988; 67:565–570.
33. Kriss JP. Studies on the pathogenesis of Graves' ophthalmopathy (with some related observations regarding therapy). Recent Prog Horm Res 1975; 31:533–566.
34. Prummel MF, Mourits MP, Berghout A, et al. Prednisone and cyclosporine in the treatment of severe Graves' disease. N Engl J Med 1989; 321:1353–1359.
35. Bahn RS, Gorman GA. Choice of therapy and criteria for assessing treatment outcome in thyroid associated ophthalmology. Endocrinol Metab Clin North Am 1987; 16:391–407.
36. Banker BQ. Other inflammatory myopathies. In: Engel AG, Frazini-Armstrong C, eds. Myology, 2nd ed. New York: McGraw-Hill, 1994:1461–1486.
37. Harris GJ, Murphy ML, Schmidt EW, Hanson GA, Dotson RM. Orbital myositis as a paraneoplastic syndrome. Arch Ophthalmol 1994; 112:380–386.
38. Slavin ML, Glaser JS. Idiopathic orbital myositis. Arch Ophthalmol 1982; 100:1261–1265.
39. Kiessling WR, Pflughaupt KW, Ricker K, Haubitz I, Mertens HG. Thyroid function and circulating antithyroid antibodies in myasthenia gravis. Neurology 1981; 31:771–774.
40. Engel AG. Thyroid function and myasthenia gravis. Arch Neurol 1961; 4:95–106.
41. Takamori M, Gutmann L, Crosby TW, Martin JD. Myasthenic syndromes in hypothyroidism: electrophysiological study of neuromuscular transmission and muscle contraction in two patients. Arch Neurol 1972; 26:326–335.
42. Muller R, Kugelberg E. Myopathy in Cushing's syndrome. J Neurol Neurosurg Psychiatry 1959; 22:314–319.
43. Urbanic RC, George JM. Cushing's disease—18 years experience. Medicine 1981; 60:14–24.
44. Lane RJM, Mastaglia FL. Drug-induced myopathies in man. Lancet 1978; 2:562–566.
45. Rothstein JM, Delitto A, Sinacore DR, Rose SJ. Muscle function in rheumatic disease patients treated with corticosteroids. Muscle Nerve 1983; 6:128–135.
46. Lacomis D, Chad DA, Aronin N, Smith TW. The myopathy of Cushing's syndrome. Muscle Nerve 1993; 16:880–881.
47. Rebuffe-Scrive M, Krotkiewski M, Elfverson J, Bjorntorp P. Muscle adipose tissue morphology and metabolism in Cushing's syndrome. J Clin Endocrinol Metab 1988; 67:1122–1128.
48. Harriman DGF, Reed R. The incidence of lipid droplets in human skeletal muscle in neuromuscular disorders. J Pathol 1972; 106:1–24.
49. Engel AG. Electron microscopic observations in thyrotoxic and corticosteroid myopathies. Mayo Clin Proc 1966; 41:785–796.
50. Trainer PJ, Grossman A. The diagnosis and differential diagnosis of Cushing's syndrome. Clin Endocrinol 1991; 34:317–330.
51. Cronin C, Igoe D, Duffy MJ, Cunningham SK, McKennar TJ. The overnight dexamethasone suppression test is a worthwhile screening procedure. Clin Endocrinol 1990; 33;27–33.
52. Khaleeli AA, Betteridge DJ, Edwards RH, et al. Effect of treatment of Cushing's syndrome on skeletal muscle structure and function. Clin Endcrinol 1983; 19:547–556.
53. Falduto MT, Czerwinski SM, Hickson RC. Glucocorticoid-induced muscle atrophy prevention by exercise in fast-twitch fibers. J Appl Physiol 1990; 69:1058–1062.
54. Ramsay DA, Zochodne DW, Robertson DM, et al. A syndrome of acute severe muscle necrosis in intensive care unit patients. J Neuropathol Exp Neurol 1993; 52:387–398.

55. Shee CD. Risk factors for hydrocortisone myopathy in acute severe asthma. Respir Med 1990; 84:229–233.
56. Phineas J, Hall R, Barwick DD, Watson AJ. Myopathy associated with pigmentation following adrenalectomy for Cushing's syndrome. Q J Med 1968; 37:63–77.
57. Conn JW, Knopf RF, Nesbit RM. Clinical characteristics of primary aldosteronism from analysis of 145 cases. Am J Surg 1964; 107:159–172.
58. Mor F, Green P, Wysenbeek AJ. Myopathy in Addison's disease. Ann Rheum Dis 1987; 46:81–83.
59. Mier A, Laroche C, Wass J, Green M. Respiratory muscle weakness in Addison's disease. Br Med J 1988; 297:457–458.
60. Dumas P, Archambeaud-Mouveroux F, Vallat JM, et al. Myasthenia gravis associated with adrenocortical insufficiency: report of two cases. J Neurol 1985; 232:354–356.
61. Vilchez JJ, Cabello A, Bendito J, Villarroya T. Hyperkalaemic paralysis, neuropathy, and persistent motor neuron discharges at rest in Addison's disease. J Neurol Neurosurg Psychiatry 1980; 43:818–822.
62. Bhatnagar D, Carey P, Pollard A. Focal myocarditis and elevated creatine kinase levels in a patient with phaeochromocytoma. Postgrad Med J 1986; 62:197–198.
63. Van Vilet PD, Burchell HB, Titus JC. Focal myocarditis associated with phaeochromocytoma. N Engl J Med 1966; 274:1102–1108.
64. Thomas JE, Rooke ED, Kvale WF. The neurologists experience with phaeochromocytoma. JAMA 1966; 197:754–758.
65. Ross EJ, Griffith DNW. The clinical presentations of phaeochromocytoma. Q J Med 1989; 71:485–496.
66. Cunliffe WJ, Hudgson P, Fulthorpe JJ, Black MM, Hall R, Johnston IDA, Shuster S. A calcitonin-secreting medullary thyroid carcinoma associated with mucosal neuromas, marfanoid features, myopathy and pigmentation. Am J Med 1970; 48:120–126.
67. Khaleeli AA, Levy RD, Edwards RHT, et al. The neuromuscular features of acromegaly: A clinical and pathological study. J Neurol Neurosurg Psychiatry 1984; 47:1009–1015.
68. Mastaglia FL. Pathological changes in skeletal muscle in acromegaly. Acta Neuropathol 1973; 24:273–286.
69. Raben MS. Growth hormone. 2. Clinical use of growth hormone. N Engl J Med 1962; 266:82–86.
70. Senior B, Gellis SS. The syndromes of total lipodystrophy and of partial lipodystrophy. Paediatrics 1964; 33:593–612.
71. Afifi AK, Bergman RA, Zaynoun ST, Bahuth NB, Kraydieh M. Partial (localised) lipodystrophy. J Am Acad Dermatol 1985; 12:198–203.
72. Orrell RW, Peatfield RC, Collins CE, Woodrow DF, Moss J, Press M, Lane RJM. Myopathy in acquired partial lipodystrophy. Clin Neurol Neurosurg 1995; 97:181–186.
73. Sasaki T, Ono H, Nakajima H, Sugimoto J. Lipoatrophic diabetes. J Dermatol 1992; 19:246–249.
74. Oseid S, Beck-Nielsen H, Pederson O, Sovid O. Decreased binding of insulin to its receptor in patients with congenital generalized lipodystrophy. N Engl J Med 1977; 296:245–248.
75. Golden MP, Charles MA, Arquilla ER, Myers GL, Lippe BM, Duckworth WC, Zuniga OF, Tanner SM, Palmer AM, Spell M, Ocariz J. Insulin resistance in total lipodystrophy: evidence for a pre-receptor defect in insulin action. Metabolism 1985; 34:330–335.
76. Barohn RJ, Sahenk Z, Warmolts JR, Mendell JR. The Bruns-Garland syndrome ("diabetic amyotrophy"): revisited 100 years later. Arch Neurol 1991; 48:1130–1135.
77. Chokroverty S, Reyes MG, Rubino FA, Tonaki H. The syndrome of diabetic amyotrophy Ann Neurol 1977; 2:181–194.
78. Barohn RJ, Kissel JT. Painful thigh mass in a young woman: diabetic muscle infarction. Muscle Nerve 1992; 15:850–855.
79. Banker BQ, Chester CS. Infarction of thigh muscle in the diabetic patient. Neurology 1973; 23:667–677.
80. Patten BM, Bilezidian JP, Mallette LE, et al. Neuromuscular involvement in primary hyperparathyroidism. Ann Intern Med 1974; 80:182–193.
81. Frame B, Heinze EG, Block MA, Manson GA. Myopathy in primary hyperthyroidism. Ann Inter Med 1968; 68:1022–1027.
82. Turken SA, Cafferty M, Silverberg SJ, et al. Neuromuscular involvement in mild, asymptomatic primary hyperparathyroidism. Am J Med 1989; 87:553–557.
83. Ljunghall S, Akerstrom G, Johansson G, et al. Neuromuscular involvement in primary hyperparathyroidism. J Neurol 1984; 231:263–265.
84. Delbridge LW, Marshman D, Reeve JS, et al. Neuromuscular symptoms in elderly patients with hyperparathyroidism: improvement with parathyroid surgery. Med J Aust 1988; 149:74–76.
85. Floyd M, Ayyar DR, Barwick DD, et al. Myopathy in chronic renal failure. Q J Med 1974; 43:509–524.
86. Savica V, Bellinghier G, Di Stefano C, et al. Plasma and muscle carnitine levels in haemodialysis patients with morphological-ultrastructural examination of muscle samples. Nephron 1983; 35:232.
87. Yamaguchi H, Okamoto K, Shooji M, et al. Muscle histology of hypocalcemic myopathy in hypoparathyroidism. J Neurol Neurosurg Psychiatry 1987; 50:817–818.
88. Cape CA. Phosphorylase A deficiency in pseudohypoparathyroidism. Neurology 1969; 19:167–172.

89. Chauhan AK, Katiyar BC, Misra S, Thacker AK, Singh NK. Muscle dysfunction in male hypogonadism. Acta Neurol Scand 1986; 73:466–471.
90. Vermeulen A. Androgens in the aging male. J Clin Endocrinol Metab 1991; 73:221–224.
91. Orrell RW, Woodrow DF, Barrett MC, Press M, Dick DJ, Rowe RCG, Lane RJM. Testosterone deficiency myopathy. J Roy Soc Med 1995; 88:454–456.
92. Bardin CW, Swerdloff RS, Santen RJ. Androgens: risks and benefits. J Clin Endocrinol Metab 1991; 73:4–7.
93. Wilson JD, Griffin JE. The use and misuse of androgens. Metabolism 1980; 29:1278–1295.
94. Wilson JD. Androgen abuse by athletes. Endocr Rev 1988; 9:181–199.

31
Toxic and Drug-Induced Myopathies

Russell J. M. Lane
Charing Cross Hospital and Charing Cross and Westminster Medical School, London, England

INTRODUCTION

Little attention had been paid to the problem of neuromuscular disorders induced by drugs and other agents prior to the original review and classification proposed in 1978 (1). Since then, dozens of drugs and chemicals have been implicated and extensive reviews on the subject have been published (for example, Refs. 2–4). Understanding of the basis of myotoxicity has increased but remains incomplete, and classification on clinical and pathological grounds is still the most useful approach for practical purposes (Table 1). For example, the clinical effects of myotoxicity of a certain drug may encompass myalgia in isolation, myalgia with raised serum creatine kinase (CK) levels, and a necrotizing myopathy with profound weakness. Yet the same drug may also cause a slowly progressive, painless weakness under different circumstances.

This chapter will be confined to the most common and clinically important drug and toxin-induced myopathies and recent developments concerning pathogenesis. Earlier detailed reviews (2–4) should be consulted for original sources where appropriate. Drug-induced myasthenia will not be considered, and malignant hyperthermia is discussed in depth in Chapter 36. The many manifestations of myotoxicity due to alcohol, the most used and abused of all drugs, deserve special consideration and are discussed at the end of the chapter.

DRUG-INDUCED MYOPATHIES

Diagnosis of Drug-Induced Myopathies

The characteristics of a drug or toxin-induced neuromuscular disorder include: chronological relationship between exposure and development of symptoms; lack of personal or family history of previous neuromuscular disease; predisposing features, such as renal impairment, which might increase plasma drug levels; and particularly, resolution of symptoms and signs on removal of the suspect agent. In addition, drugs and toxins may also induce or exacerbate underlying neuromuscular disorders, particularly myasthenia and myotonias.

Table 2 summarizes the pathological features of toxic and drug-induced myopathies and the agents most frequently implicated.

Focal Myopathy

Acute Form

Intramuscular (IM) injections or insertion of an electromyography needle may cause focal muscle fiber damage, with a rise in serum CK between 4 and 72 hr later, and may result in significant histological

Table 1 Classification of Drug-Induced Myopathies

I. Focal myopathy
   A. Acute
   B. Chronic
II. Acute and subacute painful myopathy
   A. Myalgia
   B. Myalgia with weakness and raised CK
   C. Necrotizing myopathy
   D. Vacuolar myopathy
   E. Inflammatory myopathy
   F. Mitochondrial myopathy
III. Drug-induced rhabdomyolysis
IV. Chronic, painless proximal myopathy
   A. Type 2 atrophy
   B. Vacuolar myopathy
V. Drug-induced myotonia

Source: After Lane and Mastaglia (1).

abnormalities ("*needle myopathy*"). This should be borne in mind when assessing a raised CK level in a hospitalized patient (2). These mechanical effects can be amplified by the specific chemical effects of the injected drugs and by secondary problems, including hematoma formation and infection. Such problems were reported in 0.4% of hospitalized patients receiving at least one IM injection, with *cephalothin*, *tetracycline*, and *paraldehyde* being particularly troublesome (5). The volume and concentration of drug and the nature of incipient also play a part, with small volume, dilute and oil-based preparations being less damaging (2).

Table 2 Pathological Characteristics of Drug and Toxic Myopathies

Type	Drug[a]	Pathology
Focal myopathy		
Acute	IM injections	"Needle myopathy"
Chronic	IM antibiotics	Fat, fibrous tissue, loss of muscle cells
Acute and subacute painful myopathy		
Necrotizing	Alcohol, opiates, lipid-lowering drugs, ipecac alkaloids, IV steroids, venoms, vit. E, organophosphates	Multifocal necrosis, phagocytosis, regeneration
Vacuolar	Drugs, toxins causing hypokalemia	Cytoplasmic vacuoles
Inflammatory	D-penicllamine	Lymphocytic infiltration, necrosis
Mitochondrial	AZT, germanium	Ragged red fibers, abnormal mitochondria on EM
Chronic painless proximal myopathy		
Type 2 atrophy	Alcohol, steroids	Type 2 atrophy
Vacuolar	Chloroquine, amiodarone, perhexilene, colchicine, vincristine	Lysosomes with myeloid bodies

[a]The drugs cited are commonly implicated, but many others can cause pathology of this type—see text.

Chronic Form

Repeated intramuscular injections may result in local induration and replacement of muscle and subcutaneous fat by granulation tissue, leading to fibrosis and contracture. This can lead to fixed limb deformities requiring surgical correction. The deltoid and quadriceps are affected most commonly, although multiple muscles may be involved. This type of reaction has been associated most commonly with *narcotic* abuse and with intramuscular administration of *antibiotics* in children. Peripheral nerve entrapment may occasionally complicate the picture (2).

Acute and Subacute Painful Myopathy

Myalgia

Myalgia, with or without cramp, is a common drug- related symptom and is rarely investigated unless accompanied by weakness and a significant increase in CK levels. Diverse, unrelated agents including *anticholinesterases, carbimazole, diuretics, calcium antagonists, β-agonists, danazol, enalapril, to*pi*cal minoxidil*, and *dipyridamole* have been implicated (2–4,6). *Suxamethonium*, a depolarizing muscle relaxant, causes myalgia in about 50% of patients the day following surgery and is associated with an increase in CK levels. The pathogenesis may include fiber damage due to Ca^{2+}-stimulated phospholipid hydrolysis, with generation of prostanoids, and membrane damage by free radical–mediated peroxidation (3). Aspirin 600 mg preoperatively appears to halve the incidence (7).

Myalgia with Weakness and Increased CK

Some drugs induce an acute or subacute syndrome comprising myalgia, together with tenderness and weakness, particularly affecting the proximal limb girdle and axial musculature, associated with a marked rise in CK. This typically develops within days to a few weeks of starting the drug and resolves rapidly on withdrawal. If the responsible drug is not stopped, rhabdomyolysis may ensue, with potentially life-threatening consequences. Electromyographic changes of myopathy together with evidence of muscle fiber "irritability" (increased insertional activity, spontaneous potentials, and pseudomyotonia) would be typical. The responsible agents can be grouped on the basis of the pathological changes produced, and whether these result from a direct toxic action of the drug or are secondary to hypokalemia, to inflammatory reaction, or to effects on mitochondrial structure and function.

Necrotizing Myopathy

This syndrome occurs most frequently in association with *alcohol* and *opiate* abuse, and also with the fibrinolytic inhibitor *ε-aminocaproic acid*; *ipecac alkaloids* such as *emetine* and *ipecacuanha;* β-receptor blockers (*labetolol, sotalol*); and the vitamin A analogs *etretinate* and *isotretinoin*. Histologically, there is striking fiber necrosis, phagocytosis, and marked regenerative activity (Fig. 1). Such cases are rare and have been discussed previously (2–4).

More recently, however, this syndrome has been reported rather commonly with new cholesterol- and lipid-lowering drugs, the 3-hydroxy-3-methyl-glutaryl-coenzyme A (HMG-CoA) inhibitors and fibric acid derivatives.

The first drug of this type to be implicated was a fibric acid derivative, *clofibrate*, the ethyl ester of *p*-chlorophenoxyisobutyric acid (CPIB). CPIB, the active derivative, acts on plasma proteins to lower cholesterol and triglycerides. In approximately 80% of previously reported cases, myalgia developed within 3 weeks, and weakness, often severe, occurred in over half (8). Some cases were asymptomatic but CK levels were raised. Clofibrate may also induce myotonia (see below). The drug is partly bound by albumin, and a low plasma protein level, as in cirrhosis and nephrotic syndrome, increases the risk of myotoxicity. Renal insufficiency prolongs the half-life of the drug, and most cases of clofibrate myopathy have been reported in association with uremia (9).

Myopathy has been reported only rarely with other fibric acid derivatives, such as *bezafibrate*,

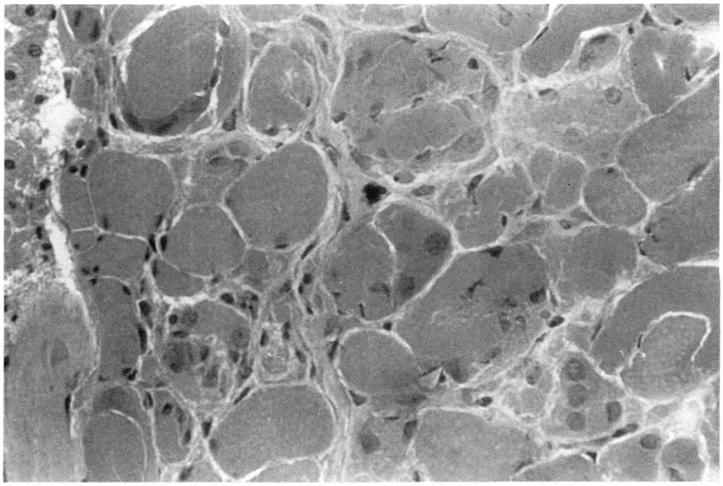

Figure 1 Necrotizing myopathy due to ε-aminocaproic acid myopathy. Multifocal necrosis with infiltration of fibers by macrophages and active regeneration. (H&E, ×250.)

fenofibrate, and *gemfibrozil*, and appears to occur largely when such drugs are used in combination with HMG-CoA reductase inhibitors, such as *lovastatin*.

Lovastatin and analogs such as *pravastatin*, *memvastatin*, and *simvastatin* inhibit HMG-CoA reductase, a microsomal enzyme that converts HMG-CoA to mevalonic acid during cholesterol synthesis, thus reducing cholesterol synthesis and lowering cholesterol levels. Myalgia with mildly increased CK levels occurs in about 0.5% of patients taking lovastatin alone, but the incidence is 10 times higher when lovastatin is combined with gemfibrozil and much higher still with cyclosporin, for example in transplant patients; necrotizing myopathy may occur in up to 30% of such cases (10). The risk is also greater in patients taking erythromycin or nicotinic acid, and with liver disease. Experimentally, lovastatin alone showed no myotoxicity, but *cyclosporin* appeared to impair its biliary excretion, resulting in much greater plasma and tissue levels of the drug (11). It is not known how the drug or its metabolites produce the muscle damage, although it might reflect drug-related effects on the composition of muscle membranes (12).

Steroids usually produce a chronic, painless proximal myopathy (see below), but parenteral administration of large doses of hydrocortisone to experimental animals will produce a necrotizing myopathy, and several cases of acute, painful necrotizing myopathy have been reported in patients with asthma treated with several grams of this drug (2–4). In addition, a syndrome resulting from high-dose steroid treatment in the presence of functional denervation due to neuromuscular blocking agents, such as pancuronium and vecuronium (13), or in association with an underlying functionally "denervating" disorder such as myasthenia (14), has recently been recognized. This comprises moderate or severe flaccid tetraparesis, occasionally with extraocular involvement, with raised CK levels and unique muscle biopsy findings of selective loss of myosin filaments (Fig. 2, a and b). Such patients are typically being treated for status asthmaticus in intensive care, and the clinical syndrome may be misdiagnosed as Guillain-Barré syndrome or critical illness neuropathy or myopathy. Withdrawal of steroids usually results in gradual improvement (Fig. 2c). The specific histopathological lesions have been reproduced experimentally in rats treated by surgical denervation plus dexamethasone (15), and it has been postulated that functional denervation might predispose to myosin degradation by up-regulation of steroid receptors (16).

Vacuolar Myopathy Associated with Hypokalemia

Vacuolar change, with relatively little necrosis, may underlie the syndrome of acute and subacute painful proximal myopathy, although in some instances, pain may be inconspicuous despite rhabdomyolysis. Severe hypokalemia (serum $K^+ < 2$ mM), induced for example by *diuretics*, *purgatives*, *amphotericin* B, or *licorice* and derivatives of glycyrrhizinic acid such as *carbenoxolone*, can induce a remitting or persistent flaccid weakness with areflexia, reminiscent of hypokalemic periodic paralysis, but often with raised CK. In rare cases, rapid progression to fiber necrosis and rhabdomyolysis can occur. Vacuoles, if present, are not membrane-bound, however, unlike those of periodic paralysis (17). The syndrome may be superimposed on a more chronic painless weakness, in which case type 2 fiber atrophy may be seen in addition.

Vacuolar change may also be seen in drug-induced, chronic painless myopathy, but the nature of the vacuoles differs (see below).

Inflammatory Myopathy

Inflammatory change is an uncommon basis for drug-induced myopathies. It may be seen as a nonspecific feature in general allergic reactions to drugs such as penicillin and phenytoin, and isolated examples have been reported with drugs such as *levodopa*, *cromoglycate*, *propylthiouracil*, and *cimetidine*, and following *vaccinations* (2–4). However, a number of instances of myositis have followed D-*penicillamine* treatment of systemic sclerosis, cystinuria, Wilson's disease, and particularly rheumatoid disease (18). Cases of dermatomyositis due to this drug have also been reported (2–4). *Procainamide*, too, has been linked to the development of an inflammatory myopathy as part of a drug-induced lupus syndrome, and this drug at high concentration may also cause an acute necrotizing myopathy (2,3).

Mitochondrial Myopathy

Zidovudine (azidothymidine, AZT), used in the treatment of HIV infection, has been shown to cause mitochondrial damage in muscle fibers, with a decrease in mitochondrial DNA and respiratory chain enzyme activities in treated cases, which was reversed when the drug was stopped (19). This probably results from the inhibitory action of the drug on mitochondrial DNA polymerase. Patients on AZT treatment may develop a syndrome akin to a subacute painful proximal myopathy, with degenerative changes in muscle fibers but little or no inflammatory reaction, and progressively increasing CK levels, with rapid resolution on withdrawing the drug. However, such symptoms usually develop only after taking about 1 g AZT daily for approximately 1 year (20), and the syndrome has become much less common since maintenance doses were halved.

In cases of AZT myopathy muscle biopsies may show "ragged red fibers," with abnormal accumulations of subsarcolemmal mitochondria on ultrastructural examination (21). However, it has been argued that such changes occur only in fibers undergoing degeneration (and not in structurally normal fibers as in naturally occurring mitochondrial myopathies), and such changes can also be seen in cases of HIV-related myopathy in patients not taking AZT (22). This matter is discussed further in Chapter 51.

Germanium myopathy, discussed below, is also characterized by accumulations of abnormal mitochondria and reduced cytochrome c oxidase activity, suggesting that mitochondrial dysfunction may play a part in pathogenesis (23).

Drug-Induced Rhabdomyolysis

Rhabdomyolysis is the most serious manifestation of drug-induced myopathy. While in most instances it merely reflects an exaggeration of the pathological processes underlying acute and subacute painful proximal myopathies, and commonly involves the same drugs, the distinctive clinical picture and the resulting serious metabolic disturbances merit particular consideration. These aspects are discussed in detail in the section on myoglobinuria (see Chapter 49).

Drug-induced rhabdomyolysis in clinical practice is encountered most frequently in the context of drug abuse, particularly with *alcohol* (see below), and with *opiates* and stimulants such as *amphetamines*, *phencyclidine*, and *ketamine* (2–4).

Initial reports of severe myopathy with rhabdomyolysis associated with *heroin* abuse were

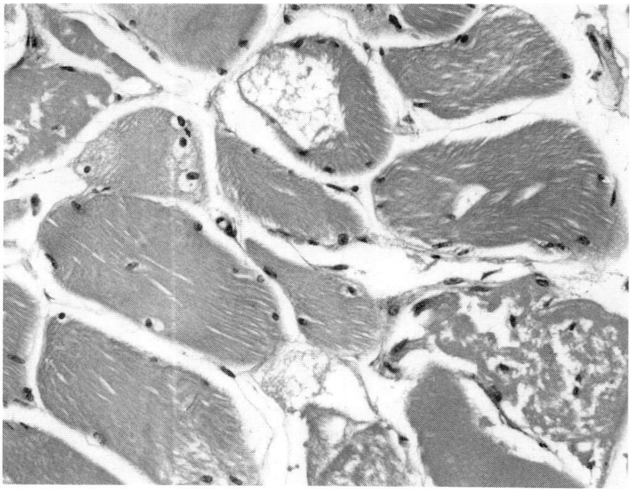

(a)

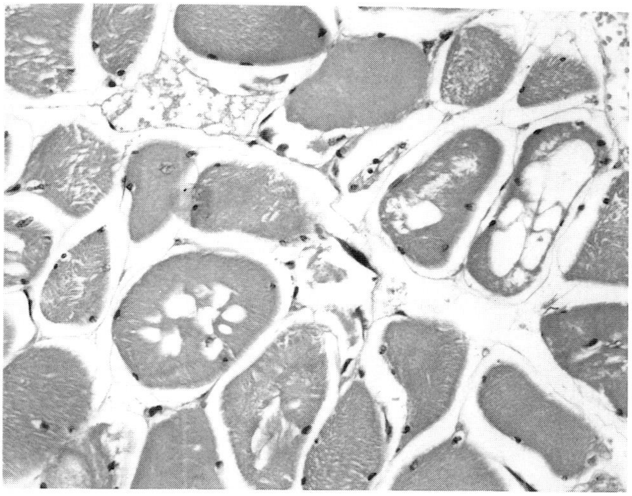

(b)

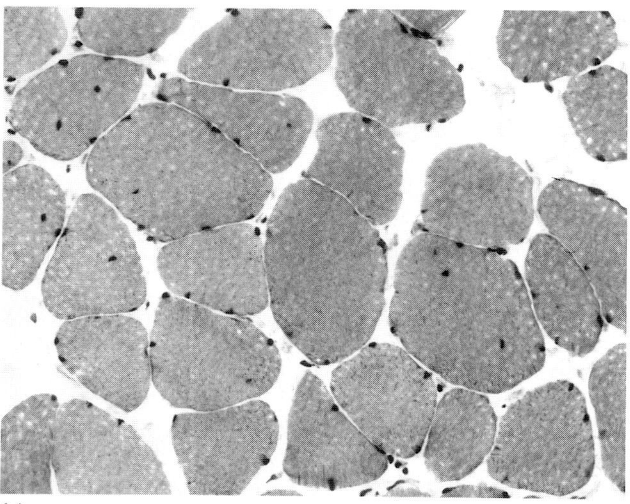

(c)

followed by others implicating *cocaine*. However, it was unclear whether the effect was mediated directly by the drug or was the result of an adulterant or factors such as trauma, coma, or infection, which are independent risk factors for myoglobinuria. It was subsequently reported that intravenous use of relatively pure heroin and self-poisoning with prescribed *dihydrocodeine* and *morphine* could precipitate rhabdomyolysis, and myopathy has also followed administration of opiates during anesthesia and following ingestion of *codeine linctus*, indicating that these drugs have direct myotoxic effects (2–4).

Phencyclidine and the related drug ketamine were widely abused drugs a decade ago, and rhabdomyolysis was reported in a number of cases as a result of extreme motor excitation. Experimentally, phencyclidine administered to restrained rats produced a myopathy characterized by extensive disruption of sarcomeres and marked increase in CK, which could be prevented by denervation or by prior administration of dantrolene (2–4).

Overdoses of other drugs such as the methylxanthine inhibitors *aminophylline* and *theophylline* and pentamidine have also occasionally induced rhabdomyolysis (2–4), but direct myotoxic actions cannot necessarily be inferred since pressure effects on muscle in comatose patients may also induce myoglobinuria.

The causes of rhabdomyolysis and myoglobinuria, including drugs, are discussed in Chapter 48.

Chronic, Painless Proximal Myopathy

This is the most common form of drug-induced myopathy. The clinical features are typically of a slowly progressive, symmetrical proximal weakness, with variable wasting, normal or depressed tendon reflexes, and normal muscle enzymes. The pathological changes on biopsy are usually of either type 2 fiber atrophy or a vacuolar myopathy.

Type 2 Atrophy

This is most frequently seen in chronic alcoholic myopathy (see below) and with use of steroids, but may also be a feature of persistent drug-induced hypokalemia, as mentioned above.

Myopathy may occur with any type of steroid, but fluorinated steroids (*triamcinolone, betamethasone*, and *dexamethasone*) are most potent in this regard. There is no clear association between dose or duration of therapy and development of myopathy, but it is unlikely to occur with daily doses equivalent to 10 mg *prednisolone* or less (2–4). The mechanism of steroid myopathy is not fully understood, but appears to relate principally to impairment of protein and carbohydrate metabolism. Experimentally, protein synthesis is inhibited through impaired RNA synthesis and reduced ribosomal function and peptide chain synthesis, particularly in type 2 fibers. At higher doses, myofibrillary degradation is increased through induction of lysosomal proteases, again largely in type 2 fibers (2–4). Steroids also profoundly inhibit myophosphorylase activity, increasing the relatively greater susceptibility of type 2 fibers to the degradative process. A degree of weakness and muscle wasting is likely to occur in any patient on steroid treatment, but it has been found that isokinetic exercise can combat these effects to some extent (24).

Vacuolar Myopathy

Chronic painless proximal myopathy can be caused by a number of drugs that induce autophagic vacuoles containing membranous bodies—a drug-induced lysosomal storage myopathy. In such cases, an associated drug-induced peripheral neuropathy is almost always present, and this may play an

Figure 2 (a and b) Acute steroid myopathy during neuromuscular paralysis. A 40-year-old woman received hydrocortisone 200 mg IV qid while being ventilated for severe asthma. Two weeks later, she remained tetraplegic and her CK was 26,000 U/L, but K^+ was normal. Biopsy showed a severe vacuolar myopathy with little inflammation or necrosis. She recovered over 6 months following withdrawal of steroids. (c) Repeat biopsy 1 year later. No vacuolar changes were observed and the biopsy was normal except for an increase in internal nuclei (11%).

important part in both the clinical picture and the pathological changes. This is most frequently seen with amphiphilic drugs, such as *chloroquine* and its derivatives, used as antimalarials and immunosuppressants in a variety of conditions (2–4). Such drugs, which have positively charged groups, interact with anionic residues of phospholipid membranes to produce curvilinear and concentric laminar structures (myeloid bodies) that are resistant to lysosomal degradation. This results in a typical histological picture of cytoplasmic vacuolation, which may be associated with various degrees of fiber necrosis, induced by markedly increased lysosomal enzyme activity. A high proportion of fibers demonstrate increased acid phosphatase activity, unlike cases of "rimmed vacuolar myopathy" (see Chapter 42), and in contrast to steroid and hypokalemic myopathy, the changes affect principally type 1 fibers. Myopathy is now much less commonly seen with lower maintenance doses of these drugs, and hydroxychloroquine appears to be less myotoxic than chloroquine.

Neuromyopathy with similar muscle biopsy changes has also been reported with *amiodarone* and *perhexilene*, although the associated neuropathy tends to dominate the symptomatology and may occur in isolation. Similarly, *colchicine* and *vincristine* produce a vacuolar myopathy with membranous inclusions, but here the effects of the drugs appear to relate to disruption of the cytoskeletal architecture with secondary effects on lysosomal function (25). With colchicine, a more acute syndrome may arise if plasma drug levels increase rapidly due to renal failure, and CK levels may increase significantly, despite lack of necrotic changes on biopsy (16). Concomitant therapy with *cyclosporin* may also increase the risk of myotoxicity, and this drug may independently also cause a mild myopathy (the role of cyclosporin in the production of necrotizing myopathy was discussed above). Finally, organic *germanium* compounds used as antitumor agents have produced a vacuolar myopathy with increased lysosomal activity, although this may be related to a primary inhibitory effect on mitochondrial function (23) (see above).

Drug-Induced Myotonia

Drug-induced myotonia in humans occurs most commonly when a drug unmasks or exacerbates the symptom in a patient with an undiagnosed myotonic disease, such as myotonic dystrophy or congenital myotonia. This has been reported with *β-blockers*, *barbiturates*, *diuretics*, and *acetazolamide*. Electromyographic myotonia and pseudomyotonia are commonly found in patients with drug-induced acute necrotizing myopathies (see above) and with myopathies due to lipid-soluble drugs that affect membrane structure, such as chloroquine and clofibrate (2–4).

Alcohol and Opiate-Related Myopathies

The principle of variability in the myotoxic manifestations of individual drugs is well illustrated by alcohol and opiates, where effects may range from asymptomatic elevation of CK levels following alcohol or opiate usage to fatal rhabdomyolysis.

Alcoholic Necrotizing Myopathy

The prevalence of alcoholic myopathy is dependent on the definition used. The most dramatic presentation of this condition is with an acute, painful myopathy, often preceded by cramps, with myoglobinuria and a very marked increase in CK level, usually following a drinking binge. Muscle swelling is often present and may be focal, for example in one leg or a calf, producing a clinical picture suggestive of deep vein thrombosis. In most patients, the condition is relatively mild, with recovery in 2–3 weeks, and a history of previous milder episodes is common in such cases. However, renal failure and other complications of myoglobinuria may follow more severe attacks (see Chapter 48) and urgent faciotomy may be required. Muscle biopsy findings in such cases can range from generalized fiber swelling, to patchy segmental necrosis and hyalinization of individual fibers, to massive necrosis, with active regenerative features. Ultrastructurally, accumulations of abnormal mitochondria and tubular aggregates may be seen (2–4).

Alcoholic Hypokalemic Myopathy

Another acute or subacute myopathy occurring in alcoholics results from severe hypokalemia, presumably resulting from recurrent vomiting and diarrhea. The clinical and pathological manifestations are similar to those described above in other forms of acute hypokalemic myopathy, with severe areflexic proximal weakness, marked increase in CK levels, and a vacuolar fiber necrosis, except that muscle pain, cramp, and swelling are inconspicuous (26).

Subclinical Alcoholic Myopathy

Such symptomatic muscle disease probably occurs in less than 1% of all alcoholics, but evidence of significant muscle disease may emerge in perhaps 10% of patients admitted with acute intoxication after drinking sprees. Such patients are specifically sensitive to rechallenge with alcohol, which results in a significant increase in CK levels 2–6 hr later; no such increase is observed in normal controls. If such elevation of CK is taken as indicative of underlying alcoholic myopathy, some 80% of alcoholics, unselected for muscle symptoms, may be affected (2–4).

Chronic Alcoholic Myopathy

Chronic alcoholic myopathy has the characteristics of a chronic painless proximal myopathy, with insidious wasting and weakness that can be reversed by abstinence from alcohol. It may be the most common of all neuromuscular disorders, since it has been estimated that approximately 10% of European and North American adults abuse alcohol, and approximately two-thirds of such chronic alcohol misusers have the histological features of chronic alcoholic myopathy: selective type 2 fiber atrophy (27).

However, the status of chronic alcoholic myopathy has been questioned, since it often arises in patients with alcoholic peripheral neuropathy, and type 2 atrophy and other pathological changes frequently encountered, such as scattered fiber necrosis, increased central nuclei, and tubular aggregates, are nonspecific and may occur in patients with peripheral neuropathy, which is frequently subclinical in alcoholics. Arguments for a primarily myopathic origin of the condition include data from quantitative electromyographic studies, the analogy with alcoholic cardiomyopathy, where denervation atrophy is not a factor, and studies on animal models (27).

The mechanisms involved in alcohol-related muscle damage are poorly understood, and several factors may be involved. Large doses of alcohol in normal subjects (28) and naïve experimental animals (29) cause ultrastructural damage to mitochondria and membrane systems and impaired ATPase activity; alcohol is intrinsically myotoxic. However, symptomatic alcoholic myopathy may result from the additional burden of malnutrition on a background of chronic alcohol abuse (29). Alcohol metabolism results in the generation of free radicals, and a diet deficient in antioxidants (such as selenium and vitamin E), common in alcoholics, could result in free radical–mediated damage (27). The main site of damage in such circumstances is probably on the integrity of cell membranes, particularly on ion homeostasis, and on inhibition of glycolytic function, resulting in critical impairment of energy metabolism, which could lead to fiber necrosis during stress (30). Type 2 fibers have significantly lower antioxidant capacities than type 1 fibers, explaining their vulnerability. In addition, alcoholics have a generalized abnormality of protein metabolism, with negative nitrogen balance and loss of amino acids that would normally be destined for incorporation into muscle (27).

TOXIC MYOPATHIES

Myopathies caused by toxins other than drugs can be similarly classified on clinical and pathological grounds.

Acute Necrotizing Myopathies

Snakebite is a common problem in tropical climates and *snake venoms* are potent myotoxins. Several components of venom appear to be myotoxic, including crotamine, myotoxin a, and the A_2 phospholipases, which have a strong predeliction for skeletal muscle (31). Severe myonecrosis with myoglobinuria may result.

Acute poisoning with *organophosphorus insecticides* results in irreversible inhibition of cholinesterases at neuromuscular junctions and severe weakness, not only through neurmuscular blockade, but also as a result of necrotizing myopathy (32).

The current trend to self-medicate with large doses of *vitamin E*, as an antioxidant, has resulted in a number of cases of necrotizing myopathy, characterized by ragged red fibers on Gomori trichrome staining. Subsarcolemmal accumulations comprise a variety of components, including lysosomes (33).

Acute and Chronic Hypokalemic Myopathy

Volatile substance abuse is common in inner-city areas, and both acute and chronic myopathies have been reported in *toluene* abusers. The myotoxic action appears to be mediated by hypokalemia, resulting from renal tubular damage (34), and the effects are usually reversible within days of withdrawal.

Inflammatory Myopathy

Although now of largely historical interest, outbreaks of *eosinophilia-myalgia syndrome* induced by contaminated *L-tryptophan*, chiefly in the United States in 1974 and 1989, and the *toxic oil syndrome*, in Spain in 1981, caused severe and persistent disabilities, including myopathy, and shared a number of characteristics (3,6).

Muscle involvement in eosinophilia-myalgia syndrome included incapacitating myalgia and perivascular and perimysial inflammatory infiltration without myonecrosis, tending to spare the endomysium, and with normal CK levels. Intramuscular nerves and muscle spindles tended to be particularly affected, and the neuromuscular symptoms and signs were probably related more to nerve than muscle involvement (35). The condition appears to have been due to contamination of L-tryptophan by analine and anthranilic acid compounds used as reactants for L-tryptophan synthesis by a single company. Similar compounds may have been adulterants of rapeseed oil in the toxic oil syndrome, which was also associated with marked myalgia and eosinophilia (6).

Finally, polymyositis has been linked to poisoning by *ciguatera* (36), a low-molecular-weight saponin derived from a coral reef dinoflagellate, *Gambierdiscus toxicus*, which enters the human food chain through seafood.

REFERENCES

1. Lane RJM, Mastaglia FL. Drug-induced myopathies in man. Lancet 1978; 2:562–565.
2. Lane RJM. Muscle in systemic and iatrogenic disease. In: Kennard C, ed. Recent Advances in Clinical Neurology, No. 5. Edinburgh: Churchill Livingstone, 1988:201–238.
3. Blain PG, Lane RJM. Neurological disorders. In: Davies DM, ed. Textbook of Adverse Drug Reactions. Oxford, England: Oxford Medical Publications, 1991:535–566.
4. Mastaglia FL. Toxic myopathies. In: Rowland LP, DiMauro S, eds. Handbook of Clinical Neurology, Vol 18. Amsterdam: Elsevier, 1992:595–622.
5. Greenblatt DJ, Allen MD. Intramuscular injection-site complications. JAMA 1978; 240:542–544.
6. Shetty HGM, Routledge PA, Davies DM. Disorders of muscle, bone and connective tissue. In: Davies DM, ed. Textbook of Adverse Drug Reactions. Oxford, England: Oxford Medical Publications, 1991:491–513.
7. McGloughlin C, Nesbitt GA, Howe JP. Suxamethonium-induced myalgia and the effect of pre-operative administration of oral aspirin. Anaesthesia 1988; 43:565–567.
8. Rimon D, Ludatsher R, Cohen L. Clofibrate-induced muscular syndrome. Isr J Med Sci 1984; 20:1082–1086.
9. Rush P, Baron M, Kaputsa M. Clofibrate myopathy: a case report and review of the literature. Semin Arthritis Rheum 1986; 15:226–229.
10. Corpier CL, Jones PH, Suki WN, Lederer D, Quinones MA, Schmidt SW, Young JB. Rhabdomyolysis and renal injury with lovastatin use: report of two cases in cardiac transplant recipients. JAMA 1988; 260:239–241.
11. Smith PF, Eydelloth RS, Grossman SJ, Stubbs RJ, Schwartz MS, Germershausen JI, Vyas KP, Kari PH, Macdonald JS. HMG-CoA reductase inhibitor-induced myopathy in the rat: cyclosporine A interaction and mechanism studies. J Pharmacol Exp Ther 1991; 257:1225–1235.
12. Nakahara K, Kuriyama M, Yoshidome H, Nagata K, Nagado T, Nakagawa M, Arimura K, Higicui I, Osame M. Experimental simvastatin-induced myopathy in rabbits. J Neurol Sci 1992; 113:114–117.
13. Hirano M, Ott BR, Raps EC, Minetti C, Lennihan L, Libbey NP, Bonilla E, Hays AP. Acute quadriplegic

myopathy; a complication of treatment with steroids, nondepolarising blocking agents, or both. Neurology 1992; 42:2082–2087.
14. Panegyres PK, Squier M, Mills KR, Newsom-Davis J. Acute myopathy associated with large parenteral doses of corticosteroids in myasthenia gravis. J Neurol, Neurosurg Psychiatry 1993; 56:702–704.
15. Massa R, Carpenter S, Holland P, Karpati G. Loss and renewal of thick myofilaments in glucocorticoid-treated rat soleus after denervation and reinnervation. Muscle Nerve 1992; 15:1290–1298.
16. Kuncl RW, Duncan G, Watson D, Alderson K, Rogawski MA, Peper M. Colchicine myopathy and neuropathy. N Engl J Med 1987; 316:1562–1568.
17. Bjorn Jensen O, Mosdal C, Reske-Nielsen E. Hypokalaemic myopathy during treatment with diuretics. Acta Neurol Scand 1977; 55/56:455–482.
18. Takahashi K, Ogita T, Okudaira H, Yoshinoya S, Yoshizawa H, Miyamoto T. D-penicillamine-induced polymyositis in patients with rheumatoid arthritis. Arthritis Rheum 1986; 29:560–564.
19. Arnaudo E, Dalakas M, Shanske S, Moraes CT, DiMauro S, Schon EA. Depletion of muscle mitochondrial DNA in AIDS patients with zidovudine-induced myopathy. Lancet 1991; 337:508–510.
20. Mhiri C, Baudimont M, Bonne G, Geny C, Degoul F, Marsac C, Roullet E, Gherardi R. Zidovudine myopathy: a distinct disorder associated with mitochondrial dysfunction. Ann Neurol 1991; 29:606–614.
21. Dalakas MC, Illa I, Pezeshkpour GH, Lautakis JP, Cohen B, Griffin JL. Mitochondrial myopathy caused by long-term zidovudine therapy. N Engl J Med 1990; 322:1098–1105.
22. Lane RJM, McLean KA, Moss J, Woodrow DF. Myopathy in HIV infection; the role of zidovudine and the significance of tubuloreticular inclusions. Neuropathol Appl Neurobiol 1993; 19:406–413.
23. Higuchi I, Izumo S, Kuriyama M, Ohtsubo S, Miyata K. Germanium myopathy: clinical and experimental pathological studies. Acta Neuropathol (Berlin) 1989; 79:300–304.
24. Horber FF, Scheidegger JR, Grunig BE, Frey FJ. Evidence that prednisone-induced myopathy is reversed by physical training. J Clin Endocrinol Metab 1985; 61:83–88.
25. Kuncl RW, Wiggins WW. Toxic myopathies. Neurol Clin 1988; 6:593–619.
26. Martin JB, Craig JW, Eckel RE, Munger J. Hypokalaemic myopathy in chronic alcoholism. Neurology 1971; 21:1160–1168.
27. Preedy VR, Peters TJ. Alcohol and muscle disease. J Roy Soc Med 1994; 87:188–189 (editorial).
28. Song SK, Rubin E, Ethanol produces muscle damage in human volunteers. Science 1972; 175:327–328.
29. Haller RG, Drachman DB. Alcoholic rhabdomyolysis: an experimental model in the rat. Science 1980; 208:412–414.
30. Trounce I, Byrne E, Dennett X. Biochemical and morphological studies of skeletal muscle in experimental chronic alcoholic myopathy. Acta Neurol Scand 1990; 82:386–391.
31. Mebs D, Ownby CL. Myotoxic components of snake venoms: their biochemical and biological activities. Pharmacol Ther 1990; 48:223–236.
32. Meshul CK, Boyne AF, Deshpande SS, Alberquerque EX. Comparison of the ultrastructural myopathy induced by anticholinesterase agents at the end-plates of rat soleus and extensor muscles. Exp Neurol 1985; 89:96–114.
33. Bardosi A, Dickmann U. Necrotising myopathy with paracrystalline inclusion bodies in hypervitaminosis E. Acta Neuropathol (Berlin) 1987; 75:166–172.
34. Streicher HZ, Gabow PA, Moss AH, Kono D, Kaehny WD. Syndromes of toluene sniffing in adults. Ann Intern Med 1981; 94:758–762.
35. Martin RW, Duffy J, Engel AG, Lie JT, Bowles CA, Moyer TP, Gleich GJ. The clinical spectrum of the eosinophilia-myalgia syndrome associated with L-tryptophan ingestion: clinical features in 20 patients and aspects of pathophysiology. Ann Intern Med 1990; 113:124–134.
36. Stommel EW, Parsonet J, Jenkyns LR. Polymyositis after ciguatera toxin exposure. Arch Neurol 1991; 50:571.

32
Myopathies in Infections and Systemic Diseases

Richard W. Orrell and Russell J. M. Lane
Charing Cross Hospital and Charing Cross Hospital and Westminster Medical School, London, England

INTRODUCTION

Skeletal muscle comprises more than 40% of the body mass and has active metabolic functions in addition to its locomotor and supportive roles. The muscle is vulnerable to involvement in many systemic infections and diseases, in addition to localized infection (1,2). This chapter covers the involvement of skeletal muscle in infection, malignant disease, connective tissue disease, vascular and blood disorders, and systemic metabolic disease. Myopathies occurring in other systemic diseases are covered more specifically in other chapters, including the inflammatory myopathies of unknown cause (where viral infection may be implicated but is as yet unproven), HIV-related myopathies, the granulomatous diseases, alcohol and opiate abuse, vitamin and nutritional deficiencies, and endocrine and drug-induced myopathies.

MUSCLE INVOLVEMENT IN INFECTIOUS DISEASES

Bacteria

Direct bacterial myositis is rare in Europe and North America, but diseases initially observed in the tropics may now be seen in temperate climates, especially in the context of immunosuppression due to chemotherapy or acquired immune deficiency syndrome (AIDS). The myositis may be generalized but is more commonly localized and related to an intramuscular abscess (pyomyositis). The infection may be due to direct invasion, for example by traumatic penetration, by hematogenous dissemination from a remote focus of sepsis, or by direct extension from adjacent tissues.

Pyomositis

Pyomositis is a bacterial infection of large skeletal muscles, usually with abscess formation. Original descriptions were from the tropics, where the condition is endemic as "tropical pyomyositis." In a review of 112 patients with tropical myositis, over a period of 4 years in one center in Nigeria (3), 90% of cases were due to *Staphylococcus aureus*, 3.5% to *Streptococcus pyogenes*, and 0.9% to *Escherichia coli*. Multiple lesions were seen in 43% of cases, most commonly in the quadriceps and other large muscles. A typical early presentation of the disease (2% of cases) is in the so-called "invasive stage." The patient will notice a cramping pain in a large muscle, followed after a week by redness, increasing pain, and low-grade fever. There may be a "wooden" stiffness on palpation of the muscle, with mild leukocytosis and eosinophilia. Between 10 and 21 days, "the suppurative stage" (90% of cases presented during this stage) commences with marked edema and tenderness of the muscle, with pyrexia. Needle aspiration yields thick yellow pus. In the "late stage" (5% of cases present at this time), the patient is significantly ill, with fever, septicemia, and possibly coma, and there is fluctuance and tenderness of the muscle.

Pyomyositis has more recently, but less commonly, been recognized in temperate climates (4–7). In a North American center (6), 18 cases of pyomyositis were identified over an 18-year period. The presentations were similar to tropical pyomyositis, although pyomyositis was often not considered in the initial diagnosis, the alternatives including muscle strain, deep venous thrombosis, synovitis, and tumor. Muscles of the hip and thigh were most commonly involved, with *S. aureus* the most common organism.

Eosinophilia is not generally found, and serum creatine kinase is not usually significantly raised (4). Computerized-tomography (CT) scan with contrast enhancement has been the most commonly used method for diagnosis and for delineating the extent of the lesion, to aid aspiration, and for distinguishing pyomyositis from other conditions such as tumor (6); magnetic resonance imaging (MRI) may also be used. In most cases a single muscle is affected, but multiple abscesses, with a "polymyositis-like" presentation, have been reported (8).

The cause of pymoyositis is uncertain, as bacteremia itself does not result in pyomyositis, and skeletal muscle is very resistant to bacterial infection. Some patients have predisposing conditions, including diabetes mellitus, steroid treatment (6), and splenectomy (9); and trauma is also felt to have a role in causation. Other theories of predisposing causes include nutritional deficiency, viruses, and spirochetes (3). Pyomyositis has been associated with many bacterial organisms, including gram-negative bacteria, in individual case reports and is increasingly recognized in patients with HIV disease and AIDS. In one center (10) six cases of pyomyositis were seen over a 20-month period. All patients had muscle injury due to exercise or trauma, and underlying HIV-related myopathy may also have been a predisposing factor in these cases. The patients had multiple abscesses, which required prolonged antibiotic treatment and repeated drainage of the abscesses. One patient developed rhabdomyolysis, which is generally rare in pyomyositis, but has been reported in HIV-related myopathies. A gram-negative organism was identified in one patient, but *S. aureus* was the most commonly defined infection.

Clostridia

Clostridia may cause a myositis, or more commonly a myonecrosis (gas gangrene). Six species of clostridia (large gram-positive anaerobic rods with a central or subterminal spore) cause gas gangrene, *Clostridium perfringens* being the most common (11). Trauma, ischemia, pyogenic infection, or foreign bodies may initiate conversion of the spore to the vegetative form that produces toxins. Clostridial myonecrosis may follow intramuscular or subcutaneous injection in otherwise healthy patients (12). In "simple contamination" the wound may heal spontaneously. In "anaerobic cellulitis" the organisms multiply in the connective tissue spaces, producing gas which does not appear in the muscle. The most serious form is "anaerobic myonecrosis" or "gas gangrene." The incubation period may be from 7 hr to 6 weeks, the patient presenting with localized acute pain and marked system illness. The muscle is initially pale and edematous, progressing to a brick red color, and gas bubbles are present between the fibers. Gas gangrene may also be associated with other, nonclostridial organisms (13).

Gas gangrene occurs spontaneously or atraumatically in about 16% of infections (14). *Clostridium septicum* is common in this condition (15). Spontaneous gas gangrene may be associated with a malignant tumor, especially carcinoma of the bowel or hematological malignancy, and with diabetes mellitus (16). Treatment requires aggressive excision of the affected tissue and antibiotics. The specific antibiotics are uncertain, but benzylpenicillin is generally used, and clindamycin, rifampicin, and tetracycline may also be effective. Hyperbaric oxygen may be used, although its benefit is not proven in clinical practice (15,16).

Less commonly an abscess, or pyomyositis, may contain clostridia. In one report, three heroin addicts with subcutaneous abscesses had pure cultures of *C. perfringens* or *septicum*, and two others had a mixed growth. There was no evidence of the myonecrosis associated with gas gangrene (17).

Muscle atrophy and weakness have been observed after recovery from the acute phase of tetanus, due to *Clostridium tetani* infection (18). This may require continued mechanical ventilation after discontinuing the muscle relaxants required for control of the tetanic spasms. In one case, features on muscle biopsy and electromyography (EMG), with raised serum muscle enzymes, suggested a myopathy due to the tetanus toxin. This predominantly proximal myopathy recovered over a period of between 2 and 15 months.

Mycobacteria

Tuberculosis of the skeletal muscle is very rare and usually due to secondary invasion from an adjacent focus or by hematogenous spread. Tuberculous granulomata may be found on muscle biopsy (19). The onset is insidious, with slow swelling in the muscle and often little pain or disability. A firm, soft, or fluctuant mass may be found in the muscle. The pathology includes a nodular form, with small tuberculous nodules with caseating centers in the connective tissues; an inflammatory tuberculous myositis, with granulomatous tissue and cold abscesses; or a sclerosing myositis, with diffuse sclerosis and loss of muscle fibers (20).

Muscle weakness in leprosy is usually neurogenic, but rarely invasion and proliferation of *Mycobacterium leprae* in muscle fibers may lead to muscle fiber degeneration, with inflammatory changes, and subsequent fibrosis (21). Other studies have described an interstital myositis (22). The clinical features associated with the myositis may be minimal, but may include pain and tenderness of the muscle and flexion deformities of the limbs due to the fibrosed muscle (23).

Legionella

Myalgia may be present in up to 50% of patients with *Legionella* pneumonia. In one patient, a necrotizing myopathy was observed with direct invasion of muscle cells by bacteria together with a striking inflammatory response, and it is suggested that cell-mediated immunity plays an important role in the myositis (24).

It is probable that muscle involvement is underdiagnosed in severe systemic bacterial infections. For example, significant increases in serum creatine kinase have been reported in patients with legionnaires' disease, other severe pneumonias, and the toxic shock syndrome (probably due to a *S. aureus* exotoxin) (25).

Viruses

Viruses have been implicated in some cases of congenital myopathy, polymyositis, and nonspecific fatigue syndromes, in addition to more specific viral myositis where the virus may be identified in the muscle (1,26,27). A wide range of DNA and RNA viruses may cause acute necrotizing inflammatory myopathies, and analogy to viral-induced muscle damage in experimental animals (28) suggests that myonecrosis is a result of direct invasion of the virus, with a secondary inflammatory response, although no direct infection of muscle fibers occurs in HIV-related myositis or myopathy. The clinical presentation ranges from a localized self-limiting myositis, to a severe acute rhabdomyolysis. Viruses have also been implicated in the etiology of more chronic muscle conditions whose pathophysiology has not been clearly defined (see below).

Benign Acute Myositis

This infection has been attributed to many viruses, in particular influenza A and B, parainfluenza, and adenovirus 2 (26,27). Postinfluenzal myositis is well recognized in epidemic childhood forms (29,30) and sporadic forms affecting children and adults (31,32). Typical symptoms are severe muscle pain, tenderness, and occasionally swelling, within the first week following an upper respiratory tract infection. The calf muscles are most commonly affected. Complete recovery usually occurs over a period of several weeks. Serum creatine kinase may be raised, with EMG features of a myopathy and a necrotizing inflammatory myopathy on muscle biopsy (30). The virus may occasionally be seen on electronmicroscopy (32) but is rarely isolated from the muscle. The age-related resistance to the epidemic childhood form may be related to several factors, including maturation of macrophages and increased effectiveness of tissue barriers to viral spread (33).

Acute Myopathy with Myoglobinuria

A more severe clinical presentation, with diffuse muscle pain, tenderness, weakness, and myoglobinuria, may occur during the febrile phase of a viral illness (26,27). Serum creatine kinase is markedly raised, and there may be severe muscle necrosis, with acute renal failure related to the myoglobinuria. Many viruses have been implicated, including influenza A and B (30), Coxsackie B5 (34), echo 9 (35), adenovirus 21 (36), herpes simplex (37), and Epstein-Barr (37), as well as *Mycoplasma pneumoniae* (38). The pathogenic role of the virus is uncertain, and the virus is rarely isolated from muscle. The attribution of an acute proximal myopathy to herpes zoster raises the possibility of antiviral therapy with acyclovir in this condition (39).

Epidemic Pleurodynia (Bornholm Disease)

This is an acute infectious disease characterized by fever, headache, and severe pain in the chest or upper abdomen. The pain is worse with respiratory movement. The affected muscles, especially the intercostals, are usually tender, and there may be muscle swelling. A pleural friction rub is sometimes heard, but the primary disease is muscular, with no specific pleural disease. Most patients are ill for 4–6 days, but there may be a recurrence of symptoms. Coxsackie B virus, especially Coxsackie B 1, 3, 4, and 5, is the commonest cause, and rarely echovirus and Coxsackie A. Treatment is with analgesics, and opiates may be necessary (40).

Postviral Fatigue Syndrome

This is a clinical syndrome of chronic fatigue, with possible, but unproven, viral cause. Some patients may have specific symptoms of myalgia, with myopathic EMG findings (41). Enteroviral mRNA has been detected in the skeletal muscle of some patients (42,43). A suggested association with Epstein-Barr virus remains unproven (44). This condition is considered further in another chapter.

Polymyositis and Dermatomyositis

Because of the association of viruses with acute myopathies, viruses have also been considered in the etiology of the idiopathic inflammatory myopathies. Results in different laboratories have been conflicting, with studies in the United Kingdom reporting enteroviral mRNA in patients with dermatomyositis, polymyositis, and dilated cardiomyopathy (45–47), while in the United States, in a study of the muscle of 44 patients, no mRNA for a number of candidate viruses could be detected, including coxsackie viruses, mumps virus, encephalomyocarditis virus, adenovirus, human T-lymphotropic virus (HTLV) types 1 and 2, and human immunodeficiency virus (HIV)) (48).

The polymyositis associated with chronic graft-versus-host disease may have a viral cause, due to the recurrent viral infections that these patients experience (49).

Retrovirus-Associated Myopathies

Myopathies have been associated with human immunodeficiency virus (HIV-1, HTLV-III), either directly (50) or through secondary opportunistic infection with organisms such as microspiridia (51), and this topic is discussed in detail in another chapter. A polymyositis has also been associated with HTLV-I infection, in addition to the more commonly recognized myelopathy (52).

Fungi

Myopathies may be associated with systemic candidiasis, especially in the context of immunosuppression. The triad of a high fever, a papular erythematous rash, and diffuse severe muscle tenderness has been described as a distinctive clinical presentation of disseminated candidiasis in six patients, presenting at one center, over a period of approximately 1 year (53). This was felt to be sufficiently distinctive to justify antifungal therapy pending confirmation of the diagnosis. All the patients had acute leukemia, and all died within 1 week despite treatment with amphotericin B. In one patient, extensive infiltration with *Candida pseudohyphae*, with myofiber necrosis and hemorrhage, was demonstrated (53). Muscle involvement in candidiasis may be more common than reported at autopsy, and antemortem muscle biopsy may aid in diagnosis, before the organism is cultured in blood or precipitating antibodies have developed (54,55).

Spirochetes

Myositis may be a feature of Lyme disease, caused by the spirochete *Borrelia burgdorferi*, which is transmitted from deer and small mammals to humans by the bite of the *Ixodes* tick. In a report from Germany, eight patients with Lyme disease had muscle biopsy evidence of myositis (56). In six patients the myositic features were localized to the affected region of the skin or neuropathic lesion, and in two the features were more generalized. On muscle biopsy, lymphoplasmocellular infiltrates were found in the region of small vessels, with spirochetes seen in six of seven muscle biopsies. Four patients responded to treatment with antibiotics.

Protozoa

Toxoplasmosis

A myositis is rarely associated with toxoplasmosis (57). Features include a recent upper respiratory tract infection with fever and lymphadenopathy and skin lesions. The features of meningoencephalitis with eosinophilia may lead to the diagnosis, which may be confirmed by an increase in *Toxoplasma* antibody titers. The muscles may be weak and tender, with elevated serum creatine kinase. Muscle biopsy shows evidence of myositis, and sometimes *Toxoplasma gondii* are found in the muscles, either isolated or in cysts (58). A patient with a diagnosis of *Toxoplasma* meningoencephalitis with myositis was successfully treated with a combination of minocycline, sulfamethoxazole, trimethoprim, piperacillin, and acetyl spiramycin (57).

Sarcocystis

Sarcocysts are cylindrical structures in the skeletal muscle, which are occasionally visible to the naked eye as minute white threads (*T. gondii* may produce similar structures). They are found in the skeletal muscles of many mammals and some birds, and occasionally in humans acting as either "prey" or "predator" host (59). (Fig. 1). The condition is generally asymptomatic in humans but, for example, a painful, localized muscle lesion of the chest wall with sarcocysts has been described (60).

Trypanosomiasis cruzi (Chagas Disease)

Trypanosomiasis infection may be clinically silent, but common clinical presentations include myocarditis and meningoencephalitis. The skeletal muscle may be involved but is generally asymptomatic (61).

Amebiasis

A local suppurative, necrotizing myositis due to *Entamoeba histolytica* has been described. The patient died despite surgical drainage (26).

Malaria

Malaria is a common infection in warm climates. Foreign travel is resulting in a significant number of imported infections to countries with temperate climates. A clinical myositis has been reported as the principal manifestation of infection with *Plasmodium falciparum* (62). In a study of 58 children with severe *P. falciparum* infection in The Gambia, there was a significantly raised serum creatine kinase in 28% and serum myoglobin in 45%, suggesting muscle involvement in the disease (63). A more severe myonecrosis, with myoglobinuria and acute renal failure, has also been described (64).

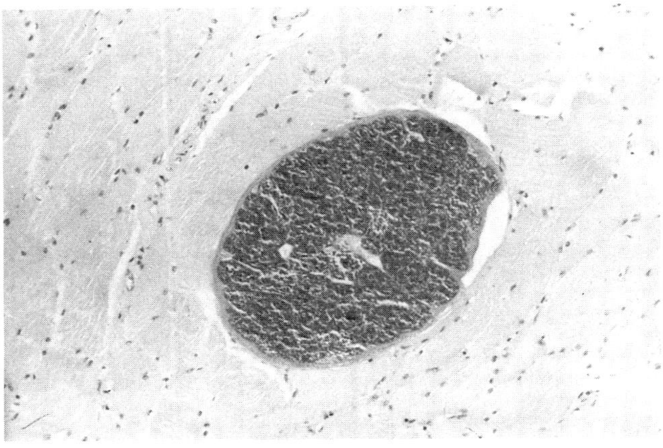

Figure 1 Transverse section of a sarcocyst in skeletal muscle, the capsule enclosing sporozoites. Hematoxylin and eosin. (Courtesy of Paul D. Lewis, Charing Cross Hospital, London.)

Cestodes

Cysticercosis

Human cysticercosis is due to infection with *Cysticercus cellulosae*, the larval form of *Taenia solium* (the pork tapeworm), through ingestion of the eggs from human feces. The central nervous system is most commonly involved by the encysted parasite, but skeletal muscle is also commonly involved, the cysts causing single or multiple nodules or a pseudohypertrophic myopathy (65). In the pseudohypertrophic form, there may be progressive muscle weakness over a period of months to years, with increase in muscle bulk. Mobile subcutaneous nodules may be noted. On muscle biopsy the cysts may be seen between the muscle fibers with the naked eye. The cause of the muscle hypertrophy is uncertain and not fully accounted for by the cysts (65). Treatment of cysticercosis is with praziquantel, although there is a possibility of complications in cysticercosis of the muscle if large numbers of larvae are rapidly destroyed (59).

Coenurosis (Multiceps Tapeworm)

Coenuri are glistening cysts of the tapeworm *Multiceps multiceps*. Humans are infected by ingestion of the eggs acquired from dogs. The clinical disease may be similar to cysticercosis. The coenuri may be as large as apricots (59) and may cause a usually solitary palpable tumor in muscle, especially on the trunk.

Hydatidosis (Echinococcosis)

Hydatid disease in humans is caused by the larva of the tapeworms *Echinococcus granulosus* and *Echinococcus multilocularis*. *E. granulosus* is the commoner, unilocular hydatid disease. Alveolar echinococcosis due to *E. multilocularis* is rare but more severe and, having no true limiting capsule, may behave as an invasive tumor. Muscle involvement is relatively common. Among 1802 patients in an Australasian Hydatid Register, 25% had lesions in the muscles (66). In some parts of the world a hydatid cyst is the most common cause of a benign muscle tumor, presenting as a deep, slowly growing, poorly mobile, spherical mass of variable size and firm consistency. Pain is rare, but if the cyst ruptures there may be severe pain and edema. The lesion may occasionally calcify (59). *E. multilocularis* presents more rarely and in one family was found in association with myotonia congenita (67).

Sparganosis

This is an uncommon infection caused by larvae of the pseudoophillidean tapeworm of the genus *Spirometra*. In most humans sparganosis is a single lesion located in the superficial muscles or subcutaneously. The lesion is painless, recurrent, and migratory (59). A 23-year-old Greek woman, living in Canada, had noted a swelling in the left arm 5 years previously, which was diagnosed as a lipoma. On surgical exploration, a white structure surrounded by connective tissue was seen beneath the biceps muscle. A live, wriggling worm 20 cm long was removed. There was a chronic inflammatory reaction of the surrounding muscle. The infection was attributed to ingestion of uncooked pork (68).

Nematodes

Trichinosis

Infection with the larvae of *Trichinella spiralis* is worldwide, the distribution depending on the eating of poorly cooked, infected pork. In striated muscle the larvae penetrate the sarcolemma, coil within the muscle fiber, and encyst (69). Calcification may occur within 6 months, but the larvae remain viable for many years. The affected muscle fiber and adjacent fibers will be destroyed by the parasite. Many infections are asymptomatic, but the condition may be fatal. There may be marked myalgia with tenderness, weakness, and swelling of the muscles. The weakness may be rapid leading to tetraparesis. A patient with a puffy face, swollen legs, and very tender muscles probably has trichinosis (59). Death usually results from nonspecific myocarditis, pneumonia, or encephalitis. Treatment is with thiabendazole, with the risk of a Herxheimer-like reaction.

This infection was much more common in the nineteenth century and was a notable cause of muscle enlargement. Duchenne de Boulogne introduced the use of "harpoon" muscle biopsy principally to help differentiate cases of trichinosis from other causes of muscle enlargement, such as pseudohypertrophic muscular dystrophy.

Toxocariasis

The ova of *Toxocara canis* (and *Toxocara catis*), present in dog (or cat) feces, may be ingested by children. The larvae form granulomatous lesions in many sites (visceral larva migrans), and motile larvae may be found in muscle (70). It has been suggested that muscle biopsy may be an appropriate diagnostic procedure for the systemic condition (70). The most usual presentation is a child with hepatosplenomegaly, pneumonitis, and eosinophilia (59).

MALIGNANT DISEASE

Direct Invasion

Direct invasion of skeletal muscle in malignant disease is rare. In a report of three patients with diffuse enlargement of the skeletal muscle of the thigh by diffuse large cell lymphoma, CT scan demonstrated diffuse infiltration of the muscle, with adjacent bone marrow involvement. It was suggested that the lymphoma spread directly from the adjacent bone or possibly in the reverse direction (71). Muscle infiltration occurs in approximately 1% of cases of lymphoma, usually in the context of widespread systemic disease. The major diagnostic problem is distinction from venous or lymphatic obstruction. CT scan of the muscle will aid in the diagnosis, which may be confirmed by muscle biopsy. The swelling may resolve with treatment of the lymphoma.

Metastatic Spread

Metastatic tumor deposits in skeletal muscle are rare, although in a detailed histological study of postmortem skeletal muscle, some metastatic tumor was found in six of 36 patients with malignant disease (72). A proximal myopathy due to metastatic invasion was described in two patients, one with squamous cell carcinoma of the bronchus, the other with carcinoma of the breast (73). EMG showed features of a myopathic process, and muscle biopsy or postmortem examination showed clumps of malignant cells between the muscle fibers and in the intermyofibrillary network. The initial clinical diagnosis in these patients had been of a nonmetastatic neuromyopathy.

Carcinomatous Neuromyopathy

A nonspecific cachexia with muscle wasting and weakness is a common feature of advanced carcinomatous disease. The specific causes are unknown, but probably include nutritional, toxic, and biochemical factors. Brain and Henson (74) distinguished a group of patients with mainly bronchial, rectal, and prostatic carcinoma, who had a proximal weakness affecting the limbs and trunk. Several of the patients had evidence of additional neural involvement—"carcinomatous neuromyopathy"—and a smaller group had evidence of muscle involvement only—"carcinomatous myopathy." This syndrome was present in no more than 5% of patients with these cancers, and there were no specific abnormalities on muscle biopsy. Carcinomatous neuromyopathy represents a heterogeneous group of conditions involving abnormalities of peripheral nerve, neuromuscular junction, and muscle, which may be associated with, and sometimes precede, a variety of carcinomas, lymphoma, and leukemia (1).

In some patients myopathy has been attributed to embolic infarction related to the thrombotic state associated with carcinoma (75). Pain may be of sudden onset, and the syndrome, described as "a migratory monomyositis multiplex," is more focal than polymyositis. Muscle histology shows features of embolic infarction. Treatment should be to prevent further systemic infarction.

Autoantibodies have been found in the sera of patients with various paraneoplastic neurological syndromes. Muscle-specific antibodies may be involved in paraneoplastic myositis; for example, a muscle-specific antibody was found in a patient with colon cancer and myositis, although its significance is uncertain (76). Finally, Lambert-Eaton myasthenic syndrome associated with neoplasia often presents with clinical features suggestive of a "myopathy." This condition appears to be due to autoantibodies directed against the presynaptic calcium channels (77).

Polymyositis and Dermatomyositis

An association between polymyositis and dermatomyositis and neoplasia has been recognized for many years, but the precise relationship has been questioned. A myositis appears to be associated with malignancy in 7–30% of cases (78), but there may be ascertainment bias in some studies, as malignancy is often actively sought in the presence of polymyositis. The association appears stronger, and with a higher rate of mortality from the cancer, in dermatomyositis (79). It has been suggested that screening for an underlying neoplastic condition should be actively pursued in patients presenting with features of myositis (especially dermatomyositis) and, in addition to detailed history and physical examination, should include mammography for women, upper and lower gastrointestinal imaging, and abdominal and thoracic CT scans in all patients, together with investigation suggested by more specific aspects of the patient's condition (78). However, such an approach is rarely rewarding in practice, and others would not initially perform gastrointestinal imaging or body CT or MRI, weighing the benefits against anxiety to the patient and preferring continued clinical follow-up (80). Polymyositis and dermatomyositis are discussed more extensively in another chapter.

Carcinoid Syndrome

A myopathy has been associated with carcinoid syndrome (81). The symptoms of carcinoid syndrome were present for more than 10 years before the development of muscle weakness. Muscle biopsy showed marked atrophy of type 2 fibers. The myopathy is probably the consequence of excess serotonin. Cyproheptadine, a serotonin and histamine antagonist, improved the muscle strength.

CONNECTIVE TISSUE DISEASE

Systemic Sclerosis

There is a clear association between systemic sclerosis and myopathy. This may be either a "simple" myopathy or a "complicated" inflammatory myopathy resembling polymyositis (82). The features are generally of a mild myopathy, with proximal muscle weakness. The EMG may be normal or show features suggestive of a myopathy, with normal or mild increase in serum creatine kinase. Muscle biopsy shows minimal lymphocytic infiltration, with scattered angulated fibers, and no necrosis. The symptoms often remit spontaneously, without steroid treatment.

There may be an associated myocarditis (83), and such patients may have more severe skeletal muscle involvement with EMG features of an inflammatory myopathy. Muscle biopsy in such cases shows an inflammatory infiltration with necrosis and degeneration and regeneration of muscle fibers. This myopathy responds to treatment with steroids.

Systemic Lupus Erythematosus (SLE)

Prominent muscle disease is present in around 5–10% of patients with SLE (84,85). The patients have mild to moderate weakness affecting especially the proximal lower limbs. The EMG usually shows myopathic features, with increased serum creatine kinase at some stage of the illness. Muscle biopsy shows features of an inflammatory myopathy. The myopathy responds to treatment with steroids and immunosuppressants, and the prognosis is similar to polymyositis in general.

Rheumatoid Disease

Muscle wasting is a well-recognized feature of rheumatoid arthritis (86). Haslock et al. (87) suggested several types of myopathy might occur based on EMG and muscle biopsy studies. The muscle may be affected by cachexia as part of the systemic disease; by polyarteritis nodosa, myositis, or chronic myopathy, as an expression of the connective tissue disease; by denervation; by steroid treatment; and by the effects of immobility. More than one of these factors may be active in an individual patient. In Haslock's 34 patients with muscle weakness or wasting and rheumatoid arthritis, five had features of an inflammatory myositis similar to polymyositis.

A nodular myositis may also be seen in rheumatoid disease. The nodules contain plasma cells and lymphocytes and are actively involved in immunoglobulin synthesis. Muscle fiber degeneration may be found in the proximity of these nodules, and local immunological effects may contribute to the pathogenesis of the myositis (86).

A proliferative myositis presented as a 4-by-6 cm tender, subcutaneous nodule near the knee of a patient with rheumatoid disease. The histology showed perimysial and perifascicular connective tissue proliferation enclosing fibroblasts and ganglion cells. It was suggested that this was a result of ischemia due to rheumatoid vasculitis (88).

Ankylosing Spondylitis

Muscle biopsy in the limbs of patients with ankylosing spondylitis may be normal, but muscle biopsy from the paraspinal muscles may show many fibers with the appearance of "targetoid cores." These fibers have an oval, central, unstained core, surrounded by normally staining concentric rings. These are seen more readily with oxidative stains and almost exclusively in type 1 fibers. On electron microscopy these fibers show disruption of organelles and myofibrillar structure. In addition, there may be atrophic angulated fibers and slight predominance of type 1 fibers. These findings appear to be unrelated to the extent and severity of involvement of the spine (89).

Sarcoidosis and Granulomatous Disease

The myopathy associated with these conditions is discussed in another chapter.

VASCULAR AND BLOOD DISORDERS

The muscle has a rich vascular network, which can respond efficiently to the changing demands of exercise. The blood supply may be compromised by disease of the blood vessels or by disorders of the blood that lead to rheological abnormalities (1).

Polyarteritis

Localised ischemia and microinfarction may occur in polyarteritis nodosa and other vasculitic disorders (1). The characteristic lesion, in larger intramuscular arteries in the perimysium and epimysium, is associated with inflammatory changes throughout the arterial wall. In lesions where the acute inflammation has resolved, the arteries may be narrowed and the walls thickened with fibrosis (2). Vasculitis is also implicated in some of the myopathies associated with connective tissue disease.

There appears to be a distinct form of childhood dermatomyositis, where the pathology includes a lymphocytic vasculitis producing temporary or permanent occlusion of small vessels and capillaries (90). Infarction and zonal myopathy may result, and ischemia is thought to be the basis of the perifasicular atrophy typical of childhood dermatomyositis. Because of the focal nature of the pathology, a generous muscle biopsy is advised. The clinical features range from a mild self-limiting illness to a multisystem disease with significant muscle weakness, which may lead to permanent disability or death.

Atherosclerosis

Ischemic myopathy and infarction may occasionally result from disease of larger vessels in atherosclerosis and diabetes. Infarction of muscle alone is relatively rare, presumably due to the rich collateral blood supply and relative resistance to ischemia. Although isolated infarction of muscles, other than that associated with gangrene, is rare in diabetics, two patients with diabetes are described who presented with sudden pain and swelling in the thigh (91). There was atherosclerosis of the medium-sized arteries and arterioles, atherosclerosis and medial sclerosis of Mönckeberg in the muscular arteries, and microangiopathy in the capillaries. The site of obstruction in these patients was probably in the large proximal vessels; more peripheral vascular occlusion leads to the more common distal ischemia. Muscle biopsy distinguished the condition from initial diagnoses of tumor and osteomyelitis.

In limbs amputated because of ischemia there is evidence of neurogenic atrophy of distal muscle. Necrosis, regeneration, and vacuolation in muscle fibers of normal size are a direct result of ischemia (92). In Volkmann's ischemic contracture, which is related to arterial damage and affects both nerves and muscles, the muscle may appear hemorrhagic initially, but in time the infarction is followed by fibrosis leading to the contracture (92).

Denervation atrophy is also found in similar conditions, and muscle damage in chronic ischemia may be primarily neurogenic (93).

Sickle Cell Disease

Muscle necrosis may be caused by massive intramuscular sickling. This may lead to rhabdomyolysis and acute renal failure (94). Two patients with homozygous sickle cell disease developed recurrent episodes of painful symmetrical swelling in the deltoid and thigh muscles (95). The symptoms settled in between 1 and 5 weeks. On some occasions surgical drainage of necrotic fluid was necessary. Acute episodes were treated with hydration, nasal oxygen, and bed rest. The muscle showed features of acute necrotizing myositis, with sickling of red cells adjacent to areas of abnormal muscle and fascia. Muscle examined after resolution of the swelling appeared normal except for a mononuclear cell infiltrate.

SYSTEMIC METABOLIC DISEASE

Weakness, myalgia, cramps, and fatigue are common symptoms in patients with renal and hepatic disease and in cachexia from other causes. Some symptoms may be related to a polyneuropathy, but disturbance of the metabolism and electrolyte balance of the muscle at a cellular level may contribute to the clinical presentation.

Hypokalemia, Hypomagnesemia, and Hypophosphatemia

Myopathy is rarely related to isolated deficiency of the cations potassium, magnesium, and phosphate, which may more often occur as part of a more generalized metabolic derangement (1).

Hypokalemic myopathy (96) is characterized by a rapid onset of skeletal muscle weakness and is associated with a decrease in both cellular and plasma potassium (unlike hypokalemic periodic paralysis where the total potassium pool is normal). Potassium depletion myopathy may occur as a result of a variety of drug treatments, steatorrhea, alcoholism, hyperaldosteronism, or renal tubular acidosis. The weakness may progress over a period of 1 week to 1 month, causing severe weakness of the limbs, especially proximally, sometimes with myalgia, polyuria, and polydipsia. Reflexes and sensation are preserved. The serum K^+ may be as low as 1.1–2.3 mmol/L, with a marked increase in serum creatine kinase. The EMG shows myopathic changes. Muscle biopsy features include fiber degeneration with phagocytosis and regeneration, especially of type 2 fibers. Full recovery, with potassium replacement, occurs within 3 weeks. The pathogenesis is uncertain, but is probably related to potassium-induced disturbance of cellular metabolism, or possibly ischaemia secondary to constriction of vascular smooth muscle.

An acute painless myopathy has been described in alcoholic patients with hypokalemia, which may respond to potassium replacement. Muscle biopsy shows necrotic muscle fibers surrounded by lymphocytes and histiocytes, with multiple small vacuoles in some of the degenerating fibers (97).

Chronic magnesium deficiency may produce structural and functional damage to muscle fibers in experimental animals (1). In experimental magnesium deficiency in humans, lethargy and generalized weakness developed, and EMG showed myopathic changes (98). Hypomagnesemia is common in chronic alcoholism, but the myopathy here is usually multifactorial (99).

Acute severe hypophosphatemia may cause rhabdomyolysis in alcoholics, patients with diabetic ketoacidosis, and following antacid treatment with aluminum hydroxide gel. Chronic phosphate depletion is associated with muscle weakness and myalgia with raised serum muscle enzyme levels (1). A myopathy may be associated with hypophosphatemic osteomalacia. Treatment with phosphate, calcium, and vitamin D leads to a significant improvement in the symptoms (100).

Uremic and Dialysis Myopathy

Fatigability, muscle weakness, and cramps are frequent symptoms in chronic renal failure. These may be due to electrolyte abnormalities, anemia, medication, or uremic polyneuropathy, but uremia may also cause a primary myopathy.

Mild proximal muscle weakness is seen in the myopathy of chronic renal failure, as opposed to the distal weakness of symmetrical sensorimotor polyneuropathy (101). The reflexes may remain brisk, with no sensory disturbance, and the degree of myopathy does not appear related to the level of uremia. Renal osteodystrophy appears to be a common feature in this myopathy. In some patients the strength may improve with subtotal parathyroidectomy (for secondary hyperparathyroidism). Selective type II fiber atrophy may be seen on muscle biopsy. In other uremic patients the appearance may be of an ischemic myopathy secondary to the arterial disease. Vitamin D administration may produce some improvement in power in patients with osteomalacia. In patients who developed myopathy while on dialysis, there was an improvement in power after renal transplantation or dialysis with deionized water (102). The myopathy of chronic maintenance dialysis can only be fully reversed by renal transplantation (103).

A negative nitrogen balance with loss of lean body mass occurs in patients with acute or chronic renal failure. This appears to be related to abnormal energy metabolism and metabolic acidosis (by increasing muscle protein degradation and increasing the requirement of essential amino acids). The low-protein diet generally used in these patients may exacerbate the loss of muscle tissue, and this may be reduced by correcting the metabolic acidosis and ensuring adequate intake of nutrients. Inadequate intake of essential amino acids in particular may limit protein synthesis (104).

Muscle cramps are a common symptom in patients on dialysis. The cause is uncertain, but calcium antagonists such as nifedipine may control the symptoms (105).

Hypermetabolism of Trauma

The metabolic response to severe trauma includes a "flow" or hypermetabolic phase, with increased nitrogen excretion and negative nitrogen balance. An adequate source of nitrogen is necessary to reduce muscle wasting, but there does appear to be some resistance to protein synthesis in skeletal muscle in this phase, despite adequate nitrogen intake. In nonstressed malnourished patients, an increase in nitrogen intake will lead to a significant increase in protein synthesis, which is not seen in the flow phase of trauma. In the recovery phase, several weeks after the trauma, an increase in nitrogen intake leads to an improvement in nitrogen balance (106).

REFERENCES

1. Lane RJM. Muscle involvement in systemic and iatrogenic disease. In: Kennard C, ed. Recent Advances in Clinical Neurology. Vol 5. Edinburgh: Churchill Livingstone, 1988:201–238.
2. Heffner RR. Inflammatory myopathies. A review. J Neuropathol Exp Neurol 1993; 52:339–350.
3. Chiedozi LC. Pyomyositis. Review of 205 cases in 112 patients. Am J Surg 1979; 137:255–259.
4. Schlech WF, Moulton P, Kaiser AB. Pyomyositis: tropical disease in a temperate climate. Am J Med 1981; 71:900–902.
5. Gibson RK, Rosenthal SJ, Lukert BP. Pyomyositis. Increasing recognition in temperate climates. Am J Med 1984; 77:768–772.
6. Hall RL, Callaghan JJ, Moloney E, Martinez S, Harrelson JM. Pyomyositis in a temperate climate. J Bone Joint Surg 1990; 72- A:1240–1244.
7. Dede H, Ozdogan H, Dumankar A, Aktugla Y, Tabak F, Yazic H. Tropical pyomyositis in a temperate climate. Br J Rheumatol 1993; 32:435–6.
8. Kallen P, Nies KM, Louie JS, Keller M, Worthen N, Bayer AS. Tropical pyomyositis. Arthritis Rheum 1982; 25:107–110.
9. Robertson-MacKay F, Al-Hillawi AH. Pyomyositis complicating pneumococcal meningitis. Postgrad Med J 1993; 69:64–65.
10. Widrow CA, Kellie SM, Saltzman BR, Mathur-Wagh U. Pyomyositis in patients with the human immunodeficiency virus: an unusual form of disseminated bacterial infection. Am J Med 1991; 91:129–136.
11. Weinstein L, Barza MA. Gas gangrene. N Engl J Med 1973; 289:1129–1131.
12. Hallagan LF, Scott JL, Horowitz BC, Feied CF. Clostridial myonecrosis resulting from subcutaneous epinephrine suspension injection. Ann Emerg Med 1992; 21:434–436.
13. Bessman AN, Wagner W. Nonclostridial gas gangrene: report of 48 cases and review of the literature. JAMA 1975; 233:958–63.
14. Hart GB, Lamb RC, Strauss MB. Gas gangrene. J Trauma 1983; 23:991–1000.
15. Stevens DL, Musher DM, Watson DA, et al. Spontaneous, nontraumatic gangrene due to *Clostridium septicum*. Rev Infect Dis 1990; 12:286–96.
16. Scully RE, Mark EJ, McNeely WF, McNeely BU. Case records of the Massachusetts General Hospital, Case 5-1993. N Engl J Med 1993; 328:340–346.
17. Gorbach SL, Thadepalli H. Isolation of clostridia in human infection: evaluation of 114 cases. J Infect Dis 1975; 131:S81–85.
18. Eyrich K, Agostini B, Schulz A, et al. Clinical and morphological studies of skeletal muscle changes in tetanus. German Med Month 1967; 12:469–474.
19. Hasan N, Baithun S, Swash M, Wagg A. Tuberculosis of striated muscle. Muscle Nerve 1993; 16:984–5.
20. Griffiths E. Tuberculosis of striated muscle. J Bone Joint Surg 1956; 38-B:855–861.
21. Job CK, Karat ABA, Karat S, Mathan M. Leprous myositis—a histopathological and electron-microscopic study. Leprosy Rev 1969; 40:9–16.
22. Jopling WH, Mehta HD. A case of leprous nodular interstitial myositis. Leprosy Rev 1972; 43:39–43.
23. Ramu G, Balakrishnan S. Arthritis in lepromatous leprosy, clinical features and biochemical findings. Leprosy India 1968; 40:62.
24. Warner CL, Fayad PB, Heffner R. *Legionella* myositis. Neurology 1991; 41:750–752.
25. Kennedy DH, Love WC, Pinkerton IW, Borland WW. Rhabdomyolysis and systemic infection. Br Med J 1983; 286:1517.
26. Mastaglia FL, Ojeda VJ. Inflammatory myopathies. Part 1. Ann Neurol 1985; 17:215–227.
27. Mastaglia FL, Ojeda VJ. Inflammatory myopathies. Part 2. Ann Neurol 1985; 17:317–323.
28. Kakulas BA. Experimental myopathies. In: Walton J, ed. Disorders of Voluntary Muscle. Edinburgh: Churchill Livingstone, 1988:427–454.
29. Middleton PH, Alexander RM, Szymanski MT. Severe myositis during recovery from influenza. Lancet 1970; 2:533–535.
30. Ruff RL, Secrist D. Viral studies in benign acute childhood myositis. Arch Neurol 1982; 39:261–263.
31. Mejlszenkier JD, Safran AP, Healy JJ, Embree L, Ouellette EM. The myositis of influenza. Arch Neurol 1973; 29:441–443.

32. Greco TP, Askanase PW, Kashgarian M. Postviral myositis: myxovirus-like structures in affected muscle. Ann Intern Med 1977; 86:193–194.
33. Seay AR, Griffin DE, Johnson RT. Experimental viral polymyositis: age dependency and immune response to Ross river virus infection in mice. Neurology 1981; 31:656–661.
34. Berlin BS, Simon NM, Bovner RN. Myoglobinuria precipitated by viral infection. JAMA 1974; 227:1414–1415.
35. Jehn UW, Fink MK. Myositis, myoglobinaemia, and myoglobinuria associated with enterovirus echo 9 infection. Arch Neurol 1980; 37:457–458.
36. Wright J, Conchonnai G, Hodges GR. Adenovirus type 21 infection: concurrence with pneumonia, rhabdomyolysis and myoglobinuria in an adult. JAMA 1979; 241:2120–2121.
37. Schlesinger JJ, Gandara D, Bensch KG. Myoglobinuria associated with herpes group viral infections. Arch Intern Med 1978; 138:422–424.
38. Rothstein TL, Kenny GE. Cranial neuropathy, myeloradiculopathy, and myositis. Arch Neurol 1979; 39:261–263.
39. Joseph TP, Chand RP, Tariq SM, Johnston WJ, Muirhead D, Buhl L. Acute proximal myopathy due to herpes zoster. J Roy Soc Med 1993; 86:360.
40. Modlin JF. Coxsackiviruses, echoviruses and newer enteroviruses. In: Mandell GL, Douglas RG, Bennett JE, eds. Principles and Practice of Infectious Diseases. New York: Churchill Livingstone, 1990:1367–1383.
41. Schwartz MS, Swash M, Gross M. Benign postinfection polymyositis. Br Med J 1978; 2:1256–1257.
42. Archard LC, Bowles NE, Behan PO, Bell EJ, Doyle D. Postviral fatigue syndrome: persistence of enterovirus RNA in muscle and elevated creatine kinase. J Roy Soc Med 1988; 81:326–9.
43. Gow J, Behan W, Clements G, Woodall C, Riding M, Behan P. Enteroviral sequences detected by polymerase chain reaction in muscle of patients with postviral fatigue syndrome. Br Med J 1991; 302:692–696.
44. Gold D, Bowden R, Sixbey J, et al. Chronic fatigue. A prospective clinical and virologic study. JAMA 1990; 264:48–53.
45. Bowles NE, Bayston TA, Zhang HY, et al. Persistence of enterovirus RNA in muscle biopsy samples suggests that some cases of chronic fatigue syndrome result from a previous inflammatory viral myopathy. J Med 1993; 24:145–160.
46. Yousef GE, Isenberg DA, Mowbray JF. Detection of enterovirus specific RNA sequences in muscle biopsy specimens from patients with adult onset myositis. Ann Rheum Dis 1990; 49:310–315.
47. Richardson PJ, Why HJF, Archard LC. Virus infection and dilated cardiomyopathy. Postgrad Med J 1992; 68(Suppl 1):S17–S20.
48. Leff RL, Love LA, Miller FW, et al. Viruses in idiopathic inflammatory myopathies: absence of candidate viral genomes in muscle. Lancet 1992; 339:1192–1195.
49. Anderson BA, Young PV, Kean WF, Ludwin SK, Galbraith PR, Anastassiades TP. Polymyositis in chronic graft vs host disease. Arch Neurol 1982; 39:188–190.
50. Levy RM, Bredesen DE, Rosenblum ML. Neurological manifestations of the acquired immune deficiency syndrome (AIDS): experience at the UCSF and review of the literature. J Neurosurg 1985; 62:475–495.
51. Ledford DK, Overman MD, Gonzalvo A, Cali A, Mester SW, Lockey RF. Microsporidiosis myositis in a patient with the acquired immunodeficiency syndrome. Ann Intern Med 1985; 102:628–630.
52. Sowa JM. Human T lymphotrophic virus I, myelopathy, polymyositis and synovitis: an expanding rheumatic spectrum. J Rheumatol 1992; 19:316–318.
53. Jarowski CI, Fialk MA, Murray HW, et al. Fever, rash and muscle tenderness, a distinctive clinical presentation of disseminated candidiasis. Arch Intern Med 1978; 138:544–546.
54. Kressell B, Szewczyk C, Tuazon C. Early clinical recognition of disseminated candidiasis by muscle and skin biopsy. Arch Intern Med 1978; 138:429–433.
55. Arena FP, Perlin M, Brahman H, Weiser B, Armstrong D. Fever, rash, and myalgias of disseminated candidiasis during antifungal therapy. Arch Intern Med 1981; 141:1233.
56. Reimers CD, deKoning J, Neubert U, et al. *Borrelia burgdorferi* myositis: report of eight patients. J Neurol 1993; 240:278–283.
57. Matsubara S, Takamori M, Adachi H, Kida H. Acute *Toxoplasma* myositis: an immunohistochemical and ultrastructural study. Acta Nueropathol 1990; 81:223–227.
58. Topi GC, D'Alessandro L, Catricala C, Zardi O. Dermatomyositis-like syndrome due to *Toxoplasma*. Br J Dermatol 1979; 101:589–591.
59. Pallis CA, Lewis PD. Involvement of human muscle by parasites. In Walton J, ed. Disorders of Voluntary Muscle. Edinburgh: Churchill Livingstone, 1988:611–627.
60. Liu CT, Roberts LM. Sarcospiridiosis in a Bantu woman. Am J Clin Pathol 1965; 44:639–641.
61. Kirchoff LV. American trypanosomiasis (Chagas' disease)—a tropical disease now in the United States. N Engl J Med 1993; 329:639–644.
62. Swash M, Schwartz MS. Malaria myositis. J Neurol Neurosurg Psychiatry 1993; 56:1328.

63. Miller MD, Whilte NJ, Lott JA, et al. Biochemical evidence of muscle injury in African children with severe malaria. J Infect Dis 1989; 159:139–142.
64. DeSilva HJ, Goonetilleke AKE, Senaratna N, et al. Skeletal muscle necrosis in severe falciparum malaria. Br Med J 1988; 296:1039.
65. Jacob JC, Mathew NT. Pseudohypertrophic myopathy in cysticercosis. Neurology 1968; 18:767–771.
66. Radford AJ. Hydatid disease. Weatherall DJ, Ledingham JGG, Warrell DA, eds. Oxford Textbook of Medicine. Oxford: Oxford University Press, 1987:5.561–5.562.
67. Schimrigk K, Emser W. Parasitic myositis caused by *Echinococcus alveolaris.* Eur Neurol 1978; 17:1–7.
68. Ali-Khan Z, Irving RT, Wignall N, Bowmer EJ. Imported sparganosis in Canada. Can Med Assoc J 1973; 108:590–593.
69. Ochoa J, Pallis C. *Trichinella* thrives in both oxidative and glycolytic human muscle fibres. J Neurol Neurosurg Psychiatry 1980; 43:281–282.
70. Dent JH, Nichols RL, Beaver PC, Carrera GM, Staggers RJ. Visceral larva migrans with a case report. Am J Pathol 1956; 32:777–803.
71. Grem JL, Neville AJ, Smith SC, Gould HR, Love RR, Trump DL. Massive skeletal muscle invasion by lymphoma. Arch Intern Med 1985; 145:1818–1820.
72. Pearson CM. The incidence and type of pathologic alterations observed in muscles in a routine autopsy survey. Neurology 1959; 9:757–766.
73. Doshi R, Fowler T. Proximal myopathy due to discrete carcinomatous metastases to muscle. J Neurol Neurosurg Psychiatry 1983; 46:358–360.
74. Brain R, Henson RA. Neurological syndromes associated with carcinoma. Lancet 1958; 2:971–975.
75. Heffner RR. Myopathy of embolic origin in patients with carcinoma. Neurology 1971; 21:840–846.
76. Ueyama H, Kumamoto T, Araki S. Circulating autoantibody to muscle protein in a patient with paraneoplastic myositis and colon cancer. Eur Neurol 1992; 32:281–284.
77. Vincent A, Lang B, Newsom-Davis J. Autoimmunity to the voltage-gated calcium channel underlies the Lambert-Eaton myasthenic syndrome, a paraneoplastic disorder. Trends Neurosci 1989; 12:496–502.
78. Schulman P, Kerr LE, Spira H. A reexamination of the relationship between myositis and malignancy. J Rheumatol 1991; 18:1689–92.
79. Sigurgeirsson B, Lindelof B, Edhag O, Allander E. Risk of cancer in patients with dermatomyositis or polymyositis. N Engl J Med 1992; 326:363–367.
80. Snaith ML. Clinical conundrum—How assiduously should one investigate for occult malignancy in an elderly or middle aged patient with dermatomyositis or polymyositis? Br J Rheumatol 1990; 29:334.
81. Swash M, Fox KP, Davidson AR. Carcinoid myopathy. Arch Neurol 1975; 32:572–574.
82. Clements PJ, Furst DE, Campion DS, et al. Muscle disease in progressive systemic sclerosis. Arthritis Rheum 1978; 21:62–71.
83. West SG, Killian PJ, Lawless OJ. Association of myositis and myocarditis in progressive systemic sclerosis. Arch Rheumatol 1981; 24:662–667.
84. Foote RA, Kimborough SM, Stevens JC. Lupus myositis. Muscle Nerve 1982; 5:65–68.
85. Tsokos GC, Moutsopoulos HM, Steinberg AD. Muscle involvement in systemic lupus erythematosus. JAMA 1981; 246:766–768.
86. Halla JT, Koopman WJ, Fallahi S, Oh SJ, Gay RE. Rheumatoid myositis. Clinical and histological features and possible pathogenesis. Arthritis Rheum 1984; 27:737–743.
87. Haslock DI, Wright V, Harriman DGF. Neuromuscular disorders in rheumatoid arthritis. Q J Med 1970; 39:335–358.
88. Chevalier X, Larget-Piet B, Gherardi R. Proliferative myositis as a complication of rheumatoid vasculitis. J Rheumatol 1993; 20:1259–1260.
89. Carraba M, Cevallard M, Colombo B, Dworzak F, Mora M, Cornelio F. Muscle pathology in ankylosing spondylitis. Clin Exp Rheumatol 1984; 2:139–144.
90. Crowe WE, Bove KE, Levinson JE, Hilton PK. Clinical and pathogenetic implications of histopathology in childhood polydermatomyositis. Arthritis Rheum 1982; 25:126–139.
91. Banker BQ, Chester CS. Infarction of the thigh muscle in the diabetic patient. Neurology 1973; 23:667–677.
92. Harriman DGF. Ischaemia of peripheral nerve and muscle. J Clin Pathol 1977; 11(Suppl):94–104.
93. Farinon AM, Marbini A, Gemignani F, et al. Skeletal muscle and peripheral nerve changes caused by chronic arterial insufficiency. Clin Neuropathol 1984; 3:240–252.
94. Devereux S, Knowles SM. Rhabdomyolysis and acute renal failure in sickle cell anaemia. Br Med J 1985; 290:1707.
95. Dorwart BB, Gabuzda TG. Symmetric myositis and fasciitis: a complication of sickle cell anaemia during vasoocclusion. J Rheumatol 1985; 12:590–595.
96. Comi G, Testa D, Cornelio F, Comola M, Canal N. Potassium depletion myopathy: a clinical and morphological study. Muscle Nerve 1985; 8:17–21.
97. Rubenstein AE, Wainapel SF. Acute hypokalaemic myopathy in alcoholics. Arch Neurol 1977; 34:553–555.

98. Shils ME. Experimental human magnesium depletion. Medicine 1969; 48:61–82.
99. Heaton FW, Pyrah LN, Beresford CC, Bryson R, Martin D. Hypomagnesaemia in chronic alcoholism. Lancet 1962; 2:802–805.
100. Schott GD, Wills MR. Myopathy in hypophosphataemic osteomalacia presenting in adult life. J Neurol Neurosurg Psychiatry 1975; 38:297–304.
101. Lazaro RP, Kirschner HS. Proximal muscle weakness in uraemia. Arch Neurol 1980; 37:555–558.
102. Floyd M, Ayyar DR, Barwick DD, Hudgson P, Weightman D. Myopathy in chronic renal failure. Q J Med 1974; 43:509–524.
103. Bautista J, Gil-Neuja E, Castilla J, Chinchon I, Rafel E. Dialysis myopathy. Report of 13 cases. Acta Neuropathol 1983; 61:71–75.
104. Mitch WE, Jurkovitz C, England BK. Mechanisms that cause protein and amino acid catabolism in uremia. Am J Kidney Dis 1993; 21:91–95.
105. Peer G, Blum M, Aviram A. Relief of haemodialysis-induced muscular cramps by nifedipine. Dial Transplant 1983; 12:180–181.
106. Larsson J, Lenmarken C, Martensson J, Sandstedt S, Vinnars E. Nitrogen requirements in severely injured patients. Br J Surg 1990; 77:413–416.

33
Glycogen Storage Diseases of Muscle

Corrado Angelini, Andrea Martinuzzi, and Lodovica Vergani
University of Padua, Padua, Italy

INTRODUCTION

Glycogen storage diseases (Table 1) are a group of disorders characterized by abnormal glycogen metabolism. Most of the enzymes that play a part in glycogen homeostasis differ in their primary structure or regulation between liver and skeletal muscle, in accordance with the fact that glycogen in the two tissues has a very different function. Glycogen in the liver is mainly utilized to keep blood glucose level constant while in the muscle it provides fuel for energy production during high-intensity acute exercise (1). Glycogen storage diseases can thus be divided in two groups: one affecting primarily liver, the other affecting primarily skeletal muscle. We will focus on the latter group. Defects of glycolytic enzymes are also treated here, since many clinical and pathophysiological aspects are shared, even though glycogen accumulation is not found in the disorders of "distal" glycolysis.

GLYCOGENOSIS TYPE II—ACID MALTASE DEFICIENCY

Clinical Presentation

Symptoms of acid maltase deficiency (AMD) or acid α-1,4-1,6 glucosidase deficiency may be seen in infants, children, or adults (2). In the infantile form (Pompe's disease) glycogen accumulates in all tissues, but especially in heart, skeletal muscle, and the central nervous system (CNS). The child is usually floppy and presents with rapid deterioration leading to death. In some children the condition mimics Werdnig- Hoffmann disease; these children present with respiratory failure and need assisted ventilation, eventually becoming ventilator-dependent. Early diaphragmatic paralysis is the cause of respiratory failure. In the childhood-and juvenile-onset variants clinical manifestations are usually limited to skeletal muscle, causing gait difficulties and simulating muscular dystrophy or polymyositis. In the adult form two different clinical patterns can be seen:

A progressive limb girdle muscular weakness
A clinical type with selective weakness of respiratory muscles presenting with episodes of respiratory insufficiency, characterized by dyspnea and low blood gases

Atypical presentations can also be seen: on computerized-tomography studies AMD patients may show posterior paraspinal muscular atrophy (3); Kretzschman et al. (4) described a late-onset AMD family with involvement of liver and skeletal muscle, aneurysms, and vacuolar degeneration of cerebral arteries. These patients showed hyperlucent white matter areas on CT and nuclear magnetic resonance (NMR) studies. Neuropathological studies showed vacuolar degeneration of small and large blood

Table 1 Glycogen Storage Disease

Type	Enzyme	Age of onset	Tissue involvement	Clinical presentation
II	Acid maltase	Infantile Childhood Adult	Generalized	Floppy baby Cardiomyopathy Limb weakness Respiratory insufficiency
III	Debrancher	Infantile Adult	Heart, liver, muscle	Hepatosplenomegaly Distal weakness
IV	Brancher	Childhood	Liver, spleen, muscle	Hepatosplenomegaly Cirrhosis
V	Myophosphorylase	Child Adult	Skeletal muscle	Myoglobinuria Weakness
VIa	Phosphorylase B kinase	Adult	Liver, muscle	Hepatomegaly Growth retardation
VII	Phosphofructokinase	Adult	RBC, muscle	Cramps, myoglobinuria
IX	Phosphoglycerokinase	Childhood	RBC, CNS, muscle	Hemolytic anemia, seizures, myopathy
X	Phosphoglyceromutase	Adult	Muscle	Cramps, myoglobinuria
XI	Lactate dehydrogenase	Adult	Muscle	Myoglobinuria

vessels that were filled with lysosomal bound glycogen. The formation of aneurysms in intracerebral arteries was explained as the result of cell damage in the vessel walls.

Genetics

Biochemical studies of family members in infantile, childhood, or adult AMD deficiency are consistent with autosomal recessive inheritance (Fig. 1). In one study (5) the asymptomatic mother of two brothers with juvenile-onset AMD had AM activity in the homozygote range. It was supposed that she was homozygous for adult AMD and the children could be genetic compounds of adult and infantile AMD.

A similar hypothesis was proposed by Koster et al. (6) in a family where the paternal grandfather was thought to be a genetic compound of adult-onset and infantile AMD, his son and daughter were in the low heterozygote range for infantile AMD, and the child was an infantile AMD homozygote. Single-base-pair missense mutation or single-codon mutation was recently reported in patients with Pompe's disease in one allele (7). Missense mutations on the other allele have been identified in infants (7a); thus, at least some infantile cases of acid maltase deficiency on compound heterozygotes have been reported. Using an acid α-glucosidase cDNA probe to human chromosomes, Halley et al. (8) found, by in situ hybridization, direct regional mapping of the α-glucosidase locus to the region q23–>q25 of chromosome 17. Studies have shown that the gene for acid maltase is located on chromosome 17, while neutral maltase is on chromosome 15 (9).

Pathology

Muscle biopsies show a vacuolar myopathy by light microscopy, with large deposits of PAS-positive material. The glycogen is lysosome-bound, but in some fibers large deposits of free glycogen are found. The histoenzymatic reaction for acid phosphatase is markedly increased in affected muscle, confirming lysosomal activation. Increased amounts of free and membrane-bound glycogen are also seen within neurons (10) and in vascular and smooth muscle of arteries and can be seen in cultured fibroblasts and muscle. Mitochondrial abnormalities have also been described. In adult AMD the disease is focal and the characteristic autophagic phenomenon can be seen in isolated fibers (Fig. 2A,B).

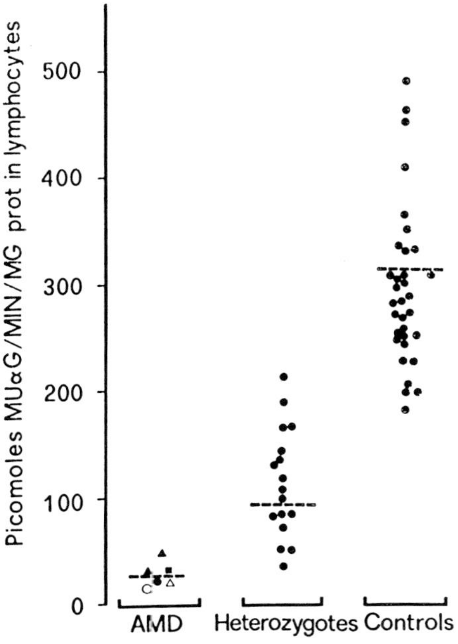

Figure 1 α1,4-1,6 glucosidase activity in white blood cells in AMD patients and heterozygote relatives. The intermediate activity in heterozygotes supports autosomal recessive inheritance.

Biochemistry

There is no detectable activity of α-glucosidase in the muscle homogenate of infantile patients. In juvenile and adult AMD a residual activity (10–20% of control) is present, but there is no correlation between residual activity and clinical presentation.

Acid maltase has been isolated from human muscle and has been shown to have a lysosomal localization and to sediment with acid phosphatase and cathepsin D (11). AMD has been observed also in Japanese quail, a useful animal model of glycogenosis. The activity of the enzyme at pH 4.0 derived from control quail pectoral muscle, sedimented with cathepsin D confirming a lysosomal origin, and on immunoblotting gave a 98-kDa band and a 110-kDa band (12). Two bands were also found in liver and both had α-glucosidase activity. The 98-kDa band however, had a sixfold greater activity compared to the other. In the affected quail only the 110-kDa band was detected, localized in a low-density fraction containing membrane soluble protein. These results indicate that while the mature lysosomal 98-kDa acid maltase (AM) was absent, its 110-kDa precursor form was present outside lysosomes. Lysosomal enzymes are synthesized as precursor forms on polysomes bound to endoplasmic reticulum, modified by proteolytic cleavage, and transported to lysosomes. In control human fibroblasts, an AM is synthesized as a precursor and then processed to the mature form. In human fibroblasts from late-onset AMD patients the enzyme defect seems to involve the early steps of enzyme maturation. Reuser et al. (13) have shown that the precursor form of acid α-glucosidase can be synthesized, but its processing to the mature form does not occur or is deficient in AMD fibroblasts.

Pathophysiology

Intralysosomal glycogen storage in the affected patients is the consequence of either total α-glucosidase deficiency (infantile cases) or the absence of the mature lysosomal form. The molecular etiology of these heterogeneous clinical presentations has not yet been worked out. Also, it is not yet well understood why in adult AMD not all muscles are affected equally and selective involvement of

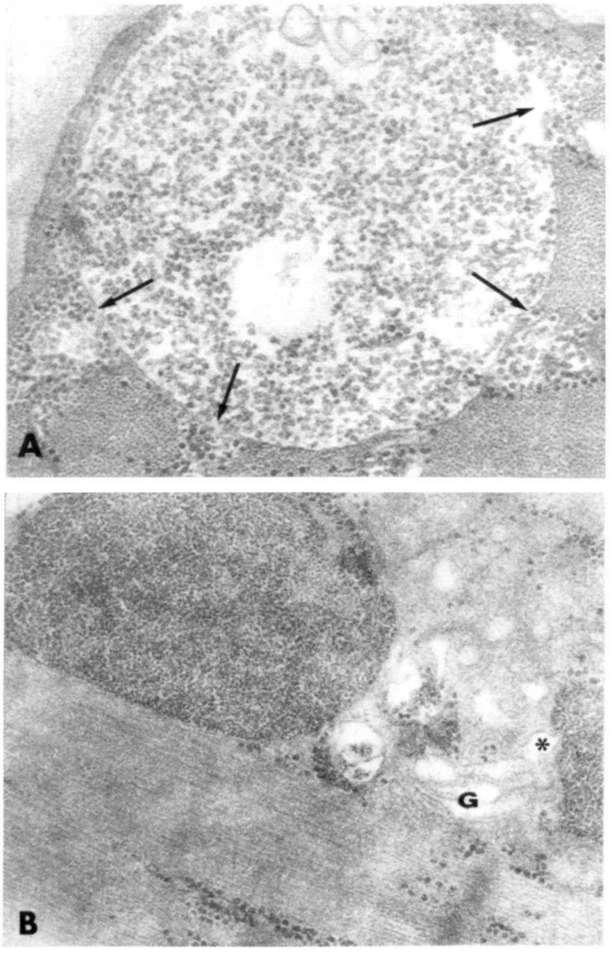

Figure 2 (A) Electron microscopy of a biopsy of AMD patient shows a vacuole with a simple or double limiting membrane filled with glycogen. Arrows indicate continuity of intra- and extralysosomal glycogen. (B) AMD biopsy shows two lysosomes filled with glycogen. Between autophagic vacuoles, G indicates Golgi complex, formed from small cisterns and dilated vescicles (*). Some vesicles are in the proximity of lysosome and may fuse with them. (From Ref. 158.)

respiratory muscles may appear. The muscle weakness has been attributed to the disruption of muscle fibers due to accumulation of glycogen in lysosomes. It is also conceivable that compensatory mechanisms coexist and that muscle protein is utilized as an alternative energy source.

Treatment

Several replacement treatments have been tried in infantile AMD without success. Enzyme replacement with an enzyme derived from *Aspergillus niger* was not successful because of immunological reactions. Pharmacological agents directed toward stabilization of lysosomes such as vitamin A or progesterone did not produce clinical benefit. Slonim et al. (14) showed moderate increase in fasting branched-chain amino acid concentrations in a 2-year-old boy with AMD. After 4 months on a high-protein diet (30% of total calories), there was a dramatic increase in strength and exercise tolerance. In adult AMD a low-carbohydrate diet (40 mg) associated with epinephrine injection has been used by Engel (15) to stabilize patients. Margolis and Hills (16) reported a case of 55-year-old man with AMD and dyspnea on exertion, who was put on a high-protein diet (47% of total calories) and had a good clinical response with improvement of respiratory parameters. Both Umpleby (1987) and Isaacs (1986) reported benefit in adult AMD with a high-protein diet. In a 43-year-old AMD patient who needed constant ventilatory support, Mobentian et al. (17) found improvement with a general diet supplemented with branched-chain amino acids (BCAA); while on this treatment, the patient gained strength and was able to be weaned from ventilator during the day. The effect of BCAA and a high-protein diet could be due to either decreased proteolysis or increased muscle protein synthesis.

DEBRANCHING ENZYME DEFICIENCY (CORI-FORBES DISEASE, GLYCOGENOSIS TYPE III, LIMIT DEXTRINOSIS)

This disease was initially described in patients presenting with liver and/or muscle dysfunction with accumulation in either tissue of an abnormal polysaccharide (18–20). The polysaccharide is composed of glycogen with a very high number of short outer chains, and it resembles the end-product of glycogenolysis in the absence of debranching enzyme (phosphorylase limit dextrin). Amylo-1-6-glycosidase (debranching enzyme) activity was shown to be defective in muscle (21), leukocytes, and fibroblasts (22). Over 120 patients have been described thus far.

Clinical Presentation

There are two main phenotypes of the disease: a predominantly hepatic form and a muscular form. The hepatic form presents in childhood with hepatomegaly, failure to thrive, and fasting hypoglycemia with high ketones and high triglycerides and cholesterol. The clinical picture is similar to a mild form of glucose-6-phosphatase deficiency (glycogenosis type I, von Gierke's disease). Symptoms tend to regress during early adulthood, but fatal cases with cardiomyopathy have been described. Fifteen percent of the patients with liver symptoms never develop extrahepatic problems. Enzymatic analysis in these cases shows debranching enzyme deficiency in liver but not in muscle (type IIIb glycogenosis), while in the majority of patients the defect is present in both tissues (type IIIa glycogenosis) (21). The muscle is usually affected later in life. Muscle involvement occurs in patients who may or may not have had liver symptoms in childhood. Patients complain of slowly progressive weakness and wasting in distal muscles (27), exercise intolerance, and, in exceptional cases, myoglobinuria (23). Cardiomyopathy is present in half the patients (both children and adults) and might be the most important prognostic parameter (28,29). Cardiomyopathy is notably absent in patients with debranching deficiency in the liver and normal activity in the muscle (29). In one case peripheral nerve involvement led to gait disturbances (30). There was no sex predominance in the larger reported series (23).

Genetics

The disease is transmitted as an autosomal recessive trait. Frequency is elevated (1/5400 live birth) in non-Ashkenazi Jews of North Africa (31). The gene is located on chromosome 1p21 (24). A single locus codes for the enzyme and no related sequences have been identified. The disease has not been associated thus far with specific mutations.

Laboratory Findings

In children presenting with liver disease, there are high serum transaminases, lactate, urate, triglycerides, and cholesterol. In all patients presenting with either liver or muscle involvement, glucagon administration (1.5 mg glucagon hydrochloride i.v.) fails to induce hyperglycemia in the fasting state; glucose tolerance test shows a "diabetic" response. Normal or even increased hyperglycemic response to galactose (1 g/kg i.v. or 1.5 g/kg orally) differentiates debranching deficiency from glucose-6-phosphate deficiency. Serum creatine kinase (CK) is moderately or even markedly elevated. The ischemic forearm exercise test (see Myophosphorylase Deficiency for details) shows low or absent postexercise lactate peak. Electrocardiographic (ECG) and echocardiographic abnormalities (ventricular hypertrophy) are present in more than 50% of patients. Electromyography (EMG) shows myopathic changes; slow conduction velocities have been described in some patients (30) but not in others (32).

Pathology

Muscle biopsy shows non-membrane-bound glycogen in vacuoles mostly in a subsarcolemmal position and mainly in type II fibers. The vacuoles are filled with granular PAS-positive material and look empty after amylase digestion. Glycogen accumulation in skeletal muscle fibers does not occur in all cases of debrancher enzyme deficiency; thus histochemical studies cannot be used to rule out this diagnosis. Ring fibers are sometimes described (33). Ultrastructural studies show large pools of glycogen in subsarcolemmal and intermyofibrillar position. Isolated glycogen-filled lysosomes are also observed. Mitochondria are normal. Liver biopsy shows intracellular glycogen accumulation. Glycogen accumulation has also been described in cultured muscle (22), skin (mostly sweat glands and smooth muscle) (32), Schwann cells (34), and erythrocytes. Glycogen content is increased in most tissues. In skeletal muscle glycogen concentration is particularly high (>3 g/100 ml). Determination of glycogen content in red blood cells is a good screening test for debranching enzyme deficiency and phosphorylase b kinase deficiency: in these disorders glycogen content in erythrocytes is higher than 10 mg/100 ml (35). Amniocytes show accumulation of abnormal glycogen when grown in glucose-free medium (36).

Biochemistry and Molecular Biology

Glycogen debranching enzyme is a monomeric protein (predicted size from cDNA sequence 172,614 D) (37) with two enzymatic activities: α-1,4-glucan transferase (E.C. 2.4.1.25) (transferring short oligoglucan chains of 3–4 glucose units from one glycogen outer chain to the top of another) and amylo-1,6-glucosidase (E.C. 3.2.1.33) (hydrolyzing the 1,6 branch point of the remaining glucose residue) (38). The two activities are present in different sites on the same molecule (39) and are coordinated with glycogen phosphorylase activity in completely degrading glycogen to glucose-1-phosphate (93%) and glucose (7%) (40). Only one enzyme is expressed in all tissues, as shown also by mRNA analysis (37), but there have to be different regulatory mechanisms to account for the tissue specific deficiency (41). Patients can be subdivided into four subgroups (21,29,41) based on detection of residual activity in muscle or liver:

Deficiency of both enzymatic activities in muscle and liver (IIIa); no (or minimal) cross-reacting material is detected by immunoblot with specific antibodies.

Deficiency of both enzymatic activities only in liver (IIIb); decreased immunoreactive material only in liver.

Deficiency of both enzymatic activities in the muscle but not in the liver (IIIc, one case reported).

Deficiency of transferase activity in both liver and muscle, normal glucosidase activity (IIId); normal or near-normal amount of cross-reacting material in both tissues.

Type IIIa is the most common form (75% of the cases), followed by type IIIb (15%), and type IIId (7%). Clinical severity is correlated with the extent of the biochemical defect (generalized to all tissues or restricted to liver only) but not with the degree of the defect (presence of residual activity) (29,42). Debranching activity can be tested in blood cells (35) and fibroblasts (43), providing good diagnostic resolution with non- or minimally invasive procedures. Northern blot analysis in 10 patients confirmed the molecular heterogeneity of the disease: four patients had a reduced amount of normal size mRNA, one had truncated mRNA, three had a normal amount of normal-size mRNA (37,44).

Therapy

Nocturnal intragastric feeding in children prevents hypoglycemia and enhances growth (45). Oral cornstarch therapy was recently shown to be effective in the initial treatment of young patients (46). No effective treatment is currently available for the adult myopathy.

BRANCHING ENZYME DEFICIENCY (ANDERSEN'S DISEASE, TYPE IV GLYCOGENOSIS, AMYLOPECTINOSIS)

This disease was described in 1952 in a family with cirrhosis and accumulation of abnormal glycogen with amylopectin-like structure. The stored glycogen has a very low frequency of branch points (3.5% vs. 6.7% of normal glycogen) and is partially resistant to amylase digestion (47). The actual defect of branching enzyme was demonstrated by Brown and Brown in 1966 (48). The disease is very rare, presents in early childhood, and usually leads to death before age 3 (48). Thirty-two patients have been reported thus far (23). Recently branching enzyme defect was described in two Ashkenazi Jewish patients with polyglucosan body disease (49,50).

Clinical Presentation

Branching enzyme deficiency typically presents in childhood, with cirrhosis leading to fatal hepatic failure. Neuromuscular involvement in these patients can be recognized as hypotonia, hyporeflexia, atrophy, and contractures, but is overshadowed by the hepatic dysfunction. Cardiopathy, when present, can be severe and dominates the clinical picture (51). Two clinical pictures affecting primarily the central nervous system and the skeletal muscle have recently been reported: a mild juvenile variant characterized by chronic progressive myopathy (52) and some cases of adult polyglucosan body disease, characterized by late-onset upper motor neuron signs and peripheral neuropathy (53–58).

Genetics

In the familial cases described the transmission seems autosomal recessive. Such mode of inheritance is confirmed by the finding of intermediate levels of enzyme activity in the leukocytes or fibroblasts of patients' relatives (49). The DNA for human glycogen branching enzyme has recently been cloned and the gene was assigned to chromosome 3 (58a).

Laboratory Findings

No specific abnormality is associated with the disease.

Pathology

The disease hallmark is the accumulation of amylopectin-like material in affected tissues. Ultrastructural studies show cytoplasmic deposits consisting of a mixture of normal and granular or filamentous diastase-resistant glycogen (54). The diastase-resistant glycogen has abnormally long outer chains that can be cleaved by isoamylase. Glycogen concentration is not usually increased in skeletal muscle, but subsarcolemmal deposits of abnormal glycogen have been described (55). In central and peripheral nervous tissue the deposits have the characteristics of polyglucosan bodies, including immunoreactivity to KM-279 antibody (52), which is shared with Lafora bodies and corpora amylacea.

Biochemistry

Branching enzyme acts by transferring chains of at least six glucose residues from the outer chains of glycogen to a 1,6 position. Purified enzyme from rabbit liver is a monomeric protein of M_r 70 kDa (56). There is no supporting evidence for multiple isoforms, even though the enzyme deficiency may occur in a tissue-specific pattern (52–58). Branching activity is measured by an indirect assay (23) in leukocytes, erythrocytes, fibroblasts, liver, muscle, or nerve. Patients with the typical hepatic form of branching enzyme deficiency have reduced or absent enzyme activity in all tissues. Prenatal diagnosis is possible (57). Patients with the mild juvenile variant have defective branching activity in muscle, but not in blood cells (52). Patients presenting with adult polyglucosan body disease have defective activity in muscle (50). The molecular basis of the tissue-specific distribution of the defects is still unclear.

Therapy

Liver transplantation can be proposed for children with the hepatic form (58). No effective treatment is otherwise available.

OTHER AMYLOID-LIKE STORAGE MYOPATHIES: POLYGLUCOSAN BODY DISEASE

In addition to the glucosan storage myopathies with defined enzymatic defect, some cases have been reported with overloading of muscle cells or peripheral nerves with an amyloyd-like substance. The clinical picture is heterogeneous, and in a collection of 15 cases reported by Cafferty et al. (53) all had an axonal peripheral neuropathy and upper motor neuron signs, and dementia was present in 8/15 cases. Sural nerve biopsy showed clusters of polyglucosan bodies. In some respects the storage material appeared as filamentous bodies surrounded by glycogen particles, resembling those seen in type IV glycogenosis, with Lafora bodies, corpora amylacea, and basophilic degeneration of the myocardium. The material was PAS-positive, stained with iodine, and was partially degraded by amylase. Branching enzyme activity in muscle extracts was normal.

MYOPHOSPHORYLASE DEFICIENCY (MCARDLE'S DISEASE, GLYCOGENOSIS TYPE V)

This was the first biochemical defect to be described in muscle (59). It is characterized by failure of muscle glycogen breakdown to lactic acid, and it is due to lack of skeletal muscle glycogen phosphorylase (M-GP) activity (60,61). Thus far approximately 140 patients have been reported.

Clinical Presentation (Table 2)

The typical presentation consists of exercise intolerance with muscle cramps, myalgia, and exercise-induced muscle stiffness, leading in some cases to myoglobinuria and possibly to renal failure. The episodes of muscle pain and stiffness are caused by exercise of short duration and high intensity or by

Table 2 Clinical Features of Myophosphorylase Deficiency

	No. of cases (literature[a])	%	No. of cases (our center)	%	Total	%
Exercise intolerance	44	91.7	11	91.7	55	91.7
Myoglobinuria	20	41.7	4	33.3	24	40.0
+ renal failure	4/20	20.0	0	0.0	4/24	16.7
Fixed weakness	17	35.4	5	41.6	22	36.6
Wasting			3	25.0		
Seizures			0	0.0		
Age at onset (years):						
<15	34	70.8	7	58.3	41	68.3
15–30	8	16.7	3	25.0	11	18.3
>50	6	12.5	2	16.7	8	13.3
Age at diagnosis (years)						
<10	2	4.2	0	0.0	2	3.3
10–30	26	54.2	6	50.0	32	53.3
30–50	10	20.8	3	25.0	13	21.7
>50	10	20.8	3	25.0	13	21.7
Positive family history	16	33.3	4	33.3	20	33.3

[a]Data from Servidei S et al, Ann Neurol, 1988; 24:774–781

more prolonged effort of intermediate intensity. The symptoms are relieved by rest. It is a common observation that the patient is able to sustain the effort for much longer if he/she is allowed to rest briefly after the first appearance of muscle pain. The patient becomes rapidly aware of this phenomenon and learns to adjust his/her activity accordingly. This "second wind" phenomenon (62,63) is probably due to activation of liver glycogenolysis and gluconeogenesis, and to mobilization of alternative fuels such as amino acids and free fatty acids (FFA).

Symptoms usually begin in childhood or early adulthood. In typical cases the disease does not show appreciable progression, and many patients remain symptom-free for most of their lives. In about a third of cases progressive muscle weakness becomes evident 5–10 years after onset of symptoms. There is a male-to-female ratio of 1.5:1 (64), which could be due to differences in muscle energetics or to hormonal influences, or to the fact that men are more often subjected to rigorous physical exercise, a condition capable of causing symptoms.

Atypical clinical presentations include:

- A fatal infantile form with early progressive muscle weakness leading to respiratory insufficiency and death in few months (65,66)
- A mild congenital form with delayed psychomotor development and proximal weakness (67)
- A late-onset form with exercise intolerance, cramps, sometimes myoglobinuria, and progressive weakness and wasting (68,69)
- A very mild form with tiredness and poor stamina but no cramps (70)

The incidence of seizures is reported to be higher than expected in patients with myophosphorylase deficiency. However, in a series of 13 patients we observed seizures in only one 43-year-old patient who had had posttraumatic epilepsy since age 23.

Since nervous tissue expresses brain-type glycogen phosphorylase (GP), the metabolic defect present in muscle should not impair glycolytic metabolism in the brain. It is still possible that the metabolic unbalance associated with severe rhabdomyolitic episodes is responsible for abnormal neuronal discharge.

Genetics

Family history is positive for 40–50% of the patients, with sibling cases being the most common observation. Transmission in these cases is autosomal-recessive; however, autosomal-dominant inheritance, and manifesting heterozygotes, have occasionally been observed (71–73). Recent work excluded deletions or insertions of >0.1 kb (73). An insertion/deletion polymorphism in the 3′ region downstream to the muscle glycogen phosphorylase (M-GP) gene, located on chromosome 11q 13 (74), and a MspI RFLP within the gene permit prenatal and postnatal diagnosis of 72% at-risk M-GP-deficient patients (73). Recently a non-sense mutation (C-to-T transition) at codon 49 of exon 1 of the M-GP gene has been found in homozygosis in about 50% of Caucasian patients (75,76). An additional 30% of patients carry the mutation of codon 49 in heterozygosis. In some of the patients heterozygous for R49X and in patients not carrying such mutation a different missense mutation (one of the following: codon 204, 291, 396, 542, 654, a splicing error at exon 14, or an insertion at codon 122) is found on one allele (76a–c). A single codon deletion (708/709) was identified in three Japanese patients (77). Ten to 20% of the patients thus far investigated do not harbor any of the described mutations. Italian patients in particular seem to fall in this latter group. These data confirm the considerable genetic heterogeneity in the etiology of M-GP deficiency.

Since most mutations can be identified by RFLP analysis of PCR products of genomic DNA purified from leukocytes, in almost 90% of the patients, molecular diagnosis can be made by simple blood sampling.

Laboratory Findings (Table 3)

High resting serum CK levels are seen in the majority of patients; the CK level may reach over 4000 U/L in some cases and rises much higher after episodes of cramps or myoglobinuria. The EMG is not characteristic and the ECG may show nonspecific abnormalities.

Table 3 Laboratory Data in Myophosphorylase Deficiency

	No. of cases (literature[a])	%	No. of cases (our center)	%	Total	%
Increased serum CK at rest	35/37	94.6	11/12	91.7	46/49	93.9
Abnormal EMG at rest	6/15	40.0	3/10	30.0	9/25	36.0
Muscle glycogen concentration						
<1.5			1/10	10	1/58	1.7
1.5–2.0%	11/48	22.9	1/10	10	12/58	20.7
2.1–3.0%	30/48	62.5	5/10	50	35/58	60.3
>3.0%	7/48	14.6	3/10	30	10/58	17.2

[a]Data from Servidei S et al, Ann Neurol, 1988; 24:774–781.

The ischemic forearm exercise test, originally described by McArdle, is still the most valuable noninvasive and inexpensive procedure by which a disorder of glycogenolysis can be demonstrated. A sphygmomanometer cuff is placed around the arm and inflated above arterial pressure; the patient is asked to squeeze energetically a rubber ball rhythmically until exhausted, or at least for 1 min, at which point the cuff is deflated. Blood samples for lactate, pyruvate, and ammonia are drawn from the antecubital vein of the same arm before and 1, 2, 5, and 10 min after the effort. The response in normal individuals (males or females) will be a >20 mg/dl rise in lactate and a >100 μg/dl rise in ammonia, with peak values between 3 and 7 min after the end of the effort (78). Patients with myophosphorylase deficiency show absent or reduced lactate peak, while there is a larger increase in ammonia (360–560 μg/dl). The observation of the ammonia rise is important in excluding false positive tests due to poor patient effort (79). Postexercise ammonia rise in M-GP deficiency patients is even more evident after incremental cycle ergometry (80,81). False negative results may be observed in patients with sizable residual M-GP activity. The test is not specific for M-GP deficiency, since blocks anywhere in the glycogenolytic or glycolytic pathway will lead to decreased lactate production after anaerobic exercise.

^{31}P NMR muscle spectroscopy, although time-consuming and expensive, is of special interest in patients with McArdle's disease; it shows no decrease in intracellular pH after exercise, low resting ATP, and no accumulation of phosphomonoesters (phosphorylated sugars) (82). ^{31}P NMR can be utilized for noninvasive diagnosis of asymptomatic patients (83).

Pathology

Light microscopy, muscle biopsy shows small deposits of PAS-positive, diastase-digested intracellular glycogen, often in subsarcolemmal "blebs." The histoenzymatic reaction for phosphorylase is negative in all striated muscle fibers. The absence of phosphorylase reaction is virtually diagnostic for the disorder.

In some biopsies small fibers with the characteristics of regenerating cells are positive with the phosphorylase stain; this can easily be seen in patients' biopsies, particularly after rhabdomyolytic episodes (Fig. 3A,B). The reappearing enzyme could be simply the fetal phosphorylase isoform, as suggested by some tissue culture studies (84,85), or could be the muscle isoform, transiently reexpressed in the regenerating cells, as other tissue culture studies suggest (86,87). Electron microscopy shows subsarcolemmal and intermyofibrillar accumulation of normal glycogen. An increased amount of free and membrane-bound glycogen was found within axons, Schwann cells, vascular smooth muscle, and endothelial cells that had been included within the skeletal muscle biopsy in a case of late-onset McArdle's disease (88). Mitochondrial abnormalities have also been described (89).

Biochemistry

There is no detectable activity of GP in the muscle homogenate of most patients. In a few cases GP activity (10–30% of normal control) is present, but there is no correlation between residual activity and severity of clinical presentation.

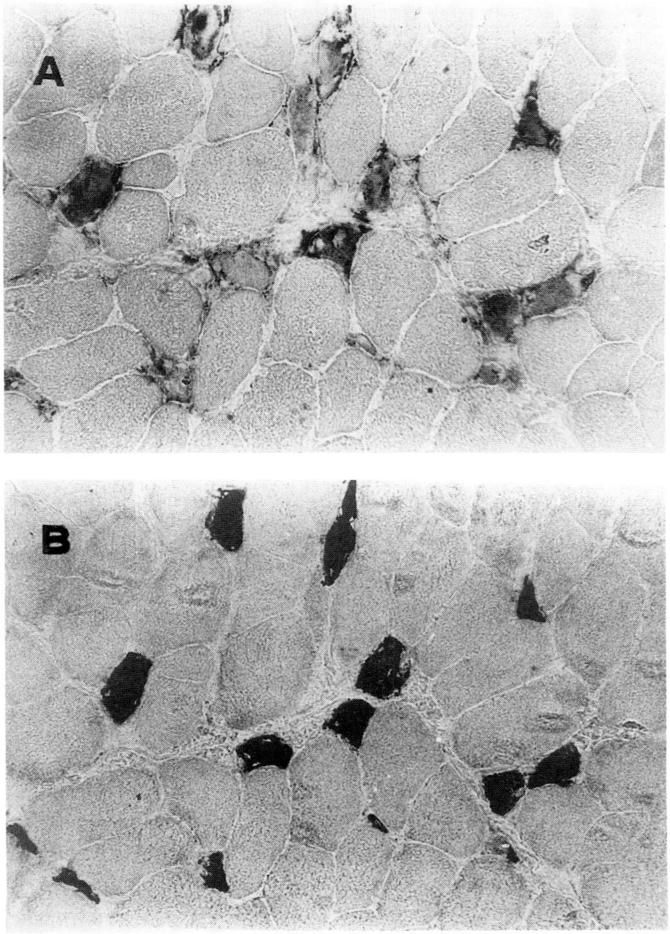

Figure 3 (A) Vacuolated basophilic fibers after a myoglobinuric episode in McArdle's disease (H&E). (B) Most regenerating fibers are positive for phosphorylase by histochemistry.

M-GP is the only isoform present in adult skeletal muscle (90). It is separately coded from the liver isozyme and from the brain isozyme (74,91,92). M-GP is also present in heart tissue together with the nonmuscle isoform and with a hybrid muscle-nonmuscle form (90). Fetal skeletal muscle has a prevalence of nonmuscle GP in early development, which shifts progressively to a M-GP-prevalent pattern during maturation (93). Identification and quantification of the M-GP isoform by immunoreaction with specific antiserum is a complementary study in muscle biopsy and muscle culture of patients with McArdle's disease. Immunological studies with antiserum against M-GP show absence of cross-reactive material in 86% of the cases, but reveal a minority of patients with low, but detectable cross-reacting inactive protein. In one case there was a normal amount of cross-reacting material (64,94) (Table 4). In two cases the concentration of cross-reacting material was studied in cardiac tissue and was found to be markedly decreased compared to controls (64).

Phosphorylase kinase activity and glycolysis downstream of glycogen breakdown are normal (95). Incubation of muscle homogenates with a high concentration of phosphorylase b kinase and ATP in four cases with detectable cross-reacting material resulted in reappearance of a normal active enzyme (96). This isolated observation was not confirmed in other cases.

Table 4 Analysis of Molecular Heterogeneity of McArdle's Disease

Phosphorylase protein/activity					Phosphorylase mRNA					
Biopsies	Method	Absent	Low	Normal	Biopsies	Absent	Low	Short	Normal	Ref.
48	ELISA	41	6	1	4	1	?	1	2	64
42	SDS-PAGE	35	6	1						
10	Activity	10	0	0						
	SDS-PAGE	10	0	0						
6	SDS-PAGE	4	2	0	—	—	—	—	—	a
11	Activity	10	1	0	9	6	1	0	2	87
12	Activity	8	4	0	10	4	—	—	6	this study
11	SDS-PAGE	10	1	0						
Total:										
87		73	13	1	31	16	4	1	10	

[a]Data from Mantle D et al, J Neurol Sci, 1987; 78:63–70.

Glycogen concentration in muscle is increased in all cases but in variable degree: between 150% and 200% of normal in 62% of cases, lower or higher in the remaining 38% of cases. Several groups have studied GP activity and M-GP immunoreaction in cultured muscle of patients with McArdle's disease. Only nonmuscle isoforms were identified in two studies (84,85), while in another two a normal amount of M-GP was reported (86,87). The possibility that M-GP is transiently expressed in patients' immature muscle raises important questions about the molecular etiology of the disease.

Molecular Biology (Table 5)

The clinical and biochemical heterogeneity shown by McArdle's disease is reflected at the molecular level. In recent years a number of studies have dealt with the analysis of M-GP mRNA in McArdle's patients (64,94,97,98) and the possible correlation of abnormalities at the transcriptional level with the biochemical and clinical picture.

M-GP mRNA was studied in the muscle biopsies of 31 patients (21 published cases, 10 studied at our center). Seventeen patients had no detectable mRNA, four had reduced amounts of normal-length (3.4 kb) transcript, nine had a normal amount of 3.4-kb transcript, and one showed a hybridization signal at 1.2 kb probably corresponding to a truncated mRNA (64). Patients with cross-reacting material identified by immunoblot had either a normal or reduced amount of 3.4-kb transcript.

Pathophysiology

The physiological consequences of the glycogenolysis block have been studied extensively using various approaches, and McArdle's disease has been a favourite model to investigate the pathophysiology of muscle fatigue (99).

Muscle fatigue in McArdle's disease is not related to ATP depletion, nor to lactic acid accumulation (100), but is associated with an increase of free ADP, Pi (101), ammonia (resulting from AMP and branched-chain amino acids (BCAA) deamination) (79,80), and potassium. High ADP concentrations inhibit myosin ATPase (102); ADP and Pi inhibit sarcolemmal Na^+/K^+-ATPase (103), K^+ activates ventilation and muscle-heart reflex (104,105); lack of pH fall inhibits muscle sympathetic reflex activation (106); ammonia rise correlates with heart rate and perceived exertion (81). There is no consensus on which is the determining factor of muscle fatigue; probably all mechanisms concur to some extent with the final manifestation of exhaustion.

The findings of reduced V_{O_2} max, increased ventilation, higher cardiac output, low respiratory rate, rise in Pi, inosine 5′-monophosphate, and ammonia, fall in acetylcarnitine, and stable level of NADH (which shows a large increase in normal subjects) (107,108) during exercise have been interpreted as

Table 5 Molecular Data in Glycogen Storage Diseases

Type	Enzyme defect	Chromosome location	Mutation	Ref.
II	Acid maltase (α-1,4-1.6 glucosidase)	17q23	M318T	7
			Δ exon 18	7
			Δ K 903	7a
			L 299 R	7a
III*a*	α-1,4-glucan transferase Amilo 1,6-glucosidase (muscle, liver)	1 p21	Unknown	
III*b*	Amilo 1,6-glucosidase (liver)			
III*c*	Amilo 1,6-glucosidase (muscle)			
III*d*	Only transferase (liver, muscle)			
IV	Branching enzyme	3	Unknown	
V	Muscle phosphorylase	11q13	R 49X	75
			G 204 S	75
			L 291 P	77
			L 396 P	76b
			L 542 T	75
			E 654 K	76b
			Δ 509 (G)	76a
			Δ F 708	77
			122 G → TT	76c
			Splice junction exon 14	76a
VI*a*	Phosphorylase B Kinase (PBK) (α,β,γ,δ subunits)	PBK-α 1 Xq12-q13 α 2 Xp22 PBK-β 16q12-q13 PBK-γ 7 p12-q21	unknown	
VII	Phosphofructokinase (PFK) (M, L, P subunits)	PFK-M 1 q 32	Δ splice site exon 13	141
		PFK-P 10p	Δ exon 5	141a
		PFK-L 21q22.3	Δ c-22	141b
			R 39P	141b
			R 606 H	141b
			D 209 G	141b
			R100 Q	141b
			Δ splice site exon 7	141c
			R 39P	141c
			D 543A	141c
IX	Phosphoglycerokinase (PGK)	Xq13	L 88P	144b
			G 157 V	144b
			A 205 P	144b
			V 265 M	144b
			D 267 N	144b
			C 315 R	144b
			I splice junction exon 4	144b
X	Phosphoglycerate mutase (PGAM)	PGAM-M 7p12-13	E 89 A	153
	(M,B subunits)	PGAM-B 10q 26.1	W 78 X	153
			R 90 W	153a
XI	Lactate dehydrogenase (LDH) (subunits M,H)	LDH-M 11p LDH-H 12p	Δ 724-42 exon 6	157

signs of impairment of oxidative metabolism, possibly due to decreased availability of substrates for the citric acid cycle (79,99). ^{31}P NMR data from other laboratories, however, seem to argue against this hypothesis, since no alteration of postexercise recovery of phosphocreatine/Pi ratio could be demonstrated (100). We also could demonstrate, in patients approaching exhaustion, a slight decrease in blood glucose, pyruvate, and lactate, a large increase in ammonia, and a postexercise rise of FFA, β-hydroxybutyrate, and medium-chain acyl-carnitines (109). These data support the view of a relative failure of readily available oxidative substrates, which is only partially and slowly compensated by an activation of β-oxidation. The importance of compensation through β-oxidation activation is suggested by the increased activity of β-OH–acyl-CoA dehydrogenase in muscle of McArdle's patients, which is not associated with an increase in the level of enzymes of Krebs cycle or high-phosphate metabolism (110).

Interesting clues to the understanding of the mechanisms of muscle exhaustion also come from electrophysiological studies. Surface EMG measurements of muscle fiber conduction velocity and of frequency spectrum during intermittent isometric contractions suggest that fatigue in McArdle's patients occurs without alterations in muscle membrane excitability and is due to failure of excitation-contraction coupling (111,112).

Therapy

No effective or established therapy is presently available for McArdle's patients. Dietary approaches have been tried with mixed results. A high-protein diet improved muscle weakness in one patient (113). The improvement was documented by ^{31}P NMR in another case (114). The effect was initially explained by suggesting that alanine and BCAA provide an alternative energy source. Further experiments, however, did not confirm this hypothesis (80,100). Glucose or fructose infusion is capable of preventing exhaustion and improving exercise capacity (99,100). Similar effects have been reported with high oral doses of ribose (116). These approaches, however, are not suitable as long-term therapy. Glucagon infusion (95,117) to promote muscle glucose utilization, norepinephrine, and heparin infusion (118) to promote fat utilization have been shown to have positive effects in some patients, but are not indicated for long-term therapy. The combination of branched-chain oxoacids (BCOA) and glucose infusion has remarkable effects on ammonia production and exercise performance, possibly because BCOA can act as glutamate precursors and can be utilized in the citric acid cycle. The same rationale guided some experiments with propionyl-L- carnitine, in which lower ammonia production and no decrease in blood pyruvate were demonstrated in McArdle's patients during aerobic exercise (119,120). A trial of long-term treatment with oral propionyl-carnitine is presently underway. Myoblast transfer therapy was attempted in two patients, but was unsuccessful (121).

PHOSPHORYLASE B KINASE (PBK) DEFICIENCY (GLYCOGENOSIS TYPE VIa)

Clinical Manifestation

Different clinical syndromes are associated with PBK deficiency (122):

- Hepatomegaly, growth retardation, delayed motor development, and fasting hypoglycemia as an X-linked disease (123)
- A static myopathy with liver disease showing autosomal-recessive transmission (124,125)
- Myopathy presenting as a childhood hypotonia or as exercise intolerance with limb weakness (126) and myoglobinuria (127)
- Fatal infantile cardiomyopathy (128)

Laboratory Findings

Some patients have abnormal serum lactate on ischemic exercise, others have normal lactate responses. CK is elevated and the EMG is usually myopathic.

Pathology

Type II fibers show subsarcolemmal blebs comprising free, normal glycogen.

Biochemistry

PBK is a hexadecamer composed of four subunits, α, β, γ, and δ, which form the whole enzyme. The γ subunit is catalytically active. The δ subunit is a calcium-sensitive calmodulin; subunits β and, to a lesser extent, subunits α and γ show calmodulin-binding sites and are involved in calcium-dependent activation, possibly through modulation of binding interaction with the δ subunit (129). PBK can be activated either by calcium that binds to the δ subunit or by epinephrine, through a cascade of reactions with activation of α and β subunits. Tissue-specific isoforms have been described for subunits α, β, and γ.

Molecular Genetics

cDNAs for the different subunits have been obtained. The genes coding for subunits α1 and α2 have been mapped to the X chromosome at region q12–q13 and p22, respectively. The gene for β subunits, which are generated by differential splicing, is on chromosome 16 at q12–q13 (130,131). The gene for the γ subunit is on chromosome 7p12–q21 (132).

The defect in X-linked liver PBK has been localized to Xp22, the same region where the gene for subunit α2 maps (133). In some patients with myopathy due to autosomal recessive PBK deficiency, Western blot studies with specific antibodies indicate a selective defect of the β subunit (134). However, there is considerable biochemical heterogeneity even among patients presenting only muscle PBK deficiency (134a). A single nucleotide insertion disrupting the reading frame of the α subunit has been identified in a mouse model of PBK deficiency (134b).

PHOSPHOFRUCTOKINASE (PFK) DEFICIENCY (GLYCOGENOSIS TYPE VII, TARUI'S DISEASE)

This disease was first described in three siblings who presented with symptoms similar to, but more severe than, McArdle's disease (135).

Clinical Manifestation

Patients present with intolerance to vigorous exercise, cramps, and sometimes myoglobinuric episodes. There is also a form that presents with scapuloperoneal atrophy (136). In most cases there is a mild hemolytic anemia and sometimes signs of gout and arthritis. Fatal cases with infantile onset, associated with limb weakness, seizures, cortical blindness, and corneal opacifications, have also been described (137).

Genetics

The disease has been described mostly in patients of Japanese and/or Ashkenazy Jewish heritage, while it is very rare in Europe. In familiar cases it is transmitted as an autosomal-recessive trait. Autosomal-recessive inheritance is also confirmed by low PFK activity in erythrocytes of parents and children of affected patients.

Laboratory Findings

There is no rise of lactate during ischemic exercise. Red blood cell count shows moderate reticulocytosis. Bilirubin and uric acid may be elevated. ^{31}P NMR spectroscopy reveals abnormal accumulation of phosphorylated monoesters (82).

Pathology

Muscle biopsy shows a vacuolar myopathy with subsacrcolemmal and intermyofibrillary glycogen accumulation. PFK activity may be visualized histochemically (138). In infantile cases, variation of fiber size, increased connective tissue, and granular aggregates are seen. PAS-positive, diastase-digestible material is found in muscle and heart.

Biochemistry

PFK is made of muscle (M), liver (L), and platelet (P) subunits. Mature muscle PFK is made of an M monotetramer. Red blood cells (RBC) express both M and L subunits. In PFK deficiency the defect of M subunits causes total lack of activity in muscle and 50% activity in RBC. The incompetent M subunit is a major component of heart and brain PFK, but clinical cardiopathy and encephalopathy are seen only in the infantile variant.

Molecular Genetics

The genes encoding for subunits M, P, and L have been localized on chromosome 1, 10, and 21, respectively (139). A full-length cDNA for the M subunit has been isolated and sequenced (140). The gene defect in the original cases of Tarui et al. (141) has been identified in a G-to-T mutation in the 5′ end of intron 13 of the gene for muscle PFK. Several other point mutations and, in some cases, splicing error have been described in non-Japanese PFK patients (141a–c).

Pathophysiology

Since PFK-deficient muscle cannot utilize glucose or glycogen, its maximal oxygen uptake is dependent on plasma FFA (142). Hyperuricemia is due to exercise-induced degradation of muscle purine nucleotides and may be seen in several types of muscle glycogenolysis disorders (phosphorylase or debrancher deficiency) or defects of terminal glycolysis.

DEFECTS OF TERMINAL GLYCOLYSIS: PHOSPHOGLYCEROKINASE (PGK) DEFICIENCY

The most common manifestation of PGK deficiency is hemolytic anemia, in some cases with signs of CNS dysfunction (seizures, mental retardation). In 1989, however, DiMauro et al. described a myopathic form of PGK deficiency (143): exercise intolerance, myalgia, and myoglobinuria were seen in three male patients (144). A fourth patient had myopathy and hemolysis (144a), a fifth had myopathy and mental retardation (144b), a sixth had myoglobinuria, hemolytic anemia and mental retardation (145).

Laboratory Findings

Forearm ischemic exercise causes no elevation of lactic acid; muscle glycogen concentration is normal. Muscle biopsy shows no specific changes. Resting CK is mildly elevated and ECG is normal.

Molecular Biology

The gene locus for PGK was shown by somatic cell hybridization to be on the X chromosome. PGK pseudogenes have been mapped on chromosomes X and 6, and a testicular isoform maps on chromosome 19. A full-length cDNA has been obtained (146). PGK deficiency has an X-linked inheritance pattern and prenatal diagnosis is possible. The enzyme deficiency is expressed in cultured fibroblasts. Using a full-length cDNA as a probe, Tonin et al. (147) have documented genetic heterogeneity in three patients, since two had normal RNA and one had abnormally large RNA. Several missense mutations and a 10-amino-acid insertion due to a splice junction mutation (144b) are responsible for the enzymatic defect.

PHOSPHOGLYCERATE MUTASE (PGAM) DEFICIENCY

PGAM deficiency (148) has been described in four patients, two men and two women. The signs and symptoms observed were intolerance to strenuous exercise, cramps, and recurrent myoglobinuria. The patients were all adults, but their clinical histories extended back into childhood.

Laboratory Findings

Forearm ischemic exercise causes a low rise of lactate. Muscle glycogen is normal. The enzyme defect is pronounced, with residual activity of 5% of control.

Biochemistry and Molecular Genetics

There are two tissue-specific isozymes for PGAM:PGAM B, characteristically found in the brain, and PGAM-M, specific for muscle (149). About 95% of total PGAM activity in muscle is represented by the MM dimer, and therefore any residual activity found is due to the B isozyme. Inheritance is autosomal-recessive. A full-length cDNA and the genomic clone have been isolated and sequenced (150,151). The gene for PGAM-M is localized on chromosome 7p12–13 (152). A G-to-A transition at codon 78 of PGAM-M gene resulting in an in-frame stop codon has been found in three black patients. The same mutation was not found in a fourth Caucasian patient from Italy (153), who showed a new C-to-T mutation at codon 90 (153a).

LACTATE DEHYDROGENASE (LDH) DEFICIENCY

Two young men with exercise intolerance, pigmenturia, and fatigability have been reported by Kanno et al. (154). One asymptomatic woman with LDH-M defect has also been reported (155).

Laboratory Findings

Forearm ischemic exercise gave a clue to the diagnosis since the lactate:pyruvate ratio fell, and the overall rise in lactate was minimal.

Biochemistry

LDH in muscle of patients is 5% of control. LDH is a tetrameric enzyme composed of two subunits: M predominant in muscle and H in the heart. In patients the M subunit is absent while the subunit H is present. The H subunit accounts for the residual activity in muscle. The defect is transmitted as an autosomal-recessive trait. The gene for muscle LDH-M has been assigned to chromosome 11 (156) and that for the heart-specific isoenzyme, to the short arm of chromosome 12. Recently a 20-bp deletion in exon 6 (subunits contact domain) of LDH-M gene has been described in two Japanese sisters with LDH-M deficiency (157).

Since Northern analysis and immunocytochemistry with anti-LDH-M antibodies revealed normal amounts of transcript and protein, the loss of activity is probably due to impaired subunit assembly.

CONCLUSIONS

In the field of metabolic myopathies remarkable advances have been made in the diagnosis and molecular biology of glycogen storage disorders. While the disorders involving the oligosaccharide pathway (acid maltase) or glycogen degradation (phosphorylase and debrancher enzyme) are characterized by weakness and a progressive course, disorders of the Embden-Meyerhof glycolytic pathway are intermittent and, apart from myoglobinuric crises, are compatible with normal life and function. This suggests a different relevance of these biochemical pathways in muscle function. Glycogen storage disorders have been among the first metabolic diseases for which biochemical and molecular basis has been investigated and, in most cases, clarified. The pathophysiology of these diseases has been extensively studied, and in many instances it represents a natural model of physiology of muscle fatigue. Therapy is now the next challenge, with the hope that, as for biochemistry, molecular biology, and physiopathology, this group of diseases may lead to the development of innovative and successful therapeutic approaches.

REFERENCES

1. Felig P, Wahren J. Fuel homeostasis in exercise. N Engl J Med 1975; 293:1078–1084.
2. Engel AG, Gomez MR, Seybold ME, Lambert EH. The spectrum of diagnosis of acid maltase deficiency. Neurology 1973; 23:95–106.
3. Cimiamon J, Slomin AE, Black KS, Gorey MT, Scuderi DM, Hyman RA. Evaluation of the lumbar spine in patients with glycogen storage disease: CT demonstration of paraspinal muscle atrophy. J Neuroradiol 1991; 12:1099–1103.
4. Kretzschman HA, Wagner H, Hubner G, Denek A, Witt TN, Martieni P. Aneurysm and vacuolar degeneration of cerebral arteries in late onset acid maltase deficiency. J Neurol Sci 1990; 98:169–183.
5. Danon MJ, DiMauro S, Shanske S, Archer FL, Miranda AF. Juvenile-onset acid maltase deficiency with unusual familial features. Neurology 1986; 36:818–822.
6. Koster JF, Busch HFM, Slee RG, Van Weerden TW. Glycogenosis type II: the infantile and late onset acid maltase deficiency observed in one family. Clin Chim Acta 1978; 87:451–453.
7. Zhong N, Martiniuk F, Tzall F, Hirschnon R. Identification of a missense mutation in one allele of a patients with Pompe disease, and use of endonuclease digestion of PCR-amplified RNA to demonstrate lack of mRNA expression from the second allele. Am J Hum Genet 1991; 49:635–641.

7a. Boerkoel N, Raben N, Exelbert R, Nicastri C, Plotz P. Mutations in the acid-maltase gene resulting phenotypically in infantile and adult glycogen storage disease type II (gsd II). Muscle Nerve 1994; (Suppl 1):s126.

8. Halley DJJ, Konings A, Hupkes P, Galiard H. Regional mapping of the human gene for lysosomal α-glucosidase by in situ hybridization. Hum Genet 1984; 67:326–328.
9. Martinuk F, Smith R, Desinck RJ, Astri K, Mitra J, Hirshon R. Assignement of the gene for the neutral alfa-glucosidase AB to chromosome 15. Cytogenet Cell Genet 1983; 35:110–116.
10. Gambetti PL, Di Mauro S, Baker L. Nervous system in Pompe's disease. J Neuropathol Exp Neurol 1971; 30:412–430.
11. Angelini C, Engel AG. Subcellular distribution of acid and neutral-glucosidase in normal, acid maltase deficient and myophosphorylase deficient human skeletal muscle. Arch Biochem Biophys 1973; 156:350–353.
12. Suhara Y, Ishiura S, Tsukahava T, Sugita H. Mature 98.000-Dalton acid α-glucosidase is deficient in Japanese quails with acid maltase deficiency. Muscle Nerve 1989; 12:670–678.
13. Reuser AJJ, Kroos M, Willunsen R, Swallow D, Tager JM, Galiciard H. Clinical diversity in glycogenosis type II biosynthesis and in situ localization and acid α-glucosidase in mutant fibroblasts. J Clin Invest 1987; 79:1689–1699.
14. Slonim AE, Coleman RA, McEligot MA, Najjar J, Hirschhorn K, Labadie GU, Mrak R, Evans OB, Shipp E, Presson R. Improvement of muscle function in acid maltase deficiency by high protein therapy. Neurology 1983; 33:34–38.
15. Engel AG. Acid maltase deficiency in adults: studies in four cases of a syndrome which may mimic muscular dystrophy or other myopathies. Brain 1970; 93:599–616.
16. Margolis ML, Hills AR. Acid maltase deficiency in an adult. Am Rev Respir Dis 1986; 134:328–331.
17. Mobentian S, Puntozzi RL, Daniele P, Fredman H. Treatment of acid maltase deficiency with a diet high in branched-chain aminoacids. J Parent Ent Nutr 1990; 14:210–212.
18. Illingworth B, Cori GT. Structure of glycogens and amylopectins. III. Normal and abnormal human glycogen. J Biol Chem 1952; 199:653–660.
19. Illingworth B, Cori GT, Cori CF. Amylo-1,6-glucosidase in muscle tissue in generalized glycogen storage disease. J Biol Chem 1956; 218:123–129.
20. Forbes GB. Glycogen storage disease. Report of a case with abnormal glycogen structure in liver and skeletal muscle. J Pediatr 1953; 42:645–650.
21. Van Hoof F, Hers HG. The subgroups of type III glycogenosis. Biochemistry 1967; 2:265–270.
22. Miranda AF, DiMauro S, Antler A, Stern LZ, Rowland LP. Glycogen debranching deficiency is reproduced in muscle culture. Ann Neurol 1981; 9:283–288.
23. Brown BI. Debranching and branching enzyme deficiencies. In: Engel AG, Banker BQ, eds. Myology. New York: McGraw-Hill, 1986:1653–1661.
24. Yang-Feng TL, Zheng K, Yu J, Yang BZ, Chen VT, Kao FT. Assignment of the human glycogen debrancher gene to chromosome 1p21. Genomics 1992; 13:931–934.

25. Miller CG, Alleyne GA, Brooks SEH. Case report: gross cardiac involvement in glycogen storage disease type III. Br Heart J 1972; 34:862–865.
26. Rossignol AM, Meyer M, Rossignol B, Palcoux MP, Raynaud EJ, Bost M. La Myocardiopathie de la glycogenose type III. Arch Fr Pediatr 1979; 36:303–309.
27. DiMauro S, Hartwig GB, Hays AP, Eastwood AB, Franco R, Olarte M, Chang M, Roses AD, Fetell M, Schoenfeldt RS, Sern LZ. Debrancher deficiency: neuromuscular disorder in 5 adults. Ann Neurol 1979; 5:422–436.
28. Smit GPA, Fernandes J, Leonard JV, Matthews EE, Moses SW, Odievre M, Ullrich K. The long term outcome of patients with glycogen storage diseases. J Inher Metab Dis 1990; 13:411–418.
29. Coleman RA, Winter HS, Wolf B, Gilchrist JM, Chen VT. Glycogen storage disease type III (glycogen debranching enzyme deficiency): correlation of biochemical defects with myopathy and cardiomyopathy. Ann Intern Med 1992; 116:896–900.
30. Ugawa J, Inoue K, Takeura T, Iwamasa T. Accumulation of glycogen in sural nerve axons in adult-onset type III glycogenosis. Ann Neurol 1986; 19:294–297.
31. Levin S, Moses SW, Chayoth R, Jagoda N, Steinitz K. Glycogen storage disease in Israel. A clinical, biochemical and genetic study. Isr J Med Sci 1967; 3:397–404.
32. Sancho S, Navarro C, Fernandez JM, Dominguez C, Ortega A, Roig M, Cervera C. Skin biopsy findings in glycogenosis III: clinical, biochemical, and electrophysiological correlations. Ann Neurol 1990; 27:480–486.
33. Pellissier JF, de Barsy TH, Faugere MC, Rebuffel P. Type III glycogenosis with multicore structures. Muscle Nerve 1979; 2:124–132.
34. Powel HC, Haas R, Hall CL. Peripheral nerve in type III glycogenosis: selective involvement of unmyelinated fiber Schwann cells. Muscle Nerve 1985; 8:667–671.
35. Shin YS. Diagnosis of glycogen storage disease. J Inher Dis 1990; 13:419–434.
36. Yang BZ, Ding JH, Brown BI, Chen VT. Definitive prenatal diagnosis for the type III glycogen storage disease. Am J Hum Genet 1990; 47:735–739.
37. Yang BZ, Ding JH, Endhild JJ, Bao Y, Chan VT. Molecular cloning and nucleotide sequence of cDNA encoding human muscle glycogen debranching enzyme. J Biol Chem 1992; 267:9294–9299.
38. Gordon RB, Brown DH, Brown BI. Studies on the structure and mechanism of action of the glycogen debranching enzyme of muscle and liver. Biochim Biophys Acta 1972; 289:97–107.
39. Bates EH, Heaton GM, Taylor C, Kernohan JC, Cohen P. Debranching enzyme from rabbit skeletal muscle: evidence for the location of two active centers on a single polypeptide chain. FEBS Lett 1975; 58:181–185.
40. Taylor C, Cox AJ, Kernohan JC, Cohen P. Debranching enzyme from rabbit skeletal muscle, purification, properties and physiological role. Eur J Biochem 1975; 51:105–115.
41. Chen VY, He JK, Ding JH, Brown BI. Glycogen debranching enzyme: purification, antibody characterization, and immunoblot analyses of type III glycogen storage disease. Am J Hum Genet 1987; 41:1002–1015.
42. Yang BZ, Stewart C, Ding JH, Chen VY. Type III glycogen storage disease: an adult case with mild disease but complete absence of debrancher protein. Neuromusc Disord 1991; 1:173–176.
43. Furutani A, Yano O, Gotoh S, Higashi K, Hasegawa E, Ohnishi A. Some Properties of fibroblasts from a patient with debrancher deficiency. Sangyo Ika Daigaku Zasshi 1990; 12:441–418.
44. Ding JH, Harris DA, Bing-Zi Y, Chen VY. Cloning of cDNA for human muscle glycogen debrancher, the enzyme deficient in type III glycogen storage disease. J Pediatr 1989; 25:140A.
45. Slonim AE, Coleman RA, Moses SW. Myopathy and growth failure in debrancher enzyme deficiency: improvement with high protein nocturnal enteral therapy. J Pediatr 1984; 105:906–911.
46. Gremse DA, Bucuvalas JC, Balistreri WF. Efficacy of cornstarch therapy in type III glycogen-storage disease. Am J Clin Nutr 1990; 52:671–674.
47. Andersen DH. Familial cirrhosis of the liver with storage of abnormal glycogen. Lab Invest 1956; 5:11–20.
48. Brown BI, Brown DH. Lack of an α-1,4- glucan: α-1,4-glucan 6- glycosyl transferase in a case of type IV glycogenosis. Proc Natl Acad Sci USA 1966; 56:725–729.
49. Lossos A, Barasch V, Soffer D, Argov Z, Gomori M, Ben-Nariah, Abramsky O, Steiner I. Hereditary branching enzyme dysfunction in adult polyglucosan body disease: a possible metabolic cause in two patients. Ann Neurol 1991; 30:655–662.
50. Bruno C, Servidei S, Shanske S, Karpati G, Carpenter S, McKee D, Barohn RJ, Hirano M, Rifai Z, DiMauro S. Glycogen branching enzyme deficiency and adult polyglucosan body disease. Ann Neurol 1993; 33:88–93.
51. servidei S, Riepe RE, Langston C, Tani L, Briker JT, Crisp-Lindgren N, Travers M, Armstrong D, DiMauro S. Severe cardiopathy in branching enzyme deficiency. J Pediatr 1987; 111:51–56.
52. Reusche E, Asku F, Goebel HH, Shin YS, Yokota T, Reichmann H. A mild juvenile variant of type IV glycogenosis. Brain Dev 1992; 14:36–43.
53. Cafferty S, Lovelace RE, Hays AP, Servidei S, DiMauro S, Rowland LP. Polyglucosan body disease. Muscle Nerve 1991; 14:102–107.
54. Peres N, DiMauro S, Roxburgh V. Adult polysaccharidosis; clinicopathological, ultrastructural and biochemical features. Arch Neurol 1979; 36:840–845.

55. Schochet SS Jr, McCormick WF, Kovarsky J. Light and electron microscopy of skeletal muscle in type IV glycogenosis. Acta Neuropathol 1971; 19:137–144.
56. Zimmermann CP, Gold AM. Isolation and characterization of glycogen branching enzyme from rabbit liver. Biochemistry 1983; 22:3387-3391.
57. Brown BI. Diagnosis of glycogen storage disease. In: Wapnir RA, ed. Congenital Metabolic Diseases. New York: Marcel Dekker, 1985:227–250.
58. Hers HG, Van Hoof F, de Barsy T. Glycogen storage diseases. In: Scriver CR, Beaudet AL, Sly WS, Valle D, eds. The Molecular Basis of Inherited Diseases. Vol 1. New York: McGraw-Hill, 1989:425–452.
58a. Thon VS, Khalil M, Cannon JF. Isolation of human glycogen branching enzyme cDNA by screening complementation in yeast. J Biol Chem 1993; 268:7509–7513.
59. McArdle B. Myopathy due to a defect in muscle glycogen breakdown. Clin Sci 1951; 10:13–35.
60. Mommaerts WFHM, Illingworth B, Pearson CM, Guillory RJ, Seraydarian K. A functional disorder of muscle associated with the absence of phosphorylase. Proc Natl Acad Sci USA 1959; 45:781–797.
61. Schmid R, Robbins PW, Traut RR. Glycogen synthesis in muscle lacking phosphorylase. Proc Natl Acad Sci USA 1959; 45:1236–1240.
62. Pearson CM, Rimer DG, Mommaerts WFHM. A metabolic myopathy due to absence of muscle phosphorylase. Am J Med 1961; 30:502–517.
63. Pernow BB, Havel RJ, Jennings DB. The second wind phenomenon in McArdle's syndrome. Acta Med Scand 1967; 472:294–307.
64. Servidei S, Shanske S, Zeviani M, Lebo R, Fletterick R, DiMauro S. McArdle's disease: biochemical and molecular genetic studies. Ann Neurol 1988; 24:774–781.
65. DiMauro S, Hartlage PL. Fatal infantile form of muscle phosphorylase deficiency. Neurology 1978; 28:1124–1129.
66. De LaMaza M, Patten BM, Williams JC, Chambers JP. Myophosphorylase deficiency: a new cause of infantile hypotonia simulating infantile muscular atrophy. Neurology 1980; 30:402.
67. Cornelio F, Bresolin N, DiMauro S, Mora M, Balestrino MR. Congenital Myopathy due to phosphorylase deficiency. Neurology 1983; 33:1383–1385.
68. Engel WK, Eyerman EL, Williams HE. Late onset type of skeletal muscle phosphorylase deficiency: a new variety with completely and partially affected subjects. N Engl J Med 1963; 268:135–137.
69. Felice KJ, Schneebaum AB, Royden Jones H Jr. McArdle's disease with late-onset symptoms: case report and review of the literature. J Neurol Neurosurg Psychiatry 1992; 55:407–408.
70. DiMauro S, Bresolin N. Phosphorylase deficiency. In: Engel AG, Banker BQ, eds. Myology. New York: McGraw-Hill, 1986:1585–1601.
71. Chui LA, Munsat TL. Dominant inheritance of McArdle's syndrome. Arch Neurol 1976; 33:636–641.
72. Schmidt B, Servidei S, Gabbai AA, Silva AC, de Sousa Bulle de Oliveira A, DiMauro S. McArdle's disease in two generations: autosomal recessive transmission with manifesting heterozygote. Neurology 1987; 37:1558–1561.
73. Lebo RV, Anderson LA, Di Mauro S, Lynch E, Hwang P, Fletterick RJ. Rare locus polymorphic site on 11q13 contains CpG sequence. Hum Genet 1990; 86:17–24.
74. Lebo RV, Gorin F, Fletterick RJ, Lao FT, Cheung MC, Bruce BD, Kan YW. High resolution chromosome sorting and DNA spot-blot analysis assign McArdle's syndrome to chromosome 11. Science 1984; 225:57–59.
75. Tsujino S, Shanske S, Di Mauro S. Molecular genetic heterogeneity of myophosphorylase deficiency (McArdle's Disease). N Engl J Med 1993; 329:241.245.
76. Bartram C, Edwards RHT, Clague J, Beynon J. McArdle's disease: a nonsense mutation in exon 1 of the muscle glycogen phosphorylase gene explains some but not all cases. Hum Mol Genet 1993; 2:1291–1293.
76a. Tsujino S, Shanske S, Goto YI, Nonaka I, DiMauro S. Two mutations, one novel and one frequently observed, in Japanese patients with McArdle's disease. Hum Mol Genet 1994; 3:1005–1006.
76b. Tsujino S, Shanske S, Martinuzzi A, Heiman-Patterson T, DiMauro S. Two novel missense mutations in Causasian patients with myophosphorylase deficiency (McArdle's disease). Hum Mutat 1995; 6:276–277.
76c. Bartram C, Edwards RHT, Clague J, Beynon RJ. McArdle's disease: a rare frameshift mutation in exon 1 of the muscle phosphorylase gene. Biochim Biophys Acta 1996; 1226:391–393.
77. Tsujino S, Shanske S, Nonaka I, Eto Y, Mendell J, Fenichel G, DiMauro S. Three new mutations in patients with myophosphorylase deficiency (McArdle's disease). Am J Hum Genet 1994; 54:44–52.
78. Coleman RA, Stajich JM, Pact VW, Pericak-Vance MA. The ischemic exercise test in normal adults and in patients with weakness and cramps. Muscle Nerve 1986; 9:216–221.
79. Fishbein WN, Foellmer JW, Davis JI. Medical implications of lactate and ammonia relationship in anaerobic exercise. Int J Sports Med 1990; 11:S91–S100.
80. Wagenmakers AJ, Coakley JH, Edvards RH. Metabolism of branched-chain amino acids and ammonia during exercise: clues from McArdle's disease. Int J Sports Med 1990; 11:S101–S113.
81. Coakley JH, Wagenmakers AJM, Edwards RTH. Relationship between ammonia, heart rate, and exertion in McArdle's disease. Am J Physiol 1992; 262:E167–E172.

82. Argov Z, Bank WJ. Phosphorus magnetic resonance spectroscopy (^{31}P MRS) in neuromuscular disorders. Ann Neurol 1991; 30:90–97.
83. Gruetter T, Kaelin P, Boesch C, Martin E, Werner B. Non invasive ^{31}P magnetic resonance spectroscopy revealed McArdle disease in an asymptomatic child. Eur J Pediatr 1990; 149:483–486.
84. Sato K, Imai F, Hatayama I, Roelofs RI. Characterization of glycogen phosphorylase isoenzymes present in cultured skeletal muscle from patients with McArdle's disease. Biochem Biophys Res Commun 1977; 78:663–667.
85. DiMauro S, Arnold S, Miranda A, Rowland P. McArdle's disease: the mystery of reappearing phosphorylase activity in muscle culture: a fetal isoenzyme. Ann Neurol 1978; 3:60–66.
86. Meinhofer C, Askanas V, Proux-Daegelen D, Dreyfus JC, Engel WK. Muscle-type phosphorylase activity present in muscle cells cultured from three patients with myophosphorylase deficiency. Arch Neurol 1977; 34:779–781.
87. Martinuzzi A, Vergani L, Fanin M, Carrozzo R, Bartoloni L, Angelini C, Askanas V, Engel WK. Muscle-type glycogen phosphorylase (M-GP) activity in innervated and non-innervated cultured muscle of McArdle patients. J Clin Invest 1993; 92:1774–1780.
88. Byard RW, Lach B, Preston DN. Peripheral nerve and vasculature involvement in myophosphorylase deficiency (McArdle's disease). Pathology 1991; 23:62–65.
89. Schotland DL, Spiro D, Rowland LP, Carmel P. Ultrastructural studies of muscle in McArdle's disease. J Neuropathol Exp Neurol 1965; 24:629–637.
90. Proux D, Dreyfus JC. Phosphorylase isoenzymes in tissues: prevalence of the liver type in man. Clin Chim Acta 1973; 48:167–172.
91. Newgard CB, Fletterick RJ, Anderson LA, Lebo RV. The polymorphic locus for glycogen storage disease VI (liver glycogen phosphorylase) maps to chromosome 14. Am J Hum Genet 1987; 351–364.
92. Newgard CB, Littman DR, van Genderen C, Smith M, Fletterick RJ. Human brain glycogen phosphorylase: cloning, sequence analysis, chromosomal mapping, tissue expression and comparison with the human liver and muscle isozymes. J Biol Chem 1988; 263:3850–3857.
93. Richter F, Bohme HJ, Hofmann E. Developmental changes of glycogen phosphorylase b isozymes in rat tissues. Biomed Biochim Acta 1983; 9:1229–1235.
94. McConchie SM, Coakley J, Edwards RTH, Beynon RJ. Molecular heterogeneity in McArdle's disease. Biochem Biophys Acta 1991; 1096:26–32.
95. Rowland LP, Fahn S, Schotland DL. McArdle's disease. Arch Neurol 1963; 9:325–342.
96. Cerri CG, Willner JH. Phosphorylation of McArdle phosphorylase induces activity. Proc Natl Acad Sci USA 1981; 78:2688–2692.
97. Daegelen D, Munnich A, Lvin MJ, Girault A, Goasguen J, Kahn A, Dreyfus JC. Absence of functional messenger RNA for glycogen phosphorylase in the muscle of two patients with McArdle's disease. Ann Hum Genet 1983; 47:107–115.
98. Gautron S, Dagelen D, Mennecier F, Dubocq D, Kahn A, Dreyfus JC. Molecular mechanisms of McArdle's disease (muscle glycogen phosphorylase deficiency). J Clin Invest 1987; 79:275–281.
99. Lewis SF, Haller RG. The pathophysiology of McArdle's disease: clues to regulation in exercise and fatigue. J Appl Physiol 1986; 61:391–401.
100. Argov Z, Bank WJ, Maris J, Chance B. Muscle energy metabolism in McArdle's syndrome by in vivo phosphorus magnetic resonance spectroscopy (^{31}P NMR). Neurology 1987; 37:1720–1724.
101. Radda GK. Control of bioenergetics: from cells to man by phosphorus nuclear-magnetic-resonance spectroscopy. Biochem Soc Trans 1986; 14:517–525.
102. Bendal JR. A study of the kinetics of the fibrillar adenosine triphosphatase of rabbit skeletal muscle. Biochem J 1961; 81:520–535.
103. Ponce-Hornos JE, Langer GA. Effects of inorganic phosphatase on ion exchange, energy state, and contraction in mammalian heart. Am J Physiol 1982; 242:H79–H88.
104. Paterson DJ, Friedland JS, Bascom DA, Clement ID, Cunningham DA, Painter R, Robbins PA. Changes in arterial K^+ and ventilation during exercise in normal subjects and subject with McArdle's syndrome. J Physiol 1990; 492:339–384.
105. Rybicki KJ, Waldrop TG, Kaufman MP. Increasing gracilis muscle interstitial potassium concentrations stimulate group III and IV afferents. J Appl Physiol 1985; 58:936–941.
106. Pryor SL, Lewis SF, Haller RG, Bertocchi LA, Victor RG. Impairment of sympathetic activation during static exercise in patients with muscle phosphorylase deficiency (McArdle's disease). J Clin Invest 1990; 85:1444–1449.
107. Haller RG, Lewis SF, Cook JD, Blomqvist CG. Myophosphorylase deficiency impairs muscle oxidative metabolism. Ann Neurol 1985; 17:196–199.
108. Sahlin K, Areskog NH, Haller RG, Henriksson KG, Jorfeldt L, Lewis SF. Impaired oxidative metabolism increases adenine nucleotide breakdown in McArdle's disease. J Appl Physiol 1990; 69:1231–1235.
109. Martinuzzi A, Pegoraro E, Cadaldini M, Schiraldi C, Costa A, Vergani L, Angelini C. Physiological responses

in McArdle's patients during maximal aerobic exercise. In: Carraro U, Salmons S, eds. Basic and Applied Myology: Perspectives for the 90's. Padua: Unipress, 1992:183–190.
110. Turk WR, Heller SL, Norris BJ, Nemeth PM. Increased muscular β-hydroxyacyl CoA dehydrogenase with McArdle's disease. Muscle Nerve 1990; 13:607–612.
111. Cooper RG, Edwards RHT, Stokes MJ. Human muscle fatigue: excitation-contraction coupling failure in McArdle's disease. J Physiol 1987; 394:142p.
112. Linsenn WH, Jacobs M, Stegeman DF, Joosten EM, Moleman J. Muscle fatigue in McArdle's disease. Muscle fiber conduction velocity and surface EMG frequency spectrum during ischaemic exercise. Brain 1990; 113:1779–1793.
113. Slonim AF, Goans PJ. Myopathy in McArdle's syndrome: improvement with a high protein diet. N Engl J Med 1985; 312:355–359.
114. Jensen KE, Jakobsen J, Thomsen C, Heriksen O. Improved energy kinetics following high protein diet in McArdle's syndrome. A ^{31}P magnetic resonance spectroscopy study. Acta Neurol Scand 1990; 81:499–503.
115. Lewis SF, Haller RG, Cook JD, Nunnally RL. Muscle fatigue in McArdle's disease studied by ^{31}P-NMR: effect of glucose infusion. J Appl Physiol 1986; 59:1991–1994.
116. Wagner DR, Zollner N. McArdle's disease: successful symptomatic therapy by high dose oral administration of ribose. Klin Wochenschr 1991; 69:92.
117. Kono N, Ineo S, Shiizu T, Kang J, Nonaka K, Tarui S. Metabolic basis of improved exercise tolerance: muscle phosphorylase deficiency after glucagon administration. Neurology 1984; 34:1471–1476.
118. Porte D Jr, Crawford DW, Jennings DB, Aber C, McIlroy MB. Cardiovascular and metabolic responses to exercise in a patient with McArdle's syndrome. N Engl J Med 1966; 275:406–412.
119. Wagenmakers AJM, Coakley JH, Edwards RTH. Propionyl-L- carnitine reduces plasma ammonia in patients with McArdle's disease. In: Benzi G, ed. Advances in Myochemistry 2. London: John Libbey Eurotext, 1989:195.
120. Pegoraro E, Martinuzzi A, Rosa M, Carrozzo R, Schiraldi C, Bevilacqua M, Angelini C. Effect of propionyl L carnitine infusion during exercise in McArdle's patients. J Neurol Sci 1990; 98 (Suppl 1):242.
121. Karpati G, Johnston W, Ajdukovic G, Arnold D, Vanasse M, Guttmann R, Crerar M, Carpenter S, Shoubridge E. Myoblast transfer (MT) in McArdle's disease. Neurology 1992; 42(Suppl 3):387.
122. Van der Berg IET, Berger R. Phosphorylase B kinase deficiency in many. J Inher Metab Dis 1990; 13:442–451.
123. Huijing F, Fernandes J. X-chromosomal inheritance of liver glycogenosis with phosphorylase kinase deficiency. Am J Hum Genet 1969; 21:275–284.
124. Abarbanel JM, Bashan N, Potashnik R, Osimani A, Moses SW, Herishanu Y. Adult muscle phosphorylase B kinase deficiency. Neurology 1986; 36:560–562.
125. Clemens PR, Yamamoto M, Engel AG. Adult phosphorylase B kinase deficiency. Ann Neurol 1990; 28:529–538.
126. Carrier M, Maire I, Vial C, Rambaud G, Flocard F, Flechaire A. Myopathic evolution of an exertional muscle pain syndrome with phosphorylase B kinase deficiency. Acta Neuropathol 1990; 81:84–88.
127. Tonin P, Lewis P, Servidei S, DiMauro S. Metabolic causes of myoglobinuria. Ann Neurol 1990; 27:181–185.
128. Eishi Y, Takemura T, Sone R, Yakamura H, Narisawa K, Ichinohasma R, Tanaka M, Hatekayama S. Glycogen storage disease confined to the heart with deficient activity of cardiac phosphorylase kinase: a new type of glycogen storage disease. Hum Pathol 1985; 16:193–198.
129. Newsholme P, Angelos KL, Walsh DA. High and intermediate affinity calmodulin bindings domains of the α and β subunits of phosphorylase kinase and their potential role in phosphorylation-dependent activation of the holoenzyme. J Biol Chem 1992; 267:810–818.
130. Francke U, Darras BT, Zander NF, Kilimann MW. Assignment of human genes for phosphorylase kinase subunits alpha (PHKA) to Xq12-q13 and beta (PHKB) to 16 q12-q13. Am J Hum Genet 1989; 45:276–282.
131. Davidson JJ, Ozcelic T, Hamacher C, Willems PJ, Francke U, Kilimann MW. cDNA cloning of a liver isoform of the phosphorylase kinase alpha subunit and mapping of the gene to Xp22.22-p22.1, the region of human X-linked liver glycogenesis. Proc Natl Acad Sci USA 1992; 89:2096–2106.
132. Jones TA, Dacruz e Silva EF, Spurr NK, Sheer D, Cohen PIW. Localization of the gene encoding the catalytic gamma subunit of phosphorylase kinase to human chromosome bands 7p12-q21. Biochem Biophys Acta 1990; 1048:24–29.
133. Hendrickx J, Coucke P, Bossuyt P, Wauters J, Raeymaekers P, Marchau F, Smit GPA, Stolte I, Sardharwalla IB, Berthelot J, Van den Bergh I, Berger R, Van Broeckhoven C, Baussan C, Wapenaar M, Fernandes J, Willems PJ. X-linked liver glycogenosis: localization and isolation of a candidate gene. Hum Mol Genet 1993; 2:583–589.
134. Clemens PR, Yamamoto M, Engel AG. Muscle phosphorylase B-kinase (PBK) deficiency in two adults: selective deficiency of the β (regulatory) PBK subunit. Neurology 1990; 40(Suppl 1):413.
134a. Wilkinson DA, Tonin P, Shanske S, Lombes A, Carlson GM, Di Mauro S. Clinical and biochemical features of 10 adult patients with muscle phosphorylase kinase deficiency. Neurology 1994; 44:461–466.

134b. Schneider A, Davidson JJ, Wulrich A, Kilimann MW. Phosphorylase kinase deficiency in I-strain mice is associated with a frameshift mutation in the alpha subunit muscle isoform. Nat Genet 1993; 5:381–385.
135. Tarui S, Okuno G, Ikua Y, Tanaka T, Suda M, Nishikawa M. Phosphofructokinase deficiency in skeletal muscle: a new type of glycogenosis. Biochem Biophys Res Commun 1965; 19:517–523.
136. Serratrice G, Monges A, Roux H, Aquaron R, Gambarelli D. Forme myopathique du déficit en phosphofructokinase. Rev Neurol 1969; 120:271–277.
137. Servidei S, Bonilla E, Diedrich RG, Kornfeld M, Oates JD, Davidson M, Vora S, DiMauro S. Fatal infantile form of phosphofructokinase deficiency. Neurology 1986; 36:1465–1470.
138. Bonilla E, Schotland DL. Histochemical diagnosis of muscle phosphofructokinase deficiency. Arch Neurol 1970; 22:8–12.
139. Vora S, Miranda AF, Hernandez E, Francke U. Regional assignment of the human gene for platelet-type phosphofructokinase (PFKP) to chromosome 10p, a novel use of polyspecific rodent anti sera to localize human enzyme genes. Hum Genet 1993; 63:374–379.
140. Nakajima H, Noguchi T, Yamasaki T, Kono N, Tonaka T, Tarui S. Cloning of human muscle phosphofructokinase cDNA. FEBS Lett 1988; 233:133–166.
141. Nakajima H, Kono N, Yamasaki T, Hotta K, Kawachi M, Kuwajiama M, Nogochi T, Tanaka T, Tarui S. Genetic defect in muscle phosphofructokinase deficiency. J Biol Chem 1991; 265:9292–9295.
141a. Raben N, Sherman J, Miller F, Mena H, Plotz P. A 5′ splice junction mutation leading to exon deletion in an Ashkenazic Jewish family with phosphofructokinase deficiency (Tarui disease). J Biol Chem 1993; 268: 4963–4967.
141b. Raben N, Sherman J, Nakajima H, Argov Z, Spiegel R, Plotz P. Genetic heterogeneity of phosphofructokinase deficiency (Tarui disease). Muscle Nerve 1994; (Suppl 1):s126.
141c. Tsujino S, Servidei S, Tonin P, Shanske S, Azan G, DiMauro S. Three novel mutations in non-Ashkenazi Italian patients with muscle phosphofructokinase deficiency (Tarui disease). Muscle Nerve 1994; (Suppl 1):s126.
142. Haller RG, Lewis SF. Glucose-induced exertional fatigue in muscle phosphofructokinase deficiency. N Engl J Med 1991; 324:364–369.
143. DiMauro S, Dalakas M, Miranda AF. Phosphoglycerate kinase deficiency: a new cause of recurrent myoglobinuria. Trans Am Neurol Assoc 1989; 106:202–205.
144. Tonin P, Shanske S, Brownell AK, Wyse JP, DiMauro S. Phosphoglycerate kinase deficiency: a third case with recurrent myoglobinuria. Neurology 1989; 39:359–360.
144a. Fujii H, Kanno H, Hirono A, Shiomura T, Miwa S. A single amino acid substitution (157 Gly to Val) in a phosphoglycerate kinase variant (PGK Shizuoka) associated with chronic hemolysis and myoglobinuria. Blood 1992; 79:1582–1585.
144b. Tsujino S, Tonin P, Shanske S, Nohria V, Boustany RM, Lewis D, Chen YT, DiMauro S. A splice junction mutation in a new myopathic variant of phosphoglycerate kinase deficiency (PGK North Carolina). Ann Neurol 1994; 35:349–353.
145. Sugie H, Sugie Y, Nishida M, Masataka I, Tarui S, Suzuki M, Miyamoto R, Igarashi Y. Recurrent myoglobinuria in a child with mental retardation: phosphoglycerate kinase deficiency. J Child Neurol 1989; 4:95–99.
146. Michelson AM, Markham AF, Orkin SH. Isolation and DNA sequence of a full length cDNA clone for human X chromosome- encloded phosphoglycerate kinase. Proc Natl Acad Sci USA 1983; 80:472–476.
147. Tonin P, Shanske S, Miranda AF, Brownell AK, Wyoe JP, Tsujino S, DiMauro S. Phosphoglycerate kinase deficiency: biochemical and molecular genetics studies in a new myopathic variant (PGK Alberta). Neurology 1993; 43:387–391.
148. DiMauro S, Miranda AF, Kahn S, Gitlin K, Friedman R. Human muscle phosphoglycerate mutase deficiency: newly discovered metabolic myopathy. Science 1981; 212:1277–1279.
149. Fundele R, Krietsch WGK. Purification and properties of the phosphoglycerate mutase isozymes from the mouse. Comp Biochem Physiol 1985; 81:965–968.
150. Shanske S, Sakoda S, Hermodson MA, DiMauro S, Schon E. Isolation of a cDNA encoding the muscle-specific subunit of human phosphoglycerate mutase. J Biol Chem 1987; 262:14612–14617.
151. Tsujino S, Sakoda S, Mizuno R, Kobayaschi T, Suzuki T, Kishimoto S, Shanske S, DiMauro S, Schon E. Structure of human phosphoglycerate mutase. J Biol Chem 1989; 264:15334–15337.
152. Edwards YH, Sakoda S, Schon E, Povey S. The gene for human muscle-specific phosphoglycerate mutase, PGAM-M mapped to chromosome 7 by polymerase chain reaction. Genomics 1989; 5:948–951.
153. Tsujino S, Shanske S, DiMauro S. Molecular genetic basis of muscle phosphoglycerate mutase (PGAM-M) deficiency. Am J Hum Genet 1993; 52:472–477.
153a. Toscano A, Tsujino S, Vita G, Shanske S, Di Mauro S, Messina C. Molecular basis of phosphoglycerate mutase (PGAM-M) deficiency in an Italian kindred. Neurology 1993; 43 (Suppl 1):a279.
154. Kanno T, Sudo K, Takeuchi I, Kanada S, Honda N, Nishimura Y, Dyama K. Hereditary deficiency of lactate dehydrogenase M subunits. Clin Chim Acta 1980; 108:267–276.

155. Bryan W, Lewis SF, Bertocci L, Gunder M, Ayyad K, Gustafson P, Haller RG. Muscle lactate dehydrogenase deficiency: a disorder of anaerobic glycogenolysis associated with exertional myoglobinuria. Neurology 1990; 40 (Suppl 1):203.
156. Boone CM, Chen TR, Ruddle FH. Assignment of three human genes to chromosomes (LDH-M to 11, TK to 17, and IDH to 20) and evidence for translocation between human and mouse chromosomes in somatic cell hybrids. Proc Natl Acad Sci USA 1972; 69:510–514.
157. Miyajima H, Takahashi Y, Suzuki M, Shimizu T, Kaneko E. Molecular characterization of gene expression in human lactate dehydrogenase-M deficiency. Neurology 1993; 43:1414–1419.
158. Angelini C, Engel AG. Riv Patol Nerv Ment 1971; 92:289–305.

34

Lipid Storage Myopathies and Inborn Errors of Fatty Acid Oxidation

Jochen Schaefer, Sandra Jackson, and D. M. Turnbull
University of Newcastle upon Tyne, Newcastle upon Tyne, England

MORPHOLOGICAL FEATURES OF LIPID STORAGE MYOPATHIES

The intracellular accumulation of lipid in skeletal muscle can be a feature of a variety of metabolic disorders, including defects of mitochondrial or peroxisomal fatty acid oxidation or of the mitochondrial respiratory chain. The definition of a lipid storage myopathy is therefore based only on morphological observations and is not pathognomonic for a special defect. Although the underlying enzyme abnormalities have been identified in comparatively few cases, it has become apparent in recent years that most lipid storage myopathies are due to defects of mitochondrial fatty acid β-oxidation. In these defects fatty acyl-CoA esters are diverted to triacylglycerol synthesis, which leads to intracellular deposition of excess lipid. Since fatty acids are predominantly oxidized in the slow-twitch oxidative type 1 fibers, most lipid storage myopathies show preferential lipid deposition in these fibers. On histological examination of muscle, lipid droplets are often visible as small vacuoles throughout the affected muscle fibers and further histochemical staining with Sudan black allows demonstration of both neutral and phospholipids.

This chapter will deal exclusively with inborn errors of mitochondrial fatty acid β-oxidation since virtually all patients with lipid storage myopathies have defects involving these enzymes. However, in these patients the amount of lipid storage is dependent on the site and severity of the metabolic defect and on the metabolic state of the patient. There may be minimal or even no lipid storage. For example, patients with carnitine palmitoyltransferase (CPT) deficiency usually have minimal lipid storage, although they may suddenly present with rhabdomyolysis and histological evidence of muscle necrosis.

BIOCHEMICAL PATHWAYS AND PHYSIOLOGICAL ROLE OF MITOCHONDRIAL β-OXIDATION

Fatty acids are released from dietary fat or mobilized from endogenous lipid stores during fasting or stress. In times of prolonged fasting, fatty acids serve several important functions (1). First, they are the primary metabolic fuel for cardiac and skeletal muscle. Second, they provide the substrate for hepatic ketogenesis, which supplies an important auxiliary fuel for many organs including the central nervous system (CNS) in the form of ketone bodies (acetoacetate, β-hydroxybutyrate). Third, the metabolic intermediates (acetyl-CoA), reducing equivalents (NADH), and energy derived from fatty acid oxidation are necessary for maintaining high rates of hepatic gluconeogenesis during fasting.

Fatty acids are transported in blood either as free fatty acids bound to albumin or as lipoproteins. After tissue uptake long-chain fatty acids are activated to the corresponding acyl-CoA ester by an

ATP-dependent acyl-CoA synthetase in the outer mitochondrial membrane (2). The acyl-CoA esters formed then enter the mitochondria through a specific carnitine-dependent transport mechanism, which is composed of three membrane-bound enzymes, carnitine palmitoyltransferase I and II and carnitine-acylcarnitine translocase (Fig. 1a). Medium- and short-chain fatty acids enter the mitochondria directly and are activated to their respective CoA esters in the mitochondrial matrix. Inside the mitochondria the acyl-CoA esters are degraded by the sequential removal of acetyl-CoA units from the carboxy-terminal end. This requires repeated cycles of four consecutive reactions (Fig. 1b) until the acyl-CoA is completely cleaved to acetyl-CoA. The initial step, oxidation of an acyl-CoA to 2-enoyl-CoA, is catalyzed by the FAD-dependent acyl-CoA dehydrogenases. Two other flavoproteins, electron transfer flavoprotein (ETF) and ETF-dehydrogenase, link this step to the respiratory chain. The three other steps involve, in sequence: enoyl-CoA hydration, NAD^+-dependent dehydrogenation, and thiolysis (Fig. 1b). The complete intramitochondrial oxidation of long-chain fatty acids requires a minimum of 10 enzymes due to the varying, but overlapping, chain-length specificity of the enzymes involved. There are four genetically distinct acyl-CoA dehydrogenases (very-long-chain, long-chain, medium-chain, short-chain) and a trifunctional enzyme bearing activities for long-chain enoyl-CoA hydratase, long-chain hydroxyacyl-CoA dehydrogenase, and long-chain ketoacyl-CoA thiolase (3). In addition, a short-chain hydratase, a short-chain hydroxyacyl-CoA dehydrogenase, and two ketoacyl-CoA thiolases (medium-chain or general and short-chain) have been described.

In contrast to the proteins of the respiratory chain, all β-oxidation enzymes are encoded on nuclear genes and all defects so far reported are inherited in an autosomal recessive fashion. Recently several enzymes have been sequenced, including CPT-II (4), the acyl-CoA dehydrogenases (5,6), ETF (7), and some short-chain enzymes (6,8). This has allowed the identification of the molecular genetic basis of the enzyme deficiency in some patients (see later) and also enabled an estimate of the incidence of these disorders. The most commonly described defect of fatty acid oxidation is medium-chain acyl-CoA dehydrogenase (MCAD) deficiency, and one specific point mutation, an A-G transition at position 985 of the cDNA, accounts for 85% of all mutant alleles. Screening for the prevalence of the mutation in a population in the United Kingdom suggests the carrier frequency to be between 1:40 and 1:68 (9). Thus the frequency of the homozygous state may be as high as 1 in 6000 births, which is more common than phenylketonuria.

PATHOPHYSIOLOGY AND CLINICAL FEATURES OF β-OXIDATION DEFECTS

Fatty acid oxidation contributes to energy homeostasis in many organs, particularly skeletal muscle, heart, liver, and kidney. The known defects share many common clinical symptoms, although there are often indications as to the probable site of the defect. However, it is important to note that β-oxidation defects do not only present with muscle involvement, but usually affect other organs as well. Furthermore, the same biochemical defect may present with "muscle" symptoms in one patient and with "hepatic" or "cardiac" signs in another. Even in identical twins the symptoms can vary, which emphasizes the importance of environmental factors for the clinical presentation of these disorders. In this section we will highlight the common characteristics of β-oxidation defects.

Two main pathophysiological processes in fatty acid oxidation defects help explain most of the clinical and laboratory features associated with these diseases: (1) buildup of potentially toxic intermediates (long-chain acyl-CoA esters) proximal to the metabolic block and (2) depletion of energy-yielding substrates (acetyl-CoA, NADH) distal to the block. Given the central role of fats for energy homeostasis during fasting, many patients with inborn errors of fatty acid oxidation develop symptoms under conditions of fasting or metabolic stress (viral illness, fever, vigorous prolonged exercise). One of the hallmarks of fatty acid oxidation disorders is *fasting hypoglycemia* accompanied by *hypoketosis* and *elevated levels of free fatty acids in plasma*. The increase in serum free fatty acids with concomitant hypoketosis indicates impaired β-oxidation of fatty acids in the liver and thus the failure of ketone body

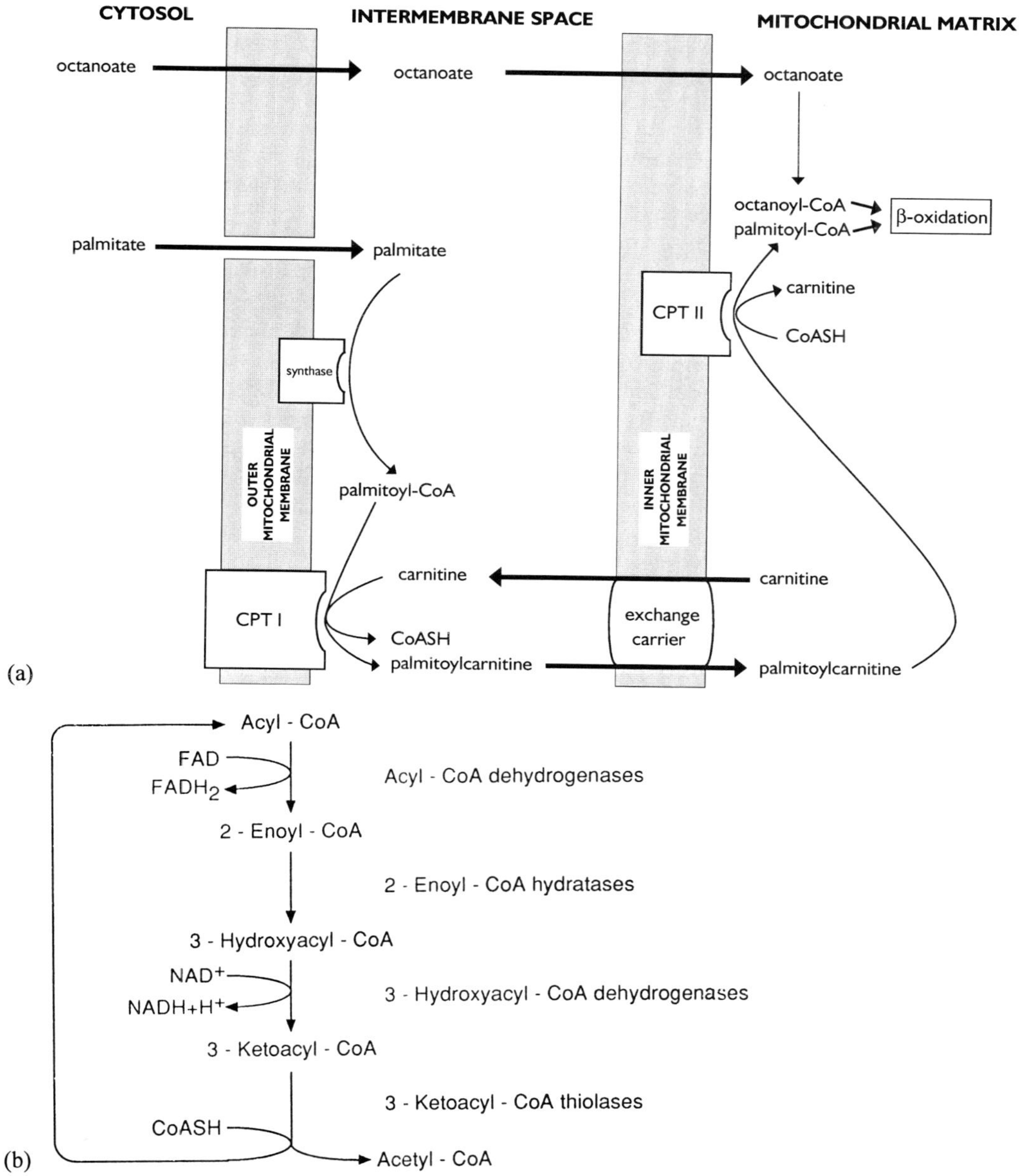

Figure 1 Long-chain fatty acids are transported into the mitochondrial matrix by the combined action of CPT-I, carnitine-acylcarnitine translocase, and CPT-II (a). Medium- and short-chain fatty acids (e.g., octanoate) enter the mitochondrial matrix directly. Once inside the mitochondria the fatty acyl-CoA esters are converted to acetyl-CoA by enzymes of β-oxidation (b).

synthesis. The hypoglycemia can be explained by an acceleration of peripheral glucose utilization, because ketone bodies are not available as alternative fuels. At the same time, hepatic gluconeogenesis is markedly reduced, which contributes to the hypoglycemia. The consequence is a rapid metabolic decompensation eventually leading to multiorgan failure, in particular encephalopathy and coma. It must be stressed that the duration and depth of coma may be unrelated to the hypoglycemia and are presumably due to an accumulation of toxic intermediary metabolites.

The muscle symptoms in these patients are predominantly muscle pain and weakness. Adults may present with a progressive proximal myopathy, neonates and infants often with a generalized profound hypotonia and failure to thrive. Typically these symptoms are provoked by prolonged exercise, fasting, or cold, when fatty acids become an essential metabolic fuel. In patients with such chronic recurrent symptoms, a lipid storage myopathy is often found on histology, whereas in other cases the energy crisis in muscle cells leads to acute muscle necrosis with rhabdomyolysis, elevated creatine kinase, and myoglobinuria, sometimes severe enough to induce acute renal failure.

In addition to abnormalities of liver and skeletal muscle function, β-oxidation defects, especially long-chain defects, frequently involve the heart and occasionally the nervous system or kidneys. Patients with a cardiomyopathy are at a permanent risk of cardiac arrest due to arrhythmias (intraventricular tachycardia, third-degree heart block) (10) and progressive heart failure. Apart from a lipid storage myopathy and encephalopathy, some long-chain deficiencies may present with specific neurological signs, i.e., trifunctional protein deficiency can be associated with a progressive peripheral neuropathy and retinitis pigmentosa (11). Extrapyramidal symptoms (progressive dyskinesia and dystonia), renal and cerebral dysplasia, and malformations have been described in ETF and ETF-dehydrogenase deficiency (12). It is likely, however, that a disturbance of amino acid metabolism rather than fatty acid oxidation is the predominant factor in the pathogenesis of CNS and renal involvement in ETF and ETF-dehydrogenase deficiency.

A summary of common clinical features is given in Table 1.

DIAGNOSTIC APPROACH TO A PATIENT WITH A LIPID STORAGE MYOPATHY OR A SUSPECTED β-OXIDATION DEFECT

In spite of all recent advances in laboratory techniques, a thorough history and physical examination are still of fundamental importance in the investigation of a patient with a lipid storage myopathy or a suspected β-oxidation defect. It must be stressed again that, while the main presentation may be with muscle signs, other organs can be affected. Appropriate screening for involvement of heart, liver, kidneys, and nervous system therefore forms an essential part of the investigation.

History and Examination

Fatty acid oxidation defects may present either with acute and episodic symptoms or with a more chronic clinical course. Although the episodic presentation is more typical, patients with β-oxidation defects may have a variety of findings ranging from neonatal hypotonia and coma to late-onset myopathy. When onset is in infancy or childhood, the patients predominantly present with acute metabolic decompensation. A child characteristically comes to hospital with a history of increasing lethargy and coma following a prolonged fast or minor viral illness. A number of patients have been described who presented for the first time in their teens or early twenties. Then the prevailing symptoms are usually exercise-induced progressive muscle weakness and/or muscle pain.

Important features that suggest an underlying β-oxidation defect are:

- Previous episodes of unexplained rhabdomyolysis or myoglobinuria, especially if they are recurrent and related to exercise or fasting
- Previous attacks of unexplained hypoglycemia, encephalopathy, or cardiorespiratory arrest
- Positive family history of myopathy, unexplained encephalopathy and psychomotor retardation, Reye's syndrome, or sudden infant death
- Evidence for simultaneous involvement of heart, liver, and skeletal muscle

Table 1 Diagnostic Approach to Fatty Acid Oxidation Disorders

I. Clinical features	
A. Acute metabolic decompensation	
Fasting-induced coma	
Reye-like syndrome with hepatic encephalopathy	
Hypoketotic hypoglycemia	
Sudden infant death	
Hepatomegaly, hepatic steatosis	All defects
Recurrent episodes triggered by fasting or minor viral illness	
B. Myopathy	
Chronic: progressive proximal muscle weakness; lipid storage in muscle cells	
Acute: exercise-induced myalgia, rhabdomyolysis, myoglobinuria	
Histological evidence for acute muscle necrosis or regenerating muscle fibers	All defects
Neonatal hypotonia	
C. Cardiomyopathy	
Progressive heart failure (hypertrophic or dilated cardiomyopathy	Long-chain defects
Arrhythmias, cardiorespiratory arrest	
Fatty infiltration of tissue on biopsy	
D. Involvement of the nervous system	
Progressive peripheral neuropathy	TFP deficiency
Retinitis pigmentosa	TFP deficiency
Malformations of the CNS	ETF or
Extrapyramidal movement disorders	ETF-DH deficiency
E. Involvement of other organs	
Renal cysts	ETF or ETF-DH deficiency
II. Laboratory findings	
A. Nonspecific abnormalities	
Hypoglycemia with inappropriately low blood and urine ketones and elevated serum free fatty acids	
Hyperammonemia	
Elevated serum transaminases with normal bilirubin	
Hyperuricemia	
Elevated serum creatine kinase	
Metabolic acidosis, lactic acidosis (in TFP deficiency)	
B. Urinary organic acids	
Dicarboxylic aciduria; pattern often suggestive, but not diagnostic, of a specific defect (glutaric aciduria type II, TFP deficiency, MCAD deficiency)	
Acyl glycines	
C. Plasma and tissue carnitines	
Decreased total plasma carnitine concentration, increased percentage of esterified carnitine (exception: CPT-I deficiency)	
Separation of individual carnitine esters (HPLC, tandem mass spectrometry) may allow localization of the defect	
D. Specific in vitro tests	
Overall β-oxidation flux (oxidation of ^{14}C- or ^{3}H-labeled fatty acids by fibroblasts or lymphocytes)	
Analysis of pathological β-oxidation intermediates (radio-HPLC, tandem mass spectrometry)	
Measurement of individual enzyme activities	
DNA studies (so far possible only for MCAD deficiency)	

TFP, trifunctional enzyme deficiency; ETF, electron transfer flavoprotein deficiency; ETF-DH, electron transfer flavoprotein dehydrogenase deficiency.

Physical examination may reveal muscle weakness, wasting, or hypotonia, cardiomegaly and/or signs of heart failure, hepatomegaly, encephalopathy, or peripheral neuropathy.

Further clinical studies must be guided by history and physical examination. They should include abdominal ultrasound, echocardiography, electrocardiography, chest X-ray, and neurophysiological tests (electromyography, nerve conduction studies).

Laboratory Investigations (Table 1)

Initial laboratory tests should include: urea and electrolytes, liver function tests, plasma glucose, ammonia, acid-base status, creatine kinase, and urinalysis. A metabolic acidosis is usually mild; a large anion gap often indicates superimposed lactic acidosis. Hypoglycemia (<3 mmol/L) with only a trace of urinary ketones, hyperammonemia, elevated transaminases with normal bilirubin, and hyperuricemia are all features of a metabolic disorder and merit further, more specific investigations.

Valuable information about the presence of a β-oxidation defect can be obtained by studying metabolic fuels and intermediary metabolite (lactate, pyruvate, ketone bodies, glycerol, alanine, free fatty acids) concentrations in plasma. A typical finding is hypoketotic hypoglycemia with elevated free fatty acids. However, the presence of ketone bodies in blood or urine does not rule out a β-oxidation defect, because the abnormal constellation of intermediary metabolites may only occur under conditions of metabolic stress. It is therefore essential to obtain blood and urine samples when the patient is ill and before any intravenous infusions (especially glucose) are given.

Measurement of Carnitine Concentrations in Plasma and Urine

Low concentrations (<30 μmol/L) of total (= free and esterified) plasma carnitine together with an increased percentage (>80%) of esterified (acylated) carnitine are a common finding in patients with β-oxidation defects. The acyl-CoA esters that cannot be metabolized accumulate within the mitochondria and are conjugated with carnitine to form acylcarnitine esters. The acylcarnitine esters are then transported out of the mitochondria and finally excreted by the kidneys. This results in a higher proportion of acylated plasma carnitines, an increased renal clearance, and consequently a secondary carnitine deficiency.

Separation and identification of the individual acylcarnitines (for example by fast atom bombardment tandem mass spectrometry) (13) may allow further localization of the defect, since specific carnitines accumulate in some deficiencies (e.g., octanoylcarnitine in MCAD deficiency). In primary carnitine deficiency, which frequently leads to a lipid storage myopathy, both free and total carnitine are extremely low with a paradoxically high renal excretion of carnitine and acylcarnitines.

Urinary Organic Acid Analysis

When the capacity for mitochondrial β-oxidation is exceeded, as is the case in inborn errors of fatty acid metabolism, alternative pathways (peroxisomal β-oxidation, ω-oxidation) may contribute significantly to total cellular fatty acid oxidation. This results in the accumulation of dicarboxylic acids, which can be identified in plasma or urine by gas chromatography–mass spectrometry. Urinary organic acids are diagnostically very useful, because many β-oxidation disorders are associated with a specific pattern of dicarboxylic aciduria. However, dicarboxylic acids are not reliably present when the patient is well and they generally disappear once normoglycemia is restored. Moreover, dicarboxylic acids are seen in a variety of other conditions, such as diabetic ketoacidosis, and during administration of medium-chain triglycerides or with certain medications (valproic acid). The third limitation is that disorders of fatty acid transport into the mitochondria (CPT deficiency, primary carnitine deficiency, carnitine-acylcarnitine translocase deficiency) do not produce a dicarboxylic aciduria. Thus, while the presence of dicarboxylic acids in excess of ketones provides strong evidence for a β- oxidation defect, the absence of dicarboxylic acids does not necessarily rule out fatty acid oxidation disorders (14). Alternatively, acyl-CoA esters accumulating in fatty acid oxidation defects may be conjugated to glycine. The glycine metabolites formed are diagnostically useful in several defects (MCAD deficiency, ETF deficiency) (15).

In Vivo Provocation Tests for the Detection of β-Oxidation Disorders

Fasting and fat-loading tests are useful in the detection of β-oxidation disorders, when there is no disturbance of metabolism in the basal state. The aims of a fasting study are to distinguish fatty acid oxidation defects from other causes of hypoglycemia and to assess the individual fasting tolerance as an aid in giving therapeutic recommendations to the patient. Hypoketotic hypoglycemia with inappropriately high levels of plasma free fatty acids is a characteristic finding in β-oxidation defects. However, fasting yields unequivocal results only if it is prolonged enough (usually at least 18 hr in adults) to produce either hypoglycemia (in β-oxidation defects) or a marked ketosis (in normal persons). This leads to obvious risks for the patients, who should therefore be closely monitored and promptly treated with intravenous glucose infusions if necessary. Fasting studies can also induce myoglobinuria in some patients, again stressing the potential risks of the procedure. Fat loading tests are sometimes useful in localizing the site of a fatty acid oxidation defect. A long-chain triglyceride load (sunflower oil) allows detection of long-chain defects, since plasma ketone bodies do not rise after the load.

In Vitro Tests for the Diagnosis of β-Oxidation Disorders

Defects in the β-oxidation pathways have been identified indirectly by measuring the products of ^{14}C-labeled fatty acids in cultured fibroblasts (16), white cells, or muscle mitochondria (17). Separation of these products by a recently developed HPLC-based technique (18) or by tandem mass spectrometry allows the detection of a pathological accumulation of β-oxidation intermediates.

In spite of these recent advances, identification of a β-oxidation defect in many cases still requires measurement of the individual enzyme activities (19), which is difficult owing to the lack of commercially available substrates.

Improved understanding of the genetic basis of each of the deficiencies should in future allow identification of β-oxidation disorders by molecular genetics techniques. To some extent this is already true for MCAD deficiency.

SPECIFIC FEATURES OF INDIVIDUAL β-OXIDATION DEFECTS

Carnitine Palmitoyltransferase Deficiency

CPT deficiency was the first β-oxidation defect to be described (20). Clinically and biochemically it can be classified into three subtypes:

The muscle form presents in adolescence or adulthood with recurrent episodes of myalgia, rhabdomyolysis, and myoglobinuria, typically induced by fasting and/or prolonged exercise. The biochemical defect involves CPT-II. There is often minimal or no lipid storage in muscle cells. The defect is inherited in an autosomal-recessive manner, although males are more commonly reported, probably because they perform more prolonged exercise. Between episodes most patients are symptom-free, but in approximately 20% a mild muscle weakness can be found. Hypoglycemia and cardiac involvement are unusual features.

In the much rarer infantile hepatic form, fasting hypoglycemia and hypoketonemia occur in the absence of muscle symptoms (21). CPT-I activity is low in liver and fibroblasts from the patients, but not in muscle. CPT-I deficiency is the only β-oxidation defect where free plasma carnitine concentrations are normal, because the transfer of acyl groups from CoA to carnitine is impaired.

Recently a hepatomuscular presentation of CPT-II deficiency has been reported with severe disease in infancy (hypoglycemia, cardiomyopathy, hypotonia, renal cysts) (22).

Primary Carnitine Deficiency

In contrast to secondary carnitine deficiency, primary carnitine deficiency is rare; it presents in childhood with progressive proximal muscle weakness, cardiomyopathy, and in some cases hypoglycemia (23). It is caused by impaired cellular carnitine uptake.

Long-Chain Acyl-CoA Dehydrogenase Deficiency

According to the clinical phenotype, the patients may be divided into two groups:

In the first group the clinical course is very similar to that of MCAD deficiency, with recurrent episodes of hypoketotic hypoglycemia, vomiting, and coma brought on by fasting. Compared to MCAD deficiency, the illness seems to have a higher mortality, an earlier onset (first 6 months of life) and is associated with lipid storage myopathy and cardiomyopathy (24,25).

The second group is characterized by later onset (18 months–20 years), often without metabolic disturbance, but with prominent muscle symptoms (exercise-induced myalgia and myoglobinuria) resembling a CPT-II deficiency. These patients often do not show cardiac abnormalities, and the lipid storage in muscle may be mild. It has now become apparent that the deficiency in this group of patients in fact involves the recently described very-long-chain acyl-CoA dehydrogenase (26).

Medium-Chain Acyl-CoA Dehydrogenase (MCAD) Deficiency

This is probably the most common fatty acid oxidation defect, with a proposed incidence of approximately 1:6000 births in the United Kingdom (9). Symptoms are often precipitated by metabolic stress and include a Reye-like syndrome with stupor, encephalopathy, hepatomegaly, hypoketotic hypoglycemia, and sudden infant death (27). The majority of patients reported are young children, but some patients present in adult life with muscle pain, lipid storage myopathy, and rhabdomyolysis. Cardiomyopathy has not been reported in patients with MCAD deficiency. The recognition of a common point mutation at position 985 of the MCAD cDNA allows a molecular diagnosis of MCAD deficiency in up to 90% of affected patients.

Short-Chain Acyl-CoA Dehydrogenase (SCAD) Deficiency

SCAD deficiency appears to be very rare, having been identified in only a few patients (28). Clinically two forms can be distinguished:

A severe infantile form with hypoglycemic episodes, vomiting, lethargy, muscle weakness, microcephaly, and psychomotor retardation. These children all have a systemic SCAD deficiency (29).

A mild late-onset form with slowly progressive proximal lipid storage myopathy and secondary carnitine deficiency (28). The enzyme deficiency is limited to skeletal muscle.

Some cases of myopathic SCAD deficiency may represent a riboflavin-responsive multiple acyl-CoA dehydrogenase deficiency (30, and below).

Glutaric Aciduria Type II (ETF or ETF-Dehydrogenase Deficiency)

In ETF or ETF-dehydrogenase deficiency, the defect affects not only fatty acid oxidation, but also amino acid metabolism, since the dehydrogenases involved in this pathway connect to the respiratory chain via ETF and ETF dehydrogenase. Clinically and biochemically these disorders mimic multiple acyl-CoA dehydrogenase (MAD) deficiency and for this reason are sometimes incorrectly referred to as MAD deficiency. The clinical picture may vary:

Patients with the severe form (31) present at birth with congenital abnormalities, hypoketotic hypoglycemia, hepatomegaly, acidosis, and a complex organic aciduria consisting of many fatty acid– and amino acid–derived metabolites. The predominant urinary metabolite, glutarate, has given this defect its name: glutaric aciduria type II. Most patients die in the neonatal period.

With the mild form patients usually have similar, but less severe symptoms, and the age of onset ranges from infancy to adolescence. In some cases muscle weakness associated with a lipid storage myopathy may be the prominent clinical feature (32).

Riboflavin-Responsive Multiple Acyl-CoA Dehydrogenase (MAD) Deficiency

A number of patients have been described in whom the activity of more than one of the acyl-CoA dehydrogenases was low. These patients often present with muscle pain and proximal muscle weakness, which responds well to large doses of riboflavin, an essential cofactor for the acyl-CoA dehydrogenases (30). Both the clinical picture and organic aciduria are similar to those seen with ETF or ETF-dehydrogenase deficiency, but the low activity of the acyl-CoA dehydrogenases and the responsiveness to riboflavin usually enable one to distinguish MAD deficiency from those disorders.

Trifunctional Enzyme (TFP) Deficiency

This recently described enzyme defect shares many similarities with other long-chain enzyme deficiencies, including hypoketotic hypoglycemia, hepatomegaly, encephalopathy, cardiomyopathy, and sudden infant death (3). Muscle weakness and myoglobinuria associated with respiratory failure have been reported in several children (3). The age of onset may be as early as 3 days, but some patients have presented as adults (personal observation). Some clinical and laboratory features distinguish TFP deficiency from other inborn errors of β-oxidation: patients often show signs of a sensorimotor polyneuropathy and in one instance pigmentary retinopathy. The majority of patients have a hydroxy-dicarboxylic aciduria, although this may be variable and depend on the metabolic state of the patient. A unique biochemical finding in TFP deficiency is lactic acidosis, which is not seen in other β-oxidation defects (3).

MANAGEMENT OF PATIENTS WITH β-OXIDATION DEFECTS

Fatty acid oxidation defects can usually be treated successfully if recognized at an early age. In addition to nonspecific measures (for example, inotropic and antiarrhythmic drugs for an associated cardiomyopathy), the specific management of these conditions includes the following:

Dietary Management

The primary treatment is the avoidance of prolonged fasting. It is vital to ensure adequate calorie intake in the form of carbohydrates with a low percentage of fat in the diet. Dietary advice is essential, because any excess carbohydrates will be stored as long-chain fats in adipose tissue. Additional carbohydrates must be given if there is an intercurrent illness. If oral feeding is not possible, intravenous glucose must be administered in amounts sufficient to suppress lipolysis completely. In patients with defects involving long-chain fatty acid oxidation, medium-chain triglycerides may be useful as a nutritional source, because medium-chain fatty acids enter β-oxidation below the metabolic block. Obviously, a defect of medium-chain fatty acid oxidation such as MCAD must be ruled out before medium-chain triglycerides are administered.

Carnitine

Carnitine supplementation (100 mg/kg/day) has been highly beneficial for patients with primary carnitine deficiency, resulting in normalization of cardiac function and resolution of clinical disease.

In other fatty acid oxidation disorders, carnitine supplementation corrects the secondary carnitine deficiency, but its clinical value is uncertain.

Riboflavin

Some patients with multiple acyl-CoA dehydrogenase deficiency have responded well to high doses of riboflavin (100–300 mg/day) (30).

OTHER CAUSES OF LIPID STORAGE MYOPATHY

Respiratory chain defects may lead to a secondary lipid storage myopathy. These defects are discussed in detail elsewhere in this book. Peroxisomal disorders (Refsum's disease, Zellweger syndrome) are a very rare cause for lipid accumulation in muscle cells; for a detailed review, see Ref. 33.

About 25 patients with neutral lipid storage disease have been reported. This disorder is characterized by massive lipid storage in multiple tissues leading to a proximal myopathy, congenital ichthyosis, cataracts, neurosensory deafness, abnormal liver function, and various other neurological manifestations (34). Lipid vacuoles in either granulocytes or cultured skin fibroblasts can be identified by simple light microscopy and are pathognomonic for the disease. Biochemical studies have shown normal oxidation of fatty acids, but revealed a specific impairment in the degradation of endogenously synthesized triglycerides (34). The exact nature of the underlying defect remains unknown, but may involve an intracellular lipase or a protein necessary for directing the substrate (triglyceride) to its enzyme (lipase). Most patients present in early to middle adult life and the course of the disease is associated with slowly progressive deterioration. No effective treatment is available.

REFERENCES

1. Cahill GF. Starvation in man. N Engl J Med 1970; 282:668–675.
2. Miyazawa S, Hashimoto T, Yokota S. Identity of long-chain acyl-CoA synthetase of microsomes, mitochondria and peroxisomes in rat liver. J Biochem 1985; 98:723–733.
3. Jackson S, Kler RS, Bartlett K, Briggs H, Bindoff LA, Pourfarzam M, Gardner-Medwin D, Turnbull DM. Combined enzyme defect of mitochondrial fatty acid oxidation. J Clin Invest 1992; 90:1219–1225.
4. Finocchiaro G, Taroni F, Rocchi M, Liras Martin A, Colombo I, Torri Tarelli G, DiDonato S. cDNA cloning, sequence analysis, and chromosomal localization of the gene for human carnitine palmitoyltransferase. Proc Natl Acad Sci USA 1991; 88:661–665.
5. Indo Y, Yang-Feng T, Glassberg R, Tanaka K. Molecular cloning and nucleotide sequence of cDNA encoding human long-chain acyl-CoA dehydrogenase and assignment of the location of its gene (ACADL) to chromosome 2. Genomics 1991; 11:609–620.
6. Naito E, Ozasa H, Ikeda Y, Tanaka K. Molecular cloning and nucleotide sequence of complementary cDNAs encoding human short-chain acyl-CoA dehydrogenase and the study of the molecular basis of short-chain acyl-CoA dehydrogenase deficiency. J Clin Invest 1989; 83:1605–1613.
7. Finocchiaro G, Archidiacono N, Gellera C, Bloisi W, Colombo I, Valdameri G, Romeo G, Tanaka K, DiDonato S. Molecular cloning and chromosomal localization of the β-subunit of human electron transfer flavoprotein (ETF). Am J Hum Genet 1989; 45:A185.
8. Mori M, Amaya Y, Arakawa H, Takiguchi M. Biosynthesis and mitochondrial import of fatty acid oxidation enzymes. In: Tanaka K, Coates PM, eds. Fatty Acid Oxidation: Clinical, Biochemical and Molecular Aspects. New York: Alan R Liss, 1990:663–672.
9. Blakemore AIF, Singleton H, Pollitt RJ, Engel PC, Kolvraa S, Gregersen N, Curtis D. Frequency of the G985 MCAD mutation in the general population. Lancet 1991; 337:298–299.
10. Pande SV, Brivet M, Slama A, Demaugre F, Aufrant C, Saudubray JM. Carnitine-acylcarnitine translocase deficiency with severe hypoglycaemia and auriculo-ventricular block. J Clin Invest 1993; 91:1247–1252.
11. Poll-The BT, Bonnefont JP, Ogier H, Charpentier C, Pelet A, Le Fur JM, Jakobs C, Kok RM, Duran M, Divry P, Scotto J, Saudubray JM. Familial hypoketotic hypoglycaemia associated with peripheral neuropathy, pigmentary retinopathy and C_6-C_{14} hydroxydicarboxylic aciduria: a new defect in fatty acid oxidation. J Inher Metab Dis 1988; 11(Suppl 2):183–185.
12. Loehr JP, Goodman SI, Frerman FE. Glutaric acidaemia type II: heterogeneity of clinical and biochemical phenotypes. Pediatr Res 1990; 27:311–315.

13. Millington DS, Terada N, Chace DH, Chen YT, Ding JH, Kodo N, Roe CR. The role of tandem mass spectrometry in the diagnosis of fatty acid oxidation disorders. Progr Clin Biol Res 1992; 375:339–354.
14. Pollitt RJ. Disorders of mitochondrial β- oxidation: prenatal and early postnatal diagnosis and their relevance to Reye's syndrome and sudden infant death. J Inter Metab Dis 1989; 12(Suppl 1):215–230.
15. Rinaldo P, O'Shea JJ, Coates PM, Hale DE, Stanley CA, Tanaka K. Medium-chain acyl-CoA dehydrogenase deficiency: diagnosis by stable isotope dilution measurement of urinary *n*-hexanoylglycine and 3-phenylpropionylglycine. N Engl J Med 1988; 319:1308–1313.
16. Saudubray JM, Coude FX, Demaugre F, Johnson C, Gibson KM, Nyhan WL. Oxidation of fatty acids in cultured fibroblasts: a model system for the detection and study of defects in oxidation. Pediatr Res 1982; 16:877–881.
17. Watmough NJ, Bhuijan AKMJ, Bartlett K, Sherratt HSA, Turnbull DM. Skeletal muscle mitochondrial β-oxidation—a study of the products of oxidation of [U-^{14}C]-hexadecanoate by h.p.l.c using continuous on-line radiochemical detection. Biochem J 1988; 253:541–547.
18. Schaefer J, Pourfarzam M, Bartlett K, Jackson S, Turnbull DM. Fatty acid oxidation in peripheral blood cells: characterisation and use for the diagnosis of defects of fatty acid oxidation. Pediatr Res 1995; 37:354–360.
19. Birch-Machin MA, Jackson S, Singh-Kler R, Turnbull DM. Study of skeletal muscle mitochondrial dysfunction. Meth Toxicol 1993; 2:51–69.
20. DiMauro S, DiMauro PP. Muscle carnitine palmitoyltransferase deficiency and myoglobinuria. Science 1973; 182:929–931.
21. Demaugre F, Bonnefont JP, Mitchell G, Nguyen-Hoang N, Pelet A, Rimoldi M, DiDonato S, Saudubray JM. Hepatic and muscular presentations of carnitine palmitoyltransferase deficiency: two distinct entities. Pediatr Res 1988; 24:308–311.
22. Demaugre F, Bonnefont JP, Colonna M, Cepanec C, Leroux JP, Saudubray JM. Infantile form of carnitine palmitoyltransferase-II deficiency with hepatomuscular symptoms and sudden death. J Clin Invest 1991; 87:859–864.
23. Stanley CA, DeLeeuw S, Coates PM, Vianey-Liaud C, Divry P, Bonnefont JP, Saudubray JM, Haymond M, Trefz FK, Breningstall GN, Wappner RS, Byrd DJ, Sansaricq C, Tein I, Grover W, Valle D, Rutledge SL, Treem WR. Chronic cardiomyopathy and weakness or acute coma in children with a defect in carnitine uptake. Ann Neurol 1991; 30:709–716.
24. Hale DE, Batshaw ML, Coates PM, Frerman FE, Goodman SI, Singh I, Stanley CA. Long-chain acyl-CoA dehydrogenase deficiency: an inherited cause of nonketotic hypoglycaemia. Pediatr Res 1985; 19:666–671.
25. Treem WR, Stanley CA, Hale DE, Leopold HB, Hyams JS. Hypoglycaemia, hypotonia and cardiomyopathy: the evolving clinical picture of long-chain acyl-CoA dehydrogenase deficiency. Pediatrics 1991; 87:328–333.
26. Ogilvie I, Pourfarzam M, Jackson S, Stockdale C, Bartlett K, Turnbull DM. Very long-chain acyl-CoA dehydrogenase deficiency presenting with exercise-induced myoglobinuria. Neurology 1994; 44:467–473.
27. Roe CR, Coates PM. Acyl-CoA dehydrogenase deficiencies. In: Scriver CR, Beaudet AL, Sly WS, Valle D, eds. The Metabolic Basis of Inherited Disease, 6th ed. New York: McGraw-Hill, 1989:889–914.
28. Turnbull DM, Bartlett K, Stevens DL, Alberti KGMM, Gibson GJ, Johnson MA, McCulloch AJ, Sherratt HSA. Short-chain acyl-CoA dehydrogenase deficiency associated with a lipid storage myopathy and secondary carnitine deficiency. N Engl J Med 1984; 311:1232–1236.
29. Amendt BA, Greene C, Sweetman L, Cloherty J, Shih V, Moon A, Teel L, Rhead WJ. Short-chain acyl-CoA dehydrogenase deficiency. J Clin Invest 1987; 79:1303–1309.
30. Gregersen N. Riboflavin-responsive defects of β-oxidation. J Inher Metab Dis 1985; 8(Suppl 1):65–69.
31. Amendt BA, Rhead WJ. The multiple acyl-CoA dehydrogenation disorders, glutaric aciduria type II and ethylmalonic-adipic aciduria. J Clin Invest 1986; 78:205–213.
32. Turnbull DM, Bartlett K, Eyre JA, Gardner-Medwin D, Johnson MA, Fisher J, Watmough NJ. Lipid storage myopathy due to glutaric aciduria type II: treatment of a potentially fatal myopathy. Dev Med Child Neurol 1988; 30:667–672.
33. Wanders RJA, Schutgens RBH, Barth PG, Tager JM, van den Bosch H. Postnatal diagnosis in peroxisomal disorders: a biochemical approach. Biochimie 1993; 75:269–279.
34. DiDonato S, Garavaglia B, Striscinglio B, Borrone C, Andria G. Multisystem triglyceride storage disease is due to a specific defect in the degradation of endocellularly synthesised triglyercides. Neurology 1988; 38:1107–1110.

35
Myoadenylate Deaminase Deficiency

Richard L. Sabina
Medical College of Wisconsin, Milwaukee, Wisconsin

INTRODUCTION

Myoadenylate deaminase (skeletal muscle AMP deaminase isoform; EC 3.5.4.6) deficiency is a common muscle enzyme defect exhibiting a heterogeneous clinical picture, presenting either as a primary, inherited myopathy or in an acquired form in a variety of neuromuscular diseases. To date, over 180 cases have been described in the literature, of which nearly one-half are classified as inherited (for a more comprehensive review see Ref. 1).

CLINICAL AND LABORATORY CRITERIA FOR DIAGNOSIS

Based on a variety of parameters such as differential clinical presentation, activity and immunological profile of residual AMP deaminase protein, and levels of other muscle-specific enzymes (see below), inherited and acquired forms of myoadenylate deaminase (mAMPD) deficiency have been proposed (2) and supported by subsequent studies (3,4).

Clinically, inherited mAMPD deficiency is characterized by exercise-induced myopathy with no other apparent neuromuscular complications. By this criterion alone, therefore, inherited mAMPD deficiency is indistinguishable from a variety of other exertional myopathies of known (e.g., glycolytic and lipid metabolism) and unknown origin. Acquired mAMPD deficiency, on the other hand, is observed secondary to numerous neuromuscular and rheumatological diseases. Although it is unclear whether any particular disease is more likely to exhibit an acquired mAMPD deficiency (or if there are any that do not display this anomaly), it most commonly occurs in their advanced stages (see below). In these patients, therefore, overwhelming clinical manifestations of the associated disease make it difficult to assess whether an acquired deficiency of mAMPD contributes to the clinical picture.

mAMPD deficiency can be diagnosed by several laboratory procedures, including ischemic forearm exercise (5–8) and direct histochemical (9) or enzymatic (10–12) assay of biopsied tissue. Ischemic forearm exercise is a simple, relatively noninvasive technique, although inadequate exercise performance may result in a false negative diagnosis. In this regard, an increase in blood lactate of at least 2.5 mM should be attained (7).

The acquisition of biopsied tissue allows for additional laboratory tests that can distinguish the inherited and acquired forms of this disease. These include determining the relative amount and immunoreactivity of residual AMP deaminase activity and levels of other muscle-specific enzymatic activities, such as M-creatine phosphokinase and adenylate kinase. Inherited mAMPD-deficient patients generally exhibit lower residual AMP deaminase activities that are less immunoreactive with anti-mAMPD serum than are observed in acquired mAMPD-deficient individuals. Acquired mAMPD-deficient patients also display reductions in other muscle-specific enzymatic activities, although at much higher levels than AMP deaminase (1,2). Because of the overlap between the two groups of patients, however, it is desirable to evaluate as many criteria as possible (1–3). Elevations in serum creatine kinase have been noted in approximately one-half of all patients with inherited mAMPD deficiency, although these are generally mild and transient (for a review, see Ref. 1).

Muscle biopsies obtained from individuals with inherited mAMPD deficiency are generally unremarkable on routine histopathological examination. Similar evaluation of material obtained from acquired mAMPD-deficient patients, however, typically reveals pathology characteristic of the associated neuromuscular or rheumatological disease. Electromyographic analyses of inherited mAMPD-deficient individuals are generally normal or indicative of a nonspecific myopathic process (for a review, see Ref. 1).

AMP deaminase histochemical staining may also be useful for distinguishing between alternate forms of mAMPD deficiency. In skeletal muscle sections taken from individuals with inherited mAMPD deficiency, histochemical staining indicates a uniform lack of AMP deaminase activity in all myofibers (9,10,13–17). These results are consistent with an inherited defect in mAMPD; i.e., all myocytes would be expected to lack this enzymatic activity. Conversely, histochemical staining in acquired mAMPD-deficient skeletal muscle sections is variable. In some cases there is a uniform lack of staining (16–20), whereas in others differential staining of myofibers is evident (2,17). The former may represent an inherited mAMPD deficiency coincidental with other neuromuscular disease. Such individuals might be anticipated based on the incidence of inherited mAMPD deficiency, recently proposed to be 1–3% in the Caucasian and Afro-American populations (21; see below). The existence of a third class of mAMPD-deficient patients, initially classified as acquired but actually harboring an inherited defect in the *AMPD1* gene, might explain the significant overlap in laboratory findings between the inherited and acquired forms. Differential histochemical staining of myofibers, on the other hand, might be characteristic of a true acquired mAMPD deficiency because of the wide variety of pathological conditions in this patient group (see below).

MYOADENYLATE DEAMINASE AND SKELETAL MUSCLE ENERGETICS: IMPLICATIONS FOR THE DISEASE

mAMPD is an integral component of a potentially important energy-related pathway in skeletal muscle, the purine nucleotide cycle (PNC). Although outside the scope of this review, the PNC has been extensively reviewed (see Ref. 22–24 for recent reviews). The PNC is illustrated in Figure 1, which also depicts how these enzymatic reactions impact on other energy-generating metabolic pathways. Proposed functions of the PNC, operative during exercise, include (1) maintenance of a high ATP/ADP ratio by the combined action of myokinase and mAMPD, (2) contribution to the expansion of tricarboxylic acid (TCA) cycle intermediates, (3) production of ammonia, and (4) activation of glycolysis through the generation of small molecule effectors. Although there is controversy over whether the enzymatic reactions of the PNC work in concert or in series, a recent study performed in human subjects supported operation of the entire cycle during exhaustive exercise (25). Overall, it is apparent that mAMPD plays a role in skeletal muscle energetics and that a deficiency of this enzymatic activity in humans could result in exercise-induced myopathic symptoms.

MOLECULAR BASIS OF INHERITED AND ACQUIRED MYOADENYLATE DEAMINASE DEFICIENCY

Significant progress has been made in the last several years regarding the molecular basis of mAMPD deficiency. Beginning with the isolation of rat mAMPD cDNA in 1987 (26), several unique AMP deaminase genes and their respective cDNAs have been cloned and partially characterized (3,27–30).

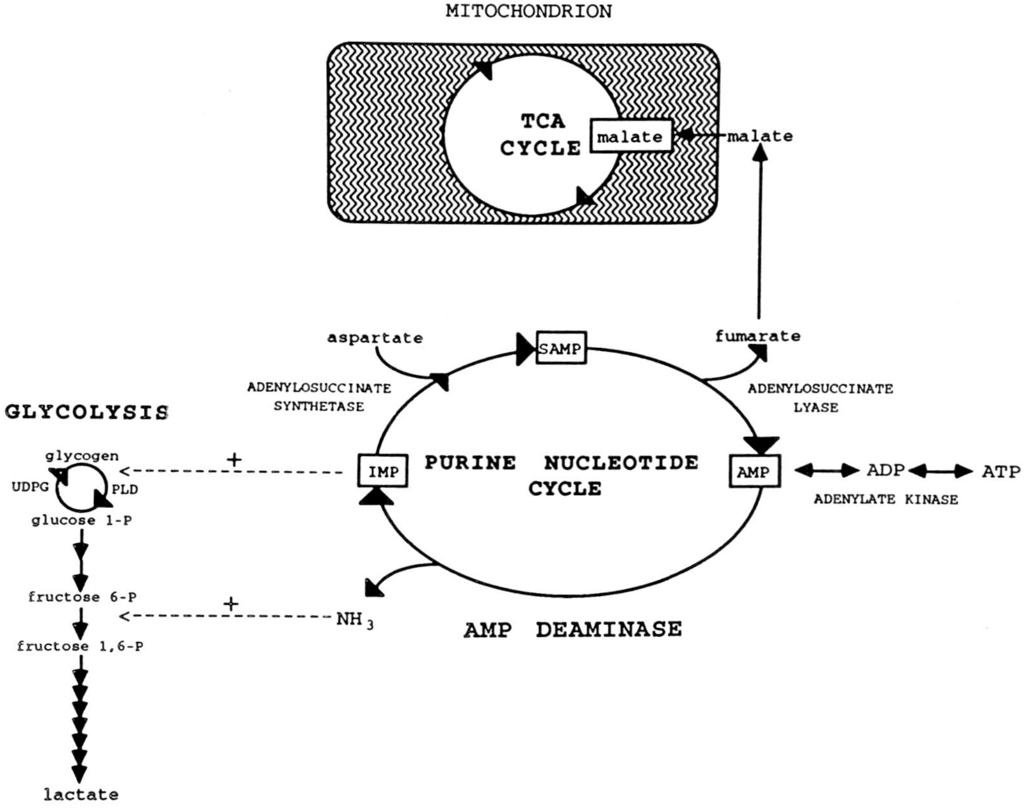

Figure 1 Purine nucleotide cycle (PNC), a series of enzymatic reactions catalyzed by AMP deaminase, adenylosuccinate synthetase, and adenylosuccinate lyase. Proposed interactions of the PNC with glycolysis [denoted by dotted lines indicating the steps that are positively (+) affected by IMP (phosphorylase b) and NH_3 (phosphofructokinase)] and the tricarboxylic acid (TCA) cycle are also illustrated. PLD, phosphorylase-limit-dextrin; UDPG, uridine diphosphoglucose; SAMP, adenylosuccinate.

Before the molecular basis of the disease is discussed, it is necessary to first consider AMP deaminase expression in normal skeletal muscle.

Molecular Basis of AMP Deaminase Expression in Normal Human Skeletal Muscle: Implications for Myoadenylate Deaminase Deficiency

The human AMP deaminase multigene family is composed of at least three genes (*AMPD1*, *AMPD2*, and *AMPD3*), each expressing transcripts encoding a unique AMP deaminase isoform (see Table 1). *AMPD1* produces a 2.5-kb transcript encoding mAMPD (26), and Northern blot analysis of multiple human tissues detects its expression only in skeletal muscle (30). This latter observation is consistent with the tissue specificity ascribed to mAMPD deficiency (2,10,31). Consistent with pedigree analyses that have indicated an autosomal pattern of inheritance for mAMPD deficiency (2,32,33), the *AMPD1* gene is located on the short arm of human chromosome 1 in the region p13–p21 (26). The human and rat *AMPD1* genes are structurally similar, each being composed of 16 exons, including a small, 12-bp exon 2 (26). A developmental and fiber type-regulated alternative splicing event that removes exon 2 from the mature *AMPD1* transcript has been documented in the rat (34,35). Although thought originally to be species-specific (35), this alternative splicing event occurs at low levels in human skeletal muscle

Table 1 Molecular Basis of AMP Deaminase Expression in Normal Human Skeletal Muscle

Gene	Size (kb)[a]	Alternatively spliced transcript	Size (bp)	Isoform	Subunit[b] mass (kDa)
AMPD1	23	Exon 2+	2341	mAMPD	86.5
		Exon 2–	2329	mAMPD	86.0
AMPD2	14	Exon1A/2	3755	lAMPD	92.1
		Exon1B/2	3970	lAMPD	100.7
		Exon1B/3	3839	lAMPD	92.9
AMPD3	60	Type 1a	3915[c]	eAMPD	89.7
		Type 1b	3879[c]	eAMPD	88.8
		Type 1b/1c	4076[c]	eAMPD	89.5
		Type 1c	4018[c]	eAMPD	89.5

[a]Estimated sizes.
[b]Predictions (based on cloned cDNA sequences).
[c]Mahnke-Zizelman DK et al: *Biochim Biophys Acta* (in press).

(35a). Production of an *AMPD1* transcript that has had exon 2 removed might have a bearing on the clinical manifestations of inherited mAMPD deficiency in human skeletal muscle (see below).

The human *AMPD2* gene produces three alternative transcripts encoding isoform L variants (35b), but Northern blot analyses demonstrate only a very low level of its expression in adult skeletal muscle (30). Immunohistochemical studies performed in adult rat skeletal muscle indicate that *AMPD2* gene expression is confined primarily to nonmyocyte components of this tissue (36).

The human *AMPD3* gene produces multiple transcripts through exon shuffling at the 5′-end of the gene (29). The processed mRNAs are proposed to encode the kinetically and immunologically similar E isoforms. Northern blot analysis demonstrates that the *AMPD3* gene is expressed in adult skeletal muscle (29), apparently at an intermediate level relative to the *AMPD1* and *AMPD2* genes. Analysis of residual AMP deaminase activity in skeletal muscle biopsies obtained from mAMPD-deficient patients, furthermore, indicates a significant component that is immunoreactive with anti-E serum (37). The existence of this second AMP deaminase in human skeletal myocytes may serve to minimize the clinical manifestations of an inherited mAMPD deficiency (see below).

Molecular Basis of Inherited Myoadenylate Deaminase Deficiency

Southern blot analyses yield normal restriction endonuclease digestion patterns of the *AMPD1* gene in patients diagnosed with inherited mAMPD deficiency (38), and Northern blot analyses demonstrate normal abundance of apparently normal-sized mAMPD transcript (3,21). Western blot analyses, however, do not detect immunoreactive protein in patients with inherited mAMPD deficiency (3,21). Together, these results imply small structural changes in the *AMPD1* gene in these individuals. Furthermore, these changes are not predicted to alter transcript stability, but rather affect immunoreactive protein production or stability. Accordingly, 11 unrelated patients with inherited mAMPD deficien- cy have been found homozygous for a mutant allele characterized by C—>T(U) transitions at nucleotides +34 (codon 12 in exon 2) and +143 (codon 48 in exon 3) (21). The C—>T(U) transition at codon 48 predicts an amino acid substitution of a leucine for a proline. This change in the mAMPD protein, however, does not appear to have a significant effect on catalytic activity (21). The transition at codon 12, on the other hand, creates a stop codon in the *AMPD1* transcript (21) (see Fig. 2), an alteration that would predict premature termination of the mAMPD polypeptide after only 11 amino acids.

In addition to explaining the molecular basis of inherited mAMPD deficiency, the identification of a C—>T(U) transition at nucleotide +34 in the mAMPD transcript also impacts on the molecular diagnosis of this genetic disease. Nucleotide +34 is positioned at the last base of miniexon 2 in the

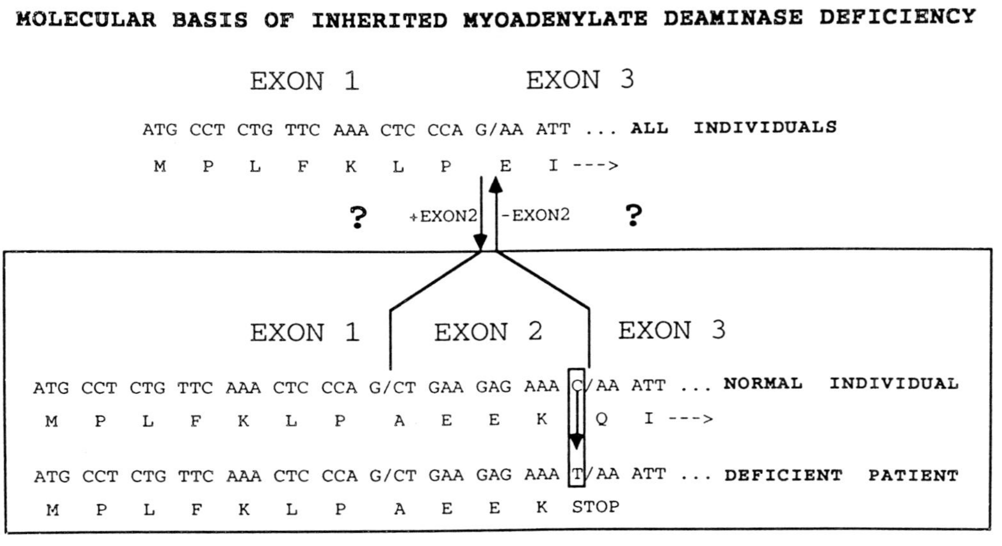

Figure 2 Identified point mutation at nucelotide +34 in the human *AMPD1* gene in 11 unrelated patients with inherited myoadenylate deaminase (mAMPD) deficiency (21). The predicted effect of this transition is premature termination of the mAMPD polypeptide in deficient patients. An alternative splicing event involving exon 2 of the *AMPD1* gene, however, would remove the mutation, resulting in a functional polypeptide. This phenomenon may be relevant to the apparent discrepancy between the frequencies of the mutant allele in the Caucasian and African-American populations and exercise-induced myopathy in these individuals.

AMPD1 gene and is part of a Mae II restriction endonuclease recognition site, AC/GT, in normal individuals. In those harboring the mutant allele, however, this restriction endonuclease site is destroyed (AT/GT). This observation serves as the basis for a diagnostic test designed to distinguish between the two alleles (21). Using oligonucleotides designed against flanking intron 2 and intron 3 sequences, the polymerase chain reaction (PCR) is employed to amplify this region of the *AMPD1* gene, the latter obtained from a processed blood sample. The region of genomic DNA selected for amplification contains the Mae II restriction endonuclease site only once in normal individuals, whereas this genetic marker is absent in those with the mutant allele. Using this approach in a limited survey, Morisaki et al. detected the mutant allele in 12% of Caucasians and 19% of African-Americans (21). These apparent allele frequencies support previous studies in which 1–3% of randomly sampled muscle biopsies were found to be deficient in AMP deaminase activity (14,20,31,32,39,40). These results also suggest that a single mutant allele may account for all cases of inherited mAMPD deficiency. A similar percentage of the *entire* population, furthermore, would be predicted to be mAMPD-deficient. However, 1–3% of these individuals clearly do not have symptoms severe enough to be classified as a metabolic myopathy.

In an attempt to resolve the discrepancy between frequency of the mutant allele and manifestation of the clinical symptoms in the general Caucasian and African-American populations, several explanations might be offered. An inherited deficiency of mAMPD may be nothing more than a harmless genetic variant. The existence of symptomatic patients who have no other known neuromuscular complications other than an inherited deficiency of mAMPD (to date 77 cases in the literature; for a review, see Ref. 1) argues against this possibility, however. Alternatively, an inherited deficiency of mAMPD could predispose an individual to develop myopathic symptoms in association with another inherited or acquired derangement in energy metabolism; however, if this is true, abnormalities other than a deficiency in mAMPD have so far gone undiagnosed in these individuals. Finally, it is relevant to consider the alternative splicing event involving exon 2 of the *AMPD1* gene. If this phenomenon was variable in human skeletal muscle even at low frequency, the removal of the point mutation at nucleotide +34 resulting in the functional expression of mAMPD, combined with that contributed by other AMP deaminase isoforms proposed to be expressed in skeletal muscle (i.e., the E isoforms), might be sufficient to render an individual homozygous for the mutant allele asymptomatic, although homozygous for the mutant allele. A quantitative evaluation of this alternative splicing event in symptomatic and asymptomatic mAMPD-deficient patients appears to be warranted.

Molecular Basis of Acquired Deficiency of Myoadenylate Deaminase

The acquisition of a deficiency of mAMPD has been related to the severity of the associated primary neuromuscular or rheumatological disease and is most often found deficient in advanced stages of several such diseases (4,41,42). In such patients, however, the clinical manifestations of the associated primary disease far outweigh any contributions from a deficiency of mAMPD. An acquired deficiency of mAMPD would perhaps be expected in diseases characterized particularly by muscle wasting/degeneration or invasion by nonmuscle cells. AMP deaminase activity is 20- to 100-fold higher in normal adult skeletal muscle compared to other tissues and cells (43), and thus a reduction in the percentage of normal, mature myocytes would be expected to produce lower levels in biopsy material. One study also reported reduced creatine kinase activities in severely affected muscle obtained from patients with a variety of neuromuscular diseases (41). Discordance between the degree of reduction in AMP deaminase (attaining a deficient state) versus creatine kinase (reduced by approximately half) activities in these patients, however, indicates that factors other than skeletal muscle pathology must play a role in the acquisition of an mAMPD deficiency.

To date, only two studies have addressed the problem of acquired mAMPD deficiency at the molecular level. In the first (3), four cases of acquired mAMPD deficiency were analyzed for relative *AMPD1* transcript abundance by Northern blot analysis. Two of these cases, one secondary to denervation and the other with Becker muscular dystrophy, exhibited normal abundance of mAMPD transcript. These results are similar to observations in patients with inherited mAMPD deficiency (see above). Although further molecular analyses are necessary, the Northern blot data might indicate that these two acquired cases belong to the third category of mAMPD-deficient patients described above, i.e., those individuals with a primary neuromuscular or rheumatological disease and a coincidental inherited deficiency of mAMPD. On the other hand, two other acquired mAMPD-deficient cases, both secondary to an inflammatory myopathy, displayed severely reduced levels of relative mAMPD transcript abundance. This Northern blot profile is unique to all mAMPD-deficient patients studied to date and may be indicative of a true acquired deficiency of mAMPD. Consistent with this interpretation, relative M-creatine kinase transcript abundance was also reduced in these two cases. These results prompted a second study, in which biopsy material obtained from patients with idiopathic inflammatory myopathy was evaluated for relative myofiber necrosis. This index of relative pathological change was compared to endogenous AMP deaminase activity and relative *AMPD1* and M-creatine kinase transcript abundances (4). The results confirmed the previously established correlation between an increase in relative pathological change (in this case relative myofiber necrosis) and a decrease in AMP deaminase activity. This study also extended available information regarding acquired deficiency of mAMPD by demonstrating that reductions in muscle-specific enzymatic activities observed in these patients are likely due to limited availability of transcript. Molecular analyses of additional cases of acquired mAMPD deficiency, however, are needed.

TREATMENT AND MANAGEMENT OF THE MYOADENYLATE DEAMINASE DEFICIENCIES

In considering potential treatments for mAMPD deficiency, it is appropriate to consider the various forms of this disease. Owing to the overwhelming clinical features attributable to the associated neuromuscular or rheumatological disease in acquired mAMPD deficiency, it seems unlikely that treating the mAMPD component in these patients would lead to a significant improvement in their prognosis. Nevertheless, the role of an mAMPD deficiency in the clinical presentation of these patients is currently poorly understood and should not be dismissed a priori. Further elucidation of factors involved in the acquisition of an mAMPD deficiency may also impact on treatments of the inherited disease by contributing, for example, to existing knowledge of *AMPD1* gene regulation.

Currently, there is no good evidence to refute the apparent causal relationship between an inherited deficiency of mAMPD and an exercise-related myopathy in those patients exhibiting both features. The development of a rational treatment for these patients might significantly enhance their life-styles. Although there is no clearly established regimen for treating an inherited deficiency of mAMPD, oral ribose has been used with mixed success in limited clinical evaluations. In two mAMPD-deficient patients, diminished exertional myalgia (44) and enhanced stamina (45) were reported, whereas two additional patients failed to respond to this intervention (46).

In light of the identified point mutation at nucleotide +34 in the mAMPD transcript found in patients with inherited mAMPD deficiency, a better appreciation of the factors regulating the alternative splicing event involving miniexon 2 of the *AMPD1* gene could ultimately improve therapy in these patients. By increasing the endogenous rate of this alternative splicing event, for example, it would theoretically be possible to restore mAMPD activity to levels that would alleviate symptoms in these individuals.

REFERENCES

1. Sabina RL. Myoadenylate deaminase deficiency. In: Rosenberg RN, Prusiner SB, DiMauro S, Barchi RL, Kunkel LM, eds. The Molecular and Genetic Basis of Neurological Disease. Boston: Butterworth-Heinemann, 1993:261–275.
2. Fishbein WN. Myoadenylate deaminase deficiency: inherited and acquired forms. Biochem Med 1985; 33:158–169.
3. Sabina RL, Fishbein WN, Pezeshkpour G, Clarke PRH, Holmes EW. Molecular analysis of the myodenylate deaminase deficiencies. Neurology 1992; 42:170–179.
4. Sabina RL, Sulaiman AR, Wortmann RL. Molecular analysis of acquired myoadenylate deaminase deficiency in polymyositis (idiopathic inflammatory myopathy). Adv Exp Med Biol 1991; 309B:203–205.
5. Sinkeler SPT, Daanen HAM, Wevers RA, Oei TL, Joosten EMG, Binkhorst RA. The relation between blood lactate and ammonia in ischemic handgrip exercise. Muscle Nerve 1985; 8:523–527.
6. Coleman RA, Stajich JM, Pact VW, Pericak-Vance MA. The ischemic exercise test in normal adults and in patients with weakness and cramps. Muscle Nerve 1986; 9:216–221.
7. Sinkeler SP, Wevers RA, Joosten EM, et al. Improvement of screening in exertional myalgia with a standardized ischemic forearm test. Muscle Nerve 1986; 9:731–737.
8. Valen PA, Nakayama DA, Veum J, Sulaiman AR, Wortmann RL. Myoadenylate deaminase deficiency and forearm ischemic exercise testing. Arthritis Rheum 1987; 30:661–668.
9. Fishbein WN, Griffin JL, Armbrustmacher VW. Stain for skeletal muscle adenylate deaminase: an effective tetrazolium stain for frozen biopsy specimens. Arch Pathol Lab Med 1980; 104:462–466.
10. DiMauro S, Miranda AF, Hays AP, et al. Myoadenylate deaminase deficiency: muscle biopsy and culture in a patient with gout. J Neurol Sci 1980; 47:191–202.
11. Fishbein WN. Indicator enzyme assays. I. Adenylate deaminase: principles and application to human muscle biopsies and blood cells. Biochem Med 1979; 22:307–322.
12. Leech AR, Newsholme EA. Radiochemical assays for adenylate kinase and AMP deaminase using polyethyleneimine-cellulose thin layers. Anal Biochem 1978; 90:576–589.
13. Sabina RL, Swain JL, Patten BM, Ashizawa T, O'Brien WE, Holmes EW. Disruption of the purine nucleotide cycle: A potential role for muscle dysfunction in myoadenylate deaminase deficiency. J Clin Invest 1980; 66:1419–1423.
14. Hayes DJ, Summers BA, Morgan-Hughes JA. Myoadenylate deaminase deficiency or not? Observations on two brothers with exercise-induced muscle pain. J Neurol Sci 1982; 53:125–136.
15. Ashwal S, Peckham N. Myoadenylate deaminase deficiency in children. Pediatr Neurol 1985; 1:185–191.
16. Goebel HH, Bardosi A, Conrad B, Kuhlendahl HD, DiMauro S, Rumpf KW. Myoadenylate deaminase deficiency. Klin Wochenschr 1986; 64:342–347.
17. Goebel HH, Bardosi A. Myoadenylate deaminase deficiency. Klin Wochenschr 1987; 65:1023–1033.
18. Mercelis R, Martin JJ, Dehaene I, de Barsy Th, Van den Berghe G. Myoadenylate deaminase deficiency in a patient with facial and limb girdle myopathy. J Neurol 1981; 225:157–166.

19. Gertler PA, Jacobs RP. Myoadenylate deaminase deficiency in a patient with progressive systemic sclerosis. Arthritis Rheum 1984; 27:586–590.
20. Mercelis R, Martin JJ, deBarsy Th, Van den Berghe G. Myoadenylate deaminase deficiency: absence of correlation with exercise intolerance in 452 muscle biopsies. J Neurol 1987; 234:385–389.
21. Morisaki T, Gross M, Morisaki H, Pongratz D, Zollner N, Holmes EW. Molecular basis of AMP deaminase deficiency in skeletal muscle. Proc Natl Acad Sci USA 1992; 89:6457–6461.
22. Van Waarde A. Operation of the purine nucleotide cycle in animal tissues. Biol Rev 1988; 63:259–298.
23. Lowenstein JM. The purine nucleotide cycle revised. Int J Sports Med 1990; 11:S37–S46.
24. Van den Berghe G, Bontemps F. Vincent MF, Van den Bergh F. The purine nucleotide cycle and its molecular defects. Prog Neurobiol 1992; 39:547–561.
25. Bangsbo J, Graham T, Johansen L, Strange S, Christensen C, Saltin B. Elevated muscle acidity and energy production during exhaustive exercise in humans. Am J Physiol 1992; 32:R891–R899.
26. Sabina RL, Marquetant R, Desai NM, Kaletha K, Holmes EW. Cloning and sequence of rat myoadenylate deaminase cDNA: evidence for tissue-specific and developmental regulation. J Biol Chem 1987; 262:12397–12400.
27. Sabina RL, Morisaki T, Clarke P, et al. Characterization of the human and rat myoadenylate deaminase genes. J Biol Chem 1990; 265:9423–9433.
28. Morisaki T, Sabina RL, Holmes EW. Adenylate deaminase: a multigene family in humans and rats. J Biol Chem 1990; 265:11482–11486.
29. Mahnke-Zizelman DK, Sabina RL. Cloning of human AMP deaminase isoform E cDNAs: evidence for a third AMPD gene exhibiting alternatively spliced 5′-exons. J Biol Chem 1992; 267:20866–20877.
30. Bausch-Jurken MT, Mahnke-Zizelman DK, Morisaki T, Sabina RL. Molecular cloning of AMP deaminase isoform L: sequence and bacterial expression of human AMPD2 cDNA. J Biol Chem 1992; 267:22407–22413.
31. Fishbein WN, Armbrustmacher VW, Griffin JL. Myoadenylate deaminase deficiency: a new disease of muscle. Science 1978; 200:545–548.
32. Keleman J, Rice DR, Bradley WG, Munsat TL, DiMauro S, Hogan EL. Familial myoadenylate deaminase deficiency and exertional myalgia. Neurology 1982; 32:857–863.
33. Sinkeler SPT, Joosten EMG, Wevers RA, et al. Myoadenylate deaminase deficiency: a clinical, genetic, and biochemical study in nine families. Muscle Nerve 1988; 11:312–317.
34. Sabina RL, Ogasawara N, Holmes EW. Expression of three stage-specific transcripts of AMP deaminase during myogenesis. Mol Cell Biol 1989; 9:2244–2246.
35. Mineo I, Clarke PRH, Sabina RL, Holmes EW. A novel pathway for alternative splicing: identification of an RNA intermediate that generates an alternative 5′ splice donor site not present in the primary transcript of AMPD1. Mol Cell Biol 1990; 10:5271–5278.
35a. Morisaki H, Morisaki T, Newby LK, Holmes EW. Alternative splicing: a mechanism for phenotypic rescue of a common inherited defect. J Clin Invest 1993; 91:2275–2280.
35b. Van den Bergh F, Sabina RL. Characterization of human AMP deaminase 2 (AMPD2) gene expression reveals alternative transcripts encoding variable N-terminal extensions of isoform L. Biochem J 1995; 312:401–410.
36. Thompson JL, Sabina RL, Ogasawara N, Riley DA. AMP deaminase histochemical activity and immunofluorescent isozyme localization in rat skeletal muscle. J Histochem Cytochem 1992; 40:931–946.
37. Fishbein WN, Sabina RL, Ogasawara N, Holmes EW. Immunologic evidence for three isoforms of AMP deaminase in mature skeletal muscle. Biochim Biophys Acta 1993; 1163:97–104.
38. Gross M, Morisaki T, Pongratz D, Holmes EW, Zollner N. Normal restriction pattern (Hind III) of the myoadenylate deaminase gene in enzyme deficient patients. Klin Wochenschr 1990; 68:1084.
39. Shumate JB, Katnik, Ruiz, M, et al. Myoadenylate deaminase deficiency. Muscle Nerve 1979; 2:213–216.
40. Kar NC, Pearson CM. Muscle adenylate deaminase deficiency: report of six new cases. Arch Neurol 1981; 38:279–281.
41. Kar NC, Pearson CM. Muscle adenylic acid deaminase activity: selective decrease in early-onset Duchenne muscular dystrophy. Neurology 1973; 23:478–482.
42. Nagao H, Habara S, Morimoto T, et al. AMP deaminase activity of skeletal muscle in neuromuscular disorders in childhood: histochemical and biochemical studies. Neuropediatrics 1986; 17:193–198.
43. Ogasawara N, Goto H, Yamada Y, Watanabe T, Asano T. AMP deaminase isozymes in human tissues. Biochim Biophys Acta 1982; 714:298–306.
44. Patten BM. Beneficial effect of D-ribose in a patient with myoadenylate deaminase deficiency. Lancet 1982; 1:1071.
45. Zollner N, Reiter S, Gross M, et al. Myoadenylate deaminase deficiency: Successful symptomatic therapy by high dose oral administration of ribose. Klin Wochenschr 1986; 64:1281–1290.
46. Lecky BRF. Failure of D-ribose in myoadenylate deaminase deficiency. Lancet 1983; 1:193.

36
Malignant Hyperthermia

Beverley A. Britt
University of Toronto and Toronto General Hospital, Toronto, Ontario, Canada

INTRODUCTION

Malignant hyperthermia (MH) is a hereditary skeletal muscle disorder of humans and pigs characterized by hypercatabolic reactions induced by anesthetic drugs or by physical or emotional stress. In this chapter, I have included statistical studies obtained from 1200 human MH reactions reported to my institution between 1966 and the present.

EPIDEMIOLOGY

About one-half of all human MH reactions have been preceded by one to as many as 13 previous apparently uneventful general anesthetics (1). A reaction appears to be more likely to occur if a potent inhalational anesthetic has been used in combination with succinylcholine for a long period, or if immediately prior to the anesthetic the patient has been exercising violently, has sustained extensive muscle trauma, or has been feverish or apprehensive. Many cases of MH go unrecognized either because they are so mild as to be undetectable without the aid of an end-tidal carbon dioxide ($ETco_2$) monitor or temperature monitor or because they are misdiagnosed as something else—for instance, as a cerebral or pulmonary embolus, infection, myocardial infarction, or a mismatched blood transfusion.

The incidence of recognized, fulminant human MH reactions is about 1:15,000 anesthetics in children and in all age groups is about 1:50,000 to 1:200,000 anesthetics (1). Kosko et al. (2) have reported that the incidence of masseter muscle rigidity (MMR) is about 1% in the pediatric population who have received halothane followed by succinylcholine. When these patients are subsequently biopsied, about half are found to be positive for MH (3,4). This would make the true gene frequency for MH about 1:200 of the human population.

The geographical and racial distribution of MH is worldwide (1,5,6).

Males are more commonly affected than females (1) (Fig. 1). In both sexes MH is rare under the age of 3, but then the incidence rises rapidly until age 30, when the incidence declines rather precipitously in both sexes (1). By old age (over 75 years), reactions are almost unknown.

MH has been described in species other than humans. Sporadic cases have occurred in dogs such as racing greyhounds, in cats, race horses, cattle, and giraffes (1). While MH in these species is relatively rare, MH in pigs is quite common (1,7).

INHERITANCE

MH in humans is inherited as an autosomal-dominant trait (1,8) (Fig. 2). Inheritance in pigs appears to be recessive (7) or codominant (9,10).

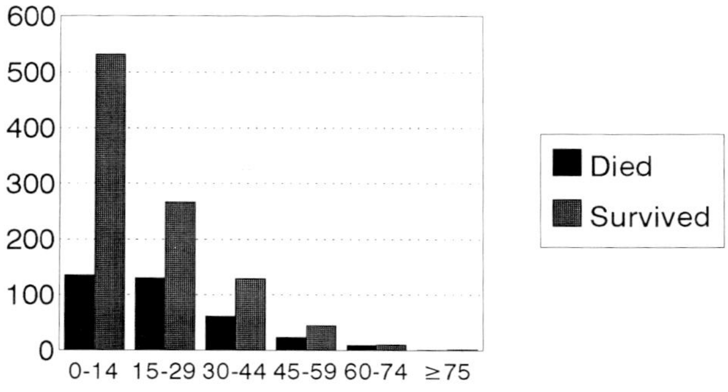

Figure 1 Age and sex distribution of malignant hyperthermic reactions.

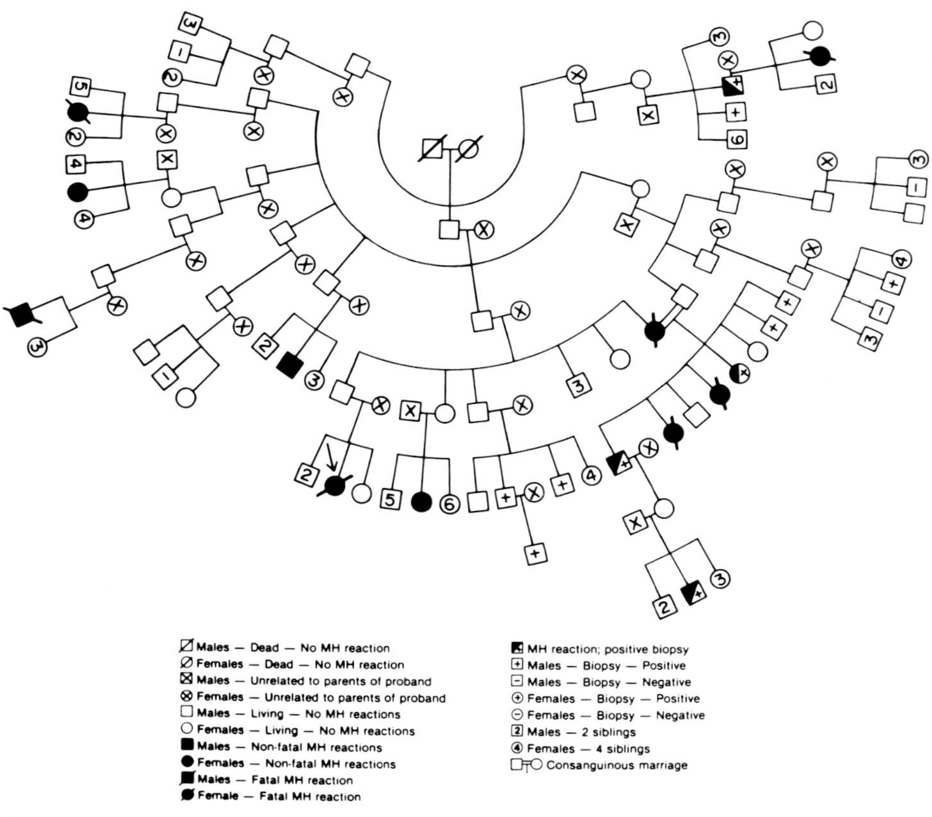

Figure 2 Inheritance of malignant hyperthermia.

It is now known, by means of restriction fragment length polymorphism (RFLP) and sequencing studies, that in all pigs so far investigated from at least five different breeds, the gene responsible for MH lies on the long arm of chromosome 6 (11–17). This gene, known as the *HAL* gene, controls the calcium release channel in the muscle sarcoplasmic reticulum (SR). Within this gene a substitution of thymine for cytosine at position 1840 of the *RYR1* (ryanodine receptor) transcript results in a substitution of cysteine for arginine at position 615 of the amino acid sequence of the *RYR1* receptor.

In humans a similar syntenic gene, known as the ryanodine gene (*RYR1*), lies on the long arm of chromosome 19 (19q) (18). In a few human MHS families, linkage and sequencing studies have also shown that this *RYR1* gene appears to be responsible for the MH defect (19,20) and the mutation within the gene appears to be the same as that present in MHS pigs (18,21). For many families, however, linkage between the MH defect and the *RYR1* gene cannot be found (22–24). In other MHS families, the candidate gene appears to be the sodium channel gene on the long arm of chromosome 17 (25,26). In still other human MHS families, neither of these genes appears to be the causative gene (27), so a third gene (or perhaps even more genes) must be postulated. This is powerful evidence in favor of the MH syndrome being heterogeneic.

Since one single mutation seems to be the cause of all cases of porcine MH, a blood test is now a practical possibility for diagnosis of MH in pigs. In humans, however, at least three, and probably more, mutations appear to be involved in causing the MH syndrome. In humans, therefore, a genetically based blood test does not appear to be possible for the foreseeable future (28).

TRIGGERING FACTORS

Drugs

In human patients succinylcholine is a potent precipitator of early and intense, but nevertheless transient, rigidity beginning with masseter muscle rigidity (MMR), but is a weak inducer of fever and biochemical derangements (1). By itself succinylcholine is not usually lethal.

Of the modern agents, halothane is the most common and strongest trigger of MH reactions (1). Other modern implicated agents include, in addition to succinylcholine or halothane: enflurane, isoflurane, desflurane (a mild trigger), sevoflurane, and methoxyflurane (1). The order of potency of these is: halothane > isoflurane > enflurane > methoxyflurane (1).

Stress

Stress immediately preceding surgery (exercise, trauma, and apprehension) appears to increase the incidence of anesthetic-induced reactions. Stress may also trigger MH reactions in MHS patients during anesthesia with nontriggering drugs, especially if the anesthesia is long (29). The commencement of pain, agitation, or shivering in the recovery room following anesthesia may also induce MH reactions (30) even when anesthesia has been conducted with nontriggering agents (29). Similarly, drug-induced reactions may be aggravated after the partial return of consciousness by the stress of pain, physical manipulation, loud noises, or glaring lights.

THE ACUTE REACTION

Clinical Appearance

Rigidity

In humans, muscle rigidity, the cardinal feature of MH, typically begins as MMR shortly after the infusion of succinylcholine. Intubation or insertion of an airway or mouth gag is thus made difficult or impossible (1). Care may be taken to ensure that the jaw stiffness is truly due to MH and not to an insufficient dose of succinylcholine, the normal tone-increasing action of succinylcholine on the jaw muscles (31,32), temperomandibular joint dysfunction, or several other causes detailed by Lerman (33).

In MH the rigidity is intense and hard, like postmortem rigor mortis, and makes opening of the jaws extremely difficult or even impossible. The incidence of MMR is high when halothane is given first, followed by succinylcholine and low when these two agents are given in the reverse order (33,34). This difference is probably partly due to the protective effect of thiopentone used with the latter, but not the former, technique (33). It is also probably due to the temporal separation of succinylcholine from halothane that occurs with the latter, but not the former, technique (33). Repeated doses of succinylcholine worsen rather than relieve the stiffness. Isolated MMR, particularly in children, may be the only abnormality observed (31,35). Sometimes, however, rigidity spreads to the remainder of the skeletal muscles after a variable period of time.

About 20% of patients never manifest any perceptible increase in muscle tone (36). Mortality is lower in nonrigid than in rigid patients. While some of these individuals may have a different (and as yet unknown) skeletal muscle defect, some other cases appear to be etiologically similar to those displaying rigidity. They may have very mild inherent defects, have had anesthesia with weak triggering agents, or have not undergone muscle exercise or injury immediately prior to anesthesia.

End-Tidal CO_2 Elevation

With the aid of an $ETCO_2$, incipient MH crisis can be detected moments after commencement of the reaction and substantially prior to any other rapidly measurable clinical sign (37).

Tachycardia and Arrhythmias

Prior to the development of $ETCO_2$ analyzers, the most consistent early feature of a MH crisis was tachycardia (1). Tachycardia is eventually present in the overwhelming majority of cases (1). It has little effect on survival and usually occurs prior to the detection of fever, although often it is not seen until after MMR has occurred and the $ETCO_2$ has risen.

Rapid multifocal ventricular arrhythmias, which have a severe influence on mortality, frequently occur in conjunction with or shortly after the onset of the tachycardia and may take the form of frequent multifocal ventricular extrasystoles, trigeminal or bigeminal rhythms, or runs of ventricular tachycardia (1). Later, tall, peaked T waves secondary to potassium elevation may develop.

Unstable Blood Pressure

Another common early sign of MH is instability of the systolic blood pressure (1). The levels vary more than usual from one reading to the next, often with an overall rise until shortly before cardiac arrest, when profound hypotension along with bradycardia supervenes. The early hypertension seems to be due to a combination of an early rise in cardiac output and intense peripheral arteriolar spasm.

Hyperventilation

If the patient is breathing spontaneously, early rapid and deep respirations (1) may occur and, if present, are associated with a rise in mortality. Sweating and excessive heat and discoloration of the soda lime canister may be observed (1). These changes represent an attempt by the body to excrete the excess CO_2 with heat being produced by the hypercatabolizing muscles and other tissues.

Flushing–Mottled Cyanosis

A peculiar mottled cyanosis of ominous prognostic import may follow an initial flushing of the skin (1). This cyanosis is partly secondary to arteriolar spasm and initially is only partly due to hypoxemia arising from increased oxygen consumption by the muscles. It may also be secondary to arrhythmias and to chest wall rigidity.

Fever

The fever of a MH reaction is due both to decreased heat loss secondary to peripheral vasospasm and to increased heat production in the muscles. Since the fever is a result, not a cause, of the various biochemical derangements that occur in the muscles during a MH reaction, it tends to be of relatively late onset (1). As a rule, the higher the maximum temperature, the higher the death rate (1). Death has, however, occurred when the maximum temperature has been relatively low or even without a temperature rise, while on the other hand, survival has occurred after fevers of 44°C. When fever is the sole abnormal sign during anesthesia, the muscle biopsy is rarely positive for MH (38).

Laboratory Observations

In patients and in pigs, marked biochemical changes occur during an MH reaction. With the exception of a rising creatine kinase (CK), abnormalities of all these parameters increase mortality. Pco_2 rises; base deficit increases; later Po_2 and Sao_2 decrease; serum Ca^{2+} rises. An early rise in serum K^+ is followed by a fall in serum K^+; serum Mg^+ and PO_4 rise; blood glucose and blood urea nitrogen increase; after several hours, large molecules leak out across the by now damaged sarcolemma, resulting in dramatic rises in serum and urine myoglobin and in serum levels of CK and other enzymes (39,40). Enzymes and myoglobin reach maximal urine or blood levels 24–96 hr after the beginning of a MH reaction. If the patient has been treated with dantrolene sodium, enzyme and myoglobin elevations may be attenuated or absent.

Late Complications

Following blood urea nitrogen (BUN) elevation, oliguria develops and leads to renal failure in some cases (1) with a significant increase in mortality.

Platelets, fibrinogen, and other clotting factors may become depleted, leading to multiple hematomas and intractable and usually lethal bleeding from body orifices and wound and needle sites (1). These changes are probably due to a primary defect in the platelets (see below).

Another potentially lethal late complication is acute pulmonary edema characterized by loud, moist rales in both lung fields and copious amounts of pink frothy sputum surging up through the endotrachial tube (1). Acute left heart failure may be secondary to increased tone of the pulmonary microcirculation arising because of the vasoconstricting action of the elevated blood catecholamine concentrations and low blood oxygen tensions (1).

With the aid of heroic treatment, some patients linger on for several hours or days before finally dying, never regaining consciousness, with fixed and dilated pupils, absent deep tendon reflexes, upward plantar reflexes, and perhaps intermittent, generalized, or localized convulsions (1). Whether such brain death is caused by a primary MH brain defect or is secondary to the many biochemical and physiological disturbances to which the brain is subjected during a MH crisis has not yet been determined. However, statistically, a cardiac arrest during a MH reaction is followed by an increased incidence of neurological damage only if death ultimately ensues (1).

Occasionally, during convalescence from severe but nonfatal reactions, generalized skeletal muscle edema develops due to a combination of increased water production within the muscle cell and accelerated influx of extracellular water into the myoplasm. Much more frequently, the patient complains of generalized muscle pain and weakness often associated with general fatigue (1), which may persist for prolonged periods postoperatively.

Late-Onset MH

Occasionally, MH reactions may commence only in the recovery room or even after return to the ward. These reactions, which are usually less fulminant than those that start in the operating room, tend to be characterized by moderate fever and tachycardia and occasionally by slight cyanosis, hyperventilation, and agitation. Sometimes, however, these late reactions may be lethal (30), particularly after cardiopulmonary bypass procedures, because of the masking effect of this operation.

Prognosis

The survival rate from MH reactions has steadily been improving since study of this condition was begun in the early 1960s (Fig. 3). Survival in the 1990s is remaining fairly constant at about 96–97%. The higher survival rates appear to be a result, in part, of more frequent and better educational programs

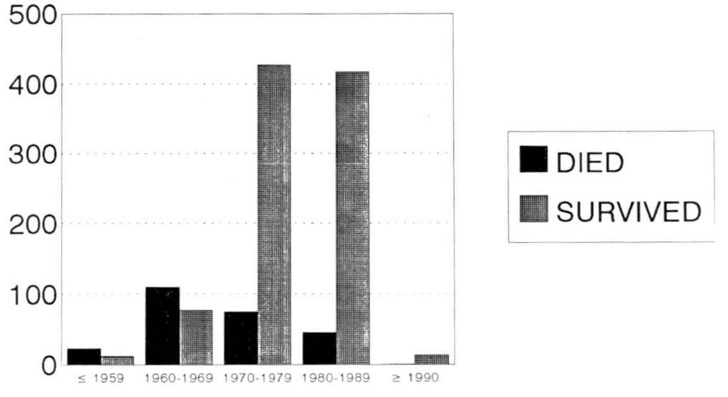

Figure 3 Mortality by year.

within hospitals; more consistent use of monitoring, for example, of temperature, blood gases, and end-tidal CO_2 during anesthesia; greater use of and familiarity with MH protocols and MH emergency trolleys; earlier cessation of triggering drugs with consequent lower maximum temperatures and serum potassium; and the early administration of dantrolene sodium.

Differential Diagnosis

MH may be confused with decreased heat loss due to hyperthermia from excessive use of heavy drapes during anesthesia, or increased external heating by a K-thermia blanket or by a malfunctioning ventilator (1). Internal heat production may be increased due to: excessive shivering in response to cold or struggling; endocrinopathies such as pheochromocytoma and thyrotoxicosis; osteogenesis imperfecta; infection, either preexisting or induced by infusion of infected blood, by intravenous fluids or by contamination of the wound site; and mismatched blood transfusion (1).

A combined decreased heat loss and increased heat production may arise from: (1) an abnormality of the central temperature-regulating mechanism in the hypothalamus secondary to an intracranial lesion (1); (2) strychnine poisoning, characterized by fulminant fever, lactic acidosis, severe convulsions and muscle spasms, profound rhabdomyolysis, and marked CK elevation; strychnine may be a contaminant of illegal cocaine used by addicts (1); or (3) neuroleptic malignant syndrome—an idiosyncratic reaction to psychotropic drugs such as haloperidol, phenothiazines, tricyclics amphetamines, lithium, and monoamine oxidase inhibitors (1,41,42).

ETIOLOGY OF MH

The most likely immediate cause of the acute hypercatabolism of rigid MH appears to be a sudden rise in the concentration of myoplasmic calcium (43) secondary to a defect in calcium release from the sarcoplasmic reticulum (Fig. 4a–c) (1,44,45). Halothane, Ca^{2+}, and caffeine further stimulate calcium release from MHS but not from normal SR (46).

The calcium release channel of the SR is a footplate lying between the terminal cisternae of the SR and the transverse tubules (TT). It is termed the ryanodine receptor because ryanodine, a plant alkaloid, exhibits binding to the receptor. Binding of ryanodine to MHS SR has been reported to be greater than to normal SR by many workers (47).

The ryanodine receptor is a homotetrameric complex (23,48) of about 980 kDa with a four-leaf-clover appearance. Caffeine opens the channel, and this effect is dramatically greater in MHS than in normal porcine SR (10). A direct interaction of dantrolene with the ryanodine receptor has not yet been demonstrated (23).

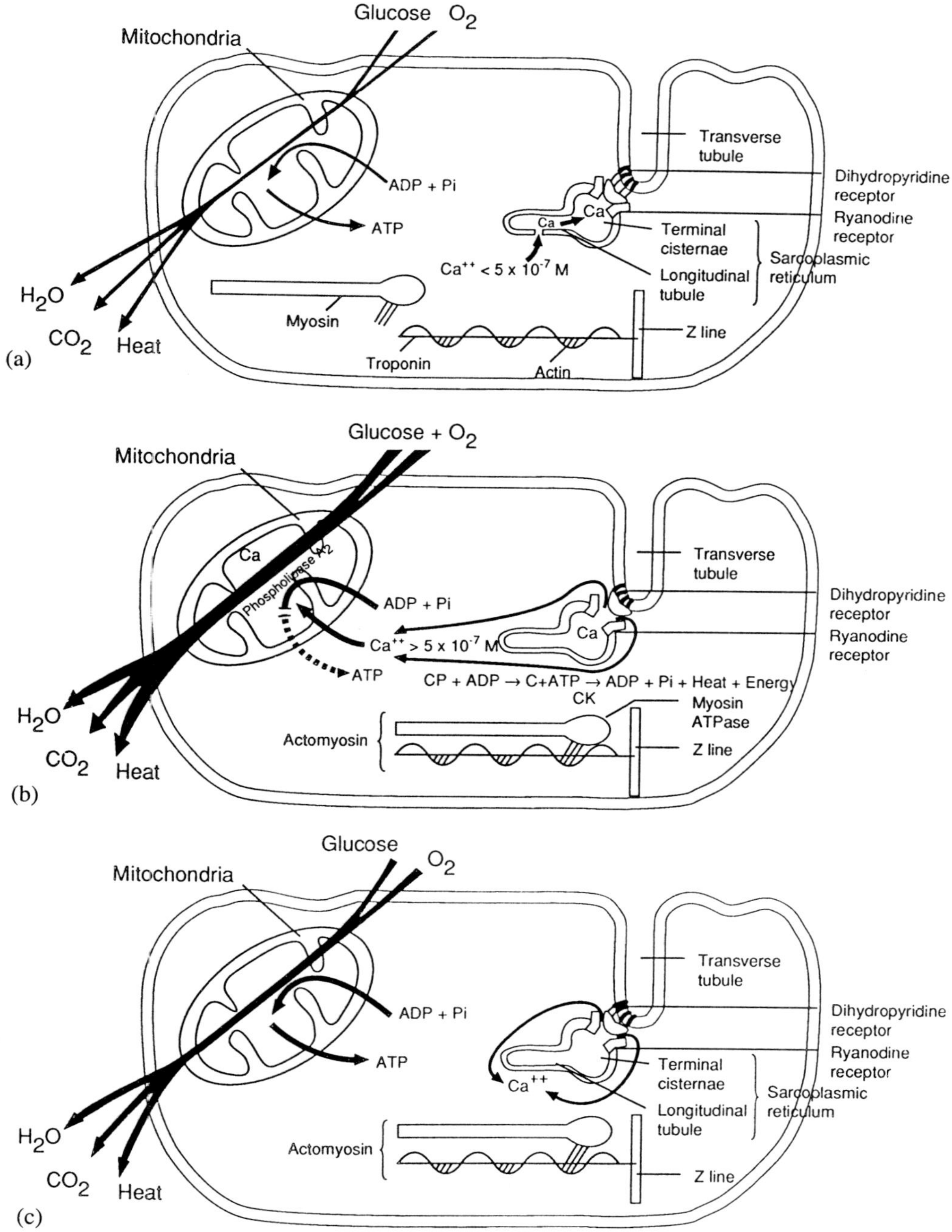

Figure 4 (a) Normal relaxation of skeletal muscle. (b) Muscle cell during a MH reaction. (c) Normal contraction of skeletal muscle.

MH—A Widespread Membrane Defect

MH appears to be a widespread membrane defect, involving different membranes within the same cell type and membranes of different cell types. For example, in skeletal muscle there is evidence for calcium-related defects not only in the sarcoplasmic reticulum, but also in the mitochondria and sarcolemma, including the catecholamine receptor area of the sarcolemma (1). There is also some evidence that similar calcium-related biochemical defects are present in other types of cells, for example, nerve cells, bone, red blood cells, lymphocytes, cardiac muscle, hormone-producing and -secreting cells, and brain (1).

DIAGNOSIS OF MH

Clinical Features

Musculoskeletal System

Some, patients, but by no means all, complain of skeletal muscle cramps, aches, pains, and tenderness that occur with variable frequency, spontaneously, during and after exercise, emotional stress, and infectious illness, or after anaesthesia (1,49). These may occur in the muscles of the limbs, back, chest, or abdominal walls. When present, they may occasionally be so severe as to be almost incapacitating.

Some other musculoskeletal and connective tissue abnormalities that are common in MHS patients are the following.

1. The King syndrome (5) is characterized by short, stocky stature, pectus carinatum, lumbar lordosis and thoracic kyphosis, and cryptorchidism, in males who have had MH reactions and demonstrate positive caffeine halothane contracture test results.
2. Duchenne muscular dystrophy (MD) occurs in about 4% of all biopsy-proven MHS patients (50). The exact relationship between these two disorders is unknown since the gene for at least some MH patients appears to be on either chromosome 17 or 19 while the DMD gene is on the X chromosome.
3. Microscopic findings similar to those of central core disease (CCD) occur in the muscle of a few MHS patients, although most, but not all, of these individuals have had none or few of the overt clinical signs of CCD (51,52). The gene of CCD is known to reside on the long arm of chromosome 19 close to the *RYR1* gene (18,53).
4. A number of myotonic dystrophy (MyD) patients also have biopsies positive for MH (52) and some MHS patients develop MyD-like signs. The MyD gene is also on chromosome 19 but 25 cm distant from the *RYR1* gene, which makes it unlikely that the two conditions are genetically linked (53).

Nervous System

Prior to anesthesia the stress levels of MHS persons are higher than are those of normal individuals (54).

Early changes in the electroencephalogram (EEG) and in cerebral blood flow are seen during halothane-induced MH reactions in pigs (55,56). In MHS pigs, when the heart rate increases but other parameters are still normal, the EEG shows a decrease in total power and a shift to lower frequencies (delta-theta activity) (55). At the same time cerebral blood flow is significantly increased in MHS brains compared to control brains (56). These changes begin when the PaO_2 is greater than 90 mmHg and $PaCO_2$ is less than 50 mmHg (55). Such results do not support the hypothesis that early EEG changes during MH occur as a result of systemic hypotension, hypoxemia, or hypercarbia (55,56).

Laboratory Studies

Electromyography

Most MHS patients have not been subjected to electromyography, but among those who have, abnormalities are found in about 68% of individuals. The abnormalities most frequently noticed are increased incidence of polyphasic action potentials and of short-duration action potentials (1). Fibrillation potentials have also been detected occasionally (1,57).

Serum Creatine Kinase Elevation

While mean values are higher in MHS probands than in control patients, many false positive and false negative values occur (39,40). Furthermore, normal subjects may have elevated serum CK levels for many reasons other than MH (1).

Nuclear Magnetic Resonance Spectroscopy (MRS)

MRS of human muscles in vivo reveals that MHS muscle is similar to normal muscle at rest (58) except that the inorganic phosphate/phosphocreatine ratios are higher than normal (59). However, these muscles exhibit a greater rise in phosphate and a faster and deeper drop in pH during rhythmical exercise and a slower rise toward baseline values during recovery from exercise (59–61). These differences are increased by ischemia (58,60).

Skeletal Muscle Biopsy—Caffeine Halothane Contracture Test (CHCT)

The CHCT is now performed in many laboratories throughout the world (1,62,63). The methodology and interpretation employed in Europe (62) are somewhat different from those utilized in North America (63).

The North American Method. In North America (63), muscle for this test is obtained from the vastus lateralis, rectus abdominus, or other large mixed type I/II muscle belly such as the gracilis, pectoralis major, or the paravertebral muscles. Prior to dissection, each end of the fascicle may be clamped and then meticulously dissected free from the surrounding muscle belly while constant neutral tension is maintained on the muscle with the clamp or sutures. The muscle samples are transported immediately to the laboratory in Krebs-Ringer (KR) solution pregassed in carbogen (95% oxygen and 5% carbon dioxide) and maintained at 22°C and pH 7.4. The sample, still in KR solution, is then divided into several strips (enough to permit studies in triplicate for each test procedure) approximately 1–5 mm by 10–20 mm.

Each muscle strip, secured by a clamp or by a silk suture to a plastic electrode frame, is immersed in a bathing chamber of KR solution at 37°C and pH 7.4 (Fig. 5). The bathing solution is aerated with 95% O_2 and 5% CO_2 (carbogen) at 20 ml/min. The upper end of the muscle is connected via a second silk suture to a force displacement transducer. Isometric tension is recorded with a polygraph. To assess viability, duplicate muscle fascicles are stimulated every 5 sec via a platinum plate or prong electrodes connected to a stimulator that is set to deliver impulses of 2–5 msec duration. The initial tension is 1.0–2.0 g resting tension or supramaximal twitch tension. Before drug additions are given, 15–60 min should elapse until stable twitch and baseline tensions are achieved.

Caffeine, an agent known to accelerate calcium release from the SR (62,64), is added directly to the bath in increments from 0.5 to 32.0 mM. Each caffeine dose is left in the bath until the contracture plateau has been reached (about 5 min after the drug addition). The caffeine is then removed and replaced by a dose double that of the previous concentration. The parameter measured is the contracture achieved 4 min after the addition of caffeine, expressed as grams of tension.

The dose of caffeine required to raise the resting tension by 1 g is termed the caffeine specific concentration (CSC) (1). This value is lower in MHS than in normal muscle (Table 1, Fig. 6). The reference value for the CSC is 4.0 mM. Another measurement made on the caffeine dose-response curves is contracture amplitude in grams at 2.0 mM caffeine (Table 2). The reference point for this parameter is 0.2 g.

In separate muscle strips, the amplitude of contractures produced by 3% halothane is measured. The concentration of the halothane in the gas line just before entry to the bathing chamber should be measured daily with an instrument such as a Reiken Anaesthetic Gas Indicator. The concentration of halothane in the KR solution should be measured periodically by gas chromatography.

Halothane alone induces larger contractures in MHS than in normal muscle (Table 3, Fig. 7)

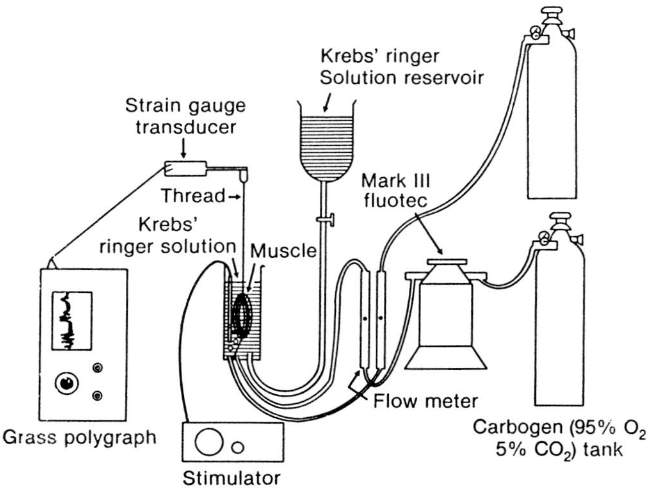

Figure 5 Experimental method for isolated isometric muscle contracture study.

(1,62,63). The reference value for 3.0% halothane-induced contractures is 0.5–0.8 g, depending on the individual laboratory.

The European Method. In the European protocol (62) halothane 0.5, 1.0, 1.5, 2.0, and 3.0 vol% is added to a single muscle fascicle. A threshold (0.2 g) contracture seen at 2 vol% (0.44 mM) or less of halothane is considered positive for MH. To another single muscle fascicle is added 0.5, 1.0, 1.5, 2.0, 3.0, and 4.0 mM of caffeine or until a threshold (0.2 g) contracture is obtained, whichever is the less. A threshold contracture observed at ≤2.0 mM caffeine is deemed positive for MH. At the termination of drug additions to each of the two muscle strips, a dose of 32 mM caffeine is added to test for continued viability of the muscle. Optionally in the European protocol the dynamic test is performed. In this test, during the addition of halothane, the muscle is progressively stretched. Tension is then removed. Normal muscle then relaxes while MHS muscle develops an increase in tension. Individuals whose muscles react in the MH range in the presence of halothane and also in the presence of caffeine are termed MHS. Those whose muscles react in the MH range in the presence of only

Table 1 Statistics for Caffeine Specific Concentration Values

	CSC	
	MH	Control
Mean	3.47	4.76
Standard error	0.16	0.23
Standard deviation	1.61	1.34
Sample size	107	34
Nonparametric probabilities of MH/MP vs. control	0.0001	
F-value for MH, MP, and control	17.95	
Probability of *F*-value	0.0001	
Duncan	B	A

MH(S) = malignant hyperthermic susceptible patients (see text); Control = normal control patients (see text); Duncan = Duncan's Multiple Range Test; CSC = caffeine-specific concentration (see text); p = probability.

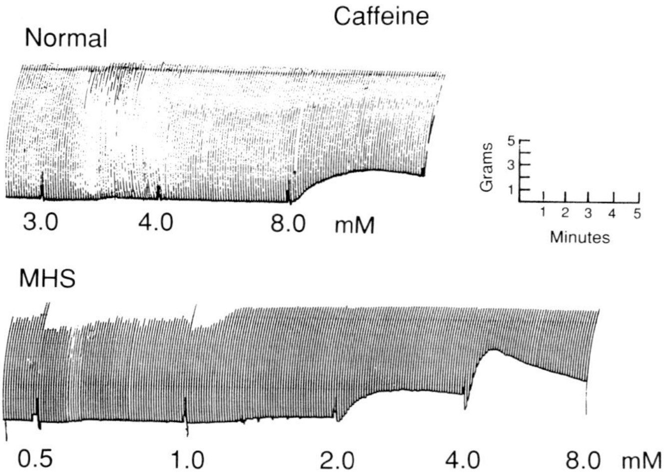

Figure 6 Caffeine specific concentration in normal muscle and in MHS.

halothane are termed malignant hyperthermic equivalent (halothane) [MHE(H)], while those whose muscles react in the MH range in the presence of only caffeine are termed MHE caffeine [MHE(C)]. Finally, those whose muscles react in the normal range in the presence of halothane and also in the presence of caffeine are called MH normal (MHN).

Comparison of the North American and European Protocols. The European method tends to produce substantially flatter halothane dose-response curves and slightly flatter caffeine dose-response curves (65) and, at least in pigs, produces a higher incidence of false positive and false negative tests than the North American method (66).

Microscopy

The muscle of most MHS patients looks normal under the microscope. However, a few MHS patients exhibit minor, nonspecific, inconsistent microscopic abnormalities of their skeletal muscle (1,40,67). The most common finding is a greater-than-normal variation in fiber diameter. There may be small, faintly basophilic angular type I and type II fibers and hypertrophied fibers that have excessive numbers

Table 2 Statistics for 2 mM Caffeine-Induced Contracture Values

	2 mM Halothane	
	MH	Control
Mean	0.47	0.02
Standard error	0.12	0.01
Standard deviation	1.11	0.05
Sample size	107	34
Nonparametric probabilities of MH/MP vs. control	0.0001	
F-value for MH, MP, and control	5.51	
Probability of *F*-value	0.0203	
Duncan	A	B

MH(S) = malignant hyperthermic susceptible patients (see text); Control = normal control patients (see text); Duncan = Duncan's Multiple Range Test; 2 mM caffeine = amplitude of contractures in the presence of 2 mM caffeine; p = *probability*.

Table 3 Statistics for 3% Halothane-Induced Contracture Values

	3% halothane	
	MH	Control
Mean	1.70	0.20
Standard error	0.24	0.04
Standard deviation	2.45	0.22
Sample size	107	34
Nonparametric probabilities of MH/MP vs. control	0.0001	
F-value for MH, MP, and control	11.79	
Probability of *F*-value	0.0008	
Duncan	A	B

MH (S) = malignant hyperthermic susceptible patients (see text); Control = normal control patients (see text); Duncan = Duncan's Multiple Range Test; 3% halothane = amplitude of contractures in the presence of 3% halothane; p = probability.

of diffusely distributed central nuclei. Rarely, frankly atrophic and even necrotic fibers are seen that may be accompanied by regenerating fibers or lymphocytic infiltration (40) and that contain dark-staining pyknotic nuclei. Cytoplasmic bodies may be seen with a central, deeply eosinophilic, homogeneous mass with a lighter peripheral halo. Tubular aggregates rarely occur (67).

On oxidative enzyme stains, localized foci containing disorganized mitochondria may rarely be seen in individual fibers, which may appear either as ill-defined, multiple "moth-eaten" foci or as larger, more sharply circumscribed "targetoid fibers" (1). Much more extensive areas of mitochondrial loss and myofibrillar disorganization, which have been defined as "central cores," have also been described (1).

Electron microscopy demonstrates increased numbers of degenerating and swollen giant mitochondria of variable shape with crystalline inclusions and only remnants of cristae. There may be areas of expansion of the cisterns of the SR and proliferation of the sarcolemma (1). Myofibrillar disarray with streaming of the Z-lines may be observed (1).

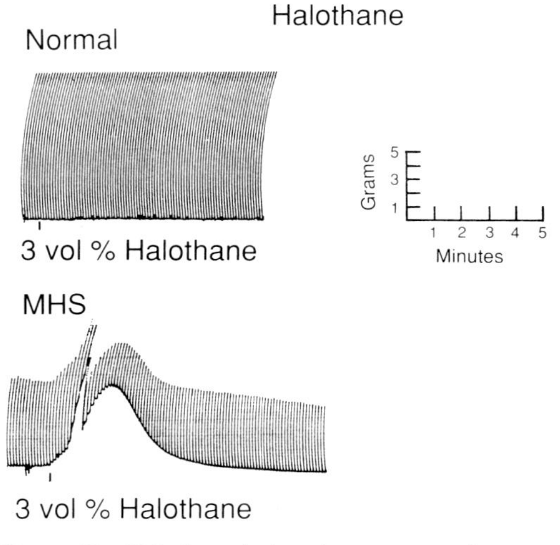

Figure 7 Halothane-induced contractures in normal muscle and in MHS.

Lymphocytes

Using the Quin-2 or Indo-1 technique and spectrofluoroscopy, Klip et al. (68) and O'Brien et al. (69) have observed that halothane induces a greater rise in cytoplasmic Ca^{2+} in MHS human and pig lymphocytes than in control lymphocytes. The same workers have shown that the halothane-induced rise in cytoplasmic calcium concentration occurs only in the presence of exogenous calcium. This suggests that the source of the extra calcium is not the intracellular calcium storage sites, but rather transport across the lympholemma from the extracellular fluid.

Platelets

Employing Quin-2 and spectrofluoroscopy, Fink et al. (70) noted that the resting platelet cytoplasmic Ca^{2+} is slightly but significantly greater in MHS than in normal human platelets. Addition of halothane induces a dose-dependent, rapid calcium release from the intracellular stores in both normal and MHS platelets, which is significantly higher in the MHS than in the normal platelets (71). Under the electron microscope, and with a manual optical photoanalyzer, MHS big platelets are observed to have substantially more dilated open caniculae than normal pig platelets (72).

Red Blood Cells

The V_{max} for Ca^{2+}-ATPase activity of normal red blood cell (RBC) ghosts is greater than that of MHS RBC ghosts (73,74). In normal RBC membranes the Arrhenius plot of spin-label probe motion is linear over 5–40°C while MHS red blood cell membranes exhibit two transsition temperatures at 24 and 33°C (74). The K_m for Ca^{2+} is greater at the high-affinity binding sites and less at the low-affinity binding sites in normal than in MHS RBC ghosts isolated from pigs (74).

Spin-labeled RBC membranes from MHS and normal humans exhibit order parameters and rotational correlation times that are significantly smaller in the MHS group than in the normal group (75). Using a similar technique, Ohnishi and Ohnishi found that halothane increases the membrane fluidity of the RBCs of pigs to a greater degree in MHS than in normal RBCs (76). However, these findings have not been confirmed by Halsall (77). Spin-labeled MHS human RBCs, on electron paramagnetic resonance spectroscopy (EPR), exhibit a much greater decrease in the rotation correlation time in the presence of halothane than do normal human RBCs (78).

MANAGEMENT OF MH

Dantrolene Sodium—General Considerations

Dantrolene sodium (Dantrium, Proctor and Gamble Pharmaceuticals, Norwich, NY) is a lipid-soluble diphenylhydantoin analog. Lerman et al. (80) have demonstrated that, in human children, the elimination half-life of 2.4 mg/kg of dantrolene administered IV over 10.2 min is 10.0 hr. They also have noted that whole blood concentrations of dantrolene decline from 6.03 μg/ml 1 min after the end of the infusion to 3.56 μg/ml at 1 hr and then fall slightly so that at 6.5 hr the concentration is still 3.0 μg/ml.

Orally administered dantrolene is slowly and completely absorbed. However, Allen et al. (81) have reported that in 10 humans given a total oral dose of 5 mg/kg in three or four divided doses every 6 hr with the last dose 4 hr preoperatively, plasma dantrolene levels were over 2.8 μg/ml at induction of anesthesia and for at least the next 6 hr and, in three of the 10 patients, for up to 18 hr after induction. Thus the protective action of oral dantrolene is at least as long, and may be longer in some, as that of IV bolus dantrolene.

Dantrolene sodium is the drug of choice in the treatment and prevention of acute MH reactions (1,82). It acts peripherally by inhibiting twitch and resting tensions of normal and more especially of MHS skeletal muscle fascicles and in some unknown, and probably indirect, way preventing release of calcium from the SR (46,82), thereby lowering myoplasmic Ca^{2+} (83).

Azumolene is a water-soluble analog of dantrolene (84). It appears to have a similar muscle relaxant action to dantrolene in the intact animal (84). In vivo azumolene reverses episodes of MH in MHS pigs (85). This agent may, therefore, prove to be beneficial in the treatment of MH since, because of its water solubility, it can be administered in high concentrations without the need for lyophilization or the addition of irritating sodium hydroxide.

MANAGEMENT OF ELECTIVE ANESTHESIA

Every institution anesthetizing MHS patients should be equipped with all drugs and equipment necessary to treat MH reactions. These should be kept together on mobile trolleys in a central part of the operating suite. A protocol outlining the management of elective and emergency MH cases should be available and familiar to all.

Controversy exists regarding the need to pretreat MHS patients with dantrolene sodium prior to elective anesthesia. Some authorities consider it to be desirable (86). However, in humans much experience has shown that dantrolene pretreatment is not usually necessary (87).

Controversy is also extant about the route by which this drug should be given. The oral dosage required is sufficiently large (82) to cause unpleasant side effects, including muscle weakness, dizziness, double vision, nausea, and diarrhea (82). To eliminate this problem, some anesthesiologists recommend pretreating known or suspected MHS patients with intravenous dantrolene for 30 min immediately prior to anesthesia.

Intravenous administration of dantrolene 2.4 mg/kg achieves MH prophylaxis in humans (88). However, 3.5 mg/kg IV dantrolene is required to prevent MH reactions in susceptible pigs. If the procedure is long, a second bolus dose of dantrolene can be given 6 hr after the first dose to maintain prophylactic blood concentrations (80). Intravenous dantrolene, however, in addition to being very expensive, causes thrombophlebitis of the veins into which it is infused. This occurs because commercial dantrolene is highly alkaline owing to its high sodium hydroxide content and because of the corrosive nature of dantrolene itself. Thrombophlebitis is particularly likely to occur if the drug is administered by continuous infusion rather than as a bolus dose (80).

Some other workers, such as Wingard (89), therefore prefer the oral route unless the surgery is urgent. Wingard (89) states that the side effects of oral dantrolene can be minimized and adequate blood levels reached if a total dose of 2.2 mg/kg of body weight is given in two divided doses as 4 and 8 hr preoperatively.

A vapor-free gas machine should be used. Preferably, such a machine should have had the vaporizers permanently removed and be kept specifically for the purpose of anesthetizing MHS patients. If such a machine is not available, the vaporizers can be temporarily removed and the machine purged with O_2 for a few minutes before induction (90). A circle system with a CO_2 absorber should be used since, with a Bain circuit without such an absorber, it may not be possible to keep the $ETCO_2$ levels within the normal range (91).

Anesthesia can be induced with any desired combination of a narcotic, barbiturate, tranquilizer (1) [particularly midazolam (92)], Althesin (93), propofol (94), ketamine (95), any nondepolarizing skeletal muscle relaxant (1), or any local anesthetic, including amide local anesthetics.

Anesthesia can be maintained with nitrous oxide and oxygen (N_2O:O_2) plus any desired narcotic, propofol, or tranquilizer. If skeletal muscle relaxation is essential, further doses of a nondepolarizing skeletal muscle relaxant may be given.

The patient should not be discharged from hospital until those in attendance are satisfied that temperatures are remaining in the normal range following the withdrawal of dantrolene. Yentis et al. (96), in an extensive study conducted at the Hospital for Sick Children in Toronto, considered it safe to send children home the same day surgery was performed.

Monitoring during anesthesia of all MHS patients must be meticulous (1) and should include end-tidal CO_2, pulse oximetry, temperature measurements from several different sites, electrocardiogram, blood pressure, skin circulation and appearance, and serial measurments of blood gases, electrolytes, and muscle enzymes.

Axillary temperature should be measured because this is the best reflection of muscle temperature and it is in muscles that the temperature first starts to rise. The tip of the probe should be firmly secured in the apex of the axilla with several strips of adhesive tape placed parallel to the probe. The arm should then be tucked closely to the patient's side. If the arm is left on an arm board, falsely low temperature values will be recorded.

Obstetrical Anesthesia for MHS Patients

MH reactions are rare, but not completely unknown, in pregnant women (97). Some factor, possibly hormonal, appears to protect pregnant women from MH reactions, but seems to put males at increased risk for MH crises. Nevertheless, patients known or suspected to have MH must not be given triggering drugs during delivery.

Dental Anesthesia for MHS Patients

Minor and intermediate dental procedures can be managed as usual with an amide local anesthetic. Temperature, pulse rate, and electrocardiogram should be checked before, during, and after minor or intermediate intervention. Major dental operations (e.g., extraction of impacted wisdom teeth) may, because of the inherent apprehensiveness of MHS patients, require general anesthesia. General anesthesia for dental surgery for MHS patients should *never* be attempted in a dentist's office unless the drugs, equipment, and anesthetic expertise are equivalent to those found in an up-to-date general hospital.

Cardiopulmonary Bypass Anaesthesia

Special changes in technique are required when a cardiopulmonary bypass is performed on an MHS patient. These have been well described in detail by Byrick et al. (98) and by Allen and Cattran (99). In particular, the halothane vaporizer must be removed from the extracorporeal membrane oxygenator before surgery. The oxygenator system is then continuously vented with oxygen before the operation. Dantrolene pretreatment may be necessary for cardiopulmonary bypass procedures because the physiological stress of the procedure could conceivably precipitate an MH episode and an MH episode may be difficult to detect in the rewarming period because of acidosis and dysrrhythmias, as well as increases in core temperature independent of an MH episode (98,99).

Anaesthesia for Skeletal Muscle Biopsies

Anesthesia for muscle biopsy for diagnosis of MH may consist of a combined "3-in-1" lumbar plexus block of the femoral-lateral femoral cutaneous nerves (100) if the amount of muscle to be removed is small. If, on the other hand, the amount of muscle to be removed is large or the patient is a child, general anesthesia, using some combination of thiopentone, fentanyl, droperidol, diazepam, nondepolarizing muscle relaxant, or nitrous oxide, may be given. Unlike barbiturates, propofol, and Althesin, these drugs do not significantly affect the tests being performed on the excised muscle. Barbiturates, propofol, and Althesin should not be used because they tend to normalize the muscle. Pretreatment with dantrolene sodium should be avoided as it may inhibit caffeine-augmented and halothane-induced contractures (101). Patients should also not be pretreated with calcium channel blockers as they also tend to normalize the muscle, and if use of dantrolene should become necessary, unacceptable falls in blood pressure and rises in serum potassium would be likely to occur.

Management of Isolated Masseter Muscle Rigidity

When isolated MMR occurs, disagreement exists as to whether to discontinue the surgical procedure entirely (102), to continue the surgical procedure but change the anesthetic agents to drugs known to be safe for MHS patients (103,104), or to continue the surgical procedure with triggering anesthetic agents. Littleford et al. (31), who retrospectively investigated 57 cases of isolated MMR, argue in favor of the last of these options provided the patient is carefully monitored for signs of an impending fulminant reaction. We strongly disagree with this position. First, they identified another 11 children who did go on to develop generalized rigidity in addition to MMR. In other words, 16.2% of these MMR patients did eventually develop overt signs of MH. Since even successfully treated MH is often associated with severe morbidity, not only during and immediately after the crisis, but also for prolonged periods afterward in the form of severe muscle pain, weakness, and fatigue, this is an unacceptably high incidence. Therefore, we and others (105–107) strenuously recommend that once true MMR has been detected, all triggering drugs be immediately discontinued. If necessary, anesthesia can be continued with nontriggering agents.

MANAGEMENT OF FULMINANT MH REACTIONS

Stop Triggering Agents

The single most important factor in the successful management of fulminant MH reactions is early diagnosis and early cessation of triggering drugs (1). Therapy must be carried out with the same urgency as therapy of a cardiac arrest as there may be only a 5-min window of opportunity during which dantrolene is efficacious. Any case in which the diagnosis is uncertain should be treated as MH until proven otherwise beyond all doubt. It is essential not only to turn off the inhalational agent vaporizer immediately, but also to renew the soda lime and change the rubber parts of the gas machine to new rubber tubing, reservoir bags, and ventilator bellows.

Lower Body Temperature

In the past, active cooling to reduce the temperature has been extensively used (1,30,99) since mortality is proportional to the maximum temperature attained. However, cooling seems to be ineffective (1), and retrospective statistical studies show that active cooling worsens survival. Perhaps active cooling exerts its deleterious effects by further exacerbating the preexisting vasoconstriction of the skin and mucosal surfaces and by stimulating shivering and nonshivering thermogenesis. Since the advent of earlier diagnosis and of dantrolene, which lowers the temperature effectively by itself, the need for cooling has largely disappeared.

Correct Oxygen and Carbon Dioxide Tensions

To bring the arterial oxygen and carbon dioxide tensions into the normal ranges, the inspired oxygen concentration should be enriched but not raised to 100%, and artificial ventilation should be increased so that the minute volume approximates three times the Radford nomogram prediction, that is, about 15–20 ml O_2/kg body weight, 15–20 times per minute—or about 201 L/min in a 70-kg man. Use of hyperventilation and 100% O_2 worsens rather than improves survival, even though many case reports describe their use (99). The reason for the adverse statistical data observed for hyperventilation and 100% O_2 is not entirely clear, but may relate to the fact that high pressures and 100% Pao_2 over several hours have a deleterious effect on the alveoli of the lungs.

Base Deficit

Metabolic acidosis has usually been treated with sodium bicarbonate (1,99) with a dose sufficient to return the pH halfway to normal. Complete correction of the base deficit seems to worsen survival statistically. This may be because, as the most recent "Guidelines for Cardiopulmonary Resuscitation and Emergency Cardiac Care" (108) have noted, during a cardiac arrest sodium bicarbonate may shift the oxyhemoglobin saturation curve to the right, may induce hyperosmolarity and hypernatremia, lower coronary perfusion pressure, cause CO_2 to diffuse into the cell membranes, and so produce "paradoxical acidosis," which can depress organ function, may cause extracellular alkalosis, and may increase lactate production and reduce its clearance.

Dantrolene (in Skeletal Muscle)

Dantrolene produces significant improvement in survival from MH reactions (1,82). It is essential to begin dantrolene therapy early. If given late, after the SR has been depleted of calcium, it may be ineffective since the major action of dantrolene is to prevent release of calcium from the SR. In spite of dantrolene, unnecessary deaths from MH still continue to be reported (109). This is because of delay in early diagnosis due to preoccupation with the symptom "hyperthermia," lack of adequate MH monitoring—$ETco_2$, Sao_2, and blood gas analysis, preoccupation with nonspecific aspects of therapy such as cooling, changing the gas machine, transfer of the patient to the intensive care unit, or administration of drugs that have been shown to be ineffective in treating MH, administration of insufficient amounts of dantrolene too late, and failure to increase minute ventilation immediately after the diagnosis is suspected (109).

Because dantrolene is marketed in an unsuitable container, it is time-consuming and difficult to prepare. To treat one patient, many vials (up to 36 in the first 15 min of treatment) must be reconstituted (110). Only bacteriostatic free water may be used, and since 60 cc is necessary for each dantrolene vial, many water vials ($36 \times 6 = 216$) may have to be on hand in addition to the dantrolene, if larger bags of bacteriostatic free water are not available. Transference of the dantrolene from its vial to an infusion bag can be considerably hastened by use of a semiautomatic fluid-dispensing syringe such as the Multi-Ad Fluid Dispensing system (product code MAC-1000) and Vented Needle (product code VN-1000) (both can be obtained from Burron Medical, Inc., Bethlehem, PA) (111).

Dantrolene is administered at the rate of 1.0 mg (2.5 ml)/kg/min until the temperature begins to fall and/or muscle stiffness starts to subside. A maximum of 10 mg/kg may be given over a 15-min period. This dosage should be repeated if the reaction fails to disappear or recurs. Repeated doses must be continued until all evidence of an active MH crisis has disappeared. As soon as possible, intravenous dantrolene should be replaced with oral dantrolene.

Calcium Channel Blockers

Verapamil, a papaverine analog, inhibits influx of calcium into the myocardial sarcoplasm by blocking the calcium-conducting channels of the sarcolemma. However, Gallant et al. (112) have reported that in pigs dantrolene and verapamil in combination produce unacceptable reductions in blood pressure.

Correct Abnormal Serum Potassium Levels

Early hyperkalemia has usually been treated with insulin (1) although not with helpful results. Fifty units of regular insulin in 50 ml of 50% glucose is the recommended dose for a 70-kg adult (1). Serum potassium should be measured frequently, since once the serum potassium starts to decline it falls rapidly through the normal range to sometimes excessively low levels. Insulin therapy must then be discontinued and replaced with infusion of potassium chloride.

Control Tachycardia

The heart rate should be kept below 100/min in adults and 120/min in small children through the use of dantrolene. Failure to do so leads to early onset of myocardial exhaustion with associated clinical signs of heart failure, particularly acute pulmonary edema.

Mannitol

Infusion of mannitol helps to reverse cerebral and muscle edema by "drawing" fluid from brain and muscle cells and interstitial spaces. Use of mannitol slightly improves survival. Mannitol, through its diuretic action, also dislodges myoglobin from the renal tubules, thereby preventing acute renal failure. The dose of mannitol should be about 15 ml of 20% mannitol/kg body weight.

Monitoring

Monitoring during acute MH reaction should include: measurement of Sao_2 by pulse oximetry O_2 and of end-tidal CO_2 levels; axillary and core temperatures; electrocardiogram; heart rate; central venous pressure, urine color, specific gravity and volume; BUN and blood glucose, lactate and pyruvate; serum electrolytes (sodium, potassium, chloride, calcium, phosphorus, and magnesium); serum enzymes (CK, LDH, AST, ALT, aldolase, and alkaline phosphatase); serum creatinine and bilirubin; arterial and venous Po_2, Pco_2, pH, and base deficit; level of consciousness, pupillary size, equality, and reaction, deep tendon and plantar reflexes, evidence of convulsions and skeletal muscle tone; and skin color and skin circulation.

REFERENCES

1. Britt BA. Malignant hyperthermia—a review. In: Schonbaum E, Lomax P, eds. Thermoregulation: Pathology, Pharmacology and Therapy. New York: Pergamon Press, 1991:179–292.
2. Kosko JR, Brandom BW, Chan KH. Masseter spasm and malignant hyperthermia: a retrospective review of a hospital-based pediatric otolaryngology practice. Int J Pediatr Otorhinolaryngol 1992; 23:45–50.
3. Rosenberg H, Fletcher JE. Masseter muscle rigidity and malignant hyperthermia susceptibility. Anesth Analg 1986; 65:161–164.
4. Christian AS, Ellis FR, Halsall PJ. Is there a relationship between masseteric muscle spasm and malignant hyperpyrexia? Br J Anaesth 1989; 62:540–544.
5. King JO, Denborough MA. Malignant hyperpyrexia in Australia and New Zealand. Med J Aust 1973; 1:525–528.
6. Nishimoto Y, Mizumoto K, Kawaguchi Y, et al. A case of enflurane induced malignant hyperthermia with normal result of calcium induced calcium release. Hiroshima J Anesth 1990; 26:309–317.
7. Eikelenboom G., Minkema D., Van Eldick P., et al. Inheritance of the malignant hyperthermia syndrome in Dutch Landrace swine. In: Aldrete JA, Britt BA, eds. Second International Symposium on Malignant Hyperthermia. New York: Grune & Stratton, 1978: 141–146.
8. Britt BA, Locher WG, Kalow W. Hereditary aspects of malignant hyperthermia. Can Anaesth Soc J 1968; 16:89–98.
9. Gallant EM, Lentz LR. Excitation-contraction coupling in pigs heterozygous for malignant hyperthermia. Am J Physiol 1992; 262:C422–C426.
10. Fill M., Stefani E., Nelson TE. Abnormal human sarcoplasmic reticulum Ca^{2+} release channels in malignant hyperthermic skeletal muscle. Biophys J 1991; 59:1085–1090.
11. Fujii J, Otsu K, Zorzato F. Identification of a mutation in porcine ryanodine receptor associated with malignant hyperthermia. Science 1991; 253:448–451.
12. Davies W, Harbitz I, Fries R, et al. Porcine malignant hyperthermia carrier detection and chromosomal assignment using a linked probe. Anim Genet 1988; 19:203–212.
13. Brenig B, Brem G. Marker-directed diagnosis of MH-susceptibility in pigs. J Neurol Sci 1990; 98:136.
14. Gellin J, Yerle M, Dalens M. Human and pig malignant hyperthermia: an homologous genetic control. J Neurol Sci 1990; 98:413 (abstract).
15. Harbitz I, Chowdhary B, Thomsen PD, et al. Assignment of the porcine calcium release channel gene, a candidate for the malignant hyperthermia locus, to the 6P11-Q21 segment of chromosome 6. Genomics 1990; 8:243–248.
16. Otsu K, Khanna VK, Archibald AL, et al. Cosegregation of porcine malignant hyperthermia and a probable causal mutation in the skeletal muscle ryanodine receptor gene in backcross families. Genomics 1991; 11:744–750.
17. Harbitz I, Kristensen T, Bosnes M, et al. DNA sequence of the skeletal muscle calcium release channel CDNA and verification of the ARG 615 to CYS 615 mutation, associated with porcine malignant hyperthermia, in Norwegian Landrace pigs. Anim Genet 1992; 23:395–402.
18. MacLennan DH, Fujii J, Zorzato F, et al. Characterization of CDNA encoding the ryanodine receptor in humans and swine predisposed to malignant hyperthermia. J Neurol Sci 1990; 98:512 (abstract).
19. MacLennan DH, Duff C, Zorzato F, et al. Ryanodine receptor gene is a candidate for predisposition to malignant hyperthermia. Nature 1990; 343:559–561.
20. McCarthy TV, Healy SJM, Heffron JJA, et al. Mapping of the gene for malignant hyperthermia susceptibility to human chromosome 19q12-13.2. J Neurol Sci 1990; 98:512 (abstract).
21. Gillard EF, Otsu K, Fujii J, et al. A substitution of cysteine for arginine 614 in the ryanodine receptor is potentially causative of human malignant hyperthermia. Genomics 1991; 11:751–755.
22. Gillard EF, Otsu K, Fujii J, et al. Polymorphisms and deduced amino acid substitutions in the coding sequence

of the ryanodine receptor (*RYR1*) gene in individuals with malignant hyperthermia. Genomics 1992; 13:1247–1254.
23. MacLennan DH, Phillips MS. Malignant hyperthermia. Science 1992; 256:789–794.
24. Deufel T, Golla A, Iles D, et al. Evidence for genetic heterogeneity of malignant hyperthermia susceptibility. Am J Hum Genet 1992; 50;1151–1161.
25. Levitt RC, Olckers A, Meyers S. Evidence for the localization of a malignant hyperthermia susceptibility locus (MHS2) to human chromosome 17q. Genomics 1992; 14:562–566.
26. Olckers A, Meyers DA, Meyers S. Adult muscle sodium channel a-subunit is a gene candidate for malignant hyperthermia susceptibility. Genomics 1992; 14:829–831.
27. Levitt RC, Nouri N, Jedlicka AE. Evidence for genetic heterogeneity in malignant hyperthermia susceptibility. Genomics 1991; 11:543–547.
28. Larach MG. Malignant hyperthermia. Science 1992; 257:11–12 (letter).
29. Grinberg R, Edelist G, Gordon A. Postoperative malignant hyperthermia episodes in patients who received "safe" anaesthetics. Can Anaesth Soc J 1983; 30:273–276.
30. Kochiyama K, In-Nami H, Matsuyama T, et al. A case of malignant hyperthermia in the immediate postoperative period. Hiroshima J Anesth 1988; 24:81–86.
31. Littleford JA, Patel LR, Bose D, et al. Isolated masseter spasm in children—implications of continuing the trigger anaesthetic. J Neurol Sci 1990; 98:525 (abstract).
32. Van der Spek AFL. Triggering agents continued after masseter spasm: there is proof in the pudding! Anesth Analg 1991; 73:364 (letter).
33. Lerman J. Controversies in paediatric anaesthesia. Can J Anaesth 1988; 35:S18–S22.
34. Rosenberg H, Fletcher J. More about masseter spasm and malignant hyperthermia. Anesthesiology 1985; 62:212 (letter).
35. Flewellen EH, Nelson TE. Masseter spasm induced by succinylcholine in children: contracture testing for malignant hyperthermia. Report of six cases. Can Anaesth Soc J 1982; 29:42.
36. Jago RH, Panne MJ. Malignant hyprepyrexia—the difficulty of diagnosis. Anaesthesia 1977; 32:74–80.
37. Baudendistel N, Goudsouzian N, Cote C. et al. End-tidal CO_2 monitoring. Anaesthesia 1984; 39:1000–1003.
38. Halsall PJ, Ellis FR. Does postoperative pyrexia indicate malignant hyperthermia susceptibility? Br J Anaesth 1992; 68:209–210.
39. Britt BA, Endrenyi L, Peters PL, et al. Screening of malignant hyperthermic susceptibile families by CPK measurement and other clinical investigations. Can Anaesth Soc J 1976; 23:263–284.
40. Isaacs H. High serum creatine phosphokinase levels in asymptomatic members of the families of patients developing malignant hyperpyrexia—a genetic study. In: Gordon RA, Britt BA, Kalow W, eds. International Symposium on Malignant Hyperthermia. Springfield, IL: Charles C Thomas 1973:331–338.
41. Pearlman AC. Neuroleptic malignant syndrome: a review of the literature. J Clin Psychopharmacol 1986; 6:257–273.
42. Ward A, Chauffman MO, Sorkin EM. Dantrolene: a review of its pharmacodynamic and pharmacokinetic properties and therapeutic use in malignant hyperthermia, the neuroleptic syndrome and disorders of muscle spasticity. Drugs 1986; 32:130–168.
43. Lopez JR, Allen PD, Alamo L, et al. Myoplasmic free (Ca^{2+}) during a malignant hyperthermia episode in swine. Muscle Nerve 1988; 11:82–88.
44. Kim DH, Sreter FA, Ikemoto N. Involvement of the 60 KDA phosphoprotein in the regulation of Ca^{2+} release from sarcoplasmic reticulum of normal and malignant hyperthermia susceptible pig muscles. Biochim Biophys Acta 1988; 945:246–252.
45. Ohta T, Endo M, Nakano T. Ca-induced Ca release in malignant hyperthermia-susceptible pig skeletal muscle. Am J Physiol 1989; 256:C358–C367.
46. Ohnishi ST. Effects of halothene, caffeine and dantrolene on the skeletal sarcoplasmic reticulum of pigs susceptible to malignant hyperthermia. Presented at the 4th International Malignant Hyperpyrexia Workshop, York, England, 1986.
47. Mickelson JR, Gallant EM, Litterer LA. Abnormal sarcoplasmic reticulum ryanodine receptor in malignant hyperthermia. J Biol Chem 1988; 263:9310–9315.
48. Fleisher S, Inui M. Biochemistry and biophysics of excitation-contraction coupling. Annu Rev Biophys Biophys Chem 1989; 18:333–364.
49. Walther JU, Zafiriou D, Jensen M. Malignant hyperthermia and Marinesco-Sjögren sundrome. Lancet 1991; 338:1603 (letter).
50. Brownell AKW, Paasuke RT, Fowlow SB. MH, abnormal contracture studies and muscle diseases. Presented at the 4th International Malignant Hyperthermia Workshop, York, 1986.
51. Harriman DGF, Ellis FR. Central core disease and malignant hyperthermia. Br Med J 1973; 1:545.
52. Lehmann-Horn F, Iaizzo PA, Klein W. Neuromuscular diseases and the association to malignant hyperthermia. J Neurol Sci 1990; 98:135.

53. Kausch K, Lehmann-Horn F, Janka M, et al. Evidence for linkage of the central core disease locus to the proximal long arm of human chromosome 19. Genomics 1991; 10:765–769.
54. Smith RJ, Newbegin HF. Anxiety, hypothalamic-pituitary-adrenal axis activity and malignant hyperpyrexia. Presented at the 4th International Malignant Hyperpyrexia Workshop, York, England, 1986.
55. Kochs E, Hoffman WE, Roewer N, et al. Alterations in brain electrical activity may indicate the onset of malignant hyperthermia in swine. Anesthesiology 1990; 73:1236–1242.
56. Kochs E, Nollen H, Schulte am Esch J. Changes in cerebral blood flow and EEG in experimental malignant hyperthermia. Anasth Intensivther Notfallmed 1990; 25:347–353.
57. LaCour D, Juul-Jensen P, Reske-Nielsen E. Anaestesi induceret malign hypertermi hos 4 patients efterundersogt neurofysiologisk og neuropathologisk. Nord Med 1970; 17:1636–1637.
58. Foster PS, Hopkinson K, Denborough MA. ^{31}P-NMR spectroscopy: the metabolic profile of malignant hyperpyrexic porcine skeletal muscle. Muscle Nerve 1989; 12:390–396.
59. Olgin J, Argov Z, Rosenberg H. Noninvasive evaluation of malignant hyperthermia susceptibility with phosphorus nuclear magnetic resonance spectroscopy. Anesthesiology 1988; 68:507–513.
60. Kozak-Reiss G, Jehenson P, Gascard JP, et al. ^{31}P NMR spectroscopy in subjects susceptible of malignant (MH) or exercise (EH) hyperthermia. Presented at the 4th International Malignant Hyperpyrexia Workshop, York, 1986.
61. Allsop P, Jorfeldt L, Rutberg H, et al. Muscle pH recovery after high intensity cycle exercise in malignant hyperthermia susceptible subjects. Biochem Soc Trans 1991; 19:1315.
62. European MH Group. A protocol for the investigation of malignant hyperpyrexia (MH) susceptibility. Br J Anaesth 1984; 56:1267–1269.
63. Larach MG. Standardization of the caffeine halothane muscle contracture test. Anesth Analg 1989; 69:511–515.
64. Frank GB, Buss WC. Caffeine-induced contractures in mammalian skeletal muscle. Arch Int Pharmacodyn 1967; 170:343.
65. Bendixen D, Ording H. Comparison of different methods of performing the in vitro muscle test for malignant hyperthermia. J Neurol Sci 1990; 98:524.
66. Fletcher JE, Conti PA, Rosenberg H. Comparison of North American and European Malignant Hyperthermia Group halothane contracture testing protocols in swine. Acta Anaesth Scand 1991; 35:483–487.
67. Harriman DGF, Ellis FR, Franks AJ, et al. Malignant hyperthermia myopathy in man: an investigation of 75 families. In: Aldrete JA, Britt BA, eds. Second International Symposium on Malignant Hyperthermia. New York: Grune & Stratton, 1978:67–88.
68. Klip A, Britt BA, Elliott ME. Increase in cytoplasmic free Ca^{2+} caused by halothane in lymphocytes from malignant hyperthermia patients and pigs. J Gen Physiol 1986; 88:33a (abstract).
69. O'Brien PJ, Britt BA, Kalow BI. Malignant hyperthermia: lymphocyte Ca test for Ca-channel hypersensitivity and compensatory increase in Ca-ATPase activity VII International Congress on Neuromuscular Diseases, 1990, (abstract).
70. Fink HS, Hofmann JG, Hentschel H, et al. Abnormalities in the regulation of blood platelet free cytosolic calcium in malignant hyperthermia. I. Human platelets. Cell Calcium 1992; 13:149–155.
71. Fink HS, Maak S. Halothane-induced changes in calcium permeability of blood platelets. J Neurol Sci 1990; 98:411 (abstract).
72. Basrur PK, Bouvet A, McDonell WN. Open canalicular system of platelets in porcine stress syndrome. Can J Vet Res 1988; 52:380–385.
73. Louis CF. Red blood cell abnormalities in porcine MH. Presented at the 4th International Malignant Hyperpyrexia Workshop, York, 1986.
74. Thatte HS, Mickelson JR, Addis PB, et al. Erythrocyte membrane ATPase and calcium pumping activities in porcine malignant hyperthermia. Biochem Med Metab Biol 1978; 38:355–365.
75. Cooper P, Kudynska J, Buckmaster HA. An EPR investigation of spin-labelled erythrocytes as a diagnostic technique for malignant hyperthermia. Biochim Biophys Acta 1992; 1139:70–76.
76. Ohnishi ST, Ohnishi T. Halothane induced disorder of red cell membranes of subjects susceptible to malignant hyperthermia. Cell Biochem Funct 1988; 6:257–261.
77. Halsall PJ. EPR measurement of membrane fluidity erythrocytes from patients tested for MH susceptibility by muscle biopsy (abstract).
78. Ohnishi ST, Katagi H, Ohnishi T, et al. Detection of malignant hyperthermia susceptibility using a spin label technique on red blood cells. Br J Anaesth 1988; 61:565–568.
79. Ellis FR. The diagnosis of MH: its social implications. Br J Anaesth 1988; 60:251–252 (editorial).
80. Lerman J, McLeod ME, Strong HA. Pharmacokinetics of intravenous dantrolene in children. Anesthesiology 1989; 70:625–629.
81. Allen GC, Cattran CB, Peterson RG, et al. Plasma levels of dantrolene following oral administration in malignant hyperthermia-susceptible patients. Anesthesiology 1988; 69:900–904.

82. Britt BA, Scott E, Frodis W, et al. Dantrolene. In vitro studies in malignant hyperthermia susceptible (MHS) and normal muscle. Can Anaesth Soc J 1984; 31:130–154.
83. Lopez JR, Lopez M, Allen P. Dantrolene reduces myoplasmic free Ca^{2+} in patients with malignant hyperthermia. Acta Cient Venezol 1990; 41:135–137.
84. Leslie GC, Part NJ. The effect of EU4093 (azumolene sodium) on the contraction of intrafusal muscle of the soleus muscle of the anaesthetized rat. Br J Pharmacol 1989; 97:1151–1156.
85. Dershwitz M, Sreter FA. Azumolene reverses episodes of malignant hyperthermia in susceptible swine. Anesth Analg 1990; 70:253–255.
86. Shindoh M, Enomoto K, Tatekawa S, et al. Prophylactic use of oral dantrolene for spinal anaesthesia in a malignant hyperthermia susceptible patient. Masui 1988; 37:861–867.
87. Hackl W, Mauritz W, Winkler M, et al. Anaesthesia in malignant hyperthermia-susceptible patients without dantrolene prophylaxis: a report of 30 cases. Acta Anaesthesiol Scand 1990; 34:534–537.
88. Flewellen EH, Nelson TE. Is theophylline, aminophylline, or caffeine (methylxanthines) contraindicated in malignant hyperthermia susceptible patients? Anesth Analg 1983; 62:115.
89. Wingard DW. Controversies regarding the prophylactic use of dantrolene for malignant hyperthermia. Anesthesiology 1983; 58:489–490 (letter).
90. McGraw TT, Keon TP. Malignant hyperthermia and the clean machine. Can J Anaesth 1989; 36:530–532.
91. Allen G. A circle system is best when anesthetizing malignant hyperthermia susceptible patients. Anesthesiology 1989; 70:1025–1026 (letter).
92. Brooks JH. Midazolam in a malignant hyperthermia-susceptible patient. Anesthesiology 1989; 70:167–168 (letter).
93. Shimonaka H, Yamamoto M, Suzuki A, et al. Althesin anesthesia for malignant yperthermia susceptible patients. Hiroshima J Anesth 1981; 17:44.
94. McKenzie AJ, Couchman KG, Pollock N. Propofol is a "safe" anaesthetic agent in malignant hyperthermia susceptible patients. Anaesth Intens Care 1992; 20:165–168.
95. Hackl W, Winkler M, Mauritz W, et al. In vitro studies of ketamine for use in human malignant hyperthermia. Anaesthesist 1989; 38:681–685.
96. Yentis SM, Levine MF, Hartley EJ. Should all children with suspected or confirmed malignant hyperthermia by admitted after surgery? A 10 year review. Anesth Analg 1992; 75:345–350.
97. Cupryn JP, Kennedy A, Byrick RJ. Malignant hyperthermia in pregnancy. Am J Obstet Gynecol 1984; 150:327–328.
98. Byrick RJ, Rose DK, Ranganathan N. Management of a malignant hyperthermia patient during cardiopulmonary bypass. Can Anaesth Soc J 1982; 29:50–54.
99. Allen GC, Cattran CB. Rewarming following hypothermic cardiopulmonary bypass in the malignant hyperthermia-susceptible patient: implications for diagnosis and peri-operative management. Can J Anaesth 1989; 36:81–85.
100. Madej TH, Ellis FR, Halsall PJ. Evaluation of "3 in 1" lumbar plexus block in patients having muscle biopsy. Br J Anaesth 1989; 62:515–517.
101. Lambert W. Dantrolene and caffeine contracture test. Can Anaesth Soc J 1980; 27:304.
102. Rosenberg H. Management of patients in whom trismus occurs following succinylcholine. Anesthesiology 1988; 687:654–655.
103. Sinclair G. Masseter spasm following suxamethonium. Br J Hosp Med 1990; 45:342.
104. Gronert GA. Management of patients in whom trismus occurs following succinylcholine. Anesthesiology 1988; 68:653–654 (letter).
105. Rosenberg H, Shutack JG. Masseter muscle spasm in children. Anesth Analg 1991; 73:361–362 (letter).
106. Ellis FR, Hopkins PM, Halsall PJ, et al. Masseter muscle spasm and the diagnosis of malignant hyperthermia susceptibility. Anesth Analg 1992; 75:143–146 (letter).
107. Kaplan R, Becker M, Broadman L. Isolated masseter muscle spasm versus generalized rigidity? Anesth Analg 1992; 75:146–147 (letter).
108. Guidelines for cardiopulmonary resuscitation and emergency cardiac care. JAMA 1992; 268:2171–2302.
109. Schulte-Sasse VU, Eberlein H-J. Reasons for the persistence of a high mortality due to malignant hyperthermia and proposals for its reduction. Anaesthesiol Reanimat 1991; 16:202–207.
110. Haag GP. Are you prepared for a malignant hyperpyrexia crisis? J Am Assoc Nurse Anaesth 1990; 58:250–251.
111. Hargrove JC. Are we really prepared for malignant hyperthermia? Part II. Anesthesiology 1988; 68:815 (letter).
112. Gallant E, Foldes F, Rempel WM. Verapamil is not a therapeutic adjunct to dantrolene in porcine malignant hyperthermia. Anesth Analg 1985; 64:601–606.

37

Sarcoplasmic Reticulum Ca^{2+}–Adenosine Triphosphatase Deficiency (Brody's Disease)

J. A. P. Hiel, P. J. H. Jongen, P. J. E. Poels, A. A. G. M. Benders, J. H. Veerkamp, N. Knoers, F. J. M. Gabreëls, and R. A. Wevers
University Hospital Nijmegen, Nijmegen, The Netherlands

HISTORICAL BACKGROUND

In 1969 Brody described a patient with an unusual disorder of muscle function, characterized by impairment of muscle relaxation worsening during exercise (1). Electromyography showed electrical silence. Biochemical studies revealed a markedly decreased uptake of calcium ions (Ca^{2+}) by isolated sarcoplasmic reticulum (SR). This finding was considered the cause of the disorder, which has since then been named Brody's disease (McKusick 108740). Twelve years earlier, Lambert and Goldstein had described a patient with "silent myotonia," who suffered from abnormal muscle relaxation (2). This patient may have had Brody's disease as well. More recently, in 1986, Karpati et al. described four patients with impaired skeletal muscle relaxation (3). Biochemical, immunochemical, and immunocytochemical studies on muscle biopsies demonstrated a deficiency of the SR Ca^{2+}-transport protein of type II fibers.

SR Ca^{2+}-ATPase deficiency appears to be a clinically, electromyographycally, and biochemically distinct metabolic myopathy. As far as we know, until now only 21 patients have been reported in the literature (1,3–8).

CLINICAL MANIFESTATIONS

The clinical picture emerging from the reported patients is rather heterogeneous (Tables 1 and 2). The symptoms usually begin in childhood and initially preferably affect the musculature of the limbs, later also that of the trunk and face. Exercise-induced stiffness, cramps, and difficulty with muscle relaxation are the common signs and symptoms. Myalgia and muscle weakness may also occur. Patients are not able physically to keep up with their peers. They experience difficulties with strenuous exercise such as running and climbing. As a rule, stiffness and cramping develop within a few minutes after exercise has started. After a short period of rest (seconds to minutes) exercise is possible again for a while, but the time required for relaxation of the muscles increases. There are problems with loosening handgrip. Activities such as typing and piano playing can lead to stiffening and "freezing up" of the fingers. Typically, symptoms may worsen in the cold (1,5).

On examination striking phenomena consist of delayed opening of a clenched fist with flexion of the wrist as the patient attempts to extend the fingers and delayed opening of closed eyes with

Table 1 Number of Patients Reported in the Literature

Author	Ref.	Number of cases
Brody	1	1
Karpati	3	4
Danon	4	4
Taylor	5	2
Wevers/Poels	6,7	1
Benders	8	10

furrowing of the brow as the eyes begin to open (3). Percussion myotonia is absent. Although muscle volume and strength are usually normal, mild atrophy and paresis may be present in advanced stages of the disease. Usually the course is slowly progressive, with shortening of the symptom-free interval and worsening of exercise-induced symptoms. In some patients, however, the course of the disease is stationary for many years. Two patients have been reported with recurrent exertional rhabdomyolysis (6–8).

ANCILLARY STUDIES

Serum creatine kinase (CK) levels are normal or slightly elevated (less than twice the upper limit of the reference range). A patient with recurrent rhabdomyolysis had episodic myoglobinuria and CK levels continuously above 500 IU/L (reference range 15–91) (6,7).

Electromyography (EMG) shows a physiological recruitment of normal motor unit potentials. Fasciculations, fibrillation potentials, or myotonic discharges are absent. The most significant finding is the electrical silence during the phase of delayed relaxation of strongly contracted muscles, in contrast to myotonia associated with sustained electrical activity. The amplitude of the muscle action potential immediately after sustained contraction is the same as that of the resting muscle. Repetitive stimulation

Table 2 Signs, Symptoms, and Relevant Technical Data on the Reported Patients

	Patient											
	1	2	3	4	5	6	7	8	9	10	11	12
Reference	1	3	3	3	3	4	4	4	4	5	5	6,7
Age	26	42	42	26	36	28	30	54	19	37	46	38
Age of onset	5	<5	<5	<5	?	<5	<5	?	<5	27	30	<5
Sex	M	M	M	M	M	F	F	F	M	M	F	M
Symptoms												
Stiffness	+	+	+	+	+	+	+	–	+	+	+	+
Cramps	–	+	+	–	?	+	–	–	–	+	+/–	+
Myalgia	–	+	–	–	?	–	–	–	–	+	+	+
Relaxation problems	+	+	+	+	?	+	+	+	+	+/–	+/–	+
Muscle weakness	–	+	–	–	?	+	+	+	?	–	–	+
Rhabdomyolysis	–	–	–	–	?	–	–	–	–	–	–	+
Influence of cold	+	–	?	?	?	?	?	?	?	+	+	?
Serum CK	n	n	n	↑	?	n	n	?	?	+	↑	↑↑

n = normal; ↑ = elevated; + = present; – = absent.

causes no decrement of the compound muscle action potential. Nerve conduction studies show normal results.

HISTOPATHOLOGY

Although Brody reported in his original case no abnormalities on transverse or longitudinal sections by light microscopy, later authors found a significant atrophy in about 50% of the histochemical type II (A and B) fibers (3,4). The cross-sectional outline of most type II fibers was angular. About 50% of the type II fibers had one to five internal nuclei, and the atrophic type II fibers had noticeably reduced glycogen content. No muscle fiber necrosis was noted (3,4).

Electron microscopy disclosed triads and tubular elements of SR without gross abnormalities (3,4). On longitudinal sections tubular elements of the SR could be found here and there between myofibrils. On transverse sections these were seen bordering the Z-disc occasionally (3). Some fibers showed a focal reduplication of the basal lamina.

BIOCHEMICAL CONSIDERATIONS

SR Ca^{2+}-ATPase is an integral membrane protein, having a transmembrane organization (9), that serves to pump calcium out of the cytoplasm into the sarcoplasmic reticulum against a large concentration gradient (Fig. 1).

Brody (1) in his original paper used $^{45}Ca^{2+}$-uptake studies in isolated SR fractions to prove the Ca-flux abnormality in his patient's muscle. Subsequent authors used various additional techniques to measure the Ca^{2+}-ATPase involved. It was demonstrated that upon incubation with gamma ^{32}P-ATP, the steady-state phosphorylated Ca^{2+}-ATPase, a 100-kDa phosphoprotein, was lacking in patients (3,4). Antibody studies both biochemically and immunohistochemically confirmed the absence of the Ca^{2+}-ATPase, in particular in type II muscle fibers of the patients (3,4). These studies indicate that Ca^{2+}-ATPase as a protein is almost absent in type II fibers in Brody's disease. This seems not to be the

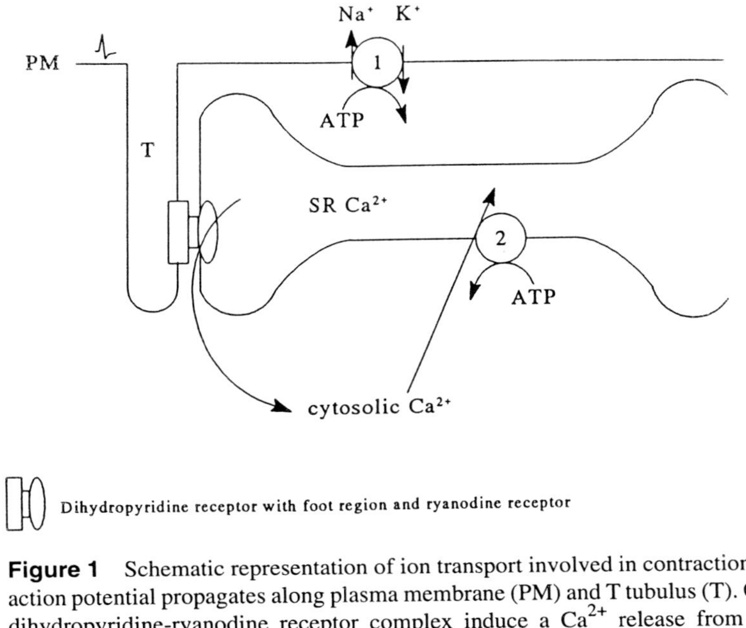

Figure 1 Schematic representation of ion transport involved in contraction and relaxation of muscle. An action potential propagates along plasma membrane (PM) and T tubulus (T). Conformational changes in the dihydropyridine-ryanodine receptor complex induce a Ca^{2+} release from the SR. Na^{+}/K^{+}-ATPase (1) repolarizes the membrane potential and SR Ca^{2+}-ATPase (2) normalizes the cytosolic Ca^{2+} concentration.

case in one of the patients of Taylor et al. (5). As the protein yield in the SR fraction was normal, where Ca^{2+}-ATPase normally forms a major contribution to the protein content of SR (10), Taylor et al. suggested that in their patient there is a normal amount of Ca^{2+}-ATPase protein present that lacks activity. Unfortunately, no immunochemical studies were included in the paper of Taylor et al. Recently Benders et al. (8) found that the concentration of the fast-twitch muscle SR Ca^{2+}-ATPase isoform (=SERCA1) in Brody's disease is normal but the activity is reduced. This implies a reduction of the molecular activity of the Ca^{2+}-ATPase in Brody's disease.

Quantitative data on Ca^{2+}-ATPase are difficult to compare, because various authors used a wide range of techniques. Some studies present quantitative data after SR isolation (5) while others give data on whole muscle biopsy extracts (3,6,7). Furthermore, the techniques used to quantify the protein or its activity varied considerably. Karpati et al. (3) used immunoquantitation, thus measuring the amount of protein, while others (5–7) measured the activity of the protein. In view of methodological discrepancies between the studies, the residual presence of the protein or its activity is hard to interpret. Roughly, the measurements indicate that between 0 and 50% residual Ca^{2+}-ATPase is found in the various studies. The existence of both a slow and a fast form of Ca^{2+}-ATPase in muscle further complicates the interpretation of the biochemical results. Only the fast form seems to be involved in Brody's disease while the slow form may have influenced the biochemical results in several studies. Typically Figure 2 demonstrates an age dependency of reference values for Ca^{2+}-ATPase activity that has not been considered in studies on Brody's disease. New developments are the possibility to study the SR Ca^{2+}-ATPase in cultured muscle cells and the additional option of continuous monitoring of intracellular calcium with fluorescence techniques (8,11).

Obviously, an unequivocal, specific, and reliable technique to confirm Brody's disease biochemically has not yet been established. Further studies are necessary to decide the method of choice in this respect.

Interestingly ^{31}P magnetic resonance spectroscopy studies revealed an unusual metabolic response to exercise (5). Taylor et al. (5) found a rapid and excessive acidification due to lactate production upon exercise. The data were consistent with increased glycolytic flux. Lactate levels during rest were normal. Until now only Brody (1) found elevated blood lactate concentrations under resting conditions in his patient.

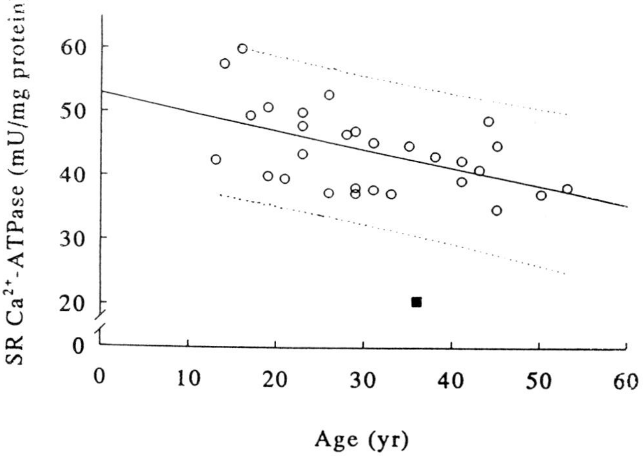

Figure 2 Age dependency (11) and residual Ca^{2+}-ATPase activity in muscle biopsies in controls (○) and patient with Brody's disease (■) (6,7).

DNA AND GENETICS

Rabbit genomic DNA contains two separate genes encoding SR Ca^{2+}-ATPases (12,13). The product of one is found in cardiac and slow-twitch muscle while that of the second is located in fast-twitch muscle fibers. The fast-twitch SR Ca^{2+}-ATPase isoform, involved in Brody's disease, was found to exist in two developmentally regulated forms in rabbit [neonatal and adult form (14)]. Alternative splicing of the fast-twitch Ca^{2+}-ATPase transcript, in which a 42-base-pair exon is excised in the neonatal transcript and retained in the adult transcript, explains the existence of a neonatal and an adult Ca^{2+}-ATPase form (14). The fast-twitch Ca^{2+}-ATPase gene is located on chromosome 16 (15). Whether this gene is directly involved in either one of the presumed dominant and recessive forms of Ca^{2+}-ATPase deficiency remains to be elucidated.

The Mendelian pattern of transmission of SR Ca^{2+}-ATPase deficiency in the families described so far indicates that the disorder is genetic. Due to the low number of patients reported in literature, however, the mode(s) of inheritance is (are) not yet clear. The available family data suggest that the disease may be heterogeneous since both families with an autosomal dominant (4) and families with a recessive (X-linked or autosomal) pattern of transmission (3,5) have been reported.

DIAGNOSIS

A myopathy caused by SR Ca^{2+}-ATPase deficiency should be considered in patients with exercise-induced muscle stiffness, cramps, and myalgia with impairment of muscle relaxation. It should also be considered in cases of exertional myalgia with postexercise muscle weakness or recurrent unexplained rhabdomyolysis. Electrical silence in the EMG is a characteristic of the disease. Definite diagnosis requires biochemical and/or immuno(histo)chemical confirmation of SR Ca^{2+}-ATPase deficiency in muscle tissue (3,4). Cases may exist where the Ca^{2+}-ATPase *protein* is present in normal quantities but the *activity* of the Ca^{2+}-ATPase is disturbed (5,8). Therefore biochemical assays measuring activity (1,3,6–8,11) should be preferred to such techniques as enzyme-linked immunosorbent assay (11) or immunoblotting (3,4). Given that clinical signs and symptoms may lack specificity, a good and noninvasive screening test would be invaluable. In a patient with SR Ca^{2+}-ATPase deficiency and recurrent episodes of exertional rhabdomyolysis, ischemic forearm testing showed an excessive K^{+}-efflux into the blood from working muscle (6,7). However, the use of an ischemic forearm test as a reliable screening test for this disease has yet to be established.

DIFFERENTIAL DIAGNOSIS

The differential diagnosis of SR Ca^{2+}-ATPase deficiency includes myotonia due to muscle cramp, tetany, myotonic disorders (myotonic dystrophy, myotonia congenita, paramyotonia congenita), and glyco(geno)lytic disorders (e.g., myophosphorylase deficiency, phosphofructokinase deficiency). Myotonia resulting from myotonic disorders is associated with sustained electrical activity of the shortened muscles. Persistent muscle action potentials on EMG can distinguish these disorders from SR Ca^{2+}-ATPase deficiency. Glyco(geno)lytic disorders are characterized by electrically silent contractures and must be ruled out by appropriate tests, the ischemic forearm test being the most well known.

TREATMENT

There is no satisfactory treatment for Brody's disease. Karpati et al. treated a patient with dantrolene sodium (acting on the dihydropyridin-ryanodine receptor complex; Fig. 1) and nifedipine (acting on voltage-operated Ca^{2+} channels) four times a day for 1 month, but without any positive effect on muscle function (3). Taylor et al. (5), however, noticed some relief from nifedipine on exertional muscle pain in their two patients. Dantrolene sodium given in one patient improved muscle pain slightly, but produced unacceptable muscle weakness. Walton (16) reported a positive effect of verapamil on exercise-induced muscle pain of undetermined cause. A positive effect of dantrolene sodium was found by Wevers and Poels in a patient with recurrent rhabdomyolysis (6,7). When the patient was given dantrolene sodium, exercise intolerance improved, muscular symptoms disappeared, and both the serum CK level and the abnormal K^+ efflux normalized. On discontinuation of dantrolene medication the symptoms returned. Single cell measurements by Benders et al. (8) in muscle cell cultures have shown that dantrolene or verapamil reduces the Ca^{2+} response. Therefore, less Ca^{2+} pumping capacity is required and Ca^{2+} overload can be prevented by the residual activity of the Ca^{2+}-ATPase in the patient's cells. This balances the excitation-relaxation cycle, which may explain the observed beneficial effect of administration of the drugs in some patients.

REFERENCES

1. Brody I. Muscle contracture induced by exercise: a syndrome attributable to decreased relaxing factor. N Engl J Med 1969; 281:187–192.
2. Lambert EH, Goldstein NP. Unusual forms of "myotonia." Physiologist 1957; 1:51.
3. Karpati G, Charuk J, Carpenter S, Jablecki C, Holland P. Myopathy caused by a deficiency of Ca^{2+}-adenosine triphosphatase in sarcoplasmic reticulum (Brody's disease). Ann Neurol 1986; 20:38–49.
4. Danon MJ, Karpati G, Charuk J, Holland P. Sarcoplasmic reticulum adenosine triphosphatase deficiency with probable autosomal dominant inheritance. Neurology 1988; 38:812–815.
5. Taylor DJ, Brosnan MJ, Arnold DL, et al. Ca^{2+}-ATPase deficiency in a patient with an exertional muscle pain syndrome. J Neurol Neurosurg Psych 1988; 51:1425–1433.
6. Wevers RA, Poels PJE, Joosten EMG, Steenbergen GGH, Benders AAGM, Veerkamp JH. Ischaemic forearm testing in a patient with Ca^{2+}-ATPase deficiency. J Inher Metab Dis 1992; 15:423–425.
7. Poels PJE, Wevers RA, Braakhekke JP, Benders AAGM, Veerkamp JH, Joosten EMG. Exertional rhabdomyolysis in a patient with Ca-ATPase deficiency. J Neurol Neurosurg Psych 1993; 56:823–826.
8. Benders AAGM, Veerkamp JH, Ooosterhof A, et al. Ca^{2+} homeostasis in Brody's disease: a study in skeletal muscle and cultured muscle cells and the effects of dantrolene and verapamil. J Clin Invest 1994; 94:741–748.
9. Toyoshima C, Sasebe H, Stokes DL. Three-dimensional cryo-electron microscopy of the calcium ion pump in the sarcoplasmic reticulum membrane. Nature 1993; 362:469–471.
10. Fleischer S, Inui M. Biochemistry and biophysics of excitation-contraction coupling. Annu Rev Biophys Biophys Chem 1989; 18:333–364.
11. Benders AAGM, Timmermans JAH, Oosterhof A, et al. Deficiency of Na^+/K^+-ATPase and SR Ca-ATPase in skeletal muscle and cultured muscle cells of myotonic dystrophy patients. Biochem J 1993; 293:269–274.
12. MacLennan DH, Brandl CJ, Korczak B, Green M. Amino-acid sequence of a Ca^{2+}+Mg^{2+}-dependent ATPase from rabbit muscle sarcoplasmic reticulum, deduced from its complementary DNA sequence. Nature 1985; 316:696–700.
13. Brandl CJ, Green NM, Korczak B, MacLennan DH. Two Ca^{2+}ATPase genes: homologies and mechanistic implications of deduced amino acid sequences. Cell 1986; 44:597–607.
14. Brandl CJ, deLeon S, Martin DR, MacLennan DH. Adult forms of the Ca^{2+}ATPase of sarcoplasmic reticulum; expression in developing skeletal muscle. J Biol Chem 1987; 262(8):3768–3774.
15. MacLennan DH, Brandl CJ, Champaneria S, Holland PC, Powers PE, Huntington FW. Fast-twitch and slow-twitch cardiac Ca^{2+}-ATPase genes map to human chromosomes 16 and 12. Somat Cell Mol Gen 1987; 13:341–346.
16. Walton J. Diffuse exercise-induced muscle pain of undetermined cause relieved by verapamil. Lancet 1981; 1:933.

38

Clinical Features of Mitochondrial Myopathies and Encephalomyopathies

Michio Hirano and Salvatore DiMauro
Columbia University College of Physicians and Surgeons, New York, New York

INTRODUCTION

The classification of neurological diseases has relied upon the acumen of clinicians. This is especially true of mitochondrial encephalomyopathies. These disorders present diagnostic challenges to the most skilled and knowledgeable physicians because of the heterogeneity of presentations and the multiplicity of organ systems affected. Fortunately, with sophisticated laboratory tools, including histopathology, biochemical assays, and molecular genetic analyses, we now possess an understanding of the pathogenesis of clinical phenotypes, and therefore we can classify the mitochondrial disorders more logically. Rather than attempting to review all of the mitochondrial diseases, we will focus on defects of the respiratory chain and oxidative phosphorylation.

MITOCHONDRIAL PHYSIOLOGY

Mitochondria are intracellular organelles that perform vital functions in the aerobic production of energy (ATP). Although it is impossible to adequately describe the current state of knowledge about mitochondria, a brief summary of mitochondrial functions might provide helpful background to place the diseases in perspective.

Each mitochondrion is about the size of a bacterium (1–2 μm by 0.5–1 μm) with two membranes: an external or outer membrane, and an internal or inner membrane (1,2). The outer membrane separates the organelle from the cytoplasm. The mitochondrial matrix is enclosed within the inner membrane. Substrates, such as fatty acids and pyruvate, are transported across the mitochondrial membranes into the matrix where they are further processed to form acetyl CoA (Fig. 1). This molecule enters the citric acid (or Krebs) cycle and is oxidized to CO_2, while NAD^+ and FAD are reduced to NADH and $FADH_2$, respectively (3). The NADH and $FADH_2$ carry electrons to the respiratory (electron transfer) chain located in the inner mitochondrial membrane (4–6). The electrons flow through four enzyme complexes (I–IV) and create a proton gradient across the inner membrane. Complex V utilizes this proton gradient to generate ATP from ADP and inorganic phosphate (7).

Defects in each of the five major mitochondrial processing steps have been identified: (1) substrate transport; (2) substrate utilization; (3) Krebs cycle; (4) oxidation-phosphorylation coupling; and (5) respiratory chain (Table 1). Unfortunately, some of the mitochondrial encephalomyopathies do not fit precisely into this biochemical classification scheme; for example, there can be variable defects in the respiratory chain with the same clinical phenotype, or even with the same genotype as described below.

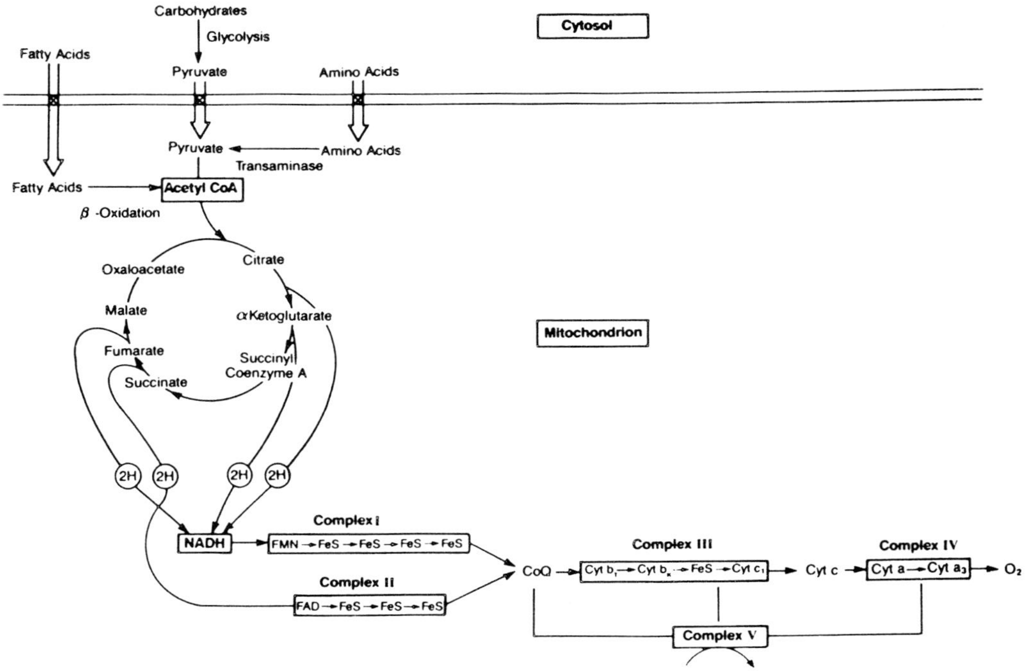

Figure 1 Mitochondrial metabolic pathways: substrate transport, Krebs cycle, and respiratory chain. NADH, nicotinamide adenine nucleotide; FMN, flavin mononucleotide; FAD, flavin adenine dinucleotide; FeS, (nonheme) iron-sulfur; CoQ, coenzyme Q; Pi, inorganic phosphate. (From: DiMauro S, DeVivo DC: Diseases of carbohydrate, fatty acid, and mitochondrial metabolism. In: Siegel GJ, et al, eds. Basic Neurochemistry: Molecular, Cellular, and Medical Aspects, 4th ed. New York: Raven Press, 1989:647.)

Mitochondrial Genetics

The mitochondria have distinct genetic properties because they are the products of two DNA sources; the nucleus and the organelle itself. The mitochondria encode 13 polypeptides, 22 transfer RNAs (tRNA), and two ribosomal RNAs (8) (Fig. 1). All of the polypeptides are subunits of the respiratory chain/oxidative phosphorylation system. Over the last 8 years, there has been an outburst of information linking mtDNA mutations to human diseases.

Four important concepts are vital in understanding the clinical expression of mtDNA mutations.

1. *Maternal inheritance*: MtDNA is transmitted exclusively from the mother through the ovum; therefore, the inheritance pattern of mitochondria does not follow the classic Mendelian model. Only women can transmit mutations to offspring; accordingly, progeny of both sexes inherit the abnormal mtDNA.
2. *Heteroplasmy*: Each mitochondrion contains several copies of mtDNA, and each cell contains multiple mitochondria; therefore, the ratio of normal to mutant mtDNA can vary widely. This coexistence of different mtDNA populations within the same cell is called heteroplasmy.
3. *Threshold effect*: In order for a mtDNA mutation to be expressed, it must be present above a certain threshold number or percent relative to the normal mtDNA. The threshold level probably varies according to the oxidative metabolism requirements of the tissue.
4. *Nuclear factors* can influence the expression of mtDNA mutations. Leber's hereditary optic neuropathy (LHON) causes loss of vision, mostly in young men. The male predominance has been attributed to an, as yet unidentified, X-linked factor.

Table 1 Biochemical Classification of Mitochondrial Encephalomyopathies

1.	Defects of transport	
	a.	CPT deficiency
	b.	Carnitine deficiency
	c.	Defect of FAD uptake (?)
2.	Defects of substrate utilization	
	a.	Pyruvate carboxylase deficiency
	b.	Pyruvate dehydrogense complex deficiency
	c.	Defects of β-oxidation
3.	Defects of the Krebs cycle	
	a.	Fumarase deficiency
	b.	α-Ketoglutarate dehydrogenase (dihydrolipoyl dehydrogenase) deficiency
4.	Defects of oxidation-phosphorylation coupling	
	a.	Luft's syndrome (loose coupling of muscle mitochondria)
5.	Defects of the respiratory chain	
	a.	Complex I deficiency
	b.	Complex II deficiency
	c.	Complex III deficiency
	d.	Complex IV deficiency
	e.	Complex V deficiency
	f.	Combined defects of respiratory chain components

Source: From DiMauro S et al. Mitochondrial encephalomyopathies. Neurol Clin 1990; 8:483–506.

DEFINING MITOCHONDRIAL ENCEPHALOMYOPATHIES

Mitochondrial encephalomyopathies with respiratory chain and oxidation-phosphorylation dysfunction can be characterized at four levels: clinical, histological, biochemical, and genetic. This chapter focuses on the clinical aspects, but the other three are discussed briefly.

Morphology

The concept of mitochondrial disease originated in the early 1960s with the report by Luft et al. (9) of a young woman with nonthyroidal hypermetabolism thought to be due to a defect coupling between respiration and phosphorylation. In the mid-1960s, Shy et al. (10) described the characteristic ultrastructural alterations of mitochondria in skeletal muscle that are used to define the mitochondrial myopathies. These changes include: (1) an overabundance of ultrastructurally normal mitochondria, "pleoconial myopathy"; (2) enlarged mitochondria with disoriented cristae, "megaconial myopathy"; (3) inclusions within mitochondria, "paracrystalline" and "osmiophilic" inclusions. Engel and Cunningham (11) detected excessive accumulations of mitochondria in skeletal muscle by light microscopy using a modified Gomori trichrome stain. The subsarcolemmal region of abnormal fibers stained red and these were called ragged-red fibers (RRF) (Fig. 2) (11). Histochemical stains for mitochondrial enzymes are also used to detect excessive proliferation of mitochondria and to demonstrate specific enzyme defects. These stains include: succinate dehydrogenase (SDH); NADH-tetrazolium reductase (NADH-TR); and cytochrome *c* oxidase (COX). Immunohistochemical techniques are used to identify defects in specific mitochondrial polypeptides and in situ hybridization is used to detect messenger RNA and DNA abnormalities.

The histological features of mitochondria are important in diagnosing and classifying mitochondrial disorders, as shown in Table 2. Unfortunately, the morphological features are neither specific nor sensitive enough to define all mitochondrial diseases. RRF and ultrastructurally abnormal mitochondria have been detected in many conditions that are not primary mitochondrial disorders (Table 3), such as: inflammatory myopathies (12,13), Duchenne muscular dystrophy (14), myotonic dystro-

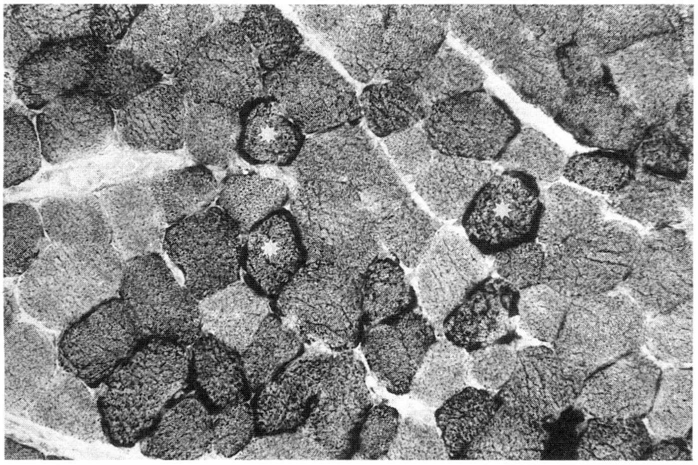

Figure 2 Succinate dehydrogenase (SDH) histochemical stain of skeletal muscle from a MELAS patient demonstrating ragged-red fibers (white stars). (Courtesy of Dr. Eduardo Bonilla.)

phy (15), and acid maltase deficiency (16), and even in healthy aged individuals (17). Conversely, some conditions with mitochondrial enzyme defects do not have morphologically abnormal mitochondria, including: carnitine palmitoyl transferase (CPT) deficiency and pyruvate dehydrogense (PDH) deficiency.

Biochemistry

Before DNA mutations were identified, many mitochondrial diseases were identified and classified by their biochemical defects (Table 1). CPT II deficiencies are examples of monoenzyme defects that are related to specific DNA mutations and distinct clinical phenotypes (18). Unfortunately for those who would like a comprehensive and precise biochemical classification scheme, defects of oxidative phosphorylation do not always correlate well with clinical phenotypes, or with specific DNA mutations. Various combinations of deficiencies and even normal respiratory chain enzyme activity levels have been reported in muscle of patients with Kearns-Sayre syndrome (KSS), mitochondrial encephalomyopathy, lactic acidosis with stroke-like episodes (MELAS), and myoclonus epilepsy with ragged-red fibers (MERRF). There are at least three possible causes for the phenotypic variability: (1) mtDNA heteroplasmy, which leads to variable proportions of mutant to wild-type mtDNA and correspondingly variable biochemical defects; (2) mutations in mtDNA-encoded transfer RNAs (tRNAs) might impair synthesis of multiple polypeptides; and (3) uneven tissue distribution of the mutant mtDNA might lead to clinical expression of the disease only in those tissues with a higher proportion of mutations.

Genetics

Since the initial discovery of large-scale mtDNA deletions in 1988 (19), there has been an explosion of information relating molecular genetic defects to biochemical dysfunction, and clinical phenotypes (Table 4). Several types of mtDNA mutations have been identified: point mutations, duplications, depletions, and multiple deletions. Some of these mutations are primary and pathogenic (point mutations in MERRF, MELAS, LHON; single deletions in KSS and progressive external ophthalmoplegia, PEO); others are secondary to another process yet might have a role in pathogenesis (mtDNA depletion in AZT myopathy). Mutations of nuclear DNA can also affect respiratory chain functions.

Table 2 Muscle Histological Features of Mitochondrial Encephalomyopathies and Mitochondrial Myopathies

Clinical syndrome	Ragged-red fibers	COX-negative fibers	Increased SDH stain	SSV[a]	Other features
Kearns-Sayre syndrome	+	+	+	–	
PEO with deletion	+	+	+	–	
PEO with 3243 mutation	+	+	+	–	
Autosomal-dominant PEO	+	+	+	?	
MELAS-3243	+	rare	+	+	
MELAS-3271[b]	+	rare	+	–	
MERRF-8344	+	+	+	–	
MERRF-8356	+	+	+	–	
LHON	–	–	–	–	
NARP-8993	–	–	–	–	
MNGIE or MEPOP	+	+	–	–	
Fatal infantile myopathy	+	+	+	–	COX VIIab defect by immunohistochemistry; some cases have been associated with mtDNA depletion.
Benign infantile myopathy (2)	+	+	+	–	COX II and VIIab defect by immunohistochemistry
AZT myopathy	+	+	–	–	Associated with mtDNA depletion

[a]Strongly SDH-positive vessels.
[b]E. Bonilla, personal communication.

Table 3 Syndromes in Which Histologically Abnormal Mitochondria Have Been Seen, but Are Not Primary Mitochondrial Encephalomyopathies, or Mitochondrial Myopathies

Inflammatory myopathies (12,13)
Aging (17)
Hypertrophic cardiomyopathy (17,148,149)
Duchenne muscular dystroph (14)
Myotonic dystrophy (15)
Acid maltase deficiency (16)
Periodic paralysis (160,161)
Clofibrate myopathy (162)
Oculopharyngeal myopathy (163)
Facioscapulohumeral myopathy (164)

GENERAL CLINICAL CONSIDERATIONS

Dysfunction of mitochondria can be expressed in virtually every human organ system (Table 5); however, certain tissues are more susceptible to respiratory chain defects because of their lower threshold levels. In general, cells with high oxidative metabolism requirements, such as muscle cells and neurons, seem most vulnerable.

For many years, there was a lively debate between clinical "lumpers" and "splitters" of human mitochondrial encephalomyopathies (20–23). The "lumpers" stated that "rational classification of mitochondrial myopathies is not possible on clinical grounds alone" (20). Rowland and colleagues steadfastly asserted that "distinct clinical syndromes do exist" (21). With our advancing understanding of the molecular basis of the mitochondrial syndromes, a consensus is emerging that both sides are partly correct (23). The "splitters" were supported by the increasing numbers of case reports that fulfilled the criteria for the acronymic syndromes and by the high correlation of specific mutations with clinical phenotypes. On the other hand, the existence of overlap cases with clinical features of more than one syndrome and the multiple clinical phenotypes associated with the same mutation reinforced the arguments of the "lumpers." Ultimately, a deeper knowledge of the molecular biological sequelae of the mtDNA mutations will probably provide us with a better understanding of the clinical expression and an improved classification scheme.

Clinical classification of the mitochondrial encephalomyopathies is important for several reasons:

1. As demonstrated by the discovery of mtDNA mutations, if the clinical criteria are sufficiently specific, there is a greater likelihood that there will be a correlation with a mutation or a narrow range of mutations.
2. A clinical diagnosis can narrow the range of known mtDNA mutations for which to screen.
3. Following the natural history of clinical phenotypes can be useful in predicting future progression and as a guide to therapy [for example, patients with mtDNA deletions and Pearson syndrome (sideroblastic anemia and exocrine pancreatic dysfunction) are likely to develop KSS and should be monitored for cardiac conduction block, which might warrant pacemaker placement].
4. Atypical clinical presentations should be identified to broaden our understanding of the syndromes and pathogenesis.

We have organized the clinical syndromes by known genetic defects in an attempt to rationally classify the mitochondrial encephalomyopathies (Table 4). Some clinical syndromes have mtDNA abnormalities that are due to a toxin (zidovudine myopathy), or nuclear DNA defect (e.g., the mitochondrial neuropathy, gastrointestinal encephalomyopathy, or MNGIE syndrome) and are categorized by the primary etiological factor. Other syndromes in which morphologically or biochemically abnormal mitochondria might be epiphenomena are mentioned (e.g., RRF in inflammatory myopathies).

Table 4 Mitochondrial Encephalomyopathy and Mitochondrial Myopathy Syndromes with Reported Associated DNA Mutations

- I. MtDNA mutations
  - A. Single deletions and duplications
    - 1. Kearns-Sayre syndrome (KSS) (19,39,66,67,75)
    - 2. Sporadic progressive external ophthalmoplegia (PEO) (19,43,75)
    - 3. Pearson syndrome (65,73,74)
    - 4. Maternally inherited diabetes mellitus and deafness (DAD) (78)
  - B. MtDNA point mutations
    - 1. MELAS
      tRNA$^{Leu(UUR)}$-3243 (119)
      tRNA$^{Leu(UUR)}$-3271 (120)
      ND4-11084 (121)
    - 2. MERRF
      tRNA^{Lys}-8344 (99)
      tRNA^{Lys}-8356 (108)
    - 3. MERRF-MELAS overlap?
      tRNA^{Lys}-8356 (109)
    - 4. Myopathy-cardiomyopathy
      tRNA$^{Leu(UUR)}$-3260 (123)
    - 5. Mitochondrial myopathy
      tRNA$^{Leu(UUR)}$-3243 (72)
      tRNA^{Lys}-8344 (107)
    - 6. PEO
      tRNA$^{Leu(UUR)}$-3243 (72)
      tRNA^{Lys}-8344 (107)
    - 7. Kearns Sayre syndrome
      tRNA$^{Leu(UUR)}$-3243 (72)
  - C. Multiple mtDNA deletions
    - 1. Familial recurrent myoglobinuria (86)
    - 2. Dilating cardiomyopathy (149)
    - 3. Aging (144–147)
- II. Nuclear DNA mutations (unidentified)
  - A. Autosomal-dominant PEO (with multple mtDNA deletions) (71)
  - B. Fatal infantile myopathy with COX deficiency (130–132)
  - C. Benign infantile myopathy with COX deficiency (129,130)
  - D. MNGIE (with or without multiple mtDNA deletions) (87)
  - E. Limb-girdle myopathy
- III. MtDNA-nDNA intercommunication defects
  - A. Fatal infantile myopathy with mtDNA depletion (125,126)

MITOCHONDRIAL ENCEPHALOMYOPATHY SYNDROMES WITH mtDNA MUTATIONS

MtDNA Deletions and Duplications

Single Deletions

Kearns-Sayre Syndrome. The history of progressive ophthalmoplegia and "ophthalmoplegia-plus syndromes," a term originated by Drachman (24), has been well described by Rowland (25,26). In 1958, Kearns and Sayre (27) described the syndrome as a combination of retinitis pigmentosa, external ophthalmoplegia, and complete heart block. In 1983, Rowland and colleagues (22,28) redefined the syndrome by the invariant triad of ophthalmoplegia, pigmentary retinopathy, and onset before

Table 5 General Clinical Features of Mitochondrial Encephalomyopathies

Muscle	Heart
External ophthalmoplegia	Cardiac conduction block
Exercise intolerance	Hypertrophic cardiomyopathy
Limb weakness	Preexcitation syndrome
Lack of muscle bulk	
Elevated creatine kinase	Renal
	Tubular acidosis (de Toni-Fanconi-Debré syndrome)
CNS	Bartter-like syndrome
Acute optic neuropathy, atrophy, or both	
Pigmentary retinopathy	Pancreas
Mental retardation or dementia	Exocrine deficiency
Seizures	
Ataxia	Liver
Sensorineural hearing loss	Elevated transaminases
Leukodystrophy	Hepatic failure
Myoclonus	Steatorrhea
Stroke-like episodes at a young age	
Migraine headaches	Hematological
Extrapyramidal signs	Pancytopenia
	Sideroblastic anemia
PNS	Acanthocytosis
Sensorimotor neuropathy	
	Gastrointestinal tract
Endocrinopathy	Dysmotility
Diabetes mellitus	Intestinal pseudo-obstruction
Hypoparathyroidism	
Hypothyroidism	Psychiatric
Infertility	Depression
Growth hormone deficiency with short stature	Schizophrenic-like episodes
Delayed puberty	
Hirsutism	Dermatological
Irregular menses	Purpuric lesions
	Lipomas
	Hirsutism

age 20 with at least one of the following three additional features: cardiac conduction block, cerebellar syndrome, or cerebrospinal fluid (CSF) protein greater than 100 mg/dl (Table 6). Using these criteria, we found 86 additional KSS patients (Table 7) (29–72) since the last review in 1982 (22). A major breakthrough in our understanding of this disorder came in 1988 with the identification of large-scale mtDNA heteroplasmic deletions in KSS patients (19,39). The majority of KSS patients have mtDNA deletions; the pathogenic significance will be discussed later. Duplications of mtDNA have also been reported (66,67).

Frequently associated symptoms, in addition to the diagnostic features listed above, include hearing loss and limb weakness. Endocrinopathies have been reported; these include: diabetes mellitus, hypoparathyroidism, irregular menses, and growth hormone deficiency. Renal tubular dysfunction is usually in the form of deToni-Fanconi-Debré syndrome; however, Bartter-like (52) and Lowe-like (57) syndromes have also been reported. Six patients with KSS had strokes (30,43,54,60,72), the clinical hallmark of MELAS, which makes the clinical distinction between the two syndromes unclear in these patients. Molecular genetic analysis was helpful in these unusual cases; four of the six patients had

Table 6 Clinical Definitions of Mitochondrial Encephalomyopathy Syndromes

I.	Kearns-Sayre syndrome	1. Onset before age 20 years 2. Pigmentary retinopathy 3. Ophthalmoplegia plus, at least one of the following 1. CSF protein greater than 100 mg/dl 2. Cardiac conduction block 3. Cerebellar syndrome
II.	MELAS	1. Stroke before age 40 2. Encephalopathy (seizures, dementia, or both) 3. Ragged-red fibers, lactic acidosis, or both plus, at least two of the following: 1. Normal early development 2. Recurrent headaches 3. Recurrent vomiting
III.	MERRF	1. Myoclonus, or myoclonic epilepsy 2. Cerebellar syndrome 3. Myopathy with ragged-red fibers
IV.	MNGIE	1. Ophthalmoparesis 2. Peripheral neuropathy 3. Gastrointestinal dysmotility 4. Morphologically abnormal mitochondria

mtDNA deletions (43,54,60), one had the $tRNA^{Leu(UUR)}$-3243 mutation typically associated with MELAS (72), and one was not tested. Only five of the 156 KSS patients had seizures (22,46,60), a common feature of MELAS and a defining symptom of MERRF; the seizures were attributed to electrolyte imbalance in four of five patients who had hypoparathyroidism.

Virtually all of the cases have been sporadic, suggesting that the deletion arises early in embryogenesis or in the ovum. One woman with KSS and a mtDNA deletion gave birth to a daughter who had no detectable mutant mtDNA in muscle by polymerase chain reaction at age 5 months, suggesting no transmission of the deleted mtDNA (61). There are only five families in which more than one member had complete or incomplete KSS. In one family, identical twins had the full syndrome while a brother and the parents were not affected (40); in another family, two brothers had KSS, while five other siblings were in good health (68). There are two reports of paternal transmission, suggesting autosomal inheritance (69,70). The affected family members might have had the autosomal dominant PEO and one or more members fulfilled the criteria for KSS. Finally, a woman with KSS and hallucinatory psychosis had a brother with ophthalmoplegia who died of a cardiopathy in adolescence and a maternal uncle with ptosis and ophthalmoplegia (41). The absence of a deletion in the Southern blot of the propositus and the inheritance pattern consistent with maternal transmission suggest the presence of a mtDNA point mutation, such as the $tRNA^{Leu(UUR)}$-3243 mutation that Moraes et al. (72) found in 15 cases of familial PEO and in one case of KSS.

Typically, patients with KSS are normal at birth and have normal early development, which raises a pathogenic problem: if the mtDNA deletion is present from conception or early embryogenesis, why do symptoms begin in childhood or adolescence rather than in infancy? One possible explanation is that the deletions accumulate over years, perhaps because the shorter mtDNA replicate more rapidly. It has been demonstrated by serial biopsies that the amount of deleted mtDNA increases in patients

Table 7 Clinical Features of 156 Kearns-Sayre Syndrome Patients

Sign/symptom	1977[a]	1982[b]	1992
Number of cases	35	35	86
Impaired intellect	14	20	25/29
Cerebellar syndrome	24	32	53/63
Babinski signs	9	9	8
Limb weakness	19	24	61/65
Hearing loss	19	24	33/34
Seizures	0	3	2
Delayed puberty	11	—	1
CSF protein			
Reported	25	30	25
Normal	1	2	0
>100 mg/dl	20	26	17
Lactic acidosis	—	—	13/17
RRF in limb muscle	5/7	28/30	65/66
mtDNA deletions	—	—	47/50
mtDNA duplications	—	—	1/50
$tRNA^{Leu(UUR)}$-3243 mutation	—	—	1/50
CT lucencies	—	4	10
White matter lucencies	—	—	5
BG lucencies	—	—	5
Cerebral atrophy	—	—	7
Pontine atrophy	—	4	0
BG calcification	—	2	1
Normal	—	4	1
Autopsies	5	8	6
Spongy degeneration	5	8	6
Diabetes mellitus	—	—	11
Neuropathy	—	—	8
Stroke	—	—	6
Hypoparathyroidism	—	—	5
deToni-Fanconi syndrome	—	—	4
Family history	—	—	3
Cardiomyopathy	—	—	2
Cataracts	—	—	2
Irregular menses	—	—	2
Hepatomegaly	—	—	1
Dystonia	—	—	1
Bartter-like syndrome	—	—	1
Lowe-like syndrome	—	—	1
Anhydrosis	—	—	1
Chorea	—	—	1

[a]From Berenberg RA, et al. Ann Neurol 1977; 1:37–43.
[b]From Rowland, et al. In: Cerri, C, Scarlato G, eds. Mitochondrial Pathology in Muscle Diseases. 1983:141–158.

(53). A few patients have presented with Pearson syndrome (sideroblastic anemia and exocrine pancreatic dysfunction) in infancy or early childhood and later developed KSS (65). MtDNA deletions have been found in blood, bone marrow, and pancreas of patients with Pearson syndrome (65,73,74), in contrast to KSS in which blood typically does not harbor detectable mutation by routine Southern blot. These findings suggest that the percentage of deleted mtDNA in these patients decreases in hematopoietic cells of the bone marrow and increases in skeletal muscle and other tissues if the child survives the anemia.

Progressive External Ophthalmoplegia. PEO and ptosis typically present in childhood and adolescence in a symmetrical fashion usually without diplopia (Fig. 3). Limb weakness is a common associated feature. About half of these patients have a heteroplasmic population of deleted mtDNA (75). Like KSS patients, patients with mtDNA deletions are usually sporadic. Patients with a family history of PEO or other manifestations typical of mitochondrial disorders are likely to have a different mtDNA defect, such as the autosomal-dominant PEO with multiple deletions (71) or the $tRNA^{Leu(UUR)}$-3243 A-to-G transition mutation found in most patients with MELAS (72).

PEO-Plus. A recurrent motif in mitochondrial disorder is the occurrence of incomplete clinical syndromes such as incomplete KSS. There is no clear distinction between the terms PEO-plus and incomplete KSS. Patients with these diagnoses may represent intermediates in the clinical spectrum, from isolated PEO to the multisystem KSS disorder, and probably have a different tissue distribution of deleted mtDNA, a lower proportion of deleted mtDNA, or both when compared to KSS patients. Often young patients will be classified as having an incomplete syndrome before all of the clinical manifestations appear.

The mtDNA deletions range in size from about 2.0 to 10.4 kb (19,43,75,79–84). These have been mainly confined to an 11-kb region that does not include the origins of replication or mtDNA promoters. Approximately 30–40% of mtDNA deletions involve an identical 4977-bp segment (75,81–84). This "common deletion" is flanked by a 13-bp direct repeat. Over 20 different deletion breakpoints have

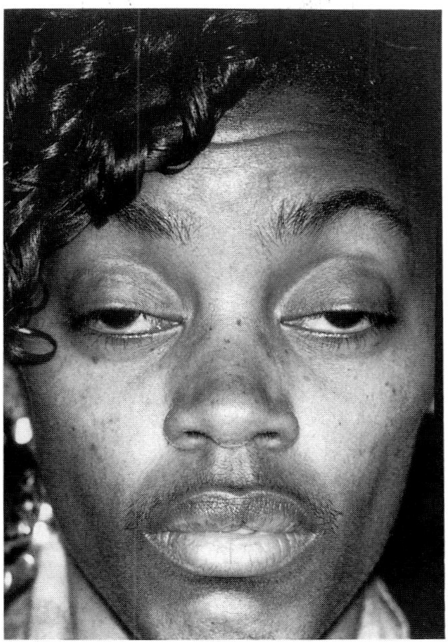

Figure 3 Patient with progressive external ophthalmoplegia and ptosis. Note also the thin habitus, which is commonly present in patients with mitochondrial myopathies and mitochondrial encephalomyopathies.

been sequenced. Mita et al. (84) found that the mtDNA deletions fall into two main classes: class I deletions are flanked by direct sequence repeats, class II deletions are not. The majority of deletions are class I, which suggests that deletions are created by homologous recombination events. The recombination mechanism, however, and the origin of class II deletions have not been identified.

Multiple DNA Deletions

Autosomal-Dominant PEO. Unlike KSS patients who are usually sporadic, families with autosomal-dominant progressive external ophthalmoplegia with mitochondrial myopathy have also been reported (68–70,85). This disorder is described in the nuclear DNA mutation section.

Recurrent Myoglobinuria. There has been only one report of familial multiple mtDNA deletions with recurrent myoglobinuria (86). Two brothers in their twenties began having recurrent myoglobinuria at age 18. The episodes were provoked by strenuous exercise, heavy alcohol intake, or prolonged fasting. RRF and COX-negative fibers were present.

Mitochondrial Neuropathy, Gastrointestinal Encephalomyopathy (MNGIE). This syndrome is described in the nuclear DNA mutation section because it seems to be an autosomal-recessive trait. In our laboratory, four of eight MNGIE patients had multiple mtDNA deletions (87).

MtDNA Point Mutations

Myoclonus Epilepsy with Ragged-Red Fibers (MERRF)

Myoclonus epilepsy (or myoclonic epilepsy) with ragged-red fibers (MERRF) was identified as a clinical syndrome in 1980 by Fukuhara (88). The defining clinical features are: myoclonus, generalized seizures, ataxia, and RRF in the muscle biopsy (88–107). Symptoms usually begin in childhood, but adult onset has been described. Other common clinical manifestations include: impaired hearing, family history of mitochondrial encephalomyopathy (although not always the full MERRF syndrome), dementia, neuropathy, short stature, exercise intolerance, optic atrophy, and lactic acidosis (Table 8). RRF were found in over 90% of the MERRF patients. Pathological studies have shown neuronal loss in the dentate nucleus, inferior olivary nucleus, diffuse gliosis of the cerebellar white matter and of the brainstem, and degeneration of the posterior columns of the spinal cord. Biochemical studies have shown variable respiratory chain defects, but complex IV is most frequently affected. Symptoms and signs typical of KSS, such as pigmentary retinopathy and ophthalmoparesis, are uncommon (Table 8). Lipomas have been reported in MERRF and seem to be unique to this syndrome. There are usually oligosymptomatic and asymptomatic maternal relatives in MERRF pedigrees. Seven individuals had only limb-girdle myopathy.

A point mutation in $tRNA^{Lys}$ has been found in the majority of MERRF patients tested; specifically, there is an adenine-to-guanine transition at position 8344 (99). Silvestri et. al. (107) studied 15 patients with MERRF and found the mutation in 11 individuals. The percent of mutant mtDNA did not correlate with the severity of the disease. A second $tRNA^{Lys}$ mutation at 8356nt (108) was found in a pedigree that was clinically indistinguishable from MERRF pedigrees with the mutation at 8344bp.

Some controversies surround the clinical syndrome of MERRF. First, there are patients with the cardinal features of MERRF who also had a stroke at a young age. These individuals have been considered to have overlap MERRF-MELAS syndrome. Molecular genetics may have resolved this controversy. Zeviani et al. (109) recently reported patients with MERRF/MELAS overlap in a description of the second pedigree with the $tRNA^{Lys}$-8356 mutation. A secondary controversy is whether Ramsay-Hunt syndrome is still a meaningful clinical diagnosis or is now obselete because all individuals with progressive myoclonus epilepsy can be classified as having MERRF, sialidosis, ceroid lipofuscinosis, or some other defined disorder (110–112).

Mitochondrial Encephalomyopathy, Lactic Acidosis, with Stroke-like Episodes (MELAS)

MELAS is clinically characterized by: (1) stroke before age 40 years; (2) encephalopathy characterized by seizures, dementia, or both; and (3) lactic acidosis, RRF, or both (113,114). For the diagnosis to be secure, at least two of the following must be present: normal early development, recurrent headaches,

Table 8 Clinical Features of 62 MERRF Patients

Sign/symptom	Present	Recorded	Percent
Myoclonus	62	62	100
Epilepsy	62	62	100
Normal early development	17	17	100
Ragged-red fibers	47	51	92
Hearing loss	41	45	91
Lactic acidosis	24	29	83
Family history	34	42	81
Exercise intolerance	8	10	80
Dementia	39	52	75
Neuropathy	17	27	63
Short stature	4	7	57
Impaired sensation	9	18	50
Optic nerve atrophy	14	36	39
Cardiomyopathy	2	6	33
Wolff-Parkinson-White syndrome	2	9	22
Pigmentary retinopathy	4	26	15
Pyramidal signs	8	60	13
Ophthalmoparesis	3	28	11
Pes cavus	5	60	8
Multiple lipomas	2	60	3

The following symptoms and signs were reported once each in the 60 patients: migraine headache, hypothyrodism, irregular menses, elevated creatine kinase, extrapyramidal signs, and hepatomegaly.
Diabetes mellitus, hypoparathyroidism, strokes, CSF protein > 100 mg/dl, and heart block were not noted in these patients.

or recurrent vomiting (114). Other frequent manifestations include: exercise intolerance, limb weakness, short stature, and elevated CSF protein. Features of MERRF, myoclonus and cerebellar signs, are sometimes seen, and seizures is one of the clinically defining symptoms in both syndromes; therefore, the major clinical distinction is stroke at a young age. The features of KSS are less often seen in MELAS; none of the 69 MELAS patients (114) fulfilled the clinical criteria for KSS. Analogous to MERRF, MELAS pedigrees often have a history of maternally related affected members.

Computerized tomography and magnetic resonance imaging studies have shown lesions consistent with infarcts, often in the posterior cerebral hemispheres. However, they typically do not conform to the distribution of a major cerebral artery. There have been reports that the lesions fluctuate and sometimes disappear after intervals of 2 weeks–2 months. Angiograms were normal in 18 of 28 cases (114). Abnormalities of major vessels were found in only two cases. In one, there was occlusion of the extracranial portion of the internal carotid artery, which was possibly iatrogenic (113). In the other, there was occlusion of the supraclinoid carotid with the typical radiological appearance of moyamoya (115). These studies have not elucidated the mechanism of stroke in this disorder; however, large vessel occlusion seems unlikely in the majority of MELAS strokes.

Biochemical analysis has shown variable respiratory chain defects, but the majority of cases (29/42) have had a defect of complex I (114). Of 14 autopsy reports (114) of MELAS patients, focal cortical lesions were seen in 10 and were characterized by necrosis, loss of neurons, gliosis, microcyst formation, and capillary proliferation. In two cases, the cortical lesions were confluent. The cerebellum was affected in seven of nine cases and the brainstem in four of six patients. Basal ganglionic calcification was noted in 11 cases with mineral deposits found predominantly in and around blood vessels. Large blood vessels were normal; however, three reports noted abnormal or excessive

Table 9 Clinical Features of 69 Patients with MELAS

Sign/symptom	Present	Recorded	Percent
1. Cardinal manifestation			
Exercise intolerance	31	31	100
Onset < age 40	66	66	100
Stroke	65	66	98
Ragged-red fibers	60	62	97
Seizures	61	65	94
Lactic acidosis	60	66	91
Normal early development	49	54	91
2. Frequent manifestations			
Dementia	44	49	90
Limb weakness	49	56	88
Onset < age 20	55	66	83
Short stature	49	61	80
Hemiparesis	47	59	80
Hemianopsia	35	46	76
Headache	33	45	73
Nausea, vomiting	41	56	73
Hearing loss	34	48	71
Learning disability	28	47	60
CSF protein > 45	16	31	52
3. Other manifestations			
Basal ganglia calcification[a]	24	62	39
Myoclonus	18	61	30
Cerebellar signs	17	63	26
Family history	14	54	26
Episodic coma	9	46	20
Optic atrophy	8	42	19
Congestive heart failure	8	51	16
Wolff-Parkinson-White syndrome	6	43	14
Pigmentary retinopathy	7	61	11
PEO	7	66	11
Cardiac conduction block	3	42	7
CSF protein > 100	1	30	3
Diabetes mellitus	1	36	3
Impaired sensation	1	44	2

[a]By CT scan.
Source: From Hirano M, et al. Neuromusc Disord 1992; 2:125–135.

mitochondria in endothelial and smooth muscle cells in small arteries, arterioles, and capillaries (114,116–118). These blood vessel abnormalities might be contributing to the pathogenesis of stroke.

Between 80 and 90% of MELAS patients have been found to have an adenine-to-guanine point mutation in $tRNA^{Leu(UUR)}$ at 3243 bp (114,119). Two other mutations have been reported in MELAS patients—one also in $tRNA^{Leu(UUR)}$, but at 3271 (120), the other in subunit 4 of complex I (121). Using mtDNA-less cells repopulated with mtDNA containing the $tRNA^{Leu(UUR)}$-3243 mutation, King et al. (122) found decreased mitochondrial protein synthesis in cells with greater than 95% mutant mtDNA. These results support the pathogenic role of the tRNA mutation.

PEO Without mtDNA Deletions

Moraes et al. (72) screened 91 patients with mitochondrial encephalomyopathies who did not fulfill the criteria for MELAS. They found 21 patients with the 3243-bp mutation (PEO^{3243}); 15 had PEO, one had KSS. Unlike the patients with PEO and large-scale mtDNA deletions (PEO^{Δ}), most of the PEO^{3243} patients had an affected relative (8/13) although not always with MELAS or PEO. The COX histochemistry in these patients' muscle had more COX-negative muscle fibers than MELAS-3243 patients' muscle.

Myopathy-Cardiomyopathy

One large pedigree (123) included five clinically affected members with exertional dyspnea, proximal limb weakness, and a hypertrophic cardiomyopathy. The cardiac disorder was severe enough to cause death in one individual at age 30 years. A second patient was awaiting cardiac transplant at age 23. An A-to-G transition mutation in the mtDNA-encoded $tRNA^{Leu(UUR)}$ at nucleotide pair 3260 was found in all of the affected individuals, as well as in 10 symptom-free maternal relatives. All family members with the mutation had RRF, COX-negative fibers, or both on muscle histology. Unlike MERRF, there appeared to be a correlation between the percent mutation in skeletal muscle and clinical expression; the symptomatic patients had 86–90% mutant mtDNA while the asymptomatic individuals had 20–78% mutant DNA.

MtDNA Depletion—Fatal Infantile Mitochondrial Myopathy

In 1983, Boustany et. al. (124) described a floppy infant with ptosis, ophthalmoplegia, and elevated serum lactate. Muscle biopsies revealed RRF and ultrastructurally abnormal mitochondria in skeletal muscle, but not in the smooth muscle of blood vessels, or in fibroblasts. This child died at age 3 months. Postmortem biochemical analysis of isolated muscle mitochondria showed deficiencies of cytochrome b and aa_3, but normal cytochrome spectra in: lung, kidney, heart, brain, spinal cord, and liver (125). COX activity was less than 1% of control in skeletal muscle, but normal in liver. A sister had nearly identical clinical, pathological, and biochemical features and died at age 2½ months of respiratory failure. A second cousin (related through a maternal grandfather) died at age 9 months of hepatic failure.

Moraes et al. (125) studied the postmortem material from the first patient and from two other unrelated infants with similar clinical and histological features and found low levels of mtDNA in muscle. In addition, these infants had renal dysfunction characterized by aminoaciduria, glycosuria, and phosphaturia (DeToni–Fanconi–Debré syndrome). A postmortem kidney sample from one of the infants also revealed mtDNA depletion. Tritschler et. al. (126) found five other infants with a later onset (age 5–12 months), but otherwise similar clinical, biochemical, morphological, and genetic features. The pathogenic mechanism of the mtDNA depletion is unknown.

NUCLEAR DNA MUTATIONS

Yeast *pet* Mutations as a Paradigm

The interactions between mammalian nDNA and mitochondria are poorly understood. In the yeast *Saccharomyces cerevisiae*, a large body of knowledge about the relationship of nDNA with mitochondria has been gathered from the study of nDNA mutations. This yeast is a facultative anaerobe, capable of meeting its energy requirements through fermentation; however, respiratory-defective strains form smaller colonies than their wild-type counterparts when grown on media containing glucose. When the genetic lesion is in a nuclear gene, the strain is called a nuclear *petite*, or *pet* mutant (127). More than 1700 *pet* mutants have been identified (128).

Infantile Myopathies with Cytochrome c Oxidase Deficiency

There are two forms of infantile myopathies with COX deficiency: benign infantile myopathy (BIM) (129) and fatal infantile myopathy (FIM) (130). The fatal form presents as generalized muscle weakness during infancy, respiratory muscle insufficiency, lactic acidosis, and death before age 1 year. Some of these patients have DeToni–Fanconi–Debré syndrome. The occurrence of this fatal myopathy in siblings and in children of consanguineous parents indicates that this disorder is hereditary and is presumably autosomal-recessive. The common biochemical feature of these patients is COX deficiency in skeletal muscle, sparing heart, kidney, liver, and brain (130,131). Histochemical studies have confirmed the decrease in COX activity, while the amounts of immunologically reactive proteins were markedly decreased in muscle extracts and in mitochondria (131,133).

In contrast to these patients with the fatal syndrome, some infants also present with severe myopathy and lactic acidosis soon after birth, but improve dramatically and are virtually normal by age 2 or 3 years (129,134). The serum lactate and muscle biopsy abnormalities also resolve. Because of this striking improvement, the syndrome has been called benign infantile myopathy, even though this disorder can be fatal if the patients are not properly managed in the critical first few months of life. The recurrence of this syndrome in more than one family member indicates that, like the fatal myopathy, it too is hereditary.

Using immunohistochemistry with COX subunit–specific antibodies, Tritschler et al. (135) found absence of the nDNA-encoded COX subunits VIIab in muscle from patients with both types of infantile myopathies and an absence of the mtDNA-encoded COX II only in the benign form.

Autosomal Dominant PEO

In 1989, Zeviani et al. (71) reported multiple deletions of mtDNA in a family with autosomal dominant PEO. The clinical features besides PEO in this pedigree included: age at onset 24–30 years, ptosis, dysphagia, dysphonia, facial and limb weakness, cataracts, exercise intolerance, and death in middle age (43–59 years) (85). Four people died during febrile illnesses causing rapid symptomatic worsening with respiratory muscle weakness and metabolic disarray. Cormier et al. (136) reported a family with multiple deletions in which the proband presented at age 5 years with ketoacidotic comas and repeated attacks of ataxia and drowsiness and later developed ptosis, PEO, and hearing loss. Suomalainen et al. (137) found severe depression and pyschomotor retardation in one member of an autosomal-dominant PEO pedigree with multiple deletions. RRF and cytochrome c oxidase–negative fibers were reported in all patients studied. Muscle biochemical analysis revealed variable or no respiratory chain complex defects.

Zeviani et al. (71) sequenced 19 deletions and found that they started at the D-loop region (3′-end of mtDNA L-strand orientation). The 5′-ends had short regions of homology with the 3′-end (3–12 nucleotides), suggesting a specific recombination event. The genetic defect causing this disorder has not been identified; however, it is probably a *trans*-acting nuclear-encoded factor.

Mitochondrial Neuropathy, Gastrointestinal Encephalomyopathy (MNGIE), or Mitochondrial, Pseudoobstruction, Ophthalmoplegia, and Polyneuropathy (MEPOP)

Another multisystem autosomal-recessive mitochondrial disorder is MNGIE, which is characterized by ophthalmoparesis, peripheral neuropathy, leukoencephalomyopathy, gastrointestinal symptoms (recurrent nausea, vomiting, or diarrhea) with intestinal dysmotility, and histologically abnormal mitochondria in muscle (87) (Table 10). Variable respiratory chain defects have been reported in muscle from these patients. In our laboratory, four of eight MNGIE patients were found to have multiple deletions in mtDNA (87). This disorder is thought to be transmitted in an autosomal-recessive manner; 20 of 25 cases had affected siblings, and three pairs of parents in two kinships were consanguineous (87).

TOXIC STATES

Zidovudine or azidothymidine (AZT) use has been associated with RRF myopathy (138). This nucleoside analog lacks a 3′-OH group and inhibits infectivity of retroviruses by being incorporated into replicating viral DNA and leading to premature termination of the elongating DNA chain (139).

Table 10 Clinical and Laboratory Features of 26 Patients with MNGIE

Age	29.1 years (range 11–51 years)		Other clinical features		
Sex	13 male, 13 female patients		Thin body	25/25	10%
Age at onset	13.9 years (range 2.5–32 years)		Family history	20/25	80%
Onset before 20	16/22	73%	Affected siblings	19/25	76%
First symptom			Consanguineous parents	5/25	20%
Gastrointestinal	16/24	67%	(3 pairs of parents in 2 kinships)		
Ocular	3/24	13%	Short stature	10/14	71%
Gastrointestinal and ocular	1/24	4%	Died (average age 30 years, range 17–51)	13/26	50%
Peripheral neuropathy	2/24	7%			
Hearing loss	1/24	4%	Laboratory features		
Gait disturbance	1/24	4%	Abnormal NCS/EMG	22/22	100%
			Neurogenic	18/22	82%
Gastrointestinal symptoms/signs			Neurogenic and myogenic	3/22	14%
Dysmotility	26/26	100%	Myogenic	1/22	5%
Diarrhea	25/25	100%	Lactic acidosis	7/11	64%
Pseudoobstruction	23/25	92%	Mitochondrial respiratory chain defect	7/9	77%
Diverticulosis	9/14	64%	Complex IV deficiency	2/9	22%
Nausea/vomiting	21/23	91%	Multiple complex defect	3/9	33%
Malabsorption	3/8	38%	Complex I deficiency	2/9	22%

Table 10 *(Continued)*

Neurological manifestations			MRI leukodystrophy	8/9	89%
Ocular	24/26	92%	Ragged-red fibers or increased SDH stain	8/10	80%
External ophthalmoparesis	22/26	85%	Heart block	3/8	38%
Ptosis	17/26	65%	(1 RBBB, 2 incomplete RBBB)		
Pigmentary retinopathy	4/17	24%			
Neuropathy	21/21	100%	Ethnic origin		
Limb weakness	21/22	95%	German	9/22	41%
Distal	7/22	32%	Jewish-Iranian	5/22	23%
Unspecified	6/22	27%	Mexican	4/22	18%
Proximal	4/22	18%	Italian	1/22	5%
Diffuse	4/25	18%	?Turkish	1/22	5%
Hearing loss	11/18	61%	Mexican-American-Indian-Philippino	1/22	5%
Areflexia	8/20	40%	English-Irish-German	1/22	5%
Ataxia	2/15	13%			
Mental retardation	1/11	9%			
Dementia	0/11	0%			

Source: From Hirano M, et al. Neurology 1993; 44: 721–727.

Patients with the RRF myopathy on AZT therapy have a reduction in mtDNA content (140). The drug has been shown to be incorporated into mtDNA by the DNA polymerase gamma in vitro (141).

IDIOPATHIC MITOCHONDRIAL DISEASES

It is ironic that the first disorder to be attributed to mitochondrial dysfunction, Luft disease, has remained pathogenically undefined. There are only two reported cases of Luft disease (9,142,143). Both patients were sporadic and without known consanguinity in either family. Symptoms began in childhood with fever, heat intolerance, profuse diaphoresis, resting tachycardia and tachypnea, polydipsia, and polyphagia. In one of the two patients (143), there was cutaneous erythema on the legs because of dilated capillaries. Basal metabolic rates were elevated without clinical or laboratory signs of hyperthyroidism. Electromyography was consistent with a myopathy. Muscle biopsy revealed many RRF and ultrastructurally enlarged mitochondria with tightly packed cristae and dark osmiophilic inclusions. Measurements of oxidative phosphorylation in isolated mitochondria showed maximal respiratory rates with or without ADP, suggesting that respiratory control was lost. The high rate of respiration probably accounted for the hypermetabolism and hyperthermia. The loss of ADP control of respiration indicated "loose coupling" of oxidation and phosphorylation.

In addition, there are patients with isolated mitochondrial myopathy, characterized by exercise intolerance and proximal limb weakness, who have RRF and variable respiratory chain defects. Some of these patients have oligosymptomatic presentations of a known mitochondrial disorder such as MERRF or MELAS; however, in other cases, the primary cause is unknown.

(NONMITOCHONDRIAL?) DISEASES OR STATES ASSOCIATED WITH MITOCHONDRIAL ABNORMALITIES

There is a rapidly expanding list of conditions and diseases that are associated with mitochondrial abnormalities (Table 4). The significance of these mitochondrial changes remains uncertain, but intriguing. There are several reports of COX-negative skeletal muscle fibers (17), as well as mtDNA deletions in low percentages in muscle and brain of aged individuals (144–147). Similarly, mtDNA deletions have been found in hypertrophic cardiomyopathy (presbycardia) (148,149). In another disorder of the elderly, Parkinson disease, complex I defects have been found in platelets (150), skeletal muscle (151), and brain (152). Further evidence supporting the pathogenic role of complex I in Parkinson disease comes from the meperidine analog MPTP, which causes a Parkinson-like disorder. The active MPTP metabolite MPP^+ has been shown to inhibit complex I activity (153).

THERAPEUTIC CONSIDERATIONS

There are two types of therapy of mitochondrial encephalomyopathies and mitochondrial myopathies. First, there is treatment for many of the symptoms and signs listed in Table 1. Clinical testing for diabetes mellitus, hypoparathyoidism, and cardiac conduction block is important in patients with Kearns-Sayre syndrome. Recognition of heart block is especially important because cardiac pacemaker placement can avert death. Seizures can be ameliorated in MERRF and MELAS with anticonvulsant medications. Blood transfusions in Pearson syndrome might allow patients to survive the severe anemia.

Second, there are treatments of the primary biochemical defects, although these have not yielded dramatic improvement. Multiple medications have been tried with anecdotal improvement in some patients. Coenzyme Q_{10} has been reported to benefit some patients with KSS, MELAS, and MERRF. The major function of coenzyme Q_{10} in mitochondria is to shuttle electrons from complexes I and II to complex III (1). It might also stabilize the respiratory chain complexes within the inner mitochondrial membrane (154,155). Idebenone, another quinone compound, has also been used and has the theoretical advantage of crossing the blood-brain barrier (118). Vitamins have been used to donate electrons directly to cytochrome c and include: phylloquinone (vitamin K_1), and menadione (vitamin K_3), with ascorbic acid (vitamin C) (156). Dichloroacetate (DCA) inhibits pyruvate dehydrogenase (PDH)-specific kinase, which activates the PDH complex; anecdotal reports have suggested improvement of patients with MELAS (157,158). Treatment of mitochondrial myopathies is discussed in depth in another chapter.

Finally, genetic counseling must be mentioned as part of the care of patients with mitochondrial encephalomyopathies. Patients with mtDNA mutations should be educated about maternal inheritance. This is described in greater detail in another chapter.

CONCLUSIONS

Histology, biochemical assays, and molecular biological analyses have added a wealth of knowledge that has deepened our understanding of the human disorders of the mitochondria. Despite all we have learned, many fundamental questions remain. How are mtDNA deletions formed? How do mtDNA point mutations cause cellular dysfunction? How can a single mtDNA mutation cause multiple clinical syndromes (for example, MELAS or PEO with the A-to-G transition in nucleotide 3243 of $tRNA^{Leu(UUR)}$)? The tools presently being used will continue to provide new insights. In addition, new techniques are being developed. King and Attardi (159) developed a human tissue cell line called ρ^0 that can grow without mtDNA. Furthermore, cells with mutant mtDNA can be fused with the ρ^0 cells, thereby establishing cybrids that provide an excellent in vitro system to study the effects of mtDNA mutations on cell function.

Although not completely specific, the delineation of clinical syndromes is useful for several purposes. If the clinical criteria are sufficiently strict, there is a high correlation with specific mutations as demonstrated by KSS, MELAS, and MERRF. Careful clinical characterization can narrow the range of genetic mutations for which to screen. By following the natural history of a syndrome, we can make predictions about the potential outcome of future patients and make more rational therapeutic decisions. For example, KSS patients should be screened regularly for cardiac conduction block, which can be treated with a pacemaker. Finally, by recognizing clinical features that are common to mitochondrial encephalomyopathies, we can identify patients who should be screened further (i.e., by muscle biopsy, mtDNA testing) because they might have a novel clinical phenotype.

ACKNOWLEDGMENTS

This work was supported by Center Grants from the Muscular Dystrophy Association and the National Institute of Neurological Diseases and Stroke (NS 11766 and RR-00645) and by a donation from Libero and Graziella Danesi. MH is supported by NIH Clinical Investigator Development Award 1K08NS01617-01.

REFERENCES

1. Tzagoloff A. Mitochondria. New York: Plenum Press, 1982.
2. Palade G. An electron microscopy study of cytoplasmic structures. J Histochem Cytochem 1953; 1:188–203.
3. Krebs HA, Kornberg HL. Energy transformations in living matter. Ergeb Physiol 1957; 49:212.
4. Bookelman H, Trijbels J, Sengers R, et al. Pyruvate oxidation in rat and human skeletal muscle mitochondria. Biochem Med 1978; 20:395–403.
5. Cooperstain S, Lazarow A. A microspectrophotometric method for the determination of cytochrome oxidase. J Biol Chem 1951; 189:665–670.
6. Ernster L, Kuylenstierna B. Outer membrane of mitochondria. In: Racker E, ed. Membranes of Mitochondria and Chloroplasts. New York: Van Nostrand Reinhold, 1970:172–210.
7. Mitchell P. Coupling of phosphorylation to electron and hydrogen transfer by a chemiosmotic type of mechanism. Nature 1961; 191:144–148.
8. Anderson S, Bankier AT, Barrell BG, et al. Sequence and organization of the human mitochondrial genome. Nature 1981; 290:457–465.

9. Luft R, Ikkos D, Palmieri G, et al. A case of severe hypermetabolism of nonthyroid origin with a defect in the maintenance of mitochondrial respiratory control: a correlated clinical, biochemical, and morphological study. J Clin Invest 1962; 41:1776–1804.
10. Shy GM, Gonatas NK, Perez M. Childhood myopathies with abnormal mitochondria. I. Megaconial myopathy–pleoconial myopathy. Brain 1966; 89:133–158.
11. Engel W, Cunningham G. Rapid examination of muscle tissue: an improved trichrome stain method for fresh frozen biopsy section. Neurology 1963; 19:919–923.
12. Carpenter S, Karpati G, Eisen A. A morphologic study of muscle in polymyositis: clues to pathogenesis of different types. In: Bradley WG, Gardner Medwin D, Walton JN, eds. Recent Advances in Myology. Amsterdam: Excerpta Medica, 1975:374–379.
13. Chew EC, Araoz C, Sun CN, White HJ. Dermatomyositis: ultrastructure of abnormal mitochondria in skeletal muscle. Ann Clin Lab Sci 1977; 7:29–34.
14. Cullen MJ, Fulthorpe JJ. Stages in fiber breakdown in Duchenne muscular dystrophy: an electronmicroscopy study. J Neurol Sci 1975; 24:179–200.
15. Fardeau M. Ultrastructural lesions in progressive muscular dystrophies: a critical study of their specificity. In: Canal N, Scarlato G, Walton JN, eds. Muscle Diseases. Amsterdam: Excerpta Medica, 1970:98–108.
16. Engel AG, Dale AJD. Autophagic glycogenosis of late onset with mitochondrial abnormalities: light and electron microscopy observations. Proc Staff Meet Mayo Clin 1968; 43:233–279.
17. Müller-Höcker J. Cytochrome c oxidase deficient fibres in the limb muscles and diaphragm of man without muscular disease. J Neurol Sci 1990; 100:14–21.
18. Taroni F, Verderio E, Fiorucci S, et al. Molecular characterization of inherited carnitine palmitoyltransferase II deficiency. Proc Natl Acad Sci USA 1992; 89:8429–8433.
19. Holt IJ, Harding AE, Morgan-Hughes JA. Deletions of mitochondrial DNA in patients with mitochondrial myopathies. Nature 1988; 331:717–719.
20. Petty RKH, Harding AE, Morgan-Hughes JA. The clinical features of mytochondrial myopathy. Brain 1986; 109:915–938.
21. Rowland LP, Hays AP, DiMauro S, et al. Diverse clinical disorders associated with morphological abnormalities of mitochondria. In: Cerri C, Scarlato G, eds. Mitochondrial Pathology in Muscle Diseases. Padua: Piccin Editore, 1983:141–158.
22. Rowland LP, Blake DM, Hirano M, et al. Clinical syndromes associate with ragged-red fibers. Rev Neurol (Paris) 1991; 147:467–473.
23. Rowland LP. Mitochondrial encephalomyopathies: lumping, splitting, and melding. In: Schapira AVH, DiMauro S, eds. Mitochondrial Disorders in Neurology. Boston: Butterworth-Heinemann, 1994; 14:116–129.
24. Drachman DA. Ophthalmoplegia plus: a classification of the disorders associated with progressive external ophthalmoplegia. Handbook Clin Neurol 1975; 22:203–216.
25. Rowland LP. Progressive external ophthalmoplegia. Handbook Clin Neurol 1975; 22:177–202.
26. Rowland LP. Progressive external ophthalmoplegia and ocular myopathies. Handbook Clin Neurol 1992; 62:287–329.
27. Kearns TP, Sayre GP. Retinitis pigmentosa, external ophthalmoplegia, and complete heart block. Arch Ophthalmol 1958; 60:280–289.
28. Berenberg RA, Pellock JM, DiMauro S. Lumping or splitting? "ophthalmoplegia plus" or Kearns-Sayre syndrome? Ann Neurol 1977; 1:37–43.
29. Egger J, Lake B, Wilson J. Mitochondrial cytopathy. A multisystem disorder with ragged red fibres on muscle biopsy. Arch Dis Child 1981; 56:741–752.
30. Byrne E, Blumbergs P, Hallpike J, Mukherjee T. Clinical features of mitochondrial myopathy. Aust NZ J Med 1983; 13:353–358.
31. Rheuban K, Ayres N, Sellers T, DiMarco J. Near-fatal Kearns-Sayre syndrome. Clin Pediatr 1983; 22:822–825.
32. Mechler F, Mastaglia F, Serena M, et al. Mitochondrial myopathies. A clinicopathological study of cases with and without extra-ocular muscle involvement. Aust NZ J Med 1986; 16:185–192.
33. Sjö O, Trojaborg W. A multimodality neurophysiological study of children with progressive ophthalmoplegia, pigmentary retinopathy, and mitochondrial myopathy. In: Gallai V, ed. Maturation of the CNS and Evoked Potentials. Amsterdam: Elsevier Science B. V. (Biomedical Division), 1986; 297–306.
34. Case records of the Massachusetts General Hospital. Case 34-1987. N Engl J Med 1987; 317:493–501.
35. Byrne E, Marzuki S, Sattayasai N, et al. Mitochondrial studies in Kearns-Sayre syndrome: normal respiratory chain function with abscence of a mitochondrial translation product. Neurology 1987; 37:1530–1534.
36. Channer K, Channer J, Campbell M, Russell Rees J. Cardiomyopathy in the Kearns-Sayre syndrome. Br Heart J 1988; 59:486–490.
37. Tanabe Y, Miyamoto S, Kinoshita Y, et al. Diabetes mellitus in Kearns-Sayre syndrome. Eur Neurol 1988; 28:34–38.

38. Martens M, Peterson P, Lee C, et al. Kearns-Sayre syndrome: biochemical studies of mitochondrial metabolism. Ann Neurol 1988; 24:630–637.
39. Zeviani M, Moraes CT, DiMauro S, et al. Deletions of mitochondrial DNA in Kearns-Sayre syndrome. Neurology 1988; 38:1339–1346.
40. Rowland LP, Hausmanowa-Petrusewicz I, Bardurska B, et al. Kearns-Sayre syndrome in twins: lethal dominant mutation or acquired disease? Neurology 1988; 38:1399–1402.
41. Desnuelle C, Pellissier J, Serratrice G, et al. Le syndrome de Kearns et Sayre: encéphalomyopathie mitochondriale par déficit de la chaine respiratoire. Rev Neurol (Paris) 1989; 145:842–850.
42. Piccolo G, Aschei M, Ricordi A, et al. Normal insulin receptors in mitochondrial myopathies with ophthalmoplegia. J Neurol Sci 1989; 94:163–172.
43. Holt IJ, Harding AE, Cooper JM, et al. Mitochondrial myopathies: clinical and biochemical features of 30 patients with major deletions of muscle mitochondrial DNA. Ann Neurol 1989; 26:699–708.
44. Gerbitz K-D, Obermaier-Kusser B, Zierz S, et al. Mitochondrial myopathies: divergences of genetic deletions, biochemical defects and the clinical syndromes. J Neurol 1990; 237:5–10.
45. Bordarier C, Duyckaerts C, Robain O, Ponsot G, Laplane D. Kearns-Sayre syndrome. Two clinico-pathological cases. Neuropediatrics 1990; 21:106–109.
46. Shanske S, Moraes CT, Lombes A, et al. Widespread tissue distribution of mitochondrial DNA deletions in Kearns-Sayre syndrome. Neurology 1990; 40:24–28.
47. Goto Y-i, Koga Y, Horai S, Nonaka I. Chronic progressive external ophthalmoplegia: a correlative study of mitochondrial DNA deletions and their phenotypic expression in muscle biopsies. J Neurol Sci 1990; 100:63–69.
48. Tveskov C, Angelo-Nielsen K. Kearns-Sayre syndrome and dilated cardiomyopathy. Neurology 1990; 40:553–554.
49. Eviatar L, Shanske S, Gauthier B, et al. Kearns-Sayre syndrome presenting as renal tubular acidosis. Neurology 1990; 40:1761–1763.
50. Oldfors A, Fyhr I-M, Larsson N-G, Tulinius M. Neuropathology in Kearns-Sayre syndrome. Acta Neupathol 1990; 80:541–546.
51. Truong D, Harding A, Scaravilli F, et al. Movement disorders in mitochondrial myopathies: a study of nine cases with two autopsy studies. Movement Disord 1990; 5:109–117.
52. Goto Y-i, Itami N, Kajii N, et al. Renal tubular involvement mimicking Bartter syndrome in a patient with Kearns-Sayre syndrome. J Pediatr 1990; 116:904–910.
53. Larsson N-G, Holme E, Kristiansson B, et al. Progressive increase of the mutated mitochondrial DNA fraction in Kearns-Sayre syndrome. Pediatr Res 1990; 28:131–136.
54. McKelvie P, Morley J, Byrne E, Marzuki S. Mitochondrial encephalomyopathies: a correlation between neuropathological findings and defects in mitochondrial DNA. J Neurol Sci 1991; 102:51–60.
55. DeGoul F, I N, Lestienne P, et al. Deletions of mitochondrial DNA in Kearns-Sayre syndrome and ocular myopathies: genetic, biochemical and morphological studies. J Neurol Sci 1991; 101:168–177.
56. Yamamoto M, Clemens P, Engel A. Mitochondrial DNA deletions in mitochondrial cytopathies: observations in 19 patients. Neurology 1991; 41:1822–1828.
57. Moraes CT, Zeviani M, Schon EA, et al. Mitochondrial DNA deletion in a girl with manifestations of Kearns-Sayre and Lowe syndromes: an example of phenotypic mimicry. Am J Med Genet 1991; 41:301–305.
58. Tulinius M, Holme E, Kristiansson B, et al. Mitochondrial encephalomyopathies in childhood. I. Biochemical and morphologic investigations. J Pediatr 1991; 119:242–250.
59. Mori K, Narahara K, Ninomiya S, et al. Renal and skin involvement in a patient with complete Kearns-Sayre syndrome. Am J Med Genet 1991; 38:583–587.
60. Zupanc M, Moraes CT, Shanske S, et al. Mitochondrial DNA deletion in patients with combined features of Kearns-Sayre and MELAS syndromes. Ann Neurol 1991; 29:680–683.
61. Larsson N-G, Eiken H, Boman H, et al. Lack of transmission of deleted mtDNA from a woman with Kearns-Sayre syndrome to her child. Am J Hum Genet 1992; 50:360–363.
62. Fischel-Ghodsain N, Bohlman C, Prezant T, et al. Deletion in blood mitochondrial DNA in Kearns-Sayre syndrome. Pediatr Res 1992; 31:557–560.
63. Quade A, Zierz S, Klingmüller D. Endocrine abnormalities in mitochondrial myopathy with external ophthalmoplegia. Clin Invest 1992; 70:396–402.
64. Gross-Jendroska M, Schatz H, MeConald H, Johnson R. Kearns-Sayre syndrome: a case report and review. Eur J Ophthalmol 1992; 2:15–20.
65. McShane MA, Hammans SR, Sweeney M, et al. Pearson syndrome and mitochondrial encephalomyopathy in a patient with a deletion of mtDNA. Am J Hum Genet 1991; 48:39–42.
66. Poulton J, Deadman ME, Gardiner RM. Tandem direct duplications of mitochondrial DNA in mitochondrial myopathy: analysis of nucleotide sequence and tissue distribution. Nucleic Acids Res 1989; 17:10223–10229.
67. Poulton J, Deadman ME, Gardiner RM. Duplications of mitochondrial DNA in mitochondrial myopathy. Lancet 1989; 1:236–240.

68. Schnitzler E, Robertson W. Familial Kearns-Sayre syndrome. Neurology 1979; 29:1172–1174.
69. Leveille A, Newell F. Autosomal dominant Kearns-Sayre syndrome. Ophthalmology 1980; 87:99–108.
70. Jankowicz E, Berger H, Kurasz S, et al. Familial progressive external ophthalmoplegia with abnormal mitochondria. Eur Neurol 1977; 15:318–324.
71. Zeviani M, Servidei S, Gellera C, et al. An autosomal dominant disorder with multiple deletions of mitochondrial DNA starting at the D-loop region. Nature 1989; 339:309–311.
72. Moraes C, Ciacci F, Silvestri G, et al. Atypical clinical presentations associated with the MELAS muation at position 3243 of human mitochondrial DNA. Neuromusc Disord 1993 (in press).
73. Rötig A, Colonna M, Bonnefont JP, et al. Mitochondrial DNA deletion in Pearson's marrow/pancreas syndrome. Genomics 1989; 10:502–504.
74. Cormier, V., A. Rötig, A. Rasore Quartino, G. et al. 1990. Widespread multitissue deletions of the mitochondrial genome in the Pearson marrow-pancreas syndrome. J Pediatr 117:599–602.
75. Moraes CT, DiMauro S, Zeviani M, et al. Mitochondrial DNA deletions in progressive external ophthalmoplegia and Kearns-Sayre syndrome. N Engl J Med 1989; 320:1293–1299.
76. Pearson H, Lobel J, Kocoshis S, et al. A new syndrome of refractory sideroblastic anemia with vacuolization of marrow precursors and exocrine pancreatic dysfunction. J Pediatr 1979; 95:976–984.
77. Stoddard R, McCurin D, Shultenover S, et al. Syndrome of refractory sideroblastic anemia with vacuolization of marrow precursors and exocrine pancreatic dysfunction presenting in the neonate. J Pediatr 1981; 99:259–261.
78. Ballinger S, Shoffner J, Hedaya E, et al. Maternally transmitted diabetes and deafness associated with a 10.4 kb mitochondrial DNA deletion. Nature Genet 1992; 1:11–15.
79. Johns DR, Rutledge SL, Stine OC, Hurko O. Directly repeated sequences associated with pathogenic mitochondrial DNA deletions. Proc Natl Acad Sci USA 1989; 86:8059–8062.
80. Shoffner JM, Lott MT, Voljavec AS, et al. Spontaneous Kearns-Sayre/chronic external ophthalmoplegia plus syndrome associated with a mitochondrial DNA deletion: a slip-replication model and metabolic therapy. Proc Natl Acad Sci USA 1989; 86:7952–7956.
81. Nelson I, Degoul F, Overmaier-Kusser B, et al. Mapping of heteroplasmic mitochondrial DNA deletions in Kearns-Sayre syndrome. Nucleic Acids Res 1989; 17:8117–8124.
82. Holt IJ, Harding AE, Morgan-Hughes JA. Deletions of muscle mitochondrial DNA in mitochondrial myopathies: sequence analysis and possible mechanisms. Nucleic Acids Res 1989; 17:4465–4469.
83. Schon EA, Rizzuto R, Moraes CT, et al. A direct repeat is a hotspot for large-scale deletions of human mitochondrial DNA. Science 1989; 244:346–349.
84. Mita S, Rizzuto R, Moraes CT, et al. Recombination via flanking direct repeats is a major cause of large-scale deletions of human mitochondrial DNA. Nucleic Acids Res 1990; 18:561–567.
85. Servidei S, Zeviani M, Manfredi G, et al. Dominantly inherited mitochondrial myopathy with multiple deletions of mitochondrial DNA: clinical, morphologic, and biochemical studies. Neurology 1991;41:1053–1059.
86. Ohno K, Tanaka M, Sahashi K, et al. Mitochondrial DNA deletions in inherited recurrent myoglobinuria. Ann Neurol 1991; 29:364–369.
87. Hirano M, Silvestri G, Blake DM, et al. Mitochondrial neuropathy, gastrointestinal encephalomyopathy (MNGIE): clinical, biochemical, and genetic features of an autosomal recessive mitochondrial disorder. Neurology 1994; 44:721–727.
88. Fukuhara N, Tokigushi S, Shirakawa K, Tsubaki T. Myoclonus epilepsy associated with ragged-red fibers (mitochondrial abnormalities): disease entity or syndrome? Light and electron microscopic studies of two cases and review of the literature. J Neurol Sci 1980; 47:117–133.
89. Tsaris P, Engel W, Park P. Familial myoclonic epilepsy syndrome associated with skeletal muscle abnormalities. Neurology 1973; 23:408.
90. Fitzsimons RB, Clifton-Bligh P, Wolfenden WH. Mitochondrial myopathy and lactic acidemia with myoclonic epilepsy, ataxia and hypothalamic infertility: a variant of Ramsay-Hunt syndrome? J Neurol Neurosurg Psychiatry 1981; 44:79–82.
91. Morgan-Hughes JA, Hayes DJ, Clark JB, et al. Mitochondrial encephalomyopathies: biochemical studies in two cases revealing defects in the respiratory chain. Brain 1982; 105:553–582.
92. Nakano T, Sakai H, Amano N, et al. An autopsy case of degenerative type myoclonus epilepsy associated with Friedreich's ataxia and mitochondrial myopathy. Brain Nerve 1982; 34:321–332.
93. Roger J, Pellissier J, Dravet C, et al. Dégénérescence spino-cérébelluese—atrophie optique épilepsie-myoclonies-myopathie mitochondriale. Rev Neurol 1982; 138:187–200.
94. Sasaki H, Kuzuhara S, Kanazawa I, et al. Myoclonus, cerebellar disorder, neuropathy, mitochondrial myopathy, and ACTH deficiency. Neurology 1983; 33:1288–1293.
95. Rosing HS, Hopkins LC, Wallace DC, et al. Maternally inherited mitochondrial myopathy and myoclonic epilepsy. Ann Neurol 1985; 17:228–237.
96. Garcia Silva M, Aicardi J, Goutiéres F, Chevrie J. The syndrome of myoclonic epilepsy with ragged-red fibers: report of a case and review of the literature. Neuropediatrics 1987; 18:200–204.

97. Federico A, Cornelio F, DiDonato S, et al. Mitochondrial encephalo-neuro-myopathy with myoclonus epilepsy, basal nuclei calcifications and hyperlactacidemia. Ital J Neurol Sci 1988; 9:65–71.
98. Berkovic SF, Carpenter S, Evans A, et al. Myoclonus epilepsy and ragged-red fibres (MERRF): a clinical, pathological, biochemical, magnetic resonance spectrographic and positron emission tomographic study. Brain 1989; 112:1231–1260.
99. Shoffner J, Lott M, Lezza A, et al. Myoclonic epilepsy and ragged-red fiber disease (MERRF) is associated with a mitochondrial DNA $tRNA^{Lys}$ mutation. Cell 1990; 61:931–937.
100. Yoneda M, Tanno Y, Horai S, et al. A common mitochondrial DNA mutation in the $tRNA^{Lys}$ of patients with myoclonus epilepsy associated with ragged-red fibers. Biochem Int 1990; 21:789–796.
101. Zeviani M, Amati P, Bresolin N, et al. Rapid detection of the A->G(8344) mutation of mtDNA in Italian families with myoclonus epilepsy and ragged-red fibers (MERRF). Am J Hum Genet 1991; 48:203–211.
102. Bindoff L, Desnuelle C, Birch-Machin M, et al. Multiple defects of the mitochondrial respiratory chain in a mitochondrial encephalopathy (MERRF): a clinical, biochemical and molecular study. J Neurol Sci 1991; 102:17–24.
103. Shih K-D, Yen T-C, Pang C-Y, Wei Y-H. Mitochondrial DNA mutation in a Chinese family with myoclonic epilepsy and ragged-red fiber disease. Biochem Biophys Res Commun 1991; 174:1109–1116.
104. Noer A, Sudoyo H, Lertrit P, et al. A $tRNA^{Lys}$ mutation in the mtDNA is the causal genetic lesion underlying myoclonic epilepsy and ragged-red fiber (MERRF) syndrome. Am J Hum Genet 1991; 49:715–722.
105. Seibel P, Degoul F, Romero N, et al. Identification of point mutations by mispairing PCR as exemplified in MERRF disease. Biochem Biophys Res Commun 1990; 173:561–565.
106. Tanno Y, Yoneda M, Nonaka I, et al. Quantitation of mitochondrial DNA carrying $tRNA^{Lys}$ mutation in MERRF patients. Biochem Biophys Res Commun 1991; 179:880–885.
107. Silvestri G, Ciafaloni E, Santorelli F, et al. Clinical features associated with the A > G transition at nucleotide 8344 of mtDNA ("MERRF" mutation). Neurology 1993; 43:1200–1206.
108. Silvestri G, Moraes C, Shanske S, et al. A new mtDNA mutation in the $tRNA^{Lys}$ gene associated with myoclonic epilepsy and ragged-red fibers (MERRF). Am J Hum Genet 1992; 51:1213–1217.
109. Zeviani M, Muntoni F, Savarese N, et al. A MERRF/MELAS overlap syndrome associated with a new point mutation in the mitochondrial DNA $tRNA^{Lys}$ gene. Eur J Hum Genet 1992; 1:80–87.
110. Marsden C, Obeso J. Viewpoints on the Ramsay Hunt syndrome. 1. The Ramsay Hunt syndrome is a useful clinical entity. Movement Disord 1989; 4:6–12.
111. Andermann F, Berkovic S, Carpenter S, Andermann E. Viewpoints on the Ramsay Hunt syndrome. 2. The Ramsay Hunt syndrome is no longer a useful diagnostic category. Movement Disord 1989; 4:13–17.
112. Harding A. Ramsay Hunt syndrome, Unverricht-Lundborg disease, or what? Movement Disord 1989; 4:18–19.
113. Pavlakis SG, Phillips PC, DiMauro S, et al. Mitochondrial myopathy, encephalopathy, lactic acidosis, and stroke-like episodes: a distinctive clinical syndrome. Ann Neurol 1984; 16:481–487.
114. Hirano M, Ricci E, Koenigsberger R. MELAS: an original case and clinical criteria for diagnosis. Neuromusc Disord 1992; 2:125–135.
115. Kotagal S, Peterson, PL Martens, ME, et al. Impaired NADH-CoQ reductase activity in a child with moyamoya syndrome. Pediatr Neurol 1988; 4:241–244.
116. Ohama E, Ohara S, Ikuta F, et al. Mitochondrial angiopathy in cerebral blood vessels of mitochondrial encephalomyopathy. Acta Neuropathol (Berl) 1987; 74:226–233.
117. Kishi M, Yamamura Y, Kurihara T, et al. An autopsy case of mitochondrial encephalomyopathy: biochemical and eletron microscopic studies of the brain. J Neurol Sci 1988; 86:31–40.
118. Ihara Y, Namba R, Kuroda S, et al. Mitochondrial encephalomyopathy (MELAS): pathological study and successful therapy with coenzyme Q_{10} and idebenone. J Neurol Sci 1989; 90:263–271.
119. Goto Y-I, Nonaka I, Horai S. A mutation in the $tRNA^{Leu(UUR)}$ gene associated with the MELAS subgroup of mitochondrial encephalomyopathies. Nature 1990; 348:651–653.
120. Goto Y-I, Nonaka I, Horai S. A new mtDNA mutation associated with mitochondrial myopathy, encephalopathy, lactic acidosis and stroke-like episodes (MELAS). Biochim Biophys Acta 1991; 1097:238–240.
121. Lertrit P, Noer A, Jean-Francois M, et al. A new disease-related mutation for mitochondrial encephalopathy, lactic acidosis, and strokelike episodes (MELAS) syndrome affects the ND4 subunit of the respiratory complex I. Am J Hum Genet 1992; 51:457–468.
122. King MP, Koga Y, Davidson M, Schon EA. Defects in mitochondrial protein synthesis and respiratory chain activity segregate with the $tRNA^{Leu(UUR)}$ mutation associated with mitochondrial myopathy, encephalopathy, lactic acidosis, and strokelike episodes. Mol Cell Biol 1992; 12:480–490.
123. Zeviani M, Gellera C, Antozzi C, et al. Maternally inherited myopathy and cardiomyopathy: association with mutation in mitochondrial DNA $tRNA^{Leu(UUR)}$. Lancet 1991; 338:143–147.
124. Boustany R, Aprille J, Halperin J, et al. Mitochondrial cytochrome deficiency presenting as a myopathy with hypotonia, external ophthalmoplegia, and lactic acidosis in an infant and as fatal hepatopathy in a second cousin. Ann Neurol 1983; 14:462–470.

125. Moraes CT, Shanske S, Tritschler H-J, et al. Mitochondrial DNA depletion with variable tissue expression: a novel genetic abnormality in mitochondrial diseases. Am J Hum Genet 1991; 48:492–501.
126. Tritschler H-J, Andreeta F, Moraes CT, et al. Mitochondrial myopathy of childhood associated with depletion of mitochondrial DNA. Neurology 1992; 42:209–217.
127. Tzagoloff A, Dieckmann C. *PET* genes of *Saccharomyces cerevisiae.* Microbiol Rev 1990; 54:211–225.
128. Myers A, Pape L, Tzagoloff A. Mitochondrial protein synthesis is required for maintenance of intact mitochondrial geneomes in *Saccharomyces cerevisiae.* EMBO J 1985; 4:2087–2092.
129. DiMauro S, Nicholson JF, Hays AP, et al. Benign infantile mitochondrial myopathy due to reversible cytochrome c oxidase deficiency. Ann Neurol 1983; 14:226–234.
130. DiMauro S, Hirano M, Bonilla E et al. Cytochrome oxidase deficiency: progress and problems. In: Schapira AVH, DiMauro S. eds. Mitochondrial Disorders in Neurology. Boston: Butterworth-Heinemann, 1994; 14:91–115.
131. Bresolin N, Zeviani M, Bonilla E, et al. Fatal infantile cytochrome c oxidase deficiency: decrease of immunologically detectable enzyme in muscle. Neurology 1985; 35:802–812.
132. Takayanagi T, Inoue M, Tomimasu K, et al. Infantile cytochrome c oxidase deficiency with neonatal death. Pediatr Neurol 1989; 5:179–181.
133. Zeviani M, Nonaka I, Bonilla E, et al. Fatal infantile mitochondrial myopathy and renal dysfunction due to cytochrome c oxidase deficiency: immunological studies in a new patient. Ann Neurol 1985; 17:414–417.
134. Zeviani M, Peterson P, Servidei S, et al. Benign reversible muscle cytochrome c oxidase deficiency: a second case. Neurology 1987; 37:64–67.
135. Tritschler HJ, Bonilla E, Lombes A, et al. Differential diagnosis of fatal and benign cytochrome c oxidase deficient myopathies of infancy: an immunohistochemical approach. Neurology 1991; 41:300–305.
136. Cormier V, Rotig A, Tardieu M, et al. Autosomal dominant deletions of the mitochondrial genome in a case of progressive encephalomyopathy. Am J Hum Genet 1991; 48:643–648.
137. Suomalainen A, Majander A, Haltia M, et al. Multiple deletions of mitochondrial DNA in several tissues of a patient with severe retarded depression and familial progressive external ophthalmoplegia. J Clin Invest 1992; 90:61–66.
138. Dalakas M, Illa I, Pezeshkpour GH, et al. Mitochondrial myopathy caused by long-term AZT (Zidovudine) therapy: management and differences from HIV-associated myopathy. N Engl J Med 1990; 322:1098–1105.
139. Byars N, Kidson C. Chain termination by 2′,3′-dideoxythymidine in replicating mammalian cells. Biochemistry 1975; 14:3159–3164.
140. Arnaudo E, Dalakas M, Shanske S, et al. Depletion of mitochondrial DNA in AIDS patients with zidovudine-induced myopathy. Lancet 1991; 337:508–510.
141. Simpson MV, Chin CD, Keilbaugh SA, et al. Studies on the inhibition of mitochondrial DNA replication by 3′ azido-3′-deoxythimidine and other dideoxinucleoside analogs which inhibit HIV-1 replication. Biochem Pharm 1989; 38:1033–1036.
142. DiMauro S, Bonilla E, Lee CP, et al. Luft's disease: further biochemical and ultrastructural studies of skeletal muscle in the second case. J Neurol Sci 1976; 27:217–232.
143. Haydar NA, Conn HL, Afif A, et al. Severe hypermetabolism with primary abnormality of skeletal muscle mitochondria. Ann Intern Med 1971; 74:548–558.
144. Linnane AW, Baumer A, Maxwell RJ, et al. Mitochondrial gene mutation: the ageing process and degenerative diseases. Biochem Int 1990; 22:1067–1076.
145. Soong NW, Hinton DR, Cortopassi G, Arnheim N. Mosaicism for a specific somatic mitochondrial DNA mutation in adult human brain. Nature Genet 1992; 2:318–323.
146. Corral-Debrinski M, Horton T, Lott MT, et al. Mitochondrial DNA deletions in human brain: regional variability and increase with advanced age. Nature Genet 1992; 2:324–329.
147. Simonetti S, Chen X, DiMauro S, Schon EA. Accumulation of deletions in human mitochondrial DNA during normal aging: analysis by quantitative PCR. Biochim Biophys Acta 1992; 1180:113–122.
148. Muller-Hocker J. Cytochrome-c-oxidase deficient cardiomyocytes in the human heart—an age-related phenomenon: a histochemical ultracytochemical study. Am J Pathol 1989; 134:1167–1173.
149. Hattori K, Tanaka M, Sugiyama S, et al. Age-dependent increase in deleted mitochondrial DNA in the human heart: possible contributory factor to presbycardia. Am Heart J 1991; 121:1735–1742.
150. Parker W, Boyson S, Parks J. Abnormalities of the electron transport chain in idiopathic Parkinson's disease. Ann Neurol 1989; 26:719–723.
151. Bindoff L, Birch-Machin M, Cartlidge N, et al. Mitochondrial function in Parkinson's disease. Lancet 1989; 2:49.
152. Cooper J, Mann V, Krieg D, Schapira A. Human mitochondrial complex I dysfunction. Biochim Biophys Acta 1992; 1101:198–203.
153. Nicklas W, Vyas I, Heikkila R. Inhibition of NADH-linked oxidation in brain mitochondria by MPP^+, a metabolite of the neurotoxin MPTP. Life Sci 1985; 36:2503–2508.

154. Lenaz G, Espoti M. Physical properties of ubiquinones in model systems and membranes. In: Lenaz G, ed. Coenzyme Q. New York: Wiley, 1985:83–105.
155. Lenaz G, De Santis A, Bertoli E. A survey of the function and specificity of ubiquinone in the mitochondrial respiratory chain. In: G Lenaz, ed. Coenzyme Q. New York: Wiley, 1985:165–199.
156. Shoffner JM, Wallace DC. Oxidative phosphorylation diseases: disorders of two genomes. Adv Hum Genet 1990; 19:267–330.
157. DeVivo DC, Jackson A, Wade C, et al. Dichloroacetate treatment of MELAS-associated lactic acidosis. Ann Neurol 1990; 28:437 (abstract).
158. Saito T, Naito E, Ito M, Takeda E, Hashimoto T, Kuroda Y. Therapeutic effects of sodium dichloroacetate on visual and auditory hallucinations in a patient with MELAS. Neuropediatrics 1991; 22:166–167.
159. King MP, Attardi G. Human cells lacking mtDNA: repopulation with exogenous mitochondria by complementation. Science 1989; 246:500–503.
160. Buruma O, Schipperheyn J. Periodic paralysis. Handbook Clin Neurol 1979; 41:147–174.
161. Guarino M, Tarateta A, DiStefano A. Mitochondrial cytopathy in familial periodic hypokalemic paralysis. Lancet 1984; 2:49.
162. Bardosi A, Scheidt P, Goebel H. Mitochondrial myopathy: a result of clofibrate/etiofibrate treatment? Acta Neuropathol (Berl) 1985; 68:164–168.
163. Julien J, Vital C, Vallat J, et al. Oculopharyngeal muscular dystrophy: a case with abnormal mitochondrial and "fingerprint" inclusions. J Neurol Sci 1974; 21:165–169.
164. Hudgson P, Bradley W, Jenkinson M. Familial mitochondrial myopathy. Part I. Clinical, electrophysiological and pathological findings. J Neurol Sci 1974; 16:343–370.

39
Genetics of Mitochondrial Myopathies

S. R. Hammans
Wessex Neurological Centre, Southampton General Hospital, Southampton, England

A. E. Harding[+]
Institute of Neurology, London, England

THE MITOCHONDRION

The mitochondrion is an intracellular organelle with the vital function of generating ATP, the energy currency of the cell. This is accomplished by oxidative phosphorylation, in which electron transfer along the mitochondrial respiratory chain is coupled to the formation of ATP. The respiratory chain comprises five enzyme complexes, assembled from polypeptide subunits (Table 1). The majority of subunits are encoded by the 3×10^9 base pair (bp) nuclear genome, translated in the cytoplasm and imported into mitochondria. The 13 remaining subunits are encoded by mitochondrial DNA, a 16,569-bp circular genome present in several hundred copies per cell with 2–10 copies per mitochondrion (Fig. 1). This genome is distinguished from its larger nuclear counterpart by a different genetic code and extremely compact organization. Mitochondrial (mt) DNA encodes its own translation apparatus, comprising two ribosomal (r) RNAs and 22 transfer (t) RNAs. mtDNA replication and transcription is largely under the control of nuclear encoded factors.

mtDNA has been shown to be almost exclusively maternally inherited in mammals. The spermatozoon contains a few mitochondria located in the midpiece behind the head, with about 50 mtDNA molecules in total (1). In contrast, the ovum contains much larger numbers of mitochondria. Paternal mtDNA molecules have been detected in backcrossed mice at a frequency of 10^{-4} relative to the maternal contributions (2). However, in the context of human disease, mtDNA can be regarded as being exclusively maternally inherited.

MYOPATHOLOGY

The term "mitochondrial myopathy" (MM) is applied to a clinically, biochemically, and genetically heterogeneous group of diseases that usually show mitochondrial structural abnormalities in skeletal muscle. The morphological hallmark of mitochondrial myopathies is the ragged red fiber (RRF), which contains large subsarcolemmal and intermyofibrillar collections of mitochondria. These may be visualized in fresh frozen muscle sections stained by the modified Gomori trichrome (mGT) technique; mitochondrial aggregates also show an intense stain for the exclusively intramitochondrial enzyme succinic dehydrogenase (SDH). Other fibers may show abnormal SDH or cytochrome oxidase activity but do not have a ragged appearance. The ragged red and histochemical changes usually affect segments rather than the entire length of individual muscle fibers.

[+]Deceased.

Table 1 Respiratory Chain Polypeptides

Complex		Number of subunits	Number encoded by mtDNA
I	NADH ubiquinone reductase (ND)	Up to 41	7 (ND1, ND2, ND3, ND4, ND4L, ND5, ND6)
II	Succinate dehydrogenase	4	0
III	Ubiquinone-cytochrome *c* reductase	11	1 (cytochrome b)
IV	Cytochrome *c* oxidase (COX)	13	3 (COX I, COX II, COX III)
V	ATP synthase (ATPase)	14	2 (subunits, 6, 8)

Although the diagnostic importance of muscle biopsy is well established, the definition of mitochondrial myopathy by morphological criteria alone is complicated by the fact that small numbers of RRF and COX-negative fibers are occasionally observed in the elderly and in other muscle diseases such as muscular dystrophies, polymyositis, and glycogen storage diseases (3,4). Conversely, the diagnosis of mitochondrial myopathy has been made in the absence of RRF but on the basis of a suggestive clinical course supported by other histochemical abnormalities (5), a family history (6,7), a biochemical defect (7), or mitochondrial DNA mutations (8).

GENOTYPE AND PHENOTYPE

Largely using histological and histochemical diagnostic criteria, specifically the presence of RRF, clinically diverse syndromes became "lumped" under the rubric of mitochondrial myopathy. Muscle

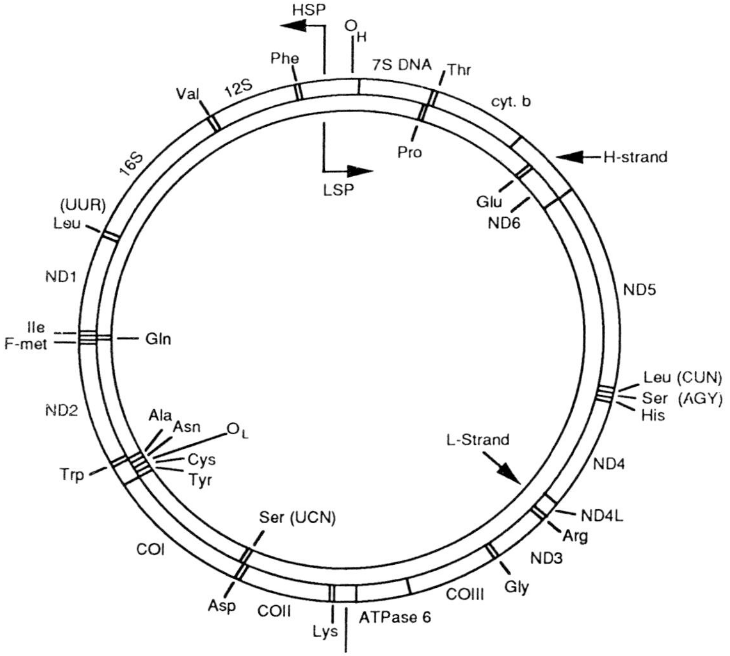

Figure 1 The mitochondrial genome. H-strand and L-strand, heavy and light strands, respectively; O_H and O_L, origins of heavy and light strand replication; HSP and LSP, heavy and light strand promoters; ND, NADH dehydrogenase; CO, cytochrome oxidase. Transfer RNAs are indicated by short transverse lines and relevant amino acid abbreviation; 16S and 12S are ribosomal RNAs.

Table 2 Mitochondrial Myopathies/Encephalopathies: Summary of Clinical Classification

Onset in infancy or early childhood
Combinations of:
Myopathy
Lactic acidosis
Renal tubulopathy
Hepatopathy
Cardiomyopathy
Encephalopathy
Pearson's marrow–pancreas syndrome (may evolve into Kearns-Sayre syndrome)
Leigh's syndrome[a] (some cases)
Onset in childhood or adulthood
Proximal limb weakness—enhanced by exercise
Progressive external ophthalmoplegia
if onset < 20 years and retinopathy is present with one of ataxia, heart block, or ↑ CSF/ protein meets criteria for Kearns-Sayre syndrome (KSS).
Encephalopathies
MERRF (myoclonic epilepsy and RRF)
MELAS (mitochondrial encephalomyopathy, lactic acidosis, strokelike episodes)
MNGIE (myo-, neuro-, gastrointestinal encephalopathy) (63,64)
Others (e.g., ataxia, deafness, dementia, seizures)
Diabetes and deafness
Neurogenic muscle weakness, ataxia, retinitis pigmentosa (NARP)[a]
Multiple symmetrical lipomatosis

[a]RRF usually absent.

and the central nervous system were the most frequently affected organs, although involvement of the heart, endocrine system, kidney, gut, and hemopoietic tissues also occurs (9,10). While clinical classification remains controversial (11), clustering of clinical features within mitochondrial myopathies allows a useful basic clinical taxonomy (Table 2).

In 1988, an association between human disease and mtDNA was established when Holt et al. described large-scale deletions of mtDNA in some patients with mitochondrial myopathy. Since then, molecular genetic studies have further increased the understanding of the etiology and pathogenesis of these disorders. These advances have implications for clinical practice in investigation, diagnosis, genetic counseling, and classification of MMs. The remainder of this chapter will follow the genetic classification outlined in Table 3.

Primary mtDNA Mutations

Single Heteroplasmic Deletions

Single heteroplasmic deletions are observed in approximately 40% of patients with MM. Almost all patients with this genetic defect and neurological disease have progressive external ophthalmoplegia (PEO) with or without other features, with some satisfying the criteria for Kearns-Sayre syndrome (KSS; Table 2). Conversely, about 80% of patients with KSS and about 60% of patients with PEO but not KSS have deletions. In muscle, two populations of mtDNA are present (heteroplasmy), one apparently normal and the other with a large deletion of mtDNA. The proportion of abnormal mtDNA varies between patients, typically between 20 and 80%. Deletions are present in much lower proportions in blood, only rarely being detectable by Southern blot. The deletion may be from 2 to 9 kb in size and does not involve either of the two origins of replication. The site and size of the deletion within the

Table 3 Summary of MTDNA Defects Associated with Human Disease

MtDNA defect	Phenotype	Tissue distribution of mtDNA defect	Inheritance	RRF?	Ref.
Primary large-scale mtDNA rearrangements					
Heteroplasmic, large, single deletions	PEO + other manifestations	Present in muscle, usually not detectable in blood by Southern blot techniques	Usually sporadic	Usually	19,65–71
	Pearson's syndrome	a	Sporadic	No	13–15,72, 73
Tandem, heteroplasmic duplications	KSS + other manifestations	a	Sporadic	Yes	29
Heteroplasmic, large deletion involing O_L (? duplication)	Diabetes and deafness	a	Matrilineal	No	12
Heteroplasmic point mutations					
$tRNA^{Leu(UUR)}$ $A \rightarrow G^{(3243)}$	MELAS and others (see text)	a	Matrilineal	Usually	8,30–32,74
$tRNA^{Leu}$ $T \rightarrow C^{(3271)}$	MELAS	a	Matrilineal	Usually	33
$tRNA^{Lys}$ $A \rightarrow G^{(8344)}$	MERRF and other features	a	Matrilineal	Usually	8,40–42,75
$tRNA^{Lys}$ $A \rightarrow G^{(8356)}$	MERRF	a	Matrilineal	Yes	76,77
$tRNA^{Leu(UUR)}$ $A \rightarrow G^{(3251)}$	PEO, early death	a	Matrilineal	Yes	78
$tRNA^{Leu(UUR)}$ $A \rightarrow G^{(3260)}$	Cardiomyopathy and myopathy	a	Matrilineal	Yes	79,80
$tRNA^{Leu(UUR)}$ $A \rightarrow G^{(3250)}$	Myopathy	a	Matrilineal	Yes	81
ND4 $A \rightarrow G^{(11084)}$	MELAS	a	Matrilineal	Yes	82
$tRNA^{Pro}$ $G \rightarrow A^{(15990)}$	Myopathy	Muscle	Not known	Yes	83
ND4 $G \rightarrow A^{(11778)}$ ND1 $G \rightarrow A^{(3460)}$	Leber's hereditary optic neuropathy	Present in blood and muscle	Matrilineal	No	84–88

ND6 $T \rightarrow C^{(14484)}$					
and others					
$T \rightarrow G^{(8993)}$	Neurogenic muscle weakness, ataxia, and retinitis pigmentosa	Present in blood and muscle	Matrilineal	No	89–91
Nuclear gene defects					
Large multiple deletions	PEO + other manifestations	Present in muscle, not detectable in blood by Southern blot	Autosomal dominant	Yes	45–47
	KSS	Present in muscle, blood not examined	Autosomal dominant	Yes	92
	Recurrent ketoacidosis, encephalomyopathy	[a]	?	Yes	51
	Ptosis, optic arophy, peripheral neuropathy	Present in muscle, blood not examined	Two brothers. ? autosomal recessive	Yes	48
	Recurrent myoglobinuria	Present in muscle, not detectable in blood by PCR	Two brothers. ? autosomal recessive	Yes	52
	Dilated cardiomyopathy	Present in muscle, blood not examined	Mother and son	No	50
MtDNA depletion	Infantile onset myopathy, cardiomyopathy, renal or hepatic failure	MtDNA depletion found in affected organs	? Autosomal recessive	Yes	54,55

Low abundance deletions demonstrated in Parkinson's disease(93), cardiomyopathy(94), and normal aging(95) are excluded from this summary.
[a]Detectable in blood in lower proportion than muscle.

mitochondrial genome do seem to have a limited correlation with the biochemical phenotype (see below), but do not appear to influence the clinical phenotype. A single family with a larger deletion (12) has many atypical features and is discussed separately. Pearson's syndrome presents in infancy with refractory sideroblastic anameia and exocrine pancreatic failure. MtDNA deletions are found in both muscle and blood. If the infant survives, then the features of KSS may evolve, suggesting that this syndrome may be an age-dependent manifestation of the same underlying defect (13–15).

In general, single deletions are sporadic. No family has been described with two symptomatic members with the same deletion. However, a symptomatic mother and daughter had deletions that differed in site and size (16). Poulton and colleagues found that in one family deletions were detected at low level by the polymerase chain reaction in blood of the asymptomatic mother and sister of a patient (17); these findings differ from another study of the same family by Brockington et al. (18).

A high proportion of deleted mtDNAs (72%) was observed in the brain of one patient with KSS, and 55% of liver mtDNA was also deleted, as compared with 85% in muscle (19). Other studies have also shown widespread tissue distribution of deleted mtDNA in KSS (20,21). No similar studies have been performed in patients with clinical features confined to PEO without central nervous system involvement. The uneven distribution of wild-type and deleted mtDNA may arise from random segregation of mtDNA and mitochondria in the oocyte or zygote, but more probably from the influence of selection pressures on mtDNA populations, which vary between tissues and cell types.

Approximately 130 patients with mtDNA deletions have been described to date. About one-third of deletions are identical, being flanked by a 13-bp direct repeat existing at bp 8470–8482 and bp 13,447–13,459, and designated the "common deletion." The high incidence of this deletion may arise from the length of the direct repeat. In a study of 28 patients with 17 different mtDNA deletions (22), 12 had the common deletion and eight had deletions flanked by other direct repeats 5–11 bp in length. The remaining eight were flanked by imperfect repeats or by nonhomologous sequences, suggesting that pathogenetic mechanisms may be heterogeneous. Models involving slipped mispairing or illegitimate elongation have been proposed. A study by Brockington et al. (18) demonstrated that many patients with deletions also have evidence of a tandem duplication in the D-loop region of mtDNA in a small proportion of molecules. Since this region is involved in the initiation of mtDNA replication, it is possible that the duplication may promote rearrangements of mtDNA, including large deletions. Alternatively, duplications and deletions may both be the result of an inherent tendency toward replicative infidelity brought about by another factor, presumably one of the nuclear encoded proteins involved in mtDNA replication.

In situ hybridization studies of muscle of patients with mtDNA deletions (23–25) have shown that deleted mtDNA appears to be confined to histochemically abnormal fiber segments and is transcribed, while normal mtDNA is depleted within these segments (Fig. 2). Mitochondria within the aggregations observed in RRF mostly contain only deleted mtDNA (25). Deletions of mtDNA create deficiencies of mRNA species derived from involved protein subunit reading frames, but all known mtDNA deletions also involve tRNA genes. Deletions would thus be expected to cause a translational defect involving all mtDNA-encoded subunits. Although the distribution of different mtDNA species between mitochondria is unclear, coexistence of deleted and normal mtDNA in the same mitochondrion could potentially ameliorate the translational defect, if normal mtDNA supplied missing tRNAs (complementation). A study of cybrid clones formed by fusing enucleated cultured fibroblasts containing deleted mtDNA with HeLa cells without mtDNA (26) showed that translational complementation between genomes occurred when the proportion of deleted mtDNA was less than 60%, while higher proportions were associated with progressive inhibition of translation and concomitant COX deficiency. Patients with deletions involving COX subunits tend to have absent COX activity in RRF, while COX activity is relatively preserved in RRF of patients with other deletions. Intramitochondrial, intergenomic complementation between deleted and normal mtDNA is the likely explanation of this phenomenon (27).

Ballinger and colleagues (12) described a family with maternally inherited diabetes and deafness

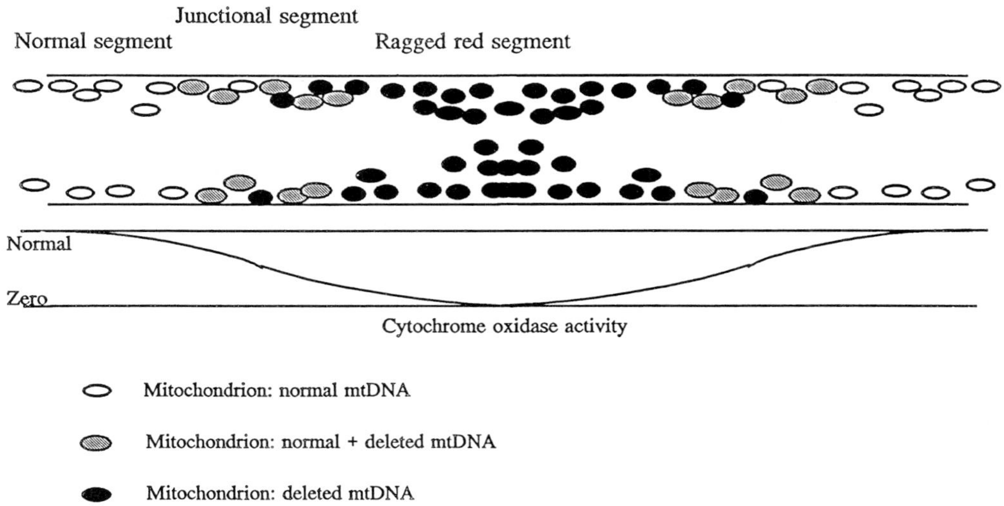

Figure 2 Distribution of normal and deleted mtDNA in ragged red fibre segments in muscle of patients with mitochondrial myopathy.

with a large deletion of mtDNA including the origin of light strand replication. This differs from all other mtDNA deletions in its inheritance, phenotype, and site of deletion. The persistence of this deletion with an absent origin of replication presents theoretical difficulties. Poulton and colleagues (28) suggest that the genetic data were also consistent with the presence of a mtDNA duplication, rather than a deletion, allowing an explanation for the persistence of the mutant species.

Mitochondrial DNA Duplications

Direct tandem duplications of mtDNA have a similar phenotype to mtDNA deletions and are often sporadic (29), although maternal transmission has been suggested in one kindred (12,28). They are heteroplasmic and may arise from deletion of mtDNA dimers. There is some evidence that duplications may be significant in the pathogenesis of deletions, forming a transient intermediate form (28).

Mitochondrial DNA Point Mutations

In 1990 Shoffner and colleagues described an $A \rightarrow G^{(8344)}$ transition in the $tRNA^{Lys}$ gene in association with the syndrome of myoclonic epilepsy and ragged red fibers (MERRF). Since this description, an increasing number of point mutations have been described (Table 3), the most prevalent being the $tRNA^{Leu(UUR)}$ $A \rightarrow G^{(3243)}$ mutation. In general, the mutations are heteroplasmic, present in blood as well as muscle, and are inherited through the maternal line. There is a correlation between the individual genetic defects and clinical features, although this is not perfect. Two well-characterized point mutations will be discussed further.

The $tRNA^{Leu(UUR)}$ $A \rightarrow G^{(3243)}$ Mutation

The $tRNA^{Leu(UUR)}$ $A \rightarrow G^{(3243)}$ mutation was initially identified in patients with the syndrome of mitochondrial myopathy, encephalopathy, lactic acidosis, and stroke-like episodes (MELAS). Further studies have demonstrated the mutation in 60 of 68 (88%) patients with MELAS (8,30–32). Some of the remainder may harbor the more recently described $tRNA^{Leu(UUR)}$ $T \rightarrow C^{(3271)}$ mutation (33). Conversely, when unselected cases are examined for the mutation, phenotypic specificity has not been confirmed (8). The mutation is harbored by patients with phenotypes ranging from the classical MELAS syndrome to patients with PEO or myopathy alone or the syndrome of diabetes and deafness (34–37).

The mutation is generally heteroplasmic in both blood and muscle, comprising 50–92% of total muscle mtDNA, with lesser proportions detectable in blood.

The A $\rightarrow$ $G^{(3243)}$ mutation alters a conserved nucleotide in the dihydrouridine loop of the $tRNA^{Leu(UUR)}$ gene. The pathogenesis of respiratory chain dysfunction is unclear. The transition may not only cause tRNA dysfunction by structural changes within the tRNA molecule, but may have other effects stemming from its position within a conserved tridecamer sequence involved in heavy strand transcription termination. Hess et al. (38) showed that the mutation inhibited transcription termination in vitro, although major quantitative effects do not occur in vivo (25) or in cybrid lines (39).

The $tRNA^{Lys}$ $A \rightarrow G^{(8344)}$ Mutation

This transitional mutation was initially described by Shoffner and colleagues (40) in association with the MERRF phenotype. The mutation is now established as a major cause of the MERRF syndrome, present in 10 of 14 families (8,41). Some of the other families may have the $tRNA^{Lys}$ T $\rightarrow$ $C^{(8356)}$ mutation (33). The core clinical features include myoclonus, ataxia, and seizures. Many additional features may be present, including optic neuropathy, peripheral neuropathy, deafness, dementia, ophthalmoplegia, stroke-like episodes, and Leigh's syndrome (8,42). The mutation is detectable in blood and muscle, with up to 99% mutant being present in muscle and less in blood. The proportion of mutant mtDNA in blood shows correlation with age of onset of disease and clinical severity (42), of potential clinical use.

The A $\rightarrow$ $G^{(8344)}$ mutation alters a conserved nucleotide in the TΨC loop of the $tRNA^{Lys}$ gene. The TΨC loop is thought to interact with the ribosomal surface, so a mutation in this region is likely to affect incorporation of lysine residues into mitochondrial proteins. A defect of translation has been demonstrated (43,44).

Primary Nuclear DNA Defects

The evidence that some defects of mtDNA are secondary to nuclear gene abnormalities lies mainly in the observation that some mtDNA abnormalities are inherited as Mendelian traits. Nonmaternal transmission precludes a primary inherited mtDNA defect and suggests a *trans*-acting nuclear encoded protein that causes secondary deleterious effects on mtDNA. The secondary mtDNA defects may be qualitative or quantitative.

Multiple mtDNA Deletion Syndromes

The syndrome of multiple mtDNA deletions was first reported by Zeviani and colleagues (45,46). The disorder was inherited in simple autosomal dominant fashion. The clinical features are relatively stereotyped, frequently including adult-onset progressive external ophthalmoplegia, proximal myopathy, and deafness. Cataracts, ataxia, and peripheral neuropathy are also commonly observed. The histological findings of ragged red and cytochrome oxidase negative fibers resemble those seen in patients with single mtDNA deletions.

Southern blot analysis of muscle shows the normal band of 16.5 kb, but also several smaller bands, giving rise to the term pleioplasmy. The deletions were not detectable in blood. Other families with similar genetic and clinical features have been described (47). In contrast to the autosomal dominant form of the disease, two siblings born to consanguineous healthy parents with similar clinical features were thought to demonstrate autosomal recessive inheritance (48,49).

Following the recognition of the typical clinical features associated with multiple deletions, a number of other phenotypes have been described. A mother and son with idiopathic dilated cardiomyopathy both had multiple mtDNA deletions (50). The authors suggested either maternal or autosomal dominant inheritance. Cormier and colleagues (51) described a patient presenting with ataxia and ketoacidotic comas, with subsequent external ophthalmoplegia, deafness, and spasticity with RRF on muscle biopsy. The patient, his asymptomatic mother, and his maternal aunt had multiple deletions of mtDNA in muscle and blood, which were not detected in other family members. These features appear to make this family unique. A further phenotype of recurrent exertional myoglobinuria was described in two brothers (52). RRF were present in muscle together with multiple mtDNA deletions. No further features were found on examination, or in the patients' mother or brother.

The origin of multiple mtDNA deletions is obscure. The most obvious candidate genes are the nuclearly encoded genes that are involved in mtDNA replication, with the implicit hypothesis that unfaithful replication leads to accumulation of mtDNA deletions. Sequencing of deleted molecules generally identifies flanking direct repeats similar to those found in most single deletions. In contrast to single deletions, repeats flanking multiple deletion more often show imperfections. One autosomal locus predisposing to multiple deletions has been assigned to chromosome 10q (53).

MtDNA Depletion with Variable Tissue Expression

Moraes et al. studied two cousins with a fatal mitochondrial disease, affecting muscle in one and liver in the other (54). Quantitative analysis demonstrated a severe depletion of mtDNA in affected tissues. Similar findings of mtDNA depletion in affected tissues (muscle, muscle and kidney) were observed in two unrelated infants. Tritschler et al. (55) studied five additional children, with mitochondrial myopathy manifesting within or soon after the first year of life, all of whom had depletion of muscle mtDNA (2–34% of normal). Two of these patients were siblings born to healthy nonconsanguineous parents. The authors speculated that the nonmaternal mode of inheritance indicated a nuclear gene defect, probably with an autosomal-recessive mode of inheritance. One possible mechanism may involve failure of the resumption of mtDNA replication after early embryogenesis.

Other Possible Genetic Defects

About 40% of adult patients with mitochondrial myopathy in our series do not have a known mtDNA defect (unpublished data). It is likely that some of these will have as yet undefined point mutations, especially as these are being reported exponentially. However, since the majority of respiratory chain subunits are nuclearly encoded, it seems probable that some will have nuclear defects that will prove more difficult to identify.

MANAGEMENT

Assessment

A provisional diagnosis of mitochondrial myopathy is usually made on the basis of suggestive clinical features supported by histological and histochemical abnormalities on muscle biopsy. Pedigree analysis is often helpful diagnostically as well as being essential for genetic assessment. Asymptomatic relatives may have mild ptosis, limb weakness or fatigue, retinopathy, or hearing loss and should be examined if possible. Detailed ophthalmological or audiological assessment may be helpful. Mitochondrial myopathies may be sporadic or show matrilineal, autosomal dominant, or recessive modes of inheritance.

Serum creatine kinase is unhelpful in diagnosis. Resting blood lactate concentrations are less than twice normal in approximately 70% of cases, but are more often abnormal following exercise. In one series peak exercise serum lactate exceeded 2 mM in 19/29 patients, compared to a maximum of 1.8 mM in normal or diseased controls (10).

Magnetic resonance imaging (MRI) or computerized tomography (CT) scan of the brain may be normal, but in encephalopathies may show a wide range of abnormalities including low-density lesions or atrophy of the cerebrum or cerebellum, or lucencies or calcification of the basal ganglia. Multifocal areas of hyperintense signal confined to the cortex of the cerebrum, cerebellum, and immediately adjacent white matter on T_2-weighted MRI scan have been suggested as a specific finding in the MELAS syndrome (56); a characteristic distribution of lesions on MRI may also aid diagnosis of Leigh's syndrome (57).

Screening leukocyte DNA for the $A \rightarrow G^{(3243)}$ and $A \rightarrow G^{(8344)}$ mutations for diagnosis may establish a genetic diagnosis in patients with mitochondrial encephalopathy without recourse to muscle biopsy (8). Absence of the mutations does not exclude the diagnosis. Other published mutations may be sought, but are less prevalent. In most cases, a muscle biopsy in necessary to confirm the diagnosis. Histochemistry is required [particularly for succinic dehydrogenase and cytochrome oxidase activity (5)] together with ultrastructural (7) or genetic studies. A negative muscle biopsy does not exclude mitochondrial disease.

Genetic counseling in mitochondrial myopathies is often difficult since similar clinical presentations may vary widely in their associated recurrence risks, depending on the genetic diagnosis. For this reason molecular genetic analysis is mandatory prior to counseling. The demonstration of a large single mtDNA deletion confers a very low recurrence risk to relatives, including the offspring of females. Autosomal-dominant inheritance of multiple deletions can be assumed if there is paternal transmission in the pedigree, and this disorder appears to be fully penetrant. The presence of a point mutation of known pathogenicity confers no risk to the offspring of affected males, but counseling affected or asymptomatic females is not straightforward. In families with the $A \rightarrow G^{(8344)}$ mutation, there is some correlation between clinical severity and the proportion of mutant mtDNA in blood, and it also appears that women with affected children have a higher proportion than those who do not (42). However, these observations cannot be reliably used in genetic counseling as there is considerable overlap, and it is not clear whether the proportion of mutant mtDNA in blood alters with age. Prenatal diagnosis is not advisable for the same reasons. The situation is more complex with the most common $A\text{->}G^{(3243)}$ mutation, as there is little correlation between proportion of mutant mtDNA and severity (32).

TREATMENT

Effective treatment remains limited. Exercise, alcohol, or intercurrent infection may precipitate lactic acidosis and patients should be counseled accordingly. Acute exacerbations of lactic acidosis may be improved by slow infusion of sodium bicarbonate. Carnitine supplements may help patients with secondary carnitine deficiency. In an open trial of ubidecarone (CoQ10), 16 of 44 patients showed a greater than 25% decrease in postexercise lactate levels (58). However, this subgroup of "responders" was studied further in a blind placebo-controlled trial for 3 months; no significant difference was observed between placebo and treated groups. Other agents reported to improve clinical status include succinate (59), menadione and ascorbic acid (60), and riboflavin (61). Headaches and stroke-like episodes seem to improve rapidly with steroid treatment (62).

REFERENCES

1. Hecht NB, Liem H, Kleene KC, Distel RJ, Ho S-M. Maternal inheritance of the mouse mitochondrial genome is not mediated by a loss or gross alteration of the paternal mitochondrial DNA or by methylation of the oocyte mitochondrial DNA. Dev Biol 1984; 102:452–461.
2. Gyllensten U, Wharton D, Josefssoon A, Wilson AC. Paternal inheritance of mitochondrial DNA in mice. Nature 1991; 352:255–257.
3. Yamamoto M, Koga Y, Ohtaki E, Nonaka I. Focal cytochrome c oxidase deficiency in various neuromuscular diseases. J Neurol Sci 1989; 91:207–213.
4. Müller-Höcker J. Cytochrome c oxidase deficient fibres in the limb muscle and diaphragm of man without muscular disease: an age related alteration. J Neurol Sci 1990; 100:14–21.
5. Turnbull DM, Johnson MA, Dick DJ, Cartlidge NEF, Sherrat HSA. Partial cytochrome oxidase deficiency without subsarcolemmal accumulation of mitochondria in chronic external ophthalmoplegia. J Neurol Sci 1985; 70:93–100.
6. Rosing HS, Hopkins LC, Wallace DC, Epstein CM, Weidenheim K. Maternally inherited mitochondrial myopathy and myoclonic epilepsy. Ann Neurol 1985; 17:228–237.
7. Berkovic SF, Carpenter S, Evans A, et al. Myoclonus epilepsy and ragged-red fibres (MERRF). 1. A clinical, pathological, biochemical, magnetic resonance spectrographic and positron emission tomographic study. Brain 1989; 112:1231–1260.

8. Hammans SR, Sweeney MG, Brockington M, Morgan-Hughes JA, Harding AE. Mitochondrial encephalopathies: molecular genetic diagnosis from blood samples. Lancet 1991; 337:1311–1313.
9. DiMauro S, Bonilla E, Zeviani M, Nakagawa M, DeVivo DC. Mitochondrial myopathies. Ann Neurol 1985; 17:521–538.
10. Petty RKH, Harding AE, Morgan-Hughes JA. The clinical features of mitochondrial myopathy. Brain 1986; 109:915–938.
11. Rowland LP, Blake DM, Hirano M, et al. Clinical syndromes associated with ragged red fibres. Rev Neurol 1991; 147:467–473.
12. Ballinger SW, Shoffner JM, Hadaya EV, et al. Maternally transmitted diabetes and deafness associated with a 10.4kb mitochondrial DNA deletion. Nature Genet 1992; 1:11–15.
13. Rotig A, Cormier V, Blanche S, et al. Pearson's marrow-pancreas syndrome: a multisystem mitochondrial disorder in infancy. J Clin Invest 1990; 86:1601–1608.
14. McShane MA, Hammans SR, Sweeney M, et al. Pearson syndrome and mitochondrial encephalomyopathy in a patient with a deletion of mtDNA. Am J Hum Genet 1991; 48:39–42.
15. Larsson N-G, Holme E, Kristiansson B, Oldfors A, Tulinius MH. Progressive increase of the mutated mitochondrial DNA fraction in Kearns-Sayre syndrome. Pediatr Res 1990; 28:131–136.
16. Ozawa T, Yoneda M, Tanaka M, et al. Maternal inheritance of deleted mitochondrial DNA in a family with mitochondrial myopathy. Biochem Biophys Res Commun 1988; 154:1240–1247.
17. Poulton J, Deadman ME, Ramacharan S, Gardiner RM. Germ-line deletions of mtDNA in mitochondrial myopathy. Am J Hum Genet 1991; 48:649–653.
18. Brockington M, Sweeney MG, Hammans SR, Morgan-Hughes JA, Harding AE. A tandem duplication in the D-loop of human mitochondrial DNA is associated with deletions in mitochondrial myopathies. Nature Genet 1993; 4:67–71.
19. Moraes CT, DiMauro S, Zeviani M, et al. Mitochondrial DNA deletions in progressive external ophthalmoplegia and Kearns-Sayre syndrome. N Engl J Med 1989; 320:1293–1299.
20. Zeviani M, Gellera C, Pannacci M, et al. Tissue distribution and transmission of mitochondrial DNA deletions in mitochondrial myopathies. Ann Neurol 1990; 28:94–97.
21. Shanske S, Moraes CT, Lombes A, et al. Widespread tissue distribution of mitochondrial DNA deletions in Kearns-Sayre syndrome. Neurology 1990; 40:24–28.
22. Mita S, Rizzuto R, Moraes CT, et al. Recombination via flanking direct repeats is a major cause of large-scale deletions of human mitochondrial DNA. Nucleic Acids Res 1990; 18:561–567.
23. Mita S, Schmidt B, Schon EA, DiMauro S, Bonilla E. Detection of "deleted" mitochondrial genomes in cytochrome-c oxidase–deficient muscle fibers of a patient with Kearns-Sayre syndrome. Proc Natl Acad Sci USA 1989; 86:9509–9513.
24. Shoubridge EA, Karpati G, Hastings KE. Deletion mutants are functionally dominant over wild-type mitochondrial genomes in skeletal muscle fiber segments in mitochondrial disease. Cell 1990; 62:43–49.
25. Hammans SR, Sweeney MG, Wicks DAG, Morgan-Hughes JA, Harding AE. A molecular genetic study of focal histochemical defects in mitochondrial encephalomyopathies. Brain 1992; 115:343–365.
26. Hayashi J, Ohta S, Kikuchi A, Takemitsu M, Goto Y, Nonaka I. Introduction of disease-related mitochondrial DNA deletions into HeLa cells lacking mitochondrial DNA results in mitochondrial dysfunction. Proc Natl Acad Sci USA 1991; 88:10614–10618.
27. Hammans SR, Sweeney MG, Holt IJ, et al. Evidence for intramitochondrial complementation between deleted and normal mitochondrial DNA in some patients with mitochondrial myopathy. J Neurol Sci 1992; 107:87–92.
28. Poulton J, Deadman ME, Bindoff L, Morten K, Land J, Brown G. Families of mtDNA re-arrangements can be detected in patients with mtDNA deletions: duplications may be a transient intermediate form. Hum Mol Genet 1993; 2:23–30.
29. Poulton J, Deadman ME, Gardiner RM. Duplications of mitochondrial DNA in mitochondrial myopathy. Lancet 1989; 1:236–240.
30. Goto Y, Nonaka I, Horai S. A mutation in the $tRNA^{Leu(UUR)}$ gene associated with the MELAS subgroup of mitochondrial encephalomyopathies. Nature 1990; 348:651–653.
31. Kobayashi Y, Momoi MY, Tominaga K, et al. A point mutation in the mitochondrial tRNA(Leu)(-UUR) gene in MELAS (mitochondrial myopathy, encephalopathy, lactic acidosis and stroke-like episodes). Biochem Biophys Res Commun 1990; 173:816–822.
32. Ciafaloni E, Ricci E, Shanske S, et al. MELAS: Clinical features, biochemistry, and molecular genetics. Ann Neurol 1992; 31:391–398.
33. Goto Y, Nonaka I, Horai S. A new mtDNA mutation associated with mitochondrial myopathy, encephalopathy, lactic acidosis and stroke-like episodes (MELAS). Biochim Biophys Acta 1991; 1097:238–240.
34. Reardon W, Ross RJM, Sweeney MG, et al. Diabetes mellitus associated with a pathogenic point mutation in mitochondrial DNA. Lancet 1992; 340:1376–79.
35. Ouweland JMW, Lemkes HHPJ, Ruitenbeek W, et al. Mutation in mitochondrial $tRNA^{Leu(UUR)}$ gene in a

large pedigree with maternally transmitted type II diabetes mellitus and deafness. Nature Genet 1992; 1:368–371.
36. Sue CM, Holmes-Walker DJ, Morris JGL, Boyages SC, Crimmins DS, Byrne E. Mitochondrial gene mutations and diabetes mellitus. Lancet 1993; 341:437–438.
37. Schulz JB, Klockgether T, Dichgans J, Seibel P, Reichmann H. Mitochondrial gene mutations in diabetes mellitus. Lancet 1993; 341:438–439.
38. Hess JF, Parisi MA, Bennett JL, Clayton DA. Impairment of mitochondrial transcription termination by a point mutation associated with the MELAS subgroup of mitochondrial encephalomyopathies. Nature 1991; 351:236–239.
39. King MP, Koga Y, Davidson M, Schon EA. Defects in mitochondrial protein synthesis and respiratory chain activity segregate with the $tRNA^{Leu(UUR)}$ mutation associated with mitochondrial encephalopathy, lactic acidosis, and strokelike episodes. Mol Cell Biol 1992; 12:480–490.
40. Shoffner JM, Lott MT, Lezza AM, Seibel P, Ballinger SW, Wallace DC. Myoclonic epilepsy and ragged-red fiber disease (MERRF) is associated with a mitochondrial DNA tRNA(Lys) mutation. Cell 1990; 61:931–937.
41. Zeviani M, Servidei S, Bresolin N, et al. Rapid detection of the $A\text{->}G^{(8344)}$ mutation in Italian families with myoclonus, epilepsy and ragged red fibres (MERRF). Am J Hum Genet 1991; 48:203–211.
42. Hammans SR, Sweeney MG, Brockington M, et al. The mitochondrial DNA transfer RNA^{Lys} $A>G^{(8344)}$ mutation: Clinical phenotype and relationship to proportion of mutant mitochondrial DNA. Brain 1993; 116:617–632.
43. Chomyn A, Meola G, Bresolin N, Lai S, Scarlato G, Attardi G. In vitro genetic transfer of protein synthesis and respiration defects to mitochondrial DNA-less cells with myopathy-patient mitochondria. Mol Cell Biol 1991; 11:2236–2244.
44. Boulet L, Karpati G, Shoubridge EA. Distribution and threshold expression of the $tRNA^{Lys}$ mutation in skeletal muscle of patients with myoclonic epilepsy and ragged red fibers (MERRF). Am J Hum Genet 1992; 51:1187–1200.
45. Zeviani M, Servidei S, Gellera C, Bertini E, DiMauro S, DiDonato S. An autosomal dominant disorder with multiple deletions of mitochondrial DNA starting at the D-loop region. Nature 1989; 339:309–311.
46. Zeviani M, Bresolin N, Gellera C, et al. Nucleus-driven multiple large-scale deletions of the human mitochondrial genome: A new autosomal dominant disease. Am J Hum Genet 1990; 47:904–914.
47. Servidei S, Zeviani M, Manfredi G, et al. Dominantly inherited mitochondrial myopathy with multiple deletions of mitochondrial DNA: clinical, morphologic, and biochemical studies. Neurology 1991; 41:1053–1059.
48. Yuzaki M, Ohkoshi N, Kanazawa I, Kagawa Y, Ohta S. Multiple deletions in mitochondrial DNA at direct repeats of non-D-loop regions in cases of familial mitochondrial myopathy. Biochem Biophys Res Commun 1989; 164:1352–1357.
49. Mizusawa H, Watanabe M, Kanazawa I, et al. Familial mitochondrial myopathy associated with peripheral neuropathy: partial deficiencies of complex I and complex IV. J Neurol Sci 1988; 86:171–184.
50. Suomalainen A, Paetau A, Leinonen H, Majander A, Peltonen L, Somer H. Inherited idiopathic dilated cardiomyopathy with multiple deletions of mitochondrial DNA. Lancet 1992; 340:1319–1320.
51. Cormier V, Rotig A, Tardieu M, Colonna M, Saudubray J-M, Munnich A. Autosomal dominant deletions of the mitochondrial genome in a case of progressive encephalopathy. Am J Hum Genet 1991; 48:643–648.
52. Ohno K, Tanaka M, Sahashi K, et al. Mitochondrial DNA deletions in inherited recurrent myoglobinuria. Ann Neurol 1991; 29:364–369.
53. Suomalainen A, Kaukonen J, Amati P, et al. An autosomal locus predisposing to deletions of mitochondrial DNA. Nat Genet 1995; 9:146–151.
54. Moraes CT, Shanske S, Tritschler HJ, et al. mtDNA depletion with variable tissue expression: a novel genetic abnormality in mitochondrial diseases. Am J Hum Genet 1991; 48:492–501.
55. Tritschler HJ, Andreetta F, Moraes CT, et al. Mitochondrial myopathy of childhood associated with depletion of mitochondrial DNA. Neurology 1992; 42:209–217.
56. Matthews PM, Tampieri D, Berkovic SF, et al. Magnetic resonance imaging shows specific abnormalities in the MELAS syndrome. Neurology 1991; 41:1043–1046.
57. Berkovic SF, Karpati G, Carpenter S, Lang AE. Progressive dystonia with bilateral putaminal hypodensities. Arch Neurol 1987; 44:1184–1187.
58. Bresolin N, Doriguzzi C, Ponzetto C, et al. Ubidecarenone in the treatment of mitochondrial myopathies: a multi-center double-blind trial. J Neurol Sci 1990; 100:70–78.
59. Shoffner JM, Lott MT, Voljavec AS, Soueidan SA, Costigan DA, Wallace DC. Spontaneous Kearns-Sayre/chronic external ophthalmoplegia plus syndrome associated with a mitochondrial DNA deletion: a slip-replication model and metabolic therapy. Proc Natl Acad Sci USA 1989; 86:7952–7956.
60. Eleff S, Kennaway NG, Buist NRM, et al. ^{31}P-NMR study of improvement in oxidative phosphorylation by

vitamins K_3 and C in a patient with a defect in electron transport at complex III in skeletal muscle. Proc Natl Acad Sci USA 1984; 81:3529–3533.
61. Arts WFM, Scholte HR, Bogaard JM, Kerrebijn KF, Luyt-Houwen IEM. NADH-CoQ reductase deficient myopathy: successful treatment with riboflavin. Lancet 1983; 2:581–582.
62. Shapira Y, Cederbaum SD, Cancilla PA, Nielsen D, Lippe BM. Familial poliodystrophy, mitochondrial myopathy, and lactate acidemia. Neurology 1975; 25:614–621.
63. Hurano M, Silvestin G, Blake DM, et al. Mitochondrial neurogastrointestinal encephalopathy (MNGIE): clinical, biochemical, and genetic features of an autosomal recessive mitochondrial disorder. Neurology 1994; 44:721–727.
64. Bardosi A, Creutzfeldt W, DiMauro S, et al. Myo-, neuro-, gastrointestinal encephalopathy (MNGIE syndrome) due to partial deficiency of cytochrome-c-oxidase. Acta Neuropathol (Berl) 1987; 748:248–258.
65. Holt IJ, Harding AE, Cooper JM, et al. Mitochondrial myopathies: clinical and biochemical features of 30 patients with major deletions of muscle mitochondrial DNA. Ann Neurol 1989; 26:699–708.
66. Nelson I, Degoul F, Obermaier Kusser B, et al. Mapping of heteroplasmic mitochondrial DNA deletions in Kearns-Sayre syndrome. Nucleic Acids Res 1989; 17:8117–8124.
67. Gerbitz KD, Obermaier Kusser B, Zierz S, Pongratz D, Müller-Höcker J, Lestienne P. Mitochondrial myopathies: divergences of genetic deletions, biochemical defects and the clinical syndromes. J Neurol 1990; 237:5–10.
68. Degoul F, Nelson I, Lestienne P, et al. Deletions of mitochondrial DNA in Kearns-Sayre syndrome and ocular myopathies: genetic, biochemical and morphological studies. J Neurol Sci 1991; 101:168–177.
69. Geny C, Cormier V, Meyrignac C, et al. Muscle mitochondrial DNA in encephalomyopathy and ragged red fibres: a Southern blot analysis and literature review. J Neurol 1991; 238:171–176.
70. Bresolin N, Martinelli P, Barbiroli B, et al. Muscle mitochondrial DNA delection and ^{31}P-NMR spectroscopy alterations in a migraine patient. J Neurol Sci 1991; 104:182–189.
71. Poulton J, Deadman ME, Turnbull DM, Lake B, Gardiner RM. Detection of mitochondrial DNA deletions in blood using the polymerase chain reaction: non-invasive diagnosis of mitochondrial myopathy. Clin Genet 1991; 39:33–38.
72. Rotig A, Colonna M, Bonnefont JP, et al. Mitochondrial DNA deletion in Pearson's marrow/pancreas syndrome. Lancet 1989; 1:902–903.
73. Rotig A, Colonna M, Blanche S, et al. Deletion of blood mitochondrial DNA in pancytopenia. Lancet 1988; 2:567–568.
74. Goto Y, Horai S, Matsuoka T, et al. Mitochondrial myopathy, encephalopathy, lactic acidosis, and stroke-like episodes (MELAS): a correlative study of the clinical features and mitochondrial DNA mutation. Neurology 1992; 42:545–550.
75. Berkovic SF, Shoubridge EA, Andermann F, Andermann E, Carpenter S, Karpati G. Clinical spectrum of mitochondrial DNA mutation at base pair 8344. Lancet 1991; 338:457.
76. Silvestri G, Moraes CT, Shanske S, Oh SJ, DiMauro S. A new mtDNA mutation in the $tRNA^{Lys}$ gene is associated with myoclonic epilepsy and ragged red fibres (MERRF). Am J Hum Genet 1992; 51:1213–1217.
77. Zeviani M, Muntoni F, Savarese N, et al. A MERRF/MELAS overlap syndrome associated with anew point mutation of mitochondrial $tRNA^{Lys}$ gene. Eur J Mol Genet 1993; 1:80–87.
78. Sweeney M, Bundey S, Brockington M, Poulton KR, Winer JB, Harding AE. Mitochondrial myopathy associated with sudden death in young adults and a novel mutation in the mitochondrial DNA transfer $RNA^{(UUR)}$ gene (submitted).
79. Zeviani M, Gellera C, Antozzi C, et al. Maternally inherited myopathy and cardiomyopathy: association with mutation in mitochondrial DNA $tRNA^{Leu(UUR)}$. Lancet 1991; 338:143–147.
80. Sweeney MG, Brockington M, Weston MJ, Morgan-Hughes JA, Harding AE. Mitochondrial DNA transfer RNA mutation $^{LEU(UUR)A\text{->}G\ 3260}$: a second family with myopathy and cardiomyopathy. Q J Med 1993; 86:435–438.
81. Goto Y, Tojo M, Tohyama J, Horai S, Nonaka I. A novel point mutation in the mitochondrial $tRNA^{Leu(UUR)}$ gene in a family with mitochondrial myopathy. Ann Neurol 1992; 31:672–675.
82. Lertrit P, Noer AS, Jean-Francois MJB, et al. A new disease-related mutation for mitochondrial encephalopathy lactic acidosis and strokelike episodes (MELAS) syndrome affects the ND4 subunit of respiratory complex I. Am J Hum Genet 1992; 51:457–468.
83. Moraes CT, Ciacci F, Bonilla E, Ionasescu V, Schon EA, DiMauro S. A mitochondrial tRNA anticodon swap associated with a muscle disease. Nature Genet (in press).
84. Wallace DC, Singh G, Lott MT, et al. Mitochondrial DNA mutation associated with Leber's hereditary optic neuropathy. Science 1988; 242:1427–1430.
85. Howell N, Kubacka I, Xu M, McCullough DA. Leber hereditary optic neuropathy: involvement of the mitochondrial NDI gene and evidence for an intragenic suppressor mutation. Am J Hum Genet 1991; 48:935–942.

86. Huoponen K, Vilkki J, Aula P, Nikoskelainen EK. A new mtDNA mutation associated with Leber hereditary optic neuroretinopathy. Am J Hum Genet 1991; 48:1147–1153.
87. Johns DR, Berman J. Alternative, simultaneous complex I mitochondrial DNA mutations in Leber's hereditary optic neuropathy. Biochem Biophys Res Commun 1991; 174:1324–1330.
88. Mackey D, Howell N. A variant of Leber hereditary optic neuropathy characterized by recovery of vision and by an unusual mitochondrial genetic etiology. Am J Hum Genet 1992, 51:1218–1228.
89. Holt IJ, Harding AE, Petty RKH, Morgan-Hughes JA. A new mitochondrial disease associated with mitochondrial DNA heteroplasmy. Am J Hum Genet 1990; 46:428–433.
90. Harding AE, Holt IJ, Sweeney MG, Brockington M, Davis MB. Prenatal diagnosis of mitochondrial $DNA^{8993T\text{->}G}$ disease. Am J Hum Genet 1992; 50:629–633.
91. Tatuch Y, Christodoulou J, Feigenbaum A, et al. Heteroplasmic mtDNA mutation (T-G) at 8993 can cause Leigh disease when the percentage of abnormal mtDNA is high. Am J Hum Genet 1992; 50:852–858.
92. Otsuka M, Niijima K, Mizuno Y, Yoshida M, Kagawa Y, Ohta S. Marked decrease of mitochondrial DNA with multiple deletions in a patient with familial mitochondrial myopathy. Biochem Biophys Res Commun 1990; 167:680–685.
93. Ikebe S, Tanaka M, Ohno K, et al. Increase of deleted mitochondrial DNA in the striatum in Parkinson's disease and senescence. Biochem Biophys Res Commun 1990; 170:1044–1048.
94. Ozawa T, Tanaka M, Sugiyama S, et al. Multiple mitochondrial DNA deletions exist in cardiomyocytes of patients with hypertrophic or dilated cardiomyopathy. Biochem Biophys Res Commun 1990; 170:830–836.
95. Cortopassi GA, Arnheim N. Detection of a specific mitochondrial DNA deletion in tissues of older humans. Nucleic Acids Res 1990; 18:6927–6933.

40

Biochemical Investigation of Mitochondrial Respiratory Chain Abnormalities

J. M. Cooper and Anthony H. V. Schapira
Royal Free Hospital School of Medicine, London, England

INTRODUCTION

Patients with mitochondrial respiratory chain defects can be classified by their clinical presentation, biochemical abnormality, or genetic defect. In this chapter we will discuss the criteria used for biochemical classification of these disorders; the clinical and genetic classification will be covered elsewhere.

The clinical suspicion that a patient may have a mitochondrial respiratory chain abnormality may be raised by a variety of symptoms and signs, including muscle weakness, fatigue, ptosis, ophthalmoplegia, and a variety of encephalopathic features including epilepsy, dementia, deafness, and ataxia. The presence of ragged red fibers in skeletal muscle is the most characteristic morphological feature in patients with mitochondrial myopathy. The site of the functional and molecular defects can be identified using a variety of biochemical and molecular analyses, which will be discussed in this chapter with reference to specific patients. However, this will not be an exhaustive analysis of all the possible experimental techniques.

EXERCISE TEST

An abnormality of the respiratory chain will lead to a decrease in aerobic ATP synthesis from glucose and fatty acids. The flux through the glycolytic pathway will increase to accelerate the rate of ATP synthesis from substrate level phosphorylation. With a decrease in the oxidation of NADH by the mitochondrial respiratory chain, to allow glycolysis to continue, NADH is oxidized as pyruvate is reduced to lactate by lactate dehydrogenase. The consequent lactic acidosis is a useful parameter to confirm the presence of a respiratory chain abnormality. In patients with severe respiratory chain defects there is usually a raised resting plasma lactate concentration, which is exacerbated by light exercise during which demands for ATP rise. The aerobic exercise test is therefore a convenient method to identify the presence of a respiratory chain abnormality. In its simplest form this involves taking blood samples for pyruvate and lactate measurements before, during, and after exercise. It is essential that the period of exercise is monitored and standardized to keep within the aerobic threshold of control subjects. In the majority of patients with respiratory chain abnormalities the lactate levels increase markedly above control values during exercise with a delayed recovery to normal levels (Fig. 1). The lactate: pyruvate ratio also rises to abnormally high levels. Both these parameters are good indicators for the presence of respiratory chain disorders. Clearly, an exercise test will not be suitable for all patients such as the very young, the ataxic, severely weak patients, or those with a cardiac defect; however, these patients often have a lactic acidosis or a raised lactate:pyruvate ratio at rest.

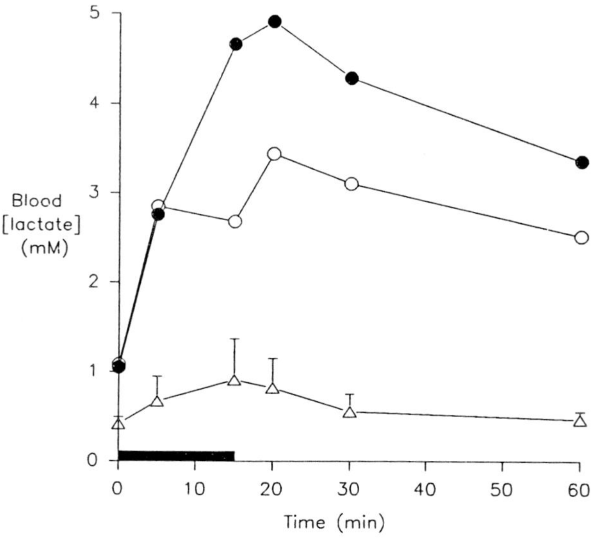

Figure 1 Blood lactate concentrations of control (open triangles), a patient (JH) with a complex I defect (open circles), and a patient (EP) with a complex I–III defect (closed circles). The filled bar represents the duration of exercise, except for JH, who could manage only 5 min of exercise. The amount of work was control—4.3 ± 1.3 kpd km; JH—0.25 kpd km; EP—0.4 km. (Reproduced with permission from Ref. 43.)

BIOCHEMICAL ANALYSES

The detailed analysis of mitochondrial function can be determined on a variety of tissue samples. Skeletal muscle is usually (although not always) affected clinically and is therefore a suitable choice for biochemical analysis. However, when the main clinical symptoms involve other tissues, the severity of the biochemical defect observed in muscle mitochondria may be less pronounced. From a small muscle biopsy (100–300 mg) several informative investigative procedures can be performed, including histochemical, electron microscopic, DNA, and limited biochemical and polypeptide analyses. From a larger biopsy (3–6 g) it is possible to perform a more extensive biochemical investigation involving mitochondrial isolation.

Characteristic morphological changes in the skeletal muscle from patients with mitochondrial respiratory chain dysfunction include: ragged red fibers observed with the modified Gomori trichrome stain; accumulations of mitochondria observed with the succinate dehydrogenase stain; cytochrome oxidase–negative fibers; and intramitochondrial paracrystalline inclusions and other structural mitochondrial abnormalities on electron microscopy. However, not all cases with a respiratory chain abnormality show all these features (1,2) and abnormal mitochondria have been observed in other disorders (3,4).

Accurate assessment of mitochondrial function is best performed on intact mitochondria isolated from freshly biopsied tissue, although it is possible to enrich mitochondria from frozen tissue samples (5).

ISOLATION OF SKELETAL MUSCLE MITOCHONDRIA

Mitochondria can be isolated from almost any tissue; however, for diagnostic purposes skeletal muscle has been the tissue of choice mainly because it is usually involved in the disease, is relatively accessible, and usually exhibits characteristic changes in patients with mitochondrial abnormalities.

Both white cells and platelets contain mitochondria and are therefore potential sources of material for mitochondrial analyses. This can involve screening for certain mtDNA abnormalities (6) or the analysis of mitochondria respiratory chain function (7). However, caution must be exercised when assessing the results from such samples because the tissue distribution of the mitochondrial abnormality may not always include these cell types. This is most clearly seen in the distribution of deletions of mtDNA, which, although found in muscle, are absent in blood in the majority of patients.

A number of techniques exist for mitochondrial isolation from various-sized skeletal muscle samples and from both fresh and frozen muscle (8–13). However, a mixture of protease digestion and mechanical disruption of the skeletal muscle followed by differential centrifugation is a method frequently adopted, although without the use of density gradient centrifugation all preparations will have some contamination from other organelles. Although high yields of pure, intact mitochondria are not realistically attainable, higher yields should not be sought at the expense of significantly lower purities. In our laboratory enough mitochondria of sufficient purity can be reproducibly obtained from 3–6 g of fresh muscle. Smaller sample sizes tend to result in higher and more variable degrees of contamination.

Mitochondrial-enriched fractions have been used, especially from smaller sample sizes, correcting for variations in purity by standardizing the results to a mitochondrial marker such as citrate synthase (CS). While this can be acceptable, it assumes all the mitochondria are intact and have not lost proportions of their CS activity, which depends on the severity and reproducibility of tissue homogenization. Using postnuclear supernatants there is potential for variability as the number of mitochondria released from the cells can vary in both absolute and relative terms. Corrections made using CS activities assumes there is uniform release of CS (a soluble enzyme) and mitochondrial membranes. The use of whole-tissue homogenates has the advantage that any analysis measures all the mitochondria present in the tissue, but has the disadvantages of high turbidity and decreased sensitivity of the analyses. In many situations the approach used is a compromise governed by sample availability. While these approaches are acceptable with the appropriate safeguards, in comparison to purified mitochondrial fractions they inevitably introduce a greater degree of data scattering. The important point is the complete validation of the technique in terms of purity and reproducibility. Contamination of the mitochondrial preparation can be assessed by electron microscopic examination of the mitochondrial pellet or the measurement of markers of other subcellular compartments, relative to the total activity in whole-cell extract (i.e., acid phosphatase–lysosomes; esterase–endoplasmic reticulum; catalase-peroxisomes; alkaline phosphatase–plasma membrane; lactate dehydrogenase–cytosol and DNA–nuclear fraction, although mitochondria also have DNA).

POLAROGRAPHIC ANALYSES

The polarographic analysis of oxygen utilization by intact mitochondria can be used to screen a large number of mitochondrial functions; however, it can be reliably performed only on either freshly isolated intact mitochondria or freshly harvested cells.

It is important to isolate the mitochondria immediately after the biopsy and to complete the analysis of the mitochondria within 3 or 4 hr to avoid variability in the results, although mitochondria can still be active and coupled to the phosphorylation of ADP more than 24 hr after their preparation when stored appropriately. The integrity of the mitochondrial preparation can be assessed by determining the respiratory control ratio (RCR). The RCR can be calculated for a given substrate by the ratio of the rate of oxygen utilization by mitochondria in the presence of ADP to the rate in the absence of ADP (state 3 respiration/state 4 respiration, Fig. 2) and is dependent on the integrity of the mitochondrial membranes. Generally, the higher the RCR, the better the mitochondrial preparation, and an RCR greater than 4 should be sought.

Using a microincubation chamber and a commercially available microelectrode (Yellow Springs Instruments, Ohio), oxygen utilization can be determined in a final volume of 200 μl, containing 100–200 μg of mitochondria. Measuring the rate of oxygen utilization by mitochondria in the presence of a variety of substrates, both natural and artificial (Fig. 2), enables a large proportion of the

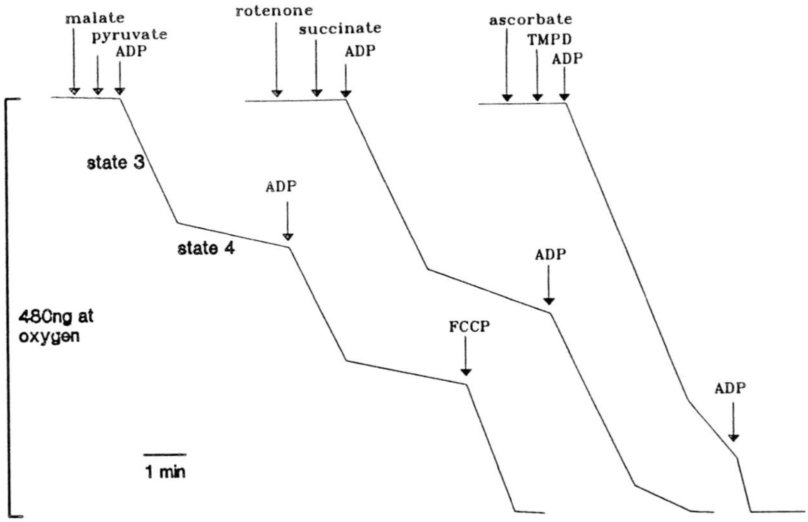

Figure 2 An example of the polarographic determination of mitochondrial function. The experimental conditions were as described (17). The respiratory control ratio (RCR) is the state 3 rate/state 4 rate of oxygen utilization.

mitochondrial functions to be analyzed (Fig. 3). As mitochondria are impermeable to nearly all molecules, the normal rate of oxygen utilization by mitochondria using pyruvate, glutamate, palmitoyl carnitine, succinate, and ascorbate plus *NNN'N'*-tetramethyl-*p*-phenlene diamine (TMPD) requires not only a normal respiratory chain and oxidative phosphorylation system, but also many other mitochondrial systems including: normal transporters (mono- and dicarboxylic acid), carnitine palmitoyl transferase II, electron-transferring flavoprotein, ADP/ATP translocator, primary dehydrogenases (pyruvate dehydrogenase, glutamate dehydrogenase/glutamate oxaloacetate transaminase), and tricarboxylic acid (TCA) cycle and β-oxidation enzymes. The classification of the functional biochemical abnormality using polarographic data alone is summarized in Table 1. Table 2 shows some polarography data from three patients with a mitochondrial respiratory chain abnormality. These data clearly confine the defects in these patients to the respiratory chain, although confirmation of the site of the abnormality needs to be determined enzymatically (see later). The polarographic data are typical of a severe complex I defect (YD), a complex I–III defect (JDM), and a complex I–IV defect (JL), and abnormalities of other mitochondrial systems can be discounted, especially if the enzyme data confirm decreased respiratory chain activities.

While it is possible to purify mitochondria from cells in culture, it is difficult to obtain enough mitochondria of sufficient quality to perform polarographic analysis. However, by selective permeabilization of the plasma membrane with a detergent (usually digitonin), the cell cytoplasm can be removed by washing the cells, and polarographic analysis can be performed. This avoids mitochondrial fractionation and therefore sample recovery is less of a problem. This approach has been used successfully (14), although care must be taken to add sufficient detergent to effectively solubilize the plasma membrane but leave the mitochondrial membranes intact.

ENZYME ANALYSES

The enzyme activities of all five complexes of the respiratory chain and oxidative phosphorylation system can be analyzed spectrophotometrically as individual or functional units (15). These analyses

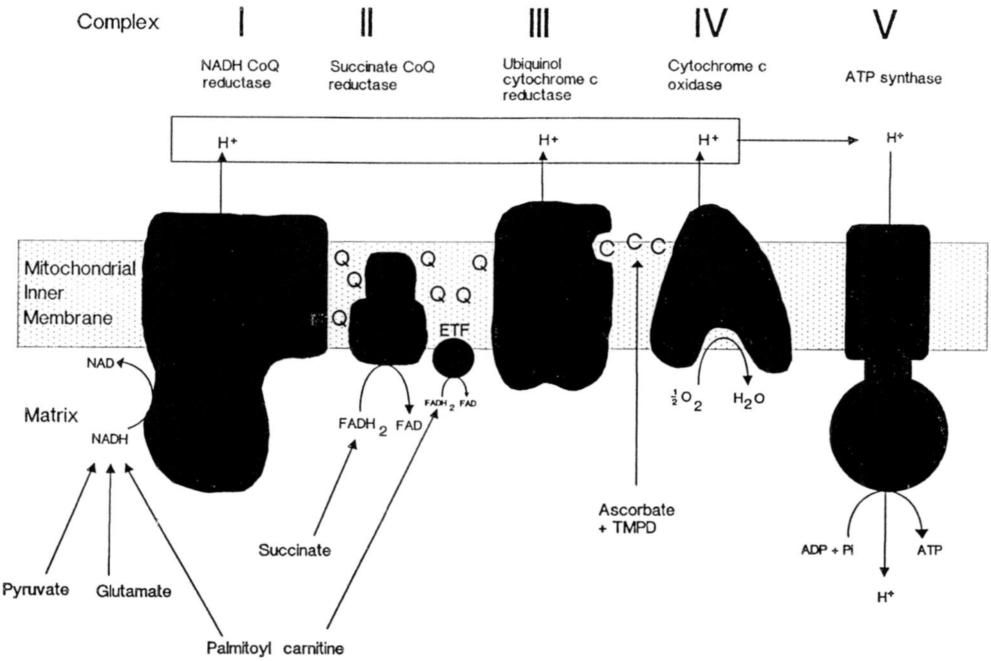

Figure 3 The mitochondrial respiratory chain and oxidative phosphorylation system. Electrons are transferred from complex I, complex II, and electron-transferring flavoprotein (ETF) to ubiquinone (Q). From Q electrons follow a common pathway to oxygen involving complex III, cytochrome c (c), and complex IV. Cytochromes b and c_1, and aa_3, are located in complexes III and IV, respectively. The proton motive force generated by the translocation of protons at complexes I, III, and IV is used to drive ATP synthesis by complex V.

Table 1 Effect of Specific Mitochondrial Abnormalities on the State 3 Rates of Oxygen Utilization in the Presence of Various Substrates Relative to Controls

	Site of mitochondrial defect							
	Complex					pyr	β	ADP/ATP
Substrate	I	II	III	IV	V	dh	oxid/ETF	transl
Pyruvate	↓	N	↓	↓	↓	↓	N	↓
Glutamate	↓	N	↓	↓	↓	N	N	↓
Palmitoyl carnitine	↓	N	↓	↓	↓	N	↓	↓
Succinate	N	↓	↓	↓	↓	N	N	↓
Ascorbate + TMPD	N	N	N	↓	↓	N	N	↓
+ FCCP	NC	NC	NC	NC	↑	NC	NC	↑

Conditions as described by Morgan-Hughes et al. (17). The effect of FCCP is relative to the state 3 rate for a given substrate. In normal mitochondria the addition of carbonyl cyanide *p*-trifluoromethoxyphenylhydrazone (FCCP) may increase the state 3 rate by up to 25%, whereas in complex V and ADP/ATP translocase defects addition of FCCP restores all state 3 rates to normal rates.

pyr dh-pyruvate dehydrogenase: β oxid-β oxidation: ETF = electron transferring flavoprotein; ADP/ATP transl = ADP/ATP translocator. ↓ = decreased; N = normal; NC = no change; ↑ = increased.

Table 2 State 3 Rates of Oxygen Utilization Measured Polarographically for Three Patients with Mitochondrial Respiratory Chain Abnormalities

	Site of Defect (complex)			
	I YD	I-III JDM	I-IV JL	Control ($n = 6$)
Pyruvate	11	36	33	106 ± 17
Glutamate	11	36	35	107 ± 25
Palmitoyl carnitine	ND	28	33	85 ± 23
Succinate	144	72	70	147 ± 49
Ascorbate + TMPD	220	469	119	305 ± 109

Values are ng at O/min/mg mitochondrial protein. The conditions of analysis are as described (17). ND—not determined.

can be performed on all types of sample preparation, although complex V (ATP synthase) activity can be difficult to differentiate from other ATPase activities in preparations other than purified mitochondria. There are also methods for the determination of the activities of other mitochondrial enzymes including TCA enzymes, primary dehydrogenases, and β-oxidation enzymes (12).

In contrast to the polarographic determination of mitochondrial respiratory chain function, the enzymatic analysis of the mitochondrial respiratory chain measures specific enzyme functions and usually involves artificial electron donors and/or acceptors. The latter are generally saturating to give maximal activities (V_{max}), although by varying substrate concentrations it is possible to determine the kinetics of the enzymes as some pathological abnormalities may affect the kinetics of the enzymes and not the V_{max} rates. With enzyme analysis, identification of any functional respiratory chain defect observed by polarographical analysis can be confirmed and localized to the defective complex(es). Taken together the polarographic and enzymatic determinations of mitochondrial function are powerful tools to identify the site of any mitochondrial defect. Table 3 shows the mitochondrial enzyme data for patients listed in Table 2. While the defect in patient YD clearly shows a defect of complex I activity in agreement with the polarographic data, the defect in patient JDM involves complexes I, II/III, and IV. The abnormality of complex IV contradicts the polarographic data. This discrepancy between the ascorbate plus TMPD data and cytochrome oxidase activity is often observed and may reflect the different mechanisms of the activities measured. While the spectrophotometric analysis of cytochrome oxidase activity is dependent on the binding and release of cytochrome c from the enzyme, the ascorbate plus

Table 3 Mitochondrial Enzyme Activities for Skeletal Muscle Mitochondrial Isolated from Three Patients with Mitochondrial Myopathy

	Patient			Control
	YD	JDM	JL	($n = 6$)
NADH CoQ1 reductase	22	24	15	182 ± 64
Succinate cytochrome c reductase	321	112	272	242 ± 134
Cytochrome oxidase	28.5	16.7	4.0	44.4 ± 14.6
Citrate synthase	1483	1053	1844	1180 ± 186

Activities are nmol/min/mg mitochondrial protein, except cytochrome oxidase activity, which is the first-order rate constant k/min/mg mitochondrial protein. The conditions of analysis are as described by Schapira et al. (42).

TMPD activity is not. Consequently, different abnormalities of the complex may differentially affect these activities depending on how they affect the function of the complex. The enzyme data for patient JL suggest the abnormality involves complexes I and IV in agreement with the polarographic data. Combined abnormalities of complex I and IV are a common observation and may reflect the involvement of mtDNA in the disease as these complexes are particularly dependent on mtDNA products.

ANALYSIS OF COMPONENTS OF THE RESPIRATORY CHAIN

The mitochondrial respiratory chain and oxidative phosphorylation system is composed of five multisubunit protein complexes (complexes I–V) and two mobile electron carriers, ubiquinone and cytochrome c. The number of protein subunits identified as part of this system has steadily risen over the years and now totals up to 83 different proteins, which include: 41 proteins in complex I; four proteins in complex II; 11 proteins in complex III; 13 proteins in complex IV; and 14 proteins in complex V (16). Associated with a number of these proteins are various redox centers, which are either iron sulfur centers or cytochromes, which can be detected by either electron spin resonance (ESR) or light spectroscopy, respectively. Unfortunately, these analyses require a relatively large amount of material, preferably purified mitochondria, which is not always available from patients.

CYTOCHROME ANALYSES

Complexes III and IV contain cytochromes b_{562}, b_{566}, c_1, and cytochromes a and a_3, respectively, and cytochrome c is a mobile electron carrier.

The cytochromes have a distinct visible spectrum when reduced. However, because the organelles produce a lot of light scattering and high levels of nonspecific absorption, the difference (reduced minus oxidized) spectrum has to be determined. The peaks at 550 nm (cytochrome c), 556 nm (cytochrome c_1), 562 nm (cytochrome b), and 602 nm (cytochromes aa_3) can be used to calculate the cytochrome levels of the mitochondria (17). The spectrum can be obtained at room temperature, but the signal is enhanced at liquid nitrogen temperatures because there is a decrease in line broadening due to molecular motion and because there is an increased effective light path due to multiple internal reflections from the ice crystals.

Samples can be reduced by either a nonspecific reducing agent (dithionite) or a physiological substrate (succinate). Nonspecific reducing agents must not be used when the sample contains hemoglobin or myoglobin because their difference spectra interfere with the cytochrome spectra (5). When a physiological reductant is being used, any functional defect of the respiratory chain will affect the reduction of the cytochromes and may reflect the ability of the respiratory chain to reduce the cytochromes rather than the absolute level of the cytochrome.

The cytochrome levels in isolated mitochondria from patients with various polarographic abnormalities are presented in Table 4. Clearly, because of the localization of the cytochromes within

Table 4 Cytochrome Levels in Mitochondria Isolated from Skeletal Muscle from Three Patients with Mitochondrial Myopathy

	Patient			Control
	YD	JDM	JL	($n = 6$)
Cytochrome cc_1	0.77	0.54	0.58	0.69 ± 0.10
Cytochrome b	0.43	0.08	0.16	0.41 ± 0.09
Cytochrome aa_3	0.10	0.09	ND	0.42 ± 0.09

Values are nmol/mg mitochondrial protein. The analysis was conducted according to the method described (17). ND, not detectable.

complexes III and IV, cytochrome levels are usually most informative in patients with abnormalities involving these complexes. However, from the data in Table 4 it can be seen that a complex I defect (YD) can also be associated with deficiencies of cytochromes aa_3 in particular. The relevance of this is not always apparent, especially in this patient where the other parameters of complex IV activity were normal. However, the 3243 bp (A to G) point mutation of mtDNA has been detected in this patient; therefore, the involvement of complex IV and cytochromes aa_3 in particular could be anticipated. In both JDM and JL, cytochromes b and aa_3 were markedly diminished, in agreement with their more delocalized defects involving, in particular, complexes III and IV, and complex IV, respectively.

IRON SULFUR CENTERS

The iron sulfur centers present in complexes I, II, and III have characteristic ESR spectra (18,19), and the presence of these centers can be detected in mitochondria. The requirement for relatively large amounts of material and specialist equipment has limited the usefulness of this analysis. However, two investigations of patients with a complex I defect identified decreased ESR signals from the ESR-visible Fe-S centers in complex I but normal signals from the other Fe-S centers in liver mitochondria obtained at postmortem (20,21). Western blot analysis in one of these patients (21) revealed a deficiency of all detectable complex I subunits in addition to a more severe decrease in the 75- and 51-kDa subunits, while in the second patient (22) there was a marked decrease in the 75- and 13-kDa subunits of complex I. In agreement with the ESR data, these subunits are associated with Fe-S centers (16).

POLYPEPTIDE ANALYSES

Specific respiratory chain subunits can be detected using antisera raised to either the whole complex or individual subunits that have been purified from mitochondria usually obtained from bovine heart. The specific antisera can be used to identify the specific components when used in conjunction with one- or two-dimensional SDS-PAGE followed by Western blotting (23); quantitate the levels of the subunits by ELISA (24); localize the cellular distribution of the subunits using immunocytochemistry (25); and purify the complex/subunit using immunoprecipitation (22). However, when antisera are being used, there are a number of drawbacks, including the specificity of the cross-reaction; the limited number of subunits that cross-react; and the difficulty in comparing data between laboratories because of the variability in the specificity and subunit recognition between antisera preparations.

Analysis of complex I, II, III, and IV subunits has been performed on a number of cases with mitochondrial myopathy. A deficiency of all cross-reactive complex I subunits has been reported in many cases (21,23,26), with the severity of the deficiency broadly correlating with the severity of the complex I dysfunction (21,27). In some cases there is a marked decrease in selective complex I subunits, which have included the 75-kDa and 13-kDa subunits (22); the 75-kDa and 51-kDa subunits (21); the 20-kDa subunit (28); and the 24-kDa, 24-kDa, and 13-kDa subunits (26). An example of these can be seen in Figure 4 where muscle mitochondria isolated from both patients 2 and 3 have a deficiency of all cross-reactive complex I subunits, in addition to which patient 3 has a specific deficiency of the 24-kDa subunit.

Analysis of complex III subunits has revealed a deficiency of all cross-reacting subunits in one patient (29) and a deficiency of the iron sulfur protein in two patients (1,15). In one of these patients the defect appears to be due to an abnormality in protein transport into the mitochondria (1).

Analysis of respiratory chain subunits has not significantly improved our understanding of the underlying abnormalities in these diseases. The main reason for this has been the selective nature of these studies. Only a small proportion of the respiratory chain subunits have been detected in the patients studied, and most of those detected are nuclear-encoded subunits. Consequently, it is difficult to draw many conclusions from the Western blots especially in the absence of information regarding the content of mtDNA-encoded subunits. This is particularly true for complex I where the number of detectable

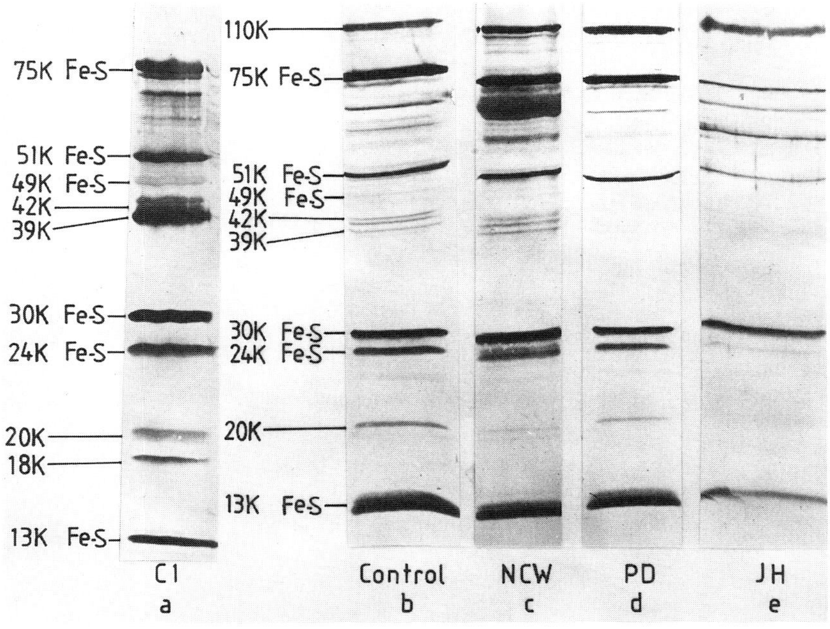

Figure 4 Western blot of human muscle mitochondrial protein with antisera against bovine complex I: (a) 25 μg control muscle mitochondria; (c) 250 μg muscle mitochondria from patient 1; (d) 250 μg muscle mitochondria from patient 2; (e) 250 μg of muscle mitochondria from patient 3. Antisera used were raised to the complex I holoenzyme in addition to serum raised to the specific subunits 30, 24, and 13 kDa. (Reproduced from Ref. 23, with permission.)

subunits is very low; only 10–15 of the 41 subunits can be routinely detected and these are probably all nuclear-encoded subunits.

Two-dimensional gel electrophoresis of mitochondrial proteins, involving isoelectric focusing or none equilibrium pH gradient gel electropheresis (NEPHGE) followed by SDS-PAGE, has been used in an attempt to identify a fingerprint of mitochondrial proteins (30). However, a system of sufficient reproducibility to identify many subunits has yet to be described.

The five complexes of the respiratory chain and oxidative phosphorylation system can be separated from one another and most other cellular components using the technique known as blue native gel electrophoresis (31). This involves separation of the complexes in the first dimension while they are still in their native state, followed by a second-dimensional SDS-PAGE separation. After Coomassie staining, the relative levels of the complexes in the sample (i.e., taking the ratio of prominent components of the individual complexes) can be determined. However, it is more difficult to determine the composition of the individual complexes (especially complex I) because the number of constituent subunits is often too large and they cannot always be distinguished from either one another or from contaminants. However, this system has the potential of allowing analysis of the five complexes in relatively small volumes of material, either purified mitochondria or whole tissue.

DNA ANALYSES

The fact that the mitochondrion contains its own DNA that codes for 13 polypeptides, which are all components of the respiratory chain and oxidative phosphorylation system, has focused much attention on this genome. Major alterations in mtDNA (large deletions and duplications) and point mutations affecting restriction endonuclease sites (either loss or gain of a site) have been detected by RFLP analysis. However, in the majority of cases this can now be performed more easily using the polymerase chain reaction (PCR). A number of methods are available for the detection of novel mutations of mtDNA, including a combination of PCR and sequencing of selected regions of the mtDNA, single-stranded conformational polymorphism (32), and chemical cleavage by mismatch (33). The genetics of mitochondrial myopathies are covered in another chapter.

MITOCHONDRIAL BIOGENESIS

The translation products of the mitochondrial DNA transcripts can be determined using either mitochondria freshly isolated from the patient's biopsy samples or in cells cultured from the patient. The results of such studies in patients with a point mutation of the $tRNA^{leu}$ or $tRNA^{lys}$ genes have indicated that there is decreased synthesis of the larger mtDNA-encoded proteins (34,35). Hayashi et al. (36) have also been able to demonstrate the translation of a predicted fusion protein resulting from a deletion of mtDNA.

TISSUE CULTURE

Cells cultured from patients with respiratory chain abnormalities have been increasingly used to study these abnormalities. Although mtDNA deletions in particular were often found to disappear in a mixed culture, if individual cells were clonally selected the mutation could be observed to persist in selected lines (37). Clonal cell culture has therefore proved to be a useful model of mtDNA abnormalities. Having a homogeneous population of cells not only makes interpretation of data a little easier, but also by providing a variety of different clones had led to a better understanding of the relationship between the molecular, biochemical, and clinical defects (34–36). However, there is still a barrier between our understanding of the molecular mechanisms in individual cell lines and the situation observed in vivo, where there is a heterogeneous population of cells.

CELL FUSION EXPERIMENTS

The development of cells devoid of mtDNA [rho° cells (38)] and their fusion with enucleated cells from patients with mitochondrial DNA abnormalities (Fig. 5) has enabled the investigation of the relationship between the mtDNA mutation and the functional abnormality and the role played by the nucleus in the mitochondrial respiratory chain abnormality. These cell fusion experiments have been successfully performed using cells expressing either the 3243-bp point mutation of the $tRNA^{leu}$ gene (34), the 8344-bp point mutation of the $tRNA^{lys}$ gene (35), or a mtDNA deletion (36). From these studies it would appear the mtDNA abnormalities alone are sufficient to account for the respiratory chain dysfunction observed in the patients. The $tRNA^{lys}$ mutation appears to exert its effect by impairing mtDNA translation, while the mechanism of the $tRNA^{leu}$ mutation may involve abnormal RNA processing. There is evidence that nuclear factors do play a role in the clinical expression of some disorders involving mtDNA mutations (39). However, cell fusion experiments are not able to identify whether nuclear factors, other than those that interact directly with the mitochondrial respiratory chain or mtDNA, are involved in the disease pathogenesis.

Rho° cells can be particularly useful in the identification of respiratory chain abnormalities due to an abnormal nuclear gene(s). A good example of this is the use of rho° cells to demonstrate that depletion of mtDNA in the fibroblasts from two siblings was due to a defect of a nuclear-encoded factor (40).

^{31}P NMR ANALYSES

A useful noninvasive analysis of disrupted muscle energetics is the use of ^{31}P nmr spectroscopy. ^{31}P nmr can monitor the levels of high-energy phosphate molecules [phosphocreatine (PCr) and ATP] and

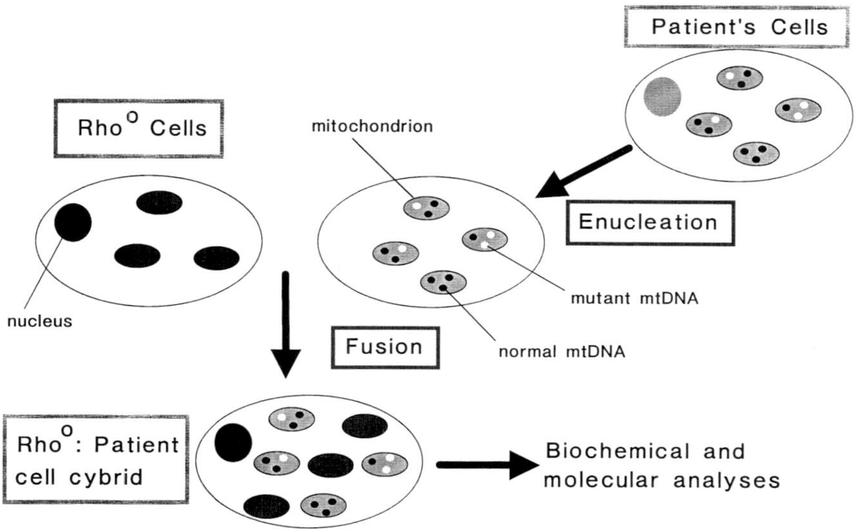

Figure 5 Generation of Rho°: patient cell cybrids. Experimental rho° cells can be fused with enucleated cells expressing a mitochondrial abnormality to produce a cybrid containing nuclear DNA from the rho° cell and mitochondrial DNA from the patient's cell line. Conditions for enucleation and cell fusion as described (40).

their response in muscle during exercise and recovery and also allow intramuscular pH and free ADP concentrations to be calculated.

Mitochondrial respiratory chain disorders give characteristic changes on nmr: an accelerated decline in phosphocreatine levels and intramuscular pH and elevated free ADP concentrations during exercise; and a delay in the recovery of PCr concentrations after exercise (41). These parameters can be used to confirm the presence of a mitochondrial respiratory chain defect, but like the exercise test, this technique cannot distinguish between patients with different molecular abnormalities. This subject is covered in depth in another chapter.

ACKNOWLEDGMENTS

We are grateful to Dr. J. A. Morgan-Hughes for allowing us to cite the biochemical data on some of his patients. We are indebited to the Muscular Dystrophy Group (UK) and the Medical Research Council for financial support.

REFERENCES

1. Schapira AHV, Cooper JM, Morgan-Hughes JA, Landon DN, Clark JB. Mitochondrial myopathy with a defect of mitochondrial protein transport. N Engl J Med 1990; 323:37–42.
2. Holt IJ, Harding AE, Petty RK, Morgan-Hughes JA. A new mitochondrial disease associated with mitochondrial DNA heteroplasmy. Am J Hum Genet 1990; 46:428–433.
3. Cullen MJ, Fulthorpe JJ. Stages in fiber breakdown in Duchenne muscular dystrophy. An electron microscopic study. J Neurol Sci 1975; 24:179–200.
4. Fardeau M. Ultrastructural lesions in progressive muscular dystrophies: a critical study of their specificity. In: Canal N, Scarlato G, Walton JW, eds. Muscle Diseases. Amsterdam: Excerpta Medica, 1970:98–108.
5. Bookelman H, Trijbels JMF, Sengers RCA, Janssen AJM. Measurement of cytochromes in human skeletal muscle isolated from fresh frozen stored muscle specimens. Biochem Med 1978; 19:366–373.
6. Hammans SR, Sweeney MG, Brockington M, Morgan-Hughes JA, Harding AE. Mitochondrial encephalopathies: molecular genetic diagnosis from blood samples. Lancet 1991; 337:1311–1313.

7. Krige D, Carroll MT, Cooper JM, Marsden CD, Schapira AHV. Platelet mitochondrial function in Parkinson's disease. Ann Neurol 1992; 32:782–788.
8. Byrne E, Trounce I. Oxygen electrode studies with human skeletal muscle mitochondria in vitro. J Neurol Sci 1985; 69:319–333.
9. Zheng X, Shoffner JM, Voljavec AS, Wallace DC. Evaluation of procedures for assaying oxidative phosphorylation enzyme activities in mitochondrial myopathy muscle biopsies. Biochim Biophys Acta 1990; 1019:1–10.
10. Fischer JC, Ruitenbeek W, Stadhouders AM, Trijbels JMF, Sengers RCA, Janssen AJM, Veerkamp JH. Investigation of mitochondrial metabolism in small human skeletal muscle biopsy specimens. Improvement of preparation procedure. Clin Chim Acta 1985; 145:89–100.
11. Spiro AJ, Moore CL, Prineas JW. Strasberg PM, Rapin I. A cytochrome-related inherited disorder of the nervous system. Arch Neurol 1970; 23:103–112.
12. Sherratt HSA, Watmough NJ, Johnson M, Turnbull DM. Methods used for the study of normal and abnormal skeletal muscle mitochondria. Meth Biochem Anal 1988; 33:245–335.
13. Holt IJ, Harding AE, Cooper JM, Schapira AHV, Toscano A, Morgan-Hughes JA, Clark JB. Mitochondrial myopathies: clinical and biochemical features of 30 patients with major deletions of muscle mitochondrial DNA. Ann Neurol 1989; 26:699–708.
14. Granger DL, Lehninger AL. Sites of inhibition of mitochondrial electron transport in macrophage-injured neoplastic cells. J Cell Biol 1982; 95:527–535.
15. Birch-Machin MA, Shepherd IM, Watmough NJ, Sherratt HS, Bartlett K, Darley-Usmar VM. Fatal lactic acidosis in infancy with a defect of complex III of the respiratory chain. Pediatr Res 1989; 25:553–559.
16. Cooper JM, Clark JB. The structural organisation of the mitochondrial respiratory chain. In: Schapira AHV. Di Mauro S, eds. Mitochondrial Disorders in Neurology. Oxford: Butterworths-Heinemann, 1994:1–30.
17. Morgan-Hughes JA, Darveniza P, Kahn SN, Landon DN, Sherratt RM, Land JM, Clark JB. A mitochondrial myopathy characterized by a deficiency in reducible cytochrome B. Brain 1977; 100:617–640.
18. Ragan CI, Ohnishi T, Hatefi Y. Iron-sulphur proteins of mitochondrial NADH-ubiquinone reductase (complex I). In: Matsubara H, et al, eds. Iron-Sulfur Protein Research. Tokyo: Springer-Verlag, 1986:220–231.
19. Ohnishi T. Structure of the succinate-ubiquinone oxidoreductase (complex I). Curr Top Bioenerg 1987; 15:37–65.
20. Moreadith RW, Batshaw ML, Ohnishi T, Kerr D, Knox B, Jackson D, Hruban R, Olson J, Reynafarje B, Lehninger AL. Deficiencies of the iron-sulfur clusters of mitochondrial reduced nicotinamide-adenine dinucleotide-ubiquinone oxidoreductase (complex I) in an infant with congenital lactic acidosis. J Clin Invest 1984; 74:685–697.
21. Ichiki T, Tanaka M, Kobayashi M, Sugiyama N, Suzuki H, Nishikimi M, Ohnishi T, Nonaka I, Wada Y, Ozawa T. Disproportionate deficiency of iron-sulphur clusters and subunits of complex I in mitochondrial encephalopathy. Pediatr Res 1989; 25:194–201.
22. Moreadith RW, Cleeter MJW. Ragan CI, Batshaw ML, Lehninger AL. Congenital deficiency of two polypeptide subunits of the iron-protein fragment of mitochondrial complex I. J Clin Invest 1987; 79:463–467.
23. Schapira AHV, Cooper JM, Morgan-Hughes JA, Patel SD, Cleeter MJW, Ragan CI, Clark JB. Molecular basis of mitochondrial myopathies: polypeptide analysis in complex I deficiency. Lancet 1988; 1:500–503.
24. Zeviani M, Peterson P, Servidei S, Bonilla E, DiMauro S. Benign reversible muscle cytochrome c oxidase deficiency: a second case. Neurology 1987; 37:64–67.
25. Johnson MA. Cytochemical and immunocytochemical studies of human mitochondrial myopathies. In: Azzi A, Drahota Z, Papa S, eds. Molecular Basis of Membrane-Associated Diseases. Berlin: Springer-Verlag, 1989:200–215.
26. Morgan-Hughes JA, Schapira AHV, Cooper JM, Clark JB. Molecular defects of NADH-ubiquinone oxidoreductase (complex I) in mitochondrial diseases. J Bioenerg Biomemb 1988; 20:365–383.
27. Cooper JM, Schapira AHV, Holt IJ, Toscano A, Harding AE, Morgan-Hughes JA, Clark JB. Biochemical and molecular aspects of human mitochondrial respiratory chain disorders. Biochem Soc Trans 1990; 18:517–519.
28. Slipetz DM, Goodyer PR, Rozen R. Congenital deficiency of a 20 kDa subunit of mitochondrial complex I in fibroblasts. Am J Hum Genet 1991; 48:1121–1126.
29. Darley-Usmar VM, Kennaway NG, Buist NRM, Capaldi RA. (1983) Defficiency in ubiquinone cytochrome c reductase in a patient with mitochondrial myopathy and lactic acidosis. Proc Natl Acad Sci USA 1983; 80:5103–5106.
30. Schapira AHV, Keir GF. Two dimension protein mapping by gold stain and immunoblotting. Anal Biochem 1988; 169:167–171.
31. Schägger H, von Jagow G. Blue native electrophoresis for isolation of membrane protein complexes in enzymatically active form. Anal Biochem 1991; 199:223–231.

32. Suomalainen A, Ciafaloni E, Koga Y. Peltonen L, DiMauro S, Schol EA. Use of single strand conformation polymorphism analysis to detect point mutations in human mitochondrial DNA. J Neurol Sci 1992; 111:222–226.
33. Morten KJ, Cooper JM, Brown G, Lake BD, Poulton J. A new point mutation associated with mitochondrial encephalomyopathy. Hum Mol Genet 1993; 2:2081–2087.
34. Chomyn A, Martinuzzi A, Yoneda M, Daga A, Hurko O, Johns D, Lai ST, Nonaka I, Angelini C, Attardi G. MELAS mutation in mtDNA binding site for transcription termination factor causes defects in protein synthesis and in respiration but no change in levels of upstream and downstream mature transcripts. Proc Natl Acad Sci USA 1992; 89:4221–4225.
35. Chomyn A, Meola G, Bresolin N, et al. In vitro genetic transfer of protein synthesis and respiration defects to mitochondrial DNA-less cells with myopathy—patient mitochondria. Mol Cell Biol 1991; 11:2236–2240.
36. Hayashi J-I, Ohta S, Kikuchi A, Takemitsu M, Goto Y-I, Nonaka I. Introduction of disease-related mitochondrial DNA deletions into HeLa cells lacking mitochondrial DNA results in mitochondrial dysfunction. Proc Natl Acad Sci USA 1991; 88:10614–10618.
37. Moraes CT, Schon EA, DiMauro S, Miranda AF. Heteroplasmy of mitochondrial genomes in clonal cultures from patients with Kearns-Sayre syndrome. Biochem Biophys Res Commun 1989; 160:765–771.
38. King MP, Attardi G. Human cells lacking mtDNA: repopulation with exogenous mitochondria by complementation. Science 1989; 246:500–503.
39. Vilkki J, Ott J, Savontaus M-L, Aula P, Nikoskelainen EK. Optic atrophy in LHON is probably determined by an X-chromosomal gene closely linked to DX57. Am J Hum Genet 1991; 48:486–491.
40. Bodnar AG, Cooper JM, Holt IJ, Leonard JV, Schapira AHV. Nuclear complementation restores mtDNA levels in cultured cells from a patient with mtDNA depletion. Am J Hum Genet 1993; 53:663–669.
41. Arnold DL, Taylor DJ, Radda GK. Investigation of human mitochondrial myopathies by phosphorus magnetic resonance spectroscopy. Ann Neurol 1985; 18:189–196.
42. Schapira AHV, Mann VM, Cooper JM, Dexter D, Daniel SE, Jenner P, Clark JB, Marsden CD. Anatomic and disease specificity of NADH CoQ_1 reductase (complex I) deficiency in Parkinson's disease. J Neurochem 1990; 55:2142–2145.
43. Schapira AHV, Cooper JM. Biochemical and molecular features of deficiencies of complexes I, II and III. In: Schapira AHV, Di Mauro S, eds. Mitochondrial Disorders in Neurology. Oxford: Butterworths-Heinemann, 1994:75–90.

41
Treatment of Mitochondrial Myopathies

R. C. A. Sengers, J. M. F. Trijbels, and W. Ruitenbeek
University Hospital Nijmegen, Nijmegen, The Netherlands

INTRODUCTION

A mitochondrial myopathy is a muscle disease characterized by structurally or numerically abnormal mitochondria and/or abnormally functioning mitochondria. In a considerable number of patients the muscle disease is part of a multisystem disorder. The clinical symptoms vary greatly and are often dominated by manifestations other than the muscular signs and symptoms, especially when the central nervous system is involved (1–4).

Biochemically the mitochondrial myopathies form a group of disorders that have the common effect of interfering with intracellular energy production. These disorders of energy metabolism are characterized by accumulation of intermediates of carbohydrate, fat, or amino acid metabolism and a disordered conservation of oxidatively derived energy and its utilization for ATP synthesis.

Mitochondria are formed by the concerted action of two genetic systems: the nucleocytoplasmic system and the intrinsic mitochondrial system (5). The genetic contribution of the mitochondria is modest because the genetic potential of mitochondrial DNA (mtDNA) is restricted to the equivalent of 16,569 base pairs, coding for 13 proteins of the oxidative phosphorylation system.

Most reported patients with a mitochondrial myopathy are sporadic cases. The diseases inherited as Mendelian traits may be classified as autosomal dominant, autosomal recessive, and X-linked recessive. Mitochondrial DNA is maternally transmitted. Therefore, one may expect a non-Mendelian pattern of inheritance in those mitochondrial myopathies in which a deficiency of a protein encoded by mitochondrial DNA is found (6–8). Mitochondrial myopathies result from an alteration of genetic material leading to an absent or abnormal protein essential for enzymatic activity, transport, assemblage, or structure. In the near future the investigator will be able to prove a primary mitochondrial pathology by demonstrating an etiological DNA abnormality.

Structural mitochondrial abnormalities are known to occur even in normal muscle tissue. They sometimes constitute a minor factor of the myopathology of dystrophic muscle and other well-defined neuromuscular diseases that are not thought to be due to primary mitochondrial dysfunction. In such diseases the cause of the structural and/or functional abnormalities may be assumed to be situated outside the mitochondria (Table 1).

Observations in this chapter relate to patients considered to suffer from a primary mitochonrial myopathy, unless an extramitochondrial cause can be inferred.

PRINCIPLES OF TREATMENT

In 1987 Przyrembel (9) reported literature data and personal experience in the treatment of patients with a mitochondrial myopathy. In this chapter we summarize the principles and add new insights. The

Table 1 Some Causes of Nonprimary Mitochondrial Pathology

	Example
Deficiency	Carnitine
	Copper
Intoxication	Drugs
Disturbed regulatory control	Hypothyroidism
Hypoxia	
Disturbed interaction with other cell organelles	Zellweger syndrome
Unidentified causes	

therapeutic approach to patients suffering from an inborn error of metabolism basically follows that of Beaudet et al. (10), with modifications as summarized in Table 2.

Treatment at the clinical phenotype level covers a variety of conventional medical practices. heavy exercise can be withheld preventing lactic acidosis or other consequences. Anticonvulsive drugs are given in case of an encephalopathy complicated by convulsions. In the treatment of patients some drugs should be withheld (vide infra).

A quite different situation occurs during an acute crisis, provoked by an intercurrent infection, exercise, prolonged fasting, or occurring spontaneously. Symptomatic treatment including buffering blood pH or removing toxic substances by dialysis is indicated (see also Application of Treatment).

Treatment at the metabolite level often involves nutritional or pharmacological approaches.

1. Dietary alteration may be effective if the pathophysiology involves accumulation of toxic precursors whose major source is nutritional. A low-fat, high-carbohydrate diet is recommended for the treatment of acylCoA dehydrogenases and carnitine deficiencies (11,12), medium-chain triglycerides for carnitine palmitoyl transferase deficiency (13), and high-fat, low-carbohydrate diet for the treatment of pyruvate dehydrogenase complex (PDHC) deficiency (9,14).
2. It is sometimes possible to promote the conversion of an offending metabolite to a readily removable substance: supplying extra carnitine in defects leading to high nonmetabolized acylCoA esters as in defects associated with organic acid accumulation. In cases with severe lactic acidosis it may be necessary to perform peritoneal dialysis (15).
3. Supplementation of a nonsynthesized product can be illustrated by administration of carnitine in case of disturbed carnitine synthesis. In a considerable number of patients the concentration of carnitine is decreased for unknown reasons. Carnitine can be supplemented. The concentration of carnitine in serum increases but in muscle tissue varies greatly during therapy. In patients with a complex I deficiency, succinate supplements can be beneficial (16).

Table 2 Strategies for the Treatment of Inherited Metabolic Diseases

1. Treatment at the clinical phenotype level
2. Treatment at the metabolite level
   a. Substrate reduction
   b. Utilization of alternative pathways
   c. Metabolic inhibitors
   d. Replacement of deficient product
3. Treatment at the protein level
   a. Activation with vitamin cofactors
   b. Protein replacement therapies
4. Radical scavengers
5. Somatic cell therapy

Treatment at the protein level implies activation or replacement of the mutant enzyme. Those mutations that either decrease the affinity of the enzyme for its cofactor or destabilize the protein may be partially overcome by substantially increasing the cofactor concentration in the surrounding milieu. Thiamine and lipoic acid has been used to stimulate PDHC activity (17).

Riboflavin, the precursor of flavin adenine dinucleotide and flavin mononueleotide, which are part of a number of enzymes (18), has proved effective in some cases of complex I, electron transfer flavoprotein, and multiple acylCoA dehydrogenase deficiency (12,19,20). Riboflavin administration can also be successful in patients with a less well-localized defect, possibly by stimulating the synthesis of mitochondrial proteins or by stabilizing mitochondria (21). Lipoic acid has been given with beneficial effect in a case of PDHC deficiency caused by dihydrolipoyl dehydrogenase deficiency (17).

Vitamin C (ascorbic acid), K3 (menadione), and CoQ are given either to bridge a defect in the electron transport chain or to supplement endogenous coenzyme Q (CoQ). Several open studies indicate that CoQ is effective in ophthalmoplegia plus and Kearns-Sayre syndrome. In a double-blind, crossover therapeutic trial with CoQ, Müller et al. (22) could not demonstrate any evident benefit, however.

Dichloroacetate (DCA) was found to inhibit PDHC kinase in experimental studies with concomitant activation of PDHC. DCA was found to be effective in decreasing blood pyruvate and lactate. Later experimental studies suggested that DCA may not readily cross the blood-brain barrier, and since toxic side effects were also reported, the therapeutic value is presumably limited (23). Nevertheless, recent studies showed beneficial effects in patients with a disordered respiratory chain (24). Suggested supplements are summarized in Table 3.

Radical scavengers. Oxygen is reduced to mono- and tetravalent forms within mitochondria. It is not unlikely that pathological amounts of oxygen radicals are formed in case of a disordered respiratory chain. Scavenging these radicals by vitamin E (α-tocopherol) or CoQ may have beneficial effects in

Table 3 Suggestions for Supplements and Drugs

Deficient enzyme	Drug	Dosage
Pyruvate dehydrogenase complex	Dichloroacetate	30–100 mg/kg/day
	Lipoic acid	5–50 mg/kg/day
	Thiamine	400–800 mg/day
Respiratory chain		
Complex I	Carnitine	50–200 mg/kg/day
	Menadione	40–500 mg/day
	Riboflavin	30–400
	Succinate	150 mg/kg/day
Complex III	Carnitine	v.s.
	Menadione	v.s.
Complex IV	Carnitine	v.s.
	Riboflavin	v.s.
Electron transfer flavoprotein	Carnitine	v.s.
	Riboflavin	v.s.
Ill-defined deficiencies incl. multiple defects	Ascorbic acid	4000 mg/day
	Carnitine	v.s.
	Coenzyme Q	60–200 mg/day
	Dicholoroacetate	v.s.
	Idebenone	90–180 mg/day
	Menadione	v.s.
	Riboflavine	v.s.
	α-tocopherol	100–300 mg/day

v.s. = vide supra.

mitochondrial myopathies with respiratory chain defect. Vitamin C can also be considered as a scavenger for free radicals. Successful use of idebenone has recently been described in patients with the MELAS syndrome (25) and in Leber's hereditary optic neuropathy (26). This drug with electron-shuttle and radical scavenger properties (27) seems to be especially suitable in encephalomyopathies, as it easily reaches the brain (28).

Somatic cell therapy. Replacing the deficient protein is the ultimate therapeutic approach, but unfortunately this has yet to be achieved. The alternative approach of gene therapy poses major hurdles. Somatic cell therapy seems to offer a possible shortcut for delivering the normal gene or its product directly to the muscle. It entails transplanting normal muscle cells directly into the diseased muscle with the aim of fusing donor to host myoblasts. Therapeutic trials of myoblast transfer into several major muscles in Duchenne dystrophy have been performed (29) but until now no such treatment of patients with mitochondrial myopathies has been reported.

APPLICATION OF TREATMENT

In general, defects of fatty acid metaolism are amenable to therapy by dietary measures, carnitine substitution, and in some cases with vitamins (9,12). Defects in pyruvate metabolism do not respond to therapy except in some special cases (30,31). Therapeutic attempts have included dietary measures and supplementation with vitamins as coenzyme precursors. Defects in the respiratory chain appear to respond to treatment only in exceptional cases.

Treatment for secondary abnormalities such as carnitine deficiency and accumulation of oxygen radicals may be beneficial.

In the case of a mitochondrial myopathy, an exact biochemical diagnosis is a prerequisite for a rational therapy. Unfortunately, in a considerable number of reported patients, while reasonable data permitting the diagnosis of mitochondrial myopathy were obtained, specific enzymatic deficiencies were not identified. Physicians should realize that muscle must be biopsied before the institution of any therapeutic regimen that may influence the dysfunctional enzyme or enzyme complex.

Apart from measures appropriate to treatment at the clinical phenotype level, cofactors can be given separately or in combination. Monotherapy, based on insight into the underlying metabolic disorder, should preferably be instituted. In emergency cases a "cocktail" of vitamins can be administered; to such a mixture, L-carnitine (50–200 mg/kg/day or more) should be added. It is wise to give such vitamins for a predetermined period. If a treatment is successful, it can generally be adjusted in a relatively short time, making a long-term trial unnecessary. Criteria for efficacy should also be defined in advance (see below).

It is advisable to avoid drugs such as valproic acid (which sequestrates carnitine and influences respiratory chain activity), barbiturates (which inhibit the respiratory chain), tetracyclines, and chloramphenicol (which inhibit mitochondrial protein synthesis).

EVALUATION OF TREATMENT

The evaluation of the therapeutic regimens used in the treatment of patients with a mitochondrial myopathy is another difficult problem. In literature the following criteria are mentioned:

1. Clinical observation including muscle power, anaerobic threshold, and exercise testing
2. Laboratory examination including estimation of serum lactate in relation to exercise, CK, organic acids in urine, and carnitine levels in plasma
3. Electrophysiological examinations
4. Magnetic resonance imaging and spectroscopy (14,32)
5. Histopathological examination
6. Biochemical examination of biopsied muscle

It must be stressed that only double-blind, placebo-controlled studies can give an unambiguous answer concerning the validity of a therapeutic approach.

CONCLUDING REMARKS

Unfortunately, in patients with a mitochondrial myopathy, all therapy is as yet palliative. Nevertheless, symptomatic and supportive aids are available, and through early diagnosis and rational therapy, comfort, functional capacity, and even life expectancy may be increased. Apart from dietetic approaches a few therapeutic manoevres may be of some value. Favorable responses justify therapeutic trials in every new patient.

REFERENCES

1. Sengers RCA, Stadhouders AM, Trijbels JMF. Mitochondrial myopathies. Clinical, morphological and biochemical aspects. Eur J Pediatr 1984; 141:192–207.
2. DiMauro S, Bonilla E, Zeviani M, Nakagawa M, DeVivo DC. Mitochondrial myopathies. Ann Neurol 1985; 17:521–538.
3. Petty PKH, Harding AE, Morgan-Hughes JA. The clinical features of mitochondrial myopathy. Brain 1986; 109:915–938.
4. Tulinius MH, Holme E, Kristiansson B, Larsson N-G, Oldfors A. Mitochondrial encephalomyopathies in childhood. II. Clinical manifestations and syndromes. J Pediatr 1991; 119:251–259.
5. Shoffner IV JM, Wallace DC. Oxidative phosphorylation diseases. Disorders of two genomes. Adv Hum Genet 1990; 19:267–330.
6. Harding AE, Petty RKH, Morgan-Hughes JA. Mitochondrial myopathy: a genetic study of 71 cases. J Med Genet 1988; 25:528–535.
7. Wallace DC. Diseases of the mitochondrial DNA. Annu Rev Biochem 1992; 61:1175–1212.
8. Zeviani M, Antozzi C. Defects of mitochondrial DNA. Brain Pathol 1992; 2:121–132.
9. Przyrembel H. Therapy of mitochondrial disorders. J Inher Metab Dis 1987; 10 (Suppl 1):129–146.
10. Beaudet AL, Scriver CR, Sly WS, et al. Genetics and biochemistry of variant human phenotypes. In: Scriver CR, Beaudet AL, Sly WS, Valle D, eds. The Metabolic Basis of Inherited Disease, 6th ed. New York:L McGraw-Hill, 1989:3–53.
11. Engel AG. Carnitine deficiency syndromes and lipid storage myopathies. In: Engel AG, Banker BQ, eds. Myology, Basic and Clinical. New York: McGraw-Hill, 1986:1663–1695.
12. Vianey-Liaud C, Divry P, Gregersen N, Mathieu M. The inborn errors of michondrial fatty acid oxidation. J Inher Metab Dis 1987; 10 (Suppl 1):159–198.
13. DiMauro S, Papadimitriou A. Carnitine palmitoyltransferase deficiency. In: Engel AG, Banker BQ, eds. Myology, Basic and Clinical. New York: McGraw-Hill, 1986:1697–1708.
14. Krägeloh-Mann I, Grodd W, Niemann G, Haas G, Ruitenbeek W. Assessment and therapy monitoring of Leigh disease by MRI and proton spectroscopy. Pediatr Neurol 1992; 8:60–64.
15. Scholte HR, De Jonge PC. Metabolism, function and transport of carnitine in health and disease. In: Gitzelman R, Baerlocher K, Steinmann B, eds. Carnitin in der Medizin. Stuttgart: Schattauer, 1987:21–59.
16. Wijburg FA, Barth PG, Ruitenbeek W, et al. Familial NADH: Q_1 oxidoreductase (complex I) deficiency: variable expression and possible treatment. J Inher Metab Dis 1989: 12 (Suppl 2):349–351.
17. Byrd DJ, Krohn H-P, Winkler L, et al. Neonatal pyruvate dehydrogenase deficiency with lipoate responsive lactic acidemia and hyperammonaemia. Eur J Pediatr 1989; 148:43–547.
18. Merrill AH, Lambeth JD, Edmondson DE, McCormick DB Formation and mode of action of flavoproteins. Annu Rev Nutr 1981; 1:281–317.
19. Arts WFM, Scholte HR, Bogaard JM, Kerrebijn KF, Luyt-Houwen IEM. NADH-CoQ reductase deficient myopathy: successful treatment with riboflavin. Lancet 1983; 2:581–582.
20. Bernsen PLJA, Gabreëls FJM, Ruitenbeek W, Sengers RCA, Stadhouders AM, Reinier WO. Successful treatment of pure myopathy, associated with complex I deficiency, with riboflavin and carnitine. Arch Neurol 1991; 48:334–338.
21. Addison R, McCormick DB. Biogenesis of flavoprotein and cytochrome components in hepatic mitochondria from riboflavin-deficient rats. Biochem Biophys Res Commun 1978; 81:133–138.
22. Müller W, Reimers CD, Berninger T, et al. Coenzyme Q_{10} in ophthalmoplegia plus—a double blind, cross-over therapeutic trial. J Neurol Sci 1990; 98–442.
23. Stacpoole PW. The pharmocology of dichloroacetate. Metabolism 1989; 38:1124–1144.
24. De Vivo DC, personal communication.

25. Ihara Y, Namba R, Kuroda S, Sato T, Shirabe T. Mitochondrial encephalomyopathy (MELAS): pathological study and successful therapy with coenzyme Q_{10} and idebenone. J Neurol Sci 1989; 90:263–271.
26. Mashima, Y, Hiida Y, Oguchi Y. Remission of Leber's hereditary optic neuropathy with idebenone. Lancet 1992; 2:368–369.
27. Imada I, Fujita T, Sugiyama Y, Okamoto K, Kobayashi Y. Effects of idebenone and related compounds on respiratory activities of brain mitochondria, and on lipid peroxidation of their membranes. Arch Gerontol Geriatr 1989; 8:323–341.
28. Zs-Nagy I. Chemistry, toxicology, pharmacology and pharmacokinetics of idebenone: a review. Arch Gerontol Geriatr 1990; 11:177–186.
29. Dubowitz V. Transferring myoblasts in Duchenne dystrophy. Br Med J 1992; 305:844–845.
30. Robinson BH, Sherwood WG. Lactic Acidaemia. J Inher Metab Dis 1984; 7 (Suppl 1):69–73.
31. Bartlett K, Ghneim HK, Stirk J-H, Dale G, Alberti KGMM. Pyruvate carboxylase deficiency. J Inher Metab Dis 1984: 7 (Suppl 1):74–78.
32. Radda GK, Rajagopalan B, Taylor DJ. Biochemistry in vivo: an appraisal of clinical magnetic resonance spectroscopy. Magnet Reson Q 1989; 5:122–151.

42
Idiopathic Inflammatory Myopathies

Russell J. M. Lane
Charing Cross Hospital and Charing Cross and Westminster Medical School, London, England

Peter Hudgson
Newcastle General Hospital and University of Newcastle upon Tyne, Newcastle upon Tyne, England

INTRODUCTION

The term "idiopathic inflammatory myopathies" (IIM) denotes a heterogeneous group of disorders having a primary autoimmune pathogenesis, but in which the nature of the antigen(s) involved and the cause of the loss of immunological tolerance are as yet unknown. The term encompasses polymyositis (PM) and dermatomyositis (DM); IIM in association with other autoimmune diseases, notably collagen vascular diseases (overlap myositis); IIM associated with malignancy; and rarer focal and localized forms of myositis. Inflammatory infiltration is also a feature of sarcoid and other idiopathic granulomatous myopathies and of muscle involvement in vasculitic disorders, which can therefore be conveniently considered with this group of conditions.

Of course, inflammatory changes in muscle may be encountered in many other disorders, notably inclusion body myositis (IBM), which probably accounts for some cases of "polymyositis" that do not respond to steroid treatment; muscle infection, by bacteria, viruses, and parasites; occasionally in dystrophies, notably the Duchenne and facioscapulohumeral (FSH) forms; and rarely in metabolic and toxic myopathies. In such conditions, however, there are usually distinctive clinical and pathological features indicating the appropriate diagnosis, or the external antigenic stimulus is evident. IIM can also result from drug hypersensitivity reactions (notably to *d*-penicillamine, phenytoin, and procainemaide, see Chapter 31) and graft-versus-host disease. Such instances, in which the responsible antigen is defined, support the concept of an autoimmune basis for IIM.

NOSOLOGY AND CLASSIFICATION

The first cases of IIM were reported in the midnineteenth century (1–3). Skin involvement was commonly but not invariably observed, and cases were subsequently described as having PM or DM in a somewhat indiscriminate manner. Debate followed concerning the relationship between childhood and adult cases, and cases with isolated muscle disease as opposed to those with associated diseases such as collagen vascular disorders and malignancy. In the first authoritative review of the subject, Walton and Adams (4) proposed four subgroups of IIM: "pure" PM, which could present in acute (rhabdomyolytic), subacute, and chronic forms; PM with minimal or transient skin involvement; PM or DM associated with other autoimmune diseases; and PM or DM with malignancy. However, the elegant and detailed immunocytochemical studies by Engel and colleagues at the Mayo Clinic have since shown that, while PM and DM share clinical and pathological features, they are fundamentally distinct entities, at least on immunopathological grounds, and several authorities also consider DM in childhood to be a separate disorder (systemic angiopathy of childhood) (5,6), since particular manifestations and complications seen in childhood cases are rarely encountered in adults. In addition, several other forms of IIM have since been recognized. Debate concerning the nosology of IIM continues therefore (7) and has been complicated more recently by the discovery of *myositis-specific autoantibodies* (MSAs), which define new clinical subgroups of PM and DM (see below). A clinically based classification of IIM is given in Table 1.

One recent multicenter survey in Japan (8) found that the presence of at least one characteristic dermal manifestation of DM, together with at least four other typical signs of myositis, resulted in a classification of DM with 94.1% sensitivity, while a diagnosis of PM was mde in the absence of such skin lesions, with a sensitivity of 98.9%. The overall specificity of these clinical features in IIM was 95.2%

EPIDEMIOLOGY OF IIM

IIM are said to be the most common of all acquired myopathies (although see discussion of alcoholic myopathy in Chapter 31), but are still relatively rare conditions. IBM has accounted for a substantial proportion of cases of inflammatory myopathy in recent prevalence studies (see Chapter 43). One large study from North America, covering 1947–1968 (9), gave an incidence for IIM of around 0.5/100,000. A more recent nationwide study of PM/DM among children up to 15 years of age, in Finland, found an annual incidence of 0.3/100,000 (10), and a study in southern New England found a similar rate of 0.4/100,000 (11). De Vere and Bradley's 1975 study in Newcastle (12) gave a prevalence of about 8/100,000, over a 20-year period. As with all autoimmune disorders, IIM is some 2–3 times more common in females than males, and black females were affected twice as commonly as white in the American study (9). A recent Australian study (13), in which ascertainment of all cases of *biopsy-proven* IIM in New South Wales appeared complete, identified 77 cases among a population of 4.5×10^6 over a 5-year period between 1982 and 1987. It should be borne in mind that infective myositis is substantially more common than IIM worldwide, because of its high incidence in Third World countries (see Chapter 32).

Table 1 Clinical Classification of IIM

Type of IIM	Distinguishing features
Polymyositis (PM)	Symptoms and signs apparently confined to skeletal muscle
Dermatomyositis (DM)	Characteristic skin manifestations (see text)
Overlap myositis	IIM in association with defined connective tissue disease
Cancer-associated myositis (CAM)	With malignancy diagnosed within 2 years before or after onset of myositis
Juvenile (childhood) myositis	Onset before 18, frequent skin and gastrointestinal vasculitis, subcutaneous calcification
Inclusion body myositis (IBM)	Mainly males, insidious progressive weakness, often distal, little response to treatment, characteristic histological and ultrastructural features (see text)
Granulomatous myositis	Characteristic pathological changes (see text)
Eosinophilic myositis	Inflammation with eosinophils on biopsy
Focal myositis	IIM confined to a restricted area, e.g., one side of body, one limb, or one muscle
Localized nodular myositis	Single or multiple painful muscle swellings with biopsy features of inflammation and infarction; may progress to more typical PM.
Ocular (orbital) myositis	Myositis confined to extraocular muscles

CLINICAL FEATURES

Polymyositis and Dermatomyositis in Adults

Wagner (1), in the first reported case of what he termed "polymyositis," described a young German woman with established tuberculosis, who developed profound generalized muscular weakness with skin rash and died within 6 days. Such acute cases of IIM, which can result in rhabdomyolysis, gross muscle swelling, and sometimes marked subcutaneous edema, are rarely encountered (14). PM and DM in adults present more commonly as a subacute, progressive proximal weakness, usually without marked wasting but frequently with pain and muscle tenderness, notably in the upper arm and shoulders. Other common symptoms include dysphagia, due to involvement of striated muscle in the pharynx and upper esophagus, Raynaud's phenomenon, which can antedate the onset of weakness by many months, and arthralgia and sometimes deforming arthritis, in about a third of cases (12, 15, 16). In overlap syndromes there will be additional symptoms and signs related to the associated disorder.

Less commonly, PM may present as an chronic, indolent proximal myopathy without other symptoms encountered in more acute cases. Females thus affected probably account for most cases of so-called "late life" or "menopausal muscular dystrophy" (17, 18). It was the recognition that such cases were forms of acquired inflammatory myopathy, rather than "genetically determined, primary muscle cell degeneration" (Walton's definition of a muscular dystrophy), which led Nattrass (19) and Walton and Nattrass (20) to distinguish inflammatory myopathies from muscular dystrophies and to propose new classifications of muscle disease (20).

On examination, a skin rash may be present (see below), but can be difficult to see, or evanescent. Other common general manifestations might include some joint pain and swelling and changes of Raynaud's phenomenon. The muscles may be tender, and weakness is chiefly proximal and symmetrical, with normal or brisk reflexes, unless the muscle involvement is very severe or chronic, in which event the reflexes may be depressed or absent. Neck flexion and extension weakness may be particularly prominent. As mentioned, severe acute disease may result in muscle swelling, due to edema consequent upon inflammation and myonecrosis; conversely, chronic or untreated disease may result in muscle wasting (12,15). The general manifestations of PM/DM in adults are given in Table 2.

Skin Lesions

The typical rash of DM is a 'heliotrope' slightly cyanotic erythema, usually seen over the light-exposed malar and supraorbital areas on the face, on the anterior chest (the "V" sign), and over the shoulders, upper back, and upper arms (the "shawl sign"). Gottron's lesions (erythematous smooth or scaly patches over the dorsal interphalangeal or metacarpophalangeal joints, the extensor surfaces of knees or elbows, and the medial malleoli) and Gottron's papules (violaceous, flat-topped papules in similar areas) may be found (see Chapter 1, Figs. 1 and 2). Rarely, the whole body may be affected, or the erythema can be curiously localized (Fig. 1). Dilated nail fold capillaries or frank telangiectases may also be found. In the acute case there may be cutaneous edema, which can be severe (14), but subsequently there is

Table 2 Symptoms and Signs of Polymyositis and Dermatomyositis

System	Symptoms and signs
General	Fatigue, malaise, weight loss, fever
Musculoskeletal	Weakness, myalgia, arthralgia, tenderness, wasting, arthritis
Cutaneous	Raynaud's, "heliotrope cyanosis," "V" sign, "shawl sign," Gottron's lesions, nailfold capillary changes, "mechanic's hands"
Pulmonary	Dyspnea, aspiration pneumonia
Cardiac	Palpitations, cardiac failure
Gastrointestinal	Dysphagia, reflux, regurgitation

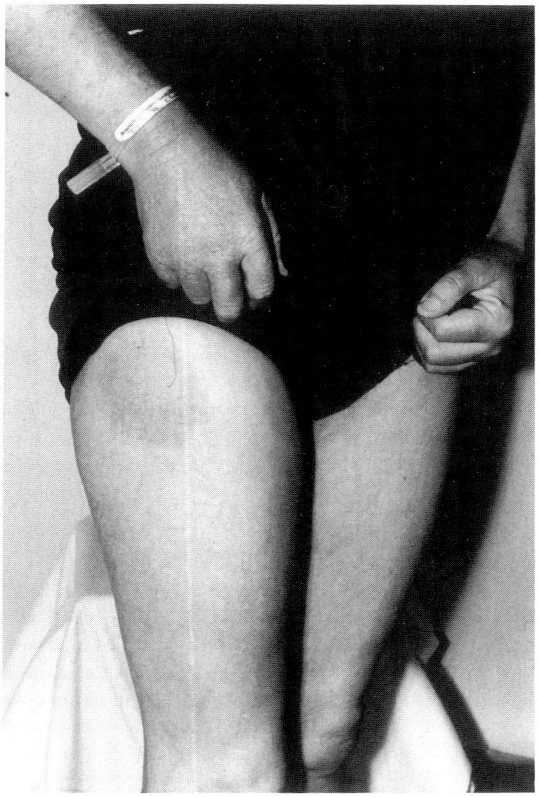

Figure 1 Dermatomyositis in a case of cancer-associated myositis. This woman, with myositis associated with underlying ovarian cancer, had several discrete areas of cyanotic cutaneous erythema.

scaling, induration, tightening, reticulate erythema, and pigmentation at the sites of the lesions. Cutaneous and subcutaneous calcification is uncommon in adult DM, but may be a feature of the disease in childhood and in patients with scleroderma. As with lupus, exposure to sunlight may worsen the rash, and also the muscle weakness.

The rash may be minimal; conversely, patients with marked DM rash but without apparent clinical or biopsy evidence of muscle involvement have been described (21). However, detailed studies of such "amyopathic DM" cases may demonstrate electromyographic, immunocytochemical, and ultrastructural characteristics of myositis (22), and phosphorus magnetic resonance spectroscopy disclosed inefficient muscle energy metabolism during exercise in most such cases in one recent study (23).

Other dermal complications encountered in IIM, such as scleroderma and calcinosis, are discussed below.

Respiratory Involvement

Subclinical restrictive defects, due to weakness of respiratory musculature, are common. Significant dyspnea is rare, although fatal respiratory failure due to respiratory muscle weakness has been reported (24). Respiratory problems are most likely to arise in overlap syndromes, since pulmonary disease occurs in many of the associated disorders. About 10% of patients with IIM have an associated interstitial lung disease (25), due to bronchiolitis, interstitial pneumonia or diffuse alveolar damage, with chest X-ray changes including reticular shadowing and patchy pneumonic consolidation, particularly at the lung bases (26). Respiratory muscle weakness usually responds well to steroid treatment, but interstitial lung disease carries a poor prognosis, with 40% mortality after 3 years in one series (22). Pulmonary involvement can rarely result in life-threatening alveolar haemorrhage (27) and pneumomediastinum with subcutaneous emphysema (28).

There is a strong relationship between interstitial lung disease in PM and the anti-Jo-1 antibody (see below).

Cardiac Disease

Cardiac involvement, particularly mild congestive cardiomyopathy, is common, but resulting symptoms and signs are generally nonspecific and only rarely pose significant management problems (29,30). However, involvement of the conducting system can lead to a variety of arrhythmias, notably complete heart block, which may require treatment by pacemaker (31). Cardiac involvement can develop at any stage of the disease, independently of treatment response or other variables, although there is a relationship with increasing age (30).

Overlap Myositis

Clinical evidence of associated collagen vascular disease is present in about 20% of IIM cases (32). Conversely, mild weakness is present in most patients presenting with primary features of these conditions. Muscle biopsy in such cases will often show inflammatory changes, but myonecrosis, associated with raised creatine kinase (CK) levels and spontaneous discharges on electromyography indicative of myositis with fiber necrosis, is relatively uncommon.

Sjögren's syndrome comprises a chronic arthritis and reduced mucous membrane secretions, leading to dry mouth and eyes (sicca syndrome) and sometimes affecting the gut and genitourinary tract. It is probably the most common associated condition in IIM overlap syndromes and is also found to various degrees in most other collagen vascular diseases and in other organ-specific autoimmune diseases, such as rheumatoid arthritis, primary biliary cirrhosis, chronic active hepatitis, and thyroiditis. Muscle weakness is a significant feature in most cases and can be severe (33), and muscle biopsy often shows inflammation and vasculitis even in patients without significant weakness. A predominantly sensory polyneuropathy is often found in addition.

Scleroderma is also associated frequently with IIM ("scleroderma myositis"). In the classical form, the condition is predominantly confined to the skin, with hardening and tightening over the face, around the mouth (restricting mouth opening), and over the fingers (sclerodactyly or acrosclerosis—see Chapter 1, Fig. 2), while additional features are seen in the CREST syndrome (calcinosis, Raynaud's phenomenon, esophageal disease, sclerodactyly, cutaneous telangiectasia). In progressive systemic sclerosis (PSS), the gut and other organs are involved, leading to severe dysphagia and malabsorption.

In one study of IIM in scleroderma, Raynaud's phenomenon and arthritis were present in virtually all cases, and pulmonary involvement was present in three-quarters, while calcinosis and sicca symptoms were found in 50% (34). Cardiac involvement was detected in 20% of cases in another study (35).

Localized scleroderma (morphea) and linear scleroderma (coup de sabre) may both be associated with myositis deep to the site of skin involvement.

Systemic lupus erythematosus (SLE), by contrast, is encountered only infrequently in patients with PM or DM (12). Involvement of other organ systems, such as the skin, joints, kidney, and nervous system, is likely to dominate the clinical picture, although dysphagia and Raynaud's phenomenon are found more commonly in overlap syndrome than in SLE alone (36).

Rheumatoid arthritis (RA) is also rarely found in association with IIM, and muscle biopsy in RA cases is more likely to demonstrate type 2 fiber atrophy from disuse or steroid treatment than inflammatory changes (12). Moreover, a number of clinical forms of neuropathy can occur that may complicate the picture.

Mixed connective tissue disease (MCTD) presents with variable features of SLE, scleroderma, and

DM and is defined by the clinical characteristics of hand edema, synovitis, myositis, Raynaud's phenomenon, and acrosclerosis. These criteria, together with a high titer of the extractable nuclear antibody anti-U_1 snRNP, are said tc provide a diagnostic sensitivity and specificity approaching 100%. This multisystem disorder can also result in hepatosplenomegaly, lymphadenopathy, and aseptic meningitis, as well as pulmonary, renal, and cardiac disease (37,38). However, the nosological status of MCTD has recently been questioned. On follow-up over 5 years, 18 patients with anti-U_1 snRNP, originally diagnosed as having MCTD, were eventually found to fulfill diagnostic criteria for SLE, PSS, or RA (39).

Juvenile Dermatomyositis (Childhood DM)

IIM without skin involvement is unusual in childhood, where the condition usually presents in an insidious, subacute fashion (often incorrectly assumed to represent a "viral infection"), with skin rash followed by weakness, myalgia, and general malaise (the "miserable, weak child with rash"). In most respects, the manifestations of childhood DM are very similar to those in the adult, in terms of the characteristics of the rash, distribution of weakness, and involvement of other systems and organs. However, complications such as muscle contractures, gastrointestinal hemorrhage (due to vasculitis with ischemic mucosal ulceration), and dystrophic subcutaneous and muscle calcification appear to be significantly more common than in adults (5,6,40). Greater awareness of this condition and the institution of effective, early treatment may limit the incidence of such manifestations (41).

PM/DM and Malignancy

The relationship between muscle disease (and inflammatory muscle disease in particular) and malignancy remains a vexed issue. The association was first suggested by Stertz in 1916, but not emphasized until the studies by Bezecny in 1935 (42).

Early surveys noted an association between DM and PM and malignancy in 6–45% and 0–28% of cases, respectively, with an increasing incidence in patients over 40 (especially males) and a low incidence in children (43). It is known that detection of a neoplasm may precede or follow the development of PM/DM, sometimes by years, so the intensity of investigation, duration of follow-up, and frequency of postmortem confirmation can confound observations on the relationship. Thus, Lakhanpal et al. (44) found 25.2% of 115 PM/DM patients had a malignancy at any time compared to 17.4% of 115 age-matched controls, which was not statistically different overall. However, the incidence was just significantly greater in the IIM patients in the first 2 years after evaluation (45) Manchul et al. (46) also found a concurrent malignancy of 21% in 76 PM/DM patients. This was substantially greater than in their controls, but the PM/DM group did not show a greater incidence of malignancy on subsequent prolonged follow-up. Two more recent epidemiological studies, from Denmark (47) and Boston (48), found that the overall risk of malignancy in DM was about 4–5 times, and in PM about 2–3 times, greater than in the general population, but noted that there was a "window" of highest risk in the first few years after the development of IIM, with a lesser, but still increased, risk 5 years before and after this time (48).

PM/DM may be associated with virtually any form of malignancy, including Hodgkin and non-Hodgkin lymphomas (49,50), hematopoietic malignancies (47), and malignant melanoma (with DM) (51). Perhaps the most compelling evidence of an association between IIM and malignancy comes from the observation that certain tumors, such as ovary (52,53) and particularly nasopharyngeal carcinoma in the Chinese (54), occur significantly more frequently in association with IIM than in the general population. Early studies also showed that some patients have serum antibodies specific to their own tumor that cross-react with muscle components (55).

Even greater controversy surrounds the issue of "carcinomatous myopathy," the term used by Henson et al. (56) and Heathfield and Williams (57) in 1954 to describe a myopathy in association with malignancy, with clinical features similar to polymyostitis, but showing little or no inflammation on muscle biopsy. Since muscle biopsy in otherwise clear-cut cases of IIM may also sometimes show a paucity of inflammatory changes, it was subsequently suggested that IIM associated with malignancy and "carcinomatous myopathy" were in fact the same condition (4). However, myopathy associated with carcinomas of the bronchus, breast, and ovary may be characterized histologically by an extremely aggressive *necrotizing* myopathy, without lymphocytic infiltration (58), and similar findings have been observed in myeloproliferative disorders (59), raising the possibility that such myopathies result from some humoral myotoxic factor, rather than primary immunological insult. This is considered further in Chapter 32, and the question of screening for malignancy in IIM is addressed below.

LABORATORY DIAGNOSIS

The diagnosis of IIM depends in large measure on fulfilling the criteria established by Bohan and Peter (15), and modified subsequently by Hudgson and Peter (7). Thus, a suggestive clinical picture should be supported by an increase in serum muscle enzymes, a "mixed" pattern of "denervation" and myopathic changes on electromyography (EMG), and inflammatory and necrotic changes on muscle biopsy.

Blood Tests

The *erythrocyte sedimentation rate* is only raised in about half of all PM/DM cases and shows no close association with disease activity in muscle (12), but is more commonly increased in active overlap syndromes.

The extent of elevation of *creatine kinase* (CK) reflects the degree of muscle fiber damage to a large degree. Conversely, significant muscle inflammation can be present in the absence of parenchymal damage, for example in scleroderma myositis (see below). Other enzyme levels (such as SGOT, aldolase, lactate dehydrogenase) may also be raised, but are generally less useful since they are not muscle-specific. However, elevation of *serum myoglobin*, a smaller molecule than CK, has been reported to be more sensitive and to correlate better with the clinical state in several disorders, including IIM (see Chapter 49) (60,61). Curiously, IIM rarely causes myoglobinuria, even in the acute form (14) (see Chapter 49). Levels of the BB and MB *isoenzymes of CK,* which are normally present only in immature skeletal muscle fibers, also rise, reflecting muscle fiber regeneration (62).

A range of *autoantibodies* may be detected in patients with IIM. Some of these are common to a number of autoimmune diseases, some are associated closely with particular collagen vascular diseases in overlap syndromes, and some are specific to myositis and have prognostic as well as diagnostic value (see below). Certain clinical forms of IIM, and particular autoantibodies, are associated with specific *human leucocyte antigen* (HLA) *phenotypes*, and some authorities have suggested that the related antigens might therefore be involved in the etiology of IIM. This concept is considered in depth in Pathogenesis and Etiology (see below).

Electromyography and Myoimaging

EMG characteristics of IIM may change as the disease progresses, but during active myositis there is typically increased muscle fiber irritability, comprising increased insertional activity (sometimes including pseudomyotonic and complex repetitive discharges), fibrillations, and positive sharp waves, together with a myopathic interference pattern on contraction. The increased insertional activity is believed to result from functional "denervation," through isolation of areas of fiber membrane from the motor end-plate, and from damage to surface membranes resulting in depolarization. In long-standing cases, only "denervation" changes may be found, or less frequently the pattern may be purely "myopathic" (12). Sandstedt et al. (63) reported that serial EMG could be used to monitor disease activity during treatment, with reduction in spontaneous activity and high-frequency polyphasia reflecting increasing muscle strength and fall in CK levels.

Patients with IIM have been studied by a variety of *myoimaging* techniques, including radionuclide, computerized tomography, and magnetic resonance imaging (MRI) scanning. These can be useful in documenting the extent and activity of disease and in monitoring response to treatment (see Chapter 8), but it is not clear that this approach is superior to careful clinical evaluation. One recent study suggested that prior localization of lesions by MRI might increase the diagnostic yield of muscle biopsy (64).

Ancillary Investigations

In view of the known pulmonary and cardiac complications of IIM, all patients should have a baseline chest X-ray (CXR), electrocardiogram, and pulmonary function tests. Echocardiography and dynamic cardiac studies should be considered on clinical grounds. Opinion differs on how assiduously underlying malignancy should be sought. It is the authors' practice to perform screening in all patients over 40 years of age, particularly in cases of DM, to include CXR, abdominal ultrasound, mammography, and colonoscopy, to detect the most commonly associated malignancies. Repeat screening might be performed in cases refractory to treatment. The search for ovarian tumors in particular needs to be assiduous (52).

Muscle Biopsy

The presence of mononuclear inflammatory cell infiltrates in muscle biopsy is the sine qua non of a confident diagnosis of IIM. However, the nature of the inflammatory cells, their distribution, and the extent of associated necrobiotic changes vary substantially between cases. Inflammation may not be evident in perhaps 10% of biopsy samples, although diagnostic yield may be improved by multiple site sampling (see below). The inflammatory cells comprise lymphocytes, plasma cells, and histiocytes (macrophages), whose relative abundance in different disorders and in different areas of a biopsy can be studied using specific immunocytochemical markers (65–71).

In *dermatomyositis*, inflammation is notably perivascular and perifascicular (Fig. 2), with relatively sparse endomysial involvement, and the cells are principally $CD4^+$ (helper) T cells and B cells, with some plasma cells (Fig. 3a–d). Perifascicular atrophy may be striking, particularly in childhood DM (Fig. 4). This appears to arise as a result of fiber necrosis or degeneration with incomplete regeneration. The distribution of the atrophy, and the relationship to areas of muscle infarction that can be seen in some cases, suggest that it might be secondary to ischemia (71); similar morphological changes were produced by intramuscular injection of serotonin in a model of experimental ischemic myopathy (72).

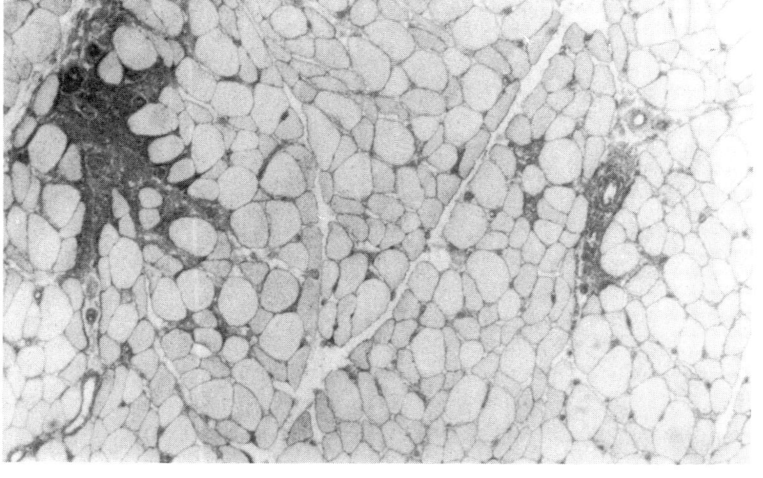

Figure 2 Dermatomyositis. The inflammation is typically perivascular and perifascicular in distribution. (H&E × 50.)

There is now good evidence that the primary lesion in DM is indeed a microangiopathy, mediated by local humoral mechanisms. Arterioles and venules show intimal and endothelial hyperplasia, and platelet and fibrin thrombi may be found occluding vessel lumens, while immune complexes and elements of the membrane attack complex (immunoglobulins and complement components) can be demonstrated on intimal surfaces of vessels (68–70). A recent ultrastructural study also demonstrated a remarkable reduction in endomysial capillary size and density in very early cases of DM, before other significant inflammatory or necrobiotic changes were evident, together with tubuloreticular inclusions (TRI) in vascular endothelial cells, macrophages, and satellite cells, indicating production of interferon (and presumably other cytokines) (69). Neither of these ultrastructural findings was observed in cases of PM. (The significance of TRI is discussed further in relation to myopathy in HIV infection in Chapter 52).

In contrast, lymphocytic infiltration in *polymyositis* is principally endomysial and comprises T cells, macrophages, virtually no natural killer (NK) cells, and few B cells (71). Unfortunately, conflicting observations on the precise nature of the T cells in the infiltrates have been made. Some studies reported a predominance of $CD4^+$ lymphocytes (73–75), while others found that $CD8^+$ cells were the main effectors (65–67, 76). Subsequent detailed analysis of T-cell-receptor protein characteristics suggests that the latter view is correct (71) (see below). One group noted that the proportion of activated cytotoxic T lymphocytes increases from perivascular to perifascicular to endomysial sites, while the distribution of macrophages tends to be fairly constant (65,66), but others could not confirm this (76). These discrepancies may reflect variation in patient selection and disease activity. It is clear, however, that the majority of the T lymphocytes bear HLA-DR antigens, indicating that they are activated, supporting the concept of cell-mediated immunopathogenesis in PM (71).

Macrophages may be particularly prominent in apparently refractory cases of IIM treated with long-term immunosuppressants (see below) and can be shown to be invading only necrotic fibers, while T lymphocytes invade both nonnecrotic and necrotic cells (65,66). There is little in the way of microvascular damage or evidence of humorally mediated disease in PM, while necrosis is found in both PM and DM, but is much more pronounced in PM. It may involve single fibers or clusters of fibers in a focal area, and it is typically patchy. Often, only a segment of a fiber will show necrosis. Regeneration occurs sequentially with degeneration, regenerating fibers being characterized by basophilic cytoplasm and internal nuclei, with prominent nucleoli, the resulting fibers having type 2c histochemical characteristics. These changes are reflected in an increased variation of fiber diameters, and there is generally an increase in connective tissue relating to the extent of fiber destruction, possibly as a result of increased fibroblast activity and collagen synthesis in response to the inflammation. The histological and immunocytochemical characteristics of DM and PM are summarized in Table 3.

Table 3 Pathological and Immunopathological Characteristics of DM and PM

	Dermatomyositis	Polymyositis
Light microscopy		
Inflammation	Perivascular, perifasicular	Endomysial
Necrosis	+	+++
Atrophy	++ perifasicular	± general
Ultrastructure		
Vessels	Fibrin, platelet thrombi	No specific changes
TRI	+	–
Immunocytochemistry		
$CD4^+$	+++	+
$CD8^+$	+	+++
Macrophages	+	+
Plasma cells	+++	+
Immunoglobulins	+++	+

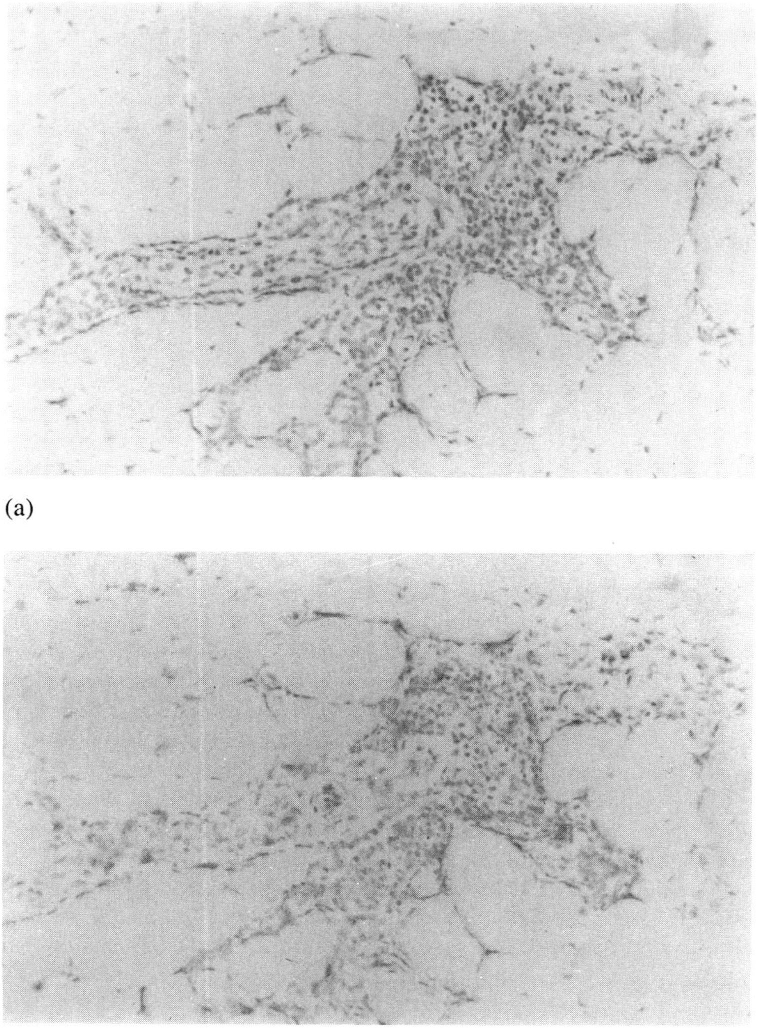

(a)

(b)

Figure 3 Immunocytochemistry in DM. (a) Perifascicular infiltrate showing adundance of $CD4^+$ helper T-cells. (b) Serial section stained for $CD8^+$; few positive cells are seen. (c) A large number of B cells are demonstrated with the B-cell marker CD22. (d) Macrophages in the infiltrate shown with CD 68. (All sections ×64.)

Histochemical analysis of biopsies in PM/DM typically shows numbers of "moth-eaten" fibers on oxidative enzyme stains, in which there is a loss of the intermyofibrillar network in the center of fibers. These are particularly notable at the periphery of fascicles in DM showing perifascicular atrophy, again possibly reflecting an ischemic insult. Morpohometric studies of ATPase preparations generally show a nonspecific type 2 atrophy, and often an increased complement of 2c fibers; fiber hypertrophy is not a common feature (77,78). "Ragged red" fibers, with abnormal succinate dehydrogenase staining characteristics suggestive of mitochondrial abnormality (Chapter 6), may be observed but are no more common than in age-matched normal controls (79).

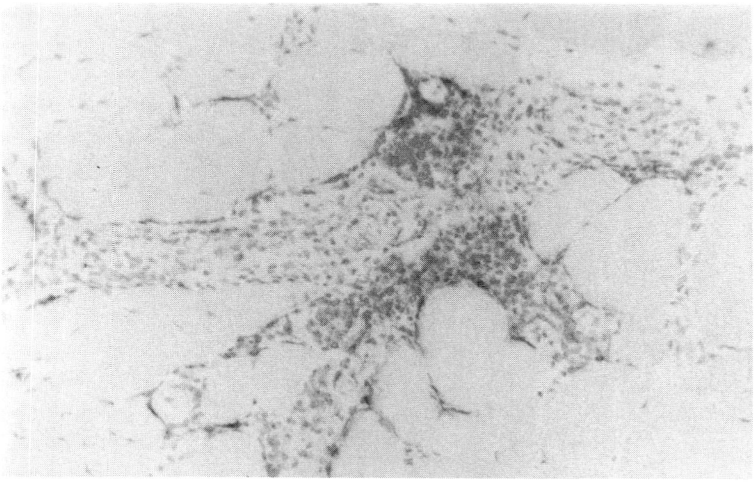

(c)

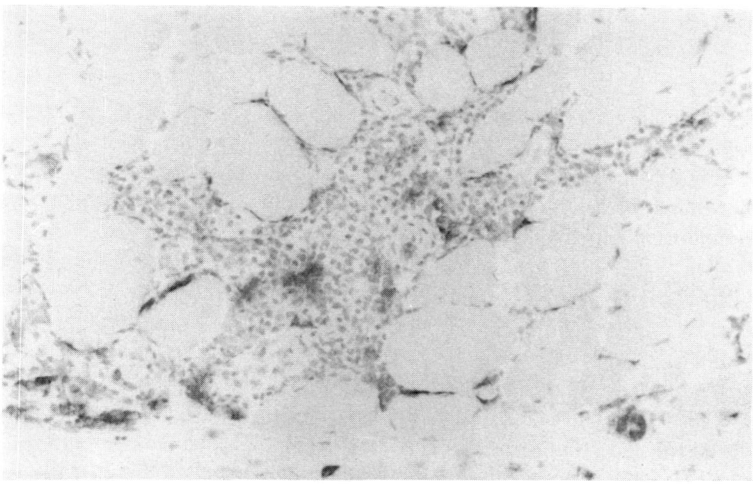

(d)

PATHOGENESIS AND ETIOLOGY

IIM is thought to be a primary autoimmune disorder. Although the antigens involved have not yet been identified, there is substantial circumstantial evidence to support this concept. As previously mentioned, IIM is associated with a number of diseases known or believed to be autoimmune in origin, including Hashimoto's thyroiditis, myasthenia gravis, primary biliary cirrhosis (80,81), ulcerative colitis (82), and celiac disease (83). IIM has also been reported in other conditions in which there is an altered immune state, including pregnancy (84,85). There are also clear associations between PM/DM and HLA markers, including HLA B8/DR3 in about half of all cases of IIM in whites, both adults and children (86,87); B7 and DRw6 in blacks with PM (88); B14 in DM associated with overlap syndromes (89); and a DQAl allele in childhood DM (87). Other HLA associations in PM and DM cases with MSAs have also been recently demonstrated (see below). In addition, inflammatory myopathy identical to spontaneously occurring disease can be induced by certain drugs, notably *d*-penicillamine, phenytoin,

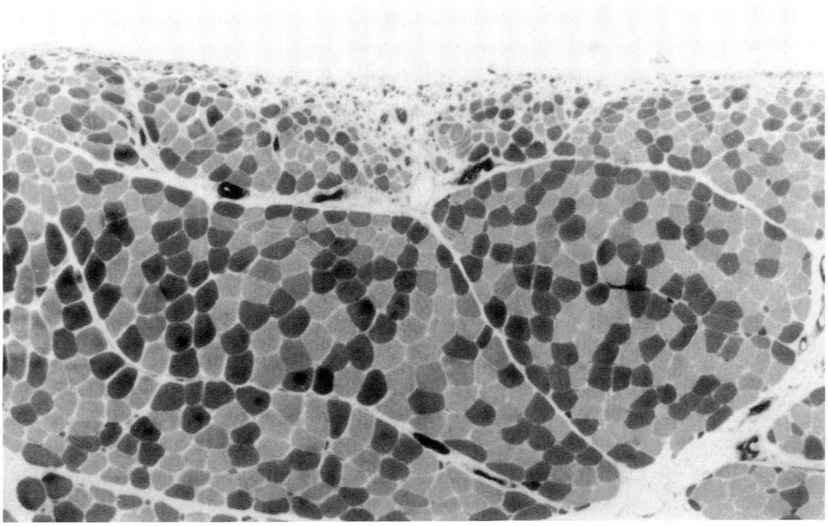

Figure 4 Dermatomyositis. Marked perifasicular atrophy. The atrophy affects both main fiber types. (Myosin ATPase pH 9.5, ×25.)

and procaineamide, through immunological mechanisms (90–94); and myositis has also been reported as a feature of chronic graft-versus-host disease (GVHD), in which muscle becomes a target for immunopathological damage by allogenic lymphocytes in the same way as the skin, liver, and gut (95–97). In all these instances, treatment with immunosupressants and/or removal of the antigenic stimulus improves the inflammatory myopathy.

Mechanisms of Muscle Cell Damage in IIM

Cell-Mediated Inflammation

Early opinion was that PM was likely to result from cell-mediated rather than humoral mechanisms (98), and the subsequent development of an animal model, through injection of crude muscle homogenate in Freund's adjuvant, provided support for this view (99). Kakulas showed that lymphocytes from such animals were cytotoxic to cultured myocytes (100), and Takayanagi demonstrated that experimental myositis was transferable by lymphocytes to other animals (101). Several early studies also demonstrated that autologous peripheral blood lymphocytes from PM/DM patients could lyse muscle cells in tissue culture (102,103) and would show proliferative responses in the presence of skeletal muscle homogenates (104,105). However, these were relatively crude experiments. More recently, it has been shown that myotubes expressing class I antigens could be lysed by autologous muscle-derived lymphocytes from PM/DM cases (106) and also by cloned $CD8^+$ lymphocytes sensitized to donors' HLA antigens (107). Furthermore, lysis was completely blocked by a monoclonal antibody against a component of the HLA antigen complex (107).

HLA antigens are known to be a crucial part of the immune mechanism, being expressed on antigen-presenting cells (APCs). Unlike most cells, however, skeletal and cardiac muscle cells do not normally express class I (HLA-ABC type) antigens, but expression can be induced by cytokines, such as interferons and tumor necrosis factor (TNF), generated in inflammatory and necrotic processes, such as IIM and muscular dystrophies (108,109). In particular, strong class I expression is seen in nonnecrotic fibers being infiltrated by T lymphocytes, whose cytotoxic effects are largely HLA-restricted (109). However, HLA class I expression on muscle cell surfaces is restricted to areas of inflammatory infiltration (78) (although clearly evident in blood vessels in other areas) using some techniques, and absence of such expression in a biopsy section cannot be construed as being evidence against a diagnosis of IIM; conversely, HLA class I expression supports this diagnosis when inflammation is sparse.

Fibers expressing class I antigens also express cell adhesion molecules, such as intercellular adhesion molecule-1 (ICAM-1) and lymphocyte function–associated antigens (LFA-1,2,3), essential for the binding of the T-cell receptor (TCR) to the antigen/HLA molecule on the APC, forming the trimolecular complex (110,111). In support of this concept, it has recently been shown that ICAM-1 is normally barely detectable on the surface of cultured human muscle cells, but that induction of ICAM by α-interferon increases T-cell adhesion in a dose-dependent fashion. This suggests that expression of such molecules may be important for the localization of T cells in IIM (112).

Inflammatory Mechanisms in DM

Despite these observations, the primary insult in DM appears to be humoral rather than cell-mediated, and directed at microvascular elements. There is good evidence that this results from the action of immune complexes (5,6,113,114): $CD4^+$ lymphocytes and B cells are abundant in perivascular locations in DM; class II antigen expression on inflammatory cells is frequent; immunoglobulin and complement components are present on vascular endothelial cells (65–71); and it was recently demonstrated that ICAM-I and LFA-1 were expressed on the endothelial cells of perimysial vessels in DM, while in other forms of IIM, expression was increased mainly in relation to areas of inflammation (115). Kissel et al. (116) found components of the membrane attack complex in the microvasculature in 10/12 children and 5/19 adults with DM, but not in a range of other muscle diseases, and these deposits were associated with features of local ischemia and infarction. These findings suggest that DM results from a primary T-helper-cell-dependent, B-cell-mediated, local humoral immune mechanism, leading to vascular occlusion and capillary obliteration and ischemic damage to muscle fibers (69). In this respect, it is of interest that Fulthorpe and Hudgson noted ultrastructural abnormalities in muscle capillaries before any changes in muscle fibers in their studies of experimental allergic myositis (117). The mechanism of complement activation is not entirely clarified, however, since it can occur in the absence of immunoglobulin (68,118). In addition, DM and other immune-mediated rheumatological disorders, have been reported in patients with hereditary deficiency of components of the classic (C1-C4-C2) complement pathway, indicating the importance of the properdin pathway in this situation (119).

As noted above, fiber necrosis in DM is often not prominent and nonnecrotic fibers are not invaded by lymphocytes. NK cells are scarce, but macrophages are common and $CD8^+$ cytotoxic T cells are present in endomysial infiltrates, so cell-mediated muscle fiber damage also plays a part in DM (71).

Inflammatory Mechanisms in Polymyositis

$CD8^+$ cytotoxic T lymphocytes expressing the α/β receptor comprise the bulk of the endomysial inflammatory cells in PM, but the immune effector system also includes macrophages, which constitute from one-third to one-half of the inflammatory cell infilatrates, occasional $CD8^-$ CTLs expressing the γ/δ receptor, and some $CD4^+$ cells. The cytotoxic T cells surround and invade *intact* myofibers, making intimate contact with the plasmalemma, by penetration of the basal lamina, but the final pathway involved in fiber destruction is unclear (71).

Recent studies have shown that the majority of both CD8 and CD4 T cells in IIM infiltrates are "primed" memory cells, rather than "virgin" cells (120). In addition, PCR studies of mRNAs coding for TCR proteins have shown that in IIM, there is a restricted repertoire of TCR gene activity and oligoclonal expanison of effector cells, while in other myopathies in which inflammation can occur, such as muscular dystrophy, and indeed in DM, TCR gene expression on inflammatory cells is more heterogeneous (121). In addition, none of the TCR sequences detected in muscle cells could be identified in blood or skin, indicating that while T cells extravasate nonselectively in muscle, only cells bearing certain TCR families proliferate after exposure to locally expressed superantigens, and autosensitization occurs within muscle itself (122,123). A particularly graphic example of oligoclonal or even monoclonal expansion of autoaggressive T cells in PM was provided by a rare case in which the inflammatory infiltrate was comprised almost exclusively of $CD4^-$ $CD8^-$ γ/δ. T cells, expressing a single, unusual

TCR phenotype (124,125) (see below). However, as a rule, γ/δ T cells are rarely encountered in PM infiltrates and TCR analysis in one study revealed considerable heterogeneity, suggesting that their appearance in an infiltrate is probably a nonspecific, secondary response (126).

PM can thus be regarded as being primarily the result of an HLA-restricted, T-cell-mediated myotoxicity, and some muscle damage in DM may also occur by this route.

Necrosis in IIM

Several mechanisms have been proposed to explain segmental muscle fiber necrosis induced by the inflammatory process. One theory is that CTLs bring about apoptosis (programmed cell death), through secretion of proteins such as tumor necrosis factor (TNF) and lymphotoxin (71). Another hypothesis is that membrane lysis occurs through the concerted action of perforins and trypsin-like serine proteases located within granules in CTL cytoplasm (granazymes) (71,127). In a recent study, perforin-positive CTLs comprised some 10% of $CD8^+$ α/β cytotoxic T cells in inflammatory infiltrates in PM, while such cells were rare in DM, in keeping with the concepts of primary cell-mediated myonecrosis in PM (128).

The Antigenic Stimulus in IIM

HLA molecules expressed on surface membranes of muscle cells bind immunogenic peptides. Recognition of such peptides by T lymphocytes results in clonal expansion, and the inflammatory and necrobiotic reaction of IIM ensues. The induction of these HLA antigens may therefore be the defining event in the immunopathogenesis of IIM (123), but the etiological factor(s) involved in their expression remain unknown.

Class I molecules on APCs present 8- to 10-mer peptides, derived from *endogenous* proteins (which could include viral transcripts, for example) and interact with $CD8^+$ cytotoxic T lymphocytes (CTLs). The ability to express such HLA molecules is probably ubiquitous. Expression of class II antigens (HLA DR, DP, DQ), which bind 8- to 25-mer peptides derived from *exogenous* antigens, is generally restricted to specialized antigen-presenting cells (APCs), such as B cells, macrophages, and dendritic cells, but can also be induced on muscle and other cells by cytokines generated in inflammatory reactions. Class II expression is necessary for presentation of the exogenous peptide antigens to $CD4^+$ (helper) T cells (71,110).

The nature of the peptide antigens in IIM is unknown, and their characterization would greatly advance our understanding of the etiology of IIM. Three lines of research have developed recently in this regard.

Autoantibodies

Three groups of autoantibodies can be identified in patients with IIM: *nonspecific antibodies*, detection of which might support an autoimmune pathogenesis for a myopathy; antibodies associated with or diagnostic of particular conditions in *overlap myositis*; and *myositis-specific autoantibodies* (MSAs), which some authorities believe may be related to the etiological antigens in IIM.

Nonspecific Autoantibodies

The recent development of immunofluorescent techniques on human epithelial (Hep) tumor cell lines has greatly increased the detection rate for antinuclear antibodies (ANA). In one series, 52% of myositis patients were ANA-positive (129). The incidence is greater in overlap syndromes and in DM, particularly when a high ANA titer is found, but a positive test can occasionally be found in normal subjects. By contrast, *rheumatoid factor* is found in only 5–8% of IIM cases and may be detected in the absence of other features of RA (129). Other antibodies of this type, such as antithyroid, antimitochondrial, antimyosin, and so forth, may also be detected. Interestingly, IIM is one of the few conditions, apart from myasthenia gravis, in which acetylcholine receptor antibodies may be detected (130), and specific involvement of the neuromuscular junction regions in the inflammatory process (131) might result not only in myasthenic fatiguability, but in autosensitization and true myasthenia gravis.

Autoantibodies in Overlap Syndromes

Three main forms of autoantibodies may be found in these conditions: *antiribonucleoprotein antibodies*, which are directed against antigens on small nuclear ribonucleoprotein particles, comprising RNA and nine associated polypeptides (132); antibodies against *nuclear DNA components* and DNA-binding proteins; and *anticentromeric antibodies*.

Sjögren's syndrome is associated with ANAs with high specificity to ribonucleoprotein antigens. *Anti-Ro* (*SS-A*) and *La* (*SS-B*) are found in about 90% of cases, although they are also found in some cases of SLE and PM (33,132). The antibody *anti-PM-1* or *anti-PM-Scl* is a marker for *scleroderma myositis* in North American patients (133), although again it is not specific to this disorder. It appears to be directed against a nucleolar component and is strongly linked to HLA-DR3 and DQw2. In Japanese cases, scleroderma myositis is associated with *anti-Ku*, which recognizes a DNA-binding protein (134), which appears to be an 80-kDa subunit of a protein that can be found in human ciliary muscle (135). Sera from patients with the *CREST* variant exhibit a centromere-binding antibody in more than 80% of cases, while 44–85% of patients with *systemic sclerosis* have anti-DNA topoisomerase I (*anti-scl-70*) (136,137).

As mentioned previously, *mixed connective tissue disease* is associated with high titers of the extractable nuclear antigen (ENA) antibody *anti-U_1snRNP* in virtually all cases (138).

In *systemic lupus erythematosus*, antibodies against a number of nuclear components are present. *Anti-ds DNA* is highly specific for SLE, while *anti-ssDNA* is less so (139). However, both antibodies appear to play a role in tissue damage. Antiribonucleoprotein antibodies are found in less than half the patients. One, *anti-Sm*, is specific for SLE (140), but others are found in other conditions (e.g., anti-U_1 snRNP in MCTD, anti-Ro and anti-La in Sjögren's syndrome), as discussed above. These associations are summarized in Table 4.

Myositis-Specific Antibodies (MSAs)

These recently described antibodies are of considerable interest, in that they are specific to IIM, have prognostic value, and are associated with clinical features that are sufficiently distinctive and coherent to constitute definable IIM syndromes (129,141,142).

MSAs are directed against cytoplasmic and nuclear proteins involved in protein translation and have been named either with reference to the initials of the first patient in whom the were detected (e.g., anti-Jo, anti-OJ), or from the number of the antibody immunodiffusion precipitin line found to be associated with a particular disease (e.g., PL-7, PL-12) (142). They are not uncommon, although

Table 4 Autoantibody Associations in Overlap Myositis

Antibody type	Class	Disease associations
Anti-RNP	Anti-Ro (SS-A) Anti-La (SS-B)	Sjögren's syndrome, SLE, PM
	PM-Scl (PM-1)	Scleroderma myositis
	Anti-Ku	Scleroderma myositis (Japanese)
	Anti-Sm	SLE
Anti DNA	Anti-Scl 70	Systemic sclerosis
	Anti-ds DNA	SLE
Anticentromeric		CREST
ENA	Anti-U_1 sn RNP	"Mixed connective tissue disease"

Table 5 Myositis-Specific Antibodies (MSAs) and Their Antigens

Autoantibody	Autoantigen	Autoantigen function
Antisynthetases		
Anti-Jo 1	Histidyl-tRNA synthetase	
Anti-PL-7	Threonyl-tRNA synthetase	
Anti-PL-12[a]	Alanyl-tRNA synthetase	Bind respective amino acids to associated tRNA
Anti-EJ	Glycyl-tRNA synthetase	
Anti-SRP	SRP proteins	Moves newly synthesized polypeptides to ER for processing/secretion
Anti-Mi2	Not known	Not known
Anti-FER	Elongation factor 1 alpha	Assists movement of aminoacyl-tRNAs to ribosomes and movement of ribosomes
Anti-KJ	Unidentified cytoplasmic protein	Not known
Anti-MAS	Unidentified cytoplasmic RNA	Not known

[a]Also directed against alanyl tRNA.
Source: After Ref. 129.

techniques to test for their presence are not widely established. One study (129) (not population-based) found MSAs in 33% of IIM cases, MSAs being the most frequent form of autoantibody after ANAs in this population. MSAs were rarely found in overlap myositis (9%) and almost never in cancer-associated myositis or IBM. Table 5 summarizes the target antigens against which MSAs are directed, together with their functions, where known.

Antisynthetase antibodies are directed against a range of cytoplasmic tRNA synthetases and can be detected in about 20–25% of all IIM cases (129,141,142). Patients with such antibodies have a nonerosive arthritis affecting chiefly the small joints of the hands, leading to subluxations, together with Raynaud's phenomenon (in 60% of cases), with characteristic cracking and thickening of the skin over the hands and fingers ("mechanic's hands"), and in some cases features of scleroderma. However, the skin rash of DM is rare. This clinical picture is now referred to as the *antisynthetase syndrome* (129,143). The most comon of these antibodies, *anti-Jo-1*, is directed against histidyl-tRNA synthetase and found in about 18% of PM sera (144). As mentioned previously, it is also strongly associated with interstitial lung disease, being present in 50–75% of patients having the antibody, against 20–25% of those who do not (144). This antibody is strongly linked to the DR3, DRw52, and one of the DQA1 haplotypes (DQA1*0501) (129). Other antisynthetase antibodies (against threonyl, alanyl, and glycine tRNA) are found in only 1–3% of cases (144).

The *signal recognition protein* (SRP) assists the transport of newly translated polypeptides into the endoplasmic reticulum. Anti-SRP antibodies are found in about 4% of PM cases, mainly black females, and are associated with an aggressive, acute myositis, with marked myalgia and cardiac involvement, but usually without skin or pulmonary disease. This antibody is also associated with HLA-DR3 and DRw52, and with DQA1*0301. The 5-year survival of patients with this autoantibody is only around 25% (141,145).

The *Mi-2* antibody is directed against a 220-kd nuclear protein of unknown function. This antibody is particularly associated with DM and was detected in 10 of 79 IIM patients with a florid skin rash and one of 13 myositis patients with carcinoma. This antibody is linked to HLA-DR7, DRw53, and DQA1*0201. Patients usually respond well to treatment (129,142,143).

The clinical associations with the MSAs are outlined in Table 6. Other forms of MSAs have been reported (Table 5), but their significance is at present unknown.

Table 6 Clinical Associations with Myositis-Specific Autoantibodies

MSA	Clinical features	Prognosis	HLA
Antisynthetase syndrome	Arthritis, fever, lung disease, Raynaud's, mechanic's hands	Moderate, 5-year survival 70%	DR3 DRw52 DQA1 0501
Anti-SRP	Severe myositis, cardiac involvement	Poor, 5-year 25%	DR5 DRw52 DQA1 0301
Anti-Mi2	Classical DM, good response to treatment	Good, 5-year >95%	DR7 DRw53 DQa1* 0201

Source: After Ref. 129.

Only one type of MSA is found in any individual case, and the type of MSA in a particular patient does not change during the course of the disease. Antibody titers also correlate with disease activity and disappear when complete remission occurs (141,142). It has therefore been suggested that they might be directed against the antigens directly involved in the autosensitization. However, they have been detected in only a minority of cases, and unlike antibodies to DNA in lupus, for example, there is no evidence that they are part of the immune effector response (141,142). Nevertheless, it is of interest that enteroviruses, which may be involved in the etiology of some cases of IIM, have an RNA genome that must interact with host cell translation machinery, and such interaction with tRNA synthetases and other proteins, which are the predominant MSA antigens, could conceivably result in their becoming targets for the autoimmune attack (142).

Heat Shock Proteins

The discovery of the rare case of PM, discussed above, in which the majority of autoaggressive T cells bore the γ/δ receptor (124,125), raised the possibility that heat shock proteins (HSPs) might be autoantigens in IIM, since γ/δ T cells in particular, and also some α/β T cells, react to HSPs (146). Indeed, naturally derived HSP peptides have been isolated from HLA class I molecules (147). However, an immunocytochemical study revealed that 65-kd HSP was widely distributed on inflammatory cells, necrotic and nonnecrotic muscle cells, and endothelial cells, not only in IIM but also in normal muscle (148), so mere expression of HSP is not sufficient to excite an inflammatory reaction. Nevertheless, the techniques used in these studies indicate the likely direction of future attempts to isolate autoantigens.

Viruses as Potential Antigens in IIM

The notion that viral infection might trigger the autoimmune response in IIM remains seductive. A number of viruses, notably influenza, enteroviruses (such as Coxsackie and ECHO), and herpes viruses, have been associated with acute myositis and sometimes rhabdomyolysis (149) (and see Chapter 32). However, viruses have rarely been recovered from muscle in such cases (150), and it is unclear whether the muscle damage results from direct cytolytic effects or secondary immune responses. For example, both HIV (HTLV-III) and HTLV-I infection can cause PM/DM, but in both instances, viral antigens are found only in the invading inflammatory cells and not in the muscle fibers (151,152) (and see Chapter 52). Hepatitis C virus (HCV) has also recently been linked to the development of polymyositis (153–155). In one case (155), chronic HCV antigenemia was demonstrated, and HCV RNA was detected in muscle extract by PCR. However, only the plus (genomic) strand of RNA could be detected;

lytic virus would normally have demonstrated both genomic and template strands, albeit with a marked excess of plus strand. This suggests that the myositis was the result of an immune response to replication-defective HCV. An additional case in a patient who also had Jo-1 antibodies supports this view (156).

One line of research has focused on the possibility that viral infection might induce immune reactions through molecular mimicry. Certain autoantibodies in overlap syndromes (such as the antiribonucleoprotein antibodies anti-Sm, anti-Ro, and anti-La) and some myositis-specific autoantibodies (such as anti-Jo-1 and other t-RNA synthetase antibodies) are directed against host antigens that are not only normally accessible to the immune system, but are critical to viral replication or insertion (157,158). It has been suggested that such antibodies are viral anti-idiotype antibodies, which therefore resemble in structure the host element with which the viral immunogen interacts. The presence of such autoantibodies might thus imply a triggering viral infection in the etiology of the immune reactions (158). For example, *Escherichia coli* histidyl- and alanyl-tRNA synthetases share amino acid sequence homology with part of the Epstein-Barr virus protein, and with myosin light chain, and tropomyosin and keratin, respectively (159). This hypothesis supports the view that autoantibodies are not necessarily implicated in the pathogenesis of IIM, but could reflect a viral etiology.

Another approach has involved attempts to detect viral genomic material in muscle in IIM. The development of molecular genetics techniques has shown that persistent, *noncytolytic* infection by viruses may occur, and such viral persistence has been demonstrated in a number of animal and human cell lines and in animal models of myocarditis and polymyositis (160). The mouse model first described by Strongwater and colleagues (161) is of particular interest. Mice infected with Coxsackie B1 Tucson strain develop a chronic inflammatory myopathy following recovery from the acute infection, which is not closely related to the humoral immune responses generated, but largely reflects cell-mediated myotoxicity (162,163). In one study, viral RNA could be recovered from some muscles, but not other tissues, up to a year later, and the time course of the inflammatory response followed that of viral persistence (163).

Viral persistence has also been linked to a number of chronic diseases in humans, including chronic meningoencephalitis in hypogammaglobulinemic patients, postpolio syndrome, motor neuron disease, myocarditis and dilated cardiomyopathy, and inflammatory myopathy (160). The possible involvement of enterovirus infection in IIM was first raised by observations on the increased incidence of Coxsackie IgM antibodies in affected patients compared to controls (164–166). Subsequent studies, notably by Archard and colleagues, stemming from work on myocarditis and dilated cardiomyopathy, utilized cDNAs specific for generic enteroviral RNA to screen muscle RNA extracts using slot-blot hybridization and revealed positive results consistent with enteroviral persistence in muscle in 5/9 cases of PM and DM, but in none of 10 Duchenne dystrophy cases and histologically normal control biopsies (167). The presence of enteroviral RNA in muscle in some cases of IIM was later confirmed by others, by in situ hybridization (168). The subsequent demonstration of enteroviral RNA in some cases of chronic fatigue syndrome (CFS), by both hybridization (169) and PCR techniques (170), raised the possibility that IIM and certain forms of CFS might represent a spectrum of myopathy resulting from virus infection (171) (see Chapter 44), and it was shown that enterovirus persists as a replication-defective mutant in such cases (172). However, several subsequent investigations, using the far more sensitive PCR reaction, failed to detect viral genomes for enterovirus, mumps, adenovirus, HTLV-1 and -2, and HIV in inflammatory myopathies (173–175). Nevertheless, involvement of virus in the initiation and possibly perpetuation of IIM remains of considerable interest.

UNCOMMON CLINICAL VARIANTS OF IIM

IIM usually presents as a subacute or chronic, symmetrical proximal myopathy, but occasionally other modes of presentation are encountered. It is generally accepted that such presentations are variants on the more common manifestations of IIM, rather than separate disease entities.

Acute Myositis

Acute presentations, with severe pain, edema of muscles and subcutaneous tissues, but no skin rash, such as those reported by Wagner (1,2) and Hepp (3) in the 19th century, are rarely seen. The cases of this type described more recently by Venables and colleagues (14) showed surprisingly modest elevations of CK and relatively little inflammation on biopsy, but quite extensive fiber necrosis was present. The pathological features of such a picture do not sit easily with the mechanisms described above, and muscle ischemia may be an important factor (14).

Distal Polymyositis

Rarely, polymyositis may present with chronic progressive distal weakness and wasting, reminiscent of some cases of IBM, but with marked improvement with steroid treatment (176,177).

Focal Myositis

Occasionally, IIM will affect only a restricted area of the body, such as the limbs on one side, a single limb (178), or even one muscle. The myositis may remain confined to this region for years (179,180), with consequent muscle wasting, before becoming more generalized, and until that time, EMG and biopsy evidence of more widespread involvement is minimal.

Localized Nodular Myositis

Cumming et al. (181) described three young to middle-aged males who each presented with an isolated painful muscle swelling that resolved within a few weeks. This was followed by similar painful mass lesions in other muscles, subsequently evolving into a more generalized myopathy. They referred to this pattern of IIM as *localized nodular myositis*. The muscle masses were several centimeters in diameter, hot and tender, resembling areas of thrombophlebitis, and function was limited by pain rather than weakness. The biopsy appearances were typical of IIM, but in addition, there were large foci of necrosis suggestive of infarction. A similar case had been reported earlier, as "relapsing myositis" (182), and additional cases have since been described (183). Heffner and Barron (184) subsequently described six cases of polymyositis presenting with a single, localized painful muscle mass, initially diagnosed as benign inflammatory pseudotumor. The mass regressed in each case, as in localized nodular myositis, but was followed by more typical IIM features over the next few months, but without the development of other muscle masses. Other examples of focal myositis (185), associated with thyroiditis (186) and showing spontaneous resolution (187), have been described more recently. It should be noted that the differential diagnosis of isolated muscle masses would include "pseudosarcoma" (proliferative myositis) (188), localized myositis ossificans (Chapter 46) metastases and rhabdomyosarcoma, in addition to nodular myositis.

"Facioscapulohumeral" Polymyositis

Two of the patients described by Cumming et al. (181) developed a facioscapulohumeral pattern of weakness. This facioscapulohumeral presentation has been noted by others (189–191), but care must be taken to distinguish this from inflammatory reaction seen in some biopsies in cases of FSH dystrophy, which should be considered if there is a family history of the condition. A marked improvement with steroids might suggest a primary autoimmune pathogenesis (191). Immunocytochemical studies of inflammatory cells in FSH dystrophy have shown recently that, unlike PM, there is no invasion of nonnecrotic fibers by $CD8^+$ perforin and granazyme-positive CTLs, and that the extent of inflammation is related to the number of necrotic fibers (192). This suggests that the inflammation in FSH dystrophy is a secondary phenomenon, as in Duchenne dystrophy. The myopathology of the FSH syndrome is discussed in detail in Chapter 23.

Eosinophilic Myositis

Eosinophilia is associated with parasitic infections, vasculitis, hypersensitivity reactions, and the hypereosinophilic syndrome (HES), which shares features with the Churg-Strauss variant of polyarteritis nodosa (PAN) (see below). HES is defined by idiopathic eosinophilia (>1500/mm^3) of at least 6 months' duration, unless additional features of organ involvement by eosinophils or related mechanisms intervene sooner (193). HES commonly causes cardiac and pulmonary disease (Loeffler's syndrome), and eosinophilic endocarditis has been associated with Balint's syndrome, due to bilateral infarction of the parietal lobes (194). Myositis can occur in conditions characterized by eosinophilia and vasculitis, and unlike classical IIM, such conditions appear to be more common in males.

Eosinophilic polymyositis in HES is rare, but resembles other forms of IIM, except that the inflammatory infiltrates in muscle contain a large proportion of eosinophils (195,196). The clinical picture is dominated by the cardiopulmonary involvement, and the prognosis is poor (193), although many cases of HES respond to treatment of the form outlined below (see Treatment and Prognosis).

Eosinophilic Fasciitis

Shulman's syndrome (197) usually presents with a short prodromal illness of fatigue, myalgia, and arthralgia, with low-grade fever, followed rapidly by a characteristic progressive thickening of subcutaneous tissues and tightening of the skin, with pitting edema, resembling scleroderma. The skin contractures result in progressive limitation of joint movement. The ESR is typically raised, and the condition is also associated with eosinophilia, which may be intermittent, and other nonspecific features seen in HES such as hypergammaglobulinemia. Unlike scleroderma, Raynaud's phenomenon is rare and other organs are not involved (198). Full-thickness skin and muscle biopsy reveals that, unlike scleroderma myositis (where dermal involvement is primary and muscle inflammation prominent), there is fascial thickening and infiltration by mononuclear cells, notably eosinophils and macrophages. Subcutaneous tissue and muscle inflammation is not invariable and occurs by extension (199–201). If muscle is involved, patients may experience weakness, with raised CK, and the diagnosis may first be suspected during electromyography, where an unusual resistance to insertion of the needle electrode can be encountered!

The condition often resolves spontaneously and usually responds to steroids, but can have a relapsing remitting course and may be complicated by hematological disorders including aplastic anemia, thrombocytopenia, and myeloproliferative and lymphoproliferative diseases (202).

The cutaneous and pathological features of this condition have much in common with the eosinophilia myalgia and toxic oil syndromes discussed in Chapter 32, and a tentative etiological link has indeed been noted with L-tryptophan ingestion (203).

Myositis in Vasculitic Disorders

Clinical evidence of muscle involvement in vasculitis (polyarteritis) is not often prominent, and muscle fibers tend to be involved either by spread of inflammation from perivascular regions or by infarction. However, muscle and nerve biopsy are frequently of diagnostic value in these conditions. Vasculitic diseases are classified on the basis of size of vessels involved: large—Takayasu's and giant cell arteritis; medium—polyarteritis nodosa (rare cases where smaller vessels are not affected) and Kawasaki's disease; and small—Wegener's granulomatosis and microscopic polyangiitis (including most cases of polyarteritis nodosa, Churg-Strauss syndrome and hypersensitivity angiitis). Muscle is affected chiefly in this last group of conditions.

Polyarteritis nodosa (PAN) may be associated with weakness and myalgia. The picture is often complicated by the presence of mononeuritis multiplex, however. Biopsy shows marked perivascular inflammation, notably comprising polymorphonuclear leukocytes, and fibrinoid necrosis in affected vessels, which leads to segmental vessel wall destruction and the aneurysm formation characteristic of the disorder. Muscle fiber necrosis resulting in raised CK is uncommon and probably results from ischemia.

The *Churg-Strauss syndrome* variant is distinguished from PAN by prominent pulmonary involvement, presenting as asthma, together with marked eosinophilia (204). Perivascular inflammatory infiltrates in muscle in such cases are markedly eosinophilic and are associated with extravascular granulomas containing macrophages, epithelioid cells, and giant cells (Fig. 5).

Wegener's granulomatosis affects small vessels and sometimes capillaries. The primary symptoms are respiratory, often presenting with upper tract involvement including sinusitis and headache, nasal discharge and ulceration, otitis media, and cough, in addition to lower respiratory tract symptoms, of chest pain, pneumonia, and hemoptysis. Renal involvement occurs in 90% of cases and may produce an aggressive glomerulonephritis. The diagnosis is usually made from biopsy of respiratory epithelium, but changes in muscle can again include perivascular inflammation and granulomata, both intra- and extravascular, with abundant giant cells. The use of steroids and cyclophosphamide has transformed the prognosis in this condition (205).

A number of vasculitides fall under the general rubric of *hypersensitivity angiitis*, including those associated with rheumatoid disease, SLE, cryoglobulinemia, and drug-induced vasculitis. Unlike the conditions discussed so far, the vasculitic lesions in such conditions are all of the same age, involve only small vessels, and are not associated with granulomas. Skin lesions are prominent, including purpura, ecchymoses, and ulcerations, particularly over the lower legs and feet, and muscle may show a nonspecific perivascular mononuclear cell infiltration.

Giant cell arteritis, and its manifestations of polymyalgia rheumatica and cranial arteritis, are discussed in Chapter 45.

Vasculitis is usually associated with a raised sedimentation rate, leukocytosis (often with eosinophilia), and anemia. Hepatitis B surface antigen can be detected in blood in about half of all cases of vasculitis (206), as can antineutrophil cytoplasmic antibodies (ANCA). ANCA subgroups have some diagnostic specificity. Thus, cANCA is directed against the cytoplasmic 29-kDa proteinase 3 enzyme. Its detection is strongly suggestive of Wegener's granulomatosis, and the titer reflects disease activity (207), with a specificity of around 90%, but its sensitivity for this condition is low (208). pANCA is directed against myeloperoxidase, which migrates to neutrophil surface membranes and can be found in various forms of microscopic polyangiitis. Interestingly, circulating DNA can be detected in vasculitides such as PAN, due to release of DNA from blood vessel walls; it is not found in cases of PM or SLE unless significant vessel damage occurs (209).

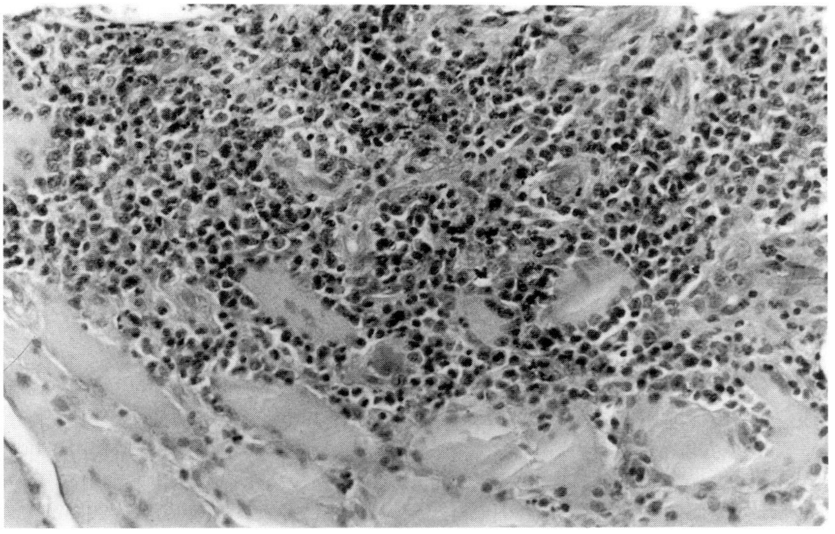

Figure 5 Myositis in Churg-Strauss syndrome. There is a florid inflammatory infiltrate, containing eosinophils and nonepithelioid giant cells.

Inflammation in Other Myopathies

Inflammatory infiltration in muscle has been reported in neonates (see Chapter 11), and the inflammation sometimes seen in muscle biopsies in cases of FSH and other dystrophies, and in other forms of myopathy, has been mentioned previously. The inflammatory changes in Duchenne dystrophy, for example, are associated with HLA class I antigen expression and $CD8^+$ preponderance in the cellular infiltrate (67), as in polymyositis, but the extent of inflammation and invasion of muscle cells by $CD8^+$ T cells is an order of magnitude less. This inflammation is presumably a secondary phenomenon and has no bearing on the mechanism of the short-lived beneficial effects of steroids in this condition; azathiaprine reduces inflammation but does not improve the condition (210).

SARCOID AND OTHER GRANULOMATOUS MYOPATHIES

Granulomatous inflammation of muscle with giant cell reaction is most often encountered in sarcoidosis, but identical lesions were also described in *granulomatous myopathy*, in which there was no evidence of systemic illness and pathology was apparently confined to muscle (211,212). Similar lesions have also been reported in association with Crohn's disease (213), and the rare syndrome first described by Namba et al. of giant cell myositis, giant cell myocarditis, and myasthenia with benign or malignant thymic tumor (214). Granulomas with giant cells are also a feature of lethal midline granuloma (Stewart granuloma) and of vasculitides such as Churg-Strauss syndrome and Wegener's granulomatosis, as discussed above.

Sarcoidosis is a multisystemic disorder of unknown cause, which affects blacks some 10–20 times more commonly than whites and is slightly more common in women. It is uncommon in childhood. The most common presentation is with arthritis and erythema nodosum, and such patients will typically show evidence of mediastinal lymphadenopathy and pulmonary involvement. The diagnosis is usually confirmed by finding characteristic changes on X-ray, increased levels of angiotensin-converting enzyme (ACE) in blood, and a positive gallium scan indicating lymph node involvement, in virtually all cases (215). A positive Kveim-Siltzbach skin test is also strongly indicative (215). Subclinical involvement of muscle, revealed by biopsy, may be found in some 80% of such cases, and muscle may yield tissue confirmation of the diagnosis as often as liver or lip gland biopsy (216). Clinical or electrophysiological evidence of polyneuropathy or mononeuropathy, in addition to myopathic changes, is also common, although since the pathology resides principally outside excitable structures, the EMG and nerve conduction in sarcoidosis may be quite normal.

Clinically, sarcoidosis affecting muscle may present in several ways. It is most commonly asymptomatic, but can cause a spectrum of myopathic conditions, from an acute or subacute myositis with myalgia, similar to PM, to chronic progressive weakness with wasting (217,218). Granuloma formation in muscle may also result in palpable nodules, comprising a form of nodular myositis.

The "noncaseating" granulomas in muscle are found in blood vessel walls and extravascularly, in connective tissue, are composed of epithelioid cells, macrophages, lymphocytes, and giant cells, including Langhans cells, surrounded by inflammatory cells. The pathology is typically focal or multifocal. Muscle fiber damage appears to occur through compressive destruction rather than immune-mediated damage, and fiber necrosis is unusual (Fig. 6).

A recent immunocytochemical study (219) found that, while macrophages predominated within the granulomas, $CD4^+$ lymphocytes constituted much of the inflammatory reaction, with a $CD4^+/CD8^+$ ratio of 2.1 in the perimysial and perivascular regions and 3.6 in the endomysium, greater than in any other form of IIM. In early granulomas, $CD8^+$ cells were noted to be uniformly distributed throughout the infiltrate, while in more mature lesions they were largely confined to the mantle. It was suggested that $CD4^+$ cells are involved with induction and formation of granulomas, while $CD8^+$ cells are involved with later maturation stages. HLA class I antigen expression was not found, unlike in IIM, but numerous T cells displayed DR antigen. The authors also pointed out that giant cells of myogenic origin could be distinguished from Langhans cells by their desmin-positive staining.

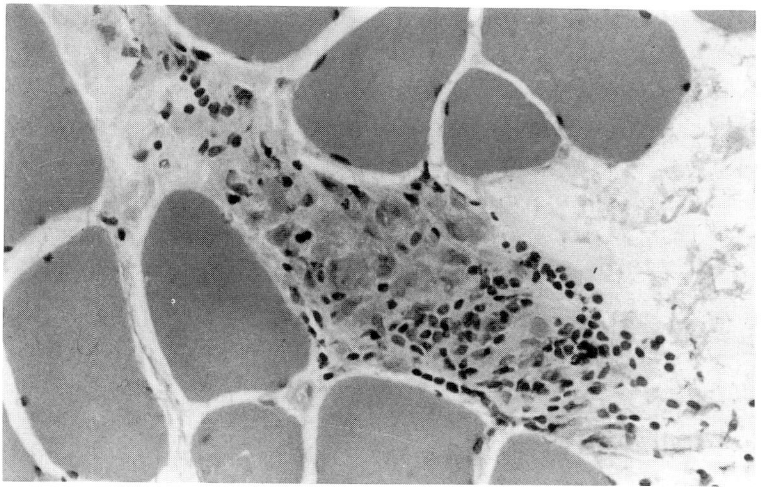

(a)

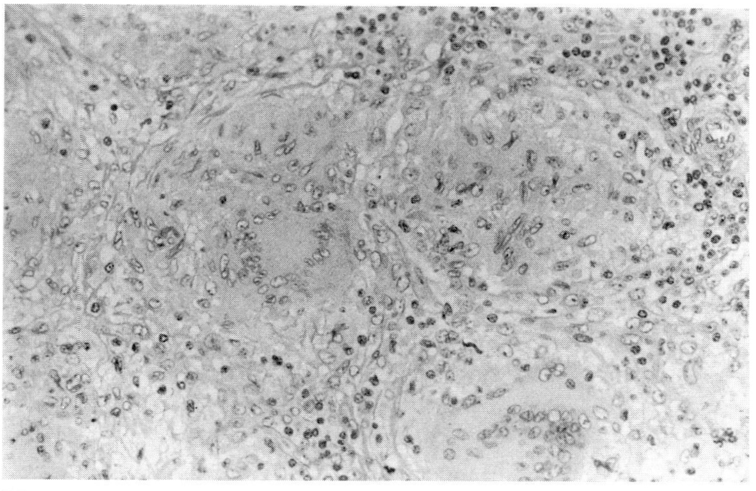

(b)

Figure 6 Muscle biopsy findings in sarcoidosis. (a) A sarcoid granuloma in the endomysium, comprising lymphocytes and giant cells. (H&E ×75.) (b) Sarcoid granuloma in muscle, showing epithelioid (Langhans) giant cells.

TREATMENT AND PROGNOSIS OF IIM

Natural History

It is difficult to assess the efficacy of current treatment strategies for IIM against the "natural history" of these disorders. Early studies of "polymyositis" may have included cases that would not have conformed to current concepts and definitions, and the management of respiratory and cardiac complications has improved sustantially over the years.

O'Leary and Waisman (220) reported 38 patients ascertained between 1926 and 1939, prior to the introduction of steroids, for whom follow-up data were available. The series included children, and only two patients had no cutaneous involvement. The overall mortality was 50%, death occurring 4 months to 6 years after diagnosis, mainly from respiratory complications. Four patients (10%) recovered spontaneously, and in the remainder, the disease was either slowly progressive, followed a relapsing and remitting course, or stabilized with residual weakness and chronic skin changes. The findings of a retrospective study by Sheard (221) were similar. Of 25 cases of PM/DM diagnosed between 1927 and 1947, 13 (52%) died after an average of 20 months.

Are Steroids Effective?

No randomized, placebo-controlled trial of steroids in IIM has ever been conducted, and there is no good evidence that steroids improve the *overall* morbidity and mortality of PM/DM. An indication that steroids might be effective came with the first report of the successful use of ACTH in 1949 (4,222). Subsequent reports of improvement with ACTH and cortisone in the early 1950s were, however, tempered by others recording lack of response (4,223).

Winkelmann et al. (224) reported outcome in 289 cases diagnosed at the Mayo Clinic to 1959, including 157 patients who had received steroids. Eighty patients had died (28%), substantially less than in the earlier studies of untreated patients noted above, but the numbers of steroid-treated and untreated patients dying or achieving remission were found to be similar. However, twice as many patients receiving "high-dose" steroids were recorded as being "better," compared to the "low-dose" treated, or "no steroids" groups. A subsequent study from the same institution, almost a decade later (225), compared 31 low-dose-treated to 31 high-dose-treated patients, matched for age and sex, and with similar disease severity; no difference in overall survival was found between these groups.

Current Treatment: Initiation and Maintenance

Despite these observations, it is common experience that high-dose oral preparations, such as prednisolone 1–2 mg/kg/day, usually produce a gratifying clinical response, with increase in strength and fall in CK over 1–2 months. In one large retrospective study, the extent of improvement was found to correlate with the mean dose of steroids received in the first 3 months of treatment, although the dose did not influence eventual outcome (226). It is common practice to change to an alternate-day regime once remission is achieved, since this seems to reduce the risk of long-term steroid-related side effects; daily and alternate day treatments appear to be equally effective (227). For example, if remission had been achieved with a dose of 60 mg/day, we would continue at a dose of 100 mg on alternate days initially. However, occasional patients experience unacceptable recrudescence of symptoms, including muscle pain, on their "off" days, while others dislike the psychomotor excitation and fluid-retention induced by large steroid doses on "on" days. Some physicians advocate the use of histamine H_2-receptor blockers to reduce the risk of dyspepsia, and calcium, vitamin D supplements, and hormone replacement therapy in postmenopausal women to prevent osteoporosis, but we do not use these drugs routinely.

We would then usually reduce the alternate-day dose by 10 mg/month until a maintenance dose of 10–20 mg on alternate days (or 5–10 mg/day) is achieved, and then continue this for 2 years before attempting withdrawal. If relapse occurs, it is heralded by an increase in CK levels [an increase in myoglobin level is more sensitive (228)], followed by evidence of increased weakness on strength testing (228). Relapses can usually be managed by a prompt increase in steroid dose, to perhaps double that currently being taken, followed by a slower withdrawal regime. However, it should be noted that steroids are themselves myotoxic (see Chapter 31), and if increasing weakness occurs without an increase in CK, steroid myopathy should be suspected (229). This may require repeat electromyography and muscle biopsy for evaluation (63,228). The incidence of steroid-related side effects reported in different studies has varied markedly, from 5–10% to almost half of all cases (12,226).

We also usually start azathioprine, at around 2 mg/kg/day, together with steroids. This drug was shown to have useful therapeutic effects in refractory cases, and in one study, azathioprine and steroids in combination appeared to be superior to steroids alone, in terms of response rate and in providing a steroid-sparing effect (230,231). Azathioprine has a low incidence of toxic side effects at this dose in our experience, but takes about 6 months to have full effect.

Monitoring Treatment Response

Response to treatment can generally be followed clinically. This usually involves the use of manual muscle strength tests (MMT, see Chapters 1 and 2), but in one prospective study (228) it was shown that functional (dynamic) muscle tests (FMT) were more sensitive to change in clinical status. For example, simple techniques, such as measuring the time to stand 10 times from a standard chair, have been shown to be a reliable index of change in leg function during treatment (232). Serial serum CK or myoglobin levels are useful (60,61,233), but these reflect the extent of muscle fiber damage rather than "disease activity" and can be misleading at times. As mentioned above, serial EMG studies (63) and myoimaging (64) have also been used to follow treatment responses, but these are expensive and time-consuming and it is not clear that they are superior to careful clinical assessments.

Prognosis

We would expect at least 80% of all IIM patients to show a good initial response to this standard regime. Overall, published data suggest that subsequently, about one-third of cases will remit completely, another third will improve but require long-term treatment, and the remainder will fail to respond adequately. The overall mortality rate in two sequential Newcastle series, reported in 1966 (234) and 1975 (12), in which ascertainment in a local population was reasonably complete, was consistent at 30% and 28%, respectively. When deaths associated with malignancy and complications of associated collagen vascular diseases were excluded, however, the mortality rate directly attributable to myositis was only about 15%. Thus the casemix obviously has an important influence, and a series from the same era, derived mainly through "tertiary" referral and including many cases with disease of long duration, found a mortality rate of 40% (235).

Data from a number of more recent studies are still more encouraging. With adequate treatment, average 5- and 10-year survival rates approach 85% and 75%, respectively, with stable improvement in around 70% and cure in one-third (71). Mortality in childhood dermatomyositis is reported to vary from 7 to 18% (71). The recent study from Victoria, noted previously (13), of all biopsy proven-cases between 1982 and 1987 found a mortality rate of only 11%, with an additional 9% of patients failing to recover strength. Associated malignancy obviously has an adverse effect on prognosis, but removal of a tumor may result in improvement of IIM, while relapse may follow regrowth (71). Involvement of the lungs and heart, impaired breathing and bulbar function, aggressive myositis, with acute onset and rhabdomyolysis, and the presence of high levels of anti-SRP [where the 5-year mortality rate may be as high as 75% (142–145)] are also poor prognostic factors. Conversely, chronic, indolent myositis may also show little response to treatment.

Refractory Cases

Careful diagnostic review should be undertaken in patients who fail to respond to treatment, and this may include repeat biopsy. It is self-evident that patients with chronic disease and marked replacement of muscle by connective tissue are less likely to respond. The mode of treatment is also critical. The Victoria study found that patients with disease onset before 50 years, duration of symptoms less than 12 months, and concomitant use of immunosuppressants with steroids had improved outlook, and that when deaths occurred, they did so within 6 months of diagnosis. Laboratory findings, and whether patients had DM or PM, had no prognostic significance (13). Another recent study emphasized the importance of early diagnosis and treatment on outcome (236).

In one study (228), clinical responses to treatment, as measured by MMT, FMT, and serum CK and myoglobin levels, were studied in relation to serial muscle biopsy findings in IIM. Inflammation cleared from biopsies within 6 weeks of introduction of steroids in all cases, but significant abnormalities of fiber architecture, and weakness on FMT, were still present at the end of the study, a year later.

Type 2 fiber atrophy was rarely observed and was also occasionally a feature of the initial diagnostic biopsy, so repeat biopsy can be of value in excluding steroid myopathy. By contrast, in another study, seven of 11 patients who underwent repeat biopsy because of poor response to treatment had evidence of ongoing inflammation up to 20 months after starting treatment (237). Immunocytochemical studies on serial biopsies have shown that steroids lower $CD8^+$ counts in infiltrates, but that MHC class 1 expression may persist, possibly explaining recrudescence of inflammation if withdrawal of immunosuppression is attempted prematurely (238).

There is no consensus regarding the management of patients who fail to respond completely, or persistently relapse during steroid withdrawal, but in general, measures having the least risk of significant side effects should be considered first. High-dose *intravenous methylprednisolone* 1 g daily for 3 days, followed by the standard regime, has been widely used in the treatment of immune-mediated disorders (239) and can be helpful in inducing remission in such cases. Several alternative immunosuppressive drugs to azathioprine have been reported to be of benefit in refractory cases. *Methotrexate* 5–10 mg/week is given in three doses 12 hr apart and can also be given intravenously. Reported response rates vary from 55 to 75% (240,241). The major toxic effects of methotrexate are liver damage and bone marrow suppression. *Cyclophosphamide* is given 1–2 mg/kg/day, orally or intravenously. It is used more frequently in the treatment of vasculitis than myositis, where experience of its efficacy has been mixed (242–244). It can cause a severe hemorrhagic cystitis when administered intravenously in high doses (>2 g/ daily), but the risks of this are significantly reduced by the concurrent administration of mesna. Pulsed intravenous cyclophosphamide has been used successfully in rheumatoid disease and multiple sclerosis but is largely untried in IIM.

Intravenous *immunoglobulin* (IVIg) therapy is being used increasingly in the management of a number of immune-mediated conditions, being widely regarded as simple and safe, with a low risk of significant side effects. Its mode of action is unknown and may be multiple, or may vary in different conditions, but most postulated mechanisms involve actions on the humoral system (245). Pooled results from three trials showed efficacy in 20/23 cases of DM and 11/14 with PM, and in DM, and in one, treatment was associated with clearance of membrane attack complexes from the microvasculature, increased capillary density, and decreased HLA class I expression on muscle fibers (246–248). Its main drawback is cost, and treatment usually has to be repeated at 3–4-monthly intervals; furthermore, a recent trial of IVIg alone, 1 g/kg daily for 2 days each month, for 4 months, showed no benefit (249).

By contrast, *cyclosporin* inhibits T-cell activation and cytotoxicity and has been reported to be effective in PM and DM refractory to other therapies (250–252). This drug is nephrotoxic at doses over 6 mg/kg day, and blood levels have to be monitored because of unpredictable absorption.

There have also been reports that plasmapheresis is beneficial in refractory IIM (253), but since such treatment was combined with immunosuppressive drug treatment, such claims are difficult to evaluate. A recent controlled study found plasmapheresis or leukopheresis to be no more effective than sham apheresis (254), and that has certainly been our experience. Similarly, we have had little success with total-body irradiation, reported anecdotally to be of benefit in some cases (255,256), but not others (257). Finally, we (258) and others (259) (and Dr. David Bates, personal communication, 1984) have had occasional success with thymectomy in refractory IIM.

Despite the problems that can arise in diagnosing, classifying, and managing these patients, the confirmation of IIM as the cause of a progressive and incapacitating muscle weakness provides an important, and often rewarding opportunity for effective intervention.

REFERENCES

1. Wagner E. Fall eines seltner Muskelkrankheit. Arch Heilkunde 1863; 4:282–283.
2. Wagner E. Ein Fall von acuter Polymyositis, Deutsches Arch Klini Med 1887; 40:241–266.
3. Hepp P. Ueber einen Fall von acuter parenchymatoser Myositis, welche Geschwulste bildete und Fluctuation, vortauschte. Klin Wochenschr 1887; 24:389–391.
4. Walton JN, Adams RD. Polymyositis. Edinburgh: E&S Livingstone, 1958:1–17.

5. Banker BQ, Victor M. Dermatomyositis (systemic angiopathy) of childhood. Medicine (Baltimore) 1966; 45:261–289.
6. Banker BQ. Dermatomyositis of childhood: ultrastructural alterations of muscle and intramuscular blood vessels. J Neuropathol Exp Neurol 1975; 34:46–75.
7. Hudgson P, Peter JB. Classification. Clin Rheum Dis 1984; 10:3–7.
8. Tanimoto K, Nakano K, Kano S, Mori S, Ueki H, Nishitani H, Sato T, Kuichi T, Ohashi Y. Classification criteria for polymyositis and dermatomyositis. J Rheumatol 1995; 22:668–674.
9. Medsger TA, Dawson WN, Masi AT. The epidemiology of polymyositis. Am J Med 1970; 48:715–723.
10. Pelkonen PM, Jalanko HJ, Lantto RK, Makela AL, Pietikainen MA, Savolainen HA, Verronen PM. Incidence of systemic connective tissue diseases in children: a nationwide prospective study in Finland. J Rheumatol 1994; 21:2143–2146.
11. Denardo BA, Tucker LB, Miller LC, Szer IS, Schaller JG. Demography of a regional pediatric rheumatology patient population. J Rheumatol 1994; 21:1552–1661.
12. De Vere R, Bradley WG. Polymyositis: its presentation, morbidity and mortality. Brain 1975; 98:637–666.
13. Lilley H, Dennett X, Byrne E. Biopsy proven polymyositis in Victoria 1982–1987: analysis of prognostic factors. J Roy Soc Med 1994; 87:323–326.
14. Venables, GS, Bates D, Cartlidge NEF, Hudgson P. Acute polymyositis with subcutaneous oedema. J Neurol Sci 1982; 55:161–164.
15. Bohan A, Peter JB. Polymyositis and dermatomyositis. N Engl J Med 1975; 292:344–347 and 403–407.
16. Schumacher H, Schimmer B, Gordon GV, Bookspan MA, Brogadir S, Dorwart BB. Articular manifestations of polymyositis and dermatomyositis. Am J Med 1979; 6:287–292.
17. Nevin S. Two cases of muscular degeneration occurring in late adult life with a review of the recorded cases of late progressive muscular dystrophy (late progressive myopathy). Q J Med 1936; 5:51–68.
18. Shy GM, McEachern D. The clinical features and responses to cortisone of "menopausal muscular dystrophy." J Neurol Neurosurg Psychiatry 1951; 14:101–107.
19. Nattrass FJ. Recovery from "muscular dystrophy." Brain 1954; 77:549–570.
20. Walton JN, Nattrass FJ. On the classification, natural history and treatment of the myopathies. Brain 1954; 77:169–231.
21. Euwer RL, Sontheimer RD. Amyopathic dermatomyositis: a review. J Invest Dermatol 1993; 124S–127S.
22. Stonecipher MR, Jorizzo JL, White WL, Walker FO, Prichard E, Cutaneous changes of dermatomyositis in patients with normal muscle enzymes: dermatomyositis sine myositis? J Am Acad Dermatol 1993; 28:951–956.
23. Park JH, Olsen NJ, King L Jr, Vital T, Buse R, Kari S, Hernanz-Schulman M, Price PR. Use of magnetic resonance imaging and P-31 magnetic resonance spectroscopy to detect and quantify muscle dysfunction in the amyopathic and myopathic variants of dermatomyositis. Arthritis Rheum 1995; 38:68–77.
24. Sano M, Suzuki M, Sato M, Sakamoto T, Uchigata M. Fatal respiratory failure due to polymyositis. Intern Med 1994; 33:185–187.
25. Arsura EL, Greenberg AS. Adverse impact of pulmonary fibrosis on prognosis in polymyositis and dermatomyositis. Semin Arthritis Rheum 1988; 18:29–37.
26. Byred SL, Case BA, Bouleware DW. Pulmonary manifestations of rheumatic disease. Postgrad Med 1993; 93:149–151.
27. Schwarz MI, Sutarik JM, Nick JA, Leff JA, Emlen JW, Tuder RM. Pulmonary capillaritis and diffuse alveolar haemorrhage: a primary manifestation of polymyositis. Am J Respir Crit Care Med 1995; 151:2037–2040.
28. De Toro Santos FJ, Verea-Hernando H, Montero C, Blanco-Aparicio M, Torres Lanzas J, Pombo Felipe F. Chronic pneumomediastinum and subcutaneous emphysema: association with dermatomyositis. Respiration 1995; 62:53–56.
29. Hochberg MC, Feldman D, Stevens MB. Adult onset polymyositis/dermatomyositis: an analysis of clinical and laboratory features and survival in 76 patients with a review of the literature. Semin Arthritis Rheum 1986; 15:168–178.
30. Taylor AJ, Wortham DC, Burge JR, Rogan KM. The heart in polymyositis: a prospective evaluation in 26 patients. Clin Cardiol 1993; 16:802–808.
31. Henderson A, Cumming WJK, Williams DO, Hudgson P. Cardiac complications of polymyositis. Neurol Sci 1980; 47:425–428.
32. Bohan A, Peter JB, Bowman RL, Pearson CM. A computer-assisted analysis of 153 patients with polymyositis and dermatomyositis. Medicine 1977; 56:255–286.
33. Ringel SP, Forstot JZ, Tan EM, Wehling C, Griggs RC, Butcher D. Sjogren's syndrome and polymyositis or dermatomyositis. Arch Neurol 1982; 39:157–163.
34. Marguerie C, Bunn C, Copier J, Bernstein RM, Gilroy JM, Black CM, So AK, Walport MJ. The clinical and immunogenetic features of patients with autoantibodies to the nucleolar antigen PM-Scl. Medicine 1992; 71:327–336.
35. Follansbee WP, Zerbe TR, Medsger TA Jr. Cardiac and skeletal muscle disease in systemic sclerosis (scleroderma): a high risk association. Am Heart J 1993; 125:194–203.

36. Foote RA, Kimbrough SM, Stevens JC. Lupus myositis. Muscle Nerve 1982; 5:65–68.
37. Sharp GC, Irvin WS, Tan E, Gould RG, Holman HR. Mixed connective tissue disease—an apparently distinct rheumatic disease syndrome associated with a specific antibody to an extractable nuclear antigen (ENA). Am J Med 1972; 52:148–159.
38. Alarcon-Segovia D, Cardiel MH. Comparison between 3 diagnostic criteria for mixed connective tissue disease: Study of 593 patients. J Rheumatol 1989; 16:328–334.
39. Van den Hoogen FH, Spronk PE, Boerbooms AM, Bootsma H, de Rooji DJ, Kallenberg CG, van de Putte LB. Long-term follow-up of 46 patients with anti-(U1) snRNP antibodies. Br. J Rheumatol 1994; 33:1117–1120.
40. Pachman LM. Juvenile dermatomyositis: a clinical overview. Pediatr Rev 1990; 12:117–125.
41. Bowyer SL, Blane CE, Sullivan DB, Cassidy JT. Childhood dermatomyositis: factors predicting functional outcome and development of dystrophic calcification. J Pediatr 1983; 103:882–888.
42. Bezecny R. Dermatomyositis. Arch Dermatol Syph (Wien) 1935; 171:242–251.
43. Callen JP. Malignancy in polymyositis/dermatomyositis. Clin Dermatol 1988; 6:55–63.
44. Lakhanpal S, Bunch TW, Nelton LJ. Polymyositis-dermatomyositis and malignant lesions: does an association exist? Mayo Clin Proc 1986; 61:645–653.
45. Masi AT, Hochberg MC. Temporal association of polymyositis/dermatomyositis with malignancy: methodological considerations. Mt Sinai J Med 1988; 55:471–478.
46. Manchul LA, Jin A, Pritchard KI, Tenenbaum J, Boyd NF, Lee P, Germanson T, Gordon DA. The frequency of malignant neoplasms in patients with polymyositis-dermatomyositis: a controlled study. Arch Intern Med 1985; 145:1835–1839.
47. Chow WH, Gridley G, Mellemkjeer L, McLaughlin JK, Olsen JH, Fraumeni JF Jr. Cancer risk following polymyositis and dermatomyositis: a nationwide cohort study in Denmark. Cancer Causes Control 1995; 6:9–13.
48. Zantos D, Zhang Y, Felson D. The overall and temporal association of cancer with polymyositis and dermatomyositis. J Rheumatol 1994; 21:855–859.
49. Bouscary D, Tulliez M, Dreyfus F, Raphael M, Hermine O, Picard F, Delsol G, Sicard D. Polymyositis associated with Ki-1 lymphoma. Leukaemia Lymphoma 1995; 17:185–188.
50. Shay M, Braester A, Cohen I. Dermatomyositis as presenting symptom of Hodgkin's disease. Ann Haematol 1991; 63:116–118.
51. Scerri L, Zaki I, Allen BR, Golding P. Dermatomyositis associated with malignant melanoma—case report and review of the literature. Clin Exp Dermatol 1994; 19:523–525.
52. Whitmore SE, Rosenheim NB, Provost TT. Ovarian cancers in patients with dermatomyositis. Medicine 1994; 73:153–160.
53. Cherin P, Piette JC, Herson S, Bletry O, Wechsler B, Frances C, Godeau P. Dermatomyositis and ovarian cancer: a report of 7 cases and literature review. J Rheumatol 1993; 20:1897–1899.
54. Lee KH, Ong B, Lim TK, Chan HL. Polymyositis associated with symptomless nasopharyngeal carcinoma. Singapore Med J 1994; 35:323–324.
55. Alexander S, Forman L. Dermatomyositis and carcinoma. Br J Dermatol 1968; 80:86–96.
56. Henson RA, Russell D, Wilkinson M. Carcinomatous neuropathy and myopathy: a clinical and pathological study. Brain 1954; 77:82–121.
57. Heathfield KWG, Williams JRB. Peripheral neuropathy and myopathy associated with a bronchogenic carcinoma. Brain 1954; 77:122–137.
58. Brownell B, Hughes JT. Degeneration of muscle in association with carcinoma of the bronchus. J Neurol Neurosurg Psychiatry 1975; 38:363–370.
59. Emslie-Smith AM, Lane RJM, Cumming WJK, Saunders P, Goff DK, Hudgson P. Necrotising inflammatory myopathy in myeloproliferative disorders (in preparation).
60. Kagen LJ. Myoglobinaemia in inflammatory myopathies. JAMA 1977; 237:1448–1452.
61. Nishikai M, Reichlin M. Radioimmunoassay of serum myoglobin in polymyositis and other conditions. Arthritis Rheum 1977; 20:1514–1518.
62. Zweig MH, Adornato B, van Steirteghem AC, Engel WK. Serum creatine kinase BB and MM concentrations determined by radioimmunoassay in neuromuscular disorders. Ann Neurol 1980; 7:324–328.
63. Sandstedt P, Henriksson KG, Larsson LE. Quantitative electromyography in polymyositis and dermatomyositis: a long-term study. Acta Neurol Scand 1982; 65:110–121.
64. Pitt AM, Fleckenstein JL, Greenlee RG Jr, Burns DK, Bryan WW, Haller R. MRI-guided biopsy in inflammatory myopathy: initial results. Magnetic Resonance Imaging 1993; 11:1093–1099.
65. Arahata K, Engel AG. Monoclonal antibody analysis of mononuclear cells in myopathies. I. Quantitation of subsets according to diagnosis and sites of accumulation and demonstration and counts of muscle fibres invaded by T cells. Ann Neurol 1984; 16:193–208.
66. Engel AG, Arahata K. Mononuclear antibody analysis of mononuclear cells in myopathies: II. Phenotypes of autoinvasive cells in polymyositis and inclusion body myositis. Ann Neurol 1984; 16:209–215.

67. Arahata K, Engel AG. Mononuclear antibody analysis of mononuclear cells in myopathies. IV. Cell-mediated cytotoxicity and muscle fibre necrosis. Ann Neurol 1988; 23:168–173.
68. Whitaker JN, Engel WK. Vascular deposits of immunoglobulin and complement in idiopathic inflammatory myopathy. N Engl J Med 1972; 286:333–338.
69. Emslie-Smith AM, Engel AG. Microvascular changes in early and advanced dermatomyositis: a quantitative study. Ann Neurol 1990; 27:343–356.
70. Kissel JT, Halterman RK, Rammohan KW, Mendell JR. The relationship of complement-mediated microvasculopathy to the histologic features and clinical duration of disease in dermatomyositis. Arch Neurol 1991; 48:26–30.
71. Engel AG, Hohlfeld R, Banker BQ. Inflammatory myopathies. The polymyositis and dermatomyositis syndromes. In: Engel AG, Franzini-Armstrong C, eds. Myology, Vol 2, 2nd ed. New York: McGraw-Hill, 1994:1335–1383.
72. Munsat TL, Hudgson P, Johnson MA. Experimental serotonin myopathy. Neurology (Minneapolis) 1979; 27:772–782.
73. Rowe DJ, Isenberg DA, Beverley PCL. Monoclonal antibodies to leucocyte antigens in polymyositis and muscular dystrophy. Clin Exp Immunol 1983; 54:327–336.
74. Behan WMH, Behan PO, Durward WF, McQueen A. The inflammatory process in polymyositis; monoclonal antibody analysis of muscle and peripheral blood immunoregulatory lymphocytes. J Neurol Neurosurg 1987: 50:1486–1474.
75. Giorno R, Barden MT, Kohler PF, Ringel SP. Immunological characterisation of the mononuclear cell infiltrating muscle of patients with inflammatory and non-inflammatory myopathies. Clin Immunol Immunopathol 1984; 30:405–412.
76. Lemoine NR, Ryan JF, Cox EL, Mayston V. Revell PA, Swash M. Immunohistochemical analysis of mononuclear cell subsets in inflammatory and non-inflammatory myopathies. J Pathol 1986; 39:271–274.
77. Dubowitz V. Inflammatory myopathies. In: Muscle Biopsy. A Practical Approach, 2nd ed. London: Balliere Tindall, 1985:570–611.
78. Cumming WJK, Fulthorpe JJ, Hudgson P, Mahon M. Inflammatory myopathies. In: Colour Atlas of Muscle Pathology. London: Mosby-Wolfe, pp 113–134.
79. Rifai Z, Welle S, Kamp C, Thornton CA. Ragged red fibres in normal aging and inflammatory myopathy. Ann Neurol 1995; 37:24–29.
80. Simpson WG, Nickl NJ. Polymyositis associated with primary biliary cirrhosis. J Kentucky Medical Assoc 1995; 93:93–94.
81. Yasuda Y, Nahano S, Akiguchi I, Tanaka M, Kameyama M. Polymyositis associated with asymptomatic primary biliary cirrhosis. Eur Neurol 1993; 33:51–53.
82. Chugh S, Dilawari JB, Sawhney IM, Dang N, Radotra BD, Chawla YK. Polymyositis associated with ulcerative colitis. Gut 1993; 34:567–569.
83. Henriksson KG, Hallert C, Norrby K, Walan A. Polymyositis and adult coeliac disease. Acta Neurol Scand 1982; 65:301–319.
84. Pinheiro G da R, Goldenberg J, Atra E, Peneira RB, Camano L, Schmidt B. Juvenile dermatomyositis and pregnancy: report and literature review. J Rheumatol 1992; 19:1798–1801.
85. Ishii N, Ono H, Kawaguchi T, Nakajima H. Dermatomyositis and pregnancy: case report and review of the literature. Dermatologia 1991; 183:146–149.
86. Plotz PH. Dalakas M, Leff RL, Love LA, Miller FW, Cronin ME. Current concepts in the idopathic inflammatory myopathies: polymyositis, dermatomyositis and related disorders. Ann Intern Med 1989; 111:143–157.
87. Reed AM, Pachman L, Ober C. Molecular genetic studies of major histocopmatibility complex genes in children with juvenile dermatomyositis: increased risk associated with HLA-DQA1*0501.
88. Hirsch TJ, Enlow RW, Bias WB, Arnett FC. HLA-D related (DR) antigens in various kinds of myositis. Hum Immunol 1981; 3:181–186.
89. Cumming WJK, Hudgson P, Latimer D. HLA and serum complement in polymyositis. Lancet 1977; 2:978–979.
90. Lane RJM, Mastaglia FL. Drug-induced myopathies in man. Lancet 1978; 2:562–565.
91. Blain P, Lane RJM. Neurological disorders. In: Textbook of Adverse Drug Reactions, 4th ed. Oxford: Oxford University Press, 1991:535–566.
92. Swartz MO, Silver RM. D-penicillamine-induced polymyositis in juvenile chronic arthritis: report of a case. J Rheumatol 1984; 11:251–252.
93. Flowers FP, Arawo OE, Hamm KA. Phenytoin hypersensitivity syndrome. J Emergency Med. 1987; 5:103–108.
94. Fontiveros ES, Cumming WJK, Hudgson P. Procainemide-induced myositis. J Neurol Sci 1980; 45:143–147.
95. Pier N, Dubowitz V. Chronic graft versus host disease presenting with polymyositis. Br. Med. J 1983; 286:2024.

96. Anderson BA, Young V, Kean WF, Ludwin SK, Galbraith PR, Anastassides TP. Polymyositis in chronic graft vs host disease. Arch Neurol 1982; 39:188–190.
97. Urbano-Marquez A, Estruch R. Grau JM, Granena A, Martin-Ortega E, Palou J, Rozman C. Inflammatory myopathy associated with chronic graft-versus-host disease. Neurology 1986; 36:1091–1093.
98. Bywaters EGL. Discussion on hypersensitivity and the collagen diseases. Proc Roy Soc Med 1956; 49:287–289.
99. Dawkins RL. Experimental myositis associated with hypersensitivity to muscle. J Pathol. Bacteriol 1965; 90:619–525.
100. Kakulas BA. Destruction of differentiated muscle cells by sensitised cells. J Pathol Bacteriol 1966; 91:495–503.
101. Takayanagi T. Immunological studies of experimental myositis in relation to human polymyositis. Folia Psychiatr Neurol Jpn 1967; 21:117–127.
102. Currie S, Saunders M, Knowles M, Brown A. Immunological aspects of polymyositis: the in vitro activity of lymphocytes on incubation with muscle antigen and with muscle cultures. Q J Med 1971; 40:63–84.
103. Haas DC, Arnason BGW. Cell-mediated immunity in polymyositis: creatine phosphokinase release from muscle cultures. Arch Neurol 1974; 31:192–196.
104. Saunders M, Knowles M, Currie S. Lymphocyte stimulation with muscle homogenates in polymyositis and other muscle wasting disorders. J Neurol Neurosurg Psychiatry 1969; 32:546–571.
105. Esiri MM, MacLennan ICM, Hazelman BL. Lymphocyte reactivity to skeletal muscle in patients with polymyositis and other disorders. Clin Exp Immunol 1973; 14:25–35.
106. Hohlfeld R, Engel AG. Coculture with autologous myotubes of cytotoxic T cells isolated from muscle in inflammatory myopathies. Ann Neurol 1991; 29:498–507.
107. Hohlfeld R, Engel AG. Lysis of myotubes by alloreactive cytotoxic T cells and natural killer cells: relevance to myoblast transplantation. J Clin Invest 1990; 86:370–374.
108. Appleyard ST, Dubowitz V, Dunn MJ, Rose ML. Increased expression of HLA ABC class I antigens by muscle fibres in Duchenne muscular dystrophy, inflammatory myopathy, and other neuromuscular disorders. Lancet 1985; 1:361–363.
109. Karpati G, Pouliot Y, Carpenter S. Major histocompatibility complex products in human skeletal muscles. Ann Neurol 1988; 23:64–72.
110. Janeway CA. How the immune system recognises invaders. Sci Am 1993; 269:72–79.
111. Goebels N, Michaelis D, Werkele H, Hohlfeld R. Human myoblasts as antigen presenting cells. J Immunol 1992; 149:661–667.
112. Kalovidouris AE, Horn CA, Plotzkin Z. The role of cytokines in polymyositis. III. Recombinant human interferon-gamma enhances T cell adhesion to cultured human muscle cells. Arthritis Rheum 1994; 37:907–914.
113. Jerusalem F, Rakusa M, Engel AG. Morphometric analysis of skeletal muscle capillary ultrastructure in inflammatory myopathies. J Neurol Sci 1974; 23:391–402.
114. Carpenter S, Karpati S, Rothman S, Watters G. The childhood type of dermatomyositis. Neurology 1976; 26:952–962.
115. De Bleeker JL, Engel AG. Expression of cell adhesion molecules in inflammatory myopathies and Duchenne dystrophy. J Neuropathol Exp Neurol 1994; 53:369–376.
116. Kissel JT, Mendell JR, Rammohan KW. Microvascular deposition of complement membrane attack complex in dermatomyositis. N Engl J Med 1985; 314:329–334.
117. Fulthorpe JJ, Hudgson P. Experimental allergic myositis. Neuropathol Appl Neurobiol 1975; 1:401 (abstract).
118. Gotoff SP, Smith RD, Sugar O. Dermatomyositis with cerebral vasculitis in a patient with agammaglobulinaemia. Am J Dis Child 1972; 123:53–56.
119. Leddy JP, Griggs RC, Klemperer MR, Frank MM. Hereditary complement (C2) deficiency with dermatomyositis. Am J Med 1975; 58:83–91.
120. De Bleecker JL, Engel AG. Immunocytochemical study of CD45 T cell isoforms in inflammatory myopathies. Am J Pathol 1995; 146:1178–1187.
121. O'Hanlon, TP, Dalakas MC, Plotz PH, Miller FW. Predominant TCR-alpha beta variable and joining gene expression by muscle-infiltrating lymphocytes in the idiopathic inflammatory myopathies. J Immunol 1994; 152:2569–2576.
122. Bender A, Ernst N, Iglesias A, Dornmair K, Wekerle H, Hohlfeld R. T cell receptor repertoire in polymyositis: clonal expansion of autoaggressive $CD8^+$ T cells. J Exp Med 1995; 181:1863–1868.
123. Tajima Y, Morikawa F, Tashiro K. Temporal alterations of immunohistochemical findings in polymyositis. Intern Med 1994; 33:263–270.
124. Hohlfeld R, Engel AG, Ii L, Harper MC. Polymyositis mediated by T lymphocytes that express the gamma/delta receptor. N Engl J Med 1991; 324:877–881.
125. Pluschke G, Ruegg D, Hohlfeld R, Engel AG. Autoaggressive myocytotoxic T-lymphocytes expressing an unusual gamma-delta T-cell receptor. J Exp Med 1992; 176:1785–1789.

126. O'Hanlon TP, Messersmith WA, Dalakas MC, Plotz PH, Miller FW. Gamma delta T cell receptor gene expression by muscle-infiltrating lymphocytes in the idiopathic inflammatory myopathies. Clin Exp Immunol 1995; 100:519–528.
127. Nakamura K, Arahata K, Ishuira S, Osame M, Sugita H. Degradative activity of granazme A on skeletal muscle proteins in vitro: a possible molecular mechanism for skeletal muscle fibre damage in polymyositis. Neuromusc Disord 1993; 3:303–310.
128. Orimo S, Koga R, Goto K, Nakamura K, Arai M, Tamki M, Sugita H, Nanaka I, Arahata K. Immunohistochemical analysis of perforin and granazyme A in inflammatory myopathies. Neuromusc Disord 1994; 4:219–226.
129. Love LA, Leff RL, Fraser DD, Targoff IN, Dalakas M, Plotz PH. A new approach to the classification of idiopathic inflammatory myopathy: myositis-specific autoantibodies define useful homogenous patient groups. Medicine; 1991: 70:360–374.
130. Pestronk A, Drachman DB. Antibody-mediated membrane abnormalities in polymyositis: reduction of acetylcholine receptors by immunoglobulin. Ann NY Acad Sci 1987; 505:357–367.
131. De Reuck J, Thiery E, de Coster W, van der Eecken H. Myasthenic syndrome in polymyositis. Eur Neurol 1976; 14:275–284.
132. Reichlin M. Significance of the Ro antigen system. J Clin Immunol 1986; 6:339–348.
133. Reichlin M, Maddison PJ, Targoff IN, Bunch T, Arnett F, Sharp G, Treadwell E, Tan EM. Antibodies to a nuclear/nucleolar antigen in patients with polymyositis-overlap syndrome. J Clin Immunol 1984; 4:40–44.
134. Mimori T. Scleroderma-polymyositis overlap syndrome. Intl J Dermatol 1987; 26:419–425.
135. Hong TJ, Escribano J, Coca-Prados M. Isolation of cDNA clones encoding the 80-kd subunit protein of the human autoantigen Ku (p70/80) by antisera raised against ciliary processes of human eye donors. Invest Ophthalmol Visual Sci 1994; 35:4023–4030.
136. Johanet C, Agostini MM, Vaysairrat M, Abuaf N. Anti-scl-70 and anticentromere autoantibodies: biological markers of two forms of systemic sclerosis. Presse Med 1989; 18:207–211.
137. Jarzabek-Chorzelska M, Blaszczyk M, Jablonska S. Scl 70 antibody—a specific marker of systemic sclerosis. Br J Dermatol 1986; 115:393–401.
138. Hoffman RW, Rettenmaier LJ, Takeda Y, Hewett JE, Petterson I, Nyman V, Luger AM, Sharp GC. Human autoantibodies against the 70 kD polypeptide of U_1 small nuclear RNP are associated with HLA-DR4 among connective tissue disease patients. Arthritis Rheum 1990; 33:666–673.
139. Eaton RB, Schneider G, Schur PH. Enzyme immunoassays for antibodies to native DNA. Arthritis Rheum 1983; 26:52–62.
140. Lerner MR, Steitz JA. Antibodies to small nuclear RNAs complexed with proteins are produced by patients with systemic lupus erythematosus. Proc Natl Acad Sci USA 1979; 76:5495–5499.
141. Targoff IN. Immune manifestations of inflammatory muscle disease. Rheum Dis Clin North Am 1994; 20:857–880.
142. Miller FW. Myositis-specific autoantibodies: touchstones for understanding the inflammatory myopathies. JAMA 1993; 270:1846–1849.
143. Marguerie C, Bunn CC, Beynon HLC, Bernstein RM, Hughes JMB, So A, Walport MJ. Polymyositis, pulmonary fibrosis and autoantibodies to aminoacyl-tRNA synthetase enzymes. Q J Med 1990; 282:1019–1038.
144. Targoff IN. Humoral immunity in polymyositis/dermatomyositis. J Invest Dermatol 1993; 100:116S.
145. Targoff IN. Johnson AE, Miller FW. Antibody to signal recognition particle in polymyositis. Arthritis Rheum 1990; 33:1361–1370.
146. Haas W, Pereira P, Tonegawa S. Gamma-delta cells. Annu Rev Immunol 1993; 11:637–685.
147. Jardetzky TS, Lane WS, Robinson RA, Madden DR, Wiley DC. Identification of self peptides bound to purified HLA-B27. Nature 1991; 353:326–329.
148. Hohlfeld R, Engel AG. Expression of 65-kd heat shock proteins in the inflammatory myopathies. Ann Neurol 1992; 32:821–823.
149. Mastaglia FL, Oeda VJ. Inflammatory myopathies: part 1. Ann Neurol 1985; 17:215–217.
150. Gamboa ET, Eastwood AB, Hays AP, Maxwell J, Penn AS. Isolation of influenza virus from muscle in myoglobinuric polymyositis. Neurology 1979; 29:1323–1335.
151. Dalakas MC, Pezeshkpour GH. Neuromuscular diseases associated with human immunodeficiency virus infection. Ann Neurology 1988; 23(Suppl):S38–S48.
152. Sherman MP, Amin RM, Rodgers-Johnson PE, Morgan OS, Char G, Mora CA, Iannone R, Collins GH, Papsideuro L, Gibbs CJ Jr et al. Identification of human T cell leukaemia virus type I antibodies, DNA, and protein in patients with polymyositis. Arthritis Rheum 1995; 38:690–698.
153. Arai H, Tanaka M, Ohata K, Kojo T, Niijima K, Imawari M. Symptomatic myopathy associated with interferon therapy for chronic hepatitis C. Lancet 1995; 345:582.
154. Horsmans Y, Geubel AP. Symptomatic myopathy in hepatitis C infection without interferon therapy. Lancet 1995; 345:1236.

155. Ueno Y, Kondo K, Kidokoro N, Kobashi R, Kanaji K, Matsumura T. Hepatitis C infection and polymyositis. Lancet 1995; 346:319–320.
156. Weidensaul D, Inam T, Holyst MM, King PD, McMurray RW. Polymyositis, pulmonary fibrosis, and hepatitis C. Arthritis Rheum 1995; 38:437–439.
157. Mathews MB, Bernstein RM. Myositis autoantibody inhibits histidyl tRNA transferase: a model for autoimmunity. Nature 1983; 304:177–179.
158. Plotz PH. Autoantibodies are anti-idiotype antibodies to antiviral antibodies. Lancet 1983; 2:824–826.
159. Walker EJ, Jeffrey PD. Polymyositis and molecular mimicry, a mechanism of autoimmunity. Lancet 1986; 2:605–607.
160. Muir P, Archard LC. Persistent enterovirus infections are causally associated with chronic medical conditions in humans. Rev Med Virol 1994; 4:245–250.
161. Strongwater SL, Dorivini-Zis K, Ball RD, Schnitzer TJ. A murine model of polymyositis induced by Coxsackie B1 (Tucson strain). Arthritis Rheum 1984; 27:433–422.
162. Jongen PJ, Heessen FW, Bergmann I, de Waal RM, ter Laak HJ, Galama JM, van Venrooij WJ, Gabreels FJ, van de Putte LB, Berden JH. Coxsackie B1 virus–induced murine myositis: a correlative study of muscular lesions and serological changes. J Autoimmun 1994; 7:727–737.
163. Tam PE, Schmidt AM, Ytterberg SR, Messner RP. Duration of virus persistence and its relationship to inflammation in the chronic phase of Coxsackievirus B1-induced murine polymyositis. J Lab Clin Med 1994; 123:346–356.
164. Travers RL, Hughes GVR, Cambridge G, Sewell JR. Coxsackie B neutralisation titres in polymyositis/dermatomyositis. Lancet 1977; 1:1268.
165. Christiansen ML, Pachman LM, Schneiderman R, Patel DC, Friedman JM. Prevalence of Coxsackie B virus antibodies in patients with juvenile dermatomyositis. Arthritis Rheum 1986; 29:1365–1370.
166. Nishikai M. Coxsackie virus infection and the development of polymyositis/dermatomyositis. Rheumatol Int 1994; 14:43–46.
167. Bowles NE, Dubowitz V, Sewry CA, Archard LC. Dermatomyositis, polymyositis and Coxsackie-B-virus infection. Lancet 1987; 1:1004–1007.
168. Yousef GE, Isenberg DA, Mowbray JF. Detection of enterovirus specific RNA sequences in muscle biopsy specimens from patients with adult onset myositis. Ann Rheum Dis 1990; 49:310–315.
169. Archard LC, Bowles NE, Behan PO, Bell EJ, Doyle D. Postviral fatigue syndrome: persistence of enterovirus RNA in muscle and elevated creatine kinase. J Roy Soc Med 1988; 81:326–329.
170. Gow JW, Behan WM, Clements GB, Woodall C, Riding M, Behan PO. Enteroviral RNA sequences detected by polymerase chain reaction in muscles of patients with postviral fatigue syndrome. Br Med J 1991; 302:692–696.
171. Bowles NE, Bayston TA, Zhang H-Y, Doyle D, Lane RJM, Cunningham L, Archard LC. Persistence of enterovirus RNA in muscle biopsy samples suggests that some cases of chronic fatigue syndrome result from a previous, inflammatory viral myopathy. J Med 1993; 24:145–160.
172. Cunningham L, Bowles NE, Lane RJM, Dubowitz V, Archard LC. Persistence of enteroviral RNA in chronic fatigue syndrome is associated with the abnormal production of equal amounts of positive and negative strands of enteroviral RNA. J Gen Virol 1990; 71:1399–1402.
173. Leff RL, Love SJ, Miller FW, Greenberg SJ, Klein EA, Dalakas MC, Plotz PH. Viruses in idiopathic inflammatory myopathies: absence of candidate viral genomes in muscle. Lancet 1992; 339:1192–1195.
174. Leon-Monzon M, Dalakas MC. Absence of persistent infection with enterovirus in muscles of patients with inflammatory myopathies. Ann Neurol 1992; 32:219–222.
175. Fox SA, Finklestone E, Robbins PD, Mastaglia FL, Swanson NR. Search for persistent enterovirus infection of muscle in inflammatory myopathies. J Neurol Sci 1994; 125:70–76.
176. Hollingrake K. Polymyositis presenting as distal muscle weakness: a case report. J Neurol Sci 1969; 8:479–484.
177. Van Kastern BJ. Polymyositis presenting with chronic progressive distal muscle weakness. J Neurol Sci 1979; 41:307–310.
178. Stark RJ. Polymyositis presenting with severe weakness involving only one arm. Aust NZ J Med 1978; 8:544–546.
179. Lugt LVD. Dermatomyositis. Acta Dermat Venereol (Stockholm) 1952; 32:27–39.
180. Lederman RJ, Salanga VD, Wilbourne AJ, Hanson MR, Dudley AW Jr. Focal inflammatory myopathy. Muscle Nerve 1984; 7:142–146.
181. Cumming WJK, Weiser R, Teoh R, Hudgson P, Walton JN. Localised nodular myostitis: a clinical and pathological variant of polymyositis. Q J Med New Ser 1977; 46:531–546.
182. McLetchie NGB, Aikens RL. Relapsing myositis. Arch Pathol 1952; 53:497–508.
183. Smith CA, Pinals RS. Localised nodular myositis J Rheumatol 1981; 8:815–819.
184. Heffner RR Jr, Barron SA. Polymyositis beginning as a focal process. Arch Neurol 1981; 38:439–442.

185. Flaisler F, Blin D, Asencio G, Lopez FM, Combe B. Focal myositis: a localised form of polymyositis? J Rheumatol 1993; 20:1414–1416.
186. Fujitake J, Ishikawa Y, Ishimaru S, Koh T, Tatsuoka Y. Localised polymyositis associated with chronic thyroiditis. J Rheumatol 1994; 21:1147–1149.
187. Fleisler F, Blin D, Asencio G, Lopez FM, Combs B. Focal myositis: a localised form of polymyositis? J Rheumatol 1993; 20:1414–1416.
188. Enzinger FM, Dulcey F. Proliferative myositis: report of 33 cases. Cancer 1967; 20:2213–2223.
189. Rothstein TL, Carlson CB, Sumi SM. Polymyositis with facioscapulohumeral distribution. Arch Neurol (Chicago) 1971; 25:313–319.
190. Munsat TL, Piper D, Cancilla P, Mednick J. Inflammatory myopathy with facioscapulohumeral distribution. Neurology 1972; 22:335–347.
191. Bates D, Stevens JC, Hudgson P. "Polymyositis" with involvement of facial and distal musculature. One form of facioscapulohumeral syndrome? J Neurol Sci 1973; 19:105–108.
192. Arahata K, Ishihara T, Fukunaga H, Orimo S, Lee JH, Goto K, Nonaka I. Inflammatory response in facioscapulohumeral muscular dystrophy (FSHD): immunocytochemical and genetic analyses. Muscle Nerve 1995; (Suppl 2):S56–66.
193. Fauci AS, Harley JB, Roberts WC, Ferrans VJ, Gralnick HR, Bjornson BH. The idiopathic hypereosinophilic syndrome: clinical, patholphysiologic and therapeutic considerations. Ann Intern Med 1982; 97:78–92.
194. Gardner-Thorpe C, Harriman DGF, Parsons MR, Rudge P. Loffler's eosinophilic endocarditis with Balint's syndrome (optic ataxia and paralysis of visual fixation). Q J Med 1971; 40:249–260.
195. Layzer RB, Shearn MA, Satya-Murti S. Eosinophilic polymyositis. Ann Neurol 1977; 1:65–71.
196. Stark RJ. Eosinophilic polymyositis. Arch Neurol 1979; 36:721–722.
197. Shulman LE. Diffuse fasciitis with eosonophilia: a new syndrome? Trans Assoc Am Physicians 1975; 88:70–86.
198. Cranmer SF, Kent L, Abramowsky C, Moskowitz RW. Eosinophilic fasciitis: immunopathology, ultrastructure, literature review, and consideration of its pathogenesis and relation to scleroderma. Arch Pathol Lab Med 1982; 106:85–91.
199. Barnes L. Rodnan GP, Medsger TA, Short D. Eosinophilic fasciitis: a pathologic study of twenty cases. Am J Pathol 1979; 96:493–517.
200. Bjelle A, Henriksson K-G, Hofer P-A. Polymyositis in eosinophilic fasciitis: review and case report. Eur Neurol 1980; 19:128–137.
201. Michet CJ Jr, Doyle JA, Ginsburg WW. Eosinophilic fasciitis, report of 15 cases. Mayo Clin Proc 1981; 56:27–34.
202. Hoffman R, Young M, Ershler WB, Mazur E, Gewirtz A. Diffuse fasciitis and aplastic anaemia: a report of four cases revealing an unusual association between rheumatological and haematological disorders. Medicine 1982; 61:373–381.
203. Martin RW, Duffy J, Lie JT. Eosinophilic fasciitis associated with use of L-tryptophan. A case control study. Mayo Clin Proc 1991; 66:892–898.
204. Chumbley LC, Harrison EG, DeRemee RA. Allergic granulomatosis and angiitis (Churg-Strauus syndrome): report and analysis of 30 cases. Mayo Clin Proc 1977; 52:477–484.
205. Fauci AS, Haynes BF, Katz P, Wolff SM. Wegener's granulomatosis: prospective clinical and therapeutic experience wih 85 patients for 21 years. Ann Intern Med 1983; 98:76–85.
206. Michalak T. Immune complexes of hepatitis B surface antigens in the pathogenesis of periarteritis nodosa. Am J Pathol 1978; 90:619–632.
207. Van der Woude FJ, Rasmussen N, Lobatto, S, Wiik A, Permin H, van Es, LA, van der Giessen M, van der Hem GK, The TH. Autoantibodies against neutrophils and monocytes: tool for the diagnosis and marker of disease activity in Wegener's granulomatosis. Lancet 1988; 1:425–429.
208. Rao JK, Allen NB, Feussner JR, Weinberger M. A prospective study of antineutrophil cytoplasmic antibody (c-ANCA) and clinical criteria in diagnosing Wegener's granulomatosis. Lancet 1995; 346:926–931.
209. Steinmann CR. Circulating DNA in polyarteritis nodosa and related syndromes. Arthritis Rheum 1982; 25:1425–1430.
210. Kissel JT, Lynn DJ, Rammohan KW, Klein JP, Griggs RC, Moxley RT III, Cwik VA, Brooke MH, Mendell JR. Mononuclear cell analysis of muscle biopsies in prednisone and azathioprine-treated Duchenne muscular dystrophy. Neurology 1993; 43:532–536.
211. Lynch PG, Bansal DV. Granulomatous polymyositis. J Neurol Sci 1973; 18:1–9.
212. Hewlett RH, Brownell B. Granulomatous myopathy: its relationship to sarcoidosis and polymyositis. J Neurol Neurosurg Psychiatry 1975; 38:1090–1099.
213. Menard DB, Haddad H, Blain JG, Beaudry R, Devroede G, Masse S. Granulomatous myositis and myopathy associated with Crohn's colitis. N Engl J Med 1976; 295:818–819.
214. Namba TJ, Grunner NG, Grob D. Idiopathic giant-cell polymyositis. Report of a case and review of the syndrome. Arch Neurol (Chicago) 1974; 31:27–30.

215. Nosal A, Schleissner LA, Mishkin FS, Lieberman J. Angiotensin-1-converting enzyme and gallium scan in noninvasive evaluation of sarcoidosis. Ann Intern Med 1979; 90:328–331.
216. Scadding JG, Mitchell DN. Saroidosis, 2nd ed. London: Chapman and Hall, 1985:246–250.
217. Stjernberg N, Cajander S, Truedsson H, Uddenfeldt P. Muscle involvement in sarcoidosis. Acta Med Scand 1981; 209:213–216.
218. Crompton MR, MacDermot V. Sarcoidosis associated with progressive muscular wasting and weakness. Brain 1961; 84:62–74.
219. Tew DS, Pongratz DE. Immunohistological analysis of sarcoid myopathy. J Neurol Neurosurg Psychiatry 1995; 59:322–325.
220. O'Leary PA, Waisman M. Dermatomyositis. A study of forty cases. Arch Dermatol Syphilol 1940; 41:1001–1019.
221. Sheard P. Dermatomyositis. Arch Intern Med 1951; 88:640–658.
222. Elkinton JR, Hunt AD, Godfrey L, McCrory WW, Rogerson AG, Stokes J. Effects of pituitary adrenocorticotropic hormome (ACTH) therapy. JAMA 1949; 141:1273–1279.
223. Kierland RR, O'Leary PA, Brunsting LA, Didcot JW. Cortisone and corticotrophin (ACTH) in dermatology. JAMA 1952; 148:23–26.
224. Winklelmann RK, Mulder DW, Lambert EH, Howard FM, Diessner GR. Course of dermatomyositis-polymyositis: comparison of untreated and cortisone-treated patients. Mayo Clin Proc 1968; 43:545–556.
225. Carpenter JR, Bunch TW, Engel AG, O'Brien P. Survival in polymyositis: corticosteroids and risk factors. J Rheumatol 1977; 4:207–214.
226. Henriksson KG, Sandstedt P. Polymyositis—treatment and prognosis. Acta Neurol Scand 1982; 65:280–300.
227. Uchino M, Araki S, Yoshida O, Uekawa K, Nagata J. High single-dose alternate-day corticosteroid regimens in treatment of polymyositis. J Neurol 1985; 232:175–178.
228. Lane RJM, Emslie-Smith A, Mosquera IE, Hudgson P. Clinical, biochemical and histological responses to treatment in polymyositis: a prospective study. J Roy Soc Med 1989; 82:333–338.
229. Edwards RHT, Wiles CM, Round JM, Jackson MJ, Young A. Muscle breakdown and repair in polymyositis: a case study. Muscle Nerve 1979; 2:223–228.
230. Bunch TW, Worthington JW, Combs JJ, Ilstrup DM, Engel AG. Azathioprine with prednisone for polymyositis. Ann Intern Med 1980; 92:365–369.
231. Bunch TW. Prednisone and azathioprine for polymyositis. Long-term follow up. Arthritis Rheum 1981; 24:45–48.
232. Csuka M, McCarty DJ. Simple method for measurement of lower extremity muscle strength. Am J Med 1985; 78:77–81.
233. Bombardieri S, Clerico A, Riente L, Grazia del Chicca M, Vitali C. Circadian variations of serum myoglobin in normal subjects and patients with polymyositis. Arthritis Rheum 1982; 25:1419–1424.
234. Rose AL, Walton JN. Polymyositis: a survey of 89 cases with particular reference to treatment and prognosis. Brain 1966; 89:747–768.
235. Riddoch D, Morgan-Hughes JA. Prognosis in adult polymyositis. J Neurol Sci 1975; 26:71–80.
236. Fafalak RG, Peterson MG, Kagen LJ. Strength in polymyositis and dermatomyositis: best outcome in patients treated early. J Rheumatol 1994; 21:643–648.
237. Schwartz HA, Slavin G, Ward P, Ansell BM. Muscle biopsy in polymyositis and dermatomyositis: a clinicopathological study. Ann Rheum Dis 1980; 39:500–507.
238. Tajima Y, Moriwaka F, Tashiro K. Temporal alterations of immunohistochemical findings in polymyositis. Intern Med 1994; 33:263–270.
239. Clements PJ, Davis J. Cytotoxic drugs; their clinical application to rheumatic diseases. Semin Arthritis Rheum 1986; 15:231–254.
240. Joffe MM, Love LA, Leff LA, Fraser DD, Targoff IN, Hicks JE, Plotz PH, Miller FW. Drug therapy of idiopathic inflammatory myopathies: predictors of response to prednisone, azathioprine, and methotrexate and a comparison of their efficacy. Am J Med 1993; 94:379–387.
241. Metzger AL, Bohan A, Goldberg LS. Polymyositis and dermatomyositis: combined methotrexate and corticosteroid therapy. Ann Intern Med 1974; 81:182–194.
242. Bombardieri S, Hughes GR, Neri R, del Bravo P, del Bono L. Cyclophosphamide in severe polymyositis. Lancet 1989; 1:1138–1139.
243. Leroy JP, Drosos AA, Yiannopoulos DI, Youniou P, Moutsopoulos HM. Intravenous pulse cyclophosphamide therapy in myositis and Sjogren's syndrome. Arthritis Rheum 1990; 33:1579–1581.
244. Cronin ME, Miller FW, Hicks JE, Dalakas M, Plotz PH. The failure of intravenous cyclophosphamide therapy in refractory idiopathic inflammatory myopathy. J Rheumatol 1989; 16:1225–1228.
245. Kaveri SV, Mouthon L, Kazatchkine MD. Immunomodulating effects of intravenous immunoglobulin in autoimmune and inflammatory diseases. J Neurol Neurosurg Psychiatry 1994; 57(Suppl):6–8.
246. Lang B, Laxer RM, Murphy G, Silverman ED, Roifman CM. Treatment of dermatomyositis with intravenous gammaglobulin. Am J Med 1991; 91:169–172.

247. Cherin P, Herson S, Wechsler B, Piette JC, Bletry O, Coutellier A, Ziga JM, Godeau P. Efficacy of intravenous gamma-globulin therapy in chronic refractory polymyositis and dermatomyositis: an open study with 20 adult patients. Am J Med 1991; 91:162–168.
248. Dalakas MC, Illa I, Dambrosia JM, Soueidan SA, Stein DP, Otero C, Dinsmore ST, McGrosky S. A controlled trial of high-dose intravenous immune globulin infusions as treatment for dermatomyositis. N Eng J Med 1993; 329:1993–2000.
249. Cherin P, Piette JC, Wechsler B, Bletry O, Ziza JM, Laraki R, Godeau P, Herson S. Intravenous gamma globulin as first line therapy in polymyositis and dermatomyositis: an open study in 11 adult patients. J Rheumatol 1994; 21:1092–1097.
250. Jones DW, Snaith ML, Isenberg DA. Cyclosporine treatment for intractable polymyositis. Arthritis Rheum 1987; 30:959–960.
251. Heckmatt J, Saunders C, Peters AM, Rose M, Hasson N, Thompson N, Cambridge G, Hyde SA. Cyclosporin in juvenile dermatomyositis. Lancet 1989; 1:1063–1066.
252. Lueck CJ, Trend P, Swash M. Cyclosporin in the management of polymyositis and dermatomyositis. J Neurol Neurosurg Psychiatry 1991; 54:1007–1008.
253. Dau PC. Plasmaphereisis in idiopathic inflammatory myopathy: experience with 35 patients. Arch Neurol 1981; 38:544–552.
254. Miller FW, Leitman SF, Cronin ME, Hicks JE, Leff RL, Wesley R, Fraser DD, Dalakas M, Plotz PH. Controlled trial of plasma exchange and leukaphereisis in polymyositis and dermatomyositis. N Engl J Med 1991; 326:1380–1384.
255. Engel WK, Lichter AS, Galdi AP. Polymyositis: remarkable response to total body irradiation. Lancet 1981; 1:658.
256. Hubbard WN, Walport MJ, Halnan KE, Beaney RP, Hughes GRV. Remission from polymyositis after total body irradiation. Br Med J 1982; 284:1915–1916.
257. Schon F, Thomas PK, Senanyake LF. Remission from polymyositis after total body irradiation. Br Med J 1982; 285:290.
258. Lane RJM, Hudgson P. Thymectomy in polymyositis. Lancet 1984; 1:626.
259. Behan P. Thymectomy in idiopathic polymyositis. Irish J Med Sci 1978; 147:159.

43
Inclusion Body Myositis and Other Rimmed Vacuolar Myopathies

Michael Rose
King's College Hospital, London, England

Russell J. M. Lane
Charing Cross Hospital and Charing Cross and Westminster Medical School, London, England

INTRODUCTION

In 1967, Chou (1) reported an unusual case of "chronic polymyositis" in which a considerable proportion of muscle fibers exhibited vacuoles rimmed by basophilic material, together with inflammatory infiltration and filamentous cytoplasmic and nuclear inclusions resembling myxovirus nucleocapsids on ultrastructural examination. Subsequently, Yunis and Samaha (2) introduced the term "inclusion body myositis" (IBM) to define a subgroup of chronic polymyositis cases having these histopathological characteristics. Successive series based on variations of these pathological criteria have retained sporadic IBM (S-IBM) as a discrete clinical entity, even though the initial defining pathological features of "rimmed vacuoles" and filamentous inclusions are not pathognomonic of IBM and can occur in other myopathic and neurogenic disorders (Table 1) (3,4). Whether this reflects a common etiological or pathogenetic relationship between IBM and these other disorders, and with other forms of idiopathic inflammatory myopathy, is the subject of current debate.

CLINICAL FEATURES OF S-IBM

S-IBM has been found to comprise 17% (5), 28% (6), or as much as 30% (7) of all cases of inflammatory myopathy. It is usually of late onset, with more than 80% of cases occurring over the age of 50 years (6), although childhood onset has been described (8). The male:female ratio is greater than 2:1, contrasting with the female predominance of polymyositis (5,6). S-IBM is a slowly progressive disorder with painless limb weakness, which can lead to considerable disability (9). One of the early series stressed the distal limb involvement compared to proximal disease (5); a later series, however, showed that while distal weakness occurred in half the cases, it exceeded proximal weakness in only one-third (6). Atrophy and weakness of the long forearm flexors is a notable feature, giving a scooped appearance to the forearm with weakness of DIP and PIP finger joint flexion (Fig. 1). The knee extensors and ankle dorsiflexors are also particularly affected (10,11). Dysphagia has been reported to occur in up to a third of cases (6) with occasional severe examples (12). As in other types of inflammatory myopathy, this is due to upper esophageal dysfunction. Neck flexion weakness and mild facial weakness may be seen, but other cranial nerves are spared. Respiratory involvement is a late feature (6,10). Depressed or absent

Table 1 Rimmed Vacuolar Myopathies

Syndrome	Interitance	Onset	CK	Pattern of involvement	Muscle biopsy features		
					Inflammation	Necrosis	Filaments
Sporadic IBM	n/a	>30 years old	N/↑	Limb, neck flexion, mild facial	+	+/–	+
Familial IBM							
Limb girdle muscular dystrophy (31)	AD	2nd-6th decade	N	Proximal limb	–	–	–
With quadriceps sparing (19, 32, 33)	AD	2nd decade	N/↑	Distal, sparing quadriceps	–	–	–
With excessive autophagy (30)	XL	Early childhood	↑	Proximal limbs esp legs	–	+	–
With dystrophin gene abnormality (34)	XL	Early childhood	↑↑	Proximal limbs	–	–	–
With electrical myotonia (35)	?AR	Adolescent	N	Limbs	–	+	+
With leukodystrophy (28)	?AR	Early childhood	↑	Proximal limbs	–	+	+
Distal myopathies (36–39)							
Late onset type I (Welander)	AD	Late adult	N/↑	Onset in hands, anterior compartment lower leg	–	–	+

Late onset type II (Nonaka)	AD	Late adult	N/↑	Onset in legs, anterior compartment lower leg	–	–	+
Early adult onset type I (Barohn)	AR/spont.	Early adult	↑↑	Onset in legs, anterior compartment lower leg	–	–	+
Oculopharyngeal dystrophies							
Adult onset (40–43)	AD	5–6h decade	N/↑	Ocular, bulbar, ocassionally limbs	–	+/–	+
Early onset (44)	AR?	Adolescence	N/↑	Ocular, bulbar and limbs	–	+/–	+
Miscellaneous (see text)							
Rigid spine	n/a				–	–	+
Postpoliomyelitis syndrome	n/a				–	–	–
Neuropathy; amyloid, vasculitic	n/a				–	–	–
Spinal muscular atrophy	n/a				–	–	–

AD = autosomal dominant; AR = autosomal recessive; XL = X-linked; n/a = not applicable; + = present; – = absent or not mentioned; N = normal; ↑ = moderate rise; ↑↑ = marked rise.

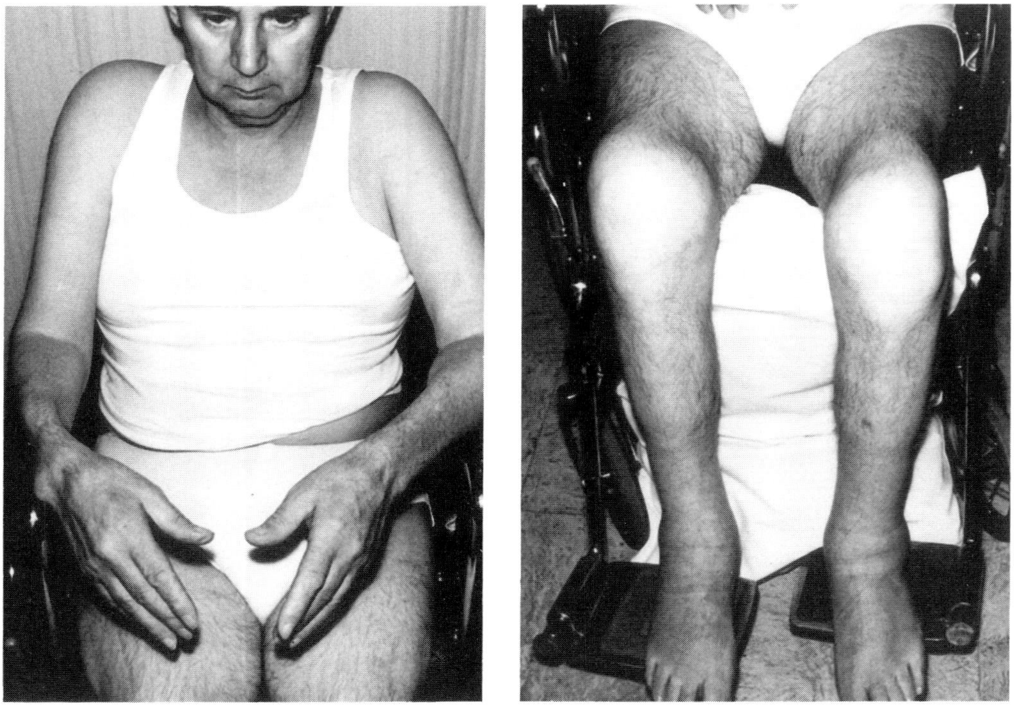

(a) (b)

Figure 1 S-IBM in a 72-year-old man. (a) Striking distal wasting, resulting in a "scooped out" appearance of muscles on the volar aspect of the forearms. (b) Severe proximal leg muscle wasting and dependent edema. (Reproduced by permission from Recent Advances in Clinical Neurology, Vol 5, Chapter 9, Churchill Livingstone, Edinburgh, 1988.)

reflexes, especially at the knee, are an early feature (unlike idiopathic polymyositis), suggesting a possible neurogenic component (6) (vide infra).

Association with other connective tissue and autoimmune diseases has been described, including lupus erythematosus, Sjögren's syndrome, scleroderma, interstitial pneumonitis, psoriasis, and dermatomyositis (13–15), with an overall incidence of 15% in one series (6). Diabetes mellitus occurred in 20% of cases (6), and cardiovascular disease, including cardiac conduction defects and rhythm disturbances, was found in more than half the patients. This incidence is similar to that seen in idiopathic inflammatory myopathies, but it may be more a consequence of coexisting hypertension in this generally older group of patients (6). Serum creatine kinase levels tend to be normal, or if raised, rarely by more than two- to fivefold. The clinical features of S-IBM, as defined by two series, are shown in Table 2.

Needle electromyography (EMG) is invariably abnormal. The most common pattern is similar to that found in polymyositis, consisting of increased insertional activity with fibrillation potentials, positive sharp waves, and short duration/small amplitude motor unit potentials (MUPs). In about a third of cases there is a mixture of short duration/small amplitude MUPs and long duration/large amplitude MUPs with a reduced interference pattern, the latter two features having been traditionally considered "neurogenic" (16). This mixed myopathic and neurogenic pattern on EMG has been regarded by some as highly specific for IBM (16), but others have described it in polymyositis (6,17), X-linked humeroperoneal neuromuscular disease (18), familial rimmed vacuolar myopathy sparing the quadriceps (19) (see below), and sporadic distal myopathy (20). A small proportion of patients with S-IBM have a purely neurogenic EMG picture of normal or long duration/large amplitude MUPs with a reduced interference pattern (16). Nerve conduction studies show mild slowing in more than a third of cases,

Table 2 Clinical Features of IBM as Defined by Carpenter et al. (5) and Lotz et al. (6)

Feature	Carpenter (1978)	Lotz (1989)
% IM	17	25
M:F	M>>F 12/14	M>F 29/11
Age of onset	15–62	28–73
Weakness	Distal > proximal	Proximal > distal but distal involvement in 50%
	Other features	
	Myalgia—rare	
	Dysphagia—40%	
	Neuropathy—25%	
	Cardiovascular disease[a]—50%	
	Diabetes—20%	
	Autoimmune diseases—10%	

[a]Chiefly hypertension, myocardial ischemia.

particularly in those with neurogenic features on EMG (6,16,21), and single-fiber EMG studies may show an increased fiber density and moderately abnormal jitter in severely affected muscles (16,22). As with the reflex abnormalities mentioned above, these neurogenic features seen neurophysiologically, as well as histologically, (grouped fiber atrophy and small angulated fibers) have generated discussion as to whether there is a neurogenic component to IBM. However, they may all only reflect the chronicity of a myopathic process in which the nerve fibers become disconnected from the muscle (23). Conversely, nerve biopsies in S-IBM have shown mild nonspecific reduction in myelinated fibers (21,24) or a mild axonal neuropathy (25).

Unlike other forms of inflammatory myopathy, S-IBM is notably resistant to steroid treatment (5,6), although exceptions have been reported (13,14).

OTHER "RIMMED VACUOLAR MYOPATHIES"

As mentioned above, a number of neuromuscular conditions share the histological features of rimmed vacuoles and filamentous inclusions as orginally seen in sporadic inclusion body myositis. As listed in Table 1, they can be divided into *hereditary inclusion body myopathies*, *distal myopathies*, *oculopharyngeal myopathies*, and a *miscellaneous group* of neurogenic and myopathic conditions. However, the clinical features of some of these groups overlap.

Hereditary Inclusion Body Myopathies (H-IBM)

IBM is generally sporadic, but there are a number of reports of cases with similar clinical and pathological features in which the disease is transmitted in an autosomal-recessive (26–28), autosomal-dominant (29), or X-linked fashion (30). They generally have an earlier onset than S-IBM, in the second or third decade. Since inflammatory infiltrates are not generally a feature of these disorders, they have been termed *hereditary inclusion body myopathies* (H-IBM) rather than *myosites*. Chutkow et al. (31) described an autosomal dominant limb-girdle myopathy with rimmed vacuoles, in which weakness was confined to proximal limb muscles without cranial, ocular, or bulbar involvement. Argov and Yarom (19) and Sadeh et al. (32) have described, in Iranian Jews, the onset usually in the second decade of an autosomal dominant limb weakness beginning in the legs and progressing mainly to the proximal upper limb and the distal lower limb muscles with striking preservation of the quadriceps even in advanced cases. Rimmed vacuoles were seen in all severely affected muscles but not in quadriceps or deltoid muscles. Hentati et al. (33) reported a similar phenotype in a Tunisian kindred associated with an asymptomatic leukoencephalopathy. An asymptomatic leukoencephalopathy was also a feature seen in four of five affected siblings from a single family reported by Cole et al. (28). These siblings had delayed motor milestones, progressive limb weakness without craniobulbar involvement, and muscle biopsies showing numerous rimmed vacuoles. An X-linked limb myopathy with excessive autophagy was described in a Finnish family by Kalimo et al. (30). The affected members had an infantile onset of a slowly progressive, mainly proximal limb weakness, especially in the legs, with moderate leg and minor shoulder atrophy. Rimmed vacuoles were seen in many fibers. de Visser et al. (34) reported five brothers with slowly progressive limb-girdle weakness, without calf hypertrophy, and with muscle biopsies showing numerous rimmed vacuoles, who had a deletion of the dystrophin gene. Juguilon et al. (35) described three patients, two of them siblings, who had limb wasting and weakness beginning in the late teens, with electrical but not clinical evidence of myotonia, and whose muscle biopsies showed rimmed vacuoles and lobulated fibers.

Distal Myopathies

Several variants of distal myopathy have been described since Welander first identified this entity and some of these share the histological feature of rimmed vacuoles (36). The original series of Welander (37) described an autosomal-dominant form with a male predominance, a mean onset in the fifties, and a slowly progressive weakness beginning in the hands and subsequently affecting the anterior compartment of the legs without cranial or pharyngeal involvement. Some of these patients' muscle biopsies showed vacuolar change. Vacuoles were a consistent morphological feature of the late-adult-onset, autosomal-dominant distal myopathy confined to the legs described by Nonaka et al. (38), and also of the early-adult-onset, autosomal-recessive form associated with high serum creatine kinase described by Barohn et al. (36). By contrast, in the Miyoshi variant of distal myopathy, the muscle biopsy appearances are those of a dystrophic process without vacuolar changes (39).

Oculopharyngeal (OPMD) and Facioscapulohumeral (FSHD) Dystrophies

OPMD is another autosomal-dominant disorder, which can be distinguished from H-IBM by its generally later onset and the prominent cranial, extraocular, and bulbar involvement (see Chapter 25). In 15–20% of cases limb weakness may occur, which can be of a proximal or distal distribution (40–42). Intrafamilial differences in the distribution of muscle involvement have been noted. Fukuhara et al. (43) reported the son of a patient with typical OPMD who at 2 years of age had developed a distal myopathy without oculopharyngeal symptoms, and recently Rose et al. (44) described two siblings with an adolescent-onset, rapidly progressive oculopharyngeal dystrophy with rimmed vacuoles. Pathological features of IBM may also be found in patients with the FSH phenotype (see Chapter 23).

Other Neuromuscular Disorders

Rimmed vacuoles have been reported in patients with polyneuropathy (3), rigid spine syndrome (3) (see Chapter 25), spinal muscular atrophy (45), and neurogenic atrophy secondary to poliomyelitis (46). Filaments have been reported in postpolio syndrome, sarcoidosis, and polyneuropathy (3–6,45,47,48).

THE PATHOLOGY OF IBM AND RIMMED VACUOLAR MYOPATHIES

In S-IBM the presence of rimmed vacuoles 2–15μm in diameter in up to 10% of fibers was considered by Carpenter et al. (5) to be the key pathological feature (Fig. 2). Rimmed vacuoles are nonstorage vacuoles with granular basophilic (on hematoxylin and eosin) or red (on Gomori trichrome) peripheral staining. These vacuoles have little acid phosphatase activity, unlike similar vacuoles seen in clinical and experimental myopathies induced by amphiphilic drugs (see Chapter 31) and in lysosomal storage diseases such as acid maltase deficiency (see Chapter 33). In the series described by Lotz et al. (6), in which S-IBM was defined pathologically by the presence of such rimmed vacuoles, 96% had scattered

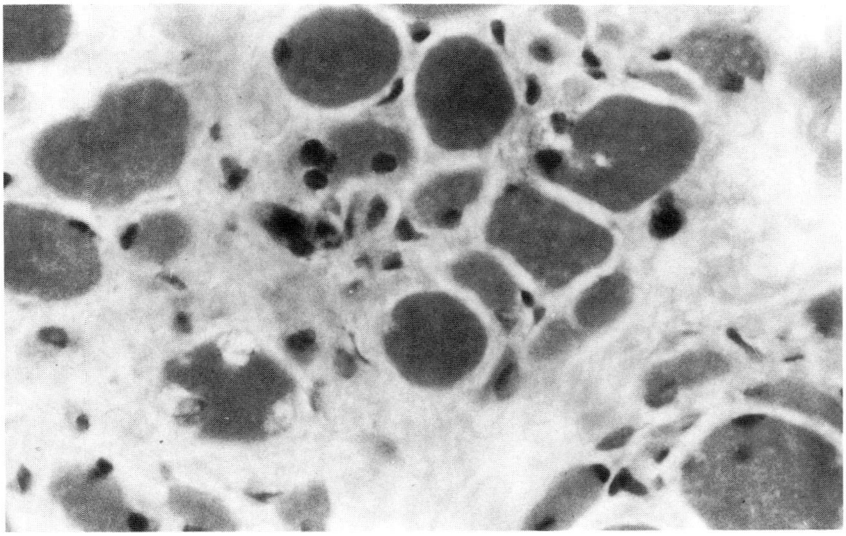

Figure 2 Needle muscle biopsy, left vastus lateralis. There is extensive replacement of muscle by fibrous tissue. Two fibers show "rimmed" vacuoles, with basophilic material in the walls of the vacuoles. (H&E ×800.)

and grouped angular atrophic fibers, 92% had inflammatory cell infiltrates, and 88% had all three features. Lotz et al. (6) also found eosinophilic inclusions in 46% of cases with the above triad. Fiber type grouping is unusual. Hypertrophied fibers are seen more frequently in S-IBM than in other inflammatory myopathies, perhaps reflecting the chronicity of this condition (49). Ragged red fibers representing subsarcolemmal accumulations of mitochondria are seen in excess of those expected for this age group (50). Muscle fiber capillary density is increased, contrasting with the decrease seen in dermatomyositis and the normal density seen in idiopathic polymyositis (5). Arahata and Engel (51–53) have shown that the inflammatory cell infiltrates, which are primarily endomysial, are comprised mainly of $CD8^+$ cytotoxic T cells and macrophages, similar to those seen in idiopathic inflammatory myopathy (Fig. 3a).

Ultrastructurally, rimmed vacuoles are not membrane-lined but contain membrane fragments as well as secondary lysosomes and other debris, which account for the basophilic staining (Fig. 3b). Closely associated with the vacuoles are cytoplasmic and intranuclear filaments, which are of several types. The filaments most characteristic of IBM are nuclear accumulations of randomly orientated tubular structures, which are 1.5 μm long, with an external diameter of 15–21 nm and an internal diameter of 3–6 nm, containing an electron-dense core (Fig. 4a,b,c). Amyloid-like 6–10 nm diameter fibrils have also been described, as have cytoplasmic tubulofilaments consisting of paired helical filaments, with twist periodicity of 45–55 nm (54). Structural mitochondrial abnormalities are frequently observed, including abnormal cristal arrangments and accumulations of crystalloids in the matrix.

Immunocytochemistry has demonstrated the presence of amyloid in vacuolated fibers in S-IBM. This is located in clusters of dense, amorphous structures and loose floccular material containing 6–10 nm amyloid-like fibrils. The precise localization of this material within the fiber is uncertain and it may also extend outside of the fiber (54). Vacuolated fibers also contain strongly ubiquitinated inclusions localized to the tubulofilaments, to amorphous floccular material, and to the amyloid-like fibrils (54). These findings are summarized in Table 3.

Specificity of Pathology

S-IBM is frequently misdiagnosed as polymyositis, particularly early in the disease when the histological features can be identical. Inflammatory cell infiltrates may be prominent in the early stage of S-IBM,

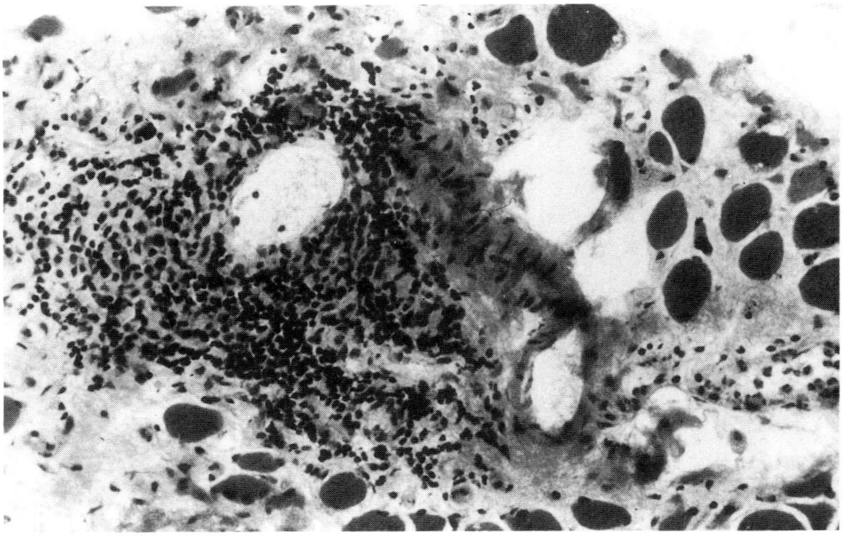

(a)

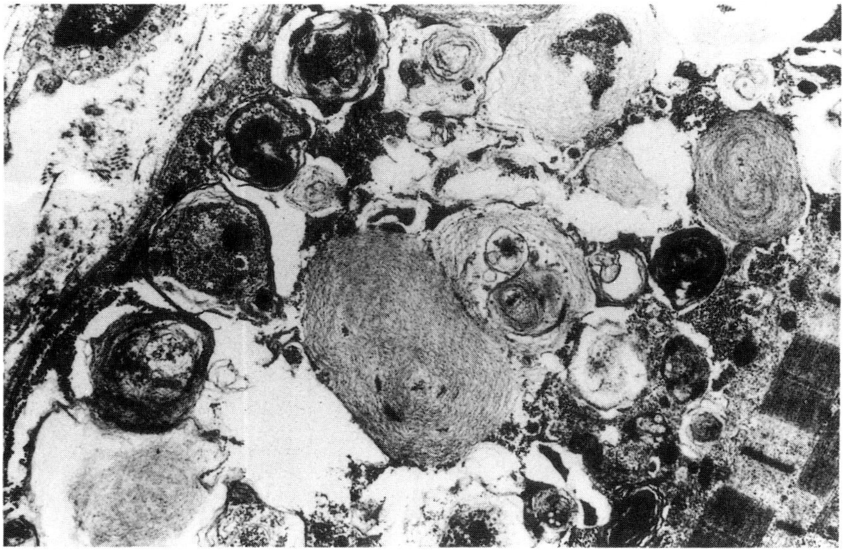

(b)

Figure 3 (a) Extensive perivascular inflammatory infiltrate from a nearby region in the biopsy. Vacuolated fibers are also visible. (H&E ×460.) (b) Electron micrograph showing whorled "myelin figures," the basophilic material in the walls of the rimmed vacuoles, lying beneath the sarcolemma. Several masses of IBM-type cytoplasmic filaments in close proximity (arrows). Normal myofibrils are seen in the lower right-hand corner. (×18,000.)

and the distinctive rimmed vacuoles of S-IBM may only appear in the later stage, and even when present can be small and sparse and therefore missed (54). The presence of angular atrophic fibers, fiber hypertrophy, and ragged red fibers favors a diagnosis of S-IBM rather than polymyositis. Acid phosphatase activity may be prominent in polymyositis in contrast to the scanty staining seen in S-IBM.

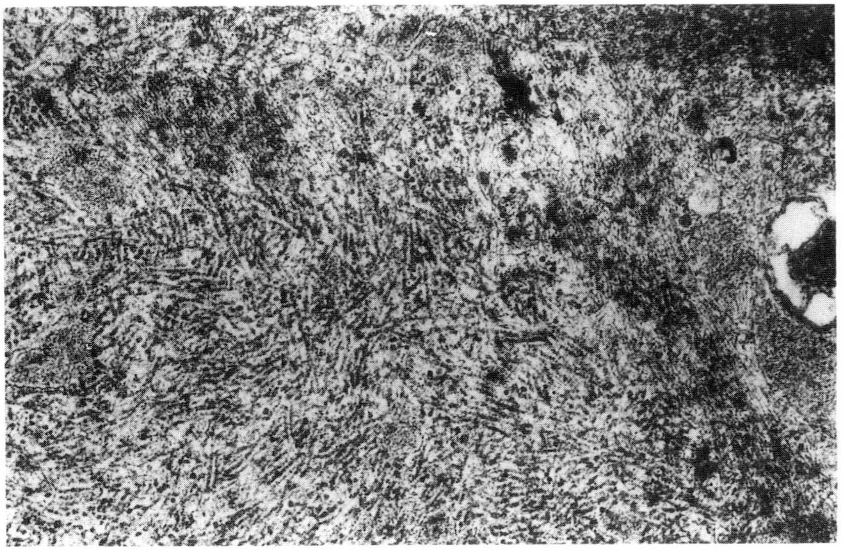

(a)

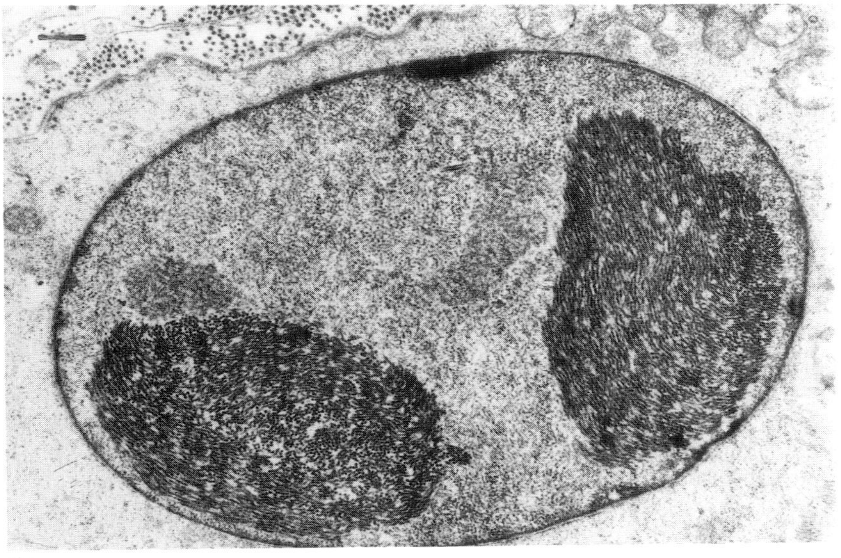

(b)

Figure 4 (a) Cytoplasmic IBM-type filaments in random orientation. Immunocytochemical studies have revealed that these filaments contain antigens recognised by antiubiquitin and β-amyloid antibodies. (×50,000.) (b) Nucleus showing groups of similar 16–18-nm diameter filaments oriented in various planes. (×22,000.) (c) Higher-power view of these nuclear filaments. (×67,000.)

Positive immunocytochemical staining for ubiquitin and fluorescence-enhanced Congo red staining for amyloid point to a diagnosis of S-IBM. The presence of filamentous inclusions on electron microscopy is helpful, but these may be present only in the later stages of S-IBM (54,55).

As mentioned, both rimmed vacuoles and filamentous inclusions can be found in other disorders

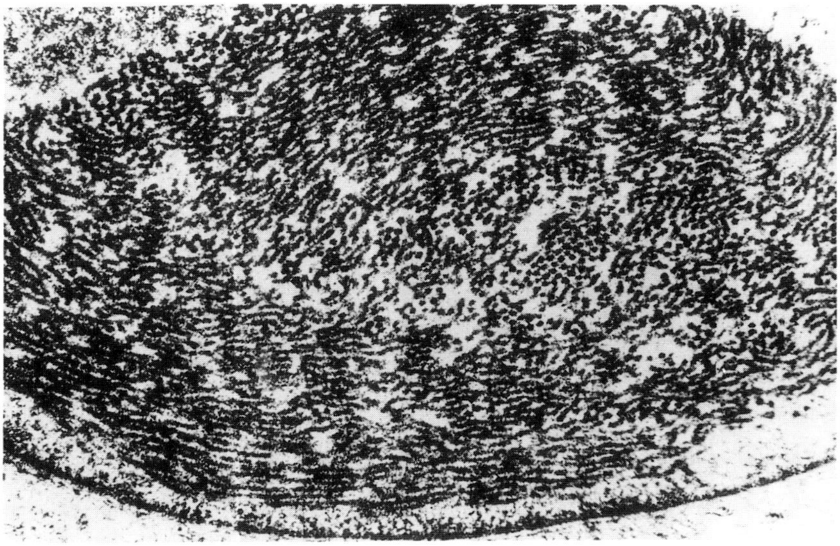

(c)

Figure 4 (*Continued*)

apart from S-IBM (Table 1). Mhiri and Gherardi (4) found nine cases of rimmed vacuole myopathy among 850 consecutive biopsies over a 7-year period, of which only five were considered to have S-IBM, the remaining four being thought to have a chronic denervating process. Filamentous inclusions were found in three of these five S-IBM cases and in one of the four "denervation" cases. Jongen et al. (3) found rimmed vacuoles in 21 of 1600 biopsies, of which only seven fulfilled criteria for the diagnosis of IBM, the remaining cases comprising patients with limb-girdle muscular dystrophy, familial OPMD with distal myopathy, distal myopathy with rimmed vacuoles, rigid spine syndrome, and miscellaneous neuropathies. Thus, the original defining pathological characteristics of S-IBM are in fact found more commonly in other disorders. However, the proportion of fibers containing rimmed vacuoles is significantly higher in S-IBM than in neurogenic disorders (3).

There have been claims for the specificity of the type of the filamentous inclusions. Thus a morphological marker for OPMD is said to be a unique type of exclusively intranuclear tubulofilamentous inclusion up to 0.25 μm long, with an external diameter of 8.5 nm and an internal diameter of 3

Table 3 Inclusions of IBM

		Immunostaining characteristics		
Structure	Dimensions	β-Amyloid	Ubiquitin	Prion protein
Nuclear inclusions	Randomly oriented, 1.5 μm long, 15–21-nm diam.	–	–	–
Cytoplasmic tubulofilaments	Twisted paired helical periodicity 45–55 nm	–	+	+
Amyloid-like fibrils	6–10-nm diam.	+	+	+
Amorphous structures	n/a	+	+	+
Floccular material	n/a	+	+	+

n/a = not applicable; + = present; – = absent.

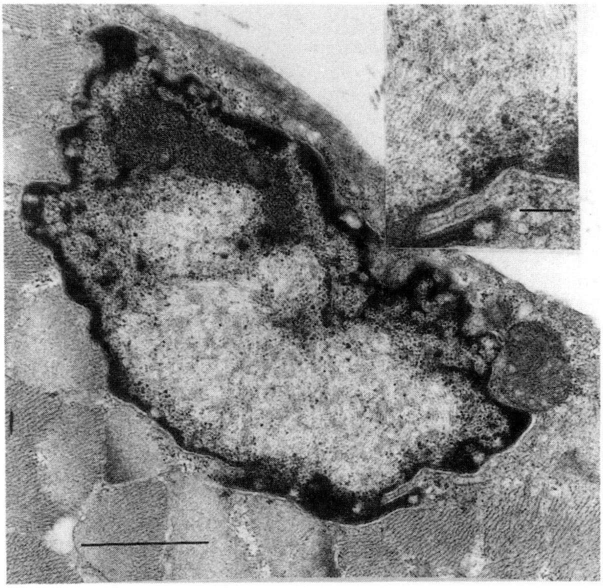

Figure 5 Oculopharyngeal muscular dystrophy (OPMD). EM showing unique 8–9-nm-diameter intranuclear filaments arranged in palisades. This patient had a 12-year history of progressive external ophthalmoplegia, followed by limb weakness. Initial diagnosis was of a mitochondrial myopathy, and it was only when he evolved dysphagia and other bulbar problems that EM scrutiny of the intramuscular nuclei indicated the correct diagnosis. (×27,000, inset ×66,000.)

nm, arranged in tangles or palisades (Fig. 5) (56–59). However, a case of OPMD has been reported with 13–18-nm diameter tubules (60). Similarly, the specificity of the immunostaining characteristics of S-IBM has been examined. Amyloid deposition is seen in virtually all cases of S-IBM if an enhanced illumination technique with Congo red is used. With the same technique only half the cases of H-IBM and virtually none of the other forms of vacuolar myopathy show similar deposition (54). Ubiquitin staining is seen in H-IBM and OPMD as well as in S-IBM (54).

ETIOLOGICAL THEORIES

The cause of IBM is unknown. Chou (1) suggested that the condition might be related to myxovirus infection based on the similarity of the filamentous inclusions to nucleocapsids, with mumps virus infection being the proposed etiology (61). However, immunological studies using antibodies to mumps and other paramyxoviruses, and in situ hybridization studies with an RNA probe to mumps nucleocapsid gene, failed to support this hypothesis (62).

The rimmed vacuoles appear to be autophagic in origin. There is a strong immunoreactivity for lysosomal cathepsins in and around rimmed vaculoes (63). Cathepsins are endopeptidases involved in breakdown of high-molecular-weight proteins. The activity of cathepsins relative to other lysosomal enzymes such as acid phosphatase and β-D-glucosidase is increased in rimmed vacuole myopathies compared to disease controls (64), and this might imply that the vacuoles were formed by the autodigestion of myofilaments (63).

It is not known whether the intranuclear inclusions (INI) are pathological per se or merely a nonpathogenic epiphenomenon. They could be a marker of a metabolic abnormality with the accumulation of a normal nuclear component, or they might represent accumulation of a cytoplasmic component entering the nucleus or the synthesis of an abnormal molecule. Muscle cultures obtained from a case of OPDM showed intranuclear inclusions of a smaller diameter than seen in the original biopsy (65). By contrast, the filamentous inclusions found in muscle cultures obtained from a case of H-IBM were identical to those seen in the original biopsy (66). Further suggestion of a primary nuclear defect in both S-IBM and H-IBM has been stimulated by the finding of mainly nuclear binding of single-stranded DNA. This disease-specific phenomenon was abolished by proteases but not by RNase, suggesting that it was mediated by a protein. This protein is as yet unidentified, but since it is present in excess of that seen in controls, it may be a foreign, e.g., viral, protein or result from nuclear accumulation of a cytoplasmic protein or the up-regulation of a nuclear protein (67).

The vacuolated fibers in S-IBM show greatly increased expression of β-amyloid precursor protein compared with normal fibers where the expression of this protein is confined to the postsynaptic neuromuscular junction. Both the vacuolated muscle fiber and the tubulofilaments in S-IBM show deposits of apolipoprotein E immunoreactive protein and the inclusions are immunoreactive for α_1-antichymotrypsin. The tubulofilaments also contain hyperphosphorylated tau protein. All of these characteristics are seen in the brain pathology of Alzheimer's disease where the senile plaques contain β-amyloid, α_1-antichymotrypsin, and apolipoprotein E, and the paired helical filaments of the neurofibrillary tangles contain apolipoprotein E as well as hyperphosphorylated tau protein (54,68). The striking parallels between these diseases may reflect similar abnormalities in the accumulation or clearance of normal or abnormal protein degradation products.

Of relevance in this regard is the demonstration of abnormal accumulation of ubiquitin in S-IBM fibers, which colocalizes with β-amyloid protein and is additionally found on the cytoplasmic twisted tubulofilaments (54). Ubiquitin is part of a highly specific proteolytic pathway possibly related to the control of gene expression and it can induce amyloidogenesis (69).

Support for the accumulation of β-amyloid precursor protein (βAPP) in S-IBM being more than an epiphenomenon comes from experiments in which normal human muscle cultures were transfected with 3-kb human 751-βAPP cDNA with resultant overexpression of βAPP and induction of features of the S-IBM phenotype (70).

In S-IBM and in H-IBM, prion protein immunonoreactive inclusions have been found within the vacuoles. This protein colocalizes with β-amyloid protein on the amorphous material and the amyloid-like fibrils, and colocalizes with ubiquitin on the cytoplasmic twisted tubulofilaments and on the amorphous material (54). The significance of this finding has yet to be determined.

The presence of ragged red fibers and ultrastructural mitochondrial changes in S-IBM has focused attention on the possibility that mitochondrial dysfunction may play an etilogical role in this disease. Carpenter et al. (5) suggested that the increase in muscle capillary density seen in S-IBM might be a reaction to defective aerobic metabolism resulting from mitochondrial abnormalities. Since morphological, biochemical, and mitochondrial DNA (mtDNA) defects are a feature of normal aging, one has to relate any such findings in S-IBM patients to age-matched controls (71,72). Oldfors et al. (73) found 2–5% cytochrome oxidase (COX)-negative muscle fibers in three cases of S-IBM compared with no COX-negative fibers in six elderly controls and less than 1% COX-negative fibers in an additional three elderly controls. Many of the COX-negative fibers in S-IBM were regenerating fibers and were near sites of inflammation. Rifai et al. (50) found an excess of ragged red equivalent muscle fibers on a modified succinate dehydrogenase stain in S-IBM compared with polymyositis and with age-matched normal controls. In two of three cases of S-IBM many of the COX-negative muscle fibers showed reduced in situ hybridization of mtDNA gene transcripts for subunit 2 of complex 1, suggesting the presence of a mtDNA deletion similar to that found in mitochondrial myopathies. In one of these two cases Southern blotting revealed that a small proportion of the total mtDNA contained two different length deletions (73). However, although none of the controls in this study showed similar mtDNA abnormalities, others have found similar abnormalities in the normal aging population (71). It therefore remains uncertain whether mitochondrial abnormalities are disease-specific, a primary event, or an epiphenomenon.

THE TREATMENT OF S-IBM

Unfortunately, treatment of S-IBM remains unsatisfactory. The lack of response of this group of inflammatory myopathies to steroids was how S-IBM was originally distinguised from polymyositis and dermatomyositis (5). Occasional cases have been reported to improve with steroid treatment (6,10,13,14) but the effects are short-lived. Uncontrolled trials of other immunosuppressant therapies such as azathioprine, cyclosporin, and methotrexate have not resulted in reversal of weakness (74,75) except in occasional cases (76). A prospective open crossover trial of combination oral azathioprine and methotrexate and of biweekly intravenous methotrexate with leucovorin rescue found stabilization or improvement of function in some patients, although treatment sessions were short, ranging from 3 to 6 months (77). Previous trials of intravenous immunoglobulin (IVIg) have shown mixed results (9) and further studies are in progress. One such randomized, double-blind, placebo-controlled, crossover study in 20 patients, using monthly infusions of 2 g/kg IVIg for 3 months, has reported some slowing in the progression of the disease with improvement in strength in some patients (78).

REFERENCES

1. Chou SM. Myxovirus like structures in a case of human chronic polymyositis. Science 1967; 158:1453–1455.
2. Yunis EJ, Samaha FJ. Inclusion body myositis. Lab Invest 1971; 25:240–248.
3. Jongen PJH, Laak HTT, Stadhouders AM. Rimmed basophilic vacuoles and filamentous onclusions in neuromuscular disorders. Neuromusc Disord 1995; 5:31–38.
4. Mhiri C, Gherardi R. Inclusion body myositis in French patients: a clinicopathological evaluation. Neuropathol Appl Neurobiol 1990; 16:333–344.
5. Carpenter S, Karpati G, Heller I, Eisen A. Inclusion body myositis: a distinct variety of idiopathic inflammatory myopathy. Neurology 1978; 28:8–17.
6. Lotz BP, Engel AG, Nishino H, Stevens JC, Litchy WJ. Inclusion body myositis: observations in 40 patients. Brain 1989; 112:727–747.
7. Karpati G, Carpenter S, Heller I, Eisen A. Idiopathic inflammatory myopathies. Curr Opin Neurol Neurosurg 1988; 1:806–814.
8. Riggs JE, Schochet SS Jr., Gutmann L, Lerfald SC. Childhood onset inclusion body myositis mimicking limb-girdle muscular dystrophy. Child Neurol 1989; 4:283–285.
9. Askanas V, Engel WK, Mirabella M. Idiopathic inflammatory myopathies: inclusion-body myositis, polymyositis, and dermatomyositis. Curr Opin Neurol 1994; 7:448–456 (review).
10. Lindberg C, Persson LI, Bjorkander J, Oldfors A. Inclusion body myositis: clinical, morphological, physiological and laboratory findings in 18 cases. Acta Neurol Scand 1994; 89:123–131.
11. Dalakas MC. Inflammatory and toxic myopathies. Curr Opin Neurol Neurosurg 1992; 5:645–654 (review).
12. Verma A, Bradley WG, Adesina AM, Sofferman R, Pendelbury WW. Inclusion body myositis with cricopharyngeus muscle involvement and severe dysphagia. Muscle Nerve 1991; 14:470–473.
13. Julien J, Vital C, Vallat JM, Lagueny A, Sapina D. Inclusion body myositis: clinical, biological and ultrastructural study. J Neurol 1982; 55:15–24.
14. Lane JM, Fulthorpe JJ, Hudgson P. Inclusion body myositis: a case with associated collagen vascular disease responding to treatment. J Neurol Neurosurg Psychiatry 1985; 48:270–273.
15. Gutmann L, Govindan S, Riggs JE, Schochet SS Jr. Inclusion body myositis and Sjögren's syndrome. Arch Neurol 1985; 42:1021–1022.
16. Joy JL, Oh SJ, Baysal AI. Electrophysiological spectrum of inclusion body myositis. Muscle Nerve 1990; 13:949–951.
17. Mechler F. Changing electromyographic findings during the chronic course of polymyositis. J Neurol Sci 1974; 23:237–242.
18. Waters DD, Nutter DO, Hopkins LC, Dorney ER. Cardiac features of an unusual X-linked humeroperoneal neuromuscular disease. N Engl J Med 1975; 293:1017–1022.
19. Argov Z, Yarom R. Rimmed vacuole myopathy sparing the quadriceps. J Neurol Sci 1984; 64:33–43.
20. Vaccario ML, Scoppetta C, Bracaglia R, Uncini A. Sporadic distal myopathy. J Neurol 1981; 224:291–295.
21. Lacy JR, Simon DB, Neville HE, Ringel SP. Inclusion body myositis: electrodiagnostic and nerve biopsy findings. Neurology 1982; 32(Suppl 2): A202 (abstract).
22. Eisen A, Berry K, Gibson G. Inclusion body myositis (IBM): myopathy or neuropathy? Neurology 1983; 33:1109–1114.
23. Desmedt JE, Borenstein S. Relationship of spontaneous fibrillation potentials to muscle fibre segmentation in human muscular dystrophy. Nature 1975; 258:531–534.

24. Ringel SP, Kenney CE, Neville HE, Giorno R, Carry MR. Spectrum of inclusion body myositis. Arch Neurol 1987; 44:1154–1157.
25. Lindberg C, Oldfors A, Hedstrom A. Inclusion body myositis: peripheral nerve involvement: combined morphological and electrophysiological studies on peripheral nerves. Neurol Sci 1990; 99:327–338.
26. Fardeau M, Askanas V, Tome FMS, et al. Hereditary neuromuscular disorder with inclusion body myositis like filamentous inclusions: clinical, pathological and tissue culture studies. Neurology 1990; 40(Suppl): 120.
27. Fardeau M, Tome FMS, Chevally M, Collin H, Lebon P, Fournier JG. Inclusion body myositis-like filamentous inclusion in several cases of hereditary neuro-muscular disorders. Muscle Nerve 1986; 9(Suppl 5):215.
28. Cole AJ, Kuzniecky R, Karpati G, Carpenter S, Andermann E, Andermann F. Familial myopathy with changes resembling inclusion body myositis and periventricular leucoencephalopathy—a new syndrome. Brain 1989; 111:1025–1037.
29. Neville HE, Baumbach LL, Ringel SP, Russo LS, Jr., Sujansky E, Garcia CA. Familial inclusion body myositis: evidence for autosomal dominant inheritance. Neurology 1992; 42:897–902.
30. Kalimo H, Savontaus ML, Lang H, et al. X-linked myopathy with excessive autophagy: a new hereditary muscle disease. Ann Neurol 1988; 23:258–265.
31. Chutkow JG, Heffner RR, Kramer AA, Edwards JA. Adult-onset autosomal dominant limb girdle muscular dystrophy. Ann Neurol 1986; 20:240–248.
32. Sadeh M, Gadoth N, Hader H, Ben-David E. Vacuolar myopathy sparing the quadriceps. Brain 1993; 116:217–232.
33. Hentati F, Ben Hamida C, Tome FMS, Oueslati S, Fardeau M. "Familial inclusion body myositis" sparing the quadriceps with asymptomatic leukoencephalopathy in a Tunisian kindred. Neurology 1991; 41:422 (abstract).
34. de Visser M, Bakker E, Defesche JC, Bolhuis PA, van Ommen GJ. An unusual variant of Becker muscular dystrophy. Ann Neurol 1990; 27:578–581.
35. Juguilon A, Chad D, Bradley WG, et al. Familial granulovacuolar lobular myopathy with electrical myotonia. J Neurol Sci 1982; 56:133–140.
36. Barohn RJ, Miller RG, Griggs RC. Autosomal recessive distal myopathy. Neurology 1991; 41:1365–1370.
37. Welander L. Myopathia distalis tarda hereditaria. Acta Med Scand 1951; 141:1–124.
38. Nonaka I, Sunohara N, Ishiura S, Satoyoshi E. Familial distal myopathy with rimmed vacuole and lamellar (myeloid) body formation. J Neurol Sci 1981; 51:141–155.
39. Nonaka I, Sunohara N, Satoyoshi E, Terasawa K, Yonemoto K. Autosomal recessive distal muscular dystrophy: a comparative study with distal myopathy with rimmed vacuole formation. Ann Neurol 1985; 17:51–59.
40. Vita G, Dattola R, Santoro M, Messina C. Familial oculopharyngeal muscular dystrophy and distal spread. J Neurol 1983; 230:57–64.
41. Peterman AF, Lillington GA, Jamplis RW. Progressive muscular dystrophy with ptosis and dysphagia. Arch Neurol 1964; 10:38–41.
42. Taylor EW. Progressive vagus-glossopharyngeal paralysis with ptosis: contribution to the group of family diseases. J Nerv Ment Dis 1915; 42:124–139.
43. Fukuhara N, Kumamoto T, Tsubaki T, Mayuzumi T, Nitta H. Oculopharyngeal muscular dystrophy and distal myopathy. Intrafamilial difference in the onset and distribution of muscular involvement. Acta Neurol Scand 1982; 65:458–467.
44. Rose MR, Landon DN, Papadimitriou A, Morgan-Hughes JA. A rapidly progressive adolescent onset oculopharyngeal somatic syndrome with rimmed vacuoles in two siblings. Ann Neurol 1996 (in press).
45. Fukuhara N, Kumamoto T, Tsubaki T. Rimmed vacuoles. Acta Neuropathol 1980; 51:229–235.
46. Borg K, Borg J, Edstrom L, Grimby L. Effects of excessive use of remaining muscle fibres in prior polio and LV lesion. Muscle Nerve 1988; 11:1219–1230.
47. Abarbanel JM, Lichtenfeld Y, Zirkin H, et al. Inclusion body myositis in post-poliomyelitis muscular atrophy. Acta Neurol Scand 1988; 78:81–84.
48. Matsubara S. Ultrastructural changes in granulomatous myopathy. Acta Neuropathol 1980; 50:91–96.
49. Verma A, Bradley WG, Soule NW, et al. Quantitative morphometric study of muscle in inclusion body myositis. J Neurol Sci 1992; 112:192–198.
50. Rifai Z, Welle S, Kamp C, Thornton CA. Ragged red fibers in normal aging and inflammatory myopathy. Ann Neurol 1995; 37:24–29.
51. Arahata K, Engel AG. Monoclonal antibody analysis of mononuclear cells in myopathies. IV. Cell-mediated cytotoxicity and muscle fiber necrosis. Ann Neurol 1988; 23:168–173.
52. Arahata K, Engel AG. Monoclonal antibody analysis of mononuclear cells in myopathies. I. Quantitation of subsets according to diagnosis and sites of accumulation and demonstration and counts of muscle fibers invaded by T cells. Ann Neurol 1984; 16:193–208.
53. Arahata K, Engel AG. Monoclonal antibody analysis of mononuclear cells in myopathies. V. Identification and quantitation of $T8^+$ cytotoxic and $T8^+$ suppressor cells. Ann Neurol 1988; 23:493–499.

54. Askanas V, Engel WK. New advances in inclusion-body myositis. Curr Opin Rheumatol 1993; 5:732–741 (review).
55. Purvis J, Fam AG, Lewis A. Clinically unsuspected inclusion body myositis. J Rheumatol 1991; 18:289–292.
56. Coquet M, Vallat JM, Vital C, et al. Nuclear inclusions in oculopharyngeal dystrophy: an ultrastructural study of six cases. J Neurol Sci 1983; 60:151–156.
57. Bouchard JP, Gagne F, Tome FM, Brunet D. Nuclear inclusions in oculopharyngeal muscular dystrophy in Quebec. Can J Neurol Sci 1989; 16:446–450.
58. Tome FMS, Leroy JP, Neville H, et al. Intranuclear tubulofilamentous inclusions as a morphological marker of oculopharyngeal muscular dystrophy. Muscle Nerve 1986; 9:212.
59. Tome FMS, Fardeau M. Nuclear inclusions in oculopharyngeal dystrophy. Acta Neuropathol 1980; 49:85–87.
60. Smith TW, Chad D. Intranuclear inclusions in oculopharyngeal dystrophy. Muscle Nerve 1984; 7:339–340 (letter).
61. Nishino H, Engel AG, Rima BK. Inclusion body myositis: the mumps virus hypothesis. Ann Neurol 1989; 25:260–264.
62. Kallajoki M, Hyypia T, Halonen P, Orvell C, Rima BK, Kalimo H. Inclusion body myositis and paramyxoviruses. Hum Pathol 1991; 22:29–32.
63. Jimi T, Satoh Y, Takeda A, Shibuya S, Wakayama Y, Sugita K. Strong immunoreactivity of cathepsin L at the site of rimmed vaculoes in diseased muscles. Brain 1992; 115 (Pt 1):249–260.
64. Leclerc A, Tome FM, Fardeau M. Ubiquitin and beta-amyloid-protein in inclusion body myositis (IBM), familial IBM-like disorder and oculopharyngeal muscular dystrophy: an immunocytochemical study. Neuromusc Disord 1993; 3:283–291.
65. Tome FM, Askanas V, Engel WK, Alvarez RB, Lee CS. Nuclear inclusions in innervated cultured muscle fibers from patients with oculopharyngeal muscular dystrophy. Neurology 1989; 39:926–932.
66. Alvarez RB, Fardeau M, Askanas V, Engel WK, McFerrin J, Tome FMS. Characteristic filamentous inclusions reproduced in cultured innervated muscle fibres from patients with familial "inclusion body myositis" (FIBM). J Neurol Sci 1990; 98(Suppl):178.
67. Nalbantoglu J, Karpati G, Carpenter S. Conspicuous accumulation of a single-stranded DNA binding protein in skeletal muscle fibers in inclusion body myositis. Am J Pathol 1994; 144:874–882.
68. Askanas V, Mirabella M, Engel WK, Alvarez RB, Weisgraber KH. Apolipoprotein E immunoreactive deposits in inclusion-body muscle diseases. Lancet 1994; 343:364–365 (letter).
69. Alizadeh-Khiavi K, Normand J, Chronopoulos S, Ali-Khan Z. Alzheimer's disease brain-derived ubiquitin has amyloid-enhancing factor activity: behavior of ubiquitin during accelerated amyloidogenesis. Acta Neuropathol 1991; 81:280–286.
70. Askanas V, McFerrin J, Baque S, Engel WK, Sarkozi E, Alvarez RB. Overexpression of b-amyloid precursor protein (bAPP) gene in cultured normal human muscle using adenovirus vector induces aspects of inclusion body myositis (IBM) phenotype. Neurology 1995; 45(Suppl 4):176S (abstract).
71. Cooper JM, Mann VM, Schapira HV. Analyses of mitochondrial respiratory chain function and mitochondrial DNA deletion in human skeletal muscle: effect of ageing. J Neurol Sci 1992; 113:91–98.
72. Trounce I, Byrne E, Marzuki S. Decline in skeletal muscle mitochondrial respiratory chain function: possible factor in ageing. Lancet 1989; 637–639.
73. Oldfors A, Larsson NG, Lindberg C, Holme E. Mitochondrial DNA deletions in inclusion body myositis. Brain 1993; 116:325–336.
74. Dalakas MC. Clinical, immunopathologic, and therapeutic considerations of inflammatory myopathies. Clin Neuropharmacol 1992; 15:327–351 (review).
75. Sayers ME, Chou SM, Calabrese LH. Inclusion body myositis: analysis of 32 cases. J Rheumatol 1992; 19:1385–1389.
76. Lueck CJ, Trend P, Swash M. Cyclosporin in the management of polymyositis and dermatomyositis. J Neurol Neurosurg Psychiatry 1991; 54:1007–1008.
77. Leff RL, Miller FW, Hicks J, Fraser DD, Plotz PH. The treatment of inclusion body myositis: a retrospective review and a randomised prospective trial of immunosuppressive therapy. Medicine 1993; 72:225–235.
78. Dalakas MC, Dambrosia JM, Sekul EA, Cupler EJ, Sivakumar K. The efficacy of high dose intravenous immunoglogulin (IVIg) in patients with inclusion body myositis (IBM). Neurology 1995; 45(Suppl 4):174S (abstract).

44
Chronic Fatigue Syndrome

Russell J. M. Lane
Charing Cross Hospital and Charing Cross and Westminister Medical School, London, England

INTRODUCTION

"Fatigue" is an extremely common complaint in industrialized societies. When this term is used to convey "tiredness," as opposed to the more exacting physiological definition of "failure to maintain force" (see Chapter 1), community-based studies have reported fatigue ("tired all the time") in up to a third of the population (1). Chronic fatigue syndrome (CFS) however, essentially comprises fatigue (tiredness) of more than 6 months' duration, present for more than 50% of the time, associated with other nonspecific symptoms including myalgia, exercise intolerance, neuropsychological problems such as mood disturbances, difficulties using appropriate words, problems with memory and concentration, and other somatic complaints, in the absence of alternative explanation in terms of definable medical or psychiatric disorders (2). The proportion of fatigued individuals fulfilling such criteria is relatively small, around 1% (3). The condition is also sometimes referred to as myalgic encephalomyelitis (ME) and postviral or postinfectious fatigue syndrome (PVFS, PIFS), while some authorities consider it to be synonymous with neurasthenia, which reached epidemic proportions in the nineteenth century (4).

ORIGIN OF CFS

The basis of the symptoms of CFS has been widely debated. The orthodox view is that CFS is essentially of psychiatric or psychological origin, and psychiatric disorder is indeed common in populations with fatigue (5). Other studies suggest that disturbances in the hypothalamic-pituitary axis (6), alterations in sleep pattern (7), and chronic hyperventilation syndrome (8) may play a part, while others hold that CFS results from persistent viral infection (9), possibly related to alterations in immune function (10). These views are not necessarily mutually exclusive, since CFS patients are almost certainly a heterogeneous group.

One point at issue is whether there is any abnormality in neurosmuscular function that may contribute to the fatigue of CFS. Quantitative studies of static muscle strength have generally shown no abnormalities (11), and interpolation of muscle contractions produced by supramaximal stimulation of a relevant nerve during attempted maximal voluntary contractions increased force generation in some cases, suggesting inadequate effort or "lack of central drive" (11–13). Similarly, CFS patients do not demonstrate fatigue in terms of inability to sustain force, when studied using repetitive stimulation techniques (13).

LABORATORY ABNORMALITIES

Incremental treadmill exercise tests, relating oxygen consumption and lactate production to work done and perceived effort, have shown that CFS patients feel more distress and reach their limits of exercise tolerance at lower work rates than control subjects, but show similar increases in lactate production and oxygen consumption, suggesting a change in "fatigue threshold" (14,15). However, data from these studies also indicated a more rapid increase in plasma lactate at low work rates in some CFS patients, compatible with a decrease in anaerobic threshold (16). Studies of blood lactate responses using the subanaerobic threshold exercise test (see Chapter 3) revealed a left shift of anaerobic threshold in about a third of cases in one study, which could not be explained in terms of muscular or cardiovascular deconditioning, suggesting a possible abnormality of muscle energy metabolism in some patients (17).

Several other laboratory abnormalities relating to neuromuscular function have been described in CFS. Serum CK levels are occasionally increased (18), and while muscle biopsy is typically normal, histological changes, including inflammation and structural mitochondrial abnormalities, have been reported (19). Cases of "benign postinfection myositis," with symptoms compatibile with CFS, showing myopathic changes on electromyogram (EMG) and mild inflammation on muscle biopsy, have also been described (20). Some patients show increased jitter on single-fiber EMG (21), although in a recent evaluation, such cases were found to be relatively rare, and the jitter was thought to be myogenic and unlikely to contribute to fatigue (22). A further study found significant changes in muscle fiber densities in CFS cases with myalgia as a prominent symptom (23).

A study using phosphorus magnetic resonance spectroscopy (MRS, Chapter 9) reported early acidification of muscle during aerobic work in one patient with postviral fatigue syndrome following herpes zoster infection (24). A subsequent study by the same group showed no overall abnormality in energy metabolism in a larger group of CFS patients, and the abnormalities observed in exceptional cases were nonspecific (25). However, another MRS study found that CFS patients reached metabolic exhaustion, with a reduction in ATP levels, more rapidly than controls, and it was concluded that such patients had defective oxidative muscle metabolism and accelerated glycolysis in working muscles, possibly reflecting the abnormal increases in lactate production found in the exercise test studies (26).

ENTEROVIRAL RNA IN MUSCLE BIOPSY EXTRACTS

Some groups have detected enteroviral RNA in muscle biopsy extracts from CFS patients (27), as also found in some cases of inflammatory myopathy (28), myocarditis, and dilated cardiomyopathy (29). The RNA detected may represent defective viral genome that is unable to replicate (30), and it has been suggested that inflammatory myopathies and PVFS/benign postinfectious polymyositis might represent extremes of a continuum, with varying immune and cytotoxic responses to the presence of virus (31). However, to date, studies using the more sensitive polymerase chain reaction in inflammatory muscle disease have been negative (32,33), and results in CFS patients have been contradictory (9,34). The debate concerning the role of possible viral persistence in cardiac and skeletal muscle continues (35,36).

REFERENCES

1. Ridsdale L. Tired all the time. Br Med J 1991; 303:1490–1491.
2. Sharpe M, Archard L, Banatvala J, et al. Guidelines for research in chronic fatigue syndromes. J Roy Soc Med 1991; 84:118–121.
3. Bates DW, Schmitt W, Buchwald D, et al. Prevalence of fatigue and chronic fatigue syndrome in a primary care practice. Arch Intern Med 1993; 153:2759–2765.
4. Wessely S. History of postviral fatigue syndrome. Br Med Bull 1991; 47:919–941.
5. David AS. Postviral fatigue syndrome and psychiatry. Br Med Bull 1991; 47:966–988.
6. Demitrack MA, Dale JK, Straus SE, et al. Evidence or impaired activation of the hypothalamic-pituitary-adrenal axis in patients with chronic fatigue syndrome. J Clin Endocrinol Metab 1991; 73:1224–1234.
7. Morriss R, Sharpe M, Sharpley AL, Cowen PJ, Hawton K, Morris J. Abnormalities of sleep in patients with the chronic fatigue syndrome. Br Med J 1993; 306:1161–1164.

8. Rosen SD, King JC, Nixon PGF. Is chronic fatigue syndrome synonymous with effort syndrome?J Royl Soc Med 1990; 83:761–764.
9. Gow J, Behan W, Clements G, Woodall C, Riding M, Behan P. Enteroviral sequences detected by polymerase chain reaction in muscle of patients with postviral fatigue syndrome. Br Med J 1991; 302:692–696.
10. Mobray JF, Yousef GE. Immunology of postviral fatigue syndrome. Br Med Bull 1991; 47:886–894.
11. Stokes MJ, Cooper RG, Edwards RHT. Normal strength and fatiguability in patients with effort syndrome. Br Med J 1988; 297:1014–1017.
12. Rutherford OM, White PD. Human quadriceps strength and fatiguability in patients with postviral fatigue. J Neurol Neurosurg Psychiatry 1991; 54:961–964.
13. Lloyd A, Gandevia S, Hales J. Muscle performance, voluntary activation, twitch properties and perceived effort in normal subjects and patients with chronic fatigue syndrome. Brain 1991; 114:85–98.
14. Riley M, O'Brien CJ, Bell N, Nicholls DP, McCloskey DR. Aerobic work capacity in patients with chronic fatigue syndrome. Br Med J 1990; 301:953–956.
15. Gibson H, Carroll N, Clague JE, Edwards RHT. Exercise performance and fatiguability in patients with chronic fatigue syndrome. J Neurol Neurosurg Psychiatry 1993; 56:993–998.
16. Lane RJM, Woodrow D, Archard L. Lactate responses to exercise in chronic fatigue syndrome. J Neurol Neurosurg Psychiatry 1994; 57:662–663.
17. Lane RJM, Burgess AP, Flint J, Riccio M, Archard LC, Lactate responses to exercise and psychiatric disorder in the chronic fatigue syndrome. Br Med J 1995; 311:544–545.
18. Archard L, Bowles N, Behan P, Bell E, Doyle D. Postviral fatigue syndrome: persistence of enterovirus in muscle and elevated creatine kinase. J Roy Soc Med 1988; 81:326–329.
19. Behan WMH, More IAR, Behan PO. Mitochondrial abnormalities in the post-viral fatigue syndrome. Acta Neuropathol (Berlin) 1991; 83:61–65.
20. Schwartz MS, Swash M, Gross M. Benign postinfection polymyositis. Br Med J 1978; 2:1256–1257.
21. Jamal G, Hansen S. Electrophysiological studies in the postviral fatigue syndrome. J Neurol Neurosurg Psychiatry 1986; 48:691–694.
22. Roberts L, Byrne E. Single fibre EMG studies in chronic fatigue syndrome: a reappraisal. J Neurol Neurosurg Psychiatry 1994; 57:375–376.
23. Connolly S, Smith DG, Doyle D, Fowler CJ. Chronic fatigue: electromyographic and neuropathological evaluation. J Neurol 1993; 240:435–438.
24. Arnold DL, Bore PJ, Radda GK, Styles P, Taylor DJ. Excessive intracellular acidosis of skeletal muscle on exercise in a patient with a post-viral exhaustion/fatigue syndrome: a ^{31}P nuclear magnetic resonance study. Lancet 1984; 1:1367–1369.
25. Barnes PRJ, Taylor DJ, Kemp GJ, Radda GK. Skeletal muscle energetics in the chronic fatigue syndrome. J Neurol Neurosurg Psychiatry 1993; 56:679–683.
26. Wong R, Lopaschuk G, Zhu G, et al. Skeletal muscle metabolism in the chronic fatigue syndrome: in vivo assessment by ^{31}P nuclear magnetic resonance spectroscopy. Chest 1992; 102:1716–1722.
27. Cunningham L, Bowles NE, Archard LC. Persistent virus infection of muscle in postviral fatigue syndrome. Br Med Bull 1991; 47:852–871.
28. Bowles NE, Dubowitz V, Sewry CA, Archard LC. Dermatomyositis, polymyositis and Coxsackie B virus infection. Lancet 1987; 1:1004–1007.
29. Bowles NE, Richardson PJ, Olsen EGC, Archard LC. Detection of Coxsackie B virus–specific RNA sequences in myocardial biopsy samples from patients with myocarditis and dilated cardiomyopathy. Lancet 1986; 1:1120–1123.
30. Cunningham L, Bowles NE, Lane RJM, Dubowitz V, Archard LC. Persistence of enteroviral RNA in chronic fatigue syndrome is associated with the abnormal production of equal amounts of positive and negative strands of enteroviral RNA. J Gen Virol 1990; 71:1399–1402.
31. Bowles NE, Bayston TA, Zhang H-Y, Doyle D, Lane RJM, Cunningham L, Archard LC. Persistence of enterovirus RNA in muscle biopsy samples suggests that some cases of chronic fatigue syndrome result from a previous inflammatory viral myopathy. J Med 1993; 24:145–160.
32. Leff RL, Love LA, Miller FW, Greenberg SJ, Klein EA, Dalakas MC, Plotz PH. Viruses in idiopathic inflammatory myopathies: absence of candidate viral genomes in muscle. Lancet 1992; 339:1192–1195.
33. Fox SA, Finklestone E, Robbins PD, Mastaglia FL, Swanson NR. Search for persistent enterovirus infection of muscle in inflammatory myopathies. J Neurol Sci 1994; 125:70–76.
34. Behan WMH, Behan PO. The role of viral infection in polymyositis, dermatomyositis and chronic fatigue syndrome. Balliere's Clin Neurol 1993; 2/3:637–657.
35. Muir P, Archard LC. Persistent enterovirus infections are causally associated with chronic medical conditions in humans. Rev Med Virol 1994; 4:245–250.
36. Melchers WJ9, Zoll GJ, van Kuppeveld FJM, Swanink CMA, Galama JMD. There is no evidence for persistent enterovirus infections in chronic medical conditions in humans. Rev Med Virol 1994; 4:235–243.

45
Polymyalgia Rheumatica

Richard W. Orrell
Charing Cross Hospital and Charing Cross and Westminster Medical School, London, England

INTRODUCTION

Polymyalgia rheumatica has to be considered in the differential diagnosis in patients complaining of myalgia. It is a clinical syndrome, for which the seven most valuable criteria for diagnosis, in hierarchy of decreasing sensitivity, are: bilateral shoulder pain or stiffness, onset of illness of less than 2 weeks' duration, initial ESR greater than 40 mm/hr, duration of morning stiffness exceeding 1 hr, age 65 years or more, depression and/or weight loss, and bilateral tenderness of the arms (1). It is proposed that a patient has probable polymyalgia if any three criteria are present, or if at least one is present with a clinical or pathological abnormality of the temporal artery. A therapeutic response to prednisolone also supports the diagnosis (1).

GIANT CELL ARTERITIS

Polymyalgia rheumatica is commonly associated with cranial arteritis, and polymyalgia and cranial arteritis may be considered manifestations of giant cell arteritis (2). In a study of 49 patients with giant cell arteritis, initial symptoms included symmetrical myalgia of the shoulders (69%), symmetrical myalgia of the thighs (34%), morning stiffiness of the joints (35%), and distal myalgias (13%) (2). In another population study in Finland, of patients with clinical features of polymyalgia or temporal arteritis, 11% of patients with polymyalgia had a positive temporal artery biopsy (showing features of an arteritis), and 7% of patients had features of both temporal arteritis and polymyalgia (3). In a Swedish study of giant cell arteritis, 53% presented with polymyalgia (31% with arteritis on temporal artery biopsy), and 18% had clinical features of both temporal arteritis and polymyalgia (4).

Arteritis causing blindness is the most significant complication of giant cell arteritis (5), demanding early diagnosis and treatment of the condition. In the Swedish study, of nine patients with major eye complications of giant cell arteritis, three presented with polymyalgia, three with temporal arteritis, one with polymyalgia and temporal arteritis, and two with general symptoms only (4).

Giant cell arteritis is diagnosed on the basis of the clinical history and findings, supported by the finding of raised ESR (6) or plasma viscosity (7), and confirmed but not specified by features of arteritis on temporal artery biopsy (8,9). A negative arterial biopsy does not exclude the diagnosis, as the pathology can be patchy. After the diagnosis of giant cell arteritis (including polymyalgia) has been made, immediate treatment should be with high-dose steroids (1 mg/kg/day), primarily to prevent ocular complications, and in cranial arteritis and polymyalgia to give additional relief of pain, which generally resolves within 24 hr. Giant cell arteritis appears to be a self-limiting disorder, but treatment may be needed for between 2 and 5 years, and relapse is possible (10,11). In a study of 66 patients with giant cell arteritis, the most common neurological complications were postural vertigo (17%), amaurosis fugax (17%), and polyneuropathy (12%) (2). Stroke is also a possible complication (12).

ETIOLOGY

The etiology and pathogenesis of polymyalgia and giant cell arteritis are uncertain. No inflammation is seen on light microscopy of skeletal muscle. In one study, skeletal muscle biopsies were examined in 13 patients with polymyalgia, and nonspecific degenerative features were seen, some of which were similar to those found in neurogenic myopathies (13). In an earlier study there were features of type 2 fiber atrophy with moth-eaten and whorled fibers (14). In a histochemical and immunohistochemical study of muscle biopsies from 11 patients with polymyalgia, standard histochemistry showed small angular fibers, pyknotic nuclear clumps, or target-targetoid fibers in 63% patients. Immunohistochemistry showed no expression of major histocompatibility complex (MHC) class 1 or 2 products on the muscle surface membrane, typically seen in inflammatory myopathies (see Chapter 41), but MHC class 2 products were seen on intramuscular vessels, and on serial sections combined deposits of IgG and C1q in the perimysial arteries were seen in 38% of cases (15).

Serological studies suggest that HLA-DR4 is an important susceptibility factor for polymyalgia (16). Hence it is possible that the myalgic symptoms, and mild myopathic features, are secondary to involvement of small-caliber arteries by an inflammatory process. The more widely recognized neuropathic features of giant cell arteritis, including mononeuropathies (especially of the median, radial, peroneal, and sciatic nerves), mononeuritis multiplex, brachial plexopathy, cervical radiculopathy, and peripheral neuropathy, are probably the result of medium- or large-vessel arteritis, with nutrient artery vasculitis rather than vasa nervorum vasculitis (12). In one report of muscle biopsy findings in 15 patients with polymyalgia, five patients had features of mitochondrial myopathy with numerous ragged red fibers. It was suggested that a mitochondrial myopathy may have predisposed to polymyalgia in these patients (19).

EMG FINDINGS

Electromyographic (EMG) findings in polymyalgia are usually normal, although occasional nonspecific features may be found, and rarely features of denervation (17). In two patients studied serially, progressive improvement in clinical and EMG features with treatment with steroids was documented, and a steroid-responsive microvascular arteritis causing ischemic damage to the axons of motor nerve terminal branches was proposed (18). In some patients it is possible that the myalgic symptoms are secondary to a synovitis of the joints, especially the shoulder joints (20).

Thus polymyalgia appears to be a clinical manifestation of giant cell arteritis, which includes temporal arteritis. The most important complication is loss of vision, which can be prevented by prompt treatment with steroids. The cause of the peripheral neurological symptoms, including myalgia and neuropathy, is uncertain, but some of the findings support an immunologically mediated arteritis of the small vessels, primarily involving the nerves, with secondary nonspecific muscle involvement.

REFERENCES

1. Bird HA, Esselinckx W, Dixon AS, Mowat AG, Wood PHN. An evaluation of criteria for polymyalgia rheumatica. Ann Rheumat Dis 1979; 38:434–439.
2. Berlit P. Clinical and laboratory findings with giant cell arteritis. J Neurol Sci 1992; 111:1–12.
3. Franzen P, Sutinen S, Knorring JV. Giant cell arteritis and polymyalgia rheumatica in a region of Finland: an epidemiologic, clinical and pathologic study, 1984–1988. J Rheumatol 1992; 19:273–276.
4. Bengtsson B, Malmvall B. The epidemiology of giant cell arteritis including temporal arteritis and polymyalgia rheumatica. Arthritis Rheum 1981; 24:899–904.
5. Hayreh SS. Ophthalmic features of giant cell arteritis. Balliere's Clin Rheumatol 1991; 5:431–459.
6. Kyle V, Cawston TE, Hazleman BL. Erythrocyte sedimentation rate and C reactive protein in the assessment

of polymyalgia rheumatica/giant cell arterits on presentation and during follow up. Ann Rheumat Dis 1989; 48:667–671.
7. Orrell RW, Johnson MH. Plasma viscosity and the diagnosis of giant cell arteritis. Br J Clin Pract 1993; 47:71–72.
8. Nadeau SE. Temporal arteritis: a decision-analytic approach to temporal artery biopsy. Acta Neurol Scand 1988; 78:90–100.
9. Paice EW. Giant cell arteritis: difficult decisions in diagnosis, investigation and treatment. Postgrad Med J 1989; 65:743–747.
10. Chuang T, Hunder GG, Ilstrup DM, Kurland T. Polymyalgia rheumatica: a 10-year epidemiologic and clinical study. Ann Intern Med 1982; 97:672–680.
11. Kyle V, Hazleman BL. Stopping steroids in polymyalgia rheumatica and giant cell arteritis. Br Med J 1990; 300:344–345.
12. Caselli RJ, Hunder GG, Whisnant JP. Neurologic disease in biopsy-proven giant cell (temporal) arteritis. Neurology 1988; 38:352–359.
13. Fassbender R, Simmling-Annefeld M. Ultrastructural examination of the skeletal muscles in polymyalgia rheumatica. J Pathol 1982; 137:181–192.
14. Brooke MH. Muscle pathology in rheumatoid arthritis, polymyalgia rheumatica, and polymyositis. Arch Pathol 1972; 94:101–118.
15. Kojima S, Takagi A, Ida M, Shiozawa R. Muscle pathology in polymyalgia rheumatica: histochemical and immunohistochemical study. Jpn J Med 1991; 30:516–523.
16. Sakkas LI, Loqueman N, Panayi GS, Myles AB, Welsh KI. Immunogenetics of polymyalgia rheumatica. Br J Rheumatol 1990; 29:331–334.
17. Gross MD, Borkin MH, Rupp S. Unusual electromyographic findings in a patient with polymyalgia rheumatica. Arthritis Rheum 1979; 22:277–280.
18. Bromberg MB, Donofrio PD. Steroid-responsive electromyographic abnormalities in polymyalgia rheumatica. Muscle Nerve 1990; 13:138–141.
19. Harle J, Pellissier J, Desnuelle C, Disdier P, Figarella-Branger D, Weiller P. Polymyalgia rheumatica and mitochondrial myopathy: clinicopathologic and biochemical studies in five cases. Am J Med 1992; 92:167–172.
20. Healey LA. Polymyalgia rheumatica and seronegative rheumatoid arthritis may be the same entity. J Rheumatol 1992; 19:270–272.

46
Myositis Ossificans

Richard W. Orrell
Charing Cross Hospital and Charing Cross and Westminster Medical School, London, England

INTRODUCTION

Two separate conditions are covered by the term "myositis ossificans": a congenital generalized form and an acquired localized form. The term "myositis" is misleading, as there is no persistent inflammatory process in the muscle (1).

CONGENITAL MYOSITIS OSSIFICANS

The congenital generalized form is a rare and progressively disabling disorder, probably better termed *fibrodysplasia ossificans progressiva*. It is usually the result of a spontaneous mutation, with autosomal dominant inheritance, complete penetrance, but variable expression (2).

The main features are progressive heterotopic calcification in the muscles and additional dysmorphic features, including short monophalangic big toes, short thumbs and other digital abnormalities (Fig. 1), exostoses, mental retardation, deafness, and baldness (3). The ossification usually begins in the first decade of life, initially with painful nodules of noninflammatory connective tissue proliferation in the fascia and dermis, progressing to form normal lamellar bone, which causes rigidity at the joints. Often this commences with a swelling in the neck mimicking congenital torticollis, progressing to the back, shoulder, and pelvic girdles. The diaphragm, extraocular, cardiac, and smooth muscles are generally spared. Premature death results from restrictive respiratory failure or aspiration pneumonia. There is no effective treatment, and attempts to excise ossified lesions lead to further ossification at the site of the trauma.

In a study of 44 patients with fibrodysplasia progressiva ossificans (4), the mean age of the patients was 27 years (range 3–69 years), with a mean age of onset of heterotopic ossification at 5 years (range birth to 25 years). Attempts are being made to identify the genetic cause of this condition.

LOCALIZED MYOSITIS OSSIFICANS

The localized form of myositis ossificans is characterized by extraosseous, localized, nonneoplastic formation of bone and cartilage, in regenerating areas of skeletal muscle (1). It is an uncommon condition, but well described, and in the typical case would be precipitated by trauma to the muscle, as in a sports or industrial injury. Less commonly the condition follows a repetitive injury such as to the adductors of the thighs in horse riders, or around the hips in patients with spastic paraplegia. Approximately one-third of cases are not associated with any recollection of injury.

In the first week following the trauma to the muscle there is active spindle cell proliferation,

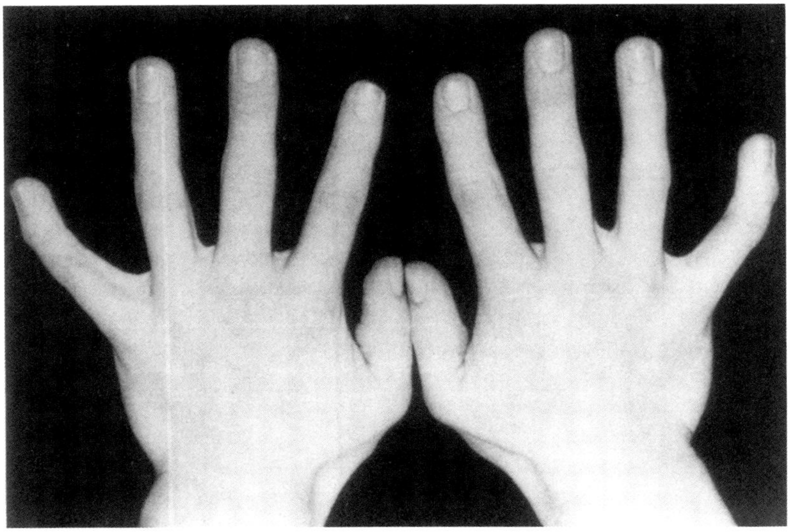

Figure 1 Short thumbs and incurving of the fingers in a patient with fibrodysplasia ossificans progressiva. (From Ref. 3, courtesy of Professor J. M. Connor.)

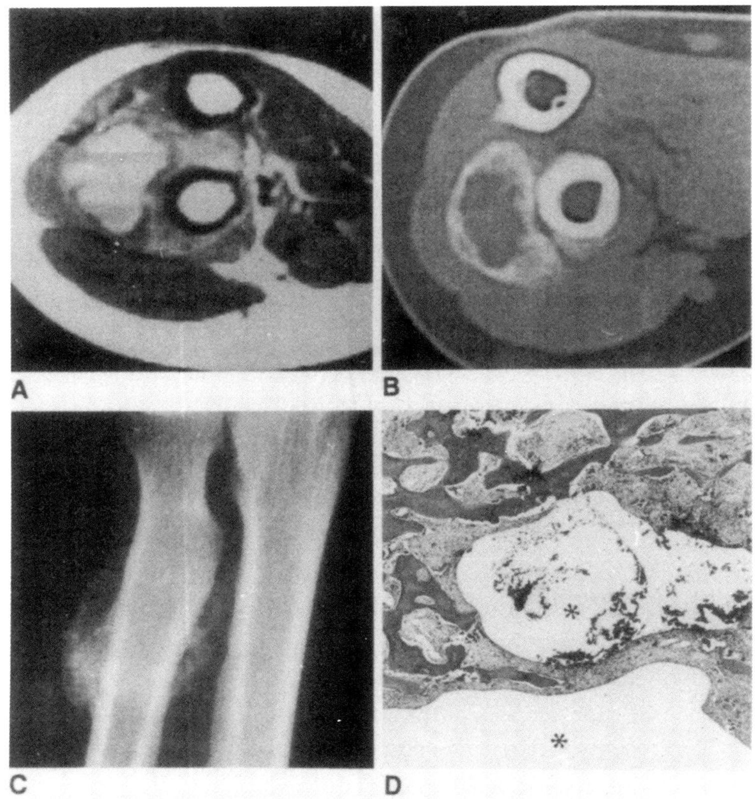

Figure 2 Magnetic resonance imaging (A), computerized tomography (B), radiographic (C), and photomicrographic (D) appearance of myositis ossificans in the forearm. (From Ref. 5, courtesy of Dr. M. J. Kransdorf.)

followed by primitive osteoid development. This progresses through primitive cartilage and bone formation to trabecular bone at 2–5 weeks. In the mature lesion there is a cellular center, surrounded by osteoid and a rim of lamellar bone. These changes can be followed radiologically, including plain films, computerized tomography, and magnetic resonance imaging (5) (Fig. 2).

The natural history of the condition is benign, although the lesion may occasionally be painful or cause functional disability, depending on the location. Excision of the mature lesion should be curative. Although very rare (6), a malignant lesion should be considered in the differential diagnosis. Generally a conservative approach should be followed, with a period of observation, but if the lesion does not follow the expected course and radiological features, biopsy or excision should be considered (7).

REFERENCES

1. Ackerman LV. Extra-osseous localised non-neoplastic bone and cartilage formation (so-called myositis ossificans). J Bone Joint Surg 1958; 40:279–298.
2. Connor JM, Evans DAP. Genetic aspects of fibrodysplasia ossificans progressiva. J Med Genet 1982; 19:35–39.
3. Connor JM, Evans DAP. Fibrodysplasia ossificans progressiva. J Bone Joint Surg 1982; 64B:76–83.
4. Cohen RP, Hahn GV, Tabas JA, et al. The natural history of heterotopic ossification in patients who have fibrodysplasia ossificans progressiva. J Bone Joint Surg 1993; 75:215–219.
5. Kransdorf MJ, Meis JM, Jelinek JS. Myositis ossificans: MR appearance with radiologic-pathologic correlation. Am J Radiol 1991; 157:1243–1248.
6. Allen A, Wetzel L, Borek D. Malignant myositis ossificans: a case report. Tumori 1992; 78:55–58.
7. Nuovo MA, Norman A, Chumas J, Ackerman LV. Myositis ossificans with atypical clinical, radiographic, or pathologic findings: review of 23 cases. Skel Radiol 1992; 21:87–101.

47

Arthrogryposis Multiplex Congenita

Richard W. Orrell
Charing Cross Hospital and Charing Cross and Westminster Medical School, London, England

Geraint N. Fuller
Gloucester Royal Hospital, Gloucester, England

INTRODUCTION

Arthrogryposis multiplex congenita (AMC) is a term originally coined by Stern to describe four children with multiple congenital contractures (1). This term has been used to describe a heterogeneous group of patients with a nonprogressive congenital syndrome characterized by poorly developed and contracted muscles, deformed joints with thickened periarticular capsules, and intact sensory system (2,3). This working definition has led to the inclusion of a wide range of different disorders within the syndrome. In one review, over 150 different entities were found to have been described as AMC (4). AMC should perhaps therefore be considered a sign rather than a syndrome (4).

All conditions that are associated with AMC have two common features: onset during fetal development; and an alteration in one of the components of the final common pathway (anterior horn cell, nerve roots, peripheral nerve, motor end-plate or muscle) (5). The causes can be divided into neurogenic and myogenic. The neurogenic causes are frequently associated with central nervous system and other abnormalities (Table 1). Common associated features include low-set ears, micrognathia, wide flat nose, short neck, high arched palate, and congenital heart defects (5).

Muscle disease was found to be the cause in one pathological study, in six of 96 patients (5), and in 11 of 21 cases in another (6). Muscle disorders associated with AMC include congenital muscular dystrophy, central core disease, nemaline myopathy, and myotonic muscular dystrophy (5,6).

A syndrome of multiple congenital contractures can be produced in animals. Drachman and Sokoloff (7) found that joints developed but no joint cavities formed, when muscles were paralyzed with d-tubocurare. This and other work [reviewed by Swinyard and Bleck (8)] has led to the view that normal intrauterine movement is essential for normal joint development and lack of intrauterine movement may lead to AMC.

MANAGEMENT

Three aspects must be considered in the management of AMC: management of the contractures, management of the underlying condition, and genetic advice regarding the risk of recurrence. Management of the contractures is primarily through passive stretching and orthopedic maneuvers (9). Long-term follow-up suggests that the patients may achieve independence, though they frequently require some support. In one series, 13 of 22 adults were employed, though seven required some additional support (10). One severely deformed patient, who had had no surgical correction, success-

Table 1 Syndromes Associated with Arthrogryposis Multiplex Congenita

Dysgenesis of the CNS—trisomy 18
Dysgenesis of the brain stem and spinal cord (Pierre Robin syndrome)
Dysgenesis of the motor nuclei of the brain stem and spinal cord (Mobius syndrome)
Dysgenesis of the CNS and hepatorenal syndrome (Zellweger syndrome)
Dysgenesis of the spinal cord and the prune-belly syndrome
Encephalocele, arrhinencephaly, and dysgenesis of the anterior horn (Meckel-Gruber syndrome)
Anencephaly and dysgenesis of the anterior horn
Myelomeningocele and dysgenesis of the CNS (Arnold-Chiari syndrome)
Agenesis of the lumbosacral cord and vertebrae
Spinal muscular atrophy
Muscle disease (see text)

Source: From Ref. 5.

fully ran his own business. In the same series there were five severely or mildly mentally retarded patients, indicating that the underlying condition will limit functional recovery.

The genetic advice will depend on the diagnosis of any underlying condition. In a series of 350 patients, 28% were found to have genetically determined syndromes, 46% had syndromes with no recurrence risk, and 20% had an unknown condition. In this last group the empirical risk was 4.7% if limbs alone were involved and 7% when there was additional central nervous system involvement (4).

REFERENCES

1. Stern WG. Arthrogryposis multiplex congenita. JAMA 1923; 81:1507.
2. Thompson GH, Bilenker RM. Comprehensive management of arthrogryposis multiplex congenita. Clin Orthopaed Related Res 1985; 194:6–14.
3. Fisher RL, Johnstone WT, Fisher WH, et al. Arthrogryposis multiplex congenita: a clinical investigation. Pediatrics 1970; 17:255.
4. Hall JG. Genetic aspects of arthrogryposis multiplex congenita. Clin Orthopaed Related Res 1985; 194:44–53.
5. Banker BQ. Arthrogryposis multiplex congenita: a spectrum of pathologic changes. Hum Pathol 1986; 17:656–672.
6. Quinn CM, Wigglesworth JS, Heckmatt J. Lethal arthrogryposis multiplex congenita: a pathological study of 21 cases. Histopathology 1991; 19:155–162.
7. Drachman DB, Sokoloff L. The role of movement in embryonic joint development. Dev Biol 1966; 14:401.
8. Swinyard CA, Bleck EE. The etiology of arthrogryposis (multiple congenital contracture). Clin Orthopaed Related Res 1985; 194:15–29.
9. Palmer PM, MacEwen GD, Bowen JR, Mathews PA. Passive motion therapy for infants with arthrogryposis. Clin Orthopaed Related Res 1985; 194:54–59.
10. Carlson WO, Speck GJ, Vicari V, Wenger DR. Arthrogryposis multiplex congenita: a long term follow up. Clin Orthopaed Related Res 1985; 194:115–124.

48
Compartment Syndromes

Richard W. Orrell
Charing Cross Hospital and Charing Cross and Westminster Medical School, London, England

INTRODUCTION

A compartment syndrome is a condition in which an increase of the interstitial pressure in a closed osseofascial compartment leads to microvascular compromise (1). Compartment syndromes occur especially in the lower leg (2) and forearm, and less commonly in the thigh (3) and upper arm. The calf, for example, contains four distinct compartments: anterior, lateral, deep posterior, and superficial posterior. The skeletal muscle may become necrotic in an acute compression or become painful in a chronic syndrome (4). The causes of both acute and chronic presentations are those that either decrease the size of the compartment, such as constrictive casts, dressings, or other external pressure, or those that increase the pressure within the compartment, such as blood (5), edematous fluid, myositis (6), or infection (7).

ACUTE COMPARTMENT SYNDROME

The acute compartment syndrome usually presents after trauma, but may also occur after nontraumatic pressure on the limbs occurring in the unconscious patient after a drug overdose, or rarely during a surgical procedure (8). Edema and hemorrhage within the compartment lead to an increase in intracompartmental pressure, and when this reaches 50 mmHg (normal resting pressure being 0–15 mmHg), the muscles and nerves within the compartment become ischemic, and if pressure is prolonged, will become necrotic. The myonecrosis will lead to further swelling and increased pressure within the compartment. An intracompartmental pressure of over 30 mmHg is an indication for fasciotomy in an acute compartment syndrome. In a patient in whom the pressure was over 70 mmHg for just over 4 hr before fasciotomy, muscle histology showed necrosis of about half the superficial fibers and complete necrosis of the deep fibers (9). An additional complication of this progressive myonecrosis is myoglobinuria and acute renal failure, the *crush syndrome* (see Chapter 49).

The classic signs and symptoms of the acute compartment syndrome are pain that is out of proportion to that expected for the injury, tenderness of the limb, decrease in muscle power, paraesthesiae in the distribution of the appropriate nerves, and pain with muscle stretch (4). In the unconscious patient these symptoms will not be available; the only signs may be a swollen limb with a tense compartment, and intracompartmental pressure monitoring will guide the management (10). Peripheral pulses will remain palpable until the pressure is very high, i.e., more than 100 mmHg. The compartment pressure may be measured by various catheter techniques (11), and treatment is surgical decompression by open fasciotomy of the affected compartment (but see also Chapter 49).

CHRONIC COMPARTMENT SYNDROME

The chronic compartment syndrome is seen especially in the lower leg of athletes. On exercise the pressure within a compartment may rise to 50 mmHg, causing ischemia and pain in the muscles. On resting the pressure returns to normal, and the pain resolves (11). Localization of the pain may point to the compartment involved. Diagnosis may be aided by intracompartmental monitoring during exercise. If symptoms have been present for less than 6 months, the initial management is rest and physiotherapy, but if the symptoms persist over 6 months or fail to respond to conservative treatment, surgical fasciotomy is indicated (4).

REFERENCES

1. Mubarak SJ, Hargens AR, Owen CA, Garetto LP, Akeson WH. The wick catheter technique for measurement of intramuscular pressure. J Bone Joint Surg 1976; 58A:1016–1020.
2. Allen MJ, Stirling AJ, Crawshaw CV, Barnes MR. Intracompartmental pressure monitoring of leg injuries. J Bone Joint Surg 1985; 67B:53–57.
3. Editorial. Compartment syndrome of the thigh. Lancet 1989; 2:486–487.
4. Allen MJ. Compartment syndromes of the lower limb. J Roy Coll Surg Edinburgh 1990; 35:S33–36.
5. Griffiths D, Jones DH. Spontaneous compartment syndrome in a patient on long-term anticoagulation. J Hand Surg 1993; 18:41–42.
6. Chow LTC, Chow W. Acute compartment syndrome: an unusual presentation of gemfibrozil induced myositis. Med J Aust 1993; 158:48–49.
7. Paletta CE, Lynch R, Knutsen AP. Rhabdomyolysis and lower extremity compartment syndrome due to influenza B virus. Ann Plastic Surg 1993; 30:272–273.
8. Owen CA, Mubarak SJ, Hargens AR, Rutherford L, Garetto LP, Akeson WH. Intramuscular pressures with limb compression. N Engl J Med 1979; 300:1169–1172.
9. Skejldal S, Stomsoe K, Alho A, Johnsen U, Torvik A. Acute compartment syndrome: for how long can muscle tolerate increased tissue pressure? Eur J Surg 1992; 158:437–438.
10. Rutgers PH, Harst E, Koumans RKJ. Surgical implications of drug-induced rhabdomyolysis. Br J Surg 1991; 78:490–492.
11. Allen MJ, Barnes MR. Exercise pain in the lower leg. J Bone Joint Surg 1986; 68B:818–823.

49
Myoglobinuria

Richard W. Orrell and Russell J. M. Lane
Charing Cross Hospital and Charing Cross and Westminster Medical School, London, England

INTRODUCTION

The term "myoglobinuria" is used to describe the appearance of cola-colored urine resulting from extensive muscle fiber damage (rhabdomyolysis). Myoglobinuria indicates a plasma myoglobin in excess of 300 ng/ml (normal range approximately 5–50 ng/ml by radioimmunoassay), and the urinary concentration is then generally greater than 250 ng/ml. Acute tubular necrosis and renal failure, the major direct complication of the phenomenon, occurs with urinary concentrations above 1000 ng/ml (1).

Myoglobin is a heme protein, which binds O_2 in a 1:1 molar ratio. It is particularly concentrated in type 1 muscle fibers and has been shown to enhance oxygen diffusion from capillaries to mitochondria and to increase the ability of muscle fibers to generate tension (2). Any pathological process leading to sarcolemmal damage results in leakage of myoglobin into the blood, and its low molecular weight of 17,500 Daltons, compared to 80,000 Daltons for creatine kinase (CK), determines its greater sensitivity to such damage. For example, in one study, serum myoglobin levels correlated more closely than CK with patients' clinical status during treatment of polymyositis (3). In most instances, however, there is a reasonably close relationship between CK and myoglobin levels (4).

CLINICAL AND LABORATORY FEATURES

The clinical presentation of rhabdomyolysis with myoglobinuria may be dramatic, with acute onset of severe myalgia, gross weakness, and stiffness and swelling of muscles, which may be focal or generalized. Large, proximal muscles are usually affected most severely. Reflexes may be depressed or absent. CK levels are often dramatically increased, in the order of 10,000 U/L, and the electromyogram typically shows severe myopathic changes together with injury potentials reflecting surface membrane damage or even frank denervation. Muscle biopsy findings can range from necrosis affecting virtually every fiber in the section, to rather inconspicuous scattered necrotic foci; or there may be no obvious histological abnormality, reflecting damage at the ultrastructural level (5). The muscle swelling may cause secondary nerve compression, and fasciotomy may be required (see Chapter 48), although some consider that such intervention can promote infection and delay healing (6).

CAUSES OF RHABDOMYOLYSIS AND MYOGLOBINURIA

The main causes of myoglobinuria are classified in Table 1.

Table 1 Causes of Myoglobinuria

Physical	
Extreme exertion	Coma (pressure)
Hyperthermia	Muscle ischemia
Compartment syndromes	Reperfusion
Crush syndrome	Involuntary movements
Inflammation	
Infections (viral, bacterial, etc.)	
Inflammatory muscle disease	
Metabolic abnormalities	
Metabolic myopathies	Systemic disorders
Defects of intermediary metabolism	Hyponatremia
Paroxysmal myoglobinuria	Hypokalemia
Mitochondrial myopathies	Hypophosphatemia
MAD deficiency	Hyperosmolar coma
Malignant hyperthermia	Acidosis
	Toxic and drug-induced myopathies
	See Table 2

Physical Causes

Extensive destruction of muscle fibers resulting in myoglobinuria can occur in crush injuries, for example in building collapse and road traffic accidents; in compartment syndromes (see Chapter 48); from muscle infarction; and with reperfusion of ischemic limbs following vascular reconstruction (7). Significant muscle crush injury can also occur in coma from any cause, particularly if the victim is lying on a hard surface, when intramuscular pressures can be 10 times greater than on soft surfaces (8).

The most common cause of life-threatening rhabdomyolysis and myoglobinuria is severe physical exertion, particularly in high ambient temperatures. This has been reported frequently in military recruits, long distance runners, and body builders (9). In such cases, damage probably results from a combination of compartment syndrome ischemia, excessive eccentric contractions, and critical depletion of energy metabolites. Warning symptoms may be masked by the elation or motivation of the exercise.

Severe, repeated muscle contractions may also induce rhabdomyolysis, for example, in status epilepticus, electroconvulsive therapy, delirium tremens, tetanus, severe dystonia, or prolonged chorea (10), and in neuroleptic malignant syndrome (11). In these conditions, the associated hyperthermia may be important since myoglobinuria may be a feature of heatstroke (12).

Muscle Inflammation

Myoglobinuria may result from muscle fiber destruction as a result of extensive infection, by viruses, bacteria, and other organisms (Chapter 31), possibly including HIV (13) (see Chapter 52), and also in cases of toxic shock syndrome, due to staphylococcal exotoxin (14). Rarely, idiopathic inflammatory myopathies are sufficiently severe to cause the syndrome (4).

Toxic and Metabolic Causes

In general, progressive myopathies such as muscular dystrophies do not cause myoglobinuria [although serum and urinary myoglobin levels are raised, corresponding to increased CK levels (4)], unless massive fiber destruction occurs over a short period, for example due to a malignant hyperthermia (MH) reaction during anesthesia. However, most metabolic myopathies involving intermediary metabolism may present in this way, and sometimes a history of previous recurrent attacks may be forthcoming. In one study of 77 consecutive patients presenting with myoglobinuria, specific enzyme defects were identified in about half the cases (15). Carnitine palmityltransferase (CPT) deficiency was the most common disorder, followed by McArdle's disease, phosphorylase kinase, myoadenylate deaminase (MAD), and phosphoglycerate kinase deficiencies. The age range of the patients was 15–65 years, with a mean of 31 years, and men were more commonly represented than women. Exercise was the most common precipitating factor. Fasting was a precipitant in patients with CPT and MAD deficiencies, but not in those with glycolytic disorders. More recently, cases of myoglobinuria associated with short- (16) and long-chain fatty acyl CoA dehydrogenase deficiencies (17) have also been reported (see Chapter 33).

Mitochondrial myopathies rarely result in myoglobinuria, but recurrent exertional myoglobinuria has been reported in patients with multiple mitochondrial DNA deletions (18), succinate dehydrogenase (19), cytochrome c oxidase (20), and coenzyme Q deficiencies (21).

Paroxysmal myoglobinuria of childhood is a potentially fatal syndrome, with a mortality of up to 35% (22). Other clinical features may include ictal bulbar signs, encephalopathy, seizures, developmental delay, and dysmorphism. Specific enzymatic defects affecting energy metabolism have been defined in some cases. Myoglobinuria resulting from malignant hyperthermia is discussed extensively in Chapter 35.

Severe abnormalities of water and electrolytes may result in muscle fiber damage and myoglobinuria. This has been reported with water intoxication and hyponatremia (23), nonketotic hyperosmolar coma (24), and hypokalemia and hypophosphatemia from a variety of causes (25). However, in several such cases, the pathogenesis may have been multifactorial, involving the toxic effects of drugs or alcohol, fever, seizures, and coma, in addition to the electrolyte disturbances. Finally, another common cause is toxic and drug-induced rhabdomyolysis. This syndrome is most often encountered in abusers of alcohol and opiates, but any drug that can cause an acute or subacute painful proximal myopathy may be implicated. The mechanisms involved are fully discussed in Chapter 30 and a list of commonly implicated drugs is given in Table 2.

Table 2 Classification of Toxic and Drug-Induced Rhabdomyolysis (Based on Pathological Categories as in Chapter 30)

Direct toxic effects on muscle		
	Drugs	*Toxins*
Necrotizing myopathy	Alcohol	Venoms (snakes, spiders, wasps bees)
	Opiates	Toluene
	Fibric acid derivatives	
	HMG-CoA reductase inhibitors	
	EACA	
	Succinylcholine	
Vacuolar myopathy secondary to hypokalemia	Glycyrrhizinic acid derivatives (licorice, carbenoxolone)	
	Diuretics	
	Amphotericin	
Secondary to motor excitation and fever		
"Ecstasy," other amphetamines	Bacterial toxins (tetanus, typhoid staphylococcal)	
Phencyclidine		
Neuroleptics (in NMS)		
Theophylline		
Unknown cause		
Haff disease (from fish)		
Quail eater's syndrome (hellbore or hemlock poisoning)		

PATHOPHYSIOLOGY AND MANAGEMENT

There are many different causes of rhabdomyolysis. Except for frank physical trauma, the fundamental mechanisms are unclear, although a critical increase in myoplasmic Ca^{2+} may be a final common pathway to muscle fiber destruction. In one recent study (26), intramuscular Ca^{2+} concentrations in external intercostal muscle fibers were found to be an order of magnitude higher than controls in 11 patients presenting with exertional rhabdomyolysis. Calcium levels fell with clinical improvement, which was hastened by dantrolene. Notably, however, 7/10 cases tested gave positive contraction tests for susceptibility to malignant hyperthermia (rhabdomyolysis is characteristic of the MH reaction, see Chapter 35), so this observation may not necessarily be pertinent to other cases.

Myoglobin is nephrotoxic in excess and causes acute tubular necrosis, possibly through toxin-induced vasoconstriction or tubular obstruction from myoglobin casts, enhanced by volume depletion (27). This may compound the hyperkalemia resulting from the release of K^+ from damaged muscle cells, which may cause fatal dysrhythmias, while phosphate release can cause a paradoxical hypocalcemia early in the oliguric phase, followed by a rebound hypercalcemia due to calcium phosphate release in the diuretic phase. Progressive muscular weakness may lead to respiratory failure.

Treatment is principally supportive, with particular regard to minimizing the risks of renal failure, monitoring respiratory function, and, where possible, eliminating or correcting the precipitating abnormality. Dantrolene should be administered, and a diuresis encouraged with careful monitoring of fluid and electrolyte balance. General strategies used in the management of the MH reaction are applicable (see Chapter 35).

REFERENCES

1. Grossman RA, Hamilton RW, Morse BM, Penn AS, Goldberg M. Non-traumatic rhabdomyolysis and acute renal failure. N Engl J Med 1974; 291:807–811.
2. Wang ZY, Noyszweski EA, Leigh JS Jr. In vivo MRS measurement of deoxymyoglobin in human forearms. Magnet Reson Med 1990; 14:562–567.
3. Lane RJM, Emslie-Smith A, Mosquera IE, Hudgson P. Clinical, biochemical and histological responses to treatment in polymyositis: a prospective study. J Roy Soc Med 1989; 82:333–338.
4. Kiessling WR, Ricker K, Pflughaupt KW, Martens HG, Haubitz I. Serum myoglobin in primary and secondary skeletal muscle disorders. J Neurol 1981; 224:229–233.
5. Blain PG, Lane RJM. Drugs and muscle. Adverse Drug React Acute Poison Rev 1983; 2:1–24.
6. Odeh M. The role of re-perfusion induced injury in the pathogenesis of the crush syndrome. N Engl J Med 1991;324:1417–1423.
7. Adeseshiah M, Round JM, Jones DA. Reperfusion injury in skeletal muscle: a prospective study in patients with acute limb ischaemia and claudicants treated by revascularisation. Br J Surg 1992; 99:1026–1029.
8. Owen CA, Mubarak SJ, Hargrens AR, Rutherford L, Garetto LP, Akeson WH. Intramuscular pressures with limb compression: clarification of the pathogenesis of the drug-induced muscle compartment syndrome. N Engl J Med 1979; 30:1169–1172.
9. Doriguzzi C, Palmucci L, Mongini T, Arnaudo E. Body building and myoglobinuria: report of three cases. Br Med J 1988; 269:826–827.
10. Jankovic J. Myoglobinuric renal failure in Huntington's chorea. Neurology 1986; 36:138–139.
11. Buckley PF, Hutchinson M. Neuroleptic malignant syndrome. J Neurol Neurosurg Psychiatry 1995; 58:271–273.
12. Vertel RM, Knochel JP. Acute renal failure due to heat injury: an analysis of ten cases associated with a high incidence of myoglobinuria. Am J Med 1967; 43:435–451.
13. Younger DS, Hays AP, Uncini A, Lange DJ, Lovelace RE, DiMauro S. Recurrent myoglobinuria and HIV seropositivity: incidental or pathogenic association. Muscle Nerve 1989; 12:842–843.
14. Saul RA, Vernon M, Roe C, Osofsky SG. Rhabdomyolysis in a patient with nonoliguric renal failure: similarities to the toxic shock syndrome. South Med J 1980; 73:261–263.
15. Tonin P, Lewis P, Servidei S, DiMauro S. Metabolic causes of myoglobinuria. Ann Neurol 1990;27:181–185.
16. Tein I, DeVivo DC, Hale DE, Clark JT, Zinman H, Laxer R, Shore A, DiMauro S. Short chain L-e-hydroxyacyl- CoA dehydrogenase deficiency in muscle: a new cause for recurrent myoglobinuria and encephalopathy. Ann Neurol 1991; 30:415–419.
17. Ogilvie I, Pourfazam M, Jackson S, Stockdale C, Bartlett K, Turnbull DM. Very long acyl-CoA dehydrogenase deficiency presenting with exercise induced myoglobinuria. Neurology 1994; 44:467–473.

18. Ohno K, Tanaka M, Sahashi K, Ibi T, Sato W, Yamamoto T, Takahashi A, Ozawa T. Mitochondrial DNA deletions in inherited recurrent myoglobinuria. Ann Neurol 1991; 29:364–369.
19. Linderholm H, Essen-Gustavsson B, Thornell LE. Low succinate dehydrogenase (SDH) activity in a patient with a hereditary myopathy with paroxysmal myoglobinuria. J Intern Med 1990; 228:43–52.
20. Johnson WSW, Penn AMW, Fuller TW, Salas VM, Hall RE, Burton MD, D'Agostino AN, Kennaway NG. Cytochrome c oxidase deficiency presenting as isolated myoglobinuria. Neurology 1993; 43:A372.
21. Ogasahara S, Engel AG, Frens D, Mack D. Muscle coenzyme Q deficiency in familial mitochondrial encephalomyopathy. Proc Natl Acad Sci USA 1989; 86:2379–2382.
22. Tein I, DiMauro S, DeVivo DC. Recurrent childhood myoglobinuria. In: Barness LA, ed. Advances in Paediatrics. St. Louis: CV Mosby, 1987:77–117.
23. Penn AS. Myoglobin and myoglobinuria. In: Vinken PJ, Bruyn GW, eds. Handbook of Clinical Neurology, Part II. North Holland: Amsterdam, 1980:259–285.
24. Singhal PC, Abramovic M, Venkatesan J. Rhabdomyolysis in the hyperosmolar state. Am J Med 1990; 88:9–12.
25. Lane RJM. Muscle involvement in systemic and iatrogenic disease. In: Kennard C, ed. Recent Advances in Clinical Neurology 5. Oxford: Churchill Livingstone, 1988:201–238.
26. Lopez JR, Rojas B, Gonzalez MA, Terzic A. Myoplasmic calcium concentration during exertional rhabdomyolysis. Lancet 1995; 345:424–425.
27. Hamilton RW, Hopkins MB, Shihabi ZK. Myoglobinuria, haemoglobinuria, acute renal failure. Clin Chem 1989; 35:1713–1720.

50
Nutritional Myopathies

Richard W. Orrell
Charing Cross Hospital and Charing Cross and Westminster Medical School, London, England

INTRODUCTION

A myopathy may be a manifestation of protein or calorie malnutrition, due to negative nitrogen balance producing muscle wasting. The contribution of specific nutrient deficiencies is less certain, especially in humans (1). Myopathy has been associated with vitamin D deficiency (in osteomalacia) and vitamin E deficiency. Ethanol excess may lead to a myopathy, which may partly be attributable to nutrient deficiency, in addition to direct toxic effects (see Chapter 30). In the clinical assessment of muscle weakness in nutritional disorders, it is necessary to establish whether the muscle weakness is primary, or secondary to a dietary deficiency neuropathy.

PROTEIN-CALORIE MALNUTRITION NEUROMYOPATHY

The muscles provide a store of amino acids, and during fasting and starvation the muscles are a major source of amino acids for other essential metabolic processes. This results in breakdown and loss of muscle protein (2). Thirty children with protein-calorie malnutrition were studied in India, of whom 23 had muscle wasting, and 17 were unable to walk. Another four had a waddling gait (3). Electromyography (EMG) demonstrated myopathic features, sometimes associated with reduced peripheral nerve conduction velocity, whereas in others the velocity was normal. Muscle biopsy showed small fibers with fragmentation and fiber type grouping, suggesting neurogenic atrophy, but other studies have described features of both myopathic and neurogenic atrophy.

A reversible neuromuscular syndrome has also been described in malnourished children in the United States (4). These children had additional preexisting neurological disease, with mental impairment, and impoverishment, which contributed to the malnutrition. The conduction in peripheral nerves was normal, but EMG showed scattered fibrillations and positive sharp waves. The children were treated with high-calorie diets and multiple vitamins, with resolution of the myopathy. Malnutrition, of protein and nutrients, may also occur in catabolic states such as trauma and chronic infection, and in some systemic diseases (Chapter 31), leading to muscle weakness and wasting.

DIETARY OSTEOMALACIA (VITAMIN D DEFICIENCY)

Proximal myopathy is a well-recognized clinical feature of osteomalacia and may be the presenting complaint. The muscle weakness is not related to the plasma calcium concentration or other hitherto defined biochemical abnormality. Muscle biopsies were examined from 19 patients in India with osteomalacia, 10 of nutritional and nine of mixed (e.g., additional hyperparathyroidism or anticonvulsant medication) etiology (5). In those with a simple nutritional etiology, a mild diffuse atrophy was seen, with electron microscopic changes including loss of myofibrils and prominence of mitrochondria and glycogen. In those with a mixed etiology, with additional endocrine or metabolic features, clear degenerative changes were seen in the myofibrils and mitochondria, in addition to the atrophic features. It was concluded that osteomalacic muscle weakness is not necessarily myopathic, but that the muscle wasting may be related to disuse and malnutrition. However, EMG studies of another similar group of patients suggested that the proximal muscle weakness of nutritional osteomalacia is a true reversible myopathy, not a disuse atrophy (6). Myopathy has also been described in patients with hypophosphatemic osteomalacia, which responded to vitamin D supplementation (7). The mechanism by which vitamin D deficiency causes a myopathy is not known.

VITAMIN E DEFICIENCY MYOPATHY

Vitamin E deficiency may be associated with a primary myopathy in humans (8), in addition to other neurological features (9). Vitamin E deficiency is rare in humans, requiring a chronic deficiency, but may be seen in malabsorption syndromes such as chronic cholestasis and cystic fibrosis, and in abetalipoproteinemia. Chronic vitamin E deficiency was described in a boy with severe malabsorption since birth, attributed to an intrahepatic syndrome with cirrhosis (8). At age 7 years he developed progressive exotropia of the left eye, bilateral ptosis, and progressive generalized weakness. EMG and nerve conduction studies were normal, but serum creatine kinase was raised to 613 U/L. Muscle biopsy showed many fibers to be small and angulated, with occasional pyknotic nuclei. Type 1 fibers were predominant, with atrophy especially in type 2 fibers. Electron microscopy was normal except for occasional areas of Z-band smearing. Serum vitamin E was markedly reduced (8 μg/dl, normal 550–1500). Following 4 months of treatment with large doses of vitamin E, the boy had a significant improvement in muscle power, with a fall in creatine kinase to 71, and his serum vitamin E rose to 350 μg/dl.

PARENTERAL NUTRITION

An acute myopathy may be associated with parenteral nutrition (10). Four patients receiving parenteral nutrition (three with celiac disease and one with a small bowel fistula) developed muscle pain, swelling, and tenderness, with a marked rise in plasma creatine kinase. Stopping the parenteral nutrition led to improvement of the symptoms within hours, with complete recovery in a few days. The symptoms were attributed to a deficiency of essential fatty acids. There were no further symptoms on recommencing parenteral nutrion with lipid supplementation. Other possible deficiencies in parenteral nutrition that are known to cause myopathies include vitamin E, phosphate, and selenium.

ALCOHOL-RELATED MYOPATHY

This is discussed in detail in Chapter 30.

REFERENCES

1. Lane RJM. Muscle involvement in systemic and iatrogenic disease. In: Kennard C, ed. Recent Advances in Clinical Neurology. Edinburgh: Churchill Livingstone 1988:201–238.
2. Pratt OE. Liver and muscle: how together they nourish the brain. In: Clifford Rose F, ed. Metabolic Diseases of the Nervous System. London: Pitman Brooks, 1981:89–101.
3. Sachdev KK, Taori GM, Pereira SM. Neuromuscular status in protein-calorie malnutrition. Neurology 1971; 21:801–805.
4. Donley DK, Evans BK. Reversible neuromuscular syndrome in malnourished children. Dev Med Child Neurol 1989; 31:797–815.

5. Dastur DK, Gagrat BM, Wadia NH, Desai MM, Bharucha EP. Nature of muscular change in osteomalacia: light- and electron-microscope observations. J Pathol 1975; 117:211–228.
6. Irani PF. Electromyography in nutritional osteomalacic myopathy. J Neurol Neurosurg Psychiatry 1976; 39:686–693.
7. Schott GD, Wills MR. Myopathy in hypophosphataemic osteomalacia presenting in adult life. J Neurol Neurosurg Psychiatry 1975; 38:297–304.
8. Tomasi LG. Reversibility of human myopathy caused by vitamin E deficiency. Neurology 1979; 29:1182–1186.
9. Hegele RA, Angel A. Arrest of neuropathy and myopathy in abetalipoproteinemia with high-dose vitamin E therapy. Can Med Assoc J 1985; 132:41–44.
10. Stewart PM, Hensley WJ. Acute polymyopathy during total parenteral nutrition. Br Med J 1981; 283:1578.

51
Amyloid Myopathy and Intermediate Filament Myopathy

David J. Dick
Norfolk and Norwich Hospital, Norwich, and Addenbrooke's Hospital, Cambridge, England

AMYLOID MYOPATHY

Amyloid appears homogeneous and amorphous under light microscopy, stains pink with hematoxylin and eosin, and metachromatically with methyl violet or crystal violet. When stained with Congo red, it produces an apple-green birefringence when viewed under polarized light. At an ultrastructural level amyloid is made up of a mat of aggregated fibrils 7.5-10-nm wide arranged in an antiparallel or cross-pleated sheet configuration.

The classification of amyloidosis is based on identification of the protein that is the major constituent of the amyloid fibril. A simplified classification is shown in Table 1. Antisera are available that allow the type of amyloid to be accurately identified.

Clinical involvement of muscle in amyloidosis appears to occur exclusively in primary amyloidosis (AL), though histological evidence of deposition may be seen in familial amyloidosis (AF). Yamada et al. (1) examined muscle biopsies from 12 cases of amyloidosis: six cases of AL, two of secondary amyloid (AA), and four cases of AF. No amyloid was found in muscle from the AA type. However, amyloid deposition was found in muscle and blood vessels in the AL and AF cases. Unsurprisingly, the most florid changes were found in those patients with a clinical myopathy, and these tend to be patients with AL amyloidosis.

Case History

A 72-year-old man was referred to a surgical team with a 6-month history of anorexia, weight loss of 7 kg, abdominal pain, and a sensation of incomplete evacuation of his rectum. His past history was unremarkable except for Menière's disease. He was a nonsmoker and nondrinker. Investigations included a gastroscopy, abdominal ultrasound, sigmoidoscopy, and barium enema. No abnormality was found. During the following 6 months he lost a further 20 kg and developed myalgia, fatigue, and symptoms of proximal limb weakness.

Examination showed a thin man with mild peripheral edema and firm hepatomegaly four finger-breadths below the costal margin. The cranial nerves were normal. In the limbs, there was a diffuse wasting and weakness with depressed tendon reflexes. Sensation was normal. Investigations revealed a mild normochromic, normocytic anemia, a urea of 12.0 mmol/L, and normal creatine kinase activity and plasma protein electrophoresis. Small pleural effusions and moderate cardiomegaly were seen on chest radiography. A bone marrow aspirate and trephine were normal. Nerve conduction studies showed normal motor studies but absent sensory nerve action potentials, and the electromyogram revealed myopathic features. An echocardiogram revealed features of a constrictive cardiomyopathy.

Table 1 Classification of Amyloidosis

Amyloid type	Clinical classification	Major protein component
AL	Primary: no evidence of associated disease except multiple myeloma	κ or λ chains
AA	Secondary: associated with other disorders such as rheumatoid disease or chronic infection	Protein A
AL	Localized—single organ involvement	κ or λ chains
AF	Familial	Transthyretin

Muscle biopsy demonstrated scattered muscle fiber atrophy and infiltration of endomysium, perimysium, blood vessels, and fibrofatty connective tissues by amyloid (Fig. 1). The atrophic fibers were of both fiber types. Immunocytochemical studies confirmed the clinical diagnosis of AL.

The typical features of amyloid myopathy are said to be an abnormal firmness of muscles, pseudohypertrophy, and palpable tumors within muscles (2,3). Macroglossia is often seen and is sometimes so severe as to interfere with closure of the mouth and mastication. There may also be dysphagia and hoarseness. Limb weakness may be generalized or proximal in distribution and some patients complain of myalgia. Occasionally proximal limb weakness may be the presenting complaint (4). Rarely, amyloid myopathy has been described as presenting with ophthalmoplegia (5) or respiratory failure (6). Despite these uncommon presentations, it is the picture of an apparently athletic but weak adult with macroglossia that should alert the clinician to the diagnosis.

The serum creatine kinase activity is normal or moderately elevated. Electromyography is not often

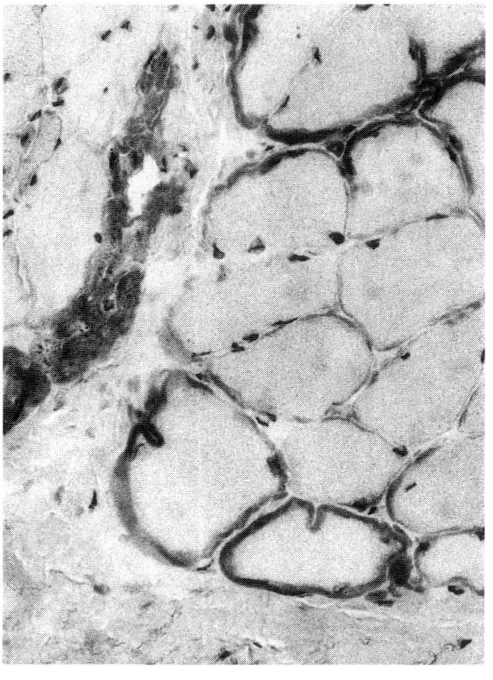

Figure 1 Amyloid myopathy. Deposition of amyloid within blood vessels and around muscle fibers (Congo red stain × 280).

helpful although it may show a mixture of myopathic features and denervation changes, which may be due to local compression and damage to fibers by amyloid deposits.

Magnetic resonance imaging of muscles shows only a slight prolongation of muscle T1 and T2 relaxation times but a striking reticulation of subcutaneous fat (7). The muscle biopsy appearances are characteristic—scattered, non-type-specific *fiber* atrophy and perimysial and endomysial deposition of amyloid. There may also be amyloid deposition in the walls of blood vessels and in connective tissue. At an ultrastructural level amyloid filaments are often found in association with the sarcolemma, sometimes invaginating it (3,4). Filaments are not seen within *fibers*.

The prognosis of primary amyloid is poor and the median survival is in the order of 2 years. Even in those cases associated with multiple myeloma where chemotherapy is given, it is rare for the myopathy to improve, although exceptions have been reported (8).

INTERMEDIATE FILAMENT MYOPATHIES

At an ultrastructural level, three main classes of filamentous structures form the cytoskeleton in muscle: *actin filaments* (6–10 nm in diameter), *intermediate filaments* (7–11 nm), and *microtubules*. The intermediate filaments are desmin, vimentin, synemin, and a neural filament–associated polypeptide (NFAP). Of these, the first two have been characterized most fully. During early development and in the early stages of muscle regeneration, vimentin is the major intermediate filament, but as the cell matures the concentration of desmin increases, and the major intermediate filament of mature skeletal muscle is desmin.

Desmin is an α-helical polypeptide (mass 53 kDa) and is coded by a single gene at the 2q35 locus in birds and mammals (9). It is located in the outer circumference of Z-bands and is thought to provide longitudinal stability by linking adjacent myofibrils and peripheral myofibrils to the plasma membrane (10). In 1978 Fared et al. (11) documented the first case of a myopathy associated with cytoplasmic aggregates of intermediate filaments. Since then over 20 additional cases have been reported and these have been recently reviewed by Helliwell and colleagues (12). Most cases are sporadic, though a few dominant or recessive families have been described. The age of onset varies from childhood to middle age.

Just over half of the cases involved only skeletal muscle, but in most of the remainder cardiac muscle was also affected. The pattern of muscle weakness varies and may be either predominantly distal or proximal. Some patients may be diffusely weak and respiratory muscle weakness has been recorded (13). Cardiac manifestations are usually those of a hypertrophic or restrictive cardiomyopathy, sometimes with conduction defects. Isolated defects of cardiac conduction are unusual. Clearly, there is a marked variation in clinical presentation, and a few patients have an associated peripheral neuropathy (14).

Serum creatine kinase activity is usually normal or moderately elevated. The electromyogram is either normal or shows myopathic features.

The histopathological features vary between cases. On light microscopy eosinophilic bodies or flattened plaques are found in both *fiber* types. They are sharply circumscribed, usually subsarcolemmal, but may occasionally be seen in the centre of *fibers* (Fig. 2). All are PAS-positive but digest with diastase, indicating the presence of glycogen. The deposits may also show increased intensity of staining with oxidative enzymes, suggesting an accumulation of mitochondria. They appear bright green with Gomori trichrome stain. Uninvolved sarcoplasm and other fibers have a normal cytoarchitecture.

The ultrastructural appearances vary. The deposits, which are often subsarcolemmal, contain glycogen granules clustered around dense granular filamentous material. This material may be associated with normal mitochondria, and occasional deposits have a close relationship with a smeared myofibrillar Z-band (Fig. 3). In some cases dense ovoid bodies may be present (15) or aggregates that resemble Mallory bodies (16). Similar appearances have been reported in cardiac muscle (17).

Immunostaining is strongly positive for desmin (Fig. 4). In some patients weaker staining is also found for vimentin or dystrophin (Fig. 2).

Clearly, there is considerable variation in the clinical and morphological expression of these

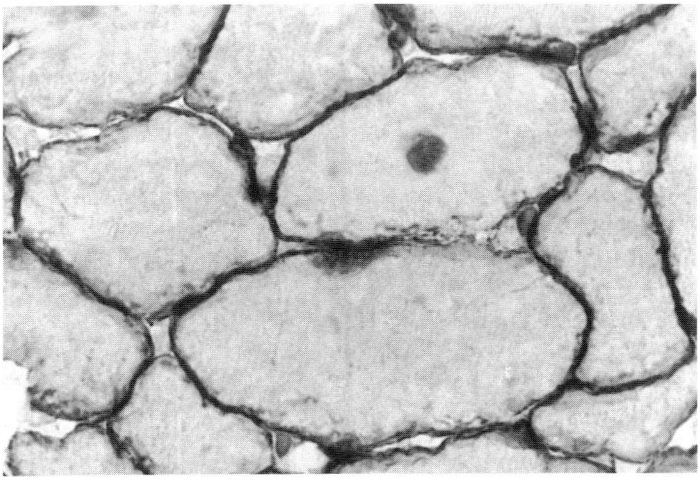

Figure 2 Desmin myopathy. Positive staining of deposits and cell membranes with antidystrophin antibody. The deposits seen in the center of one fiber and beneath the sarcolemma in the adjacent fiber were eosinophilic and PAS-positive (×400).

disorders. They have been given the name desmin storage myopathies or intermediate filament myopathies. Traditionally, storage products are energy-related compounds such as glycogen or lipid, and in metabolic myopathies it is a blockage in a biochemical pathway that leads to their accumulation. By contrast, there is no evidence that there is any defect of desmin degradation in these myopathies and their etiology is unknown. It is possible that increased synthesis may be the underlying mechanism. Presumably, suppressor genes control the synthesis of intermediate filaments at different stages of myocyte maturation, and perhaps, in these myopathies, there is a failure to switch off production.

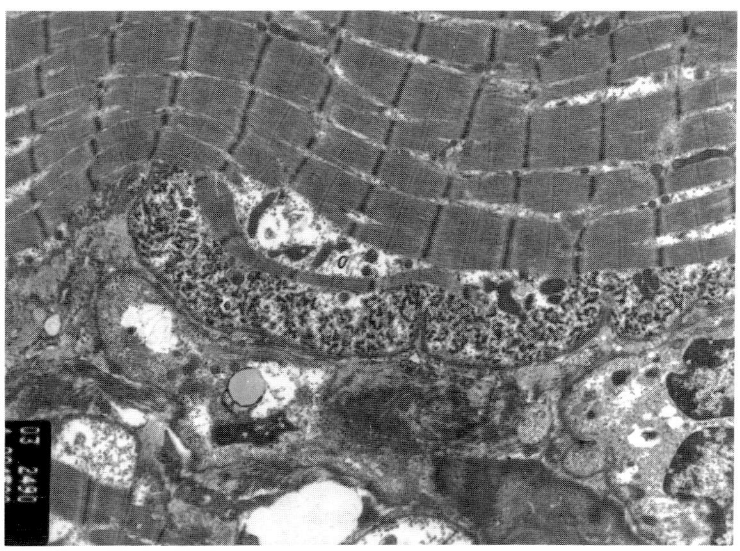

Figure 3 Irregularly shaped, electron-dense subsarcolemmal deposits associated with glycogen granules (×10,000).

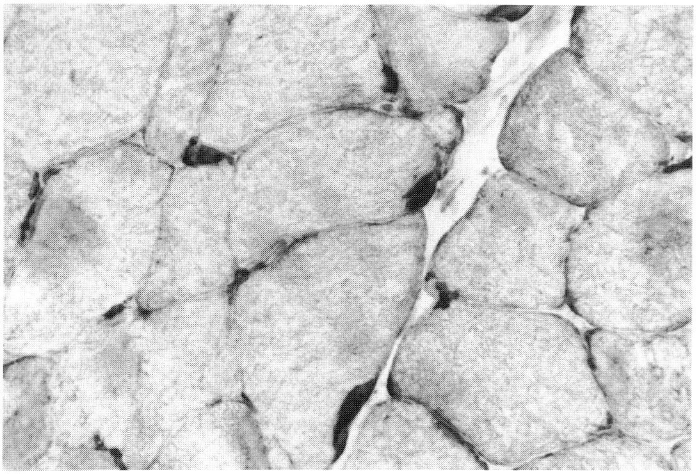

Figure 4 Deposits showing strong positive immunocytochemical staining with antidesmin antibodies (×400).

ACKNOWLEDGMENT

All figures were kindly supplied by Dr. Janice Anderson, Consultant Neurologist, Addenbrooke's Hospital, Cambridge, England.

REFERENCES

1. Yamada M, Tsukagoshi H, Hatakeyama S. Skeletal muscle deposition in AL- (primary or myeloma-associated), AA- (secondary), and prealbumin-type amyloidosis. J Neurol Sci 1988; 85:223–232.
2. Whitaker JN, Hashimoto K, Quinones M. Skeletal muscle pseudohypertrophy in primary amyloidosis. Neurology 1977; 27:47–54.
3. Ringel SP, Cleman HN. Amyloid-associated muscle pseudohypertrophy. Arch Neurol 1982; 39:413–417.
4. Jennekens FGI, Wokke JHJ. Proximal weakness of the extremities as main features of amyloid myopathy. J Neurol Neurosurg Psychiatry 1987; 50:1353–1358.
5. Katz B, Leja S, Melles RB, Press GA. Amyloid ophthalmoplegia. J Clin Neurol Ophthalmol 1989; 9:39.
6. Ashe J, O'Borel C, Hart G, Humphrey RL, Derrick DA, Kuncl RW. Amyloid myopathy presenting with respiratory failure. J Neurol Neurosurg Psychiatry 1991; 54:162–165.
7. Metzler JP, Fleckenstein JL, White CL, Haller RG, Frenkel EP, Greenlee RG. MRI evaluation of amyloid myopathy. Skeletal Radiol 1992; 21:463–465.
8. Sheehan-Dare RA, Simmons AV. Amyloid myopathy and myeloma: response to treatment. Postgrad Med J 1987; 63:141–142.
9. Capetanaki YG, Ngai J, Lazarides E. Characterization and regulation in the expression of a gene coding for the intermediate filament protein desmin. Proc Natl Acad Sci USA 1984; 81:6909–6913.
10. Tokuyasu KT, Dutton AH, Singer SJ. Immunoelectronmicroscopic studies of desmin (skeletin) localisation and intermediate filament organisation in chicken skeletal muscle. J Cell Biol 1983; 96:1727–1735.
11. Fared M, Godet-Guillain J, Tome FMS, Collin H, Gaudeau S, Boffety C, Vernant P. Une nouvelle affection musculaire familiale, definie par l'accumulation intrasarcoplasmique d'un materiel granulo-filamentaire dense en microscopie electronique. Rev Neurol 1978; 131:411–425.
12. Helliwell TR, Green ART, Green A, Edwards RHT. Hereditary distal myopathy with granulo-filamentous cytoplasmic inclusions containing desmin, dystrophin and vimentin. J Neurol Sci 1994; 124:174–187.
13. Chapon F, Viader F, Fardeau M, Tome F, Dalazeau N, Berthelin C, Thenint JP, Lechevalier B. Myopathie familiale avec inclusions de type "corps cytoplasmiques" (ou "spheroides") revelee par une insuffisance respiratoire. Rev Neurol (Paris) 1989; 145:460–465.
14. Liu HM, Gumbinas M. Axonal filamentous spheroids associated with cardiomyopathy with "targetoid fibers." Neurology 1974; 24:547–554.
15. Edstrom L, Thornell LE, Eriksson A. A new type of hereditary distal myopathy with characteristic sarcoplasmic bodies and intermediate (skeletin) filaments. J Neurol Sci 1980; 47:171–190.

16. Fidzianska A, Goebel HH, Osborn M, et al. Mallory body–like inclusions in a hereditary congenital neuromuscular disease. Muscle Nerve 1983; 6:195.
17. Takatsu T, Kawai G, Tsutsumi G, Inoue K. A case of idiopathic myocardiopathy with deposits of a peculiar substance in the myocardium: diagnosis by endomyocardial biopsy. Am Heart J 1968; 76:93.

52
HIV-Related Myopathies

Russell J. M. Lane
Charing Cross Hospital and Charing Cross and Westminster Medical School, London, England

INTRODUCTION

Myalgia and myopathic symptoms were recorded in occasional patients with acquired immunodeficiency syndrome (AIDS) in early reviews of neurological complications in this disorder (1,2), but myopathy was diagnosed in under 1% of cases prior to the introduction of zidovudine (azidothymidine, AZT). Muscle biopsy studies of HIV-infected patients with neuromuscular symptoms undertaken before the widespread use of this drug revealed a range of abnormalities, including muscle fiber necrosis with inflammatory infiltration, nemaline rod formation, neurogenic atrophy, and type 2 fiber atrophy (3). The most common histological appearance, however, proved to be an HIV-related polymyositis, similar in histological and cytological characteristics to inflammatory muscle disease in simian AIDS (3–5). It was established at an early stage that HIV antigens could not be detected within the muscle fibers in either simian or human AIDS, but were abundant in CD4 cells in the inflammatory infiltrates (3–5). In a later study (6), the nature of the inflammatory infiltrate in HIV-related polymyositis was compared with that of idiopathic inflammatory myopathy. No significant differences were observed in $CD8^+$ or macrophage cell counts, but the proportion of CD4 cells was significantly lower, as might be predicted. The inflammatory cells were noted to be surrounding and invading muscle fibers expressing MHC-1 antigens, and it was concluded that HIV-associated polymyositis was due to a T-cell-mediated, MHC-1-restricted, cytotoxic process triggered by the virus. Since the immunocytochemical features in the two forms of polymyositis were otherwise so similar, it was suggested that by analogy, idiopathic polymyositis might also be virally mediated (6).

However, a number of reports subsequently appeared, describing HIV-infected patients with proximal myopathy and other features suggestive of HIV-related polymyositis, having little or no inflammatory infiltration on muscle biopsy. Histologically, these patients had a necrotizing myopathy, often with prominent nemaline rod formation, referred to as HIV-associated myopathy, or HIV myopathy (7–9). As with the inflammatory form, some patients with this condition responded to steroids. It was apparent from these reports that HIV-associated polymyositis or myopathy could be the presenting feature of AIDS in previously asymptomatic patients (5,9), and HIV-related muscle disease has been reported, but not proven, to be a cause of rhabdomyolysis (10,11). HIV-associated myopathy has also been shown to be one cause of the common HIV wasting syndrome (12). By contrast, myopathy or myositis due to opportunistic infection appears to be relatively uncommon, although cases of pyomyositis (13,14) and muscle and nerve infection by *Mycobacterium intracellulare* (15) and microsporidiosis (16) have been reported (see also Chapter 31).

AZT MYOPATHY

The incidence of myopathic symptoms increased substantially following the introduction of AZT. Myalgia had been reported as an adverse reaction in occasional patients during trials of this drug (17), and subsequently cases of a subacute painful proximal myopathy with prominent wasting and elevation of serum CK levels were reported with increasing frequency (18–20). A sterotyped clinical presentation began to emerge (20–22). The initial symptom was myalgia, followed by symmetrical weakness and often muscle wasting, notably in the buttocks and legs; occasionally the syndrome could be complicated by additional features of HIV-related peripheral neuropathy. The symptoms generally began after about 9–12 months of treatment on a standard dose of approximately 1 g AZT daily and usually improved quite quickly after the drug was stopped. Serum CK levels were noted to be increased in all cases, typically in advance of the developing myopathy, and to fall or return to normal when the drug was stopped (22). Electromyographic findings were usually myopathic. Some studies indicated that this complication of AZT treatment might occur in up to a third of patients receiving the drug (21,22).

Muscle biopsy studies in cases of AZT myopathy typically revealed a multifocal, segmental necrotizing myopathy, but without significant inflammatory infiltration (3,18,19). In addition, it was noted that the necrotic segments of affected fibers showed a characteristic microvacuolar change in the sarcoplasm (Fig. 1). Ultrastructurally, changes noted included myofibrillar disorganisation, Z-line streaming and nemaline rod formation, vacuolation (principally due to the accumulation of swollen mitochondria), inflammatory cell infiltration, and occasional cytoplasmic body formation (Fig. 2). Several studies also commented on the frequent presence of tubuloreticular inclusions (TRI) in vascular endothelial cells (Fig. 3) (23–25).

However, similar histological and ultrastructural appearances had been noted earlier in some patients with HIV-related myopathy who had never received AZT (7–9), resulting in some controversy concerning the role of AZT in the genesis of myopathy in HIV-infected patients. Subsequently it was reported that AZT caused a specific mitochondrial myopathy, and that AZT myopathy could be distinguished from HIV-related myopathy by the presence of ragged-red fibers on trichrome staining, and by the presence of mitochondrial abnormalities on electron microscopy (26,27). It was also suggested by one group that the microvacuolar change in the sarcoplasm had different origins in the two situations, being due to aggregations of abnormal mitochondria in AZT myopathy and to dilatations of the sarcoplasmic reticulum in HIV-related myopathy (24). Indeed, in one study electron micrographs

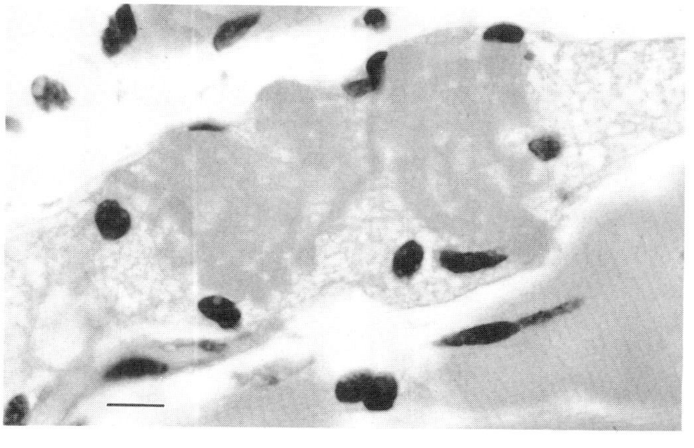

Figure 1 AZT myopathy. A segment of a muscle fiber cut longitudinally shows disintegration and clumping of the myofibrils and diffuse microvacuolation of the sarcoplasm, which is due to diffuse accumulation of swollen mitochondria. The adjacent fiber on the lower right aspect shows a normal banding pattern. H&E, bar = 10 μ.

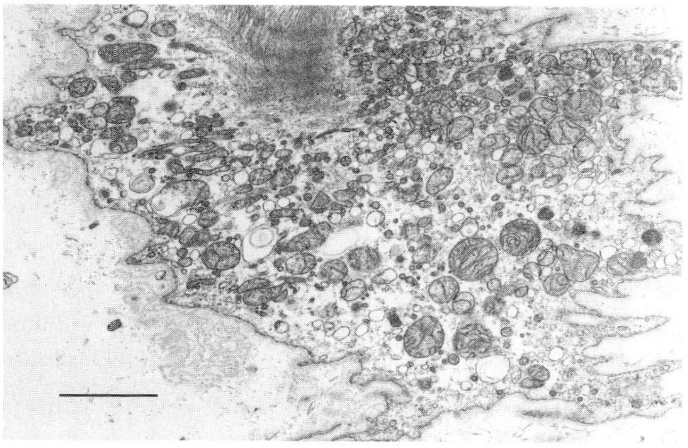

Figure 2 AZT myopathy. Electron micrograph showing gross disruption of myofibrillary components and accumulations of swollen, structurally abnormal mitochondria beneath the sarcolemma. Very similar appearances were observed in patients with HIV-related myopathy (see text). Bar = 2 μ.

of muscle from patients with AZT myopathy could be distinguished "blind" from those of untreated myopathic patients on the basis of the mitochondrial abnormalities in the former, which included subsarcolemmal proliferation, abnormalities of shape, size, and organization of the cristae, and, rarely, paracrystalline inclusions (28). It was suggested that the likely cause of mitochondrial abnormalities in AZT myopathy was inhibition of γ-DNA polymerase by the drug (26). In tissue culture studies, AZT appeared to cause an overall reduction in mitochondrial DNA synthesis and mitochondrial proliferation

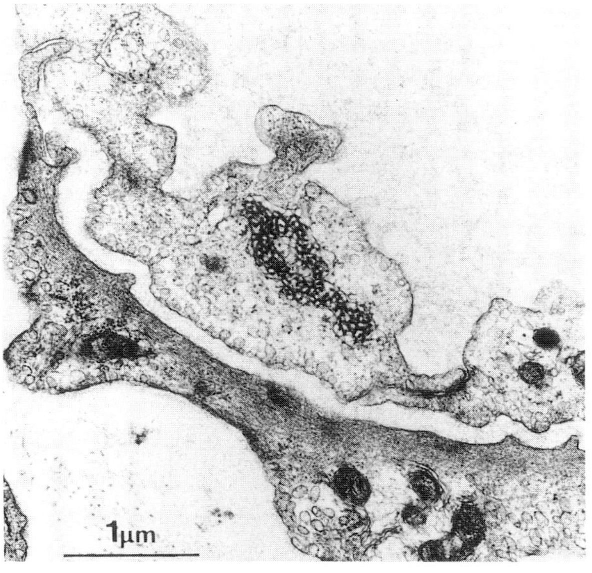

Figure 3 AZT myopathy. Electron micrograph showing tubuloreticular inclusion (TRI) in an endothelial cell. This honeycomb-like structure is produced as a response to interferon synthesis, either systemic or local. TRI are found in profusion in muscle endothelial cells in HIV patients with clinical myopathy, but less commonly in cases without myopathic signs.

(29), and the mitochondrial DNA content in muscle biopsy samples from AZT myopathy patients was found to be reduced, and to increase after the drug was stopped (30).

PATHOLOGICAL CHANGES

A number of studies subsequently examined the specificity of the pathological changes in AZT and HIV-related myopathy. Several groups noted that ragged-red changes were not seen in nonnecrotic fibers (25, 31), and one group could not distinguish the histological and ultrastructural features of patients with AZT myopathy from those of patients who had either never received the drug or had stopped it at least 5 months earlier (31). Two studies found that the proportion of ragged-red fibers seen on biopsy was directly related to the lifetime dose of AZT received (27,32), but this was not confirmed in another study (31). In one series, several patients who were found to have ragged-red fibers were entirely asymptomatic (32).

It has been suggested that the most likely relationship between HIV-related myopathy and AZT myopathy is that the latter develops only in the presence of the former (31). Thus, the drug does not appear to be strikingly myotoxic in normal subjects or in HIV-infected patients without myopathic symptoms, and the time course of AZT myopathy is not typical of other drugs that cause acute/subacute necrotizing myopathy (see Chapter 30), in that there is an unusually long lag phase from introduction of the drug to onset of symptoms. While inflammatory infiltration is typically insignificant in AZT myopathy, immunohistochemical studies show that sparse infiltration of muscle with T cells is present (26), and withdrawal of the drug may result in recrudescence of an inflammatory myopathy (personal observations). It may be, therefore, that AZT exerts its myotoxic actions principally on a background of preexisting HIV-related polymyositis, reducing the inflammatory infiltration but causing a microvacuolar necrotizing myopathy with prominent mitochondrial abnormalities. Whether it does so by a specific effect on mitochondrial DNA synthesis, causing a mitochondrial myopathy, or by a less specific effect on muscle cell growth and metabolism is unresolved (33). It has also been suggested that the appearance of TRI in endothelial cells in muscle may be due to local rather than systemic production of type 1 interferons as the first manifestation of HIV-related pathological processes in muscle, by analogy to similar observations in other tissues, since these structures were frequently not observed in patients without clinical myopathy (31).

PROGNOSIS

The prognosis of AZT myopathy is generally favorable. In one study, myalgia resolved in all cases and strength returned to normal in most when the drug was stopped, emphasising the need for early diagnosis (34). Differentiation of the early clinical features from exercise-related myalgia with raised CK may pose occasional difficulties, which can usually be resolved by needle muscle biopsy (35). However, ongoing or recrudescent inflammatory myopathy may produce persistent problems and may require serial biopsy for interpretation.

As noted above, HIV-related polymyositis and myopathy may show a gratifying response to steroids, and HIV wasting syndrome due to HIV-related myopathy may also respond (12). The recent introduction of lower maintenance doses of AZT in the treatment of HIV infection has significantly reduced the incidence of AZT myopathy.

REFERENCES

1. Snider WD, Simpson DM, Neilson S, Gold JWM, Metroka CE, Posner JB. Neurological complications of acquired immunodeficiency syndrome: analysis of 50 patients. Ann Neurol 1983; 14:403–418.
2. Levy RM, Bredensen DE, Rosenblum ML. Neurological manifestations of the acquired immunodeficiency syndrome (AIDS): experience at UCFS and review of the literature. J Neurosurg 1985; 62:475–495.
3. Dalakas MC, Pezeshkpour GH. Neuromuscular diseases associated with human immunodeficiency virus infection. Ann Neurol 1988; 23(Suppl):S38–S48.
4. Dalakas MC, Pezeshkpour GH, Gravell M, Sever JL. Polymyositis associated with AIDS retrovirus. JAMA 1986; 256:2381–2383.

5. Dalakas MC, London WT, Gravell M, Sever JL. Polymyositis in an immunodeficiency disease in monkeys induced by a type D retrovirus. Neurology 1986; 36:569–572.
6. Illa I, Nath A, Dalakas MC. Immunocytochemical and virological characteristics of HIV-associated inflammatory myopathies: similarities with seronegative polymyositis. Ann Neurol 1991; 29:474–481.
7. Stern R, Gold J, DiCarlo EF. Myopathy complicating the acquired immunodeficiency syndrome. Muscle Nerve 1987; 10:318–322.
8. Simpson DM. Myopathy associated with human immunodeficiency virus (HIV) but not with zidovudine. Ann Intern Med 1988; 109:942.
9. Simpson DM, Bender AN. Human immunodeficiency virus–associated myopathy: analysis of 11 patients. Ann Neurol 1988; 24:79–84.
10. Mahe A, Bruet A, Chabin E, Fendler J-P. Acute rhabdomyolysis coincident with primary HIV-1 infection. Lancet 1989; 2:1454–1455.
11. Younger DS, Hays AP, Uncini A, Lange DJ, Lovelace RE, DiMauro S. Recurrent myoglobinuria and HIV seropositivity: incidental or pathogenic association. Muscle Nerve 1989; 12:842–843.
12. Simpson DM, Bender AN, Farraye J, Mendelson SG, Wolfe DE. Human immunodeficiency virus wasting syndrome may represent a treatable myopathy. Neurology 1990; 40:535–538.
13. Watts RA, Hoffbrand BI, Paton DF, Davis JC. Pyomyositis associated with human immunodeficiency virus infection. Br Med J 1987; 294:1524–1525.
14. Widrow CA, Kellie SM, Saltzman BR, Mathur-Wagh U. Pyomyositis in patients with the human immunodeficiency virus: an unusual form of disseminated bacterial infection. Am J Med 1991; 91:129–136.
15. Wrzolek MA, Rao C, Kozlowski PB, Sher J. Muscle and nerve involvement in AIDS patient with disseminated *Mycobacterium avium intracellulare* infection. Muscle Nerve 1989; 12:247–248.
16. Ledford DK, Overman MD, Gonzavlo A, Cali A, Mester SW, Lockey RF. Microsporidiosis myositis in a patient with the acquired immunodeficiency syndrome. Ann Intern Med 1985; 102:628–630.
17. Richman DD, Fischl MA, Grieco MH, et al. The toxicity of azidothymidine (AZT) in the treatment of patients with AIDS and AID-related complex. N Engl J Med 1987; 317:192–197.
18. Bessen LJ, Greene JB, Louie E, Seitzman P, Weinberg H. Severe polymyositis-like syndrome associated with zidovudine therapy of AIDS and ARC. N Engl J Med 1988; 17:708.
19. Gorard DA, Henry K, Guiloff RJ. Necrotising myopathy and zidovudine. Lancet 1988; 1:1050.
20. Helbert M, Fletcher T, Peddle B, Harris JRW, Pinching AJ. Zidovudine-associated myopathy. Lancet 1988; 2:689–690.
21. Fischl M, Gagnon S, Uttamchani R, et al. Myopathy associated with long-term zidovudine therapy. Vth International Conference on AIDS, Montreal, 1989, MBP 326 (abstract).
22. Pinching AJ, Helbert M, Peddle B. Clinical experience with zidovudine in patients with acquired immune deficiency syndrome and immune deficiency-related complex. J Infect 1990; 18(Suppl):33–40.
23. Panegyres PK, Tan N, Kakulas BA, Armstrong JA, Hollingsworth P. Necrotising myopathy and zidovudine. Lancet 1988; 1:1050–1051.
24. Panegyres PK, Papdimitriou JM, Hollingsworth PN, Armstrong JA, Kakulas BA. Vesicular changes in the myopathies of AIDS. Ultrastructural observations and their relationship to zidovudine treatment. J Neurol Neurosurg Psychiatry 1990; 53:649–655.
25. Chalmers AC, Greco CM, Miller RG. Prognosis in AZT myopathy. Neurology 1991; 41:1181–1184.
26. Dalakas MC, Illa I, Pezeshkpour GH, Lautakis JP, Cohen B, Griffin JL. Mitochondrial myopathy caused by long-term zidovudine therapy. N Engl J Med 1990; 322:1098–1105.
27. Mhiri C, Baudimont M, Bonne G, et al. Zidovudine myopathy: a distinctive disorder associated with mitochondrial dysfunction. Ann Neurol 1991; 29:606–614.
28. Pezeshkpour G, Illa I, Dalakas MC. Ultrastructural characteristics and DNA immunocytochemistry in human immunodeficiency virus and zidovudine-associated myopathies. Hum Pathol 1991; 22:1281–1288.
29. Miranda AF, Sacho S, Tanji K, Walker WF, Schon EA, DiMauro S. Mitrochondrial studies in muscle culture: effects of 3′-azido-3′deoxythymidine (AZT). J Neurol Sci 1990; 98(Suppl):110 (abstract).
30. Arnaudo E, Dalakas M, Shanske S, Moraes CT, DiMauro S, Schon EA. Depletion of muscle mitochondrial DNA in AIDS patients with zidovudine-induced myopathy. Lancet 1991; 337:508–510.
31. Lane RJM, McLean KA, Moss J, Woodrow DF. Myopathy in HIV infection: the role of zidovudine and the significance of tubuloreticular inclusions. Neuropathol Appl Neurobiol 1993; 19:406–413.
32. Grau JM, Masanes F, Pedrol E, Casademont J, Fernandez-Sola J, Urbano-Marquez A. Human immunodeficiency virus type 1 infection and myopathy: clinical relevance of zidovudine therapy. Ann Neurol 1993; 34:206–211.
33. Hertzberg NH, Zorn I, Zwart R, Portegies P, Bolhuis PA. Major growth reduction and minor decrease in mitochondrial enzyme activity in cultured human muscle cells after exposure to zidovudine. Muscle Nerve 1992; 15:706–710.
34. Daniels D, McCormack SMG, McLean KA, Davies PTG, Lane RJM. Zidovudine-related myopathy. Lancet 1990; 336:1256.

53
Anesthesia and Primary Muscle Disease

Nicholas P. Hirsch
The National Hospital for Neurology and Neurosurgery, London, England

Scott H. Russell
Queen Elizabeth Hospital, Birmingham, England

INTRODUCTION

Patients with primary muscle disease are only relatively rarely encountered in general anesthetic practice. Consequently, documentation is rather sporadic and consists mainly of case reports. However, the conduct of anesthesia for this group of patients may profoundly and directly influence morbidity and mortality. This review, therefore, attempts to collate the available information concerning anesthesia for patients with myopathic disease. Malignant hyperpyrexia is covered in another chapter.

GENERAL ANESTHETIC CONSIDERATIONS

Respiratory and Bulbar Muscle Weakness

Respiratory failure, often in association with pulmonary aspiration and infection, is a common terminal event in many of the myopathies. Although in most conditions (e.g., the muscular dystrophies) this event occurs late in the disease when respiratory and bulbar muscle weakness reflects the generalized disability, in others (e.g., acid-maltase deficiency) respiratory failure may occur early while the patient is still fully ambulant. The mechanisms contributing to the development of respiratory failure include respiratory muscle weakness [especially affecting the diaphragm (1)], impaired bulbar function and cough (2), obstruction of the upper airway during sleep (3), and restrictive lung disease due to kyphoscoliosis, abdominal distension, and repeated respiratory tract infections (4 – 6). Furthermore, sleep-induced respiratory failure is not uncommon in neuromuscular disease (7).

Preoperative anesthetic assessment of pulmonary and bulbar function must be meticulous. Examination of the patient should include assessment of bulbar function, cough, careful auscultation of lung fields, and observation of chest and abdominal walls during respiration. Paradoxical inward movement of the anterior abdominal wall during early inspiration while the patient is in the supine position suggests diaphragmatic paralysis (8).

Arterial blood gas analysis and respiratory function tests (especially forced vital capacity) form an integral part of the preoperative assessment and provide a useful monitor of the condition. More reliable and sensitive assessment of respiratory muscle reserve may be obtained by measuring maximum static inspiratory and expiratory mouth pressures (9); a simple bedside meter has recently been described (10).

Preoperative preparation of the patient includes chest physiotherapy, cessation of smoking, and treatment of any pulmonary infection. Teaching the patient incentive spirometry may be useful.

Postoperatively, patients with muscle disease may prove difficult to wean from mechanical ventilation and tracheostomy may be necessary. Factors other than respiratory muscle weakness contributing to weaning problems include the respiratory depressant effects of anesthesia and opioid drugs, pain, and the presence of coexisting central or obstructive sleep apnea. Respiratory failure may occur some hours postoperatively; therefore, these patients should be managed in an intensive care setting.

Cardiac Dysfunction

Cardiac abnormalities are more commonly found in the myopathic patient than in the general population. These include conduction defects, cardiac failure, cardiomyopathy, and valve defects. These abnormalities are often asymptomatic as the patient's exercise tolerance is limited by limb muscle weakness. An unexplained persistent tachycardia may be the only indication of underlying disease. Preoperative assessment should include an electrocardiogram (ECG) in all cases and the use of 24-hr monitoring and echocardiography when indicated. Conduction defects may necessitate the insertion of a pacemaker before anesthesia is considered. Cardiac involvement in muscle disease is discussed in depth in another chapter.

Gastrointestinal Dysfunction

Smooth muscle is also affected in some myopathies, and this may result in poor intestinal mobility. Most important, gastric motility and emptying may be severely reduced, and with coexisting poor bulbar function these patients must be considered to be at high risk of aspiration of stomach contents during induction of anesthesia and in the immediate postoperative period. The anesthetic technique must therefore include the use of cricoid pressure, the passage of a nasogastric tube, and extubation of the trachea only when the patient is fully awake.

Response to Neuromuscular Blocking Drugs (11)

The indiscriminate use of neuromuscular blocking drugs in myopathic patients can result in serious complications and occasionally death. Many of these patients show an exaggerated response to the nondepolarizing muscle relaxants and careful perioperative monitoring of neuromuscular function must be performed. The response of these patients to depolarizing relaxants may be both dramatic and fatal. Complications following the administration of suxamethonium include the development of a severe increase in muscle tone in some myotonic patients, cardiac arrest, rhabdomyolysis, and an increased risk of developing malignant hyperpyrexia. It is therefore recommended that suxamethonium be avoided, if possible, in this group of patients.

Increased Susceptibility to Malignant Hyperpyrexia

The relationship between malignant hyperpyrexia (MH) and primary muscle disease is difficult to determine (12), and many features of the former (e.g., muscle rigidity, increases in body temperature, metabolic acidosis, and myoglobinuria) may be seen in the latter with or without exposure to the so-called anesthetic "triggering" agents. However, there seems to be a relatively strong link between central core disease and MH. Other, weaker associations with MH are found in the following conditions (given in decreasing order of susceptibility (13): Duchenne muscular dystrophy, King-Denborough syndrome, Fukuyama muscular dystrophy, Becker muscular dystrophy, periodic paralysis, myotonia congenita, and mitochondrial myopathy.

Regional Anesthesia

Although regional anesthesia is not extensively reported in the context of myopathic disease, local anesthetic techniques should be seriously considered. Field block has been suggested for patients undergoing muscle biopsy (14), and local blocks for dental surgery have been successful (15). Full consideration of cardiac status should be made before spinal or extradural anesthesia is performed.

SPECIFIC MYOPATHIC CONDITIONS

The Muscular Dystrophies

X-Linked Muscular Dystrophies

Respiratory insufficiency generally occurs late in the course of *Duchenne muscular dystrophy* (DMD) in association with obesity, kyphoscoliosis, and bulbar dysfunction (5). Respiratory failure with bronchopneumonia is the most common cause of death. Cardiac abnormalities are found in 70% of patients although these are of significance in only a small proportion (16). Characteristic ECG changes are seen in 70% of cases and consist of prominent R waves in leads V1 and V2 and deep, narrow Q waves over the left precordial leads (17). Atrial conduction defects (typically resulting in atrial tachycardias) are seen in 65% of patients, and echocardiography demonstrates mitral valve prolapse in 25% (18). Left ventricular hypokinesia is characteristically limited to the lateral and posterobasal walls and reflects progressive replacement of cardiac muscle with fibrous tissue (19). Smooth muscle involvement reduces gastrointestinal motility, and patients are at risk from vomiting and aspiration of stomach contents. Acute postoperative gastric dilatation has been reported (20). Although not invariable, most patients have intellectual impairment with a mean intelligence quotient of 85 (21). Patients with DMD often present for orthopedic procedures, most commonly lower limb operations (22).

Before the patient is exposed to the hazards of general anesthesia, serious consideration must be given to using regional techniques. However, these are not without difficulty. The combination of obesity and spinal deformity makes lumbar spinal and extradural anesthesia technically challenging. The sacral canal is less affected, and certainly in younger children caudal extradural blockade may be the technique of choice. If general anesthesia is unavoidable, due consideration must be given to the pulmonary and cardiac status of the patient.

Gastrointestinal dysfunction predisposes to pulmonary aspiration of stomach contents, and measures should be taken to reduce gastric acid pH and to increase the rate of gastric emptying. Cricoid pressure should be performed during induction, and following tracheal intubation, a nasogastric tube should be passed. This should be left in place in the postoperative period. Although two large early series (23,24) reported few adverse effects following the administration of a wide variety of anesthetic agents, others have commented on a number of minor or more serious sequelae. The latter include sudden cardiac arrest on induction of anesthesia and in the early postoperative period (25,26); indeed, this may be the first manifestation of the condition.

When asystole or ventricular fibrillation has occurred during induction, this followed the use of suxamethonium, although there have been reports of cardiac arrest on induction with halothane as the sole anesthetic agent (27). Postoperative asystole occurred in two patients given only halothane (28) and in one patient given isoflurane (29). The two most common mechanisms proposed to explain cardiac arrest in these patients are the occurrence of malignant hyperthermia and hyperkalemia secondary to rhabdomyolysis. Difficulties occur in distinguishing these two conditions as they share some clinical features such as metabolic acidosis, hyperkalemia, and myoglobinuria. Furthermore, a mild increase in temperature is common in patients with DMD undergoing anesthesia (14). Although it appears that DMD and MH may coexist (30,31), there is little evidence that they are transmitted together. However, it is prudent to avoid known triggering agents (especially suxamethonium) when anesthetizing these patients.

Although some authors report a normal response to nondepolarizing neuromuscular blocking agents (11,23,24,32), others report increased sensitivity (33). Neuromuscular function should therefore be monitored, and doses of relaxants titrated accordingly. Most series emphasize the frequency with which late postoperative respiratory failure may occur, and these patients should therefore be nursed in an intensive care setting.

Other X-Linked Dystrophies

Patients with *Becker muscular dystrophy* present with proximal muscle weakness in a similar distribution to DMD but the disease typically starts at the age of about 10 years and progresses more slowly.

Pseudohypertrophy is typical and ECG changes are similar but less frequent to those found in DMD (34). Cardiac manfestations are less common although cardiac arrest during induction of anesthesia has been reported (35). Anesthetic considerations are otherwise the same as for DMD.

Emery-Dreifuss muscular dystrophy is a nonhypertrophic muscular dystrophy that usually presents at the age of 4–5 with flexion contractures of the elbows, Achilles tendons, and posterior neck muscles. Initial muscle weakness occurs in the second decade and typically affects the triceps and biceps (36). Respiratory failure may occur early in the disease and is usually due to a severe and progressive restrictive deficit due to scoliosis, paraspinal contractures, and, to a lesser extent, respiratory muscle weakness. Cardiac abnormalities usually become apparent in the third decade of life and include cardiomyopathy (37) and conduction defects involving the sinoatrial and atrioventricular nodes (38). A 24-hr ambulatory ECG is essential and may indicate that preoperative pacing is necessary.

Tracheal intubation may be difficult due to restriction of neck flexion and may be further exacerbated by cervical spine abnormalities (39).

Other Muscular Dystrophies

Limb girdle muscular dystrophy and the limb girdle syndrome comprise a heterogeneous group of disorders (discussed extensively in another chapter). Anesthetic significance is related to the cardiac and pulmonary involvement that may occur in some of these disorders. There may be early cardiac failure and conduction defects resulting in sinus tachycardia and right bundle branch block. Global respiratory muscle weakness may occur early in the disease resulting in alveolar hypoventilation and respiratory failure.

Facioscapulohumeral muscular dystrophy is an autosomal dominant dystrophy characterized by variable onset of weakness of the face and scapuloperoneal muscles eventually progressing to involve the truncal and pelvic musculature. Although global respiratory muscle weakness is uncommon, diaphragmatic weakness has been described and respiratory failure may occur in the aggressive childhood form of the disease. Complete atrial inactivity has been reported and may result in severe bradyarrhythmias (40). Ventricular pacing may be necessary,

Ocular myopathies are a group of myopathies that present with a slowly progressive external ophthalmoplegia and ptosis. Facial, pharyngeal, and limb muscles are involved in some variants of the syndrome. Patients with these conditions are typically extremely sensitive to nondepolarizing muscle relaxants; these should be used with caution (41).

The Myotonic Syndromes

This group of conditions consists of a number of congenital disorders that exhibit the phenomenon of myotonia, and includes myotonic dystrophy (MyD), myotonia congenita, and paramyotonia congenita. Myotonia may also be a feature of hyperkalemic periodic paralysis. The clinical and pathogenic features of these disorders are discussed in another chapter.

Myotonic Dystrophy

The truly systemic nature of myotonic dystrophy becomes apparent when considering other systems involved in the condition. Cardiac involvement is usual and may precede other symptoms. Conduction defects include a prolonged P-R interval and interventricular conduction defects. Atrial flutter and fibrillation may occur. Complete heart block resulting in Stokes-Adams attacks has been reported (42). An ambulatory 24-hr ECG is often indicated before surgery and perioperative cardiac pacing may be required. Cardiomyopathy is common in the later stages of MyD and postmortem evidence of cardiac muscle involvement has been found in 50% of patients (43).

Respiratory involvement in MyD is multifactorial (44). Respiratory muscle weakness affects both the diaphragm and intercostal muscles, resulting in poor cough, restrictive lung defects, and alveolar hypoventilation. The occurrence of central sleep apnea suggests an additional defect of central control of respiration (45). Bulbar dysfunction contributes to the development of recurrent respiratory tract infection and obstructive sleep apnea (46). Videoradiography demonstrates swallowing defects in a high proportion of patients (47).

Smooth muscle involvement causes poor intestinal motility, the development of megacolon (48), and urinary bladder dysfunction. Endocrine features include hypothyroidism and abnormalities of insulin and glucose metabolism, which may result in diabetes mellitus (49).

Considering the systemic manifestations of the disease, it is not surprising that the patient with MyD poses many problems for the anesthetist. Induction and sedative agents produce profound apnea due to a combination of poor respiratory reserve and disorders of central respiratory drive (50,51). For these reasons it is suggested that these agents be carefully titrated against response and the patient be mechanically ventilated during surgery to avoid alveolar hypoventilation. Although inhalational agents have been widely used in patients with MyD, it would seem prudent to avoid high concentrations, which might cause a further reduction of an already reduced myocardial contractility and result in postoperative shivering.

The use of neuromuscular blocking agents during surgery poses a difficult problem. These agents do not necessarily produce the required relaxation and the muscle may still respond to surgical manipulation and electrocautery (52). However, when the nondepolarizing relaxants are effective, most reports suggest a normal duration of action (53), although increased sensitivity has been reported (11,54).

Suxamethonium produces sustained contraction in up to 50% of patients with MyD (55). This may make tracheal intubation and ventilation difficult. Furthermore, this agent may cause hyperkalemia, which further aggravates the myotonia. It should therefore be avoided.

Anticholinesterase drugs used for reversal of nondepolarizing relaxants may also cause myotonia (56), and it is recommended that the shorter-acting neuromuscular blocking drugs that do not require reversal (e.g., atracurium) be used.

During surgery temperature must be monitored and maintained as postoperative shivering may initiate myotonia (52).

The postoperative complications of MyD are related to poor respiratory reserve, bulbar dysfunction, and an abnormal sensitivity to the respiratory depressant effects of sedative and opioid analgesic agents (57).

Anesthesia for obstetric patients with MyD poses particular problems for the anesthetist, but successful general and regional anesthesia has been reported (58,59). Fifty percent of babies born to mothers with myotonic dystrophy will be affected and may require ventilatory support (60).

Myotonia Congenita

Anesthetic considerations are similar to those for patients with MyD. Hypothermia and shivering should be avoided. There has been one report of a possible association between MH and myotonia congenita and death has been reported (61).

Postoperative respiratory complications are rare as respiratory muscle strength is not affected.

Regional anesthesia may be useful.

Paramyotonia Congenita

This autosomal dominant condition, first described in 1886, is characterized by cold-induced myotonia followed by paresis. The muscles of the face, hands, and tongue are most frequently involved. It has been demonstrated that there is a temperature-dependent abnormality in the behavior of the voltage-dependent sodium channel (62). The condition thus shares features with hyperkalemic periodic paralysis, and similar anesthetic considerations apply to both conditions (vide infra). Reports of the coexistence of paramyotonia congenita and malignant hyperpyrexia have been published (61,62), and known triggering agents should be avoided.

Metabolic Myopathies

With the exception of malignant hyperpyrexia, there is little in the literature concerning these rare conditions. However, for completeness, the available information is given here.

Glycogenoses

The glycogen storage diseases result from selective enzyme deficiencies in the glycolytic pathway. Cori's classification is most commonly used and recognizes at least 10 distinct enzyme deficiencies. While many are associated with hepatic dysfunction, some (e.g, McArdle's disease) exclusively affect muscle. Patients with acid maltase deficiency (Cori's type II) often present with early respiratory failure due to diaphragmatic weakness (64).

Although anesthetic experience of the glycogenoses is extremely limited, certain generalizations can be made (65). Suxamethonium should not be used, as it may precipitate myoglobinuria. Similarly, shivering in the postoperative period should be avoided. The use of tourniquets is contraindicated.

Mitochondrial Myopathies

No firm recommendations can be made concerning anesthesia in these patients as the clinical picture varies so much. There has been a report of the uneventful use of a nitrous oxide/isoflurane anesthetic in a 6-week-old child with fumarase deficiency (66). No relaxant was used. Atracurium has been used successfully in a patient with mitochondrial myopathy (67). An earlier report (68) gave details of a possible case of MH in a patient with mitochondrial myopathy, and it therefore seems prudent to avoid known triggering agents and measure temperature peroperatively. Care must be taken to fully evaluate hepatic and renal function and clotting status. Cardiac involvement may result in cardiomyopathy.

Periodic Paralysis

Reports of anesthesia for patients with hyperkalemic periodic paralysis and paramyotonia congenita are rare. An early series (69) reported prolonged postoperative paralysis following general anesthesia administered to two of three patients; the third patient received an uneventful spinal anesthetic. Unfortunately, details concerning potassium levels were not discussed. Ashwood et al. (70) discussed the successful management of four affected members of one family. Their experience led to a number of recommendations, including a preoperative period of potassium depletion using loop or thiazide diuretics. A low-potassium diet may also be useful. Starvation should be avoided, and therefore a dextrose infusion should be started as fasting begins. Drugs that cause a rise in potassium (e.g., suxamethonium, ketamine, and potassium-containing fluids) must be avoided. In addition, the use of suxamethonium may result in generalized myotonia (71). Temperature must be monitored and maintained peri- and postoperatively. The authors did not report any sensitivity to nondepolarizing muscle relaxants. The ECG usually gives the first indication of the development of hyperkalemia.

Anesthetic recommendations with regard to the more common hypokalemic periodic paralysis include avoiding large meals the night before surgery and oral potassium supplements if there is any evidence of preexisting hypokalemia (72). Fluids containing large amounts of carbohydrate or sodium should not be used. No anesthetic data exist for the rarely encountered clinical entity of normokalemic periodic paralysis.

Benign Myopathies of Childhood

Central Core Disease

The anesthetic significance of this disease lies in the association with MH. Denborough et al. (73) gave details of a family with MH, one member of which was diagnosed as having central core disease. Furthermore, most patients with this condition test positive in the halothane contracture test (12). Agents known to trigger MH should therefore be avoided.

Nemaline Myopathy

Although nemaline myopathy is often nonprogressive, some neonates with the disease present with rapidly progressive, sometimes fatal respiratory failure (74). Anesthetic attention should be directed toward the possibility of difficult tracheal intubation due to facial dysmorphism, the existence of pulmonary and bulbar insufficiency, and the occurrence of exaggerated vagal reflexes (75). Although it has been suggested that muscle relaxants should be avoided (76), other investigators have suggested a normal response (77). There are no reports of any association with MH although there has been a report of central core disease and nemaline rod myopathy occurring in the same family (78). It therefore seems prudent to avoid the use of known triggering agents of MH.

SUMMARY OF RECOMMENDATIONS

Patients with primary muscle disease, although rarely encountered in general anesthetic practice, require careful preoperative assessment and preparation (Tables 1 and 2), and techniques of anesthesia should reflect the overall condition of the patients and the possibility of susceptibility to developing MH (Table 3).

Table 1 Preoperative Assessment

Bulbar function: gag, swallowing
Cough
Thoracic and abdominal wall movement
Blood gases
Respiratory function tests
Inspiratory and expiratory mouth pressures
Cardiac status

Table 2 Preoperative Preparation

Chest physiotherapy
Cessation of smoking
Treat preexisting chest infection
Teach incentive spirometry
Optimize cardiac function

Table 3 Diseases Associated with Malignant Hyperpyrexia

Central core disease
Duchenne muscular dystrophy
King-Denborough syndrome
Myoadenylate deaminase deficiency
Schwartz-Jampel syndrome
Fukayama muscular dystrophy
Periodic paralysis
Myotonia congenita
Mitochondrial myopathy

Source: Ref. 13.

REFERENCES

1. Laroche CM, Moxham J, Green M. Respiratory muscle weakness and fatigue. Q J Med 1992; 71:373–397.
2. Howard RS, Wiles CM, Hirsch NP, Loh L, Spencer GT, Newsom Davis J. Respiratory involvement in multiple sclerosis. Brain 1992; 115:479–494.
3. Phillipson EA, Bowes G. Control of breathing during sleep. In: Fishman AP, ed. Handbook of physiology: The Respiratory System. Bethesda: American Physiological Society, 1986:649–687.
4. Kreitzer SM, Saunders NA, Tyler HR, Ingram RH. Respiratory muscle function in amyotrophic lateral sclerosis. Am Rev Respir Dis 1978; 117:437–447.
5. Smith PEM, Calverley PMA, Edwards RHT, Evans GA, Campbell EJM. Practical problems in the care of patients with muscular dystrophy. N Engl J Med 1987; 316:1197–1205.
6. Estenne M, Heilporn A, Delhez L, Yernault YC, De Troyer A. Chest wall stiffness in patients with respiratory muscle weakness. Am Rev Respir Dis 1983; 128:1002–1007.
7. Gilmartin JJ, Cooper BG, Griffiths J, et al. Breathing during sleep with myotonic dystrophy and non myotonic respiratory muscle weakness. Q J Med 1991; 78:21–31.
8. Goldman MD. Interpretation of respiratory thoracoabdominal movements. Clin Sci 1982; 62:7–11.
9. Black LF, Hyatt RE. Maximal static respiratory pressures in generalised neuromuscular disease. Am Rev Respir Dis. 1971; 103:641–650.
10. Smith PEM, Royall RA. A portable mouth pressure meter. Anaesthesia 1992; 47:144–145.
11. Azar I. The response of patients with neuromuscular disorders to muscle relaxants: a review. Anaesthesiology 1984; 61:173–187.
12. Ellis FR. Predicting malignant hyperthermia. Br J Anaesth 1990; 64:411–412.
13. Brownell AKW. Malignant hyperpyrexia; relationship to other diseases. Br J Anaesth 1988; 60:303–308.
14. Larsen UT, Juhl B, Hein-Sorensen O, Olivarius B.deF. Complications during anaesthesia in patients with Duchenne's muscular dystrophy. Can J Anaesth 1989; 36:418–422.
15. Reich DF, Neff J. Oral-surgical management of odontogenic keratocyst in a patient with Duchenne muscular dystrophy. Pediatr Dent 1981; 3:343.
16. Griggs RC, Reeves W, Moxley RT III. The heart in Duchenne dystrophy. In: Pathogenesis of Human Muscular Dystrophies. Proceedings of the Fifth International Conference of the Muscular Dystrophy Association, Colorado, 1976. Amsterdam: Excerpta Medica, 1977:661.
17. Jellett LB, Kennedy MC, Goldblatt E. Duchenne pseudohypertrophic muscular dystrophy: a clinical and electrocardiographic study of patients and female carriers. Aust NZ Med J 1974; 4:41.
18. Berlit P, Stegaru-Hellring B. The heart in muscular dystrophy: an electrocardiographic and ultrasound study of 20 patients. Eur Arch Psychiatry Clin Neurosci 1991; 241:177–180.
19. Rubler S, Perloff K, Roberts WD. Clinical pathological conference: Duchenne muscular dystrophy. Am Heart J 1977; 94:776–784.
20. Crowe GG. Acute dilatation of the stomach as a complication of muscular dystrophy. Br Med J 1961; 1:1371.
21. Marsh GG, Munsat TL. Evidence for early impairment of verbal intelligence in Duchenne muscuar dystrophy. Arch Dis Child 1974; 49:118–122.
22. Hody JL. Regional anaesthesia in myopathic children. Acta Anaesthesiol Belg 1988; 39:209–213.
23. Cobham IG, Davis HS. Anaesthesia for muscular dystrophy patients. Anaesth Analg 1964; 43:22–29.
24. Richards WC. Anaesthesia and serum creatine phosphokinase levels in patients with Duchennes pseudohypertrophic muscular dystrophy. Anaesth Intensive Care. 1972; 1:150–153.
25. Smith CL. Anaesthesia and progressive muscular dystrophy. Br J Anaesth 1985; 57:1113–1118.
26. Genever EE. Suxamethonium induced cardiac arrest in unsuspected pseudohypertrophic muscular dystrophy. Br J Anaesth 1971; 43:984–986.
27. Sethna NF, Rockoff MA. Cardiac arrest following inhalational induction of anaesthesia in a child with Duchennes muscular dystrophy. Can Anaesth Soc J 1986; 33:799–802.
28. Boba A. Fatal postanesthetic complications in two muscular dystrophic patients. J Pediatr Surg 1970; 5:71–75.
29. Chalkiadis GA, Brach KG. Cardiac arrest after isoflurane anaesthesia in a boy with Duchenne muscular dystrophy. Anesthesia 1990; 45:22–25.
30. Brownell AKW, Fowlow SB, Paasuke RJ. Coexistence of malignant hyperpyrexia and Duchenne muscular dystrophy in the same pedigree. Neurology 1985; 35 (Suppl 1):195.
31. Brownell AKW, Paasuke RJ, Elask A. Malignant hyperpyrexia and Duchenne muscular dystrophy. Anaesthesiology 1983; 58:180.
32. Buzello W, Huttarch H. Muscle relaxation in patients with Duchenne muscular dystrophy. Br J Anaesth 1988; 60:228–231.
33. Brown JC, Charlton JE, White DJK. A regional technique for the study of sensitivity to curare in human muscle. J Neurol Neurosurg Psychiatry 1975; 38:18–26.
34. Swain M, Schwartz MS. Neuromuscular Diseases. Berlin: Springer-Verlag, 1981:170.

35. Marks WA, Bodensteiner JB, Reitz RD. Cardiac arrest during anesthetic induction in a child with Becker type muscular dystrophy. J Child Neurol 1987; 2:160–161.
36. Emery AEH, Dreifuss FE. Unusual type of benign X-linked muscular dystrophy. J Neurol Neurosurg Psychiatry 1966; 29:338–342.
37. Chakrabati A, Pearce JMS. Scapuloperoneal syndrome with cardiomyopathy: report of a family with autosomal inheritance and unusual features. J Neurol Neurosurg Psychiatry 1981; 44:1146–1152.
38. Waters DD, Nutter DO, Hopkins LC, Dorney ER. Cardiac features of an unusual X-linked humeroperoneal neuromuscular disease. N Engl J Med 1975; 293:1017–1022.
39. Morrison P, Jago RH. Emery-Dreifuss muscular dystrophy. Anaesthesia 1991; 46:33–35.
40. Baldwin BJ, Talley RC, Johnson C, Nutter DO. Permanent paralysis of the atrium in a patient with facioscapulohumeral muscular dystrophy. Am J Cardiol 1973; 31:649–653.
41. Jacob JC, Varkey GP. Curare sensitivity in ocular myopathy. Can J Anaesth 1966; 13:449–452.
42. Hawley RJ, Milner MR, Gottdeiner JS, Cohen A. Myotonic heart disease: a clinical follow up. Neurology 1991; 41:259–262.
43. Motta J, Guilleminault C, Billingham M, Barry W, Mason J. Cardiac abnormalities in myotonic dystrophy; electrophysiological and histological studies. Am J Med 1979; 67:467–473.
44. Begin R, Bureau MA, Lupien L, Bernier JP, Lemieux B. Pathogenesis of respiratory insufficiency in myotonic dystrophy. Am Rev Respir Dis. 1982; 125:312–318.
45. Gillam PMS, Heal DJD, Kaufman L, Lucas BGB. Respiration in dystrophia myotonica. Thorax 1964; 19:112–120.
46. Broughton R, Stuss D, Kates M, Roberts J, Dunham W. Neuropsychological defects in sleep and myotonic dystrophy. Can J Neurol Sci 1990; 17:410–415.
47. Josephovich RF. Myotonic dystrophy. Neurol Clin 1988; 6(3):455–472.
48. De Koninck X, Fiasse R, Jonad P, Demelenne J, Pringot T, Dive C. Digestive system manifestations in Steinerts disease: Analysis of 19 cases of which 10 with digestive system manifestations. Acta Gastroenterol Belg 1990; 53:3–15.
49. Piccardo MG, Pacini G, Rosa M, Vichi R. Insulin resistance in myotonic dystrophy. Enzyme 1991; 45:14–22.
50. Hansotia P, Frens D. Hypersomnia associated with alveolar hypoventilation in myotonic dystrophy. Neurology 1981; 31:1336–1337.
51. Cumminskey J, Lynn-Davies P, Guilleminault C. Sleep study and respiratory function in myotonic dystrophy. In: Guilleminault C, Dement WC, eds Sleep Apnoea Syndromes. New York: Alan R Liss, 1978: 295.
52. Cope DK, Miller JN. Local and spinal anaesthesia for caesarian section in a patient with myotonic dystrophy. Anaesth Analg 1986; 65:687–690.
53. Mitchell MM, Ali HH, Savarese JJ. Myotonia and neuromuscular blocking agents. Anaesthesiology 1978; 49:44–48.
54. Mudge BJ, Taylor PB, Vanderspeck AFL. Perioperative hazards in myotonic dystrophy. Anaesthesia 1980; 35:492–495.
55. Orndahl G. Starnberg K. Myotonic human musculature; stimulation with depolarising agents: mechanical registration of succinylcholine, succinylmonocholine and decamethonium. Acta Med Scand 1962; 172:S3-29.
56. Buzello W, Krieg N, Schlickwei A. Hazards of neostigmine in patients with neuromuscular disorders: report of two cases. Br J Anaesth 1982; 54:529–534.
57. Kaufman L. Anaesthesia in dystrophia myotonica: a review of the hazards of anaesthesia. Proc Roy Soc Med 1960; 53:183–188.
58. Blumgart CH, Hughes DG, Redfern N. Obstetric anaesthesia in dystrophia myotonica. Anaesthesia 1990; 45:26–29.
59. Paterson RA, Tousignant M, Shere DS. Caesarian section for twins in a patient with myotonic dystrophy. Can Anaesth Soc J 1985; 32:418–421.
60. Sarnat HB, O Connor T, Byrne PA. Clinical effects of myotonic dystrophy on pregnancy and the neonate. Arch Neurol 1976; 33:459–465.
61. Lehmann-Horn F, Iaizzo PA. Are myotonias and periodic paralyses associated with malignant hyperthermia? Br J Anaesth 1990; 65:692–697.
62. Streib EW. Paramyotonia congenita. Semin Neurol 1991; 11(3):249–25.
63. Heimann-Patterson T, Rosenberg H, Fletcher JE, Tahmoush AJ. Halothane-Caffeine contracture testing in neuromuscular disease. Muscle Nerve. 1988; 11:453–457.
64. Keunan RMW, Lambregts PCLA, Op de Coul AAW, Joosten EMG. Respiratory failure as an initial manifestation of acid maltase deficiency. J Neurol Neurosurg Psychiatry 1984; 47:549–552.
65. Smith CL, Bush GH. Anaesthesia and progressive muscular dystrophy. Br J Anaesth 1985; 57:1113–1118.
66. Burns AM, Shelly MP, Anaesthesia for patients with mitochondrial myopathy. Anaesthesia 1989;44:975–977.
67. Kelly A, O Connor M. Mitochondrial myopathy and anaesthesia. Anaesthesia 1990; 45:596.
68. Ohtani Y, Miike T, Ishitsu T, Matsuda I, Tamari H. A case of malignant hyperthermia with mitochondrial dysfunction. Brain Dev 1985; 7:245.

69. Siler JN, Discavage DW. Anaesthetic management of hypokalaemic periodic paralysis. Anaesthesiology 1975; 43:489–490.
70. Ashwood EM, Russell WJ, Burrow DD. Hyperkalaemic periodic paralysis and anaesthesia. Anaesthesia 1992; 47:579–584.
71. Ellis FR. Inherited muscle disease. Br J Anaesth 1980; 52:153–164.
72. Miller JD, Lee C. In: Katz, Bemenof, Kadis, eds. Anaesthesia and Uncommon Diseases. Saunders Philadelphia: WB Saunders, 1990:624.
73. Denborough MA, Dennett X, Anderson RM. Central core disease and malignant hyperpyrexia. Br Med J 1973; 1:272–273.
74. Schmalbruch H, Kamieniecka Z, Arroe M. Early fatal nemaline myopathy; case report and review. Dev Med Child Neurol 1987; 29:800–804.
75. Asai T, Fujise K, Uchida M. Anaesthesia for cardiac surgery in children with nemaline myopathy. Anaesthesia 1992; 47:405–408.
76. Cunliffe M, Burrows FA. Anaesthetic implications of nemaline rod myopathy. Can Anaesth Soc J 1985; 32:543–547.
77. Heard SO, Kaplan RF. Neuromuscular blockade in a patient with nemaline myopathy. Anaesthesiology 1983; 59:588–590.
78. Afifi AK, Smith JW, Zelleweger H. Congenital nonprogressive myopathy: central core disease and nemaline myopathy in one family. Neurology 1965; 15:371–381.

54

Management of Respiratory Problems in Neuromuscular Disease

Marlene Normand
University of Nebraska Medical Center, Omaha, Nebraska

John R. Bach
Kessler Institute, West Orange, New Jersey

Ann Ford Fricke, Henry D. Owens, and Richard T. Moxley III
University of Rochester Medical Center, Rochester, New York

INTRODUCTION

The purpose of this chapter is to review the management of respiratory complications that commonly occur as part of the natural history of the progressive neuromuscular diseases. We have included a brief discussion of diagnosis and treatment of both acute and chronic respiratory failure. Respiratory complications can be life-threatening and it is critical to recognize them early and treat them aggressively.

A variety of neuromuscular diseases can present with acute respiratory failure. Commonly, acute symptoms result from provocation by a concomitant process, such as aspiration pneumonia or an intercurrent respiratory tract infection. Prior to the acute deterioration, respiratory muscle weakness often develops insidiously. For this reason patients and physicians may not have discussed the options for preventing or treating acute or chronic respiratory failure (1). Unfortunately, on some occasions decisions occur in a crisis setting without adequate opportunity for the patient, the family, and the doctors to review the different options. This circumstance needs to be avoided. A discussion of the different approaches for treatment should take place on several occasions early in the clinical course, before a crisis provokes a hurried decision.

RESPIRATORY PATHOPHYSIOLOGY IN NEUROMUSCULAR DISEASE

Patients with progressive neuromuscular diseases typically develop restrictive lung disease. Vital capacity (VC) may become markedly reduced and can show a decline to 40% of predicted normal, while the patient is free of symptoms and is still able to breathe without assistance. Patients have reduced total lung capacities and functional residual capacities. However, their residual volumes are normal, as is the ratio of the forced expiratory volume in 1 sec to the total VC. There is also preservation of diffusion capacity, in contrast to patients with infiltrative lung disease.

Microatelectasis and a decrease in pulmonary compliance occur as VC decreases. Microatelectasis can begin within 1 hr of hypoventilation. Chronic ventilatory insufficiency leads to chronic microatelectasis, permanent loss of lung and chest wall elasticity, and decreased pulmonary compliance. Chest wall compliance often declines to about 70% of normal due to stiffening of the rib cage tendons and ligaments and to ankylosis of the costosternal and thoracovertebral joints. There may be an accompanying increase in stiffness of the intercostal muscles. With decreased pulmonary compliance there is an increase in physiological dead space, with blood perfusing unventilated portions of the lungs. This leads to ventilation perfusion mismatching, which, if severe enough, causes hypoxia. In patients with myotonic muscular dystrophy, myotonia of the intercostal muscles and diaphragm may contribute to the decrease in compliance and cause diminished respiratory function (2). Scoliosis, aspiration, and impaired clearing of secretions are other common problems in patients with neuromuscular diseases and cause a further increase in the work of breathing (1,3–7). These different problems exacerbate respiratory muscle dysfunction and can impair the normal neuronal regulation of respiratory muscles.

Respiratory muscle strength is variably affected in the different neuromuscular disorders. Maximal inspiratory pressure is often decreased both by loss of strength of inspiratory muscles and by instability of the chest wall. Rib cage–abdominal paradox occurs, characterized by an abnormal inward movement of the rib cage during inspiration, when the diaphragm moves caudally.

Expiratory muscle strength is probably more important to the patient. Expiratory power is necessary to cough effectively and clear the airway of secretions. Often the earliest detectable sign of respiratory weakness in patients with neuromuscular diseases is a reduction in static pressures, and maximal expiratory pressure is more affected than maximal inspiratory pressure (1,4,5). Selective appearance of greater difficulty with expiratory strength may result, in part, from the fact that during normal breathing, inspiratory muscles are trained with every breath whereas expiration is a passive event (8).

Many patients have greater impairment of respiration in the supine position, compared to sitting. Forced VC measurements should be evaluated in both positions (1). This is particularly important if the underlying disease damages the diaphragm more than the other inspiratory muscles. A lower VC when supine compared to values obtained while sitting is due to the contents of the abdominal cavity displacing the weakened diaphragm cephalad, causing a further reduction in the functional residual capacity. Obesity, acute gastric dilatation, intestinal distention due to fecal impaction, and ileus are common and treatable.

In Duchenne muscular dystrophy VC peaks between ages 10 to 14 years and averages about 1800 ml (1,7). It then declines at about 250 ml per year with the rate of loss decreasing when VC reaches 400 ml. For these patients there is very little difference between the VC in the supine and sitting positions, indicating that diaphragm function is relatively preserved. Patients with certain other myopathies or with anterior horn cell disorders often exhibit a considerable difference in supine and sitting VC, with the supine measurement being significantly lower.

A reduced respiratory response to both hypercapnia and hypoxia can appear in some neuromuscular disorders following treatment with calcium channel blockers, aminoglycosides, steroids, and benzodiazepines. This reduced response to elevations in P_{CO_2} and to low P_{O_2} may significantly worsen existing ventilatory problems. An abnormal response is especially a concern in myasthenia gravis and other neuromuscular transmission disorders. Beta-blockers not only can worsen symptoms in patients with myasthenia gravis by decreasing the efficiency of impulse transmission but can also increase the work of breathing; beta-blockers cause an increase in airway resistance, which in turn increases the ventilatory effort. This effect can worsen ventilation in almost any patient who has respiratory muscle weakness. Malnutrition, acidosis, electrolyte disturbance, infection, fatigue, and muscle disuse can also compromise ventilatory function (1,7). Oxygen therapy can depress ventilatory drive, which can lead to respiratory arrest (7,9), and it should be used only when hypoxia is due to another disease process, such as congestive heart failure or pneumonia. Even in these situations oxygen therapy should be carefully evaluated, with close monitoring of the patient for signs of hypoventilation.

Deformity of the thoracic cage, such as scoliosis, can itself cause or exacerbate respiratory compromise and lead to cardiopulmonary failure (1,6,7). Scoliosis reduces total lung capacity, VC, functional residual capacity, chest wall compliance, and the response to hypercapnia (1,6,7). The resulting chronic alveolar hypoventilation can lead to hypoxic vasoconstriction of pulmonary blood vessels and cor pulmonale (1,7). Surgical correction of scoliosis can help to ameliorate the problem. When surgery is performed early, a full correction can be achieved (1,6). Prevention of major thoracic deformity permits the patient to consider using a variety of noninvasive forms of assisted ventilation that might otherwise not be effective (10).

SLEEP IN NEUROMUSCULAR DISEASE PATIENTS

Sleep may unmask respiratory insufficiency in at-risk patients with certain neuromuscular diseases. Arterial blood gas disturbances may develop only during sleep (1,11,12). Patients may initially display short periods of hypercapnia and hypoxemia only during REM sleep, when chemotaxic drive is less sensitive to changes in blood gases (12,13). Obstructive sleep apneas may further contribute to ventilatory insufficiency. Such apnea has been described in many patients with neuromuscular disease (14–21) and is believed to be due to hypopharyngeal collapse. The incidence of apnea increases in the presence of weak oropharyngeal musculature and with the use of negative-pressure body ventilators (19,20).

The severity of hypoxemia or hypercapnia during sleep may play a major role in the development of chronic respiratory insufficiency. Patients with P_{CO_2} greater than 50 torr while awake usually have oxygen saturations of less than 85% during the night (4,11,12). Hypercapnia can itself cause a decrease in muscle power, worsen hypoxia, and lead to potentially fatal cardiac arrhythmias and heart failure. For patients with chronic hypoventilation, oxygen therapy can prolong hypopneas and apneas and may depress central nervous system–mediated reflex muscular activity, vital to the successful nocturnal use of noninvasive inspiratory muscle aids (1,9,11,22,23).

CLINICAL EVALUATION

Successful clinical management of respiratory insufficiency and other pulmonary problems in patients with progressive neuromuscular diseases depends heavily on anticipating the problems and detecting them early. Once the problems are identified, the clinician and the care team need to discuss the various management options with the patient, the family, and others involved in the ongoing care. Tables 1 and 2 outline approaches to consider in the management of patients at risk for respiratory failure and for those who have chronic ventilatory failure. Table 3 lists selected neuromuscular diseases that can cause respiratory insufficiency and indicates those diseases in which congestive heart failure can commonly accompany or precede respiratory failure.

Patients should be questioned regularly about symptoms suggesting ventilatory insufficiency. Symptoms may be difficult to recognize early in the clinical course and may be present only during intercurrent respiratory infections. They can include dyspnea, anxiety, sleep disturbances, drooling due to fatigue, nausea, vomiting, headaches, hypersomnolence, poor concentration, decreased memory,

Table 1 Evaluation of a Patient with Neuromuscular Disease at Risk for Respiratory Failure

If VC $\downarrow$ below 40% of predicted normal		
Obtain arterial blood gases		
If normal		*If abnormal*
Perfom oximetry and capnography during an overnight sleep		Use IPPV during the day and continuous ventilation during the night
If abnormal	*If normal*	Obtain oximetry at follow-up visits to assess efficacy of ventilation technique
Give overnight trials of BIPAP or IPPV	Fequent follow-ups	
	Measure VC and repeat overnight oximetry and capnography	

nightmares, and changes in body weight (1,7,12,24–34). Chronically hypercapnic patients often have resetting of chemotaxic ventilatory drive and are not tachypneic.

During each follow-up examination patients should be observed for tachypnea; shallow, irregular, asymmetrical, or paradoxical breathing; use of accessory muscles of respiration; signs of perioral or generalized cyanosis; and any evidence of ineffective swallowing with aspiration. The patient should be able to count to more than 15 with one breath. Other signs of respiratory insufficiency may be nasal flaring, hypertension, drooling, dysphagia, regurgitation of fluids through the nose, nasality of speech, signs of congestive heart failure, changes in mental status, or edema. The measurement of VC is useful at each physician visit or at least annually. As the VC approaches 1000 ml or declines to 40% of predicted normal, nocturnal oxygen saturation and noninvasive P_{CO_2} monitoring should be undertaken (see Tables 1 and 2).

Another useful measurement of ventilatory capability is the maximum insufflation capacity (MIC), which is the maximum depth of insufflation that an individual can hold. This volume of air can be delivered to the patient as one deep insufflation or the patient can air-stack 2–4 tidal volumes of air delivered by a ventilator to achieve the maximum depth. The MIC is a function of both pulmonary compliance and the competence of oropharyngeal muscles. The greater the MIC, the greater the ability to generate effective expiratory flows during both spontaneous and assisted coughing (35–37).

While peak cough expiratory flow (PCEF) is another important measure of ventilatory capacity, PCEF is often not measured in assessing the ventilatory status of patients with neuromuscular disease. The measurement of PCEF (Peak Flow Meter, Health Scan, Cedar Grove, NJ), both during unassisted coughing and with appropriately timed abdominal thrusts, is vital for determining the risk of respiratory

Table 2 Options for Ventilaion Support in Neuromuscular Patients with Chronic Respiratory Failure

For patients with fair oral strength	*For patients with poor oral strength*
In daytime consider	In daytime consider
Sipping/oral IPPV or IAPV	Nasal IPPV or IAPV
At night use	At night use
Nasal/oral IPPV	Mouth seal/Nasal IPPV or Tracheostomy

Table 3 Outline of Neuromuscular Disease that May Develop Respiratory or Cardiac Failure

Disease	Present as respiratory failure	Chronic respiratory failure	Cardiac with secondary respiratory failure
Anterior horn cell			
Amyotrophic lateral sclerosis	+	+	0
Spinal muscular atrophy	+	+	0
Neuromuscular junction			
Myasthenia gravis	+	+	0
Lambert-Eaton syndrome	+	+	0
Congenital myasthenia	+	+	0
Muscular dystrophy			
Duchenne muscular dystrophy	0	+	+
Myotonic dystrophy	+	+	+
Limb-girdle dystrophy	0	+	+
Scapuloperoneal dystrophy	0	+	0
Emery-Dreifuss syndrome	0	0	+
Becker syndrome	0	+	+
Facioscapulohumeral	0	+	0
Metabolic myopathy			
Acid maltase	+	+	+
Carnitine deficiency	+	+	+
Congenital myopathy			
Centronuclear myopathy	+	+	0
Nemaline myopathy	+	+	+
Inflammatory myopathy			
Polymositis	0	+	+
Dermtomyositis	0	+	0

complications and the potential utility of noninvasive ventilatory support methods long-term (35–37). Patients who can achieve assisted PCEF exceeding 3–4 L/sec rarely require an indwelling tracheostomy for long-term ventilatory support even when the VC is unmeasurable. When PCEF cannot exceed 3 L/sec, as for example in the presence of the severe bulbar muscle dysfunction (e.g., amyotrophic lateral sclerosis), noninvasive respiratory aids cannot be used effectively, and the patient will be at risk for aspiration and pneumonia.

ACUTE RESPIRATORY FAILURE

Patients with neuromuscular disorders may present with acute respiratory failure of unknown etiology before their neuromuscular disease has been recognized (Table 3). The differential diagnosis of respiratory failure includes conditions such as myasthenia gravis, Lambert-Eaton syndrome, myotonic dystrophy, amytrophic lateral sclerosis, and acid maltase deficiency (26–34,39,40). Most often these acute episodes are caused by intercurrent upper respiratory tract infections or exacerbations of concurrent chronic obstructive pulmonary disease. They are associated with the accumulation of secretions in the airway and lead to atelectatic pneumonitis. Certain medications can precipitate acute respiratory failure, especially in myasthenia gravis (Table 4). Respiratory arrest has occurred in a pregnant woman with myasthenia gravis after treatment with ritodrine for premature labor (39). Calcium channel blockers have caused respiratory failure in patients with Lambert-Eaton syndrome and Duchenne muscular dystrophy (40). On the other hand, it may seem that respiratory failure is due to a

Table 4 Medications that May Precipitate Respiratory Failure in Myasthenia Gravis

Antibiotic drugs	Heart disease drugs	Rheumatic disease drugs
Clindamycin	Quinidine	Penicillamine
Colistin	Propranolol (Inderal)	Cholroquine
Kanamycin	Procainamide	
Neomycin	Oxprenolol	Anesthesia drugs
Streptomycin	Practolol	Succinylocholine chloride
Tobramycin	Trimethaphan	Curace and other relaxants
Tetracyclines	Lidocaine	Decamethonium
Gentamycin	Verapamil	Gallamine
Polymyxin	Nifedipine	Ether
Bacitracin	Diltiazem	
Hormonal drugs	Psychatric drugs	Seizure drugs
ACTH	Chlorpromazine	Dilantin
Corticosteroids	Promazine	Trimethadione
Thyroid hormone	Phenelzine	
Oral contraceptives	Lithium	
		Ocular medications
		Timoptic

Pain medications
Avoid those that cause respiratory depression or use low dosage.
Check contraindications for each muscle relaxant prescribed.
Avoid Lactated Ringer's Intravenous solution if possible.
Avoid hypokalemia.

A drug alone may not cause any symptoms, but when it is taken with another drug causing neuromuscular transmission defect, the effect may become apparent.

specific neuromuscular disease, but in fact the respiratory symptoms are due entirely to conditions such as chronic obstructive airways disease (COAD), obesity, or malnutrition. Occasionally, a patient with ocular myasthenia will present with shortness of breath, which is presumed to be secondary to generalized myasthenia. In fact, COAD or congestive heart failure may be responsible, and both possibilities deserve serious consideration (41).

Patients with some neuromuscular disorders may have cardiac involvement associated with that particular disease. Ultrasound and resting gated radionuclide ventriculography (RNV or MUGA) are helpful diagnostic tests. Interestingly, there appears to be no close relationship between the severity of skeletal muscle involvement in certain of these myopathic disorders and the associated cardiac involvement (ventricular ejection fractions) (42).

NONINVASIVE INSPIRATORY AND EXPIRATORY MUSCLE AIDS

Glossopharyngeal Breathing

Glossopharyngeal breathing (GPB), or "frog breathing," is a technique used by some ventilator-supported patients with neuromuscular disease to increase maximal inspirations for assisted coughing, for raising the volume of their voice, and for increasing breathing time off the ventilator (43). Many patients learn this technique on their own although video instruction tapes and training manuals are available (44,45). With this technique the tongue and pharyngeal muscles "gulp" boluses of air past the vocal chords. Six to eight gulps of 60–100 ml may constitute one breath. A normal minute ventilation can be maintained for hours by patients who otherwise have no measurable VC. GPB may be ineffective when bulbar muscle weakness is severe.

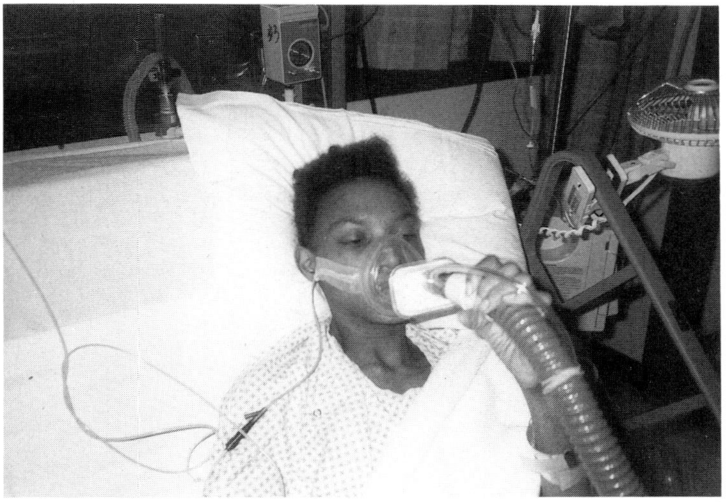

Figure 1 Patient with myasthenia gravis using mechanical insufflation-exsufflation.

Mechanical Exsufflation

Mechanical insufflation-exsufflation (In-Exsufflator, J. H. Emerson Company, Cambridge, MA) involves the use of a device to deliver a deep insufflation (Fig. 1). The deep insufflation is followed by sudden decrease in pressure, which causes a rapid exsufflation. The resulting 6–10 L/sec of expiratory flow clears debris and airway secretions (35–37,41,46,47). With this assistive device the VC and oxyhemoglobin saturation can increase markedly. The insufflation-exsufflation can be delivered via a mask or by tracheostomy tube.

Manually assisted coughing involves the use of an abdominal thrust. The thrust is timed to follow glottic opening after an inspiration that has provided an insufflation greater than 1.5 L. This maneuver can often generate a peak cough expiratory flow (PCFF) of greater than 5 L/sec and thereby effectively remove excessive accumulation of airway secretions (Fig. 2). The use of assisted coughing can be vital during intercurrent respiratory tract infections and is often necessary long before the patient develops the need for ongoing ventilatory support (35–37,41,46). Assisted coughing techniques are good alternatives to endotracheal suctioning, which can increase secretions. Suctioning may miss the left mainstem bronchus entirely, while assisted coughing is effective throughout the bronchial tree. Assisted coughing can also be more effective than chest physical therapy. However, it is generally not effective in the presence of severe scoliosis or airway obstruction.

Home Mechanical Ventilation

Noninvasive IPPV

Most patients with less than 30% of predicted VC will require nocturnal ventilation, and by the time the VC falls to less than 12% of predicted, most patients will require 24-hr ventilatory support. Each patient with ventilatory insufficiency who is a candidate for the use of noninvasive ventilatory support alternatives should have the opportunity to try several methods and choose the ones that he/she prefers (41,46–49). Introduction to these methods should begin long before the patient requires their continuous use. Mouthpiece IPPV should be introduced as a technique to provide intermittent deep insufflations to improve PCEF when necessary and provide "range of motion" to the lungs and chest wall. This approach also lessens the fatigue of inspiratory muscles. Its use in the early stages of chronic alveolar hypoventilation will decrease the likelihood of intubation during episodes of acute respiratory failure

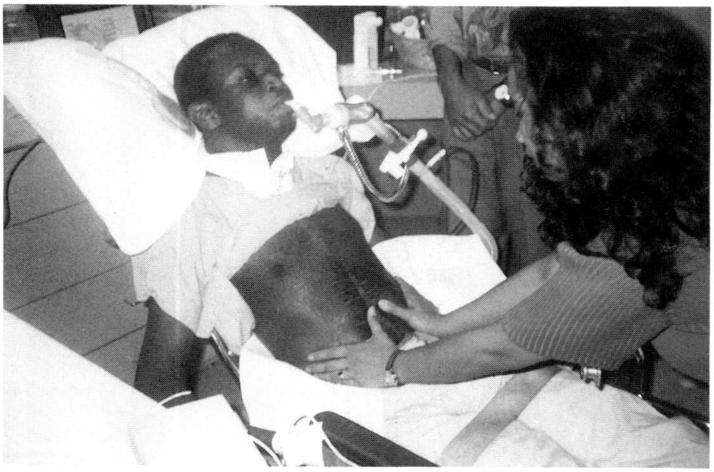

Figure 2 Manually assisted coughing—Most effective when preceded by a deep breath or mechanically assisted insufflation.

when otherwise intubation would be mandatory. Oximetry biofeedback can be very helpful in training patients to use the mouthpiece IPPV for appropriate periods. Patients can see that with just several deep insufflations their oxyhemoglobin saturations can exceed 95% within seconds. An alarm can be set to sound when the saturation decreases below an unacceptable value, e.g., 93%. This will remind the patient to take a few deep assisted breaths to correct alveolar ventilation again.

Nocturnal IPPV should be delivered via a comfortable nasal interface. The patient should have the opportunity to try at least three or four varieties to find one or two with which he/she is most satisfied (41,47–50). A large number of nasal interfaces are now commercially available (Respironics, Monroeville, PA; Puritan-Bennett Inc., Boulder, CO; Life Care International Inc., Lafayette, CO; and others). Although usually preferred only for nocturnal use or when the patient is in bed, nasal IPPV has been used up to 24 hr a day by ventilator users whose lips or neck are too weak to accommodate daytime mouthpiece IPPV (Fig. 3).

Ventilator users may choose to use mouthpiece IPPV overnight with a lipseal (Puritan-Bennett, Boulder, CO) (41). The lipseal can completely seal off the mouth from insufflation leak (Fig. 4). This technique has the advantage of offering excellent alveolar ventilation for any patient without the required central nervous system reflex muscular activity needed during sleep to prevent excessive insufflation leak. In a few patients, sealing of the nostrils with cotton pledgets has created an entirely closed system of ventilatory support similar to that of tracheostomy IPPV with an inflated cuff.

In addition to the possibility of offering mouthpiece (Fig. 5) and/or nasal IPPV as alternatives to intubation and tracheostomy, custom-made strapless acrylic interfaces that cover both the nose and mouth have been used for nocturnal support. These interfaces are retained by a biteplate that fastens the outer shell, essentially an acrylic lipseal, to the teeth by metal clasps (Fig. 6). The biteplate may be thrust out of the mouth by simple tongue movement. These interfaces are indicated for ventilator users who live alone and who have upper extremity function but not sufficient function to don the straps required to interface with the ventilator. The disadvantages are expense and need for stable dentition.

Adequate inspiratory muscle support for nasal IPPV can be delivered via bilevel positive airway pressure (BIPAP) devices, which have the advantages of being less expensive, lighter, and without annoying alarm systems. However, sometimes it is difficult to assess the extent of inspiratory muscle aid by BIPAP; the pressure gauge reads the same pressures whether or not the patient is receiving an adequate tidal volume. Unfortunately, BIPAP is often mistakenly used at inadequate inspiratory positive airway pressures. In general, clinicians with little experience in the use of noninvasive aids would be

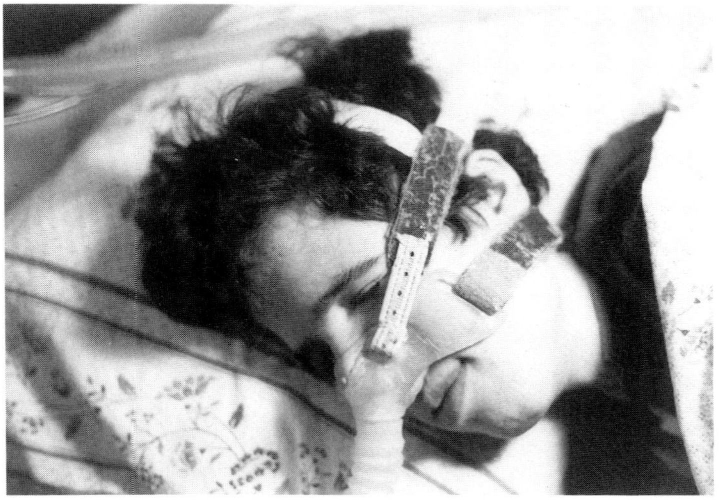

Figure 3 Nasal mask used by a patient with Duchenne dystrophy who uses nasal IPPV 24 hr/day for ventilatory support.

better advised to use portable volume ventilators and reserve BIPAP for patients with obstructive apneas for whom the expiratory positive airway pressure may be useful.

Body Ventilators

There are several types of body ventilators that have occasional use in patients with ventilatory insufficiency. There are those that act directly on the abdominal contents to produce movement of the diaphragm and air flow and those that act by creating subatmospheric pressure around the chest and

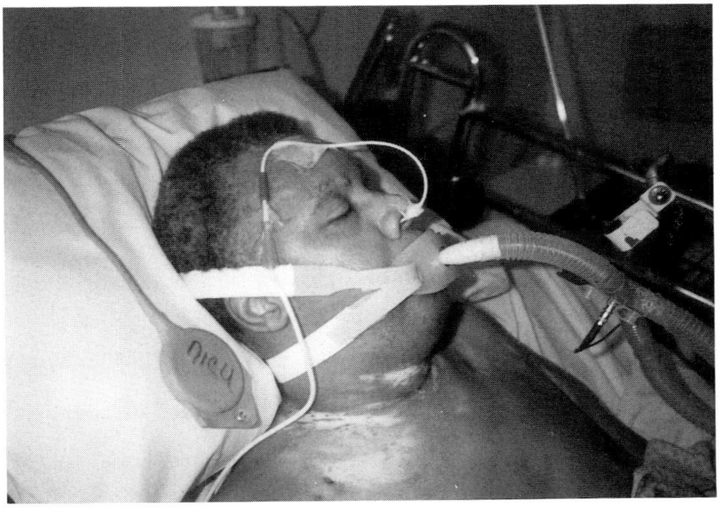

Figure 4 Ventilator-dependent patient shortly after removal of tracheostomy and using nocturnal lipseal IPPV.

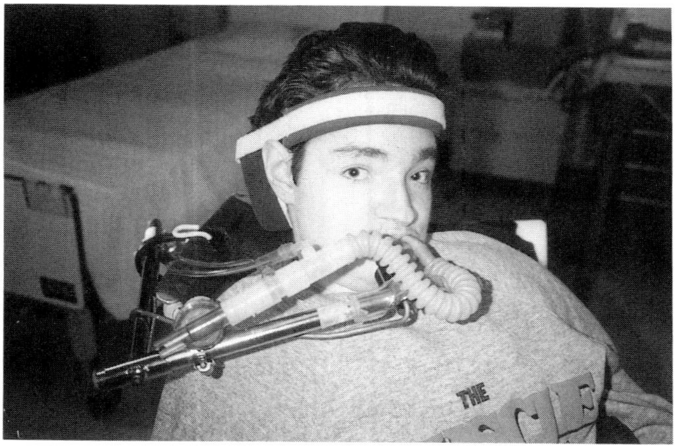

Figure 5 This individual with Duchenne muscular dystrophy has relied on 24-hr mouth intermittent positive-pressure ventilation for 7 years. The flexed mouthpiece is fixed adjacent to the tongue control switch for his motorized wheelchair.

abdomen to create alveolar airflow. The intermittent abdominal pressure ventilator (IAPV), or "pneumobelt," and the rocking bed act directly on the body. The "iron lung," Porta-lung, "wrap" style, and chest shell ventilators are negative-pressure body ventilators (NPBVs) (51–58). The IAPV optimizes patient appearance, function, and mobility. In can be used only in the sitting position, and the patient must have a straight spine in order for it to be effective. The mouth remains free, however. About 300 ml of tidal volume can be delivered with this method. The rocking bed tilts the patient 15–30° from the horizontal. This moves the abdominal contents and creates diaphragm excursion. It is often not very effective and is infrequently used today.

The Porta-lung is a lightweight version of the iron lung. Although it frees the patient's mouth and nose, it, like the use of all negative pressure body ventilator methods, is associated with obstructive apneas and oxyhemoglobin desaturations during sleep. The chest shell ventilator consists of a firm shell,

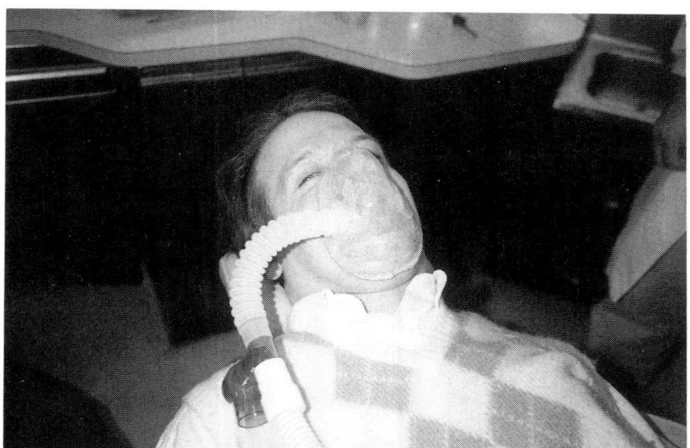

Figure 6 Patient using custom-made strapless acrylic interface, which is retained by a biteplate that fastens the outer shell (essentially an acrylic lipseal) to the teeth by metal clasps.

which fits over the chest, and a negative-pressure generator, which produces subatmospheric pressure to induce ventilation. This ventilator is less effective than an iron lung or Porta-lung, and it is ineffective in the presence of severe back deformity or poor pulmonary compliance. Wrap-style ventilators are similar to the chest shell except that a nylon or flexible plastic (Gortex) wrap covers a plastic grid that goes over the anterior chest and abdomen. The wrap is sealed around the wrists, legs, and neck. Subatmospheric pressure is created under the grid and wrap to promote alveolar ventilation (57,58). The wraps can be especially awkward when providing personal care. They can be helpful, however, to facilitate tracheostomy site closure when converting from 24-hr ventilator use via tracheostomy to 24-hr noninvasive IPPV. Body ventilators can also be useful to free the nose and mouth for secretion management during intercurrent upper respiratory infections. However, as a general rule, NPBVs are rarely if ever preferred over, or as effective as, noninvasive IPPV.

In summary, tracheostomy and intubation are rarely necessary unless assisted PCEF cannot exceed 3 L/sec. For noninvasive respiratory aids to be effective, the patient must also be able to cooperate, have adequate bulbar muscle function to avoid aspiration of saliva, be free of seizures, nasogastric tubes, supplemental oxygen use, and substance abuse, and have no orthopedic conditions that interfere with their use (59,60). Once patients begin to require nocturnal IPPV, daytime aid may not be necessary for 1–3 years or more (15).

TRACHEOSTOMY AND INTUBATION

For patients unfamiliar with noninvasive aids, who do not meet the criteria for their use, and who must be intubated, tracheostomy should not be delayed. Vocal cord damage is not uncommon with prolonged oro- or nasotracheal intubation (59,60). Upper airway obstruction can result and increase the risk of long-term pulmonary morbidity. Tracheostomy tubes may be cuffed, uncuffed, or with or without fenestrations. Cuff inflation need never be used long-term unless severe bulbar muscle involvement impairs control of upper airway secretions (61). Although cuff inflation protects to some degree against aspiration, it also causes ischemia and damage to the tracheal wall, which not infrequently lead to trachiectasis. Cuff pressures should stay below 20 mm/Hg to lessen the damage to tracheal mucosa (60).

Tracheostomies require daily care, regular tube changes, and necessitate tracheal suctioning and supplemental humidification. Frequent suctioning and inadequate humidification may lead to tracheitis. Well-humidified air and good hydration help prevent thick secretions and lead to more manageable, yet increased amounts of thin secretions. In contrast, users of noninvasive ventilatory aids generally have no difficulty with airway secretions except during respiratory tract infections. The patient with a tracheostomy should be careful to cover the site during bathing or showering or when in a dusty environment. Patients with tracheostomies lose the natural cough and ciliary clearance mechanisms. Bronchial mucus plugging is common and can be fatal. Signs of mucus plugging include appearance of gurgling, shortness of breath, oxyhemoglobin desaturation, anxiety, and restlessness. When suctioning is performed, the catheter should be no more than one size larger than one-half the size of the tracheostomy tube so that high negative pressures do not cause alveolar collapse. Some other potential risks related to a tracheostomy are aspiration, tracheoesophageal fistula, tracheoinnominate fistula, and laryngeal and tracheal stenosis (59,60).

The ability to speak is critically important for effective communication as well as for maintaining self-esteem. Cuff deflation or the use of cuffless tracheostomy tubes permits speech by allowing ventilator-delivered air to pass over the vocal cords (61). Tracheostomy tubes with double lumens with a separate compressed air source for air delivery through the glottis, or "talking trachs," are rarely effective or needed. Except for some patients with amyotrophic lateral sclerosis, virtually all neuromuscular ventilatory insufficiency can be managed without cuff inflation (61).

MEDICAL TREATMENT

Medications are not effective in the treatment of ventilatory insufficiency. Methylxanthines, although possibly useful as positive inotropic agents for the diaphragm for some patients with COAD, have no clinical effect on the paralytic/restrictive ventilatory insufficiency that occurs in progressive neuromuscular diseases. Oxygen therapy is contraindicated unless adequate oxygenation (PaO_2 greater than 60 torr) cannot be maintained by assisted ventilation alone. Although protriptyline may be useful in treating central sleep apneas in certain patients [for example, patients with myotonic dystrophy (1,7)] by suppressing REM sleep, it is of no value in treating ventilatory insufficiency from muscle weakness. Other medications, such as acetazolamide and methylprogesterone, have also been tried with questionable results. Certain electrolyte disturbances, such as hypokalemia and hypophosphatemia, may aggravate respiratory insufficiency and should be corrected.

IMPORTANCE OF DISCUSSING OPTIONS FOR RESPIRATORY CARE WITH PATIENTS

Whether a patient has a slowly progressive neuromuscular condition or is rapidly developing weakness, e.g., from Guillain-Barré syndrome, it is critical to discuss the possibility of respiratory failure early in the course of the illness (before the VC has decreased to 40% of predicted normal). Early planning often permits the patient to use noninvasive respiratory muscle aids to avoid intubation. Such discussions alleviate fear and eliminate the crisis decision-making process about whether or not to go on a respirator, be intubated, and have a tracheostomy. If noninvasive aids for providing ventilatory support have not been adequately discussed with the patient, invasive mechanical ventilation via tracheostomy or endotracheal tube is likely to be performed. Decisions as to which options are available for ventilatory support require detailed discussions with the patient and others involved in his or her care (41,62–69). Too frequently a physician will use an inaccurate assessment of a patient's circumstance and of his quality of life in recommending the best treatment for the patient's ventilatory insufficiency. In one study measuring satisfaction with health, housing, interpersonal relationships, and life in general, the satisfaction of ventilator-supported Duchenne muscular dystrophy patients was greatly underestimated by health care professionals (64). These patients reported even greater satisfaction with housing and family life than physically intact controls. All the patients stated that they would have used ventilatory assistance again if they had to. The noninvasive aids, such as intermittent positive-pressure ventilation (IPPV) via a mouthpiece and nasal IPPV, are greatly preferred by patients who had also experienced tracheostomy IPPV (62).

Despite the patients' desire to survive and their preference for and the cost advantages of using noninvasive aids (63), an elective recommendation of ventilatory assistance occurs in only 26% of Muscular Dystrophy Association clinics (64). Deciding not to provide some form of ventilatory assistance may subject patients eventually to episodes of acute and potentially life-threatening respiratory failure and tracheal intubation. Clearly, health care professionals should not impose their values and biases onto their patients. A thorough discussion of the options is necessary, followed by a care plan tailored to the individual patient.

CONCLUSION

Treatment of respiratory complications is an important concern for care providers in the management of patients with progressive neuromuscular disease. Respiratory failure is the most common cause of death in these disorders. The signs and symptoms of respiratory compromise need to be recognized early. Early diagnosis of ventilatory insufficiency allows the initiation of preventive therapy to guard against secondary complications. The natural history of ventilatory insufficiency and the noninvasive treatment options should be thoroughly discussed with the patient and his family. Different forms of noninvasive ventilatory support should be tried by each patient when appropriate. This will prevent the situation in which the patient arrives in an emergency department and is intubated for acute respiratory failure without having had an opportunity to discuss options. The prevention of respiratory complications requires a team approach and should include the patient, his physician, his family, and a host of ancillary health care providers.

REFERENCES

1. Edwards RHT, Griggs RC. Medical and psychological management of neuromuscular disease. In: Walton, Karpati, Jones, eds. Disorders of Voluntary Muscle. 1994:837–850.
2. Fitting J-W, Leuenberger P. Procainamide for dyspnea in myotonic dystrophy. Am Rev Respir Dis 1988; 140:1442–1445.
3. Schmidt-Nowara WW, Altman AR. Atelectasis and neuromuscular respiratory failure. Chest 1984; 85(6): 792–795.
4. Tobin MJ. Respiratory muscles in disease. Clin Chest Med 1988; 9(2):263–286.
5. Szeinberg A, Tabachnik E, Rashed N, et al. Cough capacity in patients with muscular dystrophy. Chest 1988; 94(6):1232–1235.
6. Kurz LT, Murbarak ST, Schultz P, Park SM, Leach J. Correlation of scoliosis and pulmonary function in Duchenne muscular dystrophy. J Pediatr Orthop 1983; 3:347–353.
7. Smith PEM, Calverley PMA, Edwards RHT, et al. Practical problems in the respiratory care of patients with muscular dystrophy. N Engl J Med 1987; 316:1197–1205.
8. Rardy RL, Reid WD, Belman MJ. Respiratory muscle training. Clin Chest Med 1988; 9(2):287–296.
9. Smith PEM, Edwards RHT, Calverley PMA. Oxygen treatment of sleep hypoxaemia in Duchenne muscular dystrophy. Thorax 1989; 44:997.
10. Bach JR, Alba AS. Total ventilatory support by the intermittent abdominal pressure ventilator. Chest 1991; 99:630–636.
11. Manni R, Zucca C, Galimberti CA, et al. Nocturnal sleep and oxygen balance in Duchenne muscular dystrophy: a clinical and polygraphic 2-year follow-up study. Eur Arch Psych Clin Neurosci 1991; 240(4–5):225–257.
12. Bye PT, Ellis ER, Issa FG, et al. Respiratory failure and sleep in neuromuscular disease. Thorax 1990; 45(4):241–247.
13. Bach, JR. Pulmonary rehabilitation considerations for Duchenne muscular dystrophy: the prolongation of life by respiratory muscle aids. Crit Rev Phys Rehabil Med 1992; 3(3):239–269.
14. Carre PC, Didier AP, Tiberge YM, et al. Amyotrophic lateral sclerosis presenting with sleep hypopnea syndrome. Chest 1988; 93(6):1309–1312.
15. Heckmatt JZ, Loh L, Dubowitz V. Nocturnal hypoventilation in children with nonprogressive neuromuscular disease. Pediatrics 1989; 83(2):250–255.
16. Kryger MH, Steljes DG, Yee WC, et al. Central sleep apnoea in congenital muscular dystrophy. J Neurol Neurosurg Psychiatry 1991; 54(8):710–712.
17. Shintani S, Shiozawa Z, Shindo K, et al. Sleep apnea in well-controlled myasthenia gravis (Japanese). Rinsho Shinkeigaku-Clin Neurol 1989; 29(5):547–553.
18. Cadieux RJ, Kales A, McGlynn TJ Jr, et al. Sleep apnea precipitated by pharyngeal surgery in a patient with myotonic dystrophy. Arch Otolaryngol 1984; 110(9):661–613.
19. Bach JR, Penek J. Obstructive sleep apnea complicating negative pressure ventilatory support in patients with chronic paralytic/restrictive ventilatory dysfunction. Chest 1991; 99:1386–1393.
20. Levy RD, Bradley TD, Newman SL, et al. Negative pressure ventilation: effects on ventilation during sleep in normal subjects. Chest 1989; 65:95.
21. Cirignotta F, Mondini S, Zucconi M, et al. Sleep-related breathing impairment in myotonic dystrophy. J Neurol 1987; 235(2):80–85.
22. Bay PC, Westbrook PR, Daube JR, Litchey WJ, Windebank AJ, Iverson R. Effects of alterations in pulmonary function and sleep variations on survival in patients with amyotrophic lateral sclerosis. Mayo Clin Proc 199; 66:686–694.
23. Smith PEM, Edward RHT, Calverley PMA. Ventilation and breathing pattern during sleep in Duchenne muscular dystrophy. Chest 1989; 96(6):1346–1351.
24. Bach JR, Alba AS. Management of chronic alveolar hypoventilation by nasal ventilation. Chest 1990; 97:52.
25. Bach JR, Campagnolo DI, Hoeman S. Life satisfaction of individuals with Duchenne muscular dystrophy using long term mechanical ventilation support. Am J Phys Med Rehabil 1991; 70:3:129–135.
26. Massey EW, Harrell LE. Motor neuron disease presenting with respiratory failure: report of two cases. Postgrad Med 1984; 76(8):216–218.
27. Kelly BJ, Luce JM. The diagnosis and management of neuromuscular diseases causing respiratory failure. Chest 1991; 99(6):1485–1494.
28. Schmidt-Nowara WW, Marder EJ, Feil PA. Respiratory failure in myasthenia gravis due to vocal cord paresis. Arch Neurol 1984; 41(5):567–568.

29. Jordan CB 2nd, Morse HG, Atkinson LS. Myasthenia gravis presenting as respiratory failure: confusion with a psychiatric illness. SC Med Assoc 1989; 85(6):292–294.
30. Marlowe FI, D'Angelo AJ Jr. Respiratory failure as the initial presentation of myasthenia gravis. Ear Nose Throat J 1989; 68(6):472–473.
31. Mier A, Laroche C, Green M. Unsuspected myasthenia gravis presenting as respiratory failure. Thorax 1990; 45(5):422–423.
32. Dushay KM, Zibrak JD, Jensen WA. Myasthenia gravis presenting as isolated respiratory failure. Chest 1990; 97(1):232–234.
33. Hannon VM, Cunningham AJ, Hutchinson M, McNicholas W. Aspiration pneumonia and coma—an unusual presentation of dystrophic myotonia. Can Anaesth Soc J 1986; 33(6):803–806.
34. Daras M, Spiro AJ, Swerdlow M. Respiratory failure in amyotrophic lateral sclerosis. NY State J Med 1984; 84(11):570–572.
35. Barach AL, Beck GJ, Bickerman HA, Seanor HE. Physical methods simulating cough mechanisms. JAMA 1952; 150:1380–1385.
36. Sortor S, McKenzie M. Toward Independence: Assisted Cough. Dallas: Bioscience Communications of Dallas, 1986 (video).
37. Barach AL, Beck GJ, Smith RH. Mechanical production of expiratory flow rate surpassing the capacity of human coughing IM. J Med Sci 1953; 226:241–248.
38. Godley PJ, Morton TA, Karboski JA, Tami JA. Procainamide-induced myasthenic crisis. Ther Drug Monit 1990; 12(4):411–414.
39. Catanzarite VA, McHargue AM, Sandberg EC, Dyson DC. Respiratory arrest during therapy for premature labor in a patient with myasthenia gravis. Obstet Gynecol 1984; 64(6):819–822.
40. Krendel DA, Hopkins LC. Adverse effect of verapamil in a patient with the Lambert-Eaton syndrome. Muscle Nerve 1986; 9(6):519–522.
41. Bach JR. Pulmonary rehabilitation. In: Delisa JD, ed. Rehabilitation Medicine: Principles and Practice. Philadelphia: JB Lippincott, 1993:952–972.
42. Stewart CA, Gilgoff I, Baydur A, Prentice W, Applebaum D. Gated radionuclide ventriculography in the evaluation of cardiac function in Duchenne's muscular dystrophy. Chest 1988; 94:1245–1248.
43. Bach JR, Alba AS, Bodofsky E, Curran FJ, Schultheiss M. Glossopharyngeal breathing and noninvasive aids in the management of post-polio respiratory insufficiency. Birth Defects 1987; 23:99–113.
44. Dail CW, Affeldt JE. Glossopharyngeal Breathing. Los Angeles: Los Angeles Dept. of Visual Education, College of Medical Evangelists, 1954 (video).
45. Dail C, Rogers M, Guess V, Atkins HV. Glossopharyngeal Breathing Manual. Downy, CA: Professional Staff Association of Rancho Amigos Hospital, 1979.
46. Bach JR. Mechanical exsufflation, non-invasive ventilation, and new strategies for pulmonary rehabilitation and sleep disorder breathing. Pulm Rehabil 1992; 68(2):321–340.
47. Bach JR, Alba AS, Saporito LR. Intermittent positive pressure ventilation via the mouth as an alternative to tracheostomy for 257 ventilator users. Chest 1993; 103:174–182.
48. Baydur A, Gilgoff I, Prentice W, et al. Decline in respiratory function and experience with long-term assisted ventilation in advanced Duchenne's muscular dystrophy. Chest 1990; 97(4):884–889.
49. McDermott I, Bach JR, Parkers V, Sortors. Custom fabricated interfaces for intermittent positive pressure ventilation. Int J Prosthodont 1989; 2:224.
50. Blitzer B. Directory of Sources for Ventilation Face Masks. St. Louis: Gazette International Networking Institute Press, 1991.
51. Curran FJ, Colbert AP. Ventilator management in Duchenne muscular dystrophy and postpoliomyelitis syndrome: twelve years' experience. Arch Phys Med Rehabil 1989; 70(3):180–185.
52. Miller HJ, Thomas E, Wilmot CB. Pneumobelt use among high quadriplegic population. Arch Phys Med Rehabil 1988; 69:369–372.
53. Braun SR, Sufit RL, Giovannoni R, et al. Intermittent negative pressure ventilation in the treatment of respiratory failure in progressive neuromuscular disease. Neurology 1987; 37(12):1874–1875.
54. Mohr Chester H, Hill Nicholas S. Long term follow-up of nocturnal ventilatory assistance in patients with respiratory failure due to Duchenne-type muscular dystrophy. Chest 1990; 97(1):91–96.
55. Splaingard ML, Frates RC, Jr., Jefferson LS, Rosen CL, Harrison GM. Home negative pressure ventilation: report of 20 years of experience in patients with neuromuscular disease. Arch Phys Med Rehabil 1985; 66:239–242.
56. Bach JR, Alba AS. Intermittent abdominal pressure ventilator in a regimen of noninvasive ventilatory support. Chest 1991; 99:630–636.
57. Bauer LE, Martin LL. How do you use a poncho respirator? Am J Nurs 1988; Dec:1705–1706.
58. Verchken JB, Rapheal JC, deLattre J, et al. Adult maltase acid deficiency myopathy: treatment with long-term home mechanical ventilation. Biomed Pharmacol 1988; 42(5):343–349.
59. Carol PF. Artificial airways equal real risk. J Nurs 1986; 16(8):56–59.

60. Mapp CS. Trach care: are you aware of all the dangers? J Nurs 1988; 18(7):34–43.
61. Bach JR, Alba AS. Tracheostomy ventilation: a study of efficacy with deflated cuffs and cuffless tubes. Chest 1990; 97:679–683.
62. Bach JR. A comparison of long-term ventilatory support alternatives from the perspective of the patient and care giver. Chest (in press).
63. Back JR, Intintola P, Alba AS, Holland I. The ventilator-assisted individual: cost analysis of institutionalization versus rehabilitation and in-home management. Chest 1992; 101:26–30.
64. Bach JR. Ventilator use by Muscular Dystrophy Association patients: an update. Arch Phys Med Rehabil 1992; 73:179–183.
65. Freed MM. Quality of life: The physician's dilemma. Arch Phys Med Rehabil 1984; 65:109–111.
66. Miller JR, Colbert AP, Osberg JS. Ventilator dependency: decision-making, daily functioning and quality of life for patients with Duchenne muscular dystrophy. Dev Med Child Neurol 1990; 32(12):1078–1086.
67. Gilgoff I, Prentice W, Baydur A. Patient and family participation in the management of respiratory failure in Duchenne's muscular dystrophy. Chest 1989; 95(3):519–524.
68. Mohr CH, Hill NS. Long-term follow-up of nocturnal ventilatory assistance in patients with respiratory failure due to Duchenne-type muscular dystrophy. Chest 1990; 97(1):91–96.
69. Splaingard ML, Frates RC, Harrison GM, Carter ER, Jefferson LS. Home positive-pressure ventilation: twenty years experience. Chest 1983; 85(4):376–382.

55
Cardiac Involvement in Neuromuscular Disease

Michael Swash
St. Bartholomew's and Royal London School of Medicine and Dentistry, Royal London Hospital, London, England

Martin S. Schwartz
Atkinson Morley's Hospital, London, England

INTRODUCTION

Involvement of the heart in neuromuscular disorders is often overlooked. Cardiac involvement may be subclinical or present as a life-threatening manifestation early or late in the natural history of the underlying disorder. In some neuromuscular disorders, e.g., Becker X-linked muscular dystrophy, myocardial involvement is unrelated to the severity or duration of skeletal muscle weakness and may even precede the onset of skeletal muscle weakness. In myotonic dystrophy 86% of patients show electrocardiographic (ECG) abnormalities, but only 16% have cardiac symptoms (1). In acute, acquired disorders, e.g., in Guillain-Barré syndrome (acute postinfectious polyradiculoneuropathy) and in acute polymyositis, cardiac complications may be simultaneous with the neuromuscular disorder.

To the cardiologist cardiomyopathy is a disorder of the heart in which myocardial disease occurs without evidence of myocardial ischemia or volume/pressure overload. The functional classification of the primary cardiomyopathies, which recognizes congestive, hypertrophic, and restrictive disturbances and disorders of atrioventricular conduction, is not easily applicable to the cardiomyopathies that complicate neuromuscular disorders.

CLASSIFICATION OF NEUROMUSCULAR DISORDERS (TABLE 1)

Cardiac involvement is not a feature of all neuromuscular disorders, although in some it is a major problem (2). In recent years a metabolic basis has been recognized for many of these diseases, especially those that have a genetic background. Increasing knowledge of the molecular genetics of these diseases is beginning to delineate them further, particularly in understanding the variations in their clinical features.

Primary Diseases of Muscle and Cardiomyopathy

The muscular dystrophies are relatively uncommon inherited disorders of muscle, characterized by a progressive course and by degenerative changes in skeletal muscle fibers. Classification of these disorders depends on clinical, genetic, and histological criteria.

Duchenne Muscular Dystrophy

This X-linked disorder (Xp21) is the most severe form of muscular dystrophy (2). The gene product, dystrophin, is a 400-kDa protein that comprises only 0.01% of protein in normal skeletal muscle cells (3). It is an essential component of the muscle cell plasma membrane. Most affected boys lose the ability to walk by age 11 years, and only 25% survive beyond age 25 years.

Table 1 Primary Diseases of Muscle Associated with Cardiac Involvement

Genetically determined myopathies
1. Muscular dystrophies
Duchenne (severe X-linked dystrophinopathy)
Becker (benign X-linked dystrophinopathy)
Limb-girdle
Facioscapulohumeral
Emery-Dreifuss (X-linked)
2. Myotonic dytrophy
3. Metabolic myopathies
Mitochondrial cytopathies
Glycogen storage diseases
Periodic paralyses
Malignant hyperpyrexia
4. Myopathies of childhood
Nemaline myopathy
Myotubular myopathy
Central core disease
Acquired myopathies
1. Inflammatory myopathies
2. Endocrine myopathies
3. Drug-induced myopathies
4. Infections and infestations

Cardiac involvement is common, occurring in 80% of boys with the disease (4), usually manifested as tachycardia, arrhythmia, and cardiac failure (5). The cardiac problems become more prominent and frequent as the disease progresses, but are often masked by the severity of the disability caused by the skeletal muscle weakness and, especially, by ventilatory difficulties related to kyphoscoliosis and intercostal muscle weakness. These thoracic deformities may result in a prominent left sternal impulse and a loud pulmonary second sound. In the preclinical stage, even in early childhood, the ECG is often abnormal (6). The precordial R waves are tall, with increased R/S wave amplitude ratio and deep Q waves in the lateral precordial leads and in leads I and aV_L (7,8). Some female carriers of the gene also show ECG abnormalities (9). Conduction defects may also occur. Hypertrophic cardiomyopathy has been described in the early stages of the disorder. The latter consist of ectopic atrial rhythms, short PR interval, and paroxysmal rapid heart action. Ventricular tachycardias also occur, and, rarely, sudden death may occur (10). In the later clinical stages a dilated cardiomyopathy commonly develops. In a third of cases this is associated with pulmonary insufficiency with nocturnal oxygen desaturation and hypercarbia. The cardiomyopathy is associated with predilection for involvement of the posterobasal and lateral walls of the left ventricle (11). In about 10% of patients death is due to cardiomyopathy.

Histopathological studies (10,12) of the heart show fatty infiltration and mild fibrosis in the sinus and AV nodes. Papillary muscle degeneration may lead to mitral valve prolapse. Fibrous replacement and vacuolar degeneration of muscle fibers, with fatty infiltration of the heart muscle and of the peripheral conduction system, occur. There are abnormalities in the distribution of dystrophin in cardiac myocytes (13). In a family with fatal X-linked dilated cardiomyopathy, Muntoni et al. (14) found that the first exon in the dystrophin gene and the muscle promoter region were deleted. In this family muscular involvement consisted only of a raised CK level and mildly increased variability in fiber size, with increased central nucleation in a needle muscle biopsy. There was no muscular weakness or hypertrophy and the electromyogram was normal.

Becker Muscular Dystrophy

This mild form of dystrophinopathy causes less disability and presents later than Duchenne muscular dystrophy, and patients may have a nearly normal life expectancy. Cardiac involvement is therefore clinically important and may be the presenting feature of the disease (15). Dilated cardiomyopathy, bundle branch block, tachyarrhythmia, or complete heart block may develop. Subclinical cardiomyopathy is common in Becker muscular dystrophy, and the severity of cardiac involvement is unrelated to age or to the severity of skeletal muscle disease (16). The prognosis of Becker dystrophy is variable, some patients having a normal life expectancy. About 30% of patients develop dilated cardiomyopathy and congestive failure (13). Steare et al. (17) found that although 89% of their patients were symptom-free, 74% had abnormal ECGs. In 42% there was right bundle branch block, and echocardiography showed left ventricular dilatation and hypokinesis. These features are similar to those found in Duchenne disease. Dystrophin staining of the myocardium in Becker dystrophy reveals abnormally patchy staining (18). Cardiac transplantation has been used in five patients with Becker dystrophy; two of these patients were able to return to work (13).

Emery-Dreifuss Syndrome

This is a rare X-linked (Xq28) disorder in which muscular weakness and wasting develop in a predominantly humeroperoneal distribution during childhood (2). There are contractures of the elbows, posterior neck muscles, and Achilles tendons. Cardiac involvement is invariable. The early cardiac features consist of premature atrial contractions and sinus bradycardia. Later, permanent atrial standstill, atrial fibrillation, or flutter develops. There is a risk of A-V block, and sudden death occurs in as many as 50% of patients (19). Cerebral emboli and stroke may also occur. Since most patients have no cardiac symptoms prior to sudden death, cardiac monitoring and follow-up are indicated, and implantation of a pacemaker has been recommended if the heart rate falls beliw 50 beats/min (13). Histologically, the heart shows hypertrophy, fibrosis, and fatty replacement (20).

Myotonic Dystrophy

This is probably the most common of the genetically determined neuromuscular disorders, affecting 3–5/100,000. The disease may present with cardiomyopathy or with the typical manifestations of myotonia, distal weakness and atrophy, and other features. Visceral smooth muscle is also involved and there may be mental instability. The severity of the clinical manfestations is related to the number of CTG repeats at the genetic locus on chromosome 19 (19q13).

ECG abnormalities, particularly impaired conduction in the bundle of His, occur in 90% of patients (1). The most common ECG abnormalities are increased PR and ST intervals. Pathologically, the heart may show little abnormality apart from patchy fibrosis and fatty infiltration in the conducting system (21). These changes lead to delayed and reentrant conduction and arrhythmias (22). It has been recommended that patients with myotonic dystrophy should be reviewed annually, including ECG studies. Patients with cardiac symptoms, e.g., dizziness or syncope, should have 24-hr monitoring, and if abnormalities are revealed, intracardiac electrophysiological studies may be useful (23). As many as 30% of patients die suddenly, with Stokes-Adams syncope, conduction block, and ventricular arrhythmias (24). Pacing may be required (23).

In addition to the cardiac manifestations, sleep apnea may develop, related to the progressive involvement of respiratory muscles in the disease, and this complication may be clinically associated with symptomatic cardiomyopathy (25).

Mitochondrial Cytopathies

Mitochondrial disorders may present with skeletal muscle syndromes, but there are often features of multiorgan involvement, including encephalopathy, external ophthalmoplegia, pigmentary retinopathy, deafness, peripheral neuropathy, small stature, and heart block (26). The most common neuromuscular disorder is the Kearns-Sayre syndrome, with progressive external ophthalmoplegia, atypical pigmentary retinal degeneration, and heart block (27). These disorders are associated with deletions of mitochondrial DNA, resulting in abnormalities of mitochondrial respiratory chain enzymes, leading to poor exercise tolerance with lactic acidosis. Most patients present before the age of 20 years. Subsarcolemmal accumulations of abnormal mitochondria may be found in cardiac muscle fibers. The cardiac conduction disturbance is slowly progressive, potentially fatal, and may necessitate pacing. Dilated cardiomyopathy may develop terminally. Careful and regular cardiological assessment is indicated in these cases (28).

A family has been described in which inherited dilated cardiomyopathy was associated with multiple deletions in mitochondrial DNA, but without neurological disease (29).

Periodic Paralysis

In the periodic paralyses, recurrent episodes of flaccid weakness develop, lasting minutes or hours, in association with hyperkalemia or hypokalemia. During hyperkalemia, due to potassium release from muscle cells, the ECG shows high peaked T waves, and during hypokalemia the converse ECG change is seen. In either form of periodic paralysis ventricular arrhythmias may occur, which are believed to originate in the left ventricle. These may revert with mild exercise but not necessarily with potassium supplements (30,31).

Inflammatory Myopathies

Cardiac involvement, including congestive heart failure, pericarditis, myocarditis, and arrhythmia, has been reported in polymyositis and dermatomyositis. Although it has been suggested that as many as 50% of patients with polymyositis have serious cardiac manifestations such as congestive failure, clinically significant features of cardiac disease are uncommon in unselected series of patients (32). Asymptomatic ECG abnormalities occur in many cases, perhaps as many as 50% (33), consisting of tachyarrhythmias and conduction defects, but the most sensitive index of cardiac involvement is an increased blood level of CK_{MB}. Serum cardiac troponin T levels may also be a sensitive index of cardiac involvement (34). Most patients with clinically significant cardiac involvement are elderly or have a history of several months' muscular weakness. The cardiomyopathy usually improves with treatment with steroids and immunosuppressant drugs, but occasionally it may be fulminant, acute, and fatal (35). Transplantation may be the only form of treatment feasible in these cases.

Autopsy studies of the heart have revealed widespread myocardial necrosis with focal areas of relative sparing and evidence of regeneration. In some cases infiltration with mononuclear cells has been recorded (36).

Inclusion Body Myositis

This disorder is a nonvasculitic form of inflammatory myopathy recognized as distinct from other forms of myositis only in the last decade. Lotz et al. (37) found cardiac abnormalities in 22 of their series of 40 patients. These consisted of ECG evidence of rhythm disturbances, conduction defects, and ischemic change. Two patients were in cardiac failure and one had hypertrophic cardiomyopathy. Eighteen were hypertensive. The average age at diagnosis was 64 years, and the specificity of these features in this disorder is uncertain.

Neurogenic Disorders with Cardiomyopathy

Neurogenic weakness may result from genetic or acquired disease of anterior horn cells or from genetic or acquired disease of peripheral nerves (Table 2). The spinal muscular atrophies, most of which are inherited as autosomal recessive disorders, do not involve the heart, and the most common inherited type of peripheral neuropathy, the hereditary motor and sensory neuropathies (HMSN; Charcot-Marie-Tooth disease), also spare the heart.

Friedreich's Ataxia

In his original description in 1863 Friedreich noted cardiac abnormalities at autopsy in three of his six subjects (38). The incidence of cardiac signs and symptoms in life is low, however, despite the finding of ECG changes in about 80% of patients (39). These ECG abnormalities consist of inverted T waves, right-axis deviation, and abnormal ST segments. The most common cardiac symptom is exertional dyspnea, a symptom that develops late in the course of this autosomal recessive neurological disorder

Table 2 Neurogenic Disorders with Cardiomyopathy

Genetically determined polyneuropathies
1. Spinocerebellar degenerations
Friedreich's ataxia
2. Peripheral neuropathies
Refsum's disease
Familial amyloid polyneuropathy (FAP)
Acute intermittent porphyria
Acquired polyneuropathies
Guillain-Barré syndrome
Diphtheria
Diabetic neuropathy
Alcoholic neuropathy

and that is limited in its expression by the immobility that usually develops as the disorder progresses. Most patients are restricted to a wheelchair by age 25 years. Hewer reported death from heart failure in 46 of his 82 fatal cases and evidence of cardiac dysfunction during life in 60 patients (40). The mean age at death is 37 years. Echocardiography commonly reveals concentric left ventricular hypertrophy and asymmetrical septal hypertrophy (41). In contrast to idiopathic hypertrophic cardiomyopathy, malignant ventricular arrhythmias are rare. Dilated cardiomyopathy is uncommon and may represent a different form of the disease, in which atrial and ventricular arrhythmias occur.

Refsum's Disease

This rare autosomal recessive disorder is characterized by peripheral neuropathy, ataxia, deafness, ichthyosis, night blindness, pes cavus, and cardiomyopathy, with storage of a lipid, phytanic acid, in the tissues. Congestive heart failure, arrhythmias, and acute left ventricular failure tend to develop in relation to relapses in the neuropathy and may lead to sudden death, perhaps associated with autonomic involvement. The heart shows patchy fibrosis (42).

Familial Amyloid Polyneuropathy (FAP)

In this dominantly inherited group of disorders, amyloid fibrillar protein is deposited in many tissues, including the heart. The main feature in the Portuguese (type I) form of the disease is progressive peripheral neuropathy with distal pain and sensory loss. There are prominent autonomic disturbances, including impotence, diarrhea, incontinence, hypotension, and pupillary abnormalities. In other forms, e.g., type II FAP, carpal tunnel syndrome is the mean presenting feature. Cardiac manifestations consist of constrictive or congestive cardiomyopathy. The importance of cardiomyopathy varies in different mutations of the trans-thyretin gene (43). In one such mutation in Denmark (MET 111) cardiomyopathy is the major feature of the disease (44).

Guillain-Barré Syndrome (GBS)

The autonomic nervous system is frequently affected in this acute postinfective polyneuropathy. This may lead to sudden death, probably due to cardiac arrhythmia (45). Most patients show sinus tachycardia, due to relative overactivity of the sympathetic innervation of the heart. Vagal involvement is shown by loss of normal sinus arrhythmia and by fluctuations of blood pressure, leading to hypertension and arrhythmias that may presage sudden death (46). The most common problem is sinus bradycardia or asystole occurring after tracheal suction is ventilated patients. The ECG shows atrial and ventricular arrhythmias, often with inverted T waves (47). Cardiac monitoring is advisable in patients with GBS sufficiently severe to warrant assisted ventilation.

Diphtheria

This is a rare disorder in developed countries. Two-thirds of patients with diphtheria develop myocarditis, 3–5 weeks after infection, but this is clinically apparent in only 10% of cases. Sudden death may occur, but congestive heart failure is rare. Myocardial fibrosis sometimes develops and permanent pacing may be needed (48).

Skeletal Muscle in Primary Cardiomyopathies

Skeletal muscle function may be abnormal in patients with primary cardiomyopathies. Cafario et al. studied skeletal muscle in patients with idiopathic dilated and hypertrophic cardiomyopathies (49). They found abnormalities of myopathic type in electromyographic studies in 35% of patients in both groups. Nine patients had muscle biopsies and all were abnormal, showing mild myopathic changes, including central cores in type 1 fibers.

These findings must be assessed in relation to the observation that in patients with congestive heart failure, of any cause, skeletal muscle is abnormal, probably as a result of disuse, hypoxia, underperfusion, and increased venous pressure. There is atrophy of type 1, slow-twitch fibers and of type 2B fibers, features typically associated with disuse atrophy (50).

CONCLUSIONS

There is marked variation in the clinical features and in the severity of cardiomyopathies associated with neuromuscular disease. This complication in this group of disorders has usually been neglected by clinicians. There is a place for regular assessment of vulnerable patients (Table 3) to prevent sudden

Table 3 Need for Cardiological Investigation and Intervention

Diagnosis	Presentation	Risk	Cardiological management
Duchenne dystrophy	Weakness	Sudden death Cardiac failure Tachyarrhythmia	Diuretics, vasodilators, and glycosides
Becker dystrophy	Weakness Cardiac failure	Dilated cardiomyopathy Tachyarrhythmia RBBB	Diuretics, vasodilators, and glycosides. Cardiac transplantation
Emery-Dreifuss syndrome	Muscular syndrome and cardiac failure	Sudden death A-V block	Regular assessment Pacemaker Transplantation
Myotonic dystrophy	Muscular syndrome and cardiac failure	Stokes-Adams syncope and ventricular arrhythmia	Annual ECG and echocardiogram; intracardiac physiology, pacemaker
Mitochondrial cytopathy	Neuromuscular syndrome and heart block or cardiac failure	Heart block	Annual ECG Pacemaker ? Transplantation
Polymyositis	Muscular syndrome	Sudden death Cardiac failure	Glycosides and diuretics Regular cardiac assessment ? Transplantation
Friedreich's ataxia	Neurological syndrome	Cardiac failure Heart block Sudden death	Glycosides and diuretics
Refsum's disease	Neurological syndrome	Sudden death Left heart failure	Glycosides and diuretics
Guillain-Barré syndrome	Acute neuropathy	Tachycardia Hypo- and hypertension Sudden death	Cardiac monitoring during ventilation

death from arrhythmia and to treat cardiac failure, especially in myotonic dystrophy, Becker dystrophy, and the mitochondrial disorders. Cardiac transplantation has begun to be used in Becker and Emery-Dreifuss dystrophies.

The molecular pathogenesis of these cardiomyopathies is currently poorly understood, but there are sufficient similarities to some of the inherited idiopathic cardiomyopathies to suggest that investigation of the former may lead to better understanding of the latter.

REFERENCES

1. Church SC. The heart in myotonia atrophica. Arch Intern Med 1967; 119:176–181.
2 Swash M, Schwartz MS. Neuromuscular Diseases: A Practical Approach to Diagnosis and Management, 2nd ed. London: Springer-Verlag, 1988.
3. Kunkel LM. Analysis of deletions in DNA from patients with Becker and Duchenne muscular dystrophy. Nature 1986; 322:73–77.
4. Hunsacker RH, Fulkerson PK, Baary FJ, Lewis RP, Leier CV, et al. Cardiac function in Duchenne muscular dystrophy: results of 10 year follow up study and non-invasive tests. Am J Med 1982; 73:235–238.
.5 Silverman L, Mendell J, Sahenk Z. Significance of creatine phosphokinase isoenzymes in Duchenne muscular dystrophy. Neurology 1967; 26:561–564.
6. Slucka C. The electrocardiogram in Duchenne's progressive muscular dystrophy. Circulation 1968; 38:933–940.
7. Fitch CW, Anger LE. The Frank vectorcardiogram and the electrocardiogram in Duchenne's progressive muscular dystrophy. Circulation 1967; 35:1124–1131.
8. Nigro G, Comi LI, Politano L, Bain RJI. The incidence and evolution of cardiomyopathy in Duchenne muscular dystrophy. Int J Cardiol 1990; 26:271–277.
9. Lane RJM, Gardner-Medwin D, Roses AD. Electrocardiographic abnormalities in carriers of Duchenne muscular dystrophy. Neurology 1980; 30:499–501.
10. Perloff JK. Cardiac rhythm and conduction in Duchenne's muscular dystrophy. J Am Coll Cardiol 1984; 3:1263–1267.
11. Sanyal SK, Johnson WW, Dische MR, et al. Dystrophic degeneration of papillary muscle and ventricular myocardium: a structural basis for mitral valve prolapse in Duchenne's muscular dystrophy. Circulation 1980; 62:430–439.
12. Wakai S, Minami R, Kameda K, et al. Electron microscopic study of the biopsied cardiac muscle in Duchenne's progressive muscular dystrophy. J Neurol Sci 1988; 84:167–176.
13. Schmidt-Achert M, Fischer P, Pongratz D. Myocardial evidence of dystrophin mosaic in a Duchenne muscular dystrophy carrier. Lancet 1992; 340:1235–1236.
14. Muntoni F, Cau M, Ganau A, Congiu R, Arvedi G, et al. Brief report: deletion of the dystrophin muscle promoter region associated with X-linked dilated cardiomyopathy. N Engl J Med 1993; 329:921–925.
15. Quinlivan RM, Dubowitz V. Cardiac transplantation in Becker muscular dystrophy. Neuromusc Dis 1992; 2:165–167.
16. Palmucci L, Doraguzzi C, Mangini T, et al. Dilating cardiomyopathy as the expression of Xp-21 Becker-type muscular dystrophy. J Neurol Sci 1992; 111:218–221.
17. Steare SE, Dubowitz V, Benatar A. Subclinical cardiomyopathy in Becker muscular dystrophy. Br Heart J 1992; 68.
18. Anan R, Higuchi I, Ichinari K, et al. Myocardial patchy staining of dystrophin in Becker's muscular dystrophy associated with cardiomyopathy. Am Heart J 1992; 123:1088–1089.
19. Merlini L, Granata C, Domenici P, Bonfiglioli S. Emery-Dreifuss muscular dystrophy. Muscle Nerve 1986; 9:481–485.
20. Hara H, Hagara H, Matawatari S, Kodo A, Sato H. Emery-Dreifuss muscular dystrophy. J Neurol Sci 1987; 79:23–31.
21. Nguyen HH, Wolfe JT, Holmes DR Jr, Edwards WD. Pathology of the cardiac conducting system in myotonic dystrophy. J Am Coll Cardiol 1988; 11:662–671.
22. Perloff JK. Stevenson WG, Roberts NK, Cabeen W, Weiss J. Cardiac involvement in myotonic dystrophy: a prospective study of 25 patients. Am J Cardiol 1984; 54:1074–1081.
23. Hawley RJ, Milner MR, Gottdiener JS, Cohen A. Myotonic heart disease: a clinical follow-up. Neurology 1991; 41:259–262.
24. Editorial. The heart in myotonic dystrophy. Lancet 1992; 339:528–529.
25. Chang L, Anderson T, Migneco A, Boone K, Mehringer CM, et al. Cerebral abnormalities in myotonic dystrophy. Arch Neurol 1993; 50:917–923.
26. Petty RKH, Harding AE, Morgan-Hughes JA. The clinical features of mitochondrial myopathy. Brain 1986; 109:915–938.

27. Berenberg RA, Pollock JM, Di Mauro S, et al. Lumping or splitting? "Ophthalmoplegia plus" or Kearns-Sayre syndrome. Ann Neurol 1988; 1:37–39.
28. Coulter DL, Allen RJ. Abrupt neurological deterioration in children with Kearns-Sayre syndrome. Arch Neurol 1981; 38:247–250.
29. Suomalainen A, Paltan A, Leinonen H, et al. Inherited dilated cardiomyopathy with multiple deletions of mitochondrial DNA. Lancet 1992; 340:1319–1320.
30. Lisak RP, LeBeau J, Tucker SH, Rowland LP. Hyperkalaemic periodic paralysis and cardiac arrhythmias. Neurology 1972; 22:810–813.
31. Karpawich PP, Hart ZH, Perry BL, et al. Childhood periodic paralysis with dysrhythmias. Am Heart J 1987; 114:186–187.
32. Haupt HM, Hutchins GM. The heart and cardiac conducting system in polymyositis/dermatomyositis. Am J Cardiol 1982; 50:998–1006.
33. Askari AD, Huettner GM. Cardiac abnormalities in polymyositis/dermatomyositis. Semin Arthritis Rheumatol 1982; 12:208–219.
34. Kobayashi S, Tanaka M, Tamura N, Hashimoto H, Hirose SJ. Serum cardiac troponin T in polymyositis/dermatomyositis. Lancet 1992; 340:726.
35. Bhan A, Baithun SI, Kopelman P, Swash M. Fatal myocarditis with acute polymyositis in a young adult. Postgrad Med J 1990; 66:229–231.
36. Gottdiener JS, Sherber HS, Hawley CJ, Engel WK. Cardiac manifestations in polymyositis. Am J Cardiol 1967; 41:1141–1149.
37. Lotz BP, Engel AG, Nishimo H, Stevens JC, Litchy WJ. Inclusion body myositis. Brain 1989; 112:727–747.
38. Friedreich N. Ueber degenerative Atrophie der Spinalen Hinterstraenge. Virchows Arch Pathol Anat Physiol Klin Med 1863; 26:391–419, 26:433–459, 27:1–26.
39. Harding AE. Friedreich's ataxia. Brain 1981; 104:589–620.
40. Hewer RL. Study of fatal cases of Friedreich's ataxia. Br Med J 1968; 3:639–652.
41. Pentland B, Fox KAA. The heart in Friedreich's ataxia. J Neurol Neurosurg Psychiatry 1983; 46:1138–1142.
42. Skjeldaal OH, Stolce O, Refsum S, Thomas PK. Phytanic acid storage disease. In: Dyck PJ, Thomas PK, eds. Peripheral Neuropathy, 3rd ed. Philadelphia: WB Saunders, 1992:1149–1160.
43. Kyle A, Dyck PJ. Amyloidosis and neuropathy. In: Dyck PJ, Thomas PK, eds. Peripheral Neuropathy, 3rd ed. Philadelphia: WB Saunders, 1992:1294–1309.
44. Husby G, Ranlov PJ, Sletten K, Marhan G. The amyloid in familial amyloid cardiomyopathy of Danish origin is related to prealbumin. Clin Exp Immunol 1985; 60:207–216.
45. Lichtenfeld P, Autonomic dysfunction in Guillain-Barré syndrome. Am J Med 1971; 50:772–780.
46. Winer JB, Hughes RAC. Identification of patients at risk of arrthythmia in Guillain-Barré syndrome. Q J Med 1988; 68:735–739.
47. Hughes RAC. Guillain-Barré Syndrome. London:Springer-Verlag, 1991:133.
48. Kurdi A, Abdul-Kader M. Clinical and electrophysiological studies of diphtheritic neuritis in Jordan. J Neurol Sci 1979; 42:243–250.
49. Cafario ALP, Rossi B, Risaliti R, et al. Type 1 fibre abnormalities in the skeletal muscle of patients with hypertrophic and dilated cardiomyopathy. J Am Coll Cardiol 1989; 14:1464–1473.
50. Swash M, Schwartz MS. Muscle Biopsy Pathology, 2nd ed. London: Chapman and Hall, 1991.

56
Physical Therapy of Muscle Diseases

Sylvia Hyde
Muskelsvindfonden, Arhus, Denmark

INTRODUCTION

Impairment of function, or, in the case of children, failure to achieve functional competence, is the main problem for patients with muscle disease. The elegance of movement and high level of physical performance of the unaffected is almost never seen in those with muscle disease. Function is dependent on the interrelationship of muscle strength, range of motion, physical ability, and the anthropometric parameters of height and weight. Clearly, in muscle disease, where muscle strength is often profoundly compromised and where the secondary effects of contracture and deformity affect the range of motion available in the spine and peripheral joints, this interrelationship is markedly disturbed. Further, the integrity of the architecture of the spine and peripheral joints, which provide the fulcrum for movement, play an important role in function, and the strength to body mass ratio, together with skeletal lever length, is a critical factor in the biomechanical status of the individual.

Change in the normal balance of muscle strength between agonist and antagonist muscle groups may force the patient into adopting unusual or compensatory postures, leading to the development of contractures and deformity, the pathomechanics of the disease.

Further considerations in the physical management of muscle disease include the age of onset of the specific disease and whether the disease is progressive or nonprogressive. Many of the disease entities discussed in this book are congenital; therefore, in planning strategies of physical management, the physical therapist must be aware of normal maturational processes and motor development.

The role of the therapist is essentially supportive. This does not imply passivity, but rather that the emphasis in physical treatment is firmly placed on the secondary or consequential aspects of preventing disuse atrophy, contractures, and deformity, while maintaining function through the use of compensatory mechanisms, appropriate orthoses, and aids to mobility.

In summary, the physical therapist's role in the general management of muscle diseases may be said to be:

To maintain or improve muscle strength
To prevent/retard the secondary complications of contractures and deformity
To promote or prolong ambulation
To promote and maintain function
To facilitate psychosocial development and interaction

ASSESSMENT

The cornerstone of successful management is assessment since this will provide the baseline data against which progression of the disease process and/or the effect of therapeutic intervention may be evaluated.

The main advantages of a comprehensive, well-documented assessment is that it allows the team of professionals managing the patient to identify, well in advance, critical times for intervention and, by doing so, largely avoids crisis management, the latter being second only to the diagnosis in order of distress factors for the patient and his/her family. The clinician is able to discuss potential difficulties and their solutions and thus give the patient and family members the time and opportunity to make informed decisions about choices in management strategies. This is particularly important in relation to decisions concerning prolongation of ambulation and the use of ventilator support in advanced respiratory insufficiency.

The assessment must be comprehensive, accurate, standardized, quantitative, and sequential because of the global involvement and often progressive natural history of the disease.

Various methodologies for assessment are cited in the literature (1–3). Most include measures of muscle strength, range of motion, measures of physical performance, such as timed task activities, and an overall measure of function. Other specific tests include those for respiratory function and gait analysis, but the inclusion of such tests in the assessment protocol is usually disease-specific. There are assessment protocols that include the use of sophisticated equipment, such as the Cybex range of dynamometers or the Lido, but this equipment is not universally available and therefore is less relevant in the clinical situation (4).

Recognition of the need for standardized protocols for measuring/monitoring the physical parameters has assumed greater importance as the prospect for gene and cell therapy making the transition from a laboratory technique to the clinical trial stage draws closer to reality. Standardized methods of measuring the individual parameters of muscle strength, range of motion, and function have been rigorously investigated, and the studies are robust in their statistical treatment to explore validity, reliability, and inter- and intratester reliability (5–7).

There can be no doubt of the importance of such studies for the purposes of research, but for the physical therapist involved in everyday management of patients with muscle disease, the interrelationship of the various parameters assumes greater importance in decision making. Lord et al. (8) further make the point that tests for function do not necessarily reflect the patient's everyday activity level.

METHODS OF TREATMENT

The methods of treatment used in managing the patient with one of the muscle diseases are exercise, passive stretching, the provision of night splints, and the techniques of rehabilitation associated with the provision of orthoses or, where applicable, aids to mobility (9–12). In addition, there are the specialized techniques necessary to manage reduced respiratory function; these are the use of mechanical adjuncts, such as intermittent positive-pressure breathing (IPPB) or periodic continuous positive airways pressure (PCPAP), together with modified postural drainage positions, to prevent atelectasis with small airways collapse and to assist pulmonary hygiene. Respiratory muscle training using the Threshold Inspiratory Muscle Trainer or similar device is beneficial in improving respiratory muscle endurance, although its role in improving strength remains controversial (13–15).

In those diseases that present at birth with marked hypotonia, for example nemaline myopathy and congenital muscular dystrophy (CMD), the physical therapist uses neurodevelopmental techniques to assist the child achieve the normal motor milestones. In such infants the additional problems of the development of postural contractures must be addressed, and this is done by the provision of appropriate use of supports, such as rolls or wedges, and advice to nursing staff and/or parents.

Bipedal stance and the functional task of walking are complex motor activities requiring the integration of a host of sensory and motor inputs, together with sufficient muscle strength to overcome the force of gravity. Profound proximal muscle weakness, often aggravated by the presence of lower extremity contractures, may make independent ambulation unsustainable or, where the onset of the weakness is at birth, unobtainable without surgical and/or orthotic intervention to prolong or initiate ambulation. Duchenne muscular dystrophy is an example of the former and congenital muscular dystrophy of the latter.

The physical therapist has a wider role to fulfill in respect to patient education, ensuring that the patient and his or her family members understand thoroughly the why, when, and how of the management strategy and that they fully understand the implications of the findings of assessment so that they are able to make informed choices.

Hospital attendance should be kept to a minimum and a carefully constructed critically monitored home program of exercises based on the findings of the assessment should be devised. It is important in instituting programs of physical management to ascertain the individual's and the family's normal cycle and rhythm of daily activities to discover times in the day when training programs will not conflict with school, work, recreation, and social activities, as this will result in greater patient compliance.

Successful management of the patient with muscle disease is a team endeavor, so the physical therapist must ensure that the findings of assessment, changes in treatment regimes, and patient/family anxieties are known to all members of the team. It is essential with children, for example, that school teachers are kept informed and given practical help on how to facilitate the child's progress in school and how to encourage his full participation in class activities leading to integration with his or her peers.

Advice on adaptive sports and recreational activities, together with the expert prescription of wheelchairs for mobility, will help with the patient's psychosocial development. Self-image is an important part of this, and the prevention of disabling, and often cosmetically unacceptable, deformities of the hands and spine makes a positive contribution.

In the later stages of the progressive diseases and in patients with onset in adult life, discomfort of mechanical origin, as a result of postural strain on the spine and peripheral joints, responds well to techniques of mobilization or manipulative therapy.

Lifting and handling techniques for transferring, lifting, and turning should be carefully taught to those caring for muscle disease patients in order to avoid back injury, while making the patient feel safe. Simple advice on back care is also helpful for caregivers.

Where corrective surgery is performed in patients with muscle disease, for example, in arthrogryposis or spinal surgery for neurogenic scoliosis, the postsurgical physiotherapy will be as directed by the orthopedic surgeon, but the physical therapist must be vigilant in preventing disuse atrophy and postural contractures. After spinal surgery for correction of scoliosis, it is essential to review the patient's seating position and chair requirements. It is usually necessary to work intensively at hand and arm function at this stage because many patients will have relied on compensatory movements for fulfilling activity of daily living needs and these may have been compromised by the changed spinal posture.

TREATMENT OF SPECIFIC DISEASES

Duchenne Muscular Dystrophy (DMD)

The areas of emphasis in treatment at arbitrary stages are given in Table 1, and only attempts to prolong ambulation and intermediate-stage respiratory treatment are described.

Loss of independent ambulation usually occurs when muscle strength has fallen to 50% and the child is scoring 18–22 on the Motor Ability Score (3). At this time the locus of center of gravity has moved forward to lie outside his base and balance is precarious (16). Growth spurts or concomitant illness at this time may be precipitative factors.

When the method of Siegel (17) is followed for prolongation of ambulation, percutaneous tenotomies of the tendo Achillis, knee flexors, and iliotibial band are performed prior to fitting lightweight polypropylene knee-ankle-foot orthoses (KAFOs). Sometimes only the TA requires release. Long-leg plasters (POPs) are applied at the time of surgery, and the following day the physiotherapist stands the child in the plasters and begins rehabilitation. The emphasis is on training the child to make the postural adaptation, particularly of the pelvic tilt, required by the plantigrade foot position. This is achieved through resisted trunk work in the parallel bars and by mat work. In addition, weight acceptance and transference, forward, sideways, and backward, is practiced as preparation for walking. Where the child has severe equinovarus contractures prior to surgery, it may be necessary to use a small

Table 1 The Emphasis in Physical Therapy Treatment at Different Stages in Duchenne Muscular Dystrophy

Stage	Emphasis	Method
Early	Maintenance of strength	Exercise
	Prevention of contractures at hip, knee, ankle.	Passive stretching, night splints
	+	+
Intermediate	Prevention of hand and wrist contractures	Passive stretching, night splints
	Maintenance of vital capacity	Inspiratory muscle training
		Breathing exercises, postural drainage
	Prolongation of ambulation	Orthoses reeducation, ambulation
	Aids to mobility	Wheelchair
	Wheelchair and seating	Advice
	+	+
Late	Control of contractures	Splinting
	Prevention of atelectasis	PCPAP, postural drainage, assisted cough

felt wedge under the heel of the plaster during the first days to assist in the transition from the equinus to plantigrade foot position. Extension of the upper trunk is needed to achieve a sitting position over the posteromedial lip of the orthoses. Three days after surgery the POPs are bivalved so that the orthotist can make the mold for the definitive orthoses, which are usually ready for fitting 3–5 days later. Rehabilitation continues until the child is walking satisfactorily, normally within a week.

Studies have shown that the use of KAFOs may extend the effective walking period by an average of 3 years. A major criterion for successful outcome has been found to be the strength of the hip abductors at the time of application (18,19).

In DMD deterioration in respiratory function is the result of reduced strength of the respiratory muscles and decreased pulmonary and chest wall compliance, causing progressive restrictive functional impairment.

The physical therapist is involved in three situations:

1. Maintenance of muscle strength and vital capacity
2. Prophylaxis with respect to pulmonary hygiene
3. Assistance in clearing secretions in the presence of intercurrent chest infection or in the event of aspiration

The therapist managing patients in the later stages of the disease should be alert for any signs of silent respiratory insufficiency, for example difficulty in concentration, change in emotional response, irritability, loss of appetite with resultant loss of weight. It is also advisable to monitor the vital capacity more closely, since it is not uncommon for this to be less than 1 L at this stage, so small changes assume magnitude. Prophylactic use of PCPAP night and morning has been shown to be helpful in clearing secretions.

Late-stage disease also tends to cause the child to reduce his fluid intake, because of the difficulty of transfer to the toilet, and it is therefore important to persuade him to take sufficient fluids to avoid dehydration.

Becker Dystrophy

This condition requires the same general pattern of physical management, but because of its more benign course, the period of emphasis on strengthening exercise is often much longer, and in many cases orthotics may not be required until adult life, if at all.

Limb-Girdle Dystrophy

Advances in diagnosis based on improved understanding of the specific genetic deletion present are likely to change this diagnostic category, but for the purposes of this chapter comments relate to those patients who present with an autosomal recessive or dominant dystrophy, of variable progression and severity. Physical therapy follows the same program for preventing contractures, maintaining muscle strength, and prolonging ambulation as outlined for DMD, based on the assessment. These patients tend to develop even more marked hyperlordosis and will not infrequently complain of debilitating back pain. This is presumably mechanical and postural; it responds well to the mobilization techniques of manipulative therapy.

Congenital Muscular Dystrophy (CMD)

The physical therapist has four concerns with the infant presenting with CMD, namely the delay in achieving motor milestones, prevention of further contractures due to abnormal postures adopted as a result of the hypotonia, improvement in muscle strength, and management of contractures present at birth.

Although the hypotonia present at birth resolves, delay in the attainment of normal milestones may further compromise the infant's later function by allowing the development of further postural contractures. This will be particularly important if the child has deformities as a result of associated arthrogryposis. It is therefore essential that the therapist works with the child and the parents using neurodevelopmental techniques to help the attainment of motor milestones and provides the stimulation necessary for interaction. Full-range passive movements of all joints are done daily, and the baby's resting posture may need to be controlled using rolls or wedges to prevent the development of contractures and early scoliosis. If the hypotonia resolves so that the infant is sitting at an age-appropriate time, the emphasis in treatment is on strengthening exercises leading to standing and functional activities. However, if severe truncal hypotonia persists, it is necessary to use molded jackets to bring the child to stable sitting without prejudicing the spine. The child is then able to use his/her arms and visually explore the environment, thus beginning the interaction necessary for social development. Passive stretching of any contractures continues.

The use of serial plasters, or splinting for correction of contractures, although beneficial in some cases, must be approached with caution because of the risk of causing further atrophy and weakness. It is also essential to ensure that there is no structural malformation underlying the contracture. In children particular care must be taken to ensure that the skin is not compromised, and overcorrection must be avoided.

Once the child has sufficient trunk and pelvic control, weight bearing is encouraged. If there is insufficient power in the knee extensors, small gaiters are used to support the knees and the child is encouraged to stand and play in a flexistand with table.

When the child is accepting weight through the hips and lower extremities, it is usually possible in all but the severest cases to fit orthoses for ambulation.

Medical opinion on the timing and extent of surgical intervention for the correction of deformities varies considerably; the physical therapy treatment postsurgically must respond to the orthopedic surgeon's prevailing prescription. The therapist, however, must also be even more vigilant in preventing disuse atrophy than in a patient unaffected by muscle disease.

Congenital Myopathies

The infant with myopathy presents the physical therapist with a great challenge with respect to helping him or her reach adult life, functioning and performing to the optimum within the constraints of the disease process. There are several disease entities, each with its own natural history; it is therefore essential that there is a definitive diagnosis as this provides the global framework within which to work. The aims and objectives of treatment and the methods of achieving them are essentially the same as those previously described, the variation in emphasis being dictated by the natural history.

Nemaline Myopathy

This condition presents at birth with often marked hypotonia; therefore, the therapist concentrates on helping the infant initiate movement, to overcome the delay in achieving motor milestones. These children often have particularly floppy heads, and strategies must be devised to encourage the development of truncal and limb control while working to gain head control. It is important to prevent the development of contractures that arise from postural adaptation.

Myotubular Myopathy

Patients with this condition usually respond well to physical therapy, but the major problem is the severe hyperlordosis that many develop as they enter adolescence. This is particularly disabling if asymmetrical contractures have been allowed to develop at the hips and will lead to loss of independent ambulation. Emphasis in this group of patients is therefore on preventing contractures.

Dermatomyositis

Dermatomyositis of childhood requires the physical therapist to perform two distinct tasks:

1. To provide accurate quantitative force measurements of the major proximal muscle groups for monitoring the effect of drug therapy (20)
2. To provide treatment during the acute and chronic phases of the disease

In the acute phase, when the child complains of pain and is generally very weak, passive range of movement is done on all the joints involved, and resting splints are provided to prevent contractures and reduce discomfort. Postural advice is provided while the patient is on bed rest; exercise is limited by pain and should as far as possible be static work. As drug therapy affects remission, a carefully controlled, supervised program of strengthening exercises is begun and maintained until full function is restored or the condition plateaus. It is usually necessary to continue to monitor the patients over a long period as small changes in muscle strength usually herald an exacerbation of the disease process.

Chronic dermatomyositis, where contractures have developed and where calcinosis is often present, requires an aggressive physical therapy program of strengthening exercises, mobilization, and the use of serial splinting in order to realize the patient's full potential for function.

SUMMARY

The use of simple physical therapy methods to maintain function make an important contribution to the management of muscle disease patients. Interpretation of the assessment and understanding the interrelationship of the many variables is the key to designing successful strategies of management.

REFERENCES

1. Brooke MH, Griggs RC, Mendell JR, Fenichel GM, Shumate JB, Pellegrino RJ. Clinical trial in Duchenne dystrophy. I. The design of the protocol. Muscle Nerve 1981; 4:186–197.
2. Vignos PJ, Spencer GE, Archibald KC. Management of progressive muscular dystrophy in childhood. JAMA 1963; 184:89–110.
3. Scott OM, Hyde SA, Goddard C, Dubowitz V. Quantitation of muscle function in children: a prospective study in Duchenne muscular dystrophy. Muscle Nerve 1982; 5:291–301.
4. Merlini L, Dell'Accio D, Holzl A, Granata C. Isokinetic muscle testing in neuromuscular diseases: preliminary report. Neuromusc Disord 1992; 3:201–207.
5. Barr AE, Diamond BE, Wade CK, Harashima T, Pecorella A, Potts CC, Rosenthal H, Fleiss JL, McMahon DJ. Reliability of testing measures in Duchenne or Becker muscular dystrophy. Arch Phys Rehabil 1991; 72:315–319.
6. Florence JM, Pandya S, King WM, Ribison JD, Signore LC, Wentzell M, Province MA. Clinical trials in Duchenne dystrophy: standardisation and reliability of evaluation procedures. Phys Ther 1984; 64(1):41–45.
7. Pandya S, Florence JM, King WM, Robison JD, Oxman M, Province MA. Reliability of goniometric measurements in patients with Duchenne muscular dystrophy. Phys Ther 1985; 65:1339–1342.

8. Lord JP, Lieberman JS, Portwood MM, Fowler WM, Carson R. Functional ability and equipment use among patients with neuromuscular disease. Arch Phys Med Rehabil 1987; 68:348–352.
9. Fowler WM. Jr. Taylor M. Rehabilitation management of muscular dystrophy and related disorders. 1. The role of exercise. Arch Phys Med Rehabil 1982; 63:319–321.
10. Scott OM, Hyde SA, Goddard C, Dubowitz V. Prevention of deformity in Duchenne muscular dystrophy: a prospective study of passive stretching and splintage. Physiotherapy 1981; 67(6):177–180.
11. Brooke MH, Fenichel GM, Griggs RC, et al. Duchenne muscular dystrophy: patterns of clinical progression and effects of supportive therapy. Neurology 1989; 39:475–481.
12. Vignos PJ. Jr. Physical treatment in neuromuscular diseases. Semin Ortho 1987; 2(3):162–172.
13. Martin AJ, Stern L, Yeates J, Lepp D, Little J. Respiratory muscle training in Duchenne muscular dystrophy. Dev Med Chil Neurol 1986; 28:314–318.
14. DiMarco AF, Kelling JS, DiMarco MS, Jaccobs I, Shields R, Altose MD. The effects of inspiratory muscle training on respiratory muscle function in patients with muscular dystrophy. Muscle Nerve 1985; 8:284–290.
15. Estrup C, Lyager S, Noeraa N, Olsen C. Effect of respiratory muscle training inpatients with neuromuscular diseases and in normals. Respiration 1986; 50:36–43.
16. Barrett R, Hyde SA, Scott OM, Dubowitz V. Changes in the centre of gravity in boys with Duchenne muscular dystrophy. Muscle Nerve 1988; 11:1157–1163.
17. Siegel IM. Plastic-molded knee-ankle-foot orthosis in the management of Duchenne muscular dystrophy. Arch Phys Med Rehabil 1975; 56:322.
18. Heckmatt JZ, Dubowitz V, Hyde SA, Florence J, Gabain AC, Thompson N. Prolongation of walking in Duchenne muscular dystrophy with lightweight orthoses: review of 57 cases. Dev Med Child Neurol 1985; 27:149–154.
19. Hyde SA, Scott OM, Goddard CM, Dubowitz V. Prolongation of ambulation in Duchenne muscular dystrophy by appropriate orthoses. Physiotherapy 1982; 68:105–108.
20. Heckmatt J, Saunders C, Peters AM, et al. Cyclosporin in juvenile dermatomyositis. Lancet 1989; 1:1063–1066.

57
Surgical Care and Orthotics in Muscle Disease

John D. Hsu and Yuriko Lynn Yasuda
Rancho Los Amigos Medical Center, Downey, and University of Southern California, School of Medicine, Los Angeles, California

GENERAL PRINCIPLES

Surgical procedures need to be planned and tailored to the exact needs of the neuromuscular disease patient. Used properly, they can relieve pain and discomfort, improve function, and allow a greater amount of independence for a physically disabled individual. Although many surgical procedures on the spine and extremities have been developed, few have been used for the neuromuscular patient. The nature and course of the neuromuscular disorder must be known. In a patient with a stable course, such as a child with congenital myopathy or a minimally progressive neuropathy, standard orthopedic procedures developed for a normal child or adult may be applicable. However, when there is progression of the disorder, whether slow or rapid, special orthopedic surgery procedures or modification of commonly used and successful methods may need to be considered (1,2).

Orthoses, when used properly, provide support. They are especially welcomed by patients and their therapists when they can augment or improve function. They are readily discarded by the user (1) when they are cumbersome, (2) when they interfere with normal or adaptive activities, and (3) if they break down and require frequent repairs. The prescriber of an orthotic device needs (1) to be totally knowledgeable about the diagnosis and clinical course of the disease, (2) to have some idea of the prognosis, (3) to understand the results of the manual muscle test and functional evaluation, and (4) to be able to work with the patient and his/her family. Success occurs when the patient is comfortable, at ease in using the well-fitting orthotic device, and can take advantage of it. Because a patient with a neuromuscular disorder presents with muscle weakness, the use of the orthotic device is to support, rather than prevent or correct active deforming forces. Thus, the orthoses must not be made out of heavy material and should be as light as possible, using state-of-the-art, space-age materials if possible, with components that do not break or joints that do not become stuck with repeated use.

SURGERY AND ORTHOSES USE

In many instances, orthoses are used in conjunction with a surgical procedure. If a joint is to be fused and made immobile, it is reasonable to use an orthosis as a trial to fix that joint and see if the desired result can be accomplished by the surgical procedure. On the other hand, after a surgical correction, an orthosis can be used to provide postoperative support either temporarily or permanently, such as in the early phases of healing of a tendon transfer where maintenance of the corrected position postcasting needs to be made orthotically.

SPECIFIC SURGICAL PROCEDURES

Lower Extremity

Hip Flexion Contracture

Fixed flexion contractures of the hips greater than 30 degrees become undesirable for standing and efficient walking. Flexion contractures can occur because of positioning, prolonged all-day seating, or muscle imbalance. Joint contractures can also occur after an inflammatory insult. Prior to surgery, stretching and active ranging are necessary. Contracture release can be accomplished via muscle and tendon lengthenings and hip joint capsule incision and maintained through positioning, ranging, and active motion. In the permanently seated patient, up to 90 degrees of hip flexion contracture can be acceptable; however, proning for comfort and pulmonary toileting is no longer possible, and if the patient desires to sleep supine, the legs would need to be supported.

Knee Flexion Contracture

Tightness of the iliotibial band, tightness of the hamstrings, weakness of the knee extensors, and gravitational forces all contribute to the development of a knee flexion contracture. Mild contractures can be released by resection of the distal iliotibial band (Yount procedure) with or without hamstring releases or fractional lengthening at the musculotendinone junction, followed by casting in the corrected position. Severe contractures may require posterior capsular release. Serial casting without joint capsule release can improve the knee position but may cause posterior knee joint subluxation.

Ankle and Foot Deformities

In the patient with a neuromuscular disorder, muscle imbalance around the ankle and foot occurs because of selective muscle weakness or muscle wasting (Fig. 1). Abnormal muscle pull in one direction is the result of the imbalance between agonist and antagonist, influenced by positioning and gravity. Dropfoot occurs when the dorsiflexors of the ankle are overpowered by the plantar flexors. If the gastrocnemius-soleus muscles becomes weakened or nonfunctioning and the anterior tibial is left unopposed, a calcaneal foot can develop. Intrinsic weakness with strong long flexors may lead to a high arch, or pes cavus. If the peroneals become inactive in the face of normal foot inverters, a varus position results. When this situation occurs with a weak ankle dorsiflexor, the foot is pulled into equinovarus. If efforts are not made to correct and maintain proper alignment passively, the deformity can become fixed. In the growing child, the bones can develop with their shapes and joint surfaces distorted. Surgical

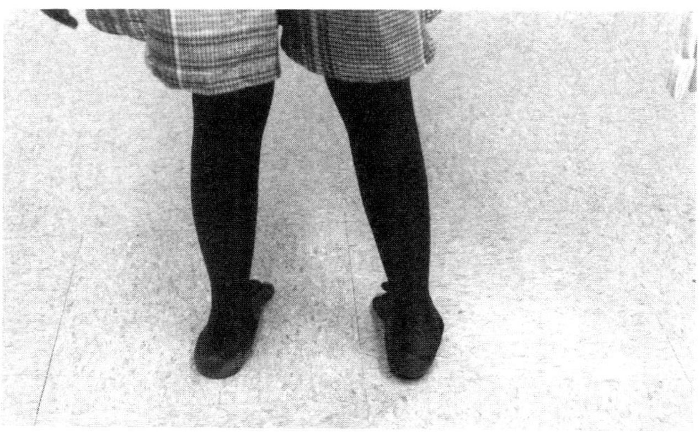

Figure 1 Child with congenital myopathy. Left foot is in valgus with pes planus (flatfoot). Right heel is in varus.

correction for foot and ankle muscle imbalance (1) relieves pain, (2) corrects the deformity, and, if possible, (3) restores function.

Standard orthopedic procedures have been developed to correct muscle imbalance via muscle and tendon releases or tendon transfers (Fig. 2). If bony deformity is also present, this also needs to be addressed in the overall preoperative planning. Osteotomies and bone fusions (arthrodesis) including triple arthrodesis (fusion of the talocalcaneus, calcaneocuboid, and talonavicular joints) stabilize the hindfoot in the mature foot (Fig. 3). It is recommended that after the bony deformities have been corrected, muscle rebalancing operations may need to be considered to avoid recurrence or loss of correction.

In the older patient, well-planned, staged procedures consisting of bony correction, tendon transfers, and postoperative bracing become the treatment of choice. In the patient whose feet are still growing and immature, initial tendon transfers to rebalance a foot may lead to more normal bone growth and maturing of the skeletal structure of the foot and ankle. They can also prevent the formation of fixed joint contractures with bony deformities. Fixed foot and ankle deformities can lead to a loss of ambulation in the child with Duchenne muscular dystrophy, and if left these deformities are uncorrected, loss of walking can occur leading to secondary complications (3–5).

Spine

Mild spinal curvatures need to be recognized. Principles developed for the treatment of idiopathic scoliosis cannot be followed in the patient with neuromuscular disorder because weak muscular support causes spinal column collapse. Mild spinal curves can be supported by orthoses or seating devices (Fig. 4), but in the ambulatory patient encumbering the spine with a corset interferes with balance and walking. A patient with a spinal curvature needs to be followed closely, and if there is any doubt about the nature and degree of the spinal curve, radiographic examination of the entire spine should be made.

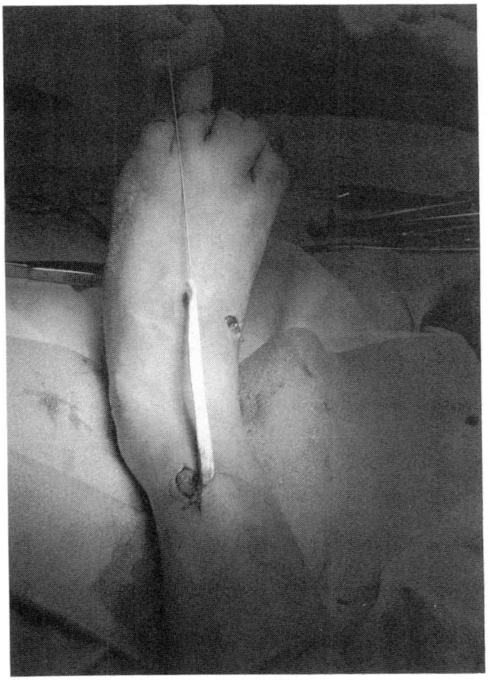

Figure 2 Tendon transfer—anterior tibial tendon exposed on medial side of the foot, detached and brought externally. Tendon will be tunnelled through the skin and reattached to the midlateral portion of the foot.

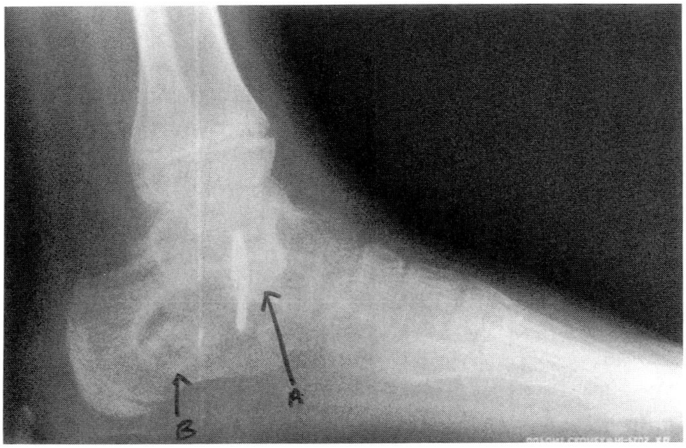

Figure 3 Radiograph showing subtalar fusion with staple between talus and calcaneus. (A) Subtalar joint; (B) donor site for bone graft used for fusion.

If there is progression of the curve, further measures are needed before the curve becomes fixed and difficult to correct. When collapse progresses to the degree affecting standing, walking, and sitting, generally a curve over 35 degrees, spinal fusion should be considered (6). A neuromuscular curve of 35–40 degrees can be corrected fully. More severe curves can only be corrected partially, and if insufficient numbers of segments are fused, a "falloff" with progression of the curve either proximally or distally to the fusion mass can occur (Fig. 5a,b). Spinal curve for the young child with much growth left is very difficult to treat. If spinal fusion is done too early, secondary problems can occur. Posterior spinal fusion arrests posterior growth, but the anterior part of the spine continues to grow, resulting in severe lordosis, which presents with difficulty in seating as the head and neck will tilt back into extension more and more with increasing age.

Upper Extremity

Functional needs are to be carefully assessed. A certain degree of shoulder, elbow, wrist, and hand contracture can be acceptable. If the elbow becomes fixed in a nonfunctional position, surgical releases of the tight structures and capsule may be indicated. Wrist and hand contractures can be reduced by proper positioning. If finger contractures become severe and interfere with proper hygiene, the increasing moisture and pressure will cause skin changes leading to ulcerations. Appropriate flexor tendon release or fractional lengthening is indicated. Opponensplasty can be used to strengthen the thumb if there is a muscle strong enough and available for this tendon transfer.

ORTHOSES (BRACES, CALIPERS)

Spinal Curvature Support and Control

A spinal curvature that can be overcome or improved by stretching or positioning against gravitational forces is known as a correctable or flexible curve; this frequently occurs in the neuromuscular patient due to the weakness of the supporting muscle. If the curvature is ignored or untreated, muscle contractures will develop, resulting in a permanent deformity and the development of a fixed curve. Spinal curves are generally never uniplanar. Scoliosis is generally associated with kyphosis, forming kyphoscoliosis. A three-dimensional analysis of the spinal curvature will show rotational changes.

Because the nature of the spinal curve in neuromuscular scoliosis is different than that of a patient with idiopathic scoliosis, orthoses made for the latter cannot be directly applicable. For neuromuscular

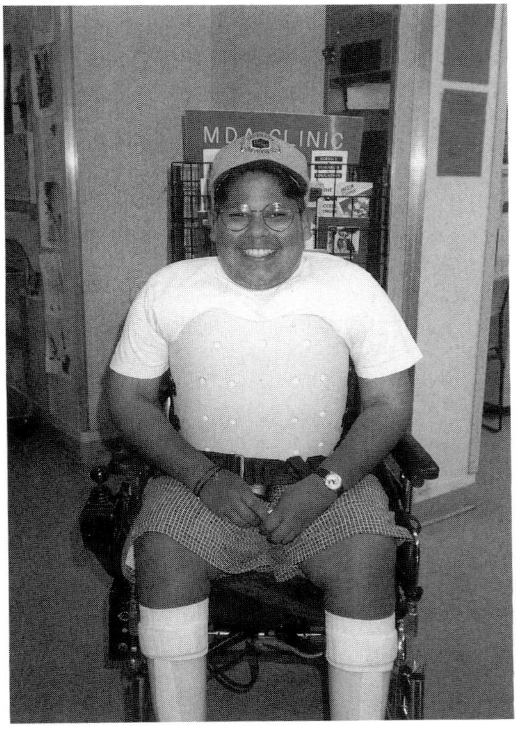

Figure 4 Ten-year-old boy with SMA—TLSO (body jacket) used for spinal support.

scoliosis, some of the techniques and materials developed for the Milwaukee brace and its subsequent modification have proven useful in the care of the neuromuscular spine (7).

Thoracolumbar Spinal Orthoses (TLSO)

The goal for the use of a spinal orthoses is to:

1. Support the spinal column
2. Allow for independent sitting
3. Free up the arms to maintain use and function
4. Provide support for the head

In dealing with a flexible curve, a high axillary TLSO can be designed to support the spine. This orthosis needs to be custom-fit and molded to the back in the corrected position (seated patient). Gravitational forces and further changes in the muscle tissue may cause further collapse; therefore, frequent inspection of the skin is necessary. Areas of redness on the skin caused by pressure should not be ignored. Minor changes to the orthosis can give relief, and if the orthosis does not fit well, it needs to be replaced. Radiographs taken with and without the orthosis can show its effectiveness.

A TLSO may restrict chest expansion, breathing, and eating. If this is a problem, the orthosis can be designed to allow for an anterior chest and/or "belly hole" to be cut out. With modern plastic and high-technology materials available, the intrinsic strength of the orthosis can be maintained even though large portions of the frame have been removed.

If a spinal orthosis cannot hold the spine and no longer serves the function it was ordered for, a spinal fusion operation should be considered to prevent further collapse if the patient does not have other adverse conditions such as respiratory failure or the inability to bring up secretions.

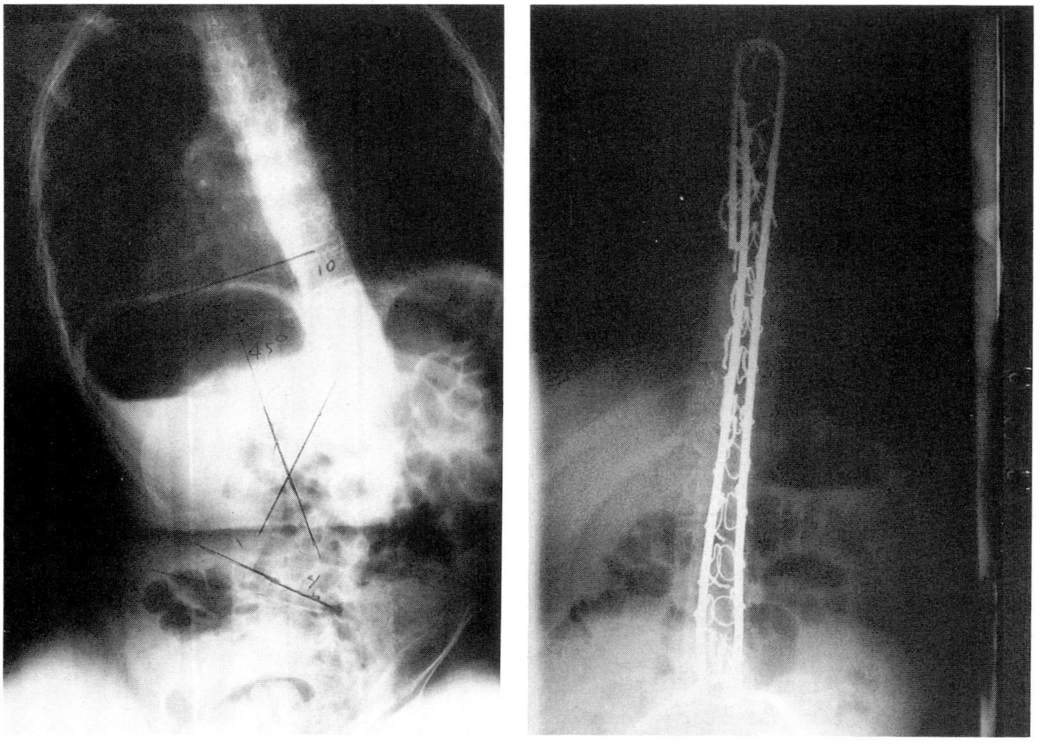

(a) (b)

Figure 5 SMA patient with 45-degree scoliosis (between T10 and L4), corrected to almost neutral (straight) by spinal fusion with segmental spinal instrumentation.

Outriggers

Frequently, neuromuscular patients lack good neck musculature. Support may be needed in either flexion or extension of the head. If head control is lacking because of weak extensors, a headband can be attached to an outrigger.

Lower Extremity Orthoses

Ankle-Foot Orthoses (AFO)

Support of the ankle is indicated when the musculature to dorsiflex the ankle is insufficient or is overpowered by the plantar flexors. This would (1) enable the remaining musculature to function better, (2) prevent the formation of a fixed equinus contracture at the ankle, and (3) allow support of the foot in walking in the ambulatory patient. In the nonambulatory neuromuscular patient, an AFO would position the feet to be placed in the wheelchair footrests in the proper supported position (12).

Although the double upright brace connected to a shoe via a movable or nonmovable joint at the ankle has been used successfully, the author feels it would be more appropriate for ultra-lightweight polypropylene orthosis to be prescribed for the neuromuscular patient, especially a walking patient (Fig. 6a,b). These types of orthoses can fit into store-bought shoes and encumber the limb minimally by weight. Frequently, there may be an objection to the use of a plastic orthosis in warmer climates because of sweating, but this secondary problem can be alleviated by cutting out segments of the orthosis

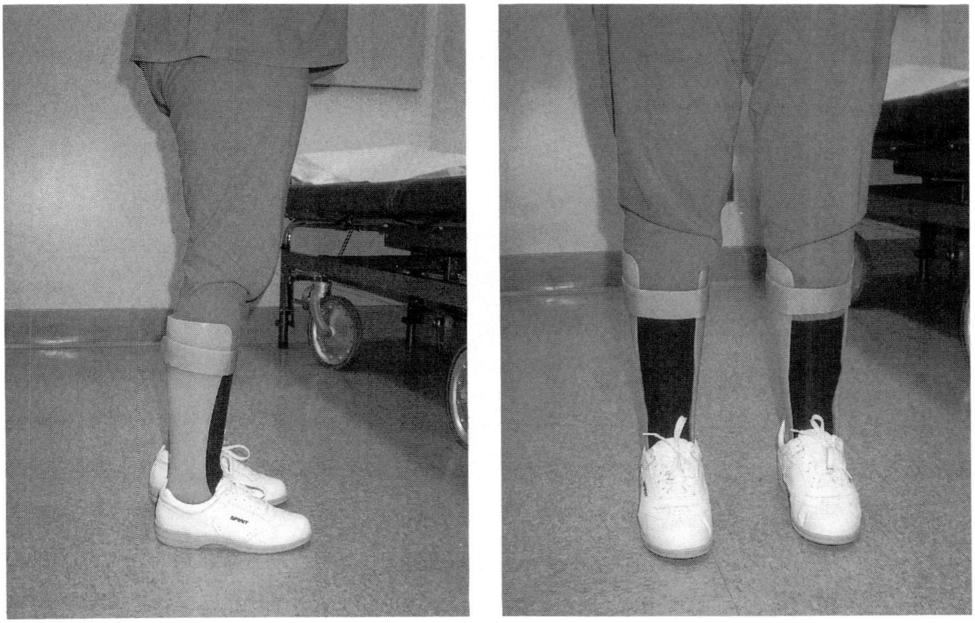

(a) (b)

Figure 6 CMT patient—polypropylene ankle-foot orthoses (AFO) fitted to the patient, using running shoes.

to allow for ventilation or by utilizing mositure-absorbing stockings. A skilled orthotist can usually provide appropriate support to the orthosis by using additional material or carbon fiber at key points.

The foot should be checked frequently when an AFO is used. If persistent muscle imbalance causes the foot to turn into varus or valgus and if abnormal callosities form at the bottom of the foot or at the skin-orthosis interface; then rebalancing the foot, either surgically or by dynamic devices, hinging the orthosis at the ankle in the correct plane, can give pressure relief.

Knee-Ankle-Foot Orthoses (KAFO)

Long leg braces are used to support a weakened quadriceps muscles and are frequently prescribed for ambulatory neuromuscular patients with proximal weakness when further standing and walking is still possible and desirable (8,9).

KAFO cannot be used successfully for these purposes if there is excessive hip flexion contracture (over 35 degrees). Knee flexion contractures may need to be reduced by serial casting or surgical release, so that the braces can be locked with the knee in extension. Ankle and foot deformities, when severe, need to be surgically treated. The droplocks need to be reliable and usable by the patient. Special locks at the knee joint have to be incorporated into the brace to allow for release from the extended position (Fig. 7). For instance, if the hands and fingers are weak, as in a Charcot-Marie-Tooth patient, and if the droplocks take too much effort to release the brace, the patient will find it difficult to sit in a wheelchair.

When KAFO is prescribed for the growing child, particularly close follow-up is required. Adjustments for growth may need to be made in many areas. The size of the plastic components, the length of the supporting bars and position of the knee and ankle joints, and the height of the brace must be periodically checked.

The author has found that when KAFO is used to enable a child who has weakened lower extremity musculature to continue ambulation, the KAFO has to "fit" almost perfectly. This is certainly possible and an experienced orthotist can make minor modifications and adjustments. Unfortunately, as more

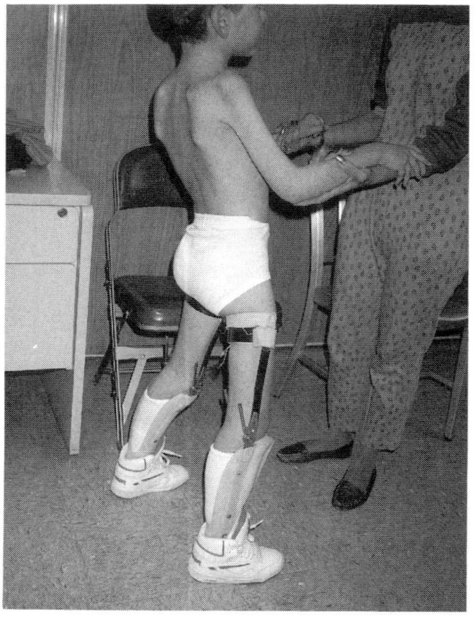

Figure 7 Knee-ankle-foot orthoses (KAFO) worn by boy with DMD. Note height of brace for ischial support, Also lightweight leg shell, carbon fiber uprights, and running shoes to minimize weight of orthoses.

plastic (polypropylene) is used and custom-molded, changes can be very difficult to make. If the orthosis does not fit well and does not allow the child to balance, the child becomes apprehensive and may not cooperate in the postbracing training, which is very important to the overall success of an ambulation program.

Additionally, with patients who have progressive weakness, frequent reevaluation of the use of the orthosis needs to be made. When function becomes impossible, the orthosis can be modified, for instance, cut down to an AFO for continued support of the ankle and feet in the nonambulatory person.

The knee and the ankle joints and shoes should be as lightweight as possible. Frequently, joints can be made out of aluminum instead of steel, and running shoes can be substituted for the heavier leather shoes.

Upper Extremity Orthoses

Wrist-Hand Orthoses (WHO)

Prevention of contractures is important for continued hand and finger function. Braces (WHO) can support the wrist to allow for the maintenance of finger function (10). If finger contractures are present, surgical releases, stretching, and attachments in the form of outriggers to the orthotic device may help minimize, improve, or prevent fixed contractures to the fingers.

In the severely physically disabled neuromuscular patient who has minimal hand function, it is important to position the wrist and hand properly so they can operate the controls to drive an electric wheelchair or to operate an environmental control system. Splinting the wrist in an appropriate position can be accomplished orthotically. Many custom-designed components may need to be used. However, when the desired result can be accomplished, the patient's activities can improve significantly.

Mobile Arm Supports (MAS)

With proximal weakness, positioning of the hands and fingers for function may become difficult. In the patient with spinal muscular atrophy, when MAS are properly fitted, function can be improved (11).

For the patient with a progressive problem, especially a spinal curvature that allows the patient to lean toward the side away from the apex of the curve, MAS allow the elbow to be propped at the table edge. By bending the trunk forward, these patients can feed themselves. However, with spinal fusion this independence is lost. Becoming dependent on others to be fed at this time can bring about an emotional crisis. Thus, MAS should be considered and may be used successfully with appropriate training (Fig. 8). For the DMD patient with progressive proximal weakness, there is a window of time during which MAS can be used before further weakness occurs.

SUMMARY

Surgical procedures to correct fixed deformities, remove deforming forces where there is imbalance, and augment weakened structures have been especially designed and adapted for the neuromuscular patient. They are used to preserve structural integrity, to maintain or augment function, and to alleviate pain and discomfort. Orthoses can be used separately, in conjunction with surgery, or as a trial prior to surgery to achieve these goals.

The neuromuscular patient who is severely physically disabled requires a team skilled in diagnosis, treatment, and rehabilitation. Diagnosis is important to understand the disease course, for prognosis, and for family counseling and genetic counseling. Accurate assessment of muscle strength performed by skilled physical and occupational therapists not only allows the treating team to know the area of weakness, but, when coupled with a functional evaluation, can help assess the overall needs of the patient and guide the patient and his/her family on the use of assistive devices. The team needs to include a rehabilitation engineer and orthotist so that proper orthoses and assistive devices can be designed and adapted for the patient's specific needs. In this way, the goals of maintaining or improving the activity and living conditions of a severely physically disabled neuromuscular patient can be achieved.

CASE STUDIES

Case 1

A 15-year-old Duchenne muscular dystrophy patient, wheelchair-dependent since age 11, started to show a mild spinal curvature at age 13. An initial attempt to control the spinal curve was made using

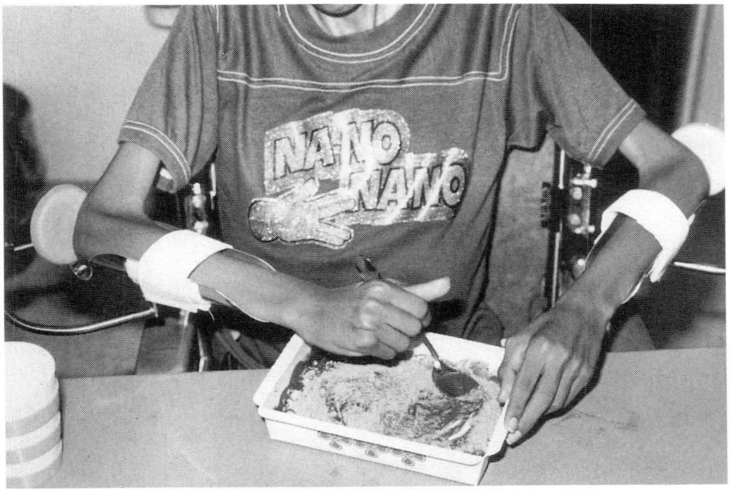

Figure 8 DMD patient fitted with mobile arm supports.

a TLSO, but the curve progressed despite faithful use of the orthosis. When the patient was 15, with a vital capacity of 55% and a spinal curvature of 45 degrees, spinal fusion was made using the Luque rod fusion technique from T5 to the sacrum with nearly full correction of the spinal curvature. Postoperatively, he was out of bed within the first week of surgery and has been able to maintain his activities. There was no loss of function.

Case 2

A 9-year-old girl with Charcot-Marie-Tooth disease, had a strong family history of this disease. She had evident weakness in the peroneal musculature. Her foot and ankle turned in at a very early age. Initially, a short leg brace (AFO) was used. The muscle imbalance progressed such that there was heel pain because of the varus position. At age 8, she had a posterior tibial tendon transfer anteriorly to interosseous membrane, correcting the deformity. The transfer acted as a tenodesis and an AFO was used for approximately 6 months to protect the transfer. The brace was discontinued because the patient had sufficient active dorsiflexion of the ankle via the anterior tibial to maintain a plantigrade foot and could walk normally.

Case 3

A 19-year-old boy with Charcot-Marie-Tooth disease had progressive foot and ankle deformity with pes cavus, turning in of the foot, and lateral callosity. Initial treatment was symptomatic and the callosities were trimmed and appropriate shoe modifications made. However, because the deformity was persistent and progressive, with pain leading to loss of function, and with permanent bony deformity with fixed contractures of the foot, a triple arthrodesis was performed. This was followed by anterior tibial tendon transfers to the middle of the foot and posterior tibial tendon transfers to the dorsum of the foot, to rebalance the foot. Because of persistent footdrop, AFOs were used as a "permanent" measure.

Case 4

An 11-year-old boy with spinal muscular atrophy did not stand until he was about 2 years old. He was a full-term baby, first born of a family whose parents did not have any other affected family member. His mother noticed that he was weak at birth and he was slow in his motor developmental milestones. Muscle biopsy showed type 1 fiber predominance and grouping compatible with chronic neuropathy. When the patient was 2 years old, he could pull himself up to standing and stand with support. Using a walker, he "back-kneed." An AFO was fitted, allowing him to walk. Because of the planovalgus heel position bilaterally, he could not be comfortably fitted into the braces and developed painful callosities on the medial side of the foot and under the medial malleolus.

Bilateral extra-articular talocalcaneal fusions were made at age 3 and 3/4 using a fibula graft. The heel valgus was corrected and the grafts became incorporated within 8 months. His heels are at neutral and he can fit into his AFOs without trouble. He can take some steps independently and walk around the house. The fibula, which had a 3-cm defect, filled in and healed completely within 6 months.

Case 5

A 20-year-old spinal muscular atrophy patient, presently a student at a major university, manages a motorized wheelchair independently. Surgical spinal fusion from T3 to sacrum was performed at age 9 secondary to correction of progressive spinal collapse. The patient was fitted with bilateral mobile arm supports in the same year. His shoulder flexion and abduction strength were zero to poor-minus, the remaining upper extremity strength ranging from poor-minus to fair-plus. He has continued to use the right mobile arm support for over 10 years. He uses his left hand to operate the motorized wheelchair, and the right mobile arm support is used full-time while he is in the wheelchair, about 12 hr/day. He reports that he uses the MAS for self-feeding, for oral hygiene, and for turning the pages of his books; he also uses it for balance and to enable him to shift his weight in the wheelchair.

ACKNOWLEDGMENT

The authors thank Ms. Leslie Sakioka for typing the manuscript.

REFERENCES

1. Hsu JD. Surgical treatment of the child with Duchenne muscular dystrophy (DMD). In: Noble J, Galasko CSB, eds. Recent Developments in Orthopaedic Surgery. Manchester: Manchester University Press, 1987: 135–142.
2. Hsu JD, Furumasu J. Gait and posture changes in the Duchenne muscular dystrophy (DMD) child. Clin Orthop Rel Res 1993; 288:122–125.
3. Hsu JD. Management of musculoskeletal complications: spinal deformity and the role of bracing and surgery. State-of-the-Art Reviews—Phys Med Rehabil 1988; 2:537–552.
4. Hsu JD. Management of foot deformity in Duchenne's pseudohypertrophic muscular dystrophy (DMD). Ortho Clin North Am 1976; 7:979–984.
5. Rodillo EB, Fernandez-Bermejo E, Heckmatt JZ. Prevention of rapidly progressive scoliosis in Duchenne muscular dystrophy by prolongation of walking with orthoses. J Child Neurol 1988; 3:269–274.
6. Hsu JD. The natural history of spine curvature progression in the non-ambulatory Duchenne muscular dystrophy patient. Spine 1983; 7:771–775.
7. Hsu JD. Spine care of the patient with Duchenne muscular dystrophy. State-of-Art Rev—Spine, 1990; 4:161–172.
8. Hyde SA, Scott OM, Gaddard CM, Dubowitz V. Prolongation of ambulation in Duchenne muscular dystrophy by appropriate orthoses. Physiotherapy 1982; 68:105–108.
9. Merritt JL. Knee-ankle-foot orthotics. State-of-Art Rev—Phys Med Rehabil 1987, 1:67–82.
10. Irani KD. Wrist and hand orthoses. State-of-Art Rev—Phys Med Rehabil 1987; 1:137–160.
11. Yasuda YL, Bowman K, Hsu JD. Mobile arm supports: criteria for successful use in muscle disease patients. Arch Phys Med Rehabil 1986; 67:253–256.
12. Hsu JD, Jackson R. Treatment of symptomatic foot and ankle deformities in the nonambulatory neuromuscular patient. Foot Ankle 1985; 5:238–244.

58

Genetic Counseling and Prenatal Diagnosis

Elizabeth M. Thompson
Women's and Children's Hospital, North Adelaide, South Australia, Australia

INTRODUCTION

Is Genetic Counseling Relevant to Muscle Disorders?

Many muscle diseases are genetically determined and several specific entities are common. For example, Duchenne muscular dystrophy affects 1 in 3500 liveborn males and myotonic dystrophy occurs in about 1 in 8000 individuals. Like the majority of genetically determined disorders, hereditary muscle diseases are a great burden to affected individuals and their families. At present, all are incurable. Relatives of affected individuals often want to know the risk of developing the particular disorder or the chance of having an affected child. For many, prenatal diagnosis provides the opportunity to choose whether or not to have affected children. It is the role of the clinical geneticist to provide information on these issues and set it into a meaningful context for the inquirer, who can then make informed choices. Genetic counseling is therefore an important aspect in the management of most muscle diseases, for the patient and the whole family. The clinical geneticist's work is becoming easier, thanks to the successful efforts of colleagues in molecular biology in identifying an increasing number of disease gene loci and specific mutations (Tables 1–3). Accurate presymptomatic tests, carrier detection, and prenatal diagnosis can now be performed for many diseases using molecular techniques.

What Is Genetic Counseling?

Genetic counseling provides information about an individual's risk of developing a particular disorder or transmitting it to offspring and then helping that person make informed reproductive choices. In order to make such decisions, the patient needs an understanding of (1) the nature of the disorder, including its age of onset, range of severity, progression, complications, and treatment (2) the mode of inheritance, (3) the tests that can be done to clarify the risks, and (4) the tests that can be done in pregnancy. Since the aim of genetic counseling is not to dictate what should be done, the information should be imparted in an unbiased way. In other words, counseling should be as nondirective as possible. However, it is difficult to be completely impartial, since one's own beliefs and views will inevitably color the way in which one conveys the information.

What Does Genetic Counseling Involve?

Once a referral is received, it is necessary to confirm the diagnosis in the index case. For example, the clinical geneticist is often asked to see a young woman who has a male relative with "muscular dystrophy." As all types of Mendelian inheritance are represented among the dystrophies, it is essential to establish the precise clinical diagnosis. It may be necessary to trace old records or even to see the index case in person. Referral letters that contain results of investigations or other information confirming the diagnosis in the index case are very useful.

Patients are often uncertain as to why they are being referred for genetic counseling and may even

Table 1 Localization of Gene(s) Causing Various Muscle Disorders

Disease	Inheritance	Chromsome location	Symbol (gene product)
Duchenne/Becker muscular dystrophy	XL	Xp21.2	DYS[a] (dystrophin)
Severe childhood AR muscular dystrophy (SCARMD)	AR	13q12 17q21	SCARMD1 SCARMD2 (adhalin)
Myotonic dystrophy	AD	19q13	DM[a] (myotonin-protein kinase)
Emery-Dreifuss muscular dystrophy	XL	Xq28	EMD
Facioscapulohumeral muscular dystrophy	AD	4q35 ?	FSHD FSHD2
Limb-girdle muscular dystrophy	AD AR AR	51 15q 2p	LGMD1 LGMD2 LGMD3

AD = autosomal dominant; AR = autosomal recessive; XL = X-linked recessive.
[a]The gene has been cloned.

be fearful about what it will entail. Many genetic units have specially trained health visitors or genetic nurses who carry out home visits before clinic appointments, which help reduce anxiety and may improve attendance rates.

In the clinic, the pedigree is drawn, usually encompassing at least three generations. The patient is examined for signs of the disease in question, appropriate information about the disease is given, and its mode of inheritance explained, often with the use of diagrams. Risks to the patient and offspring are discussed and appropriate investigations arranged. Often it is necessary to examine and obtain blood samples from other family members for DNA studies.

Table 2 Muscle Diseases Caused by Abnormalities in Ion Channels

Ion channel	Disease[a]	Chromosome location	Gene symbol (product)
Calcium	Malignant hyperthemia[b] Central core disease	19q13.1	MHS1[c] (ryanodine receptor)
Calcium	Hypokalemic periodic paralysis	1q31–32	HypoPP/ CACNL1A3[c] (DHP receptor)
Sodium	Hyperkalemic periodic paralysis Paramyotonia congenita and some variants	17q13.1–13.3	SCN4A[c] (adult skeletal muscle sodium channel α-subunit)
Chloride	Myotonia contenita: Thomsen's disease Becker's variant	7q35	CLC-1[c] (human skeletal muscle chloride channel)

[a]All autosomal dominant, except Becker variant of myotonia congenita, which is autosomal recessive.
[b]Fifty percent of families linked to this locus; some families linked to 7q or 3q (see text).
[c]The gene has been cloned.

Table 3 Congenital Myopathies in Which the Causative Gene Has Been Mapped

Disease	Inhertiance	Chromosome location	Gene symbol (product)	Ref.
Central core disease	AD	19q13.1	CCD-MH	See text
Fukuyama congenital muscular dystrophy	AR	9q31–33	FCMD	117,118
Myotublar (centronuclear) myopathy	XL	Xq28	MTM1, MTMX	119–121
Nemaline myopathy	AD	1q22–25	NEM1[a] (α tropomyosin, TPM3)	122,123
Congenital muscular dystrophy (some)	AR	6q	CMD (merosin)	31

AD = autosomal dominant; AR = autosomal recessive; XL = X-lnked recessive.
[a]The gene has been cloned.

Once all the information is at hand, which may take several weeks or even months if a large family linkage study is required, the risks to the patient can be calculated and conveyed in meaningful terms. The reproductive options are discussed. These include ignoring the risk, not having (more) children, insemination by donor or ovum donation when appropriate (ovum donation is not widely available), and prenatal diagnosis with a view to terminating the pregnancy if the fetus is affected with the disorder in question. The aim of prenatal diagnosis is to provide reassurance to mothers whose babies are found to be at low risk and to offer termination of pregnancy if the fetus is affected or at high risk. Occasionally, a woman at risk requests prenatal testing "just to know" the result without planning to terminate a high risk fetus, but this is uncommon when invasive tests are involved. Counseling before a prenatal test should include a discussion as to how the test is done, and when (for disorders of muscle, most are done by DNA analysis of chorion villus samples taken transcervically or transabdominally at around 11 weeks of gestation), the risks of miscarrying because of the test (around 1 in 40 for chorion villus sampling), how long it takes to get the results (usually up to 2 weeks depending on the DNA technique used), and the method of termination.

Preimplantation diagnosis is being developed for a number of disorders but, again, is not widely available at present because of resource limitations. Briefly, it involves in vitro fertilization, taking a single cell (blastomere) from the very early embryos, testing the blastomeres for the mutation in question, and placing unaffected or low-risk embryos into the mother. DNA testing of single cells has become possible because of a technique called "polymerase chain reaction" (PCR), which greatly amplifies the relevant sequence of DNA in a single cell. In X-linked disorders, such as Duchenne muscular dystrophy, an affected pregnancy can be avoided by placing into the mother only female embryos, whose sex has been determined by DNA probes, again using PCR on single cells (1).

Support and follow-up are an important part of genetic counseling and are often provided by specialist health visitors and genetic nurses attached to clinical genetic units. Clearly, it is highly desirable for the referral to take place *before* a pregnancy. If a couple first face their genetic risk once a pregnancy has begun, it is difficult for them to make decisions and there may be insufficient time to carry out the preliminary family DNA studies that make prenatal diagnosis possible.

Genetic Registers

One way to ensure that all adult individuals at risk of a particular disorder are aware of their options regarding carrier or presymptomatic tests and prenatal diagnosis is to maintain registers documenting all affected families in a given geographical area. The register needs continual updating and a mechanism for offering counseling to individuals at risk as they reach adulthood. Many regional genetic centers maintain registers for a variety of disorders, employing staff dedicated to the task of running the register.

Testing of Children

It is generally thought that carrier and presymptomatic tests should be deferred until individuals reach adult life, perhaps around 16–18 years. This allows the individual to make his or her own informed decision as to whether or not to have such tests. One argument against this view is that it would be better for a person carrying a harmful gene to grow up with the knowledge rather than face possible sudden disappointment after taking a test as a young adult. It could also be argued that, if a child at risk of inheriting a disorder such as myotonic dystrophy is diagnosed presymptomatically, then appropriate surveillance could begin rather than waiting for problems to develop. Conversely, a child at risk who is shown not to have inherited the gene need not be the subject of unnecessary anxiety of parents and others. If a child does develop symptoms, then clearly it is reasonable to use DNA tests to confirm the diagnosis, in the same way that any other diagnostic test would be employed.

DNA Storage

In many families seen for genetic counseling, the affected individuals have died without DNA being stored. It is then often difficult to perform carrier, presymptomatic, and prenatal tests for relatives at risk, particularly if linkage studies are required. It is extremely important to arrange for storage of DNA from individuals with fatal disorders such as Duchenne muscular dystrophy. Arguably, this applies to all Mendelian and mitochondrial disorders, even if the gene locus is not yet known, because the chances are it soon will be!

The accounts of the genetics of dystrophies and myopathies that follow should be considered with reference to other chapters dealing with the clinical aspects of these conditions. The genetics of mitochondrial disorders and metabolic myopathies are covered in other chapters.

DUCHENNE AND BECKER MUSCULAR DYSTROPHIES

Duchenne muscular dystrophy (DMD) is a progressive disorder affecting about 1 in 3500 newborn boys. It is caused by mutations in the dystrophin gene on the X chromosome (at Xp21), and so affects males whose female relatives can be carriers. Female carriers are usually healthy but occasionally may manifest symptoms. In about two-thirds of sporadic cases, the mother is a carrier; in the other one-third, the mutation has arisen at gametogenesis and the mother is not a carrier and is not at increased risk of having another affected son.

Becker muscular dystrophy (BMD), a similar but more benign disorder affecting about 1 in 30,000 male births, is caused by mutations in the same gene.

The gene encodes dystrophin, a very large and rare component of the membrane skeleton of skeletal muscle cells. It accounts for only 0.002% of total striated muscle protein (2). The abnormality in dystrophin correlates with the severity of the phenotype; in DMD, dystrophin is absent or markedly reduced in quantity, whereas in BMD, dystrophin is present but abnormal.

Female relatives of males with DMD or BMD related through the maternal line should be offered genetic counseling and carrier tests.

Carrier Testing for DMD and BMD

Determination of a woman's risk of being a carrier takes into account the pedigree, creatine kinase estimations, and DNA studies.

The Pedigree

A woman is an "obligate" carrier if she has two or more affected sons or an affected son and another affected relative. A daughter of a known carrier has a 50% chance of being a carrier. Risks for women in a DMD family can be assessed using Bayesian calculation. This modifies the risk according to the closeness in the pedigree of a female to the affected male(s) and to the number of normal male relatives she has. For example, a woman with one affected son and four healthy sons is statistically less likely to be a carrier than if she had only one healthy son.

Creatine Kinase Estimations

Plasma creatine kinase (CK) levels are elevated in some carrier women because of uneven Lyonization. (Lyonization refers to the random inactivation of one of the two X chromosomes in each cell of a woman's body. Occasionally, more cells with the normal X chromosome are inactivated than are ones carrying the DMD mutation.) If a woman has markedly elevated CK levels, she is probably a carrier. However, about one-third of carriers have a result in the normal range. Using samples from obligate carriers, some laboratories have constructed curves that give a probability that the woman is a carrier for a given CK level. For example, the CK result might give a woman a 1 in 4 chance of being a carrier. This result does not stand alone but is incorporated into the pedigree-derived risk and the risk derived from molecular studies (see below), using Bayesian calculation, to give the woman an overall probability figure for being a carrier.

There are many pitfalls in CK testing. The sample must not be subjected to undue sunlight or prolonged storage, and it must be tested in a reliable laboratory with a properly defined normal range. An average of three CK levels is taken, and the patient must avoid strenuous exercise before collection of the samples. The timing is important; CK levels are usually tested during the reproductive years, as results are unreliable before puberty and after the menopause. CK levels are markedly reduced during pregnancy, when a result in the normal range may be unreliable.

It may be argued that CK testing has been superseded by DNA tests, but it is still useful especially when it is high.

DNA Studies

Before the introduction of DNA testing, if a woman was not an obligate carrier, she might be left with an uncertain carrier risk of say 10–20% based on assessment of the pedigree and CK results. She then faced a 5–10% chance of having a son with DMD, which is a moderate risk. The only way that females at significant carrier risk could avoid having affected sons was to terminate all male pregnancies.

Linkage Studies. In 1983, the gene causing DMD was localized to the short arm of the X chromosome at Xp21, and it became possible to perform linkage studies in families to refine female carrier risks. Using restriction fragment length polymorphisms (RFLPs) as markers, a woman's two X chromosomes can be distinguished and then the one cosegregating with the disease in that particular family can be identified (Fig. 1). RFLPs can be linked (i.e., just outside the gene) or intragenic. There is an inherent error rate because the marker and the mutation can become separated at meiosis by recombination. In many diseases, when intragenic markers are used, the risk of recombination between the marker and the mutation is negligible, and the results of marker studies are then very accurate. The dystrophin gene, however, is so large that there is a 12% risk of recombination right across the gene (3), leading to an average risk of around 5% that the mutation and an intragenic marker will become separated, and hence a wrong prediction be made.

A limitation of RFLPs is that they tend to be two-allele systems. In other words, for one RFLP, there are only two possible forms, e.g., allele 1 or allele 2. If a woman is homozygous for the markers, in other words, both of her X chromosomes share the same marker and so cannot be distinguished, then tracking of the mutation through the family cannot be done. Intragenic CA repeats are highly polymorphic repeats of the cytosine and adenosine bases within the DNA. The number of alleles is much greater, so the chance of being able to "track" a given X chromosome through a family is much higher.

Detection of Mutations. In DMD, 55% of mutations are deletions, 5% are duplications, and the other 40% are presumed to be point mutations (reviewed in Ref. 4). In Becker dystrophy, about 65% of mutations are deletions (5). Using linked markers, it is obvious when an affected male has a deletion;

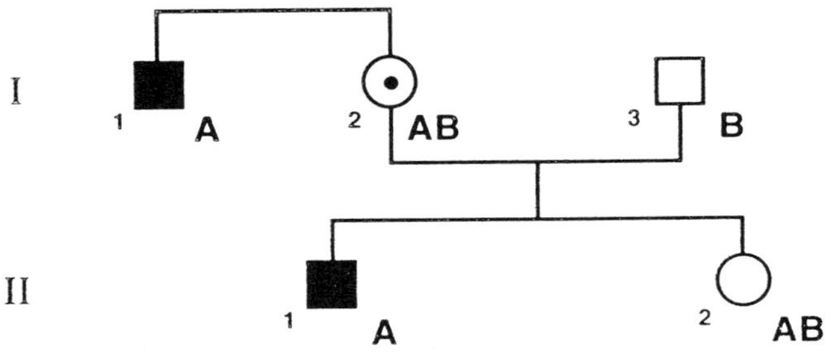

Figure 1 Principle of a genetic linkage study using DNA markers. A woman (individual II2 on the pedigree) wishes to know if she is a carrier for Duchenne muscular dystrophy (DMD). Using DNA markers, the X chromosome of her affected brother (II1) and uncle (I1) can be identified as having allele A. In this family, the DMD mutation is "tracking" with allele A. Individual I2, an obligate carrier, has alleles A and B. Individual II2 therefore inherited allele B from her father and allele A from her mother and so is at *high risk* of being a carrier. The chance of recombination between the mutation and the DNA marker has to be taken into account. If this is 5%, then her carrier risk is 95%. (Males have one X chromosome and females have two; hence they have one and two alleles, respectively.)

there is no allele (Fig. 2). For females, the situation is complicated by the fact that the allele on the normal X chromosome masks the presence of a deletion on the other X. In other words, if the woman has allele 1, it is unclear under routine circumstances whether she is homozygous (that is, 1,1) or hemizygous (that is, 1, delta, where delta indicates a deletion). Sometimes it is possible to determine the "dose" of DNA in the allele, which might help distinguish the two scenarios, but dosage estimations are difficult and often unreliable. Occasionally, it becomes clear fortuitously that a female carries a deletion. For example, it sometimes appears that a woman cannot be her mother's daughter! (Fig. 3). This is because the daughter has inherited from the mother a deletion that removes one allele of the marker. The two women are then hemizygous for the marker (that is, having only one allele), which cannot be distinguished routinely from their being homozygous (that is, having the same allele on both X chromosomes). This apparent nonmaternity is proof of the presence of a deletion (provided there is no evidence for true nonmaternity!). Occasionally a deletion falls within an RFLP, producing an extra fragment of DNA called a junction fragment, which is clearly visible on DNA studies (Fig. 4). This makes it easy to detect all carrier females in these families and to offer accurate prenatal tests to carriers.

Various other molecular methods of detecting mutations in females are theoretically possible (e.g., pulsed field gel electrophoresis to detect deletions and duplications; studies of messenger RNA; single-strand conformational polymorphism (SSCP) analysis to detect point mutations; fluorescence in situ hybridization (FISH) to detect deletions). Many of these are expensive and time-consuming and may be available only in some centers.

Dystrophin Testing

Dystrophin immunofluorescence can be performed on thin sections of frozen muscle tissue to detect female carrier status if a mutation cannot be identified. Because of random X-chromosome inactivation, a female carrier will show a mosaic pattern of dystrophin immunofluorescence, that is, a mixture of dystrophin-positive and dystrophin-negative muscle fibers from the normal and dystrophin-deficient fibers, respectively. However, a normal result does not exclude carrier status because known carriers have produced normal patterns. Muscle cell cultures can also be established to carry out dystrophin immunofluorescence or immunoblotting. If the woman has two populations of cells, one that does produce dystrophin and one that does not, then this indicates that she is a carrier. Neither of these techniques is widely available, particularly the latter, which is very time-consuming.

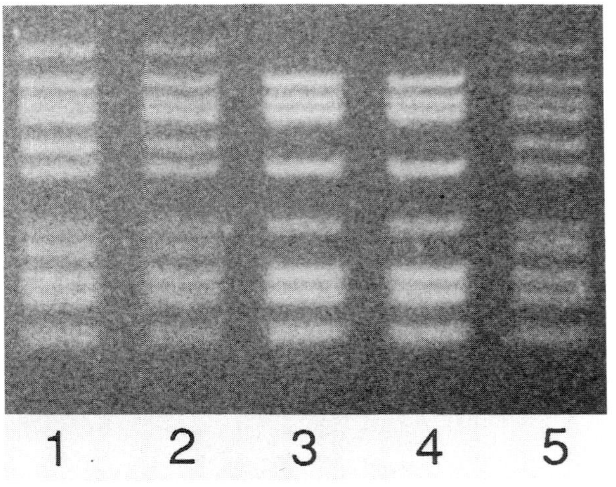

Figure 2 A deletion in the dystrophin gene in Duchenne muscular dystrophy. This multiplex PCR ethidium bromide gel shows normal control males whose DNA is in lanes 1, 2, and 5 and a male with Duchenne muscular dystrophy in lanes 3 and 4. The bands represent various exons (coding regions) in the dystrophin gene. It can be seen that the patient lacks bands 1, 5, and 8, which corresponds to a deletion of exons 45–48, inclusive. (The term "multiplex" refers to the fact that multiple exons can be visualized on the one gel.) (Courtesy of Ms. Jenni Bell, Kennedy-Galton Centre, Harrow, Middlesex, England.)

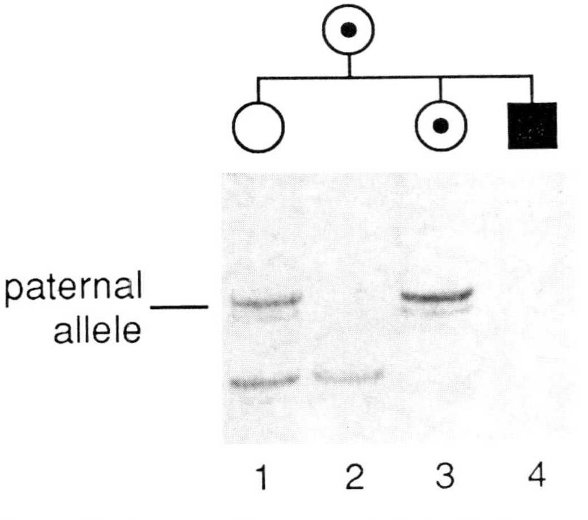

Figure 3 Apparent "nonmaternity" in Duchenne muscular dystrophy. Shown are a mother, her two daughters, and her affected son. The affected male has a deletion of exons 51–53 (no band is visible using this particular probe, a CA repeat called STR50). The mother appears to be homozygous for the lower band and it appears that individual 3 shares no band in common with her mother ("nonmaternity"). The correct interpretation, however, is that the mother and individual 3 both have the deletion and both must be carriers. (The mother's lower band and individual 3's upper band are their normal bands.) Individual 1 has the mother's normal lower band, so she is not a carrier. Correct maternity was established using other DNA probes. (Courtesy of Ms. Jenni Bell, Kennedy-Galton Centre, Harrow, Middlesex, England.)

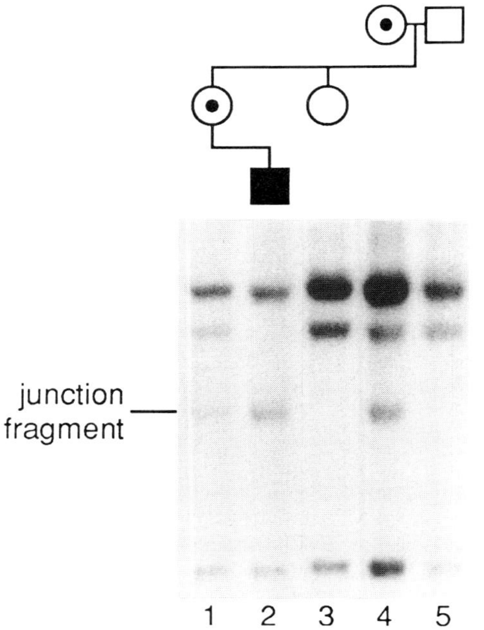

Figure 4 Carrier testing when a junction fragment is visible. Southern blot digested with the restriction enzyme EcoR1 and probed with CaIa. The affected male (lane 2) has a deletion that falls within the RFLP, generating a junction fragment. His mother and maternal grandmother also demonstrate the junction fragment, which proves that they are carriers of the deletion. (Courtesy of Ms. Jenni Bell, Kennedy-Galton Centre, Harrow, Middlesex, England.)

Gonadal Mosaicism

Although it is possible to prove that a woman *is* a carrier by the various means described above, it is presently impossible to *exclude* the mother of an isolated case as being a carrier. This is because she could be a gonadal mosaic, that is, carrying the mutation not in somatic cells but in the gonad.

Females with a Dystrophinopathy

It has recently been shown, using dystrophin protein analysis, that many females with a primary myopathy have an underlying dystrophinopathy, despite a negative family history of Duchenne dystrophy (6–8). In a retrospective study of 505 muscle biopsies from female myopathy patients, as many as 10% were diagnosed as Duchenne dystrophy carriers, using the dystrophin immunofluorescence assay (9). It has been hypothesized that these patients must manifest symptoms because of skewed (rather than the normal random) X-inactivation, such that more X chromosomes bearing the normal dystrophin gene are inactivated, leaving the X chromosome with the mutated gene active. Evidence in favor of this theory was provided in a study that demonstrated skewed X-inactivation at a DNA level in 13 female dystrophinopathy patients (10). A fascinating finding was that eight of nine of these patients had inherited the mutation from their father, either as a new mutation or because the father is a gonadal mosaic, which has implications for genetic counseling in these families.

Prenatal Diagnosis

Women with a significant risk of being carriers for DMD and BMD frequently request prenatal diagnosis. It is important to note, however, that women's perception of the significance of the risk varies widely. For example, one woman may feel reassured that her risk of having an affected son is "only" 5% whereas another may be so anxious that she is unprepared to take any chance at all. Risk perception involves not only the magnitude of the actual figures. Other factors, such as the closeness of the woman to the affected individual and her experience of the illness, also play a major role.

Prenatal tests involve the same kinds of DNA analyses as described for the carrier tests. The first requirement is to determine the sex of the fetus by chromosome analysis and/or DNA techniques. If the fetus is female, the parents can be reassured. If the fetus is male, further tests on the CVS material are performed. If the mutation is identifiable, it can be sought, or linked markers may be used. All too often, a woman at risk does not present for genetic counseling until she is pregnant. All is not lost, however, as it is usually possible to make some estimation of her carrier risk, and if she is unwilling to take that risk, some form of assessment of the fetus can be made. If the fetus is male, a deletion screen can be carried out. As 55% of affected boys have deletions, the risks will improve if a deletion is absent. "Prenatal exclusion" can then also be undertaken, provided it is possible to collect the relevant family blood samples and find informative polymorphisms in the family (Fig. 5). In this example, if the fetus has inherited his mother's father's X chromosome markers, he is at low risk of being affected; that is, the diagnosis has been excluded. If the baby has the grandmaternal X, he may still be at risk and the mother may opt for termination of pregnancy. There are, however, many traps for the unwary. One is the "grandfather trap," illustrated in Figure 6. For example, a woman's sister has a son with DMD and he is the only affected male in the family (Fig. 6). If the affected boy has his grandpaternal allele, it could either mean that the affected boy or his mother represents a fresh mutation or that the grandfather is a gonadal mosaic. The two sisters could have inherited the mutation from their father and be carriers. It would therefore be inappropriate to offer either woman a "prenatal exclusion test," as described in Figure 5, in which it is clear that the maternal grandmother is a carrier. The exact incidence of gonadal mosaicism is unknown but it is a well-documented phenomenon.

As mentioned above, preimplantation diagnosis is now possible for Duchenne and Becker dystrophies by determining the sex of embryos produced by in vitro fertilization and replacing only female embryos in the mother. However, this procedure is not widely available because of resource limitations. Laboratory procedures are also being developed that will allow specific preimplantation diagnosis of Duchenne dystrophy and fetal sex, simultaneously (11). Preimplantation diagnosis offers hope to couples who are ethically opposed to termination of pregnancy, or who have had several terminations following prenatal diagnosis. There are, however, a number of potential drawbacks to this

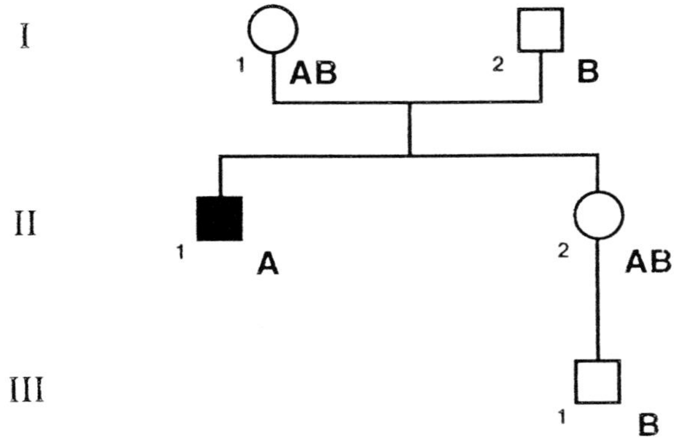

Figure 5 Prenatal exclusion test. A woman (II2) has a brother (II1) with DMD. Using a two-allele DNA marker system, the affected male has allele A, their mother (I1) has alleles A and B, and their father (I2) has allele B. Analysis of the fetus (III1) by testing DNA from a chorion villus sample shows that the baby, a male, has inherited the allele that his mother received from her healthy father. Therefore, the baby is at low risk of DMD and the diagnosis has been "excluded."

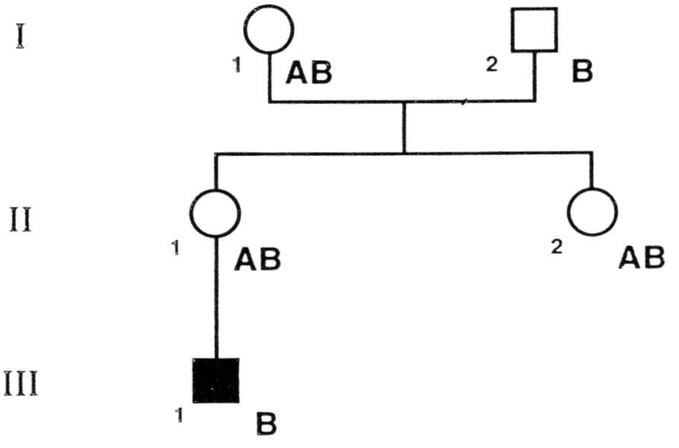

Figure 6 The "grandfather trap." A woman (II2) has a sister (II1) with a son with DMD, and he is an isolated case. Using a two-allele DNA marker system, the affected male (III1) has allele B, which his mother inherited from her father (I2). Although the latter is healthy, he could be a "gonadal mosaic" (carrying the mutation in the gonads but not in somatic cells) and could have passed the mutation to his daughters with allele B. The other explanation is that III1 represents a fresh mutation in the dystrophin gene with a low risk of recurrence in the family (if the actual mutation cannot be identified). Therefore, the "prenatal exclusion test" described above could not be offered to the two sisters (II1 and II2). (In the "prenatal exclusion" in Fig. 5, we do not suspect the father of the affected boy of being a gonadal mosaic, because he passed a Y chromosome not an X, to his affected son.)

approach, such as the time involved, both for the mother and for the laboratory, the expense, the need to confirm the diagnosis with a conventional prenatal test, the risk of multiple pregnancy, and the possibility of failure to achieve a viable pregnancy.

Gene Therapy

At present, research is being directed to gene therapy by cell implantation of normal muscle precursor cells into dystrophin-deficient muscles or by direct DNA injection (12). The former involves implantation of normal myoblast precursor cells into the muscle of affected boys. Preliminary results in mice have been encouraging, but human trials have had little success. Direct injection of dystrophin DNA has again yielded hopeful preliminary results in mammalian cells in culture and in mice.

The Future

DMD/BMD were the first disorders whose gene products were identified by "positional cloning"; that is, the position of the gene on the chromosome was pinpointed, the gene was identified, and then its product was characterized. In this way, molecular analysis has contributed enormously to our understanding of these disorders. It is expected that this trend will continue and will lead to a variety of therapeutic strategies for many diseases in future.

SEVERE CHILDHOOD AUTOSOMAL RECESSIVE MUSCULAR DYSTROPHY (SCARMD)

Severe autosomal recessive Duchenne-like muscular dystrophy is well recognized in Africa and Brazil, especially in populations with a high parental consanguinity rate (13–15), and it also occurs in other populations (16). The disease begins at around 5 years; there is predominantly proximal muscle wasting, pseudohypertrophy of the calves, progression with confinement to a wheelchair around 12–15 years, and death between 20 to 30 years if respiratory insufficiency is not controlled. Dystrophin is normal in these patients (17,18). In these high-risk populations, a considerable number of sporadic cases may also have autosomal recessive disease, rather than X-linked Duchenne dystrophy (19).

Three highly inbred families from Tunisia with Duchenne-like autosomal recessive muscular dystrophy showed linkage to the pericentromeric region of chromosome 13q (20). This gene is now called severe childhood autosomal recessive muscular dystrophy (SCARMD).

Dystrophin is associated with several novel sarcolemmal proteins, including a transmembrane glycoprotein of 50 kDa (50 DAG, or adhalin), which was recently shown to be deficient in SCARMD (21). Adhalin was subsequently shown to be deficient in children who had a severe Duchenne or Becker-like dystrophy in five European (22), six Brazilian (23), and two Japanese (24) cases, confirming that SCARMD does occur outside Northern Africa and demonstrating the usefulness of the im-muinocytochemical analysis of adhalin in establishing the correct diagnosis (22). Three other Brazilian SCARMD patients did not, however, have deficiency of adhalin (23). Data on 13 Tunisian (25) and six Moroccan (26) families suggested that the adhalin locus would be on chromsome 13, but other authors showed that the SCARMD phenotype is probably caused by more than one recessive gene and that the adhalin gene is not on chromsome 13q (27). Indeed, the adhalin gene was later mapped to chromosome 17q21, and one family showed mutations in this gene (28,29). A locus on chromosome 15q, which is linked to milder forms of autosomal recessive limb-girdle muscular dystrophy (see below), has been excluded as the SCARMD locus in four Brazilian families (30). At present, therefore, there is evidence that at least two genes cause the SCARMD phenotype: the adhalin gene on 17q and another gene on chromosome 13q.

CONGENITAL MUSCULAR DYSTROPHY

Thirteen of 20 cases of classical congenital muscular dystrophy studied by Tomé et al. (30a) lack a protein called merosin (laminin-M), a subgroup of laminin, which is a group of extracellular matrix proteins. Merosin acts as a ligand for one of the extracellular components of the dystrophin-glycoprotein complex. Some of the merosin-negative families show linkage to the locus for merosin on chromosome 6q (31). The *dy* (dystrophic) mouse is also deficient in merosin, and the murine merosin gene maps to chromosome 10, which corresponds to human chromosome 6 (32).

Those patients lacking merosin generally have severe neonatal hypotonia, contractures, delayed motor development, respiratory insufficiency, and normal mental development. Patients without merosin deficiency have a milder clinical condition (31).

MYOTONIC DYSTROPHY

Genetic counseling in myotonic dystrophy (DM) has become considerably simpler since the mutation causing the disorder was identified in early 1992 (33–35). It has long been observed that "anticipation" (worsening of clinical severity down the generations) occurred in DM, but this was sometimes attributed to biased ascertainment. Now there is an explanation for anticipation at a molecular level. The mutation is one of a new family of mutations that involves an unstable expansion of a trinucleotide repeat (CTG in the case of DM). Normal individuals may have from five to 40 CTG repeats in the DM gene, which is found on chromosome 19. Patients with DM have more than 50 copies of the triplet. As the DM gene (an autosomal dominant gene) is transmitted to offspring, however, the number of CTG repeats can increase, with correspondingly greater clinical severity (Fig. 7). Expansion can occur whether the DM gene is being transmitted by the mother or father, but greater increases are seen through the mother. The greatest expansion (2000 or more CTG copies) occur in babies with the severe congenital form who virtually always inherit the disorder from the mother. Other disorders involving unstable expansion of trinucleotide repeats are fragile X (both FRAXA and FRAXE), X-linked spinobulbar muscular

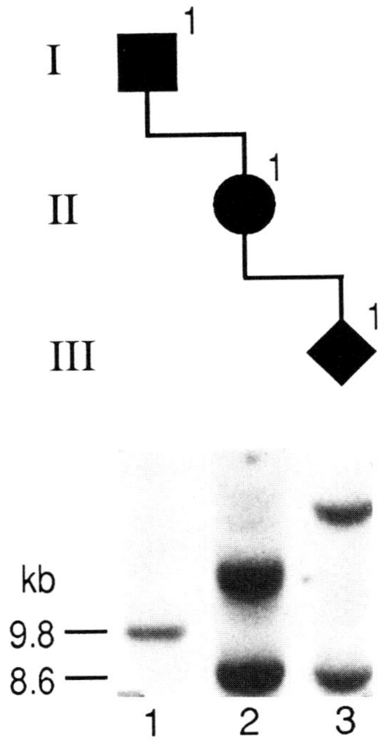

Figure 7 Anticipation in myotonic dystrophy. Shown is a three-generation family with increasing clinical severity of myotonic dystrophy down the generations. I1 has cataracts only, II1 has classical myotonic dystrophy, and III1 has severe congenital myotonic dystrophy. Beneath each individual is a lane showing the DNA result (i.e., a Southern blot using the probe p5B1.4 and restriction enzyme EcoR1). The 8.6- and 9.8-kb bands are normal. Individual I1 has a normal 9.8-kb band and a very small enlargement of the other 9.8-kb band. (The two bands are so close that they are indististinguishable on this blot.) His daughter II1 has a normal 8.6-kb band (inherited from her mother) and an obvious enlargement of her paternal 9.8-kb band. The child III1 has a normal 8.6-kb band and a further enlargement of his maternal 9.8-kb band. This family clearly shows increasing numbers of copies of the CTG repeat down the generations. (Courtesy of Mrs. Kokila Chotai, Kennedy-Galton Centre, Harrow, Middlesex, England.)

atrophy, Huntington's disease, spinocerebellar ataxia type 1, dentatorubropallidoluysian atrophy (DRPLA), and Machado-Joseph disease.

Before the mutation was discovered, individuals at risk of DM wanting presymptomatic testing would be examined clinically, undergo specialist ophthalmological examination using the slit lamp looking for cataracts, and would have an electromyogram to look for evidence of myotonia. Negative results did not, however, guarantee freedom from the illness later in life. When the gene was localized to chromosome 19, DNA linkage studies were added to the armamentarium of tests, usually providing more helpful results but occasionally causing anxiety when conflicting with clinical findings. If key affected individuals and their spouses were unavailable to give samples, then linkage analysis was impossible. With new knowledge, however, came a new battery of questions, often unanswerable! Now well-informed patients want to know exactly how many repeats they have, how much these will expand in their children, and precise clinical/molecular correlations that would be useful for prenatal diagnosis. It is possible to make limited clinical correlations with the size of the expansion only in families in which there has been a precedent. For example, if a severely congenitally affected baby has been born and the CTG repeat number in the baby is known, it is likely that future babies in the family with the same size expansion will also be severely affected. If, however, an affected mother has not yet had a baby, prediction of severity of an affected fetus might be difficult at the time of prenatal diagnosis.

It has been demonstrated that there is variation in the CTG repeat length in the tissues of patients with DM (somatic mosaicism). More recently, there is also evidence that somatic instability may lead to further expansion in the numbers of CTG repeats in the tissues with time (36). Whether this has any relationship to the onset of symptoms remains to be seen.

There are occasional families with an illness similar to DM, but in which affected individuals lack the abnormal CTG expansion. Two such families were described with apparently typical DM, including frontal balding, cataracts, cardiac conduction defects, and testicular atrophy, together with muscle weakness and myotonia (37). There was no apparent anticipation in the families. These patients may have a different type of mutation in the DM gene, or perhaps another gene is involved. Three other families were described with dominantly inherited myotonia, proximal muscle weakness, and cataracts, which the authors called proximal myotonic myopathy (PROMM) (38). The clinical course was mild with slow progression, and again, there was no evidence of anticipation down the generations. DNA studies showed normal numbers of CTG repeats and no linkage to the DM locus on chromosome 19, nor to loci for the muscle sodium and chloride channel disorders (see below). The genetic aspects of DM are discussed further in another chapter.

EMERY-DREIFUSS MUSCULAR DYSTROPHY (EDMD)

This rare X-linked disorder is characterized by early contractures of the Achilles tendons, elbows, and posterior cervical muscles, slowly progressive muscle wasting and weakness, which is predominantly humeroperoneal initially, cardiomyopathy, and conduction defects with risk of sudden death. Inheritance is usually X-linked but can be autosomal dominant. Family linkage studies have mapped the X-linked form to Xq28 (39,40). The localization of the gene was later refined (41), which improves the accuracy of prenatal diagnosis and carrier detection using DNA linkage. Two genes that encode isoforms of ABP-280, a dystrophin-like protein, were mapped to Xq28 and chromosome 7, suggesting that these are candidate genes for the X-linked and dominant forms of EDMD (42). However, the authors were unable to demonstrate any mutations in several patients with the X-linked form.

It is important to be aware that female carriers are at risk of sudden death and they may have cardiac pathology similar to that of affected males (43).

FACIOSCAPULOHUMERAL MUSCULAR DYSTROPHY (FSHD)

As its name suggests, this disorder causes weakness primarily in the muscles of the face, shoulders, and upper arms. The onset is usually in late childhood or adolescence, progression is slow, and the weakness is often asymmetrical, unlike other forms of muscular dystrophy. Late in the illness, peroneal muscles may be affected. Hearing loss is part of the disease. A retinal vasculopathy has been reported but it is unclear whether this is a specific association (44). Cardiomyopathy is not part of the illness.

FSHD is autosomal dominantly inherited but sporadic cases representing fresh mutations occur frequently. It is often difficult to determine clinically whether a relative has inherited FSHD because of the great variation in age of onset and severity of symptoms. A study of 168 individuals in 41 families found that 95% of FSHD patients showed some evidence of the illness by 20 years of age, but over the age of 40 years, one-third were still mildly affected while 19% were in wheelchairs (45). Creatine kinase levels were also unreliable indicators; only 80% of affected males and 48% of affected females over 40 years had elevated levels.

A study of 46 Brazilian patients showed that most would be interested in a preclinical test, so as to seek better help, avoid strenuous activities, prepare themselves emotionally, or choose an appropriate profession. Most would also be interested in prenatal diagnosis, but only two would terminate an affected pregnancy (46). Reproductive fitness in these patients did not seem reduced.

A gene for FSHD was mapped to 4q (47,48). Using a closely linked highly polymorphic DNA marker (D4S139), it seemed likely that presymptomatic and prenatal screening would be possible in FSHD families with 98% accuracy, using linkage studies (49). However, two families with typical FSHD were reported in which the disease is not linked to 4q, suggesting that there may be a second gene locus for the FSHD phenotype (50). Another study found that 5% of families show no linkage to chromosome 4q35 (51). This complicates the use of linkage studies, especially in small families in which it is difficult to establish linkage.

Recently, probe p13E-11 was shown to identify a de novo DNA rearrangement on 4q that is associated with FSHD (52). The rearrangements are deletions that generate a shortened p13E-11 allele (53), and their discovery will assist in the characterization of the gene, but their exact relationship to the gene is unclear. The presence of a shortened p13E-11 allele in a sporadic patient whose parents have normal DNA studies is diagnostic of FSHD. In familial cases, it is necessary to show that the shortened p13E-11 allele is indeed segregating with the FSHD, as about 10% of normal individuals have a shortened p13E-11 allele located at chromosome 10q (51), which has nothing to do with FSHD! The first prenatal test using the shortened p13E-11 allele was reported in 1994 (51); the index patient was a de novo case with the shortened allele and the fetus had inherited this allele.

At present, presymptomatic and prenatal tests within families are usually done using linkage studies, but care is needed in the interpretation of the results.

LIMB-GIRDLE MUSCULAR DYSTROPHY (LGMD)

Individuals with LGMD have slowly progressive proximal muscular weakness, usually beginning in childhood or adolescence. Caution in making the diagnosis is essential, especially in isolated cases, as there is clinical overlap with a number of disorders, including spinal muscular atrophy, metabolic myopathies, and inflammatory myopathies. These, however, can all be differentiated from LGMD on muscle biopsy. More important is the clinical similarity of LGMD to the dystrophinopathies, as the two are often indistinguishable histopathologically. A study of 41 cases previously diagnosed clinically as LGMD found that five males had abnormal dystrophin patterns (using immunochemistry and immunoblotting) diagnostic of Becker muscular dystrophy and two females had dystrophin patterns consistent with being manifesting Duchenne muscular dystrophy carriers (6). Misclassification of isolated cases had occurred in 31% (4/13) of males and in 13% (2/15) of females. The authors concluded that as many as 40% of isolated LGMD/Becker muscular dystrophy cases cannot be distinguished by routine clinical or histopathological means and that analysis of lymphocyte DNA and dystrophin analysis of muscle are important in establishing the diagnosis in any patient with muscular dystrophy involving proximal limb weakness, especially in sporadic cases.

A dystrophin-like gene identified on chromsome 6 (54) was a possible candidate gene for LGMD but linkage analysis excluded this locus in the Brazilian population (55). However, a study of an inbred population of recessive LGMD families from Reunion, a small island in the Indian Ocean, showed tight linkage to a DNA marker on the proximal long arm of chromosome 15 (56). This locus is called LGMD 2. Patients included in the DNA study had an onset before 20 years (usually between 8 and 15 years), muscle weakness and atrophy that was usually symmetrical, beginning and predominating in the shoulder and pelvic girdles, involvement of muscles of the neck and trunk but not the face, and slow progression with loss of ambulation at around 30 years. Creatine kinase levels were raised and muscle biopsies showed necrosis and regeneration. The population are descended from French settlers, and the disease probably started with one individual in the middle of the 17th century (founder effect) and was maintained in this inbred, closed community. Studies in the Old Order Amish in Indiana confirmed linkage to 15q (57). In these families, the age of onset was between 4 and 15 years and confinement to a wheelchair was from 12 to 44 years. This population is also a highly inbred isolated group, descended from immigrants from Switzerland, where there is an increased incidence of recessive muscular dystrophy. Other recessive LGMD families have been tested; only some showed linkage to the chromosome 15 locus (58,59). The proportion of families linked to this locus is unknown.

A second gene locus for autosomal recessive LBMD was reported on chromosome 2p, in two unrelated families from Palestine and Sicily, with onset in the pelvic girdle in the late teens and usually a slow progression (60).

Dominant forms of LGMD are also well described (61–63) but appear to be less common than the recessive forms. Dominant LGMD usually begins in adulthood and is generally mild and only slowly progressive. Prominent contractures, as in Emery-Dreifuss myopathy, have been described in some families, and the name "Bethlem myopathy" has been suggested for this subgroup (64). The large North American family with autosomal dominant LGMD originally described by Gilchrist et al. in 1988 (62) has shown linkage to chromosome 5q (65). This locus is called LGMD1.

The LGMD group of disorders is clearly heterogeneous in age of onset, rate of progression, severity, mode of transmission, and at a molecular level. An international consortium supported by the European Neuromuscular Centre has been established, in an attempt to clarify the range of conditions diagnosed as LGMD through the worldwide collection of such patients (60).

ION CHANNEL MUSCLE DISEASES

Recently, several muscle diseases have been found to be caused by defects in various ion channels in skeletal muscle (Table 2).

CALCIUM CHANNEL DISORDERS

Skeletal Muscle Calcium Release Channel (Ryanodine Receptor) Disorders

Malignant Hyperthermia (MH)

Susceptibility to MH is the most common cause of death from general anesthesia (66). The condition is inherited as an autosomal dominant with incomplete penetrance, meaning that not all gene carriers manifest the clinical phenotype. In susceptible individuals, halogenated volatile anesthetics or depolarizing muscle relaxants such as succinylcholine may trigger hyperthermia, generalized muscle rigidity, accelerated muscle metabolism, metabolic acidosis, and myoglobinuria. If the full syndrome develops, death from cardiac arrest may follow, unless the patient is treated with dantrolene. Many susceptible individuals present with only some of the classical signs after exposure and others have been given anesthetics with the triggering agents previously without adverse effects.

The incidence is variously reported as 1 in 50,000–75,000 anesthetized patients in the United Kingdom, to 1 in 10 000, but the latter includes childhood cases, so there may be reduced penetrance with age.

The main method in Europe of screening for susceptibility is to carry out a standardized in vitro contraction test. Muscle is exposed to halothane or caffeine in vitro to look for abnormal contraction. Drawbacks of this test are that it is highly invasive and a specialized laboratory is needed to produce reliable results. In addition, it is only feasible to study relatives of susceptible individuals, rather than to screen all patients preoperatively. The subclinical response in some susceptible individuals can make early diagnosis difficult. Clearly, there is a need for a simpler test.

Certain breeds of pigs have a disorder known as porcine stress syndrome (PSS), which is similar to human MH (67). Studies in halothane-sensitive pigs implicated the calcium release channel (CRC) of the sarcoplasmic reticulum in the pathophysiology of PSS (68); the calcium release channel binds the plant alkaloid ryanodine and is often called the ryanodine receptor (RYR). The gene for PSS is on porcine chromosome 6, which is homologous with human chromosome 19q. Studies in human families showed linkage between MH and markers in the human 19q12–13.2 region (69), and this enabled the human ryanodine receptor (called the (CRC)RYR1) locus to be mapped to 19q13.1–13.2 (70). These and further linkage data supported the notion that a defect in the calcium release channel underlies human and porcine MH. The same mutation (Arg615Cys)* in the ryanodine gene of the pig has been found in all susceptible strains (71). Studies in humans, however, showed that less than 50% of human MH is attributable to this gene (66). Various mutations in the human ryanodine gene in MH families have been identified, including the same mutation as seen in pigs (Arg615Cys), which occurs in 3–5% of human MH families (72,73). The most common human ryanodine mutation (Gly341Arg) occurs in only 10% of caucasian MH cases (74).

Data presented as evidence for a second MH locus on chromosome 17q (75) were not substantiated (76–78). One family showed linkage to a locus on 7q (79) and another to one on 3q (80).

At present, susceptibility screening using DNA tests can be offered only to those families with a documented mutation or proven linkage to one of these loci. The heterogeneity of loci and mutations means that it is often impossible to use DNA testing for these families, especially if they are small, in which case, muscle biopsy and in vitro muscle contraction testing is the only option (66).

It is important to note that the MH phenotype as demonstrated in vitro occurs in other myopathies, such as myotonic dystrophy, myotonia congenita, Duchenne and Becker dystrophies, central core disease, and carnitine palmitoyl transferase deficiency (81). Thus it is imperative that any underlying myopathy be excluded before chromosome 19q markers are used to investigate families.

Central Core Disease (CCD)

This congenital myopathy is closely associated with MH. Inherited as an autosomal dominant, CCD is characterized by hypotonia and proximal muscle weakness in infancy. Skeletal muscle biopsy shows amorphous central areas (cores) in type 1 fibers, best demonstrated by oxidative enzyme histochemistry. Associated clinical features include kyphoscoliosis, congenital hip dislocation, foot deformities, and joint contractures, which are not necessarily related to the degree of weakness. The clinical association with MH prompted linkage studies on CCD families using DNA markers from chromosome 19q12–13.2 and showed tight linkage (82–85). Two novel mutations in the ryanodine receptor gene (RYR1) were reported in two different CCD pedigrees and one of these mutations was also detected in an unrelated MH pedigree (without CCD) (86). Another mutation in the ryanodine receptor gene in a large CCD family has been described (87).

Skeletal Muscle Calcium Channel (Dihydropyridine Receptor) Disorder: Hypokalemic Periodic Paralysis

Hypokalemic periodic paralysis (HypoPP) is characterized by transient attacks of muscle weakness of varying duration and severity, associated with a fall in serum potassium levels. HypoPP is the most common of the periodic paralyses and is usually transmitted as an autosomal dominant disease. The onset is usually in the first two decades and the frequency of episodes may increase to weekly or daily, but attacks become less frequent after the age of 30 years. Permanent muscle weakness may develop as a result of a vacuolar myopathy, which may be disabling in older patients.

The HypoPP gene locus was assigned to chromosome 1q, following a genome-wide search (88). The authors suggested the calcium channel gene [dihydropyridine (DHP) receptor α_1 subunit, also called the CACNL1A3 gene] as a candidate gene, and mutations in this gene have been described subsequently (89,90). Linkage to chromosome 1q was recently confirmed in the largest known HypoPP family, a five-generation Dutch family, and a mutation in the calcium channel (CACNL1A3) gene was demonstrated (91).

---

*Arg615Cys indicates that in the ryanodine receptor protein, the amino acid arginine has been replaced by cysteine at amino acid number 615, due to a point mutation in the gene.

SODIUM CHANNEL DISORDERS

Hyperkalemic Periodic Paralysis (HPP) and Paramyotonia Congenita (PC)

These disorders are all autosomal dominantly inherited and caused by sodium channel gene defects (for review, see Ref. 92).

HPP is characterized by recurrent paralysis associated with elevations in serum potassium. Three main clinical subtypes exist: HPP with or without myotonia and HPP with paramyotonia.

In PC, cold triggers myotonia and may lead to muscle paralysis. In most myotonias, muscle activity improves myotonia, but in PC, exercise aggravates myotonia. Potassium levels are typically normal or raised during attacks, but occasionally, weakness is precipitated by potassium loading. In both HPP and PC, weakness eventually develops between attacks. Clinical variants have been described, including pure PC, myotonia fluctuans, and myotonia permanens, but there is clinical variability within families.

Electrophysiological studies in HPP show abnormalities in skeletal muscle membrane sodium conductance. The suggestion that HPP might result from a defect in a skeletal muscle sodium channel gene was supported by genetic linkage of the gene for the adult form of the skeletal muscle sodium channel (SCN4A on chromosome 17q) in families with HPP, both with and without myotonia (93–96). The clinical and electrophysiological similarities between HPP and PC prompted linkage studies in PC and confirmed linkage to the SCN4A locus (96–98). A form of myotonia congenita with potassium-induced myotonia but without periodic paralysis is also linked to this locus (99).

The presence of abnormal sodium channels in affected skeletal muscle has been confirmed (100), and several mutations in the SCN4A gene have been identified in HPP and PC and in some of the clinical variants (101–106). A mutation in the SCN4A gene was also found in a family with acetazolamide-responsive myotonia congenita (107). All these mutations result in the substitution of an amino acid in highly conserved positions in the sodium channel (92).

Most HPP and PC patients can now be characterized at a molecular level and phenotype-genotype correlations are possible (107). There may be differences in the frequencies of the various mutations in Europe and North America (107,108). However, one particular mutation (Thr704Met) is common in both populations in HPP. It occurred in five of seven HPP French families and was associated with permanent weakness over the age of 30 years, but myotonia and hyperkalemia were not always present (108). Similarly, five of 11 North American patients had this mutation, and permanent weakness between attacks was more common and severe in this group (107). A large multicenter trial is in progress to correlate the therapeutic response to various drugs in the molecular variants of HPP and PC (107).

CHLORIDE CHANNEL DISORDERS

Dominant Myotonia Congenita (Thomsen's Disease) and Recessive Myotonia Congenita (Becker's Variant)

Thomsen's disease (dominant myotonia congenita, DMC) presents with generalized nonprogressive myotonia often from birth, most prominent in the leg muscles, and often associated with muscular hypertrophy. The condition remains stable over many years. This disorder is not linked to the SCN4A gene (109), nor to the myotonic dystrophy gene (110). Muscle from patients with DMC shows abnormal chloride conductance. A murine myotonia model called "arrested development of righting response" (ADR mouse) mapped to mouse chromosome 6 (111) and the gene responsible was found to be a muscle-membrane chloride channel gene (112). The homologous human region is on chromosome 7q35, and linkage to the T-cell receptor β (TCRB) locus in this region in four Thomsen families was demonstrated (113). A human skeletal muscle chloride channel gene (CLC-1) is present in this area and is tightly linked to Thomsen's disease (114). A single nucleotide change in the human CLC-1 gene was demonstrated in three unrelated Thomsen pedigrees (115), making it highly likely that the CLC-1 gene is responsible for the disease.

In the Becker's variant of myotonia congenita, the inheritance is autosomal recessive. Onset is later than in the dominant form, usually between 6 and 12 years. Transient muscle weakness in the arms and hands is typical and a progressive myopathy may develop, signs that often allow clinical differentiation from the dominant form of the disease. As with the dominant form, tight linkage of the CLC-1 and TCRB loci in German families was shown and a mutation in the *CLC-1* gene was identified in several families (114,116). Thus the dominant and recessive forms of this disease are allelic and can now be described under the heading of muscle chloride channel disorders. It might be thought unusual that mutations in the same gene can cause recessive or dominant disease, but this has been demonstrated in other disorders, for example, osteogenesis imperfecta, and relates to the type of mutation, or its position in the gene rather than the locus itself.

OTHER CONGENITAL MYOPATHIES

Several congenital myopathies have been mapped, as summarized in Table 3.

NOTE ADDED IN PROOF

Since this chapter was submitted, much has been published on the molecular genetics of muscle disorders. A useful overview of the muscular dystrophies is given in Ref. 124. There are now five forms of recessive limb-girdle muscular dystrophy (LGMD-2A, 2B, 2C, 2D, 2E) mapping to 15q, 2p, 13q, 17q, and 4q, respectively (cited in Ref. 124). Further mutations in the adhalin gene on chromosome 17 in severe childhood autosomal recessive muscular dystrophy have been described and some families have a milder course (125). Mutations in the laminin alpha2-chain gene (LAMA2) which cause merosin-deficient muscular dystrophy have been documented (126). Diagnosis of merosin-deficient congenital muscular dystrophy by skin biopsy has been made in two patients (127). A summary of molecular tests for facioscapulohumeral muscular dystrophy is given in Ref. 128. Linkage of a family with autosomal dominant distal myopathy to chromosome 14 is reported in Ref. 129. A review of the skeletal muscle sodium and chloride channel diseases is given in Ref. 130.

REFERENCES

1. Handyside AH, Pattinson JK, Penketh RJA, Delhanty JDA, Winston RML, Tuddenham EGD. Biopsy of human preimplantation embryos and sexing by DNA amplification. Lancet 1989; 1:347–349.
2. Hoffman EP, Brown RH Jr, Kunkel LM. Dystrophin: the protein product of the Duchenne muscular dystrophy locus. Cell 1987; 51:919–928.
3. Abbs S, Roberts RG, Mathew CG, Bentley DR, Bobrow M. Accurate assessment of intragenic recombination frequency within the Duchenne muscular dystrophy gene. Genomics 1990; 7:602–606.
4. Hoffman EP. Genotype/phenotype correlations in Duchenne/Becker muscular dystrophy. In: Partridge TA, ed. Molecular and Cell Biology of Muscular Dystrophy. London: Chapman & Hall, 1993:12–36.
5. Beggs AH, Hoffman EP, Snyder JR, et al. Exploring the molecular basis for variability among patients with Becker muscular dystrophy: dystrophin gene and protein studies. Am J Hum Genet 1991; 49:54–67.
6. Arikawa E, Hoffman EP, Kaido M, Nonaka I, Sugita H, Arahata K. The frequency of patients having dystrophin abnormalities in a limb-girdle patient population. Neurology 1991; 41:1491–1496.
7. Minetti C, Chang HW, Medori R, et al. Dystrophin deficiency in young girls with sporadic myopathy and normal karyotype. Neurology 1991; 41:1288–1291.
8. Sewry CA, Sansome A, Clerk A, et al. Manifesting carriers of Xp21 muscular dystrophy: lack of correlation between dystrophin expression and clinical weakness. Neuromusc Disord 1993; 3:141–148.
9. Hoffman EP, Arahata K, Minetti, C, et al. Dystrophinopathy in isolated cases of myopathy in females. Neurology 1992; 42:967–975.
10. Pegoraro E, Schimke RN, Arahata K, et al. Detection of new paternal dystrophin gene mutations in isolated cases of dystrophinopathy in females. Am J Hum Genet 1994; 54:989–1003.

11. Holding C, Bentley D, Roberts R, Bobrow M, Mathew C. Development and validation of laboratory procedures for preimplantation diagnosis of Duchenne muscular dystrophy. J Med Genet 1993; 30:903–909.
12. Morgan JE. Cell and gene therapy in Duchenne muscular dystrophy. Hum Gene Ther 1994; 5:165–173.
13. Ben Hamida M, Fardeau M, Attia N. Severe childhood muscular dystrophy affecting both sexes and frequent in Tunisia. Muscle Nerve 1983; 6:469–480.
14. Salih MAM, Ibrahim M, Omer A, Bayoumi RA, Karrar O, Johnson M. Severe autosomal recessive muscular dystrophy in an extended Sudanese kindred. Dev Med Child Neurol 1983; 25:43–52.
15. Zatz M, Passos-Bueno MR, Rapaport D. Estimate of the proportion of Duchenne muscular dystrophy with autosomal recessive inheritance. Am J Med Genet 1989; 32:407–410.
16. Somer H, Voutilainen A, Knuutila S, Rapola J, Leinonen H. Duchenne-like muscular dystrophy in two sisters with normal karyotypes: evidence for autosomal recessive inheritance. Clin Genet 1985; 28:151–156.
17. Ben Jelloun-Dellagi S, Chaffey P, Hentati F, et al. Presence of normal dystrophin in Tunisian severe childhood autosomal recessive muscular dystrophy. Neurology 1990; 40:1903.
18. Passos-Bueno MR, Vainzof M, Pavanello RdeCM, Pavanello-Filho I, Lima MABO, Zata M. Limb-girdle syndrome: a genetic study of 22 large Brazilian families: comparison with X-linked Duchenne and Becker dystrophies. J Neurol Sci 1991; 103:65–75.
19. Vainzof M, Pavanello RCM, Pavanello-Filho I, et al. Screening of male patients with autosomal recessive Duchenne dystrophy through dystrophin and DNA studies. Am J Med Genet 1991; 39:38–41.
20. Ben Othmane K, Ben Hamida M, Pericak-Vance MA, et al. Linkage of Tunisian autosomal recessive Duchenne-like muscular dystrophy to the pericentromeric region of chromosome 13q. Nature Genet 1992; 2:315–317.
21. Matsumura K, Tome FMS, Collin H, et al. Deficiency of the 50 K dystrophin-associated glycoprotein in severe childhood autosomal recessive muscular dystrophy. Nature 1992; 359:320–322.
22. Fardeau M, Matsumura K, Tome FM, et al. Deficiency of the 50 kDa dystrophin associated glycoprotein (adhalin) in severe autosomal recessive muscular dystrophies in children native from European countries. CR Acad Sci III 1993; 316:799–804.
23. Zatz M, Matsumura K, Vainzof M et al. Assessment of the 50kDa dystrophin-associated glycoprotein in Brazilian patients with severe childhood autosomal recessive muscular dystrophy. J Neurol Sci 1994; 123:122–128.
24. Higuchi I, Yamada H, Fukunaga H, et al. Abnormal expression of laminin suggests disturbance of sarcolemma-extracellular matrix interaction in Japanese patients with autosomal recessive muscular dystrophy deficient in adhalin. J Clin Invest 1994; 94:601–606.
25. Azibi K, Bachner L, Beckmann JS, et al. Severe childhood autosomal recessive muscular dystrophy with the deficiency of the 50 kDa Dystrophin-associated glycoprotein maps to chromsome 13q12. Hum Mol Genet 1993; 2:1423–1428.
26. El Kerch F, Sefiani A, Azibi K, et al. Linkage analysis of families with severe childhood autosomal recessive muscular dystrophy in Morocco indicates genetic homogeneity of the disease in north Africa. J Med Genet 1994; 31:342–343.
27. Passos-Bueno MR, Oliveira JR, Bakker E, et al. Genetic heterogeneity for Duchenne-like muscular dystrophy (DLMD) based on linkage and 50 DAG analysis. Hum Mol Genet 1993; 2:1945–1947.
28. Roberds SL, Leturcq F, Allamand V, et al. Missense mutations in the adhalin gene linked to autosomal recessive muscular dystrophy. Cell 1994; 78:625–633.
29. McNally EM, Yoshida M, Mizuno Y, Ozawa E, Kunkel LM. Human adhalin is alternatively spliced and the gene is located on chromosome 17q21. Proc Natl Acad Sci USA 1994; 91:9690–9694.
30. Passos-Bueno MR, Bakker E, Marie SKN, et al. Exclusion of the 15q locus as a candidate gene for severe childhood autosomal recessive Duchenne-like muscular dystrophy in Brazilian families. Hum Mol Genet 1993; 2:201–201.
30a. Tomé FMS, Evangelista T, Leclerc A, et al. Congenital muscular dystrophy with merosin deficiency. CR Acad Sci Paris Sci vie/Life Sci 1994; 317:351–357.
31. Dubowitz V. Workshop report. Proceedings of the 27th ENMC sponsored workshop on congenital muscular dystrophy held in Holland, 22–24 April 1994. Neuromusc Disord 1994; 4:75–81.
32. Sunada Y, Bernier SM, Kozak CA, Yamada Y, Campbell KP. Deficiency of merosin in dystrophic dy mice and genetic linkage of laminin M chain gene to dy locus. J Biol Chem 1994; 269:13729–13732.
33. Harley HG, Brook JD, Rundle SA, et al. Expansion of an unstable DNA region and phenotypic variation in myotonic dystrophy. Nature 1992; 355:545–546.
34. Buxton J, Shelbourne P, Davies J, et al. Detection of an unstable fragment of DNA specific to individuals with myotonic dystrophy. Nature 1992; 355:547–548.
35. Aslanidis C, Jansen G, Amemiya C, et al. Cloning of an essential myotonic dystrophy region and mapping of the putative defect. Nature 1992; 355:548–550.
36. Monckton DG, Wong L-J C, Ashizawa T, Caskey CT. Somatic mosaicism, germline expansions, germline reversions and intergenerational reductions in myotonic dystrophy males: small pool PCR analyses. Hum Mol Genet 1995; 4:1–8.

37. Thornton CA, Griggs RC, Moxley RT. Myotonic dystrophy with no trinucleotide repeat expansion. Ann Neurol 1994; 35:269–272.
38. Ricker K, Koch MC, Lehmann-Horn F, et al. Proximal myotonic myopathy: a new dominant disorder with myotonia, muscle weakness, and cataracts. Neurology 1994; 44:1448–1452.
39. Hodgson S, Boswinkel E, Cole C, et al. A linkage study of Emery-Dreifuss muscular dystrophy. Hum Genet 1986; 74:409–416.
40. Romeo G, Roncuzzi L, Sangiorgi S, et al. Mapping of the Emery-Dreifuss gene through reconstruction of crossover points in two Italian pedigrees. Hum Genet 1988; 80:59–62.
41. Yates JR, Warner JP, Smith JA, et al. Emery-Dreifuss muscular dystrophy: linkage to markers in distal Xq28. J Med Genet 1993; 30:108–111.
42. Maestrini E, Patrosso C, Mancini M, et al. Mapping two genes encoding isoforms of the actin binding protein ABP-280, a dystrophin like protein, to Xq28 and to chromosome 7. Hum Mol Genet 1993; 2:761–766.
43. Fishbein MC, Siegel RJ, Thompson CE, Hopkins LC. Sudden death of a carrier of X-linked Emery-Dreifuss muscular dystrophy. Ann Intern Med 1993; 119:900–905.
44. Padberg GW, Lunt PW, Koch M, Fardeau M. Workshop report. Diagnostic criteria for facioscapulohumeral muscular dystrophy. Neuromusc Dis 1991; 1:231–234.
45. Lunt PW, Harper PS. Genetic counseling in facioscapulohumeral muscular dystrophy. J Med Genet 1991; 28:655–664.
46. Eggers S, Passos-Bueno MR, Zatz M. Facioscapulohumeral muscular dystrophy: aspects of genetic counselling, acceptance of preclinical diagnosis, and fitness. J Med Genet 1993; 30:589–592.
47. Wijmenga C, Frants RR, Brouwer OF, Moerer P, Weber JL, Padberg GW. Location of facioscapulohumeral muscular dystrophy gene on chromosome 4. Lancet 1990; 336:651–653.
48. Upadhyaya M, Lunt PW, Sarfarazi M, et al. DNA marker applicable to presymptomatic and prenatal diagnosis of facioscapulohumeral disease. Lancet 1990; 2:1320–1321.
49. Upadhyaya M, Lunt PW, Sarfarazi M, et al. A closely linked DNA marker for facioscapulohumeral disease on chromosome 4q. J Med Genet 1991; 28:665–671.
50. Gilbert JR, Stajich JM, Wall S, et al. Evidence for heterogeneity in facioscapulohumeral muscular dystrophy (FSHD). Am J Hum Genet 1993; 53:401–408.
51. Bakker E, van der Wielen MJR. Losekoot M, Padberg GW, Wijmenga C, Frants RR. Post- and prenatal testing for FSHD: diagnostic approach for sporadic and familial cases. Am J Hum Genet 1994; 55 (Suppl):A211 (abstract 1226).
52. Wijmenga C, Hewitt JE, Sandkuijl LA, et al. Chromosome 4q DNA rearrangements associated with facioscapulohumeral muscular dystrophy. Nature Genet 1992; 2:26–30.
53. Wijmenga C, van Deutekom JCT, Hewitt JE, et al. Pulsed-field gel electrophoresis of the D4F104S1 locus reveals the size and the parental origin of the facioscapulohumeral muscular dystrophy (FSHD)-associated deletions. Genomics 1994; 19:21–26.
54. Love DR, Hill D, Dickson G, et al. An autosomal transcript in skeletal muscle with homology to dystrophin. Nature 1989; 339:55–58.
55. Passos-Bueno MR, Terwilliger J, Ott J, et al. Linkage analysis in families with autosomal recessive limb-girdle muscular dystrophy (LGMD) and 6q probes flanking the dystrophin-related sequence. Am J Med Genet 1991; 38:140–146.
56. Beckmann JS, Richard I, Hillaire D, et al. A gene for limb-girdle muscular dystrophy maps to chromosome 15 by linkage. CR Acad Sci III 1991; 312:141–148.
57. Young K, Williams P, Foroud T, et al. Confirmation of linkage of limb-girdle muscular dystrophy to chromosome 15. Cytogenet Cell Genet 1991; 58:1996 (abstract).
58. Passos-Bueno MR, Richard I, Vainzof M, et al. Evidence of genetic heterogeneity in the autosomal recessive adult forms of limb-girdle muscular dystrophy following linkage analysis with 15q probes in Brazilian families. J Med Genet 1993; 30:385–387.
59. Gilchrist JM, Jackson C, Beckmann J, et al. Linkage analysis in autosomal recessive limb-girdle muscular dystrophy. Am J Hum Genet 1991; 53:1694 (abstract).
60. Bashir R, Strachan T, Keers S et al. A gene for autosomal recessive limb-girdle muscular dystrophy maps to chromosome 2p. Hum Mol Genet 1994; 3:455–457.
61. Chutkow JG, Heffner RR, Kramer AA, Edwards JA. Adult-onset autosomal dominant limb girdle muscular dystrophy. Ann Neurol 1986; 20:240–248.
62. Gilchrist JM, Pericak-Vance M, Silverman L, Roses AD. Clinical and genetic investigation in autosomal dominant limb-girdle muscular dystrophy. Neurology 1988; 38:5–9.
63. Nielsen JF, Jakobsen J. A Danish family with limb-girdle muscular dystrophy with autosomal dominant inheritance. Neuromusc Disord 1994; 4:139–142.
64. Mohire MD, Tandan R, Fries TJ, Little BW, Pendlebury WW, Bradley DM. Early-onset benign autosomal dominant limb-girdle myopathy with contractures (Bethlem myopathy). Neurology 1988; 38:573–580.
65. Speer MC, Yamaoka LH, Gilchrist JH, et al. Confirmation of genetic heterogeneity in limb-girdle muscular dystrophy: linkage of an autosomal dominant form to chromosome 5q. Am J Hum Genet 1992; 50:1211–1217.

66. Ball SP, Johnson KJ. Review article. The genetics of malignant hyperthermia. J Med Genet 1993; 30:89–93.
67. Hall LW, Woolf N, Bradley JWP, Jolly DW. Unusual reaction to suxamethonium chloride. Br Med J 1966; 2:1305.
68. Nelson T. Abnormality in calcium release from skeletal sarcoplasmic reticulum of pigs susceptible to malignant hyperthermia. J Clin Invest 1983; 72:862–872.
69. McCarthy T, Healy J, Heffron J, et al. Localisation of the malignant hyperthermia susceptibility locus to human chromosome 19q12-13.2. Nature 1990; 343:562–564.
70. MacKenzie A, Korneluk R, Zorzato F, et al. The human ryanodine receptor gene: its mapping to 19q13.1, placement in a chromosome 19 linkage group and exclusion as the gene causing myotonic dystrophy. Am J Hum Genet 1990; 46:1082–1089.
71. Fujii J, Otsu K, Zorzato F, et al. Identification of a mutation in porcine ryanodine receptor associated with malignant hyperthermia. Science 1991; 253:448–451.
72. Gillard EF, Otsu K, Fujii J, et al. A substitution of cysteine for arginine 614 in the ryanodine receptor is potentially causative of human malignant hyperthermia. Genomics 1991; 11:751–755.
73. Gillard EF, Otsu K, Fujii J, et al. Polymorphisms and deduced amino acid substitutions in the coding sequence of the ryanodine receptor (RYR1) gene in individuals with malignant hyperthermia. Genomics 1992; 13:1247–1254.
74. Quane KA, Keating KE, Manning BM, et al. Detection of a novel common mutation in the ryanodine receptor gene in malignant hyperthermia: implications for diagnosis and heterogeneity studies. Hum Mol Genet 1994; 3:471–476.
75. Levitt RC, Olckers A, Meyers S, et al. Evidence for the localisation of a malignant hyperthermia susceptibility locus (MHS2) to human chromosome 17q. Genomics 1992; 14:562–566.
76. Sudbrak R, Golla A, Hogan K, et al. Exclusion of malignant hyperthermia susceptibility (MHS) from a putative MHS2 locus on chromosome 17q and of the α1, β1, and δ subunits of the dihydropyridine receptor calcium channel as candidates for the molecular defect. Hum Mol Genet 1993; 2:857–862.
77. Iles DE, Segers B, Sengers RC, et al. Genetic mapping of the β1- and δ-subunits of the human skeletal muscle L-type voltage-dependent calcium channel on chromosome 17q and exclusion as candidate genes for malignant hyperthermia. Hum Mol Genet 1993; 2:863–868.
78. Olckers A, Jedlicka AE, Powers P, Hogan K, Gregg RG, Levitt RC. G to A polymorphism in the CACNLG gene. Hum Mol Genet 1993; 2:2198.
79. Iles DE, Lehmann-Horn F, Scherer SW. Localization of the gene encoding the α2/δ-subunits of the L-type voltage-dependent calcium channel to chromosome 7q and analysis of flanking markers in malignant hyperthermia susceptible families. Hum Mol Genet 1994; 3:969–975.
80. Sudbrak R, Procaccio V, Klausnitzer M, et al. Mapping of a further malignant hyperthermia susceptibility locus to chromosome 3q13.1. Am J Hum Genet 1995; 56:684–691.
81. Vladutiu GD, Hogan K, Saponara I, Tassini L, Conroy J. Carnitine palmitoyl transferase deficiency in malignant hyperthermia. Muscle Nerve 1993; 16:485–491.
82. Haan EA, Freemantle CJ, McCure JA, Friend KL, Mulley JC. Assignment of the gene for central core disease to chromosome 19. Hum Genet 1990; 86:187–190.
83. Kausch K, Lehmann-Horn F, Janka M, Wieringa B, Grimm T, Muller CR. Evidence for linkage of the central core disease locus to the proximal long arm of human chromosome 19. Genomics 1991; 10:765–769.
84. Mulley JC, Kozman HM, Phillips HA, et al. Refined localization for central core disease. Am J Hum Genet 1993; 52:398–405.
85. Schwemmle S, Wolff K, Palmucci LM, et al. Multipoint mapping of the central core disease locus. Genomics 1993; 17:205–207.
86. Quane KA, Healy JMS, Keating KE, et al. Mutations in the ryanodine receptor gene in central core disease and malignant hyperthermia. Nature Genet 1993; 5:51–55.
87. Zhang Y, Chen HS, Khanna VK, et al. A mutation in the human ryanodine receptor gene associated with central core disease. Nature Genet 1993; 5:46–50.
88. Fontaine B, Vale-Santos J, Jurkat-Rott K, et al. Mapping of the hypokalemic periodic paralysis (HypoPP) locus to chromsome 1q31-32 in three European families. Nature Genet 1994; 6:267–272.
89. Ptacek LJ, Tawil R, Griggs RC, et al. Dihydropyridine receptor mutations cause hypokalemic periodic paralysis. Cell 1994; 77:863–868.
90. Jurkat-Rott K, Lehmann-Horn F, Elbaz A, et al. A calcium channel mutation causing hypokalemic periodic paralysis. Hum Mol Genet 1994; 3:1415–1419.
91. Boerman RH, Ophoff RA, Links TP, et al. Mutation in DHP receptor α1 subunit (CACNL1A3) gene in a Dutch family with hypokalemic periodic paralysis. J Med Genet 1995; 32:44–47.
92. Ptacek LJ, Johnson KJ, Griggs RC. Genetics and physiology of the myotonic disorders. N Engl J Med 1993; 328:483–488.
93. Fontaine B, Khurana TS, Hoffman EP, et al. Hyperkalemic periodic paralysis and the adult muscle sodium channel α-subunit gene. Science 1990; 250:1000–1002.

94. Ptacek LJ, Tyler F, Trimmer JS, Agnew WS, Leppert M. Analysis in a large hyperkalemic periodic paralysis pedigree supports tight linkage to a sodium channel locus. Am J Hum Genet 1991; 49:378–382.
95. Koch MC, Ricker K, Otto M, et al. Confirmation of linkage of hyperkalaemic periodic paralysis to chromosome 17. J Med Genet 1991; 28:583–586.
96. Ebers GC, George AL, Barchi RL, et al. Paramyotonia congenita and hyperkalemic periodic paralysis are linked to the adult muscle sodium channel gene. Ann Neurol 1991; 30:810–816.
97. Ptacek LJ, Trimmer JS, Agnew WS, Roberts JW, Petajan JH, Leppert M. Paramyotonia congenita and hyperkalemic periodic paralysis map to the same sodium-channel gene locus. Am J Hum Genet 1991; 49:851–854.
98. Koch MC, Ricker K, Otto M, et al. Linkage data suggesting allelic heterogeneity for paramyotonia congenita and hyperkalemic periodic paralysis. Hum Genet 1991; 88:71–74.
99. Ptacek LJ, Tawil R, Griggs RC, Storvick D, Leppert M. Linkage of atypical myotonia congenita to a sodium channel locus. Neurology 1992; 42:431–433.
100. Cannon SC, Brown RH Jr, Corey DP. A sodium channel defect in hyperkalemic periodic paralysis: potassium-induced failure of inactivation. Neuron 1991; 6:619–626.
101. Ptacek LJ, George Al Jr, Griggs RC, et al. Identification of a mutation in the gene causing hyperkalemic periodic paralysis. Cell 1991; 67:1021–1027.
102. Rojas CV, Wang J, Schwartz LS, Hoffman EP, Powell BR, Brown RH Jr. A Met-to Val mutation in the skeletal muscle Na+ channel α-subunit in hyperkalemic periodic paralysis. Nature 1991; 354:387–389.
103. McClatchey AI, Van den Bergh P, Pericak-Vance MA, et al. Temperature-sensitive mutations in the III-IV cytoplasmic loop region of the skeletal muscle sodium channel gene in paramyotonia congenita. Cell 1992; 68:769–774.
104. Ptacek LJ, George AL Jr, Barchi RL, et al. Mutations in an S4 segment of the adult skeletal muscle sodium channel cause paramyotonia congenita. Neuron 1992; 8:891–897.
105. Ptacek LJ, Gouw L, Kwiecinski H, et al. Sodium channel mutations in paramyotonia congenita and hyperkalemic periodic paralysis. Ann Neurol 1993; 33:300–307.
106. Heine R, Pika U, Lehmann-Horn F. A novel SCN4A mutation causing myotonia aggravated by cold and potassium. Hum Mol Genet 1993; 2:1349–1353.
107. Ptacek LJ, Tawil R, Griggs RC, et al. Sodium channel mutations in acetazolamide-responsive myotonia congenita, paramyotonia congenita, and hyperkalemic periodic paralysis. Neurology 1994; 44:1500–1503.
108. Plassart E, Reboul J, Rime C-S, et al. Mutations in the muscle sodium channel gene (SCN4A) in 13 French families with hyperkalemic periodic paralysis and paramyotonia congenita: phenotype to genotype correlations and demonstration of the predominance of two mutations. Eur J Hum Genet 1994; 2:110–124.
109. Ptacek LJ, Ziter FA, Roberts JW, Leppert MF. Evidence of genetic heterogeneity among the nondystrophic myotonias. Neurology 1992; 42:1046–1048.
110. Koch M, Harley H, Sarfarazi M, et al. Myotonia congenita (Thomsen's disease) excluded from the region of the myotonic dystrophy locus on chromosome 19. Hum Genet 1989; 82:163–166.
111. Rudel R. The myotonic mouse—a realistic model for the study of human recessive generalised myotonia. Trends Neurosci 1990; 13:1–3.
112. Steinmeyer K, Klocke R, Ortland C. Inactivation of muscle chloride channel by transposon insertion in myotonic mice. Nature 1991; 354:304–308.
113. Abdalla JA, Casley WL, Cousin HK, et al. Linkage of Thomsen disease to the T-cell-receptor beta (TCRB) locus on chromosome 7q35. Am J Hum Genet 1992; 51:579–584.
114. Koch MC, Steinmeyer K, Lorenz C, et al. The skeletal muscle chloride channel in dominant and recessive human myotonia. Science 1992; 257:797–800.
115. George AL, Crackower MA, Abdalla JA, Hudson AJ, Ebers GC. Molecular basis of Thomsen's disease (autosomal dominant myotonia congenita). Nature Genet 1993; 3:305–310.
116. Koch MC, Ricker K, Otto M, et al. Evidence for genetic homogeneity in autosomal recessive generalised myotonia (Becker). J Med Genet 1993; 30:914–917.
117. Toda T, Segawa M, Nomura Y, et al. Localization of a gene for Fukuyama type congenital muscular dystrophy to chromosome 9q31-33. Nature Genet 1993; 5:283–286.
118. Toda T, Ikegawa S, Okui K, et al. Refined mapping of a gene responsible for Fukuyama-type congenital muscular dystrophy: evidence for strong linkage disequilibrium. Am J Hum Genet 1994; 55:946–950.
119. Thomas NST, Sarfarazi M, Roberts K, et al. X-linked myotubular myopathy (MTM1): evidence for linkage to Xq28 DNA markers. Cytogenet Cell Genet 1987; 46:704 (abstract).
120. Thomas NST, Williams H, Cole G, et al. X linked neonatal centronuclear/myotubular myopathy: evidence for linkage to Xq28 DNA marker loci. J Med Genet 1990; 27:284–287.
121. Dahl N, Samson F, Thomas NST et al. X linked myotubular myopathy (MTM1) maps between DXS304 and DXS305, closely linked to the DXS455 VNTR and a new, highly informative microsatellite marker (DXS16884). J Med Genet 1994; 31:922–924.

122. Laing NG, Majda BT, Akkari PA et al. Assignment of a gene (NEM1) for autosomal dominant nemaline myopathy to chromosome 1. Am J Hum Genet 1992; 50:576–583.
123. Laing NG, Wilton SD, Akkari PA, et al. A mutation in the α tropomyosin gene TPM3 associated with autosomal dominant nemaline myopathy. Nature Genet 1995; 9:75–79.
124. Worton R. Muscular dystrophies: diseases of the dystrophin-glycoprotein complex. Science 1995; 270:755–756.
125. Piccolo F, Roberds SL, Jeanpierre M, et al. Primary adhalinopathy: a common cause of autosomal recessive muscular dystrophy of variable severity. Nature Genet 1995; 10:243–245.
126. Hebling-Leclerc A, Zhang X, Topaloglu H, et al. Mutations in the laminin alpha-2 chain gene (LAMA2) cause merosin-deficient congenital muscular dystrophy. Nature Genet 1995; 11:216–218.
127. Sewry CA, Philpot J, Sorokin LM, et al. Diagnosis of merosin (laminin-2) deficient congenital muscular dystrophy by skin biopsy. Lancet 1996; 347:582–584.
128. Bakker E, Van der Wielen MJR, Voorhoeve E, et al. Diagnostic, predictive, and prenatal testing for facioscapulohumeral muscular dystrophy: diagnostic approach for sporadic and familial cases. J Med Genet 1996; 33:29–35.
129. Laing NG, Laing BA, Meredith C, et al. Autosomal dominant distal myopathy: linkage to chromosome 14. Am J Hum Genet 1995; 56:422–427.
130. Hudson AJ, Ebers GC, Bulman DE. Review article. The skeletal muscle sodium and chloride channel diseases. Brain 1995; 118:547–563.

Appendix: Diagnostic Index of Neuromuscular Disorders

Russell J. M. Lane and Richard W. Orrell
Charing Cross Hospital and Charing Cross and Westminster Medical School London, England

Hiba Tappouni
University of North Carolina, Chapel Hill, North Carolina

INTRODUCTION

This classification is compiled in two parts. Part I is a *categorical classfication*, based on the anatomical site and pathogenesis of neuromuscular disorder. The categories and primary diagnostic terms used are generally those specified by the Neurosciences Specialty Working Group of the Centre for Coding and Classification (the Clinical Terms Project), which was commissioned by the National Health Service Management Executive of the United Kingdom to expand the thesaurus of Read Terms and Codes for the documentation of all aspects of health care in the United Kingdom. These Read terms (from Version 3 of the thesaurus) are supplemented as neccessary by alternative descriptors relating to clinical features, pathological characteristics, or current understanding of etiology, and by commonly used eponyms, and are cross-referenced to the International Classfication of Diseases 10th Revision (ICD-10, three- and four-character categories)* and, where appropriate, to catalogue numbers from McKusick's *Mendelian Inheritance in Man*. The chromosomal location for genetically determined disorders is given where the gene or locus for a disease has been established, together with details of the relevant genes and gene products, if known. The classification includes the most commonly encountered disorders but not those of dubious significance or doubtful standing, and it is therefore simpler than the current World Federation of Neurology Classification.† Read Codes are Crown Copyright, and copyright material is included by kind permission of the Controller of Her Majesty's Stationery Office.

Part II is an *elementary alphabetical index* of the main neuromuscular disorders, using the same key terms, with the appropriate range of Read and ICD-10 codes, for quick reference and ease of coding in the clinical setting. Only the most common disorders or categories are included in this alphabetical index, and if the disorder to be coded is not found, the categorical classification in Part I should be consulted.

We recognize that the nosology and molecular genetics of the neuromuscular diseases is changing rapidly, and the current classification will inevitably require frequent updating. The journal *Neuromuscular Diseases* (Pergamon Press) provides regular information on genetic aspects.

---

*International Statistical Classification of Diseases and Related Health Problems, 10th rev. Geneva: World Health Organization, 1992.
†Walton JN, Rowland LP, McLeod JG. Classification of neuromuscular disorders. J Neurol Sci 1994; 124:109–130.

PART I. CATEGORICAL CLASSIFICATION*

Diseases of the Anterior Horn Cells

Genetically Determined

1. Spinal Muscular Atrophy (SMA)

Condition	Alternative names	Mode of inheritance	Read	ICD-10	Gene location	Gene product symbols	MIM
SMA Type I	Severe Infantile form (Werdnig-Hoffman disease)	AR	F150	G12.0	5q11–q13	SMN NAIP	253300
SMA Type II	Arrested, chronic, or intermediate infantile form	AR	F151	G12.1	5q11–q13	SMN NAIP	253300
SMA Type III	Juvenile SMA, Kugelberg-Welander disease	AR	F151	G12.1	5q11–q13	SMN NAIP	253400
SMA Type IV	Proximal form, adult onset, limb-girdle form	AR	F151	G12.9	5q11–q13	SMN NAIP	
Progressive bulbar palsy of childhood	Fazio-Londe disease	AR	X002a	G12.1			211500
with deafness	Brown-Vialetto-van Laere syndrome						211530
Distal SMA		AR	X002b F151	G12.9			271120
Scapuloperoneal SMA		AR	F151, X002c	G12.1			271220
Facioscapulo humeral SMA		AR	F151	G12.9			182970
Scapulohumeral SMA		AR	F151, X002f	G12.9			
Oculopharyngeal SMA		AR	F151	G12.9			
X-linked bulbospinal neuronopathy	Kennedy syndrome	XR	X002h	G12.8	Xq21–22	SBMA (Androgen receptor	313200

SMN, survival motor neuron protein; NAIP, neuronal apoptosis inhibitory protein.

2. Motor Neuron Disease (MND)

Condition	Alternative names	Mode of inheritance	Read	ICD-10	Gene location	Gene product symbols	MIM
Familial ALS I	FALS-I	AD	F152	G12.2	21q22	SOD-1 Cu/Zn superoxide dismutase in 10%	105400
Familial ALS II	FALS-II	AR	F152	G12.2	2q33–q35	ALS2	205100

*Abbreviations: MIM, Reference number in McKusick VA. *Mendelian Inheritance in Man: Catalogue of Autosomal Dominant, Autosomal Recessive, and X-Linked phenotypes*, 11th ed. Baltimore: Johns Hopkins University Press, 1994; AD, autosomal dominant; AR, autosomal recessive; XR, X-linked recessive.

Sporadic

Condition	Alternative names	Read	ICD-10
Amyotrophic lateral sclerosis (ALS)	Motor neuron disease (MND)	F152	G12.2
Primary lateral sclerosis	PLS	F152.	G12.2
Progressive bulbar palsy	PBP	F152.	G12.2
Progressive muscular atrophy	PMA	F152	G12.2
ALS with dementia		F152	G12.2
Madras-type motor neurone disease		F152	G12.2
Paraneoplastic MND			G13.0?
Post-polio syndrome		X002p	B91
Amyotrophy due to herpes zoster		X002q	B02.2, G13.8

Nerve, Plexus, and Root Disorders

Genetically Determined

Sporadic

1. Charcot-Marie-Tooth Disease (CMT), Hereditary Motor and Sensory Neuropathy (HMSN)

Condition	Alternative names	Mode of inheritance	Read	ICD-10	Gene location	Gene product symbols	MIM
CMT Ia	CMT of demyelinating type Chromosome 17-linked (with prominent tremor constitutes Roussy-Levy syndrome)	AD	F3610	G60.0	17p11.2	PMP-22 (Peripheral myelin protein P22)	118220
CMT Ib	CMT of demyelinating type Chromosome 1 and Duffy blood group linked	AD	F3610	G60.0	1q21–q23	PMP0 (Peripheral myelin protein Po)	118200
CMT II	CMT of axonal type	AD	F3610	G60.0	1p35–p36	CMT 2A	118210
CMT III	Hypertrophic CMT of infancy (Déjerine-Sottas disease)	AD	F3610	G60.0	17p11.2	PMP-22 (peripheral mylein protein 22)	145900
		AD	F3610	G60.0	1q21–q23	PMPO	145900
HMSN-IV	Refsum's disease (infantile type)*see also* Refsum's disease	AR	F3610	G60.0	8q	CMT 4A	214400
HMSN-V	HMSN with pyramidal signs, Silver syndrome, Troyer syndrome	AR/AD	F3610	G60.0			
HMSN-VI	HMSN with optic atrophy	AD/AR	F3610	G60.0	3q		
HMSN-VII	HMSN with retinitis pigmentosa	XR	F3610	G60.0			302800
X-linked HMSN		XD	F3610	G60.0	Xq13	CMTX (connexin)	302801/2
		XR			Xp22, q26		310490

2. Hereditary Sensory and Autonomic Neuropathy (HSAN)

Condition	Alternative names	Mode of inheritance	Read	ICD-10	Gene location	Gene product symbols	MIM
HSAN-I	Autosomal dominant sensory neuropathy	AD	X00AF	G60.8			162380
HSAN-II	Autosomal recessive sensory neuropathy	AR	X00AF	G60.8			201300
HSAN-III	Familial dysauton--omia, Riley-Day syndrome	AR	X00AF	G60.8	9q31–q33	HSAN3	223900
	with motor neuropathy	AD	X00AF	G80.8			
HSAN-IV	Congenital sensory neuropathy with anhidrosis and mental retardation	AR	X00AF	G60.8			256800

3. Other Genetic or Inherited Neuropathies

Condition	Alternative names	Mode of inheritance	Read	ICD-10	Gene location	Gene product symbols	MIM
Familial amyloid polyneuropathy	FAP Type I, Portuguese type	AD	C3732	G63.3, E85.1	18q11.2–q12.1	PALB (prealbumin, transthyretin)	176300
Amyloid neuropathy Type IV	Iowa type	AD	C3732	G63.3, E85.1	11q23–qter	APOA1 (apolipoprotein A1)	107680
Amyloid neuropathy Type V	Finnish type	AD	C3732	G63.3, E85.1	9q33	GSN (gelsoline)	105120
Other heredofamilial amyloid neuropathy		AD	C3732	G63.3, E85.1			
Porphyric neuropathy	Neuropathy in porphyria Porphyria, AIP; Porphyria variegata)	AD	C3712/ Xa01K	G63.3, E80.2	11q24.1–q24.2	AIP-PBG deaminase	176000
						Porphyria variegata-protoporhyrinogen oxidase	176200
Familial hypoalphalipo- proteinaemia	Tangier disease A-alphalipoproteinaemia	AR	X40Vt	G63.3, E78.6			205400
A-beta-lipoproteinaemia	Bassen-Kornzweig disease	AR	C3253	G63.3, E78.6			200100
Metachromatic leucodystrophy	Arylsuphatase A deficiency, sulphatide lipidosis	AR	F1003	G63.3, E75.2	22q13	arylsulphatase, sphingolipid activator protein 1	250100

(continued)

Condition	Alternative names	Mode of inheritance	Read	ICD-10	Gene location	Gene product symbols	MIM
Globoid cell leucodystrophy	Krabbe's disease	AR	F1000	G63.3, E75.2	14q21–q31	sphingolipid activator protein 1, galactosylceramidase beta-galactosidase	245200
Niemann-Pick disease	Sphyngomyelin lipidosis	AR	C3272	G63.3, E75.2	11p15.1–p15.4	sphingomyelinase	257200
Fabry's disease	Anderson-Fabry's disease, glycosphingolipidosis	XR	No code	G63.3, E75.2	Xq22–qter	alpha-galactosidase A	310500
Refsum's disease	Heredopathia atactica polyneuretiformis, phytanic acid storage disease	AR	F363	G60.1		phytanic acid oxidase	266500
X-linked adrenoleucodystrophy	Adrenomyeloneuropathy	XR	X40Wn	E71.3	Xq28	perioxosomal lignoceroyl-CoA ligase	300100
Hereditary liability to pressure palsies	Tomaculous neuropathy	AD	X00AM	G60.8	17p11.2	PMP-22 (peripheral myelin protein 22)	162500
Giant axonal neuropathy		AD	X00AS	G58.8			256850
Congenital hypomyelinating polyneuropathy	Lyon's hypomyelinating polyneuropathy, congenital polyneuropathy	AD	X00AS	G60.8			

Sporadic

1. Inflammatory Polyneuropathy

Condition	Alternative names	Read	ICD-10
Guillain-Barré syndrome (GBS)	Acute postinfective polyneuropathy	F3700	G61.0
Miller Fisher variant of GBS	Miller Fisher variant of GBS	F3700, X00AV	G61.0
Chronic inflammatory demyelinating polyneuropathy (CIDP)		X00AW	G61.8

2. Metabolic Neuropathy and Neuromyelopathy

a. Neuropathy due to nutritional factors	Alternative names	Read	ICD-10
Sub-acute combined degeneration	B_{12} deficiency neuromyelopathy	F162	G32.0, D51
Folate deficiency		C2620	E53.8, G63.4
Beri-beri neuropathy	Thiamine (B_1) deficiency neuropathy	E37415	E51.1, G63.4
B_6 (pyridoxine) deficiency		C261	E53.1, G63.4
Vitamin E deficiency		X00Ad	E56.0, G63.4
Burning feet syndrome		X00Ae	E53.9

b. Neuropathy associated with endocrine disorder

Condition	Alternative names	Read	ICD-10
Diabetic polyneuropathy		F372	G63.2, E14.4
mononeuropathy		X00Ag	G59.0, E14.4
radiculopathy		X00Ag	G59.0, E14.4
diabetic amyotrophy	Diabetic femoral mononeuropathy	X00Ag	G59.0, E14.4
autonomic neuropathy		X00Ag	G99.0, E14.4
Hypoglycemic neuropathy		F3745	G63.3, E16.2
Myxedema neuropathy	Neuropathy in hypothyroidism, hypothyroid neuropathy	X00Ao	G63.3, E03.9
Neuropathy in acromegaly		X00Ap	G63.3, E22.0

c. Neuropathy in systemic illness

Neuropathy in renal failure	Uremic polyneuropathy, uremic neuropathy	F374A	G63.8, N18.8
Neuropathy in liver disease		X00Aq	G58.8, G62.8
Sarcoid neuropathy		F3749	G63.8, D86.8
Neuropathy in vasculitis/connective tissue disease	Vasculitic neuropathy	F371	G63.5, M35.9
Anoxic neuropathy	Neuropathy in chronic obstructive airways disease	X00B1	G58.8, G62.8
Critical illness polyneuropathy		X00B2	G62.8

3. Neuropathy in Malignant Disease, Blood Dyscrasia, and Dysproteinemia

Paraneoplastic neuropathy	Carcinomatous neuropathy, paraneoplastic/carcinomatous neuromyopathy	F373	G63.1
Neuropathy in reticulosis/lymphoma		X00Ar	No code
Neuropathy in blood dyscrasias		X00As	G63.1, D75.9
Neuropathy associated with dysproteinaemias	Paraproteinemic neuropathy	X00At	G63.1, E88.0
Neuropathy with multiple myeloma		X00At	G63.1, E88.0
Neuropathy in macroglobulinaemia	Waldenstrom's macroglobulinaemia	X00At	G63.1, E88.0
Neuropathy in cryoglobulinaemia		X00At	G63.1, E88.0
Neuropathy with benign monoclonal gammopathy		X00At	G63.1, E88.0
Motor neuropathy with multiple conduction block	Multifocal motor neuropathy with conduction block	X00Ay	G58.8, G62.8
Polyneuropathy, organomegaly, endocrinopathy, monoclonal gammopathy and skin changes (POEMS)		X00B0	G62.8
Neuropathy in secondary amyloidosis	Amyloid neuropathy	F1710	G63.3, G99.0 E85.4

4. Neuropathy due to Infective Agents

Lepromatous neuropathy		X00B3	G63.0, G58.9
Diphtheritic neuropathy		X00B3	G63.0, G58.9
Post-herpetic neuralgia		X75sv	B02.2, G53.0
Neuropathy due to human immunodeficiency virus	HIV-related neuropathy	X00B3	B24, G63.0

5. Ischemic Neuropathy

Neuropathy in vasculitis/connective tissue disease		F371	M35.9
Neuropathy in arteriosclerotic occlusive disease		X00B9	G58.9, 170.9, G63.8
Neuropathy in thrombangitis obliterans	Buerger's disease	G7310	G58.9, G63.8, 173.1
Nerve trunk infarction	Embolic infarction of nerve trunk, Nerve infarction	X00BB	G64
Hemorrhage into nerve trunk		X00BC	G54.8
Compression neuropathy	Tourniquet palsy	X00Bm	G58.9
Mononeuritis multiplex		F345	G58.7

6. Toxic Damage to Nerves, Nerve Roots, and Motor Neurons

Drug-related neuropathy	Drug-induced polyneuropathy, drug-induced neuropathy	F376.	G62.0
Alcohol-related polyneuropathy		F375.	G62.1
Toxic neuropathy	Chemical/toxin-induced neuropathy	X00BG	G62.2

7. Physical Damage to Nerves, Nerve Plexuses, and Roots

a. *Nerve root lesions*

Cervical spondylosis		X703	M47.8
Cervical root lesion	Cervical radiculopathy		
Disc prolapse		N12C0	M50.1
Spondylotic compression		N119	M47.2
Transection of nerve root		XA051	S14.2
Partial division of nerve root		XA056	S14.2
Contusion of nerve root		XA05B	S14.2
Traction injury		XA05G	S142.
Avulsion of nerve root		XA05L	S14.2
Thoracic root lesion	Thoracic radiculopathy		
Disc prolapse		N12C1	M51.1, M54.1
Spondylotic compression		N11B	M47.2
Transection of nerve root		XA052	S24.2
Contusion of nerve root		XA05C	S42.2
Traction injury		XA05H	S24.2
Avulsion of nerve root		XA05M	S24.2
Lumbar spondylosis		X703k	M47.8
Lumbar root lesion	Lumbar radiculopathy		

(continued)

a. *Nerve root lesions*

Condition	Alternative names	Read	ICD-10
Lumbar root lesion	Lumbar radiculopathy		
Disc prolapse		M12C2	M51.1, M54.1
Transection of nerve root		XA053	S34.2
Partial division of nerve root		XA058	S34.2
Contusion of nerve root		XA05D	S34.2
Traction injury		XA051	S34.2
Avulsion of nerve root		XA05N	S34.2
Radiculopathy syndromes			
Lumbar canal stenosis		No code	No code
Sciatica		N143	M54.3
Arachnoiditis		F023	G03.9
Cauda equina syndrome		F246	G83.4

b. *Plexus lesions*

a. Brachial plexus			
Thoracic outlet syndrome	Cervical rib syndrome	F3303	G54.0
Neuralgic amyotrophy	Brachial plexus neuritis, Parsonage-Turner syndrome	F335	G54.5
Erb's paralysis	Erb's palsy, birth injury to brachial plexus	Q2061	P14.0
Klumpke's paralysis	Klumpke's palsy, birth injury to brachial plexus	Q2062	P14.1
Brachial plexus lesion due to neoplasm	Pancoast's syndrome	F330	G54.0
Brachial plexus neuropathy due to radiation			
Familial recurrent brachial plexus neuropathy	(*see also* Tomaculous neuropathy)	X00AN	G54.0
b. Lumbosacral plexus lesion, unspecified aetiology		F331.	G54.1
Lumbosacral plexus injury		SJ35	S34.4
Inflammatory lumbosacral plexus lesion		X00AY	G54.1

c. *Peripheral nerve lesions*

Entrapment neuropathies in the upper extremity

Median nerve entrapment			
Carpal tunnel syndrome		F3411	G56.0
Proximal median neuropathy	Pronator syndrome, supracondylar process entrapment Struther's ligament entrapment, anterior interosseous entrapment	F3411	G56.1
Ulnar nerve entrapment		Xa0Ks	G56.2
at the elbow (cubital tunnel)	Tardy ulnar nerve palsy		
at the wrist (Guyon's canal)			
deep palmar branch			
Radial nerve compression		X00BB	G56.3
Posterior interosseous nerve compression		F343	G56.8
Suprascapular nerve compression		X00Bv	G58.8
Axillary nerve compression		X00Bw	G54.0

Entrapment neuropathies in the lower extremity

Sciatic nerve compression		X00By	G57.0
Obturator nerve compression		X00Bz	G57.8
Femoral nerve compression		X00CO	G57.2
Meralgia paraesthetica	Compression of the lateral cutaneous nerve of the thigh	F351.	G57.1
Common peroneal nerve compression	Lesion of the lateral popliteal nerve, lateral popliteal palsy	X00C1	G57.3
Tibial nerve compression	Medial popliteal nerve palsy, medial popliteal palsy	X00C2.	G57.4
Tarsal tunnel syndrome		F355	G57.5
Medial plantar nerve compression		X00C3	G57.6
Morton's metatarsalgia	Morton neuroma, interdigital neuralgia	F360	G57.6
Compression neuropathy of the trunk			
Intercostal neuropathy		X00C5.	G58.0
Abdominal neuropathy	Abdominal cutaneous entrapment syndrome	X00C6	G58.8
Multiple entrapment syndrome		X00C8	G62.9

d. Nerve injury from other causes

Electric shock injury of peripheral nerve		X00B	T14.4
Trench Foot	Immersion injury	X2054	T69.0
Vibration injury		X00B1	T14.4
Neuroma of amputation stump	Traumatic neuroma	SP161	T87.3

8. Chronic Neuropathy with No Known Cause or Association

Demyelinating sensorimotor polyneuropathy	Demyelinating polyneuropathy/neuropathy	F33	G62.9
Axonal sensorimotor polyneuropathy	Axonal polyneuropathy/neuropathy	F33	G62.9
Polyneuropathy (unclassified)	Neuropathy ?cause	F33	G62.9

Disorders of Spinal Neurons Manifest by Hyperactivity

Muscle cramps	Cramps	X00CB	R25.2
Tetany	Carpopedal spasm, spasmophilia	X00CB	R29.0
Benign fasciculation-cramp syndrome	Denny-Brown and Foley syndrome	X00CC	R25.3
Neuromyotonia	Isaac's syndrome, continuous muscle fiber activity, quantal squander	X00CD	G71.1
Stiff-man syndrome	Moersch-Woltman syndrome	F13z1	G25.8
Painful legs and moving toes	Spillane's syndrome	X00CE	G95.8
Restless legs	Ekbom's syndrome	F13z2	G25.8

Diseases of the Neuromuscular Junction

Genetically Determined

1. Congenital Myasthenia

Condition	Alternative names	Read	ICD-10
Familial infantile myasthenia	Acetylcholine resynthesis deficiency	X00Ce	G70.2
Congenital myasthenia	Congenital end-plate acetylcholine receptor deficiency	X00Ce	G70.2
Slow-channel syndrome (SCS)	Pseudomyopathic myasthemia, limb-girdle myasthenia	X00Ce	G70.2
Congenital end-plate acetylcholinesterase deficiency		X00Ce	G70.2
Other types (specify)		X00Ce	G70.2

Sporadic

2. Toxic Disorders of the Neuromuscular Junction

Cholinergic paralysis	Depolarizing neuromuscular blockade	X00Cy	G70.8
Botulism		F3812	A05.1
Tick paralysis		X00Cn	T63.4
Puffer fish paralysis	Tetrodotoxin poisoning	X00Co	T63.5
Paralytic shell fish poisoning	Saxitoxin poisoning	SM80	T61.9
Drug-induced myasthenia	e.g., D-penicillamine	X00Cq	G70.0

3. Autoimmune Disorders of the Neuromucular Junction

Myasthenia gravis (MG)		F380.	G70.0
	Transient neonatal MG	F380.	G70.0
	Ocular MG	F380.	G70.0
	Generalized MG	F380.	G70.0
	with thymoma	F380.	G70.0
	malignant	F380.	G70.0
	benign	F380.	G70.0
	Thymic hyperplasia	F380.	G70.0
	Atrophic thymus	F380.	G70.0
	Myasthenic myopathy	F380.	G70.0
Lambert-Eaton myasthenic syndrome (LEMS)	Eaton-Lambert syndrome	F3810	G73.1

Disorders of Muscle

1. Muscular Dystrophy (MD)

a. MD with predominantly limb-girdle distribution

Condition	Alternative names	Mode of inheritance	Read	ICD-10	Gene location	Gene product symbols	MIM
i. X-linked recessive							
Duchenne muscular dystrophy (DMD)	X-linked dystrophinopathy, severe form	XR	F3910	G71.0	Xp21.2	Dystrophin	310200
Becker muscular dystrophy (BMD)	X-linked dystrophinopathy, benign form	XR	F3918	G71.0	Xp21.2	Dystrophin	310200
Intermediate X-linked muscular dystrophy		XR	F3918	G71.0	Xp21.2	Dystrophin	310200
Manifesting carrier of DMD/BMD			X708L	G71.0	Xp21.2	Dystrophin	310200
ii. Autosomal dominant							
Limb-girdle muscular dystrophy (LGMD), dominant, adult onset	5q-linked LGMD	AD	X708H	G71.0	5q22–q35	LGMD1A LGMD1B	159000
iii. Autosomal recessive							
Limb-girdle muscular dystrophy (LGMD), recessive, adult onset	Reunion/Indiana, 15q-linked	AR	X708H	G71.0	15q15.1–q21.1	LGMD2A (calpain 3)	253600
	Amish, non 15q-linked	AR	X708H	G71.0	2p	LGMD2B	253600
Severe childhood autosomal recessive muscular dystrophy (SCARMD)	AR muscular dystrophy with abnormal dystrophin associated glycoproteins (DAGS)	AR	Xa0ON	G71.0	13q12 17q12–q21.33 4q11–q12	LGMD2C (γ-sarcoglycan) LGMD2D (adhalin = α-sarcoglycan) LGMD2E (β-sarcoglycan)	253700

b. Muscular dystrophy not predominantly limb-girdle in distribution

Condition	Alternative names	Mode of inheritance	Read	ICD-10	Gene location	Gene product symbols	MIM
i. X-linked							
Emery-Dreifuss muscular dystrophy		XR	X708i	G71.0	Xq28	EMD (emerin)	310300
ii. Autosomal dominant							
Facioscapulohumeral muscular dystrophy (FSHMD)	FSH muscular dystrophy	AD	F3914	G71.0	4q35	FSHD	158900

(*continued*)

Condition	Alternative names	Mode of inheritance	Read	ICD-10	Gene location	Gene product symbols	MIM
Oculopharyngeal muscular dystrophy (OPMD)		AD	F3917	G71.0	14q11.2–q13	OPMD	164300
Distal muscular dystrophy	Autosomal dominant distal myopathy	AD	X708m	G71.0	14	MPD1	160500
Welander muscular dystrophy		AD	X708r	G71.0			160500
Scapuloperoneal muscular dystrophy		AD		G71.0			160500
benign			X708o				
benign with cardiomyopathy			X708i				
severe			X708p				
iii. Autosomal recessive							
Scapulohumeral muscular dystrophy		AR	X7081	G71.0			
Distal MD with juvenile onset		AR	F3915	G71.0			

2. Congenital Muscular Dystrophies and Related Disorders

Congenital muscular dystrophy without CNS involvement	Ulrich (hypotonic-sclerotic) type Eichsfeld type	AR	X708V	G71.2			254100
Congenital muscular dystrophy with CNS involvement		AR	X708V	G71.2		FCMD	253800
Fukuyama congenital MD	Congenital MD with DAG deficiency		X708S		9q31–q33		
Congenital MD with merosin deficiency	Walker-Warburg syndrome, Santavouri muscle-eye-brain (MEB) syndrome		X708V		6q2	LAMM (merosin)	
Rigid spine syndrome			No code	No code			
Bethlem myopathy			No code	No code			

3. Congenital Myopathies

Condition	Alternative names	Mode of inheritance	Read	ICD-10	Gene location	Gene product symbols	MIM
Congenital fiber type disproportion	Congenital myopathy with abnormality of fiber type, Benign congenital hypotonia (cause of)		X708t	G71.2			
Core diseases							
Central core disease			F3901	G71.2	19q13.1	CCD (RYR1) (ryanodine receptor)	117000
Minicore disease	Multicore disease, focal loss of cross striations		X7092	G71.2			
Nemaline myopathy	Rod body myopathy	AD	X708t	G71.2	1q21–q23	NEM1 (TPM3) (alpha tropomyosin)	161800
Centronuclear myopathy		AR	F3902	G71.2			
Myotubular myopathy		XR	X708t	G71.2	Xq28	MTMI (MTMX)	310400
Myopathy with cytoplasmic inclusions			X708t	G71.2			
Sarcotubular myopathy			X708t	G71.2			
Myopathy with tubular aggregates			X708t	G71.2			
Hereditary myositis ossificans			X7095	M61.5			

4. Myotonic Disorders

a. Dystrophic myotonias

Condition	Alternative names	Mode of inheritance	Read	ICD-10	Gene location	Gene product symbols	MIM
Myotonic dystrophy (DM)	Dystrophia myotonica, Steinert's disease	AD	F3920	G71.1	19q13	DM (?DMPK)	160900
Congenital DM							
Proximal myotonic myopathy	PROMM	AD	No code	No code			
Schwartz-Jampel syndrome	Chondrodystrophic myotonia	AR	PG421	Q78.8	1p34–p36.1	SJS	255800
Rippling muscle disease		AD	No code	No code	1q41	RMD	

(continued)

b. Nondystrophic myotonias

Condition	Alternativenames	Mode of inheritance	Read	ICD-10	Gene location	Gene product symbols	MIM
Myotonia congenita, autosomal dominant	Thomsen's disease	AD	F3921	G71.1	7q35	CLC-1muscle chloride channel	160800
Myotonia congenita, autosomal recessive	Becker's disease, generalised myotonia, recessive	AR	X709c	G71.1	7q35	CLC-1 muscle chloride channel	255700
Paramyotonia congenita	Eulenberg's disease	AD	F3922	G71.1	17q13.1–q13.3	SCN4A muscle sodium channel alpha subunit	168300

c. Familial Periodic Paralysis

Condition	Alternativenames	Mode of inheritance	Read	ICD-10	Gene location	Gene product symbols	MIM
Hyperkalaemic periodic paralysis	Adynamia episodica hereditaria	AD	F393	G72.3	17q13.1–q13.3	SCN4A (sodium channel alpha subunit	170500
Normokalemic periodic paralysis		AD	F393	G72.3		SCN4A	
Hypokalaemic periodic paralysis		AD	F393	G72.3	1q31–q32	CACNL1A3 (calcium channel, dihydropyridine receptor)	170400

5. Physical Injury to Muscle

Condition	Alternative names	Read	ICD-10
Crush syndrome	Muscle crush syndrome, traumatic rhabdomyolysis	X70A9	M62.8
Ischaemia/infarction of muscle	Muscle infarction, ischaemic myopathy	X70AA	M62.2
Volkmann contracture	Volkmann's ischaemic contracture	SK06	T79.6
Compartment syndromes		XA013	T79.8

6. Drug-and Toxin-Induced Muscle Disease

Drug-induced myopathy, unspecified		F394	G72.0
Steroid myopathy		X709Y	G72.0
Alcoholic myopathy		X709V	G72.0
Toxic muscle disease	Toxic myopathy	F394	G72.2

7. Myopathy Associated with HIV Infection

Human immunodeficiency virus (HIV) myopathy			
	HIV polymyositis	X709a	B23.8
	HIV myopathy	X709a	B23.8
Myopathy due to secondary infection in HIV infection (etiology-Section 9a)		X709a	B23.8

8. Myopathy in Systemic Disease (*for Myopathy in Infection and Connective Tissue Disorders—see Inflammatory Myopathies, Section 9.*)

Condition	Alternative names	Read	ICD-10
a. Endocrine myopathies			
Thyrotoxic myopathy	Myopathy in thyrotoxicosis	F3954	G73.5, E05.9
Thyrotoxic periodic paralysis		F393	G72.3
Dysthyroid ophthalmopathy	Exophthalmic Graves' disease	X00ZW	H05.2
Myxedema myopathy	Hypothyroid myopathy	F3953	G73.5
Myopathy in—hypopituitarism		F3952	G73.5
Cushing's syndrome		C150	G73.5, E24.9
Conn's syndrome	Hyperaldosteronism	C1510	G73.5, E26.0
Addison's disease	Adrenocortical insufficiency	C1541	G73.5, E27.1
Nelson's syndrome		X40MC	G73.5, E24.1
Acromegaly		X7097	G73.5, E22.0
Hyperparathyoidism		X7099	G73.5, E21.3
Hypoparathyroidism		X709A	G73.5, E20.9
Osteomalacic myopathy		X709B	G73.5, M83.1
Hypogonadism	Testosterone deficiency myopathy	X400v	G73.5, E29.1
Hypersecretion of calcitonin	Medullary cell carcinoma of thyroid	C060	G73.5, E07.0
Carcinoid syndrome		C1z2.	G73.5, E34.0
Diabetic myopathy			G73.5
Lipodystrophic myopathy		X50E8	G72.8, E88.1
b. Malignancy			G72.9
Paraneoplastic myopathy	Carcinomatous myopathy	No code	No code
Embolic/direct invasion of muscle by malignant tissue		No code	No code

9. Inflammatory Disorders of Muscle

a. Infective myositis	Infection of muscle, Viral myositis	N230	M60.0
Viral myositis		N230	M60.0
Pyomyositis	Tropical pyomyositis	A3Ay0	M60.0
Fungal myositis		N230	M60.0
Protozoal myositis		N230	M60.0
Cestode myositis		N230	M60.0
Cystercerosis myositis		N230	M60.0
Hydatid myositis		N230	M60.0
Nematode myositis		N230	M60.0
Trichinosis myositis		N230	M60.0
Toxocara myositis		N230	M60.0
b. Idiopathic Inflammatory Myopathies			
Dermatomyositis (DM)		N003	M33.1
childhood DM	Juvenile DM	N003	M33.0
with malignancy		N003	M33.1
Polymyositis (PM)		N004	M33.2
with other autoimmune disease		N004	M33.2
with malignancy		N004	M33.2
Localised nodular myositis		N004	M33.2
Inclusion body myositis (IBM)		X709w	M60.8
c. Other inflammatory myopathies			
Sarcoid myopathy	Granulomatous myopathy	F3965	G73.7, D86.8
Vasculitic myopathy	Myopathy in vasculitis	N004	M33.2

10. Metabolic Myopathies

Condition	Alternative names	Mode of inheritance	Read	ICD-10	Gene location	Gene/product symbols	MIM
a. Glycogen storage diseases Type I (glucose 6 phosphatase deficiency)-hypotonia but no myopathy	von Gierke's disease	AR	C3102	E74.0			232200
Type II acid maltase (amylo-1, 4-glucosidase) deficiency infantile form juvenile form adult-onset form	Pompe's disease	AR	X40TW	E74.0	17q23	GAA	232300
Type III debrancher enzyme (amylo-1,6-glucosidase) deficiency involving muscle and liver, involving muscle alone	Cori-Forbes disease	AR	C3103	E74.0			232400
Type IV brancher enzyme (1,4-glucan; 1,4-glucan-6-glycosyl tranferase) deficiency	Brancher enzyme deficiency	AR	C310	E74.0			232500
Type V myophosphorylase deficiency	McArdle's disease	AR	C310	E74.0	11q13	PYGM (muscle-type glycogen phosphorylase)	232600
Type VII phosphofructokinase (PFK) deficiency infantile myopathic form myopathic form due to lack of M-subunit hemolysis without myopathy due to lack of L-subunit	Tarui's disease	AR	C310	E74.0	1cenq32	PFKM (muscle-type phosphofructokinase)	232800
Type IX phosphoglycerate kinase deficiency			C310	E74.0	Xq13	PGK1 (phosphoglycerate kinase)	311800
Type X phosphoglycerate mutase deficiency			C310	E74.0	7p12–p13	PGAMM (muscle phosphoglycerate mutase)	261670

(*continued*)

Condition	Alternative names	Mode of inheritance	Read	ICD-10	Gene location	Gene/product symbols	MIM
Type XI lactate dehydrogenase deficiency			C310	E74.0	11p15.4	LDHA (lactate dehydrogenase)	150000
b. Myoadenylate deaminase deficiency homozygous heterozygous (carrier)	MADD deficiency	AD?	X40Um	G73.6			
c. Lipid Storage myopathies carnitine deficiency short-chain acyl-CoA dehydrogenase deficiency long chain acyl-CoA dehydro genase deficiency multiple acyl-CoA deficiency carnitine palmityl transferase (CPT) defciency trifunctional enzyme deficiency lipid storage myopathy due to undetermined metabolic defect			X709H	G73.6, E75.6			
d. Mitochondrial cytopathies							
MELAS	Mitochondrial encephalopathy, lactic acidosis and stroke-like episodes		X40U0	E74.4	mt DNA point mutations		
MERRF	Myoclonic epilepsy and ragged red fibers		X006Y	G40.3	mt DNA point mutations		
Chronic progressive external ophthalmoplegia (and variants)	Kearns-Sayre syndrome CPEO and CPEO plus, oculocraniosomatic syndrome		X00fb	G71.3	mt DNA deletions		
Mitochondrial cytopathy, other			X709I	G71.3			
e. Other Metabolic Myopathies							
Malignant hyperthermia (MH)	Malignant hyperpyrexia	AD	X709O	T88.3	19q13.1	MHS1 (ryanodine receptor)	145600
Brody's disease	Sarcolemmal calcium ATPase deficiency						

11. Miscellaneous Muscle Disorders/Laboratory Abnormalities

Condition	Alternative names	Read	ICD-10
Myalgia	Muscle pain, cause unknown	X75rs	No code
Myopathy ?cause		X708s	G72.9
Myopathic change on muscle biopsy		No code	No code
Neurogenic change on muscle biopsy		No code	No code
CK increased ?cause	HyperCKemia	No code	No code
Chronic fatigue syndrome (CFS)	Myalgic encephalomyelitis (ME) Postviral fatigue syndrome, neurasthenia	Xa01F	G93.3
Polymyalgia rheumatica		N20	M35.3
Myositis ossificans		X40R4	M61
Arthrogryposis multiplex congenita	(should also be classified by etiology)	PFy0	Q74.3
Myoglobinuria	Paroxysmal myoglobinuria	X709S	R82.1
Amyloid myopathy		F3960	G73.7
familial	Primary familial amyloidosis		
sporadic	Primary amyloidosis		
in myelomatosis	Secondary amyloidosis		
Disuse atrophy		X70AF	M62.5
Cachexia	Cachexia of muscle	X70AD	M62.5
Rhabdomyolysis		X70AI	M62.8
Tumor of muscle	Muscle tumor	X78V1	D48

PART II. ALPHABETICAL INDEX OF COMMON NEUROMUSCULAR DISORDERS WITH READ CODINGS

Disorder	Read	ICD-10
Acid maltase deficiency	X40TW	E74.0
Adrenoleucodystrophy	X40Wn	E71.3
Alcoholic myopathy	X709V	G72.1
Alcoholic polyneuropathy	F375	G62.1
Amyloid myopathy	F3960	G73.7
Amyloid neuropathy, secondary	F1710	E85.4
Amyloid neuropathy, familial (FAP)	C3732	G63.3/E85.1
Amyotrophic lateral sclerosis (ALS)	F152	G12.2
Amyotrophic lateral sclerosis, familial (FALS)	F152	G12.2
Anterior interosseous palsy	F341	G56.1
Arthrogryposis multiplex congenita (also classify under etiology, if known)	PFy0	Q74.3
Axonal polyneuropathy (also classify under etiology, if known)	F33	G62.9
B_{12} deficiency neuropathy neuromyelopathy (*see Subacute Combined Degeneration*)	F162	G32.0
Becker muscular dystrophy (BMD)	F3918	G71.0
Benign fasciculations, fasciculation/cramp syndrome (*see also* Cramp)	X00CC	R25.3
Beri-beri polyneuropathy (*see also* Thiamine Deficiency Polyneuropathy)	F3741	G63.4/E51.1
Bethlem myopathy	No code	No code
Burning feet syndrome	X00Ae	E53.9
Cachexia of muscle		

(*continued*)

Disorder	Read	ICD-10
Calcium ATPase deficiency (*see also* Brody's Disease)	No code	No code
Carcinomatous myopathy (*see also* Paraneoplastic Myopathy)	No code	No code
Carcinomatous polyneuropathy (*see also* Paraneoplastic Neuropathy)	F373	G63.1
Carnitine deficiency (*see* Lipid Storage Myopathy)	X709H	G73.6/E75.6
Carpal tunnel syndrome (*see also* Median Nerve Compression)	F3411	G56.0
Cauda equina syndrome	F246	G83.4
Central core disease	F3901	G71.2
Centronuclear myopathy	F3902	G71.2
Cervical root lesion (requires subclassification—see Part I)	See Part I	See Part I
Charcot-Marie-Tooth disease	F3610	G60.0–60.5 (see part I)
Chronic fatigue syndrome (CFS) (*see also* Myalgic Encephalomyelitis, Post-Viral Fatigue Syndrome, Neurasthenia)	Xa01F	G93.3
Chronic inflammatory demyelinating polyneuropathy (CIDP)	X00AW	G61.8
Chronic progressive external ophthalmoplegia (CPEO)/CPEO-plus syndrome (*see also* Kearns-Sayre Syndrome)	X00fb	G71.3
Common peroneal nerve palsy	X00C1	G57.3
Compartment Syndrome	XA013	T79.8
Congenital Fiber type abnormality/ disproportion	X708t	G71.2
Congenital muscular dystrophy	X708V	G71.0
Congenital myasthenia	X00Ce	G70.2
Congenital hypomyelinating polyneuropathy	X00AQ	G60.9
Cramp (*see also* Benign Fasciculations/Faciculation-Cramp Syndrome)	X00CB	R25.2
Critical illness polyneuropathy	X00B2	G62.8
Cubital tunnel syndrome (*see* Ulnar Nerve Lesion)	Xa0Ks	G56.2
Demyelinating polyneuropathy (also classify etiology, if known)	F33	G62.9
Dermatomyositis	N003	M33
Dermatomyositis with malignancy	N0034	M33
Diabetic amyotrophy	X00Ag	G59.0/E14.4
Diabetic mononeuropathy	X00Ag	G59.0/E14.4
Diabetic polyneuropathy	F372	G63.2
Distal myopathy/Dystrophy (*see also* Welander Distal Muscular Dystrophy)	X708m	G71.0
Distal spinal muscular atrophy (SMA)	X002b, F151	G12.9
Disuse atrophy	X70AF	M62.5
Drug-induced myopathy (*see* Toxic Myopathy)	F394	G72.0
Drug-induced myasthenia	X00Cq	G70.0
Duchenne muscular dystrophy (DMD)	F3910	G71.0
Dystrophia myotonica (*see* Myotonic Dystrophy)	F3920	G71.1
Ekbom's syndrome (*see* Restles Legs)	F13z2	G25.8
Emery-Dreifuss muscular dystrophy (EDMD)	X708i	G71.0
Fascioscapulohumeral muscular dystrophy (FSHMD)	F3914	G71.0
Familial motor neuron disease (*see* Amyotrophic Lateral Sclerosis, Familial)	F152	G12.2-
Familial periodic paralysis	F393	G72.3
Fazio-Londe disease	X002a	G12.1
Fukuyama muscular dystrophy (*see* Congenital Muscular Dystrophy)	X708S	G71.2
Glycogen storage disease	C310	E74.0
Granulomatous myopathy (*see* Sarcoid Myopathy)	F3965	G73.7/D86.8
Guillain-Barré syndrome (GBS)	F3700	G61.0
Hereditary liability to pressure palsies	X00AM	G60.8
Hereditary motor and sensory neuropathy (HMSN) (*see* Charcot-Marie-Tooth Disease)	F3610	G60.0–60.5 (see Part I)
Hereditary sensory and autonomic neuropathy	X00AF	G60.8
Hyperkalemic periodic paralysis (*see* Familial Periodic Paralysis)	F393	G72.3
Hypokalemic periodic paralysis (*see* Familial Periodic Paralysis)		

(*continued*)

Disorder	Read	ICD-10
Hypothyroid myopathy (*see* Myedema Myopathy)	F3953	G73.5
Hypothyroid neuropathy (*see* Myxedema Neuropathy)	X00Ao	G63.3, E03.9
HIV-related myopathy/myositis	X709a	B23.8
HIV-related neuropathy	X00B3	G63.0/B24
Inclusion body myositis (IBM)	X709w	M60.8
Infection of muscle/infective myositis (*see also* Pyomyositis)	N230	M60.0
Isaac's syndrome (*see* Neuromyotonia)	X00CD	G71.1
Ischemic neuropathy (requires subclassification–see Part I)	See Part I	See Part I
Ischemic infarction of muscle/ ischemic myopathy/muscle infarction	X70AA	M62.2
Kearns-Sayre syndrome (*see also* Chronic Progressive External Opthalmoplegia, CPEO)	X00fb	G71.3
Kennedy's disease (*see* Bulbopinal Neuropathy)	F151	G12.8
Kugelberg-Welander disease (*see* Spinal Muscular Atrophy)	F151	G12.1
Lambert-Eaton myasthenic syndrome (LEMS)	F3810	G73.1
Lateral popliteal nerve palsy (*see* Common Peroneal Nerve Palsy)	X00C1	G57.3
Limb-girdle muscular dystrophy (LGMD)	X708H	G71.0
Lipid storage myopathy (see Part I)	X709H	G73.6/E75.6
Lipodystrophic myopathy	X50E8	G72.8/E88.1
Lumbar spondylosis	X703k	M47.8
Lumbosacral plexus lesion (requires subclassification–see Part I)	See Part I	See Part I
Lumbosacral radiculopathy (*see also* Sciatica)	N11C	M47.2
Malignant hyperthermia/hyperpyrexia	X7090	T88.3
McArdle's disease (*see* Glycogen Storage Disease)	C310	E74.0
MELAS (mitochondrial myopathy, encephalopathy, lactic acidosis, and stroke-like episodes)	X40U0	E74.4
Meralgia paresthetica	F351	G57.1
MERRF (myoclonus, epilepsy, ragged red fibers)	X006Y	G40.3
Miller Fisher syndrome (*see also* Guillain Barré syndrome)	X00AV, F3700	G61.0
Minicore/multicore disease	X7092	G71.2
Mitochondrial cytopathy, other	X709I	G71.3
Mononeuritis multiplex (also classify under etiology, if known)	F345	G58.7
Morton's metatarsalgia	F3560	G57.6
Motor neuron disease (*see* Amyotrophic Lateral Sclerosis)	F152	G12.2
Multifocal motor neuropathy with conduction block	X00Ay	G58.8, G62.8
Myalgia	X57rs	No code
Myalgic encephalomyelitis (*see* Chronic Fatigue Syndrome)	Xa01F	G93.3
Myasthenia gravis	F380	G70.0
Myoadenylate deaminase deficiency (MADD)	X40Um	G72.8
Myoglobinuria (also classify under etiology, if known)	X709s	R82.1
Myopathy, unclassified	X708s	G72.9
Myophosphorylase deficiency (*see* McArdle's Disease)	C310	E74.0
Myositis ossificans	X40R4	M61
Myotonia congenita, dominant (Thomsen's)	F3921	G71.1
Myotonia congenita, recessive (Becker')	X709c	G71.1
Myotonic dystrophy (DM)	F3920	G71.1
Myotubular myopathy	X708t	G71.2
Myxedema myopathy	F3953	G73.5
Myxedema neuropathy	X00Ao	G63.3, E34.9
Nemaline myopathy	X708t	G71.2
Nerve infarction (*see* Ischemic Neuropathy)	X00BB	G58.9, G64.0
Neurasthenia (*see* Chronic Fatigue Syndrome)	Xa01F	G93.3
Neuralgic amyotrophy (*see also* Brachial Plexus Neuritis)	F335	G54.5
Neuromyotonia	X00CD	G71.1
Neuropathy asociated with dysproteinemia (*see* Paraproteinemic Neuropathy)	X00At	G58.9/E88.0

(*continued*)

Disorder	Read	ICD-10
Neuropathy in liver disease	X00Aq	G58.8/G62.8
Neuropathy in vasculitis/connective tissue disease (also classify by etiology, if known)	F371	G63.5/M35.9
Neuropathy in renal failure (*see also* Uremic Neuropathy)	F374A	G63.8/N18.8
Neuropathy, unclassified	F33	G62.9
Nerve root lesion, cervical (requires subclassification–see Part I)	See Part I	See Part I
Nerve root lesion, lumbosacral (requires subclassification–see Part I)	See Part I	See Part I
Normokalemic periodic paralysis (*see* Familial Periodic Paralysis)	F393	G72.3
Oculocraniosomatic syndrome (*see* Chronic Progressive External Ophthalmoplegia)	X00fb	G71.3
Oculopharyngeal muscular dystrophy (OPMD)	F3917	G71.0
Painful legs and moving toes	X00CE	G95.8
Paramyotonia congenita	F3922	G71.1
Paraneoplastic myopathy (*see* Carcinomatous Myopathy)	No code	No code
Paraneoplastic neuropathy (*see* Carcinomatous Neuropathy)	F373	G63.1
Paraproteinaemic neuropathy (*see* Neuropathy Associated with Dysproteinemia)	X00At	G58.9/E88.0
Periodic paralysis (*see* Part I)	F393	G72.3
Peroneal muscular atrophy (*see* Charcot-Marie-Tooth Disease)	F3610	G60.0–60.5 (see Part I)
Plantar nerve compression	X00C3	G57.6
Polymyalgia rheumatica	N20	M35.3
Polymyositis (PM)	N004	M33.2
Polymyositis with connective tissue disease (overlap myositis)	N004	M33.2
Polymyositis with malignancy	N004	M33.2
Polyneuropathy, unclassified	F33	G62.9
Pompe's disease (*see* Glycogen Storage Disease)	X40TW	E74.0
Postherpetic neuralgia	X75sv	G53.0/B02.2
Postpolio syndrome	X002p	B91
Postviral fatigue syndrome (*see* Chronic Fatigue Syndrome)	Xa01F	G93.3
Primary lateral sclerosis (PLS) (*see* Amyotrophic Lateral Sclerosis)	F152	G12.2
Progressive bulbar palsy (PBP) (*see* Amyotrophic Lateral Sclerosis)	F152	G12.2
Progressive muscular atrophy (PMA) (*see* Amyotrophic Lateral Sclerosis)	F152	G12.2
Proximal myotonic myopathy (PROMM)	No code	No code
Posterior interosseous nerve lesion	F3432	G56.8
Pyomyositis	A3Ay0	M60.0
Radial nerve lesion	X008B	G56.3
Restless legs (*see also* Ekbom's Syndrome)	F13z2	G25.8
Rigid spine syndrome	No code	No code
Sarcoid myopathy	F3965	G73.7/D86.8
Scapulohumeral SMA (spinal muscular atrophy)	X002f, F151	G12.9
Scapulohumeral muscular dystrophy	X708I	G71.0
Scapuloperoneal SMA (spinal muscular atrophy)	X002c, F151	G12.9
Scapuloperoneal muscular dystrophy (*see* Part I)	X708	G71.0
Sciatica (*see also* Lumbosacral Radiculopathy)	N143	M54.3
Sciatic nerve lesion	X00By	G57.0
Schwartz-Jampel syndrome (*see* Chondrodystrophic Myotonia)	PG421	Q78.8
Severe autosomal recessive muscular dystrophy of childhood (SCARMD)	Xa0ON	G71.0
Spinal muscular atrophy (SMA) (*see* Part I)	F151	G12.9
Stiff man syndrome	F13z1	G25.8
Subacute combined degeneration (*see* B_{12} Deficiency Polyneuropathy)	F162	G32.0
Thiamine (B_1) deficiency neuropathy (*see* Beri-Beri Polyneuropathy)	F3741	G63.4/E51.1
Thoracic outlet syndrome	F3303	G54.0
Thyrotoxic myopathy	F3954	G73.5/E05.9
Thyrotoxic periodic paralysis (*see also* Familial Periodic Paralysis)	F393	G72.3
Tomaculous neuropathy (*see* Herditary Liability to Pressure Palsies)	X00AM	G60.8
Toxic myopathy	F394	G72.2

(*continued*)

Disorder	Read	ICD-10
Toxic neuropathy	X00BG	G62.2
Tropical myositis (Pyomyositis)	A3Ay0	M60.0
Tumor of muscle	X78V1	D48
Ulnar nerve lesion	Xa0Ks	G56.2
Uremic neuropathy (*see* Neuropathy in Renal Failure)	F374A	G63.8/N18.8
Vasculitic myopathy (*see* Polymyositis)	N004	M33.2
Vasculitic neuropathy (*see* Neuropathy in Vasculitis/Connective Tissue Disease)	F371	G63.5/M35.9
Viral myositis	N230	M60.0
Volkmann's contracture	SK06	T79.6
Welander distal muscular dystrophy (*see also* Distal Myopathy/Muscular Dystrophy)	X708r	G71.0
Werding Hoffman disease (*see* Spinal Muscular Atrophy)	F150	G12.0

Index

A-band, 83, 86
Abscess, of muscle, 392
Acetazolamide, in periodic paralysis, 333, 350
Acetazolamide-responsive myotonia, 343, 348
Acetylcholine receptor, 82
Acid maltase deficiency, 407–411
  histochemistry, 74
  and respiratory failure, 629
Acid phosphatase staining, 63
Acromegaly, and myopathy, 372–373
ACTH myopathy, 368, 371, 372
Actin, 83
Actinin, structure, 224, 235 (*see also* Costameres)
Actinin, 83, 190
  immunostaining, 71
Acute necrotising myopathy of intensive care (*see* Critical illness myopathy)
Acyl-CoA dehydrogenase, 432
  deficiency, 75, 438–439
Acyl-CoA synthetase, 432
Addison's disease, myopathy in, 372
Adhalin deficiency, 693 (*see also* Severe childhood autosomal recessive muscular dystrophy, SCARMD)
ADP/ATP translocator, 522
Adrenal insufficiency, and myopathy, 372
Adrenocorticotrophic hormone (*see* ACTH)
AIDS, and muscle disease, 623–627 (*see also* HIV-related myopathies)
Alcoholic myopathy, 386–387, 401
Alligator sign, 153
α-glucosidase deficiency (*see* Acid maltase deficiency)
ALS (*see* Amyotrophic lateral sclerosis)
Amebiasis, 395
Amiodarone myopathy, 386
Amish type limb-girdle muscular dystrophy, 246, 261, 696
Ammonia production, in ischemic exercise test, 35 (*see also* Myoadenylate deaminase deficiency)
AMPD genes, 444, 445–448
Amyloid:
  in inclusion body myositis, 582, 585
  myopathy, 617–619
Amylopectinosis (*see* Branching enzyme deficiency)
Amyopathic dermatomyositis, 542 (*see also* Dermatomyositis)
Amyotrophic lateral sclerosis, 26, 28
Anaesthesia, and muscle diseases, 629–638
ANCA (*see* Antineutrophil cytoplasmic antibodies)
Andersen's disease (*see* Branching enzyme deficiency)
Ankylosing spondylitis, 399
Ankyrin, 84
Anti-Ku, 553
Anti-U_1 sn RNP, 553
Anticentromeric antibodies, 553 (*see also* CREST syndrome)
Anticipation, in myotonic dystrophy, 322
Antineutrophil cytoplasmic antibodies, 559
Antisynthetase syndrome, 554
Apamin binding, and myotonia, 312
Apoptosis, in inflammatory myopathies, 552
Arthrodesis, 673
Arthrogryposis, 142, 603–604
  and congenital myotonic dystrophy, 142, 321
  and myotubular myopathy, 183
Artifacts, muscle biopsy, 57
Ascorbic acid, in treatment of mitochondrial cytopathy, 535
ATPases, and myotonia, 312
Atrophy, neurogenic, 13
Autoantibodies, in inflammatory myopathies, 552–555

Autoimmune diseases, and inflammatory myopathies, 543
Autophagic vacuoles, 90
Azidothymidine (*see* AZT)
AZT myopathy, 120, 383, 482, 484, 494–497, 624–626

β-adrenergic agonists, in periodic paralysis, 350
β-oxidation (*see* Fatty acid oxidation)
B-lymphocytes, immunocytochemistry, 71
Bacterial myositis, 391–392 (*see also* Pyomyositis)
Bartter syndrome, in mitochondrial cytopathy, 486
Basement membrane, 81
  reduplication of, 85
Batten's disease, 63
Becker muscular dystrophy, 203–204, 209–212, 238–239
  magnetic resonance spectroscopy in, 121
  myoimaging, 105
  pathology, 67, 209
Becker's disease (*see* Myotonia congenita, recessive)
Benign congenital hypotonia, 195
Benign infantile myopathy (BIM), 493–494
Benign post-infection myositis, 592
Bethlem myopathy, 159–162, 168, 289, 697
BMD (*see* Becker muscular dystrophy)
Bornholm disease, 394
Borrelia myositis, 395
Branching enzyme deficiency, 74, 413–414
Brody's disease (sarcoplasmic reticulum calcium ATPase deficiency), 8, 473–478
  magnetic resonance spectroscopy, 122

Cachexia, 8
Caffeine-halothane contracture tests, 459–461
Calcaneal foot, 672
Calcification, subcutaneous in dermatomyositis, 542
Calcification, myoimaging of, 103
Calcifying epithelioma of Malherbe (*see* Pilomatrixoma)
Calcitonin, and myopathy, 372
Calcium ATPase deficiency (*see* Brody's disease)
Calcium release channel disorder (*see* Ryanodine receptor)
Calmodulin, 224
cAMP-dependent protein kinase (*see* DMPK)
Candidate genes, 128
Candidiasis, and myositis, 394–395
Carcinoid syndrome, 398
Carcinomatous myopathy, 544–545
Carcinomatous neuromyopathy, 398
Cardiac disease, and neuromuscular disorders, 655–662
Cardiac disease:
  in branching enzyme deficiency, 413
  in debranching enzyme deficiency, 411
  in desmin storage myopathy, 304, 305
  in dystrophinopathies, 204, 210, 240, 656–657
  in Emery Dreifuss muscular dystrophy, 216
  in facioscapulohumeral muscular dystrophy, 282
  in hyokalaemic periodic paralysis, 329
  in inclusion body myositis, 578
  in inflammatory myopathy, 543
  in lipid storage myopathies, 434
  in malignant hyperthermia, 454
  in minicore disease, 181
  in mitochondrial cytopathy, 493, 497, 512
  in myotonic dystrophy, 318, 657
  in nemaline myopathy, 190
  in rigid spine syndrome, 152
  in systemic sclerosis, 399
Carnitine, 436
  deficiency, 91, 374, 438
  in treatment of myopathy, 439–440, 514
Carnitine palmitoyl transferase (CPT), 432
  deficiency, 74, 437
  measurement of, 522
Carpopedal spasm, 7
Carriers, of X-linked muscular dystrophy, 68, 202, 214, 238, 265–269, 686–690
  of Emery Dreifuss muscular dystrophy, 216
  germline mosacism in, 266
  magnetic resonance spectroscopy in, 121
  manifesting, 696
Cathepsins, and rimmed vacuoles, 585
Caveolae, 81
CD4/CD8 cells (*see* T-cells)
cDNA libraries, 133
Central core disease, 177–180, 698
  histology, 63, 70
  and malignant hyperthermia, 458, 462
  myoimaging, 103
  ultrastructure, 86
Centronuclear myopathy, 181, 185–187
  histology, 70–71
  myoimaging, 104
Cerebro-ocular-dysplasia syndromes (*see* Muscle eye-brain syndromes)
CFS (*see* Chronic fatigue syndrome)
Chagas disease, 395
Channelopathies, 3, 339–349 (*see also* Periodic paralysis, Congenital myotonia)

[Channelopathies]
differential diagnosis, 349–350
genetics, 347–349
Charcot-Marie-Tooth disease, 127, 134
Childhood dermatomyositis, 544 (*see also* Dermatomyositis)
Chloride channel, 348–349
disorders of, 699–670 (*see also* Myotonia congenita)
Chloroquine myopathy, 386
Chondrodystrophic myopathy (*see* Schwartz-Jampel syndrome)
Choree fibrillaire de Morvan, 356
Chronic fatigue syndrome, 394, 591–593
and magnetic resonance spectroscopy, 122
and viral persistence, 556
Chronic progressive external ophthalmoplegia (CPEO) (*see* Progressive external ophthalmoplegia)
Churg-Strauss syndrome, myositis in, 558–559
Ciguatera poisoning, and myositis, 388
CK (*see* Creatine kinase)
Clenbuterol, 15
Clofibrate myopathy, 381
Clostridial infection, and myositis, 392
Coenurosis, 396
Coenzyme Q, 514
in treatment of mitochondrial cytopathy, 535
Colchicine myopathy, 386
Compartment syndromes, 605–606, 608
Complement, in inflammatory myopathies, 551
Complementation, of mitochondrial DNA, 510
Complex repetitive discharges (CRDs), 44
Compound muscle action potential (CMAP), 41
Computed tomography of muscle (*see* CT scanning)
Concentric contraction, 3, 20
Concentric laminated bodies, 88
Concentric needle electromyography (CNEMG), 42
Conchotome biopsy, 54
Conduction block, 46
Congenital fiber type disproportion, 65, 171, 195–198
Congenital muscular dystrophy (CMD),
with absence of dystrophin, 147
with cerebral hypomyelination, 144
classification, 140
and dystrophinopathy, 240
and limb-girdle muscular dystrophy, 251
myoimaging in, 104
Congenital myopathies, classification, 165–174
histology, 70–71 (*see also relevant specific diseases*)
Conn's syndrome (*see* Primary hyperaldosteronism)
Continuous muscle activity, 335
Contraction, concentric and eccentric, 3, 20 (*see also* Eccentric contractions)
Contraction, eccentric, 3, 20
Contracture:
in chronic drug-induced focal myopathy, 381
in exertional myalgia, 8
in glycogen storage diseases
Contracture tests, in diagnosis of malignant hyperthermia, 459–461
Contractures, treatment of in muscular dystrophy, 663, 672
in neuromyotonia, 356
Cores, structured and unstructured, 180
Cori-Forbes disease (*see* Debranching enzyme deficiency)
Cosmids, 132
Costameres, 228
Coxsackie, virus infection, 394 (*see also* Chronic fatigue syndrome, Enteroviral persistence)
CPT deficiency (*see* Carnitine palmitoyl transferase deficiency)
Cramp, 7–9
dialysis-induced, 401
and dystrophinopathy, 240
familial, 7
Cranial arteritis, and polymyalgia, 595–596
Creatine kinase, in diagnosis, 14
equilibrium reaction, 117
CREST syndrome, 543,
autoantibodies in, 553
Critical illness myopathy, 371, 382
Crush syndrome, 605, 608
CT scanning of muscle, 95–113
Curvilinear bodies, 91
Cushing's syndrome, and myopathy, 368–370
Cyclosporin myopathy, 382, 386
Cysticercosis, 396
myoimaging in, 107
Cytochrome analysis, 524, 525–526
Cytochrome c oxidase (COX), deficiency, 75, 493–494
staining, 481
Cytokines, and inflammatory myopathies, 550–551
Cytoplamic bodies, 88
in malignant hyperthermia, 462
myopathy, 88, 167,169
Cytoskeleton, 226
ultrastructure, 83

DAGs (*see* Dystrophin-associated glycoproteins)
Dantrolene, in treatment of malignant hyperthermia, 455, 463–464, 467
De Toni-Fanconi-Debre syndrome, in mitochondrial cytopathy, 75, 486, 493
Debrancher enzyme deficiency, 37, 411–412
Delta lesions, 85, 207, 212
Denny-Brown and Foley syndrome, 7
Dermatomyositis (DM), 9, 541–542 (*see also* Inflammatory myopathies)
  histology, 71–72
  immunocytochemistry, 546
  inflammatory mechanisms in, 551
Desmin, 82–84, 92
  storage myopathy, 169, 303–305, 619–621
Diabetes, and muscle infarction, 400
  in myotonic dystrophy, 317
Diabetic amyotrophy, 373–374
Dichloroacetate, in treatment of mitochondrial cytopathy, 535
Diabetic neuromyopathy, 373–374
Dihydropyridine receptor, 475, 698 (*see also* Ryanodine receptor, Hypokalaemic periodic paralysis)
Distal muscular dystrophy, 303 (*see also* Welander's disease)
  myoimaging in, 106
Distal myopathy, 300–303 (*see also* Welander's disease)
  with rimmed vacuoles, 580, 584
Distal polymyositis, 557
Disuse atrophy, 663
Dive-bomber potentials (*see* Myotonic runs)
DMD (*see* Duchenne muscular dystrophy)
DMPK gene, 324
DNA fingerprinting, 131
Drug-induced myopathies, 379–387
Duchenne muscular dystrophy, 201–215, 237–238
  anesthesia in, 214–215, 631
  clinical features of, 202–203
  in females, 269–270 (*see also* Carriers)
  in fetus and neonate, 205
  laboratory diagnosis of, 213
  magnetic resonance spectroscopy in, 121
  malignant hyperthermia reaction in, 458
  mental retardation in, 204, 210–212
  myoimaging in, 104
  neonatal screening, 213
  and normal dystrophin, 238
  pathogenesis of, 212, 236
  pathology, 66, 206
Duchenne muscular dystrophy:
  prenatal diagnosis in, 213, 237, 265
  management of, 214–215
Dynamic mutations, 321, 693–694
Dynamic strength (force), 20, 23–24
Dynamometry, 22, 24–26
Dystroglycan, 260
Dystrophia myotonica (DM) (*see* Myotonic dystrophy)
Dystrophin, 82, 91, 250 (*see also* Dystrophinopathies)
  in cardiac muscle, 228
  function of, 229–231
  immunoblotting, 68
  immunocytochemistry, 66
  localisation of, 225–228
  structure of, 224
  tissue distribution, 228–229
Dystrophin gene, 202
  mutations, 687–688
Dystrophin-associated glycoproteins (DAGS), 144, 224, 229,
  and limb-girdle muscular dystrophy, 250–251
Dystrophin-related protein (DRP, *see* Utrophin)
Dystrophinopathies 201 (*see also* Dystrophin)
  animal models of, 237
  asymptomatic, 241
  cardiac disease in, 656–657
  in females, 690
  hyperCKemia in, 204
  and limb-girdle syndrome, 248
  myalgia in, 8, 204
  phenotypic mainfestations, 204, 239–241

Eccentric contractions, 3, 20 (*see also* Contraction, eccentric)
  and myoglobinuria, 608
  and myalgia, 5
Echinococcosis (*see* Hydatidosis)
Edema of muscle, 98, 103
EDMD (*see* Emery-Dreifuss muscular dystrophy)
Effort syndrome, 3
Ekbom's syndrome (*see* Restless legs)
Electron microscopy (*see* Ultrastructure)
Electron transfer chain, 479
Electron transport flavoprotein, 432
  deficiency, 75, 438–439
  measurement of, 522
Emery-Dreifuss muscular dystrophy (EDMD), 156, 215–217, 251, 695
  histology of, 67–68
Emetine myopathy, 381
End plate noise, 43
Endocrine myopathies, 365–378

Enteroviral persistence, 592 (*see also* Chronic fatigue syndrome, Inflammatory myopathies)
Eosinophilia-myalgia syndrome, 388, 558
Eosinophilic fasciitis and myositis, 558
Eosinophilic inclusions, in inclusion body myositis, 581
Epidemic pleurodynia (*see* Bornholm disease)
Epinephrine, and myopathy, 372
Epsilon-amino caproic acid (EACA) myopathy, 381
Epstein-Barr virus infection, 394
Equinovarus deformity, 672
ETF (*see* Electron transport flavoprotein)
Excitation-contraction coupling, 3
Exercise intolerance, 3
Exercise tests, 35–39, 519 (*see also* Ischemic exercise tests)
Exertional muscle pain syndrome (EMPS), 5
Exophthalmic Graves' disease (*see* Thyroid-associated ophthalmopathy)

Facioscapulohumeral muscular dystrophy (FSHD), 11, 276–280, 695–696
  dystrophinopathy in, 241
  genetics of, 282
  myoimaging in, 105
  pathology of, 276
  rimmed vacuoles in, 580
  treatment of, 282
Facioscapulohumeral syndrome, causes of, 276–277
  histology, 69
  in inflammatory myopathy, 557
  myoimaging in, 276
Fasciculations, 7
  myoimaging of, 102–103
Fatal infantile myopathy, 493–494 (*see also* Mitochondrial depletion)
Fatigue, 3, 591–592
  central, 4
  high frequency, 3
  low frequency, 3
Fatty acid oxidation, 431–433
Fetal myositis, 144
Fiber density, 48
Fiber type disproportion (*see* Congenital fiber type disproportion)
Fiber type grouping, 65
Fiber typing, 62
Fibrodysplasia ossificans progressiva (*see* Myositis ossificans)
Fibromyalgia, 4
Fibronectin, 81, 82
Filamentous bodies, 88
Filamentous inclusions:
  in inclusion body myositis, 582–585
  in oculopharyngeal muscular dystrophy, 584
Fingerprint bodies, 91
Focal loss of cross striations, 181
Focal myopathy, 379–381
Focal myositis, 557
Fukuyama-type congenital muscular dystrophy, 141–143
Functional (dynamic) strength tests, 563
Fungal myositis, 394

Gas gangrene, 392
Gene therapy, 215, 692
Genetic counseling, in neuromuscular diseases, 683–701
Genetic mapping, 129–130
Germanium myopathy, 383, 386
Giant cell arteritis, 74, 595–596
Giant cell myositis, 560
Gilchrist myopathy, 697
Glossopharyngeal breathing, 644
Glutaric aciduria (*see* Electron transport flavoprotein deficiency)
Glycerol kinase deficiency (*see* Phosphoglycerokinase deficiency)
Glycogen storage diseases (*see* Glycogenoses)
Glycogenoses, 407–430 (*see also specific disorders*)
  histochemistry, 64
  magnetic resonance spectroscopy in, 120–121
  ultrastructure, 85, 90
Goldflam myopathy, 294
Golgi apparatus, 84
Gomori trichrome stain, 64, 71
  in mitochondrial myopathies, 481, 505
Gottron's lesion, 541–542
Gower's sign, 14, 203, 206
Graft-versus-host disease, 394, 550
Granazymes, 552
Granulomatous myopathies, 560
Granulomatous myositis, myoimaging in, 108
Graves' ophthalmopathy (*see* Thyroid-associated ophthalmopathy)
Growth hormone (*see* Acromegaly, Hypopituitarism)

H-zone, 83
HAL gene, 453
Halothane, and malignant hyperthermia reaction, 454
Hardy-Weinberg equilibrium, 130
Heat shock proteins, and myositis, 555
Heatstroke, and myoglobinuria, 608

Hemolytic anaemia, in glycogenoses, 421, 422
Herpes zoster, and myositis, 394
Histochemistry techniques, 62–64
HIV-related myopathies, 394, 623–627 (*see also* AIDS)
HIV wasting syndrome, and muscle disease, 623, 626
HLA antigens, and inflammatory myopathies, 549–550
Hoffman rule, 209, 236
Hoffman's syndrome, 366–367
Honeycomb structures, 90
Hounsfield Units, definition of, 95
HTLV-I infection, and myopathy, 394
HUMCLC (*see* Chloride channel protein)
Hyaline fibers, 86, 209
Hydatidosis, 396
Hyperaldosteronism, 401
HyperCKaemia, 14
  drug-induced, 381
  and dystrophinopathy, 241
Hyperkalaemic periodic paralysis, 340 (*see also* sodium channel diseases)
  in Addison's disease, 372
Hyperparathyroidism, and myopathy, 374, 401
Hypersensitivity angiitis, 559
Hypertrophy of muscle, in myotonia congenita, 345, 346
  in neuromyotonia, 356
Hyperuricaemia, in glycogenoses, 422
Hypogonadism (*see* Testosterone deficiency myopathy)
Hypokalaemic myopathy, 400–401
Hypokalaemic periodic paralysis, 329–337, 698
  in Bartter's syndrome, 330
  diagnosis of, 330–331
  in hyperaldosteronism, 330
  pathophysiology of, 331–332
  in renal tubular acidosis, 330
  treatment of, 332–334
HypoKPP (*see* Hypokalaemic periodic paralysis)
Hypomagnesaemia, and myopathy, 401
Hypoparathyroidism, myopathy in, 374
Hypophosphataemia, and myopathy, 401, 614
Hypopituitarism, myopathy in, 373
Hypothyroid myopathy, 8, 366–367
  magnetic resonance spectroscopy in, 122
Hypotonia, 13

I-band, 83
IBM (*see* Inclusion body myositis)
Idebenone, in treatment of mitochondrial cytopathy, 497, 536
Iliotibial band, 672

Inclusion body myositis (IBM), 292, 575–593,
  and autoimmune diseases, 578
  and cardiac disease, 658
  and diabetes, 578
  histology of, 73
  and mitochondrial disorder, 586
  myoimaging in, 108
Infarction of muscle, 374, 608
  in atherosclerosis, 400
  in diabetes, 400
  in rheumatoid disease, 399
Inflammatory myopathies, idiopathic, 539–573 (*see also* Myositis, Polymyositis, Dermatomyositis)
  acute, 557
  and malignancy, 398, 544–545
  and viral infection, 394
  cardiac disease in, 658
  classification of, 539–540
  congenital, 148
  drug-induced, 383, 549
  histology of, 71–74
  myoimaging in, 99, 107
  treatment of, 562–564
Innervation zone, 46
Insertional activity, 43
Insulin resistance, 373
  in myotonic dystrophy, 317
Intermediate filament myopathies, 619–621
Intermediate filaments:
  in oculopharyngeal muscular dystrophy, 299
  ultrustructure of, 83
Intranuclear rods, 189, 190
Iron sulfur centers, 526
Isaac's syndrome (*see* Neuromyotonia)
Ischemic exercise test, 35
  in glycogenoses, 416
  in myoadenylate deaminase deficiency, 443
Ischemic myopathy, experimental, 546 (*see also* Muscle infarction)
Isokinetic dynamometry, 26, 664

Jitter, 48
Juvenile dermatomyositis, 544 (*see also* Childhood dermatomyositis)

Kearns-Sayre syndrome, 482, 485–489, 507
  histochemistry, 77
Kennedy's disease, 322
King (-Denborough) syndrome, 179, 458
Kocher-Debre-Semelaigne syndrome, 367
Krebs cycle, 479
Kugelberg-Welander disease (*see* Spinal muscular atrophy)

Kyphoscoliosis (*see* Scoliosis)

Lactate, in diagnosis of mitochondrial cytopathy, 36, 519
Lactate dehydrogenase (LDH) deficiency, 423
Lambert-Eaton syndrome, 42
Lamina densa, 81
Lamina lucida, 81
Laminin, 81, 82, 91, 224, 260 (*see also* Merosin)
Latent myotonia, 340
Lateral sacs, 84
LDH (*see* Lactate dehydrogenase)
Leber's hereditary optic neuropathy (LHON), 119, 480
Legionella myositis, 393
Leigh's disease, 75, 507, 512
Leprosy, and myositis, 393
Lid-lag myotonia, 345,
  in hypokalaemic periodic paralysis, 329
Limb-girdle muscular dystrophy (LGMD):
  autosomal dominant, 287–294
  autosomal recessive, 245–264, 696–697
  classification of, 253
  differential diagnosis of, 245
  gene frequency in, 258, 259
  myoimaging in, 106
Limb-girdle syndrome:
  and dystrophinopathy, 204, 239–240
  histology of, 68
Limit dextrinosis (*see* Debranching enzyme deficiency)
Lipid storage myopathies, 431–441
  hitochemistry, 64
  ultrastructure, 90
Lipodystrophy, 373
Lipofuscin, 91
Lipoic acid, 535
Lipomatosis, 98, 105
Liquorice-induced myopathy, 383
Localised nodular myositis, 557
Lod scores, 129
Lovastatin (and analogues) and myopathy, 382
Lowe syndrome, in mitochondrial cytopathy, 486
Luft's disease, 481, 497
Luque rods, 680
Lyme disease (*see Borrelia* myositis)
Lymphoma, muscle involvement in, 397
Lyonization, 202, 265, 687
Lysosomal glycogen storage disease, 63, 167

M-line, 83
Macro EMG, 49
Macroglossia:
  in acid maltase deficiency, 11
  in amyloid myopathy, 618
Macrophages, 63
MADD (*see* Myoadenylate deaminase deficiency)
Magnesium (*see* Hypomagnesaemia)
Magnetic resonance imaging of muscle (*see* MRI of muscle)
Magnetic resonance spectroscopy (MRS):
  of normal muscle, 115–119
  in myopathies in systemic diseases, 122–124
  in primary muscle diseases, 119–122, 416, 528
Malaria myositis, 395
Malignancy and inflammatory myopathy, 544–545 (*see also* Carcinomatous myopathy)
Malignant hyperthermia, 451–471, 630, 697
  and central core disease, 179
  magnetic resonance spectroscopy in, 122
  management of, 466–468
  myoglobinuria in, 608, 609
  raised creatine kinase in, 241
Malnutrition, and myopathy, 613
mAMPD (*see* Myoadenylate deaminase)
Manifesting carriers, 696 (*see also* Carriers of X-linked myscular dystrophy)
Manual muscle strength testing, 24–25, 563
Masseter muscle rigidity, in malignant hyperthermia, 451, 453
  management of, 465–466
Mathieu scale, 314
MCAD (*see* Medium-chain acyl-CoA dehydrogenase)
McArdle's disease, 3, 35, 414–420 (*see also* Glycogenoses)
  histochemistry, 74
  magnetic resonance spectroscopy in, 120
  myoimaging in, 107
McLeod syndrome, 202, 241
Medium-chain acyl-CoA dehydrogenase deficiency, 432
Medullary carcinoma of the thyroid, 372
Megaconial myopathy, 481
Meiotic drive, 323
MELAS, 482, 490–492
Membrane attack complex (MAC), 71
Membranous whorls, 91
Menadione, 514, 535
Menopausal muscular dystrophy, 541
Merosin deficiency, 146, 693 (*see also* Congenital muscular dystrophy)
MERRF, 48, 490
Metastases, in muscle, 397
MH (*see* Malignant hyperthermia)
MHC (major histocompatibility complex) immunocytochemistry, 71
Mi-2 antibody, 554

Microtubules, 82
Miller Dicker syndrome, 143
Minicore disease, 63, 70, 86, 181 (*see also* Multicore disease)
Mitochondria, isolation of, 522
 ultrastructure, 84
  in mitochondrial cytopathies, 88–89
Mitochondrial depletion, 493 (*see also* Fatal infantile myopathy)
Mitochondrial myopathies (cytopathies), 10, 11, 37, 479–504
 and AZT, 624–625
 and cardiac disease, 657–658
 drug-induced, 383
 and inclusion body myositis, 586
 magnetic resonance spectroscopy in, 119–120
 polymyalgia in, 596
 histochemistry of, 63–64
 treatment of, 119, 497–498, 514, 533–538
Mitochondrial, pseudoobstruction, ophthalmoplegia, and polyneuropathy (MEPOP) (*see* MNGIE)
Mitochondrial DNA, 505, 506
 duplications, 511
Mitochondrial DNA, point mutations, 511–512
Mitochondrial encephalomyopathy, lactic acidosis, with stroke-like episodes (*see* MELAS)
Mitochondrial neuropathy, gastrointestinal encephalomyopathy syndrome (*see* MNGIE)
Mixed connective tissue disease, 543–544
 autoantibodies in, 553
Miyoshi myopathy, 580
MNGIE syndrome, 484, 490, 494
Monaco's reading frame hypothesis, 236, 239
Morphea, 543
Mosaicism, 324
 gonadal, 690, 691
 somatic, 695
Moth-eaten fibers, 68, 71, 74
Motor unit action potentials (MUAPs), 45–46
 analysis of, 47–48
Magnetic resonance imaging (MRI) of muscle, 97–113
MRS (*see* Magnetic resonance spectroscopy)
Multicore disease (*see* Minicore disease)
Muscle aplasia, myoimaging, 103
Muscle biopsy, techniques of, 53–59
 tissue preparation, 56–57, 61–62
Muscle calcium channel disorder (*see* Dihydropyridine receptor)
Muscle pain (*see* Myalgia)
Muscle-eye-brain (MEB) syndromes, 141–144
Myalgia, 3–9
 drug-induced, 381
[Myalgia]
 in dystrophinopathy, 240
 exertional, 3
 pathogenesis of, 5
 treatment of, 15
Myalgic encephalomyelitis (*see* Chronic fatigue syndrome)
Myasthenia gravis, 3, 11, 42
 and thyroid dysfunction, 368
Mycoplasma myositis, 394
Myelin-like figures, 91
Myoadenylate deaminase deficiency (MADD), 35, 443–450,
 acquired, 48
 histochemistry, 77
Myoblast transfer, 215
Myocarditis (*see* Cardiac disease)
Myoclonus epilepsy with ragged red fibers (*see* MERRF)
Myoedema, 8, 14, 367
Myofibrils, ultrastructure, 83, 86
Myoglobin, 28, 607
Myoglobinuria, 6, 9, 607–611 (*see* Rhabdomyolysis)
Myokinase, 444
Myokymia, 7, 13,
 in Neuromyotonia, 355, 359
Myophosphorylase deficiency (*see* McArdle's disease)
Myosin, 83
Myositis: (*see also* Inflammatory myopathies)
 bacterial, 392
 drug-induced, 549
Myositis ossificans, 599–601
Myositis-specific autoantibodies, 553–554
Myosonography, 95–113
Myotendinous junction, ultrastructure, 82
Myotonia, 9
 "silent," 473
 drug-induced, 386
 experimental, 317
 pathophysiology of, 312, 346–347
 percussion, 14
Myotonia congenita:
 dominant (Thomsen's), 344–345
 recessive (Becker's), 345–346
Myotonia fluctuans, 343
Myotonia levior, 345
Myotonic discharges (*see* Myotonic runs)
Myotonic dystrophy, 10, 311–328
 anesthesia in, 632–633
 congenital, 316, 320–321
 disease severity classification, 314
 genetics of, 321–324, 693–695

[Myotonic dystrophy]
magnetic resonance spectroscopy in, 121–122
malignant hyperthermia in, 458
myoimaging in, 106
pathology of, 69, 314, 319–320
Myotonic runs, 9, 44, 312, 340
Myotubular myopathy, 183–185
histology, 71
Myxoedema myopathy (*see* Hypothyroid myopathy)

NADH dehydrogenase, 62
NADH-tetrazolium reductase (NADH-TR) staining, 481
Nebulin, 83
Needle muscle biopsy, 54
Needle myopathy, 380
Nelson's syndrome, myopathy in, 371
Nemaline myopathy, 187–190
histology, 71
myoimaging, 103
Nemaline rods, 64, 88
in HIV-related myopathy, 623
Nerve conduction measurement, 41
Neurasthenia (*see* Chronic fatigue syndrome)
Neurocristopathy, and myotonic dystrophy, 317
Neurofilaments, 82
Neuroleptic malignant syndrome, 456
and myoglobinuria, 608
Neuromuscular junction, ultrastructure, 3, 82
Neuromyotonia, 9, 14, 340, 355–363
and autoimmune disorders, 359–360
myalgia in, 5
neurophysiological diagnosis of, 357–358
pathogenesis of, 360
pathology of, 358
pathophysiology of, 356–357
secondary, 358–359
and thymoma, 359
treatment of, 360
Neutral lipid storage disease, 440
Nodular myositis, in rheumatoid disease, 399
Nondystrophic myotonias, 344–347
Normokalaemic periodic paralysis, 343, 348 (*see also* Periodic paralysis)
Nuclear inclusions:
in dermatomyositis, 85
in inclusion body myositis, 85
in oculopharyngeal muscular dystrophy, 85
Nuclear rods, 88
Nucleus, ultrastructure, 85

Oculopharyngeal muscular dystrophy (OMPD), 11, 169, 297–300
[Oculopharyngeal muscular dystrophy (OMPD)
filamentous nuclear inclusions in, 585
magnetic resonance spectroscopy in, 121
myoimaging in, 106
pathology of, 299–300
rimmed vacuoles in, 580, 584
Oil Red O stain, 64
Open muscle biopsy technique, 56
Opiate-induced myopathy, 383–385
Orbital myositis (pseudotumor), 367
Organic aciduria, in lipid storage myopathies, 436
Organophosphate poisoning, 387–388
Orthoses, in treatment of muscle disease, 674–679
Osteomalacic myopathy, 401, 613–614
Osteotomy, 673
Overlap myositis 543 (*see also* Inflammatory myopathies)
Oxidative phosphorylation, 479–480
biochemical analysis of, 522–525
magnetic resonance studies in, 117–118

Painful legs and moving toes, 7
Paradoxical myotonia, 340
Paralysis periodica paramyotonica, 340, 343
Paramyotonia congenita, 8, 9, 343, 348 (*see also* Sodium channel disorders)
magnetic resonance spectroscopy in, 122
Paraneoplastic myopathy (*see* Carcinomatous neuromyopathy)
Parenteral nutrition, and myopathy, 614
Paroxysmal myoglobinuria of childhood, 609
PAS (periodic acid-Schiff) reaction, 64
Pearson syndrome, 484, 489, 507
Pelger-Huet anomally, 288
Pennation angle, 21
PEO (*see* Progressive external ophthalmoplegia)
Percussion myotonia, 340
Perforins, 552
Perhexilene myopathy, 386
Perifascicular atrophy, in dermatomyositis, 71, 400
Periodic paralyses, 90, 339–344 (*see also* Sodium channel disorders, Hypokalemic periodic paralyis)
myoimaging in, 107
pathology of, 77
Permenant myotonia, 343
Pes cavus, treatment of, 672,
in neuromyotonia, 356
Pes planus, 672
Pet mutant model, 493
Peters syndrome, 143
PFK deficiency (*see* Phosphofructokinase deficiency)

Phaeochromocytoma, and myopathy, 372
Phosphglycerate kinase deficiency, 422
  and dystrophin deficiency, 241
  magnetic resonance spectroscopy in, 120
Phosphocreatine, 28
Phosphofructokinase (PFK) deficiency, 421–422
  histochemistry, 74
Phosphoglucomutase deficiency, 74
Phosphoglycerate mutase deficiency, 423
  magnetic resonance spectroscopy in, 120
Phosphorylase b kinase deficiency, 37, 420–421
Phosphorylation potential, 117
Physiotherapy in muscle diseases, 663–669
Pilomatrixoma, 318, 323
Pleoconial myopathy, 481
Polarography, in diagnosis of mitochondrial myopathy, 521
Polyarteritis nodosa, and vasculitic myopathy, 74, 400, 558
Polyglucosan body disease, 414
Polymyalgia rheumatica, 4, 73–74, 595–597
  mitochondrial disorder in, 596
Polymyositis (PM), histology, 73, 546–548 (*see also* Inflammatory myopathies)
  bacterial, 392 (*see also* Pyomyositis)
  and HIV infection, 623
  immunocytochemistry, 547
  inflammatory mechanisms in, 551–552
Pompe's disease (*see* Acid maltase deficiency)
Porcine stress syndrome, 697
Positional cloning, 128–129
Post-polio syndrome, myoimaging, 103
Postviral fatigue syndrome (*see* Chronic fatigue syndrome)
Pre-natal diagnosis, of muscle disease, 685, 690–692
Primary hyperaldosteronism, myopathy in, 371–372
Progressive external ophthalmoplegia (PEO), 77, 482, 489, 492–493, 494, 507
Progressive systemic sclerosis, 543
PROMM (*see* Proximal myotonic myopathy)
Proteoglycans, 81
Proximal myotonic myopathy (PROMM), 311, 316, 695
Pseudohyperparathyroidism, myopathy in, 374
Pseudohypertrophy:,
  in amyloid myopathy, 618
  in cysticerosis, 396
  in limb-girdle muscular dystrophy, 292
[Pseudohypertrophy]
  in trichinosis, 397
  in X-linked muscular dystrophies, 203
Pseudomyotonia, in hypothyroid myopathy, 366
  in neuromyotonia, 355
Purine nucleotide cycle, 44
Pyomyositis, 391–392 (*see also* Bacterial myositis)
  myoimaging in, 107

Quadriceps myopathy, 294
  and dystrophinopathy, 204, 240
Quantal squander, 355 (*see also* Neuromyotonia)
Quokka, 212, 236

Ragged red fibers (RRF), 75, 481, 505
Reducing bodies, 91
Reducing body myopathy, 171
Refsum's disease, and lipid storage in muscle, 440
Renal failure (*see* Uremic myopathy)
Renal tubular acidosis:
  and hypokalemic myopathy, 401
  and hypokalemic periodic paralysis, 330
Repetitive nerve stimulation (RNS), 42
Respiratory chain complexes, 522–525,
  histochemistry, 75–77
  immunoquantitation, 526–527
Respiratory control ratio, 521
Respiratory failure:
  in acid maltase deficiency, 407
  in inflammatory myopathy
  in myopathies, 542–543, 640–644
  in nemaline myopathy, 187, 189
Restless legs, 7
Restriction fragment length polymorphissm (RFLPs), 129
Reunion Island form of limb-girdle muscular dystrophy, 246, 261, 696
Reverse genetics, 128–129
Revertant fibers, 209, 237
RFLPs (*see* Restriction fragment length polymorphims)
Rhabdomyolysis, 9, 607–611 (*see also* Myoglobinuria)
  drug–induced, 383–385, 609
  HIV infection, 623
  in hyperthyroid myopathy, 365
  in hypophosphatemia, 401
  in inflammatory myopathies, 541
  myoimaging in, 109
  in sickle cell disease, 400
Rheumatoid disease, myopathy in, 399, 543
Rho^0 cells, 528

Riboflavin, in treatment of mitochondrial cytopathy, 440, 514, 535
Rigid spine syndrome, 141, 151–156, 216
and rimmed vacuoles, 580, 584
in minicore disease, 181
myoimaging in, 153
Rimmed vacuoles, 73
in denervating disorders, 584
in distal myopathy, 302, 303 (*see also* Welander's disease)
myoimaging in, 106
in myopathies, 579–580
in nemaline myopathy, 190
in oculopharyngeal muscular dystrophy, 298–299
pathology of, 580–584, 586
Ringbinden (ring fibers), 87, 315
Rippling muscle syndrome, 8
Rod-body myopathy (*see* Nemaline myopathy)
Ryanodine receptor, 122, 179, 453, 456, 697–698

Santavuori congenital muscular dystrophy, 142–144 (*see also* Congenital muscular dystrophy)
Sarcocystis, 395
Sarcoid myopathy, 560–561
myoimaging in, 108
Sarcolemma, 81, 226
Sarcomeres, 83
Sarcoplasmic reticulum, 3, 84, 90
Sarcotubular myopathy, 169
Satellite cells, 82
Scanning EMG, 47
Scapuloperoneal syndrome, 216, 276
histology, 69
in phosphfructokinase deficiency, 421
SCARMD (*see* Severe childhood autosomal recessive muscular dystrophy)
Schwartz-Jampel syndrome, 311, 348
Scleroderma myositis, 543,
autoantibodies in, 553
SCN4A (*see* Sodium channel protein)
Scoliosis, 14, 673–675
and respiratory failure, 641
Second wind phenomenon, 415
Secondary myopathic change, 65
Selenium deficiency myopathy, 614
Serotonin myopathy, 398
Severe childhood autosomal recessive muscular dystrophy (SCARMD), 213, 249, 692–693
histology of, 67
Shin splints, 605
Sickle cell disesase, 400 (*see also* Rhabdomyolysis)
Signal recognition protein, 554
Silent myotonia, 473
Single fibre electromyography (SFEMG), 48–49
Sjogren's syndrome, 543
autoantibodies in, 553
Skeletin (*see* Desmin)
SLE (*see* Systemic lupus erythematosus)
Sleep apnoea:
in myopathies, 641
in myotonic dystrophy, 657
Snake-bite, and myonecrosis, 387
Sodium channel disorders 343–344, 699 (*see also* Hyperkalemic periodic paralysis, Paramyotonic congenita)
and malignant hyperthermia, 453
Sodium channel myotonia, 343, 348
Sodium channel protein, 347–348
Southern blotting, 130
Sparganosis, 396
Spasmophilia, 7
Spectrin, 82, 84, 91,
immunohistochemistry, 68–69
structure, 224, 228, 235
Spheroid body myopathy, 169
Spillane's syndrome (*see* Painful legs and moving toes)
Spinal bracing, 675
Spinal fusion, 674
Spinal muscular atrophy (SMA), 65
Becker phenotype, 239
myoimaging in, 103
Spontaneous activity, 44
Stahlberg's cloud, 48
Starvation, and myopathy, 613
Steinert's disease, 311 (*see also* Myotonic dystrophy)
Steroid myopathy, 368–370, 382, 385–386
Stick man dystrophy, 141
Stiff-man syndrome, 5, 9, 359, 360
Strength measurement, 14, 19–34
Sub-anaerobic threshold exercise test (SATET), 37–38, 592
Succinate, in treatment of mitochondrial myopathy, 514
Succinate dehydrogenase (SDH), 62, 481
Sudan Black stain, 64
Supernumery muscles, myoimaging, 103
Suxamethonium myalgia, 381
Synaptic vesicles, 82
Systemic angiopathy of childhood, 540
Systemic lupus erythematosus, and myositis, 399, 543
autoantibodies in, 553
Systemic sclerosis, 398

T cells, in inflammatory myopathies, 551–552
immunocytochemistry, 71

T-tubular system, 81, 84, 90
Tandem repeats, 129–130
Tapeworm infection of muscle (*see* Coenurosis, Sparganosis)
Target fibers, 65, 86, 87
Tarui's disease (*see* Phosphofructokinase deficiency)
Terminal cisternae, 84
Testosterone deficiency myopathy, 374–375
Tetanus, 392–393
Tetany, 7, 372, 374
Thiamine, in treatment of pyruvate dehydrogenase deficiency, 535
Thomsen's disease (*see* Myotonia congenita, dominant)
Thyroid-associated ophthalmopathy, 367
Thyrotoxic myopathy (*see* Hyperthyroid myopathy)
Thyrotoxic periodic paralysis, 330, 367
Titin, 83, 92
Tolosa Hunt syndrome, 367
Toluene myopathy, 388
Toxic myopathies, 387–388 (*see also* Drug-induced myopathies)
Toxic oil syndrome, 388, 558
Toxocariasis, 397
Toxoplasmosis myositis, 395
Tracheostomy, in myopathic respiratory failure, 649
Transverse tubular system (*see* T-tubular system)
Trauma, and muscle wasting, 401
Trendelenberg test, 14
TRI (*see* Tubuloreticular inclusions)
Triads, 84, 226
Trichinosis, 397
Trifunctional enzyme, 432, 439
Trilaminar fibers, 87
Tropical myositis, 391–392 (*see also* Pyomyositis)
Tropomyosins, 83
Troponins, 83
Tuberculous myositis, 393
Tubular aggregates, 77, 90
  in central core disease, 180
  in malignant hyperthermia, 462
Tubuloreticular inclusions (TRI), 624, 626,
  in inflammatory myopathies, 547
  in HIV-related myopathies, 626
Tunisian form of limb-girdle muscular dystrophy, 249, 253, 261
Turner's syndrome, 238
Type I and II fibers, histochemistry, 62

Ubiquinone, 514
Ubiquitin, and inclusion body myositis, 585–586
Ullrich form of congenital muscular dystrophy, 140
Ultrasound imaging (*see* Myosonography)
Ultrastructure, of muscle, 81–93
Uraemic myopathy, 374, 401
Utrophin, 69, 82, 223, 260

Vacuolar myopathy:
  in acid maltase deficiency, 408
  in drug-induced myopathies, 382, 385
  in hypokalemia, 382–383
  in hypokalemic periodic paralysis, 331
  in thyrotoxic periodic paralysis, 367
Vasculitic myopathy, 400, 558–559
Ventilatory aids, 645–649, 664, 666
Vimentin, 84
Vincristine myopathy, 386
Vinculin, 84, 228
Viral myositis, 393–394
Viral persistence, 556 (*see also* Chronic fatigue syndrome, Inflammatory myopathies)
Viruses, and inflammatory myopathies, 555–556
Vitamin D deficiency, and myopathy, 613
Vitamin E:
  deficiency myopathy, 388, 614
  in treatment of mitochondrial cytopathy, 535
Volkmann's ischaemic contracture, 400
Voltage-gated potassium channels, 360

Walker-Warburg syndrome, 142–144
Warm-up phenomenon, 9, 312, 340
Weakness, pathogenesis of, 2
Wegener's granulomatosis, 559
Welander's disease, 300–303, 580
Werdnig-Hoffman disease (*see* Spinal muscular atrophy)
Wheelchairs, 665

X-inactivation (*see* Lyonization)
X-linked dominant myopathy, 288
Xanthine oxidase deficiency, 77
Xp21.2 myopathies (*see* Dystrophinopathies)

Yeast artificial chromosomes (YACS), 132

Z-line, 83, 86
  streaming, 71, 88
Zebra bodies, 91
Zebra body myopathy, 171
Zellweger's syndrome, and lipid storage in muscle, 440
Zidovudine (*see* AZT)

About the Editor

RUSSELL J. M. LANE is a Consultant Neurologist at the Regional Neurosciences Centre, Charing Cross Hospital, London, England, and a Senior Lecturer in Neurology in the Department of Clinical Neuroscience at Charing Cross and Westminster Medical School. Formerly, Dr. Lane was a Muscular Dystrophy Association Research Fellow at Duke University, Durham, North Carolina, and a Wellcome Senior Research Fellow in Clinical Sciences at the University of Newcastle upon Tyne, England. A Fellow of the Royal College of Physicians (UK) and the Royal Society of Medicine, Dr. Lane received the B.Sc. degree (1970), the M.B.B.S. degree (1973), and the M.D. degree (1984) from the University of Newcastle upon Tyne.